PRINCIPLES OF NAVAL ENGINEERING

An Introduction to the Theory and Design of Engineering Machinery and Equipment Aboard Ship

Principles of Naval Engineering Hardcover

ISBN: 978-0-9858282-7-1

Digitally Reproduced by:
CONVERPAGE
23 Acorn Street
Scituate, MA 02066

www.converpage.com

PREFACE

This text provides an introduction to the theory and design of engineering machinery and equipment aboard ship. Primary emphasis is placed on helping the student acquire an overall view of shipboard engineering plants and an understanding of basic theoretical considerations that underlie the design of machinery and equipment. Details of operation, maintenance, and repair are not included in this text.

The text is divided into five major parts. Part I deals with the development of naval ships, ship design and construction, stability and buoyancy of ships, and preventive and corrective damage control. Certain theoretical considerations that apply to virtually all engineering equipment are discussed in part II. Part III takes up the major units of machinery in the main propulsion cycle of the widely used steam turbine propulsion plant. Auxiliary machinery and equipment are discussed in part IV. Other types of propulsion machinery, together with a brief survey of new developments in naval engineering, are considered in part V. In addition to these five major parts, the text includes an appendix which surveys and briefly describes a number of references that should be of value to engineering officers.

This text was prepared by the Training Publications Division, Naval Personnel Program Support Activity, Washington, D. C., for the Bureau of Naval Personnel. Review of the manuscript and technical assistance were provided by Officer Candidate School at Newport, Rhode Island; Naval Development and Training Center, San Diego, California; Service School Command, Great Lakes, Illinois; and Naval Ship Systems Command.

Stock Ordering No.
0502-LP-053-9405

First Edition 1958
Revised, 1966, 1970
Reprinted 1987

For sale by the Superintendent of Documents, U.S. Government Printing Office
Washington, D.C. 20402

CONTENTS

CREDITS

The illustrations indicated below are included in this edition of <u>Principles of Naval Engineering</u> through the courtesy of the designated publishers, companies, and associations. Permission to reproduce these illustrations must be obtained from the original sources.

<u>Source</u>	<u>Figures</u>
American Engineering Co.	21-7
American Society of Mechanical Engineers	22-43
Babcock & Wilcox Co.	10-15, 10-19, 11-6
Bert A. Shields	7-33
Buffalo Meter Co.	7-28
Cooper-Bessemer Corp.	22-11, 22-13, 22-23, 22-26, 22-27, 22-33
Crane Co.	14-12, 14-14
Crosby Valve and Gage Co.	7-25
Cutler-Hammer, Inc.	20-25
De Laval Steam Turbine Co.	16-7, 16-8
Fairbanks, Morse and Co.	22-9
General Motors Corp.	
Cleveland Diesel Engine Division	22-14, 22-35
Detroit Diesel Engine Division	22-19, 22-20, 22-24, 22-25, 22-34
Gray Marine Motor Co.	22-2, 22-3
James G. Biddle Co.	7-35, 7-36
John Wiley & Sons, Inc.	8-11
From: Joseph H. Keenan and Frederick G. Keyes, <u>Thermodynamic Properties of Steam</u>, New York: John Wiley & Sons, Inc. 1937. (Excerpts from Table III.)	
Kingsbury Machine Works, Inc.	5-10, 5-11
Leeds and Northrup Co.	7-12
Leslie Co.	14-23, 14-24, 14-25
Manning, Maxwell & Moore, Inc.	7-17, 7-18, 7-20 7-23, 7-24
Moeller Instrument Co.	18-13
Packard Motor Car Co.	22-27, 22-28
Shell Oil Co.	5-12
Society of Naval Architects and Marine Engineers	5-19, 5-20, 5-21
Stewart Warner	7-34
U. S. Naval Institute	
<u>Naval Auxiliary Machinery</u>	7-8, 7-16, 7-29 7-32, 11-35, 13-4 14-2, 14-5, 14-6, 14-9, 14-10, 14-16,

Source	Figures
Naval Auxiliary Machinery (Cont'd)	14-20, 14-27, 14-29 15-20, 17-5, 21-5, 21-6
Naval Boilers	11-15, 11-17
Naval Turbines	12-6, 12-7, 12-9, 12-10, 12-11, 12-12, 12-13, 12-14, 12-15, 12-18, 12-20, 12-21, 16-1, 16-2, 16-3, 16-4, 16-5, 16-6
Velan Engineering Companies	14-8
Westinghouse Electric Corp.	15-32, 15-33, 15-34, 15-36, 24-7, 24-8, 24-9, 24-10, 24-11

PART I—THE NAVAL SHIP

The four chapters included in this part of the text deal primarily with the ship as a whole, rather than with specific items of engineering equipment. Most of the information given in this part applies to naval ships in general, without regard to the type of ship or the type of propulsion plant employed.

Chapter 1 provides a brief historical survey of the development of naval ships. Chapter 2 takes up basic design considerations, ship flotation, ship structure, compartmentation, and the geometry of the ship. Chapter 3 deals with the basic principles of stability and buoyancy; although this information is largely theoretical, it is essential for a true understanding of the naval ship and for an understanding of many aspects of damage control. Chapter 4 is concerned with preparations to resist damage, the damage control organization, material conditions of readiness, the investigation of damage, the control of damage, and certain aspects of nuclear, biological, and chemical defense.

CHAPTER 1

THE DEVELOPMENT OF NAVAL SHIPS

The story of the development of naval ships is the story of prime movers: oars, wind-filled sails, reciprocating steam engines, steam turbines, internal combustion engines, gas turbine engines. It is also the story of the conversion and utilization of energy: mechanical energy, thermal energy, chemical energy, electrical energy, nuclear energy. Seen in broader context, the development of naval ships is merely one fascinating aspect of man's long struggle to control and utilize energy and thereby release himself from the limiting slavery of physical labor.

We have come a great distance in the search for the better utilization of energy, from the muscle power required to propel an ancient Mediterranean galley to the vast reserves of power available in a shipboard nuclear reactor. No part of this search has been easy; progress has been slow, difficult, and often beset with frustrations. And the search is far from over. Even within the next few years, new developments may drastically change our present concepts of energy utilization.

This chapter touches briefly on some of the highlights in the development of naval ships. In any historical survey, it is inevitable that a few names will stand out and a few discoveries or inventions will appear to be of crucial significance. We may note, however, that our present complex and efficient fighting ships are the result not only of brilliant work by a relatively small number of well known men but also of the steady, continuing work of thousands of lesser known or anonymous contributors who have devised small but important improvements in existing machinery and equipment. The primitive man who invented the wheel is often cited as an unknown genius; we might do well to remember also the unknown genius who discovered that wheels work better when they turn in bearings. Similarly, the basic concepts

involved in the design of steam turbines, internal combustion engines, and gas turbine engines may be attributed to a few men; but the innumerable small improvements that have resulted in our present efficient machines are very largely anonymous.

THE DEVELOPMENT OF STEAM MACHINERY

One of the earliest steam machines of record is the aeolipile developed about 2000 years ago by the Greek mathematician Hero. This machine, which was actually considered more of a toy or novelty than a machine, consisted of a hollow sphere which carried four bent nozzles. The sphere was free to rotate on the tubes that carried steam from the boiler, below, to the sphere. As the steam flowed out through the nozzles, the sphere rotated rapidly in a direction opposite to the direction of steam flow. Thus Hero's aeolipile may be considered as the world's first reaction turbine.[1]

Giovanni Battista della Porta's treatise on pneumatics (1601) describes and illustrates a device which utilizes steam pressure to force water up from a separate vessel. In the same treatise, the author suggests that the condensation of steam could be used to create a vacuum, and that the vacuum could be utilized to draw water upward from a lower level—a remarkably sophisticated concept, for the time.

Throughout the 17th century, many other devices were suggested (and some of them built) which attempted to utilize the motive power of steam. In many instances the scientific principles were sound but the technology of the day did not permit full development of the devices.

---

[1] Hero's aeolipile is illustrated in chapter 12 of this text.

In 1698, Thomas Savery patented a condensing steam engine which was designed to raise water. This machine consisted of two displacement chambers (or one, in some models), a main boiler, a supplementary boiler, and appropriate piping and valves. The operating principles are simple, though most ingenious for the time. When steam is admitted to one of the displacement vessels, it displaces the water and forces it upward through a check valve. When the displacement vessel has been emptied of water by this method, the supply of steam is cut off. The steam already in the displacement vessel is condensed as cold water is sprayed on the outside surface of the vessel. The condensation of the steam creates a vacuum in the displacement vessel, and the vacuum causes more water to be drawn up through suction piping and a check valve. When the displacement vessel is again full of water, steam is again admitted to the vessel and the cycle is repeated. In a model with two displacement chambers, the cycles are alternated so that one vessel is discharging water upward while the other is being filled with water drawn up through the suction pipe.

Although technological difficulties prevented Savery's engine from being used as widely as its inventor would have liked, it was successfully used for pumping water into buildings, for filling fountains, and for other applications which required a relatively low steam pressure. The machine was originally designed as a device for removing water from mines, and Savery was convinced that it would be suitable for this purpose. It was never widely used in mines, however, because very high steam pressures would have been required to lift the water the required distance. The metalworking skills of the time were simply not up to producing suitable pressure vessels for containing steam at high pressures.[2]

Although Savery's machine was used throughout the 18th century and well into the 19th century, two new steam engines had meanwhile made their appearance. The first of these, Newcomen's "atmospheric engine," represents a real breakthrough in steam machinery. Like

Savery's device, the Newcomen engine was originally designed for removing water from mines, and in this it was highly successful. However, the significance of the Newcomen engine goes far beyond mere pumping. The second was the Watt engine, which brought the reciprocating steam engine to the point where it could be used as a prime mover on land and at sea.

The Newcomen engine was the first workable steam engine to utilize the piston and cylinder. As early as 1690, Denis Papin[3] had suggested a piston and cylinder arrangement for a steam engine. The piston was to be raised by steam pressure from steam generated in the bottom of the cylinder. After the piston was raised, the heat would be removed and condensation of steam in the bottom of the cylinder would create a vacuum. The downward stroke of the piston would thus be caused by atmospheric pressure acting on top of the piston. Papin's theory was good but his engine turned out to be unworkable, chiefly because he attempted to generate the steam in the bottom of the cylinder. When Papin heard of Savery's engine, he stopped working on his own piston and cylinder device and devoted himself to improving the Savery engine.

The Newcomen engine, shown in figure 1-1, was built by Thomas Newcomen and his assistant, John Cawley, in the early part of the 18th century.[4] The Newcomen engine differs from the engine suggested by Papin in several important respects. Most important, perhaps, is the fact that Newcomen separated the boiler from the cylinder of the engine.

As may be seen in figure 1-1, the boiler is located directly under the cylinder. Steam is admitted through a valve to the bottom of the cylinder, forcing the piston up. The piston is connected by a chain to the arch on one side of a large, pivoted, working beam. The arch on the other side of the beam is connected by a chain to the rod of a vertical lift pump.

[2]It is reported that Savery attempted to use steam pressures as high as 8 or 10 atmospheres. When one considers the weakness of his pressure vessels and the total lack of safety values, it appears somewhat remarkable that he survived.

[3]Papin is also credited with the invention of the safety valve.

[4]The year 1712 is frequently given as the date of the Newcomen engine, and it is probably the year in which the engine was first demonstrated to a large public. It is likely, however, that previous versions of the engine were built at a considerably earlier date, and some authorities give the year 1705 as the date of the Newcomen engine.

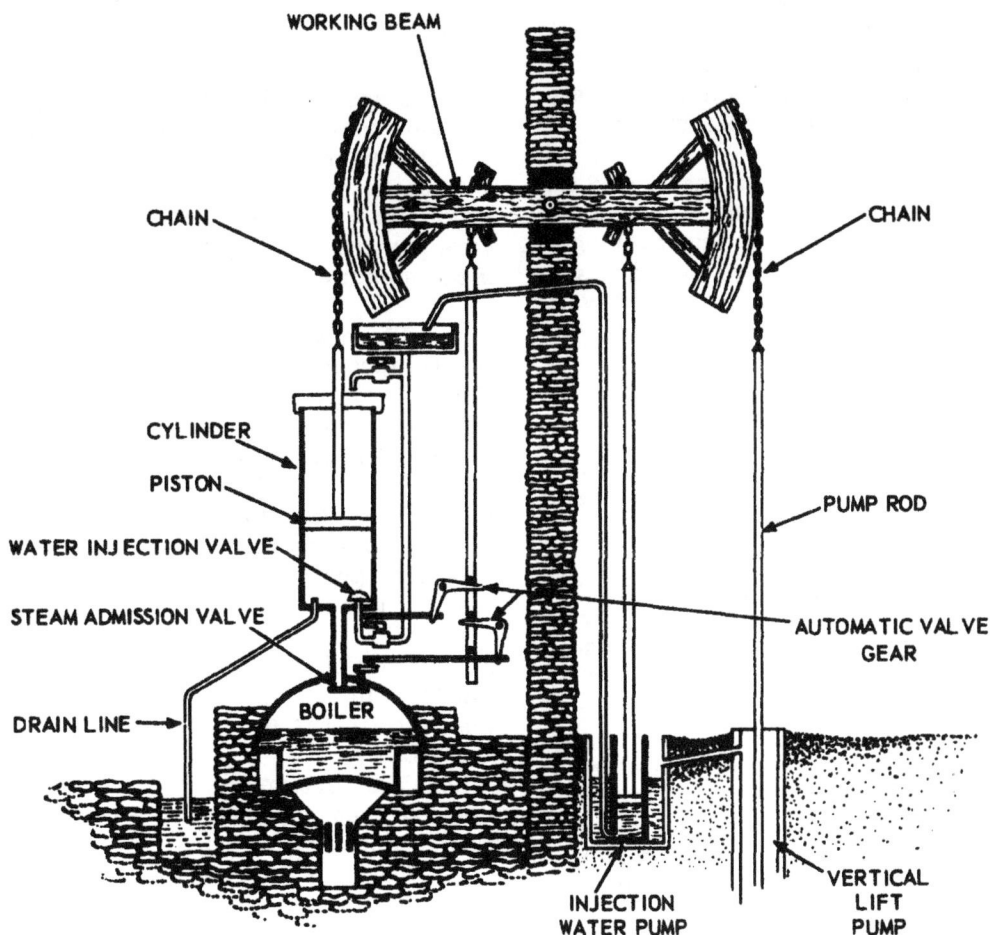

147.1

Figure 1-1.—The Newcomen engine.

After the steam has forced the piston to its top position, the steam valve is shut and a jet of cold water enters the cylinder, condensing the steam and creating a partial vacuum. Atmospheric pressure then causes the down stroke (work stroke) of the piston.

As the piston comes down, the working beam is pulled down on the cylinder side. As the beam rises on the pump side, the pump rod also rises and water is lifted upward. As soon as the pressure in the cylinder equals atmospheric pressure, an escape valve in the bottom of the cylinder opens and the condensate is discharged through a drain line into a sump.

The use of automatic valve gear to control the admission of steam and the admission of cold water made the Newcomen engine the first self-acting mechanism since the invention of the clock. In the earliest versions of the Newcomen

engine, it is most likely that the admission of steam and cold water was controlled by the manual operation of taps rather than by automatic gear. The origin of the automatic gear is a matter of some dispute. One story has it that a young boy named Humphrey Potter, who was hired to turn the taps, invented the valve gear so that he could go fishing while the engine tended itself. This story, although persistent, is considered "absurd" by some serious historians of the steam engine.[5]

James Watt, although often given credit for inventing the steam engine, did not even begin working on steam engines until some 50 years

_____

[5]See, for example, Eugene S. Ferguson, "The Origins of the Steam Engine," Scientific American, January 1964, pp. 98-107.

or so after the Newcomen engine was operational. However, Watt's brilliant and original contributions were ultimately responsible for the utilization of steam engines in a wide variety of applications beyond the simple pumping of water.

In 1799 Watt was granted a patent for certain improvements to "fire-engines" (Newcomen engines). Since some of these improvements represent major contributions to steam engineering, it may be of interest to see how Watt himself described the improvements in a specification:

"My method of lessening the consumption of steam, and consequently fuel, in fire-engines, consists of the following principles:—

"First, That vessel in which the powers of steam are to be employed to work the engine, which is called the cylinder in common fire-engines, and which I call the steam vessel, must, during the whole time the engine is at work, be kept as hot as the steam that enters it; first by inclosing it in a case of wood, or any other materials that transmit heat slowly; secondly, by surrounding it with steam or other heated bodies; and thirdly, by suffering neither water nor any other substance colder than the steam to enter or touch it during that time.

"Secondly, In engines that are to be worked wholly or partially by condensation of steam, the steam is to be condensed in vessels distinct from the steam-vessels or cylinders, although occasionally communicating with them; these vessels I call condensers; and, whilst the engines are working, these condensers ought at least to be kept as cold as the air in the neighbourhood of the engines, by application of water or other cold bodies.

"Thirdly, Whatever air or other elastic vapour is not condensed by the cold of the condenser, and may impede the working of the engine, is to be drawn out of the steam-vessels or condensers by means of pumps, wrought by the engines themselves or otherwise.

"Fourthly, I intend in many cases to employ the expansive force of steam to press on the pistons, or whatever may be used instead of them, in the same manner in which the pressure of the atmosphere is now employed in common fire-engines. In cases where cold water cannot be had in plenty, the engines may be wrought by

this force of steam only, by discharging the steam into the air after it has done its office."

As a result of these and other improvements, the Watt engine achieved an efficiency (in terms of fuel consumption) which was twice that of the Newcomen engine at its best. Among the other major contributions made by Watt, the following were particularly significant in the development of the steam engine:

1. The development of devices for translating reciprocating motion into rotary motion. Although Watt was not the first to devise such arrangements, he was the first to apply them to the task of making a steam engine drive a revolving shaft. This one improvement alone opened the way for the application of steam engines to many uses other than the pumping of water; in particular, it paved the way for the use of steam engines as propulsive devices.

2. The use of a double-acting piston—that is, one which is moved first in one direction and then in the opposite direction, as steam is admitted first to one end of the cylinder and then to the other.

3. The development of parallel-motion linkages to keep a piston rod vertical as the beam moved in an arc.

4. The use of a centrifugal "flyball" governor to control the speed of the steam engine. Although the centrifugal governor had been used before, Watt brought to it the completely new—and very significant—concept of feedback. In previous use, the centrifugal governor had been capable of making a machine automatic; by adding the feedback principle, Watt made his machines self-regulating.[6]

Neither Newcomen nor Watt were able to utilize the advantages of high pressure steam, largely because a copper pot was about the best that could be done in the way of a boiler. The first high pressure steam engines were built by

---

[6]The distinction between automatic machines and self-regulating machines is of considerable significance. An automatic pump, for example, can operate without a human attendant but it cannot change its mode of operation to fit changing requirements. A self-regulating pump, on the other hand, operates automatically and can change its speed (or some other characteristic) to meet increased or decreased demands for the fluid being pumped. To be self-regulating, a machine must have some type of feedback information from the output side of the machine to the operating mechanism.

Chapter 1—THE DEVELOPMENT OF NAVAL SHIPS

Oliver Evans, in the United States, and Richard Trevethick, in England. The Evans engine, which was built in 1804, had a vertical cylinder and a double-acting piston. A boiler, made of copper but reinforced with iron bands, provided steam at pressures of several atmospheres. The boiler was one of the first "fire-tube" boilers; the "tubes" were actually flues which were installed in such a way as to carry the combustion gases several times through the vessel in which the water was being heated. This type of boiler, with many refinements and variations, became the basic boiler design of the 19th century. Trevethick, using a similar type of boiler, built a successful steam carriage in 1801; in 1804, he built what was probably the first modern type of steam locomotive.

Continuing efforts by many people led to steady improvements in the steam engine and to its eventual application as a prime mover for ships. For many years, the major effort was to improve the reciprocating steam engine. However, the latter half of the 19th century saw the introduction of the first practicable steam turbines. Sir Charles Parsons, in 1884, and Dr. Gustaf de Laval, in 1889, made major contributions to the development of the steam turbine. The earliest application of a steam turbine for ship propulsion was made in 1897, when a 100-ton vessel was fitted with a steam turbine which was directly coupled to the propeller shaft. After the installation of the steam turbine, the vessel broke all existing speed records for ships of any size. In 1910, Parsons introduced the reduction gear, which allowed both the steam turbine and the screw propeller to operate at their most efficient speeds—the turbine at very high speeds, the propeller at much lower speeds. With this improvement, the steam turbine became the most significant development in steam engineering since the development of the Watt engine. With further refinements and improvements, the steam turbine is today the primary device for utilizing the motive power of steam.

THE DEVELOPMENT OF MODERN NAVAL SURFACE SHIPS

The 19th century saw the application of steam power to naval ships. The first steam-driven warship in the world was the Demologos (voice of the people) which was later renamed the Fulton in honor of its builder, Robert

Fulton. The ship, which is shown in figure 1-2, was built in the United States in 1815. The ship had a displacement of 2475 tons. A paddle wheel 16 feet in diameter, was mounted in a trough or tunnel inside the ship, for protection from gunfire. The paddle wheel was driven by a one-cylinder steam engine with a 48-inch cylinder and a 60-inch stroke.

The next large steam-driven warship to be built in the United States was the Fulton 2nd. This ship was built in 1837 at the Brooklyn Navy Yard. The Fulton 2nd, like the Fulton (or Demologos) before it, was fitted with sails as well as with a steam engine. The plant efficiency of the Fulton 2nd has been calculated[7] to be about 3 percent. Its maximum speed was about 15 knots, with a shaft horsepower of approximately 625.

The Fulton 2nd was rebuilt in 1852 and named the Fulton 3rd. The Fulton 3rd had a somewhat different kind of steam engine, and its operating steam pressure was 30 psi, rather than the 11 psi of the Fulton 2nd. Several other significant changes were incorporated in the Fulton 3rd—but the ship still had sails as well as a steam engine. The Navy was still a long ways away from abandoning sails in favor of steam.

The Mississippi and the Missouri, built in 1842, are sometimes regarded as marking the beginning of the steam Navy—even though they, too, still had sails. The two ships were very much alike except for their engines. The Missouri had two inclined engines. The Mississippi had two side-lever engines of the type shown in figure 1-3. Three copper boilers were used on each ship. Operating steam pressures were approximately 15 psi.

The Michigan, which joined the steam Navy about 1843, had iron boilers rather than copper ones. These boilers lasted for 50 years. The Michigan operated with a steam pressure of 29 psi.

The Princeton, which joined the steam Navy in 1844, was remarkable for a number of reasons. It was the first warship in the world to use screw propellers, although they had been tried

_____

[7]See Morris Welling, Gerald M. Boatwright, and Maurice R. Hauschildt, "Naval Propulsion Machinery," Naval Engineers Journal, May 1963, pp. 339-348.

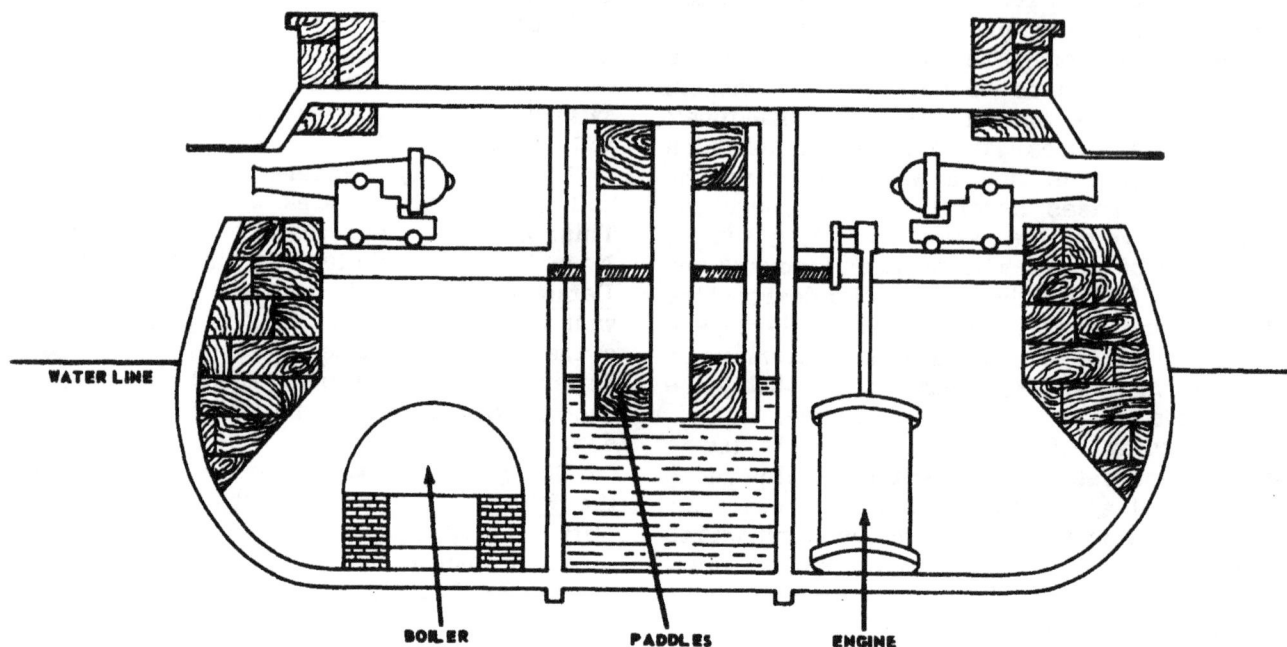

147.2

Figure 1-2.—The Demologos: the first steam-driven warship.

out more than forty years before.[8] The Princeton had an unusual oscillating, rectangular-piston type of engine (fig. 1-4). The piston rod was connected directly to the crankshaft, and the cylinder oscillated in trunnions. This ship was also noteworthy for being the first warship to have all machinery located below the waterline, the first to burn hard coal, and the first to supply extra air for combustion by having blowers discharge to the fireroom. But even the Princeton still had sails.

Almost twenty steam-driven warships joined the steam Navy between 1854 and 1860. One of

these was the Merrimac. Another was the Pensacola, which was somewhat ahead of its time in several ways. The Pensacola had the first surface condenser (as opposed to a jet condenser) to be used on a ship of the U.S. Navy. It also had the first pressurized firerooms.

It was not until 1867 that the U.S. Navy obtained a completely steam-driven ship. In the Navy's newly created Bureau of Steam Engineering, a brilliant designer, Benjamin Isherwood, conceived the idea for a fast cruiser. One of Isherwood's ships, the Wampanog, attained the remarkable speed of 17.75 knots during her trial runs, and maintained an average of 16.6 knots for a period of 38 hours in rough seas.

The Wampanog had a displacement of 4215 tons, a length of 335 feet, and a beam of 45 feet. The engines consisted of two 100-inch single-expansion cylinders turning one shaft. The engine shaft was geared to the propeller shaft, driving the propeller at slightly more than twice the speed of the engines. Steam was generated by four boilers at a pressure of 35 psi and was superheated by four more boilers. The Wampanog propulsion plant, shown in figure 1-5, was a remarkable power plant for its time.

---

[8]In 1802, Colonel John Stevens applied Archimedes' screw as a means of ship propulsion. The first ship that Stevens tried the screw on was a single-screw ship, which unfortunately ran in circles. The second application—a twin-screw ship with the screws revolving in opposite directions—was more successful. John Ericsson, who designed the Princeton, developed the forerunner of the modern screw propeller in 1837. It is interesting to note that the original problem with screw propellers was that they were inefficient at the slow speeds provided by the large, slow engines of the time. This is just the opposite of the present-day problem. Both then and now, the solution is gears: step-up gears, in the old days, and reduction gears, at the present time.

Sad to relate, the Wampanog came to an ignominious end. A board of admirals concluded that the ship was unfit for the Navy, that the four-bladed propeller was an interference to good sailing, and that the four superheater boilers were merely an unnecessary refinement. As a result of this expert opinion, two of the four propeller blades and all four of the superheater boilers were removed. The Wampanog was thus reduced from a superior steam-driven ship to an inferior sailing vessel, with steam used merely as an auxiliary source of power.

The modern U.S. Navy may be thought of as dating from 1883, the year in which Congress appropriated funds for the construction of the first steel warships. The major type of engine was still the reciprocating steam engine; however, the latter part of the 19th century saw increasing interest in the development of internal combustion engines and steam turbines.

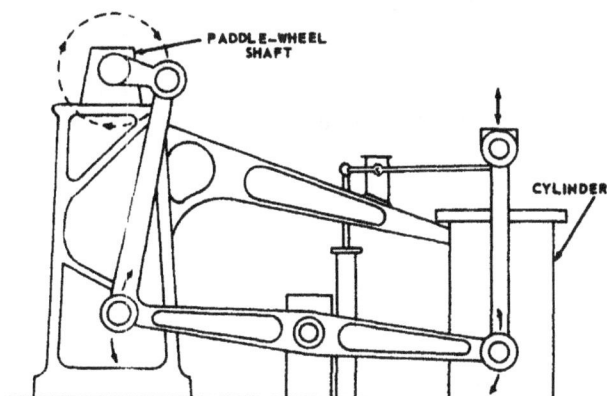

147.3

Figure 1-3.—Side-lever engine, USS Mississippi (1842).

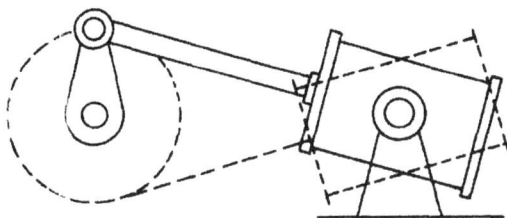

147.4

Figure 1-4.—Oscillating engine, USS Princeton (1844).

Ship designers approached the close of the 19th century with an intense regard for speed. Shipbuilders were awarded contracts with bonus and penalty clauses based on speed performance. In the construction of the cruisers Columbia and Minneapolis, a speed of 21 knots was specified. The contract stipulated a bonus of $50,000 per each quarter-knot above 21 knots and a penalty of $25,000 for each quarter-knot below 21 knots. The Columbia maintained a trial speed of 22.8 knots for 4 hours, and thereby earned for her builders a bonus of $350,000. Her sister ship, the Minneapolis, made 23.07 knots on her trials, earning $414,600 for that performance. Other shipbuilders profited in similar fashion from the speed race. And some, of course, were penalized for failure. The builders of the Monterey, for example, lost $33,000 when the ship failed to meet the specified speed.

By the early part of the 20th century, steam was here to stay; the ships of all navies of the world were now propelled by reciprocating steam engines or by steam turbines. Coal was still the standard fuel, although it had certain disadvantages that were becoming increasingly apparent. One of the problems was the disposal of ashes. The only practicable way to get rid of them was to dump them overboard, but this left a telltale floating line on the surface of the sea, easily seen and followed by the enemy. Furthermore, the smoke from the smokestacks was enough to reveal the presence of a steam-driven ship even when it was far beyond the horizon. The military disadvantages of coal were further emphasized by the fact that it took at least one day to coal the ship, another day to clean up—a minimum of two days lost, and the coal would only last for another two weeks or so of steaming.

Then came oil. The means for burning oil were not developed until the early part of the 20th century. Once the techniques and equipment were perfected, the change from coal to oil took place quite rapidly. Our first oil-burning battleships were the Oklahoma and the Nevada, which were laid down in 1911. All coal-burning ships were later altered to burn oil.

While the coal-to-oil conversion was in progress, a tug-of-war was going on in another area. The reciprocating steam engine and the steam turbine each had its proponents. To settle the matter, the Bureau of Engineering made the decision to install reciprocating engines in the Oklahoma and steam turbines in the Nevada. Although there were still many problems to be solved, the steam turbine was well on its way to becoming the major prime mover for naval ships.

With the advent of the steam turbine, the problem of reconciling the speed of the prime mover and the speed of the propeller became critical. The turbine operates most efficiently at high speed, and the propeller operates most efficiently at low speed. The obvious solution was to use reduction gears between the shaft of the prime mover and the shaft of the propeller; and, basically, this is the solution that was adopted and that is still in use on naval ships today. However, other solutions are possible; and one—the use of turboelectric drive—was tried out on a fairly large scale.

During World War I, the collier Jupiter (later converted to the aircraft carrier Langley) was fitted with turboelectric drive. The high speed turbines drove generators which were electrically connected to low speed motors. The "big five" battleships—the Maryland, the Colorado, the West Virginia, the California, and the Tennessee—were all built with turboelectric drive. Ultimately, however, starting with the modernization of the Navy in 1934, the turbo-electric drive gave way to the geared-turbine

drive; and today there are relatively few ships of the Navy that have turboelectric drive.

The period just before, during, and after World War II saw increasing improvement and refinement of the geared-turbine propulsion plant. One of the most notable developments of this period was the increase in operating steam pressures—from 400 psi to 600 psi and finally, on some ships, to 1200 psi. Other improvements included reduction in the size and weight of machinery and the use of a variety of new alloys for high pressure and high temperature service.

Although the development of naval surface ships, unlike the development of submarines, has been largely dependent upon the development of steam machinery, we should not overlook the importance of an alternate line of work—namely, the development of internal combustion engines. In the application of diesel engines to ship propulsion, Europe was considerably more advanced than the United States; as late as 1932, in fact, the United States was in the embarrassing position of having to buy German plans for diesel submarine engines. A concerted effort was made during the 1930's

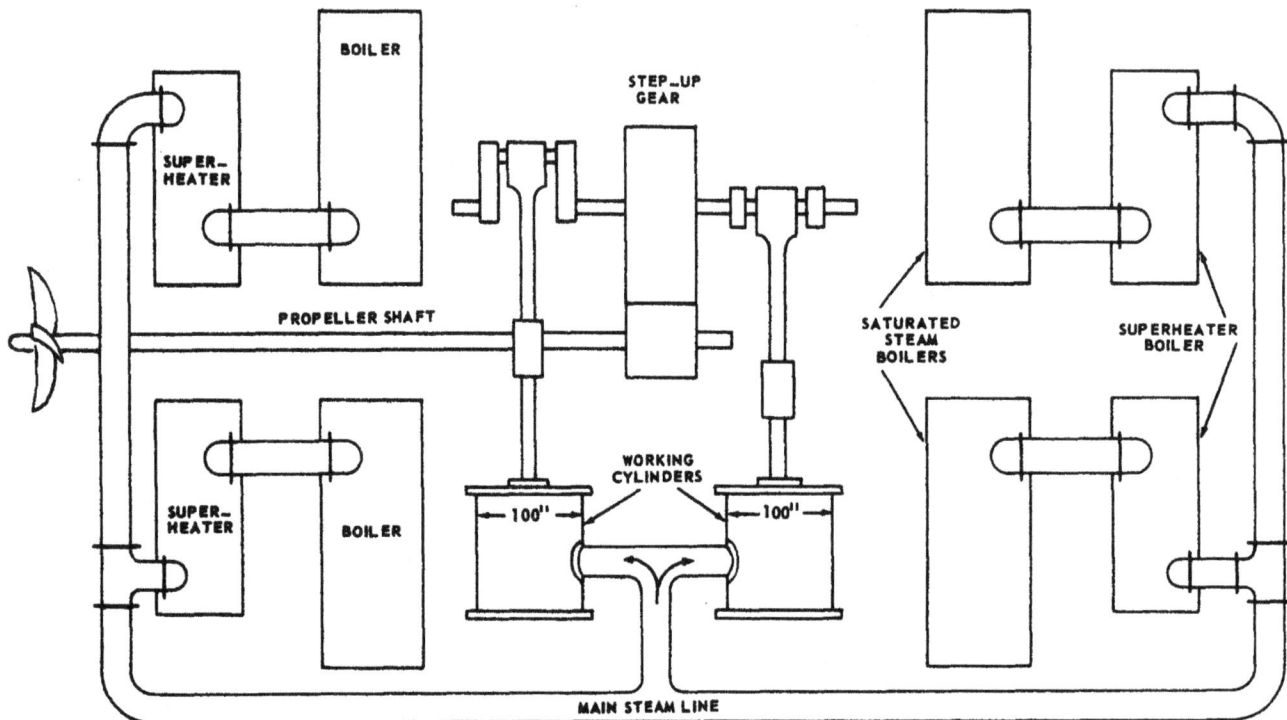

147.5

Figure 1-5.—Propulsion plant of the Wampanog.

147.6

Figure 1-6.—The nuclear surface fleet, 1965: USS Long Beach, CGN 9 (top); USS Enterprise, CVAN 65 (middle); and USS Bainbridge, DLGN 25 (bottom).

to develop an American diesel industry, and the U.S. Naval Engineering Experiment Station (now the Marine Engineering Laboratory) at Annapolis undertook the testing and evaluation of prototype diesel engines developed by American manufacturers. The success of this effort may be seen in the fact that by the end of World War II the diesel horsepower installed in naval vessels exceeded the total horsepower of naval steam plants.

Since World War II, the gas turbine engine has come into increasing prominence as a possible prime mover for naval ships. It is considered likely that the next few years will see enormously increased application of the gas turbine engine for ship propulsion, either singly or in combination with steam turbines or diesel engines.

One of the most dramatic events in the entire history of naval ships is the application of nuclear power, first to submarines and then to surface ships. The first three ships of our nuclear surface fleet are shown in figure 1-6. The middle ship is the aircraft carrier USS Enterprise, CVAN 65; on the flight deck, crew members are shown forming Einstein's famous equation which is the basis of controlled nuclear power. The other two ships are (top) the guided missile cruiser USS Long Beach, CGN 9, and (bottom) the guided missile frigate USS Bainbridge, DLGN 25. The fourth nuclear surface ship to join the fleet was the USS Truxton, DLGN 35. A fifth nuclear surface ship soon to join the fleet is the USS Nimitz, CVAN 68. Although the full implications of nuclear propulsive power may not yet be fully realized,

11

one thing is already clear: the nuclear-powered ship is virtually free of the limitations on steaming radius that apply to ships using other forms of fuel. Because of this one fact alone, the future of nuclear propulsive power seems assured.

THE DEVELOPMENT OF SUBMARINES

Although ancient history records numerous attempts of varying degrees of success to build underwater craft and devices, the first successful submersible craft—and certainly the first to be used as an offensive weapon in naval warfare—was the Turtle, a one-man submersible invented by David Bushnell during the American Revolutionary War. The Turtle, which was propelled by a hand-operated screw propeller, attempted to sink a British man-of-war in New York Harbor. The plan was to attach a charge of gunpowder to the ship's bottom with screws and to explode it with a time fuse. After repeated failures to force the screws through the copper sheathing of the hull of the British ship, the submarine gave up, released the charge, and withdrew. The powder exploded without any result except to cause the British man-of-war to shift to a berth farther out to sea.

closed automatically when the water reached a certain level.

In 1798, Robert Fulton built a small submersible which he called the Nautilus. This vessel, which is shown in figure 1-8, had an overall length of 20 feet and a beam of 5 feet. The craft was designed to carry three people and to stay submerged for about an hour. The first Nautilus carried sails for surface propulsion and a hand-driven screw propeller for submerged propulsion. The periscope had not yet been invented, but Fulton's craft had a modified form of conning tower which had a porthole for underwater observation. In 1801, Fulton tried to interest France, Britain, and America in his idea, but no nation was willing to sponsor the development of the craft, even though this was the best submarine that had yet been designed.

Interest in the development of the submarine was great during the period of the Civil War, but progress was limited by the lack of a suitable means of propulsion. Steam propulsion was attempted, but it had many drawbacks, and hand propulsion was obviously of limited value. The first successful steam-driven submarine was built in 1880 in England. The submarine had a coal-fired boiler and a retractable smokestack.

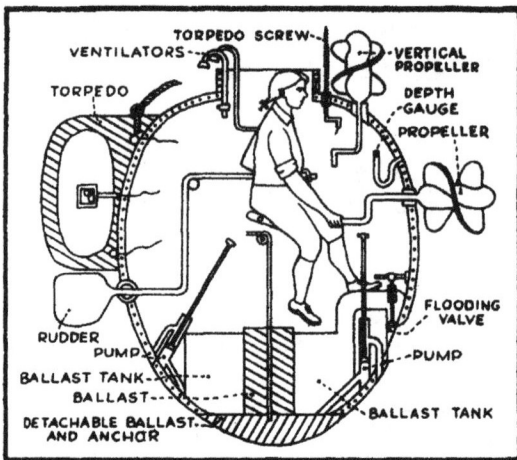

110.104

Figure 1-7.—The Turtle—the first submersible used in naval warfare.

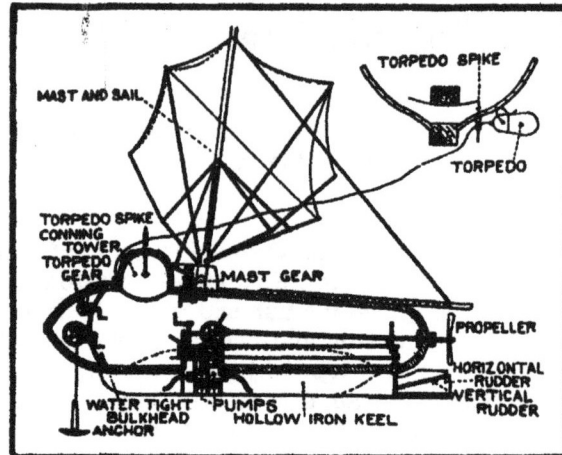

147.7

Figure 1-8.—The first Nautilus (Robert Fulton, 1798).

The Turtle, shown in figure 1-7, looked somewhat like a lemon standing on end. The vessel had a water ballast system with hand-operated pumps, as well as the hand-operated propeller. It also had a crude arrangement for drawing in fresh air from the surface. The vent pipes even

In 1886, an all-electric submarine was built by two Englishmen, Campbell and Ash. Their boat was propelled at a surface speed of 6 knots by two 50-horsepower electric motors operated from a 100-cell storage battery. However, this craft suffered from one major defect: its batteries had to be recharged and overhauled at such short intervals that its effective range never exceeded 80 miles.

In 1875, in New Jersey, John P. Holland built his first submarine. Twenty-five years and nine boats later, Holland finally built the U.S. Navy's first submarine, the USS Holland (fig. 1-9). Although Holland's early models had features which were later discontinued, many of his initial ideas, perfected in practice, are still in use today. The Holland had a length of 54 feet and a displacement of 75 tons. A 50-horsepower gasoline engine provided power for surface propulsion and for battery charging; electric motors run from the storage batteries provided power for underwater running.

Just before Holland delivered his first submarine to the Navy, his company was reorganized into the Electric Boat Company, which continued to be the chief supplier of U.S. Navy submarines until 1917. After the acceptance of the Holland, new contracts for submarines came rapidly. The A-boats, of which there were seven, were completed in 1903. These were improved versions of the Holland; they were 67 feet long and were equipped with gasoline engines and electric motors. This propulsion combination persisted through a series of B-boats, C-boats, and D-boats turned out by the Electric Boat Company.

The E-boat type of submarine was the first to use diesel engines. Diesel engines eliminated much of the physical discomfort that had been caused by fumes and exhaust gases of the old gasoline engines. The K-boats, L-boats, and O-boats of World War I were all driven by diesel engines.

There was little that was spectacular about submarine development in the United States between 1918 and 1941. The submarines built just before and during World War II ranged from 300 to 320 feet in length and displaced approximately 1500 tons on the surface. These included such famous classes as Balao, Gato, Tambor, Sargo, Salmon, Perch, and Pike.

In the latter part of World War II, the Germans adopted a radical change in submarine design known as the "schnorkel." The spelling was reduced to "snorkel" by the Americans and to "snort" by the British. The snorkel is a breathing tube which is raised while the submarine is at periscope depth. With the snorkel in the raised position, air for the diesel engines can be obtained from the surface.

The snorkel was developed and improved by the U.S. Navy at the end of World War II and was installed on a number of submarines. Another post-war development was the Guppy submarine. The Guppy (Greater Underwater Propulsion Power) was a conversion of the fleet-type submarine of World War II. The main change was in the superstructure of the hull; this was changed by reducing the surface area, streamlining every protruding object, and enclosing the periscope shears in a streamlined metal fairing.

With the advent of nuclear power, a new era of submarine development has begun. The first nuclear submarine was the USS Nautilus, SSN 571, which was commissioned on 30 September 1954. At 1100 on 17 January 1955, the Nautilus sent its historic message: "Underway on nuclear power."

The Nautilus broke all existing records for speed and submerged endurance, but even these

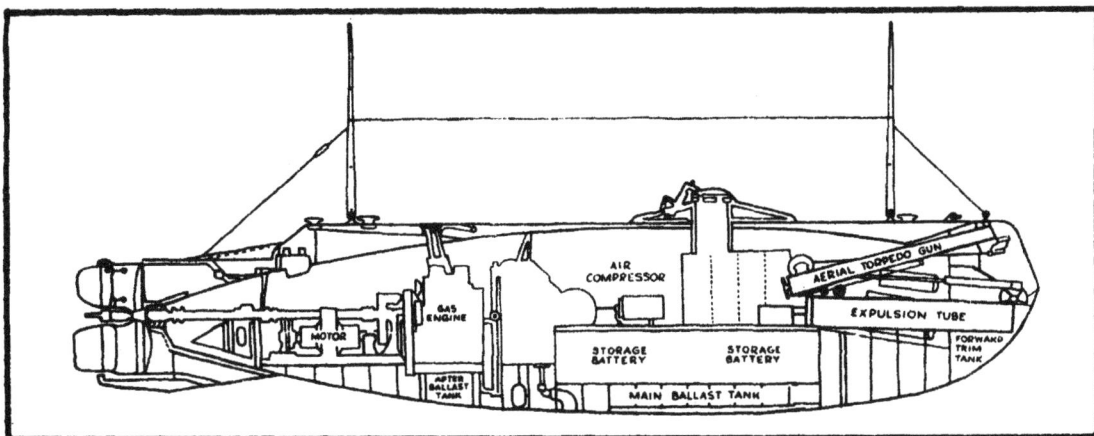

Figure 1-9.—The USS Holland; first U.S. Navy submarine.   71.1

records were soon broken by subsequent generations of nuclear-powered submarines. The modern nuclear-powered submarine is sometimes considered "the first true submarine" because it is capable of staying submerged almost indefinitely.

With the development of the modern missile firing submarine—a nuclear-powered submarine which is capable of submerged firing of the Polaris/Poseidon missile—the submarine has become one of the most vital links in our national defense. The modern nuclear-powered missile firing submarine is a far cry from David Bushnell's hand-propelled Turtle, but an identical need led to the development of both types of vessels: the need for better and more effective fighting ships.

CHAPTER 2

SHIP DESIGN AND CONSTRUCTION

As ships have increased in size and complexity, plans for building them have become more detailed and more numerous. Today only meticulously detailed plans and well conceived organization, from the designers to the men working in the shops and on the ways, can produce the ships required for the Navy.

After intensive research, many technical advances have been adopted in the design and construction of warships. These changes were brought about by the development of welding techniques, by the rapid development in aircraft, submarines, and weapons, and by developments in electronics and in propulsion plants.

This chapter presents information concerning basic ship design considerations, ship flotation, basic ship structure, ship compartmentation, and the geometry of the ship.

BASIC DESIGN CONSIDERATIONS

Combat efficiency is the prime requisite of warships. Some important factors contributing to combat efficiency are sea-keeping capabilities, maneuverability, and ability to remain in action after sustaining combat damage.

Basic considerations involved in the design of naval ships include the following:

1. Cost.—The initial cost is important in warship design, but it is not the only cost consideration. The cost of maintenance and operation, as well as the cost and availability of the required manning, are equally important considerations.

2. Life Expectancy.—The life expectancy of a ship is limited by ordinary deterioration in service and also by the possibility of obsolescence due to the design of more efficient ships.

3. Service.—The service to be performed substantially affects the design of any ship.

4. Port Facilities.—The port facilities available in the normal operating zone of the ship affect the design to some extent. Dockyard facilities available for drydocking and maintenance work must be taken into consideration.

5. Prime Mover.—The type of propelling machinery to be used must be considered from the point of view of the required speed of the ship, the location in the ship, space and weight requirements, and the effect of the machinery on the center of gravity of the ship.

6. Special Considerations.—Special considerations such as the fuel required, the crew to be carried, and special weapons are factors which restrict the designer of a naval ship.

Naval ships are designed for maximum simplicity that is compatible with the requirements of service. Naval ships are designed as simply as possible in order to lower building and operating costs and in order to ensure greater availability of construction facilities.

The following operating considerations affect the size of a naval ship:

1. Width and Length of Canal Locks and Dock Facilities. These considerations obviously have an effect on the size of ships that must use the canals or dock facilities.

2. Effect of Speed.—For large ships, speed may be maintained with a smaller fraction of displacement devoted to propulsion machinery than is the case for smaller ships. Also, large ships lose proportionately less speed through adverse sea conditions.

3. Effect of Radius of Action.—An increasing cruising radius may be obtained by increasing displacement without increasing the fraction of displacement allotted to fuel and stores. If the fraction of displacement set aside for fuel and stores is increased, some other weight must be decreased. Since the hull

weight is a constant percentage of the displacement, an increase in the fraction of the displacement assigned to one military characteristic involves the reduction of other fractions of displacement. By increasing the displacement of the ship as a whole, it is possible to increase the speed and the radius of action without adversely affecting the other required characteristics.

4. <u>Effect of Seagoing Capabilities.</u>—Larger ships are more seaworthy than smaller ships. However, smaller ships are more maneuverable because they have smaller turning circle radii than larger ships, where all other factors are proportional. The maneuverability of large ships may be increased somewhat by the use of improved steering gear and large rudders.

In general, larger ships have the advantage of greater protection because of their greater displacement. From the point of view of underwater attack, larger ships also have an advantage. If compartments are of the same size, the number of compartments increases linearly with the displacement. It is apparent, then, that protection against both surface and subsurface attacks may be more effective on larger ships without impairing other military characteristics.

Many compromises must be made in designing any ship, since action which improves one feature may degrade another. For example, in the design of a conventionally powered ship there is the problem of choosing the hull line for optimum performance at a cruising speed of 20 knots and at a trial speed of 30 knots or more. One may select a hull type which would minimize resistance at top speed, and thus keep the weight of propulsion machinery to a minimum. When this is done, however, resistance at cruising speed may be high and the fuel load for a given endurance may be relatively great, thus nullifying some of the gain from a light machinery plant. On the other hand, one may choose a hull type favoring cruising power. In this case, fuel load will be lighter but the shaft horsepower required to make trial speed may be greater than before. Now the machinery plant is heavier, cancelling some of the weight gain realized from the lighter fuel load. A compromise based on the interrelationship of these considerations must usually be adopted. The need for compromise is always present, and the manner in which it is made has an important bearing on the final design of any naval ship.

SHIP FLOTATION

When a body floats in still water, the force which supports the body must be equal to the weight. Assume that an object of given volume is placed under water. If the weight of this object is greater than the weight of an equal volume of water, the object will sink. It sinks because the force which buoys it up is less than its own weight. However, if the weight of the submerged object is less than the weight of an equal volume of water, the object will rise. It rises because the force which buoys it up is greater than its own weight. The object will continue to rise until part of it is above the surface of the water. Here it floats at such a depth that the submerged part of the object displaces a volume of water, the weight of which is equal to the weight of the object.

The principle implied in this discussion is known as Archimedes' law: <u>the weight of a floating body is equal to the weight of the fluid displaced.</u>

The cube of steel shown in part A of figure 2-1 is a solid cube of the dimensions shown. If this cube is dropped into salt water, it will sink because it weighs approximately 490 pounds and the weight of the salt water it displaces is approximately 64 pounds. If the cube is hammered out into a watertight flat plate of the dimensions shown in part B of figure 2-1, with the edges bent up one foot all around, the box thus formed will float. This box could be made from the same volume of steel as that of the cube. The box will not only float; in calm water, it will carry an additional 1800 pounds of weight before sinking. As a box, the metal displaces a greater volume than the same amount of metal does as a 1-foot cube.

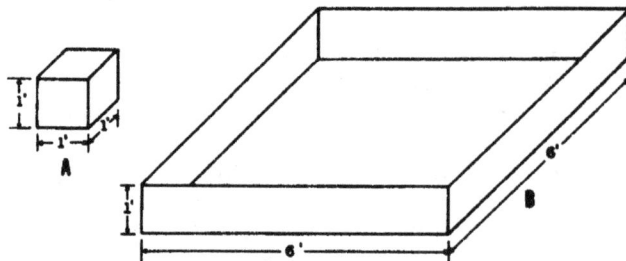

8.44

Figure 2-1.—Steel cube and box made from same volume of steel.

BASIC SHIP STRUCTURE

In considering the structure of a ship, it is common practice to liken the ship to a box girder. Like a box girder, a ship may be subjected to tremendous stresses. The magnitude of stress is usually expressed in pounds per square inch (psi).

When a pull is exerted on each end of a bar, as in part A of figure 2-2, the bar is under the type of stress called tension. When a pressure is exerted on each end of a bar, as in part B of figure 2-2, the bar is under the type of stress called compression. If an equal but opposite pull is exerted on the upper and lower bars, as shown in part C of figure 2-2, the pins connecting these bars are subjected to a stress at right angles to their length. This stress is called shear. When a shaft, bar, or other material is subjected to a twisting motion, the resulting stress is known as torsional stress. Torsional stress is not illustrated in figure 2-2.

When a material is compressed, it is shortened. When it is subjected to tension, it is lengthened. This change in shape is called strain. The change of shape (strain) may be regarded as an effect of stress.

If a simple beam is supported at its two ends and various vertical loads are applied over the center of the span, the beam will

bend (fig. 2-3). As the beam bends, the upper section of the beam compresses and the lower part stretches. Somewhere between the top and bottom of the beam, there is a section which is neither in compression nor in tension; this is known as the neutral axis. The greatest stresses in tension and compression occur near the middle of the length of the beam, where the loads are applied.

LONGITUDINAL BENDING AND STRESSES

In an I-beam, the greater mass of structural material is placed in the upper and lower flanges to resist compression and tension. Relatively little material is placed in the web which holds the two flanges so that they can work together; the web, being near the neutral axis, is less subject to tension and compression stresses than are the flanges. The web does take care of shearing stresses, which are sizeable near the supports.

A ship in a seaway can be considered similar to this I-beam (or, more correctly, it can be likened to a box girder) with supports and distributed loads. The supports are the buoyant forces of the waves; the loads are the weight of the ship's structure and the weight of everything contained within the ship.

The ship shown in figure 2-4 is supported by waves, with the bow and stern each riding a crest and the midship region in the trough. This ship will bend with compression at the top and tension at the bottom. A ship in this condition is said to be sagging. In a sagging ship, the weather deck tends to buckle under compressive stress and the bottom plating tends to stretch under tensile stress. A sagging ship is undergoing longitudinal bending—that is, it is bending in a fore-and-aft direction.

When the ship advances half a wave length, so that the crest is amidships and the bow and stern are over troughs, as shown in figure 2-5,

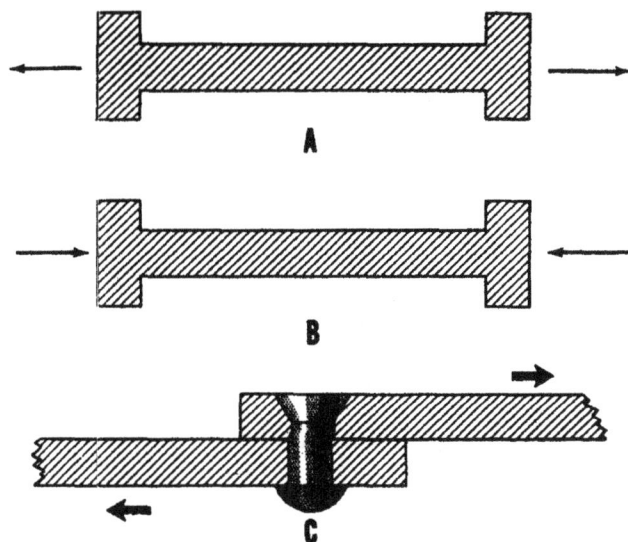

147.8
Figure 2-2.—Stresses in metal: (A) tension; (B) compression; (C) shear.

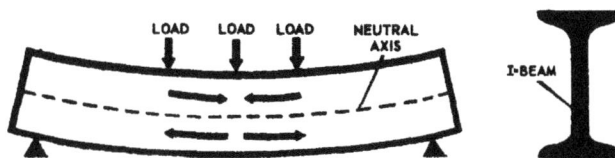

147.9
Figure 2-3.—I-beam with load placed over center.

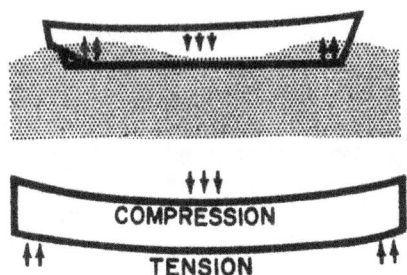

147.10

Figure 2-4.—Sagging.

the stresses are reversed. The weather deck is now in tension and the bottom plating is in compression. A ship in this condition is said to be hogging. Hogging, like sagging, is a form of longitudinal bending. The effects of longitudinal bending must be considered in the design of the ship, with particular reference to the overall strength that the ship must have.

In structural design, the terms hull girder and ship girder are used to designate the structural parts of the hull. The structural parts of the hull are those parts which contribute to its strength as a girder and provide what is known as longitudinal strength. Structural parts include the framing (transverse and longitudinal), the shellplating, the decks, and the longitudinal bulkheads. These major strength members enable the ship girder to resist the various stresses to which it is subjected.

The ship girder is subjected to rapid reversal of stresses when the ship is in a seaway and is changing from a hogging condition to a sagging condition (and vice versa), since these changes occur in the short time required for the wave to advance half a wave length. Other dynamic stresses are caused by pressure loads forward due to the ship's motion ahead, by panting[1] of forward plating due to variations of pressure, by the thrust of the propeller, and by the rolling of the ship.

Transverse stress results from the pressure of the water on the ship's sides which subjects the transverse framing, deck beams, and shellplating below water to a hydrostatic load. Local stresses occur in the vicinity of masts, windlasses, winches, and heavy weights. These areas are strengthened by thicker deck plating or by deeper or reinforced deck beams.

[1]Panting is a small in-and-out working of the plating at the bow.

HULL MEMBERS

The principal strength members of the ship girder are at the top and bottom, where the greatest stresses occur. The top flange includes the main deck plating, the deck stringers, and the sheer strakes of the side plating. The bottom flange includes the keel, the outer bottom plating, the inner bottom plating, and any continuous longitudinals in way of the bottom. The side webs of the ship girder are composed of the side plating, aided to some extent by any long, continuous fore-and-aft bulkheads. Some of the strength members of a destroyer hull girder are indicated in figure 2-6.

Keel

The keel is a very important structural member of the ship. The keel, shown in figure 2-7, is built up of plates and angles into an I-beam shape. The lower flange of this I-beam structure is the flat keel plate, which forms the center strake of the bottom plating.[2]

The web of the I-beam is a solid plate which is called the vertical keel. The upper flange is called the rider plate; this forms the center strake of the inner bottom plating. An inner vertical keel of two or more sections, consisting of I-beams arranged one on top of the other, is found on many large combatant ships.

Framing

Frames used in ship construction may be of various shapes. Figure 2-8 illustrates frames of the angle, I-beam, tee, bulb angle, and channel shapes. Figure 2-9 shows two types of

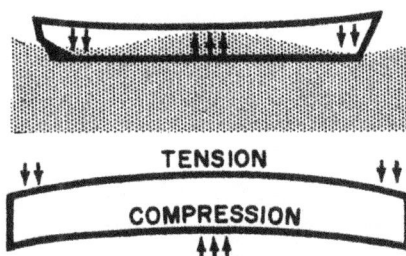

147.11

Figure 2-5.—Hogging.

[2]On large ships, an additional member is attached to this flange to serve as the center strake.

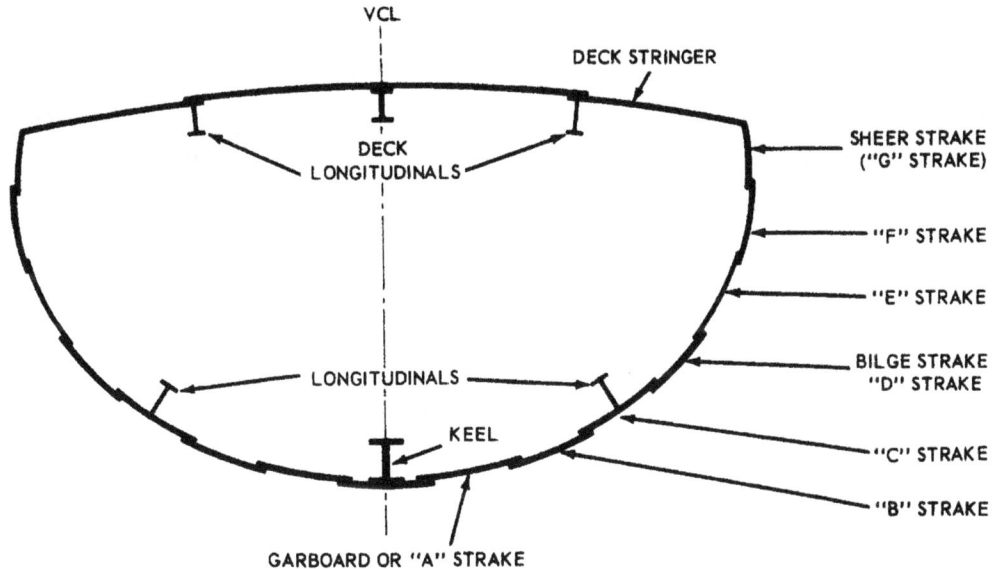

147.12

Figure 2-6.—Destroyer hull girder, showing some strength members.

built-up frames, one of welded construction and the other of riveted construction.

Frames are strength members. They act as integral parts of the ship girder when the ship is exposed to longitudinal or transverse stresses. Frames stiffen the plating and keep it from bulging or buckling. They act as girders between bulkheads, decks, and double bottoms, and transmit forces exerted by load weights and water pressures. The frames also support the inner and outer shell locally and protect against unusual forces such as those caused by underwater explosions. As may be inferred, frames are called upon to perform a variety of functions, depending upon the location of the frames in the ship. Figure 2-10 shows a web frame used in wing tank construction.

There are two important systems of framing in current use: the transverse system and the longitudinal system. The transverse system provides for continuous transverse frames with the longitudinals intercostal between them. Transverse frames are closely spaced and a small number of longitudinals are used. The longitudinal system of framing consists of closely spaced longitudinals which are continuous along the length of the ship, with transverse frames intercostal between the longitudinals.

Transverse frames are attached to the keel and extend from the keel outward around the turn of the bilge and up to the edge of the main deck. They are closely spaced along the length of the ship, and they define the form of the ship.

Longitudinals (fig. 2-11) run parallel to the keel along the bottom, bilge, and side plating. The longitudinals provide longitudinal strength, stiffen the shellplating, and tie the transverse frames and the bulkheads together. The longitudinals in the bottom (called side keelsons) are of the built-up type.

Where two sets of frames intersect, one set must be cut to allow for the other set. The frames which are cut, and thereby weakened, are known as intercostal frames; those which continue through are called continuous frames.

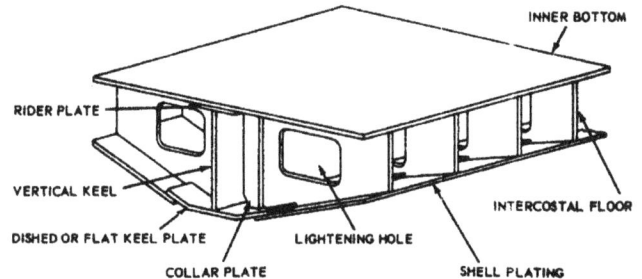

147.13

Figure 2-7.—One type of keel structure.

19

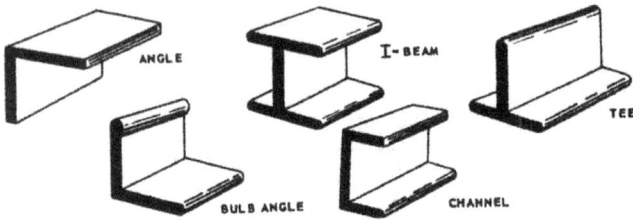

11.30(147)A

Figure 2-8.—Angle, I-beam, tee, bulb
angle, and channel frames.

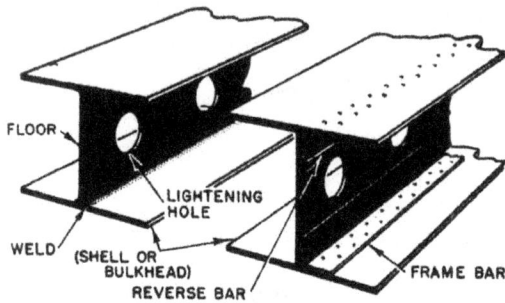

11.30(147)B

Figure 2-9.—Built-up frames.

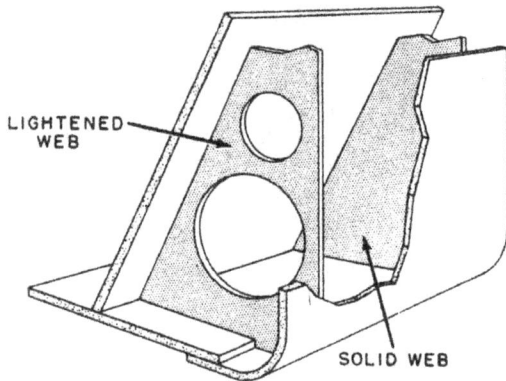

147.14

Figure 2-10.—Web frame used in wing
tank construction.

Both intercostal and continuous frames are
shown in figure 2-12.

A cellular form of framing results from a
combination of longitudinal and transverse
framing systems utilizing closely spaced deep
framing. Cellular framing is used on most
naval ships.

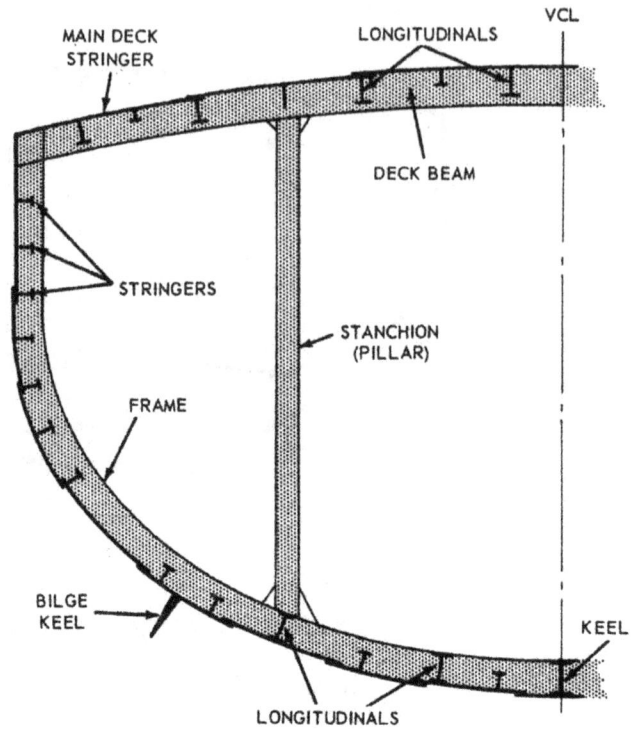

147.15

Figure 2-11.—Basic frame section
(longitudinal framing).

In the bottom framing, which is probably
the strongest part of a ship's structure, the
floors and keelsons are integrated into a rigid
cellular construction (fig. 2-13). Heavy loads
such as the ship's propulsion machinery are
bolted to foundations which are built directly
on top of the bottom framing (fig. 2-14).

Double Bottom

In many naval ships, the inner bottom plating is a watertight covering laid on top of the
bottom framing. The shellplating, framing,
and inner bottom plating form the space known
as the double bottom. This space may be used
for stowage of fresh water or fuel oil or it
may be used for ballasting.

The inner bottom plating is a second skin
inside the bottom of the ship. It prevents
flooding in the event of damage to the outer
bottom, and it also acts as a strength member.

Stem and Bow Structure

The stem assembly, which is the forward
member of the ship's structure, varies in form

20

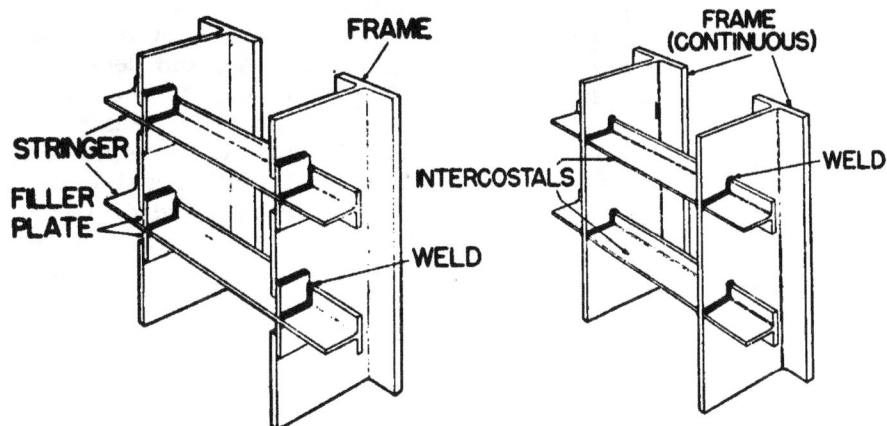

147.16

Figure 2-12.—Intercostal and continuous frames.

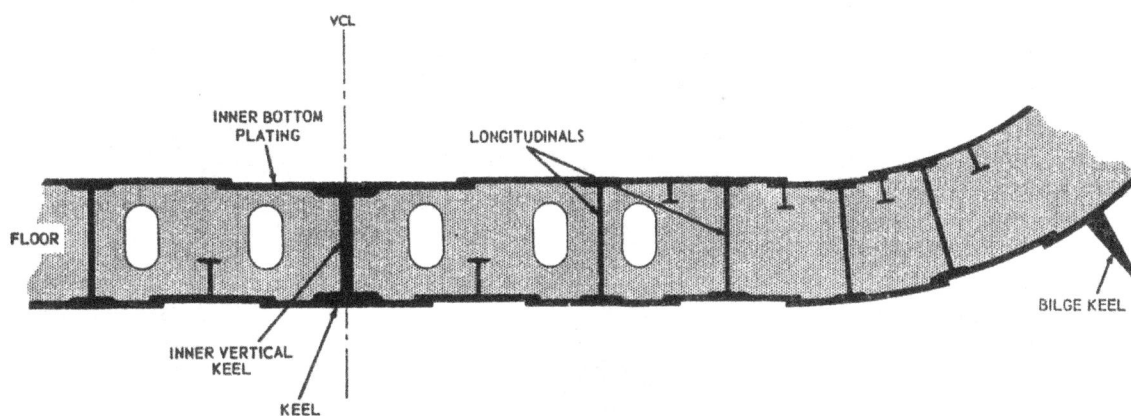

147.17

Figure 2-13.—Bottom structure.

from one type of ship to another. The external shape shown in figure 2-15 is commonly used on combatant ships. This form is essentially bulbous at the forefoot, tapering to a sharp entrance near the waterline and again widening above the waterline. Figure 2-16 shows the relationship between the stem assembly and the keel. Internally, the stem assembly has a heavy centerline member which is called the stem post (not illustrated). The stem post is recessed[3] along its after edge to receive the

shellplating, so that the outside presents a smooth surface to cut through the water. The keel structure is securely fastened to the lower end of the stem by welding. The stem maintains the continuity of the keel strength up to the main deck. The decks support the stem at various intermediate points along the stem structure between the keel and the decks.

Triangular plates known as breast hooks are fitted parallel to and between the decks or side stringers in the bow for the purpose of rigidly fastening together the peak frames, the stem, and the outside plating.

---

[3] This recess is called a rabbet.

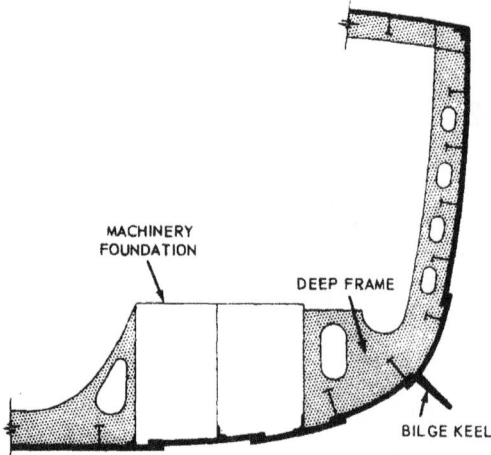

Figure 2-14.—Deep floor assembly for machinery foundations.

147.18

Stern Structure

The aftermost section of the ship's structure is the stern post, which is rigidly secured to the keel, shellplating, and decks. On single-screw ships, the stern post is constructed to accommodate the propeller shaft and rudder stock bosses. Because of its intricate form, the stern post is usually either a steel casting or a combination of castings and forgings. In modern warships having transom sterns, multiple screws, and twin rudders, the stern post as such is difficult to define, since it has been replaced by an equivalent structure of deep framing. This structure (fig. 2-17) consists of both longitudinal and transverse framing that extends throughout the width of the bottom in the vicinity of the stern. In order to withstand the static and dynamic loads imposed by the rudders, the stern structure is strengthened in the vicinity of the rudder post by a structure known as the rudder post weldment.

147.19

Figure 2-15.—Bulbous-bow configuration.

Plating

The outer bottom and side plating forms a strong, watertight shell. Shellplating consists of approximately rectangular steel plates arranged longitudinally in rows or courses called strakes. The strakes are lettered, beginning with the A strake (also called the garboard strake) which is just outboard of the keel and working up to the uppermost side strake (called the sheer strake).

The end joint formed by adjoining plates in a strake is called a butt. The joint between the edges of adjoining strakes is called a seam. Butts and seams in side plating are illustrated in figure 2-18.

Since the hull structure is composed of a great many individual pieces, the strength and tightness of the ship as a whole depend very much upon the strength and tightness of the connections between the individual pieces. In modern naval ships, welded joints are used to a

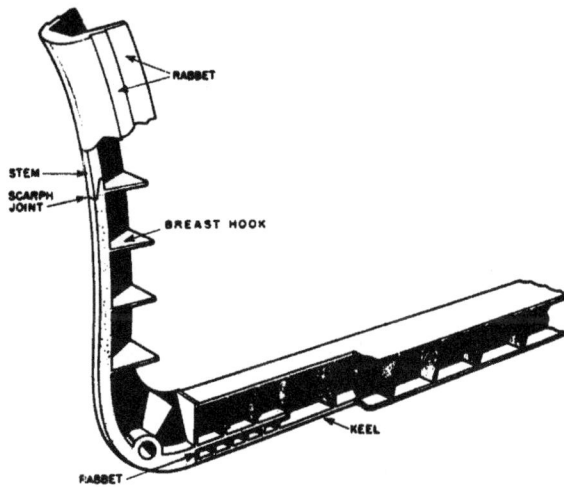

147.20

Figure 2-16.—Relationship of stem assembly to keel.

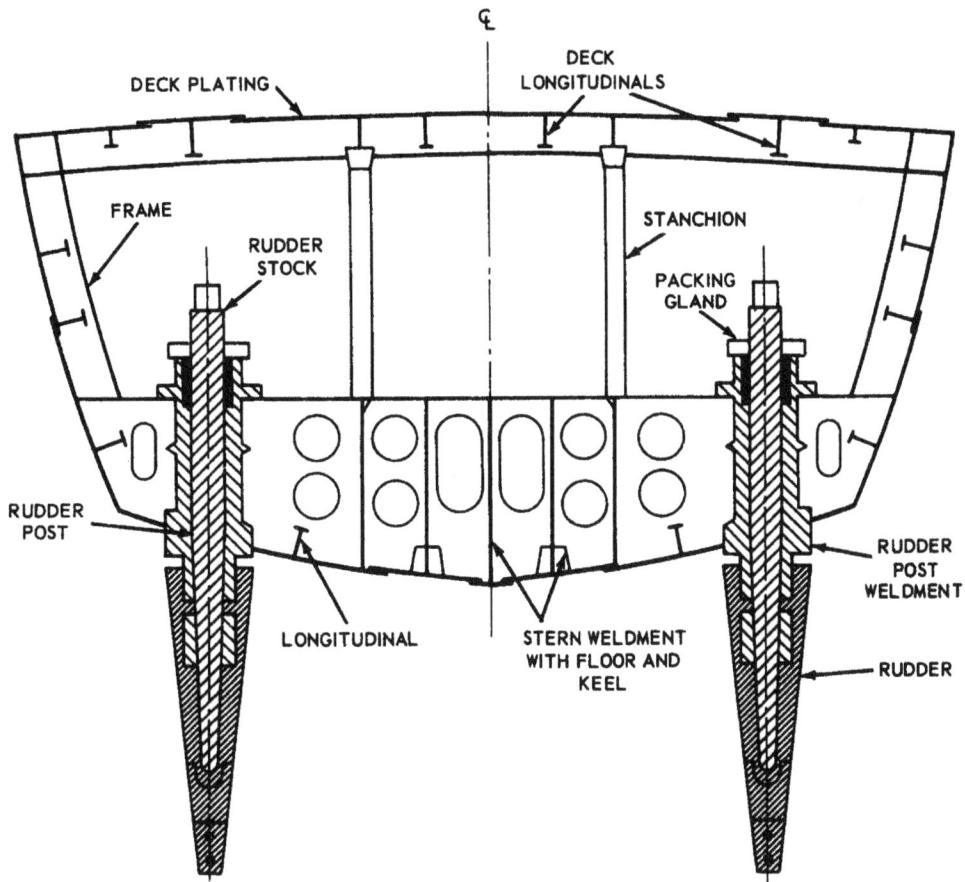

147.21

Figure 2-17.—Stern structure.

23

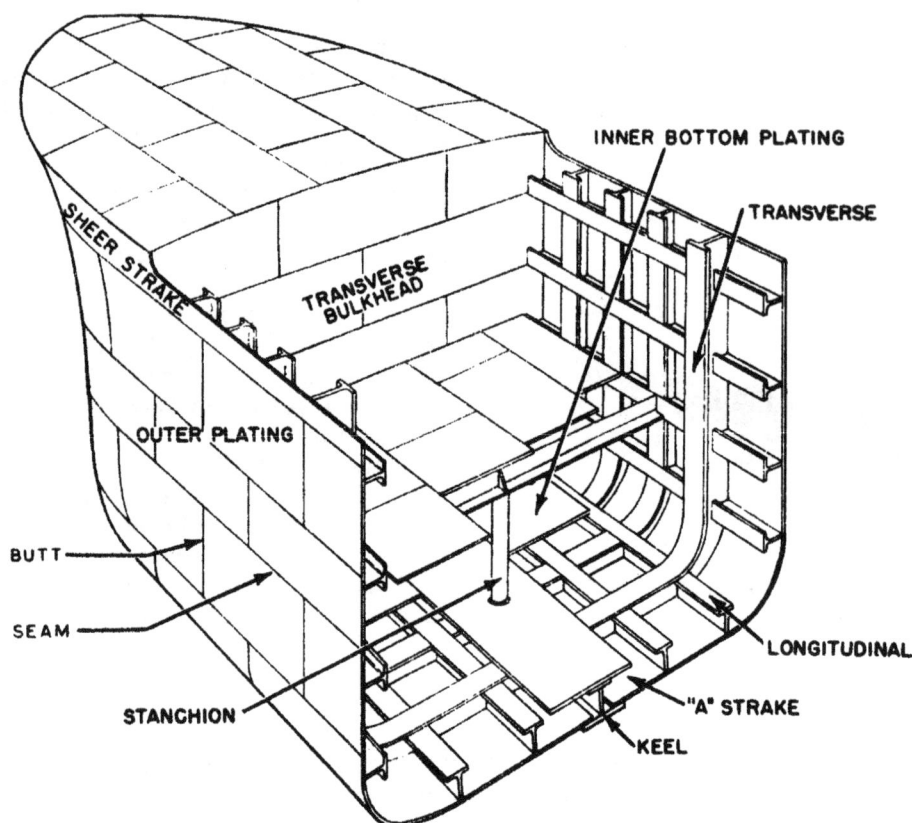

3.92

Figure 2-18.—Section of ship, showing plating and framing.

very great extent. However, riveted joints are still used for some applications.

Bilge Keels

Bilge keels, which may be seen in figures 2-11, 2-13, and 2-14, are fitted in practically all ships at the turn of the bilge. The bilge keels extend fifty to seventy-five percent of the length of the hull. A bilge keel usually consists of a plate about 12 inches deep, standing at right angles to the shellplating and secured to the shellplating by double angles. On more recent ships, bilge keels consist of two plates forming a Vee shape welded to the hull and on large ships may extend out from the hull nearly three feet. Bilge keels serve to reduce the extent of the ship's rolling.

Decks

Decks provide both longitudinal and transverse strength to the ship. Deck plates, which are similar to the plates used in side and bottom shellplating, are supported by deck beams and deck longitudinals.

The term strength deck is generally applied to the deck which acts as the top flange of the hull girder. It is the highest continuous deck— usually the main or weather deck. However, the term strength deck may be applied to any continuous deck which carries some of the longitudinal load. On destroyers and similar ships in which the main deck is the only continuous high deck, the main deck is the strength deck. The flight deck is the strength deck on recent large aircraft carriers (CVAs) and helicopter support ships (LPH), but the main or hangar deck is the strength deck on older types of carriers.

The main deck is supported by deck beams and deck longitudinals. Deck beams are the transverse members of the framing structure. The beams are attached to and supported by the frames at the sides, as shown in figure 2-19.

In most naval construction, light deck beams are interspaced at regular intervals with deep deck beams. Deck longitudinals are used to provide longitudinal strength. When possible, the heaviest longitudinals are located at the center and near the outboard edges.

The outboard strake of deck plating which connects with the shellplating is called the deck stringer (fig. 2-11). The deck stringer, which is heavier than the other deck strakes, serves as a continuous longitudinal stringer, providing longitudinal strength to the ship's structure.

Upper Decks and Superstructure

The decks above the main deck are not strength decks on most ships other than CVAs. The upper decks are usually interrupted at intervals by expansion joints. The expansion joints keep the upper decks from acting as strength decks (which they are not designed to be) and thus prevent cracking and buckling of deck houses and superstructure.

Stanchions

In order to reinforce the deck beams and to keep the deck beam brackets and side frames from carrying the total load, vertical stanchions or columns are fitted between decks. Stanchions are constructed in various ways of various materials. Some are made of pipe or rods; others are built up of various plates and shapes, welded or riveted together. The stanchion shown in figure 2-20 is in fairly common use; this pipe stanchion consists of a steel tube which is fitted with special pieces for securing it at the upper end (head) and at the lower end (heel).

Bulkheads

Bulkheads are the vertical partitions which, extending athwartships and fore and aft, provide compartmentation to the interior of the ship. Bulkheads may be either structural or nonstructural. Structural bulkheads, which tie the shell-plating, framing, and decks together, are capable of withstanding fluid pressure; these bulkheads usually provide watertight compartmentation. Nonstructural bulkheads are lighter; they are used chiefly for separating activities aboard ship.

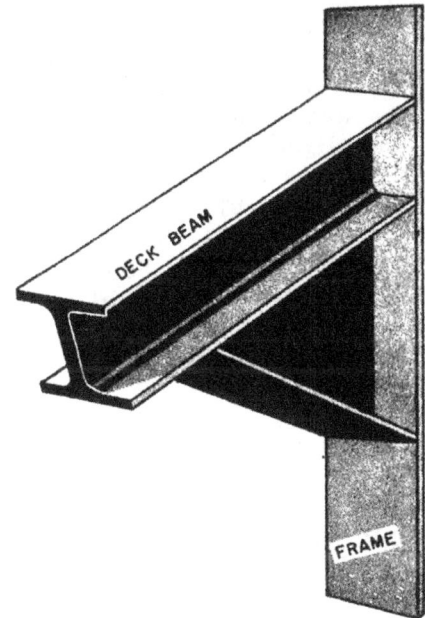

147.22
Figure 2-19.—Deck beam and frame.

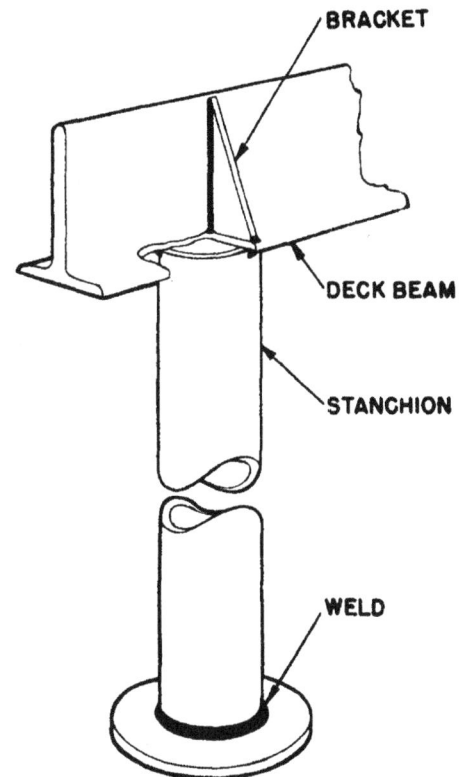

147.23
Figure 2-20.—Pipe stanchion.

Bulkheads consist of plating and reinforcing beams. The reinforcing beams are known as bulkhead stiffeners. Two types of bulkhead stiffeners are shown in figure 2-21. Bulkhead stiffeners are usually placed in the vertical plane and aligned with deck longitudinals; the stiffeners are secured at top and bottom to any intermediate deck by brackets attached to deck plating. The size of the stiffeners depends upon their spacing, the height of the bulkhead, and the hydrostatic pressure which the bulkhead is designed to withstand.

Bulkheads and bulkhead stiffeners must be strong enough to resist excessive bending or bulging in case of flooding in the compartments which they bound. If too much deflection takes place, some of the seams might fail.

In order to form watertight boundaries, structural bulkheads must be joined to all decks, shellplating, bulkheads, and other structural members with which they come in contact. Main transverse bulkheads extend continuously through the watertight volume of the ship, from the keel to the main deck, and serve as flooding boundaries in the event of damage below the waterline.

In general, naval ships are divided into as many watertight compartments, both above and below the waterline, as are compatible with the missions and functions of the ships. The compartmentation provided by transverse and longitudinal bulkheads is illustrated in the bow section shown in figure 2-22.

SHIP COMPARTMENTATION

Every space in a naval ship (except for minor spaces such as peacoat lockers, linen lockers, cleaning gear lockers, etc.) is considered as a compartment and is assigned an identifying letter-number symbol. This symbol is marked on a label plate secured to the door, hatch, or bulkhead of the compartment.

There are two systems of numbering compartments, one for ships built prior to March 1949 and the other for ships built after March 1949. In both of these systems, compartments on the port side end in an even number and those on the starboard side end in an odd number. In both systems, a zero precedes the deck number for all levels above the main deck.

Figure 2-23 illustrates both systems of numbering decks. The older system identifies decks by the numbers 100, 200, 300, etc., with the number 900 always being used for the double

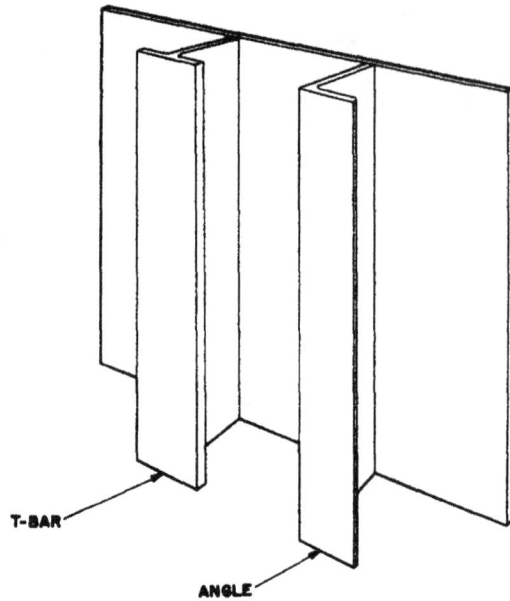

11.30(147)C

Figure 2-21.—Bulkhead stiffeners.

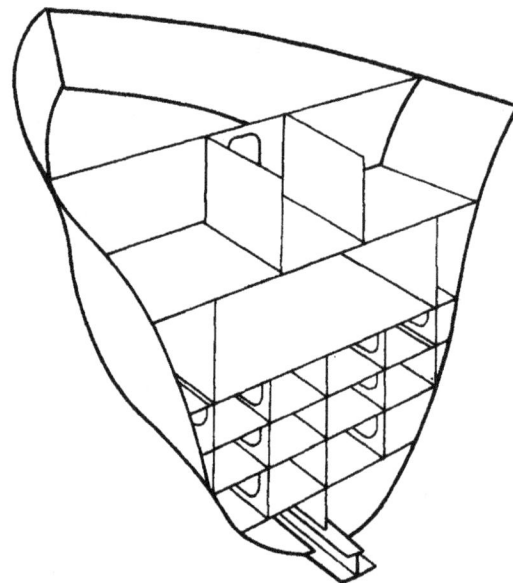

147.24

Figure 2-22.—Compartmentation provided by transverse and longitudinal bulkheads.

bottoms. In the newer system, decks are identified as 1, 2, 3, 4, etc., and the double bottoms are given whatever number falls to them.

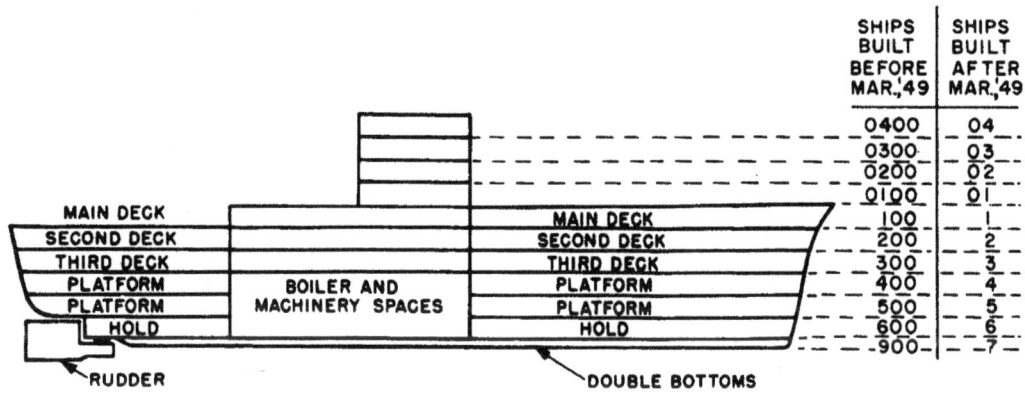

		SHIPS BUILT BEFORE MAR. '49	SHIPS BUILT AFTER MAR. '49
		0400	04
		0300	03
		0200	02
		0100	01
MAIN DECK	MAIN DECK	100	1
SECOND DECK	SECOND DECK	200	2
THIRD DECK	THIRD DECK	300	3
PLATFORM	PLATFORM	400	4
PLATFORM	PLATFORM	500	5
HOLD	HOLD	600	6
		900	7

BOILER AND MACHINERY SPACES

RUDDER   DOUBLE BOTTOMS

3.106

Figure 2-23.—Deck symbols for naval ships.

SHIPS BUILT BEFORE MARCH 1949

For ships built prior to March 1949, the first letter of the identifying symbol is A, B, or C, and indicates the section of the ship in which the compartment is located. The A section extends from the bow of the ship aft to the forward bulkhead of the engineering spaces. The B section includes the engineering spaces, while the C section extends from the after bulkhead of the engineering spaces aft to the stern. The divisions of the ship are indicated in figure 2-24. The lower half of the diagram shows the numbering of compartments, beginning at the forward end of each section. The even numbers are on the port side and the odd numbers are on the starboard side.

After the division letter, the deck designation comes next in the symbol. Main deck compartments are indicated by numbers from 101 to 199.

Second deck compartments run from 201 through 299, third deck compartments form a 300 series, etc. A zero preceding the number indicates a location above the main deck. The double bottoms always form the 900 series on any ship built before March 1949, regardless of the number of decks above.

The use of the compartments is indicated by the following letters:

A—Supply and storage
C—Control
E—Machinery
F—Fuel
L—Living quarters
M—Ammunition
T—Trunks and passages
V—Voids
W—Water

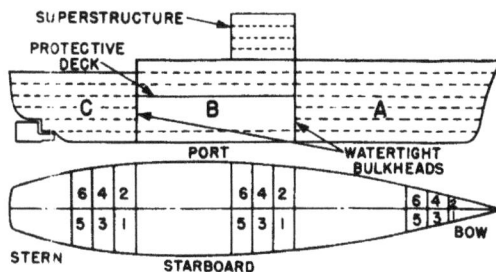

3.107

Figure 2-24.—Divisions of a ship built prior to March 1949.

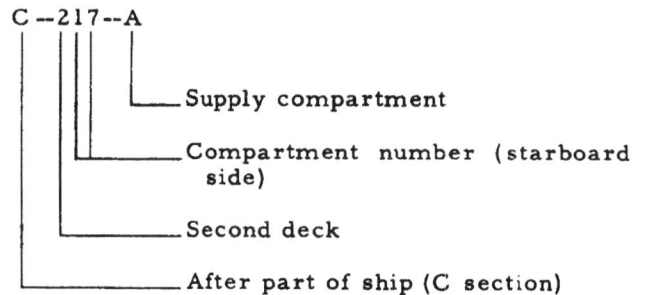

C —217—A

Supply compartment

Compartment number (starboard side)

Second deck

After part of ship (C section)

147.25

Figure 2-25.—Example of compartment symbol on ship built prior to March 1949.

27

Letter	Type of compartment	Examples
A	Stowage spaces	Storerooms; issue rooms; refrigerated compartments.
AA	Cargo holds	Cargo holds and cargo refrigerated compartments.
C	Control centers for ship and fire-control operations (normally manned).	CIC room; plotting rooms, communication centers; radio, radar, and sonar operating spaces; pilot house.
E	Engineering control centers (normally manned).	Main propulsion spaces; boiler rooms; evaporator rooms; steering gear rooms; auxiliary machinery spaces; pumprooms; generator rooms; switchboard rooms; windlass rooms.
F	Oil stowage compartments (for use by ship).	Fuel-oil, diesel-oil, lubricating-oil, and fog-oil compartments.
FF	Oil stowage compartments (cargo).	Compartments carrying various types of oil as cargo.
G	Gasoline stowage compartments (use by ship).	Gasoline tanks, cofferdams, trunks, and pumprooms.
GG	Gasoline stowage compartments (cargo).	Gasoline compartments for carrying gasoline as cargo.
K	Chemicals and dangerous materials (other than oil and gasoline).	Chemicals, semisafe materials, and dangerous materials carried for ship's use or as cargo.
L	Living spaces	Berthing and messing spaces; staterooms, washrooms, heads, brigs; sickbays, hospital spaces; and passageways.
M	Ammunition spaces	Magazines; handling rooms; turrets; gun mounts; shell rooms; ready service rooms; clipping rooms.
Q	Miscellaneous spaces not covered by other letters.	Shops; offices; laundry; galley; pantries; unmanned engineering, electrical, and electronic spaces.
T	Vertical access trunks	Escape trunks or tubes.
V	Void compartments	Cofferdam compartments (other than gasoline); void wing compartments; wiring trunks.
W	Water compartments	Drainage tanks; fresh water tanks; peak tanks; reserve feed tanks.

147.26

Figure 2-26.—Compartment letters for ships built after March 1949.

An example of a compartment symbol on a ship built prior to March 1949 is given in figure 2-25.

SHIPS BUILT AFTER MARCH 1949

For ships constructed after March 1949, the compartment numbers consist of a deck number, frame number, relation to centerline of ship, and letter showing use of the compartment. These are separated by dashes. The A, B, C divisional system is not used.

The main deck is always numbered 1. The first deck or horizontal division below the main deck is numbered 2; the second below is numbered 3; etc., consecutively for subsequent lower division boundaries. Where a compartment extends down to the bottom of the ship, the number assigned the bottom compartments is used. The first horizontal division above the main deck is numbered 01, the second above is numbered 02, etc., consecutively, for subsequent upper divisions. The deck number becomes the first part of the compartment number and indicates the vertical position within the ship.

The frame number at the foremost bulkhead of the enclosing boundary of a compartment is its frame location number. Where these forward boundaries are between frames, the frame number forward is used. Fractional numbers are not used. The frame number becomes the second part of the compartment number.

Compartments located so that the centerline of the ship passes through them carry the number 0. Compartments located completely to starboard of the centerline are given odd numbers and those completely to port of the centerline are given even numbers. Where two or more compartments have the same deck and frame number and are entirely to port or entirely to starboard of the centerline, they have consecutively higher odd or even numbers, as the case may be, numbering from the centerline outboard. In this case, the first compartment outboard of the centerline to starboard is 1; the second is 3, etc. Similarly, the first compartment outboard of the centerline to port is 2; the second 4, etc. When the centerline of the ship passes through more than one compartment, the compartment having that portion of the forward bulkhead through which the centerline of the ship passes carries the number 0, and the others carry the numbers 01, 02, 03, etc., in any sequence found desirable. These numbers indicate the relation to the centerline,

and are the third part of the compartment number.

The fourth and last part of the compartment number is the capital letter which identifies the assigned primary usage of the compartment. A single capital letter is used, except that on dry and liquid cargo ships a double letter designation is used to identify compartments assigned to cargo carrying. The compartment letters for ships built after March 1949 are shown in figure 2-26. An example of a compartment symbol on a ship built after March 1949 is given in figure 2-27.

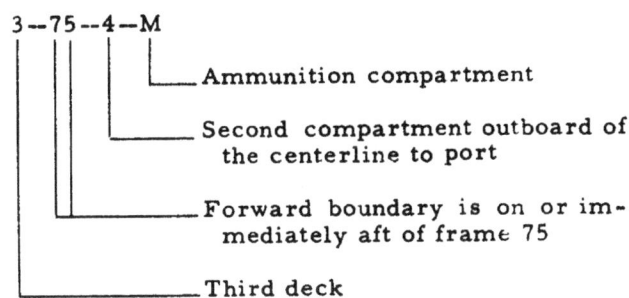

3—75—4—M

Ammunition compartment

Second compartment outboard of the centerline to port

Forward boundary is on or immediately aft of frame 75

Third deck

147.27
Figure 2-27.—Example of compartment symbol on ship built after March 1949.

GEOMETRY OF THE SHIP

Since a ship's hull is a three-dimensional object having length, breadth, and depth, and since the hull has curved surfaces in each

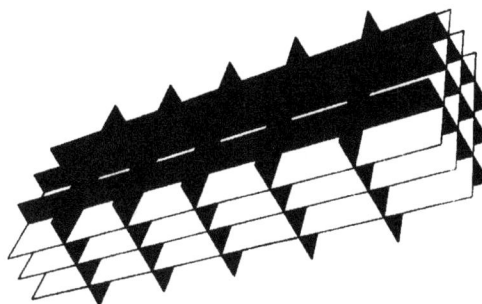

23.191
Figure 2-28.—Transverse, horizontal, and vertical planes.

dimension, no single drawing of a ship can give an accurate and complete representation of the lines of the hull. In naval architecture, a hull shape is shown by means of a lines drawing (sometimes referred to merely as the lines of the ship). The lines drawing consists of three views or projections—a body plan, a half-breadth plan, and a sheer plan—which are obtained by cutting the hull by transverse, horizontal, and vertical planes (fig. 2-28). The use of these three planes to produce the three projections is illustrated in figure 2-29.

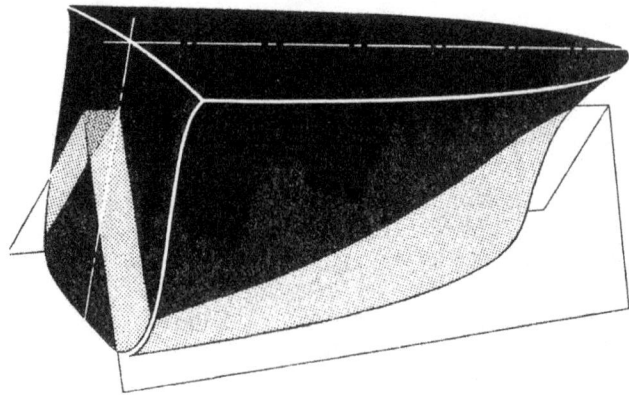

23.195

Figure 2-30.—Diagonal plane.

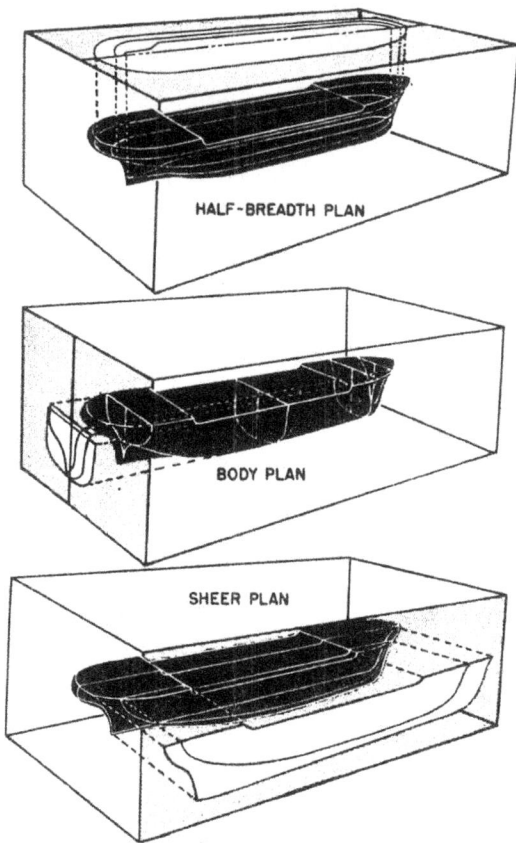

147.28

Figure 2-29.—Half-breadth plan, body plan, and sheer plan.

In addition to using transverse, horizontal, and vertical planes, ship designers frequently use a set of planes known as diagonals. A diagonal plane is illustrated in figure 2-30. As a rule, three diagonals are used; these are identified as diagonal A, diagonal B, and diagonal C. Diagonals are frequently shown as projections on the body plan and on the half-breadth plan.

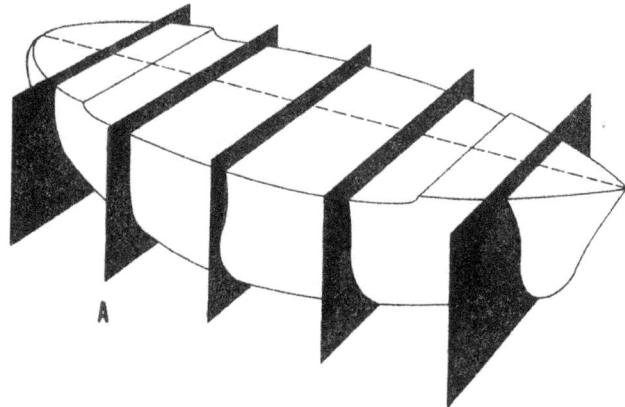

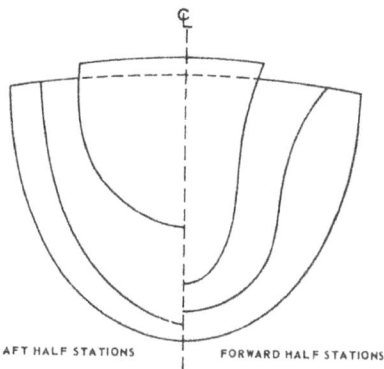

23.192

Figure 2-31.—Transverse planes and body plan.

30

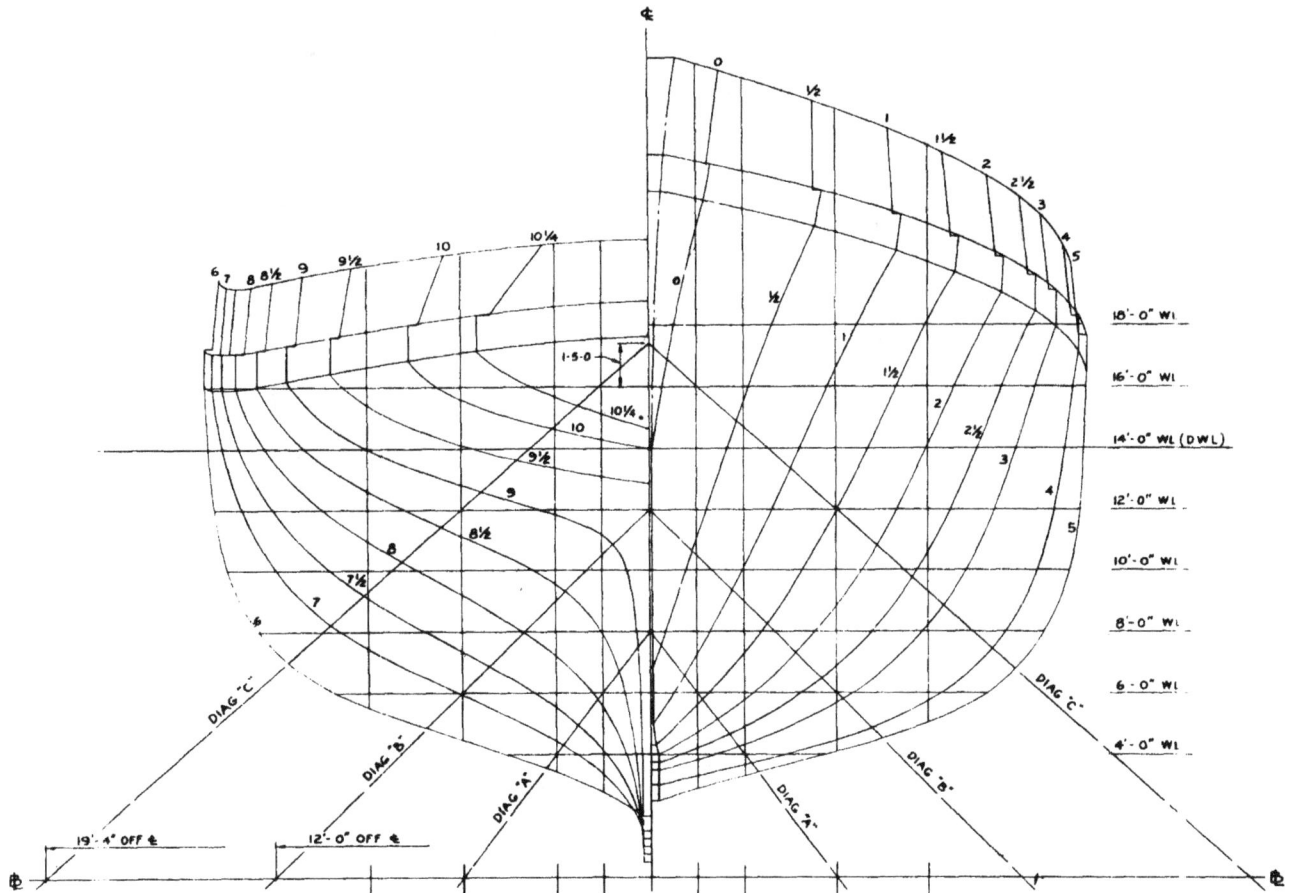

147.29

Figure 2-32.—Body plan of a YTB.

BODY PLAN

To visualize the projection known as the body plan, we must imagine the ship's hull cut transversely in several places, as shown in part A of figure 2-31. The shape of a transverse plane intersection of the hull is obtained at each cut; when the resulting curves are projected onto the body plan (part B of fig. 2-31) they show the changing shape of transverse sections of the hull.

Since the hull is symmetrical about the centerline of the ship, only one-half of each curve obtained by a transverse cut is shown on the body plan. Each half curve is called a half station. The right-hand side of the body plan shows the forward half stations—that is, the half stations resulting from transverse cuts between the bow and the middle of the ship. The left-hand side of the body plan shows the aft half stations—that is, the half stations

resulting from transverse cuts between the stern and the middle of the ship.

As may be inferred, a station is a complete curve such as would be obtained if each transverse cut were projected completely, rather than as half a curve, onto the body plan. The stations are numbered from forward to aft, dividing the hull into equally spaced transverse sections. The station where the forward end of the designer's waterline[4] and the stem contour intersect is known as the forward perpendicular, or station O. The station at the intersection of the stern contour and the designer's waterline is known as the after perpendicular. The station midway between the

---

[4]The waterline at which the ship is designed to float is known as the designer's waterline.

31

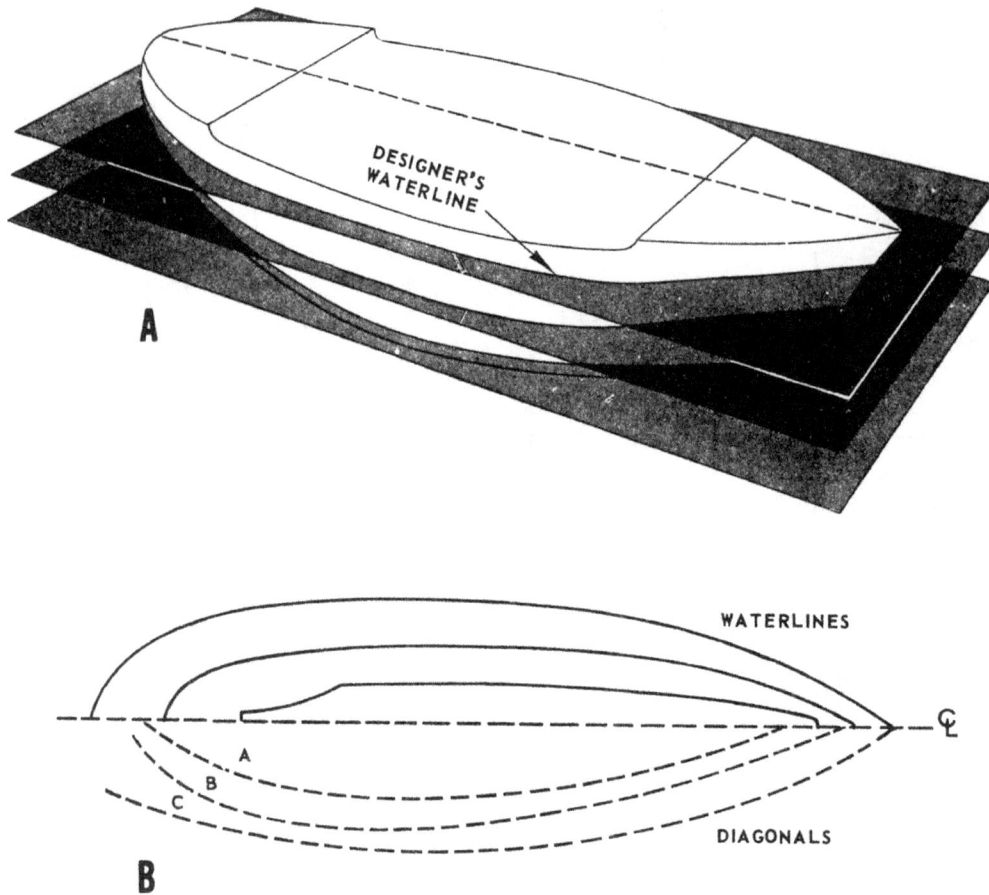

23.194

Figure 2-33.—Horizontal planes and half-breadth plan.

forward and after perpendiculars is known as the middle perpendicular.

An actual body plan for a YTB is shown in figure 2-32. Note the projection of the diagonals on this body plan.

HALF-BREADTH PLAN

To visualize the half-breadth plan, we must imagine the ship's hull cut horizontally in several places, as shown in part A of figure 2-33. The cuts are designated as waterlines, although the ship could not possibly float at many of these lines. The base plane which serves as the point of origin for waterlines is usually the horizontal plane that coincides with the top of the flat keel. Waterlines are designated according to their distance above the base plane; for example, we may have a

6-foot waterline, an 8-foot waterline, a 10-foot waterline, and so forth.

The waterlines are projected onto the half-breadth plan, as shown in part B of figure 2-33. Since the hull is symmetrical, only half of the waterlines are shown in the half-breadth plan. Diagonals are frequently shown on the other half of the half-breadth plan.

SHEER PLAN

To visualize the sheer plan, we must imagine the ship's hull cut vertically in several places, as shown in part A of figure 2-34. The resultant curves, known as buttocks or as bow and buttock lines, are projected onto the sheer plan, as shown in part B of figure 2-34. The centerline plane is designated as zero buttock. The other buttocks are designated according to their

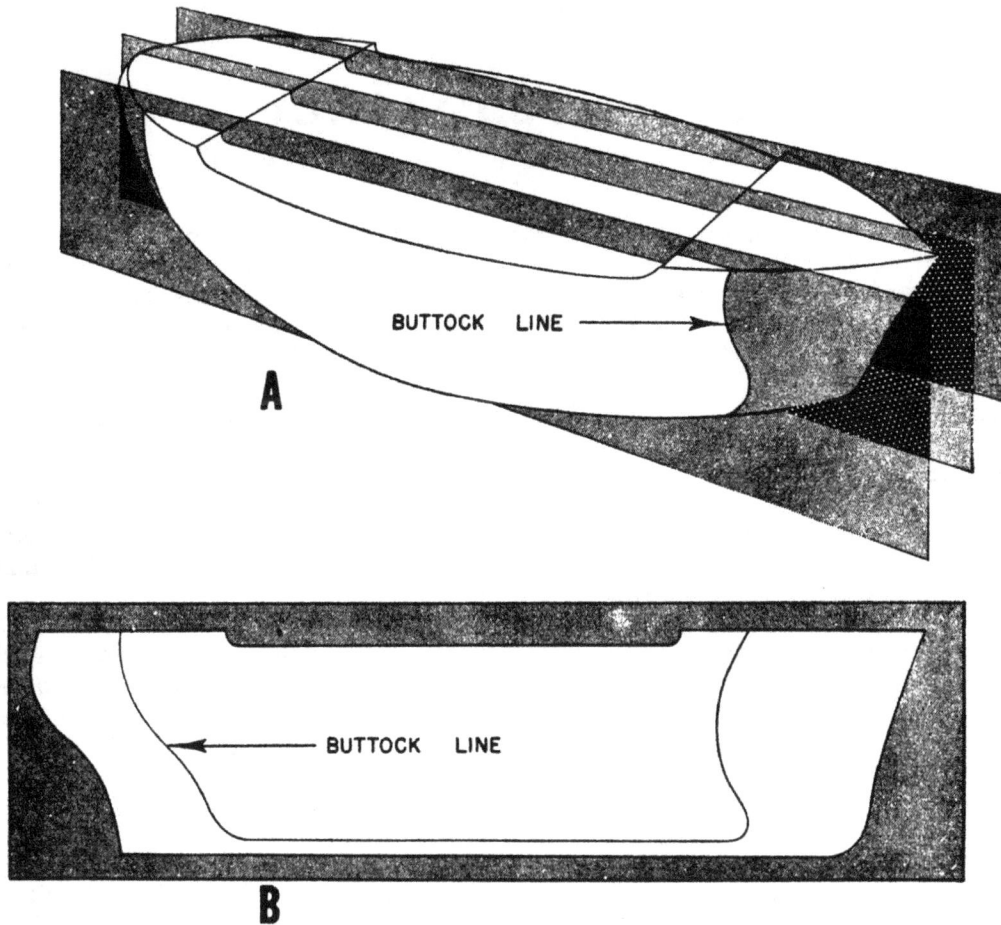

Figure 2-34.—Vertical planes and sheer plan.

23.193

distance from zero buttock. The spacing of the vertical cuts is chosen to show the contours of the forward and after quarters of the ship.

CHAPTER 3

STABILITY AND BUOYANCY

This chapter deals with the principles of stability, stability curves, the inclining experiment, effects of weight shifts and weight changes, effects of loose water, longitudinal stability and effects of trim, and causes of impaired stability. The damage control aspects of stability are discussed in chapter 4 of this text.

PRINCIPLES OF STABILITY

A floating body is acted upon by forces of gravity and forces of buoyancy. The algebraic sum of these forces must equal zero if equilibrium is to exist.

Any object exists in one of three states of stability: stable, neutral, or unstable. We may illustrate these three states by placing three cones on a table top, as shown in figure 3-1. When cone A is tipped so that its base is off the horizontal plane, it tends, up to a certain angle of inclination, to assume its original position again. Cone A is thus an example of a stable body—that is, one which tries to attain its original position through a specified range of angles of inclination.

Cone B is an example of neutral stability. When rotated, this cone may come to rest at any point, reaching equilibrium at some angle of inclination.

Cone C, balanced upon its apex, is an example of an unstable body. Following any slight inclination by an external force, the body will come to rest in a new position where it will be more stable.

From Archimedes' law, we know that an object floating on or submerged in a fluid is buoyed up by a force equal to the weight of the fluid it displaces. The weight (displacement) of a ship depends upon the weight of all parts, equipment, stores, and personnel. This total weight represents the effect of gravitational force. When a ship is floated, she sinks into the water until the weight of the fluid displaced by her underwater volume is equal to the weight of the ship. At this point, the ship is in equilibrium—that is, the forces of gravity (G) and the forces of buoyancy (B) are equal, and the algebraic sum of all forces acting upon the ship is equal to zero. This condition is shown in part A of figure 3-2. If the underwater volume of the ship is not sufficient to displace an amount of fluid equal to the weight of the ship, the ship will sink (part B of fig. 3-2) because the forces of gravity are greater than the forces of buoyancy.

The depth to which a ship will sink when floated in water depends upon the density of the water, since the density affects the weight per unit volume of a fluid. Thus we may expect a ship to have a deeper draft in fresh water than in salt water, since fresh water is less dense (and therefore less buoyant) than salt water.

Although gravitational forces act everywhere upon the ship, it is not necessary to attempt to consider these forces separately. Instead, we may regard the total force of gravity as a single resultant or composite force which acts vertically downward through the ship's center of gravity (G). Similarly, the force of buoyancy may be regarded as a single resultant force which acts vertically upward through the center of buoyancy (B) located at the geometric center of the ship's underwater body. When a ship is at rest in calm water, the center of gravity and the center of buoyancy lie on the same vertical line.

DISPLACEMENT

Since weight (W) is equal to the displacement, it is possible to measure the volume of the underwater body (V) in cubic feet and multiply this volume by the weight of a cubic foot of sea

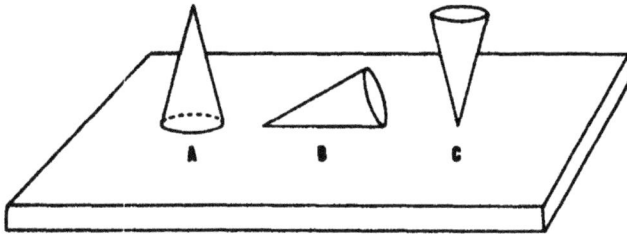

147.30
Figure 3-1.—Three states of stability.

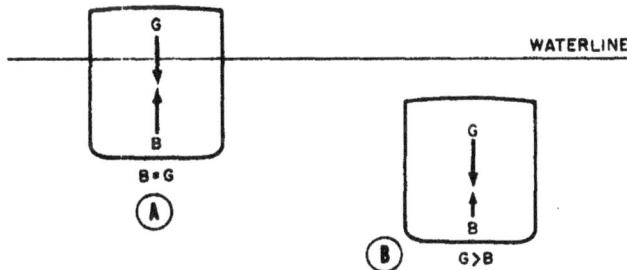

147.31
Figure 3-2.—Interaction of force of gravity and force of buoyancy.

water, in order to find what the ship weighs. This relationship may be written as:

$$(1) \quad W = V \times \frac{1}{35}$$

$$(2) \quad V = 35W$$

where

V = volume of displaced sea water, in cubic feet

W = weight, in tons

35 = cubic feet of sea water per ton (When dealing with ships, it is customary to use the long ton of 2240 pounds.)

It is also obvious, then, that displacement will vary with draft. As the draft increases, the displacement increases. This is indicated in figure 3-3 by a series of displacements shown for successive draft lines on the midship section of a cruiser.

The volume of an underwater body for a given draft line can be measured in the drafting room by using graphic or mathematical means. This is done for a series of drafts throughout the probable range of displacements in which a ship is likely to operate. The values obtained are plotted on a grid on which feet of draft are measured vertically and tons of displacement horizontally. A smooth line is faired through the points plotted, providing a curve of displacement versus draft, or a displacement curve as it is generally called. The result is shown in figure 3-4 for a cruiser.

To use the curve shown in figure 3-4 for finding the displacement when the draft is given, locate the value of the mean draft on the draft scale at left and proceed horizontally across the diagram to the curve. Then drop vertically downward and read the displacement from the scale. For example, if the mean draft is 24 feet, the displacement found from the curve is approximately 14,700 tons.

KB VERSUS DRAFT

As the draft increases, the center of buoyancy (B) rises with respect to the keel (K). Figure 3-5 shows how different drafts result in different values of KB, the height of the center of buoyancy from the keel (K). A series of values for KB is obtained and these values are plotted on a curve to show KB versus draft. Figure 3-6 illustrates a typical KB curve.

To read KB when the draft is known, start at the proper value of draft on the scale at the left and proceed horizontally to the curve. Then drop vertically downward to the baseline (KB).

Thus, if a ship were floating at a mean draft of 19 feet, the KB found from the chart would be approximately 10.5 feet.

WATERLINE	DISPLACEMENT
28 FEET	17,900 TONS
24 FEET	14,800 TONS
20 FEET	11,800 TONS
16 FEET	8,800 TONS
12 FEET	5,900 TONS

8.45
Figure 3-3.—Displacement data.

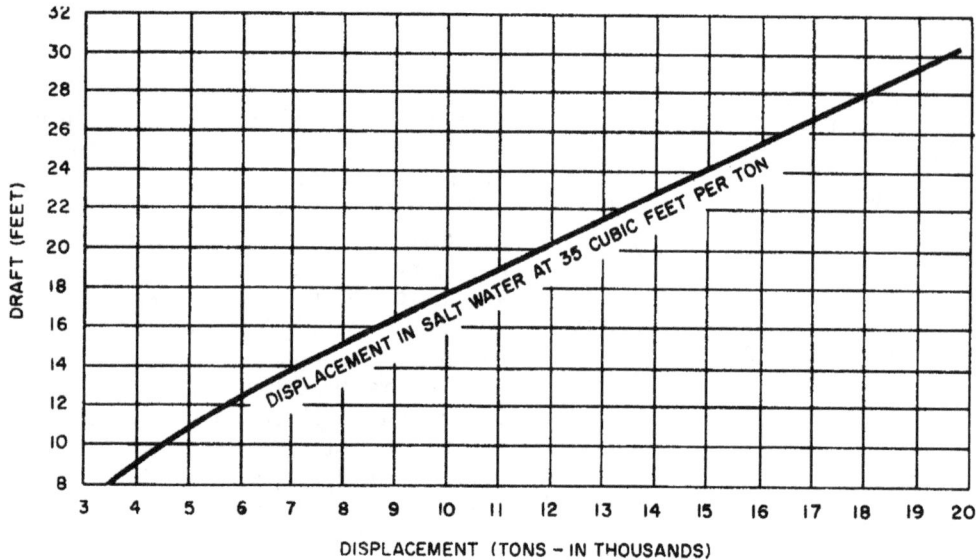

8.46

Figure 3-4.—Displacement curve of a cruiser.

RESERVE BUOYANCY

The volume of the watertight portion of the ship above the waterline is known as the ship's reserve buoyancy. Freeboard, a rough measure of the reserve buoyancy, is the distance in feet from the waterline to the main deck. Freeboard is calculated at the midship section. As indicated in figure 3-7, freeboard plus draft is equal to the depth of the hull in feet.

When weight is added to a ship, draft and displacement increase in the same amount that freeboard and reserve buoyancy decrease. Reserve buoyancy is an important factor in a ship's ability to survive flooding due to damage. It also contributes to the seaworthinesss of the ship in very rough weather.

INCLINING MOMENTS

The moment of a force is the tendency of the force to produce rotation or to move an object about an axis. The distance between the point at which the force is acting and the axis of rotation is called the moment arm or the lever arm of moment.[1] To find the value of a moment, we multiply the magnitude of the force by the distance between the force and the axis of rotation. The magnitude of the force is expressed in some unit of weight (pounds, tons, etc.) and the distance is expressed in some unit of length (inches, feet, etc.); hence the unit of the moment is the foot-pound, the foot-ton, or some similar unit.

When two forces of equal magnitude act in opposite and parallel directions and are separated by a perpendicular distance, they form a couple. The moment of a couple is found by multiplying the magnitude of one of the forces by the perpendicular distance between the lines of action of the two forces.

When a disturbing force exerts an inclining moment on a ship, causing the ship to heel over to some angle, there is a change in the shape of the ship's underwater body and a consequent relocation of the center of buoyancy. Because of this shift in the location of B, B and G no longer act in the same vertical line. Instead of acting as separate equal and opposite forces, B and G now form a couple.

The newly formed couple produces either a righting moment or an upsetting moment, depending upon the relative locations of B and G. The ship illustrated in figure 3-8 develops a

---

[1]The significance of the distance between the force and the axis of rotation may be seen if we consider a simple see-waw. If two persons of equal weight sit on opposite ends, equally distant from the center support, the see-saw balances. But if one person moves closer to or farther away from the center, the person farthest away from the support moves downward because the effect of his weight is greater.

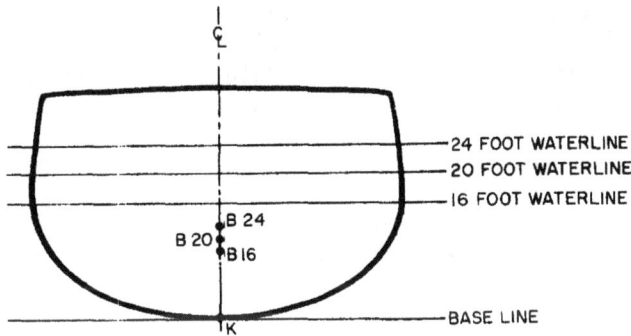

8.50

Figure 3-5.—Successive centers of buoyancy
(B) for different drafts.

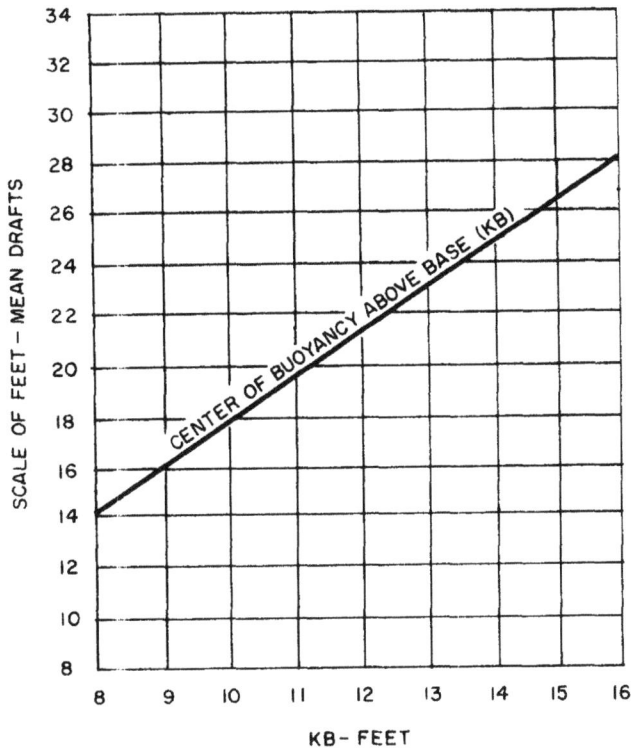

8.51

Figure 3-6.—KB curve.

righting moment, the magnitude of which is
equal to the magnitude of one of the forces (B
or G) times the perpendicular distance (GZ)
which separates the lines of action of the forces.
The distance GZ is known as the righting arm
of the ship. Mathematically,

$$RM = W \times GZ$$

where

RM = righting moment (in foot-tons)
W = displacement (in tons)
GZ = righting arm (in feet)

For example, a ship which displaces 10,000
tons and has a 2-foot righting arm at a certain
angle of inclination has a righting moment of
10,000 tons times 2 feet, or 20,000 foot-tons.
This 20,000 foot-tons represents the moment
which in this instance tends to return the ship
to an upright position.

Figure 3-9 shows the development of an upsetting moment resulting from the inclination
of an unstable ship. In this case, it is apparent
that the high location of G and the new location
of B contribute to the development of an upsetting moment rather than a righting moment.

THE METACENTER (M)

A ship's metacenter is the intersection of
two successive lines of action of the force of
buoyancy as the ship heels through a very small
angle. Figure 3-10 shows two lines of buoyant
force. One of these represents the ship on an
even keel, the other is for a small angle of
heel. The point where they intersect is the
initial position of the metacenter. When the
angle of heel is greater than the angle used to
compute the metacenter, M moves off the centerline and the path of movement is a curve.
However, it is the initial position of the metacenter that is most useful in the study of
stability. In the discussion which follows, the

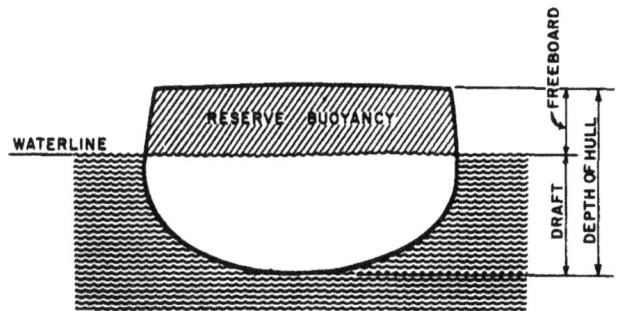

8.47

Figure 3-7.—Reserve buoyancy, freeboard,
draft, and depth of hull.

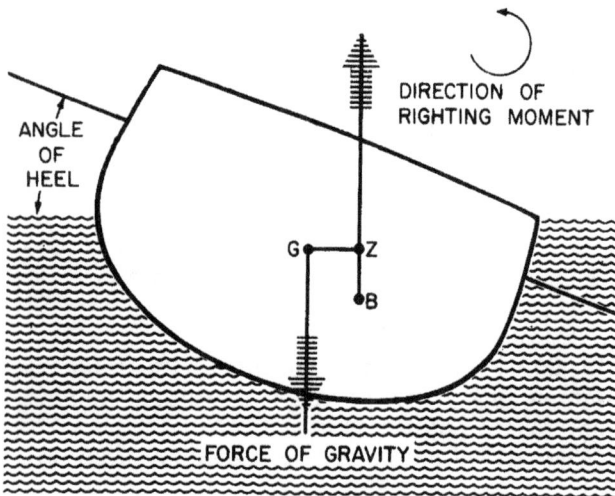

8.52

Figure 3-8.—Development of righting moment when a stable ship inclines.

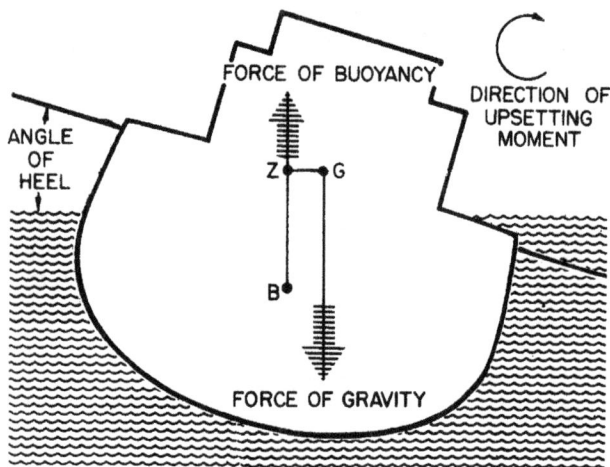

8.53

Figure 3-9.—Development of upsetting moment when unstable ship inclines.

initial position is referred to as M. The distance from the center of buoyancy (B) to the metacenter (M) when the ship is on even keel is the metacentric radius.

METACENTRIC HEIGHT (GM)

The distance from the center of gravity (G) to the metacenter is known as the ship's metacentric height (GM). Figure 3-11 shows a ship heeled through a small angle (the angle

is exaggerated in the drawing), establishing a metacenter at M. The ship's righting arm GZ is one side of the triangle GZM. In this triangle GZM, the angle of heel is at M. The side GM is perpendicular to the waterline at even keel, and ZM is perpendicular to the waterline when the ship is inclined.

It is evident that for any angle of heel not greater than 7°, there will be a definite relationship between GM and GZ because GZ = GM sin θ. Thus, GM acts as a measure of GZ, the righting arm.

GM is also an indication of whether the ship is stable or unstable at small angles of inclination. If M is above G, the metacentric height is positive, the moments which develop when the ship is inclined are righting moments, and the ship is stable (part A of fig. 3-11). But if M is below G, the metacentric height is negative, the moments which develop are upsetting moments, and the ship is unstable (part B of fig. 3-11).

INFLUENCE OF METACENTRIC HEIGHT

When the metacentric height of a ship is large, the righting arms that develop at small angles of heel are also large. Such a ship resists roll and is said to be stiff. When the

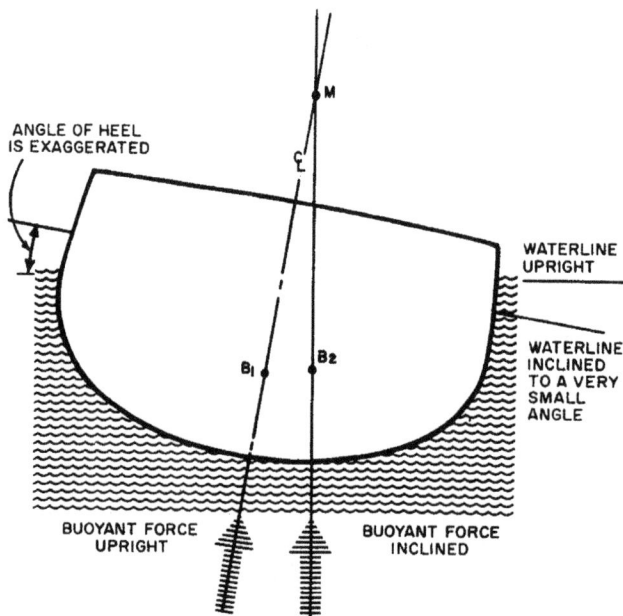

8.54

Figure 3-10.—The metacenter.

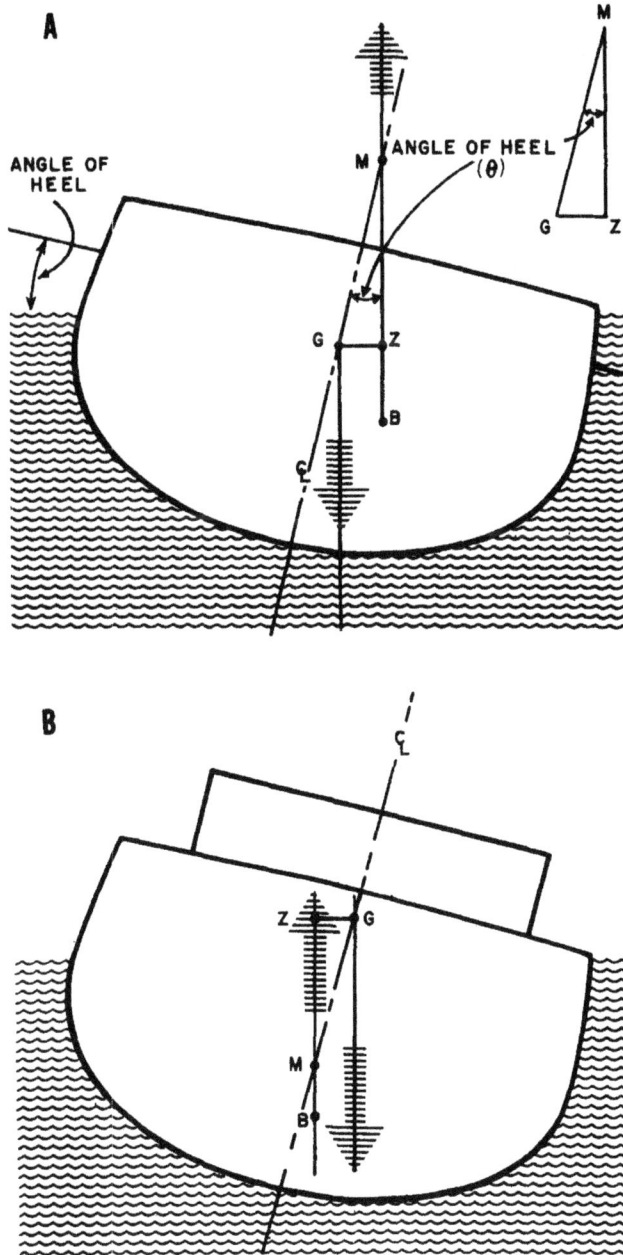

8.55

Figure 3-11.—(A) Stable condition, G is below M. (B) Unstable condition, G is above M.

metacentric height is small, the righting arms are also small. Such a ship rolls slowly and is said to be tender. Some GM values for various naval ships are: CLs, 3 to 5 feet; CAs, 4 to 6 feet; DDs, 3 to 4 feet; DEs, 3 to 5 feet; and AKs, 1 to 6 feet.

Large GM and large righting arms are desirable for resistance to the flooding effects of damage. However, a smaller GM is sometimes desirable for the slow, easy roll which makes for more accurate gunfire. Thus the GM value for a naval ship is the result of compromise.

STABILITY CURVES

When a series of values for GZ at successive angles of heel are plotted on a graph, the result is a stability curve. The stability curve shown in figure 3-12 is called a curve of static stability. The word static indicates that it is not necessary for the ship to be in motion for the curve to apply; if the ship were momentarily stopped at any angle during its roll, the value of GZ given by the curve would still apply.[2]

To understand the stability curve, it is necessary to consider the following facts:

1. The ship's center of gravity does not change position as the angle of heel is changed.
2. The ship's center of buoyancy is always at the center of the ship's underwater hull.
3. The shape of the ship's underwater hull changes as the angle of heel changes.

Putting these facts together, we see that the position of G remains constant as the ship heels through various angles, but the position of B changes according to the angle of inclination. Initial stability increases with increasing angle of heel at an almost constant rate; but at large angles the increase in GZ begins to level off and gradually diminishes, becoming zero at very large angles of heel.

EFFECT OF DRAFT ON RIGHTING ARM

A change in displacement will result in a change of draft and freeboard; and B will shift to the geometric center of the new underwater body. At any angle of inclination, a change in draft causes B to shift both horizontally and vertically with respect to the waterline. The horizontal shift in B changes the distance between B and G, and thereby changes the length of the righting arm, GZ. Thus, when draft is increased, the righting arms are reduced throughout the entire range of stability. Figure 3-13 shows how the righting arm is reduced

---

[2]Design engineers usually use GM values as a measure of stability up to about 7° heel. For angles beyond 7°, a stability curve is used.

ANGLE OF HEEL IN DEGREES

GZ = 1.4 FEET   GZ = 2.0 FEET   GZ = 1 FOOT

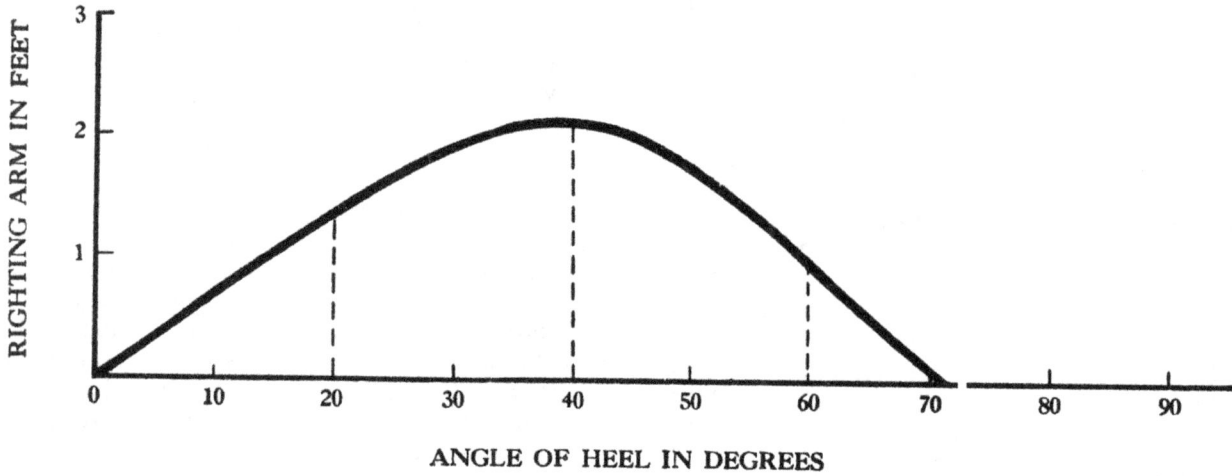

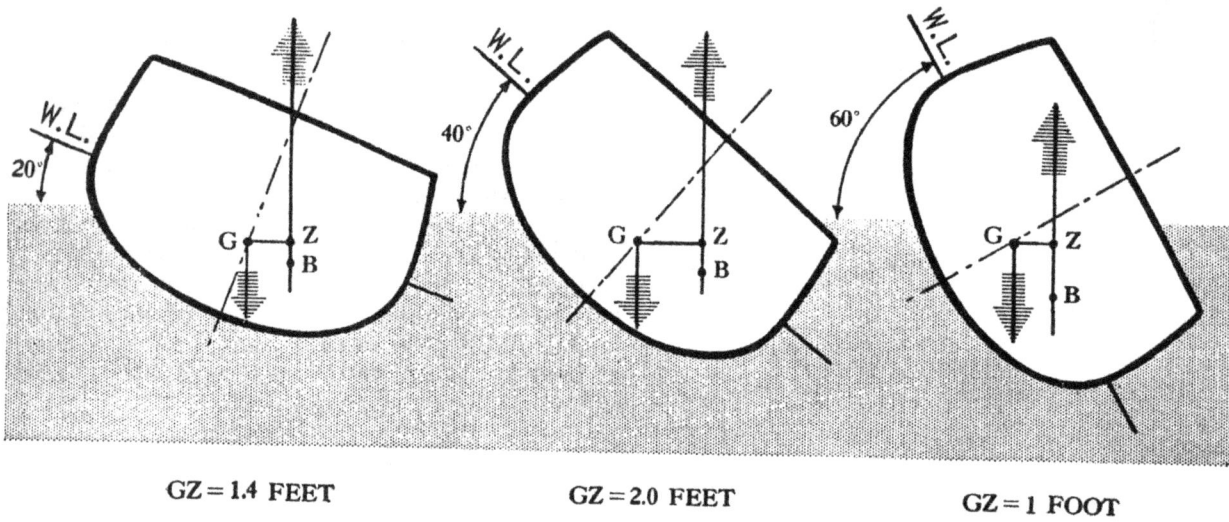

Figure 3-12.—Righting arms of a ship inclined at successively larger angles of heel.

8.56

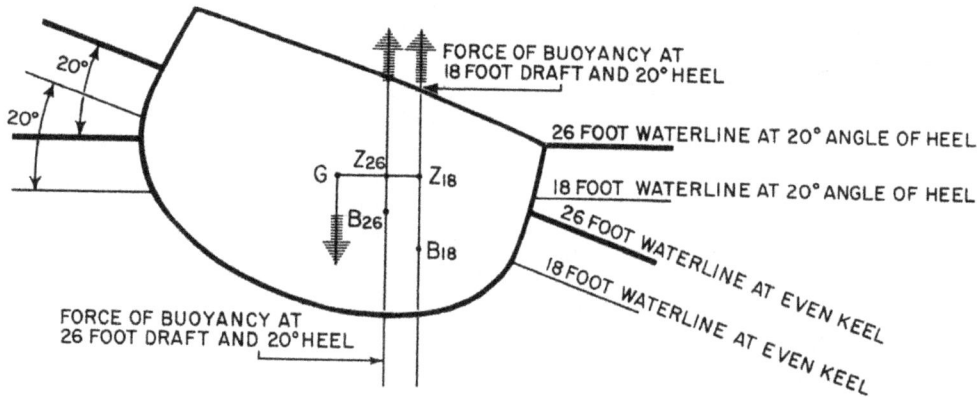

Figure 3-13.—Effect of draft on righting arm.

8.57

when the draft is increased from 18 feet to 26 feet, when the ship is inclined at an angle of 20°. At smaller angles up to 30°, certain hull types show flat or slightly increasing righting arm values with an increase in displacement.

A reduction in the size of the righting arm usually means a decrease in stability. When the reduction in GZ is caused by increased displacement, however, the total effect on stability is more difficult to evaluate. Since the righting moment is equal to W times GZ, the righting moment will be increased by the gain in W at the same time that it is decreased by the reduction in GZ. The gain in the righting moment, caused by the gain in W, does not necessarily compensate for the reduction in GZ.

In brief, there are several ways in which an increase in displacement affects the stability of a ship. Although these effects occur at the same time, it is best to consider them separately. The effects of increased displacement are:

1. Righting arms (GZ) are decreased as a result of increased draft.
2. Righting moments (foot-tons) are decreased as a result of decreased GZ (for a given displacement).
3. Righting moments may be increased as a result of the increased displacement (W), if (GZ x W) is increased.

CROSS CURVES OF STABILITY

To facilitate stability calculations, the design activity inclines a lines drawing of the ship at a given angle, and then lays off on it a series of waterlines. These waterlines are chosen at evenly spaced drafts throughout the probable range of displacements. For each waterline the value of the righting arm is calculated, using an assumed center of gravity rather than the true center of gravity. A series of such calculations is made for various angles of heel—usually 10°, 20°, 30°. 40°, 50°, 60°, 70°, 80°, and 90°—and the results are plotted on a grid to form a series of curves known as the cross curves of stability (fig. 3-14). Note that, as draft and displacement increase, the curves all slope downward, indicating increasingly smaller righting arms.

The cross curves are used in the preparation of stability curves. To take a stability curve from the cross curves, a vertical line (such as line MN in fig. 3-14) is drawn on the cross curve sheet at the displacement which corresponds to the mean draft of the ship. At the intersection of

this vertical line with each cross curve, the corresponding value of the righting arm on the vertical scale at the left can be read. Then this value of the righting arm at the corresponding angle of heel is plotted on the grid for the stability curve. When a series of such values of the righting arms from 10° through 90° of heel have been plotted, a smooth line is drawn through them and the uncorrected stability curve for the ship at that particular displacement is obtained. The curve is not corrected for the actual height of the ship's center of gravity, since the cross curves are based on an assumed height of G. However, the stability curve does embody the effect on the righting arm of the freeboard for a given position of the center of gravity.

Figure 3-15 shows an uncorrected stability curve (A) for the ship operating at 11,500 tons displacement, taken from the cross curves shown in figure 3-14. This stability curve cannot be used in its present form, since the cross curves are made up on the basis of an assumed center of gravity. In actual operation, the ship's condition of loading will affect its displacement and, therefore, the location of G. To use a curve taken from the cross curves, therefore, it is necessary to correct the curve for the actual height of G above the keel (K)—that is, it is necessary to use the distance KG. As far as the new center of gravity is concerned, when a weight is added to a system of weights, the center of gravity can be found by taking moments of the old system plus that of the new weight and dividing this total moment by the total final weight. Detailed information concerning changes in the center of gravity of a ship can be obtained from chapter 9880 of the Naval Ships Technical Manual.

Assume that the cross curves are made up on the basis of an assumed KG of 20 feet, and the actual KG, which includes the added effects of Free Surface, for the particular condition of loading, is 24 feet. This means that the true G is 4 feet higher than the assumed G, and that the righting arm (GZ) at each angle of inclination will be smaller than the righting arm shown in figure 3-15 (curve A) for the same angle. To find the new value of GZ for each angle of inclination, the increase in KG (4 feet) is multiplied by the sine of the angle of inclination, and the product is subtracted from the value of GZ shown on the cross curves or on the uncorrected stability curve. In order to facilitate the correction of the stability curves, a table showing the necessary sines of the angles of inclination is

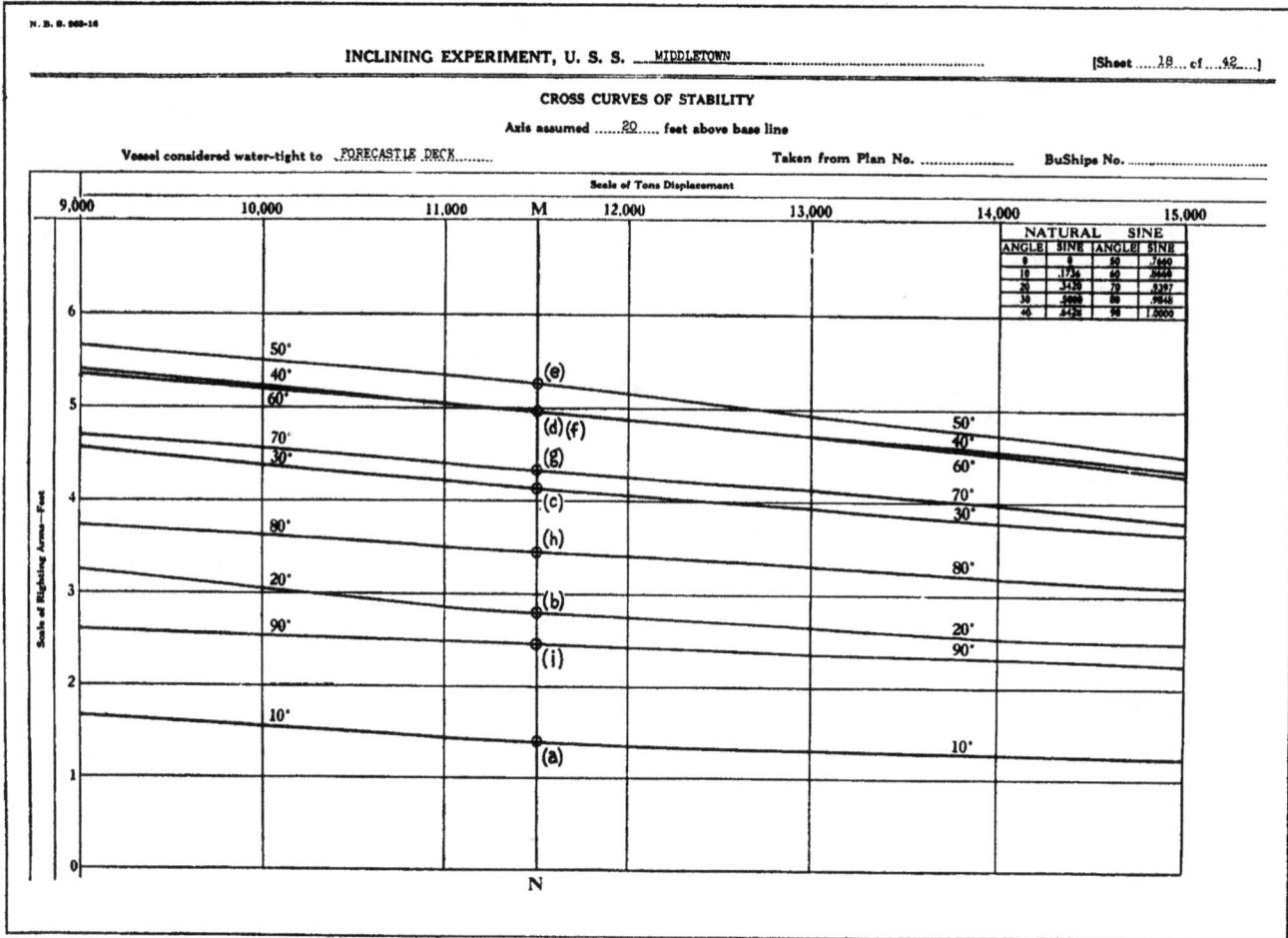

8.58

Figure 3-14.—Cross curves of stability.

included on the cross curves form (fig. 3-14).

Next, the corrected values of GZ for the various angles of heel shown on the stability curve (A) in figure 3-15 should be found and plotted on the same grid to make the corrected stability curve (B) shown in figure 3-15.

When the values from 10° through 80° are plotted on the grid and joined with a smooth curve, the corrected stability curve (B) shown in figure 3-15 results. The corrected curve shows maximum stability to be at 40°; it also shows that an upsetting arm, rather than a righting arm, generally exists at angles of heel in excess of 75°.

THE INCLINING EXPERIMENT

The vertical location of the center of gravity must be known in order to determine the stability characteristics of a ship. Although the position of the center of gravity as estimated by calculation is sufficient for design purposes, an accurate determination is required to establish the ship's stability. Therefore, an inclining experiment is performed to obtain a precise measurement of KG, the vertical height of G above the keep (base line), when the ship is completed. An inclining experiment consists of moving one or more large weights across the ship and measuring the angle of list produced. (See fig. 3-16). This angle of list, produced by the weight movement and measured by means of a pendulum

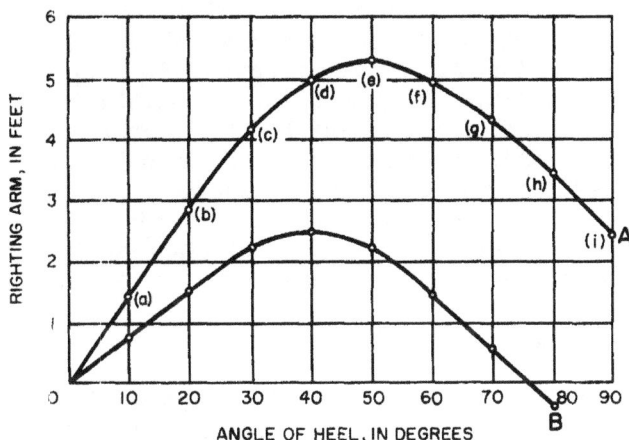

8.59

Figure 3-15.—(A) Uncorrected stability curve taken from cross curves. (B) Corrected stability curve.

and a horizontal batten or an inclinometer device designed for this purpose, usually does not exceed two degrees. The metacentric height is calculated from the formula

$$GM = \frac{wd}{W \tan \theta}$$

where

w = inclining weight, in tons

d = distance weight is moved athwartships, in feet

W = displacement of ship, including weight w, in tons

tan θ = tangent of angle of list

The results of this experiment are calculated and tabulated in the Inclining Experiment Data Booklets, which consist of two parts. Part 1, Report of Inclining Experiment, contains the observations and calculations that determine the displacement and location of the center of gravity of the ship in the light condition. Part 2, Stability Data for surface ships and Stability and Equilibrium Data for submarines, contains data relative to the characteristics of the ship in operating condition. These booklets are prepared by the inclining activity, and Part 2 is issued to the ships for their information.

The KG obtained from the inclining experiment is accurate for the particular condition of loading in which the ship was inclined. This is known as Condition A, or the "As-Inclined" condition. The ship may have been in any condition of loading at the time of the experiment, and this may not have been in operating condition. In order to convert the data thus obtained to practical use, KG must be determined for various operating conditions. The standard loading conditions as found in the Inclining Experiment Data Booklets are as follows:

Condition A—Light ship
Condition A1—Light, without permanent ballast[3]
Condition B—Minimum operating condition[4]
Condition C—Optimum battle condition[5]
Condition D—Full load

Other special conditions, including special low stability operating conditions, conditions of light load with water ballast, and similar conditions may be included.

Condition A—Light Condition assumes that the ship is complete and in all respects ready for sea, but with no load aboard—no fuel oil, stores, crew and effects, ammunition, water, gasoline, JP-5, or water or oil in machinery. Although not an operating condition, Condition A is the basic condition from which other conditions are calculated.

After obtaining the displacement and locating the center of gravity for the ship in Condition A, corresponding values may be computed for other standard conditions of loading. The weights and vertical moments of all consumables to go aboard are determined and, starting with the displacement and KG for Condition A, a new displacement, KG, and GM are calculated for each of the other conditions of loading. The GM thus obtained is in each case corrected for the free surface assumed to exist in the ship's tanks for that particular condition of loading. (Free surface is discussed later in this chapter.)

Having determined displacement and KG, it is possible to draw a curve of stability for each condition of load. Additional information concerning inclining experiment data can be obtained from chapters 9290 and 9880 of the Naval Ships Technical Manual.

---

[3]This condition is listed only when ships have permanent ballast.

[4]For ships without underwater defense systems.

[5]For ships with underwater defense systems.

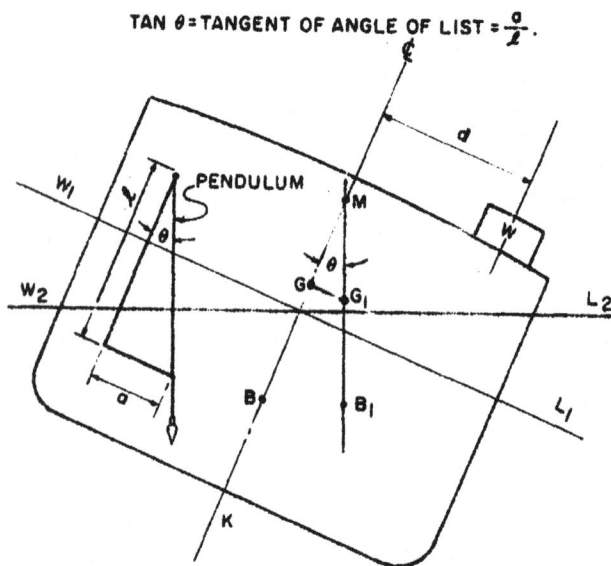

147.32

Figure 3-16.—Measuring the angle of list produced in performing the inclining experiment.

EFFECTS OF WEIGHT SHIFTS

If one weight in a system of weights is moved, the center of gravity of the whole system moves along a path parallel to the path of the component weight. The distance that the center of gravity of the system moves may be calculated from the formula.

$$GG_1 = \frac{ws}{W}$$

where

w = component weight, in tons

s = distance component weight is moved, in feet

W = weight of entire system, in tons

GG_1 = shift in center of gravity of system, in feet

Weight movements in a ship can take place in three possible directions—athwartships, fore and aft, and vertically (perpendicular to the decks). The most general type of movement is inclined with respect to all three of these. Such a diagonal movement can be divided into components in each of the three directions, and one component can be studied at a time without reference to the others. For example, if a weight is moved from the main deck, starboard side, aft, to a storeroom on the 4th deck, port side, forward, this movement may be regarded as taking place in three steps, as follows:
1. from main deck to 4th deck (down)
2. from starboard side to port side (across)
3. from stern to bow (forward)

VERTICAL WEIGHT SHIFT

If a weight is moved straight up a vertical distance on a ship, the ship's center of gravity will move straight up on the centerline (fig. 3-17). The vertical rise in G (explained later in the chapter) can be computed from the formula mentioned previously.

Example: A ship is operating with a displacement of 11,500 tons. Her ammunition, totaling 670 tons, is to be moved from the magazines to the main deck, a distance of 36 feet. Find the rise in G.

$$GG_1 = \frac{670 \times 36}{11,500} = 2.1 \text{ feet}$$

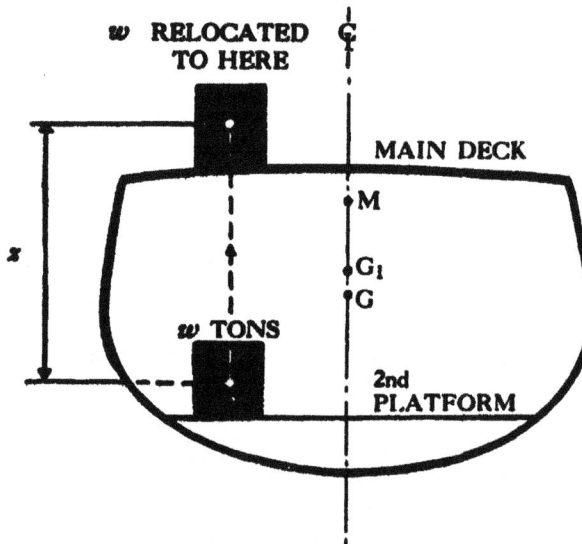

147.33

Figure 3-17.—Shift in G due to vertical weight shift.

Since moving a weight which is already aboard will cause no change in displacement, there can be no change in M, the metacenter. If M remains fixed, then the upward movement of the center of gravity results in a loss of metacentric height:

$$G_1M = GM - GG_1$$

where

G_1M = new metacentric height (after weight movement), in feet

GM = old metacentric height (before weight movement), in feet

GG_1 = rise in center of gravity, in feet

If the ammunition on the main deck is moved down to the 6th deck, the positions of G and G_1 will be reversed. The shift in G can be found from the same formula as before, the only difference being that GG_1 now becomes a gain in metacentric height instead of a loss (fig. 3-18).

If a weight is moved vertically downward, the ship's center of gravity, G, will move straight down on the centerline and the correction is additive. In this case the sine curve is plotted below the abscissa. The final stability curve is that portion of the curve above the sine correction curve.

A vertical shift in the ship's center of gravity changes every righting arm throughout the entire range of stability. If the ship is at any angle of heel, such as θ in figure 3-18, the righting arm is GZ with the center of gravity at G. But if the center of gravity shifts to G_1 as the result of a vertical weight shift upward, the righting arm becomes G_1Z_1, which is smaller than GZ by the amount of GR. In the right triangle GRG_1, the angle of heel is at G_1; hence the loss of the righting arm may be found from

$$GR = GG_1 \times \sin \theta$$

This equation may be stated in words as: The loss of righting arm equals the rise in the center of gravity times the sine of the angle of heel. The sine of the angle of heel is a ratio which can be found by consulting a table of sines.

If the loss of GZ is found for 10°, 20°, 30°, and so forth by multiplying GG_1 by the sine of the proper angle, a curve of loss of righting arms can be obtained by plotting values of

$GG_1 \times \sin \theta$ vertically against angles of heel horizontally, which results in a sine curve. When plotted, the curve is as illustrated in figure 3-19.

The sine curve may be superimposed on the original stability curve to show the effect on stability characteristics of moving the weight up in a ship. Inasmuch as displacement is unchanged, the righting arms of the old curve need be corrected for the change of G only, and no other variation occurs. Consequently, if $GG_1 \times \sin \theta$ is deducted from each GZ on old stability curve, the result will be a correct righting arm curve for the ship after the weight movement.

In figure 3-20 a sine curve has been superimposed on an original stability curve. The dotted area is that portion of the curve which was lost due to moving the weight up, whereas the lined area is the remaining or residual portion of the curve. The residual maximum righting arm is AB and occurs at an angle of about 37°. The new range of stability is from 0° to 53°.

The reduced stability of the new curve becomes more evident if the intercepted distances between the old GZ curve and the sine curve are transferred down to the base, thus forming a new curve of static stability (fig. 3-21). Where the old righting arm at 30° was AB, the new one has a value of CB, which is plotted up from the base to locate point D (CB = AD) and thus a point is established at 30° on the new curve. A series of points thus obtained by transferring intercepted distances down to the base line delineates the new curve, which may be analyzed as follows:

GM is now the quantity represented by EF.
Maximum righting arm is now the quantity represented by HI.
Angle at which maximum righting arm occurs is 37°.
Range of stability is from 0° to 53°.

Total dynamic stability is represented by the shaded area.

HORIZONTAL WEIGHT SHIFT

When the ship is upright, G lies in the fore and aft centerline, and all weights on board are balanced. Moving any weight horizontally will result in a shift in G in an athwartship direction, parallel to the weight movement. B and G are no longer in the same vertical line and an upsetting moment exists at 0° inclination, which will cause the ship to heel until B moves under the new position of G. In calm water the ship will

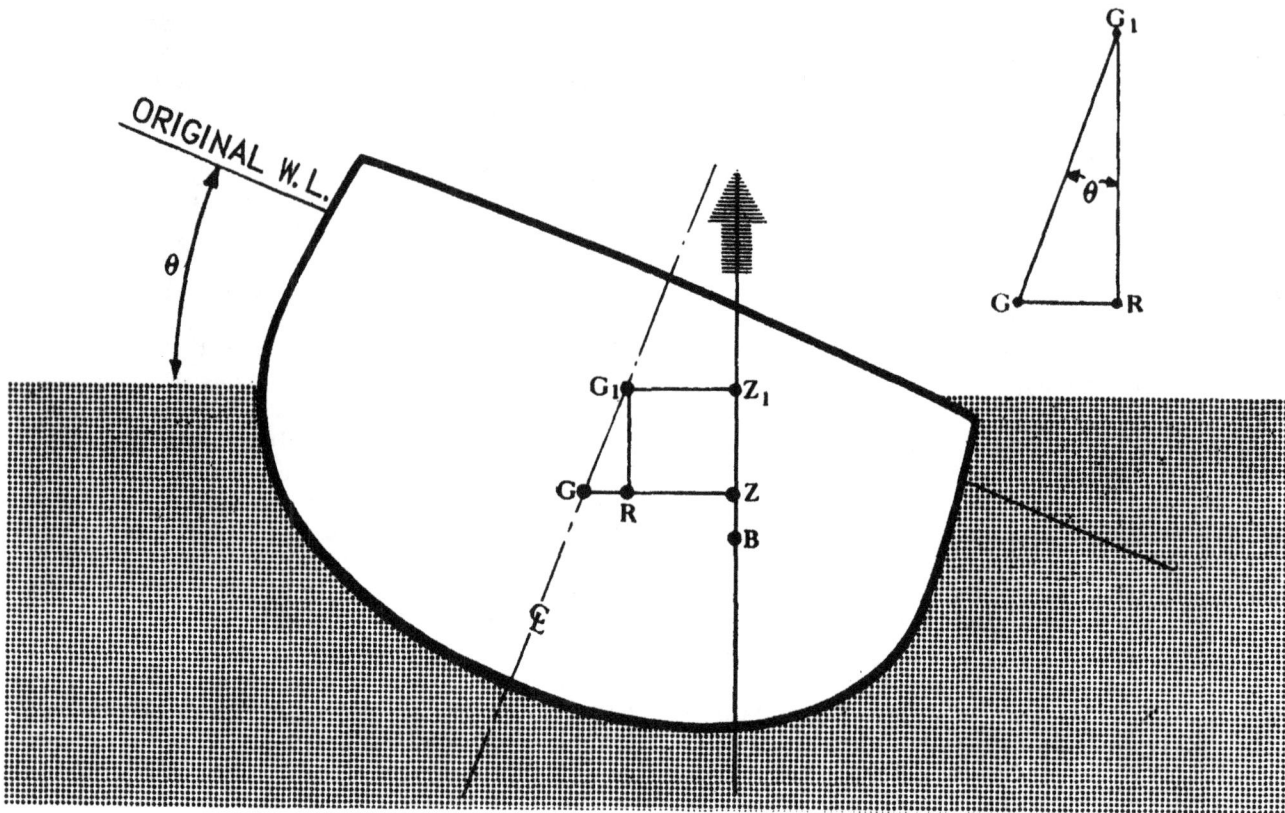

147.34

Figure 3-18.—Loss of righting arm due to rise in center of gravity.

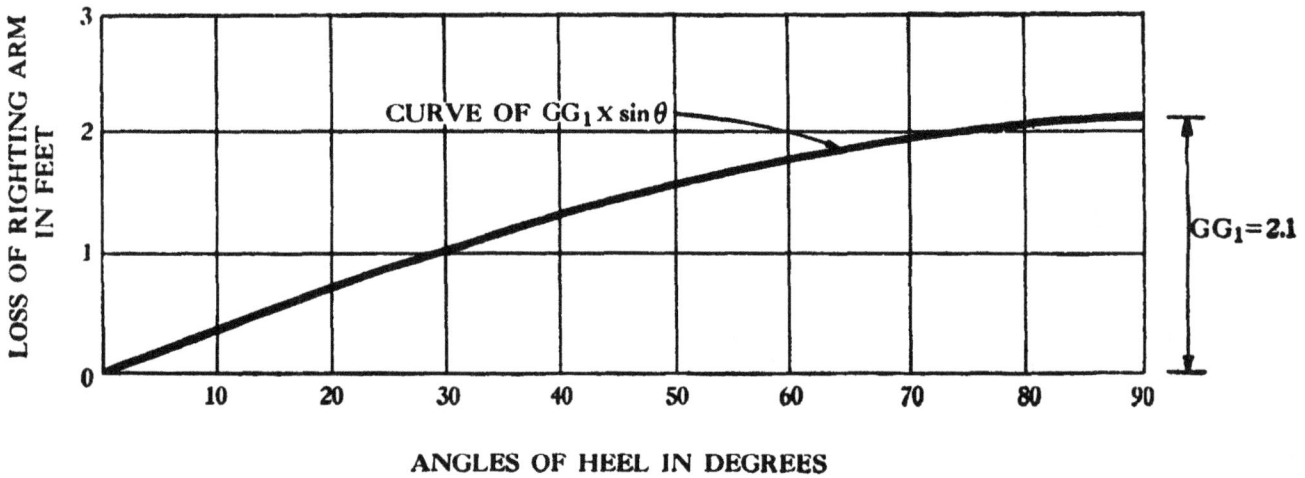

147.35

Figure 3-19.—Sine curve showing the loss of righting arm at various angles of heel.

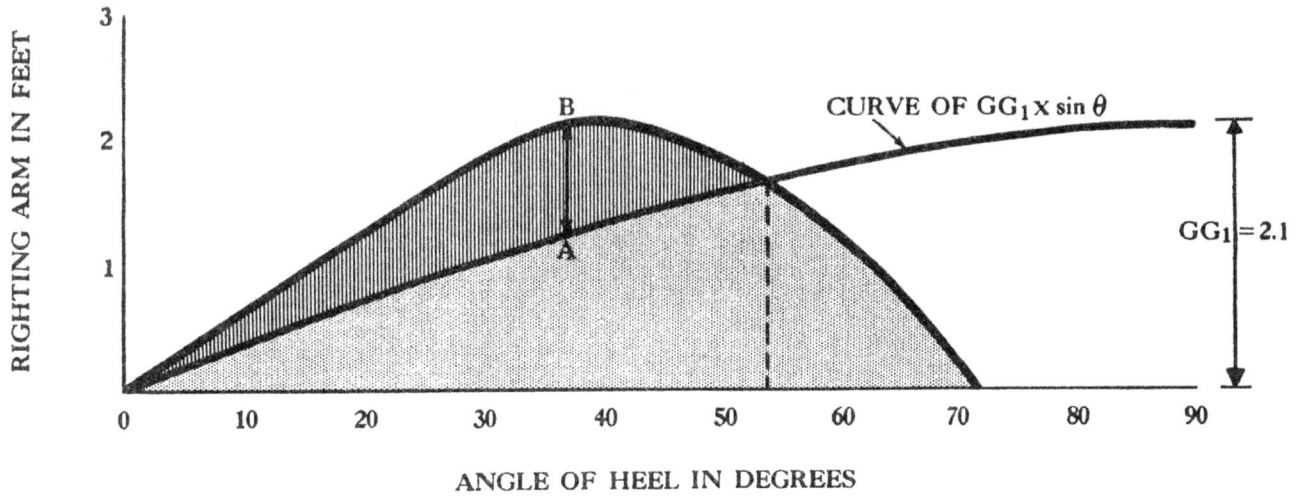

147.36

Figure 3-20.—Sine curve superimposed on original stability curve.

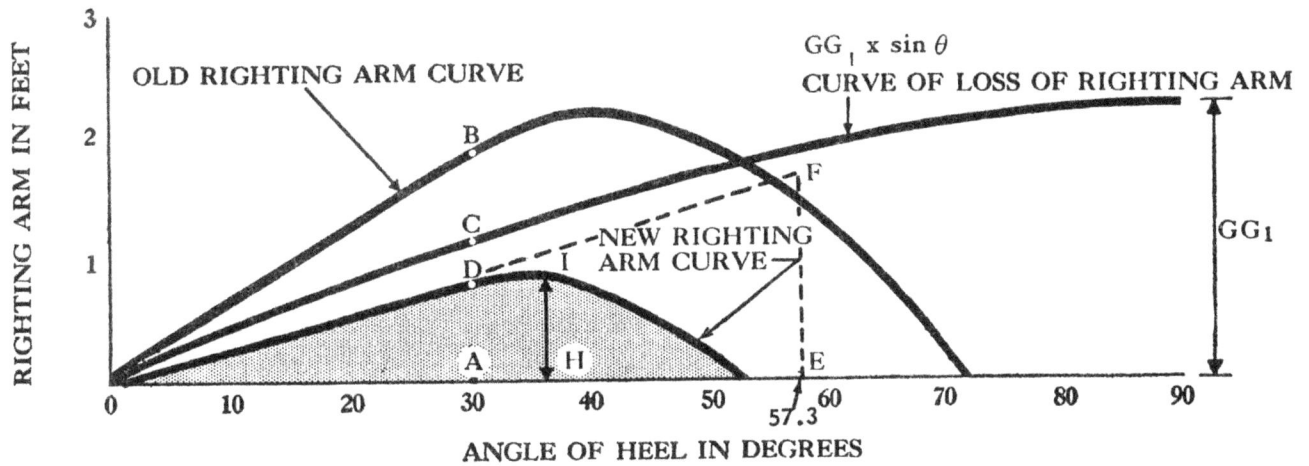

147.37

Figure 3-21.—Curve of static stability as corrected for loss of
stability due to a vertical weight shift.

remain at this angle and in a seaway it will roll about this angle of permanent list. This shift of G can be computed from the formula

$$G_1G_2 = \frac{wd}{W}$$

where

G_1G_2 = athwartship shift of G, in feet

w = weight moved over, in tons

d = distance w moved, in feet

w = displacement of ship, in tons

Going back to our original problem, let us further assume that ship's stores totaling 185 tons are shifted from port storerooms to starboard storerooms, a horizontal distance of 56 feet. Using the formula:

$$G_1G_2 = \frac{185 \times 56}{11,500} = 0.90 \text{ foot}$$

In figure 3-22 the righting arm has been reduced from G_1Z_1 to G_2Z_2 by this weight shift. G_2Z_2 is smaller than G_1Z_1. However, the distance G_1T is equal to $G_1G_2 \times \cos \theta$. Thus, the

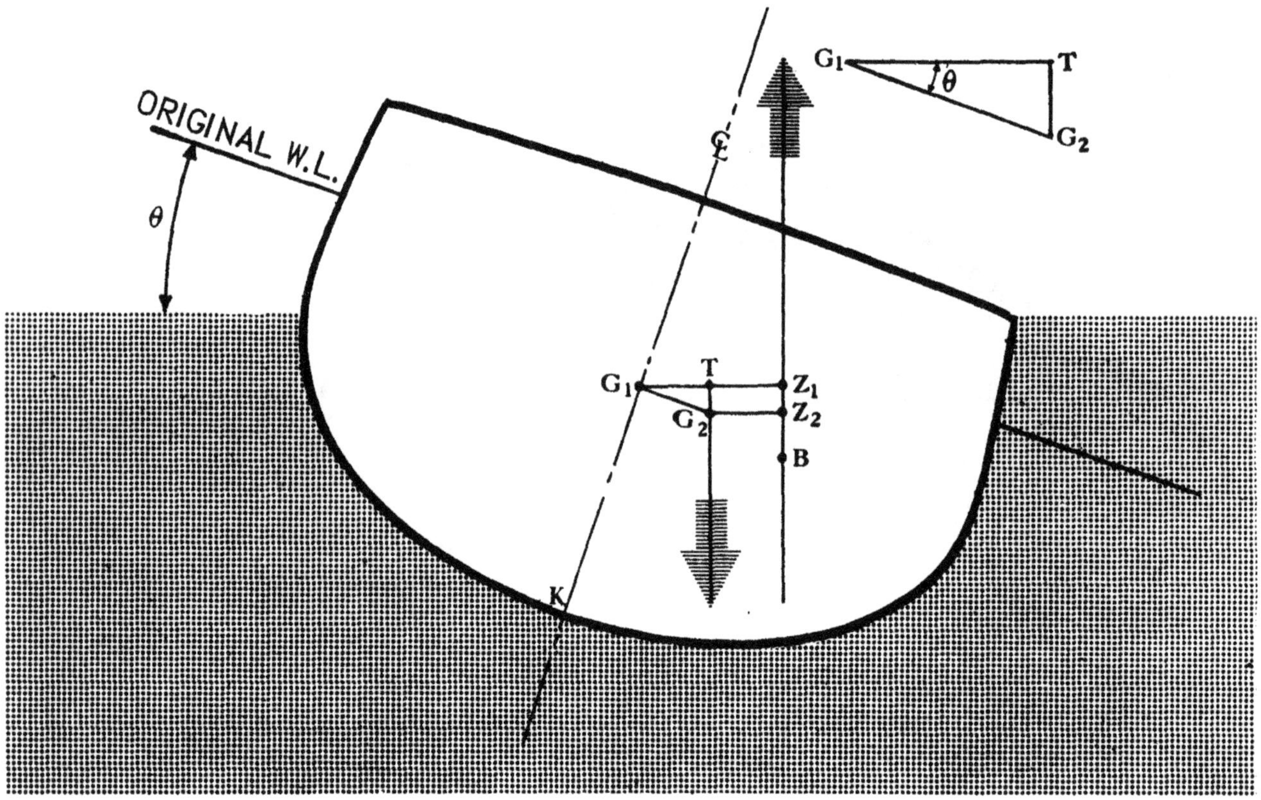

147.38

Figure 3-22.—Loss of righting arm when center of gravity is moved off the center line.

loss of righting arm involved in an athwartship movement of G is equal at any angle of heel to $G_1G_2 \times \cos \theta$. This variable distance ($G_1G_2 \times \cos \theta$) is called the ship's inclining arm; when this value is multiplied by the displacement, W, the product is the ship's inclining moment.

The expression $G_1T = G_1G_2 \times \cos \theta$ may be stated as: The inclining arm is equal to the athwartship shift in the center of gravity times the cosine of the angle of heel. The cosine of the angle of heel is a ratio which can be found by consulting a table of cosines. If the inclining arm is computed for $10°$, $20°$, $30°$, etc. by multiplying G_1G_2 by the cosine of the proper angle, a curve of inclining arms can be obtained by plotting values of $G_1G_2 \times \cos\theta$ vertically against angles of heel horizontally, which results in a cosine curve. Note that the cosine curve (fig. 3-23) is just the opposite of the sine curve (fig. 3-20) but is otherwise identical in shape.

Just as the sine curve was superimposed on the GZ curve, so may the cosine curve be superimposed on the stability curve to show the effect on stability of moving a weight athwartship. The cosine curve has been placed on the original stability curve, corrected for the actual height of the center of gravity. The dotted area (fig. 3-23) is that portion of the curve which was lost due to the weight shift, and the lined area is the remaining or residual portion of the curve. The residual maximum righting arm is AB which develops at an angle of about $37°$. The new range of stability is from $20°$ to $50°$.

The new curve of static stability can be plotted on the base by transferring down the intercepted distances between the cosine curve and the old GZ curve. For example, in figure 3-24 the old righting arm at $37°$ was AD, the loss of righting arm (inclining arm) at this angle is AC, leaving a residual GZ of CD. This value has been plotted up from the base as AB to provide one point on the final curve. The residual stability may be analyzed on the new curve as follows:

48

1. Maximum righting arm AB.
2. Angle of maximum righting arm at A.
3. Range of stability 20° to 50°.
4. Total dynamic stability is represented by the lined area.

The ship will have a permanent list at 20° which is the angle where B is under G, inclining arm equals original righting arm, cosine curve crosses original GZ curve, and residual righting arm is zero. In a seaway the ship will roll about this angle of list. If it rolls farther to the listed side, a righting moment develops which tends to return it toward the angle of list. If it rolls back towards the upright, an upsetting moment develops which tends to return it toward the angle of list. The upsetting moment (between 0° and the angle of list) is the difference between the inclining and righting moments.

DIAGONAL WEIGHT SHIFT

A weight may be shifted diagonally, so that it moves up or down and athwartship at the same time, or by moving one weight up or down and another athwartship. A diagonal shift should be treated in two steps; first by finding the effect on GM and stability of the vertical shift, and second, by finding the effect of the horizontal movement. The corrections are applied as previously described.

EFFECTS OF WEIGHT CHANGES

The additional removal of any weight in a ship may affect list, trim, draft, displacement,

and stability. Regardless of where the weight is added (or removed), when determining the various effects it should be considered first to be placed in the center of the ship, then moved up (or down) to its final height, next moved outboard to its final off-center location, and finally shifted to its fore or aft position.

Assume that a weight is added to a ship so that the list or trim is not changed, and G will not shift. The first thing to do is find the new displacement, which is the old displacement plus the added weight:

$$\text{New displacement} = W + w \text{ tons}$$

where

W = old displacement (tons)
w = added weight (tons)

With the new value of displacement, enter the curves of form and on the displacement curve find the corresponding draft, which is the new mean draft. Figure 3-25 shows typical displacement and other curves generally referred to as curves of form.

If the change in draft is not over 1 foot, the procedure can be reversed. Find the tons-per-inch immersion for the old mean draft from the curves of form, divide the added weight (in tons) by the tons-per-inch immersion in order to get the bodily sinkage in inches, and add this bodily sinkage to the old mean draft to get the new mean draft. Using the new mean draft, enter the curves of form and find the new displacement.

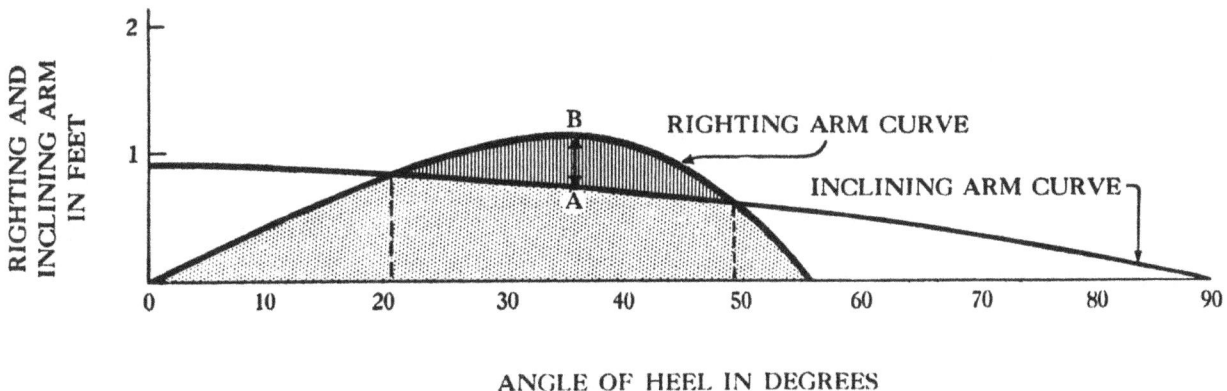

Figure 3-23.—Cosine curve superimposed on original stability curve.

147.39

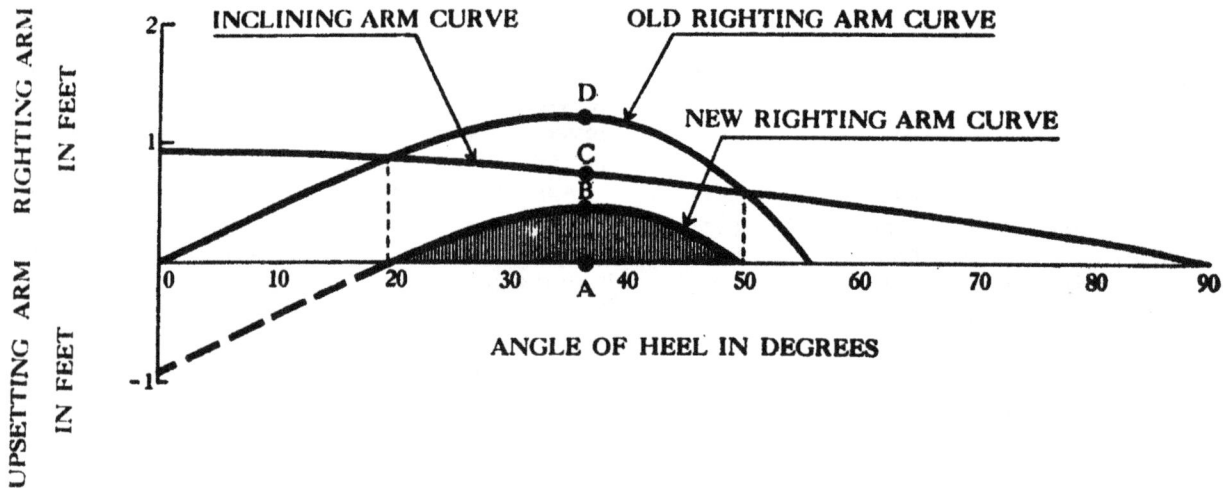

147.40

Figure 3-24.—New curve of static stability after correction for horizontal weight movement.

VERTICAL WEIGHT CHANGES

Assume that the weight added above is shifted vertically on the ship's centerline to its final height above the keel. This movement will cause G to shift up or down. To compute the vertical shift of G use the formula

$$GG_1 = \frac{wz}{(W + w)}$$

where

GG$_1$ = shift of G up or down, in feet
w = added weight, in tons
z = vertical distance w is added above or below original location of G, in feet
W = old displacement, in tons
(W + w) = new displacement, in tons

This vertical shift must be added to or subtracted from the original height of the center of gravity above the keel.

To do this, the original height KG must be known:

$$KG_1 = KG + GG_1$$

where

KG$_1$ = new height of G above keel (in feet)
KG = old height of G above keel (in feet)

GG_1 = shift of G from formula $GG_1 = \dfrac{wz}{(W + w)}$

If the final position of the added weight is below the original position of G, then GG$_1$ is minus; if it is above, then GG$_1$ is plus.

To find the new metacentric height, enter the curves of form with the new mean draft and find the height of the transverse metacenter above the base line. This is KM$_1$. The new metacentric height is determined by the formula,

$$G_1M_1 = KM_1 - KG_1$$

where

G$_1$M$_1$ = new metacentric height (in feet)
KM$_1$ = new KM
KG$_1$ = new KG

With the new displacement (W + w), enter the cross curves and pick out a new, uncorrected curve of stability. Correct this curve for the new height of the ship's center of gravity above the base line. This is accomplished by finding AG$_1$ (which is KG$_1$ minus KA) and subtracting AG$_1$ x sin θ from every vertical on the stability curve, provided G$_1$ is above A. If G$_1$ is below A, the values of AG$_1$ x sin θ must be added to the curve, as previously explained. The resulting curve of righting arms is now correct for the loss of freeboard due to the added weight and for the final height of the ship's center of gravity resulting from weight addition.

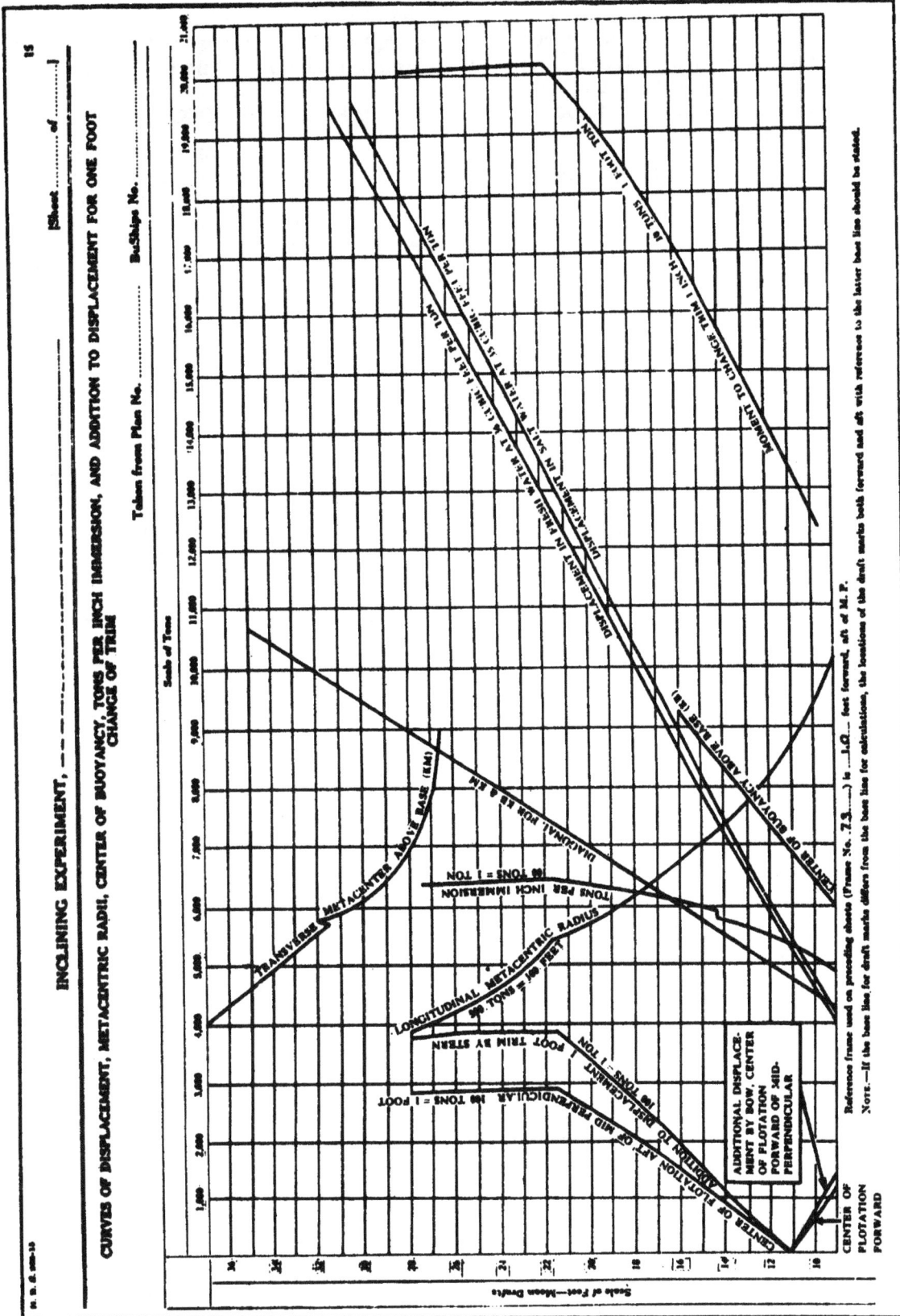

Figure 3-25.—Curves of form.

147.41

<u>Example</u>: Add four gun mounts topside to a ship with the curves of form shown in figure 3-25. Assume an initial KG of 24.5 feet. Assume that the gun mounts weigh 28 tons each and that their center of gravity is located 48 feet above the keel. What is the effect on stability?

1. New displacement = W + w = 11,500 + (4 x 28) = 11,612 tons.

2. New mean draft = 19.7 feet (fig. 3-25).

3. $GG_1 = \dfrac{wz}{W + w}$

 w = 4 x 28 = 112 tons

 z = 48 − 24.5 = 23.5 feet

 $GG_1 = \dfrac{112 \times 23.5}{11,612} = 0.23$ feet

4. $KG_1 = 24.50 + 0.23 = 24.73$ feet.

5. New $KM_1 = 28.4$ feet (fig. 3-25).

6. New $G_1M_1 = KM_1 - KG_1 = 28.4 - 24.7 = 3.7$ feet.

7. The values for the angles (0°− 70°) are taken from the cross curves for 11,612 tons displacement (fig. 3-14). KA is 20 feet. Corrections are made for $AG_1 \times \sin\theta$ = (24.73 − 20) $\sin\theta$ = 4.73 $\sin\theta$. The corrections are applied to the curve (fig. 3-26) as previously explained. Figure 3-26 shows the curve of righting arms corrected for weight addition.

HORIZONTAL WEIGHT CHANGES

In the previous example of weight addition, suppose the gun mounts are located with their center of gravity 29 feet to starboard of the centerline and the weight is moved athwartship to its final off-center location. The shift in G may be found by using the proper formula, making the required corrections, and applying the corrections to the curve in figure 3-26. This gives a correct curve of righting arms. To obtain a curve of righting moments, the righting arms are multiplied by the new displacement (W+w) =11,612 tons, and plotted in figure 3-27.

WEIGHT REMOVAL

The results of a weight removal are computed by using the previous procedure, the only

difference being that most of the operations and results will be found just the reverse of those which relate to adding a weight.

EFFECTS OF LOOSE WATER

When a tank or a compartment in a ship is partially full of liquid that is free to move as the ship heels, the surface of the liquid tends to remain level. The surface of the free liquid is referred to as <u>free surface</u>. The tendency of the liquid to remain level as the ship heels is referred to as <u>free surface effect</u>. The term <u>loose water</u> is used to describe liquid that has a <u>free surface</u>; it is not used to describe water or other liquid that completely fills a tank or compartment and thus has no free surface.

FREE SURFACE EFFECT

Free surface in a ship always causes a reduction in GM with a consequent reduction of stability, superimposed on any additional weight which would be caused by flooding. The flow of the liquid is an athwartship shift of weight which varies with the angle of inclination. Wherever free surface exists, a free surface correction must be applied to any stability calculation. This effect may be considered to cause a reduction in a ship's static stability curve in the amount of

$\dfrac{i}{V} \times \sin\theta$, due to a virtual rise in G

where

 i = the moment of inertia of the surface of water in the tank about a longitudinal axis through the center of area of that surface (or other liquid in ratio of its specific gravity to that of the liquid in which the ship is floating)[6]

 V = existing volume of displacement of the ship in cubic feet. For a rectangular compartment, i may be found from

$$i = \dfrac{b^3 l}{12}$$

where

 b = athwartship breadth of the free surface (with the ship upright) in feet

 l = fore-and-aft length of the free surface in feet

[6] It is usual to assume all liquids are salt water, and thus neglect density, unless very accurate determinations are required.

To understand what is meant by a virtual rise in G, refer to figure 3-28. This figure shows a compartment in a ship partially filled with water, which has a free surface, fs, with the ship upright. When the ship heels to any small angle, such as θ, the free surface shifts to $f_1 s_1$, remaining parallel to the waterline. The result of the inclination is the movement of a wedge of water from $f_0 f_1$ to $s_0 s_1$. Calling g_1 the center of gravity of this wedge when the ship was upright, and g_2 its center of gravity with the ship inclined, it is evident that a small weight has been moved from g_1 to g_2.

Point G is the center of gravity of the ship when upright, and G would remain at this posi-tion if the compartment contained solids rather than a liquid. As the ship heels, however, the shift of a wedge of water along the path $g_1 g_2$ causes the center of gravity of the ship to shift from G to G_2. This reduces the righting arm, at this angle, from GZ to $G_2 Z_2$.

To compute GG_2 and the loss of GZ for each angle of heel is a laborious and com-plicated task. However, an equivalent righting arm, $G_3 Z_3$ (which equals $G_2 Z_2$), can be ob-tained by extending the line of action of the force of gravity up to intersect the ship's centerline at point G_3. Raising the ship's center of gravity from G to G_3 would have the same effect on stability at this angle as shifting it from G to G_2.

① MINUS ② EQUALS ③
③ MINUS ④ EQUALS ⑤

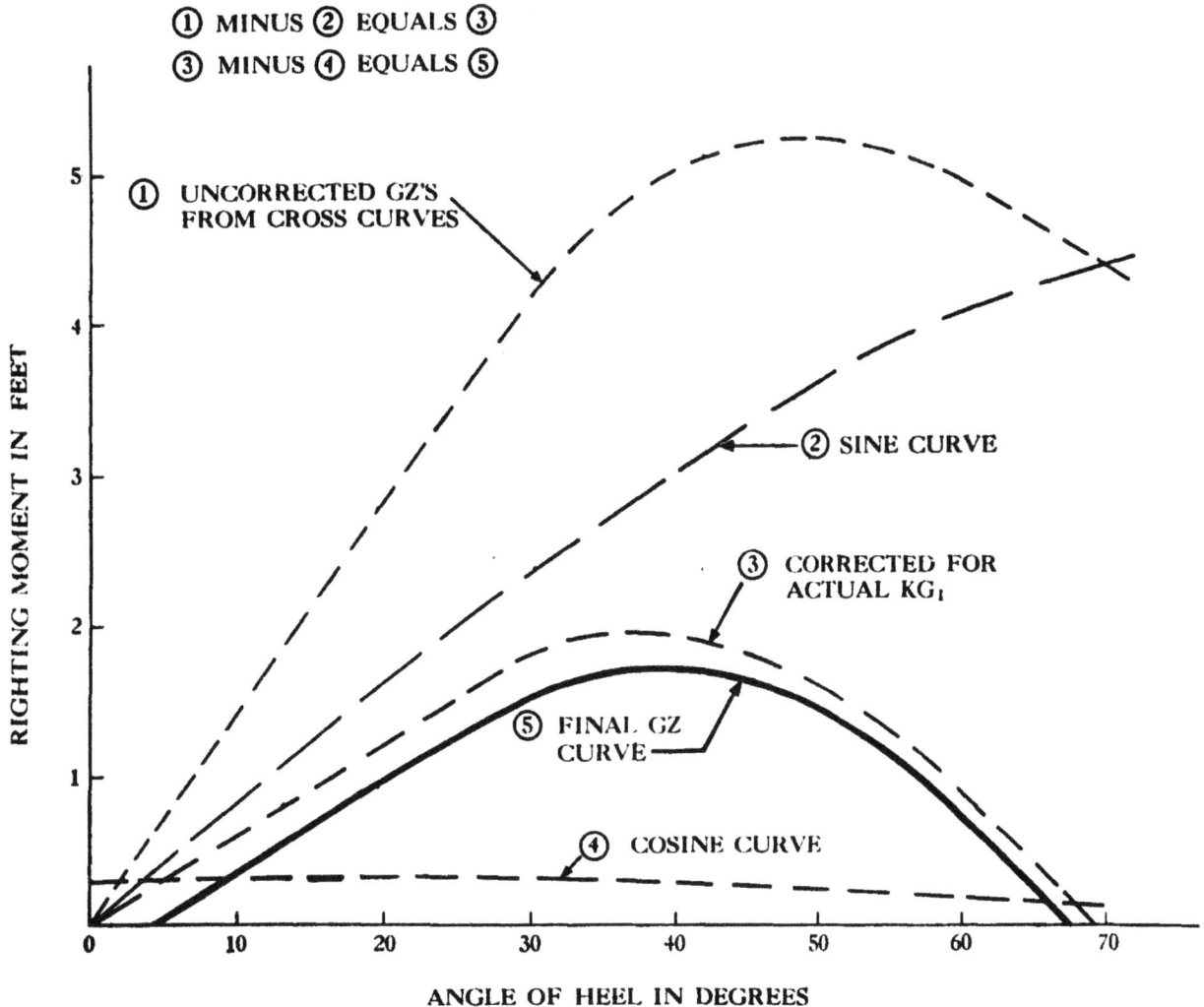

① UNCORRECTED GZ'S FROM CROSS CURVES

② SINE CURVE

③ CORRECTED FOR ACTUAL KG_1

⑤ FINAL GZ CURVE

④ COSINE CURVE

RIGHTING MOMENT IN FEET

ANGLE OF HEEL IN DEGREES

147.42

Figure 3-26.—Curve of righting arms corrected for weight addition.

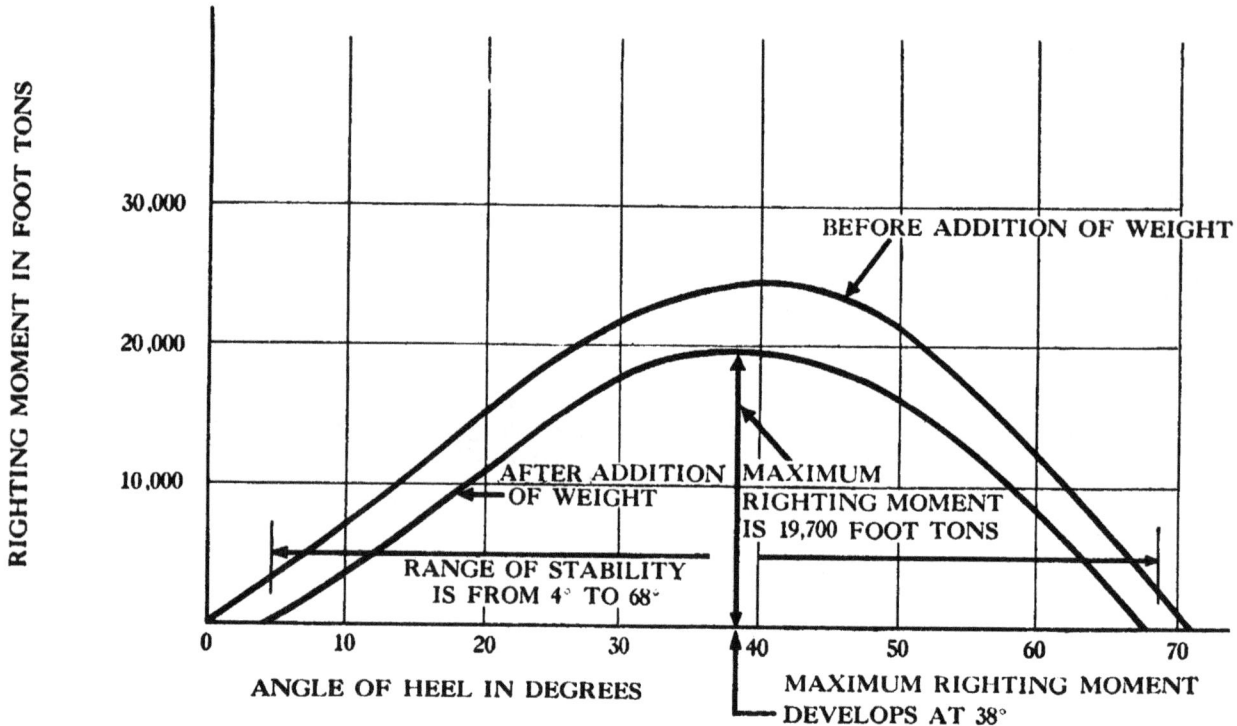

147.43

Figure 3-27.—Curve of righting moments.

The distance G_3Z_3 is the righting arm the ship would have if the center of gravity had risen from G to G_3, and this underline virtual rise of G may be computed from the formula:

$$GG_3 = \frac{i}{V}$$

Referring to the formula, loss in $GZ = \frac{i}{V} x$ sin θ. This formula is accurate for small angles of heel only, due to the pocketing effect as the angle increases. In case several compartments or tanks have free surface, their surface moments of inertia are calculated individually and their sum used in the correction for free surface. The effect of a given area of loose liquid at a given angle of heel is entirely independent of the depth of the liquid in the compartment, as is apparent in the formula,

$$i = \frac{b^3 l}{12}$$

where the only factors are the dimensions of the surface and the displacement of the ship.

The free surface effect is also independent of the free surface location in the ship, whether it is high or low, forward or aft, on the centerline or off, as long as the boundaries remain intact.

The loss of metacentric height can obviously can be reduced by reducing the breadth of the free surface, as by the installation of longitudinal bulkheads. However, off-center flooding after damage then becomes possible, causing the ship to take on a permanent list and usually bringing about a greater loss in stability than if the bulkhead were not present.

The loss of GZ due to free surface is always lessened to some extent by underline pocketing. This is the contact of the liquid with the top of the compartment or the exposure of the bottom surface of the compartment, either of which takes place at some definite angle and reduces the breadth of the free surface area. To understand how pocketing of the free surface reduces the free surface effect, study figure 3-29. Part A shows a compartment in which the free surface effect is not influenced by the depth of the loose water. The compartment shown in part B, however, contains only a small amount of water; when the ship heels sufficiently to reduce the waterline

147.44

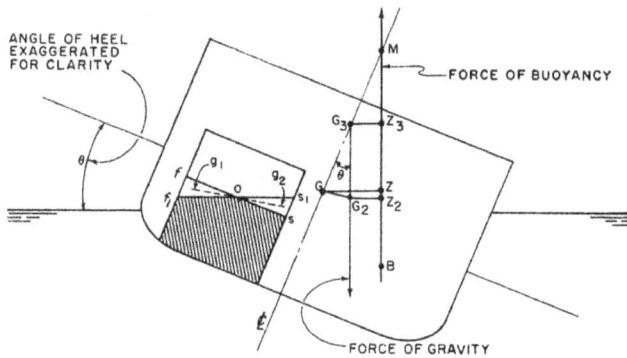

Figure 3-28.—Diagram showing virtual rise in G.

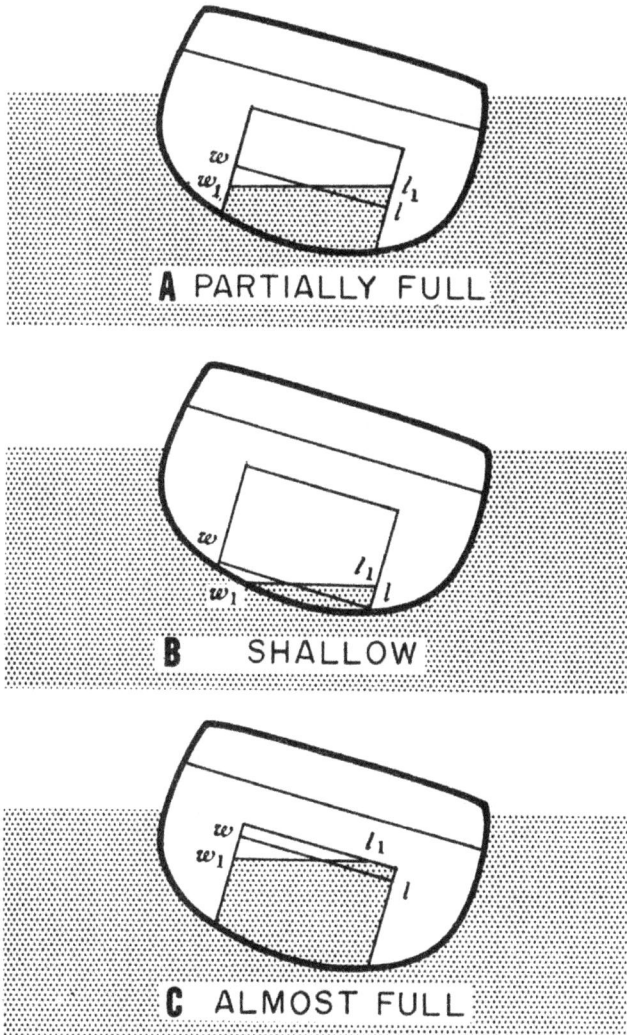

A PARTIALLY FULL

B SHALLOW

C ALMOST FULL

8.61

Figure 3-29.—Pocketing of free surface.

in the compartment from wl to w_1l_1, the breadth of the free surface is reduced and the free surface effect is thereby reduced. A similar reduction in free surface effect occurs in the almost full compartment shown in part C, again because of the reduction in the breadth of the free surface. As figure 3-29 shows, the beneficial effect of pocketing is greater at larger angles of heel.

The effect of pocketing in reducing the over all free surface effect is extremely variable and not easily determined. In practice, therefore, it is usually ignored and tends to provide a margin of safety when computing stability.

Most compartments of a ship contain some solid objects, such as machinery and stores which would project through and above the surface of any loose water. If these objects are secured so that they do not float or move about, and if they are not permeable, then the free surface area and the free surface effect is reduced by their presence. The actual value of the reduction (surface permeability effect) is difficult to calculate and, like the value of pocketing, if ignored when calculating stability will provide a further margin of safety.

Swash bulkheads (nontight bulkheads pierced by drain holes) are fitted in deep tanks and double bottoms to hinder the flow of liquid in its attempt to remain continuously parallel to the waterline as the ship rolls. They diminish the free surface effect if the roll is quick, but they have no effect when the roll is slow. A ship taking on a permanent list will incline just as far as if the swash plate were not there. When a fore-and -aft bulkhead separating two adjacent compartments is holed (ruptured) so that any flooding water present in one is free to flow athwartship from one compartment to the other, a casualty duplicating the effect of a swash bulkhead has occurred. In this case, it is incorrect to add the free surface effects of the two compartments together; an entirely new figure for the flooding effect must be computed, regarding the two as one large compartment.

In summary, the addition of a liquid weight with a free surface has two effects on the metacentric height of a ship. First, there is the effect on GM and GZ of the weight addition (considered as a solid) which influences the vertical position of the ship's center of gravity, and the location of the transverse metacenter, M. Secondly, there is a reduction in GM and GZ due to the free surface effect.

FREE COMMUNICATION EFFECT

If one or more of the boundaries of an off-center compartment are ruptured so that the sea may flow freely in and out with a minimum of restriction as the ship rolls, a condition of partial flooding with free communication with the sea exists. The added weight of the flooding water and the virtual rise in G due to the free surface effect cause what is known as free communication effect. With an off-center space flooded, a ship will assume a list which will be further aggravated by the free surface effect. As the ship lists, more water will flow into the compartment from the sea and will tend to level off at the height of the external waterline. The additional weight causes the ship to sink further allowing more water to enter, causing more list until some final list is reached. The reduction of GM due to free communication effect is approximately equal in magnitude to

$$\frac{ay^2}{V}$$

where

    a = area of the free surface in square feet

    y = perpendicular distance from the geometric center of the free surface area to the fore-and-aft centerline of the intact waterline plane in feet

    V = new volume of ship's displacement after flooding, in cubic feet. Thus reduction in GM is additional to and separate from the free surface effect.

The approximate reduction in GZ may be computed from

$$GZ = \frac{ay^2}{V} \sin \theta$$

This may be considered as a virtual rise in G, superimposed upon the virtual rise in G due to the free surface effect.

If two partially filled tanks on opposite sides of an intact ship are connected by an open pipe at or near their bottoms allowing a free flow of liquid between them, the effect on GM is the same as if both tanks were in free communication with the sea. Hence, valves in cross connections between such tanks should never be left open without anticipating the accompanying

decrease in stability. Such free flow is known as sluicing.

SUMMARY OF EFFECTS OF LOOSE WATER

The addition of loose water to a ship alters the stability characteristics by means of three effects that must be considered separately: (1) the effect of added weight; (2) the effect of free surface; and (3) the effect of free communication.

Figure 3-30 shows the development of a stability curve with corrections for added weight, free surface, and free communication. Curve A is the ship's original stability curve before flooding. Curve B represents the situation after flooding; this curve shows the effect of added weight (increased stability) but it does not show the effects of free surface or of free communication. Curve C is curve B corrected for free surface effect only. Curve D is curve B corrected for both free surface effect and free communication effect. Curve D, therefore, is the final stability curve; it incorporates corrections for all three effects of loose water.

LONGITUDINAL STABILITY AND EFFECTS OF TRIM

The important phases of longitudinal inclination are changes in trim and longitudinal stability. A ship pitches longitudinally in contrast to rolling transversely and it trims for-and-aft, whereas it lists transversely. The difference in forward and after draft is defined as trim.

CENTER OF FLOATATION

When a ship trims, it inclines about an axis through the geometric center of the waterline plane. This point is known as the center of flotation. The position for the center of flotation aft of the mid-perpendicular for various drafts may be found from a curve on the curves of form (fig. 3-25). When a center of flotation curve is not available, or when precise calculations are not required, the mid-perpendicular may be used in lieu of the center of flotation.

CHANGE OF TRIM

Change of trim may be defined as the change in the difference between the drafts forward and aft. If in changing the trim, the draft forward

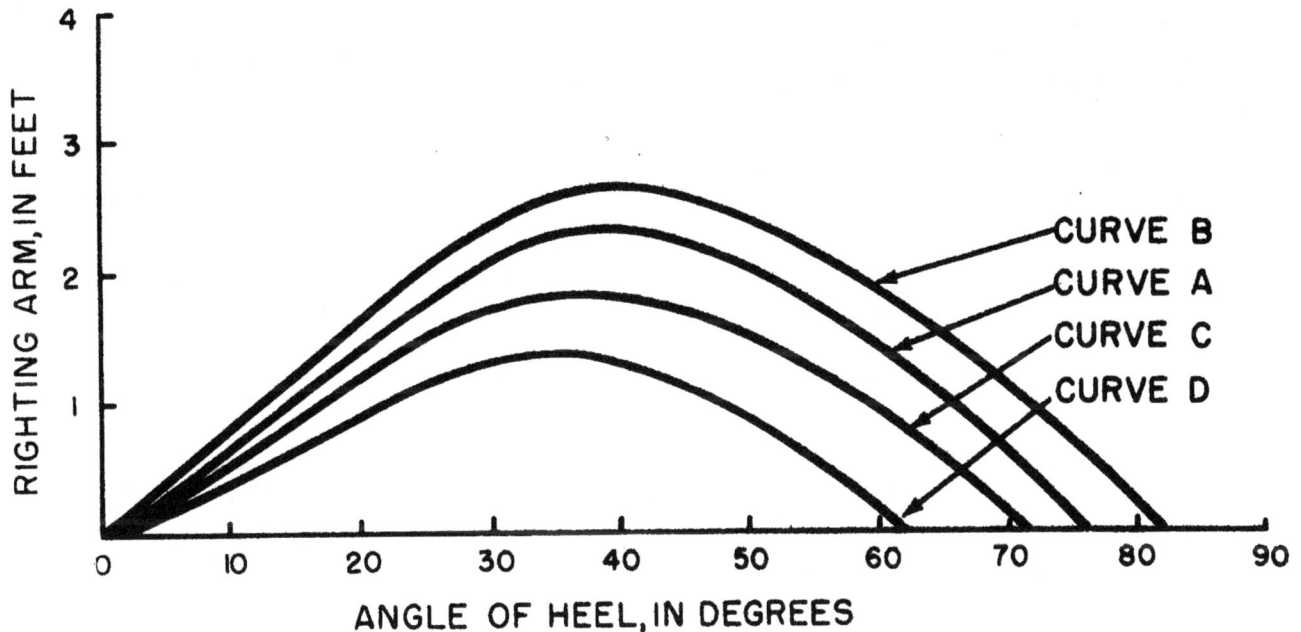

8.64

Figure 3-30.—Stability curve corrected for effects of added weight,
free surface, and free communication.

becomes greater, then the change is said to be
by the bow. Conversely, if the draft aft becomes
greater, the change of trim is by the stern.

Changes of trim are produced by shifting
weights forward or aft or by adding or subtracting weights forward of or abaft of the center of
flotation.

LONGITUDINAL STABILITY

Longitudinal stability is the tendency of a
ship to resist a change in trim. For small angles
of inclination, the longitudinal metacentric height
multiplied by the displacement is a measure of
initial longitudinal stability. The longitudinal
metacentric height is designated GM' and is
found from

$$GM' = KB + BM' - KG$$

where

KB and KG are the same as for transverse
stability BM' (the longitudinal metacentric
radius) is equal to

$$BM' = \frac{I'}{V}$$

where

I' = the moment of inertia of the ship's
waterline plane about an athwartship
axis through the center of flotation

V = the ship's volume of displacement

The value of BM' is very large—sometimes
more than a hundred times that of BM. The
values of BM' for various drafts may be found
from the curves of form (fig. 3-25).

MOMENT TO CHANGE TRIM ONE INCH

The measure of a ship's ability to resist a
change of trim is the moment required to produce a change of trim of a definite amount, such
as one inch. The value of the moment to change
trim one inch is obtained from

$$MTI = \frac{GM' \times W}{12 L}$$

where

GM' = longitudinal metacentric height (feet)

W = displacement (tons)

L = length between forward and after
perpendiculars (feet)

57

For practical work, BM' is usually substituted for GM' since they are both large and the difference between them is relatively small. When this is done, howver, MTI is called the approximate moment to change trim one inch. This value may often be found as a curve in the curves of form (fig. 3-25). If not, the approximate moment to change trim one inch may be calculated from

$$MTI = \frac{BM' \times W}{12\ L}$$

CALCULATION OF CHANGE OF TRIM

The movement of weight aboard ship in a fore-and-aft direction produces a trimming moment. This moment is equal to the weight multiplied by the distance moved. The change of trim in inches may be calculated by dividing the trimming moment by the moment to change trim one inch:

$$\text{change of trim} = \frac{w \times t}{MTI}$$

The direction of change of trim is the same as that of weight movement. If we are using midships as our axis of rotation, the change in draft forward equals the change in draft aft. This change of draft forward or aft is one-half the change of trim; for example, for a change of trim by the stern the after draft increases the same amount the forward draft decreases, that is, one-half the change of trim. The reverse holds true for a change of trim by the bow.

Example: If 50 tons of ammunition are moved from approximately 150 feet forward of the center of flotation to approximately 150 feet aft of the center of flotation (300 feet), what are the new drafts?

draft fwd = 19 feet 9 inches
draft aft = 20 feet 3 inches
mean draft = 20 feet
trimming moment = 50 x 300 = 15,000 foot-tons

moment to change trim one inch = 1940 foot-tons, from the following calculation:

$$MTI\ \frac{BM' \times W}{12\ L} = \frac{1150 \times 11,800}{12 \times 582} = 1940 \text{ foot-tons}$$

(BM' from curve in figure 3-25)

$$\text{change of trim} = \frac{15,000}{1940} = 8 \text{ inches by the stern}$$

change of draft — 4 inches fwd, 4 inches aft

new draft fwd = 19 feet 5 inches
new draft aft = 20 feet 7 inches

LONGITUDINAL WEIGHT ADDITION

The addition of a weight either directly above or below the center of flotation will cause an increase in mean draft but will not change trim. All drafts will change by the same amount as the mean draft. The reverse is true when a weight is removed at the center of flotation.

To determine the change in drafts forward and aft due to adding a weight on the ship, the computation is in two steps. First, the weight is assumed to be added at the center of flotation. This increases the mean draft and all the drafts by the same amount. The increase is equal to the weight added, divided by the tons-per-inch immersion. With the ship at its new drafts, the weight is assumed to be moved to its ultimate location. Moving the weight fore and aft produces a trimming moment and therefore a change in trim which is calculated as previously described.

FLOODING EFFECT DIAGRAM

From the flooding effect diagram of the ship's Damage Control Book it is possible to obtain the change in draft fore and aft due to solid flooding of a compartment. The weight of water to flood specific compartments is given and trimming moment produced may be computed, as well as list in degrees which may be caused by the additional weight. Additional information on the flooding effect diagram can be obtained from Chapter 9880 of the Naval Ships Technical Manual.

EFFECT OF TRIM ON TRANSVERSE STABILITY

The curves of form prepared for a ship are based on the design conditions, i.e., with no trim. For most types of ship, so long as trim does not become excessive, the curves are still applicable, and may be used without adjustment.

When a ship trims by the stern, the transverse metacenter is slightly higher than indicated by the KM curve, because both KB and BM increase. The center of buoyancy rises because of the movement of a wedge of buoyancy upward. The increased BM is the result of an enlarged waterplane as the ship trims by the stern.

Trim by bow usually means a decreased KM. The center of buoyancy will rise slightly, but this is usually counteracted by the decreased BM caused by the lower moment of inertia of the trimmed waterplane.

CAUSES OF IMPAIRED STABILITY

The stability of a ship may be impaired by several causes, resulting from mistakes or from enemy action. A summary of these causes and their effects follows:

ADDITION OF TOPSIDE WEIGHT

The addition of appreciable amounts of topside weight may be occasioned by unauthorized alterations; icing conditions; provisions, ammunition, or stores not struck down; deck cargo; and other conditions of load. Whenever a weight of considerable magnitude is added above the ship's existing center of gravity the effects are:

1. Reduction of reserve buoyancy.
2. Reduction of GM and righting arms due to raising G.
3. Reduction of GM and righting arms due to loss of freeboard (change of waterplane).
4. Reduction of righting arms if G is pulled away from the centerline.
5. Increase in righting moment due to increased displacement.

The net effect of added high weight is always a reduction in stability. The reserve buoyancy loss is added weight in tons. The new metacentric height can be obtained from:

$$G_1 M_1 \doteq KM_1 - KG_1$$

Stability is determined by selecting a new stability curve from the cross curves and correcting it for $AG_1 \sin \theta$ and $G_1 G_2 \cos \theta$.

LOSS OF RESERVE BUOYANCY

Reserve buoyancy may be lost due to errors, such as poor maintenance, failure to close fittings properly, improper classification of fittings, and overloading the ship; or it may be lost as a result of enemy action such as fragment or missile holes in boundaries, blast which carries away boundaries or blows open or warps fittings, and flooding which overloads the ship. When the above-water body is holed, some reserve buoy-ancy is lost. The immersion of buoyant volume is necessary to the development of a righting arm as the ship rolls; if the hull is riddled it can no longer do this on the damaged side, toward which it will roll. In effect, the riddling of the above-water hull is analogous to losing a part of the freeboard, thus reducing stability. When this happens, if the ship takes water aboard on the roll, the combined effects of high added weight and free surface operate to cut down the righting moment. Therefore, the under-water hull and body should be plugged and patched, and every effort should be made to re-store the watertightness of external and internal boundaries in the above-water body.

FLOODING

Flooding may take place because of under-water damage, shell or bomb burst below decks, collision, topside hit near the waterline, fire-fighting water, ruptured poping, sprinkling of magazines, counterflooding, or leakage. Regard-less of how it takes place it can be classified in three general categories, each of which can be further broken down, as follows:

1. with respect to boundaries

   a. solid footing
   b. partial flooding
   c. partial flooding in free communication with the sea

2. with respect to height in the ship

   a. center of gravity of the flooding water is above G
   b. center of gravity of the flooding water is below G

3. with respect to the ship's centerline

   a. symmetrical flooding
   b. off-center flooding

Solid Flooding

The term solid flooding designates the situation in which a compartment is completely filled from deck to overhead. In order for this to occur the compartment must be vented as by an air escape, an open scuttle or vent fitting, or through fragment holes in the overhead. Solid flooding water behaves exactly like an added weight and has the effect of so many tons placed exactly at the center of gravity of the flooding water. It is more likely to occur below the waterline, where it has the effect of any added low weight. Inasmuch as G is usually a little above the waterline in warships, the net effect of solid flooding below the waterline is most frequently a gain in stability, unless a sizeable list or a serious loss of freeboard results in a net reduction of stability. The reserve buoyancy consumed is the weight of flooding water in tons, and the new GM and stability characteristics are found as previously explained.

Partial Flooding with Boundaries Intact

The term partial flooding refers to a condition in which the surface of the flooding water lies somewhere between the deck and the overhead of a compartment. The boundaries of the compartment remain watertight and the compartment remains partially but not completely filled. Partial flooding can be brought about by leakage from other damaged compartments or through defective fittings, seepage, shipping water on the roll, downward drainage of water, loose water from firefighting, sprinklers, ruptured piping, and other damage.

Partial flooding of a compartment that has intact flooding boundaries affects the stability of the ship because of (1) the effect of added weight, and (2) the effect of free surface. The effect of the added weight will depend upon whether the weight is high in the ship or low, and whether it is symmetrical about the centerline or is off-center. The effect of free surface will depend primarily upon the athwartship breadth of the free surface. Unless the free surface is relatively narrow and the weight is added low in the ship, the net effect of partial flooding in a compartment with intact boundaries is likely to be a very definite loss in overall stability.

Partial Flooding in Free Communication with the Sea

Free communication can exist only in partially flooded compartments in which it is possible for the sea to flow in and out as the ship rolls. Partial flooding with free communication is most likely to occur when there is a large hole that extends above and below the waterline. It may also occur in a waterline compartment when there is a large hole in the shell below the waterline, if the compartment is vented as the ship rolls. Where free communication does exist, the water level in the compartment remains at sea level as the ship rolls.

When a compartment is partially flooded and in free communication with the sea, the ship's stability is affected by (1) added weight, (2) free surface effect, and (3) free communication effect. In general, net effect of partial flooding with free communication is a decided loss in stability.

CHAPTER 4

PREVENTIVE AND CORRECTIVE DAMAGE CONTROL

Aboard ship, the overall damage and casualty control function is composed of two separate but related phases: the engineering casualty control phase and the damage control phase. The engineering officer is responsible for both phases.

The engineering casualty control phase is concerned with the prevention, minimization, and correction of the effects of operational and battle casualties to the machinery, electrical systems, and piping installations, to the end that all engineering services may be maintained in a state of maximum reliability under all conditions of operation. Engineering casualty control is handled almost entirely by personnel of the engineering department.

The damage control phase, on the other hand, involves practically every person aboard ship. The damage control phase is concerned with such things as the preservation of stability and watertight integrity, the control of fires, the control of flooding, the repair of structural damage, and the control of nuclear, biological, and chemical contamination. Although under the control of the engineer officer, damage control is an all-hands responsibility.

This chapter presents some basic information on the principles of the damage control phase of the damage and casualty control function. Information on engineering casualty control is not included here; any such information would be relatively meaningless without a considerable background knowledge of the normal operating characteristics of shipboard machinery and equipment.

PREPARATIONS TO RESIST DAMAGE

Naval ships are designed to resist accidental and battle damage. Damage resistant features include structural strength, watertight compartmentation, stability, and buoyancy. Maintaining these damage resistant features and maintaining a high state of material and personnel readiness before damage is far more important for survival than are any damage control measures that can be taken after the ship has been damaged. It has been said that 90 percent of the damage control needed to save a ship takes place before the ship is damaged and only 10 percent can be done after the damage has occurred. In spite of all precautions and all preparatory measures, however, the survival of a ship sometimes depends upon prompt and effective damage control measures taken after damage has occurred. It is essential, therefore, that all shipboard personnel be trained in damage control procedures.

The maintenance of watertight integrity is a vital part of any ship's preparations to resist damage. Each undamaged tank or compartment aboard ship must be kept watertight if flooding is not to be progressive after damage. Watertight integrity can be lost in a number of ways. Failure to secure access closures and improper maintenance of watertight fittings and compartment boundaries, as well as external damage to the ship, can cause loss of watertight integrity, a thorough system of tests and inspections is prescribed. The condition of watertight boundaries, compartments, and fittings is determined by visual observation and by various tests, including chalk tests and air tests. All defects discovered by any test or inspection must be remedied immediately.

For most ships, a mandatory schedule of watertight integrity tests and inspections is prepared. This schedule informs each ship of the compartments subject to test and/or inspection, specifying which type of test or inspection shall be applied. Ships not provided with such a schedule are nevertheless required to make inspections of important watertight boundaries as required by chapter 9290 of the of Naval Ships Technical Manual.

DAMAGE CONTROL ORGANIZATION

In order to ensure damage control training and to provide prompt control of casualties, a damage control organization must be set up and kept active on all ships.

As previously noted, the engineer officer is responsible for damage control. The damage control assistant (DCA), who is under the engineer officer, is responsible for establishing and maintaining an effective damage control organization. Specifically, the DCA is responsible for the prevention and control of damage, the training of ship's personnel in damage control, and the operation, care, and maintenance of certain auxiliary, machinery, piping, and drainage systems not assigned to other departments or divisions.

Although naval ships may be large or small, and although they differ in type, the basic principles of the damage control organization are more or less standardized. Some organizations are larger and more elaborate than others, but they all function on the same basic principles.

A standard damage control organization, suitable for large ships but followed by all ships as closely as practicable, includes damage control central and repair stations. Damage control central is integrated with propulsion and electrical control in a Central Control Station on new large ships and is a separate Station on older and small ships. Repair parties are assigned to specifically located repair stations. Repair stations are further subdivided into unit patrols to permit dispersal of personnel and a wide coverage of the assigned areas.

DAMAGE CONTROL CENTRAL

The primary purpose of damage control central is to collect and compare reports from the various repair stations in order to determine the condition of the ship and the corrective action to be taken. The commanding officer is kept posted on the condition of the ship and on important corrective measures taken. The damage control assistant, at his battle station in damage control central, is the nerve center and directing force of the entire damage control organization. He is assisted in damage control central by a stability officer, a casualty board operator, and a damage analyst. In addition, representatives of the various divisions of the engineering department are assigned to damage control central.

In damage control central, repair party reports are carefully checked so that immediate action can be taken to isolate damage and to make emergency repairs in the most effective manner. Graphic records of the damage are made on various damage control diagrams and status boards, as the reports are received. For example, reports concerning flooding are marked up, as they come in, on a status board which indicates liquid distribution before damage. With this information, the stability and buoyance of the ship can be estimated and the necessary corrective measures can be determined.

If damage control central is destroyed or is for other reasons unable to retain control, the repair stations, in designated order, take over these same functions. Provisions are also made for passing the control of each repair station down through the officers, petty officers, and nonrated men, so that no group will ever be without a leader.

REPAIR PARTIES

A standard damage control organization on large ships includes the following repair stations:

Repair 1 (deck or topside repair).
Repair 2 (forward repair).
Repair 3 (after repair).
Repair 4 (amidship repair).
Repair 5 (propulsion repair).
Repair 6 (ordnance repair).

On carriers, there are two additional repair stations—Repair 7 (gallery deck and island structure repair) and Repair 8 (electronics repair party). Carriers also have special organized teams such as Aviation Fuel Repair, Crash and Salvage, and Ornance Disposal. On small ships, there are usually three repair stations—Repair 2, Repair 3, and Repair 5.

The organization of repair stations is basically the same on all types of ships; however, more men are available for manning repair stations on large ships than on small ships. The number and the ratings of men assigned to a repair station, as specified in the battle bill, are determined by the location of the station, the portion of the ship assigned to that station, and the total number of men available.

Each repair party has an officer in charge, who may in some cases be a chief petty officer. The second in charge is usually a chief petty officer who is qualified in damage control and who is capable of taking over the supervision of the repair party.

Many repair stations have unit patrol stations at key locations in their assigned areas to supplement the repair station. Operating instructions should be posted at each repair station. In general, instructions should include the purpose of the repair station; the specific assignments of space for which that station is responsible; instructions for assigning and stationing personnel; methods and procedures for damage control communications; instructions for handling machinery and equipment located in the area; procedures for nuclear, biological, and chemical (NBC) defense; sequence and procedure for passing control from one station to another; a list of current damage control bills; and a list of all damage control equipment and gear provided for the repair station.

MATERIAL CONDITIONS OF READINESS

Material conditions of readiness refers to the degree of access and system closure to limit the extent of damage to the ship. Maximum closure is not maintained at all times because it would interfere with the normal operation of the ship. For damage control purposes, naval ships have three material conditions of readiness, each condition representing a different degree of tightness and protection. The three material conditions of readiness are called X-RAY, YOKE, and ZEBRA. These titles, which have no connection with the phonetic alphabet, are used in all spoken and written communications concerning material conditions.

Condition X-RAY, which provides the least protection, is set when the ship is in no danger from attack, such as when it is at anchor in a well protected harbor or secured at a home base during regular working hours.

Condition YOKE, which provides somewhat more protection than condition X-RAY, is set and maintained at sea. It is also maintained in port during wartime and at other times in port outside of regular working hours.

Condition ZEBRA is set before going to sea or entering port, during wartime. It is also set immediately, without further orders, when manning general quarters stations. Condition ZEBRA is also set to localize and control fire and flooding when not at general quarters stations.

The closures involved in setting the material conditions of readiness are labeled as follows:

X-RAY, marked with a black X. These closures are secured during conditions X-RAY, YOKE, and ZEBRA.

YOKE, marked with a black Y. These closures are secured during conditions YOKE and ZEBRA.

ZEBRA, marked with a red Z. These closures are secured during condition ZEBRA.

Once the material condition is set, no fitting marked with a black X, a black Y, or a red Z may be opened without permission of the commanding officer (through the damage control assistant or the officer of the deck.) The repair party officer controls the opening and closing of all fittings in his assigned area during general quarters.

Additional fitting markings for specific purposes are modifications of the three basic conditions, as follows:

CIRCLE X-RAY fittings, marked with a black X in a black circle, are secured during conditions X-RAY, YOKE, and ZEBRA. CIRCLE YOKE fittings, marked with a black Y in a black circle, are secured during conditions YOKE and ZEBRA. Both CIRCLE X-RAY and CIRCLE YOKE fittings may be opened without special authority when going to or securing from general quarters, when transferring ammunition, or when operating vital systems during general quarters; but the fittings must be secured when not in use.

CIRCLE ZEBRA fittings, marked with a red Z in a red circle, are secured during condition ZEBRA. CIRCLE ZEBRA fittings may be opened during prolonged periods of general quarters, when the condition may be modified. Opening these fittings enables personnel to prepare and distribute battle rations, open limited sanitary facilities, ventilate battle stations, and provide access from ready rooms to flight deck. When open, CIRCLE ZEBRA fittings must be guarded for immediate closure if necessary.

DOG ZEBRA fittings, marked with a red Z in a black D, are secured during condition ZEBRA and during darken ship condition. The DOG ZEBRA classification applies to weather accesses not equipped with light switches or light traps.

WILLIAM fittings, marked with a black W, are kept open during all material conditions. This classification applies to vital sea suction valves supplying main and auxiliary condensers, fire pumps, and spaces that are manned during conditions X-RAY, YOKE, and ZEBRA; it also applies to vital valves that, if secured, would impair the mobility and fire protection of the ship. These items are secured only as necessary to control damage or contamination and to effect repairs to the units served.

CIRCLE WILLIAM fittings, marked with a black W in a black circle, are normally kept open (as WILLIAM fittings are) but must be secured as defense against nuclear, biological, or chemical attack.

INVESTIGATION OF DAMAGE

The DCA must be given all available information concerning the nature and extent of damage so that he will be able to analyze the damage and decide upon appropriate measures of control. The repair parties that are investigating the damage at the scene are normally in the best position to give dependable information on the nature and extent of the damage. All repair party personnel should be trained to make prompt, accurate, and complete reports to damage control central. Items that should normally be reported to damage control central include:

1. Description of important things seen, heard, or felt by personnel.

2. Location and nature of fires, smoke, and toxic gases.

3. Location and nature of progressive flooding.

4. Overall extent and nature of flooding.

5. Structural damage to longitudinal strength members.

6. Location and nature of damage to vital piping and electrical systems.

7. Local progress made in controlling fire; halting flooding; isolating damaged systems; and rigging jury piping, casualty power, and emergency communications.

8. Compartment-by-compartment information on flooding, including depth of liquid in each flooded compartment.

9. Condition of boundaries (decks, bulkheads, and closures) surrounding each flooded compartment.

10. Local progress made in reclaiming compartments by plugging, patching, shoring, and removing loose water.

11. Areas in which damage is suspected but cannot be reached or verified.

The DCA must ascertain just what information the commanding officer desires concerning the extent of the damage incurred and the corrective measures taken. The DCA must also find out how detailed the information to the CO should be and when it is to be furnished. With these guidelines in mind, the DCA must sift all information coming into damage control central and pass along to the bridge only the type of information that the CO wants to have.

CORRECTIVE MEASURES

Measures for the control of damage may be divided into two general categories: (1) overall ship survival measures, and (2) immediate local measures.

OVERALL SHIP SURVIVAL MEASURES

Overall ship survival measures are those actions initiated by damage control central for the handling of list, trim, buoyance, stability, and hull strength. Operations in this category have five general objectives: improving GM and overall stability, correcting for off-center weight, restoring lost freeboard and reserve buoyancy, correcting for trim, and relieving stress in longitudinal strength members.

Improving GM
and Overall Stability

The measures used to improve GM and overall stability in a damaged ship include (1) suppressing free surface, (2) jettisoning topside weights, (3) ballasting, (4) lowering liquid or solid weights, and (5) restoring boundaries.

Correcting for
Off-Center Weight

Off-center weight may occur as the result of unsymmetrical flooding or as the result of an athwartship movement of weight. Correcting for off-center weight may be accomplished by (1) pumping out off-center flooding water, (2) pumping liquids across the ship, (3) counterflooding, (4) jettisoning topside weights from the low side of the ship, (5) shifting solid weights athwartships, and (6) pumping liquids overboard from intact wing tanks on the low side.

Restoring Lost Freeboard
and Reserve Buoyancy

Restoring lost freeboard and reserve buoyancy requires the removal of large quantities of weight. In general, the most practicable way of accomplishing this is to restore watertight boundaries and to reclaim compartments by pumping them out. Any corrective measure which removes weight from the ship contributes to the restoration of freeboard.

Correcting for Trim

The methods used to correct for trim after damage include (1) pumping out flood water, (2) pumping liquids forward or aft, (3) counter-flooding the high end, (4) jettisoning topside weights from the low end, (5) shifting solid weights from the low end to the high end, and (6) pumping liquids over the side from intact tanks at the low end. The first of these methods— that is, pumping out flood water—is in most cases the only truly effective means of correcting a severe trim.

The correction of trim is usually secondary to the correction of list, unless the trim is so great that there is danger of submerging the weather deck at the low end.

Relieving Stress in
Longitudinal Strength Members

When a ship is partially flooded, the longitudinal strength members are subject to great stress. In cases where damage has carried away or buckled the strength members amidships, the additional stress imposed by the weight of the flooding water may be enough to cause the ship to break up. The only effective way of relieving stress caused by flooding is to remove the water. Other measures, such as removing or shifting weight, may be helpful but cannot be completely effective. In some instances, damaged longitudinals may be strengthened by welding.

IMMEDIATE LOCAL MEASURES

Immediate local measures are those actions taken by repair parties at the scene of the damage. In general, these measures include all on-scene efforts to investigate the damage, to report to damage control central, and to accomplish the following:

1. Establish flooding boundaries by selecting a line of intact bulkheads and decks to which the flooding may be held and by rapidly plugging, patching, and shoring to make these boundaries watertight and dependable.

2. Control and extinguish fires.

3. Establish secondary flooding boundaries by selecting a second line of bulkheads and decks to which the flooding may be held if the first flooding boundaries fail.

4. Advance flooding boundaries by moving in toward the scene of the damage, plugging, patching, shoring, and removing loose water.

5. Isolate damage to machinery, piping, and electrical systems.

6. Restore piping systems to service by the use of patches, jumpers, clamps, couplings, etc.

7. Rig casualty power.

8. Rig emergency communications and lighting.

9. Rescue personnel and care for the wounded.

10. Remove wreckage and debris.

11. Cover or barricade dangerous areas.

12. Ventilate compartments which are filled with smoke or toxic gases.

13. Take measures to counteract the effects of nuclear, biological, and chemical contamination or weapons.

Immediate local measures for the control of damage are of vital importance. It is not necessary for damage control central to decide on these measures; rather, they should be carried out automatically and rapidly by repair parties. However, damage control central should be continuously and accurately advised of the progress made by each party so that the efforts of all repair parties may be coordinated to the best advantage.

PRACTICAL DAMAGE CONTROL

Both the immediate local measures and the overall ship survival measures have, of course, the common aim of saving the ship and restoring it to service. The following subsections deal with the practical methods used to achieve this aim: controlling fires, controlling flooding, repairing structural damage, and restoring vital services.

It should be noted that controlling the effects of nuclear, biological, and chemical warfare weapons or agents may in some situations take precedence over other damage control measures. Because of the complex nature of NBC defense, this subject is treated separately in a later section of this chapter.

CONTROL OF FIRES

Fire is a constant potential hazard aboard ship. All possible measures must be taken to prevent its occurrence or to bring about its rapid control and extinguishment. In many cases, fire occurs in conjunction with other damage, as a result of enemy action, weather, or accident. Unless fire is rapidly and effectively extinguished, it may cause more damage than the initial casualty and it may, in fact, cause the loss of a ship even after other damage has been repaired or minimized.

Fires are classified according to the nature of the combustible material. Class A fires are those which involve ordinary combustible material such as wood, paper, mattresses, canvas, etc. Class B fires are those which involve the burning of oils, greases, gasoline, and similar materials. Class C fires are those which occur in electrical equipment. Class D fires are those which involve certain metals such as magnesium, potassium, powdered aluminum zinc, sodium, titanium, zirconium and others.

Class A fires are extinguished by the use of water. Class B fires are extinguished chiefly by smothering with foam, fog, steam, or purple K powder dry chemical agent (as appropriate for the particular fire). Class C fires are preferably extinguished by the use of carbon dioxide. Because of the danger of electric shock, a solid stream of water must never be used to extinguish a class C fire. Class D fires are presently extinguished by using large amounts of water. Personnel safety is of prime concern when fighting this class fire; toxic gasses, possible hydrogen explosions, splattering of molten metal, and intense heat are prime characteristics of this type of fire. Presently, intensive research is being conducted on better methods of attack and more suitable extinguishing agents.

The organization of a firefighting party depends on the number of men available. Figure 4-1 shows the basic organization of a small firefighting party, and figure 4-2 shows the basic organization of a large firefighting party. At times it is necessary for one person to perform more than one of the indicated duties, and this fact is taken into consideration in organizing firefighting parties.

One man in the firefighting party must be designated as the group or scene leader (investigator). His first duty is to get to the fire quickly; he investigates the situation, determines the

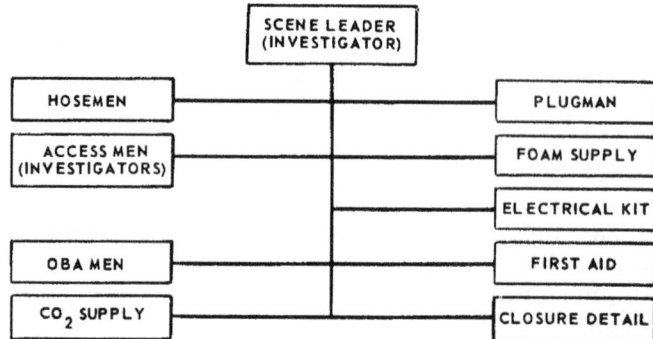

8.80

Figure 4-1.—Organization of small firefighting party.

nature of the fire, decides what type of equipment should be used, and informs damage control central. Later developments may require the use of different or additional equipment, but the scene leader must decide what to use first.

The number of hosemen assigned to a firefighting party varies in accordance with the number of men available and the size of the firehose. At least three men are required for a 1 1/2-inch hose, and four or five men are required for a 2 1/2-inch hose. The hosemen lead out the hose, remove kinks and sharp bends, and stand by the nozzles. Nozzlemen should wear oxygen breathing apparatus (OBA) while fighting fires.

The plugman stands by to operate the fireplug valve, when so ordered. He rigs and stands by jumper lines, assists on the hose lines, and clears the fireplug strainer when necessary.

The access men open doors, hatches, scuttles, and other openings and clear routes as necessary to gain access to the fire. These men carry equipment to open jammed fittings and locked doors. Once they have gained access to the fire, they make a detailed investigation of the fire area.

The foam supply man sets up the foam equipment for operation and operates it as required. He obtains spare foam cans from racks and prepares them for use.

The electrical kit man (or electrician) de-energizes all electrical equipment in the fire area, both to protect personnel and to prevent explosions or flashbacks. When necessary, he rigs power cables for portable tools, lights, and blowers.

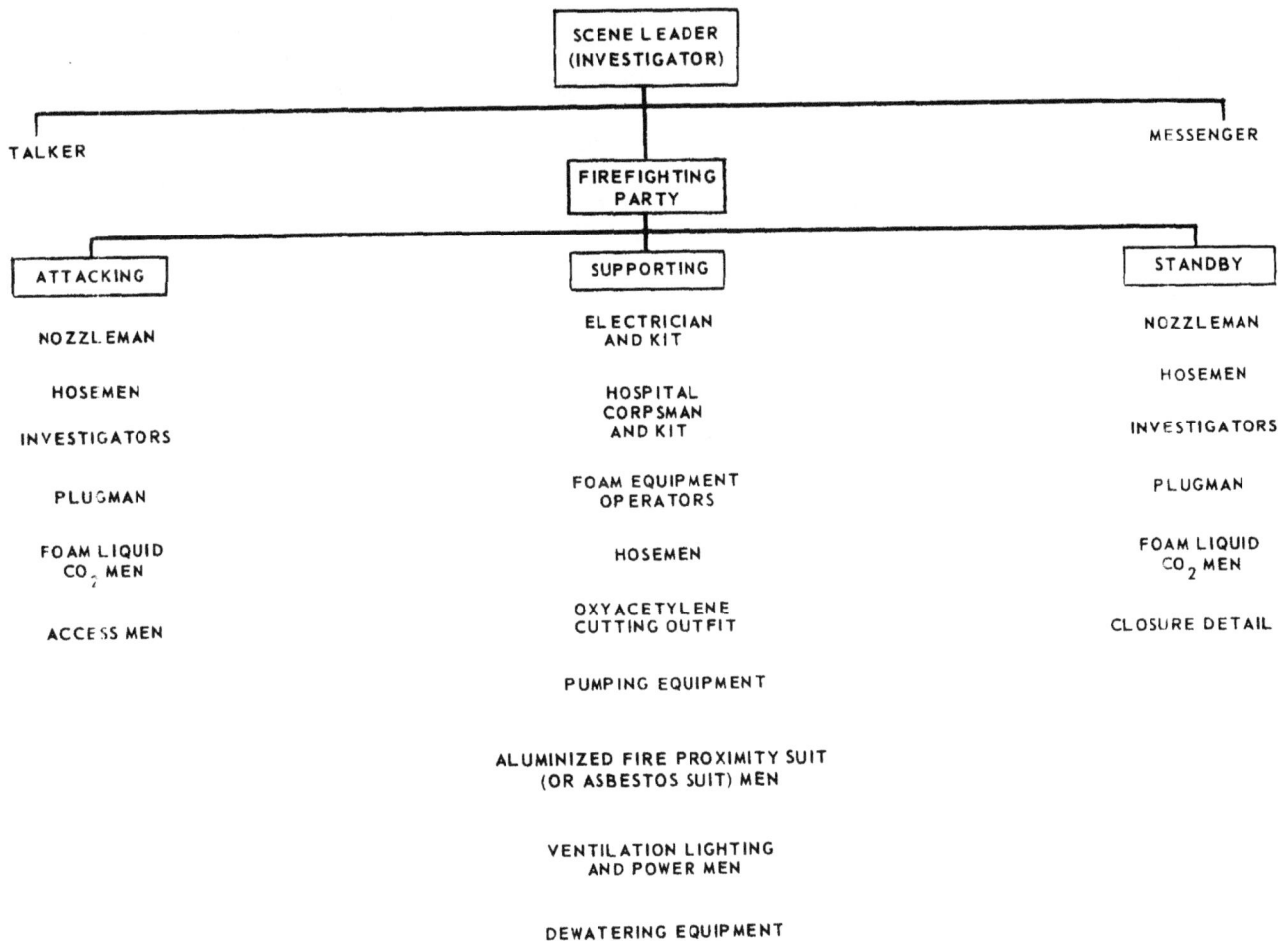

```
                              ┌─────────────────┐
                              │  SCENE LEADER   │
                              │ (INVESTIGATOR)  │
                              └─────────────────┘
          ┌──────────────────────────┼──────────────────────────┐
      TALKER                         │                        MESSENGER
                              ┌─────────────────┐
                              │  FIREFIGHTING   │
                              │     PARTY       │
                              └─────────────────┘
      ┌───────────────────────────────┼───────────────────────────────┐
 ┌───────────┐                 ┌─────────────┐                  ┌───────────┐
 │ ATTACKING │                 │ SUPPORTING  │                  │  STANDBY  │
 └───────────┘                 └─────────────┘                  └───────────┘
```

ATTACKING	SUPPORTING	STANDBY
NOZZLEMAN	ELECTRICIAN AND KIT	NOZZLEMAN
HOSEMEN	HOSPITAL CORPSMAN AND KIT	HOSEMEN
INVESTIGATORS		INVESTIGATORS
PLUGMAN	FOAM EQUIPMENT OPERATORS	PLUGMAN
FOAM LIQUID CO_2 MEN	HOSEMEN	FOAM LIQUID CO_2 MEN
ACCESS MEN	OXYACETYLENE CUTTING OUTFIT	CLOSURE DETAIL
	PUMPING EQUIPMENT	
	ALUMINIZED FIRE PROXIMITY SUIT (OR ASBESTOS SUIT) MEN	
	VENTILATION LIGHTING AND POWER MEN	
	DEWATERING EQUIPMENT	

8.81

Figure 4-2.—Organization of large firefighting party.

The CO_2 dry powder supply men take extinguishers to the fire and operate them as necessary.

The OBA men have their gear on and ready for immediate use throughout the firefighting operation. OBA tenders are in charge of tending lines, when used, and keeping spare canisters readily available. The OBA men assist in making the investigation in situations where oxygen breathing apparatus is necessary for entry. The OBA men also work with hoses and perform other duties in spaces containing toxic gases.

The closure detail secures all doors, hatches, and openings around the fire area to isolate the fire area. This group secures all ventilation closures and fans in the area of smoke and heat, establishes secondary fire boundaries by cooling down nearby areas, and assists in fighting the fire as necessary.

Hospital Corpsmen render first aid to the injured. The JZ talker establishes and maintains communications with damage control central, either directly or through the local repair party.

Since no two fires are exactly alike, the deployment of men and equipment is not always the same. In most situations, however, the following general rules are observed:

1. The attacking party, which is the first line of defense, must have a sufficient number of men. The attacking party makes the initial investigation and moves in to contain the fire.

2. The supporting party should not have any more men than are actually required to bring up auxiliary equipment, assist with foam and

67

CO_2 supply, and fight the fire. Men at the fire who are not actually engaged in fighting the fire should be sent to the standby party. The standby party makes the closures necessary to isolate the fire area, cools the surrounding areas, supplies foam and CO_2, and assists in fighting the fire when necessary.

3. The firefighting party must be quiet and orderly. There should be only two men talking at a fire: the leader of the firefighting party and the messenger or phone talker.

4. All orders issued at the scene must be clear, concise, and accurate.

5. All reports to damage control central must be clear, concise, and accurate.

6. All possible safety precautions must be observed.

7. Precautions must be taken to see that the fire does not spread. Fire boundaries must be established, and men must be stationed in adjacent compartments to see that the fire does not spread. Even distant compartments may require checking, since fire can spread a great distance through ventilation ducts.

Figures 4-3 and 4-4 show members of a shipboard firefighting party in action.

CONTROL OF FLOODING

Flooding[1] may occur from a number of different causes. Underwater or waterline damage, ruptured water piping, the use of large quantities of water for firefighting or counterflooding, and the improper maintenance of boundaries are all possible causes of flooding aboard ship. It should be noted that ballasting fuel oil tanks with sea water after the oil has been removed is not considered a form of flooding; ballasting merely consists of replacing one liquid with another in order to maintain the ship in a condition of maximum resistance to damage.

If a ship suffers such extensive damage that it never stops listing, trimming, and settling in the water, the chances are that it will go down within a very few minutes. If, on the other hand, a ship stops listing, trimming, and settling shortly after the damage occurs, it is not likely to sink at all unless progressive flooding is allowed to occur. Thus there is an excellent chance of saving any ship that does not sink immediately. There is no case on record of a ship sinking suddenly after it has stopped listing, trimming, and settling, except in cases where progressive flooding occurred.

The control of flooding requires that the amount of water entering the hull be restricted or entirely stopped. The removal of flooding water cannot be accomplished until flooding boundaries have been established. Pump capacity should never be wasted on compartments which cannot be quickly and effectively made tight. If a compartment fills rapidly, it is a sign that pumping capacity will be wasted until the openings have been plugged or patched. The futility of merely circulating sea water should be obvious.

Once flooding boundaries have been established, the removal of the flooding water should be undertaken on a systematic basis. Loose water—that is, water with free surface—and water that is located high in the ship should be removed first. Compartments which are solidly flooded and which are low in the ship are generally dewatered last, unless the flooding is sufficiently off-center to cause a serious list. Compartments must always be dewatered in a sequence that will contribute to the overall stability of the ship. For example, a ship could be capsized if low, solidly flooded compartments were dewatered while water still remained in high, partially flooded compartments.

In order to know which compartments should be dewatered first, it is necessary to know the effect of flooding on all ship's compartments. This information is given in the flooding effect diagram in the ship's Damage Control Book. The flooding effect diagram consists of a series of plan views of the ship at various levels, showing all watertight, oiltight, airtight, fumetight, and fire-retarding subdivisions. Compartments on the flooding effect diagram are colored in the following way:

1. If flooding the compartment results in a decrease in stability because of high weight, free surface effect, or both, the compartment is colored pink.

2. If flooding the compartment improves stability even though free surface exists, the compartment is colored green.

3. If flooding the compartment improves stability when the compartment is solidly flooded but impairs stability when a free surface exists, the compartment is colored yellow.

4. If flooding the compartment has no very definite effect on stability, the compartment is left uncolored.

---

[1]The effects of various types of flooding on stability are discussed in chapter 3 of this text.

8.120

Figure 4-3.—Members of firefighting party cooling hatch.

The flooding effect diagram also shows the weight of salt water (in tons) required to fill the compartment; this is indicated by a numeral in the upper left-hand corner. In addition, the transverse moment of the weight (in foot-tons) about the centerline of the ship is indicated for all compartments which are not symmetrical about the centerline.

Facilities for dewatering compartments consist of the fixed drainage systems of the ship and portable equipment such as electric submersible pumps, P-500 pumps, P-250 pumps, and eductors. On a large combat ship, the fixed drainage systems have a total pumping capacity of about 12,200 gallons per minute—less, it might be noted, than the amount of

8.122

Figure 4-4.—Members of firefighting party cooling
entrance to compartment.

water admitted by a hole 1 square foot in size in an area located 15 feet below the waterline.

Portable submersible pumps used aboard naval ships are centrifugal pumps driven by a water-jacketed, constant-speed a-c or d-c motor. When a submersible pump is being used to dewater a compartment, the pump is lowered into the water and a discharge hose is led to the nearest point of discharge. Since the delivery of the pump increases as the discharge head decreases, dewatering can be accomplished faster if the water is discharged at the lowest practicable point and if the discharge hose is short and free from kinks. When it is necessary to dewater against a high discharge head, two submersible pumps can be used in tandem, as shown in figure 4-5. The pump at the lower level lifts water to the suction side of the pump at the higher level.

The P-500 portable pump, originally developed for firefighting, is also used for dewatering flooded spaces. This pump is of the centrifugal

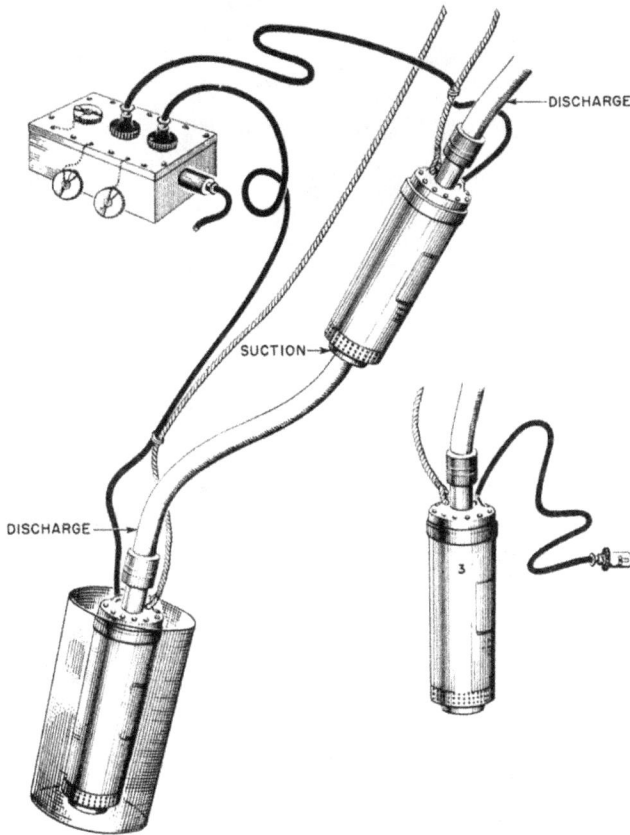

11.359

Figure 4-5.—Tandem connections for
submersible pumps.

type; it is driven by a water-cooled gasoline engine of special design. The pump delivers 500 gallons per minute at 100 pounds per square inch pressure, with a suction lift of 16 feet. The capacity may be increased by decreasing the discharge pressure.

The P-250 pump, which is similar to the P-500 pump except for capacity, is scheduled to replace the P-500 pump aboard ship. A P-250 pump is shown in figure 4-6.

In order to estimate the number of pumps required to handle a flooding situation, it is necessary to consider the amount and location of the water to be removed, the capacity and availability of the installed drainage systems, and the capacity of the available portable pumps. It is also necessary to know whether the leaks are completely plugged, partially plugged, or not plugged at all—in short, it is necessary to know how much water is coming in while water is being pumped out.

REPAIR OF STRUCTURAL DAMAGE

The kinds of damage that may have to be repaired while a ship is still in the battle area include holes above and below the waterline; cracks in steel plating; punctured, weakened, or distorted bulkheads; warped or sprung doors and hatches; weakened or ruptured beams, supports, and other strength members; ruptured or weakened decks; ruptured or cracked piping; severed electrical cables; broken or distorted foundations under machinery; broken or pierced machinery units; and a wide variety of miscellaneous wreckage that may interfere with the functioning of the ship.

One of the most important things to remember in connection with the repair of structural damage is that a ship can sink just as easily from a series of insignificant-looking small holes as it can from one larger and more dramatic-looking hole. A natural enough tendency—and one which can lead to the sinking of a ship—is to attack the large, obvious damage first and to overlook the smaller holes through interior bulkheads. Men sometimes waste hours trying to patch large holes in already flooded compartments, disregarding the smaller holes through which progressive flooding is gradually taking place. In many situations, it would be better to concentrate on the smaller interior holes; as a rule, the really large holes in the underwater hull cannot be repaired anyway until the ship is drydocked.

Holes in the hull at or just above the waterline should be given immediate attention. Although holes in this location may appear to be relatively harmless, they are actually extremely hazardous. As the ship rolls or loses buoyancy, the holes become submerged and admit water at a level that is dangerously high above the ship's center of gravity.

The methods and materials used to repair holes above the waterline are also used, for the most part, for the repair of underwater holes. The greatest difficulty encountered in repairing underwater damage is usually the inaccessibility of the damage. If an inboard compartment is flooded, opening doors or hatches to get to the damage would result in further flooding of other compartments. In such a case, it is usually necessary to send a man wearing a shallow-water diving apparatus down

3.164

Figure 4-6.—P-250 portable pump.

into the compartment. His repair work is likely to be hampered by tangled wreckage in the water, by the absence of light to work by, and by the difficulties of trying to keep buoyant repair materials submerged.

Shoring is often used aboard ship to support ruptured decks, to strengthen weakened bulkheads and decks, to build up temporary decks and bulkheads against the sea, to support hatches and doors, and to provide support for equipment which has broken loose.

The basic materials required for shoring are shores, wedges, sholes, and strongbacks. A shore is a portable beam. A wedge is a block,

triangular on the sides and rectangular on the butt end. A shole is a flat block which may be placed under the end of a shore for the purpose of distributing the pressure. A strongback is a piece used to distribute pressure or to serve as an anchor for a patch.

When to shore is a problem that cannot be solved by the application of any one set of rules. Sometimes the need for shoring is obvious, as in the case of damaged hatches; but sometimes dangerously weakened supports under guns or machinery may not be so readily noticed. Although shoring is sometimes done when it is not really necessary, the best general rule to follow is this: in case of doubt, it is always better to shore than to gamble on the strength of an important deck, bulkhead, hatch, or other member.

Some examples of shoring are illustrated in figure 4-7.

RESTORATION OF VITAL SERVICES

Thus far we have considered practical damage control operations from the point of view of combatting fires, getting rid of flooding water,

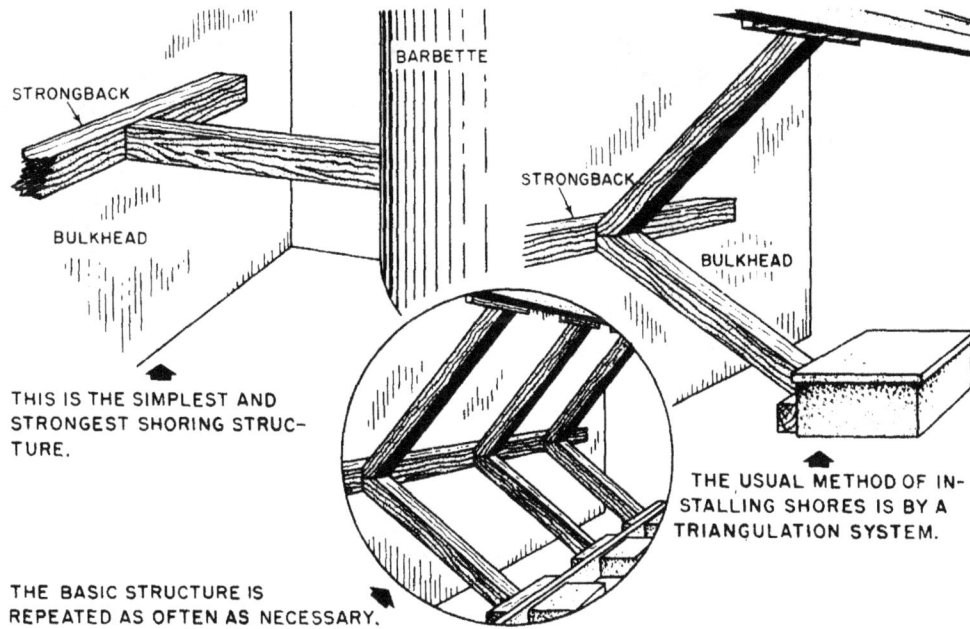

STRONGBACK
BARBETTE
BULKHEAD
STRONGBACK
BULKHEAD

THIS IS THE SIMPLEST AND STRONGEST SHORING STRUCTURE.

THE BASIC STRUCTURE IS REPEATED AS OFTEN AS NECESSARY.

THE USUAL METHOD OF INSTALLING SHORES IS BY A TRIANGULATION SYSTEM.

WHEN OBSTRUCTIONS PREVENT USE OF THE TRIANGULATION SYSTEM THIS METHOD MAY BE USED.

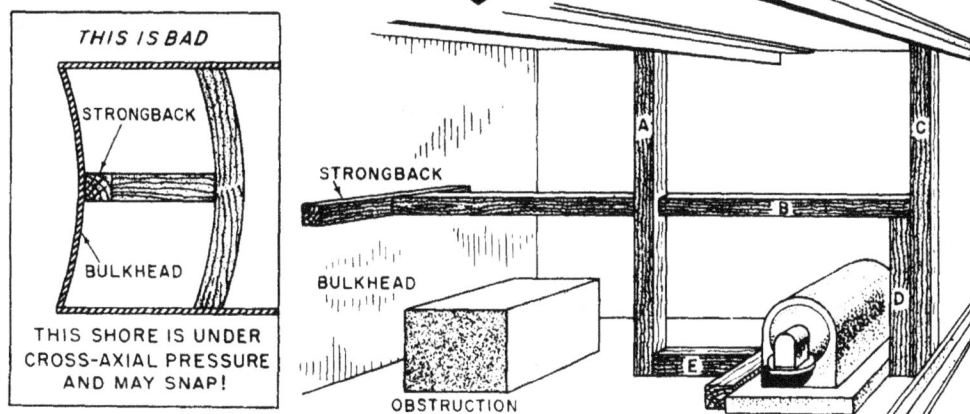

THIS IS BAD

STRONGBACK

BULKHEAD

THIS SHORE IS UNDER CROSS-AXIAL PRESSURE AND MAY SNAP!

STRONGBACK

BULKHEAD

OBSTRUCTION

ADDITIONAL STRENGTH IS AFFORDED BY SHORES B AND C. HORIZONTAL SHORE B IS SUPPORTED BY D AND A, AND IS BRACED AGAINST A UNIT OF MACHINERY BY MEANS OF E.

17.10

Figure 4-7.—Examples of shoring.

repairing structural damage, and in general restoring the ship to a stable and seaworthy condition. To function as a fighting unit, however, a ship must be more than stable and seaworthy—it must also be able to move. The restoration of vital services is therefore an integral part of damage control, even though it must often be accomplished after fires and flooding have been controlled.

The restoration of vital services includes making repairs to machinery and piping systems and reestablishing a source of electrical power. The casualty power system, developed as a result of war experience, has proved to be one of the most important damage control devices. The casualty power system is a simple electrical distribution system used to maintain a source of electrical supply for the most vital machinery. It is used to supply power only in emergencies. The casualty power system is discussed in chapter 20 of this text.

DEFENSE AGAINST NBC ATTACK

The basic guidelines for defensive and protective actions to be taken in the event of nuclear, biological, or chemical (NBC) attack are set forth in the Nuclear, Biological, and Chemical Defense Bill contained in Shipboard Procedures, NWP 50 (effective edition). Aboard ship, the engineer officer is responsible for maintaining this bill and ensuring that it is current and ready for immediate execution.

NBC defense measures may be divided into two phases: (1) preparatory measures taken in anticipation of attack, and (2) active measures taken immediately following an attack.

Preparatory measures to be taken before an attack include the following:

1. Thorough indoctrination and training of ship's force.
2. Removal of material that may constitute contamination hazards.
3. Masking of personnel who may be exposed (and of other personnel, as ordered).
4. Establishment of ship closure, including closing of CIRCLE WILLIAM fittings.
5. Donning of protective clothing by exposed personnel, as ordered.
6. Evasive action by the ship.
7. Activation of water washdown systems.

Active measures to be taken immediately following an attack include the following:

1. Evasive and self-protective action by personnel.

2. Evacuation and remanning of exposed stations, as ordered.
3. Decontamination of personnel.
4. Detection and prediction of contaminated areas.
5. Ventilation of contaminated spaces, as soon as the ship is in a clean atmosphere.

It is obvious that NBC defense is an enormously complex and wide-ranging subject, and one in which policies and procedures are subject to constant change. The present discussion is limited to a few aspects of NBC defense that are of primary practical importance aboard ship. More detailed information on all aspects of NBC defense may be obtained from chapters 9770 and 9900 of the Naval Ships Technical Manual and from Disaster Control (Ashore and Afloat), NavPers 10899-B.

PROTECTIVE CLOTHING

There are three types of clothing that are useful in NBC defense: permeable, impregnated, protective clothing, foul weather clothing, and ordinary work clothing.

Permeable protective clothing is supplied to ships in quantities sufficient to outfit 25 percent or more of the ship's compliment. Permeable clothing is olive green in color. A complete outfit includes impregnated socks, gloves, trousers with attached suspenders, and jumper (parka) with attached hood. Permeable clothing is treated with a chemical agent that neutralizes chemical agents; a chlorinated paraffin is used as a binder. The presence of these chemicals gives the permeable clothing a slight odor of chlorine and a slightly greasy or clammy feel. It is believed that the impregnation treatment should remain effective from 5 to 10 years (or possibly longer) if the clothing is stowed in unopened containers in a dry place with cool or warm temperatures and if it is protected from sunlight or daylight.

Permeable protective clothing should not be worn longer than necessary, especially in hot weather; prolonged wearing may cause a rash to develop where the skin comes in contact with the impregnated material.

Foul weather clothing of stock issue serves to protect ordinary clothing and the skin against penetration by liquid chemical agents and radioactive particles. It also reduces the amount of vapor that penetrates to the skin. Foul weather clothing, which includes a parka, trousers, rubber boots, and gloves, is easily decontaminated.

Ordinary work clothing (including long underwear, field socks, coverall, field boots, and watch cap) is partially effective in preventing droplets of liquid chemical agents and vapors from reaching the skin. However, ordinary work clothing is not as effective as the other types of clothing in preventing contamination. Under some conditions, personnel may wear two layers of ordinary work clothing to achieve greater protection than can be obtained with one layer.

PROTECTIVE MASKS

The protective mask is a very important item of protective equipment, since it protects such vulnerable areas as the eyes, the face, and the respiratory tract. The protective mask provides protection against NBC contamination by filtering the air before it is inhaled.

In general, all protective masks operate on the same principles. As the wearer inhales, air is drawn into a filtering system. This system consists of a mechanical filter which clears the air of solid or liquid particles and a chemical filling (usually activated charcoal) which absorbs or neutralizes toxic and irritating vapors. The purified air then passes to the region of the mask, where it can be inhaled. Exhaled air is expelled from the mask through an outlet valve which is so constructed that it opens only to permit exhaled air to escape.

Protective masks do not afford protection against ammonia or carbon monoxide, nor are they effective in confined spaces where the oxygen content of the atmosphere is too low (less than about 16 percent) to sustain life. When it is necessary to enter spaces where there is a deficiency of oxygen, the Navy oxygen breathing apparatus (OBA) is used.

DETECTION OF NBC CONTAMINATION

The very nature of NBC contamination makes detection and identification difficult. Nuclear radiation cannot be seen, heard, felt, or otherwise perceived through the senses. Biological agents are small in size and have no characteristic color or odor to help in identification. Although some chemical agents do have a characteristic color and odor, recently developed nerve agents are usually colorless and odorless.

It is obvious, then, that with contamination which cannot be seen, smelled, felt, tasted, or heard, specialized methods of detection are required. Mechanical, chemical, and electronic devices are available or under development for the detection of NBC contamination.

Detection of Nuclear Radiation

The instruments used for detecting radiological contamination are known as radiacs, the name being an abbreviation of radiation, detection, indication, and computation. Various types of radiacs are used aboard ship and at shore stations, since no single type of radiac can make all the radiological measurements that may be required.

The radiacs used aboard ship include (1) intensity meters for measuring gamma radiation; (2) intensity meters for measuring beta and gamma radiation; (3) survey meters for measuring alpha radiation; and (4) dosimeters for measuring accumulated doses of radiation received by individuals. These basic types of radiacs are described briefly here. Specific information on operating principles and detailed instructions for operating the instruments may be obtained from the manufacturer's technical manual furnished with each instrument.

GAMMA METERS.—Intensity meters for measuring gamma radiation include both portable instruments and fixed systems installed aboard ship. The intensity of gamma radiation is measured in roentgens per hour (r/hr) or in milliroentgens per hour (mr/hr). The roentgen is a unit of measurement for expressing the amount of gamma radiation or X-ray radiation. A milliroentgen is 1/1000 of a roentgen. Radiacs used for measuring large amounts of gamma radiation are called high-range intensity meters; these instruments are usually calibrated in roentgens per hour. Radiacs designed for measuring smaller amounts of gamma radiation are called low-range intensity meters; they are usually calibrated in milliroentgens per hour. Both high-range and low-range instruments are likely to have several scales; a range selector switch allows selection of the appropriate scale for each monitoring survey.

BETA AND GAMMA METERS.—Intensity meters which measure gamma radiation and also detect or measure beta radiation are usually of the Geiger-Mueller type. These instruments can measure gamma radiation alone or they can measure combined gamma radiation and beta radiation; an indirect measure of beta radiation can be obtained by subtracting the gamma radiation from the gamma-beta radiation.

ALPHA SURVEY METERS.—Meters for measuring alpha radiation are usually calibrated to give a meter reading in counts per minute (c/m, or cpm). However, some alpha survey meters give a reading in a unit called disintegrations per minute (d/m). The two units are not the same numerically.

DOSIMETERS.—There are two basic types of dosimeters. Self-reading dosimeters can be read by the person wearing the instrument. Nonself-reading dosimeters cannot be read directly by the wearer but must be read with the aid of special instruments. Some dosimeters are calibrated in roentgens, others in milliroentgens. Both self-reading and nonself-reading dosimeters measure exposure to radiation over a period of time—in other words, they measure accumulated radiation exposure.

Self-reading dosimeters are provided in various ranges for use by personnel aboard ship. Some of these self-reading dosimeters indicate accumulated gamma radiation from 0 to 200 mr; others indicate doses from 0 to 100 r; others from 0 to 200 r; and still others from 0 to 600 r. The dosimeter selected for any particular use will depend on the radiological situation existing at the time. Self-reading dosimeters must be charged before they are used. A special charging unit is furnished for shipboard use.

High-range nonself-reading dosimeters of the DT-60/PD type are furnished for use aboard ship. A dosimeter of this type consists of a special phosphor glass between lead filters, encased in a bakelite housing. The dosimeter, which is small, lightweight, and rugged, is worn on a chain around the neck. This dosimeter will measure accumulated doses of gamma radiation from 25 r to 600 r. A special instrument, the CP-95/PD computer-indicator, is required to read the DT-60/PD dosimeter.

Film badge dosimeters are nonself-reading devices for measuring both gamma radiation and beta radiation in low or moderate ranges. A film badge uses a special photographic film which is surrounded with moisture-proof and light-proof paper and shielded with lead, cadmium, plastic, or other shielding material. By the use of different shielding materials, the badge can be made to differentiate between gamma radiation and beta radiation. Laboratory techniques are required for the development and reading of the film.

Detection of Biological Agents

Basically, there are two possible approaches to the problem of detecting biological agents. Physical detection is based on the measurement of particles within a specified size range (and possibly the simultaneous measurement of other physical properties of the particles). Research is currently being done with a view to developing effective methods of physical detection. Biological detection involves growing the organisms, examining them under a microscope, and subjecting them to a variety of biochemical and biological tests. Although positive identification can frequently be made by biological detection methods, the procedure is difficult, exacting, and relatively slow. By the time a biological agent has been detected and identified in this fashion, personnel may well be showing symptoms of illness.

Biological detection may be divided into two phases: the sampling phase and the laboratory phase. The sampling phase may be a joint responsibility of damage control personnel and of the medical department. The laboratory phase is obviously a medical department responsibility.

Detection of Chemical Agents

Various detection devices have been developed for the detection and identification of chemical agents. Most of these devices indicate the presence of chemical agents by color changes which are chemically produced. To date, no single detector has been developed which is effective under all conditions for all chemical agents. A number of devices, including air sampling kits, papers, crayons, silica gel tubes, and indicator solutions, are in naval use. Some of these devices are also useful in establishing the completeness of decontamination and in estimating the hazards of operating in contaminated areas.

MONITORING AND SURVEYING

The monitoring and surveying of any area contaminated with NBC contamination is a vital part of NBC defense. In general, monitoring and surveying are done for the purposes of locating the hazards, isolating the contaminated areas, recording the results of the survey, and reporting the findings through the appropriate chain of command.

Specifically, the purpose of a radiological monitoring survey is to determine the location, type, and intensity of radiological contamination. The type of monitoring survey made at any given time depends on the radiological situation and on the tactical situation. Gross or rapid surveys are made as soon as possible after a nuclear weapon has been exploded, to get a general idea of the extent of contamination. Detailed surveys are made later, to obtain a more complete picture of the radiological situation.

Aboard ship, two main types of radiological surveys would be required after a nuclear attack. Ship surveys (first gross, then detailed) include surveys of all weather decks, interior spaces, machinery, circulating systems, equipment, and so forth. Personnel safety surveys (usually detailed) are concerned with protecting personnel from skin contamination and internal contamination. Personnel safety surveys include the monitoring of skin, clothing, food, and water, and the measurement of concentrations of radioactive material in the air (aerosols). Both ship surveys and personnel safety surveys are made aboard ship by members of the damage control organization. The medical department makes clinical tests, maintains dosage records, and makes specific recommendations concerning the monitoring of food, water, air, etc.; but the actual surveys are made by damage control personnel of the engineering department.

Detailed instructions for making monitoring surveys cannot be specified for all situations, since a great many factors (type ship, distance from blast, extent of damage, tactical situation, etc.) must necessarily be considered before monitoring procedures can be decided upon. However, certain basic guidelines that apply to monitoring situations may be stated as follows:

1. Monitors must be thoroughly trained before the need for monitoring arises. Learning to operate radiacs takes time. Simulated practice—as, for example, walking through a drill using a block of wood to represent a radiac—may teach a man something about the general movements made by a monitoring team, but it will not prepare him for actually using the instruments. All personnel who may be required to perform monitoring operations must be given adequate instruction and training in the use of the available radiacs.

2. Standard measuring techniques must be used. A measurement of radiation is meaningless unless the distance between the source of radiation and the point of measurement is known. For example, a radiac held 2 feet away from a source of radiation will indicate only one-fourth as much radiation as the same instrument would indicate if it were held 1 foot away from the same source. A radiac held 3 feet from the source will indicate only one-ninth as much radiation as when it is held 1 foot from the source. As may be seen, therefore, the distance between the source of radiation and the radiac must be known before the radiac reading can have any significance.

3. All necessary information must be recorded and reported. The information obtained by monitoring parties is forwarded to damage control central, where the measurements are plotted according to location and time. In order to develop an accurate overall picture of the radiological condition of the ship, damage control central must have precise and complete information from all monitoring parties. Each monitoring party must record and report the object or area monitored, the location of the object or area in relation to some fixed point, the intensity and type of radiation, the distance between the radiac and the source of radiation, the time and date of the measurements, the name of the man in charge of the monitoring party (or other identification of the party), and the type and serial number of the instrument used.

CONTAMINATION MARKERS

A standard system for marking areas contaminated by nuclear, biological, or chemical contamination has been adopted by nations included in the North Atlantic Treaty Organization. These standard survey markers are illustrated in figure 4-8.

NBC DECONTAMINATION

The basic purpose of decontamination is to minimize NBC contamination through removal or neutralization so that the mission of the ship or activity can be carried out without endangering the life or health of assigned personnel. The purpose of radiological decontamination is to remove contamination and shield personnel who

NATO COUNTRIES INCLUDING UNITED STATES

FRONT   BACK

GAS

CHEMICAL

DATE AND TIME
OF DETECTION
NAME OF AGENT,
IF KNOWN

BIO

BIOLOGICAL

DATE AND TIME
OF DETECTION
NAME OF AGENT,
IF KNOWN

ATOM

RADIOLOGICAL

DOSE RATE
DATE AND TIME
OF READING
DATE AND TIME
OF BURST,
IF KNOWN

USE OF TREFOIL
IS OPTIONAL

CURRENTLY USED WITHIN UNITED STATES

DANGER ATOM

RADIOLOGICAL

DOSE RATE
DATE AND TIME
OF READING
DATE AND TIME
OF BURST
IF KNOWN

C3.178

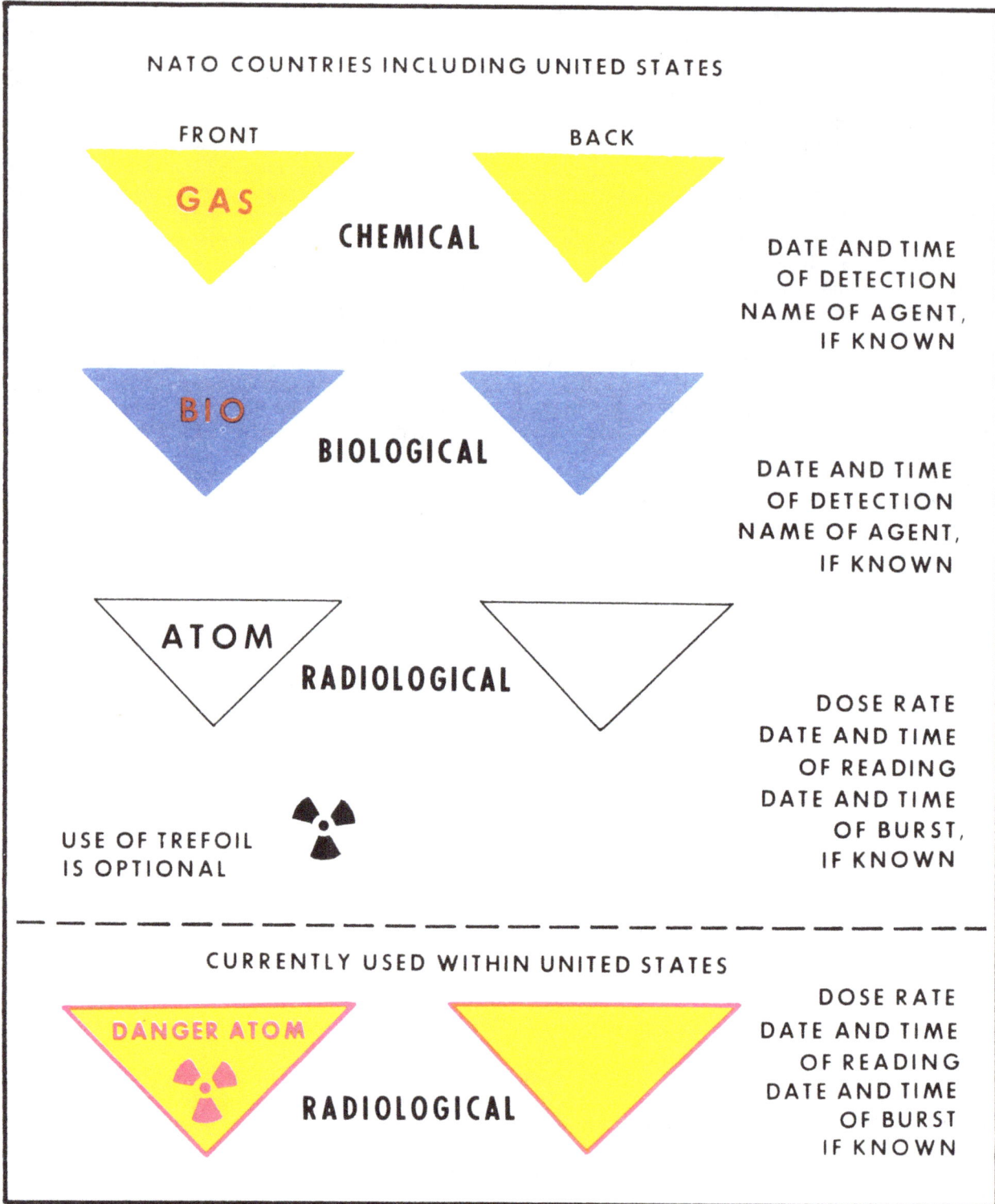

Figure 4-8.—NBC contamination markers.

are required to work in contaminated areas. The purpose of biological decontamination is to destroy the biological agents. The purpose of chemical decontamination is to remove or neutralize the chemical agents so that they will no longer be a hazard to personnel.

Decontamination operations may be both difficult and dangerous, and personnel engaged in these operations must be thoroughly trained in the proper techniques. Certain operations, such as the decontamination of food and water, should be done only by experts qualified in such work. However, all members of a ship's company should receive adequate training in the elementary principles of decontamination so that they can perform emergency decontamination operations.

After an attack, data from NBC surveys will be used to determine the extent and degree of contamination. Contaminated personnel must be decontaminated as soon as possible. Before decontamination of installations, machinery, and gear is undertaken, appraisals of urgency must be made in light of the tactical situation.

Radiological Decontamination

Radiological decontamination neither neutralizes nor destroys the contamination; instead, it merely removes the contamination from one particular area and transfers it to an area in which it presents less of a hazard. At sea, radioactive waste is disposed of directly over the side. At shore installations, the problem is more difficult.

Several methods of radiological decontamination have been developed; they differ in effectiveness in removing contamination, in applicability to given surfaces, and in the speed with which they may be applied. Some methods are particularly suited for rapid gross decontamination; others are better suited for detailed decontamination.

GROSS DECONTAMINATION.—The purpose of gross decontamination is to reduce the radiation intensity as quickly as possible to a safe level—or at least to a level which will be safe for a limited period of time. In gross decontamination, speed is the major consideration.

Flushing with water, preferably water under high pressure, is the most practicable way of accomplishing gross decontamination. Aboard ship, a water washdown system is used to wash down all the ship's surfaces, from high to low

and from bow to stern. The washdown system consists of piping and a series of nozzles which are specially designed to throw a large spray pattern on weather decks and other surfaces. The washdown system is particularly effective if it is activated before the ship is exposed to contamination; a film of water covering the ship's surfaces keeps the contaminating material from sticking to the surfaces. Figure 4-9 shows a water washdown system in operation.

Manual methods may be used to accomplish gross decontamination, but they are slower and less effective than the ship's washdown system. Manual methods that may be used by ship's force include (1) firehosing the surfaces with salt water, and (2) scrubbing the surfaces with detergent, firehosing the surfaces, and flushing the contaminating material over the side. Figure 4-10 shows men performing gross decontamination operations by manual scrubbing.

Steam is also a useful agent for gross decontamination, particularly where it is necessary to remove greasy or oily films. Steam decontamination is usually followed by hosing with hot water and detergents.

DETAILED DECONTAMINATION.—As time and facilities permit, detailed decontamination is carried out. The main purpose of detailed decontamination is to reduce the contamination to such an extent that only a minimum of radiological hazard to personnel would persist.

Three basic methods of detailed decontamination may be used—surface decontamination, aging and sealing, and disposal. Each of these methods has a specific purpose; one method can often be used to supplement another. Surface decontamination reduces the contamination without destroying the utility of the object. In aging and sealing, radioactivity is allowed to decrease by natural decay and any remaining contamination is then sealed onto the surface. The disposal method merely consists of removing contaminated objects and materials to a place where they can do little or no harm.

Biological Decontamination

The methods available for biological decontamination include scrubbing, flushing, heating, and the use of disinfectant sprays, disinfectant vapors, and sterilizing gases. The method to be used in any particular case depends upon the nature of the area or equipment to be decontaminated and upon the nature of the agent (if this is known).

3.193

Figure 4-9.—Water washdown system in operation.

Chemical Decontamination

The major problem in chemical decontamination is to decontaminate successfully after an attack by any of the blister or nerve agents. The general methods used in chemical decontamination include natural weathering, chemical action, the use of heat, the use of sealing, and physical removal.

Natural weathering relies on the effects of sun, rain, and wind to dissipate, evaporate, or decompose chemical agents. Weathering is by far the simplest and most widely applicable method of chemical decontamination; in some cases, it offers the only practicable means of neutralizing the effects of chemical agents, particularly where large areas are contaminated.

Decontamination by chemical action involves a chemical reaction between the chemical agent and the chemical decontaminant. The reaction usually results in the formation of a harmless new compound or a compound which can be removed more easily than the original agent. Neutralization of chemical agents can result from chemical reactions of oxidation, chlorination, reduction, or hydrolysis.

Expendable objects or objects of little value may be burned if they become contaminated. This procedure should not be used except as an emergency measure or as a means of disposing of material which has been highly contaminated. If this method is used, a very hot fire must be used. Intense heat is necessary for destruction of chemical agents; moderate or low heat may serve only to volatilize the agent and spread it by means of secondary aerosols. When a large amount of highly contaminated material is being burned, downwind areas may contain a dangerous

8.100

Figure 4-10.—Decontamination by manual scrubbing.

80

103.123

Figure 4-11.—Decontamination team hosing down a gun mount.

concentration of toxic vapors; personnel should be kept away from such areas.

Hot air may be blown over a contaminated surface to decontaminate it. Steam, especially high pressure steam, is also a useful decontaminating agent; the steam hydrolyzes and evaporates chemical agents and flushes them from the surfaces. Chemical decontamination may also be accomplished by sealing off porous surfaces to prevent the absorption of chemical agents or to prevent volatilization of agents already on the surface.

Decontamination can also be effected by physically removing the toxic agents from the contaminated surfaces. This can be done by washing or flushing the surfaces with water, steam, or various solvents. Figure 4-11 shows a decontamination party hosing down a gun mount in order to physically remove toxic agents.

DAMAGE CONTROL PRECAUTIONS

The urgent nature of damage control operations can lead to a dangerous neglect of necessary safety precautions. Driven by the need to act rapidly, men sometimes take chances they would not even consider taking in less hazardous situations. This is unfortunate, since there are few areas in which safety precautions are as important as they are in damage control. Failure to observe safety precautions can lead—and, in fact, has led—to the loss of ships.

Because damage control includes so many operations and involves the use of so many items of equipment, it is not feasible to list all the detailed precautions that must be observed. Some of the basic precautions that apply to practically all damage control work are noted briefly in the following paragraphs.

No one should be allowed to take any action to control fires, flooding, or other damage until the situation has been investigated and analyzed. Although speed is essential for effective damage control, correct action is even more important.

The extent of damage must not be underestimated. It is always necessary to remember that hidden damage may be even more severe than visible damage. Very real dangers may exist from damage which is not giving immediate trouble. For example, small holes at or just above the waterline may appear to be relatively minor; but they have been known to sink a ship.

It is extremely dangerous to assume that damage has been permanently controlled merely because fires have been put out, leaks plugged, and compartments dewatered. Fires may flare up again, plugs may work out of holes, and compartments may spring new leaks. Constant checking is required for quite some time after the damage appears to be controlled.

Doors, hatches, and other accesses should be kept open only as long as necessary while repairs are being made. Wartime records of naval ships show many cases of progressive flooding which were the direct result of failure to close doors or hatches.

No person should attempt to be a one-man damage control organization. All damage must be reported to damage control central or to a repair party before any individual action is taken. The damage control organization is the key to successful damage control. Separate, uncoordinated actions by individual men may actually do more harm than good.

Many actions taken to control damage can have a definite effect on ship's characteristics such as watertight integrity, stability, and weight and moment. The dangers involved in pumping large quantities of water into the ship to combat fires should be obvious. Less obvious, perhaps, is the fact that the repair of structural damage may also affect the ship's characteristics. For example, the addition of high or off-center weight produces the same general effect as high or off-center solid flooding.

While most repairs made in action would not amount to much in terms of weight shifts or additions, it is possible that a number of relatively small changes could add up sufficiently to endanger an already damaged and unstable ship. The only way to control this kind of hazard is by making sure that all damage control personnel report fully and accurately to damage control central. Ship stability problems are worked out in damage control central, but the information must come from repair personnel.

In all aspects of damage control, it is important to make full use of all available devices for the detection of hazards. Several types of instruments are available on most ships for detecting dangerous concentrations of explosive, flammable, toxic, or asphyxiating gases. Personnel should be trained to use these devices before entering potentially hazardous compartments or spaces.

PART II—BASIC ENGINEERING THEORY

We cannot proceed very far in the study of naval engineering without realizing the need for basic theoretical knowledge in many areas. To understand the functioning of the machinery and equipment discussed in later parts of this text, we must know something of the principles of mechanics, the laws of motion, the structure of matter, the behavior of molecules and atoms and subatomic particles, the properties and behavior of solids and liquids and gases, and other principles and concepts derived from the physical sciences.

Chapter 5 takes up the fundamentals of resistance, the development and transmission of propulsive power, and the principles of steering. The remaining three chapters of part II deal with basic scientific theory and engineering principles that have wide—indeed, almost universal—application in the field of naval engineering. Chapter 6 is concerned with lubrication, a subject of vital importance in practically all machinery and equipment. Chapter 7 takes up the principles of measurement and discusses basic types of measuring devices. Chapter 8 provides an introduction to some of the most fundamental concepts of energy and energy transformations, thus establishing a theoretical basis for much of the subsequent discussion of shipboard machinery and equipment. Theoretical considerations of a more specialized nature are discussed in other chapters throughout the text, as they are required for an understanding of the particular machinery or equipment under discussion.

CHAPTER 5

FUNDAMENTALS OF SHIP PROPULSION AND STEERING

The ability to move through the water and the ability to control the direction of movement are among the most fundamental of all ship requirements. Ship propulsion is achieved through the conversion, transmission, and utilization of energy in a sequence of events that includes the development of power in a prime mover, the transmission of power to the propellers, the development of thrust on the working surfaces of the propeller blades, and the transmission of thrust to the ship's structure in such a way as to move the ship through the water. Control of the direction of movement is achieved partially by steering devices which receive their power from steering engines and partially by the arrangement, speed, and direction of rotation of the ship's propellers.

This chapter is concerned with basic principles of ship propulsion and steering and with the propellers, bearings, shafting, reduction gears, rudders, and other devices required to move the ship and to control its direction of movement. The prime movers which are the source of propulsive power are discussed in detail in other chapters of this text, and are therefore mentioned only briefly in this chapter.

RESISTANCE

The movement of a ship through the water requires the expenditure of sufficient energy to overcome the resistance of the water and, to a lesser extent, the resistance of the air. The components of resistance may be considered as (1) skin or frictional resistance, (2) wave-making resistance, (3) eddy resistance, and (4) air resistance.

Skin or frictional resistance occurs because liquid particles in contact with the ship are carried along with the ship, while liquid particles a short distance away are moving at much lower velocities. Frictional resistance is therefore the result of fluid shear between adjacent layers of water. Under most conditions, frictional resistance constitutes a large part of the total resistance.

Wave-making resistance results from the generation and propagation of wave trains by the ship in motion. Figure 5-1 illustrates bow, stern, and transverse waves generated by a ship in motion. When the crests of the waves make an oblique angle with the line of the ship's direction, the waves are known as diverging waves. These waves, once generated, travel clear of the ship and give no further trouble. The transverse waves, which have a crest line at a 90° angle to the ship's direction, do not have visible, breaking crests. The transverse waves are actually the invisible part of the continuous wave train which includes the visible divergent waves at the bow and stern. The wave-making resistance of the ship is a resistance which must be allowed for in the design of ships, since the generation and propagation of wave trains requires the expenditure of a definite amount of energy.

Eddy resistance occurs when the flow lines do not close in behind a moving hull, thus creating a low pressure area in the water behind the stern of the ship. Because of this low pressure area, energy is dissipated as the water eddies. Most ships are designed to minimize the separation of the flow lines from the ship, thus minimizing eddy resistance. Eddy resistance is relatively minor in naval ships.

Air resistance, although small, also requires the expenditure of some energy. Air resistance may be considered as frictional resistance and eddy resistance, with most of it being eddy resistance.

THE DEVELOPMENT AND TRANSMISSION OF PROPULSIVE POWER

Figure 5-2 illustrates the general principles of ship propulsion and shows the functional

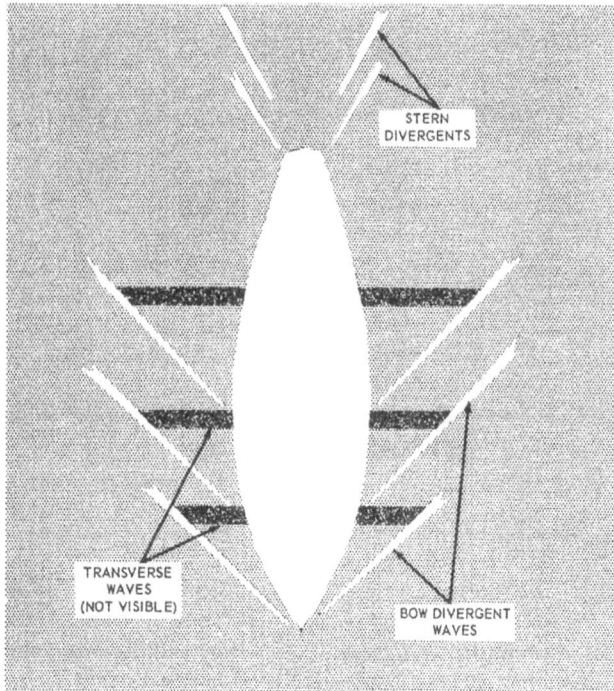

147.45

Figure 5-1.—Bow, stern, and transverse waves.

relationships of the units required for the development and transmission of propulsive power. The geared-turbine installation is chosen for this example because it is the propulsion plant most commonly used in naval service today. The same basic principles apply to all types of propulsion plants.

The units directly involved in the development and transmission of propulsive power are the prime mover, the shaft, the propelling device, and the thrust bearing. The various bearings used to support the shaft and the reduction gears (in this installation) may be regarded as necessary accessories.

The prime mover provides the mechanical energy required to turn the shaft and drive the propelling device. The steam turbines shown in figure 5-2 constitute the prime mover of this installation; in other installations the prime mover may be a diesel engine, a gas turbine engine, or a turbine-driven generator.

The propulsion shaft provides a means of transmitting mechanical energy from the prime mover to the propelling device and transmitting thrust from the propelling device to the thrust bearing.

The propelling device imparts velocity to a column of water and moves it in the direction opposite to the direction in which it is desired to move the ship. A reactive force (thrust) is thereby developed against the velocity-imparting device; and this thrust, when transmitted to the ship's structure, causes the ship to move through the water. In essence, then, we may think of propelling devices as pumps which are designed to move a column of water in order to build up a reactive force sufficient to move the ship. The screw propeller is the propelling device used on practically all naval ships.

The thrust bearing absorbs the axial thrust that is developed on the propeller and transmitted through the shaft. Since the thrust bearing is firmly fixed in relation to the ship's structure, any thrust developed on the propeller must be transmitted to the ship in such a way as to move the ship through the water.

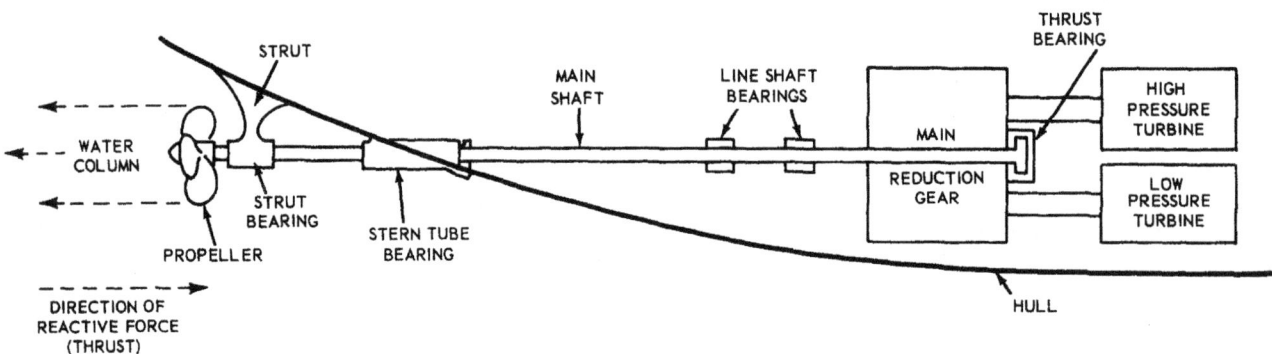

47.42A

Figure 5-2.—Principles of ship propulsion.

The purpose of the bearings which support the shaft is to absorb radial thrust and to maintain the correct alignment of the shaft and the propeller.

The reduction gears shown in figure 5-2 are used to allow the turbines to operate at high rotational speed while the propellers operate at lower speeds, thus providing for most efficient operation of both turbines and propellers.

The propellers, bearings, shafting, and reduction gears which are directly or indirectly involved in the development and transmission of propulsive power are considered in more detail following a general discussion of power requirements for naval ships.

POWER REQUIREMENTS

The power output of a marine engine is expressed in terms of horsepower. One horsepower is equal to 550 foot-pounds of work per second or 33,000 foot-pounds of work per minute. Different types of engines are rated in different kinds of horsepower. Steam reciprocating engines are rated in terms of indicated horsepower (IHP); internal combustion engines are usually rated in terms of brake horsepower (BHP); and steam turbines are rated in terms of shaft horsepower (SHP).

Indicated horsepower is the power measured in the cylinders of the engine.

Brake horsepower is the power measured at the crankshaft coupling by means of a mechanical, hydraulic, or electric brake.

Shaft horsepower is the power transmitted through the shaft to the propeller. Shaft horsepower can be measured with a torsionmeter; it can also be determined by computation. Shaft horsepower may vary from time to time within the same plant; for example, a plant that develops 10,000 shaft horsepower at 100 rpm on one occasion may develop 12,000 shaft horsepower at the same rpm on another occasion. The difference occurs because of variations in the condition of the bottom, the draft of the ship, the state of the sea, and other factors. Shaft horsepower may be determined by the formula

$$\text{SHP} = \frac{2\pi NT}{33,000}$$

where

SHP = shaft horsepower
N = rpm
T = torque (in foot-pounds) measured with torsionmeter

The amount of power which the propelling machinery must develop in order to drive a ship at a desired speed may be determined by direct calculation or by calculations based on the measured resistance of a model having a definite size relationship to the ship.

When the latter method of calculating power requirements is used, ship models are towed at various speeds in long tanks or basins. The most elaborate facility for testing models in this way is the Navy's David W. Taylor Model Basin at Carderock, Maryland. The main basin is 2775 feet long, 51 feet wide, and 22 feet deep. A powered carriage spanning this tank and riding on machine rails is equipped to tow an attached model directly below it. The carriage carries instruments to measure and record the speed of travel and the resistance of the model. From the resistance, the effective horsepower (EHP) (among other things) may be calculated. Effective horsepower is the horsepower required to tow the ship. Therefore,

$$\text{EHP} = \frac{R_T \frac{6080V}{60}}{33,000}$$

where

R_T = tow rope resistance, in pounds
V = speed, in knots

The speed in knots is multiplied by 6080 to convert it to feet per hour, and is divided by 60 to convert this to feet per minute. We have then

$$\text{EHP} = \frac{\frac{6080\,R_T V}{60}}{33,000} = \frac{6080\,R_T V}{60.33,000}$$

$$= \frac{608\,R_T V}{6.33,000} = \frac{608\,R_T V}{98 \times 10^3}$$

$$= 3.0707 \times 10^{-3}\,R_T V$$

$$= 0.0030707\,R_T V$$

The relationship between effective horsepower and shaft horsepower is called the propulsive efficiency or the propulsive coefficient of the ship. It is equal to the product of the propeller efficiency and the hull efficiency.

Variation of hull resistance at moderate speeds of any well-designed ship is approximately proportional to the square of the speed. The power required to propel a ship is proportional to the product of the hull resistance and speed. Therefore, it follows that under steady running conditions, the power required to drive a ship is approximately proportional to the cube of propeller speed. While this relationship is not exact enough for actual design, it does serve as a useful guide for operating the propelling plant.

Since the power required to drive a ship is approximately proportional to the cube of the propeller speed, 50 percent of full power will drive a ship at about 79.4 percent of the maximum speed attainable when full power is used for propulsion, and only 12.5 percent of full power is needed for about 50 percent of maximum speed.

The relation of speed, torque, and horsepower to ship's resistance and propeller speed under steady running conditions can be expressed in the following equations:

$$S = k_1 \times (rpm)$$

$$T = k_2 \times (rpm)^2$$

$$shp = \frac{2\pi k2}{33,000} \times (rpm)^3$$

where

S = ship's speed, in knots

T = torque required to turn propeller, in foot-pounds

shp = shaft horsepower

rpm = propeller revolutions per minute

k_1, k_2 = proportionality factors

The proportionality factors depend on many conditions such as displacement, trim, condition of hull and propeller with respect to fouling, depth of water, sea and wind conditions, and the position of the ship. Conditions that increase the resistance of the ship to motion cause k_1 to be smaller and k_2 to be larger.

In a smooth sea, the proportionality factors k_1 and k_2 can be considered as being reasonably constant. In rough seas, however, a ship is subjected to varying degrees of immersion and wave impact which cause these factors to fluctuate over a considerable range. It is to be expected, therefore, that peak loads in excess of the loads required in smooth seas will be imposed on the propulsion plant to maintain the ship's rated speed. Thus, propulsion plants are designed with sufficient reserve power to handle the fluctuating loads that must be expected.

There is no simple relationship for determining the power required to reverse the propeller when the ship is moving ahead or the power required to turn the propeller ahead when the ship is moving astern. To meet Navy requirements, a ship must be able to reverse from full speed ahead to full speed astern within a prescribed period of time; the propulsion plant of any ship must be designed to furnish sufficient power for meeting the reversing specifications.

PROPELLERS

The propelling device most commonly used for naval ships is the screw propeller, so called because it advances through the water in somewhat the same way that a screw advances through wood or a bolt advances when it is screwed into a nut. With the screw propeller, as with a screw, the axial distance advanced with each complete revolution is known as the pitch. The path of advance of each propeller blade section is approximately helicoidal.

There is, however, a difference between the way a screw propeller advances and the way a bolt advances in a nut. Since water is not a solid medium, the propeller slips or skids; hence the actual distance advanced in one complete revolution is less than the theoretical advance for one complete revolution. The difference between the theoretical and the actual advance per revolution is called the slip. Slip is usually expressed as a ratio of the theoretical advance per revolution (or, in other words, the pitch) and the actual advance per revolution. Thus,

$$\text{Slip ratio} = \frac{E - A}{E}$$

where

E = shaft rpm x pitch = engine distance per minute

A = actual distance advanced per minute

Screw propellers may be broadly classified as fixed pitch propellers or controllable pitch propellers. The pitch of a fixed pitch propeller cannot be altered during operation; the pitch of a controllable pitch propeller can be changed continuously, subject to bridge or engineroom control. Most propellers in naval use are of the fixed pitch type, but some controllable pitch propellers are in service.

A screw propeller consists of a hub and several (usually three or four) blades spaced at equal angles about the axis. Where the blades are integral with the hub, the propeller is known as a solid propeller. Where the blades are separately cast and secured to the hub by means of studs and nuts, the propeller is referred to as a builtup propeller.

Solid propellers may be further classified as having constant pitch or variable pitch. In a constant pitch propeller, the pitch of each radius is the same. On a variable pitch propeller, the pitch at each radius may vary. Solid propellers of the variable pitch type are the most commonly used for naval ships.

Propellers are classified as being right-hand or left-hand propellers, depending upon the direction of rotation. When viewed from astern, with the ship moving ahead, a right-hand propeller rotates in a clockwise direction and a left-hand propeller rotates in a counterclockwise direction. The great majority of single-screw ships have right-hand propellers. Multiple-screw ships have right hand propellers to port. Reversing the direction of rotation of a propeller reverses the direction of thrust and consequently reverses the direction of the ship's movement.

Some of the terms used in connection with screw propellers are identified in figure 5-3. The term face (or pressure face) identifies the after side of the blade, when the ship is moving ahead. The term back (or suction back) identifies the surface opposite the face. As the propeller rotates, the face of the blade increases the pressure on the water near it and gives the water a positive astern movement. The back of the blade creates a low pressure or suction area just ahead of the blade. The overall thrust is derived from the increased water velocity which results from the total pressure differential thus created.

The tip of the blade is the point most distant from the hub. The root of the blade is the area where the blade arm joins the hub. The leading edge is the edge which first cuts the water when the ship is going ahead. The trailing edge (also called the following edge) is opposite the leading edge. A rake angle exists when there is a rake either forward or aft—that is, when the blade is

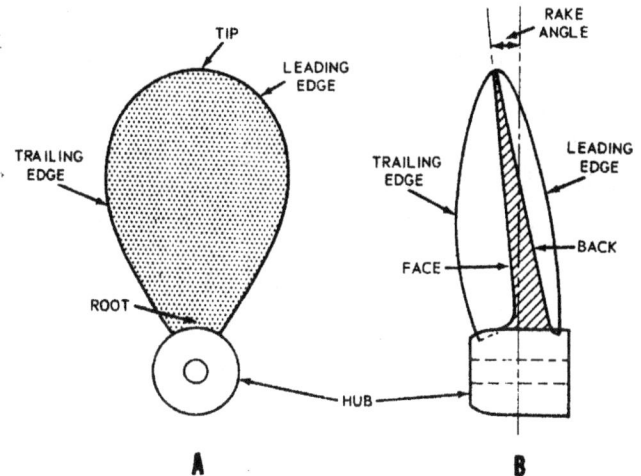

147.46

Figure 5-3.—Propeller blade.

not precisely perpendicular to the long axis of the shaft.

Blade Angle

The blade angle (or pitch angle) of a propeller may be defined as the angle included between the blade and a line perpendicular to the shaft centerline. If the blade angle were 0°, no pressure would be developed on the blade face. If the blade angle were 90°, the entire pressure would be exerted sidewise and none of it aft. Within certain limits, the amount of reactive thrust developed by a blade is a function of the blade angle.

Blade Velocity

The sternward velocity imparted to the water by the rotation of the propeller blades is partially a function of the speed at which the blades rotate. In general, the higher the speed, the greater the reactive thrust.

However, every part of a rotating blade does not give equal velocity to the water unless the blade is specially designed to do this. For example, consider the flat blade shown in figure 5-4. Points A and Z move about the shaft center with equal angular velocity (rpm) but with different instantaneous linear velocities. Point Z must move farther than point A to complete one revolution; hence the linear velocity at point Z must be greater than at point A. With the same pitch angle, therefore, point Z will exert more

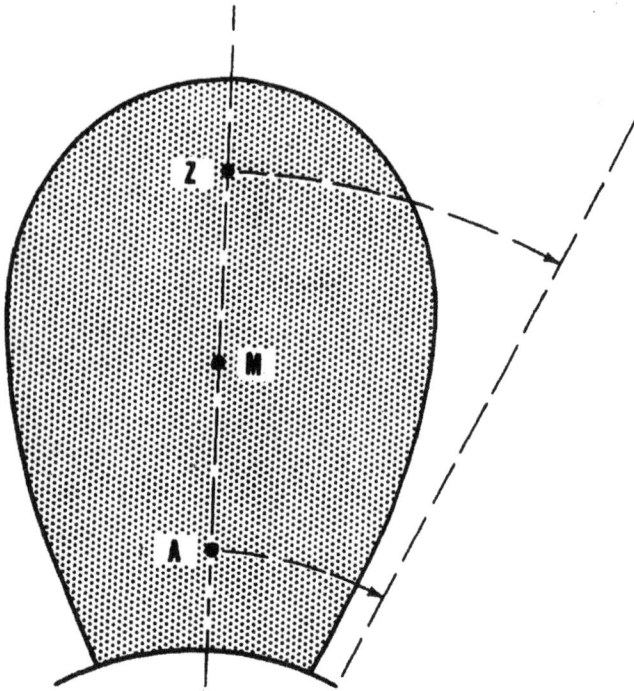

147.47

Figure 5-4.—Linear velocity
and reactive thrust.

pressure on the water and so develop more re-active thrust than point A. The higher the linear velocity of any part of a blade, the greater will be the reactive thrust.

Real propeller blades are not flat but are de-signed with complex surfaces (approximately helicoidal) to permit every infinitesimal area to produce equal thrust. Since point Z has a higher linear velocity than point A, the thrust at point Z must be decreased by decreasing the pitch angle at point Z. Point M, lying between points Z and A, would have (on a flat blade) a linear velocity less than Z but greater than A. In a real propeller, then, point M must be set at a pitch angle which is greater than the pitch angle at point Z but less than the pitch angle at point A. Since the linear velocity of the parts of a blade varies from root to tip, and since it is desired to have every infinitesimal area of the blade pro-duce equal thrust, it is apparent that a real pro-peller must vary the pitch angle from root to tip.

Propeller Size

The size of a propeller—that is, the size of the area swept by the blades—has a definite ef-fect on the total thrust that can be developed on the propeller. Within certain limits, the thrust that can be developed increases as the diameter and the total blade area increase. Since it is impracticable to increase propeller diameter beyond a certain point, propeller blade area is usually made as great as possible by using as many blades as are feasible under the circum-stances. Three-bladed and four-bladed marine propellers are commonly used.

Thrust Deduction

Because of the friction between the hull and the water, water is carried forward with the hull and is given a forward velocity. This movement of adjacent water is called the wake. Since the propeller revolves in this body of forward mov-ing water, the sternward velocity given to the propeller is less than if there were no wake. Since the wake is traveling with the ship, the speed of advance over the ground is greater than the speed through the wake.

At the same time, a propeller draws water from under the stern of the ship, thus creating a suction which tends to keep the ship from going ahead. The increase in resistance that occurs because of this suction is known as thrust de-duction.

Number and Location
of Propellers

A single propeller is located on the ship's centerline as far aft as possible to minimize the thrust deduction factor. Vertically, the propeller must be located deep enough so that in still water the blades do not draw in air but high enough so that it can benefit from the wake. The propeller must not be located so high that it will be likely to break the surface in rough weather, since this would lead to racing and perhaps a broken shaft.

A twin-screw ship has the propellers located one on each side, well aft, with sufficient tip clearance to limit thrust deduction.

A quadruple-screw ship has the outboard propellers located forward of and above the in-board propellers, to avoid propeller stream in-terference.

Controllable Pitch Propellers

As previously noted, controllable pitch pro-pellers are in use on some naval ships. Con-trollable pitch propellers give a ship excellent maneuverability and allow the propellers to de-velop maximum thrust at any given engine rpm.

A ship with controllable pitch propellers requires much less distance for stopping than a ship with fixed pitch propellers. The controllable pitch propellers are particularly useful for landing ships because they make it possible for the ships to hover offshore and because they make it easier for the ships to retract and turn away from the beach.

Controllable pitch propellers may be controlled from the bridge or from the engineroom as shown in figure 5-5. Hydraulic or mechanical controls are used to apply a blade actuating force to the blades.

A hydraulic system as shown in figure 5-6, is the most widely used means of providing the force required to change the pitch of a controllable pitch propeller. In this type of system, a valve positioning mechanism actuates an oil control valve. The oil control valve permits hydraulic oil, under pressure, to be introduced to either side of a piston (which is connected to the propeller blade) and at the same time allows for the controlled discharge of hydraulic oil from the other side of the piston. This action repositions the piston and thus changes the pitch of the propeller blades.

Some controllable pitch propellers have mechanical means for providing the blade actuating force necessary to change the pitch of the blades. In these designs, a worm screw and crosshead nut are used instead of the hydraulic devices for transmitting the actuating force to the connecting rods. The torque required for rotating the worm screw is supplied either by an electric motor or by the main propulsion plant through

pneumatic brakes. The mechanically operated actuating mechanism is usually controlled by simple mechanical or electrical switches.

Propeller Problems

One of the major problems encountered with propellers is known as cavitation. Cavitation is the formation of a vacuum around a propeller which is revolving at a speed above a certain critical value (which varies, depending upon the size, number, and shape of the propeller blades). The speed at which cavitation begins to occur is different in different types of ships; the turbulence increases in proportion to the propeller rpm. Specifically, a propeller rotating at a high speed will develop a stream velocity that creates a low pressure. This low pressure is less than the vaporization point of the water, and from each blade tip there appears to develop a spiral of bubbles (fig. 5-7). The water boils at the low pressure points. As the vapor bubbles of cavitation move into regions where the pressure is higher, the bubbles collapse rapidly and produce a high-pitched noise.

The net result of cavitation is to produce: (1) high level of underwater noise; (2) erosion of propeller blades; (3) vibration with subsequent blade failure from metallic fatigue; and (4) overall loss in propeller efficiency, requiring a proportionate increase in power for a given speed.

In naval warfare, the movements of surface ships and submarines can be plotted by sonar bearings on propeller noise. Because of the high

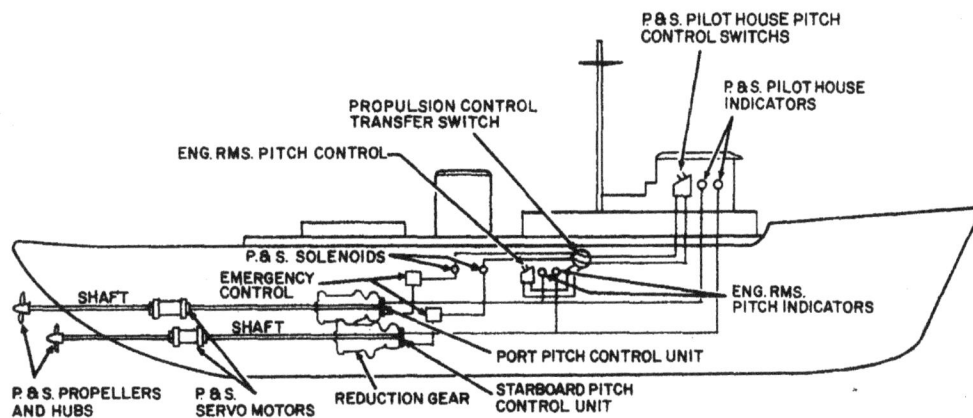

Figure 5-5.—General arrangement of pitch propeller control.

121.24

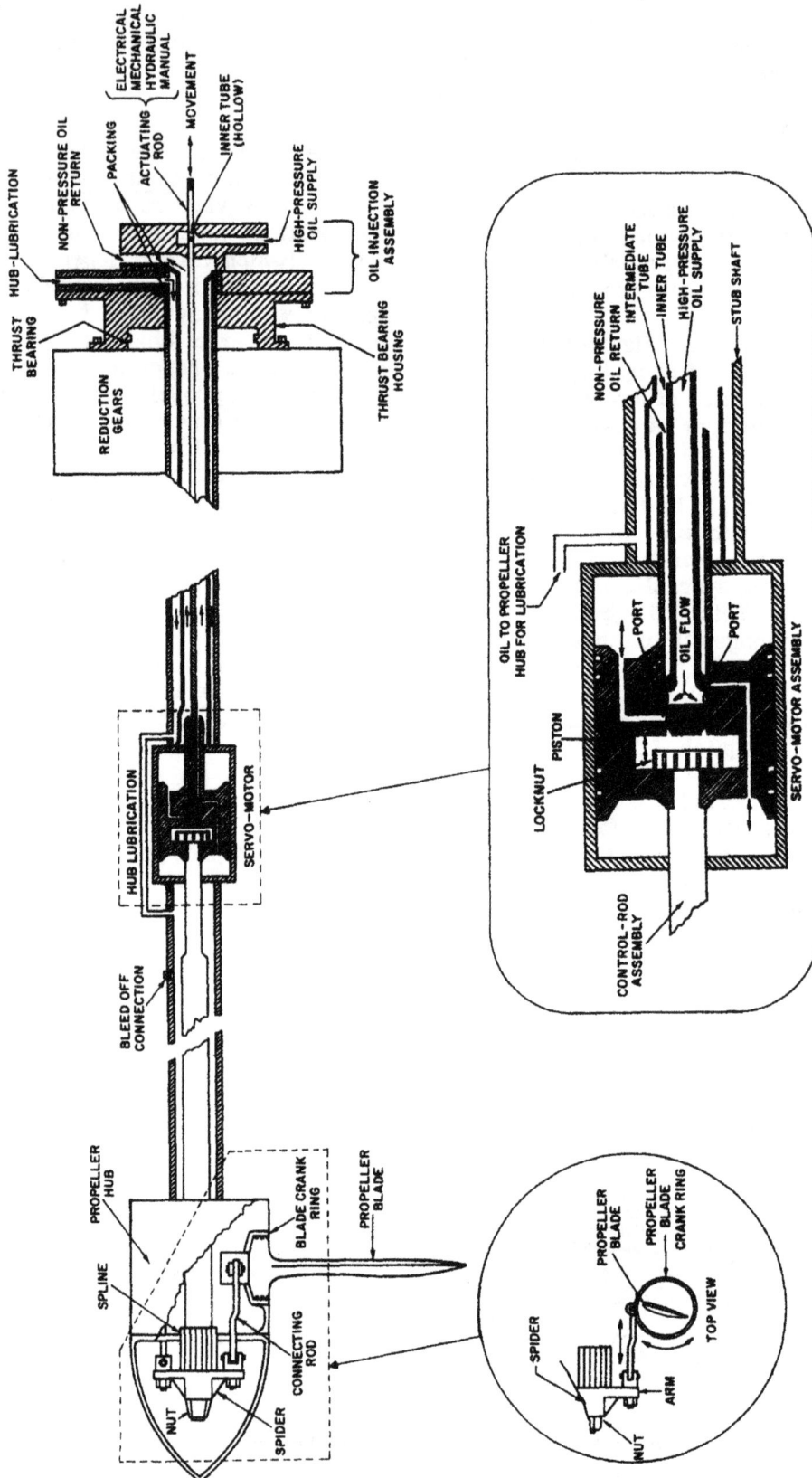

121.25

Figure 5-6.—Hydraulic controllable pitch propeller.

71.23(147B)

Figure 5-7.—Cavitating propeller.

static water pressure at submarine operational depths, cavitation sets in when a submarine is operating at a much higher rpm than when near the surface. For obvious reasons, a submarine that is under attack will immediately dive deep so that it can use high propeller rpm with the least amount of noise.

A certain amount of vibration is always present aboard ship. Propeller vibration, however, may also be caused by a fouled blade or by seaweed. If a propeller strikes a submerged object, the blades may be nicked.

Another propeller phenomen is the "singing" propeller. The usual cause of this noise is that the trailing edges of the blades have not been properly prepared before installation. The flutter caused by the flow around the edges may induce a resonant vibration. A "singing" propeller can be heard for a great distance.

BEARINGS

From the standpoint of mechanics, the term bearing may be applied to anything which supports a moving element of a machine. However, this section is concerned only with those bearings which support or confine the motion of sliding, rotating, and oscillating parts on revolving shafts or movable surfaces of naval machinery.

In view of the fact that naval machinery is constantly exposed to varying operating conditions, bearing material must meet rigid standards. A number of nonferrous alloys are used as bearing metals. In general, these alloys are tin-base, lead-base, copper-base, or aluminum-base alloys. The term babbitt metal is often used for lead-base and tin-base alloys.

Bearings must be made of materials which will withstand varying pressures and yet permit the surfaces to move with minimum wear and friction. In addition, bearings must be held in position with very close tolerances permitting freedom of movement and quiet operation. In view of these requirements, good bearing materials must possess a combination of the following characteristics for a given application.

1. The compressive strength of the bearing alloy at maximum operating temperature must be such as to withstand high loads without cracking or deforming.

2. Bearing alloys must have high fatigue resistance to prevent cracking and flaking under varying operating conditions.

3. Bearing alloys must have high thermal conductivity to prevent localized hot spots with resultant fatigue and seizure.

4. The bearing materials must be capable of retaining an effective oil film.

5. The bearing materials must be highly resistant to corrosion.

Classification

The reciprocating and rotating elements or members, supported by bearings, may be subject to external loads which can be resolved into components having normal, radial, or axial directions, or a combination of the two. Bearings are generally classified as sliding surface (friction) or rolling contact (antifriction) bearings.

Sliding surface bearings may be defined broadly as those bearings which have sliding contact between their surfaces. In these bearings, one body slides or moves on the surface of another and sliding friction is developed if the rubbing surfaces are not lubricated. Examples of sliding surface bearings are thrust bearings and journal bearings (fig. 5-8), such as the spring or line shaft bearings installed aboard ship.

Journal bearings are extensively used aboard ship. Journal bearings may be subdivided into different styles or types, the most common of which are solid bearings, half bearings, two-part or split bearings. A typical solid style journal bearing application is the piston bearing (part A of fig. 5-8), more commonly called a bushing. An example of a solid bearing is a piston rod wristpin bushing such as found in compressors. Perhaps the most common application of the half bearing in marine equipment is the propeller shaft bearing. Since the load is exerted only in one direction, they obviously are less costly than a full bearing of any type. Split bearings are used more frequently

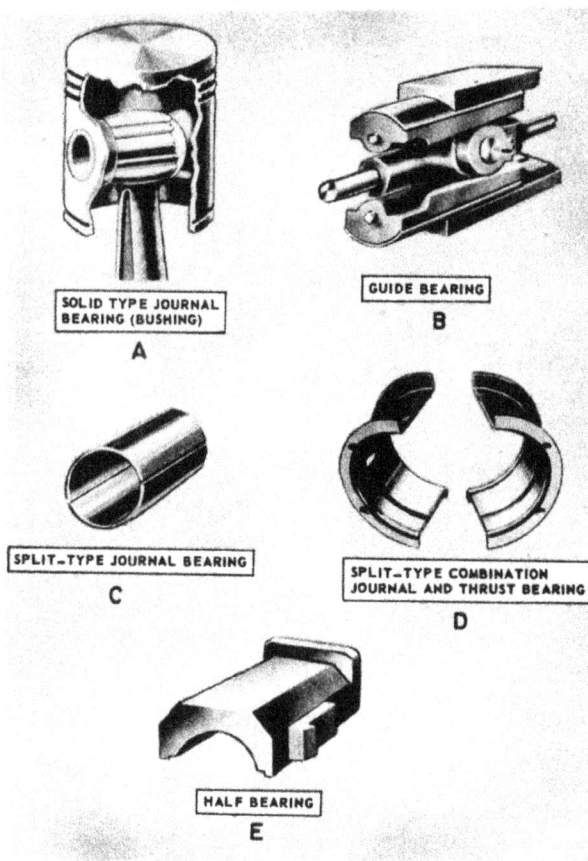

147.50
Figure 5-8.—Various types of friction bearings.

than any other friction-type bearing. A good example is the turbine bearing. Split bearings can be made adjustable to compensate for wear.

Guide bearings (part B of fig. 5-8), as the name implies, are used for guiding the longitudinal motion of a shaft or other part. Perhaps the best illustrations of guide bearings are the valve guides in an internal combustion engine.

Thrust bearings are used to limit the motion of, or support a shaft or other rotating part longitudinally. Thrust bearings sometimes are combined functionally with journal bearings.

Antifriction-type, or rolling contact, bearings are so-called because their design takes advantage of the fact that less energy is required to overcome rolling friction than is required to overcome sliding friction. These bearings may be defined broadly as bearings which have rolling contact between their surfaces. These bearings may be classified as roller bearings or ball bearings according to shape of the rolling elements. Both roller and ball bearings are made in different types, some being arranged to carry both radial and thrust loads. In these bearings, the balls or rollers generally are assembled between two rings or races, the contacting faces of which are shaped to fit the balls or rollers.

The basic difference between ball and roller bearings is that a ball at any given instant carries the load on two tiny spots diametrically opposite while a roller carries the load on two narrow lines (fig. 5-9). Theoretically, the area of the spot or line of contact is infinitesimal. Practically, the area of contact depends on how much the bearing material will distort under the applied load. Obviously, rolling contact bearings must be made of hard materials because if the distortion under load is appreciable the resulting friction will defeat the purpose of the bearings. Bearings with small, highly loaded contact areas must be lubricated carefully if they are to have the antifriction properties they are designed to provide. If improperly lubricated, the highly polished surfaces of the balls and rollers soon will crack, check, or pit, and failure of the complete bearing follows.

Both sliding surface and rolling contact bearings may be further classified by their function as follows: radial, thrust, and angular-contact (actually a combination of radial and thrust) bearings. Radial bearings, designed primarily to carry a load in a direction perpendicular to the axis of rotation, are used to limit motion in a radial direction. Thrust bearings can carry only axial loads; that is, a force parallel to the axis of rotation, tending to cause endwise motion of the shaft. Angular-contact bearings can support both radial and thrust loads.

The simplest forms of radial bearings are the integral and the insert types. The integral type is formed by surfacing a part of the machine frame with the bearing material, while the insert bearing is a plain bushing inserted into and held in place in the machine frame. The insert bearing may be either a solid or a split bushing, and may consist of the bearing material alone or be enclosed in a case or shell. In the integral bearing there is no means of compensating for wear, and when the maximum allowable clearance is reached the bearing must be resurfaced. The insert solid bushing bearing, like the integral type, has no means for adjustment due to wear, and must be replaced when maximum clearance is reached.

The pivoted shoe is a more complicated type of radial bearing. This bearing consists of a shell containing a series of pivoted pads or shoes, faced with bearing material.

The plain pivot or single disk type thrust bearing consists of the end of a journal extending into a cup-shaped housing, the bottom of which holds the single disk of bearing material.

The multi-disk type thrust bearing is similar to the plain pivot bearing except that several disks are placed between the end of the journal and the housing. Alternate disks of bronze and steel are generally used. The lower disk is fastened in the bearing housing and the upper one to the journal, while the intermediate disks are free.

The multi-collar thrust bearing consists of a journal with thrust collars integral with or fastened to the shaft; these collars fit into recesses in the bearing housing which are faced with bearing metal. This type bearing is generally used on horizontal shafts carrying light thrust loads.

The pivoted shoe thrust bearing is similar to the pivoted shoe radial bearing except that it has a thrust collar fixed to the shaft which runs against the pivoted shoes. This type bearing is generally suitable for both directions of rotation.

Angular loading is generally taken by using a radial bearing to restrain the radial load and some form of thrust bearing to handle the load. This may be accomplished by using two separate bearings or a combination of a radial and thrust (radial thrust). A typical example is the multi-collar bearing which has its recesses entirely surfaced with bearing material; the faces of the collars carry the thrust load and the cylindrical edge surfaces handle the radial load.

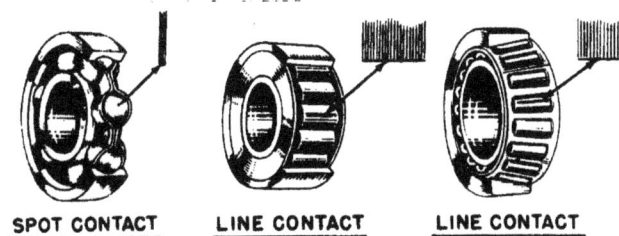

SPOT CONTACT   LINE CONTACT   LINE CONTACT

77.66

Figure 5-9.—Load-carrying areas of ball and roller bearings.

Main Reduction Gear and
Propulsion Turbine Bearings

Reduction gear bearings of the babbitt-lined split type are rigidly mounted and dowelled into the bearing housings. These bearings are split in halves, but the split is not always in a horizontal plane. On many pinion and bull gear bearings, the pressure is against the cap and not always in a vertical direction. The bearing shells are so secured in the housing that the point of pressure on both ahead and astern operation is as nearly midway between the joint faces as practicable.

Turbine bearings are pressure lubricated by means of the same forced-feed system that lubricates the reduction gear bearings.

Main Thrust Bearings

The main thrust bearing, which is usually located in the reduction gear casing, serves to absorb the axial thrust transmitted through the shaft from the propeller.

Kingsbury or pivoted segmental shoe thrust bearings of the type shown in figure 5-10 are commonly used for main thrust bearings. This type of bearing consists of pivoted segments or shoes (usually six) against which the thrust collar revolves. Ahead or astern axial motion of the shaft, to which the thrust collar is secured, is thereby restrained by the action of the thrust shoes against the thrust collar. These bearings operate on the principle that a wedge-shaped film of oil is more readily formed and maintained than a flat film and that it can therefore carry heavier loads for any given size.

In a segmental pivoted-shoe thrust bearing, upper leveling plates upon which the shoes rest and lower leveling plates equalize the thrust load among the shoes (fig. 5-11). The base ring, which supports the lower leveling plates, holds the plates in place and transmits the thrust on the plates to the ship's structure. Shoe supports (hardened steel buttons or pivots) located between the shoes and the upper leveling plates enable the shoe segments to assume the angle required to pivot the shoes against the upper leveling plates. Pins and dowels hold the upper and lower leveling plates in position, allowing ample play between the base ring and the plates to ensure freedom of movement of the leveling plates. The base ring is kept from turning by its notched construction, which secures the ring to its housing.

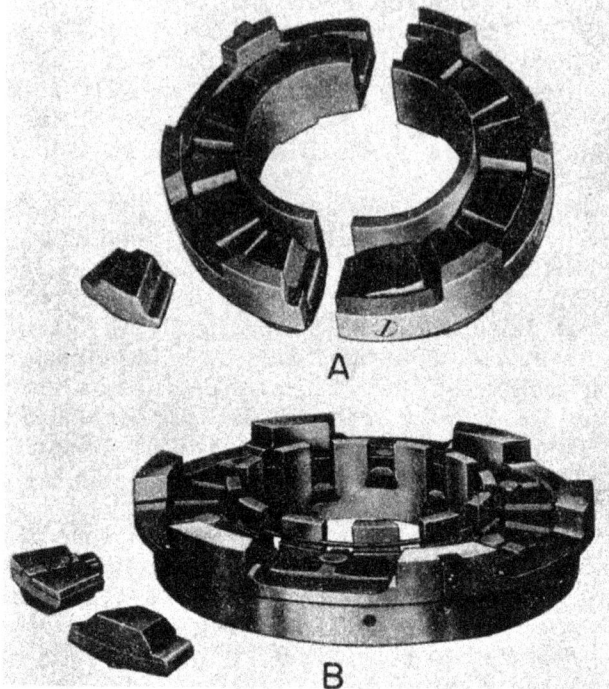

147.51X
Figure 5-10.—Kingsbury pivoted-shoe thrust bearing.

Main Line Shaft Bearings

Bearings which support the propulsion line shafting and which are located inside the hull are called line shaft bearings, spring bearings, or line bearings. These bearings are of the ring-oiled, babbitt-faced, spherical-seated, shell type. Figure 5-12 illustrates the arrangement of a line shaft bearing. The bearing is designed to align itself to support the weight of the shafting. The spring bearings of all modern naval ships are provided with both upper and lower self-aligning bearing halves.

Stern Tube and Strut Bearings

The stern tube is a steel tube built into the ship's structure for the purpose of supporting and enclosing the propulsion shafting where it pierces the hull of the ship. The section of the shafting enclosed and supported by the stern tube is called the stern tube shaft. The propeller shaft is supported at the stern by two bearings, one at each end of the stern tube. These bearings are called stern tube bearings. A packing gland known as the stern tube gland

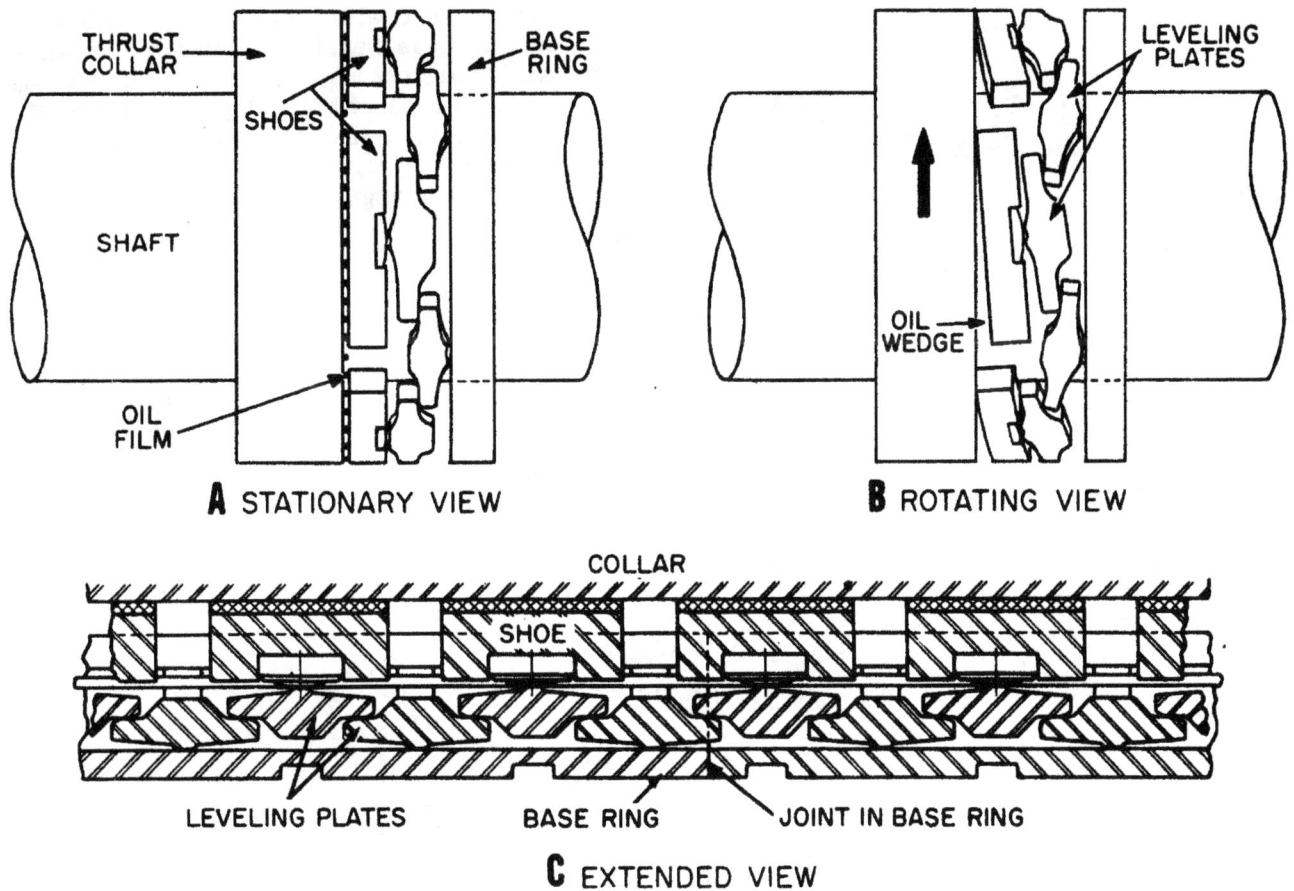

A STATIONARY VIEW

B ROTATING VIEW

C EXTENDED VIEW

38.86X

Figure 5-11.—Diagrammatic arrangement of Kingsbury thrust bearing.

is located at the inner end of the stern tube. This gland, which is shown in figure 5-13, seals the area between the shaft and the stern tube, but still allows the shaft to rotate.

The stuffing box of the stern tube gland is flanged and bolted to the stern tube. The casting is divided into two annular compartments. The forward space is the stuffing box proper; the after space has a flushing connection for providing a positive flow of water through the stern tube for lubricating, cooling, and flushing. The flushing connection is supplied by the fire and flushing system. A drain connection may be provided.

A strut bearing is shown in figure 5-14; The strut bearing has a composition bushing which is split longitudinally into two halves. The outer surface of the bushing is machined with steps to bear on matching landings in the bore of the strut. One end is bolted to the strut.

The shells of both stern tube and strut bearings are of bronze lined with a suitable bearing wearing material. The shells are normally grooved longitudinally to receive strips of

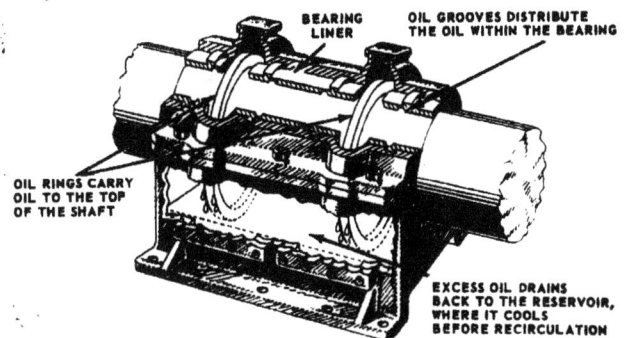

47.39X

Figure 5-12.—Line shaft bearing.

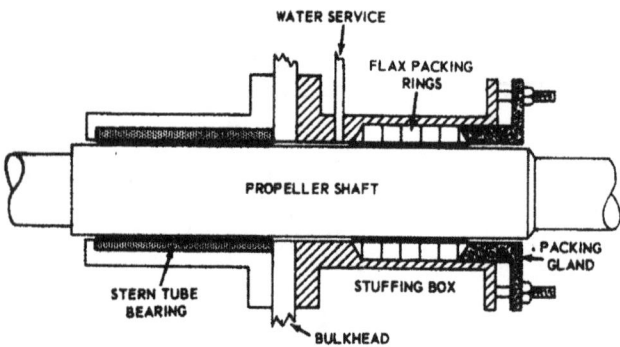

47.40

Figure 5-13.—Stern tube stuffing box and gland.

laminated resin bonded composition or strips of composition faced with rubber or synthetic rubber compounds as wearing materials. The laminated strips are cut and installed in the bearing shell so as to present the end grain to the shaft. In naval craft other than major combatant ships, resin bonded composition bearings or full molded rubber faced bearings are used.

PROPULSION SHAFTING

The propulsion shafting, which ranges in diameter from 18 to 21 inches for small twin-screw destroyers to approximately 30 inches for large four-screw carriers, is divided into four functional sections: the thrust shaft, the line

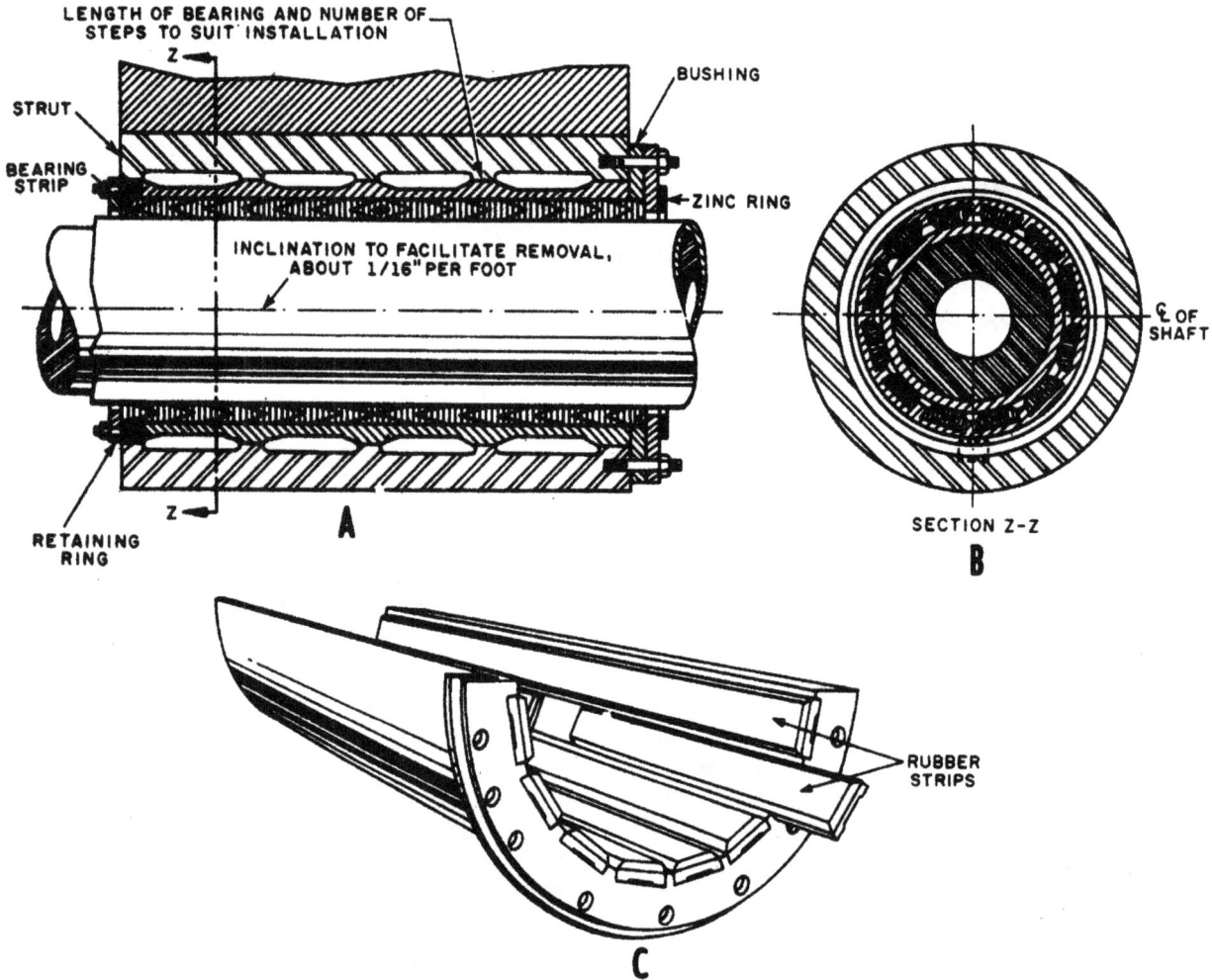

47.41

Figure 5-14.—Strut bearing. (A) Longitudinal cutaway view. (B) Cross-sectional view.
(C) Arrangement of rubber stripping in the bearing bushing.

shaft, the stern tube shaft, and the propeller or tail shaft. These portions of the shafting may be seen in figure 5-15.

Segments of the line shaft and the thrust shaft are joined together with integral flange-type couplings. The stern tube shaft is joined to the after end of the line shaft with an inboard stern tube coupling which has a removable after-sleeve flange. The tail shaft is joined to the stern tube shaft by a muff-type outboard coupling.

On single-screw ships, the portion of the outboard shaft which turns in the stern tube bearing is normally covered with a shrunk-on composition sleeve. This is done to protect the shaft from corrosion and to provide a suitable journal for the water-lubricated bearings. On multiple-screw ships, these sleeves normally cover only the bearing areas; on such ships, the exposed shafting between the sleeves is covered with synthetic sheet rubber to protect the shafting from sea water corrosion.

On carriers and cruisers, the wet shafting—that is, the shafting outboard in the sea—is composed of three sections: a tail shaft, an intermediate or dropout section, and a stern tube section. Integral flanged ends of these sections are usually used for joining the sections together.

Circular steel or composition shields known as fairwaters are secured to the bearing bushings of the stern tube and strut bearings and to both the forward and the after ends of the under-

water outboard couplings. These are intended primarily to reduce underwater resistance. The coupling fairwaters are secured to both the shaft and coupling flanges and are filled with tallow to protect the coupling from corrosion.

REDUCTION GEARS

Reduction gears are used in many propulsion plants to allow both the prime mover and the propeller to operate at the most efficient speed. Reduction gears are also used in many kinds of auxiliary machinery, where they serve the same purpose. Some of the gear forms commonly used in shipboard machinery are shown in figure 5-16.

Reduction gears are classified by the number of steps used to bring about speed reduction and by the general arrangement of the gearing. A single reduction gear consists of a small pinion gear which is driven by the turbine shaft and a large main gear (or bull gear) which is driven by the pinion. In this type of arrangement, the ratio of speed reduction is proportional to the diameters of the pinion and the bull gear. In a 2 to 1 single reduction gear, for example, the diameter of the driven gear is twice that of the driving pinion. In a 10 to 1 single reduction gear, the diameter of the driven gear is ten times that of the pinion.

All main reduction gearing in current combatant ships makes use of double helical gears (sometimes referred to as herringbone gears).

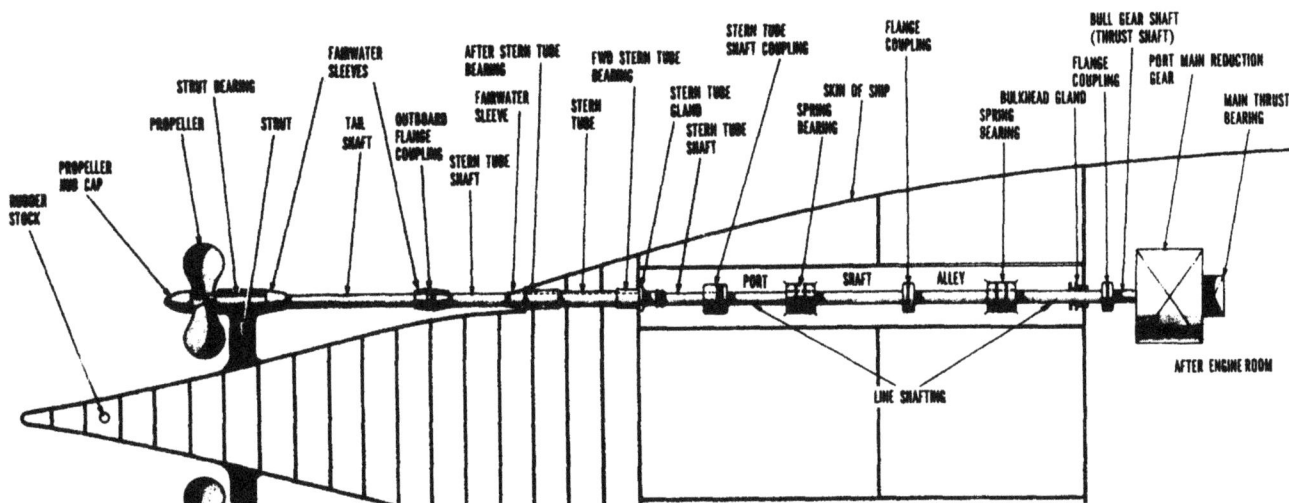

47.42B

Figure 5-15.—Propulsion shafting, twin-screw ship.

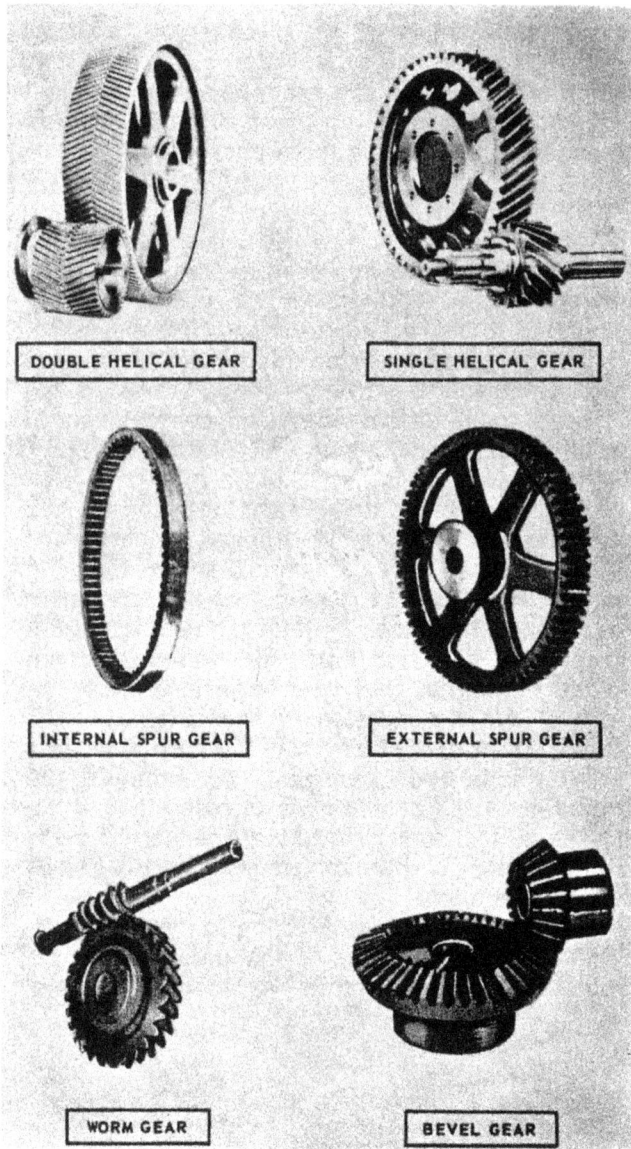

5.22

Figure 5-16.—Gear forms used in shipboard machinery.

Double helical gears have smoother action and less tooth shock than single reduction gears. Since the double helical gears have two sets of teeth at complementary angles, end thrust (such as is developed in single helical gears) is prevented.

In the double reduction gears used on most ships, a high speed pinion which is connected to the turbine shaft by a flexible coupling drives an intermediate (first reduction) gear. The first reduction gear is connected by a shaft to the low speed pinion which in turn drives the bull (second reduction) gear mounted on the propeller shaft. If we suppose a 20 to 1 speed reduction is desired, this could be accomplished by having a ratio of 2 to 1 between the high speed pinion and the first reduction gear and a ratio of 10 to 1 between the low speed pinion on the first reduction gear shaft and the second reduction gear on the propeller shaft.

A typical double reduction gear installation for a DD 692 class destroyer is shown in figure 5-17. In this type of installation, the cruising turbine is connected to the high pressure turbine through a single reduction gear. The cruising turbine rotor carries with it a pinion which drives the cruising gear, coupled to the high pressure turbine shaft. The cruising turbine rotor and pinion are supported by three bearings, one at the forward end of the turbine and one on each side of the pinion in the cruising reduction gear case.

The high pressure turbine and the low pressure turbine are connected to the propeller shaft through a locked train double reduction gear of the type shown in figure 5-18. First reduction pinions are connected by flexible couplings to the turbines. Each of the first reduction pinions drives two first reduction gears. Attached to each of the first reduction gears by a quill shaft and flexible couplings (fig. 5-19) is a second reduction pinion (low speed pinion). These four pinions drive the second reduction gear (bull gear) which is attached to the propeller shaft.

Locked train reduction gears have the advantage of being more compact than other types, for any given power rating. For this reason, all high powered modern combatant ships have locked train reduction gears. Another type of reduction gearing, known as nested gearing, is illustrated in figure 5-20. Nested gearing is used on most auxiliary ships but is not used on combatant ships. As may be seen, the nested gearing is relatively simple; it employs no quill shafts and uses a minimum number of bearings and flexible couplings.

FLEXIBLE COUPLINGS

Propulsion turbine shafts are connected to the reduction gears by flexible couplings which are designed to take care of very slight misalignment between the two units. Most flexible couplings are of the gear type shown in figure 5-21. The coupling consists of two shaft rings having internal gear teeth and an internal

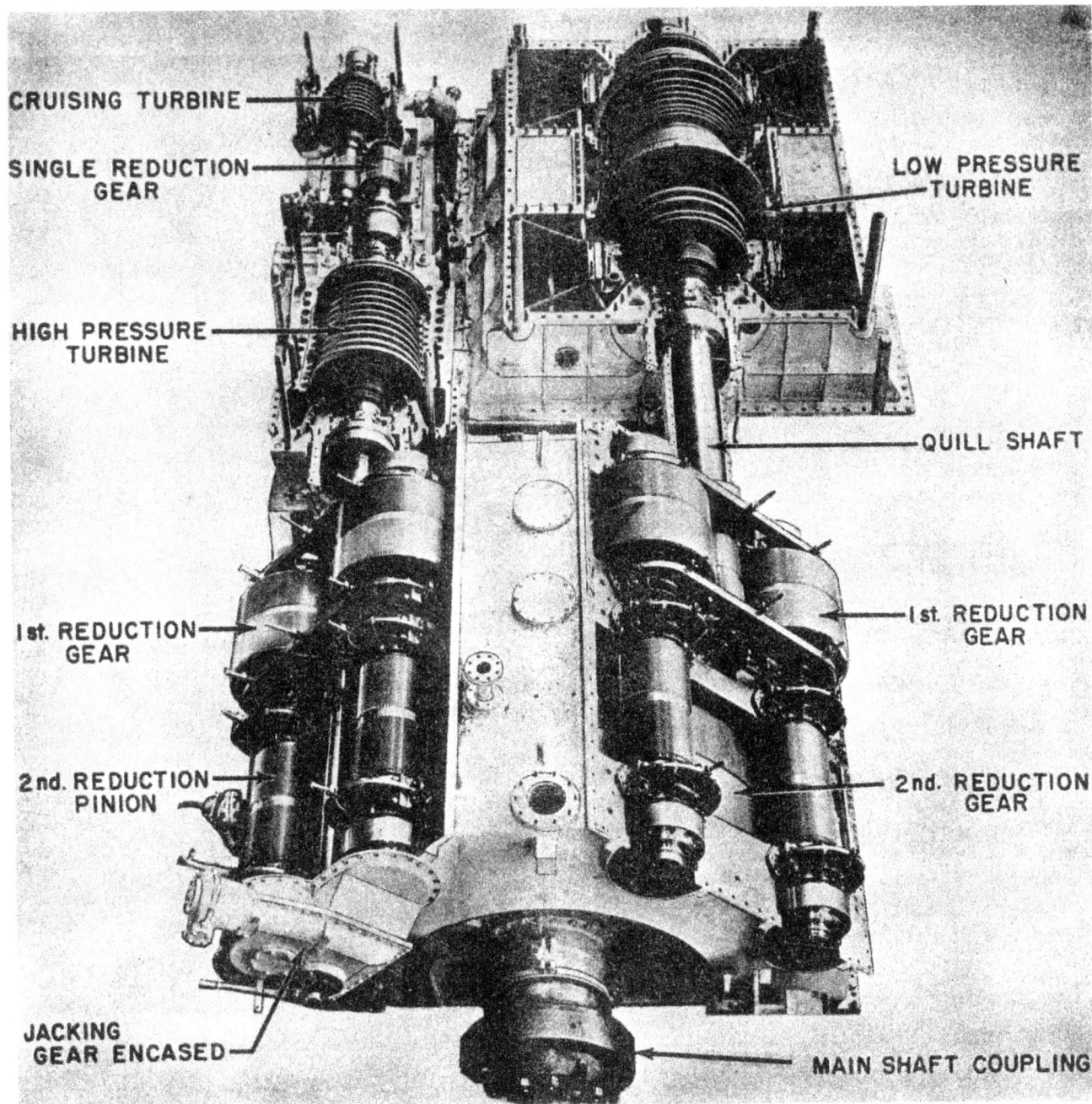

47.30

Figure 5-17.—Turbines and locked train double reduction gearing of DD 692 class destroyer.

floating member (or distance piece) which has external teeth around the periphery at each end. The shaft rings are bolted to flanges on the two shafts to be connected; the floating member is placed so that its teeth engage with those of the shaft rings.

Cruising turbine couplings which transmit lower powers may use external floating members with internal teeth. With this design, the shaft rings become spur gears with external teeth on the ends of the pinion shaft and turbine shaft. Figure 5-22 shows a flexible coupling of

101

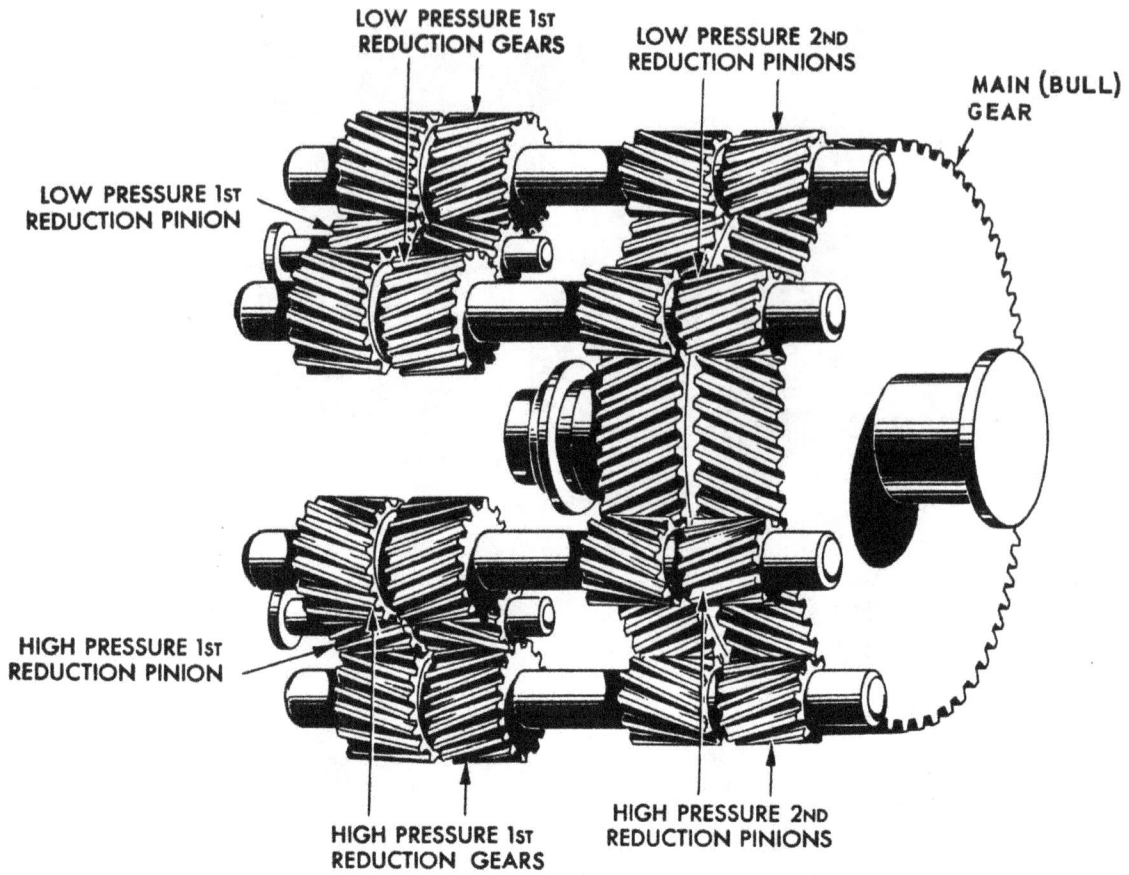

47.27

Figure 5-18.—Locked train double reduction gearing.

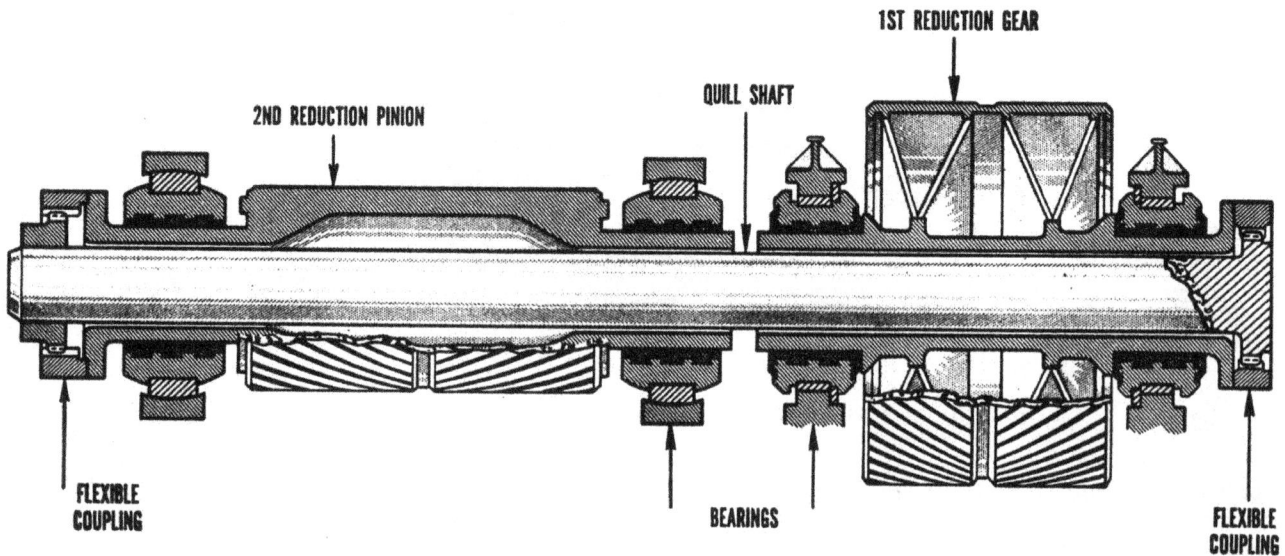

47.32X

Figure 5-19.—Quill shaft assembly.

2ND REDUCTION PINION

1ST REDUCTION PINION

1ST REDUCTION GEAR

2ND REDUCTION PINION

2ND REDUCTION PINION

1ST REDUCTION PINION

1ST REDUCTION GEAR

2ND REDUCTION PINION

2ND REDUCTION OR MAIN GEAR

1ST REDUCTION GEAR

H.P. PINION

I.P. PINION

1ST REDUCTION GEAR

L.P. PINION

2ND REDUCTION PINION

2ND REDUCTION PINION

1ST REDUCTION GEAR

1ST REDUCTION GEAR

49.29X

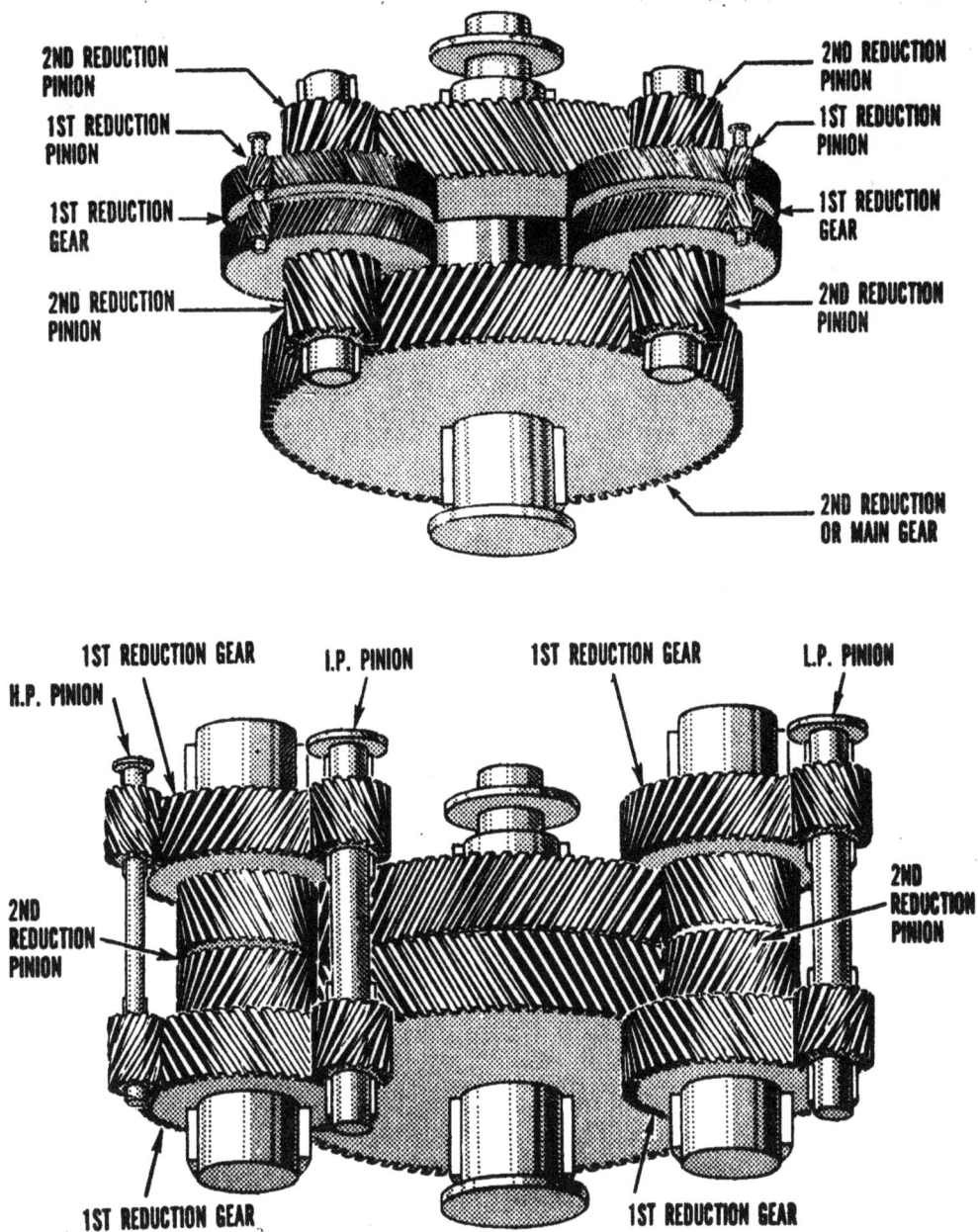

Figure 5-20.—Nested reduction gearing.

this type which is used on destroyers. The coupling is installed between the cruising turbine reduction gear and the high pressure turbine. In this coupling, the floating member is a transversely split sleeve having internal teeth which mesh completely with the external teeth of the spur gears mounted on the connected shaft ends.

CARE OF REDUCTION GEARS, SHAFTING AND BEARINGS

The main reduction gear is one of the largest and most expensive units of machinery found in the engineering department. Main reduction gears that are installed properly and operated

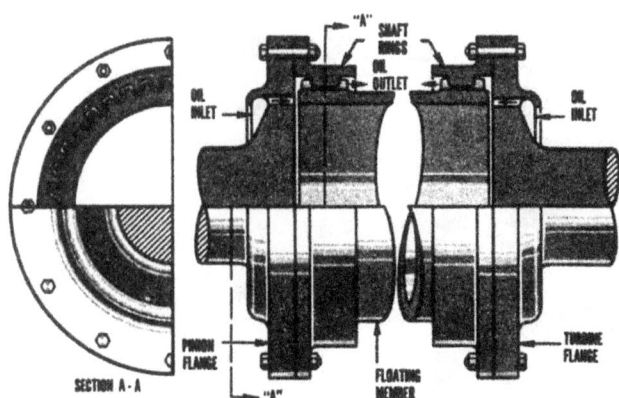

47.31X

Figure 5-21.—Gear-type flexible coupling.

properly will give years of satisfactory service. However, a serious casualty to main reduction gears, will either put the ship out of commission or force it to operate at reduced speed. Extensive repairs to the main reduction gear can be very expensive because they usually have to be made at a shipyard.

Some things are essential for the proper operation of reduction gears. Proper lubrication includes supplying the required amount of oil to the gears and bearings, plus keeping the oil clean and at the proper temperature. Locking and unlocking the shaft must be done in accordance with the manufacturer's instructions. Abnormal noises and vibrations must be investigated and corrective action taken. Gears must be inspected in accordance with the current instructions issued by NavShips, the type commander, or other proper authority. Preventive and corrective maintenance must be conducted in accordance with the 3-M System.

PROPER LUBRICATION.—Lubrication of reduction gears and bearings is of the utmost importance. The correct quantity and quality of lubricating oil must, at all times, be available in the main sump. The oil must be CLEAN; and it must be supplied to the gears and bearings at the pressure and temperature specified by the manufacturer.

In order to accomplish proper lubrication of gears and bearings, several conditions must be met. The lube oil service pump must deliver the proper discharge pressure. All relief valves in the lube oil system must be set to function at their designed pressure. On most older ships,

each bearing has a needle valve to control the amount of oil delivered to the bearing. On newer ships, the quantity of oil to each bearing is controlled by an orifice in the supply line. The needle valve setting or the orifice opening must be in accordance with the manufacturer's instructions or the supply of oil will be affected. Too small a quantity of oil will cause the bearing to run hot. If too much oil is delivered to the bearing, the excessive pressure may cause the oil to leak at the oil seal rings. Too much oil may also cause a bearing to overheat.

Lube oil must reach the bearing at the proper temperature. If the oil is too cold, one of the effects is insufficient oil flow for cooling purposes. If the oil supply is too hot, some lubricating capacity is lost.

For most main reduction gears, the normal temperature of oil leaving the lube oil cooler should be between 120°F and 130°F. For full power operation, the temperature of the oil leaving the bearings should be between 140°F and 160°F. The maximum TEMPERATURE RISE of oil passing through any gear or bearing, under any operating conditions, should not exceed 50°F; and the final temperature of the oil leaving the gear or bearing should not exceed 180°F. This temperature rise and/limitation may be determined by installed thermometers or resistance temperature elements.

Cleanliness of lubricating oil cannot be overstressed. Oil must be free from impurities, such as water, grit, metal, and dirt. Particular care must be taken to clean out metal flakes and dirt when new gears are wearing in or when gears have been opened for inspection. Lint or dirt, if left in the system may clog the oil spray nozzles. The spray nozzles must be kept open at all times. Spray nozzles must never be altered without the authorization of the Naval Ship Systems Command.

The lube oil strainers perform satisfactorily under normal operating conditions, but they cannot trap particles of metal and dirt which are fine enough to pass through the mesh. These fine particles can become embedded in the bearing metal and cause wear on the bearings and journals. These fine abrasive particles passing through the gear teeth act like a lapping compound and remove metal from the teeth.

LOCKING AND UNLOCKING THE MAIN SHAFT.—In an emergency, or in the event of a casualty to the main propulsion machinery of a turbine-driven ship, it may be necessary to stop

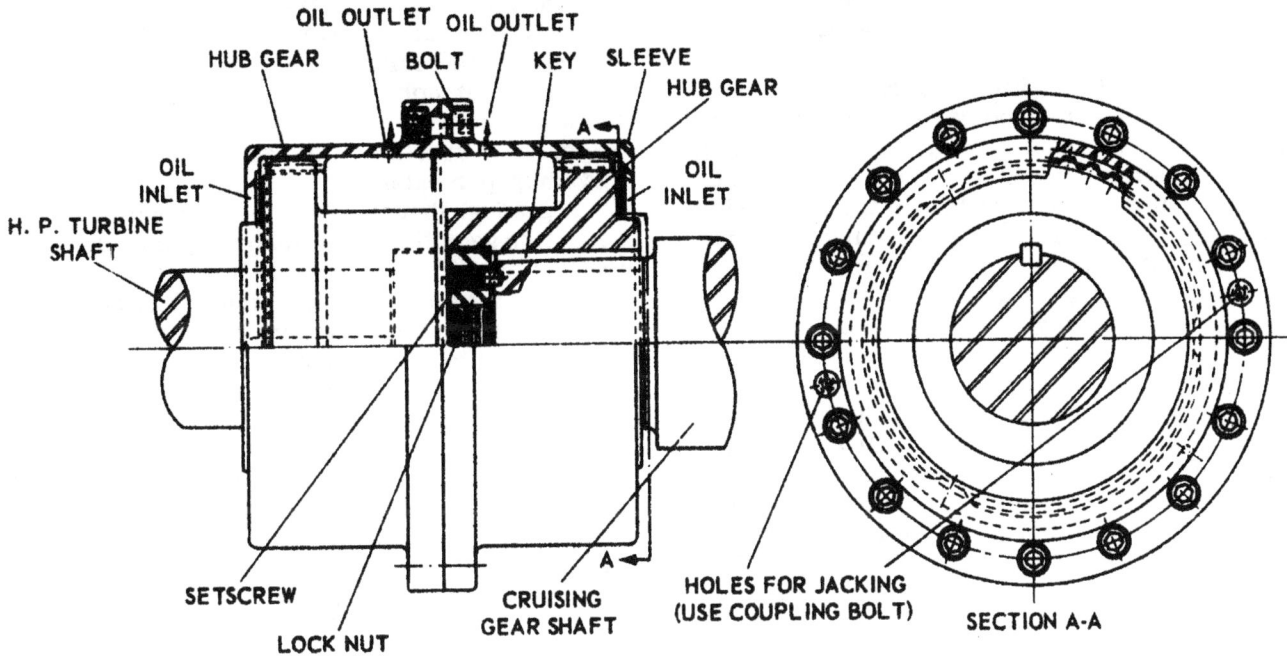

47.33

Figure 5-22.—Flexible coupling between cruising gear and high pressure turbine.

and lock a propeller shaft to prevent damage to the machinery. When the shaft is stopped, engaging the turning gear and then applying the brake is the most expeditious means of locking a propeller shaft while under way.

By carrying out actual drills, engineroom personnel should be trained to safely lock and unlock the main shaft. Each steaming watch should have sufficient trained personnel available to stop and lock the main shaft.

CAUTION: During drills the shaft should not be locked more than 5 minutes, if possible. The ahead throttle should NEVER be opened when the turning gear is engaged. The torque produced by the ahead engines is in the same direction as the torque of the locked shaft; to open the ahead throttle would result in damage to the turning gear.

The maximum safe operating speed of a ship with a locked shaft can be found in the manufacturer's technical manual. Additional information on the safe maximum speed that your ship can steam with a locked shaft can be found in Nav-Ships Technical Manual, chapter 9410. If the shaft has been locked for 5 minutes or more, the turbine rotors may have become bowed, and special precautions are recommended. Before the shaft is allowed to turn, men should be sta-

tioned at the turbines to check for unusual noises and vibration. When the turning gear is disengaged, the astern throttle should be slowly closed, the torque produced by the propeller passing through the water will start the shaft rotating. If, when the propeller starts to turn, vibration indicates a bowed rotor, the ship's speed should be reduced to the point where little or no vibration of the turbine is noticeable and this speed should be maintained until the rotor is straightened. If operation at such a slow speed is not practicable, the turbines should be slowed by use of the astern throttle, to the point of least vibration but with the turbines still operating in the ahead direction. When the turbines are slowed to the point of little or no vibration, the shaft should be operated at that speed and the ahead throttle should be opened slightly to permit some steam flow through the affected turbine. The heat from the steam will warm the shaft and aid in straightening it. Lowering the main condenser vacuum will add additional heat to the turbines; this will increase the exhaust pressure and temperature.

As the vibration decreases, the astern throttle can be closed gradually, allowing the speed of the shaft to increase. The shaft speed should be increased slowly and a check for vibration should be maintained. The turbine is not ready for normal operation until vibration has disappeared at all possible speeds.

NOISES AND VIBRATION.—On steam-turbine driven ships, noises may occur at low speeds or when maneuvering, or when passing through shallow water. Generally, these noises do not result from any defect in the propulsion machinery and will not occur during normal operation. A rumbling sound which occurs at low shaft rpm is generally due to the low pressure turbine gearing floating through its backlash. This condition has also been experienced with cruising reduction gears. The rumbling and thumping noises which may occur during maneuvering or during operation in shallow water, are caused by vibrations initiated by the propeller. These noises referred to are characteristic only of some ships and should be regarded as normal sounds for these units. These sounds will disappear with a change of propeller rpm or when the other causes mentioned are no longer present. These noises can usually be noticed in destroyers when the ship is backing, especially in choppy seas or in ground swells.

A properly operating reduction gear has a definite sound which an experienced watch-stander can easily learn to recognize. At different speeds and under various operating conditions, the operator should be familiar with the normal operating sound of the reduction gears on his ship.

If any abnormal sounds occur, an investigation should be made immediately. In making an investigation, much will depend on how the operator interprets the sound or noise.

The lube oil temperature and pressure may or may not help an operator determine the reasons for the abnormal sounds. A badly wiped bearing may be indicated by a rapid rise in oil temperature for the individual bearing. A certain sound or noise may indicate misalignment or improper meshing of the gears. If unusual sounds are caused by misalignment of gears or foreign matter passing through the gear teeth, the shaft should be stopped and a thorough investigation should be made before the gears are operated again.

For a wiped bearing, or any other bearing casualty that has caused a very high temperature, this procedure should be followed: If the temperature of the lube oil leaving any bearing has exceeded the permissible limits, slow or stop the unit and inspect the bearing for wear.

The bearing may be wiped only a small amount and the shaft may be operated at a reduced speed until the tactical situation allows sufficient time to inspect the bearing.

The most common causes of vibration in a main reduction gear installation are: faulty alignment, bent shafting, damaged propellers, and improper balance.

A gradual increase in the vibration in a main reduction gear that has been operating satisfactorily for a long period of time can usually be traced to a cause outside of the reduction gears. The turbine rotors, rather than the gears, are more likely to be out of balance.

When reduction gears are built, the gears are carefully balanced (both statically and dynamically). A small amount of unbalance in the gears will cause unusual noise, vibration, and abnormal wear of bearings.

When the ship has been damaged, vibration of the main reduction gear installation may result from misalignment of the turbine, the main shafting, the main shaft bearings, or the main reduction gear foundation. When vibration occurs within the main reduction gears, damage to the propeller should be one of the first things to be considered. The vulnerable position of the propellers makes them more liable to damage than other parts of the plant. Bent or broken propeller blades will transmit vibration to the main reduction gears. Propellers can also become fouled with line or cable which will cause the gears to vibrate. No reduction gear vibration is too trivial to overlook. A complete investigation should be made, preferably by a shipyard.

MAINTENANCE AND INSPECTION.—Under normal conditions, major repairs and major items of maintenance on main reduction gears should be accomplished by a shipyard. When a ship is deployed overseas and at other times when shipyard facilities are not available, emergency repairs should be accomplished, if possible, by a repair ship or an advanced base. Inspections, checks, and minor repairs should be accomplished by ship's force.

Under normal conditions, the main reduction gear bearings and gears will operate for an indefinite period. If abnormal conditions occur, the shipyard will normally perform the repairs. Spares are carried aboard sufficient to replace 50 percent of the number of bearings installed in the main reduction gear. Usually each bearing is interchangeable for the starboard or port installation. The manufacturer's technical manual must be checked to determine interchangeability of gear bearings.

Special tools and equipment needed to lift main reduction gear covers, to handle the quill shaft when removing bearings from it, and to take required readings and measurements, are normally carried aboard. The special tools and equipment should always be aboard in case emergency repairs have to be made by repair ships or bases not required to carry these items.

The manufacturer's technical manual is the best source of information concerning repairs and maintenance of any specific reduction gear installation. Chapters 9420, 9430, and 9440 of NavShips Technical Manual gives the inspection requirements for reduction gears, shafting, bearings, and propellers.

The inspections mentioned here are the minimum requirements only. Where defects are suspected, or operating conditions so indicate, inspections should be made at more frequent intervals.

To open any inspection plates or other fittings of the main reduction gears, permission should first be obtained from the engineer officer. Before replacing an inspection plate, connection, or cover which permits access to the gear casing, a careful inspection shall be made by an officer of the engineering department to ensure that no foreign matter has entered or remains in the casing or oil lines. If the work is being done by a repair activity, an officer from the repair activity must also inspect the gear casing. An entry of the inspections and the name of the officer or officers must be made in the Engineering Log. The inspections required on the main engine reduction gears are shown on the Maintenance Index Page, figure 5-23.

The importance of proper gear tooth contact cannot be overemphasized. Any abnormal condition which may be revealed by operational sounds or by inspections should be corrected as soon as possible. Any abnormal condition which is not corrected will cause excessive wear which may result in general disintegration of the tooth surfaces.

If proper tooth contact is obtained when the gears are installed, little wear of teeth will occur. Excessive wear cannot take place without metallic contact. Proper clearances and adequate lubrication will prevent most gear tooth trouble.

If proper contact is obtained when the gears are installed, the initial wearing, which takes place under conditions of normal load and adequate lubrication, will smooth out rough and uneven places on the gear teeth. This initial wearing-in is referred to as NORMAL WEAR or RUNNING IN. As long as operating conditions remain normal, no further wear will occur.

Small shallow pits starting near the pitch line, will frequently form during the initial stage of operation; this process is called INITIAL PITTING. Often the pits (about the size of a pinhead or even smaller) can be seen only under a magnifying glass. These pits are not detrimental and usually disappear in the course of normal wear.

Pitting which is progressive and continues at an increasing rate is known as DESTRUCTIVE PITTING. The pits are fairly large and are relatively deep. Destructive pitting is not likely to occur under proper operating conditions, but could be caused by excessive loading, too soft material, or improper lubrication. It is usually found that this type of pitting is due to misalignment or to improper lubrication.

The condition in which groups of scratches appear on the teeth (from the bottom to the top of the tooth) is termed abrasion, or scratching. It may be caused by inadequate lubrication, or by the presence of foreign matter in the lubricating oil. When abrasion or scratching is noted, the lubricating system and the gear spray fixtures should immediately be examined. If it is found that dirty oil is responsible, the system must be thoroughly cleaned and the whole charge of oil centrifuged.

The term "scoring" denotes a general roughening of the whole tooth surface. Scoring marks are deeper and more pronounced than scratching and they cover an area of the tooth, instead of occurring haphazardly, as in scratching or abrasion. Small areas of scoring may occur in the same position on all teeth. Scoring, with proper alignment and operation, usually results from inadequate lubrication, and is intensified by the use of dirty oil. If these conditions are not corrected, continued operation will result in a general disintegration of the tooth surfaces.

Under normal conditions all alignment inspections and checks, plus the necessary repairs, are accomplished by naval shipyards. Incorrect alignment will be indicated by abnormal vibration, unusual noise, and wear of the flexible couplings or main reduction gears. When misalignment is indicated, a detailed inspection should be made by shipyard personnel.

Two sets of readings are required to get an accurate check of the propulsion shafting. One

PRINCIPLES OF NAVAL ENGINEERING

System, Subsystem, or Component					Reference Publications				
Reduction Gears									

Bureau Card Control No.					Maintenance Requirement	M.R. No.	Rate Req'd.	Man Hours	Related Maintenance
MB	ZZZFGE5	35	5025	Q	1. Inspect the reduction gear including spray nozzles.	Q-1	EO MM1 MM3	1.0 1.0	None
MB	ZZ2FSC1	65	4290	Q	1. Measure main shaft thrust clearance.	Q-2	EO MM1 MM2	0.3 0.3	None
MB	ZZZFGE1	84	5064	S	1. Inspect and clean oil sump and reduction gear casing.	S-1	EO MM1 MM3 2FN	5.0 6.0 12.0	None
MB	ZZ1FCW4	65	A188	A	1. Inspect flexible couplings. Measure clearances.	A-1	MMC MM1 2FN	2.0 8.0 16.0	None
MB	ZZZFGE5	78	6669	A	1. Sound and tighten foundation bolts.	A-2	FN	1.0	None

Figure 5-23.—Maintenance Index Page.

98.171

set of readings is taken with the ship in drydock and another set of readings is taken with the ship waterborne—under normal loading conditions. The main shaft is disconnected, marked, and turned so that a set of readings can be taken in four different positions. Four readings are taken (top, bottom, and both sides). The alignment of the shaft can be determined by studying the different readings taken. The naval shipyard will decide whether or not corrections in alignment are necessary.

NOTE: During shipyard overhauls, the following inspections should be made:

a. Inspect condition and clearance of thrust shoes to ensure proper position of gears. Blow out thrusts with dry air after the inspection. Record the readings. Inspect the thrust collar, nut, and locking device.

b. If turbine coupling inspection has indicated undue wear, check alignment between pinions and turbines.

c. Clean oil sump.

When conditions warrant or if trouble is suspected, a work request may be submitted to a naval shipyard to perform a "seven year" inspection of the main reduction gears. This inspection includes clearances and condition of bearings and journals; alignment checks and readings; and any other tests, inspections, or maintenance work that may be considered necessary.

Naval Ship Systems Command authorization is not necessary for lifting reduction gear covers. Covers should be lifted when trouble is suspected. An open gear case is a serious hazard to the main plant, therefore, careful consideration of the dangers of uncovering a gear case must be balanced against the reasons for suspecting internal trouble, before deciding to lift the gear case. The seven-year interval may be extended by the type commander if conditions indicate that a longer period between inspections is desirable.

The correction of any defects disclosed by regular tests and inspections, and the observance of the manufacturers' instructions, should ensure that the gears are ready for full power at all times.

In addition to inspections which may be directed by proper authority, open the inspection plates, examine the tooth contact, the condition of teeth, and the operation of the spray nozzles. It is not advisable to open gear cases, bearings, and thrusts immediately BEFORE full power trials.

In addition to the inspections which may be directed by proper authority, open the inspection plates, and examine the tooth contact and the condition of the teeth to note changes that may have occurred during the full power trials. Running for a few hours at high power will show any possible condition of improper contact or abnormal wear that would not have shown up in months of operation at lower power. Check the clearance of the main thrust bearing.

SAFETY PRECAUTIONS

The following precautions must be observed by personnel operating or working with propulsion equipment.

1. If there is churning or emulsification of oil and water in the gear case, the gear must be slowed down or stopped until the defect is remedied.

2. If the supply of oil to the gear fails, the gears should be stopped until the cause can be located and remedied.

3. When bearings have been overheated, gears should not be operated, except in extreme emergencies, until bearings have been examined and defects remedied.

4. If excessive flaking of metal from the gear teeth occurs, the gears should not be adjusted, except in an emergency, until the cause has been determined.

5. Unusual noises should be investigated at once, and the gears should be operated cautiously until the cause for the noise has been discovered and remedied.

6. No inspection plate, connection, fitting, or cover which permits access to the gear casing should be removed without specific authorization by the engineer officer.

7. The immediate vicinity of an inspection plate should be kept free from paint and dirt.

8. When gear cases are open, precautions should be taken to prevent the entry of foreign matter. The openings should never be left unattended unless satisfactory temporary closures have been installed.

9. Lifting devices should be inspected carefully before being used and should not be overloaded.

10. When ships are anchored in localities where there are strong currents or tides,

precautions should be taken to lock the main shaft.

11. Where the rotation of the propellers may result in injury to a diver over the side, or in damage to the equipment, propeller shafts should be locked.

12. When a ship is being towed, the propellers should be locked, unless it is permissible and advantageous to allow the shafts to trail with the movement of the ship.

13. When a shaft is allowed to turn or trail, the lubrication system must be in operation. In addition, a careful watch should be kept on the temperature within the low pressure turbine casing to see that windage temperatures cannot be built up to a dangerous degree. This can be controlled either by the speed of the ship or by maintaining vacuum in the main condenser.

14. The main propeller shaft must be brought to a complete stop before the clutch of the turning gear is engaged. (If the shaft is turning, considerable damage to the turning gear will result.)

15. When the turning gear is engaged, the brake must be set quickly and securely to prevent the shaft turning and damaging the turning gear.

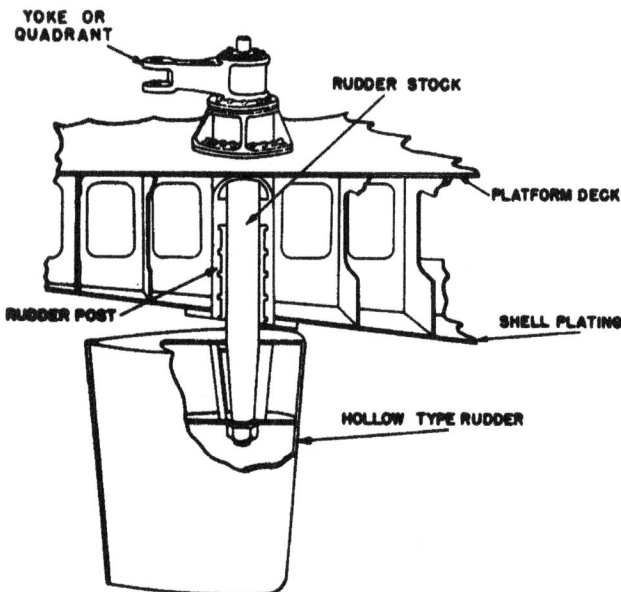

3.99
Figure 5-24.—Rudder assembly.

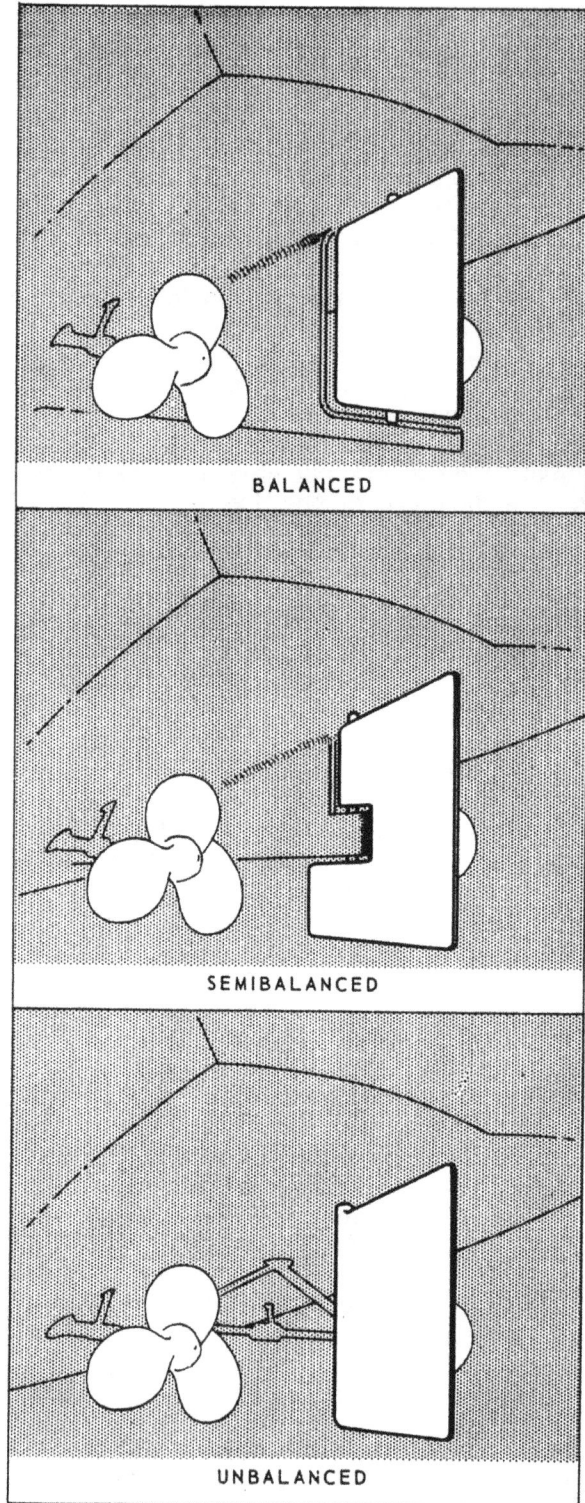

147.52
Figure 5-25.—Balanced, semibalanced, and unbalanced rudders.

16. When a main shaft is to be unlocked, precautions must be taken to disengage the jacking gear clutch before releasing the brake. If the brake is released first, the main shaft may begin to rotate and cause injury to the turning gear and to personnel.

17. In an emergency, where the ship is steaming at a high speed, the main shaft can be stopped and held stationary by the astern turbine until the ship has slowed down to a speed at which the main shaft can be safely locked.

18. Where there is a limiting maximum safe speed at which a ship can steam with a locked propeller shaft, this speed should be known and should not be exceeded.

19. Before the turning gear is engaged and started, a check should be made to see that the turning gear is properly lubricated. Some ships have a valve in the oil supply line leading to the turning gear. The operator should see that a lube oil service pump is in operation and that the proper oil pressure is being supplied to the turning gear before the motor is started.

20. It should be definitely determined that the turning gear has been disengaged before the main engines are turned over.

21. While working on or inspecting open main reduction gears, the person or persons performing the work should not have any article about their person which may accidentally fall into the gear case.

22. Tools, lights, mirrors, etc. used for working on or inspecting gears, bearings, etc. should be lashed and secured to prevent accidental dropping into the gear case.

STEERING

As noted at the beginning of this chapter, the direction of movement of a ship is controlled partly by steering devices which receive their power from steering engines and partially by the arrangement, speed, and direction of rotation of the ship's propellers.

The steering device is called a rudder. The rudder is a more or less rectangular metal blade (usually hollow on large ships) which is supported by a rudder stock. The rudder stock enters the ship through a rudder post and a watertight fitting, as shown in figure 5-24. A yoke or quadrant, secured to the head of the rudder stock, transmits the motion imparted by the steering mechanism.

Basically, a ship's rudder is used to attain and maintain a desired heading. The force necessary to accomplish this is developed by dynamic pressure against the flat surface of the rudder. The magnitude of this force and the direction and degree to which it is applied produces the rudder effect which controls stern movement and thus controls the ship's heading.

In order to function most effectively, a rudder should be located aft of and quite close to the propeller. Many modern ships have twin rudders, each set directly behind a propeller to receive the full thrust of water. This arrangement tends to make a ship highly maneuverable.

Three types of rudders are in general use— the unbalanced rudder, the semibalanced rudder, and the balanced rudder. These three types are illustrated in figure 5-25. Other types of rudders are also in naval use. For example, some ships have a triple-blade rudder which provides an increased effective rudder area.

CHAPTER 6

THEORY OF LUBRICATION

Lubrication reduces friction between moving parts by substituting fluid friction for solid friction. Without lubrication, it is difficult to move a hundred-pound weight across a rough surface; with lubrication, and with proper attention to the design of bearing surfaces, it is possible to move a million-pound load with a motor that is small enough to be held in the hand. By reducing friction, thereby reducing the amount of energy that is dissipated as heat, lubrication reduces the amount of energy required to perform mechanical actions and also reduces the amount of energy that is dissipated as heat.

Lubrication is a matter of vital importance throughout the shipboard engineering plant. Moving surfaces must be steadily supplied with the proper kinds of lubricants, lubricants must be maintained at specified standards of purity, and designed pressures and temperatures must be maintained in the lubrication systems. Without adequate lubrication, a good many units of shipboard machinery would quite literally grind to a screeching halt.

The lubrication requirements of shipboard machinery are met in various ways, depending upon the nature of the machinery. This chapter deals with lubrication in general—with basic principles of lubrication, with lubricants used aboard ship, and with the shipboard devices used to maintain lubricating oils in the required condition of purity. The separate lubrication systems that are installed for many shipboard units are discussed in other chapters of this text.

FRICTION

The friction that exists between a body at rest and the surface upon which it rests is called static friction. The friction that exists between moving bodies (or between one moving body and a stationary surface) is called kinetic friction. Static friction, which must be overcome to put any body in motion, is greater than kinetic friction, which must be overcome to keep the body in motion.

There are three types of kinetic friction: sliding friction, rolling friction, and fluid friction. Sliding friction exists when the surface of one solid body is moved across the surface of another solid body. Rolling friction exists when a curved body such as a cylinder or a sphere rolls upon a flat or curved surface. Fluid friction is the resistance to motion exhibited by a fluid.

Fluid friction exists because of the cohesion between particles of the fluid and the adhesion of fluid particles to the object or medium which is tending to move the fluid. If a paddle is used to stir a fluid, for example, the cohesive forces between the molecules of the fluid tend to hold the molecules together and thus prevent motion of the fluid. At the same time, the adhesive forces of the molecules of the fluid cause the fluid to adhere to the paddle and thus create friction between the paddle and the fluid. Cohesion is the molecular attraction between particles that tends to hold a substance or a body together; adhesion is the molecular attraction between particles that tends to cause unlike surfaces to stick together. From the point of view of lubrication, adhesion is the property of a lubricant that causes it to stick (or adhere) to the parts being lubricated; cohesion is the property which holds the lubricant together and enables it to resist breakdown under pressure.

Cohesion and adhesion are possessed by different materials in widely varying degrees. In general, solid bodies are highly cohesive but only slightly adhesive. Most fluids are quite highly adhesive but only slightly cohesive; however, the adhesive and cohesive properties of fluids vary considerably.

FLUID LUBRICATION

Fluid lubrication is based on the actual separation of surfaces so that no metal-to-metal contact occurs. As long as the lubricant film remains unbroken, sliding friction and rolling friction are replaced by fluid friction.

In any process involving friction, some power is consumed and some heat is produced. Overcoming sliding friction consumes the greatest amount of power and produces the greatest amount of heat. Overcoming rolling friction consumes less power and produces less heat. Overcoming fluid friction consumes the least power and produces the least amount of heat.

LANGMUIR THEORY

A presently accepted theory of lubrication is based on the Langmuir theory of the action of fluid films of oil between two surfaces, one or both of which are in motion. Theoretically, there are three or more layers or films of oil existing between two lubricated bearing surfaces. Two of the films are boundary films (indicated as I and V in part A of fig. 6-1), one of which clings to the

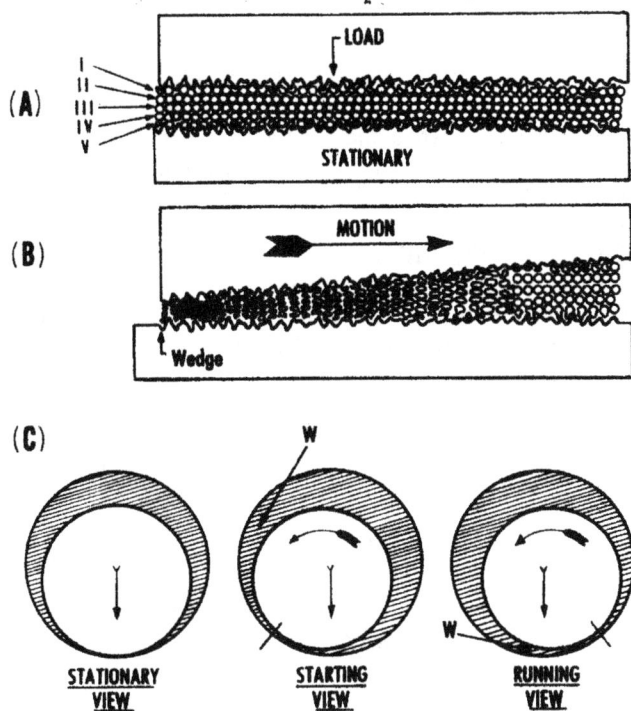

47.78

Figure 6-1.—Oil film lubrication. (A) Stationary position, showing several oil films; (B) surface set in motion, showing principle of oil wedge; (C) principle of (A) and (B) shown in a journal bearing.

surface of the rotating journal and one of which clings to the stationary lining of the bearing. Between these two boundary films are one or more fluid films (indicated as II, III, and IV in part A of fig. 6-1). The number of fluid films shown in the illustration is arbitrarily selected for purposes of explanation.

When the rotating journal is set in motion (part B of fig. 6-1), the relationship of the journal to the bearing lining is such that a wedge of oil is formed. The oil films II, III, and IV begin to slide between the two boundary films, thus continuously preventing contact between the two metal surfaces. The principle is again illustrated in part C of figure 6-1, where the position of the oil wedge W is shown with respect to the position of the journal as it starts and continues in motion.

The views shown in part C of figure 6-1 represent a journal or shaft rotating in a solid bearing. The clearances are exaggerated in the drawing in order to illustrate the formation of the oil film. The shaded portion represents the clearance filled with oil. The film is in the process of being squeezed out while the journal is at rest, as shown in the stationary view. As the journal slowly starts to turn and the speed increases, oil adhering to the surfaces of the journal is carried into the film, increasing the film thickness and tending to lift the journal as shown in the starting view. As the speed increases, the journal takes the position shown in the running view. Changes in temperature, with consequent changes in oil viscosity, cause changes in the film thickness and in the position of the journal.

If conditions are correct, the two surfaces are effectively separated, except for a possible momentary contact at the time the motion is started.

FACTORS AFFECTING LUBRICATION

A number of factors determine the efficacy of oil film lubrication, including such things as pressure, temperature, viscosity, speed, alignment, condition of the bearing surfaces, running clearances between the bearing surfaces, starting torque, and the nature and purity of the lubricant. Many of these factors are interrelated and interdependent. For example, the viscosity of any given oil is affected by temperature and the temperature is affected by running speed; hence the viscosity is partially dependent upon the running speed.

A lubricant must be able to stick to the bearing surfaces and support the load at operating speeds. More adhesiveness is required to make a lubricant adhere to bearing surfaces at high speeds than at low speeds. At low speeds, greater cohesiveness is required to keep the lubricant from being squeezed out from between the bearing surfaces.

Large clearances between bearing surfaces require high viscosity and cohesiveness in the lubricant to ensure maintenance of the lubricating oil film. The larger the clearance, the greater must be the resistance of the lubricant to being pounded out, with consequent destruction of the lubricating oil film.

High unit load on a bearing requires high viscosity of the lubricant. A lubricant subjected to high loading must be sufficiently cohesive to hold together and maintain the oil film.

LUBRICANTS

Although there is a growing use of synthetic lubricants, the principal source of the oils and greases used in the Navy is still petroleum. By various refining processes, lubricating stocks are extracted from crude petroleum and blended into a multiplicity of products to meet all lubrication requirements. Various compounds or additives are used in some lubricants (both oils and greases) to provide specific properties required for specific applications.

Types of Lubricating Oils

Lubricating oils approved for shipboard use are limited to those grades and types deemed essential to provide proper lubrication under all anticipated operating conditions.

For diesel engines, it is necessary to use a detergent-dispersant type of additive oil in order to keep the engines clean. In addition, these lubricating oils must be fortified with oxidation inhibitors and corrosion inhibitors to allow long periods between oil changes and to prevent corrosion of bearing materials.

For steam turbines, it is necessary to have an oil of high initial film strength. This oil is then fortified with anti-foaming additives and additives that inhibit oxidation and corrosion. In addition, it is necessary to use extreme pressure (EP) additives to enable the oil to carry the extremely high loading to which it is subjected in the reduction gears.

For the hydraulic systems in which petroleum lubricants are used, and for general lubrication use, the Navy uses a viscosity series

of oils reinforced with oxidation and corrosion inhibitors and anti-foam additives. The compounded oils, which are mineral oils to which such products as rape seed, tallow, or lard oil are added, are still used in deck machinery and in the few remaining steam plants that utilize reciprocating steam engines.

A great many special lubricating oils are available for a wide variety of services. These are listed in the Federal Supply Catalog. Among the more important specialty oils are those used for lubricating refrigerant compressors. These oils must have a very low pour point and be maintained with a high degree of freedom from moisture.

The principal synthetic lubricants currently in naval use are (1) a phosphate ester type of fire-resistant hydraulic fluid, used chiefly in the deck-edge elevators of carriers (CVAs); and (2) a water-base glycol hydraulic fluid used chiefly in the catapult retracting gear.

Classification of Lubricating Oils

The Navy identifies lubricating oils by symbols. Each identification number consists of four digits (and, in some cases, appended letters). The first digit indicates the class of oil according to type and use; the last three digits indicate the viscosity of the oil. The viscosity digits are actually the number of seconds required for 60 milliliters of the oil to flow through a standard orifice at a specified temperature. The symbol 3080, for example, indicates that the oil is in the 3000 series and that a 60-ml sample flows through a standard orifice in 80 seconds when the oil is at a specified temperature (210°F, in this instance). To take another example, the symbol 2135 TH indicates that the oil is in the 2000 series and that a 60-ml sample flows through a standard orifice in 135 seconds when the oil is at a specified temperature (130° F, in this case).

The letters H, T, TH, or TEP added to a basic symbol number indicate that the oil contains additives for special purposes.

Lubricating Oil Characteristics

Lubricating oils used by the Navy are tested for a number of characteristics, including viscosity, pour point, flash point, fire point, autoignition point, neutralization number, demulsibility, and precipitation number. Standard test methods are used for making all tests.

The <u>viscosity</u> of an oil is its tendency to resist flow. An oil of high viscosity flows very slowly. Raising the temperature of an oil lowers its viscosity; lowering the temperature increases the viscosity. The measurement of viscosity is discussed in chapter 7 of this text.

The <u>viscosity index</u> of an oil is a number indicating the effect of temperature changes on viscosity. A low viscosity index signifies a relatively large change of viscosity with changes of temperature. An oil which becomes thin at high temperatures and thick at low temperatures is said to have a low viscosity index; a high viscosity index signifies that the viscosity changes relatively little with changes of temperature.

The <u>pour point</u> of an oil is the lowest temperature at which the oil will barely flow from a container. The pour point is closely related to the viscosity of the oil. In general, an oil of high viscosity will have a higher pour point than an oil of low viscosity.

The <u>flash point</u> of an oil is the temperature at which enough vapor is given off to flash when a flame or spark is applied under standard test conditions.

The <u>fire point</u> (higher than the flash point) of an oil is the temperature at which the oil will continue to burn when it is ignited.

The <u>auto-ignition point</u> of an oil is the temperature at which the flammable vapors given off from the oil will burn without the application of a spark or flame.

The <u>neutralization number</u> of an oil is a measure of the acid content; it is defined as the number of milligrams of potassium hydroxide (KOH) required to neutralize one gram of the oil. All petroleum products oxidize in the presence of air and heat, and the products of oxidation include organic acids. The acids, if present in sufficient concentration, have harmful effects on alloy bearings at high temperatures. The presence of acids also may result in the formation of sludge and emulsions too stable to be broken down. An increase in acidity is an indication that lubricating oil is deteriorating.

The <u>demulsibility</u> of an oil is the ability of the oil to separate cleanly from any water present. Demulsibility is an important characteristic of lubricating oils used in forced-feed lubrication systems.

The <u>precipitation number</u> of an oil is a measure of the amount of solids classified as asphalts or carbon residue contained in the oil. The precipitation number is reached by diluting a known quantity of oil with naphtha and separating the precipitate by centrifuging. The volume of the separated solids equals the precipitation number. An oil with a high precipitation number is not suitable for certain applications because it may leave deposits in an engine or plug up valves and pumps.

Lubricating Greases

Some lubricating greases are simple mixtures of soaps and lubricating oils. Others are more exotic liquids such as silicones and di-basic acid esters, thickened with metals or inert materials to provide adequate lubrication. Requirements for oxidation inhibition, corrosion prevention, and extreme pressure performance are met by incorporating special additives.

Lubricating greases are supplied in three grades: soft, medium, and hard. The soft greases are used for high speeds and low pressures; the medium greases are used for medium speeds and medium pressures; the hard greases are used for slow speeds and high pressures.

CARE OF LUBRICATING OIL

Lubricating oils may be kept in service for long periods of time, provided the purity of the oils is maintained at the required standard. The simple fact is that lubricating oil does not wear out,[1] although it can become unfit for use when it is robbed of its lubricating properties by the presence of water, sand, sludge, fine metallic particles, acid, and other contaminants.

Proper care of lubricating oil requires, then, that the oil be kept as free from contamination as possible and that, once contaminated, the oil must be purified before it can be used again.

PREVENTING CONTAMINATION

Strainers or filters are used in many lubricating systems to prevent the passage of grit, scale, dirt, and other foreign matter. Duplex strainers are used in lubricating systems in which an uninterrupted flow of lubricating oil <u>must</u> be maintained; the flow may be diverted from one strainer basket to the other while one is being cleaned. Filters may be installed directly in pressure lubricating systems or they may be installed as bypass filters.

---

[1] The additive content of an oil may be exhausted as the additive combats the special conditions for which it was included in the oil; but this is a gradual process and is never catastrophic.

The use of strainers and filters does not solve the problem of water contamination of lubricating oil. Even a very small amount of water in lubricating oil can be extremely damaging to machinery, piping, valves, and other equipment. Water in lubricating oil can cause widespread pitting and corrosion; also, by increasing the frictional resistance, water can cause the oil film to break down prematurely. Every effort must be made to prevent the entry of water into any lubricating system.

REMOVING CONTAMINATION

In spite of all efforts, a certain amount of contamination of lubricating oil is to be expected. Aboard ship, centrifugal purifiers are used to remove impurities from lubricating oil and settling tanks are provided to permit used oil to stand while water and other impurities settle out.

Centrifugal Purifiers

A centrifugal purifier is essentially a bowl or hollow cylindrical container which is rotated at high speed while contaminated oil is forced through and rotated with the container. The centrifugal force imposed on the oil by the high rotational speed of the container causes the suspended foreign matter to separate from the oil.

Materials that are soluble in each other cannot be separated by centrifugal force. For example, salt cannot be removed from sea water by centrifugal force because the salt and water

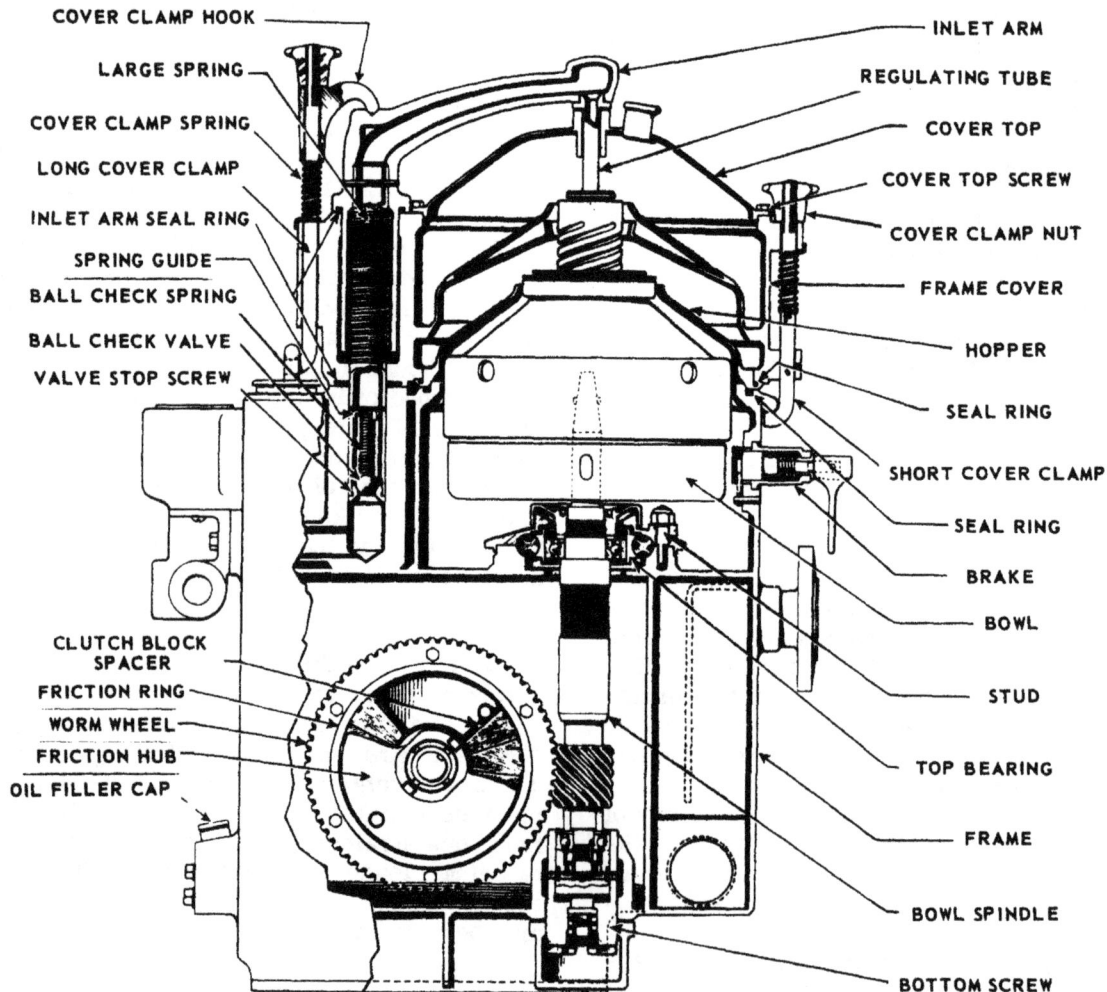

75.231

Figure 6-2.—Disk-type centrifugal purifier.

116

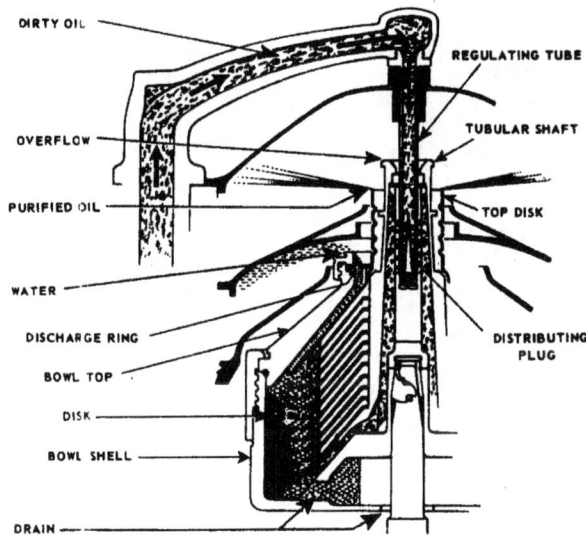

75.233

Figure 6-3.—Path of contaminated oil through disk-type purifier bowl (DeLAVAL).

are in solution. However, water can be separated from lubricating oil because water and oil do not form a solution when mixed. For separation to take place by centrifugal force, there must be a difference in the specific gravity of oil and the specific gravity of water.

When a mixture of oil, water, and sediment is allowed to stand undisturbed, gravity tends to cause the formation of an upper layer of oil, an intermediate layer of water, and a lower layer of sediment. The layers form because of differences in the specific gravities of the various substances. If the oil, water, and sediment mixture is placed in a rapidly revolving centrifugal purifier, the effect of gravity is negligible in comparison with the effect of centrifugal force. Centrifugal force, acting at right angles to the axis of rotation of the container, forces the sediment into an outer layer, the water into an intermediate layer, and the oil into an innermost layer. Centrifugal purifiers are so designed that the separated water is discharged as waste and the oil is discharged for use. The solids remain in the rotating unit and are cleaned out after each purification operation.

Two types of centrifugal purifiers are used aboard ship. The main difference between the two types is in the design of the rotating units. In the disk-type purifier, the rotating element is a bowl-like container which encases a stack of disks. In the tubular-type purifier, the rotating element is a hollow tubular rotor.

A disk-type centrifugal purifier is shown in figure 6-2. The bowl is mounted on the upper end of the vertical bowl spindle, which is driven by a worm wheel and friction clutch assembly. A radial thrust bearing is provided at the lower end of the bowl spindle to carry the weight of the bowl spindle and to absorb any thrust created by the driving action.

Contaminated oil enters the top of the revolving bowl through the regulating tube. The oil then passes down the inside of the tubular shaft and out at the bottom of the stack of disks. As the dirty oil flows up through the distribution holes in the disks, the high centrifugal force exerted by the revolving bowl causes the dirt, sludge, and water to move outward and the purified oil to move inward toward the tubular shaft. The disks divide the space within the bowl into many separate narrow passages. The liquid confined within each passage is restricted so that it can only flow along that passage. This arrangement prevents excessive agitation of the liquid as it passes through the bowl and creates shallow settling distances between the disks. The path of contaminated oil passing through a disk-type purifier is shown in figure 6-3.

Most of the dirt and sludge remains in the bowl and collects in a more or less uniform layer on the inside vertical surface of the bowl shell. Any water that may be present, together with some dirt and sludge, is discharged through the discharge ring at the top of the bowl. The purified oil flows inward and upward through the disks, discharging from the neck of the top disk.

A tubular-type centrifugal purifier is shown in figure 6-4. This type of purifier consists essentially of a hollow rotor or bowl which rotates at high speeds. The rotor has an opening in the bottom through which the dirty lubricating oil enters; two sets of openings at the top allow the oil and water (or the oil alone) to discharge. (See insert, fig. 6-4.) The bowl or hollow rotor of the purifier is connected by a coupling unit to a spindle which is suspended from a ball bearing assembly. The bowl is belt-driven by an electric motor mounted on the frame of the purifier.

The lower end of the bowl extends into a flexibly mounted guide bushing. The assembly, of which the bushing is a part, restrains movement of the bottom of the bowl but allows enough movement so that the bowl can center itself about its axis of rotation when the purifier is in operation. Inside the bowl is a device which consists of three flat plates equally spaced radially. This device is commonly referred to as the three-wing device

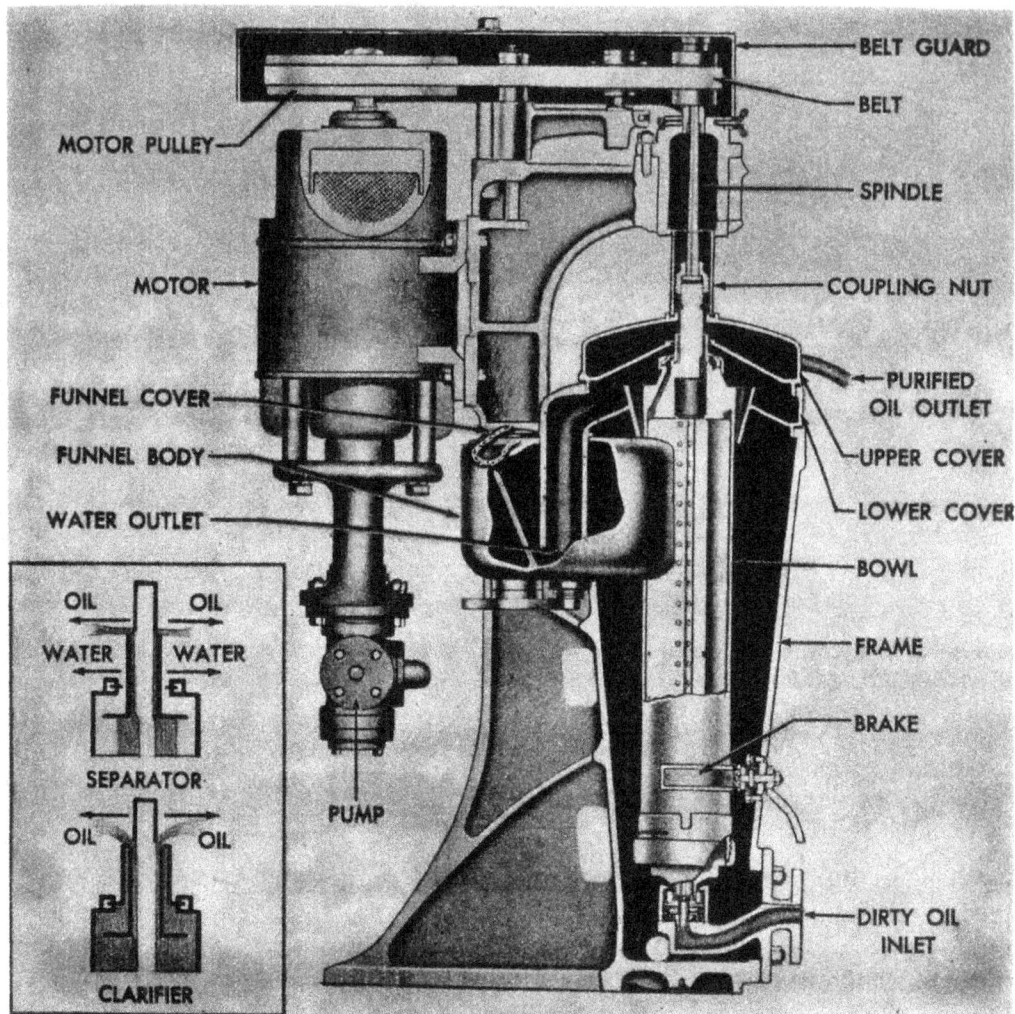

Figure 6-4.—Tubular-type centrifugal purifier.   75.234

or as the three-wing. The three-wing rotates with the bowl and forces the liquid in the bowl to rotate at the same speed as the bowl. The liquid to be centrifuged is fed into the bottom of the bowl through the feed nozzle, under pressure, so that the liquid jets into the bowl in a stream.

The process of separation is basically the same in the tubular-type purifier as in the disk-type purifier. In both types, the separated oil assumes the innermost position and the separated water moves outward. Both liquids are discharged separately from the bowl, and the solids separated from the liquid are retained in the bowl.

Settling Tanks

Lubrication systems aboard ship include settling tanks in which used oil is allowed to stand while water and other impurities settle out. Lubricating oil piping is generally arranged to permit two methods of purification: batch purification and continuous purification.

In the batch process, the lubricating oil is transferred from the sump to a settling tank by means of a purifier or a transfer pump. In the settling tank, the oil is heated to approximately 160° F and allowed to settle for several hours. Water and other impurities are removed from the settling tanks. The oil is then centrifuged and returned to the sump from which it was taken.

In the continuous purification process, the centrifugal purifier takes suction from a sump tank and, after purifying the oil, discharges it back to the same sump. The continuous method of purification is used while a ship is underway.

118

CHAPTER 7

PRINCIPLES OF MEASUREMENT

Measurement is, in a very real sense, the language of engineers. The shipboard engineering plant contains an enormous number of gages and instruments that tell operating personnel whether the plant is running properly or whether some abnormal condition—excessive speed, high pressure, low pressure, high temperature, low water level—requires corrective action. The gages and instruments also provide essential information for the hourly, daily, and weekly entries for station operating logs and for other engineering records and reports.

This chapter describes some of the basic types of gages and instruments used in shipboard engineering plants for the measurement of important variables such as temperature, pressure, fluid flow, liquid level, and rotational speed. Because of the wide variety of gages and instruments used in connection with shipboard engineering equipment, no attempt is made to cover all types that might possibly be encountered; instead, basic principles of measurement and commonly used types of gages are emphasized. Unusual or highly specialized measuring devices, or ones that have particular application to some one type of machinery or equipment aboard ship, are in general discussed in the chapters of this text that deal with the particular equipment; where an unusual type of measuring device is discussed in this chapter, it is included chiefly as a means of bringing out some interesting or important aspect of measurement. Detailed information on most gages and instruments used aboard ship can be obtained from manufacturers' technical manuals and other instructional materials furnished with shipboard engineering equipment.

THE CONCEPT OF MEASUREMENT

One of the primary ways in which we extend our knowledge and understanding of the universe and of the world around us is by the measurement of various quantities. Because we live in a world in which practically everything seems to be in some way measured or counted, we often tend to assume that measurement is basically simple. In reality, however, it may be quite difficult to develop an appropriate mode of measurement even after we have recognized the need; and, without an appropriate mode of measurement, we may even fail to recognize the significance of the phenomena we observe. Thus the development of scientific and engineering principles has been, and undoubtedly will continue to be, inextricably tied to the concept of measurement.

Many of our views on the nature of things are profoundly influenced by the procedures we devise for measurement. It is interesting to note how often in the history of science the application of a new instrument or the refinement of a measuring technique has led to new ideas about the universe or about the nature of the thing being measured.[1] Until approximately the middle of the seventeenth century, it was commonly believed that water rose in a suction pump because "nature abhors a vacuum."[2] The concept of a "sea of air" surrounding the

---

[1] As Sir Humphry Davy (1778-1829) stated, "Nothing tends so much to the advancement of knowledge as the application of a new instrument." (Quoted in Harvard Case Histories in Experimental Science, James Bryant Conant, general editor, and Leonard K. Nash, associate editor, Cambridge, Massachusetts, 1957. Vol. 1, page 119.)

[2] The explanation that "nature abhors a vacuum" persisted for quite some time in spite of the observed fact that water would not rise more than about 32 feet in the suction pumps of the time and in spite of Galileo's observation that "Evidently nature's horror of a vacuum does not extend beyond 32 feet."

earth and exerting pressure upon it was very closely related to Torricelli's experiments with a column of mercury in a glass tube. The notion that the air above us exerts a pressure was not fully accepted until after Pascal had arranged an experiment to test the hypothesis. Pascal suggested using Torricelli's new instrument, the barometer, at the base of a mountain and then again at the top of the mountain. If the air exerts a pressure, Pascal reasoned, the mercury should stand higher in the glass column at the base of the mountain than it should at the top. The experiment was performed by Pascal's brother-in-law in 1648, and the prediction was confirmed. Further experimentation and measurement by Robert Boyle and others led to the development of many important concepts concerning the nature of air and other gases, and led eventually to an understanding of the relationship between the volume and the pressure of a gas (Boyle's law).

Perhaps an even more striking example of the effects of measurement upon our basic concepts of the nature of things is to be found in the study of heat. Quantitative studies of heat were not possible before the invention of the thermometer.[3] It was not until the middle of the nineteenth century that the concept of heat as a form of energy, rather than as an invisible, weightless fluid called "caloric," was firmly established. The persistence of the caloric theory to such a late date was due partly to faulty interpretations of experimental results; but these faulty interpretations were at least in part the result of difficulties of measurement. The downfall of the caloric theory was necessary before we could conceive of heat as energy, rather than as a nebulous kind of matter, and before we could understand the relationship between heat and work.[4] In summary, it would

not be unreasonable to say that the thermometer had to be invented before we could arrive at an understanding of the nature of heat, the relationship between heat and mechanics, and the principle of the conservation of energy.

SYSTEMS, UNITS, AND STANDARDS OF MEASUREMENT

Practically all units of measurement are derived from a few basic quantities or fundamental dimensions, as they are sometimes called. In all commonly used systems of measurement, length and time are taken as two of the fundamental dimensions. A third is MASS in some systems and force (or weight) in others. In all systems, temperature is the fourth fundamental dimension.

The first three fundamental dimensions—length, time, and either mass or force—are sometimes called mechanical quantities or dimensions. All other important mechanical quantities can be defined in terms of these three fundamentals. Temperature, the fourth fundamental dimension, is in a different category because it is not a mechanical quantity. By using the three mechanical fundamental quantities and the quantity of temperature, practically all quantities of any importance may be derived.

It is often said that there are two systems of measurement—a metric system and a British system. Actually, however, there are several metric systems and several British systems. A more meaningful classification of systems of measurement can be made by saying that some systems are gravitational and others are absolute. In gravitational systems, the units of force are defined in terms of the effects of the force of gravity upon a standard sample of matter at a specified location on the surface of the earth. In absolute systems, the units of force are defined in terms that are completely independent of the effects of the force of gravity. Thus a metric system could be either gravitational or absolute, and a British system could be either gravitational or absolute, depending upon the terms in which force is defined in the particular system.

MASS AND WEIGHT

To understand what is meant by gravitational and absolute systems of measurement, it is necessary to have a clear understanding of the difference between mass and weight. Mass, a measure of the total quantity of matter in an

---

[3] Joseph Black (1728-1799), commenting on the discovery that heat tends to flow from hotter to colder bodies until a state of thermal equilibrium is reached, stated: "No previous acquaintance with the peculiar relation of each body to heat could have assured us of this, and we owe the discovery entirely to the thermometer." (From Black's Lectures on the Elements of Chemistry, assembled from notes and published in 1803 by John Robinson. Quoted in Harvard Case Histories in Experimental Science, op. cit., vol. 1, page 128.)

[4] The relationship between heat and work is, of course, basic to the entire field of engineering. Chapter 8 of this text deals with this topic in considerable detail.

object or body, is completely independent of the force of gravity, so the mass of any given object is always the same, no matter where it is located on the surface of the earth; indeed, the body would have the same mass even if it were located at the center of the earth, on the moon, in outer space, or anywhere else. Weight, on the other hand, is a measure of the force of attraction between the mass of the earth and the mass of another body or object. Since the force of attraction between the earth and another body is not identical in all places, the weight of a body depends upon the location of the body with respect to the earth.

The relationship between mass and weight can be understood from the equation

$$w = mg$$

where

    w = weight
    m = mass
    g = acceleration due to gravity

The value for acceleration due to gravity (normally represented by the letter g) is almost constant for bodies at or near the surface of the earth. This value is approximately 32 feet per second per second in British systems of measurement, 9.8 meters per second per second in one metric system, and 980 centimeters per second per second in another metric system. More precise values of g, including variations that occur with changes in latitude and changes in elevation, may be obtained from physics and engineering textbooks and handbooks.

BASIC MECHANICAL UNITS

Table 7-1 shows the basic mechanical quantities of length, mass or force, and time, together with a number of derived units, used in several systems of measurement. By examining some of the units, we may see how force is defined and thus see why each system is called "absolute" or "gravitational," as the case may be.

In the metric absolute meter-kilogram-second (MKS) system of measurement, the unit of mass is the kilogram, the unit of length is the meter, the unit of time is the second, and the unit of acceleration is meters per second per second. (This is sometimes written as m/sec^2.) The unit of force is called a Newton.

By definition, 1 newton is the force required to accelerate a mass of 1 kilogram at the rate of 1 meter per second per second. In other words, the unit of force is defined in such a way that unit force gives unit acceleration to unit mass.

The same thing holds true in the other metric absolute system shown in table 7-1. In the metric absolute centimeter-gram-second (CGS) system of measurement, the gram is the unit of mass, the centimeter is the unit of length, the second is the unit of time, and centimeters per second (cm/sec^2) is the unit of acceleration. In this system, the unit of force is called a dyne. By definition, 1 dyne is the force required to accelerate a mass of 1 gram at the rate of 1 centimeter per second per second. Again, force is defined in such a way that unit force gives unit acceleration to unit mass.

The same applies to the British absolute foot-pound-second (FPS) system of measurement, where the pound is the unit of mass, the foot is the unit of length, the second is the unit of time, and feet per second per second is the unit of acceleration. In this system, the unit of force is called a poundal. By definition, 1 poundal is the amount of force required to give a mass of 1 pound an acceleration of 1 foot per second per second. Again, force is defined in such a way that unit force gives unit acceleration to unit mass.

Now let's look at a British gravitational system—the foot-pound-second (FPS) gravitational system that we use in the United States for most everyday measurements. The foot is the unit of length, the pound is the unit of mass, the second is the unit of time, and feet per second per second is the unit of acceleration. In this system, the unit of force is called the pound. (Actually, it should be called the pound-force; but this usage is rarely followed.) In this system, a force of 1 pound acting upon a mass of 1 pound produces an acceleration of 32 feet per second per second. Note that unit force does not produce unit acceleration when acting on unit mass; rather, unit force produces unit acceleration when acting on unit weight. Since force is defined in gravitational terms, rather than in absolute terms, we say that this is a gravitational system of measurement.

The gravitational system that is usually called the British Engineering System also uses the pound (or, more precisely, the pound-force) as the unit of force. But this system has its own unit of mass: the slug. By definition, 1 slug is the quantity of mass that is accelerated

Table 7-1.—Units of measurement in Several Common Systems.

	MKS Metric Absolute System	CGS Metric Absolute System	British FPS Absolute System	British Engineering Gravitational System	British FPS Gravitational System
Length	meter (m)	centimeter (cm)	foot (ft)	foot (ft)	foot (ft)
Area	square meter (m^2)	square centimeter (cm^2)	square foot (ft^2)	square foot (ft^2)	square foot (ft^2)
Volume	cubic meter (m^3) or liter (l)	cubic centimeter $(cm^3$ or cc) or milliliter (ml)	cubic foot (ft^3)	cubic foot (ft^3)	cubic foot (ft^3)
Mass	kilogram (kg)	gram (g)	pound (lb)	slug	pound (lb)
Force (Weight)	newton (n, new, or nt)	dyne	poundal (pdl)	pound (lb) or pound-force (lbf)	pound (lb) or pound-force (lbf)
Time	second (sec)	second (sec)	second (sec)	second (sec)	second (sec)
Velocity	m/sec	cm/sec	ft/sec	ft/sec	ft/sec
Acceleration	m/sec^2	cm/sec^2	ft/sec^2	ft/sec^2	ft/sec^2
Pressure	nt/m^2	$dynes/cm^2$	pdl/ft^2	lb/ft^2 or lbf/ft^2	lb/ft^2 or lbf/ft^2
Energy	joule	erg	foot-poundal (ft-pdl)	foot-pound (ft-lb) or foot-pound-force (ft-lbf)	foot-pound (ft-lb) or foot-pound-force (ft-lbf)
Power	watt	ergs per second (erg/sec)	ft-pdl/sec	ft-lb/sec or ft-lbf/sec	ft-lb/sec or ft-lbf/sec

at the rate of 1 foot per second per second when acted on by a force of 1 pound. In other words, 1 slug equals 32 pounds, 2 slugs equals 64 pounds, and so forth. By using the slug as the unit of mass, the British engineering system sets up consistent units of measurement in which unit force acting upon unit mass produces unit acceleration. Note, however, that this is still a gravitational system rather than an absolute system.

By this time it should be obvious that the relationships expressed in one system of measurement do not necessarily hold when a different system is used. When using any particular system, it is essential to understand the precise meaning of all terms used in that system. This is not always a simple matter, since there are an enormous number of possible combinations of units and in many cases the same word is used to express quite different ideas. Another source of confusion is the way in which the various systems of measurement are used. In everyday life we use the British gravitational FPS system. In scientific work we use one of the metric systems. In engineering and other technical fields we use a British system or a metric system, depending upon the field involved. The only way to avoid total confusion in the use of measurement terms is to make sure that you understand the precise meaning of each term, as it relates to the particular system being used.

The units shown in table 7-1 are only a few of the units that may be derived in each system. For example, the unit of pressure shown for both of the British gravitational systems is pounds per square foot (lb/ft^2, or psf). However, the unit pounds per square inch (lb/in^2, or psi) is equally acceptable and is very commonly used. Similar conversions can be made for any of the other units, as long as the basic relationships of the system are accurately maintained.

When converting values from a metric system to a British system (or vice versa) it is necessary to understand the units used in each system. Most of us know quite a bit about the units commonly used in British systems, but less about the units used in metric systems. A description of the basic structure of the metric systems follows.

All metric systems of measurement are decimal systems—that is, the size of the units vary by multiples of 10. This makes computations very simple. Another handy thing about the metric systems is that the prefixes for the names of the units tell you the relative size of the units. Take the prefix kilo-, for example. Kilo-indicates 1000; so a kilogram is 1000 grams, a kilometer is 1000 meters, and so forth. Or take the prefix milli-, for another example; it indicates a thousandth. So 1 millimeter is 1 thousandth of a meter, 1 milligram is 1 thousandth of a gram, and so forth. Perhaps the best way to become familiar with the units in the metric systems is to associate the more commonly used prefixes with the positive and negative powers of 10, as shown in table 7-2.

Table 7-2.—Metric System Prefixes and Corresponding Positive and Negative Powers of 10.

METRIC SYSTEM PREFIXES	POSITIVE AND NEGATIVE POWERS OF 10	
DEKA- or DECA	10^1	= 10
HECTO-	10^2	= 100
KILO-	10^3	= 1000
MEGA-	10^6	= 1,000,000
DECI-	10^{-1}	= 0.1 (or 1/10)
CENTI-	10^{-2}	= 0.01 (or 1/100)
MILLI-	10^{-3}	= 0.001 (or 1/1000)
MICRO-	10^{-6}	= 0.000001 (or 1/1,000,000)

Table 7-3 gives some selected values for mechanical units in British systems of measurement. Table 7-4 gives some selected values for mechanical units in metric systems of measurement. Table 7-5 gives some British-metric and metric-British equivalents. The examples given in these tables are chosen primarily to help you develop an understanding of the relative sizes of the mechanical units. More complete tables are available in many physics and engineering textbooks and handbooks.

STANDARDS OF MEASUREMENT

The importance of having precise and uniform standards of measurement is recognized by all the major countries of the world, and international conferences on weights and measures are held from time to time. The International

Bureau of Weights and Measures is in France. Each major country has its own bureau or office charged with the responsibility of maintaining the required measurement standards, including the basic standards of length, mass, and time. In the United States, the National Bureau of Standards (NBS) is responsible for maintaining basic standards and for prescribing precise measuring techniques.

Length

Until quite recently, the international standard of length was a platinum-iridium alloy bar kept at the International Bureau of Weights and Measures in France. By definition, the standard meter was the distance between two parallel lines marked on this bar, measured at 0°C. Copies of this international standard were maintained by other countries; the United States

Table 7-3. —Selected Values of Mechanical Units
in British Systems of Measurement.

TYPE OF MECHANICAL UNIT	SELECTED VALUES
LENGTH	12 inches (in.) = 1 foot (ft) 3 ft = 1 yard (yd) 5280 ft = 1 mile (mi) 1760 yd = 1 mi
AREA	144 square inches = 1 square foot (sq in. or in.^{2}) (sq ft or ft^2) 9 ft^2 = 1 yd^2
VOLUME	1728 cubic inches = 1 cubic foot (cu in. or in.^{3}) (cu ft or ft^3) 27 ft^3 = 1 yd^3
FORCE (WEIGHT)	16 ounces (oz) = 1 pound (lb) 2000 lb = 1 ton
VELOCITY	60 miles per hour = 88 feet per second (ft/sec) (mi/hr or mph)

Table 7-4. —Selected Values of Mechanical Units in
Metric Systems of Measurement.

TYPE OF MECHANICAL UNIT	SELECTED VALUES
LENGTH	10 millimeters (mm) = 1 centimeter (cm) 10 cm = 1 decimeter (dm) 10 dm = 1 meter (m) 100 cm = 1 meter 1000 mm = 1 meter 10 m = 1 dekameter (dkm) 100 m = 1 hectometer (hm) 1000 m = 1 kilometer (km)
AREA	100 square millimeters (mm^2) = $1\ cm^2$ $100\ cm^2$ = $1\ dm^2$ $100\ dm^2$ = $1\ m^2$
VOLUME	1000 cubic millimeters (mm^3) = $1\ cm^3$ $1000\ cm^3$ = $1\ dm^3$ $1000\ dm^3$ = $1\ m^3$ 1 milliliter (ml) = $1\ cm^3$ 1000 ml = 1 liter (l) 100 centiliters (cl) = 1 liter 10 deciliters (dl) = 1 liter
MASS	1000 milligrams (mg) = 1 gram (g) 100 centigrams (cg) = 1 gram 1000 grams = 1 kilogram (kg)

standard meter bar was maintained at the National Bureau of Standards in Washington, D.C.

Note that the standard of length was the meter even for countries that were not on the metric system. The yard was defined in terms of the meter, 1 yard being equal to 0.9144 meter.

In 1960, the standard of length was changed by international agreement to an atomic constant: the wavelength of the orange-red light emitted by individual atoms of krypton-86 in a tube filled with krypton gas in which an electrical discharge is maintained. By definition, 1 meter is equal to 1,650,763.73 wavelengths of the orange-red light of krypton-86, and 1 inch is equal to 41,929.399 wavelengths. A device called an optical interferometer is used to determine the number of the wavelengths of the orange-red light of krypton-86 in an unknown length.

Mass

The standard of mass is the mass of a cylinder of platinum-iridium alloy defined as having a mass of 1 kilogram. The international standard kilogram mass is kept at the International Bureau of Weights and Measures in France. The

Table 7-5.—Selected British-Metric and Metric-British Conversions.

TYPE OF MECHANICAL UNIT	BRITISH-METRIC CONVERSIONS	METRIC-BRITISH CONVERSIONS
LENGTH	1 inch = 2.540 centimeters 1 foot = 0.3048 meter 1 yard = 0.9144 meter 1 mile = 1.6093 kilometers 1 mile = 1609.3 meters	1 centimeter = 0.3937 inch 1 meter = 39.37 inches 1 kilometer = 0.62137 mile
AREA	$1 \text{ in.}^2 = 6.452 \text{ cm}^2$ $1 \text{ ft}^2 = 929 \text{ cm}^2$ $1 \text{ yd}^2 = 0.8361 \text{ m}^2$ $1 \text{ mi}^2 = 2.59 \text{ km}^2$	$100 \text{ mm}^2 = 0.15499 \text{ in.}^2$ $100 \text{ cm}^2 = 15.499 \text{ in.}^2$ $100 \text{ m}^2 = 119.6 \text{ yd}^2$ $1 \text{ km}^2 = 0.386 \text{ mi}^2$
VOLUME	$1 \text{ in.}^3 = 16.387 \text{ cm}^3$ $1 \text{ ft}^3 = 0.0283 \text{ m}^3$ $1 \text{ yd}^3 = 0.7646 \text{ m}^3$ $231 \text{ in.}^3 = 3.7853 \text{ liters}$	$1000 \text{ mm}^3 = 0.06102 \text{ in.}^3$ $1000 \text{ cm}^3 = 61.02 \text{ in.}^3$ $1 \text{ m}^3 = 35.314 \text{ ft}^3$ 1 liter = 1.0567 liquid quarts
WEIGHT	1 grain = 0.0648 gram 1 ounce = 23.3495 grams 1 pound = 453.592 grams 1 pound = 0.4536 kilograms	1 gram = 15.4324 grains 1 gram = 0.03527 ounce 1 gram = 0.002205 pound 1 kilogram = 2.2046 pounds

United States standard kilogram mass is kept at the National Bureau of Standards.

The standard of mass is kept in a vault. Not more than once a year, the standard is removed from the vault and used for checking the values of smaller standards. The United States standard kilogram mass has been taken to France twice in the last seven years for comparison with the international standard. Every precaution is taken to keep the kilogram standard mass in perfect condition, free of nicks, scratches, and corrosion. The standard is always handled with forceps; it is never touched by human hands.

When the national standard is compared on a precision balance with high precision copies, the copies are found to be accurate to within one part in 100 million.

Time

Before 1960, the standard of time was the mean solar second—that is, 1/86,400 of a mean solar day, as determined by successive appearances of the sun overhead, averaged over a year. In 1960, the standard of time was changed to the tropical year 1900, which is the time it took the sun to move from a designated point back to the same point in the year 1900. By definition, 1 second was equal to 1/31,556,925.9747 of the tropical year 1900. The subdivision of the tropical year 1900 into smaller time intervals was accomplished by means of laboratory-type pendulum clocks, together with observation of natural phenomena such as the nightly movement of the stars and the moon.

Although the standard of time was changed to the tropical year 1900 in 1960, the same General Conference of Weights and Measures that approved this change also urged that work go forward on the development of an atomic clock. In 1967, the General Conference of Weights and Measures adopted as the basic standard of time the time required for the transition between two energy states of the cesium-133 atom. In accordance with this standard, 1 second is defined as 9,192,631,770 cycles of this particular transition in the cesium-133 atom.

The United States standard of time is maintained by a cesium clock which is kept at the National Bureau of Standards laboratories in Boulder, Colorado. The time signals that are broadcast by four radio stations operated by the National Bureau of Standards are based on this cesium clock.

MEASUREMENT OF TEMPERATURE

Temperature is measured by bringing a measuring system (such as a thermometer) into contact with the system in which we need to measure the temperature. We then measure some property of the measuring system--the expansion of a liquid, the pressure of a gas, electromotive force, electrical resistance, or some other mechanical, electrical, or optical property that has a definite and known relationship with temperature. Thus we infer the temperature of the measured system by the measurement of some property of the measuring system.

But the measurement of a property other than temperature will take us only so far in utilizing the measurement of temperature. For convenience in comparing temperatures and in noting changes in temperature, we must be able to assign a numerical value to any given temperature. For this we need temperature scales.

Until 1954, temperature scales were constructed around the boiling point and the freezing point of pure water at atmospheric pressure. These two fixed and reproducible points were used to define a fairly large temperature interval which was then subdivided into the uniform smaller intervals called degrees. The two most familiar temperature scales constructed in this manner are the Celsius scale and the Fahrenheit scale.

The Celsius scale is often called the centigrade scale in the United States and Great Britain. By international agreement, however, the name was changed from centigrade to Celsius in honor of the eighteenth-century Swedish astronomer, Anders Celsius. The symbol for a degree on this scale (no matter whether it is called Celsius or centigrade) is °C. The Celsius scale takes 0° C as the freezing point and 100° C as the boiling point of pure water at atmospheric pressure. The Fahrenheit scale takes 32° F as the freezing point and 212° F as the boiling point of pure water at atmospheric pressure. The interval between freezing point and boiling point is divided into 100 degrees on the Celsius scale and divided into 180 degrees on the Fahrenheit scale.

Since the actual value of the interval between freezing point and boiling point is identical, it is apparent that numerical readings on Celsius and Fahrenheit thermometers have no absolute significance and that the size of the degree is arbitrarily chosen for each scale. The relationship between degrees Celsius and degrees Fahrenheit is given by the formulas

$$°F = \frac{9}{5} \ °C + 32$$

$$°C = \frac{5}{9} \ (°F - 32)$$

Many people have trouble remembering these formulas, with the result that they either get them mixed up or have to look them up in a book every time a conversion is necessary. If you concentrate on trying to remember the basic relationships given by these formulas, you may find it easier to make conversions. The essential points to remember are these:

1. Celsius degrees are larger than Fahrenheit degrees. One Celsius degree is equal to 1.8 Fahrenheit degrees, and each Fahrenheit degree is only 5/9 of a Celsius degree.

2. The zero point on the Celsius scale represents exactly the same temperature as the 32-degree point on the Fahrenheit scale.

3. The temperatures 100° C and 212° F are identical.

In some scientific and engineering work, particularly where heat calculations are involved, an absolute temperature scale is used. The zero point on an absolute temperature scale is the point called absolute zero. Absolute zero is determined theoretically, rather than by actual measurement. Since the pressure of a gas at constant volume is directly proportional to the temperature, it is logical to assume that the pressure of a gas is a valid measure of its temperature. On this assumption, the lowest possible temperature (absolute zero) is defined as the temperature at which the pressure of a gas would be zero.

Two absolute temperature scales have been in use for many years. The rankine absolute scale is an extension of the Fahrenheit scale; it is sometimes called the Fahrenheit absolute scale. Degrees on the Rankine scale are the same size as degrees on the Fahrenheit scale, but the zero point on the Rankine scale is at -459.67° Fahrenheit. In other words, absolute zero is zero on the Rankine scale and -459.67 degrees on the Fahrenheit scale.

A second absolute scale, the kelvin, is more widely used than the Rankine. The Kelvin scale was originally conceived as an extension of the Celsius scale, with degrees of the same size but with the zero point shifted to absolute zero. Absolute zero on the Celsius scale is -273.15° C.

In 1954, a new international absolute scale was developed. The new scale was based upon one fixed point, rather than two. The one fixed point was the triple point of water--that is, the point at which all three phases of water (solid, liquid, and vapor) can exist together in equilibrium. The triple point of water, which is 0.01° C above the freezing point of water, was chosen because it can be reproduced with much greater accuracy than either the freezing point or the boiling point. On this new scale, the triple point was given the value 273.16 K. Note that neither the word "degrees" nor the symbol ° is used; instead, the unit is called a "kelvin" and the symbol is K rather than ° K.

In 1960, when the triple point of water was finally adopted as the fundamental reference for this temperature scale, the scale was given the name of International Practical Temperature Scale. However, you will often see this scale referred to as the Kelvin scale.

Although the triple point of water is considered the basic or fundamental reference for the International Practical Temperature Scale, five other fixed points are used to help define the scale. These are the freezing point of gold, the freezing point of silver, the boiling point of sulfur, the boiling point of water, and the boiling point of oxygen.

Figure 7-1 is a comparison of the Kelvin (International-Practical), Celsius, Fahrenheit, and Rankine (Fahrenheit-Absolute) temperature scales. All of the temperature points listed above absolute zero are considered as fixed points on the Kelvin scale except for the freezing point of water. The other scales, as previously mentioned, are based on the freezing and boiling points of water.

TEMPERATURE MEASURING DEVICES

Since temperature is one of the basic engineering variables, temperature measurement is essential to the proper operation of a shipboard engineering plant. The temperature of steam, water, fuel oil, lubricating oil, and other vital fluids must be measured at frequent intervals and the results of this measurement must in many cases be entered in engineering records and logs.

Devices used for measuring temperature may be classified in various ways. In this discussion we will consider the two major categories of (1) expansion thermometers, and (2) pyrometers.

Expansion Thermometers

Expansion thermometers operate on the principle that the expansion of solids, liquids, and gases has a known relationship to temperature changes. The types of expansion thermometers discussed here are (1) liquid-in-glass thermometers, (2) bimetallic expansion thermometers, and (3) filled-system expansion thermometers.

LIQUID-IN-GLASS THERMOMETERS.—Liquid-in-glass thermometers are probably the oldest, the simplest, and the most widely used devices for measuring temperature. A liquid-in-glass thermometer (fig. 7-2) consists of a bulb and a very fine bore capillary tube containing mercury, mercury-thallium, alcohol, toluol, or some other liquid which expands uniformly as the temperature rises and contracts uniformly as the temperature falls. The selection of liquid is based on the temperature

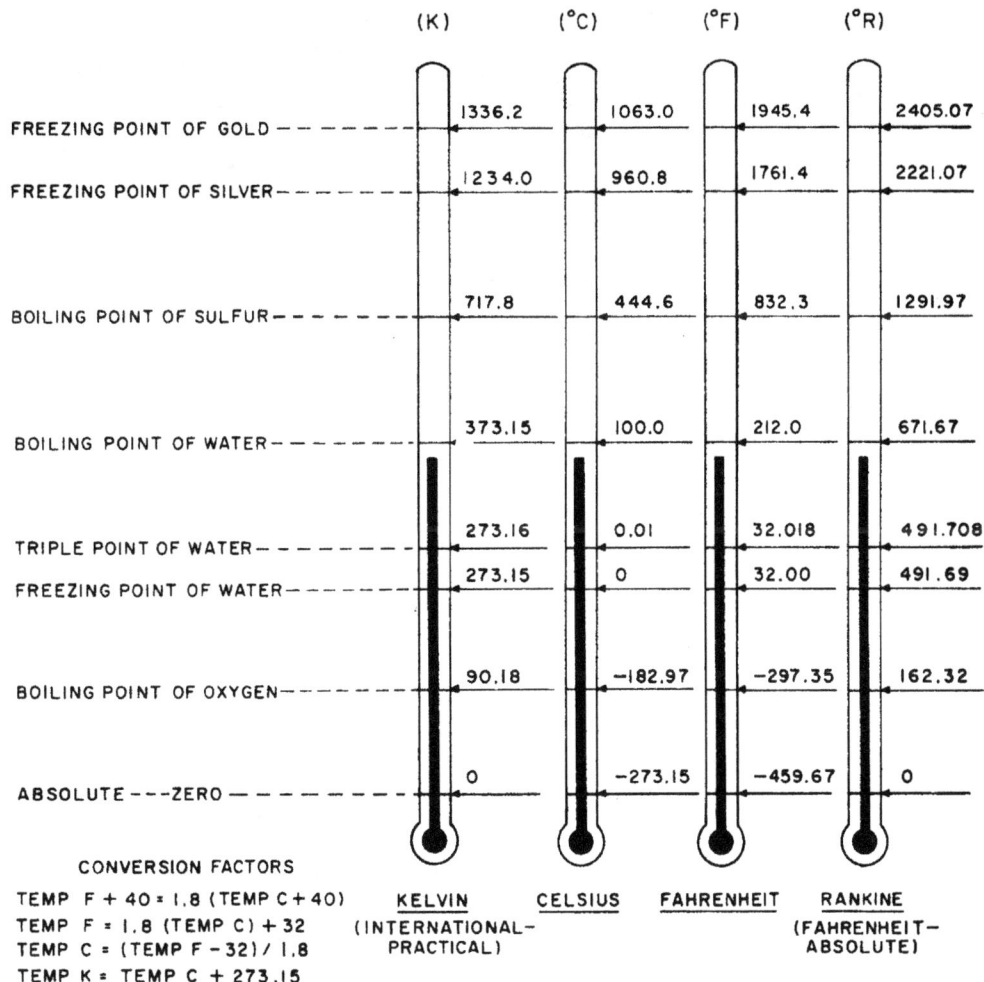

	(K)	(°C)	(°F)	(°R)
FREEZING POINT OF GOLD	1336.2	1063.0	1945.4	2405.07
FREEZING POINT OF SILVER	1234.0	960.8	1761.4	2221.07
BOILING POINT OF SULFUR	717.8	444.6	832.3	1291.97
BOILING POINT OF WATER	373.15	100.0	212.0	671.67
TRIPLE POINT OF WATER	273.16	0.01	32.018	491.708
FREEZING POINT OF WATER	273.15	0	32.00	491.69
BOILING POINT OF OXYGEN	90.18	−182.97	−297.35	162.32
ABSOLUTE ---ZERO	0	−273.15	−459.67	0

CONVERSION FACTORS

TEMP F + 40 = 1.8 (TEMP C + 40)
TEMP F = 1.8 (TEMP C) + 32
TEMP C = (TEMP F − 32)/ 1.8
TEMP K = TEMP C + 273.15

KELVIN
(INTERNATIONAL-
PRACTICAL)

CELSIUS

FAHRENHEIT

RANKINE
(FAHRENHEIT-
ABSOLUTE)

33.11(147B)

Figure 7-1.—Comparison of Kelvin, Celsius, Fahrenheit, and Rankine temperature.

range in which the thermometer is to be used. Mercury (or mercury-thallium) is commonly used because it is a liquid over a wide range of temperatures (−60° to 1200°F) and because it has a nearly constant coefficient of expansion.[5]

Almost all liquid-in-glass thermometers are sealed so that atmospheric pressure will not affect the reading. The space above the liquid in this type of thermometer may be a vacuum or it may be filled with an inert gas such as nitrogen, argon, or carbon dioxide.

The capillary bore may be either round or elliptical. In any case, it is very small so that a relatively small expansion or contraction of the liquid will cause a relatively large change in the position of the liquid in the capillary tube. Although the capillary bore itself is very small in diameter, the walls of the capillary tube are quite thick. Most liquid-in-glass thermometers are made with an expansion chamber at the top of the bore to provide a margin of safety for the instrument if it should accidentally be overheated.

[5]Not all liquids are suitable for use in thermometers. Water, for example, would be an almost impossible choice as a thermometric liquid at ordinary temperatures because its coefficient of expansion varies enormously at temperatures near 0° C. In the temperature range between 0° C and 4° C, water expands when cooled and contracts when heated; thus it actually has a negative coefficient of expansion in this range.

Liquid-in-glass thermometers may have graduations etched directly on the glass stem or the graduations may be carried on a separate strip of material which is placed behind the stem. Many thermometers used in shipboard engineering plants have the graduations marked on a separate strip, since this type is in general easier to read than the type which has the graduations marked directly on the stem.

Liquid-in-glass thermometers are made in various designs. The stem may be straight or it may be angled in various ways, depending upon the requirements of service. The thermometers may be armored or they may be partially enclosed by a metal case, if such protection is necessary. Several types of angle-stem liquid-in-glass thermometers of the type commonly used aboard ship are shown in figure 7-3. These thermometers are used in 5-inch, 7-inch, and 9-inch scale lengths. However, bimetallic thermometers are currently being substituted aboard

ship for the 5-inch scale liquid-in-glass thermometers.

Most liquid-in-glass thermometers used aboard ship are provided with wells or separable sockets. The well is installed in the piping system or equipment where the temperature is to be measured, and the thermometer glass bulb and part of the glass stem are fitted into a thin metal protection tube, packed with a heat-transfer material, and fastened in place in the well. The well is made of metals that will withstand the temperatures, pressures, and fluid velocities without damage; it protects the glass sensing bulb against damage and also eliminates the need for closing down a system or securing a piece of machinery merely in order to replace a thermometer.

One disadvantage of the well type of installation is that a certain amount of time is required for the thermometer to reach thermal equilibrium with the system in which the temperature is being measured. To some extent, the time lag can be decreased by filling the space around the bulb in the well with a heat transfer medium such as graphite. Where rapid response to temperature changes is a vital requirement, however, bare bulb thermometers are used instead of the well type of installation. Bare bulb thermometers have very much faster response to changes in temperature, but they cannot be removed for replacement or servicing while the machinery is operating or the line is under pressure.

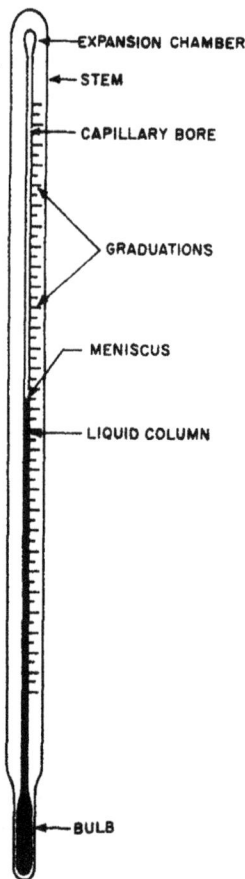

33.11(147A)

Figure 7-2.—Liquid-in-glass thermometer.

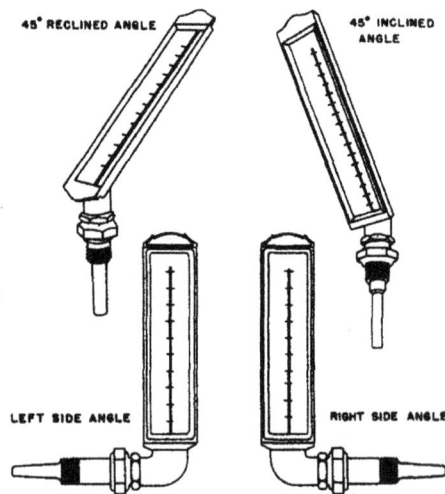

61.26

Figure 7-3.—Angle-stem liquid-in-glass thermometers.

Where special requirements exist, special types of liquid-in-glass thermometers are used. For example, maximum and minimum indicating thermometers are used in magazines aboard ship, for weather observations, and for various other applications where it is necessary to know the highest and the lowest temperatures that have occurred during a certain interval of time. One type of maximum indicating thermometer is shown in figure 7-4, and one type of minimum indicating thermometer is shown in figure 7-5.

The maximum indicating thermometer shown in figure 7-4 is a mercury thermometer with a special constriction in the bore. When the temperature rises, the pressure of the expand- ing mercury in the bulb forces mercury past the constriction in the bore. When the temperature falls, the mercury does not return to the bulb. Why? Even if the thermometer were in an upright position, the constriction in the bore would prevent the normal return flow of mercury by gravity; in addition, the maximum indicating thermometer is mounted with the bulb a few degrees above the horizontal position, so that the mercury column slopes downward from the con- striction. Thus the thermometer always indi- cates the highest temperature that has been reached since the instrument was last set. Ex- pansion and contraction of the mercury in the bore above the constriction does occur with temperature changes, but it is so slight as to

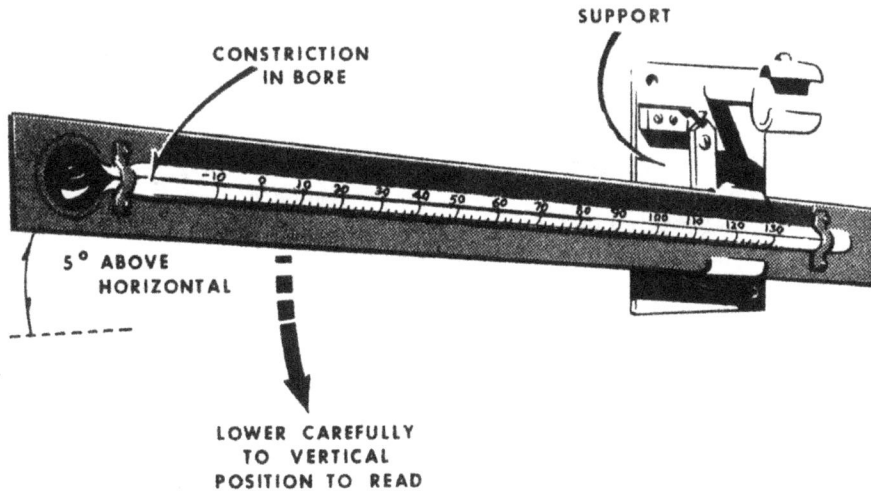

5.65A

Figure 7-4.—Maximum indicating thermometer.

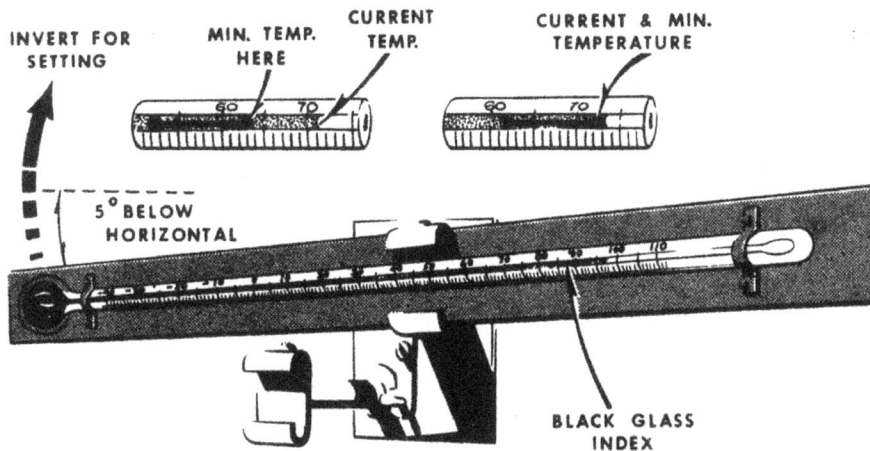

5.65B

Figure 7-5.—Minimum indicating thermometer.

be negligible for most purposes because only a very small amount of mercury is contained in the very narrow bore.

The minimum indicating thermometer shown in figure 7-5 is an alcohol-in-glass thermometer with an unusually large bore. The upper part of the bore is filled with air under pressure to help prevent evaporation of the alcohol. The thermometer is mounted with the bulb a few degrees below the horizontal position. A dumbbell-shaped piece of black glass (called an index) is the device that makes possible a reading of the minimum temperature that has occurred since the thermometer was last set. As the temperature increases, the alcohol readily flows upward past the index without moving it. As the temperature decreases, the retreating alcohol column flows past the index until the top of the column touches the upper end of the index. With a further decrease in temperature, the alcohol retreats still more and surface tension causes the index to be carried along down with the column. If the temperature increases again, the index is left undisturbed at its lowest point while the alcohol column rises again. Thus the top of the index always indicates the lowest temperature that has occurred since the thermometer was last set.

BIMETALLIC EXPANSION THERMOMETER.—Bimetallic expansion thermometers make use of the fact that different metals have different coefficients of linear expansion.[6] The essential element in a bimetallic expansion thermometer is a bimetallic strip consisting of two layers of different metals fused together. When such a strip is subjected to temperature changes, one layer expands or contracts more than the other, thus tending to change the curvature of the strip.

The basic principle of a bimetallic expansion thermometer is illustrated in figure 7-6. When one end of a straight bimetallic strip is fixed in place, the other end tends to curve away from the side that has the greater coefficient of linear expansion when the strip is heated.

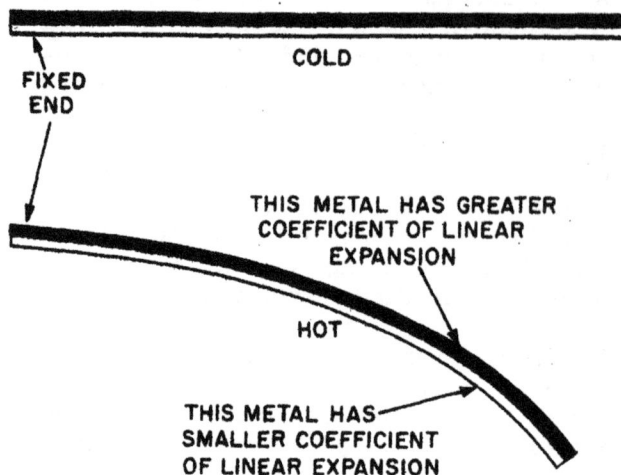

147.53

Figure 7-6.—Effect of unequal expansion of bimetallic strip.

For use in thermometers, the bimetallic strip is normally wound into a flat spiral (fig. 7-7), a single helix, or a multiple helix. The end of the strip that is not fixed in position is fastened to the end of a pointer which moves over a circular scale. Bimetallic thermometers are easily adapted for use as recording thermometers; a pen is attached to the pointer and is positioned in such a way that it marks on a revolving chart.

Bimetallic thermometers used aboard ship are normally used in thermometer wells. The wells are interchangeable with those used for mercury-in-glass thermometers.

FILLED-SYSTEM THERMOMETERS.—In general, filled-system thermometers are designed for use in locations where the indicating part of the instrument must be placed some distance away from the point where the temperature is to be measured.[7] For this reason they are often called distant-reading thermometers.

A filled-system thermometer (fig. 7-8) consists essentially of a hollow metal sensing bulb

---

[6]The coefficient of linear expansion is defined as the change in length per unit length per degree change in temperature. As is apparent from this definition, the numerical value of the coefficient of linear expansion is independent of the units in which the length is expressed but is not independent of the temperature scale chosen.

[7]This is not true of all filled-system thermometers. In a few designs the capillary tubing is extremely short and in a few it is nonexistent. In general, however, filled-system thermometers are designed to be distant-reading thermometers, and most of them do in fact serve this purpose. Some distant-reading thermometers may have capillaries as long as 125 feet.

at one end of a small-bore capillary tube, connected at the other end to a Bourdon tube or other device which responds to volume changes or to pressure changes. The system is partially or completely filled with a fluid which expands when heated and contracts when cooled. The fluid may be a gas, mercury, an organic liquid, or a combination of liquid and vapor.

The device usually used to indicate temperature changes by its response to volume changes or to pressure changes is called a Bourdon tube.[8] A Bourdon tube is a curved or twisted tube which is open at one end and sealed at the other. The open end of the tube is fixed in position and the sealed end is free to move. The tube is more or less elliptical in cross section; it does not form a true circle. The cross section of a noncircular tube which is sealed at one end tends to become more circular when there is an increase in the volume or in the internal pressure of the contained fluid, and this tends to straighten the tube. Opposing this action, the spring action of the tube metal tends to coil the tube.[9] Since the open end of the Bourdon tube is rigidly fastened, the sealed end moves as the volume or pressure of the contained fluid changes. When a pointer is attached to the sealed end of the tube through appropriate linkages, and when the assembly is placed over an appropriately calibrated dial, the result is a Bourdon-tube gage that may be used for measuring temperature or pressure, depending upon the design of the gage and the calibration of the scale.

Bourdon tubes are made in several shapes for various applications. The C-shaped Bourdon tube shown in figure 7-9 is perhaps the most commonly used type; spiral and helical Bourdon tubes are used where design requirements include the need for a longer length of Bourdon tube.

There are two basic types of filled-system thermometers: those in which the Bourdon tube responds primarily to changes in the volume of

___

[8] Bourdon tubes are sometimes called Bourdon springs, Bourdon elements, or simply Bourdons. Other devices such as bellows or diaphragms are used in some filled-system thermometers, but they are by no means as common as the Bourdon tube for this application.

[9] The precise nature of Bourdon-tube movement with pressure and volume changes is extremely complex and not completely describable in purely analytical terms. Bourdon-tube instruments are designed for specific applications on the basis of a series of empirical observations and tests.

147.54

Figure 7-7.—Bimetallic thermometer (flat spiral element).

the filling fluid and those in which the Bourdon tube responds primarily to changes in the pressure of the filling fluid. Obviously, there is always some pressure effect in volumetric thermometers and some volumetric effect in

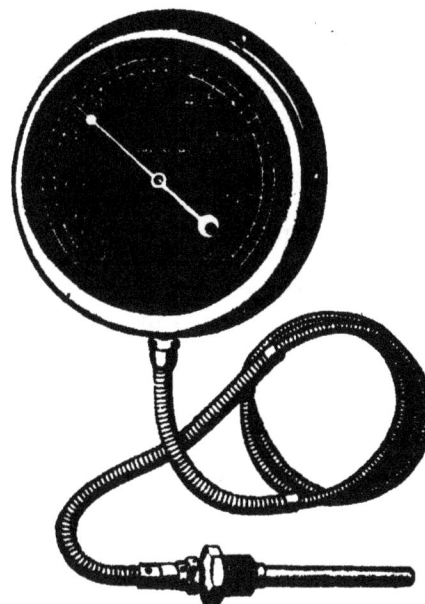

61.28X

Figure 7-8.—Distant-reading Bourdon-tube thermometer.

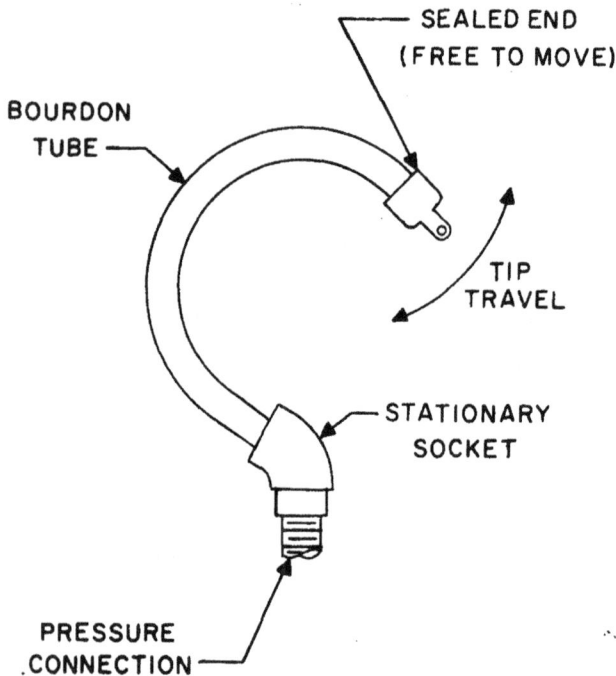

38.211(147B)
Figure 7-9.—C-shaped Bourdon tube.

pressure thermometers; the distinction deals with the major response of the Bourdon tube.

Pyrometers

The term pyrometer is used to include a number of temperature measuring devices which, in general, are suitable for use at relatively high temperatures; some pyrometers, however, are also suitable for use at low temperatures. The types of pyrometers we are concerned with here include thermocouple pyrometers, resistance thermometers, radiation pyrometers, and optical pyrometers.

THERMOCOUPLE PYROMETERS.—The operation of a thermocouple pyrometer (sometimes called a thermoelectric pyrometer) is based on the observed fact that an electromotive force (emf)[10] is generated when the two junctions of two dissimilar metals are at different temperatures. A simple thermocouple is illustrated in figure 7-10. Since the electromotive force generated is proportional to the temperature difference between the measuring junction (hot

---

[10]Basic information on electricity is given in chapter 20 of this text.

junction) and the reference junctions (cold junctions), the indicating instrument can be marked off to indicate degrees of temperature even though it is actually measuring emf's. The indicating instrument is a millivoltmeter or some other electrical device capable of measuring and indicating small direct-current emf's. The strips or wires of dissimilar metals are welded, twisted, fused, or otherwise firmly joined together. The extension leads are usually of the same metals as the thermocouple itself.

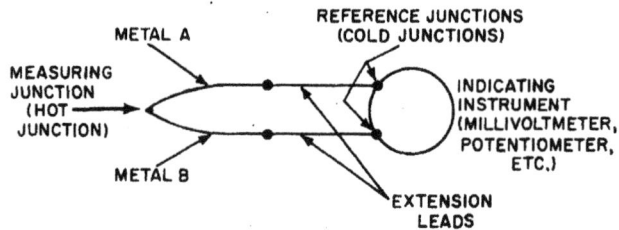

147.55
Figure 7-10.—Simple thermocouple.

RESISTANCE THERMOMETERS.—Resistance thermometers are based on the principle that the electrical resistance of a metal changes with changes in temperature. A resistance thermometer is thus actually an instrument which measures electrical resistance but which is calibrated in degrees of temperature rather than in units of electrical resistance.

The sensitive element in a resistance thermometer is a winding of small diameter nickel, platinum, or other metallic wire. The resistance winding is located in the lower end of a bulb (sometimes called a stem); it is electrically but not thermally insulated from the stem. The resistance winding is connected by two, three, or four leads to the circuit of the indicating instrument. The circuit is a Wheatstone bridge or some other simple circuit which contains known resistances with which the resistance of the thermometer winding is compared.

RADIATION AND OPTICAL PYROMETERS.—Radiation and optical pyrometers are used to measure very high temperatures. Both types of pyrometers measure temperature by measuring the amount of energy radiated by the hot object. The main difference between the two types is in their range of sensitivity; radiation pyrometers are (theoretically, at least) sensitive to the entire spectrum of radiant energy, while optical pyrometers are sensitive to only one wavelength or to a very narrow band of wavelengths.

Figure 7-11 illustrates schematically the general operating principle of a simple radiation pyrometer. Radiant energy from the hot object is concentrated on the detecting device by means of a lens or, in some cases, a conical mirror or a combination of mirror and lens. The detecting device may be a thermocouple, a thermopile (that is, a group of thermocouples in series), a photocell, or some other element in which some electrical quantity (emf, resistance, etc.) varies as the temperature of the hot object varies. The meter or indicated part of the instrument may be a millivoltmeter or some similar device.

An optical pyrometer measures temperature by comparing visible light emitted by the hot object with light from a standard source. A common type of optical pyrometer is shown in figure 7-12. This instrument consists of an eyepiece, a telescope which contains a filament similar to the filament of an electric light bulb and a potentiometer.

The person operating the optical pyrometer looks through the eyepiece and focuses the telescope on the hot object, meanwhile also observing the tin glowing filament across the field of the telescope. While watching the hot object and the filament, the operator adjusts the filament current (and consequently the brightness of the filament) by turning a knob on the potentiometer until the filament seems to disappear and to merge with the hot object. When the filament current has been adjusted so that the filament just matches the hot object in brightness, the operator turns another knob

slightly to balance the potentiometer. The potentiometer measures filament current but the dial is calibrated in degrees of temperature. As may be noted from this description, this type of optical pyrometer requires a certain amount of skill and judgment on the part of the operator. In some other types of optical pyrometers, automatic operation is achieved by use of photoelectric cells arranged in a bridge network.

MEASUREMENT OF PRESSURE

Pressure, like temperature, is one of the basic engineering variables and one that must frequently be measured aboard ship. Before taking up the devices used to measure pressure, let us consider certain definitions that are important in any discussion of pressure measurement.

PRESSURE DEFINITIONS

Pressure is defined as force per unit area.

The simplest pressure units are ones that indicate how much force is applied to an area of a certain size. These units include pounds per square inch, pounds per square feet, ounces per square inch, newtons per square millimeter, and dynes per square centimeter, depending upon the system being used.

You will also find another kind of pressure unit, and this type appears to involve length. These units include inches of water, inches of mercury (Hg), and inches of some other liquid of known density. Actually, these units do not involve length as a fundamental dimension. Rather, length is taken as a measure of force or weight. For example, a reading of 1 inch of water (1 in. H_2O) means that the exerted pressure is able to support a column of water 1 inch

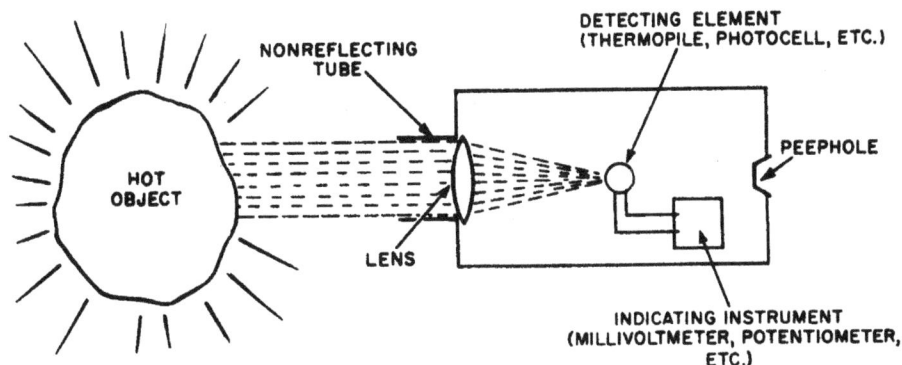

147.56

Figure 7-11.—Simple radiation pyrometer.

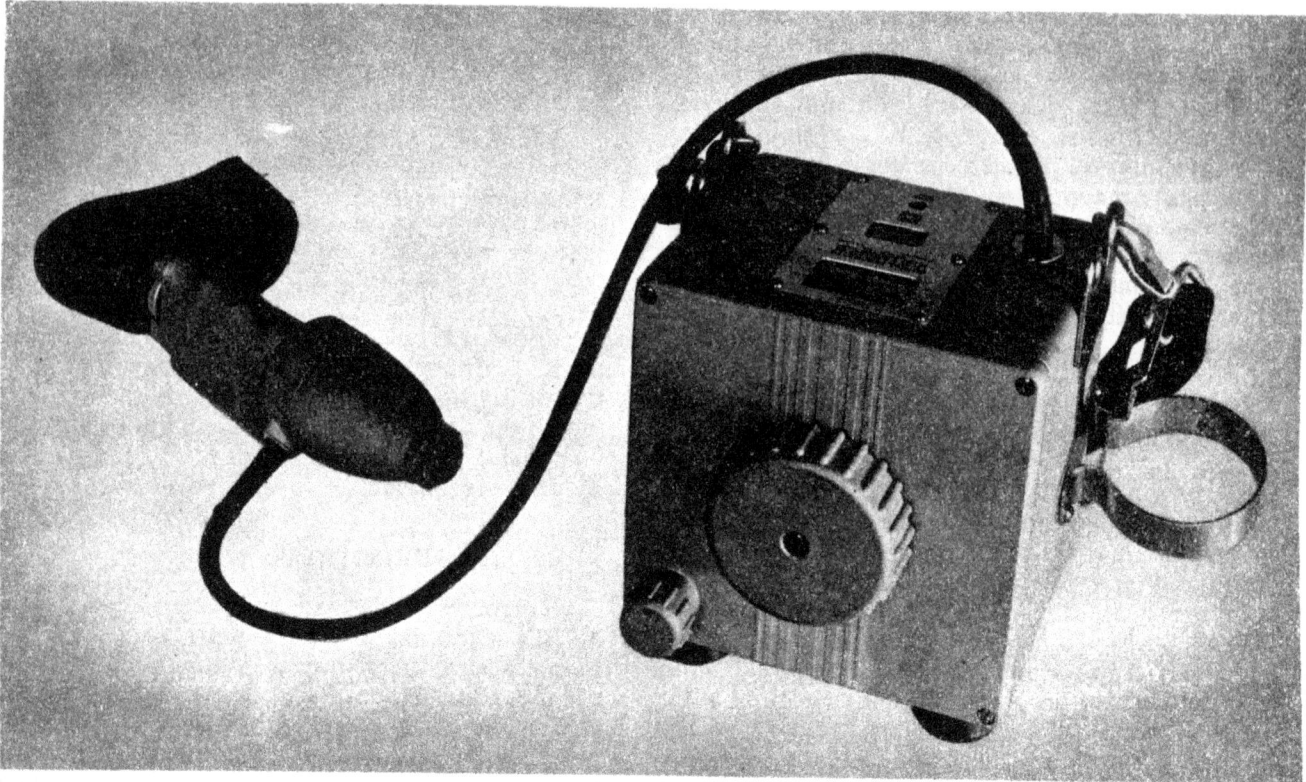

102.20X

Figure 7-12.—Optical pyrometer.

high, or that a column of water in a U-tube would be displaced 1 inch by the pressure being measured. Similarly, a reading of 12 inches of mercury (12 in. Hg) means that the measured pressure is sufficient to support a column of mercury 12 inches high. What is really being expressed (even though it is not mentioned in the pressure unit) is the fact that a certain quantity of material (water, mercury, etc.) of known density will exert a certain definite force upon a specified area. Pressure is still force per unit area, even if the pressure unit refers to inches of some liquid.

It is often necessary to convert from one type of pressure unit to another. Complete conversion tables may be found in many texts and handbooks. Conversion factors for pounds per square inch, inches of mercury, and inches of water are:

$$1 \text{ in. Hg} = 0.49 \text{ psi}$$
$$1 \text{ psi} = 2.036 \text{ in. Hg}$$

$$1 \text{ in. } H_2O = 0.036 \text{ psi}$$
$$1 \text{ psi} = 27.68 \text{ in. } H_2O$$

$$1 \text{ in. } H_2O = 0.074 \text{ in. Hg}$$
$$1 \text{ in. Hg} = 13.6 \text{ in. } H_2O$$

In interpreting pressure measurements, a great deal of confusion arises because the zero point on most pressure gages represents atmospheric pressure rather than zero absolute pressure. Thus it is often necessary to specify the kind of pressure being measured under any given conditions. To clarify the numerous meanings of the word pressure, the relationships among gage pressure, atmospheric pressure, vacuum, and absolute pressure, is illustrated in figure 7-13.

Gage Pressure is the pressure actually shown on the dial of a gage that registers pressure at or above atmospheric pressure. An ordinary pressure gage reading of zero does not mean that there is no pressure in the absolute sense; rather, it means that there is no pressure in excess of atmospheric pressure.

Atmospheric pressure is the pressure exerted by the weight of the atmosphere. At sea level, the average pressure of the atmosphere is

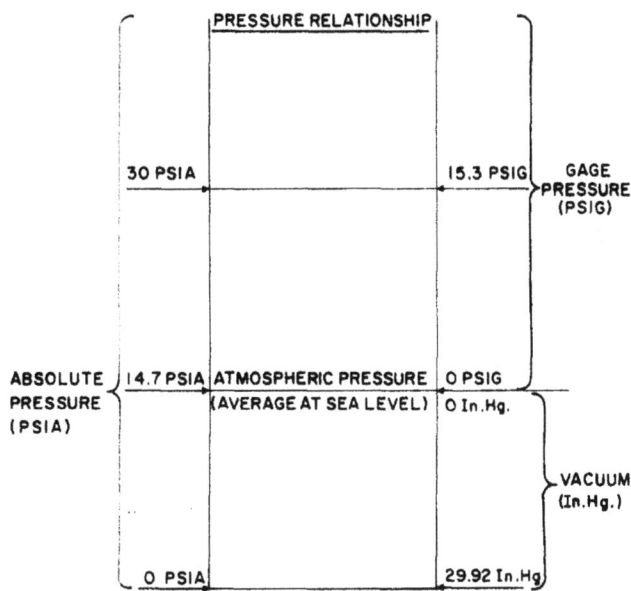

147.181

Figure 7-13.—Relationships among gage pressure, atmospheric pressure, vacuum, and absolute pressure.

sufficient to hold a column of mercury at the height of 76.0 millimeters or 29.92 inches of mercury. Since a column of mercury 1 inch high exerts a pressure of 0.49 pound per square inch, a column of mercury 29.92 inches high exerts a pressure that is equal to 29.92 x 0.49, or approximately 14.7 psi. Since we are dealing now in absolute pressure, we say that the average atmospheric pressure at sea level is 14.7 pounds per square inch absolute. It is zero on the ordinary pressure gage.

Notice, however, that the figure of 14.7 pounds per square inch absolute (psia) represents the average atmospheric pressure at sea level, and does not always represent the actual pressure being exerted by the atmosphere at the moment that a gage is being read.

Barometric pressure is the term used to describe the actual atmospheric pressure that exists at any given moment. Barometric pressure may be measured by a simple mercury column or by a specially designed instrument called an aneroid barometer.

A space in which the pressure is less than atmospheric pressure is said to be under vacuum. The amount of vacuum is expressed in terms of the difference between the absolute

pressure in the space and the pressure of the atmosphere. Most commonly, vacuum is expressed in inches of mercury, with the vacuum gage scale marked from 0 to 30 inches of mercury. When a vacuum gage reads zero, the pressure in the space is the same as atmospheric pressure—or, in other words, there is no vacuum. A vacuum gage reading of 29.92 inches of mercury would indicate a perfect (or nearly perfect) vacuum. In actual practice, it is impossible to obtain a perfect vacuum even under laboratory conditions.

Absolute pressure is atmospheric pressure plus gage pressure or minus vacuum. For example, a gage pressure of 300 psig equals an absolute pressure of 314.7 psia (300 + 14.7). Or, for example, consider a space in which the measured vacuum is 10 inches of mercury vacuum; the absolute pressure in this space must then be 19.92 or approximately 20 inches of mercury absolute. It is important to note that the amount of pressure in a space under vacuum can only be expressed in terms of absolute pressure.

You may have noticed that sometimes we say psig to indicate gage pressure and other times we merely say psi. By common convention, gage pressure is always assumed when pressure is given in pounds per square inch, pounds per square foot, or similar units. The "g" (for gage) is added only when there is some possibility of confusion. Absolute pressure, on the other hand, is always expressed as pounds per square inch absolute (psia), pounds per square foot absolute (psfa), and so forth. It is always necessary to establish clearly just what kind of pressure we are talking about, unless this is very clear from the nature of the discussion.

To this point, we have considered only the most basic and most common units of measurement. It is important to remember that hundreds of other units can be derived from these units, and that specialized fields require specialized units of measurement. Additional units of measurement are introduced in appropriate places throughout the remainder of this training manual. When you encounter more complicated units of measurement, you may find it helpful to review the basic information given here previously.

PRESSURE MEASURING DEVICES

Most pressure measuring devices used aboard ship utilize mechanical pressure elements.[11] There are two major classes of mechanical pressure elements: (1) liquid-column elements, and (2) elastic elements.

Liquid-Column Elements

Liquid-column pressure measuring elements include the devices commonly referred to as barometers and manometers. Liquid-column elements are simple, reliable, and accurate. They are used particularly (although not exclusively) for the measurement of relatively low pressures or small pressure differentials. Liquids commonly used in this type of pressure gage include mercury, water, and alcohol.

One of the simplest kinds of liquid-column elements is the fixed-cistern barometer (fig. 7-14) which is used to measure atmospheric pressure. Mercury is always used as the liquid in this type of instrument. Atmospheric pressure acts upon the open surface of the mercury in the cistern. Since the tube is open at the cistern end, and since there is a vacuum above the mercury in the tube, the height of the mercury in the tube is at all times an indication of the existing atmospheric (barometric) pressure.

A simple U-tube liquid-column element for measuring absolute pressure is shown in figure 7-15. The liquid used in this device is mercury. There is a vacuum above the mercury at the closed end of the tube; the open end of the tube is exposed to the pressure to be measured. The absolute pressure is indicated by the difference in the height of the two mercury columns.

Manometers are available in many different sizes and designs. Some are installed in such a way that the U-tube is readily recognizable, as in part A of figure 7-16; but in some designs the U-tube is inverted or inclined at an angle. The so-called single-tube or straight-tube manometer (part B of fig. 7-16) is actually a U-tube in which only one leg is made of glass.

Elastic Elements

Elastic elements used for pressure measurement include Bourdon tubes, bellows, and diaphragms. All three types of elastic elements are suitable for use in pressure gages, vacuum gages, and compound (both pressure and vacuum)

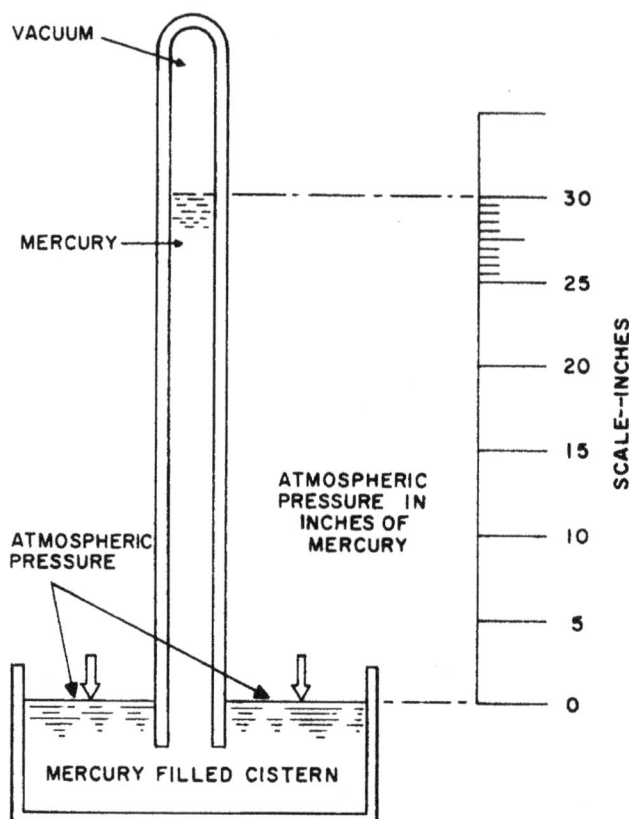

69.86(147B)
Figure 7-14.—Simple Barometer (fixed cistern) for measuring atmospheric pressure.

gages. Bourdon-tube elements are suitable for the measurement of very high pressures, up to 100,000 psig. The upper limit for bellows elements is about 800 psig and for diaphragm elements about 400 psig. Diaphragm elements and bellows elements are commonly used for the measurement of very high vacuum (or very low absolute pressure) but Bourdon tubes can be used for such applications.

BOURDON-TUBE ELASTIC ELEMENTS.— Bourdon-tube elements used in pressure gages are essentially the same as those described for use in filled-system thermometers. Bourdon tubes for pressure gages are made of brass, phosphor bronze, stainless steel, beryllium-copper, or other metals, depending upon the requirements of service.

Bourdon-tube pressure gages are often classified as simplex or duplex, depending upon whether they measure one pressure or two. A simplex gage such as the one shown in figures

_____
[11]Strain gages and other electrical pressure measuring devices are not included in this discussion; they are rarely used aboard ship.

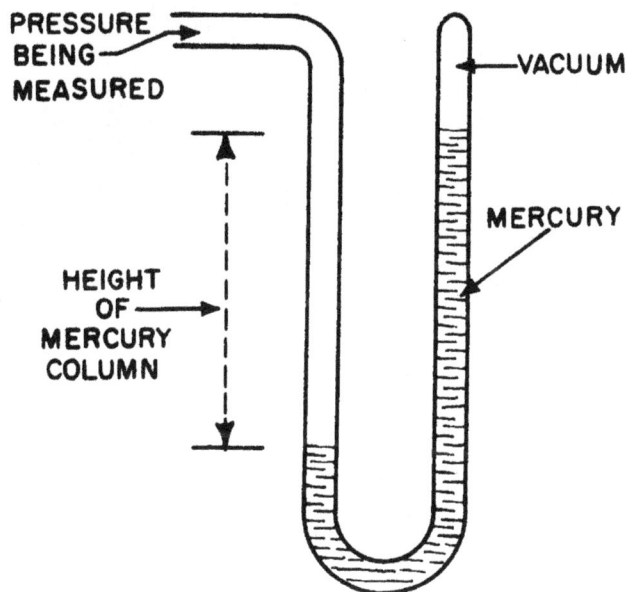

147.57

Figure 7-15.—U-tube liquid-column element for measuring absolute pressure.

7-17, 7-18, and 7-19 has only one Bourdon tube and measures only one pressure. (The pointer marked RED HAND in figure 7-17 is a manually positioned hand that is set to the normal working pressure of the machinery or equipment on which the gage is installed; the hand marked POINTER is the only hand that moves in response to pressure changes.)

When two Bourdon tubes are mounted in a single case, with each mechanism acting independently but with the two pointers mounted on a common dial, the assembly is called a duplex gage. The dial of a duplex gage is shown in figure 7-20. The two Bourdon tubes and the operating mechanism are shown in figure 7-21, and the gear mechanism is shown in figure 7-22. Note that each Bourdon tube has its own pressure connection and its own pointer. Duplex gages are used to give simultaneous indication of the pressure at two different locations.

Bourdon-tube vacuum gages are marked off in inches of mercury, as shown in figure 7-23.

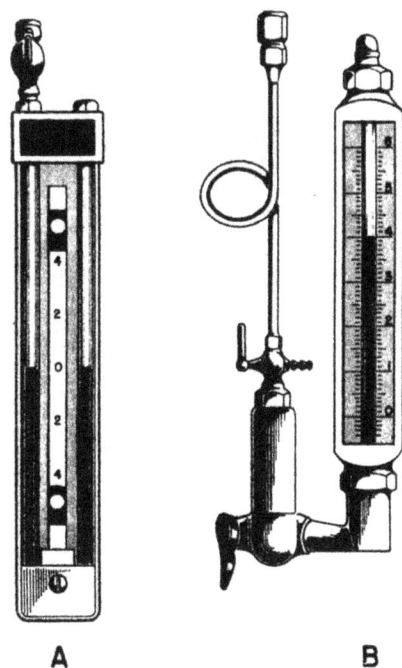

61.4X

Figure 7-16.—Two types of manometers.
(A) Standard U-tube. (B) Single-tube.

38.211BX

Figure 7-17.—Dial of a simplex Bourdon-tube pressure gage.

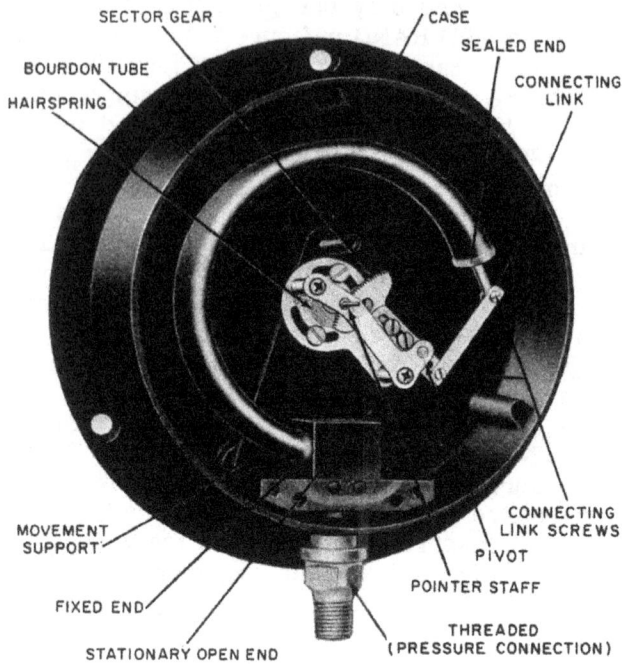

38.211AX

Figure 7-18.—Operating mechanism of simplex Bourdon-tube pressure gage.

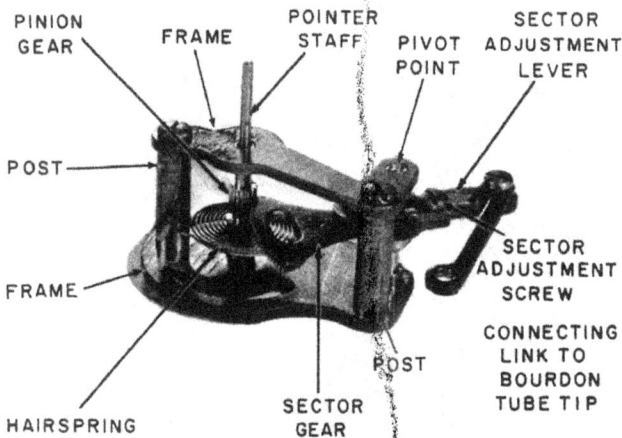

38.211C

Figure 7-19.—Gear mechanism of simplex Bourdon-tube pressure gage

38.211FX

Figure 7-20.—Dial of a duplex Bourdon-tube pressure gage.

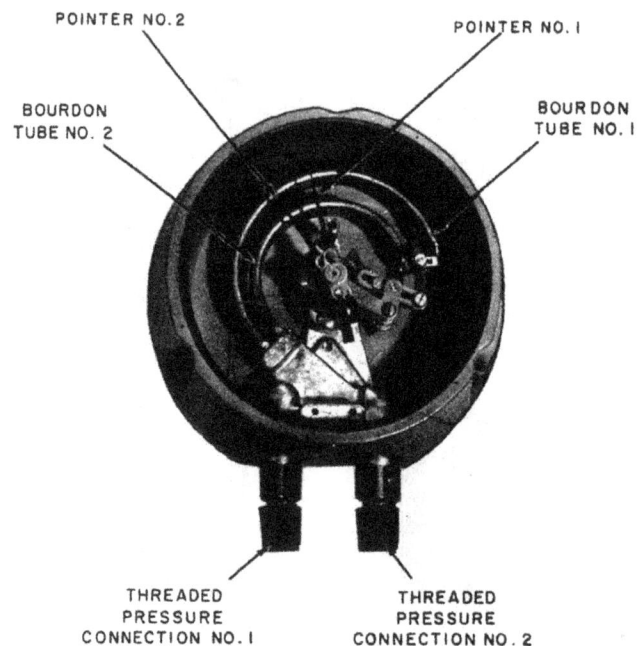

38.211G

Figure 7-21.—Two Bourdon tubes and operating mechanism of duplex Bourdon-tube pressure gage.

When a gage is designed to measure both vacuum and pressure, it is called a compound gage and is marked off both in inches of mercury and in psig, as shown in figure 7-24.

140

38.211CA

Figure 7-22.—Gear mechanism of duplex
Bourdon-tube pressure gage.

38.211DX

Figure 7-23.—Bourdon-tube vacuum gage.

Differential pressure may also be measured with Bourdon-tube gages. One kind of Bourdon-tube differential pressure gage is shown in figure 7-25. This gage has two Bourdon tubes

38.211EX

Figure 7-24.—Compound Bourdon-tube gage.

but only one pointer. The Bourdon tubes are connected in such a way that it is the pressure difference, rather than either of the two actual pressures, that is indicated by the pointer.

BELLOWS ELASTIC ELEMENTS.—A bellows elastic element is a convoluted unit that expands and contracts axially with changes in pressure. The pressure to be measured can be applied to the outside or to the inside of the bellows; in practice, most bellows-type measuring devices have the pressure applied to the outside of the bellows, as shown in figure 7-26. Bellows elastic elements are made of brass, phosphor bronze, stainless steel, beryllium-copper, or other metal suitable for the intended service of the gage.

Most bellows-type gages are spring-loaded—that is, a spring opposes the bellows and thus prevents full expansion of the bellows. Limiting the expansion of the bellows in this way protects the bellows and prolongs its life. In a spring-loaded bellows-type element, the deflection is the resultant of the force acting on the bellows and the opposing force of the spring.

141

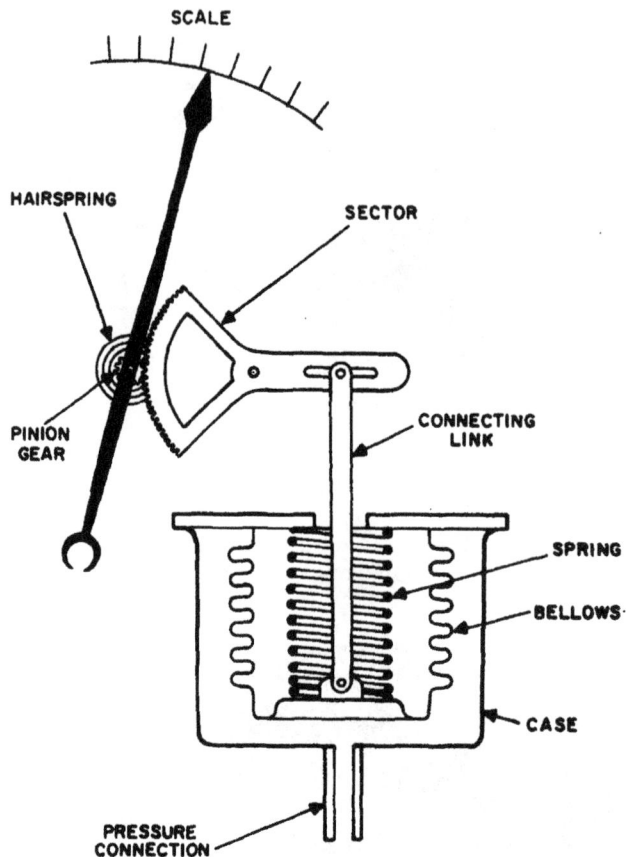

38.211KX

Figure 7-25.—Bourdon-tube differential
pressure gage.

61.3(147B)A

Figure 7-26.—Simple bellows gage.

Although some bellows-type instruments can be designed for measuring pressures up to 800 psig, their primary application aboard ship is in the measurement of quite low pressures or small pressure differentials. For example, bellows elements are widely used in boiler control systems[12] because the air pressures of the control systems are generally very low.

Many differential pressure gages are of the bellows type. In some designs, one pressure is applied to the inside of the bellows and the other pressure is applied to the outside. In other designs, a differential pressure reading is obtained by opposing two bellows in a single case.

Bellows elements are used in various applications where the pressure-sensitive device must be powerful enough to operate not only the indicating pointer but also some type of recording device.

DIAPHRAGM ELASTIC ELEMENTS.—Diaphragm elastic elements are used for the measurement of relatively low pressures or

---

[12]Boiler control systems are discussed in chapter 11 of this text.

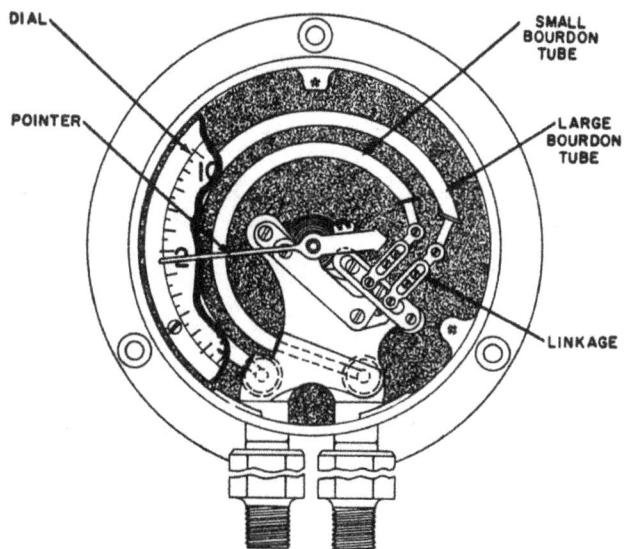

small pressure differences. Both metallic and nonmetallic diaphragms are in common use.

Metallic diaphragms are made from stainless steel, phosphor bronze, brass, or other metal. A metallic diaphragm element may consist of one or more capsules. Each capsule consists of two diaphragm shells (flat or corrugated circular disks) which are welded, brazed, or otherwise firmly fastened together to form the capsule. The capsules are all rigidly connected so that the application of pressure causes all capsules to deflect. The amount of deflection of a diaphragm gage depends upon the number of capsules, the design and the number of the corrugations, and other factors.

Nonmetallic diaphragms, also called slack or limp diaphragms, are made of leather, treated cloth, neoprene, or some other soft material. Nonmetallic diaphragms are spring-loaded. One common type of nonmetallic diaphragm pressure gage is shown in figure 7-27. When pressure is applied to the underside of the slack diaphragm, the diaphragm moves upward, although it is opposed by the action of the calibrating spring. As the spring moves, the linkage system causes the pointer to move to a higher reading. Thus the reading on the scale is proportional to the amount of pressure exerted on the diaphragm, even though the movement of the diaphragm is opposed by the calibrating spring.

PRESSURE GAGE INSTALLATION

Bourdon tube pressure gages used for steam service are always installed in such a way that the steam cannot actually enter the gage. This type of installation is necessary to protect the Bourdon type from very high temperatures. An exposed uninsulated coil is provided in the line leading to the gage, and the steam condenses into water in this exposed coil. Thus there is always a condensate seal between the gage and the steam line.

Pressure gage connections are normally made to the top of the pressure line or to the highest point on the machinery in which the pressure is to be measured. Pressure gages are usually mounted on flat-surfaced gage boards in such a way as to minimize vibration; this is a matter of considerable importance, since some ships experience very great structural vibration from screws and machinery. Efforts are currently being made to design gages capable of withstanding any vibration that may

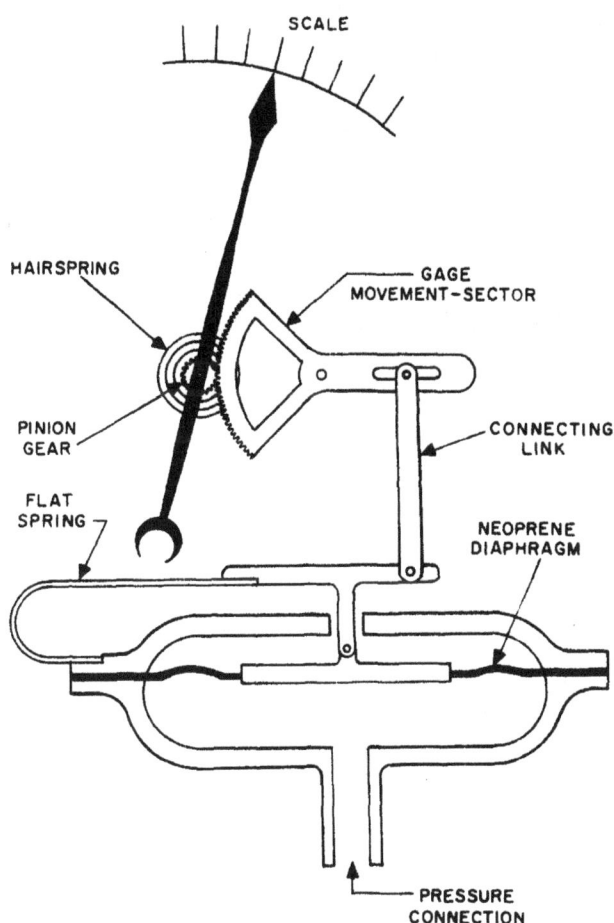

38.212(147B)

Figure 7-27.—Nonmetallic diaphragm pressure gage.

be expected from machinery. Pressure gages designed to withstand shock and vibration frequently use small size capillary tubing between the connections and the elastic elements to protect the gage mechanism and the pointer; small size tubing is used between the test connection or gage valve and the gage so that piping deflections will not cause errors in the gage readings.

MEASUREMENT OF FLUID FLOW

A great many devices, many of them quite ingenious, have been developed for the measurement of fluid flow. The discussion here is concerned primarily with the types of fluid flow measuring devices that find relatively wide application in shipboard engineering. These devices may be classified as (1) positive-displacement meters, (2) head meters, and (3) area meters.

POSITIVE-DISPLACEMENT METERS

Positive-displacement meters are used for measuring liquid flow. In a meter of this type, each cycle or complete revolution of a measuring element displaces a definite, fixed volume of liquid. Measuring elements used in positive-displacement meters include disks, pistons, lobes, vanes, and impellers. The motion of these devices may be classified as reciprocating, rotating, oscillating, or nutating, depending upon the type of measuring element used and the general design of the meter read-out or register.

A positive-displacement meter of the nutating-piston type is shown in figure 7-28. The flow of oil through the meter causes the piston (also called a disk) to move with a nutating motion. Understanding the nature of this motion is the key to understanding the operation of the meter. A nutating motion (fig. 7-29) might be described as a "rocking around" motion; it is similar to the motion of a spun coin just before the coin settles flat on its side. The piston in the meter cannot settle flat on its side like a spun coin, since the piston is nutating (or rocking around) on a lower spherical bearing surface.

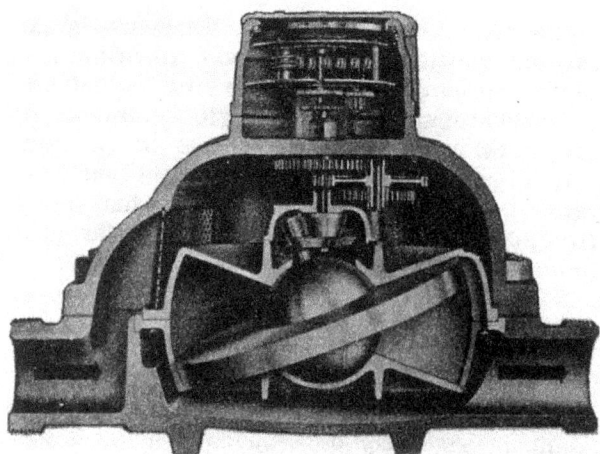

38.66X

Figure 7-29.—Diagram showing nutating motion of piston and rotary motion of pin in nutating-piston meter.

The piston cannot rotate because it is held in place by a fixed vane or guide that runs vertically through a slot in the piston. However, the nutating motion of the piston imparts a rotary motion to the pin that projects from the upper spherical surface. The rotary movement of the pin rotates the gears, and the movement of the gears actuates a counting device or register at the top of the meter.

Although the action of the nutating piston is smooth and continuous, there is nevertheless a definite cycle involved in the measurement of liquid flow through this meter. The nutating action of the piston seals the measuring chamber off into separate compartments, and these compartments are alternately filled and emptied. The meter is properly classed as a positive-displacement meter, since each compartment holds a definite volume of the liquid.

Totalizing meters have a read-out in gallons or pounds of liquid; however, they may also

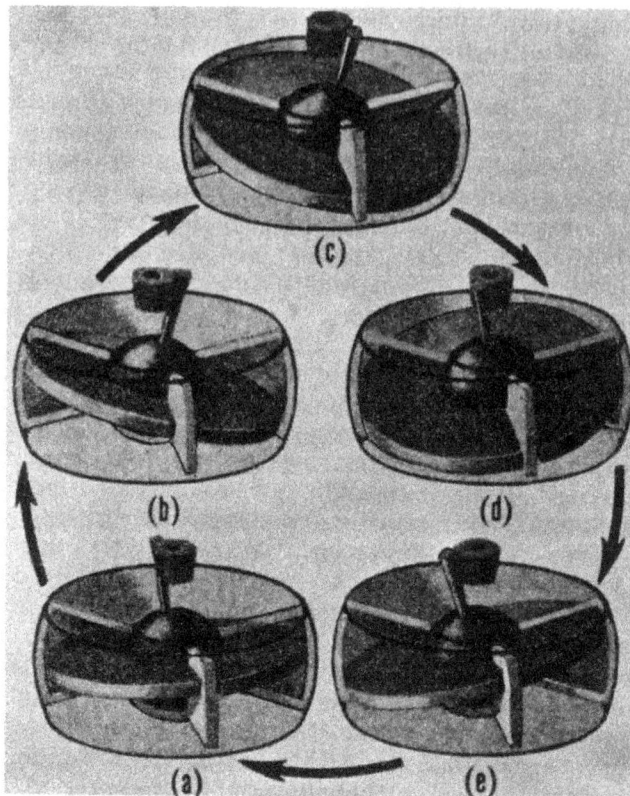

38.64X

Figure 7-28.—Nutating-piston meter for measuring liquid flow.

indicate rate of flow in gallons per minute (gpm) or in other flow units.

HEAD METERS

Head meters measure fluid flow by measuring the pressure differential across a specially designed restriction in the flow line. The restriction may be an orifice plate, a flow nozzle, a venturi tube, an elbow, a pitot tube, or some similar device. As the fluid flows toward the restriction, the velocity decreases and the pressure (or "head") increases; as the fluid flows through the restriction, the velocity increases and the pressure decreases. Figure 7-30 illustrates the pressure changes that occur as a fluid flows through a line that contains an orifice plate or similar restriction. Note that there is a slight increase in pressure just ahead of the restriction and then a sudden drop in pressure at the restriction. The point of minimum pressure and maximum velocity is slightly downstream from the restriction; this point is called the vena contracta. Beyond the vena contracta, the velocity decreases and the pressure increases until eventually normal flow is reestablished.

The flow nozzle, orifice plate, or other restriction in the line is called the primary element of the head meter. A high pressure tap upstream from the restriction and a low pressure tap downstream from it are connected to a differential bellows, a diaphragm, or some other device for measuring differential pressure.

The pressure drop occurring in a fluid flowing through a restriction varies as the square of the fluid velocity; or, to put it another way, the square root of the pressure differential is proportional to the rate of fluid flow. Because of the square-root relationship between the pressure differential and the rate of fluid flow, a square-root extracting device is usually included so that the scale can be graduated in even steps or increments. Without a device for extracting the square root of the pressure differential, the scale would have to be unevenly divided, with wider divisions at the top of the scale than at the bottom.

AREA METERS

An area meter indicates the rate of fluid flow by means of an orifice that is varied in area by variations in the fluid flow. The variations in the area of the orifice are produced by some type of movable device which is positioned by the pressure of the flowing fluid. Since the fluid itself positions the movable device and thus varies the area of the orifice, there is no significant pressure drop between the upstream side and the downstream side of the variable orifice. Since there is an essentially linear relationship between the area of the orifice and the rate of flow, there is no need for a square-root extracting device in an area meter.

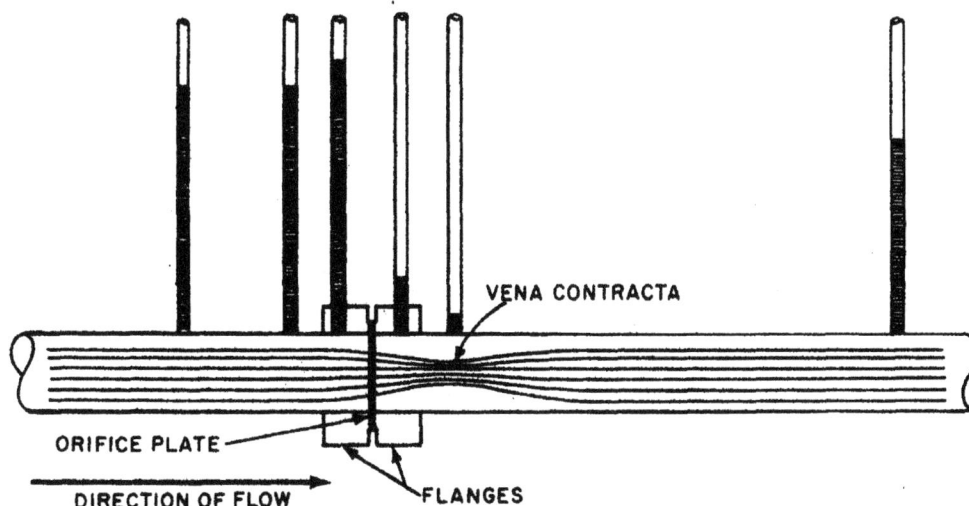

147.59

Figure 7-30.—Pressure changes in fluid flowing through restriction in line.

Most area meters are classified as (1) rotameters, or (2) piston-type meters, depending upon the type of device used to vary the area of the orifice.

Figure 7-31 shows a simple rotameter of a type used in some shipboard distilling plants. A tapered glass tube is installed vertically, with the smaller end at the bottom. Water flows in at the bottom, upward through the tube, and out at the top. A rod, supported by the end fittings of the tube, is centered in the tube. A movable rotor rides freely on the rod and is positioned by the fluid. Variations in the pressure of the fluid lead to variations in the position of the rotor; and, since the glass tube is tapered, the size of the annular orifice between the rotor and the tube is different at each position of the rotor. An increase in flow is thus indicated by the rotor rising on the rod. The glass tube is so calibrated that the flow may be read directly, with the reading being taken at the top of the rotor.

In a piston-type area meter, a piston is lifted by fluid pressure. As the piston is lifted, it uncovers a port through which the fluid flows. The port area uncovered by the lifting of the piston is directly proportional to the rate of fluid flow. Therefore, the position of the piston provides a direct indication of the rate of flow. The means by which the position of the piston is transmitted to an indicating dial varies according to the design of the particular meter and the service for which it is intended.

MEASUREMENT OF LIQUID LEVEL

In the engineering plant aboard ship, it is frequently necessary for operating personnel to know the level of various liquids in various locations. The level of the water in the ship's boilers is a prime example of a liquid level that must be known at all times, but there are other liquid levels that are also important—the level of fuel oil in service and stowage tanks, the level of water in deaerating feed tanks, the level of lubricating oil in the oil sumps of main and auxiliary machinery, and drains in various drain tanks, to name but a few.

A wide variety of devices, some of them simple and some complex, are available for measuring liquid level. Some measure liquid level quite directly by measuring the height of a column of liquid. Others measure pressure, volume, or some other property of the liquid from which we may then infer liquid level.

The gage glass is one of the simplest kinds of liquid level measuring devices and one that is very commonly used. Gage glasses are used on boilers, on deaerating feed tanks, on inspection tanks, and on other shipboard machinery. Basically, a gage glass is just one leg of a U-tube, with the other leg being the tank, drum, or other vessel in which the liquid level is to be measured. The liquid level in the gage glass is thus the same as the liquid level in the tank or drum, and the reading can be made by direct visual observation. Gage glasses vary in details of construction, depending upon the pressure, temperature, and other service conditions they must withstand.

The measurement of liquid level in tanks aboard ship may be accomplished by simple devices such as direct-reading sounding rules

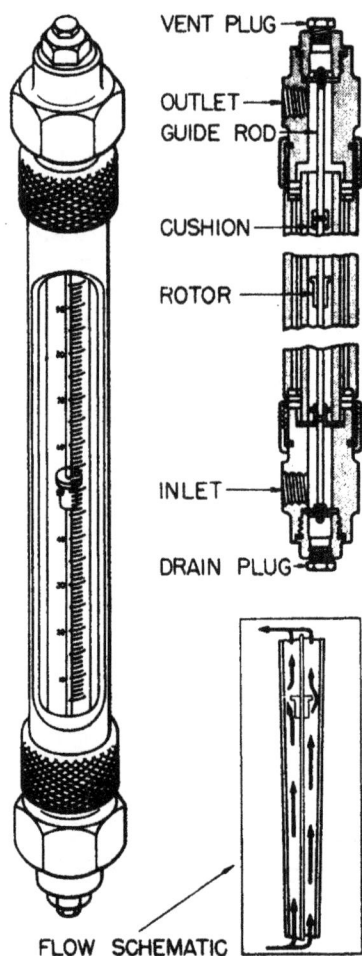

VENT PLUG

OUTLET
GUIDE ROD

CUSHION

ROTOR

INLET

DRAIN PLUG

FLOW SCHEMATIC

75.290
Figure 7-31.—Area meter (rotameter type) for measuring fluid flow.

and gaging tapes or by some form of permanently installed, remote reading gaging system. (Although remote reading gaging systems are often referred to as "tank level indicators," it should be noted that the scale may be calibrated to show level, volume, or weight; frequently there are two scales—one to show volume and one to show level.)

A static head gaging system of a type commonly used for measuring liquid level in fuel oil tanks aboard ship is shown in figure 7-32. This system balances a head of liquid in the tank against a column of liquid in a manometer or against a bellows or diaphragm differential pressure unit; the system illustrated uses a mercury manometer. The balance chamber is located so that its orifice is near the bottom of the tank; a line connects the top of the balance chamber to the mercury-filled bulb of the indicator gage, and another line connects the space above the mercury column to the top of the tank. Since the height of the liquid in the tank bears a definite relationship to the pressure exerted by the liquid, the scale can be calibrated to show height (or liquid level). When the size of the tank is known, the measurement of height can readily be converted to measurement of volume; and, when the volume of the tank and the specific weight of the liquid are known, the scale can be calibrated to indicate weight. The reading on this type of tank gaging system is always taken after compressed air has been admitted to the balance chamber through the control valve; to ensure proper readings, a sufficient amount of compressed air must be admitted to force the liquid down to the level of the bottom of the standpipe.

MEASUREMENT OF ROTATIONAL SPEED

The rotational speed of propeller shafts, turbines, generators, blowers, pumps, and other kinds of shipboard machinery is measured by means of tachometers. For most shipboard machinery, rotational speed is expressed in revolutions per minute (rpm). The tachometers most commonly used aboard ship are of the centrifugal type, the chronometric type, and the resonance type. Stroboscopic tachometers are also used occasionally.

Some types of machinery are equipped with permanently mounted tachometers, but portable tachometers are used for checking the rpm of many units. A portable tachometer of the centrifugal type or of the chronometric type is applied manually to a depression or a projection at the center of a moving shaft. Each portable tachometer is supplied with several hard rubber tips; to use the instrument, the operator selects a tip of the proper shape, fits it over the end of the tachometer drive shaft, and holds the tip against the center of the moving shaft. Some tachometers are also supplied with a small wheel which can be fitted to the end of the drive shaft and used to measure the linear velocity (in feet per second) of a wheel or a journal; with this type of instrument, the wheel is held against the outer surface of the moving object. Portable tachometers are used only for intermittent reading, not for continuous operation.

CENTRIFUGAL TACHOMETERS

As the name implies, a centrifugal tachometer utilizes centrifugal force for its operation. The main parts of a centrifugal tachometer are shown in figure 7-33 and the dial of the instrument is shown in figure 7-34. Centrifugal force acts upon weights or flyballs which are connected by linkage to an upper and a lower collar. The upper collar is fixed to the drive shaft, but the

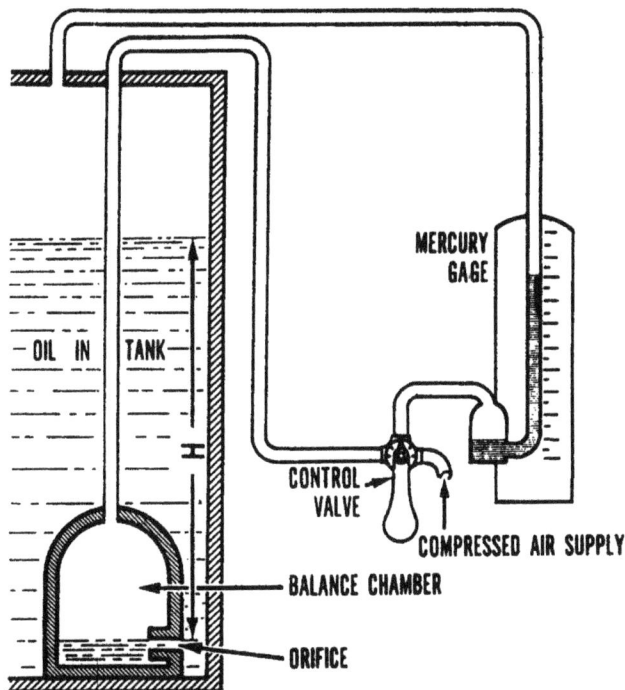

61.5X
Figure 7-32.—Tank gaging system.

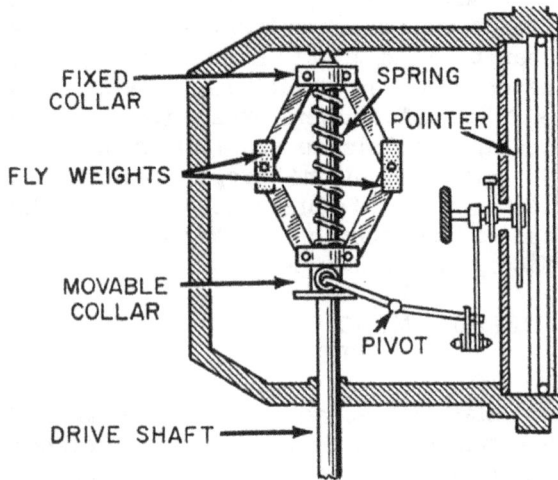

2.66X

Figure 7-33.—Main parts of centrifugal
tachometer.

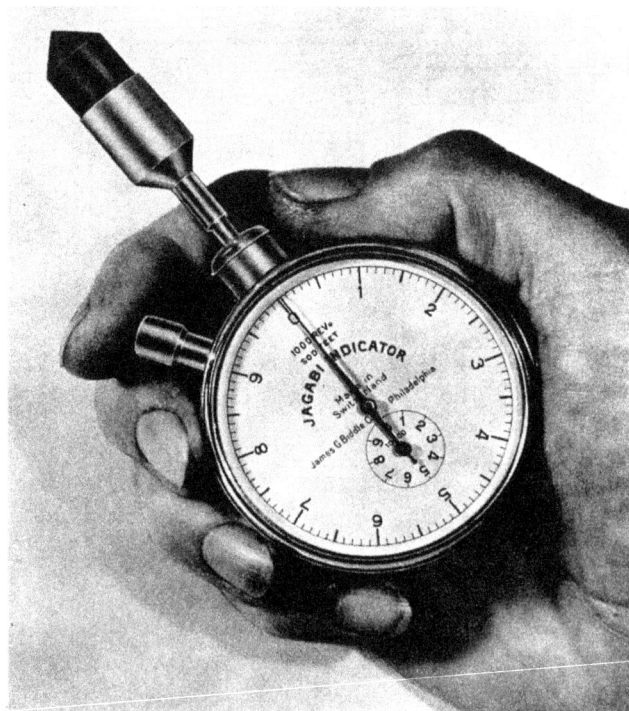

2.66X

Figure 7-34.—Dial of centrifugal tachometer.

lower collar is free to move up and down the
shaft. A spring which fits over the drive shaft
connects the upper collar and the lower collar.
As the drive shaft begins to rotate, the flyballs
spin around with it. Centrifugal force tends to
pull the flyballs away from the center, thus
raising the lower collar and compressing the
spring. The lower collar is connected to the
pointer, and the upward movement of the collar
causes the pointer to move to a higher rpm reading on the dial. The centrifugal tachometer registers rpm of a rotating shaft as long as it is in
contact with the shaft. For this reason it is called
a constant-reading tachometer.

CHRONOMETRIC TACHOMETERS

A chronometric tachometer (fig. 7-35) is a
combination of a watch and a revolution counter
which measures the average number of revolutions per minute of a rotating shaft. The device
is not a constant-reading instrument; the outer
drive shaft runs free when the instrument is
applied to a rotating shaft until a starting button
is depressed to start the timing element. After
the drive shaft has been disengaged from the
rotating shaft, the pointer remains in position on
the dial until it is returned to the zero position
by the operation of a reset button (which may be
the same as the starting button.)

RESONANCE TACHOMETERS

A resonance tachometer (fig. 7-36) consists
of a number of steel reeds, each one of which

2.66X

Figure 7-35.—Chronometric tachometer.

61.16X

Figure 7-36.—Resonance tachometer mounted on rotating machine.

vibrates at a different frequency. The reeds are fastened in a row, in order of frequency; the row is mounted with the reeds at right angles to the back of the instrument. The unattached ends of the reeds extend through a horizontal slit in the face of the instrument; the scale is stamped along the slit. When the instrument is solidly attached to the foundation or casing of a rotating machine, the reeds which are nearest in frequency to the rpm of the machine begin to vibrate. In figure 7-36, notice that five reeds are out of line with the rest of the reeds; these five are vibrating noticeably more than the others, and the one in the middle is vibrating more than the other four. To read the rpm of the machine, then, it is only necessary to read the scale marking underneath the reed that is vibrating the most—that is, the one which is most out of line with the others in the horizontal slit.

Resonance tachometers are particularly useful for measuring high rotational speeds such as those that occur in turbines, generators, and forced draft blowers. They are also particularly useful in applications where it is difficult or impossible to get at the moving ends of shafts.

The instruments give continuous readings and make very rapid—almost instantaneous—adjustments to changes in rotational speed.

STROBOSCOPIC TACHOMETERS

A stroboscopic tachometer is a device which allows rotating, reciprocating, or vibrating machinery to be viewed intermittently, under flashing light, in such a way that the movement of the machinery appears to be slowed, stopped, or reversed. Because the illumination is intermittent, rather than steady, the eye receives a series of views rather than one continuous view.

When the speed of the flashing light coincides with the speed of the moving machinery, the machinery appears to be motionless. This effect occurs because the moving object is seen each time at the same point in its cycle of movement. If the flashing rate is decreased slightly, the machinery appears to be moving slowly in the true direction of movement; if the flashing rate is increased slightly, the machinery appears to be moving slowly in the reverse direction. To measure the speed of a machine, therefore, it is only necessary to find the rate of intermittent illumination at which the machinery appears to be motionless. To observe the operating machinery in slow motion, it is necessary to adjust the stroboscope until the machinery appears to be moving at the desired speed.

The stroboscopic tachometer furnished for shipboard use is a small, portable instrument. It is calibrated so that the speed can be read directly from the control dial. The flashing rate is determined by a self-contained electronic pulse generator which can be adjusted, by means of the direct-reading dial, to any value between 600 and 14,400 rpm. The relationship between rotational speed and flashing rate may be illustrated by an example. If an electric fan is operating at a rate of 1800 rpm, it will appear to be motionless when it is viewed through a stroboscopic tachometer which is flashing at the rate of 1800 times per minute.

Because the stroboscopic tachometer is never used in direct contact with moving machinery, it is particularly useful for measuring the speed or observing the operation of machinery which is run by a relatively small power input. It is also very useful for measuring the speed of machinery which is installed in relatively inaccessible places.

MEASUREMENT OF SPECIFIC GRAVITY

The specific gravity of a substance is defined as the ratio of the density of the substance to the density of a standard substance. The standard of density for liquids and solids is pure water; for gases, the standard is air. Each standard (water or air) is considered to have a specific gravity of 1.00 under standard conditions of pressure and temperature.[13] For a solid or a liquid substance, then, we may say that

$$\text{Sp. gr. of solid or liquid} = \frac{\text{density of substance}}{\text{density of water}}$$

Density is sometimes defined as the mass per unit volume of a substance and sometimes as weight per unit volume. In engineering, fortunately, this difference in defintions rarely causes confusion because we are usually interested in relative densities—or, in other words, in specific gravity. Since specific gravity is the ratio of two densities, it really does not matter whether we use mass densities or weight densities; the units cancel out and give us a pure number which is independent of the system of units used.

Aboard ship, it is sometimes necessary to measure the specific gravity of various liquids. This is usually done by using a device called a hydrometer. A hydrometer measures specific gravity by comparing the buoyancy (or loss of weight) of an object in water with the buoyancy of the same object in the liquid being measured. Since the buoyancy of an object is directly related to the density of the liquid, then

$$\text{Sp. gr. of liquid} = \frac{\text{buoyant force of liquid}}{\text{buoyant force of water}}$$

A hydrometer is merely a calibrated rod which is weighted at one end so that it floats in a vertical position in the liquid being measured. Hydrometers are calibrated in such a way that the specific gravity of the liquid may be read directly from the scale; in other words, the comparison between the density of the liquid being measured and the density of water is

[13] For most engineering purposes, the standard pressure and temperature conditions for water as a standard of specific gravity are atmospheric pressure and 60° F.

"built in" by the calibration of the hydrometer.

For fuel oil and other petroleum products, it is customary to measure degrees API, rather than specific gravity, in accordance with a scale developed by the American Petroleum Institute. The relationship between specific gravity and API gravity is given by the formula

$$\text{Sp. gr.} = \frac{141.5}{131.5 + \text{degrees API}}$$

A hydrometer of the type normally used aboard ship to measure the degrees API of fuel oil is shown in figure 7-37. The major difference between this hydrometer and others used aboard ship is that this one is calibrated to read degrees API rather than specific gravity.

MEASUREMENT OF VISCOSITY

The viscosity of a liquid is a measure of its resistance to flow. A liquid is said to have high viscosity if it flows sluggishly, like cold molasses. It is said to have low viscosity if it flows freely, like water. The viscosity of most liquids is greatly affected by temperature; in general, liquids are less viscous at higher temperatures.

4.135

Figure 7-37.—Hydrometer.

Aboard ship, it is necessary to measure the viscosity of fuel oil and of lubricating oil. The viscosity of an oil is usually expressed as the number of seconds required for a given amount of oil to flow through an orifice of a specified size when the oil is at a specified temperature. Devices used to measure the rate of flow (and hence the viscosity) are called viscosimeters or viscometers.

The viscosimeter furnished for shipboard use is a Saybolt viscosimeter with two orifices. The larger orifice is called the Saybolt Furol orifice; the smaller one is called the Saybolt Universal orifice. The Furol orifice is used for measuring the viscosity of relatively heavy oils; the Universal orifice is used for measuring the viscosity of relatively light oils.

A Saybolt viscosimeter consists of an oil tube, a constant-temperature oil bath which maintains the correct temperature of the sample in the tube, a 60-cc (cubic centimeter) graduated receiving flask, thermometers for measuring the temperature of the oil sample and of the oil bath, and a timing device. A Saybolt viscosimeter is shown in figure 7-38; figure 7-39 shows details of the viscosimeter oil tube.

The oil to be tested is strained and poured into the oil tube. The tube is surrounded by the constant-temperature oil bath. When the oil sample is at the correct temperature, the cork is pulled from the lower end of the tube and the sample flows through the orifice and into the graduated receiving flask. The time (in seconds) required for the oil to fill the receiving flask to the 60-cc mark is noted.

The viscosity of the oil is expressed by indicating three things: first, the number of seconds required for 60 cubic centimeters of oil to flow into the receiving flask; second, the type of orifice used; and third, the temperature of the oil sample at the time the viscosity determination is made. For example, suppose that a sample of Navy Special fuel oil is heated to 122° F and that 132 seconds are required for 60 cc of the sample to flow through a Saybolt Universal orifice and into the receiving flask. The viscosity of this oil is said to be 132 seconds Saybolt Universal at 122° F. This is usually expressed in shorter form as 132 SSU at 122° F.

Saybolt Furol viscosities are obtained at 122° F. The same temperature (122° F) is used for obtaining Saybolt Universal viscosities of fuel oil, but various other temperatures are used for obtaining Saybolt Universal viscosities of oils other than fuel oil. Thus it is important that the temperature be included in the statement of viscosity.

OTHER TYPES OF MEASUREMENT

Thus far in this chapter, we have been largely concerned with basic principles of measurement and with widely used kinds of measuring devices. We have taken up many of the devices used to measure the fundamental variables of temperature, pressure, fluid flow, liquid level, and rotational speed, and we have considered the measurement of the properties of specific gravity and viscosity. For the most part, we have dealt with measuring devices that might be considered as basically mechanical in nature.

Before concluding this chapter, it might be well to point out that many other kinds of measurement are required in the shipboard engineering plant. While it is true that many of the principles of measurement discussed in this chapter apply to measuring devices other than those described here, it is also true that a specific application may require a measuring device that is not precisely the same as any device we have considered. Where appropriate, other types of measuring devices are discussed in other chapters of this text, as they relate to some particular kind of machinery or equipment. In some instances, the student may find it helpful to come back to the present chapter to renew his understanding of the basic principles of measurement we have considered here.

NAVY CALIBRATION PROGRAM

The calibration of all measuring devices begins with and is dependent upon the basic international and national standards of measurement just discussed. Obviously, however, we can't rush off to the National Bureau of Standards every time we need to measure a length, a mass, a weight, or an interval of time. Therefore, the National Bureau of Standards prepares and calibrates a great many practical standards that can be used by government and industry. Government and industry, in turn, prepare and calibrate their own practical standards. Thus there is a continuous linkage of measurement standards that begins with the international standards, comes down through the national standards, and works all the way on down to the rulers, weights, clocks, gages, and other devices that we use for everyday measurement.

38.213

Figure 7-38.—Saybolt viscosimeter.

As may be seen (fig. 7-40) in the structure of this program (Navy Calibration Program), the National Bureau of Standards is the highest level standards agency in the United States and that it has custody of this Nation's basic physical standards. The National Bureau of Standards provides the common reference for all measurements and certifies the Navy Standards that are maintained by the Navy Type I Standards Laboratories.

The Navy Type I Standards Laboratories maintain the highest standards within the Navy Calibration Program. The Type I Standards Laboratories obtain calibration services from the National Bureau of Standards and provide calibration of standards and associated

152

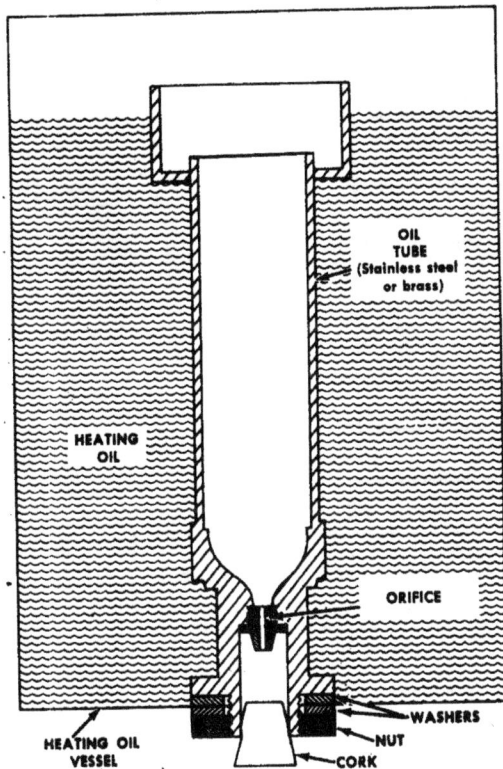

38.214

Figure 7-39.—Details of viscosimeter tube.

measuring equipment received from Type II Standards Laboratories. There are only two Type I Standards Laboratories: the Eastern SL, in Washington D.C., and the Western SL, in San Diego, California.

Navy Standards Laboratories designated as Type II furnish the second highest level of calibration services to assigned geographical areas within the Naval Establishment. The Type II Standards Laboratories obtain standards calibration services from the cognizant Type I Standard Laboratory and calibrate standards and associated measuring equipment received from lower level laboratories. There are half a dozen Type II Navy Standards Laboratories, located in various shore activities throughout the United States.

Navy Calibration Laboratories furnish the third highest level of calibration services in the Navy Calibration Program. The Navy Calibration Laboratories obtain calibration services from the Type II Standards Laboratories and they calibrate test equipment received from ships and from shore activities. There are two

basic types of Navy Calibration Laboratories. Fleet Calibration Laboratories, which are located on repair ships and tenders (MIRCS) receive and calibrate fleet equipment only. Shore Calibration Laboratories, which are located in various shore activities of the Navy, receive and calibrate shore equipment and also handle the overflow from Fleet Calibration Laboratories.

As indicated in figure 7-40 equipment to be calibrated may go directly to a Navy Calibration Laboratory or it may go to a shop or repair facility for "qualification." Qualification is not the same as calibration, and the two terms should be clearly distinguished.

Calibration is the process by which Calibration Laboratories and Standards Laboratories compare a standard or a measuring instrument with a standard of higher accuracy in order to ensure that the item being compared is accurate within specified limits throughout its entire range. The calibration process involves the use of approved instrument calibration procedures; it may also include any adjustments or incidental repairs necessary to bring the standard or instrument being calibrated within specified limits. Calibration of standards is considered mandatory.

Qualification is the process by which an activity other than officially designated Standards Laboratories or Calibration Laboratories compares a test or measuring instrument with one of higher accuracy in order to determine the need for calibration. Qualification may be performed by ships or stations that have been furnished with approved measurement standards and procedures. However, the instruments used to qualify the test or measuring equipment should be calibrated periodically by a Navy Standards Laboratory or a Navy Calibration Laboratory in order for the qualification to be valid.

Several additional terms used in connection with the Navy Calibration Program are defined in the following paragraphs. It is important to understand the precise meaning of these terms and to use them correctly.

Calibration Procedure is the term used for a document that outlines the steps and operations to be followed by standards and calibration laboratory personnel in the performance of instrument calibration.

Calibration cycle is the length of time between calibration services during which each test equipment is expected to maintain reliable measurement capability. The Metrology

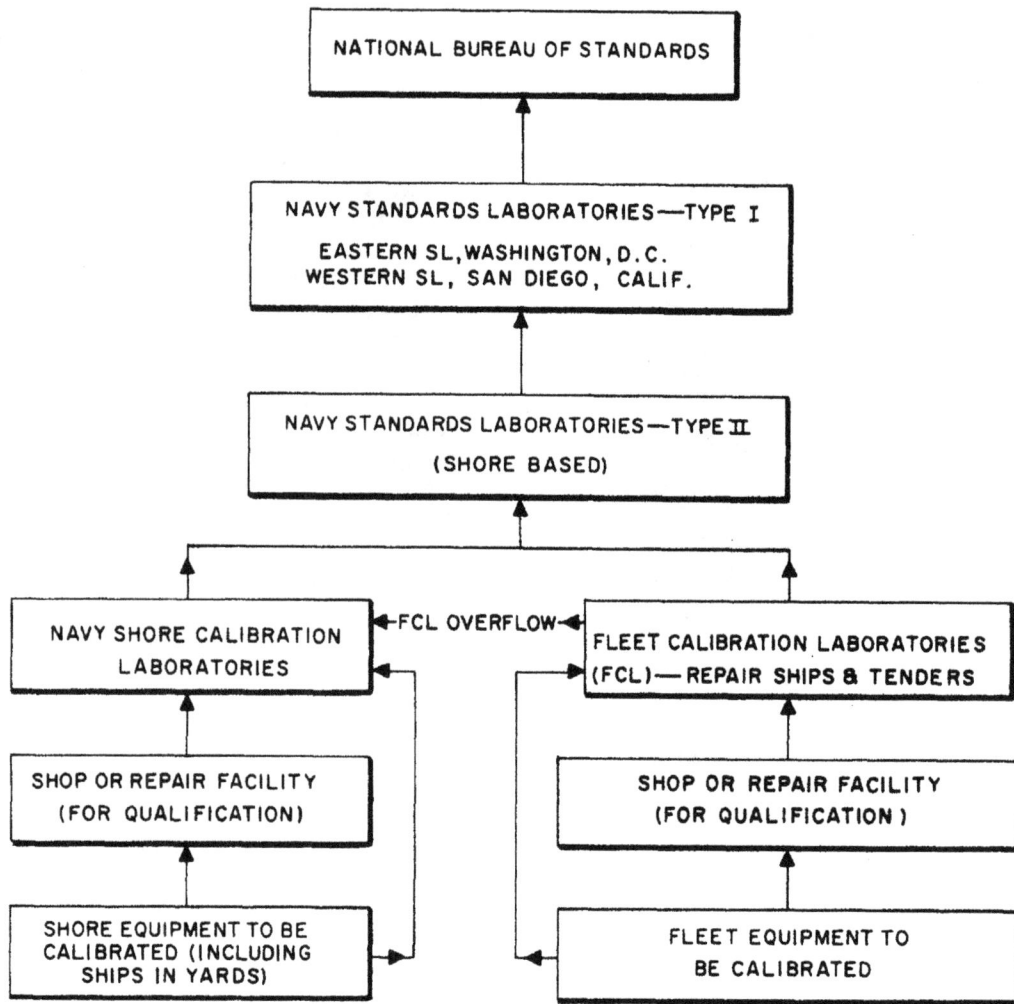

170.49

Figure 7-40.—Navy Calibration Program Structure.

Requirements List (NAVAIR 17-35MTL-1 NAV-SHIPS) is available and used to find the following information relating to a particular instrument:

(1) Instrument Calibration Recall Intervals
(2) Applicable Procedure Numbers
(3) Related Technical Numbers

(Note: The Metrology Requirements List is where you can find calibration cycles.)

Cross-checks involve the comparison of two or more instruments of equal or near equal accuracy for the sole purpose of determining if the values of any of the instruments have shifted significantly. The cross-check is used as an interim measure until a standard or instrument of sufficiently high accuracy can be used to calibrate or qualify an instrument.

Incidental repair is the term used to describe those repairs found necessary during the calibration of an operable equipment to bring it within its specified tolerances. Incidental repair includes the replacement of parts which have changed value sufficiently to prevent calibration but not enough to render the equipment inoperative. Incidental repair is normally performed in the laboratories in conjunction with the calibration of test equipment or standards.

Laboratory standard is the term used to identify a laboratory-type device that is used to maintain the continuity of values in units of measurement by periodic comparison with standards of Navy laboratories or with standards maintained by the National Bureau of Standards. A laboratory standard is used to calibrate a standard of lesser accuracy.

An acceptance for limited use indicates that an instrument which has failed certain tests or which has not been tested against all acceptance criteria is nevertheless suitable for certain specified (limited) usage. In such a situation, a limited use label (rather than a CALIBRATED or QUALIFIED label) is placed on the instrument to draw attention to the conditional acceptance. In addition to the label, a limited use tag is attached to the instrument. This tag is filled in by the servicing activity; it includes a full description of the reservations or precautions which should be observed in using the instrument. The label and the tag indicating that the instrument is suitable only for limited use must remain on the instrument until the next calibration or qualification.

The term rejected is used when an instrument fails to meet the acceptance criteria during calibration or qualification and when it cannot be made to meet these criteria by incidental repair. Under these conditions, a REJECTED label is placed on the instrument and all other servicing labels are removed. In addition to the label, a REJECTED tag is attached to the instrument. The tag, which is filled in by the servicing activity, gives the reason for rejection and such other information as may be required.

Repair is defined as the repair and/or replacement of malfunctioning parts of a measuring instrument or standard to the degree required to restore the instrument or standard to an operating condition.

Traceability is when the accuracy of a measurement made by the fleet can be directly traceable through the echelons of calibration to Reference Standards maintained by the National Bureau of Standards (unbroken chain of properly conducted calibrations.)

INSTRUMENT ACCURACY

As we have seen, all measurement is subject to a certain amount of error. The international and national standards have some error, even though it is almost unbelievably small. The error in secondary standards, while still extremely small, is somewhat greater than the error in the primary standards. When we get down to the actual measuring devices used even for precision measurement, the error is larger still.

Although it may sound backwards, the accuracy of an instrument is expressed by giving the amount of error of the instrument. For example,

an instrument with an accuracy of 1 percent is said to have an error of ±1 percent.

The error of an instrument is the difference between the reading shown on the instrument and the true value of the variable being measured. Error may be expressed in scale units, in percent of scale span, in percent of range, or in percent of indicated value (iv). By agreement among instrument manufacturers, error in instruments with uniform scales is most commonly expressed as a percentage of the full scale length, regardless of where the measurement is made on the scale. The exception to this general rule is that the measurement is not made at the extreme top or the extreme bottom of the scale, since an instrument is almost sure to be less accurate in these areas than in the working range of the scale.

Using the full scale length as a basis for determining instrument error can lead to some confusion. For example, consider several pressure gages, each one of which has a guaranteed accuracy of 1 percent. If the scale reads 0 to 30 psi, the allowable error is ± 0.3 psi. If the scale reads 0 to 100 psi, the allowable error is ±1 psi. If the scale reads 0 to 500 psi, the allowable error is ±5 psi. If the scale reads 0 to 1000, the allowable error is ±10 psi.

As far as accuracy ratios are concerned, there are recommended low and high accuracy ratios that should exist between the test and measuring equipment, and the measuring system or Standard, also between echelons of Standards.

The lower limit ratio should be at least 4 to 1; a ratio below this limit is impracticable for technical reasons. The upper limit ratio should not be more than 10 to 1; if the ratio is higher, equipment costs will become excessive.

Calibration error is taken care of by linearity and range errors. Linearity error is when the lowest and the highest indications are correct, and there is an error in between these indications. Range error occurs when the lowest indication is on and the highest indication is off, above or below the true value.

If an instrument does not give the same reading when it comes from the top of the scale down to the point of measurement as it does when it goes from the bottom of the scale up to the point of measurement, the error is called hysteresis. Hysteresis occurs from a variety of factors that cause loss of energy within the instrument; it might occur because of friction or binding of parts, fatigue of a spring, excessive play in gears, or other mechanical difficulties.

Because there will always be some error of measurement, and because the error may actually be considerably greater than that indicated by the percentage of guaranteed accuracy, the calibration of any measuring device requires the highest possible precision.

Instrument sensitivity is sometimes confused with instrument accuracy. This is a big mistake. The sensitivity of an instrument refers to the ability of the instrument to respond to changes in the value of the measured variable. If an instrument can respond to very small changes in the measured variable, it is a very sensitive instrument; if larger changes in the measured variable are required to produce effective motion of the measuring element, the instrument is less sensitive.

Sensitivity is quite directly related to friction within an instrument. An instrument that has relatively small energy losses because of friction will, all other things being equal, be more sensitive than an instrument with relatively large friction losses.

CHAPTER 8

INTRODUCTION TO THERMODYNAMICS

The shipboard engineering plant may be thought of as a series of devices and arrangements for the exchange and transformation of energy. The energy transformation of greatest importance in the shipboard plant is the production of mechanical work from thermal energy, since we depend largely upon this transformation to make the ship move through the water. On steam-driven ships, steam serves the vital purpose of carrying energy to the engines. The source of this energy may be the combustion of a conventional fuel oil or the fission of a radioactive material. In either case, the steam that is generated is the medium by which thermal energy is carried to the ship's engines, where it is converted into mechanical energy which propels the ship. In addition, energy transformations related directly or indirectly to the basic propulsion plant energy conversion provide power for many vital services such as steering, lighting, ventilation, heating, refrigeration and air conditioning, the operation of various electrical and electronic devices, and the loading, aiming. and firing of the ship's weapons.

In order to acquire a basic understanding of the design of shipboard engineering plants, it is necessary to have some understanding of certain concepts in the field of thermodynamics. In the broadest sense of the term, thermodynamics is the physical science that deals with energy and energy transformations. The branch of thermodynamics which is of primary interest to engineers is usually referred to as applied thermodynamics or engineering thermodynamics; it deals with fundamental design and operational considerations of boilers, turbines, internal combustion engines, air compressors, refrigeration and air conditioning equipment, and other machinery in which energy is exchanged or transferred in order to produce some desired effect.

This chapter deals with certain thermodynamic concepts that are particularly necessary

as a basis for understanding the shipboard engineering plant. The information given here is introductory in nature; obviously, it is not in any sense a complete or thorough exploration of the subject. Insofar as possible, we will depend upon verbal description rather than mathematical analysis to develop our understanding of the laws and principles of energy exchanges and transformations.[1]

It should perhaps be noted that many of the terms used in this chapter—including such basic terms as energy and heat—have more specialized and more precise meanings in the study of thermodynamics than they do in everyday life or even in the study of general physics. This is only to be expected; thermodynamics is a highly specialized branch of physics and, like any other specialty, it requires a certain refinement of terminology. If any difficulty arises from the fact that familiar terms are used in a somewhat unfamiliar sense, the difficulty can be largely minimized by paying particular attention to the exact meaning of each term, as defined here, rather than depending upon a general knowledge for an understanding of the terms.

ENERGY

Although energy has a general meaning to almost everyone, it is not easy to define the word in a completely satisfactory way. Energy is intangible and is largely known through its effects. Because energy is so often manifested by the production of work, energy is commonly defined as "the capacity for doing work." However, this is not entirely adequate as a definition, since work is not the only effect that is

---

[1]The student who has the mathematical background required for further study of thermodynamics will find it profitable to consult thermodynamics texts to amplify the information given in this chapter.

produced by energy. For example, heat can flow from one body to another without doing any work at all, but the heat must still be considered as energy and the process of heat transfer must be recognized as a process that has produced an effect. A broader definition, then, and one which satisfies more of the conditions under which we know energy to exist, is "the capacity for producing an effect."

Energy exists in many forms. For convenience, we usually classify energy according to the size and nature of the bodies or particles with which the energy is associated. Thus we say that mechanical energy is the energy associated with large bodies or objects—usually, things that are big enough to see. Thermal energy is energy associated with molecules. Chemical energy is energy that arises from the forces that bind the atoms together in a molecule. Chemical energy is demonstrated whenever combustion or any other chemical reaction takes place. Electrical energy, light, X-rays, and radio waves are examples of energy associated with particles that are even smaller than atoms.

Each of these types of energy must be further classified as (1) stored energy, or (2) energy in transition. Stored energy can be thought of as energy that is actually "contained in" or "stored in" a substance or system. There are two kinds of stored energy: (1) potential energy, and (2) kinetic energy. When energy is stored in a system because of the relative positions of two or more objects or particles, we call it potential energy. When energy is stored in a system because of the relative velocities of two or more objects or particles, we call it kinetic energy. It should be emphasized that all stored energy is either potential energy or kinetic energy.

Energy in transition is, as the name implies, energy that is in the process of being transferred from one object or system to another. All energy in transition begins and ends as stored energy.

In order to understand any form of energy, then, we need to know the relative size of the bodies or particles in the energy system and we need to know whether the energy is stored or in transition. Bearing in mind these two modes of classification, let us now examine mechanical energy and thermal energy—the two forms of energy which are of particular interest in practically all aspects of shipboard engineering.

MECHANICAL ENERGY

Energy associated with a system composed of relatively large bodies is called mechanical energy. The two forms of stored mechanical energy are (1) mechanical potential energy, and (2) mechanical kinetic energy.[2] Mechanical energy in transition is manifested by work.

Mechanical potential energy is stored in a system by virtue of the relative positions of the bodies that make up the system. The mechanical potential energy associated with the gravitational attraction between the earth and another body provides us with many everyday examples. A rock resting on the edge of a cliff in such a position that it will fall freely if pushed has mechanical potential energy. Water at the top of a dam has mechanical potential energy. A sled that is being held at the top of an icy hill has mechanical potential energy. Note that in each of these examples the energy resides neither in the earth alone nor in the other object alone but rather in an energy system of which the earth is merely one component.

Mechanical kinetic energy is stored in a system by virtue of the relative velocities of the component parts of the system. Push that rock over the edge of the cliff, open the gate of the dam, or let go of the sled—and something will move. The rock will fall, the water will flow, the sled will slide down the hill. In each case the mechanical potential energy will be changed to mechanical kinetic energy. Since it is customary to ascribe zero velocity to an object which is at rest with respect to the earth, it is also customary to think of kinetic energy as though it pertained only to the object which is in motion with respect to the earth. It should be remembered, however, that kinetic energy, like potential energy, is properly assigned to the system rather than to any one component of the system.

In these examples of mechanical potential energy and mechanical kinetic energy, we have used an external source of energy to get things started. Energy from some outside source is required to push the rock, open the gate of the dam, or let go of the sled. All real machines and processes require this kind of a boost from an energy source outside of the system; similarly, the energy from any one system is bound

---

[2] Although all forms of energy may be stored as potential energy or as kinetic energy, these terms refer, in common usage, to mechanical potential energy and mechanical kinetic energy, unless some other form of energy (thermal, chemical, etc.) is specified.

to affect other energy systems, since no one system can be completely isolated as far as energy is concerned. However, it is easier to understand the basic energy concepts if we disregard all the other energy systems that might be involved in or affected by each energy process. Hence we will generally consider one system at a time, disregarding energy boosts that may be received from an outside source and disregarding the energy transfers that may take place between the system we are considering and any other system.

It should be emphasized that mechanical potential energy and mechanical kinetic energy are both stored forms of energy. Some confusion arises because mechanical kinetic energy is often referred to as the "energy of motion," thus leading to the false conclusion that "energy in transition" is somehow involved. This is not the case, however. Work—mechanical work—is the only form of mechanical energy which can properly be considered as energy in transition.

Mechanical potential energy and mechanical kinetic energy are mutually convertible. To take the example of the rock resting on the edge of the cliff, let us suppose that some external force pushes the rock over the edge so that it falls. As the rock falls, the system loses potential energy but gains kinetic energy. By the time the rock reaches the ground at the base of the cliff, all the potential energy of the system has been converted into kinetic energy. The sum of the potential energy and the kinetic energy is identical at each point along the line of fall, but the proportions of potential energy and kinetic energy are constantly changing as the rock falls.

To take another example, consider a baseball that is thrown straight up into the air. The ball has kinetic energy while it is in upward motion, but the amount of kinetic energy is decreasing and the amount of potential energy is increasing as the ball travels upward. When the ball has just reached its uppermost position, before it starts to fall back toward the earth, it has only potential energy. Then, as the ball falls back toward the earth, the potential energy is converted into kinetic energy again.

The magnitude of the mechanical potential energy stored in a system by virtue of the relative positions of the bodies that make up the system is proportional to (1) the force of attraction between the bodies, and (2) the distance between the bodies. In the case of the rock which is ready to fall from the edge of the cliff, we are concerned with (1) the force of attraction

between the earth and the rock—that is, the force of gravity acting upon the rock, or the weight of the rock, and (2) the linear separation between the two objects. If we measure the weight in pounds and the distance in feet, the amount of mechanical potential energy stored in the system by virtue of the elevation of the rock is measured in the unit called the foot-pound. Specifically,

$$E_p = W \times D$$

where

E_p = mechanical potential energy, in foot-pounds

W = weight of body, in pounds

D = distance between earth and body, in feet

The magnitude of mechanical kinetic energy is proportional to the mass and to the square of the velocity of an object which has velocity with respect to another object, or

$$E_k = \frac{MV^2}{2}$$

where

E_k = mechanical kinetic energy, in foot-pounds

M = mass of body, in pounds

V = velocity of body relative to the earth, in feet per second

Where it is more convenient to use the weight of the body, rather than the mass, the equation becomes

$$E_k = \frac{WV^2}{2g}$$

where W is the weight of the body, in pounds, and g is the acceleration due to gravity, generally taken as 32.2 feet per second per second.

Work, as we have seen, is mechanical energy in transition—that is, it is a transitory form of mechanical energy which occurs only between two or more other forms of energy. Work is done when a tangible body or substance is moved through a tangible distance by the action of a tangible force. Thus we may define work as the energy which is transferred by the action of a force through a distance, or

$$E_{wk} = F \times D$$

where

E_{wk} = work, in foot pounds

F = force, in pounds

D = distance (or displacement), in feet

In the case of work done against gravity, the force is numerically the same as the weight of the object or body that is being displaced.

It is important to note that no work is done unless something is displaced from its previous position. When we lift a 5-pound weight from the floor to a table that is 3 feet high, we have done 15 foot-pounds of work. If we merely stand and hold the 5-pound weight, we do not perform any work in the technical sense of the term, even though we may feel like we are working. In this case, actually, all we are doing is exerting force in order to support the weight against the action of the force of gravity. The forces are balanced; there is no motion or displacement of the weight, so no work is done.

If the force and the displacement are neither acting in the same direction nor acting in total opposition, work is done only by that component of the force which is acting in the direction of the displacement of the body or object. A man pushing a lawnmower, for example, is exerting some force that acts in the direction in which the lawnmower is moving; but he is also exerting some force which acts downward, at right angles to the direction of displacement. In this case, only the forward component of the exerted force results in work—that is, in the forward motion of the lawnmower.

Suppose that we move an object in such a way that it returns to its original position. Have we done work or haven't we? Let us consider again the example of lifting a 5-pound weight to the top of a 3-foot table. By this action we have performed 15 foot-pounds of work. Now suppose that we let the weight fall back to the floor, so that it ends up in the same position it had originally. Displacement is zero, so work must be zero. But what has happened to the 15 foot-pounds of work we put into the system when we lifted the weight to the top of the table? By doing this work, we gave the system 15 foot-pounds of mechanical potential energy. When the weight

fell back to the floor, the mechanical potential energy was converted into mechanical kinetic energy. In one sense, therefore, we say that our work was "undone" and that no net work has been done.

On the other hand, we may choose to regard the two actions separately. In such a case, we say that we have done 15 foot-pounds of work by lifting the weight and that the force of gravity acting upon the weight has done 15 foot-pounds of work to return the weight to its original position on the floor. However, we must regard one work as positive and the other as negative. The two cancel each other out, so there is again no net work. But in this case we have recognized that 15 foot-pounds of work were performed twice, in two separate operations, by two different agencies.

This example has been elaborated at some length because we may draw several important inferences from it. First, it may help to clarify the concept of work as a form of energy that must be accounted for. Also, it may help to convey the real meaning of the statement that work is mechanical energy in transition. Work is energy in transition because it occurs only temporarily, between other forms of energy, and because it must always begin and end as stored energy. And finally, the example suggests the need for arbitrary reference planes in connection with the measurement of potential energy, kinetic energy, and work. The quantitative consideration of any form of energy requires a frame of reference which defines the starting point and the stopping point of any particular operation; the reference planes are practically always relative rather than absolute.

Note that mechanical potential energy, mechanical kinetic energy, and work are all measured in the same unit, the foot-pound. One foot-pound of work is done when a force of 1 pound acts through a distance of 1 foot. One foot-pound of mechanical kinetic energy or 1 foot-pound of mechanical potential energy is the amount of energy that would be required to accomplish 1 foot-pound of work.

The amount of work done has nothing to do with the length of time required to do it. If a weight of 1 pound is lifted through a distance of 1 foot, 1 foot-pound of work has been done, regardless of whether it was done in half a second or half an hour. The rate at which work is done is called power. In the field of mechanical engineering, the horsepower (hp) is the common unit of measurement for power. By definition, 1 horsepower is equal to 33,000 foot-pounds of work per minute or 550 foot-pounds of work per second. Thus a machine that is capable of doing 550 foot-pounds of work per second is said to be a 1-horsepower machine.

THERMAL ENERGY

Energy associated primarily with systems of molecules is called thermal energy. Like other kinds of energy, thermal energy may exist in stored form (in which case it is called internal energy) or as energy in transition (in which case it is called heat).

In common usage, the term heat is often used to include all forms of thermal energy. However, this lack of distinction between heat and the stored forms of thermal energy can lead to serious confusion. In this text, therefore, the term internal energy is used to describe the stored forms of thermal energy, and the term heat is used only to describe thermal energy in transition.

Internal Energy

Internal energy, like all stored forms of energy, exists either as potential energy or as kinetic energy.

Internal potential energy is the energy associated with the forces of attraction that exist between molecules. The magnitude of internal potential energy is dependent upon the mass of the molecules and the average distance by which they are separated, in much the same way that mechanical potential energy depends upon the mass of the bodies in the system and the distance by which they are separated. The force of attraction between molecules is greatest in solids, less in liquids and yielding substances, and least of all in gases and vapors. Whenever something happens to change the average distance between the molecules of a substance, there is a corresponding change in the internal potential energy of the substance.

Internal kinetic energy is the energy associated primarily with the activity of molecules, just as mechanical kinetic energy is the energy associated with the velocities of relatively large bodies. It is important to note that the temperature of a substance arises from and is proportional to the molecular activity with which internal kinetic energy is associated.

For most purposes, we will not need to distinguish between the two stored forms of internal energy. Instead of referring to internal potential energy and internal kinetic energy, therefore, we may often simply use the term internal energy. When used in this way, without qualification, the term internal energy should be understood to mean the sum total of all internal energy stored in the substance or system by virtue of the motion of molecules or by virtue of the forces of attraction between molecules.

Heat

Although the term heat is more familiar than the term internal energy, it may be more difficult to arrive at an accurate definition of heat. Heat is thermal energy in transition. Like work, heat is a transitory energy form existing between two or more other forms of energy.

Since the flow of thermal energy can occur only when there is a temperature difference between two objects or regions, it is apparent that heat is not a property or attribute of any one object or substance. If a person accidentally touches a hot stove, he may understandably feel that heat is a property of the stove. More accurately, however, he might reflect that his hand and the stove constitute an energy system and that thermal energy flows from the stove to his hand because the stove has a higher temperature than his hand.

As another example of the difference between heat and internal energy, consider two equal lengths of piping, made of identical materials and containing steam at the same pressure and temperature. One pipe is well insulated, one is not. From everyday experience, we expect more heat to flow from the uninsulated section of pipe than from the insulated section. When the two pipes are first filled with steam, the steam in one pipe contains exactly as much internal energy as the steam in the other pipe. We know this is true because the two pipes contain equal volumes of steam at equal pressures and temperatures. After a few minutes, the steam in the uninsulated pipe will contain much less internal energy than the steam in the insulated pipe, as we can tell by reading the pressure and temperature gages on each pipe. What has happened? Stored thermal energy—internal energy—has moved from one place to another, first from the steam to the pipe, then from the uninsulated pipe to the air. It is this movement, or this flow, of energy that should properly be called heat. Temperature is a reflection of the amount of internal kinetic energy possessed by an object or a substance, and it is therefore an attribute or property of the substance. The movement or flow of thermal energy—or, in other words, heat—is an attribute of the energy system rather than of any one component of it.[3]

Units of Measurement

In engineering, heat is commonly measured in the unit called the British thermal unit (Btu). Originally, 1 Btu was defined as the quantity of heat required to raise the temperature of 1 pound of water through 1 degree on the Fahrenheit scale. A similar unit called the calorie (cal) was originally defined as the quantity of heat required to raise the temperature of 1 gram of water through 1 degree on the Celsius scale. These units are still in use, but the original definitions have been abandoned by international agreement. The Btu and the calorie are now defined in terms of the unit of energy called the joule.[4] The

following relationships have thus been established by definition or derived from the established definitions:

$$1 \text{ calorie} = \frac{1}{860} \text{ watt-hour}$$

$$= 4.18605 \text{ joules}$$

$$= 3.0883 \text{ foot-pounds}$$

$$1 \text{ Btu} = 251.996 \text{ calories}$$

$$= 778.26 \text{ foot-pounds}$$

$$= 1054.886 \text{ joules}$$

The values given here are, of course, considerably more precise than those normally required in engineering calculations.

When large amounts of thermal energy are involved, it is often more convenient to use multiples of the Btu or the calorie. For example, we may wish to refer to thousands or millions of Btu, in which case we would use the unit kB (1 kB = 1000 Btu) or the unit mB (1 mB = 1,000,000 Btu). Similarly, the kilocalorie may be used when we wish to express calories in thousands (1 kilocalorie = 1000 calories). The kilocalorie, also called the "large calorie," is the unit normally used for indicating the thermal energies of various foods. Thus a portion of food which contains "100 calories" actually contains 100 kilocalories or 100,000 ordinary calories.

Heat Transfer

Heat flow, or the transfer of thermal energy from one body, substance, or region to another, takes place always from a region of higher temperature to a region of lower temperature.[5] In thermodynamics, the high temperature region may be called the source or the emitting region; the low temperature region may be called the sink, the receiver, or the receiving region.

---

[3]The correct definition of heat is emphasized here in order to avoid subsequent misunderstanding in the study of thermodynamic processes. It is obvious that "heat" and related words are sometimes used in a general way to indicate temperature. For example, we have no simple way of referring to an object with a large amount of internal kinetic energy except to say that it is "hot." Similarly, a reference to "the heat of the sun" may mean either the temperature of the sun or the amount of heat being radiated by the sun. Even "heat flow" or "heat transfer"—the terms quite properly used to describe the flow of thermal energy—are sometimes used in such a way as to imply that heat is a property of one object or substance rather than an attribute of an energy system. To a certain extent, such inaccurate use of "heat" and related words is really unavoidable; we must continue to "add heat" and "remove heat" and perform other impossible operations, verbally, unless we wish to adopt a very stuffy and long-winded form of speech. It is essential, however, that we maintain a clear understanding of the true nature of heat and of the distinction between heat and the stored forms of thermal energy.

[4]Several reasons contributed to the abandonment of the original definitions of the Btu and the calorie. For one thing, precise measurements indicated that the quantity of heat required to raise a specified amount of water through 1 degree on the appropriate scale was not constant at all temperatures. Second—and perhaps even more important—the recognition of heat as a form of energy makes the Btu and the calorie unnecessary. Indeed, it has been suggested that the calorie and the Btu could be given up entirely and that heat could be expressed directly in joules, ergs, foot-pounds, or other established energy units. Some progress has been made in this direction, but not much; the Btu and the calorie are still the units of heat most widely used in engineering and in the physical sciences generally.

[5]This statement, although entirely true for all practical engineering applications, should perhaps be qualified. Energy exchanges between molecules may be thought of as being random, in the statistical sense; therefore, some exchanges of thermal energy may indeed "go in the wrong direction"—that is, from a colder region to a warmer region. On the average, however, the flow of heat is always from the higher to the lower temperature.

Although three modes of heat transfer—conduction, radiation, and convection—are commonly recognized, we will find it easier to understand heat transfer if we make a distinction between conduction and radiation, on the one hand, and convection, on the other. Conduction and radiation may be regarded as the primary modes of heat flow. Convection may best be thought of as a related but basically different and special kind of process which involves the movement of a mass of fluid from one place to another.

CONDUCTION.—Conduction is the mode by which heat flows from a hotter to a colder region when there is physical contact between the two regions. For example, consider a metal bar which is held so that one end of it is in boiling water. In a very short time the end of the bar which is not in the boiling water will have become too hot to hold. We say that heat has been conducted from molecule to molecule along the entire length of the bar. The molecules in the layer nearest the source of heat become increasingly active as they receive thermal energy. Since each layer of molecules is bound to the adjacent layers by cohesive forces, the motion is passed on to the next layer which, in turn, sets up increased activity in the next layer. The process of conduction continues as long as there is a temperature difference between the two ends of the bar.

The total quantity of heat conducted depends upon a number of factors. Let us consider a bar of homogeneous material which is uniform in cross-sectional area throughout its length. One end of the bar is kept at a uniformly high temperature, the other end is kept at a uniformly low temperature. After a steady and uniform flow of heat has been established, the total quantity of heat that will be conducted through this bar depends upon the following relationships:

1. The total quantity of heat passing through the conductor in a given length of time is directly proportional to the cross-sectional area of the conductor. The cross-sectional area is measured normal to (that is, at right angles to) the direction of heat flow.

2. The total quantity of heat passing through the conductor in a given length of time is proportional to the thermal gradient—that is, to the difference in temperature between the two ends of the bar, divided by the length of the bar.

3. The quantity of heat is directly proportional to the time of heat flow.

4. The quantity of heat depends upon the thermal conductivity of the material of which the bar is made. Thermal conductivity (k) is different for each material.

These relationships may be expressed by the equation

$$Q = kTA \frac{t_1 - t_2}{L}$$

where

Q = quantity of heat, in Btu or calories

k = coefficient of thermal conductivity (characteristic of each material)

T = time during which heat flows

A = cross-sectional area, normal to the path of heat

t_1 = temperature at the hot end of the bar

t_2 = temperature at the cold end of the bar

L = distance between the two ends of the bar

This equation, which is sometimes called the general conduction equation, applies whether we are using a metric system or a British system. Consistency in the use of units is, of course, vital.

The quantity $\frac{t_1 - t_2}{L}$ is called the thermal gradient or the temperature gradient. In the metric CGS system, the temperature gradient is expressed in degrees Celsius per centimeter of length; the cross-sectional area is expressed in square centimeters; and the time is expressed in seconds. In British units, the temperature gradient is expressed in Btu per inch (or sometimes per foot) of length; the cross-sectional area is expressed in square feet; and the time is expressed in seconds or in hours. (As may be noted, some caution is required in using the British units; we must know whether the temperature gradient indicates Btu per inch or Btu per foot, and we must know whether the time is expressed in seconds or in hours.)

From the general conduction equation, we may infer that the coefficient of thermal conductivity (k) represents the quantity of heat which will flow through unit cross section and unit length of a material in unit time when there is unit temperature difference between the hotter and the colder faces of the material.

Thermal conductivity is determined experimentally for various materials. We may perhaps visualize the process of conduction more clearly and understand its quantitative aspects more fully by examining an apparatus for the determination of thermal conductivity and by setting up a problem.

Figure 8-1 shows a device that could be used for determining thermal conductivity. Assume that we have a bar of uniform diameter, made of an unknown metal. (If we knew the kind of metal, we could look up the thermal conductivity in a table; since we do not know the metal, we shall find k experimentally.) One end of the bar is inserted into a steam chest in which a constant temperature is maintained; the other end of the bar is inserted into a water chest. The quantity of water flowing through the water chest and the entrance and exit temperature of the water are measured. Also, the temperature of the bar itself is measured at two points by means of thermometers inserted into holes in the bar; we may choose any two points along the bar, provided they are reasonably far apart and provided they are some distance away from the steam chest and the water chest.

We will assume that the entire apparatus is perfectly insulated so that the temperature difference between t_1 and t_2 is an accurate reflection of the heat conducted along the bar and so

that the amount of heat absorbed by the circulating water in the water chest is a true indication of the heat conducted from the hotter end of the bar to the colder end. We will assume that the following data are known at the outset or learned by measurement or determined in the course of the experiment:

Specific heat of water = 1.00

Temperature of water entering water chest = 20° C

Temperature of water leaving water chest = 30° C

Mass of water passing through water chest = 1300 grams

t_1 (temperature at hotter end of bar) = 80°C

t_2 (temperature at cooler end of bar) = 60° C

A (cross-sectional area of bar) = 20 square centimeters

L (distance between points of temperature measurement on bar) = 10 centimeters

T (time of heat flow) = 6 minutes = 360 seconds

To determine the thermal conductivity, k, of our unknown metal, we will use two equations.

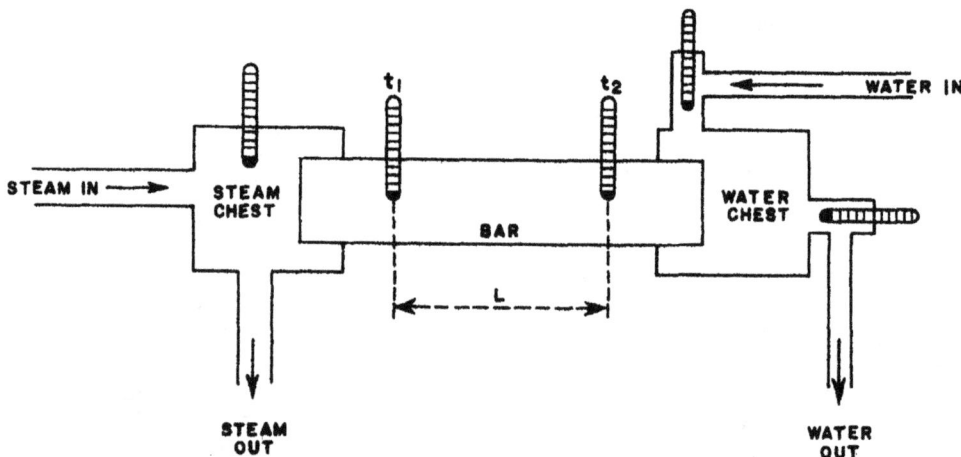

147.60

Figure 8-1.—Device for measuring thermal conductivity.

One is the general conduction equation

$$Q = kTA \frac{t_1 - t_2}{L}$$

where, as we have seen, Q may be expressed in calories or in Btu. In this example, we are using the metric CGS system and must therefore express Q in calories. The second equation we will use gives us a second way of calculating Q—that is, by determining the amount of heat absorbed by the circulating water. Thus,

Q = mass of water x temperature change of water x specific heat of water

Substituting some of our known values in this second equation, we find that

Q = (1300) (10) (1) = 13,000 calories

Using this value of Q and substituting other known values in the general conduction equation, we find that

$$13,000 = k (360) (20) \left(\frac{80 - 60}{10}\right)$$

$$= k (360) (20) (2)$$

$$= 14,400 \, k$$

$$0.9 = k$$

It should be noted that the general conduction equation applies only when there is a steady-state thermal gradient—that is, after a uniform flow of heat has been established. It should be noted also that $\underline{k}$ varies slightly as a function of temperature, although for many purposes the rise in $\underline{k}$ that goes with a rise in temperature is so slight that it can safely be disregarded.

In considering the experimental determination of thermal conductivity, why do we include "specific heat of water = 1.00" as one of the known data? What is specific heat, and what is its utility? Specific heat (also called heat capacity or specific heat capacity) is, like thermal conductivity, a thermal property of matter that must be determined experimentally for each substance. In general, we may say that specific heat is the property of matter that explains why the addition of equal quantities of heat to two different substances will not necessarily produce the same temperature rise in the two substances. We may define the specific heat of any substance as the quantity of heat required to raise the temperature of unit mass of that substance 1 degree.[6] In the metric CGS system, specific heat is expressed in calories per gram per degree Celsius; in the metric MKS system, it is expressed in kilocalories per kilogram per degree Celsius; and in British systems, it is expressed in Btu per pound per degree Fahrenheit. The specific heat of water is 1.00 in any system, and the numerical value of specific heat for any given substance is the same in all systems (although the units are, of course, different).

Specific heat is determined experimentally by laboratory procedures which are extremely complex and difficult in practice, although basically simple in theory. One of the commonest methods of determining specific heat is known as the method of mixtures. In this procedure, a known mass of finely divided metal is heated and then mixed with a known mass of water. The temperatures of the metal before mixing, of the water before mixing, and of the mixture just as it reaches thermal equilibrium are measured. Then, on the simple premise that the heat lost by one substance must be gained by the other substance, the specific heat of the metal can be found by using the equation

$$m_1 c_1 (t_1 - t_3) = m_2 c_2 (t_3 - t_2)$$

where

m_1 = mass of metal

m_2 = mass of water

c_1 = specific heat of metal

c_2 = specific heat of water (known to be 1.00)

t_1 = temperature of metal before mixing

t_2 = temperature of water before mixing

t_3 = temperature at which water and metal reach thermal equilibrium

---

[6] Specific heat as defined here should not be confused with the relatively useless concept of specific heat ratio, by which the heat capacity of each substance is compared to the heat capacity of water (taken as 1.00). The specific heat ratio is, obviously, a pure number without units.

In words, then, we may say that the mass times the specific heat times the temperature change of the first substance must equal the mass times the specific heat times the temperature change of the second substance. In this equation and in this verbal statement, we are ignoring the thermal energy absorbed by the apparatus, by the stirring rods, and by the thermometers. In actually determining specific heats, it is often necessary to account for all thermal energy, even that relatively minute quantity which is absorbed by the equipment. In such a case, the heat absorbed by the equipment is merely added to the right-hand side of the equation.

Specific heat is primarily useful in that it allows us to determine the quantity of heat added to a substance merely by observing the temperature rise, when we know the mass and the specific heat of the substance. And this, in fact, is precisely what we did in the thermal conductivity problem, where we calculated the amount of heat that had been absorbed by the water in the water chest by using the equation

$$Q = \text{mass} \times \text{temperature change} \times \text{specific heat}$$

Specific heat varies, in greater or lesser degree, according to pressure, volume, and temperature. Specific heat values quoted for solids and liquids are obtained through experimental procedures in which the substance is kept at constant pressure. The specific heat of any gas may vary tremendously, having in fact an almost infinite variety of values because of the almost infinite variety of processes and states during which energy is transferred to or by a gas. For convenience, specific heats of gases are given as specific heat at constant volume (c_v) and specific heat at constant pressure (c_p).

RADIATION.—Thermal radiation is a mode of heat transfer that does not involve any physical contact between the emitting region and the receiving region. A person sitting near a hot stove is warmed by thermal radiation from the stove, even though the air in between remains relatively cold. Thermal radiation from the sun warms the earth without warming the space through which it passes. Thermal radiation passes through any transparent substance—air, glass, ice—without warming it to any extent because transparent materials are very poor absorbers of radiant energy.

All substances—solids, liquids, and gases—emit radiant energy at all times. We tend to think of radiant energy as something that is emitted only by extremely hot objects such as the sun, a stove, or a furnace, but this is a very limited view of the nature of radiant energy. The earth absorbs radiant energy emitted by the sun, but the earth in turn radiates energy to the stars. A stove radiates energy to everything surrounding it, but at the same time all the surrounding objects are radiating energy to the stove. A child standing near a snowman may well believe that the snowman is "radiating cold" rather than emitting radiant energy; actually, however, both the child and the snowman are emitting radiant energy. The child, of course, is radiating far more energy than the snowman, so the net effect of this energy exchange is that the snowman grows warmer and the child grows colder. We are literally surrounded by—and a part of—such energy exchanges at all times. As we consider these energy exchanges, we may arrive at a new view of thermal equilibrium: when objects are radiating precisely as much thermal energy as they are receiving, in any given period of time, they are in thermal equilibrium.

Thermal radiation is an electromagnetic wave phenomenon, differing from light, radio waves, and other electromagnetic phenomenon merely in the wavelengths involved. When the wavelengths are in the infrared part of the electromagnetic spectrum—that is, when they are just below the range of visible light waves—we refer to the radiated energy as thermal radiation. It should be noted, however, that all electromagnetic waves transport energy which can be absorbed by matter and which can in many cases result in observable thermal effects. For example, one energy unit of light absorbed by a substance produces the same temperature rise in that substance as is produced by the absorption of an equal amount of thermal (infrared) energy.

When radiant energy falls upon a body that can absorb it, some of the energy is absorbed and some is reflected. The amount absorbed and the amount reflected depend in large part upon the surface of the receiving body. Dark, opaque bodies absorb more thermal radiation than shiny, bright, white, or polished bodies. Shiny, bright, white, or polished bodies reflect more thermal radiation than dark, opaque bodies. Good radiators are also good absorbers and poor radiators are poor absorbers. In general, good reflectors are poor radiators and poor absorbers.

In considering thermal radiation, the concept of black body radiation is frequently a useful construct. A black body is conceived of as an ideal or theoretical body which, being perfectly black, is a perfect radiator, a perfect absorber, and a perfect nonreflector of radiant energy. The thermal radiation emitted by such a perfect black body is proportional to T^4—that is, to the absolute temperature raised to the fourth power. Because of the fourth power relationship, doubling the absolute temperature increases the radiation 16 times, tripling the absolute temperature increases the radiation 81 times, and so forth. The thermal radiation emitted by real bodies is also proportional to the fourth power of the absolute temperature, although the total radiation emitted by a real body depends also upon the surface of the body. Consideration of the relationship between the thermal radiation of a body and the fourth power of the absolute temperature of that body explains why the problem of thermal insulation against radiation losses increases so enormously as the temperature increases.

CONVECTION.—Although convection is often loosely classified as a mode of heat transfer, it is more accurately regarded as the mechanical transportation of a mass of fluid (liquid or gas) from one place to another. In the process of this transportation, all the thermal energy stored within the fluid remains in stored form unless it is transferred by radiation or by conduction. Since convection does not involve thermal energy in transition, we cannot in the most fundamental sense regard it as a mode of heat transfer.

Convection is the transportation or the movement of some portions within a mass of fluid. As this movement occurs, the moving portions of the fluid transport their contained thermal energy to other parts of the fluid. The effect of convection is thus to mix the various portions of the fluid. The part that was at the bottom of the container may move to the top or the part that was at one side may move to the other side. As this mixing takes place, heat transfer occurs by conduction and radiation from one part of the fluid to another and between the fluid and its surroundings. In other words, convection transports portions of the fluid from one place to another, mixes the fluid, and thus provides an opportunity for heat transfer to occur. But convection does not, in and of itself, "transfer" thermal energy.

Convection serves a vital purpose in bringing the different parts of a fluid into close contact with each other so that heat transfer can occur. Without convection, there would be little heat transfer from, to, or within fluids, since most fluids are very poor at transferring heat except when they are in motion.

Two kinds of convection may be distinguished. Natural convection occurs when there are differences in the density of different parts of the fluid. The differences in density are usually caused by unequal temperatures within the mass of fluid. As the air over a hot radiator is heated, for example, it becomes less dense and therefore begins to rise. Cooler, heavier air is drawn in to replace the heated air that has moved upward, and convection currents are thus set up. Another example of natural convection, and one that may be quite readily observed, may be found in a pan of water that is being heated on a stove. As the water near the bottom of the pan is heated first, it becomes less dense and moves upward. This displaces the cooler, heavier water and forces it downward; as the cooler water is heated in turn, it rises and displaces the water near the top. By the time the water has almost reached the boiling point, a considerable amount of motion can be observed in the water.

Forced convection occurs when some mechanical device such as a pump or a fan produces movement of a fluid. Many examples of forced convection may be observed in the shipboard engineering plant: feed pumps transporting water to the boilers, fuel oil pumps moving fuel oil through heaters and meters, lubricating oil pumps forcing lubricating oil through coolers, and forced draft blowers pushing air through boiler double casings, to name but a few.

The mathematical treatment of convection is extremely complex, largely because the amount of heat gained or lost through the convection process depends upon so many different factors. Empirically determined convection coefficients which take account of these many factors are available for most kinds of engineering equipment.

Sensible Heat and Latent Heat

The terms sensible heat and latent heat are often used to indicate the effect that the transfer of heat has upon a substance. The flow of heat) from one substance to another is normally reflected in a temperature change in each substance—that is, the hotter substance becomes cooler and the cooler substance becomes hotter. However, the flow of heat is not reflected in a temperature change in a substance which is in process of changing from one physical state[7] to another. When the flow of heat is reflected in a temperature change, we say that sensible heat has been added to or removed from a substance. When the flow of heat is not reflected in a temperature change but is reflected in the changing physical state of a substance, we say that latent heat has been added or removed.

Since heat is defined as thermal energy in transition, we must not infer that sensible heat and latent heat are really two different kinds of heat. Instead, the terms serve to distinguish between two different kinds of effects produced by the transfer of heat; and, at a more fundamental level, they indicate something about the manner in which the thermal energy was or will be stored. Sensible heat involves internal kinetic energy and latent heat involves internal potential energy.

The three fundamental physical states of all matter are solid, liquid, and gas (or vapor). The physical state of a substance is closely related to the distance between molecules. The molecules are closest together in solids, farther apart in liquids, and farthest apart in gases. When the flow of heat to a substance is not reflected in a temperature change, we know that the energy is being used to increase the distance between the molecules of the substance and thus change it from a solid to a liquid or from a liquid to a gas. In other words, the addition of heat to a substance that is in process of changing from solid to liquid or from liquid to gas results in an increase in the amount of internal potential

energy stored in the substance, but it does not result in an increase in the amount of internal kinetic energy. Only after the change of state has been fully accomplished does the addition of heat result in a change in the amount of internal kinetic energy stored in the substance; hence, there is no temperature change until after the change of state is complete.

In a sense, we may think of latent heat as the energy price that must be paid for a change of state from solid to liquid or from liquid to gas. But the energy is not lost; rather, it is stored in the substance as internal potential energy. The energy price is "repaid," so to speak, when the substance changes back from gas to liquid or from liquid to solid; during these changes of state, the substance gives off heat without any change in temperature.

The amount of latent heat required to cause a change of state—or, on the other hand, the amount of latent heat given off during a change of state—varies according to the pressure under which the process takes place. For example, it takes about 970 Btu to change 1 pound of water to steam at atmospheric pressure (14.7 psia) but it takes only 62 Btu to change 1 pound of water to steam at 3200 psia.

Figure 8-2 shows the relationship between sensible heat and latent heat for one substance, water, at atmospheric pressure.[8] If we start with 1 pound of ice at 0°F, we must add 16 Btu to raise the temperature of the ice to 32°F. We call this adding sensible heat. To change the pound of ice at 32°F to a pound of water at 32°F, we must add 144 Btu (the latent heat of fusion). There will be no change in temperature while the ice is melting. After all the ice has melted, however, the temperature of the water will be raised as additional heat is supplied. Again, we are adding sensible heat. If we add 180 Btu— that is, 1 Btu for each degree of temperature between 32°F and 212°F—the temperature of the water will be raised to the boiling point. To change the pound of water at 212°F to a pound of steam at 212°F, we must add 970 Btu (the latent heat of vaporization). After all the water has been converted to steam, the addition of more heat will cause an increase in the temperature of the steam. If we add 42 Btu to the

[7] In thermodynamics, the physical state of a substance (solid, liquid, or gas) is usually described by the term phase, while the term state is used to describe the substance with respect to all of its properties—phase, pressure, temperature, specific volume, and so forth. Thus the phase of a substance may be considered as merely one of the several properties that fix the state of the substance. While the precision of this usage has some obvious advantages, it is not in standard use among engineers. In this text, therefore, the term physical state (or sometimes state) is used to denote the molecular condition of a substance that determines whether the substance is a solid, a liquid, or a gas.

[8] The same kind of chart could be drawn up for other substances, but different amounts of thermal energy would of course be required for each change of temperature or of physical state.

pound of steam which is at 212°F, we can superheat[9] it to 300°F.

The same relationships apply when heat is being removed. The removal of 42 Btu from the pound of steam which is at 300°F will cause the temperature to drop to 212°F. As the pound of steam at 212°F changes to a pound of water at 212°F, 970 Btu are given off. When a gas or vapor is changing to a liquid, we usually use the term latent heat of condensation; numerically, of course, the latent heat of condensation is exactly the same as the latent heat of vaporization. The removal of another 180 Btu will lower the temperature of the pound of water from 212°F to 32°F. As the pound of water at 32°F changes to a pound of ice at 32°F, 144 Btu are given off without any accompanying change in temperature. Further removal of heat causes the temperature of the ice to decrease.

Heat Transfer Apparatus

Any device or apparatus designed to allow the flow of thermal energy from one fluid to another is called a heat exchanger. The shipboard engineering plant contains an enormous number and variety of heat exchangers, ranging from large items such as boilers and main condensers to relatively small items such as fuel oil heaters and lubricating oil coolers.

As a basis for understanding something about heat transfer in real heat exchangers, it is necessary to visualize the general configuration of the most commonly used type of heat exchanger. With few exceptions,[10] heat exchangers used aboard ship are of the indirect or surface type— that is, heat flows from one fluid to another through some kind of tube, plate, or other "surface" that separates the two fluids and keeps them from mixing. Most surface heat exchangers

are of the shell-and-tube types, consisting of a bundle of metal tubes that fit inside a shell. One fluid flows through the inside of the tubes and the other flows through the shell, around the outside of the tubes.

The exchanges of thermal energy that take place in even a simple heat exchanger are really quite complex. The processes of conduction, radiation, and convection are involved in practically all heat exchangers. Processes involving latent heat—that is, the processes of evaporation, condensation, melting, and solidification— may contribute to the heat transfer problem. In all cases, heat transfer is affected by physical properties of the fluids which are exchanging thermal energy and by physical properties of the metal through which the change is being effected. The temperature differences involved, the extent and nature of the fluid films, the thickness and nature of the metals through which heat transfer takes place, the length and area of the path of heat flow, the types of surfaces involved, the velocity of flow, and other factors also determine the amount of heat transferred in any heat exchanger.

Because heat transfer is such a complex phenomenon, heat transfer calculations are necessarily complex. For some purposes, heat transfer problems are simplified by the use of an overall coefficient of heat transfer (U) which may be determined experimentally for any specific set of conditions. Tabulated values of U are available for various kinds of heat exchanger metal tubes, for building materials, and for other materials; in most cases the values of U are approximate, since various conditions such as temperature, velocity of flow, condition of the heat transfer surfaces, and the physical properties of the fluids have a profound effect upon the amount of heat transferred.

The transfer of heat in a heat exchanger involves the flow of heat from the hot fluid to the tube metal and from the tube metal to the cold fluid. In addition, heat must also be transferred through two layers of fluid (one on the inside and one on the outside of the tube) which are not flowing with the remainder of the fluid but are almost motionless. These relatively stagnant layers, known as boundary layers or fluid films, are extremely small in size but have an extremely important effect on heat transfer.

As previously noted, most fluids are very poor transferrers of heat. As a fluid is flowing, however, convection and mechanical mixing of

---

[9] A vapor or gas is said to be superheated when its temperature has been raised above the temperature of the liquid from which the vapor or gas is being generated. As may be inferred from the discussion, it is impossible to superheat a vapor or gas as long as it is in contact with the liquid from which it is being generated.

[10] A notable exception is the deaerating feed tank, discussed in chapter 13 of this text. Deaerating feed tanks are basically described as direct-contact heat exchangers, rather than surface heat exchangers, because heat transfer is accomplished by the actual mixing of the hotter and the colder fluids.

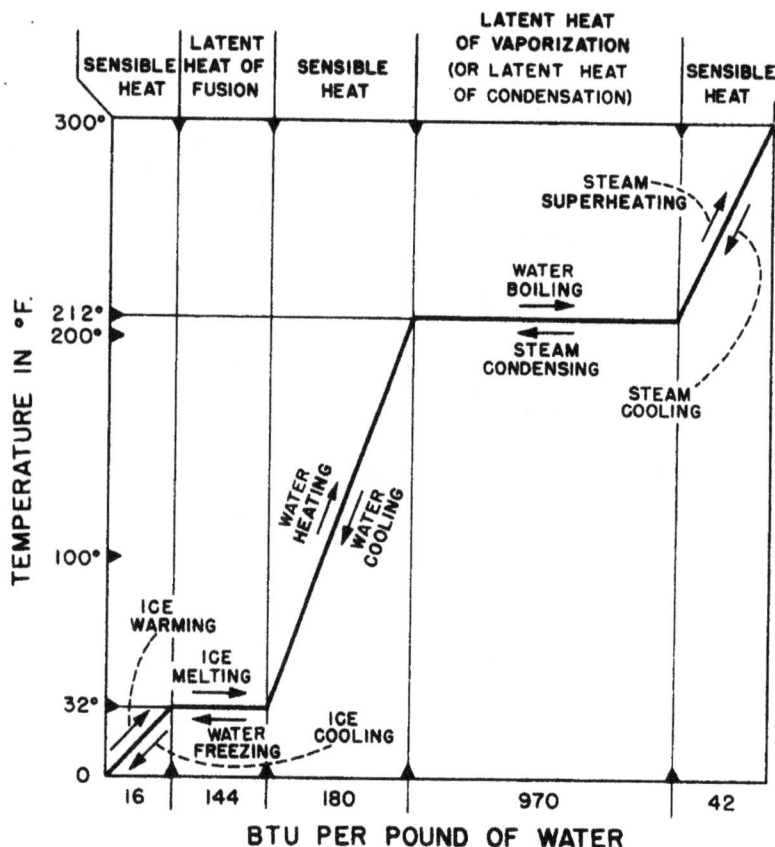

38.1

Figure 8-2.—Relationship between sensible heat and latent heat for water at atmospheric pressure.

the fluid bring the molecules into such intimate contact that heat transfer can and does occur. Other things being equal, increasing the velocity of fluid flow increases heat transfer.[11]

Since the fluid film is almost motionless, heat transfer through the film is very poor. The effect of fluid films on heat transfer is shown in figure 8-3. The temperature line indicates the changes in temperature that occur as heat is transferred from the hot fluid to the fluid film,

from this fluid film to the tube metal, from the tube metal to the other fluid film, and from this fluid film to the cold fluid. As may be seen, the major part of the temperature drop occurs in the fluid films rather than in the tube metal. Note, also, that the thicker fluid film is more resistant to heat transfer than the thinner fluid film.

The velocity of flow and the amount of turbulence in the flow affect heat transfer by altering the thickness of the fluid film. Increasing the velocity of flow diminishes the thickness of the fluid film and thus increases heat transfer. Turbulent flow breaks up the fluid film and thus increases heat transfer. Although there are some obvious disadvantages to excessive turbulence, many heat exchangers are designed to operate with a certain amount of turbulence so that the fluid films will be kept to a minimum.

In real heat exchangers, the accumulation of deposits of scale, soot, or dirt on the inside or the outside of the tubes has a profound and

---

[11]It is important here to maintain the distinction, previously established, between heat and temperature. Increasing the velocity of flow increases the amount of heat that is transferred, but decreasing the velocity increases the temperature of the fluid. This fact is of considerable practical importance in the design and operation of heat exchangers. In a heat exchanger designed for high velocity flow, stagnation of the flow is likely to cause severe overheating of the heat exchanger metal.

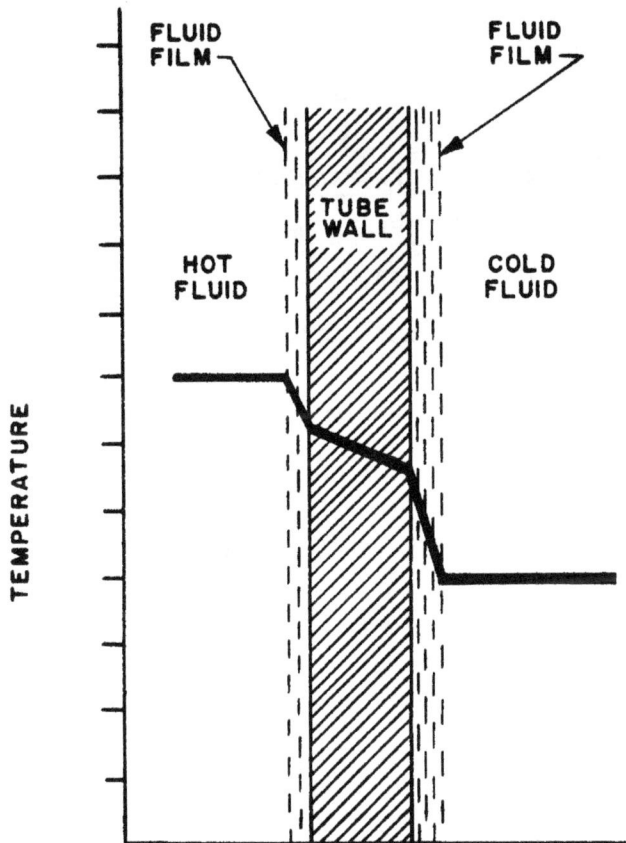

147.61

Figure 8-3.—Effect of fluid film on heat transfer.

detrimental effect upon heat transfer. Such deposits not only reduce the efficiency of the heat exchanger but also tend to cause overheating of the tube metal.

In surface heat exchangers, the components may be arranged so as to provide parallel flow, counter flow, or cross flow of the two fluids. In parallel flow (fig. 8-4) both fluids flow in the

same direction. Parallel-flow heat exchangers are rarely used for naval service, largely because they would require an impossibly long heat transfer surface to achieve the required amount of heat transfer. In counter flow (fig. 8-5) the two fluids flow in opposite directions. Many heat exchangers used aboard ship are of the counter-flow type. In cross flow (fig. 8-6) one fluid flows at right angles to the other. Cross flow is used particularly where the purpose of the heat exchanger is to remove latent heat and thus change the physical state of a substance. Main and auxiliary condensers are typically of the cross-flow type, as are several other smaller shipboard condensers.

Surface heat exchangers are referred to as single-pass units, if each fluid passes the other only once, or as multipass units, if one fluid passes the other more than once. Multipass flow may be obtained by the arrangement of the tubes and of the fluid inlets and outlets, or it may be obtained by using baffles to guide a fluid so that it passes the other fluid more than once before it leaves the heat exchanger.

THE FIRST LAW OF THERMODYNAMICS

In the previous discussion of energy, we have occasionally assumed a general principle which must now be stated. This principle is called the principle of the conservation of energy. The principle may be stated in several ways. Most commonly, perhaps, it is stated as energy can be neither destroyed nor created, but only transformed. Another statement is that energy may be transformed from one form to another, but the total energy of any body or system of bodies is a quantity that can neither be increased nor diminished by the action of the body or bodies. Still another way of stating this principle is by saying that the total quantity of energy in the

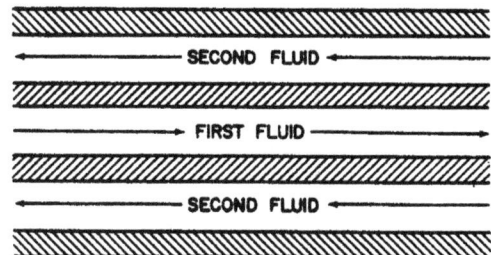

98.30

Figure 8-4.—Parallel flow in heat exchanger.

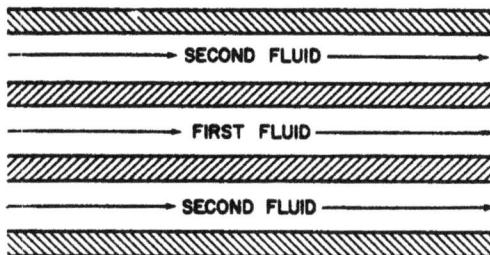

98.31

Figure 8-5.—Counter flow in heat exchanger.

171

universe is always the same. Regardless of the mode of expression, the principle of the conservation of energy applies to all kinds of energy.[12]

Energy equations for many thermodynamic processes are based directly upon the principle of the conservation of energy. When the principle of the conservation of energy is written in equation form, it is known as the general energy equation and is expressed as:

$$\text{energy in} = \text{energy out}$$

or, in more detail, it may be stated that the energy entering a system equals the energy leaving the system plus any accumulation and minus any dimunition in the amount of energy stored within the system.

The first law of thermodynamics, a special statement of the principle of the conservation of energy, deals with the transformation of mechanical energy to thermal energy and of thermal energy to mechanical energy. The first law is commonly stated as follows: Thermal energy and mechanical energy are mutually convertible, in the ratio of 778 foot-pounds to 1 Btu.

The ratio of conversion between mechanical energy and thermal energy is known as the mechanical equivalent of heat, or Joule's equivalent. It is symbolized by the letter J and, in

------

[12]The principle of the conservation of energy and the principle of the conservation of mass have been basic to the development of modern science. Until the establishment of the theory of relativity, with its implication of the mutual convertibility of energy and mass, the two principles were considered quite separate. According to the theory of relativity, however, they must be considered merely as two phases of a single principle which states that mass and energy are interchangeable and the total amount of matter and energy in the universe is constant. Nuclear fission, a process in which atomic nuclei split into fragments with the release of enormous quantities of energy, is a dramatic example of the actual conversion of matter into energy. Even in the familiar process of combustion, modern techniques of measurement have led to the discovery that a very minute quantity of matter is converted into energy; for example, about 0.00007 ounce of matter is converted into energy when 6 tons of carbon are burned with 16 tons of oxygen.

In spite of the mutual convertibility of energy and mass, the principle of the conservation of energy may still be regarded separately as the cornerstone of the science of thermodynamics. Machinery designed under this principle alone still functions in an orderly and predictable fashion.

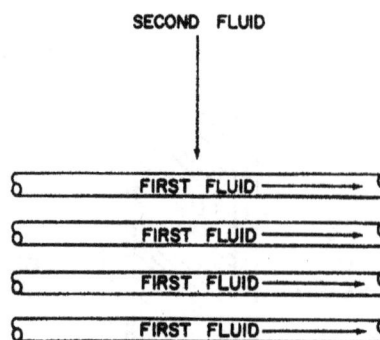

98.32
Figure 8-6.—Cross flow in heat exchanger.

accordance with the first law of thermodynamics, it is expressed as

$$J = 778 \text{ ft-lb per Btu}$$

or

$$J = \frac{778 \text{ ft-lb}}{1 \text{ Btu}}$$

The mechanical equivalent of heat provides us, directly or by extension, with a number of useful numerical values relating to heat, work, and power. Some of the most widely used values are given here; others may be obtained from engineering handbooks and similar publications.

1 Btu = 778 ft-lb
1 hp = 33,000 ft-lb per min = 550 ft-lb per sec
1 kw = 1.341 hp
1 hp = 2545 Btu per hr = 42.42 Btu per min
1 kw = 3413 Btu per hr
1 kw = 44,256 ft-lb per min
1 hp-hr = 2545 Btu
1 kw-hr = 3413 Btu

The first law of thermodynamics is often written in equation form as

$$U_2 - U_1 = Q - W$$

where

U_1 = internal energy of a system at the beginning of a process

U_2 = internal energy of the system at the end of the process

Q = net heat flowing into the system during the process

W = net work done by the system during the process

Another common statement of the first law of thermodynamics is that a perpetual motion machine of the first class is impossible. To understand the significance of this statement, it is necessary to understand the classification of perpetual motion machines. Although no perpetual motion machine exists—or, indeed, has ever been constructed—it is possible to conceive of three different categories. A perpetual motion machine of the first class is one which would put out more energy in the form of work than it absorbed in the form of heat. Since such a machine would actually create energy, it would violate the first law of thermodynamics and the principle of the conservation of energy. A perpetual motion machine of the second class would permit the reversal of irreversible processes and would thus violate the second law of thermodynamics, as discussed presently. A machine of the third class would be one in which absolutely no friction existed. Interestingly enough, there are no theoretical grounds for declaring that a machine of the third class is completely impossible; however, such a machine would be entirely contrary to our experience and would violate some of our profoundest convictions about the nature of energy and matter.

THERMODYNAMIC SYSTEMS

A thermodynamic system may be defined as a bounded region which contains matter. The boundaries may be fixed or they may vary in shape, form, and location. The matter within a system may be matter in any form—solid, liquid, or gas—or in some combination of forms. For some purposes, devices such as engines, pumps, boilers, and so forth may be regarded as being matter included within a thermodynamic system; for other purposes, each such device may be considered as a system in itself. A thermodynamic system may be entirely real, entirely imaginary, or a mixture of real and imaginary. A thermodynamic system may be capable of exchanging energy, in the form of heat and/or work, with its environs; or it may be an isolated system, in which case no heat can flow to or from the system and no work can be done on or by the system.

If a thermodynamic system appears to be a flexible thing, consider the further statement that " . . . a system may be said to be whatever one is talking about, and its environs are everything else."[13] Such flexibility of definition is entirely reasonable for most purposes. When we must account for energy, however, we will find it necessary to rigidly define and limit the system or systems under consideration. It is in terms of energy accounting, then, that the concept of a thermodynamic system is most useful.

A thermodynamic system requires a working substance to receive, store, transport, and deliver energy. In most systems, the working substance is a fluid—liquid, vapor, or gas.[14] The state of a thermodynamic system is specified by giving the values of two or more properties. These properties, which are called state variables or thermodynamic coordinates, include such common properties as pressure, temperature, volume, and mass, as well as more complex properties such as enthalpy and entropy (discussed later). Although some systems are adequately described by giving the value of only two variables, many systems require the specification of three or more variables.

THERMODYNAMIC PROCESSES

A thermodynamic process may be defined as any physical occurrence during which an effect is produced by the transformation or redistribution of energy. The occurrence of a thermodynamic process is evidenced by changes in some or all of the state variables of the system. The processes of most interest in engineering are those involving heat and work.

In connection with any process, it is usually necessary to consider the physical character of the process; the manner in which energy is transformed or redistributed as the process takes place; the kind and amount of energy that is stored in the system before and after the process, and the location of such energy; and the changes which are brought about in the system

---

[13]Kiefer, Kinney, and Stuart, Principles of Engineering Thermodynamics, 2nd ed., John Wiley & Sons, New York, 1954 (p. 32).

[14]Some writers use the term gas to indicate a gaseous substance that can be liquefied only by very large changes in pressure or temperature, reserving the term vapor for a gaseous substance that can be liquefied more easily, by slight changes of pressure or temperature. Other writers define a vapor as a gas which is in equilibrium with its liquid. For a great many purposes, the properties of a vapor are essentially the same as the properties of real gases; hence the distinction is not always important.

as the result of the process. It is also necessary to consider the energy exchanges that occur between the system and its surroundings during the process, since such energy exchanges will have an effect on the final state of the system.

The lifting of an object—as, for example, the lifting of a rock from the base of a cliff to the top of the cliff—is a simple example of a process involving work against gravity. Before the process begins, the energy which will be required to lift the rock is stored in some form in some other energy system. While the process is occurring, energy in the form of work flows from the external system to the earth-rock system. At the end of the process, the energy is stored in the earth-rock system in the form of mechanical potential energy. The change which has been brought about by this process is manifested by the separation of the rock and the earth.

Now suppose we push the rock off the top of the cliff and allow it to fall freely toward the base of the cliff. Disregarding the push (which is actually an input of energy from some external system), the process which now takes place is an example of work done by gravity. The work done by gravity converts the mechanical potential energy of the system into mechanical kinetic energy. Thus it is clear that energy in transition—work, in this case—begins and ends as stored energy.

When the rock hits the earth, other processes occur. Some work will be expended in compressing the earth upon which the rock falls, and some energy will then be stored as internal kinetic energy in the rock and in the earth. The increase in internal kinetic energy will be manifested by a rise in the temperature of the rock and of the earth, and still another process will then take place as heat flows from the rock and from the earth. Some energy may also be stored as internal potential energy because of molecular displacements in the rock and the earth.

The compression of a spring provides an example of a process involving elastic deformation. As force is applied to compress the spring, work is done. The major effect of the energy thus supplied as work is to decrease the distance between molecules in the spring, thus increasing the amount of internal potential energy stored in the spring. If we suddenly release the spring, the stored internal potential energy is suddenly released and the spring shoots away.

The turning of a shaft—as, for example, a propeller shaft of a ship—is another example of a process involving elastic deformation. Suppose that a strong twisting force is applied to a shaft at rest. The first part of this process will cause an elastic deformation of the shaft. The distance between molecules in the shaft is changed, and there is a storage of internal potential energy before the shaft begins to turn. When the applied force becomes great enough to turn the shaft, there will also be a storage of mechanical kinetic energy. As long as the applied force remains constant and the shaft continues to turn, these stored forms of energy will remain stored in unchanging amount. Meanwhile, a great deal of mechanical energy in transition (work) will continuously flow through the shaft to some other system.

When a solid body is dragged across a rough horizontal surface, the process is one of work against friction. The work done in moving the object will be equal to the force required to overcome the friction multiplied by the distance through which the object is moved. In this process, the energy supplied as work is transformed very largely into internal kinetic energy, as evidenced by an increase in temperature. Some of the energy may be transformed into internal potential energy because of molecular displacements in the object and in the surface over which it is being moved.

A propeller rotating in water is an example of a process in which work causes fluid turbulence. The first effect of the movement of the propeller is to impart various motions to the water, thus causing turbulence. For a short time this movement of the water represents mechanical kinetic energy, but the energy is rapidly transformed into internal kinetic energy, as evidenced by a rise in the temperature of the water.

The addition of thermal energy to a piece of metal is a simple example of a process involving heat. As the metal is heated, the temperature rises, indicating a storage within the metal of internal kinetic energy. Also, the metal expands; thus we know that some part of the energy delivered as heat is transformed into work as the metal expands against the resistance of its surroundings. If we continue heating the metal to its melting point, we will note a process in which the flow of heat results in a change in the physical state of the substance but does not, at this point, result in a further rise in temperature.

Because of the enormous number and variety of processes that may occur, some basic classification of processes involving heat and work is desirable. We will consider first a classification of processes according to the type of flow and then consider a classification according to the type of state change. Discussion of processes as "reversible" or "irreversible" is reserved for a later section.

Type of Flow

When classified according to type of flow of the working fluid, thermodynamic processes may be considered under the general headings of (1) non-flow processes, and (2) steady-flow processes.

A non-flow process is one in which the working fluid does not flow into or out of its container in the course of the process. The same molecules of the working fluid that were present at the beginning of the process are therefore present at the end of the process. Non-flow processes occur in reciprocating steam engines, air compressors, internal combustion engines, and other kinds of machinery. Since a piston-and-cylinder arrangement is typical of most non-flow processes, let us examine a non-flow process such as might occur in the cylinder shown in figure 8-7.

Suppose that we move the cylinder from position 1 to position 2, thereby compressing the fluid contained in the cylinder above the piston. Suppose, further, that we imagine this to be a completely ideal process, and one which

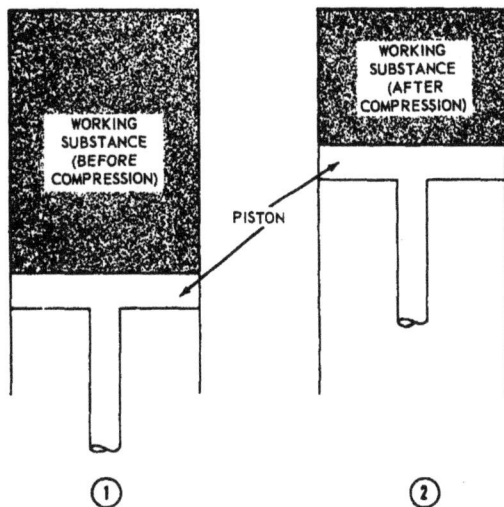

147.16.0
Figure 8-7.—Piston-and-cylinder arrangement for non-flow process.

is thus entirely without friction. The aspects of this process that we might want to know about are (1) the heat added or removed in the course of the process; (2) the work done on the working fluid or by the working fluid; and (3) the net change in the internal energy of the working substance.

From the general energy equation, we know that energy in must equal energy out. For the non-flow process, the general energy equation may be written as

$$Q_{12} = (U_2 - U_1) + \frac{Wk_{12}}{J} \text{ Btu}$$

where

Q_{12} = total heat transferred, in Btu (positive if heat is added during process, negative if heat is removed during process)

U_1 = total internal energy, in Btu, at state 1

U_2 = total internal energy, in Btu, at state 2

$U_2 - U_1$ = net change in internal energy from state 1 to state 2

Wk_{12} = work done between state 1 and state 2, in ft-lb (positive if work is done by the working substance, negative if work is done on the working substance)

J = the mechanical equivalent of heat, 778 ft-lb per Btu

$\frac{Wk_{12}}{J}$ = total work done by or on the working substance, in Btu (positive if work is done by the substance, negative if work is done on the substance)

This equation deals with total heat, total work, and total internal energy. If it is more convenient to make calculations in terms of 1 pound of the working substance, we would write the equation as

$$q_{12} = (u_2 - u_1) + \frac{wk_{12}}{J} \text{ Btu per lb}$$

where the value of J remains the same and where q, u, and wk have the general meanings

175

noted above but refer to the values for 1 pound of the working substance rather than to the values for the total quantity of the working substance. In both equations, it should be noted that the subscripts 1 and 2 refer to a separation in time rather than to a separation in space.

EXAMPLE: Four pounds of working substance are compressed in the cylinder shown in figure 8-7. The process is accomplished without the addition or removal of any heat but with a net increase in total internal energy of 120 Btu. Find the work done on or by the working substance, in Btu per pound and in foot-pounds per pound.

SOLUTION: First arrange the equation to fit the problem, as follows:

$$\frac{wk_{12}}{J} = (u_2 - u_1) + q_{12}$$

Since no heat is added or removed, $q_{12} = 0$. Since $U_2 - U_1$, or the net increase in total internal energy, is equal to 120 Btu, and since we are dealing with 4 pounds of the working substance, $u_2 - u_1 = \frac{120}{4} = 30$ Btu per pound.

The work done on or by the working substance, in Btu per pound, is given by the expression $\frac{wk_{12}}{J}$. Thus,

$$\frac{wk_{12}}{J} = (-30) + 0 \text{ Btu per lb}$$

$$= -30 \text{ Btu per lb}$$

The answer is negative, indicating that the work is done on the working substance rather than by the working substance.

To find the work done on the working substance in foot-pounds per pound, we merely solve the equation for wk_{12} rather than for $\frac{wk_{12}}{J}$ and substitute. Thus,

$$wk_{12} = (-30) (778) = -23,340 \text{ ft-lb per lb}$$

Again, the negative answer indicates that work is done on the working substance rather than by the working substance.

A steady-flow process is one in which a working substance flows steadily and uniformly through some device. Boilers, turbines, condensers, centrifugal pumps, blowers, and many other actual machines are designed for steady-flow processes. In an ideal steady-flow process, the following conditions exist:

1. The properties—pressure, temperature, specific volume, etc.—of the working fluid remain constant at any particular cross section in the flow system, although the properties obviously must change as the fluid proceeds from section to section.

2. The average velocity of the working fluid remains constant at any selected cross section in the flow system, although it may change as the fluid proceeds from section to section.

3. The system is always completely filled with the working fluid, and the total weight of the fluid in the system remains constant. Thus, for each pound of working fluid that enters the system during a given period of time, there is a discharge of 1 pound of fluid during the same period of time.

4. The net rate of heat transfer and the work performed on or by the working fluid remain constant.

In actual machinery designed for steady-flow processes, some of these conditions are not entirely satisfied at certain times. For example, a steady-flow machine such as a boiler or a turbine is not actually going through a steady-flow process until the warming-up period is over and the machine has settled down to steady operation. For most practical purposes, minor fluctuations of properties and velocities caused by load variations do not invalidate the use of steady-flow concepts. In fact, even such piston-and-cylinder devices as air compressors and reciprocating steam engines may be considered as steady-flow machines if there are enough cylinders or if some other arrangement is used to smooth out the flow so that it is essentially uniform at the inlet and the outlet.

The equations for steady-flow processes are based on the general energy equation—that is, energy in must equal energy out. Steady-flow equations are written in various ways, depending upon the forms of energy that are involved in the process under consideration. The forms of energy which, to greater or lesser degree, enter into any general equation for steady-flow processes are (1) internal energy, (2) heat, (3) mechanical potential energy, (4) mechanical kinetic energy, (5) work, and (6) flow work.

The first five of these energy terms are familiar, but the last one may be new. Flow work, sometimes called displacement energy, is the mechanical energy necessary to maintain the steady flow of a stream of fluid. The numerical value of flow work may be calculated by finding the product of the absolute pressure (in pounds per square feet) and the volume of the fluid (in cubic feet). Thus,

$$\text{flow work} = pV \text{ ft-lb}$$

or, more conveniently, using specific volume rather than total volume,

$$\text{flow work} = pv \text{ ft-lb per lb}$$

The product pv will, of course, have a numerical value even when there is no flow of fluid. However, this value represents flow work only when there is a steady, continuous flow of fluid. Flow work may also be expressed in terms of Btu per pound, as

$$\text{flow work} = \frac{pv}{J} \text{ Btu per lb}$$

As mentioned before, the steady-flow equations take various forms, depending upon the nature of the process under consideration. However, the terms for internal energy and flow work almost invariably appear in any steady-flow process. For convenience, this combination of internal energy and flow work has been given a name, a symbol, and units of measurement. The name is enthalpy (accent on second syllable). The symbol is H for total enthalpy or h for specific enthalpy—that is, enthalpy per pound. Total enthalpy, H, may be measured in Btu or in foot-pounds. Specific enthalpy (enthalpy per pound), h, may be measured in Btu per pound or in foot-pounds per pound. The enthalpy equation may be written as

$$H = \frac{pV}{J} + \text{Btu}$$

where

H = total enthalpy, in Btu
U = total internal energy, in Btu
p = absolute pressure, in pounds per square foot

V = total volume, in cubic feet
J = the mechanical equivalent of heat, 778 ft-lb per Btu

Since it is frequently more convenient in thermodynamics to make calculations in terms of 1 pound of the working substance, we should note also the equation for specific enthalpy:

$$h = u + \frac{pv}{J} \text{ Btu per lb}$$

here h, u, and v are specific enthalpy, specific internal energy, and specific volume, respectively. When it is desired to calculate enthalpy in foot-pounds, rather than in Btu, it is only necessary to drop the J from the equations.

The terms heat content and total heat are sometimes used to describe this property which we have designated as enthalpy. However, the terms heat content and total heat tend to be misleading because the change in enthalpy of a working fluid does not always measure the amount of energy transferred as heat, nor is it necessarily caused by the transfer of energy in the form of heat. Also, the transferred energy that causes a change in enthalpy is not entirely "contained" in the working fluid, as the terms heat content and total heat tend to imply; although the internal energy, u, is stored in the working fluid, the pv cannot in any way be considered as "contained" in the fluid.

Type of State Change

Thus far we have considered processes classified as non-flow or steady-flow. The nature of the state changes undergone by a working fluid provides us with another useful way of classifying processes. The terms used to identify certain common types of state changes are defined briefly in the following paragraphs.

ISOBARIC STATE CHANGES.—An isobaric state change is one in which the pressure of and on the working fluid is constant throughout the change. In other words, an isobaric change is a constant-pressure change. Isobaric changes occur in some piston-and-cylinder devices in which the piston operates in such a fashion as to maintain a constant pressure. Isobaric state changes are not typical of most steady-flow processes, but they are approximated in some steady-flow processes in which friction and shaft work are of insignificant magnitude.

An isobaric state change involves changes of enthalpy. One equation which has frequent application to isobaric state changes is written as

$$(q_{12})_p = h_2 - h_1$$

where

$(q_{12})_p$ = heat transferred between state 1 and state 2, with subscript p indicating constant pressure

h_1 = enthalpy of working fluid at state 1

h_2 = enthalpy of working fluid at state 2

ISOMETRIC STATE CHANGES. — A state change is said to be isometric when the volume (and the specific volume) of the working fluid is maintained constant. In other words, an isometric change is a constant-volume change. Isometric changes involve changes in internal energy, in accordance with the equation

$$q_v = u_2 - u_1$$

where

q_v = heat transferred, with subscript v indicating constant volume

u_1 = specific internal energy of working substance at state 1

u_2 = specific internal energy of working substance at state 2

ISOTHERMAL STATE CHANGES.—An isothermal change is one in which the temperature of the working fluid remains constant throughout the change.

ISENTHALPIC STATE CHANGES.—When the enthalpy of the working fluid does not change during the process, the change is said to be isenthalpic. Throttling processes are basically isenthalpic—that is, $h_1 = h_2$.

ISENTROPIC STATE CHANGES.—An isentropic state change is one in which there is no change in the property known as entropy. The significance of entropy and of isentropic state changes is discussed in a later section of this chapter.

ADIABATIC STATE CHANGES.—An adiabatic state change is one which occurs in such a way that there is no transfer of heat to or from the system while the process is occurring. In many real processes, adiabatic changes are produced by performing the process rapidly. Since heat transfer is relatively slow, any rapidly performed process can approach being adiabatic. Compression and expansion of working fluids are frequently achieved adiabatically. For an adiabatic process, the energy equation may be written as

$$U_2 - U_1 = W$$

where

U_1 = internal energy of working fluid at state 1

U_2 = internal energy of working fluid at state 2

W = work performed on or by the working fluid

In words, we may say that the net change of internal energy is equal to the work performed in an adiabatic process. The work term may be either positive or negative, depending upon whether work is done on the working substance, as in compression, or by the working substance, as in expansion.

THERMODYNAMIC CYCLES

A thermodynamic cycle is a recurring series of thermodynamic processes through which an effect is produced by the transformation or redistribution of energy. In other words, a cycle is a series of processes repeated over and over again in the same order.

All thermodynamic cycles may be classified as being open cycles or closed cycles. An open cycle is one in which the working fluid is taken in, used, and then discarded. A closed cycle is one in which the working fluid never leaves the cycle, except through accidental leakage; instead, the working fluid undergoes a series of processes which are of such a nature that the fluid is returned periodically to its initial state and is then used again.

The open cycle is exemplified by the internal combustion engine, in which atmospheric air supplies the oxygen for combustion and in which the exhaust products are returned to the atmosphere. In fact, another way to describe an open cycle is to say that it is one which includes the atmosphere at some point.

The closed cycle is exemplified by the condensing steam power plant used for ship propulsion on many naval ships. In such a cycle, the working substance (water) is changed to steam in the boilers. The steam performs work as it expands through the turbines, and is then condensed to water again in the condenser. The water is returned to the boilers as boiler feed, and is thus used over and over again.

Thermodynamic cycles are also classified as heated-engine cycles or as unheated-engine cycles, depending upon the point in the cycle at which heat is added to the working substance.[15] In a heated-engine cycle, heat is added to the working substance in the engine itself. An internal combustion engine has a heated-engine cycle. In an unheated-engine cycle, the working substance receives heat in some device which is separate from the engine. The condensing steam power plant has an unheated-engine cycle, since the working substance is heated separately in the boilers and then piped to the engines (steam turbines).

There are five basic elements in any thermodynamic cycle: (1) the working substance, (2) the engine, (3) a heat source, or high-temperature region, (4) a heat receiver, or low-temperature region, and (5) a pump.

The working substance is the medium by which energy is carried through the cycle. The engine is the device which converts the thermal energy of the working substance into useful mechanical energy in the form of work. The heat source supplies heat to the working substance. The heat receiver absorbs heat from the working substance. The pump moves the working substance from the low pressure side of the cycle to the high pressure side.

The essential elements of a closed, unheated-engine cycle are shown in figure 8-8. This is the basic plan of the typical condensing steam power plant.

---

[15]The terms heated engine and unheated engine should not be confused with the term heat engine. Any machine which is designed to convert thermal energy to mechanical energy in the form of work is known as a heat engine. Thus, both internal combustion engines and steam turbines are heat engines; but the first has a heated-engine cycle and the second has an unheated-engine cycle.

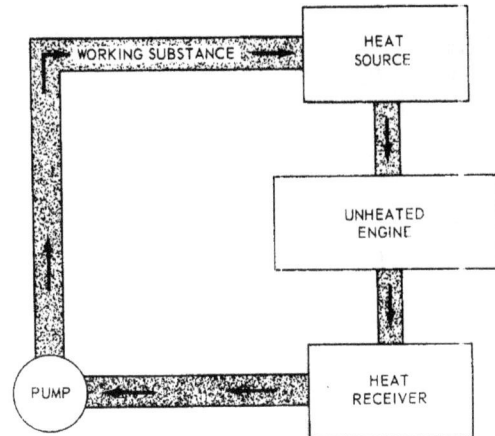

147.62

Figure 8-8.—Essential elements of closed, unheated-engine cycle.

In an open, heated-engine cycle such as that of an internal combustion engine, the essential elements are all present but are arranged in a somewhat different order. In this type of cycle, atmospheric air and fuel are both drawn into the cylinder of the engine. Combustion takes place in the cylinder, either by compression or by spark, and the resulting internal energy of the working substance is transformed into work by which the piston is moved. Since the space above the piston is a high pressure area when the piston is near the top of its stroke and a low pressure area when the piston is near the bottom, the piston may be thought of as a pump in the sense that it "pumps" the working fluid from the low pressure to the high pressure side of the system. Thus, in terms of function, the piston-and-cylinder arrangement may be thought of as including the heat source, the engine, and the pump. An open, heated-engine cycle might therefore be represented as shown in figure 8-9.

THE CONCEPT OF REVERSIBILITY

When we put a pan of water on the stove and turn on the heat, we expect the water to boil rather than to freeze. After we have mixed hot and cold water, we do not expect the resulting mixture to resolve itself into two separate batches of water at two different temperatures. When we open the valve on a cylinder of compressed air, we expect compressed air to rush out; we would be quite surprised if atmospheric air rushed into the cylinder and compressed

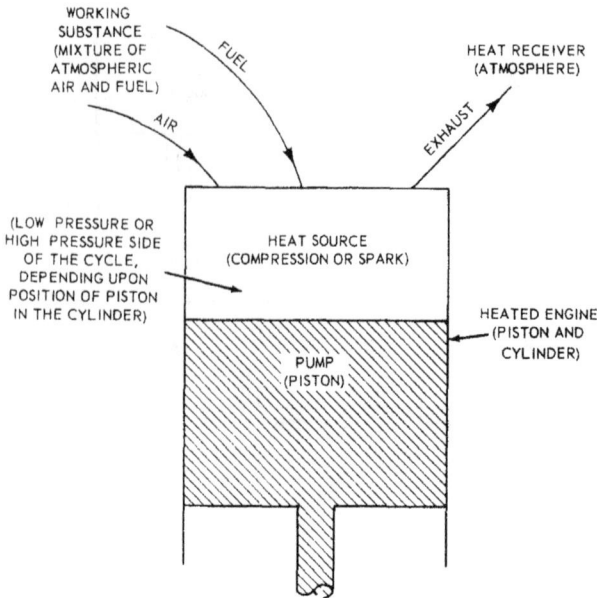

147.63

Figure 8-9.—Essential elements of open, heated-engine cycle.

itself. When a shaft is rotating, we expect a temperature rise in the bearings; when the shaft has been stopped, we would be truly amazed to observe internal energy from the bearings flowing to the shaft and causing it to start rotating again. When we drag a block of wood across a rough surface, we expect some of the mechanical energy expended in this act to be converted into thermal energy—that is, we expect a storage of internal energy in the wooden block and the rough surface, as evidenced by temperature rises in these materials. But if this stored internal energy should suddenly turn to and move the wooden block back to its original position, our incredulity would know no bounds.

All of which merely goes to show that we have certain expectations, based on experience, as to the direction in which processes will move. The reasonableness of our expectations is attested by the fact that in all recorded history there is no report of water freezing instead of boiling when heat is applied; there is no report of a lukewarm fluid unmixing itself and separating into hot and cold fluids; there is no report of a gas compressing itself without the agency of some external force; there is no report of the heat of friction being spontaneously utilized to perform mechanical work.

Are these actions really impossible? The first law of thermodynamics says that mechanical

energy and thermal energy are mutually convertible, but it says nothing about the direction of such conversions. If we consider only the first law, all the improbable actions just mentioned are perfectly possible and all processes could be thought of as being reversible. In an absolute sense, perhaps, we cannot guarantee that water will never freeze instead of boil when it is placed on a hot stove; but we are certainly safe in saying that this or any other completely reversible thermodynamic process is at the outer limits of probability. For all practical purposes, then, we will say that there is no such thing as a completely reversible process.

Nevertheless, the concept of reversibility is extremely useful in evaluating real thermodynamic processes. At this point, therefore, let us define a reversible thermodynamic process as one which would have the following characteristics: (1) the process could be made to occur in precisely reverse order, so that the energy system and all associated systems would be returned from their final condition to the conditions that existed before the process started; and (2) all energy that was transformed or redistributed during the process would be returned from its final to its original form, amount, and location.

THE SECOND LAW OF THERMODYNAMICS

Since the first law of thermodynamics does not deal with the direction of thermodynamic processes, and since experience indicates that actual processes are not reversible, it is apparent that the first law must be supplemented by some statement of principle that will limit the direction of thermodynamic processes. The second law of thermodynamics is such a statement. Although the second law is perhaps more empirical than the first law, and perhaps something less of a "law" in an absolute sense, it is of enormous practical value in the study of thermodynamics.[16]

The second law of thermodynamics may be stated in various ways. One statement, known as the Clausius statement, is that no process is possible where the sole result is the removal of heat from a low temperature reservoir and the absorption of an equal amount of heat by a high temperature reservoir. Among other things, this

[16]The interested student will find an excellent discussion of the second law of thermodynamics in Max Planck, Treatise on Thermodynamics, Dover Publications, New York, 1945. (A. Ogg, trans.)

statement indicates that water will not freeze when heat is applied. Note that the Clausius statement includes and goes somewhat beyond the common observation that heat flows only from a hotter to a colder substance.

The statement that no process is possible where the sole result is the removal of heat from a single reservoir and the performance of an equivalent amount of work is known as the Kelvin-Planck statement of the second law. Among other things, this statement says that we cannot expect the heat of friction to reverse itself and perform mechanical work. More broadly, this statement indicates a certain one-sidedness that is inherent in thermodynamic processes. Energy in the form of work can be converted entirely to energy in the form of heat; but energy in the form of heat can never be entirely converted to energy in the form of work.

A very important inference to be drawn from the second law is that no engine, actual or ideal, can convert all the heat supplied to it into work, since some heat must always be rejected to a receiver which is at a lower temperature than the source. In other words, there can be no heat flow without a temperature difference and there can be no conversion to work without a flow of heat. A further inference from this inference is sometimes given as a statement of the second law: No thermodynamic cycle can have a thermal efficiency of 100 percent.

We must say, then, that the first law of thermodynamics deals with the conservation of energy and with the mutual convertibility of heat and work, while the second law limits the direction of thermodynamic processes and the extent of heat-to-work energy conversions.

THE CONCEPT OF ENTROPY

The concept of reversibility and the second law of thermodynamics are closely related to the concept of entropy. In fact, the second law may be stated as: No process can occur in which the total entropy of an isolated system decreases; the total entropy of an isolated system can theoretically remain constant in some reversible (ideal) processes, but in all irreversible (real) processes the total entropy of an isolated system must increase.

From other statements of the second law, we know that the transformation of heat to work is always dependent upon a flow of heat from a high temperature region to a low temperature region. The concept of the unavailability of a certain portion of the energy supplied as heat to any thermodynamic system is clearly implied in the second law, since it is apparent that some heat must always be rejected to a receiver which is at a lower temperature than the source, if there is to be any conversion of heat to work. The heat which must be so rejected is therefore unavailable for conversion into mechanical work.

Entropy is an index of the unavailability of energy. Since heat can never be completely converted into work, we may think of entropy as a measure or an indication of how much heat must be rejected to a low temperature receiver if we are to utilize the rest of the heat for the production of useful work. We may also think of entropy as an index or measure of the reversibility of a process. All real processes are irreversible to some degree, and all real processes involve a "growth" or increase of entropy. Irreversibility and entropy are closely related; any process in which entropy has increased is an irreversible process.

The entropy of an isolated system is at its maximum value when the system is in a state of equilibrium. The concept of an absolute minimum—that is, an absolute zero—value of entropy is sometimes referred to as the third law of thermodynamics (or Nernst's law). This principle states that the absolute zero of entropy would occur at the absolute zero of temperature for any pure material in the crystalline state. By extension, therefore, it should be possible to assign absolute values to the entropy of pure materials, if such absolute values were needed. For most purposes, however, we are interested in knowing the values of the changes in entropy rather than the absolute values of entropy. Hence an arbitrary zero point for entropy has been established at 32° F.

Entropy changes depend upon the amount of heat transferred to or from the working fluid, upon the absolute temperature of the heat source, and upon the absolute temperature of the heat receiver. Although actual entropy calculations are complex beyond the scope of this text, one equation is given here to indicate the units in which entropy is measured and to give the relationship between entropy and heat and temperature. Note that this equation applies only to a reversible isothermal process in which $T_1 = T_2$.

$$S_2 - S_1 = \frac{Q}{T}$$

where

S_1 = total entropy of working fluid at state 1, in Btu per ° R

S_2 = total entropy of working fluid at state 2, in Btu per °R

Q = heat supplied, in Btu

T = absolute temperature at which process takes place, in ° R

The fact that the total entropy of an isolated system must always increase does not mean that the entropy of all parts of the system must always increase. In many real processes, we find increases in entropy in some parts of a system and, at the same time, decreases in entropy in other parts of the system. But the important thing to note is that the increases in entropy are always greater than the decreases; therefore, the total entropy of an isolated system must always increase.

Each increase in entropy is permanent. In a universal sense, entropy can be created but it can never be destroyed or gotten rid of, although it may be transferred from one system to another. Every natural process that occurs in the universe increases the total entropy of the universe, and this increase in entropy is irreversible. The concept of the universe eventually "running down" might be expressed in terms of entropy by saying that the entropy of the universe is constantly "building up." The so-called "heat death of the universe" is envisioned as the ultimate result of all possible natural processes having taken place and the universe being in total equilibrium, with entropy at the absolute maximum. Such a statement need not imply a total lack of energy remaining in the universe; but any energy that might remain would be completely unavailable and therefore completely useless.

THE CARNOT PRINCIPLE

According to the second law of thermodynamics, no thermodynamic cycle can have a thermal efficiency of 100 percent—that is, no heat engine can convert into work all of the energy that is supplied as heat. The question now arises as to how much heat must be rejected to a receiver which is at a lower temperature than the source? Or, looking at it another way, what is the maximum thermal efficiency that could

theoretically be achieved by a heat engine operating without friction and without any other of the irreversible processes that must occur in all real machines?

To answer this question, Carnot, a French engineer, developed an imaginary and completely reversible cycle. In the Carnot cycle, all heat is supplied at a single high temperature and all heat that must be rejected is rejected at a single low temperature. The cycle is fully reversible. When proceeding in one direction, the Carnot cycle takes in a certain amount of heat, rejects a certain amount of heat, and puts out a certain amount of work. When the cycle is reversed, the quantity of work that was originally the output of the cycle is now put into the cycle; the amount of heat that was originally taken in is now the amount rejected; and the amount of heat that was originally rejected is now the amount taken in. When thus reversed, the cycle is called a Carnot refrigeration cycle.

Obviously, no real machine is capable of such complete reversibility, but the concept of the Carnot cycle is nonetheless an extremely useful one. By analysis of the Carnot cycle, it can be proved that no engine, actual or ideal, can be more efficient than an ideal, reversible engine operating on the ideal, reversible Carnot cycle. The thermal efficiency of the Carnot cycle is given by the equation

$$\text{thermal efficiency} = \frac{\text{work output}}{\text{heat input}} = \frac{T_s - T_r}{T_s}$$

where T_s equals the absolute temperature at which heat flows from the source to the working fluid and T_r equals the absolute temperature at which heat is rejected to the receiver.

The implications of this statement are of profound importance, since it establishes the fact that thermal efficiency depends only upon the temperature difference between the heat source and the heat receiver. Thermal efficiency does not depend upon the properties of the working fluid, the type of engine used in the cycle, or the nature of the process—combustion, nuclear fission, etc.—that produces the heat at the heat source. The basic principle thus established by analysis of the Carnot cycle is called the Carnot principle, and may be stated as follows: The motive power of heat is independent of the agents employed to realize it, its quantity being fixed solely by the temperatures of the bodies between which the transfer of heat occurs.

WORKING SUBSTANCES

As previously noted, a thermodynamic system requires a working substance to receive, store, transport, and deliver energy. The working substance is almost always a fluid and is therefore frequently referred to as the working fluid. Water (together with its vapor, steam) is one of the most commonly used working fluids, although air, ammonia, carbon dioxide, and a wide variety of other fluids are used in certain kinds of systems. A working substance may change its physical state during the course of a thermodynamic cycle or it may remain in one state, depending upon the nature of the cycle and the processes involved.

To understand the behavior of working fluids, we should have some understanding of the laws of perfect gases, of the relationships between liquids and their vapors, and of the ways in which the properties of working fluids may be represented and tabulated. These topics are discussed in the following sections.

Laws of Perfect Gases

The relationships of the volume, the absolute pressure, and the absolute temperature in the hypothetical substances known as "perfect gases" were stated by the physicists Boyle and Charles in the form of various gas laws. The laws thus established may be combined and summarized in the general statement: For a given weight of any gas, the product of the absolute pressure and the volume, divided by the absolute temperature, is a constant. Or, in equation form,

$$\frac{pV}{T} = \frac{p_1V_1}{T_1} = \frac{p_2V_2}{T_2} = R$$

where

    p = absolute pressure
    V = total volume
    T = absolute temperature
    R = the gas constant

Although the laws of perfect gases were developed on the basis of experiments made with air and other real gases, later experiments showed that these relationships do not hold precisely for real gases over the entire range of pressures and temperatures. However, air and other gases used as working fluids may be treated as perfect gases over quite a wide range of pressures and temperatures without any appreciable error being introduced. Values of the gas constant for some common gases are:

    Air. 53.3
    Oxygen 48.3
    Nitrogen 55.0
    Hydrogen. 766.0
    Helium 386.0

Liquids and Their Vapors

When heat is transferred to a liquid, the average velocity of the molecules is increased and the amount of internal kinetic energy stored in the liquid is increased. As the average velocity of the molecules increases, some molecules which are at or near the surface of the liquid momentarily achieve unusually high velocities; and some of these escape from the liquid and enter the space above, where they exist in the vapor state. As more and more of the molecules escape and come into the vapor state, the probability increases that some of the vapor molecules will momentarily have unusually low velocities; these molecules will be captured by the liquid. As a result of this exchange of molecules between the liquid and the vapor, a condition of equilibrium is reached and an equilibrium pressure is established. The equilibrium pressure depends upon the molecular structure of the fluid and upon its temperature. For any given fluid, therefore, there is a definite relationship between the temperature and the pressure at which a liquid and its vapor may exist in equilibrium contact with each other.

As long as the vapor is in contact with the liquid from which it is being generated, the liquid and the vapor will remain at the same temperature. If the liquid and the vapor are in a closed container (such as a boiler with all steam stop valves closed) both the temperature and the pressure of the liquid and its vapor will increase as heat is added. If the vapor is permitted to leave the steam space at a rate equal to the evaporation rate, an equilibrium will be established at the equilibrium pressure for the particular temperature.

The pressure and the temperature which are related in the manner just described are known as the saturation pressure and the saturation temperature. Thus, for any specified pressure there is a corresponding temperature of vaporization known as the saturation temperature;

and for any specified temperature there is a corresponding saturation pressure.

A liquid which is under any specified pressure and at the saturation temperature for that particular pressure is called a saturated liquid. A liquid which is at any temperature below its saturation temperature is said to be subcooled liquid. For example, the saturation temperature which corresponds to atmospheric pressure (14.7 psia) is 212°F for water. Therefore, water at 212°F and under atmospheric pressure is said to be a saturated liquid. Water flowing in a river or standing in a pond is also under atmospheric pressure, but it is at a much lower temperature; hence, this water is said to be subcooled.

A vapor which is under any specified pressure and at the saturation temperature corresponding to that pressure is said to be a saturated vapor. Thus, water at 14.7 psia and 212°F produces a vapor known as saturated steam. As previously noted, it is impossible to raise the temperature of a vapor above the temperature of its liquid as long as the two are in contact. If the vapor is drawn off into a separate container, however, and additional heat is supplied to the vapor, the temperature of the vapor is raised. A vapor which has been raised to a temperature that is above its saturation temperature is called a superheated vapor, and the vessel or container in which the saturated steam is superheated is called a superheater. The elementary boiler and superheater illustrated in figure 8-10 show the general principle of generating and superheating steam. Practically all naval propulsion boilers have superheaters for superheating the saturated steam generated in the generating sections of the boiler; the steam is then called superheated steam. The amount by which the temperature of a superheated vapor

exceeds the temperature of a saturated vapor at the same pressure is known as the degree of superheat. For example, if saturated steam at a pressure of 600 psia and a corresponding saturation temperature of 486°F is superheated to 786°F, the degree of superheat is 300°F.

For any substance there is a critical point at which the properties of the saturated liquid are exactly the same as the properties of the saturated vapor. For water, the critical point is reached at 3206.2 psia (critical pressure) and 704.40°F (critical temperature). At the critical point, the vapor and the liquid are indistinguishable. No change of physical state occurs when the pressure is increased or when additional heat is supplied; the vapor cannot be made to liquefy and the liquid cannot be made to vaporize as long as the substance is at or above its critical pressure and critical temperature. At this point, we could no longer refer to "water" and "steam", since we cannot tell the water and the steam apart; instead, the substance is now merely called a "fluid" or a "working fluid." Boilers designed to operate above the critical point are called supercritical boilers. Supercritical boilers are not used at present in the propulsion plants of naval ships; however, some boilers of this type are used in stationary steam power plants.

Representation of Properties

The condition of a working fluid at any point within a thermodynamic cycle or system is established by the properties of the substance at that point. The properties that are of special interest in engineering thermodynamics include pressure, temperature, volume, enthalpy, entropy, and internal energy. These properties

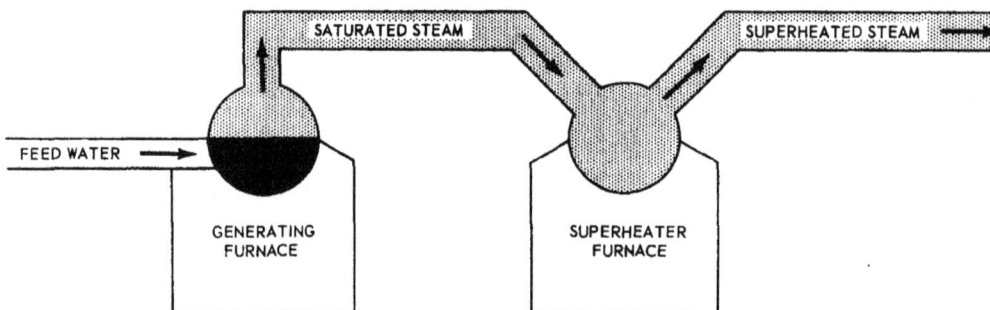

Figure 8-10.—Elementary boiler and superheater.

38.3

PROPERTIES OF SATURATED STEAM

ABS. PRESS. (PSIA) p	TEMP.°F t	SPECIFIC VOLUME		ENTHALPY			ENTROPY			INTERNAL ENERGY	
		SAT. LIQUID v_f	SAT. VAPOR v_g	SAT. LIQUID b_f	EVAP. b_{fg}	SAT. VAPOR b_g	SAT. LIQUID s_f	EVAP. s_{fg}	SAT. VAPOR s_g	SAT. LIQUID u_f	SAT. VAPOR u_g
190	377.51	0.01833	2.404	350.79	846.8	1197.6	0.5381	1.0116	1.5497	350.15	1113.1
200	381.79	0.01839	2.288	355.36	843.0	1198.4	0.5435	1.0018	1.5453	354.68	1113.7
250	400.95	0.01865	1.8438	376.00	825.1	1201.1	0.5675	0.9588	1.5263	375.14	1115.8
(1)	(2)	(3)	(4)	(5)	(6)	(7)	(8)	(9)	(10)	(11)	(12)

147.64X

Figure 8-11.—Excerpts from Keenan and Keyes steam tables.

have been discussed at some length in this chapter and in the chapter dealing with principles of measurement; at this point we are concerned less with the properties themselves than with the way in which they are tabulated and the way in which they are represented graphically.

STEAM TABLES.—In the region near a change of physical state, the behavior of a gaseous substance becomes too complex for the relatively simple energy calculations that apply to perfect gases and to many real gases over a wide range of pressures and temperatures. Because of the complicated equations needed to describe the properties of vapors, engineers customarily depend upon tables of vapor properties for information concerning the properties of liquids and their vapors. The vapor tables that are perhaps most commonly used are those which give the thermodynamic properties of steam. The most authoritative tables of thermodynamic properties of steam are those prepared by Keenan and Keyes under the title of Thermodynamic Properties of Steam.[17] Figure 8-11 is excerpted from Table III of the Keenan and Keyes steam tables; it is included here chiefly to show the general arrangement of information in these tables, rather than to provide any significant amount of data concerning the thermodynamic properties of saturated steam.

The information given in each column of the Keenan and Keyes table for the properties of saturated steam is described briefly below. Note that the subscript f is commonly used to denote properties of the saturated liquid, g to denote properties of the saturated vapor, and fg to denote property changes between the two states.

Column 1 gives the saturation pressure of the saturated water and the saturated steam at the temperature given in column 2. Note that the pressure is absolute pressure, not gage pressure.

Column 2 gives the saturation temperature for the pressure shown in column 1. This temperature is what is commonly referred to as "the boiling point" of the liquid at the pressure shown in column 1. In the steam tables, the temperature is usually given in degrees Fahrenheit rather than in degrees of an absolute temperature scale. However, the absolute temperature can always be obtained by simple computation if it should be needed.

Column 3 gives the specific volume of the saturated liquid (water) at the pressure shown in column 1 and the temperature shown in column 2. Specific volume is expressed in cubic feet per pound.

Column 4 gives the specific volume of the saturated vapor (steam) at the pressure and temperature shown.

It should be noted that some portions of the Keenan and Keyes steam tables have another column for the increase in specific volume that occurs during evaporation. This column is labelled Specific Volume, Evapo., symbolized by v_{fg}.

[17]Joseph H. Keenan and Frederick G. Keyes, Thermodynamic Properties of Steam, New York, John Wiley & Sons, Inc., 1937. Excerpts from Table III are reprinted through the courtesy of the publisher.

Column 5 gives the enthalpy per pound of the saturated liquid at the pressure and temperature shown. The enthalpy of saturated water at 32°F and the corresponding saturation pressure of 0.08854 psis is taken as zero; hence, all enthalpy figures indicate enthalpy with respect to this arbitrarily assigned zero point. For example, the enthalpy of 1 pound of saturated water at 190 psia and 377.51°F is 350.79 Btu more than the enthalpy of 1 pound of saturated water at 0.08854 psia and 32°F.

Column 6 gives the enthalpy of evaporation, per pound of working fluid—that is, the change in enthalpy that occurs during evaporation. This column is of particular significance since it indicates the Btu per pound that must be supplied to change the saturated liquid (water) to the saturated vapor (steam) at the pressure and temperature shown. In other words, the enthalpy of evaporation is what we formerly described as the latent heat of vaporization.

Column 7 gives the enthalpy per pound of the saturated vapor at the pressure and temperature shown. Note that this is the sum of the enthalpy of the saturated liquid and the enthalpy of evaporation.

Column 8 gives the entropy per pound of the saturated liquid at the pressure and temperature shown. The zero point for entropy, like the zero point for enthalpy, is arbitrarily established at 32°F and the corresponding saturation pressure of 0.08854 psia.

Column 9 gives the entropy of evaporation per pound of working fluid at the indicated pressure and temperature. In other words, this column shows the change of entropy that occurs during evaporation.

Column 10 gives the entropy per pound of the saturated vapor at the pressure and temperature shown. Note that this is the sum of the entropy of the saturated liquid and the entropy of evaporation.

Columns 11 and 12 give the internal energy of the saturated liquid and the internal energy of the saturated vapor, respectively.

In addition to giving properties of the saturated liquid and the saturated vapor, the Keenan and Keyes steam tables include data on the superheated vapor and other pertinent information.

GRAPHICAL REPRESENTATION OF PROPERTIES.—It is frequently useful to show the relationship between two or more properties of a working fluid by means of thermodynamic graphs or diagrams.

The relationships among pressure, volume, and temperature of a perfect gas are sometimes represented by a three-dimensional diagram of the type shown in figure 8-12. The p-v-T surface of a real substance may also be represented in this way, but the diagrams become much more complex because the relationships among the properties are more complex. A simplified p-v-T surface for water is shown in figure 8-13.

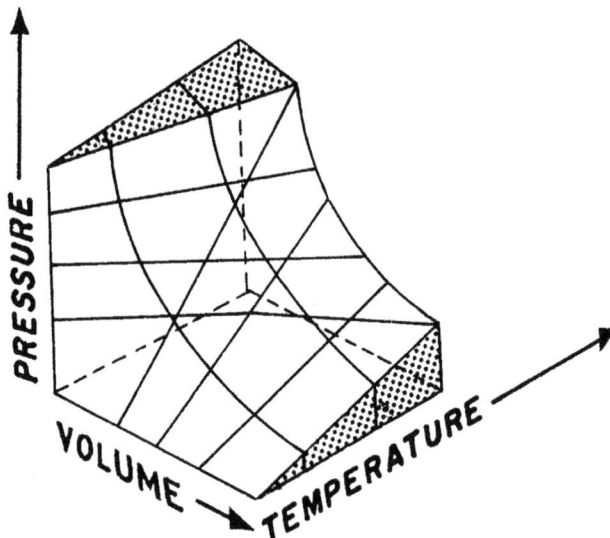

147.65

Figure 8-12.—Three-dimensional representation of p-v-T surface for perfect gas.

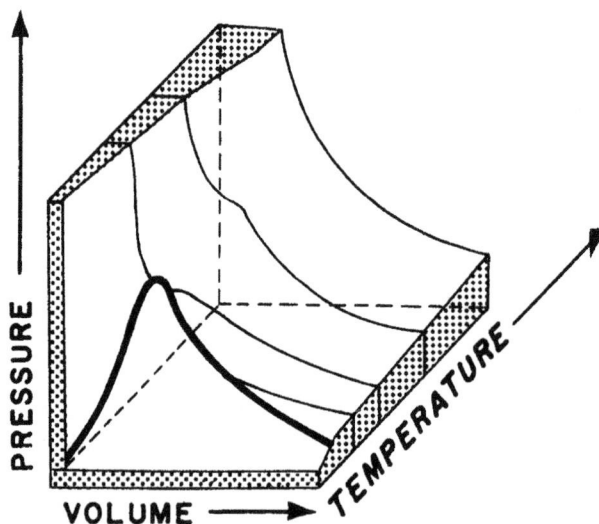

147.66

Figure 8-13.—Three-dimensional representation of p-v-T surface for water.

The significance of some of the lines on this diagram will become clearer as we consider some of the related two-dimensional diagrams.

Three-dimensional diagrams are extremely useful in giving an overall picture of the p-v-T relationships, but they are difficult to construct and are somewhat difficult to use for detailed analysis. Two-dimensional graphs are frequently projected from the three-dimensional p-v-T surfaces. Even on a two-dimensional diagram, a great many relationships of properties can be indicated by means of contour lines or superimposed curves.

The p-v diagram is made by plotting known values of pressure (p) along the ordinate and values of specific volume (v) along the abscissa.[18] To illustrate the construction of a p-v diagram, let us consider the isothermal compression of 1 pound of air from an initial pressure of 1000 pounds per square foot absolute to a final pressure of 6000 psfa. Let us assume that the air is at a temperature of 90°F, or 550° R. Since we may treat air as a perfect gas under these conditions of pressure and temperature, we may use the laws of perfect gases and the equation

$$pv = RT$$

where

p = absolute pressure, psfa
v = specific volume, cu ft per lb
R = gas constant (53.3 for air)
T = absolute temperature, °R

Since the compression is isothermal, T is constant and the expression RT is equal to 53.3 x 550, or 29,315. It is apparent from the equation that p and v must vary inversely—that is, as p goes up, v goes down. Hence, for any given value of p we may find a value of v merely by dividing 29,315 by p. Choosing six values of p and computing the values of v, we obtain the following values:

STATE A: p = 1000, v = 29.3
STATE B: p = 2000, v = 14.7
STATE C: p = 3000, v = 9.8
STATE D: p = 4000, v = 7.3
STATE E: p = 5000, v = 5.9
STATE F: p = 6000, v = 4.9

---

[18] The p-v diagram, as it applies to internal combustion engines, is discussed further in chapter 22 of this text.

By plotting these values on graph paper, we obtain the p-v diagram shown in figure 8-14. The curve applies only to the indicated temperature—that is, it is an isothermal curve. The values of p and v may be calculated for the same process at other temperatures, and plotted as before; in this case we obtain a series of isothermal curves (or isotherms) such as those shown in figure 8-15.

A p-v diagram for water and steam is shown in figure 8-16. This diagram—and, in fact, most diagrams for real substances in the region of a state change—is not drawn to scale because of the very great difference in the specific volume of the liquid and the specific volume of the vapor. Even though it is not drawn to scale, the p-v diagram serves a useful purpose in indicating the general configuration of the saturated liquid line and the saturated vapor line. These lines, which are called process lines, blend smoothly at the critical point. The shape formed by the process lines is characteristic of water and will be observed on all p-v diagrams of this substance.

A two-dimensional pressure-temperature (p-T) diagram of the type shown in figure 8-17 is useful because it indicates the way in which the phase of a substance depends upon pressure and temperature. The solid-liquid curve, for example, indicates the effects of pressure on

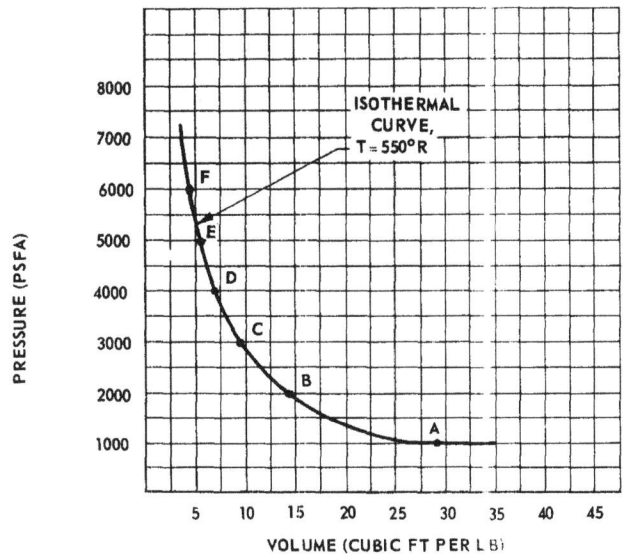

147.67

Figure 8-14.—Constant temperature (isothermal) line on p-v diagram.

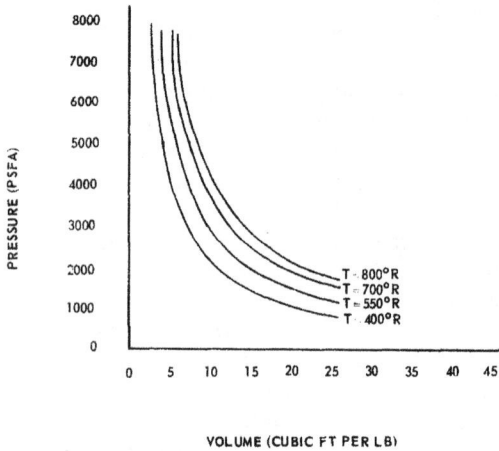

147.68

Figure 8-15.—Group of isothermal curves on p-v diagram.

the melting (or freezing) point; the liquid-vapor curve indicates the effects of pressure on the boiling point; and the solid-vapor indicates the effects of pressure on the sublimation point.

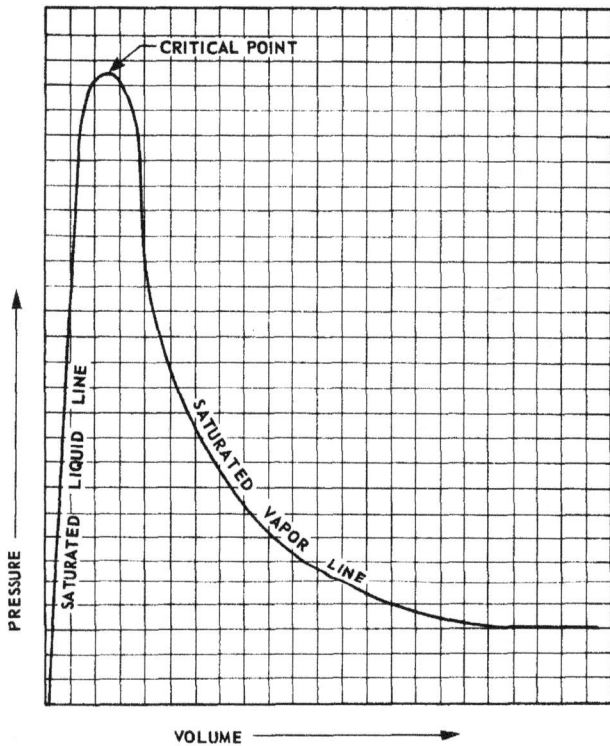

147.69

Figure 8-16.—A p-v diagram for water.

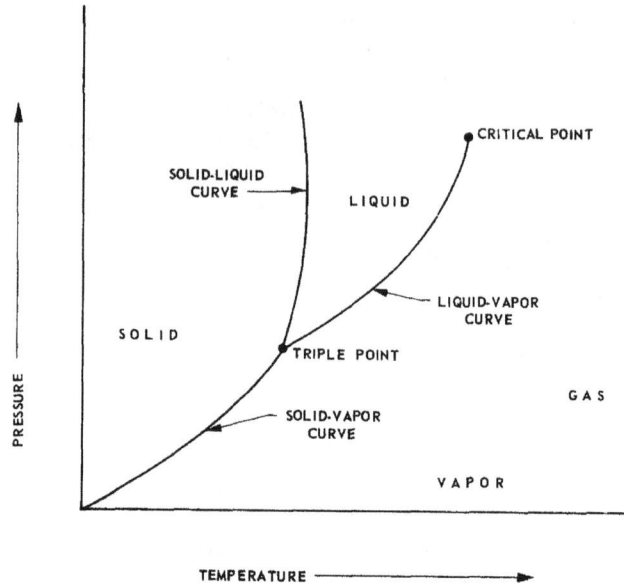

147.70

Figure 8-17.—A p-T diagram.

The intersection of these three equilibrium curves shows the triple point—that is, the single pressure and temperature at which all three phases can coexist. The termination of the liquid-vapor equilibrium curve indicates the critical point—that is, the point at which the liquid and the vapor are no longer distinguishable because their properties are identical.

Other two-dimensional diagrams that find application in engineering include the temperature-entropy (T-s) diagram; the enthalpy-entropy (h-s) diagram, also called the Mollier diagram; the pressure-enthalpy (p-h) diagram; and the enthalpy-volume (h-v) diagram. Of these, the Mollier diagram is probably of major importance in the study of steam engineering. Mollier diagrams are included in many steam tables and are also available in engineering handbooks and some thermodynamics texts.

ENERGY RELATIONSHIPS IN THE
SHIPBOARD PROPULSION CYCLE

At the beginning of this chapter it was stated that the shipboard engineering plant may be thought of as a series of devices and arrangements for the exchange and transformation of energy. Many of these transformations and energy exchanges have been discussed in this chapter, but they have not been taken up in sequence. Figure 8-18 illustrates the basic

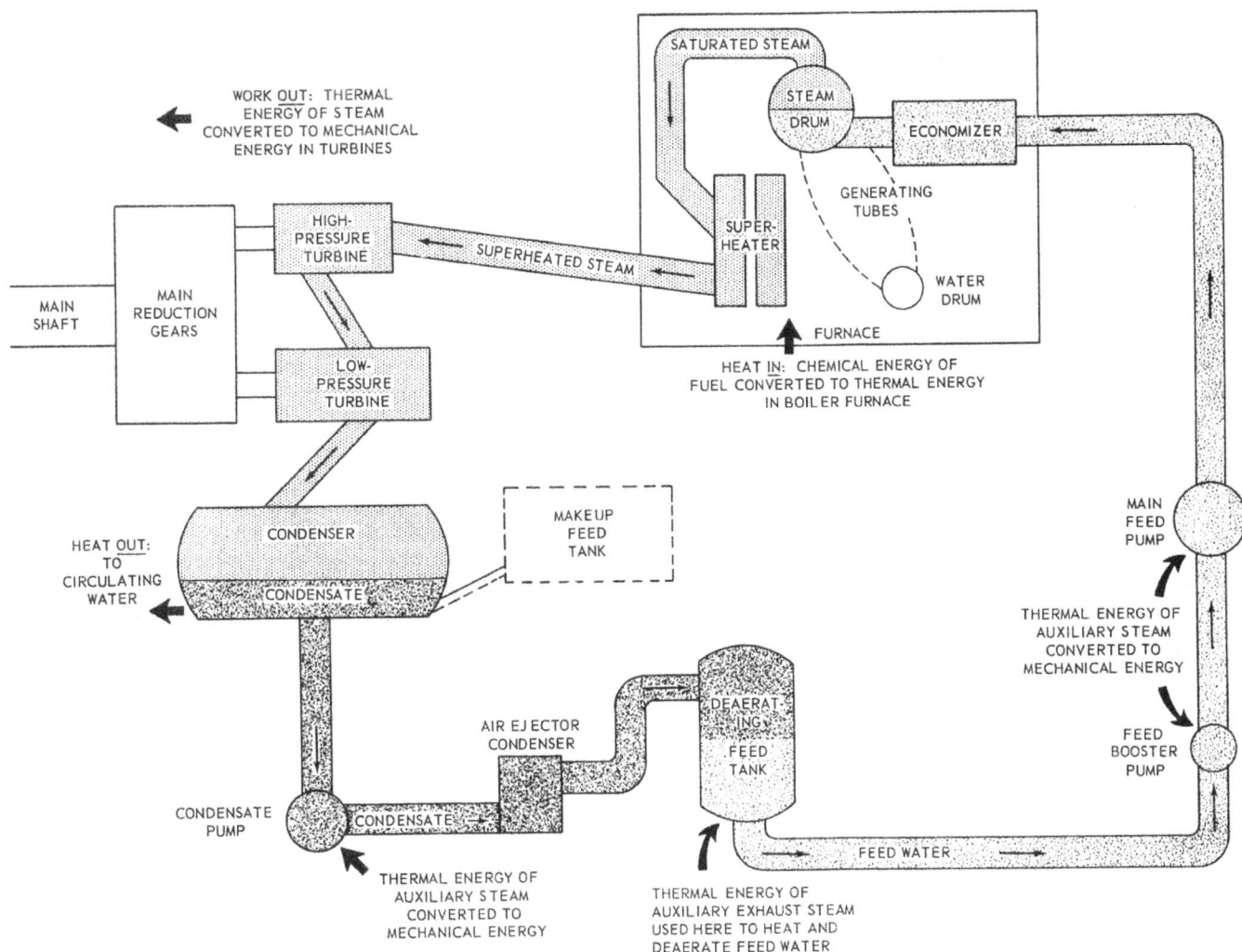

38.2

Figure 8-18.—Energy relationships in the basic propulsion cycle of conventional steam-driven ship.

propulsion cycle of a conventional steam-driven ship with geared turbine drive and shows some of the major energy transformations that take place.

The first energy transformation occurs when fuel oil is burned in the boiler furnace. By the process of combustion, the chemical energy stored in the fuel oil is transformed into thermal energy. Thermal energy flows from the hot combustion gases to the water in the boiler. While the boiler stop valves are still closed, steam begins to form in the boiler; the volume of the steam remains constant but the pressure and temperature increase, indicating a storage of internal energy. When operating pressure is reached and the steam stop valves are opened,

the high pressure of the steam causes it to flow to the turbines. The pressure of the steam thus provides the potential for doing work; the actual conversion of heat to work takes place in the turbines. The changes in internal energy between the boiler and the condenser (as evidenced by changes in pressure and temperature) indicate that heat has been converted to work in the turbines. The work output of the turbines turns the shaft and so drives the ship.

Two main energy transformations are involved in converting thermal energy to work in the turbines. First, the thermal energy of the steam is transformed into mechanical kinetic energy as the steam flows through one or more nozzles. And second, the mechanical kinetic energy of the steam is transformed into work as the steam impinges upon the projecting blades of the turbine and thus causes the turbine to turn. The turning of the turbine rotor causes the propeller shaft to turn also, although at a slower speed, since the turbine is connected to the propeller shaft through reduction gears. The steam exhausts from the turbine to the condenser, where it gives up its latent heat of condensation to the circulating sea water.

For the remainder of this cycle, energy is required to get the water (condensate and feed water) back to the boiler where it will again be heated and changed into steam. The energy used for this purpose is generally the thermal energy of the auxiliary steam. In the case of turbine-driven feed pumps, the conversion of thermal energy to mechanical energy occurs in the same way as it does in the case of the propulsion turbines. In the case of motor-driven pumps, the energy conversion is from thermal energy to electrical energy (in a turbogenerator) and then from electrical energy to mechanical energy (work) in the pumps.

ENERGY BALANCES

From previous discussion, it should be apparent that putting 1 Btu in at the boiler furnace does not mean that 778 foot-pounds of work will be available for propelling the ship through the water. Some of the energy put in at the boiler furnace is used by auxiliary machinery such as pumps and forced draft blowers to supply the boiler with feed water, fuel oil, and combustion air. Distilling plants, turbogenerators, steering gears, steam catapults, heating systems, galley and laundry equipment, and many other units throughout the ship use energy derived directly or indirectly from the energy put in at the boiler furnace.

In addition, there are many "energy losses" throughout the engineering plant. As we have seen, energy cannot actually be lost. But when it is transformed into a form of energy which we cannot use, we say there has been an energy loss. Since no insulation is perfect, some thermal

energy is always lost as steam travels through piping. Friction losses occur in all machinery and piping. Some heat must be wasted as the combustion gases go up the stack. Some heat must be lost at the condenser as the steam exhausted from the turbines gives up heat to the circulating sea water. We cannot expect all of the heat supplied to be converted into work; even in the most efficient possible cycle, we know that some heat must always be rejected to a receiver which is at a lower temperature than the source. Thus, each Btu that is theoretically put in at the boiler furnace must be divided up a good many ways before the energy can be completely accounted for. But the energy account will always balance. Energy in must always equal energy out.

Designers of engineering equipment use energy balances to analyze energy exchanges and to compute the energy requirements for proposed equipment or plants. Operating engineers use energy balances to evaluate plant performance. The engineer officer of a naval ship may find it necessary to make energy balances in order to find out whether the plant is operating at designed efficiency or whether defects are causing unnecessary waste of steam, fuel, and energy.

An energy balance for an entire engineering plant is usually made up in the form of a flow diagram similar to (but more detailed than) the one shown in figure 8-18. A number of numerical values are entered on the flow diagram, the most important of which are the quantities of the working fluid flowing per hour at various points and the thermodynamic states of the working fluid at various points. The quantity of fluid flowing per hour may be obtained by direct measurement of flow through flow meters or nozzles or by calculation; in some instances, it is necessary to estimate steam consumption of pumps and other units on the basis of available test data. Data on the state of the working fluid is obtained from pressure and temperature readings. Enthalpy calculations are made and noted at various points on the diagram. The complete energy balance includes tabular data as well as the data shown on the flow diagram.

PART III—THE CONVENTIONAL STEAM TURBINE PROPULSION PLANT

This part of the text deals with the major units of machinery in the conventional steam turbine propulsion plant—a type of plant which is at present widely used in naval ships. For the most part, the discussion is concerned with geared-turbine drive; but some of the information is also applicable to those few ships with turboelectric drive. The term "conventional" is used here to indicate that the plants under discussion utilize conventional boilers, rather than nuclear reactors, as the source of heat for the generation of steam.

Chapter 9 introduces the conventional steam turbine propulsion plant by taking up the arrangement of propulsion machinery and the major engineering piping systems found aboard conventional steam-driven ships. Chapters 10 and 11 deal with propulsion boilers and their fittings and controls. Chapter 12 describes propulsion steam turbines. Chapter 13 discusses the condensers and other heat transfer apparatus used in the condensate and feed system of the conventional steam turbine propulsion plant.

As may be noted, the sequence of presentation follows the sequence of the thermodynamic cycle. The boiler is the heat source, or high-temperature region; the turbine is the engine in which the thermal energy of the steam is converted into mechanical energy which drives the ship; and the condenser is the heat receiver to which some heat must always be rejected in order to allow the conversion of heat to work.

CHAPTER 9

MACHINERY ARRANGEMENT AND PLANT LAYOUT

To understand a shipboard propulsion plant, it is necessary to visualize the general configuration of the plant as a whole and to understand the physical relationships among the various units. This chapter provides general information on the distribution and arrangement of propulsion machinery in conventional steam turbine propulsion plants and on the arrangement of the major engineering piping systems that connect and serve the various units of machinery.

It is important to note that the information given in this chapter is general rather than specific. No two ships—not even sister ships—are exactly alike in their arrangement of machinery and piping. The examples given in this chapter are based on the arrangements used in various kinds of ships, large and small, old and new. The examples give some idea of the variety of arrangements that may be found on steam-driven surface ships, and they indicate the basic functions of the machinery and piping; but the examples cannot provide an exact picture of the machinery and piping on any one ship. For detailed information concerning the arrangements on any particular ship, it is necessary to consult the ship's blueprints, various ship's manuals, and the manufacturers' technical manuals that cover the engineering equipment and piping systems installed in the ship.

ARRANGEMENT OF PROPULSION MACHINERY

The propulsion machinery on conventional steam-driven surface ships includes (1) the propulsion boilers, (2) the propulsion turbines, (3) the condensers, (4) the reduction gears, and (5) the pumps, forced draft blowers, deaerating feed tanks, and other auxiliary machinery units which directly serve the major propulsion units.

On most steam-driven surface ships other than oilers, tankers, and certain auxiliaries, the propulsion machinery is located amidships. Turbogenerators and their auxiliary condensers are usually located in the propulsion machinery spaces; other engineering equipment that is not directly associated with the operation of the major propulsion units may be located in or near the propulsion machinery spaces or in other parts of the ship, as space permits.

A word about terminology may be helpful at this point. The boilers in a propulsion plant may be identified as propulsion boilers (or occasionally as main boilers) when it is necessary to distinguish between propulsion boilers and the auxiliary boilers that are installed on some ships. The turbines are identified as propulsion turbines when it is necessary to distinguish between them and the many auxiliary turbines that are used on all steam-driven ships to drive pumps, forced draft blowers, and other auxiliary units. The propulsion turbines are also sometimes referred to as the main engines, although this usage is not considered particularly desirable. The term propulsion unit is correctly used to identify the combination of propulsion turbines, main reduction gears, and main condenser in any one propulsion plant; however, the term propulsion unit may also be used in a more general sense to indicate any major unit in the propulsion plant.

Each propulsion shaft has an identifying number which is based on the location of the shaft, working from starboard to port. The shaft nearest the starboard side is the No. 1 shaft, the one next inboard is the No. 2 shaft, and so forth. On recent ships, the propulsion machinery that serves each shaft is given the same number as that shaft. For example, the No. 2 shaft is served by the No. 2 propulsion unit and the No. 2 boiler. Where two similar units serve one shaft, the identifying number is followed by a letter. If two boilers serve the No. 3 propulsion unit and the No. 3 shaft, for example, the boilers would be identified as No. 3A and No. 3B. Where letters are used, they are used in sequence going from starboard to port and then from forward to aft.

On older ships, the practice of identifying propulsion units by the number of the shaft they serve is slightly different. In general, each propulsion unit is numbered to correspond with the number of the shaft it serves; but the numbering of the boilers is generally not the same as the numbering of the propulsion units and the shafts. On an older ship, for example, the No. 1 boiler and the No. 2 boiler might serve the No. 1 propulsion unit and the No. 1 shaft, while the No. 3 boiler and the No. 4 boiler would serve the No. 2 propulsion unit and the No. 2 shaft.

The functional relationships of the major propulsion units and of many auxiliaries are shown in figure 9-1. This illustration does not indicate the actual location of the machinery units; indeed, the physical location is often surprisingly different from the location that might be assumed from a diagram of this type. In considering the physical arrangement of machinery, however, we must keep the functional relationships clearly in mind. The three major piping systems shown in figure 9-1 are the main steam system, the auxiliary steam system, and the auxiliary exhaust system; again, a functional rather than a physical relationship is indicated. The three systems are discussed in more detail later in this chapter; at this point it is only necessary to note the relationships of these vital systems to the propulsion units and auxiliaries.

The propulsion machinery spaces may be physically arranged in several ways. Some ships have firerooms, containing boilers and the stations for operating them, and enginerooms, containing propulsion turbines and the stations for operating them. On some ships, one fireroom serves one engineroom; on others, two firerooms serve one engineroom. Instead of firerooms and enginerooms, many large ships of recent design have spaces which are called machinery rooms. Each machinery room contains both the boilers and the propulsion turbines that serve a particular shaft. On some recent ships that have certain automatic controls, the propulsion machinery is very largely operated from separate enclosed operating stations located within the machinery room.

No matter what arrangement of machinery spaces is used, the propulsion machinery is usually on two levels. The condensers and the main reduction gears are on the lower level. The propulsion turbines and the high speed pinion gears to which they are connected are on the upper level with the low pressure turbine exhaust directly over the condenser. The boilers occupy both the lower level and the upper level; the stations for firing the boilers (sometimes referred to as "the firing aisle") are on the lower level, while the stations for operating the valves that admit feed water to the boilers are on the upper level. The boilers are usually located on the centerline of the ship or else they are distributed symmetrically about the centerline. The long axis of the boiler drums runs fore and aft rather than athwartship. Other machinery, including the propulsion auxiliaries, is arranged in various ways as space and weight considerations permit.

Figure 9-2 shows the general arrangement of propulsion machinery on destroyers of the DD 445 and DD 692 classes. The machinery is arranged so that the forward fireroom and the forward engineroom can be operated together as one completely independent plant, while the after fireroom and the after engineroom can be operated together as another completely independent plant. All propulsion machinery, including auxiliaries, is duplicated in each plant. The arrangement shown in figure 9-2 is typical of most destroyers, even the newer ones; however, the newer destroyers contain a non-machinery separation space between the forward and after machinery plants.

Figures 9-3, 9-4, 9-5, and 9-6 show the arrangement of machinery in the No. 1 fireroom and the No. 1 engineroom of the frigates DLG 14 and DLG 15. The arrangement shown in these illustrations is also typical of that in the frigates DLG 6-13. The forward (No. 1) fireroom and engineroom may be operated together as a separate plant, as may the after (No. 2) fireroom and engineroom.

Figure 9-7 shows the general arrangement of propulsion machinery on the CA 68 class of heavy cruisers. This arrangement is typical of cruisers commissioned during World War II. The two forward firerooms and the forward engineroom constitute one plant; the two after firerooms and the after engineroom constitute the other plant. Cross-connections make it possible for other operational arrangements to be used.

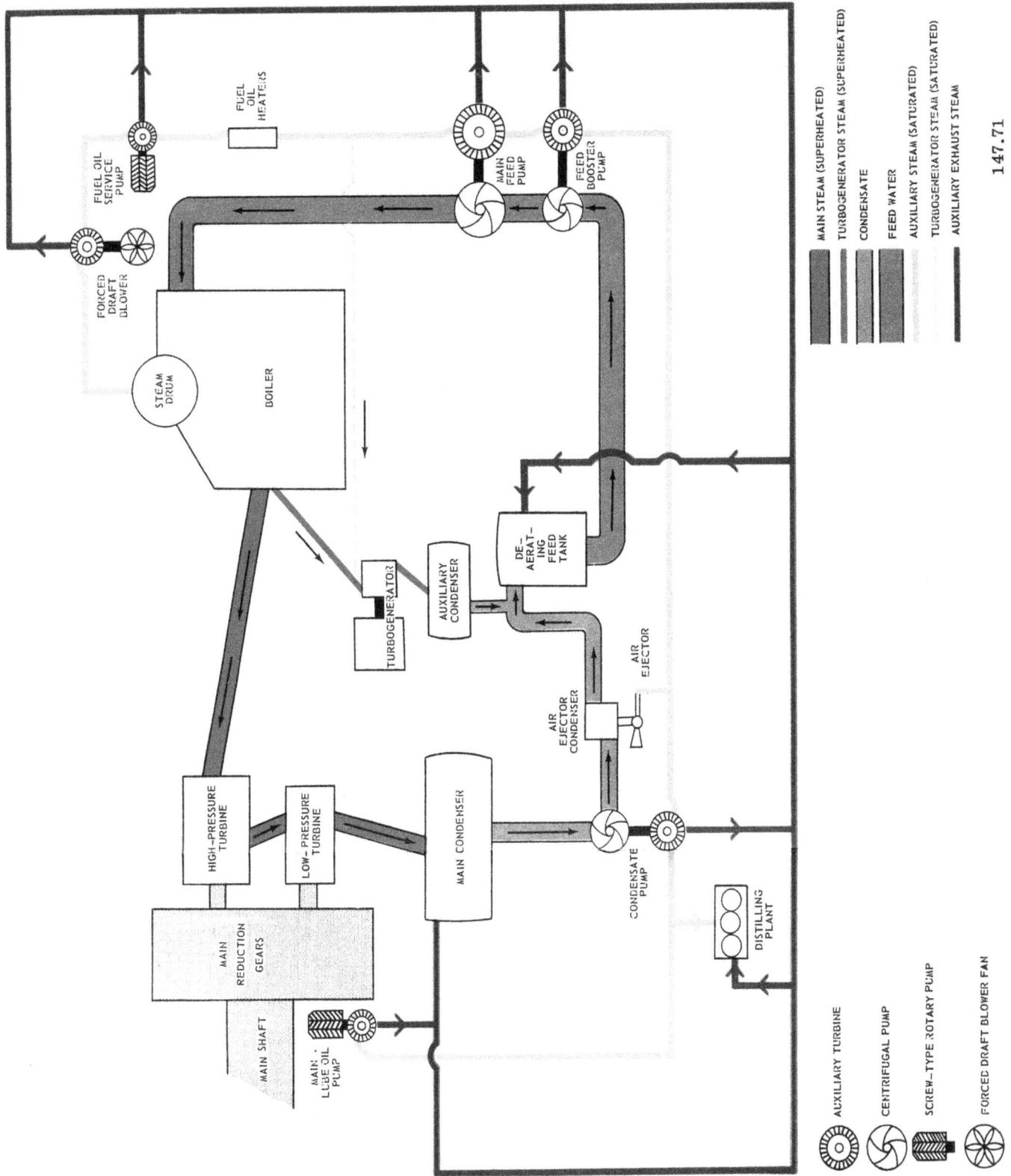

Figure 9-1.— Functional relationships of propulsion units, auxiliaries, main steam system, auxiliary steam system, and auxiliary exhaust system (Facing page 194).

147.71

MAIN STEAM (SUPERHEATED)

TURBOGENERATOR STEAM (SUPERHEATED)

CONDENSATE

FEED WATER

AUXILIARY STEAM (SATURATED)

TURBOGENERATOR STEAM (SATURATED)

AUXILIARY EXHAUST STEAM

AUXILIARY TURBINE

CENTRIFUGAL PUMP

SCREW-TYPE ROTARY PUMP

FORCED DRAFT BLOWER FAN

FUEL OIL SERVICE PUMP

FUEL OIL HEATERS

MAIN FEED PUMP

FEED BOOSTER PUMP

FORCED DRAFT BLOWER

STEAM DRUM

BOILER

TURBOGENERATOR

AUXILIARY CONDENSER

DE-AERAT-ING FEED TANK

AIR EJECTOR

AIR EJECTOR CONDENSER

HIGH-PRESSURE TURBINE

LOW-PRESSURE TURBINE

MAIN CONDENSER

CONDENSATE PUMP

DISTILLING PLANT

MAIN REDUCTION GEARS

MAIN SHAFT

MAIN LUBE OIL PUMP

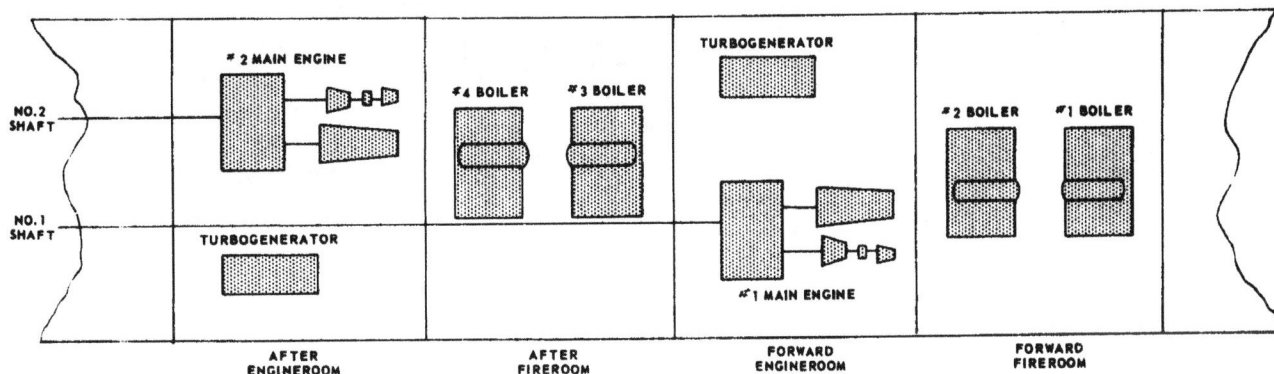

38.4

Figure 9-2.—Propulsion machinery arrangement, DD 445 and DD 692 classes.

Figure 9-8 shows the arrangement of propulsion machinery on the USS Coral Sea, CVA 43. In some ways, this arrangement of machinery represents the ultimate in designed segregation of propulsion equipment. The major units of machinery are duplicated, spread out, and compartmented to provide for maximum resistance to damage from explosion, fire, or flooding.

In recent years there has been a trend toward using machinery rooms, rather than firerooms and enginerooms, on many of the larger combatant ships. An example of the machinery room type of arrangement is shown in figure 9-9. Each machinery room contains a separate propulsion plant which is capable of independent operation. The arrangement shown in figure 9-9 is that of a heavy cruiser of the CA 139 class; the same general arrangement is used on many newer ships, including aircraft carriers.

ENGINEERING PIPING SYSTEMS

The various units of machinery and equipment aboard ship are connected by miles of piping. Each piping system consists of sections of pipe or tubing, fittings for joining the sections, and valves for controlling the flow of fluid. Most piping systems also include a number of other fittings and accessories such as vents, drains, traps, strainers, relief valves, gages, and instruments. Piping system components are discussed in chapter 14 of this text; in the present chapter, we are concerned with piping system standard symbols, piping system markings, general arrangement and layout of the major engineering piping systems aboard ship.

Piping system standard symbols are used to indicate machinery units, piping connections, valves, gages, strainers, steam traps, and other items on engineering blueprints and drawings. Figure 9-10 illustrates some of the standard symbols specified by the governing Military Standard (MIL-STD-17). In some cases, deviation from these symbols occurs on blueprints and drawings; but the basic principles of representation are usually followed. Most plans or drawings that utilize special symbols include a legend or list of symbols.

Standard piping system markings are used to mark each shipboard piping system at suitable intervals along the entire length of the system. The markings may be applied with paint and stencils or prepainted vinyl cloth markers may be used. The markings are in black letters on a white background for all systems except oxygen; oxygen systems are marked with white letters on a dark background.

The piping identification markings must include the functional name of the system and, where necessary, the specific service of the system. Markings must also include arrows to show the direction of flow.

The piping identification markings are not required for piping in tanks, voids, cofferdams, bilges, and other unmanned spaces. All other piping must be marked at least once in each manned space and at least twice in each machinery space. Systems serving propulsion plants and systems conveying flammable or toxic fluids must be marked at least twice in each space. When feasible, piping identification markings are placed near the entry and near the exit to any space and at the

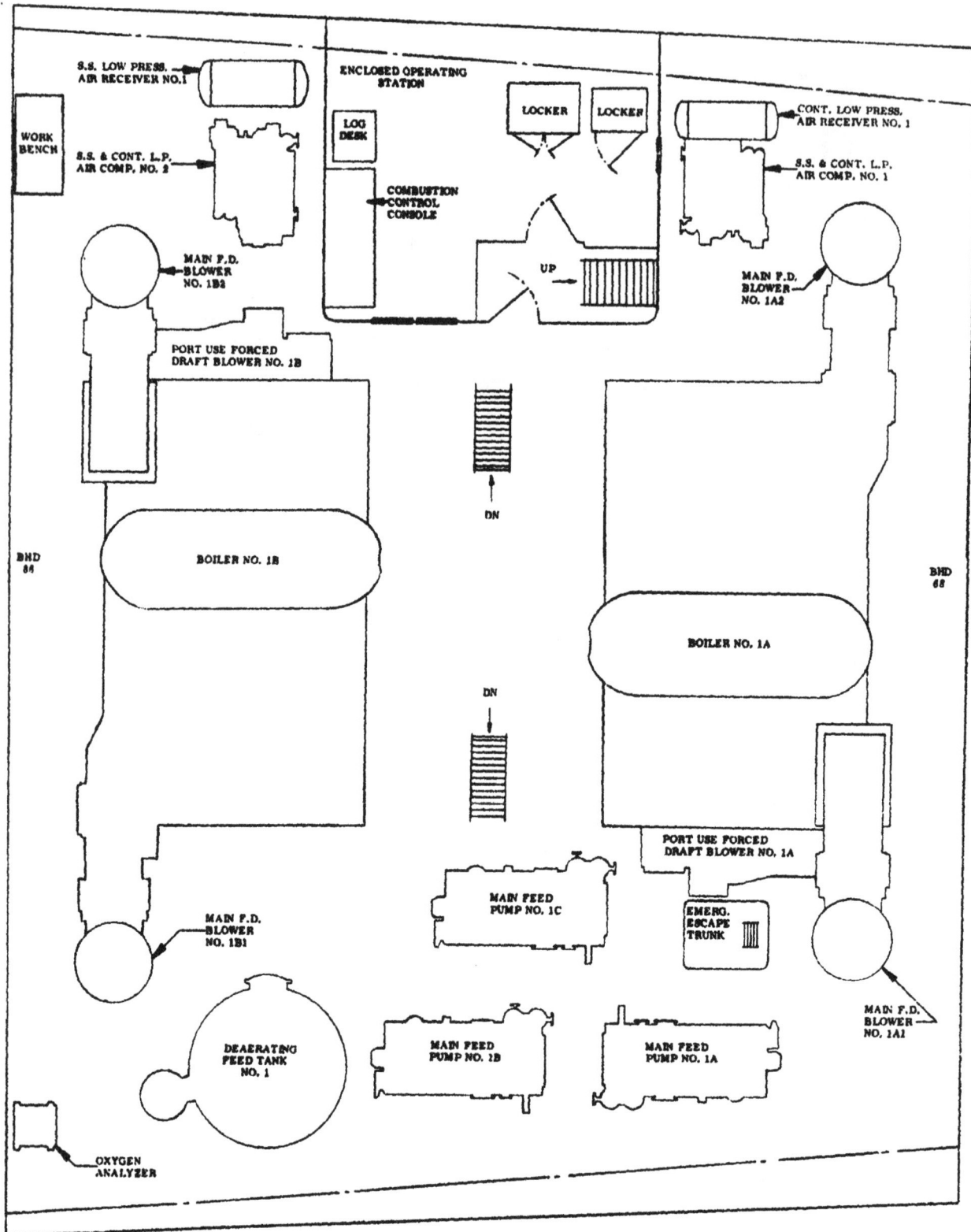

Figure 9-3.—Arrangement of machinery on upper level of No. 1 fireroom, 147.72
DLG 14 and DLG 15.

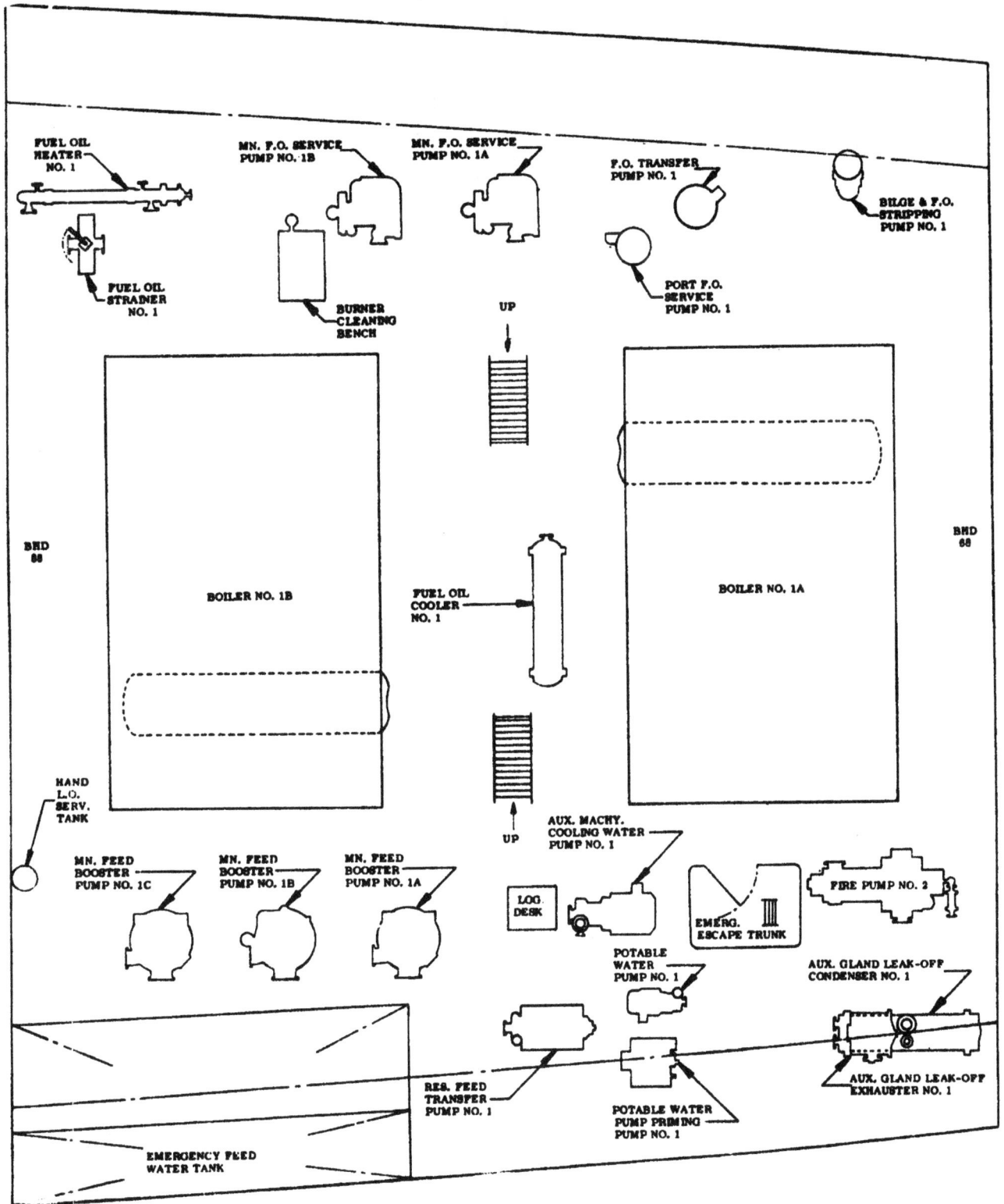

147.73

Figure 9-4.—Arrangement of machinery on lower level of No. 1 fireroom, DLG 14 and DLG 15.

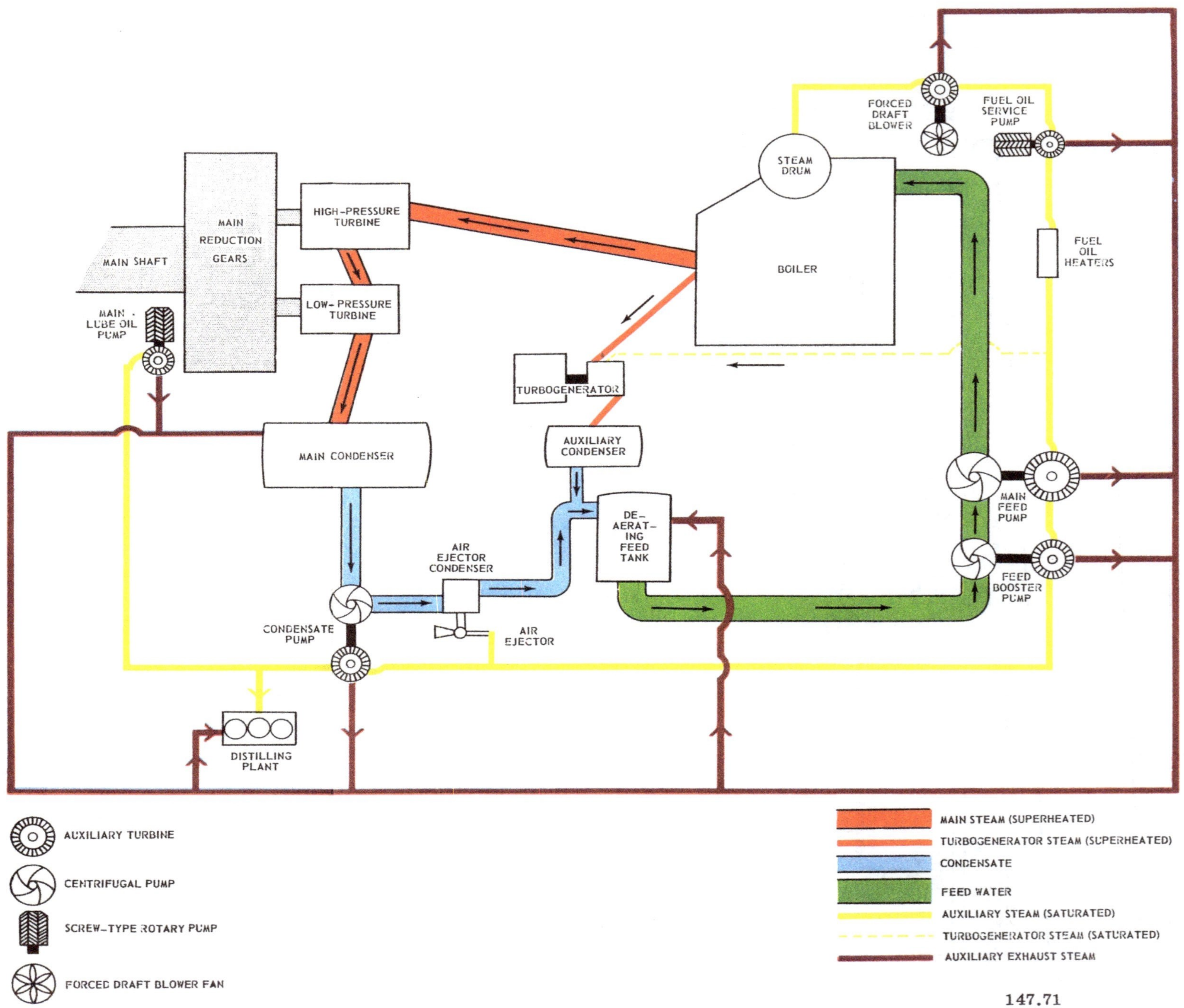

FUEL OIL HEATERS

FUEL OIL SERVICE PUMP

FORCED DRAFT BLOWER

STEAM DRUM

BOILER

MAIN FEED PUMP

FEED BOOSTER PUMP

TURBOGENERATOR

AUXILIARY CONDENSER

DE-AERAT-ING FEED TANK

AIR EJECTOR

AIR EJECTOR CONDENSER

HIGH-PRESSURE TURBINE

LOW-PRESSURE TURBINE

MAIN CONDENSER

CONDENSATE PUMP

DISTILLING PLANT

MAIN REDUCTION GEARS

MAIN SHAFT

MAIN LUBE OIL PUMP

147.71

MAIN STEAM (SUPERHEATED)

TURBOGENERATOR STEAM (SUPERHEATED)

CONDENSATE

FEED WATER

AUXILIARY STEAM (SATURATED)

TURBOGENERATOR STEAM (SATURATED)

AUXILIARY EXHAUST STEAM

AUXILIARY TURBINE

CENTRIFUGAL PUMP

SCREW-TYPE ROTARY PUMP

FORCED DRAFT BLOWER FAN

Figure 9–1. — Functional relationships of propulsion units, auxiliaries, main steam system, auxiliary steam system, and auxiliary exhaust system (Facing page 194).

198

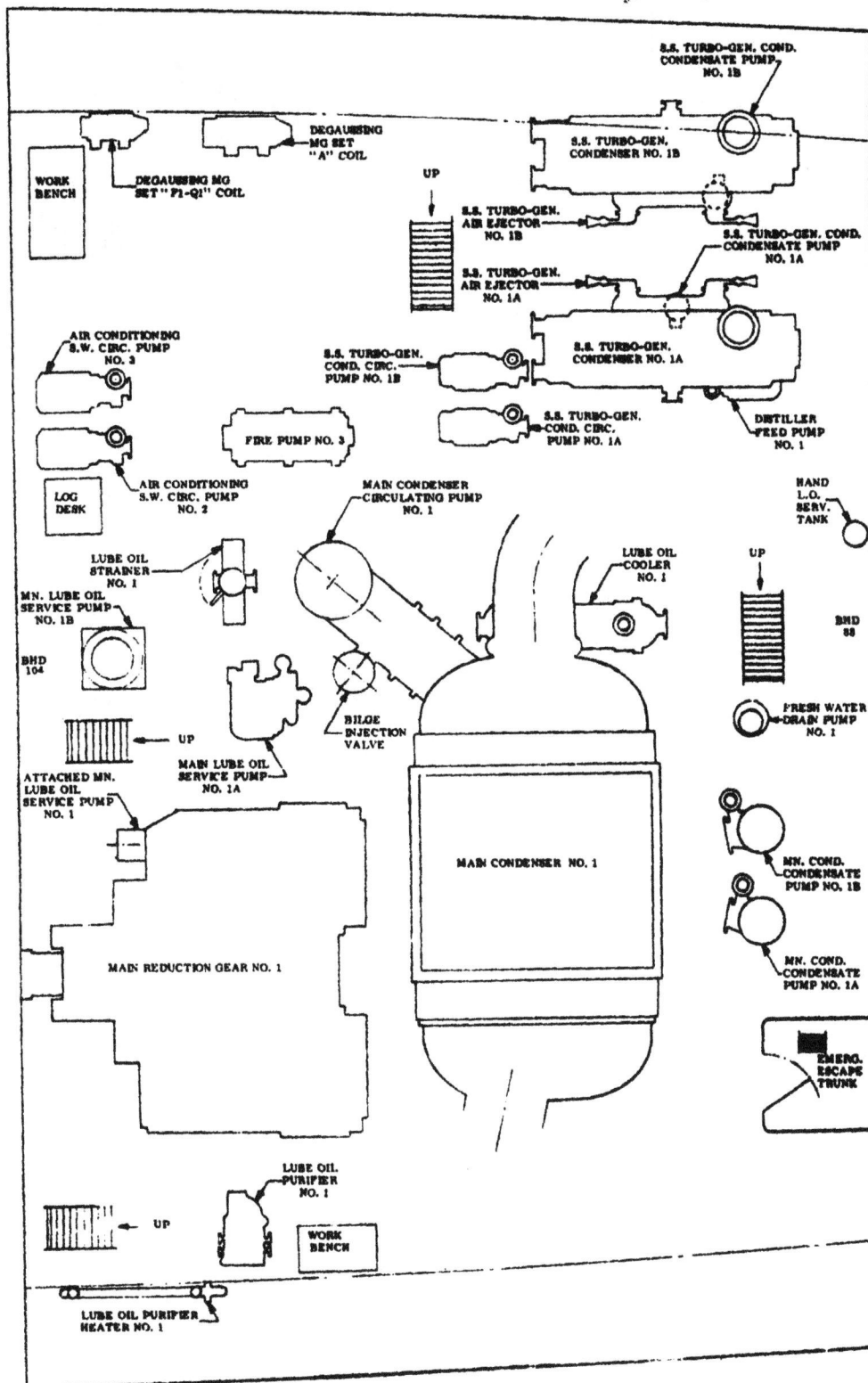

Figure 9-6.—Arrangement of machinery on lower level of No. 1 engineroom, DLG 14 and DLG 15.

147.75

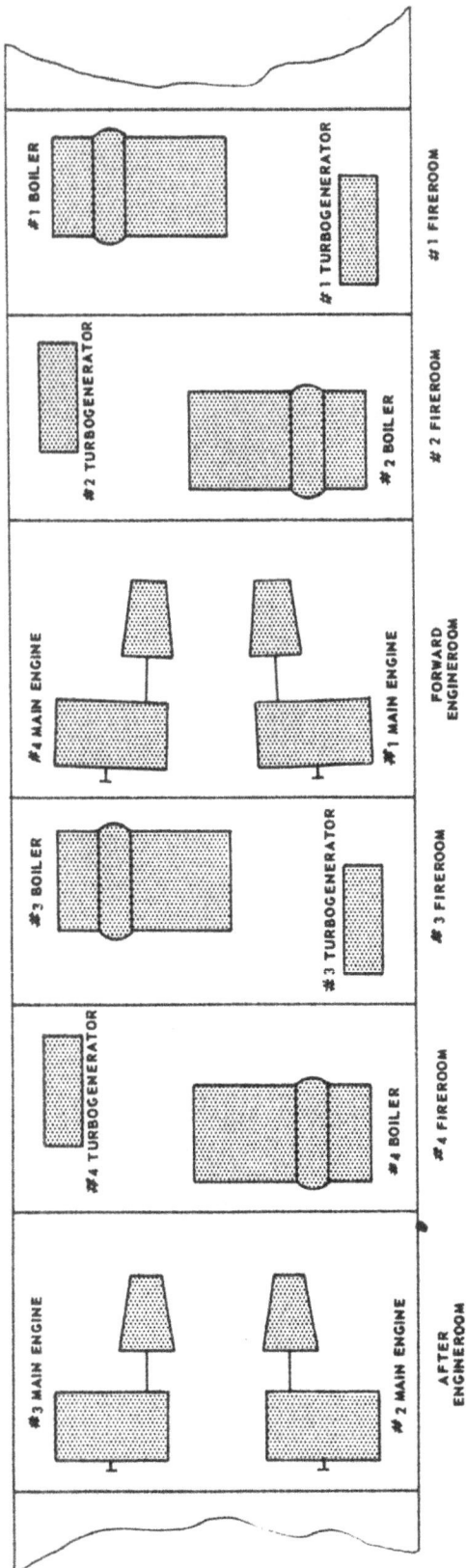

Figure 9-7.—Propulsion machinery arrangement, CA 68 class.

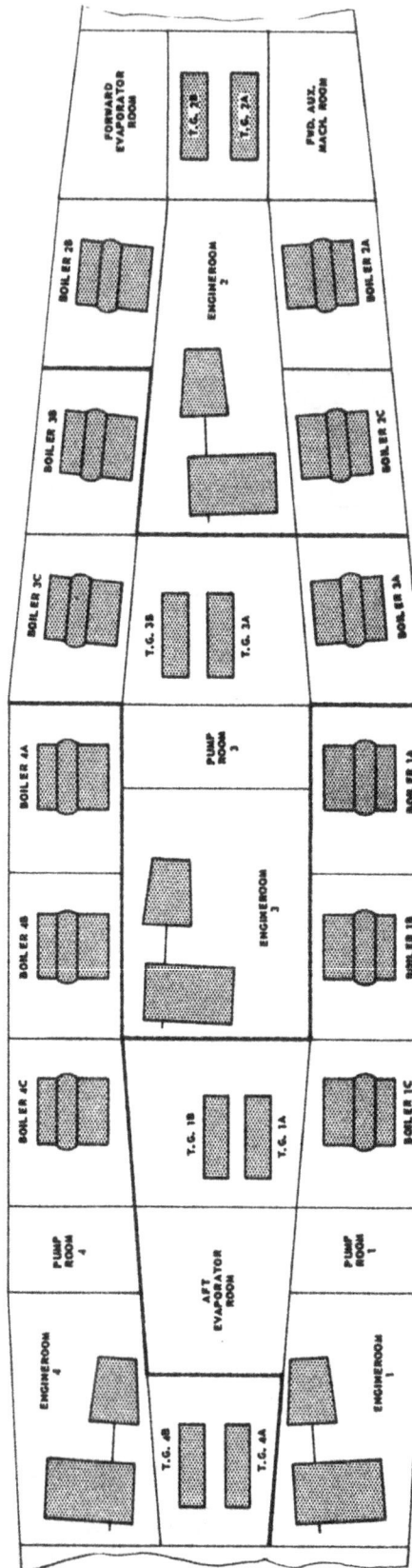

38.5

Figure 9-8.—Propulsion machinery arrangement, CVA 43.

38.6

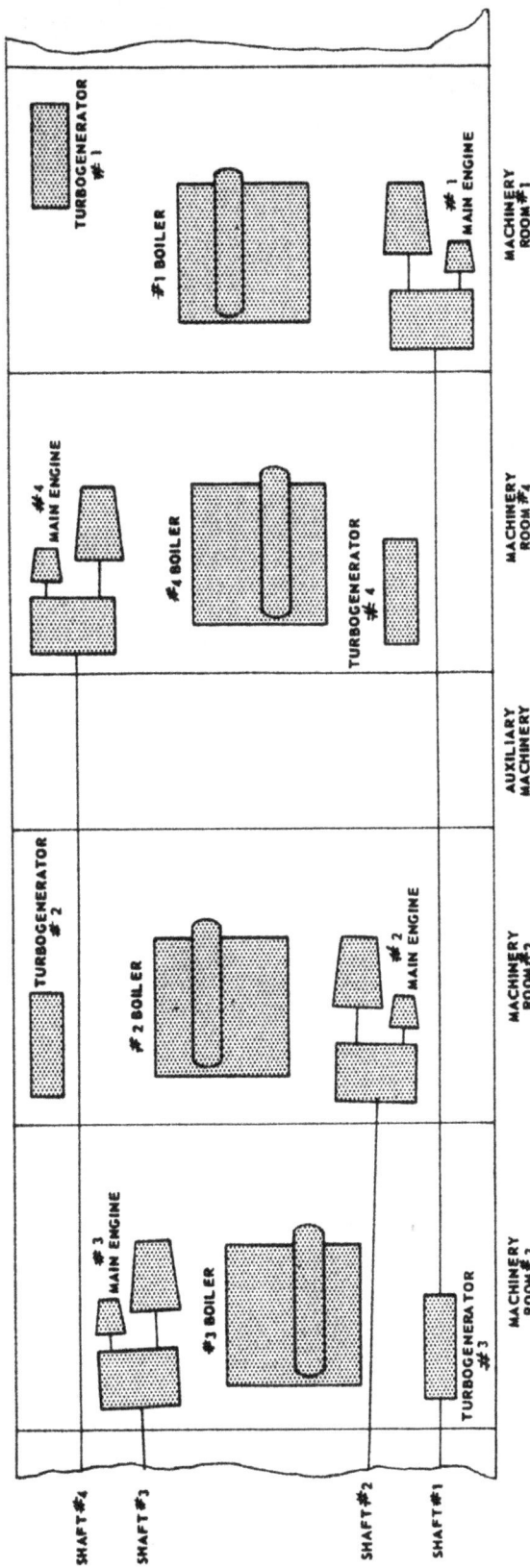

Figure 9-9.—Propulsion machinery arrangement, CA 139 class.

38.7

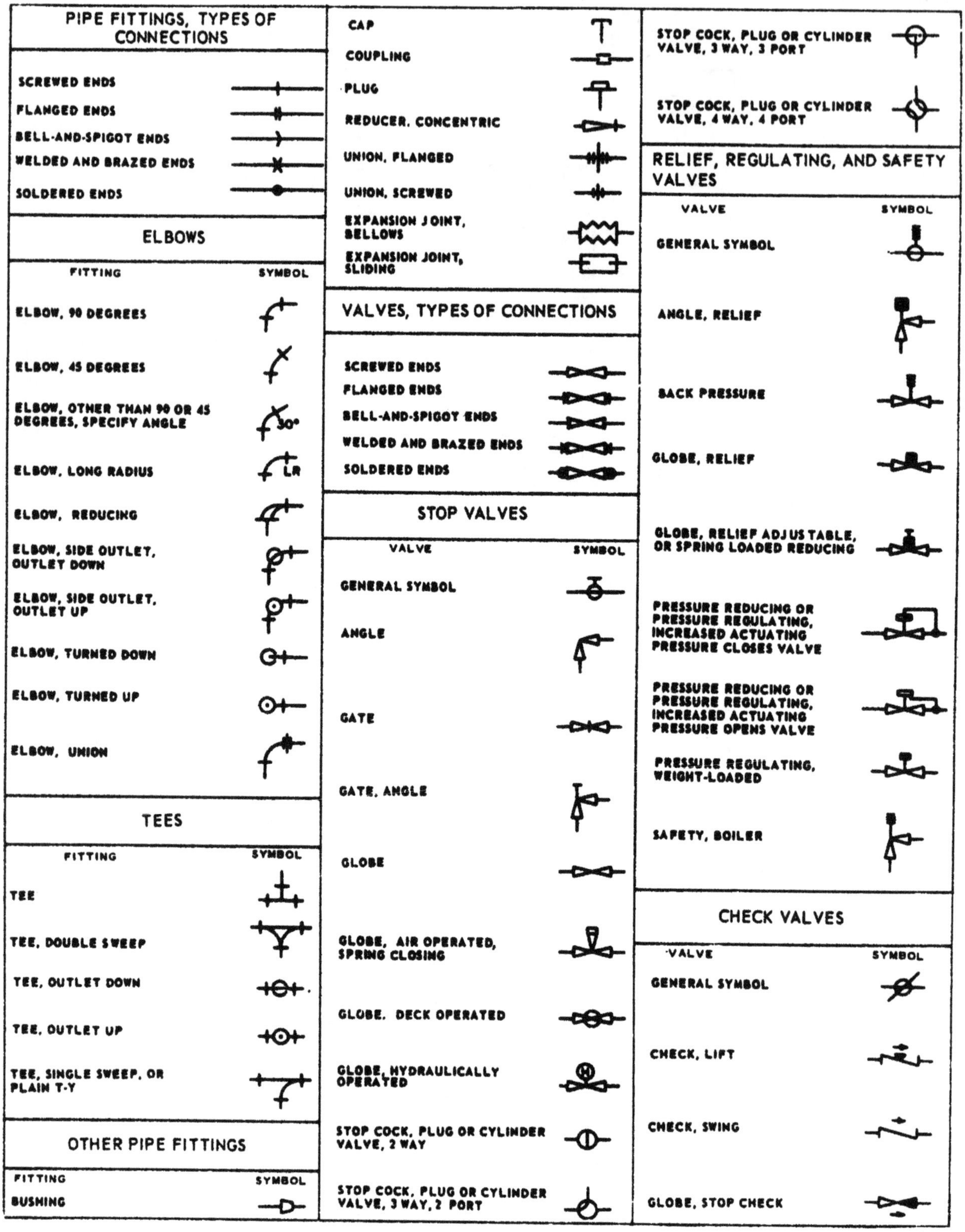

PIPE FITTINGS, TYPES OF CONNECTIONS		
SCREWED ENDS		
FLANGED ENDS		
BELL-AND-SPIGOT ENDS		
WELDED AND BRAZED ENDS		
SOLDERED ENDS		

ELBOWS	
FITTING	SYMBOL
ELBOW, 90 DEGREES	
ELBOW, 45 DEGREES	
ELBOW, OTHER THAN 90 OR 45 DEGREES, SPECIFY ANGLE	
ELBOW, LONG RADIUS	
ELBOW, REDUCING	
ELBOW, SIDE OUTLET, OUTLET DOWN	
ELBOW, SIDE OUTLET, OUTLET UP	
ELBOW, TURNED DOWN	
ELBOW, TURNED UP	
ELBOW, UNION	

TEES	
FITTING	SYMBOL
TEE	
TEE, DOUBLE SWEEP	
TEE, OUTLET DOWN	
TEE, OUTLET UP	
TEE, SINGLE SWEEP, OR PLAIN T-Y	

OTHER PIPE FITTINGS	
FITTING	SYMBOL
BUSHING	

CAP	
COUPLING	
PLUG	
REDUCER, CONCENTRIC	
UNION, FLANGED	
UNION, SCREWED	
EXPANSION JOINT, BELLOWS	
EXPANSION JOINT, SLIDING	

VALVES, TYPES OF CONNECTIONS	
SCREWED ENDS	
FLANGED ENDS	
BELL-AND-SPIGOT ENDS	
WELDED AND BRAZED ENDS	
SOLDERED ENDS	

STOP VALVES	
VALVE	SYMBOL
GENERAL SYMBOL	
ANGLE	
GATE	
GATE, ANGLE	
GLOBE	
GLOBE, AIR OPERATED, SPRING CLOSING	
GLOBE, DECK OPERATED	
GLOBE, HYDRAULICALLY OPERATED	
STOP COCK, PLUG OR CYLINDER VALVE, 2 WAY	
STOP COCK, PLUG OR CYLINDER VALVE, 3 WAY, 2 PORT	

STOP COCK, PLUG OR CYLINDER VALVE, 3 WAY, 3 PORT	
STOP COCK, PLUG OR CYLINDER VALVE, 4 WAY, 4 PORT	

RELIEF, REGULATING, AND SAFETY VALVES	
VALVE	SYMBOL
GENERAL SYMBOL	
ANGLE, RELIEF	
BACK PRESSURE	
GLOBE, RELIEF	
GLOBE, RELIEF ADJUSTABLE, OR SPRING LOADED REDUCING	
PRESSURE REDUCING OR PRESSURE REGULATING, INCREASED ACTUATING PRESSURE CLOSES VALVE	
PRESSURE REDUCING OR PRESSURE REGULATING, INCREASED ACTUATING PRESSURE OPENS VALVE	
PRESSURE REGULATING, WEIGHT-LOADED	
SAFETY, BOILER	

CHECK VALVES	
VALVE	SYMBOL
GENERAL SYMBOL	
CHECK, LIFT	
CHECK, SWING	
GLOBE, STOP CHECK	

Figure 9-10.—Engineering symbols.

11.330.1(11A)

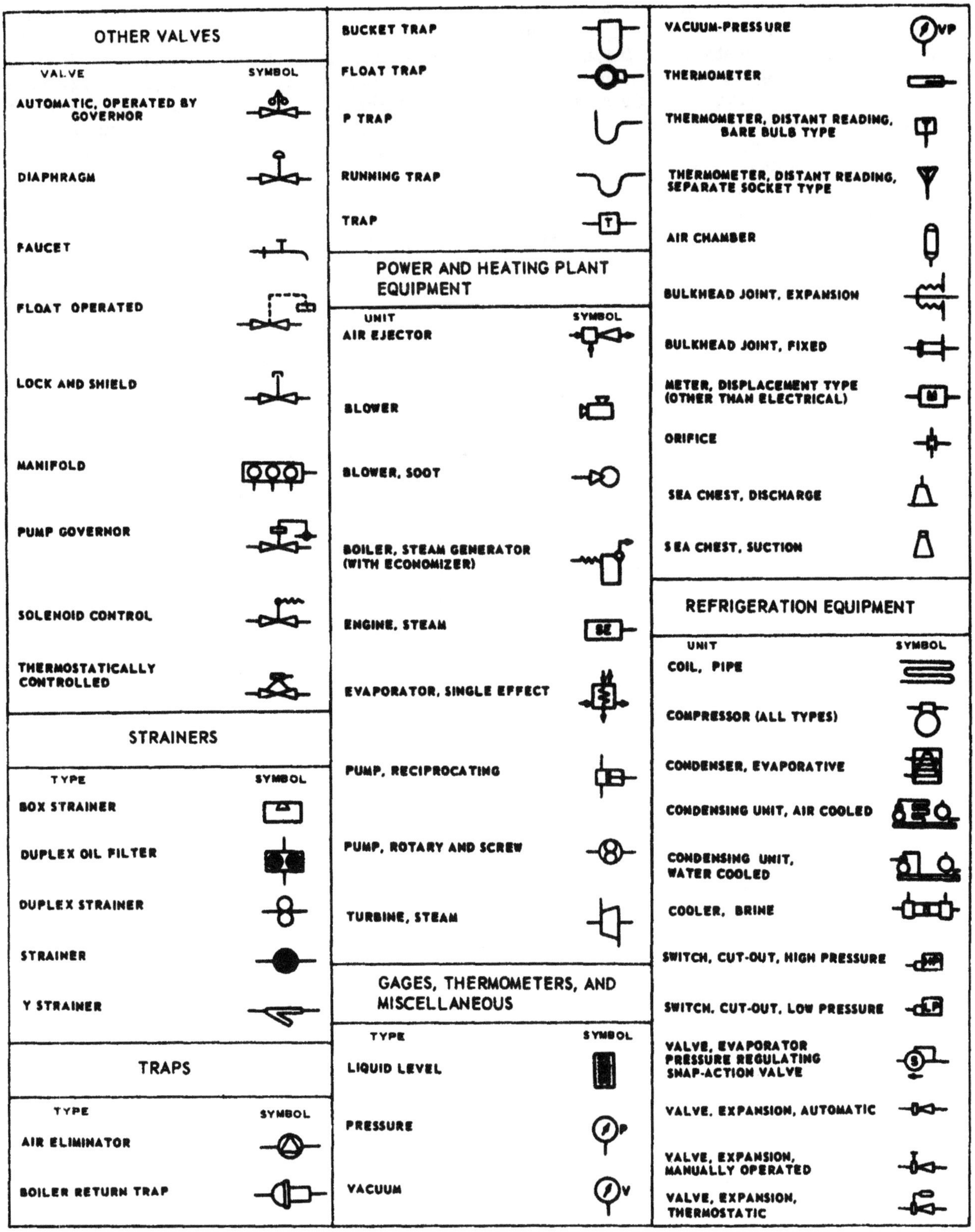

OTHER VALVES

VALVE	SYMBOL
AUTOMATIC, OPERATED BY GOVERNOR	
DIAPHRAGM	
FAUCET	
FLOAT OPERATED	
LOCK AND SHIELD	
MANIFOLD	
PUMP GOVERNOR	
SOLENOID CONTROL	
THERMOSTATICALLY CONTROLLED	

STRAINERS

TYPE	SYMBOL
BOX STRAINER	
DUPLEX OIL FILTER	
DUPLEX STRAINER	
STRAINER	
Y STRAINER	

TRAPS

TYPE	SYMBOL
AIR ELIMINATOR	
BOILER RETURN TRAP	
BUCKET TRAP	
FLOAT TRAP	
P TRAP	
RUNNING TRAP	
TRAP	

POWER AND HEATING PLANT EQUIPMENT

UNIT	SYMBOL
AIR EJECTOR	
BLOWER	
BLOWER, SOOT	
BOILER, STEAM GENERATOR (WITH ECONOMIZER)	
ENGINE, STEAM	
EVAPORATOR, SINGLE EFFECT	
PUMP, RECIPROCATING	
PUMP, ROTARY AND SCREW	
TURBINE, STEAM	

GAGES, THERMOMETERS, AND MISCELLANEOUS

TYPE	SYMBOL
LIQUID LEVEL	
PRESSURE	
VACUUM	
VACUUM-PRESSURE	
THERMOMETER	
THERMOMETER, DISTANT READING, BARE BULB TYPE	
THERMOMETER, DISTANT READING, SEPARATE SOCKET TYPE	
AIR CHAMBER	
BULKHEAD JOINT, EXPANSION	
BULKHEAD JOINT, FIXED	
METER, DISPLACEMENT TYPE (OTHER THAN ELECTRICAL)	
ORIFICE	
SEA CHEST, DISCHARGE	
SEA CHEST, SUCTION	

REFRIGERATION EQUIPMENT

UNIT	SYMBOL
COIL, PIPE	
COMPRESSOR (ALL TYPES)	
CONDENSER, EVAPORATIVE	
CONDENSING UNIT, AIR COOLED	
CONDENSING UNIT, WATER COOLED	
COOLER, BRINE	
SWITCH, CUT-OUT, HIGH PRESSURE	
SWITCH, CUT-OUT, LOW PRESSURE	
VALVE, EVAPORATOR PRESSURE REGULATING SNAP-ACTION VALVE	
VALVE, EXPANSION, AUTOMATIC	
VALVE, EXPANSION, MANUALLY OPERATED	
VALVE, EXPANSION, THERMOSTATIC	

Figure 9-10.—Engineering symbols—Continued.

11.330.2(11A)

junction of interconnecting systems. Short runs of piping which serve an immediately obvious purpose, such as short vents or drains, need not be marked. As a rule, piping on the weather decks does not require marking; if it does require marking, label plates (rather than stenciled paint or prepainted vinyl labels) are used.

Each valve is marked on the rim of the handwheel, on a circular label plate secured by the handwheel nut, or on a label plate attached to the ship's structure or to adjacent piping. The valve label gives the name and purpose of the valve, if this information is not immediately apparent from the piping system marking, and it gives the location of the valve. The location is indicated by three numbers which give, in order, the vertical level, the longitudinal position, and the transverse position. Consider, for example, a drain bulkhead stop valve that is labelled:

2-85-1

The location of this valve is indicated by these numbers. The first number indicates the vertical position—in this case, the second deck. The second number indicates the longitudinal position by giving the frame number—in this case, frame 85. The third number indicates the transverse position—starboard side if the number is odd, port side if the number is even. The numbers indicating transverse position begin at the centerline of the ship and progress out toward the sides. For example, a second drain bulkhead stop installed on the same level and at the same frame, but farther to starboard, would be indentified as

2-85-3

In either case, of course, the valve would also be identified as to system (DRAIN BULKHEAD STOP, in these examples) if the piping system identification did not make the system obvious.

A slightly different system of marking is used for identifying main line valves, cross-connection or split-plant valves, and remote-operated valves in vital engineering piping systems. Instead of being identified by location, these valves are assigned casualty control identification numbers, by system, as

Main steam............MS1, MS2, MS3, etc.
Auxiliary steam.......AS1, AS2, AS3, AS4, etc.
Auxiliary condensate...ACN1, ACN2, etc.
Auxiliary exhaust......AE1, AE2, AE3, etc.
Fuel oil service ......FOS1, FOS2, etc.

On newer ships, the system for marking valves in the vital engineering systems is slightly different, consisting of a three-part designation in the following sequence: (1) a number designating the shaft or plant number; (2) letters designating the system; and (3) a number, or a combination of a number and a letter, indicating the individual valve. Individual valve numbers are assigned in sequence, beginning at the origin of a system and going in order to the end of the system, excluding branch lines. In other words, the first valve in the main line is No. 1, the second is No. 2, and so forth. Since parallel flow paths frequently exist, it is often necessary to assign a shaft number and a system designation to the parallel flow paths as well as to the basic main line of the system. The valves in the parallel flow paths are then numbered in sequence; identical numbers are used for valves which perform like functions in each of the parallel flow paths, but a letter suffix is added to distinguish between the similar valves. This system of identification is illustrated for part of a main steam system in figure 9-11.

It is of utmost importance that all engineering personnel (officer and enlisted) become familiar with the valve markings used in the vital engineering systems. Use of the identification numbers tends to prevent confusion and error when the plant is being split or cross-connected and when damaged sections are being isolated, since it provides a means of ordering any particular valve to be opened or closed without taking time to describe the actual physical location of the valve. However, the identification markings cannot serve their intended purpose unless all engineering personnel are throroughly familiar with the physical location and the identification number of each valve they may be required either to operate themselves or to order opened or closed.

Most shipboard piping is painted to match and blend in with its surrounding bulkheads, overheads, or other structures. In a very few systems, color is used in a specified manner to aid in the rapid identification of the systems. For example, JP-5 piping in interior spaces is painted purple. Gasoline valves in interior spaces are painted yellow, except for moving parts of the valves; in exterior locations, part of the valve handwheel or the operating lever is painted yellow. Green is similarly used to identify oxygen, and red is used for fireplugs and foam discharge valves.

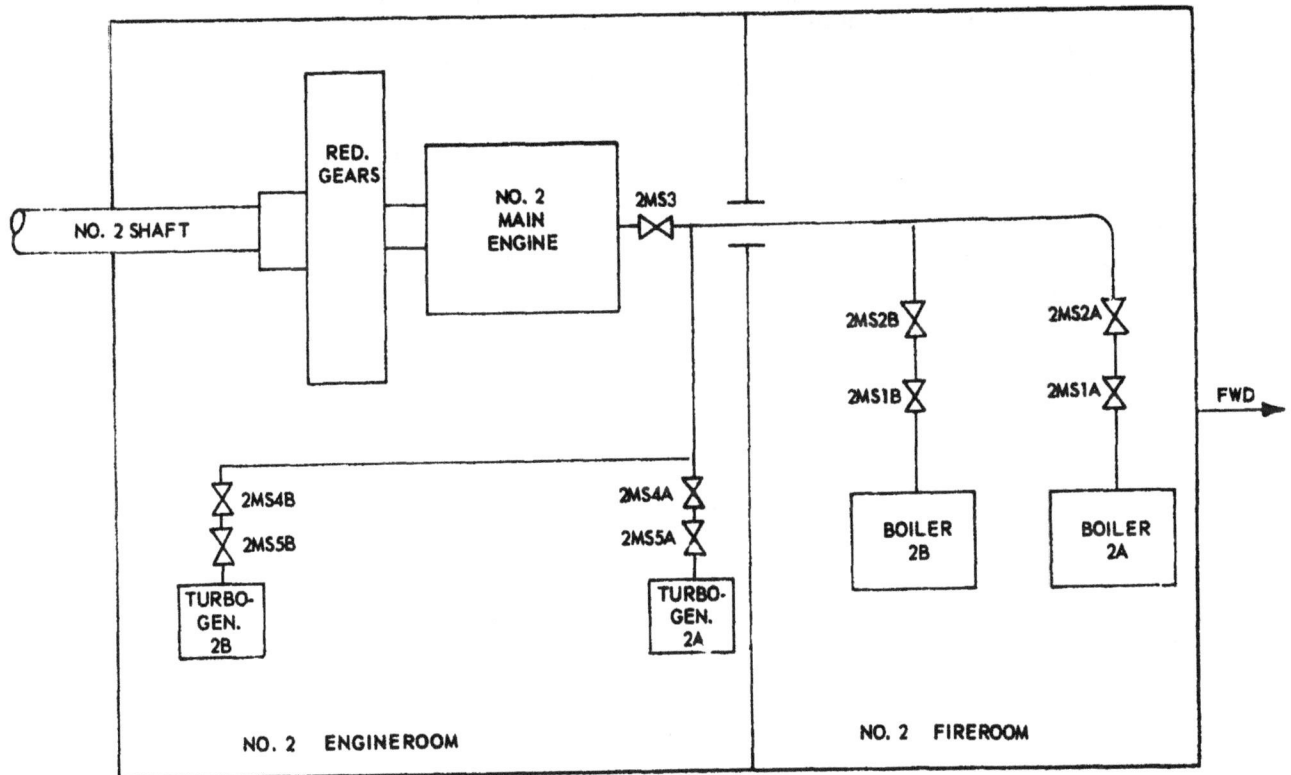

147.82

Figure 9-11.—Principle of valve identification in engineering piping systems.

MAIN STEAM SYSTEMS

The main steam system is the shortest and simplest of all the major engineering piping systems aboard ship. This statement is true regardless of the steam pressures involved. With the recent advent of the 1200-psi main steam system,[1] there is a tendency to regard high pressure main steam systems as basically different from (and mysteriously more complex than) the lower pressure systems. In reality, a 1200-psi main steam system serves the same basic purpose as a lower pressure system, and differs only in minor details, as noted in subsequent discussion. The major difference between high pressure main steam systems and lower pressure systems is in the materials used for piping and fittings; in general, the metals for 1200-psi systems must be designed to withstand operating temperatures approximately 100° to 200° F higher than the operating temperatures of the lower pressure systems.[2]

On most ships, any piping which carries superheated steam is considered as part of the main steam system. On many ships, the main steam system includes only the piping that carries superheated steam from the boilers to the propulsion turbines, the turbogenerators, and the boiler

_____

[1]Classification of main steam systems according to pressure is based on the operating pressure of the boilers. Boiler operating pressures are discussed in chapter 10 of this text; at this point, it is merely necessary to note that main steam systems are frequently referred to as 1200-psi systems, 600-psi systems, or 400-psi systems, depending upon the operating pressure of the boilers. It should be noted, also, that such pressure classifications are approximate rather than exact.

_____

[2]Although the main steam systems are approximately the same for a 1200-psi system and for a lower pressure system, it should not be inferred that the plant as a whole is identical. Important differences between 1200-psi plants and lower pressure plants are noted in appropriate places throughout this text.

soot blowers.[3] On some recent ships (both 600-psi and 1200-psi) the main steam system supplies superheated steam to several other units as well. For example, some carriers use superheated steam to supply steam catapult systems; also, some carriers and other ships use superheated steam to operate forced draft blowers, main feed pumps, main circulating pumps, and other auxiliaries. The soot blowers are not supplied from the main steam system on some ships that have 1200-psi main steam systems; instead, steam for the soot blowers is taken from the 1200-psi auxiliary steam system, as discussed later in this chapter.

Figure 9-12 illustrates the main steam system for the forward plant (No. 1 fireroom and No. 1 engineroom) of a steam-driven destroyer escort. The after plant (No. 2 fireroom and No. 2 engineroom) main steam system is very similar.

There is one boiler in each fireroom. Each boiler is provided with a boiler stop valve which can be operated either locally from the fireroom or remotely from the main deck. A second line stop valve in each fireroom provides two-valve protection for the boiler when it is not in use, and permits effective isolation in case of damage. This type of two-valve protection is standard for all boilers installed in U.S. Navy ships.

For ahead operation, the superheated steam passes through a main steam strainer, a guarding valve, and a throttle valve before entering the high pressure turbine. From the high pressure turbine, the steam passes through a crossover pipe to the low pressure turbine; then it exhausts to the condenser. For astern operation, the superheated steam passes through the steam strainer and through a stop valve; then it goes to the steam chest of the astern element, which is located at one end of the low pressure turbine.

The forward and after main steam systems are connected by cross-connection piping between the forward engineroom and the after fireroom. By means of this piping, either boiler can be used with either or both propulsion units and turbogenerators. Thus the two propulsion plants can be operated either independently (split-plant) or together (cross-connected).

Note that superheated steam for the soot blowers goes from the superheater outlet piping into a soot blower steam header.[4] Branches go from the header to the individual soot blowers.

A 600-psi main steam system is shown in figure 9-13. This is the main steam system for the two forward plants (No. 1 and No. 4) on a heavy cruiser of the CA 139 class. Although this drawing is more complicated, the system itself is still basically simple.

A 600-psi main steam system for destroyers of the DD 445 and DD 692 classes is shown in figure 9-14. A later modification was made on these ships to provide a separate superheated steam supply to the turbogenerators. With this modification, this main steam system is typical of most destroyers, even those that are considerably more recent than the DD 445 and DD 692 classes.

For comparison, figure 9-15 shows a 1200-psi main steam system for the forward plant of the frigates DLG 14 and DLG 15. Note that the 1200-psi main steam system does not supply steam to the soot blowers but that it does supply steam to the main feed pumps. In both of these respects, the 1200-psi system differs from the DD 445 and DD 692 main steam system described above.

AUXILIARY STEAM SYSTEMS

Auxiliary steam systems supply steam at the pressures and temperatures required for the operation of many systems and units of machinery, both inside and outside the engineering spaces. Although auxiliary steam is often called "saturated" steam, it has some degree of superheat in some auxiliary steam systems. Constant and intermittent service steam systems, steam smothering systems, whistles and sirens, fuel oil heaters, fuel oil tank heating coils, air ejectors, forced draft blowers, and a wide variety of pumps are typical of the systems and machinery

[3]Soot blowers are devices for removing soot from the boiler firesides while the boiler is steaming. Soot blowers are discussed in chapter 11 of this text.

[4]The term header is commonly used in engineering to describe any tube, chamber, drum, or similar piece to which a series of tubes or pipes are connected in such a way as to permit a flow of fluid from one tube (or group of tubes) to another. In essence, a header is a kind of manifold. In common usage, a distinction is made between drums and headers on the basis of size: a large piece of this kind is likely to be called a drum, a smaller one a header.

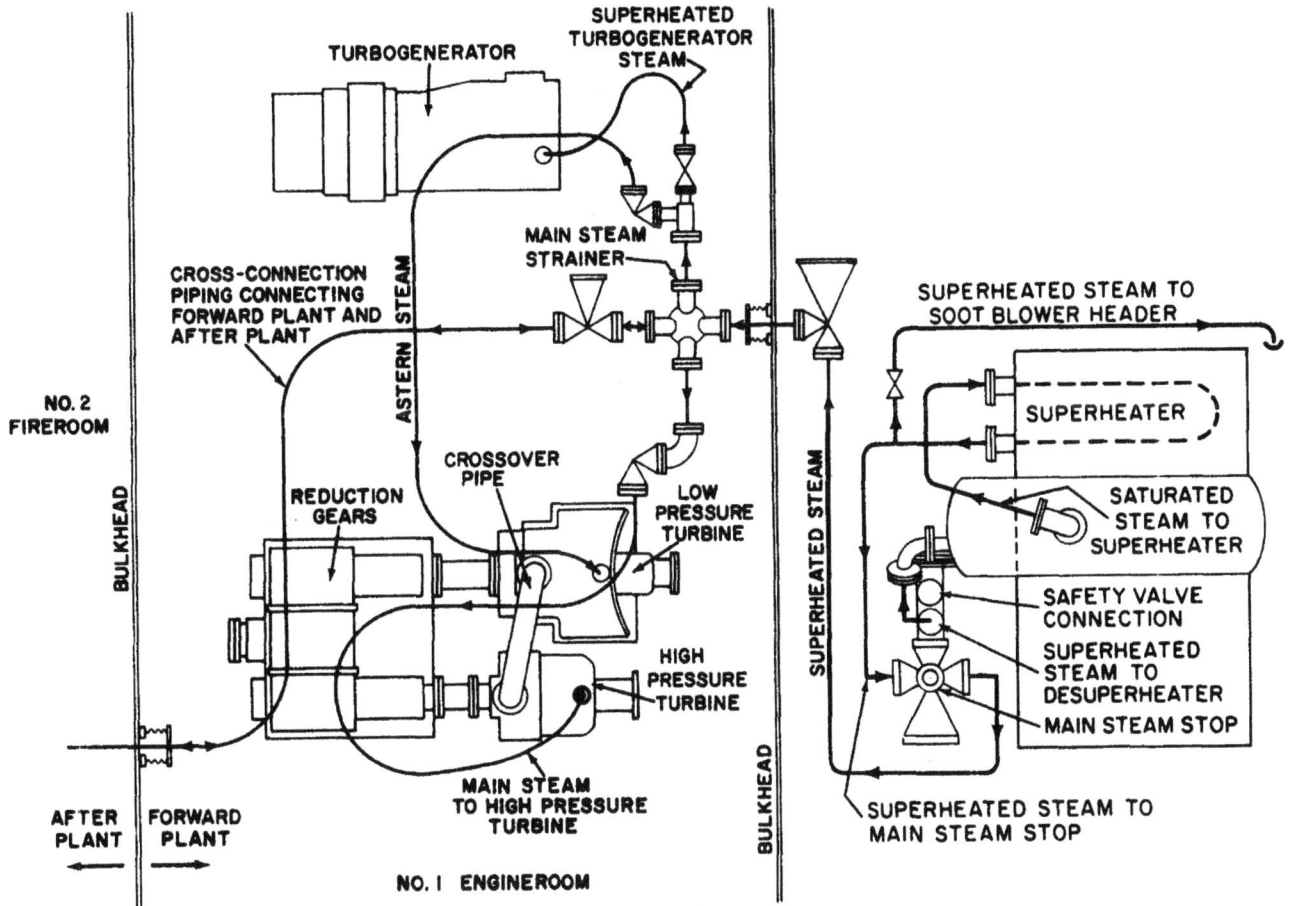

38.8

Figure 9-12.—Main steam system, destroyer escort.

that receive their steam supply from auxiliary steam systems on most steam-driven ships. As previously noted, the units are not the same on all ships. Some recent ships use main steam instead of auxiliary steam for the forced draft blowers and for some pumps. On some ships, turbine gland sealing systems receive their steam supply from an auxiliary steam system; on other ships, the source of supply is the auxiliary exhaust system. In general, an increasing use of electrically driven (rather than turbine driven) auxiliaries has led to the simplification of auxiliary steam systems on recent ships.

On ships having double-furnace boilers, auxiliary steam is taken directly from the steam drum at steam drum pressure and temperature. Since this steam is not superheated, it does not require desuperheating before it can be used as auxiliary steam. On ships having single-furnace boilers, all steam generated in the boiler goes through the superheater; the steam required for auxiliary steam systems is then desuperheated to some extent. On ships having 600-psi main steam systems, auxiliary steam is desuperheated so that it is approximately at steam drum temperature (or very slightly above). On ships having 1200-psi main steam systems, the desuperheated auxiliary steam may still have quite a bit of superheat—that is, it may be at a considerably higher temperature than the water and steam in the steam drum.

MACHINERY ROOM NO. 4

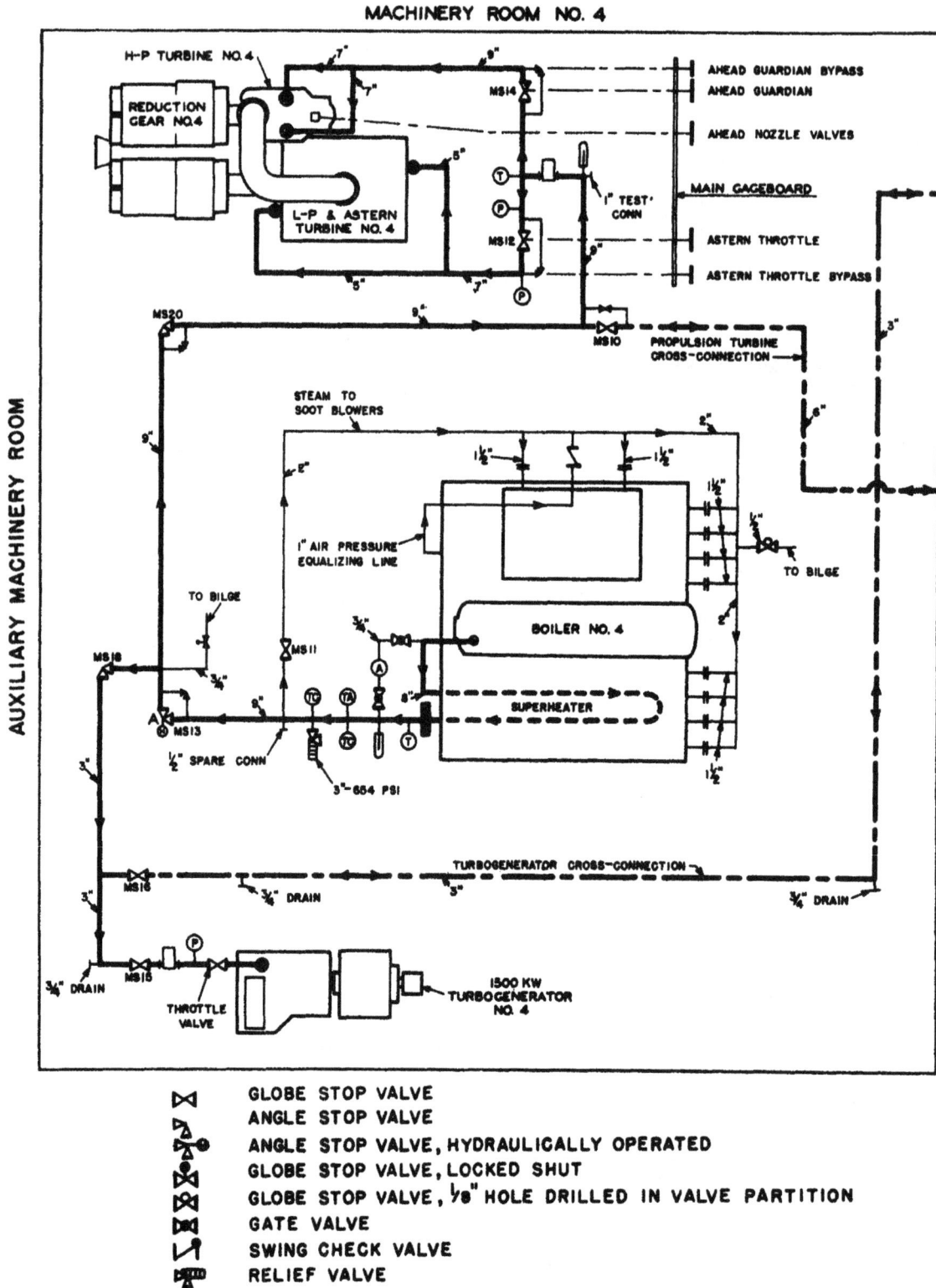

GLOBE STOP VALVE
ANGLE STOP VALVE
ANGLE STOP VALVE, HYDRAULICALLY OPERATED
GLOBE STOP VALVE, LOCKED SHUT
GLOBE STOP VALVE, ⅛" HOLE DRILLED IN VALVE PARTITION
GATE VALVE
SWING CHECK VALVE
RELIEF VALVE

38.9.1

Figure 9-13.—Main steam system, CA 139 class.

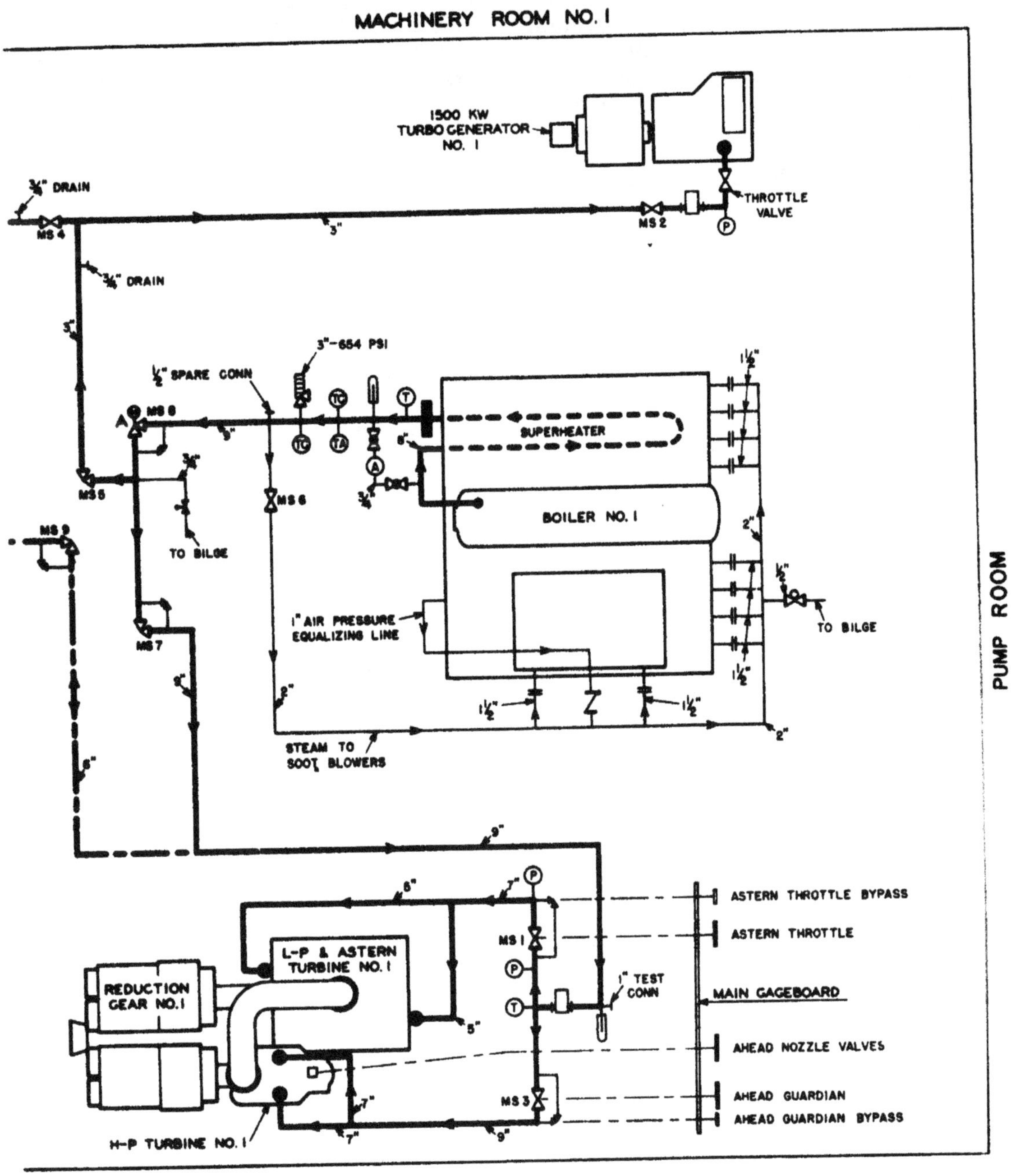

MACHINERY ROOM NO. 1

Symbol	Meaning
P	PRESSURE GAGE
A	SUPERHEATER PROTECTION DEVICE
TC	THERMOCOUPLE
TA	THERMAL ALARM
T	THERMOMETER, DISTANT READING
⊏	THERMOMETER, DIRECT READING
⊣⊢	ORIFICE
⊓	STRAINER

38.9.2

Figure 9-13.—Main steam system, CA 139 class—Continued.

Most ships that have 600-psi main steam systems have a 600-psi auxiliary steam system and a 150-psi auxiliary steam system, plus some lower pressure service systems. The 600-psi auxiliary steam system serves some machinery directly and also supplies the 150-psi system through reducing valves or reducing stations. The 150-psi auxiliary steam system serves some units directly and also provides auxiliary steam for units or systems that require auxiliary steam at even lower pressures.

Figure 9-16 shows part of a 600-psi auxiliary steam system for the two forward plants (No. 1 and No. 4) on a heavy cruiser of the CA 139 class. Note that the system is arranged in loop form, with cross connections at required intervals and with branch lines serving the various units and systems. Note, also, that the auxiliary steam system is, like the main steam system, basically rather simple.

Ships that have a 1200-psi main steam system have a 1200-psi auxiliary steam system, a 600-psi auxiliary steam system, a 150-psi auxiliary steam system, and several constant and intermittent steam service systems. The auxiliary steam systems of the DLG 14 and DLG 15 are described here in some detail as examples of auxiliary steam systems on ships having 1200-psi main steam systems.

The 1200-psi and the 600-psi auxiliary steam systems for the forward plant of the DLG 14 and DLG 15 are shown in figure 9-17. A similar arrangement exists in the after plant. The 1200-psi auxiliary steam system for each plant is entirely separate and independent; the 600-psi systems can be cross-connected but are not normally operated that way. Each plant has two boilers, both of which supply steam to the 1200-psi auxiliary steam system of that plant. The steam comes from the desuperheater outlet of each boiler; it is desuperheated from approximately 950° F (the operating temperature at the superheater outlet) to approximately 700° F. Note that the steam in this auxiliary steam system still has something more than 200° F of superheat, so it is not strictly "saturated" steam. The 1200-psi auxiliary steam lines from each boiler are interconnected so that either boiler can provide steam for everything served by this system.

The 1200-psi auxiliary steam system supplies steam directly to the soot blowers, the forced draft blowers, and the reducing stations that reduce the pressure from 1200 to 600 psig;

it also supplies augmenting steam at 12 psig to the auxiliary exhaust system, when necessary.

The 600-psi auxiliary steam system supplies steam at 600 psig and approximately 650° F to both fireroom and engineroom equipment. In the fireroom, the 600-psi system supplies steam to the fuel oil service pumps, the main feed booster pump, the fire pump, and the reducing stations that reduce the pressure from 600 to 150 psig. In the engineroom, the 600-psi system supplies steam to the standby lube oil service pump, the main condensate pump, the main circulating pump, and a reducing station that reduces the pressure from 600 to 150 psig.

The 150-psi and the 50-psi auxiliary steam systems for the after plant of the DLG 14 and DLG 15 are shown in figure 9-18. The forward plant has similar systems.

The 150-psi auxiliary steam system in each plant provides all machinery, equipment, and connections which require 150-psi steam. This system also supplies steam to other reduced pressure systems, via reducing stations, and may deliver steam to other ships or receive steam from outside sources through special piping and deck connections. Another function of the 150-psi system is to augment the auxiliary exhaust system; in fact, this function is normally performed by the 150-psi system, although it may be performed directly by the 1200-psi auxiliary steam system when necessary.

Steam for the 150-psi system in the fireroom is supplied from the reducing stations that reduce the pressure from 600 to 150 psig. There are two such stations in each fireroom. A spray-type desuperheater reduces the temperature of the fireroom 150-psi system to 400° F. Services and auxiliaries operated from the 150-psi system in the fireroom include superheater protection steam,[5] service steam systems, oil heating systems, boiler casing steam smothering systems, fireroom bilge steam smothering system, bilge and fuel oil tank stripping pumps (in No. 1 fireroom and No. 2 engineroom only), steam for burner cleaning service, and hose connections for boiling out boilers. In emergencies, the fireroom 150-psi auxiliary steam system can also supply steam for some units that are normally supplied by the engineroom 150-psi system.

---

[5] Superheater protection steam is discussed in chapter 10 of this text.

The reducing station that reduces steam from 600 to 150 psig in the engineroom supplies steam at 150 psig and 610°F to the main and auxiliary air ejectors, the distilling plant air ejectors, and the turbine gland seal systems. Line desuperheaters are not installed in the 150-psi system in the engineroom.

In the No. 2 engineroom, a reducing station reduces steam pressure from 150 psig to 100 psig and supplies steam at 100 psig and 385°F to the ship's laundry and tailor shop equipment. This 100-psi auxiliary steam system is called the 100-psi constant service system.

The 150-psi system also supplies two 50-psi systems—one a constant service system, one an intermittent service system. Both of these systems are shown in figure 9-18.

AUXILIARY EXHAUST SYSTEMS

The auxiliary exhaust system receives exhaust steam from pumps, forced draft blowers, and other auxiliaries which do not exhaust directly to a condenser. Auxiliary exhaust steam is used in various units, including deaerating feed tanks, distilling plants, and (on many ships) turbine gland seal systems.

The pressure in the auxiliary exhaust system is maintained at about 15 psig. If the pressure becomes too high, automatic unloading valves (dumping valves) allow the excess steam to go to the main or auxiliary condensers; in the event of failure of these unloading valves, relief valves allow the steam to escape to atmosphere. If the pressure in the auxiliary exhaust system drops too low, makeup steam is supplied from an auxiliary steam system (usually the 150-psi system) through augmenting valves.

The auxiliary exhaust system must be clearly distinguished from the various auxiliary steam systems. Even though the auxiliary exhaust system is a steam system, it is not considered an auxiliary steam system. A reexamination of figure 9-1 may be helpful at this point to clarify

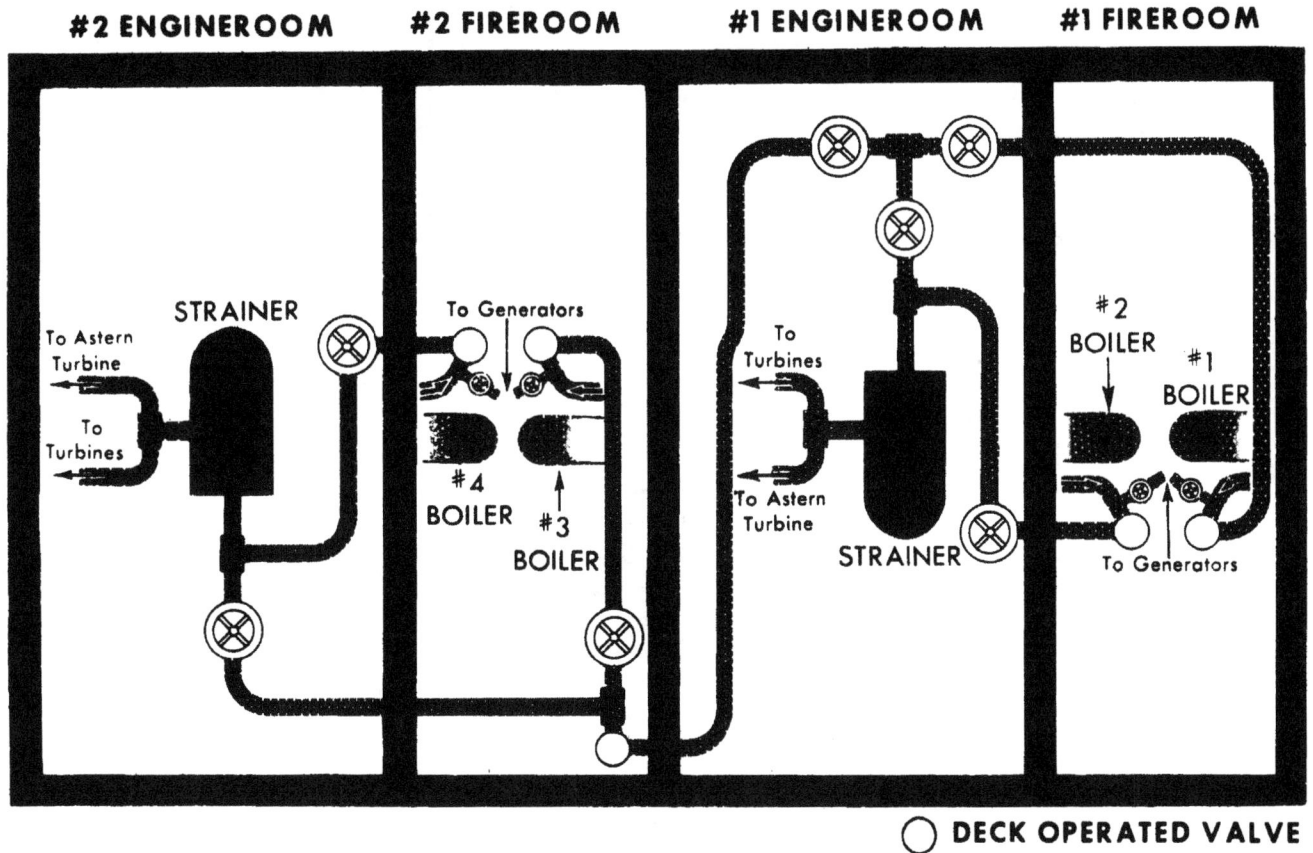

#2 ENGINEROOM   **#2 FIREROOM**   **#1 ENGINEROOM**   **#1 FIREROOM**

○ **DECK OPERATED VALVE**

147.76

Figure 9-14.—Main steam system, DD 445 and DD 692 classes.

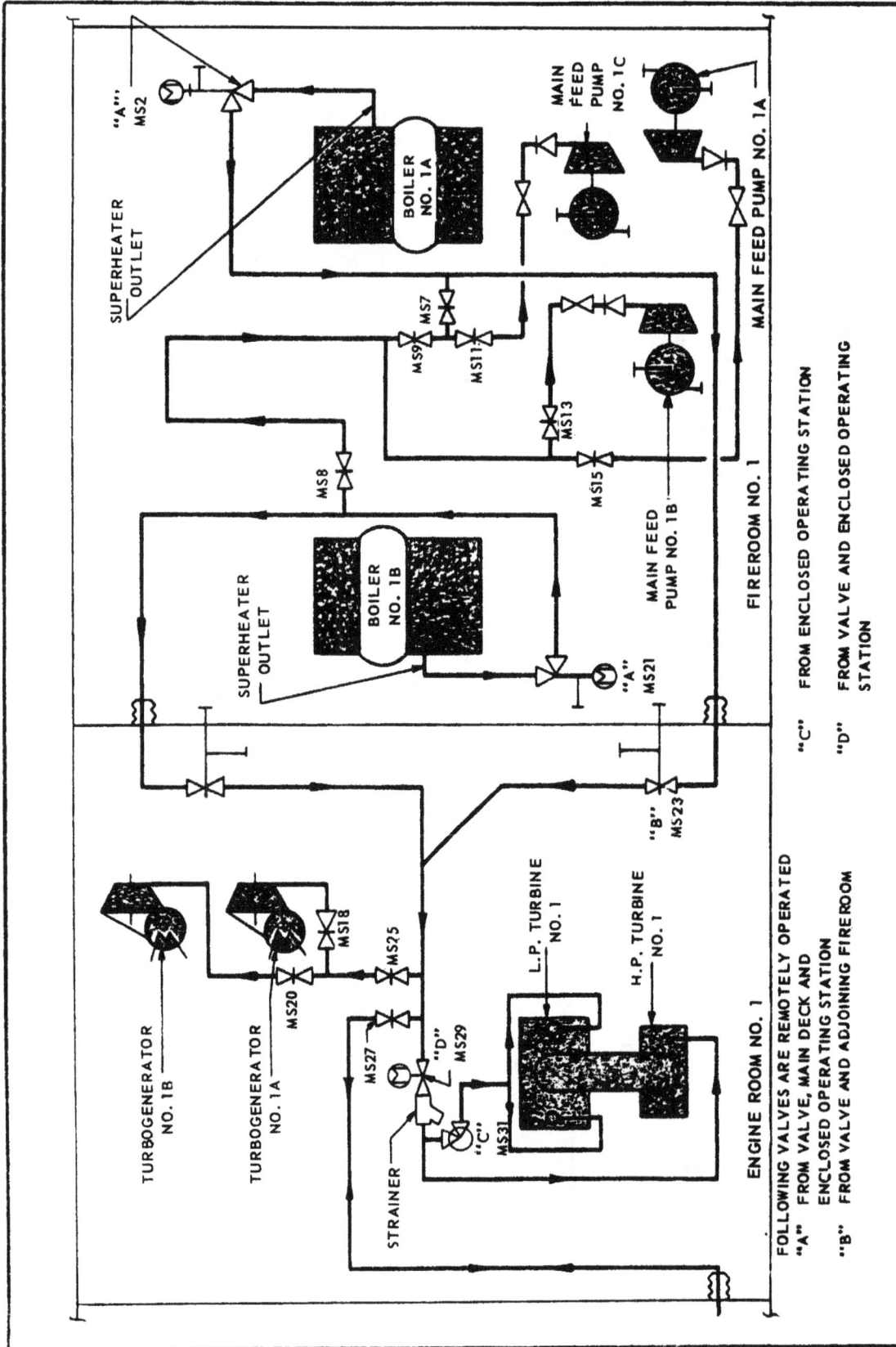

147.77

SUPERHEATER OUTLET

"A'''" MS2

BOILER NO. 1A

MAIN FEED PUMP NO. 1C

MAIN FEED PUMP NO. 1A

MS7

MS9   MS11

MS13

MS15

MS8

MAIN FEED PUMP NO. 1B

SUPERHEATER OUTLET

BOILER NO. 1B

"A" MS21

FIREROOM NO. 1

"B" MS23

TURBOGENERATOR NO. 1B

TURBOGENERATOR NO. 1A

MS18

MS25

MS20

L.P. TURBINE NO. 1

H.P. TURBINE NO. 1

MS27

"D" MS29

STRAINER

"C" MS31

ENGINE ROOM NO. 1

FOLLOWING VALVES ARE REMOTELY OPERATED

"A" FROM VALVE, MAIN DECK AND ENCLOSED OPERATING STATION

"B" FROM VALVE AND ADJOINING FIREROOM

"C" FROM ENCLOSED OPERATING STATION

"D" FROM VALVE AND ENCLOSED OPERATING STATION

Figure 9-15.—Main steam system, DLG 14 and DLG 15.

212

the relationships between the auxiliary exhaust system and the main and auxiliary steam systems.

STEAM ESCAPE PIPING

Steam escape piping is installed to provide an unobstructed passage for the escape of steam from boiler safety valves and from the relief valves installed on steam-driven auxiliaries. A line is also provided from the auxiliary exhaust system to the escape piping to allow the auxiliary exhaust to unload to atmosphere if the pressure becomes excessively high. Steam escape piping is usually shown on the same plans or drawings as the ones that show the auxiliary exhaust piping.

GLAND SEAL AND GLAND EXHAUST SYSTEMS

Gland sealing steam is supplied to the shaft glands[6] of propulsion turbines and turbogenerator turbines to seal the shaft glands against two kinds of leakage: (1) air leakage into the turbine casings, and (2) steam leakage out of the turbine casings. These two kinds of leakage may seem contradictory; however, each kind of leakage could occur under some operating conditions if the shaft glands were not sealed.

Pressures in the gland seal system are low, ranging from about 3/4 psig to 2 psig, depending upon the conditions of operation. Gland exhaust piping carries the steam and air from the turbine shaft glands to the gland exhaust condenser, where the steam is condensed and returned to the condensate system.

On most ships, gland sealing steam is supplied from the auxiliary exhaust system, although on some ships it is supplied from the 150-psi auxiliary steam system. In either case, the steam is supplied through reducing valves or reducing stations. Figure 9-19 illustrates a typical gland seal and gland exhaust system for propulsion turbines on an older type of destroyer.

CONDENSATE AND FEED SYSTEMS

Condensate and feed systems include all the piping that carries water from the condensers

to the boilers and from the feed tanks to the boilers. The condensate system includes the main and auxiliary condensers, the condensate pumps, and the piping. The boiler feed system includes the feed booster pump, the main feed pump, and the piping required to carry water from the deaerating feed tank to the boilers. Together, the condensate and feed systems begin at the condenser and end at the economizer of the boiler.

It is a little hard to say whether the deaerating feed tank is part of the condensate system or part of the boiler feed system, since the tank is generally taken as the dividing line between the two systems. The water is called condensate between the condenser and the deaerating feed tank. It is called feed water or boiler feed between the deaerating feed tank and the economizer of the boiler. Since the condensate and feed systems actually form one continuous system, the terms feed system and feed water system are quite commonly used to include both the condensate system and the boiler feed system.

Four main types of feed systems have been used on naval ships: (1) the open feed system, (2) the semiclosed feed system, (3) the vacuum-closed system, and (4) the pressure-closed system. The development of these systems, in the sequence listed, has gone along with the development of boilers. As boilers have been designed for higher operating pressures and temperatures, the removal of dissolved oxygen from the feed water has become increasingly important, since the higher pressures and temperatures accelerate the corrosive effects of dissolved oxygen. Each new type of feed system represents an improvement over the one before in reducing the amount of oxygen dissolved or suspended in the feed water.

Since practically all modern naval ships have pressure-closed feed systems, this is the only type discussed here. Pressure-closed systems are used on all naval ships having boilers operating at 600-psi and above; they are also used on some ships that have lower boiler operating pressures.

In a pressure-closed system, all condensate and feed lines throughout the system (except for the very short line between the condenser and the suction side of the condensate pump) are under positive pressure. The system is closed to prevent the entrance of air. A pressure-closed system is shown in figure 9-20.

---

[6]Shaft glands are devices for holding various kinds of packing at the point where the shaft extends through the turbine casing. Shaft glands and shaft gland packing are discussed in chapter 12 of this text.

MACHINERY ROOM NO. 4

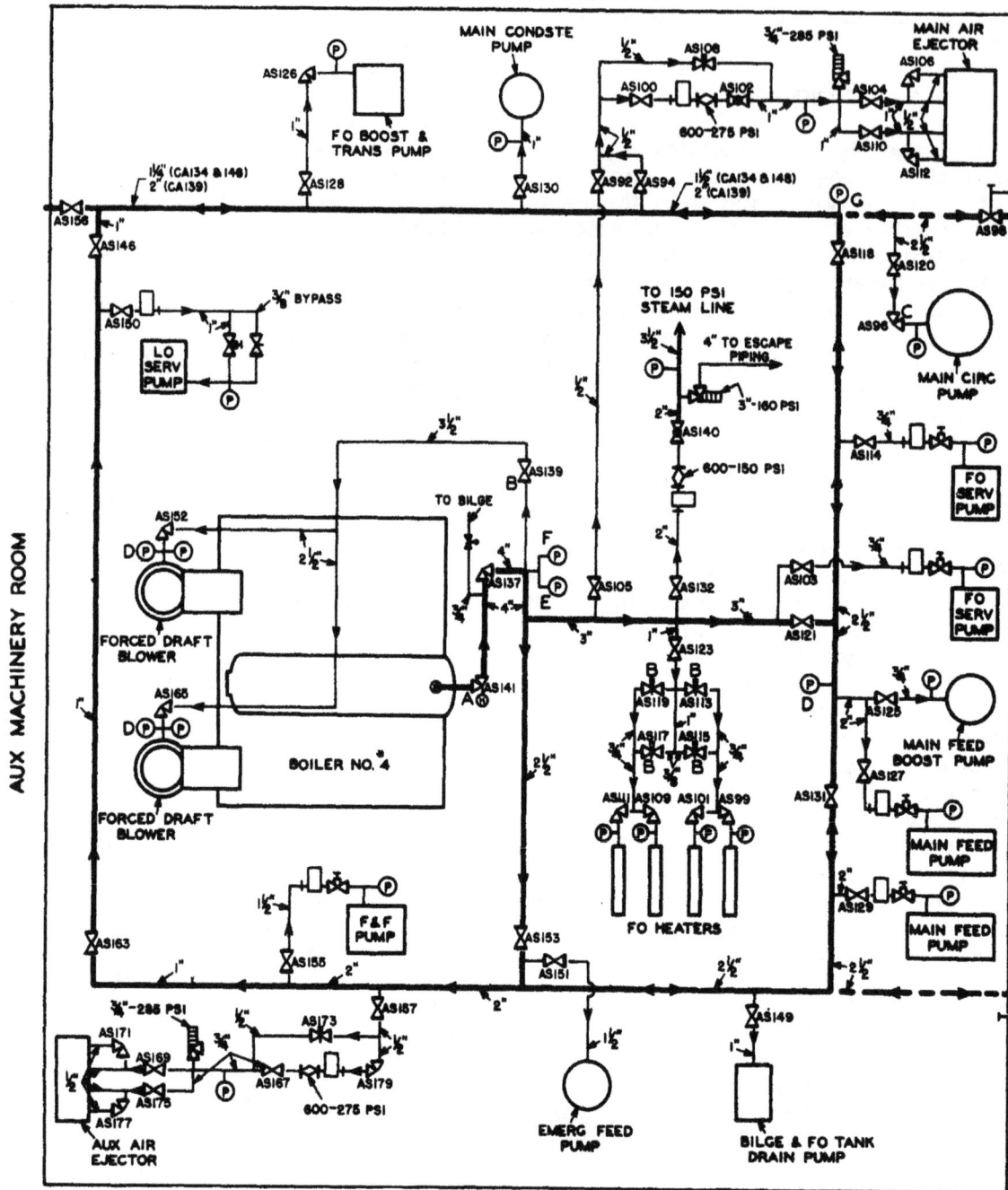

⋈	GLOBE STOP VALVE
⋈	GLOBE STOP VALVE, LOCKED SHUT
◁	ANGLE STOP VALVE
⋈	GATE VALVE
⋈	NEEDLE VALVE

Figure 9-16.—Part of 600-psi auxiliary steam system, CA 139 class.　　38.10.1

MACHINERY ROOM NO. 1

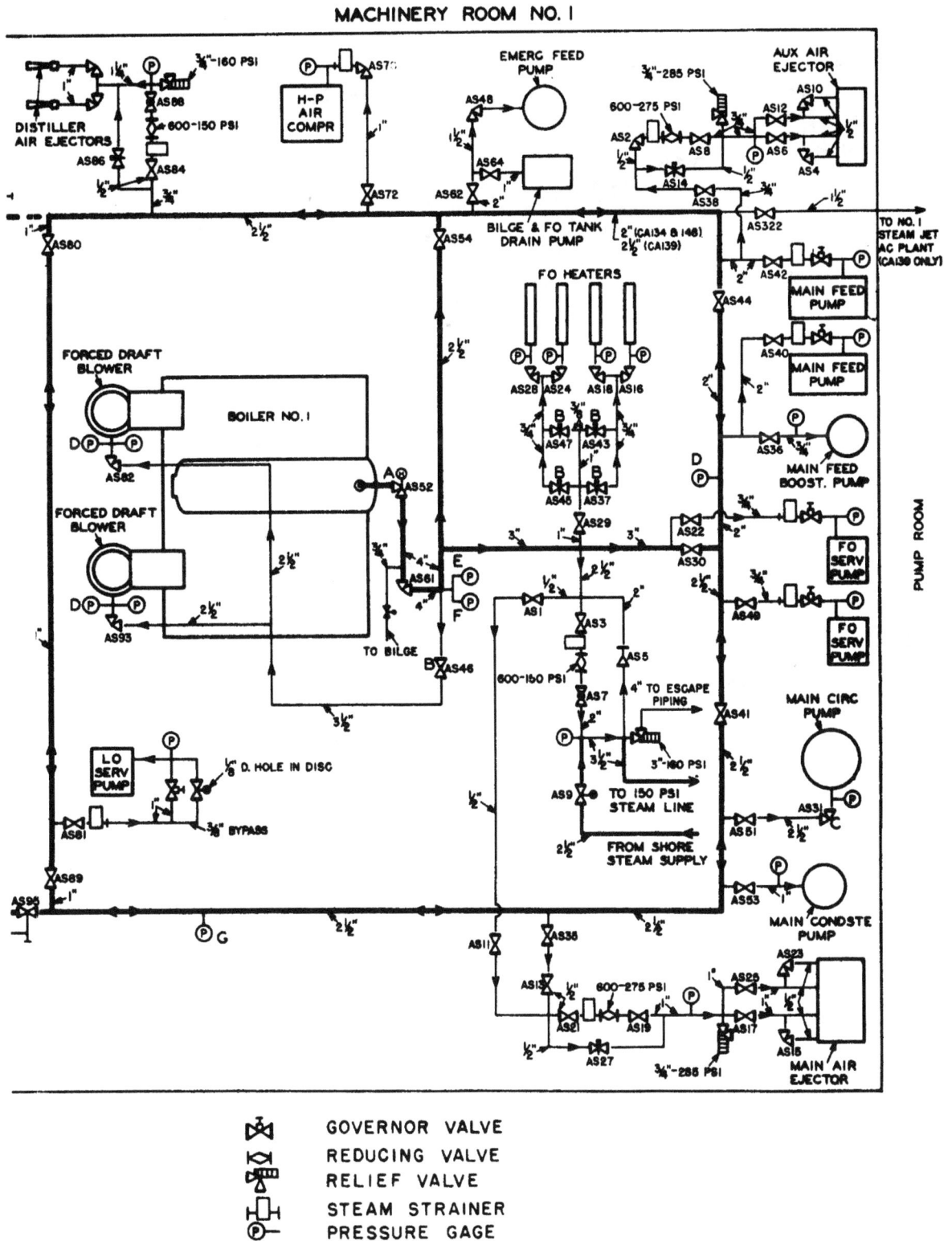

GOVERNOR VALVE
REDUCING VALVE
RELIEF VALVE
STEAM STRAINER
PRESSURE GAGE

38.10.2

Figure 9-16.—Part of 600-psi auxiliary steam system, CA 139 class--Continued.

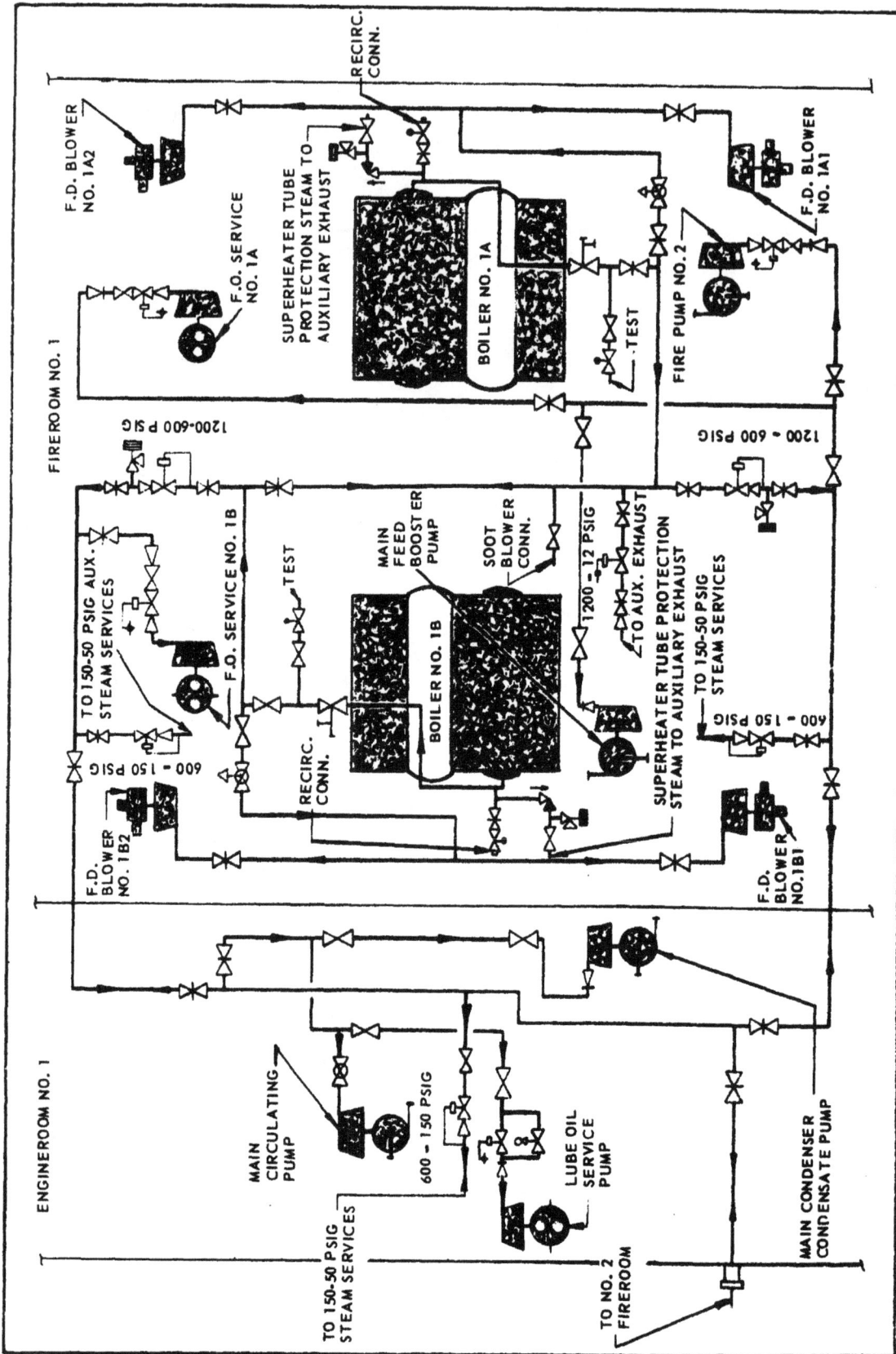

147.78

Figure 9-17. — 1200-psi and 600-psi auxiliary steam system, forward plant, DLG 14 and DLG 15.

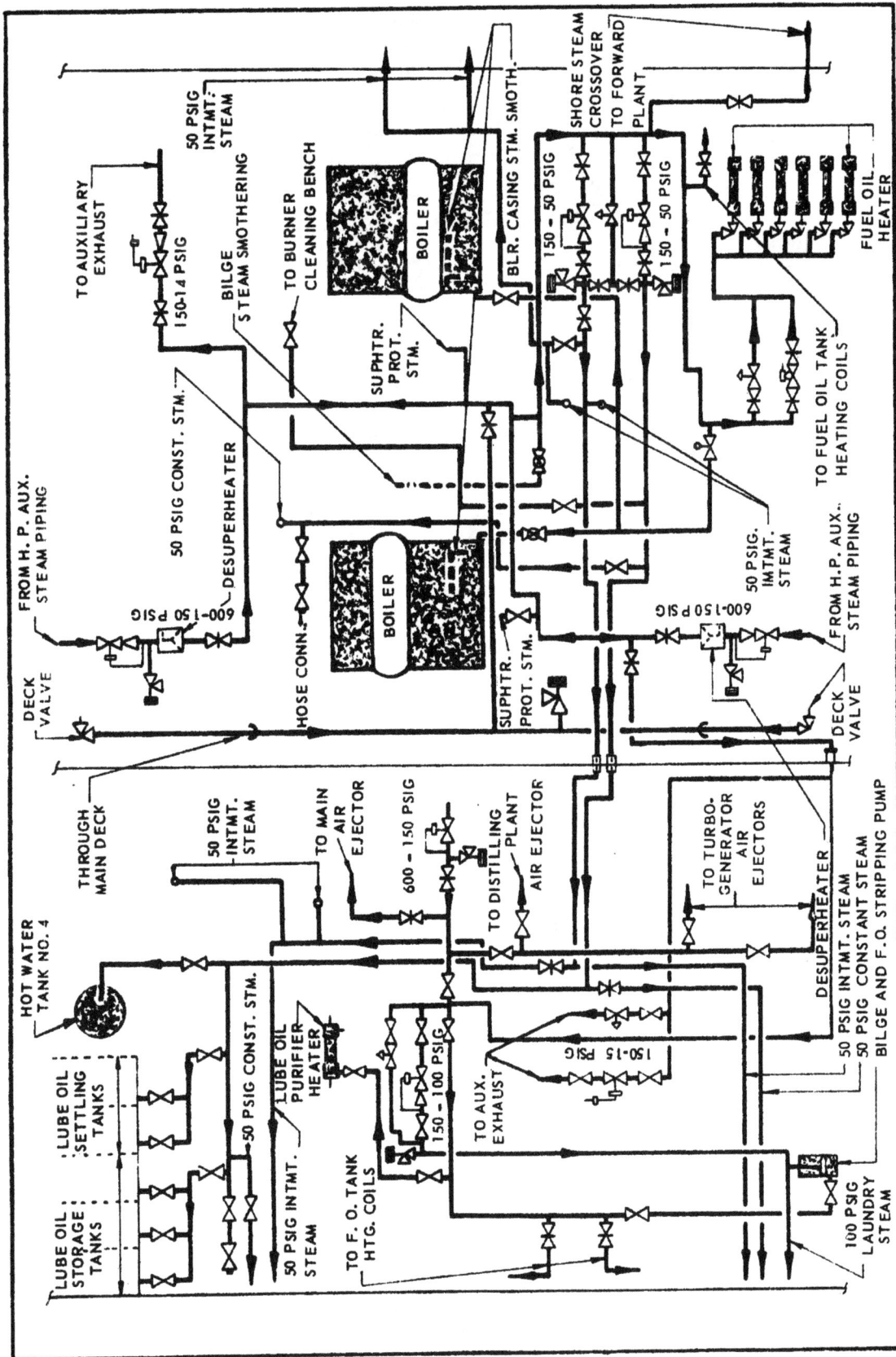

147.79

Figure 9-18.—150-psi and 50-psi auxiliary steam system, afterplant, DLG 14 and DLG 15.

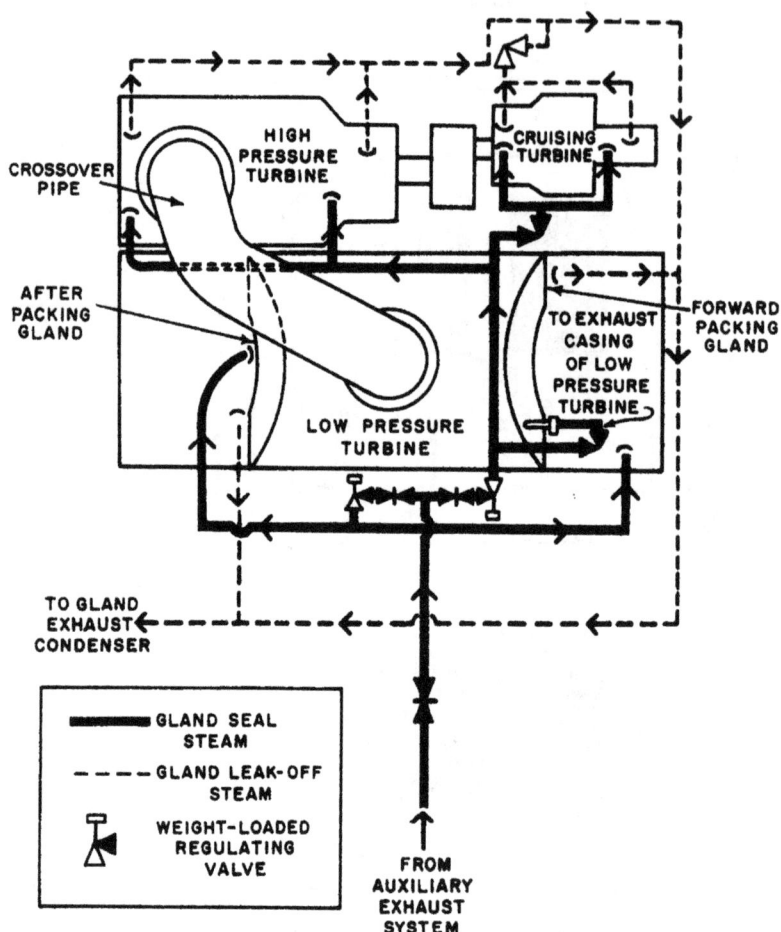

38.12

Figure 9-19.—Gland seal and gland exhaust system for propulsion turbines (destroyer).

Following this illustration, let us trace the condensate and feed system.[7]

The main condenser is the beginning of the condensate system. The main condenser is a heat exchanger in which exhaust steam from the propulsion turbines is condensed as it comes in contact with tubes through which cool sea water is flowing. The condenser is maintained under vacuum. Condensate is pumped from the condenser to the deaerating feed tank by the condensate pump. In the deaerating feed tank, the

water is heated by direct contact with auxiliary exhaust steam and is deaerated; the water (now called feed water) is pumped to the boiler by the main feed pump, with the feed booster pump providing a positive suction for the main feed pump.

Meanwhile, the air ejectors are being used to remove air and other noncondensable gases from the condenser. Condensate, on its way from the main condenser to the deaerating feed tank, is used in the air ejector condensers and in two other exchangers (the gland exhaust condenser and the vent condenser) to cool and condense the steam from steam—air mixtures and return the resulting water to the feed system. Note that the air ejectors remove air only from the condenser, not from this condensate which passes through the air ejector condensers, the gland exhaust condenser, and the vent condenser.

Makeup feed water from reserve feed tanks or from a makeup feed tank is brought into the

---

[7] The main condenser, the air ejectors, the deaerating feed tank, and other major units in the condensate and feed system are discussed in detail in chapter 13 of this text. The description given in the present chapter is intended merely to provide an overall view of the condensate and feed system.

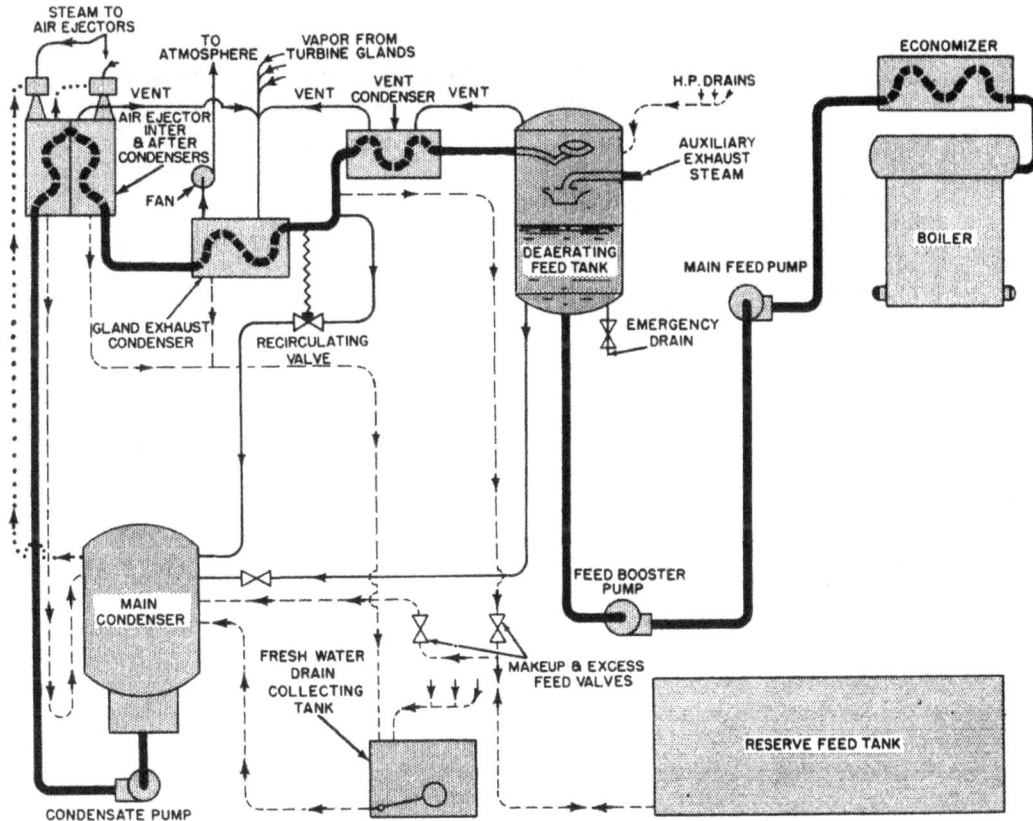

38.16

Figure 9-20.—Pressure-closed feed system.

system when necessary. A manually operated makeup feed valve is provided for this purpose. Makeup feed is brought into the condenser by vacuum drag. Another manually operated valve allows excess condensate to be discharged from the condensate line to the reserve feed tanks.

Practically all naval ships have more than one feed water system, with cross-connecting lines and valves arranged so that the systems may be operated either split-plant or cross-connected. When warming up or securing one plant, it is often necessary to transfer feed water from one plant to another. For example, it might be necessary to transfer feed water from one plant to another so that one plant will not have to take on cold makeup feed while another plant is discharging hot excess feed.

Since reserve feed tanks are normally filled by discharge from the distilling plant, it is seldom necessary to transfer feed water from one reserve feed tank to another. However, piping

system arrangements do permit this transfer to be made when necessary.

In discussing the condensate and feed system, we have not included the auxiliary condenser and its associated equipment. It should be noted that the auxiliary condenser functions in the same way as the main condenser and returns water to the condensate and feed system. The chief difference between main and auxiliary condensers is that main condensers have larger capacity.

STEAM AND FRESH WATER DRAINS

Most of the feed water in a shipboard steam plant is recovered so that it can be used over and over again for the generation of steam. As we have seen, steam is condensed in the main and auxiliary condensers and the condensate is returned to the feed system. Also, the auxiliary exhaust steam is used in the deaerating feed tank and thus becomes part of the feed system.

But steam is used throughout the ship in a good deal of machinery, equipment, and piping which does not exhaust either to a condenser or to the auxiliary exhaust system. Therefore, steam and fresh water drain systems are provided so that water can be recovered and put back into the feed system after it has been used (as steam) in fuel oil heaters, distilling plants, steam catapult systems, water heaters, whistles, and many other units and systems throughout the ship. The systems of piping which carry the water to the feed systems, and also the water carried in the systems, are known as drains.

On ships built to Navy specifications, there are four steam and fresh water drain systems which recover feed water from machinery and piping: (1) the high pressure steam drainage system, (2) the service steam drainage system, (3) the oil heating drainage system, and (4) the fresh water drain collecting system. In addition, a fifth system is provided for collecting contaminated drains which cannot be returned to the feed system. These five systems are described in the following paragraphs.

The high pressure steam drainage system generally includes drains from superheater headers, throttle valves, main and auxiliary steam lines, steam catapults (on carriers), and other steam equipment or systems which operate at pressures of 150 psi or above. On many ships, the high pressure drains are led directly into the deaerating feed tank. On some newer ships, the high pressure drains go into the auxiliary exhaust line just before the auxiliary exhaust steam enters the deaerating feed tank. In either case, of course, the high pressure drains end up in the same place—that is, in the deaerating feed tank.

The service steam drainage system collects uncontaminated drains from low pressure (below 150 psi) steam piping systems and steam equipment outside of the machinery spaces. Space heaters and equipment used in the laundry, the tailor shop, and the galley are typical sources of drains for the service steam drainage system. On some ships, these drains are discharged into the most convenient fresh water drain collecting tank. On other ships, particularly on large combatant ships such as carriers, the service steam drains discharge to special service steam drain collecting tanks located in the machinery spaces. The contents of the service steam drain collecting tanks are discharged to the condensate system; in addition, each tank has gravity drain connections to the fresh water drain collecting tank

and to the bilge sump tank located in the same space.

Note that the service steam drainage system collects only clean drains which are suitable for use as boiler feed. Contaminated service steam drains (such as those from laundry presses, for example) are discharged overboard.

The oil heating drainage system collects drains from the steam side of fuel oil heaters, fuel oil tank heating coils, lubricating oil heaters, and other steam equipment used to heat oil. Since leakage in the heating equipment could cause oil contamination of the drains, and so eventually cause oil contamination of the boilers, these drains are collected separately and are inspected before being discharged to the feed system.

The oil heating drains are collected in oil heating drain mains and are then discharged to inspection tanks. In ships that have separate enginerooms and firerooms, there is one inspection tank in the fireroom and one in the engineroom. On ships that have machinery rooms, rather than firerooms and enginerooms, each machinery room has one or more inspection tanks for the oil heating drains. The inspection tanks have small gage glasses or glass strips along the side to permit inspection of the drains. The inspection tanks normally discharge to the deaerating feed tank, but they have connections which allow the drains to be discharged to the fresh water drain collecting tank.

The fresh water drain collecting system, often called low pressure drain system, collects drains from various piping systems, machinery, and equipment which operate at steam pressures of less than 150 psi. As previously noted, both the service steam drainage system and the oil heating drainage system can discharge to the fresh water drain collecting tank, although they normally discharge more directly to the feed system. In general, the fresh water drain collecting system collects gravity drains (open-funnel or sight-flow drains), turbine gland seal drains, auxiliary exhaust drains, air ejector after condenser drains, and a variety of other low pressure drains that result from the condensation of steam during the warming up or operating of steam machinery and piping.

Fresh water drains are collected in fresh water drain collecting tanks located in the machinery spaces. The contents of these tanks may enter the feed system in two ways: they may be drawn into the condenser by vacuum drag, or in some installations they may be pumped to the condensate system just ahead of the deaerating feed tank.

A contaminated drainage system is installed in each main and auxiliary machinery space where dry bilges must be maintained. The contaminated drainage system collects oil and water from machinery and piping which normally has some leakage, and also collects drainage from any other services which may at times be contaminated. The contaminated drains are collected in a bilge sump tank located in the machinery space from which the drains are being collected. The contents of the bilge sump tank are removed by the bilge drainage system; they do not go to the feed system.

FUEL OIL SYSTEMS

Boiler fuel oil systems aboard ship include fuel oil tanks, fuel oil piping, fuel oil pumps, and the equipment used for heating, straining, measuring, and burning fuel oil.

Three main kinds of tanks are used for holding boiler fuel oil: (1) storage tanks, (2) service tanks, and (3) contaminated oil settling tanks.

The main fuel oil storage tanks are an integral part of the ship's structure. They may be located forward and aft of the machinery spaces, abreast of these spaces, and in double-bottom compartments. However, fuel oil storage tanks are never located in double-bottom compartments directly under boilers. Some fuel oil storage tanks, called fuel oil storage or ballast tanks, have connections that allow them to be filled either with fuel oil or with sea water from the ballasting system. Other fuel oil storage tanks are designated as fuel oil overflow tanks; these tanks receive the overflow from fuel oil storage tanks which are not fitted with independent overboard overflows. Overflow tanks which can also be filled with sea water from the ballasting system are called fuel oil overflow or ballast tanks.

Fuel oil is taken aboard by means of fueling trunks or special connections and is piped into the storage tanks. From the storage tanks, oil is pumped to the fuel oil service tanks. All fuel oil for immediate use is then drawn from the service tanks. The fuel oil service tanks are considered part of the fuel oil service system.

Contaminated oil settling tanks are used to hold oil which is contaminated with water or other impurities. After the oil has settled, the unburnable material such as water and sludge is pumped out through low suction connections. The burnable oil remaining in the tanks is then transferred to a storage tank or a service tank.

The contaminated oil settling tanks also serve to receive and store oil or oily water until it can be discharged overboard without violation of the Oil Pollution Acts.[8] These Acts prohibit the overboard discharge of oil and of water containing oil in port and in prohibited zones in oceans and seas throughout the world. It is standard practice, therefore, to empty the contaminated oil settling tanks before coming into port or into a prohibited zone so that the tanks will be available for storing oil and oily water until such time as it can be discharged overboard or to barges.

Fuel oil tanks are vented to atmosphere by pipes leading from the top of the tank to a location above decks. The vent pipes allow the escape of vapor when the tank is being filled and allow the entrance of air when the tank is being emptied. Most fuel oil tanks are equipped with manholes, overflow lines, sounding tubes, liquid level indicators, heating coils, and lines for filling, emptying, and cross-connecting.

The fuel oil piping system includes (1) the fuel oil filling and transfer system, (2) the fuel oil tank stripping system, and (3) the fuel oil service system. The fuel oil systems are arranged in such a way that different fuel oil pumps take suction from the tanks at different levels. Stripping system pumps have low level suction connections. Fuel oil service pumps have high suction connections from the fuel oil service tanks. Fuel oil booster and transfer pumps take suction above the stripping system pumps.

The fuel oil filling and transfer system is used for receiving fuel oil and filling the fuel oil storage tanks; filling the fuel oil service tanks; changing the list of the ship by transferring oil between port tanks and starboard tanks; changing the trim of the ship by transferring oil between forward tanks and after tanks; discharging oil for fueling other ships; and, in emergencies, transferring fuel oil directly to the suction side of the fuel oil service pumps.

The filling system on small ships such as destroyers consists of a trunk filling and tank sluicing arrangement. Larger steam-driven ships have pressure filling systems which are

---

[8]The Oil Pollution Act of 1924 (as amended) and the Oil Pollution Act of 1961 are both in effect. The 1961 Act broadens and extends the 1924 Act.

connected to the transfer mains so that the filling lines and deck connections can be used both for receiving and for discharging fuel oil. Pressure filling systems operate with a minimum pressure of approximately 40 psi at the deck connections.

In general, the filling and transfer system consists of large mains running fore and aft; transfer mains; cross-connections; risers for taking on or discharging fuel oil; fuel oil booster and transfer pumps; and lines and manifolds arranged so that the fuel oil booster and transfer pumps can transfer oil from one tank to another and, when necessary, can deliver fuel oil to the suction side of the fuel oil service pumps.

The fuel oil tank stripping system serves to clear fuel oil storage tanks and fuel oil service tanks of sludge and water before oil is pumped from these tanks by fuel oil booster and transfer pumps or by fuel oil service pumps. The stripping system is connected through manifolds to the bilge pump or, in some installations, to special stripping system pumps. The stripping system discharges the contaminated oil, sludge, and water overboard or to the contaminated oil settling tanks.

The fuel oil service system includes the fuel oil service tanks, a service main, manifolds, piping, fuel oil service pumps, meters, heaters, strainers, burner lines, and other items needed to deliver fuel oil to the boiler fronts at the required pressures and temperatures. The fuel oil service system used on any ship depends partly on the type of fuel oil burners[9] installed on the boilers. Figure 9-21 illustrates schematically a fuel oil service system typically found on ships having double-furnace boilers and straight-through-flow atomizers in the fuel oil burners. Figure 9-22 shows the fuel oil service system for the forward plant of the frigates DLG 14 and DLG 15, which use return-flow atomizers in the fuel oil burners. As may be seen in figure 9-22, a system of this type requires fuel oil return lines as well as fuel oil supply lines. Also, the use of return-flow atomizers in these burners requires a fuel oil cooler to cool the oil returned from the burners. The cooler (which is not part of the fuel oil service system on ships that do not have return-flow atomizers) serves to keep the temperature of the returned fuel oil below the flash point.

In any type of fuel oil service system, the suction arrangements for oil service pumps allow rapid changes of pump suction from one service tank (or one tank group manifold) to another. The pump suction piping is arranged to minimize contamination that might result from one service pump taking suction from a service tank that is contaminated with water.

Three classes of fuel oil service pumps are commonly used: main fuel oil service pumps, port and cruising fuel oil service pumps, and hand or emergency fuel oil service pumps.

Main fuel oil service pumps are usually screw-type rotary pumps[10] that are driven by steam turbines. However, other types of pumps are used for this purpose on some ships.

Port and cruising fuel oil service pumps on recent ships are very similar to the main fuel oil service pumps except that they are driven by two-speed electric motors. The capacity of these pumps can be adjusted by selecting the required speed of the motor and also by using a bypass arrangement to recirculate unused oil from the pump discharge to the pump suction. On older ships, the port and cruising fuel oil service pumps may be rotary pumps or they may be axial-piston variable-stroke pumps; in either case, they are normally driven by electric motors rather than by steam turbines.

Hand or emergency fuel oil service pumps are used on some ships when boilers must be lighted off and neither steam nor electric power is available. Most hand or emergency fuel oil service pumps are herringbone gear pumps. On recent ships, other means of lighting off without steam or power are used, and the hand or emergency fuel oil service pump is not required.

The fuel oil service system contains a number of valves, all of which are important to the safe and efficient operation of the boiler. The major valves in the fuel oil service system shown in figure 9-20 are listed here both to give some idea of the complexity of the fuel oil service system and to indicate the degree of precision required of operating personnel in lining up, operating, securing, and controlling casualties in the fuel oil service system.

---

[9]Fuel oil burners are discussed in chapter 10 of this text.

[10]Basic types of pumps are discussed in chapter 15 of this text.

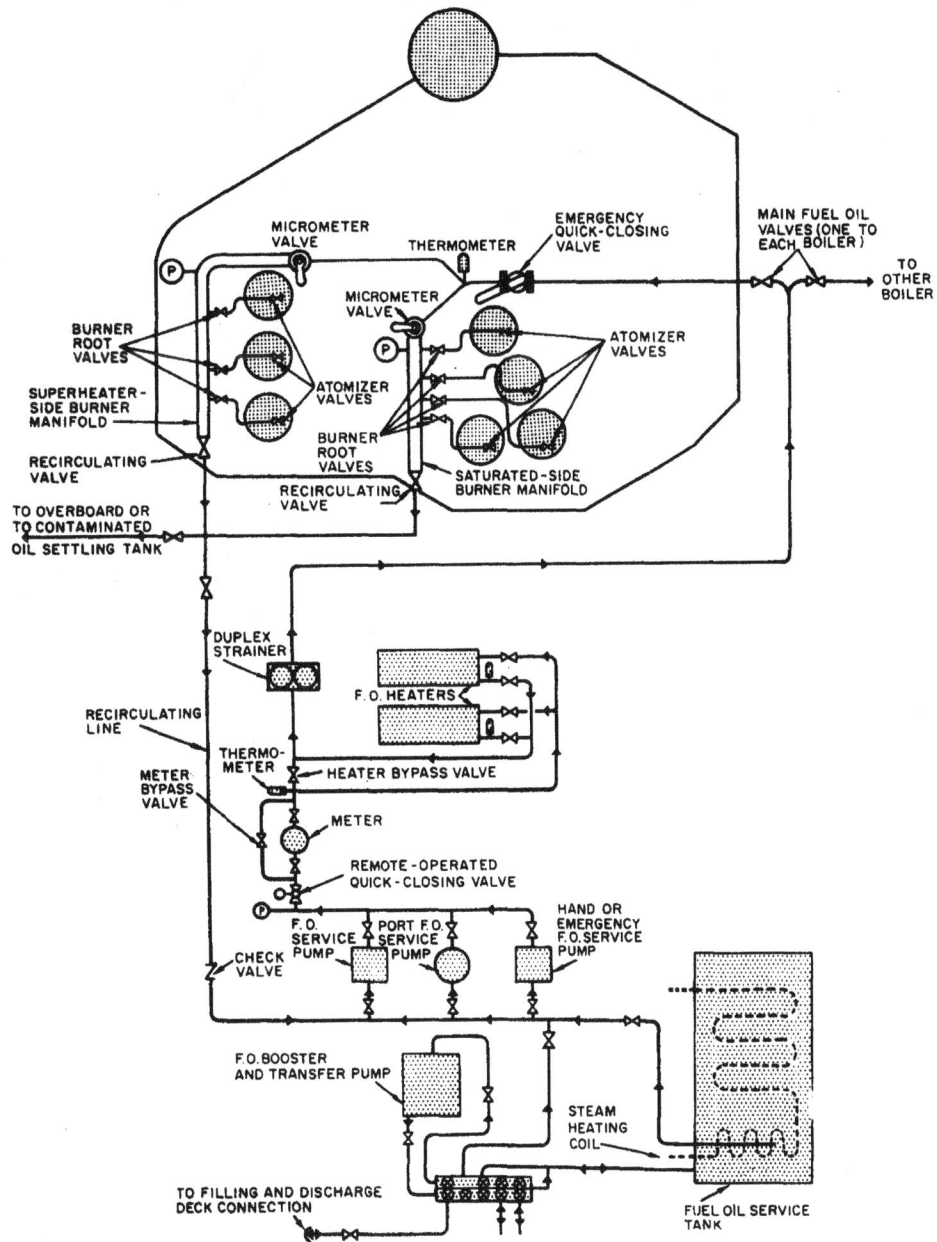

38.63

Figure 9-21.—Fuel oil service system on ship with double furnace boilers
having straight-through-flow atomizers.

Suction and discharge valves allow the pumps to be lined up for the delivery of fuel oil. The remote-operated quick-closing valve in the supply main on the discharge side of the fuel oil service pump provides a means for rapidly shutting off the fuel oil from a remote location. The remote operating gear for this valve is arranged so that the valve may be operated from two places: (1) from the fuel oil pump itself, and (2) from the fireroom escape trunk or from the deck above and near the access to the space. Fuel oil meter and meter bypass valves allow the fuel oil meter to be used or to be bypassed, as the situation requires; the fuel oil meter is bypassed when oil is being recirculated. A fuel oil heater bypass valve

223

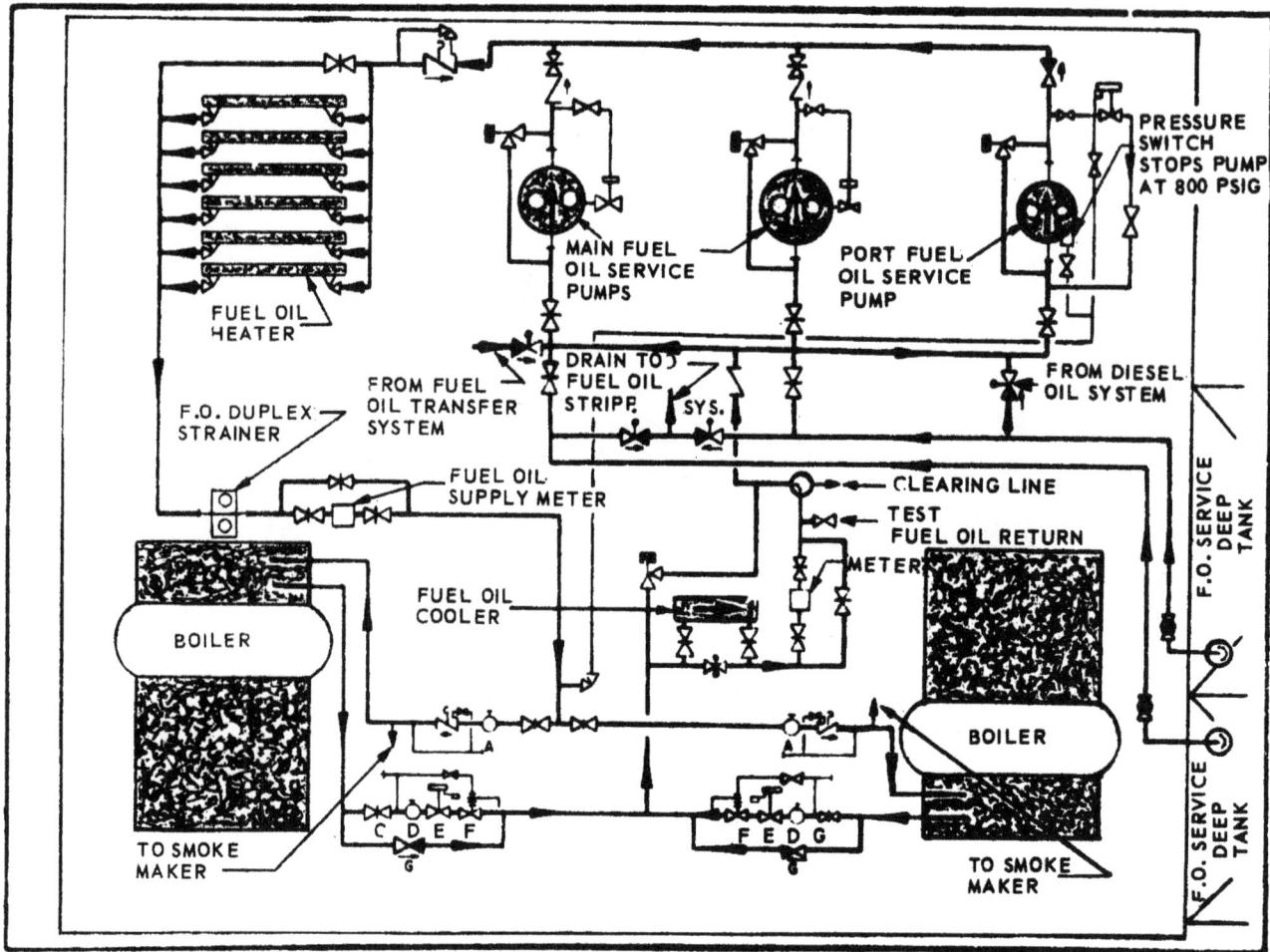

147.80

Figure 9-22.—Fuel oil service system for forward plant, DLG 14 and DLG 15.

is installed to permit bypassing the fuel oil heaters in unusual operating situations. Fuel oil heater valves control the flow of oil into the heaters and permit shifting from one heater to another.

The main fuel oil valve controls the flow of fuel oil in the line leading to each boiler. The emergency quick-closing valve can be operated from both the upper level and the lower level at the boiler front. In some installations, a latched-open solenoid valve, arranged for local tripping, is installed adjacent to each burner supply manifold; where a solenoid valve of this type is installed, it takes the place of the emergency quick-closing valve.

A micrometer valve is installed at the top of each burner manifold. The micrometer valve is used for the manual control of fuel oil pressure; thus it is the valve that controls the amount

of oil being burned in the boiler furnace. From the burner manifold, a small flexible line goes to each burner. A small valve called a burner root valve is installed in each burner line to permit shutting off the supply of oil to any burner that is not in use. And finally, an atomizer valve is installed on each burner at the atomizer connection. This valve allows oil to go through the atomizer and be sprayed out into the boiler furnace in such a way that combustion takes place.

At the lower end of each burner manifold, a recirculating valve is installed. By means of these valves and the recirculating line, fuel oil can be returned from the burner manifold to the suction side of the fuel oil service pump. The recirculating line is used to circulate oil through the fuel oil heaters and thus bring the oil up to the proper temperature for lighting off. A clearing line branches off from the recirculating line. Valves in the recirculating line and in the clearing line permit fuel oil to be discharged to the suction side of the fuel oil service pump or to the contaminated oil settling tank (or overboard). A check valve in the recirculating line prevents the back flow of oil from the fuel oil suction main; another check valve at the connection of the clearing line and the contaminated oil settling tank prevents back flow from the contaminated oil settling tank through the clearing line and into the recirculating line.

The valves just listed are typically found in fuel oil service systems on ships which use straight-through-flow atomizers. Where return-flow atomizers are used, additional valves are required in the fuel oil service system to control the return flow of oil and (in some installations) to control the flow of oil through a cooler. Where automatic boiler controls are installed, still more valves are required in the fuel oil service system; these include fuel oil supply and return valves which are operated by the boiler control system.

BALLASTING SYSTEMS

The ballasting system allows the controlled flooding of certain designated tanks, when such flooding is required for stability control. All tanks that are designated as fuel oil and ballast tanks (and also certain voids) may be flooded by the ballasting system. Sea water is used for ballasting; it may be taken from the firemain or it may be taken directly from sea chests.

Combined ballasting and drainage systems are arranged so that all designated compartments and tanks can be ballasted either separately or together and drained either separately or together. Drainage pumps or eductors are used to remove the ballast water.

DIESEL OIL AND JP-5 SYSTEMS

Diesel oil systems are found even on steam-driven ships. Ships that carry large supplies of diesel oil have fairly complex diesel oil systems which are quite similar to the boiler fuel oil systems already described. Although the diesel oil systems are separate from the boiler fuel oil systems, they are arranged so that the diesel oil can be discharged to the fuel oil service system and burned in the boiler furnace in case of emergency.

On aircraft carriers, JP-5 aviation fuel can also be used as boiler fuel in case of emergency. The JP-5 system is separate from the boiler fuel oil system but can be connected so as to discharge JP-5 to the fuel oil service system.

DISTILLATE FUEL SYSTEM

The Department of Defense has authorized the Navy to convert to an all-distillate marine type diesel fuel (Navy Special Distillate Fuel) (NSDF) to replace the Navy Special Fuel Oil (NSFO) now in use on steam-driven ships.

Testing is now being conducted on gas turbines and diesel engines for the feasibility of converting the Navy to a "one fuel" Navy for logistic simplicity and reduction of overall operating costs.

Piping system conversion and changes will require the upgrading and validation of the existing systems on all ships using NSFO. Therefore, all instructions issued by NavShips and NavSec shall be followed in upgrading and validation of the existing NSFO systems before NSDF can be introduced into the system.

Stability and buoyancy will also be affected due to the variation in specific gravity of NSFO (7.9 lbs/gal average) verses NSDF (7.2 lbs/gal average), therefore, solid ballast will be required in those ships which are now near the naval architectural limits for stability. The existing liquid loading instructions which specify sea water ballasting of empty fuel tanks will still remain in effect.

MAIN LUBRICATING OIL SYSTEMS

Main lubricating oil systems on steam-driven ships provide lubrication for the turbine bearings and the reduction gears. The main lube oil system usually includes a filling and transfer system, a purifying system, and separate service systems for each propulsion plant. On most ships, each lube oil service system includes three positive-displacement lube oil service pumps: (1) a shaft-driven pump, (2) a turbine-driven pump, and (3) a motor-driven pump. The shaft-driven pump, attached to and driven by either the propulsion shaft or the quill shaft of the reduction gear, is used as the regular lube oil service pump when the shaft is turning fast enough so that the pump can supply the required lube oil pressure. The turbine-driven pump is used while the ship is getting underway and is then used as standby at normal speeds. The motor-driven pump serves as standby for the other two lube oil service pumps.

Figure 9-23 illustrates the lube oil supply and lube oil drain piping of the service system on the frigates DLG 14 and DLG 15.

COMPRESSED AIR SYSTEMS

Completely independent compressed air systems with individual compressors include the high pressure air system, the ship's service air system, the aircraft starting and cooling air system, the combustion control air system, the air deballasting system, and the oxygen-nitrogen producer air system. For other services, air is taken from the high pressure system or from the ship's service air system, as required. Air is provided by high pressure, medium pressure, or low pressure air compressors, as appropriate.

The high pressure air system is designed to provide air above 600 psi and up to 5000 psi for charging air banks and, at required pressures, for services such as missiles, diesel engine starting and control, torpedo charging, and torpedo workshops. When air is required for these services at less than the system pressure, the outlet from the high pressure air system is equipped with a reducing valve.

Air for diesel engine starting and control is provided on some ships by a medium range compressure at a pressure of 600 psi or from the high pressure system, through appropriate reducing valves.

The ship's service compressed air system is a low pressure system that is installed on practically all surface ships. This system provides compressed air at the required pressure for the operation of pneumatic tools, the operation of oil-burning forges and furnaces, the charging of pump air chambers, the cleaning of equipment, and a variety of other uses. The ship's service air system is normally designed for a working pressure of 100 psi; on ships such as tenders and repair ships, however, where there is a greater demand for air, the system is designed for a higher working pressure (usually about 125 psi). The ship's service air system is normally supplied from a low pressure air compressor; on some ships, however, the system may be supplied from a higher pressure system, through reducing valves.

An aircraft starting and cooling air system is installed on aircraft carriers. This system is designed to provide air at various temperatures (50° to 500° F) and pressures (48 psia to 62 psia) by gas turbine. The system supplies compressed air to meet the conditions of starting and cooling aircraft being served.

Combustion control air systems (more properly called boiler control air systems) are installed on some ships to provide supply air for the pneumatic units in automatic boiler control systems. A boiler control air system usually consists of an air compressor, an air receiver, and the piping required to supply air to all units of the boiler control system On some older ships, compressed air for the operation of the boiler controls is taken from the ship's service air system, through reducing valves.

An air deballasting system is provided on some ships for deballasting by air. This system is designed to provide large quantities of air (7500 cubic feet per minute) at low pressure (200 psi). All compressors discharge to a common air loop distribution which feeds all ballast tanks.

Oxygen-nitrogen producer air systems are installed on aircraft carriers and submarine tenders. The air is supplied by high pressure air compressors, via oil filters and moisture separators, directly to the oxygen-nitrogen producer.

FIREMAIN SYSTEMS

The firemain system receives water pumped from the sea and distributes it to fireplugs, sprinkling systems, flushing systems, auxiliary machinery cooling water systems, washdown systems, and other systems as required.

There are three basic types of firemain systems used on naval ships: the single main system, the horizontal loop system, and the vertical loop system. The type of firemain system installed in any particular ship depends upon the characteristics and functions of the ship. Small ships generally have single main firemain systems; large ships usually have one of the loop systems or a composite system which is some combination or variation of the three basic types.

The single main firemain system consists of one main which extends fore and aft. The main is generally installed near the centerline of the ship, extending as far forward and as far aft as necessary. The horizontal loop firemain system consists of two single fore-and-aft

LUBE OIL SERVICE PUMP (MOTOR)

DUPLEX MAGNETIC STRAINER

600 PSIG STEAM

LUBE OIL COOLER

SUCTION FROM STORAGE AND SETTLING TANKS

LUBE OIL SERVICE PUMP (TURBINE)

LUBE OIL SERVICE PUMP (ATTACHED)

L.P. TURBINE

LUBE OIL SUMP

REDUCTION GEARS

H.P. TURBINE

LUBE OIL SUPPLY PIPING

TO PURIFIER SUCTION

DRAIN TO BUCKET

DUPLEX MAGNETIC STRAINER

DRAIN TO SUMP

LUBE OIL COOLER

DRAIN TO SUMP

FROM JOURNAL AND THRUST

LUBE OIL SERVICE PUMP (ATTACHED)

L.P. TURBINE

H.P. TURBINE

LUBE OIL SUMP TANK

MAIN REDUCTION GEAR

FROM JOURNAL AND THRUST

FROM SPRAY AND JOURNAL

LEAKAGE ALONG SHAFTS

LUBE OIL DRAIN PIPING

147.81

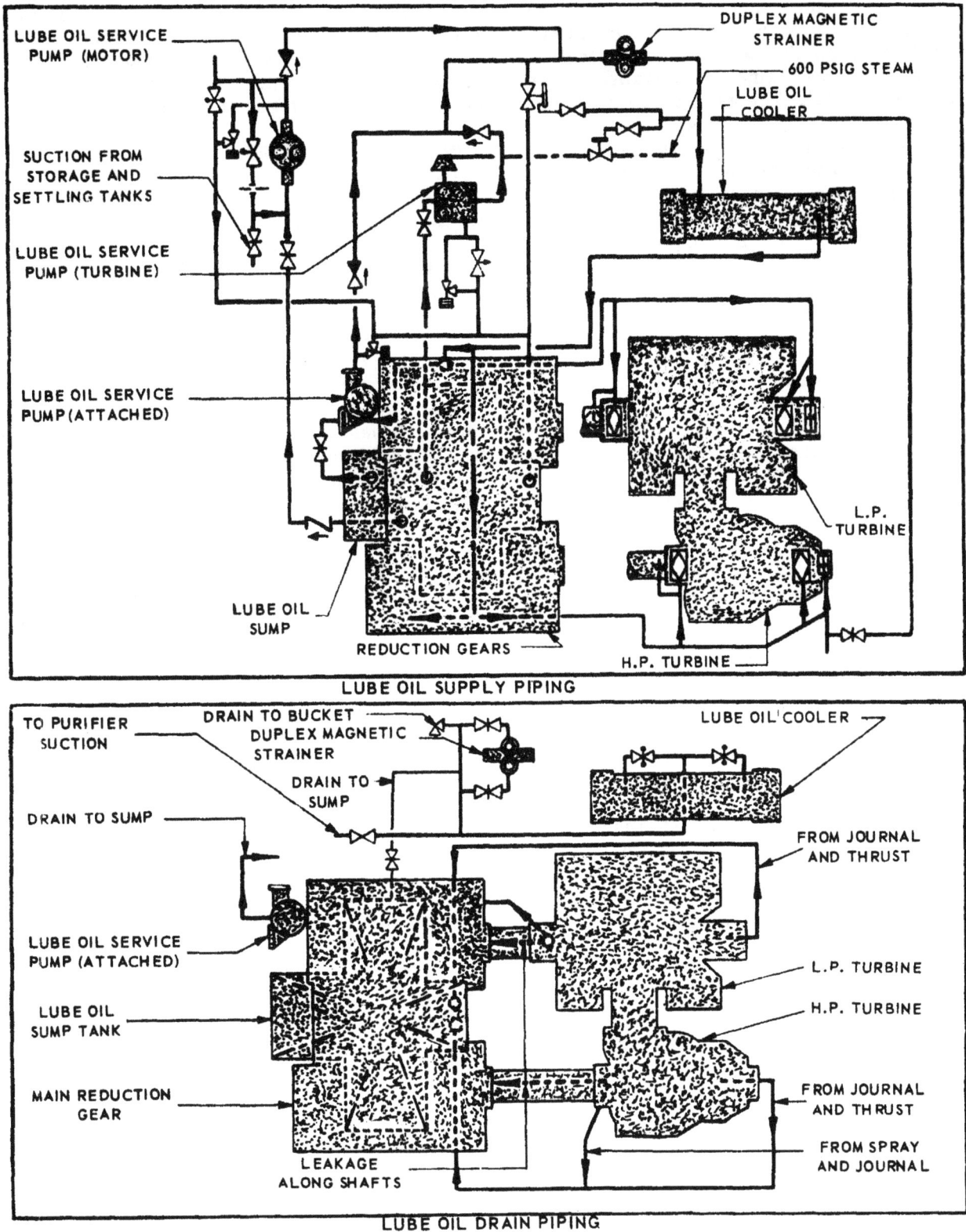

Figure 9-23.—Lubricating oil service system, DLG 14 and DLG 15.

cross-connected mains. The two mains are installed in the same horizontal plane but are separated athwartships as far as practicable. In general, the two mains are installed on the damage control deck. The vertical loop firemain system consists of two single fore-and-aft cross-connected mains. The two mains are separated both athwartship and vertically. As a rule, the lower main is located below the lowest complete watertight deck and the upper main is located below the highest complete watertight deck.

FLUSHING SYSTEMS

The shipboard flushing system is supplied with sea water by a branch from the firemain. On very small ships, a separate sanitary and flushing pump is provided which takes suction from the sea. When the flushing system is supplied from the firemain, the branch is taken as near the top of the main as possible so that sediment from the firemain will not enter the flushing system. Since the firemain pressure is too high for a flushing system, the water is led through a strainer to a reducing valve which reduces the pressure to 35 psi. Air chambers are installed in the flushing system where it runs to urinals and water closets; the air chambers absorb water hammer caused by the quick closing of the flush valves and spring-closing faucets.

DRAINAGE SYSTEMS

The drainage system aboard ship is divided into two parts: (1) the main and secondary systems, and (2) the plumbing and deck drains. Between them, these systems collect and dispose overboard all the shipboard waste fluids.

The main drainage system consists of piping installed low in the ship, with suction branches to spaces to be drained and direct connections to eductors or drainage pumps. This system generally serves the main machinery spaces and a few other spaces.

The secondary drainage system supplements the main drainage system wherever the main drainage system cannot be extended because of interference of spaces through which the passage of piping is prohibited or because the length of piping would be too great for efficient drainage. Each secondary drainage system is independent of the main drainage system and has its own pumps or eductors and its own sea connections.

Plumbing and deck drains are divided into two groups—soil drains and waste drains. Soil drains convey fluids from urinals and water closets. Waste drains convey fluids from all other plumbing fixtures and deck drains.

SPRINKLING SYSTEMS

Sprinkling systems are installed aboard ship in magazines, turret handling rooms, hangar decks, missile spaces, and other spaces where flammable materials are stowed. Water for these systems is supplied from the firemain through branch lines.

Most sprinkling systems aboard ship are of the dry type—that is, they are not charged with water beyond the sprinkling control valves except when they are in use. Sprinkling systems in magazines which contain missiles are of the wet type. The sprinkling control valves in magazine sprinkling systems are operated automatically by heat-actuated devices. Other sprinkling control valves are operated manually or hydraulically, either locally or from remote stations. In those areas of the ship in which major flammable liquid fires could occur, such as in aircraft hangars, foam sprinkling systems are provided.

WASHDOWN SYSTEMS

Washdown systems are installed aboard ship for the purpose of removing radioactive contamination from the topside surfaces of a ship. Essentially, a washdown system is a dry-pipe sprinkler system, with nozzles especially designed to throw a large spray pattern on all weather decks. For ships under construction or conversion, a permanent washdown system is installed; for ships already in service, interim washdown system kits are provided for installation by ship's force. In either case, water for the washdown system is supplied from the firemain.

POTABLE WATER SYSTEMS

Potable water systems are designed to provide a constant supply of potable water for all ship's service requirements. Potable water is stored in various tanks throughout the ship. The system is pressurized either by a pump and pressure tank or by a continuously operating circulating pump. The potable water system supplies scuttlebutts, sinks, showers, scullery, and galley, as well as providing makeup water for various fresh water cooling systems.

HYDRAULIC SYSTEMS

Hydraulic systems are used aboard ship to operate steering gear, anchor windlasses, hydraulic presses, remote control valves, and other units.[11] Hydraulic systems operate on the principle that, since liquids are noncompressible, force exerted at any point on an enclosed liquid is transmitted equally in all directions. Hence a hydraulic system permits the accomplishment of a great amount of work with relatively little effort on the part of shipboard personnel.

The medium used to transmit and distribute forces in hydraulic systems may be a petroleum-base product (hydraulic oil) or a pure phosphate ester fluid. Phosphate ester fluid is more resistant to fire and explosion than the petroleum-base oil that was used in all hydraulic systems until fairly recently. Phosphate ester fluid is now used in aircraft carrier elevators, surface ship missile systems, jet blast deflectors, seaplane servicing booms, high pressure submarine systems, and all hydraulic systems operating at pressures of more than 500 psi in new construction and conversion surface ships.

METHODS OF PROPULSION PLANT OPERATION

The major engineering systems on most naval ships are provided with cross-connections which allow the engineering plants to be opera-

ted either independently (split-plant) or together (cross-connected). In cross-connected operation, boilers may supply steam to propulsion turbines which they do not serve when the plant is split. In split-plant operation, the boilers, turbines, pumps, blowers, and other machinery are so divided that there are two or more separate and complete engineering plants.

Cross-connected operation was formerly standard for peacetime steaming, and split-plant operation was used only when maximum reliability was required—as, for example, when a ship was operating in enemy waters in time of war, operating in heavy seas, maneuvering in restricted waters, or engaged in underway fueling. However, the greater reliability of split-plant operation has led to its increasing use. At the present time, split-plant operation is the standard method of underway operation for most naval ships; cross-connected operation is used for in-port steaming but is rarely used for underway steaming.

On some ships the engineering plants can be operated by a method known as group operation. For example, the USS Forrestal, CVA 59, has four separate propulsion plants. The two forward plants (No. 1 and No. 4) constitute the forward group and the two after plants (No. 2 and No. 3) constitute the after group. Although each of these four plants is normally used for the independent (split-plant) operation of one shaft, the boilers in any one plant can be cross-connected to supply steam to the turbines in the other plant in the same group. While underway, therefore, the boilers in the No. 1 plant can be cross-connected to supply steam to the No. 4 plant, although they cannot be cross-connected to supply steam to the two plants in the other group. For in-port operation, any boiler can be cross-connected to supply steam to any turbogenerator and to all other steam-driven auxiliaries.

[11]Many hydraulically operated units are discussed in chapter 21 of this text.

CHAPTER 10

PROPULSION BOILERS

In the conventional steam turbine propulsion plant, the boiler is the source or high temperature region of the thermodynamic cycle. The steam that is generated in the boiler is led to the propulsion turbines, where its thermal energy is converted into mechanical energy which drives the ship and provides power for vital services.

In essence, a boiler is merely a container in which water can be boiled and steam generated. A teakettle on a stove is basically a boiler, although a rather inefficient one. In designing a boiler to produce a large amount of steam, it is obviously necessary to find some means of providing a larger heat transfer surface than is provided by a vessel shaped like a teakettle. In most modern boilers, the steam generating surface consists of between one and two thousand tubes which provide a maximum amount of heat transfer surface in a relatively small space. As a rule, the tubes communicate with a steam drum at the top of the boiler and with water drums and headers at the bottom of the boiler. The tubes and part of the drums are enclosed in an insulated casing which has space inside it for a furnace. As we will see presently, a boiler appears to be a fairly complicated piece of equipment when it is considered with all its fittings, piping, and accessories. It may be helpful, therefore, to remember that the basic components of a saturated-steam boiler are merely the tubes in which steam is generated, the drums and headers in which water is contained and steam is collected, and the furnace in which combustion takes place.

Practically all boilers used in the propulsion plants of naval ships are designed to produce both saturated steam and superheated steam. To our basic boiler, therefore, we must now add another component: the superheater. The superheater on most boilers consists of headers, usually located at the back or at the bottom of the boiler, and a number of superheater tubes which communicate with the headers. Saturated steam from the steam drum is led through the superheater; since the steam is now no longer in contact with the water from which it was generated, the steam becomes superheated without any appreciable increase in pressure as additional heat is supplied. In some boilers, there is a separate superheater furnace; in others, the superheater tubes project into the same furnace that is used for the generation of saturated steam.

Some question may arise concerning the need for both saturated steam and superheated steam. Many steam-driven auxiliaries—particularly if they have reciprocating engines—require saturated steam for the lubrication of the moving parts of the driving machine. The propulsion turbines, on the other hand, and many auxiliaries as well, perform much more efficiently when superheated steam is used. There is more available energy in superheated steam than in saturated steam at the same pressure, and the use of higher temperatures vastly increases the thermodynamic efficiency of the propulsion cycle since the efficiency of a heat engine depends upon the absolute temperature at the source (boiler) and at the receiver (condenser). In some instances, the gain in efficiency resulting from the use of superheated steam may be as much as 15 percent for 200 degrees of superheat. This increase in efficiency is particularly important for naval ships because it allows substantial savings in fuel consumption and in space and weight requirements. A further advantage in using superheated steam for propulsion turbines is that it causes relatively little erosion or corrosion since it is free of moisture.

BOILER DEFINITIONS

In order to ensure accuracy and uniformity in the use of boiler terms, the Naval Ship Systems

Command has established a number of standard definitions relating to boilers. Since these terms are quite widely used, the student will find it helpful to understand the following terms and to use them correctly.

BOILER FULL-POWER CAPACITY.—The total quantity of steam required to develop contract shaft horsepower of the ship, divided by the number of boilers installed in the ship, gives boiler full-power capacity. Boiler full-power capacity is expressed as the number of pounds of steam generated per hour at a specified pressure and temperature. Boiler full-power capacity is listed in the design data section of the manufacturer's technical manual for the boilers on each ship; it may be listed as capacity at full power or as designed rate of actual evaporation per boiler at full power.

BOILER OVERLOAD CAPACITY.—Boiler overload capacity is usually 120 percent of boiler full-power capacity. Boiler overload capacity is listed in the design data section of the manufacturer's technical manual for the boilers; it may be listed as boiler overload capacity or as full power plus 20 percent.

SUPERHEATER OUTLET PRESSURE.—Superheater outlet pressure is the actual steam pressure carried at the superheater outlet.

STEAM DRUM PRESSURE.—Steam drum pressure is the pressure actually carried in the boiler steam drum.

OPERATING PRESSURE.—Operating pressure is the constant pressure at which the boiler is operated in service. Depending upon various factors, chiefly design features of the boiler, the constant pressure may be carried at the steam drum or at the superheater outlet. Operating pressure is specified in the design of the boiler and is given in the manufacturer's technical manual. Operating pressure is the same as superheater outlet pressure or steam drum pressure (depending upon which is used as the controlling pressure) only when the boiler is operating at full-power capacity, for combatant ships, or some other specified rate, for other ships. When the boiler is operating at less than full-power capacity (or other specified rate), the actual pressure at the steam drum or at the superheater outlet will vary from the designated operating pressure.

DESIGN PRESSURE.—Design pressure is the pressure specified by the boiler manufacturer as a criterion for boiler design. It is often approximately 103 percent of steam drum pressure. Operating personnel seldom have occasion to be concerned with design pressure; the term is noted here because there is a good deal of confusion between design pressure and operating pressure. The two terms do not mean the same thing.

DESIGN TEMPERATURE.—Design temperature is the intended maximum operating temperature at the superheater outlet, at some specified rate of operation. The specified rate of operation is normally full-power capacity for combatant ships.

OPERATING TEMPERATURE.—Operating temperature is the actual temperature at the superheater outlet. As a rule, operating temperature is the same as design temperature only when the boiler is operating at the rate specified in the definition of design temperature.

BOILER EFFICIENCY.—The efficiency of a boiler is the ratio of the Btu per pound of fuel absorbed by the water and steam to the Btu per pound of fuel fired. In other words, boiler efficiency is output divided by input, or heat utilized divided by heat available. Boiler efficiency is expressed as a percentage.

FIREROOM EFFICIENCY.—Boiler Efficiency corrected for blower and pump steam consumption is called fireroom efficiency. Note: Fireroom efficiency is NOT boiler plant efficiency or propulsion plant efficiency.

STEAMING HOURS.—The term steaming hours is used to include all time during which the boiler has fires lighted for raising steam and all time during which steam is being generated. Time during which fires are not lighted is not included in steaming hours.

HEATING SURFACES.—The total heating surface of a boiler includes all parts of the boiler which are exposed on one side to the gases of combustion and on the other side to the water and steam being heated. Thus the total heating surface equals the sum of the generating surface, the superheater surface, and the economizer surface. All heating surfaces are measured on the combustion gas side.

The generating surface is that part of the total heating surface in which water is being heated and steam is being generated. The generating surface includes the generating tubes, the water wall tubes, the water screen tubes, and any water floor tubes that are not covered by refractory material.

The superheater surface is that part of the total heating surface in which the steam is superheated after leaving the boiler steam drum.

The economizer surface is that portion of the total heating surface in which the feed water is heated before it enters the generating part of the boiler.

DESUPERHEATERS.—On boilers with non-controlled superheaters, all steam is superheated but a small amount is redirected through a desuperheater line. The desuperheater can be located in either the water drum or the steam drum; most generally, the desuperheater will be found in the steam drum below the normal water level. The purpose of the desuperheater is to lower the superheated steam temperature back to or close to saturated steam temperature for the proper steam lubrication of the auxiliary machinery. The desuperheater is most generally an "S" shaped tube bundle that is flanged to the superheater outlet on the inlet side and the auxiliary steam stop on the outlet side.

BOILER CLASSIFICATION

Although boilers vary considerably in details of design, most boilers may be classified and described in terms of a few basic features or characteristics. Some knowledge of these methods of classification provides a useful basis for understanding the design and construction of the various types of modern naval boilers.

Location of Fire and Water Spaces

One basic classification of boilers is made according to the relative location of the fire and water spaces. By this method of classification, all boilers may be divided into two classes: fire-tube boilers and water-tube boilers. In fire-tube boilers, the gases of combustion flow through the tubes and thereby heat the water which surrounds the tubes. In water-tube boilers, the water flows through the tubes and is heated by the gases of combustion that fill the furnace and heat the outside metal surfaces of the tubes.

All boilers used in the propulsion plants of modern naval ships are of the water-tube type.

Fire-tube boilers[1] were once used extensively in marine installations and are still used in the propulsion plants of some older merchant ships. However, fire-tube boilers are not suitable for use as propulsion boilers in modern naval ships because of their excessive weight and size, the excessive length of time required to raise steam, and their inability to meet demands for rapid changes in load. The only fire-tube boilers currently in naval use are some small auxiliary boilers.[2]

Type of Circulation

Water-tube boilers are further classified according to the cause of water circulation. By this mode of classification, we have natural circulation boilers and controlled circulation boilers.

In natural circulation boilers, the circulation of water depends on the difference between the density of an ascending mixture of hot water and steam and a descending body of relatively cool and steam-free water. The difference in density occurs because the water expands as it is heated and thus becomes less dense. Another way to describe natural circulation is to say that it is caused by convection currents which result from the uneven heating of the water contained in the boiler.

Natural circulation may be either free or accelerated. Figure 10-1 illustrates free natural circulation. Note that the generating tubes are installed at a slight angle of inclination which allows the lighter hot water and steam to rise and the cooler and heavier water to descend. When the generating tubes are installed at a greater angle of inclination, the rate of water circulation is definitely increased. Therefore, boilers in which the tubes slope quite steeply from steam drum to water drum are said to have accelerated natural circulation. This type of circulation is illustrated in figure 10-2.

Most modern naval boilers are designed for accelerated natural circulation. In such boilers, large tubes (3 or more inches in diameter) are

___

[1] As, for example, the old "Scotch marine boiler."
[2] Auxiliary boilers (some water-tube, some fire-tube) are installed in diesel-driven ships and in many steam-driven combatant ships. They are used to supply steam or hot water for galley, and other "hotel" services and for other auxiliary requirements in port.

147.83

Figure 10-1.—Natural circulation (free type).

installed between the steam drum and the water drums. These large tubes, called <u>downcomers</u>, are located outside the furnace and away from the heat of combustion, thereby serving as pathways for the downward flow of relatively cool water. When a sufficient number of downcomers are installed, all small tubes can be generating tubes, carrying steam and water upward, and all downward flow can be carried by the downcomers. The size and number of downcomers installed varies from one type of boiler to another, but some are installed on all modern naval boilers.

Controlled circulation boilers are, as their name implies, quite different in design from the boilers that utilize natural circulation. Controlled circulation boilers depend upon pumps, rather than upon natural differences in density, for the circulation of water within the boiler. Because controlled circulation boilers are not limited by the requirement that hot water and steam must be allowed to flow upward while cooler water flows downward, a great variety of arrangements may be found in controlled circulation boilers.

Controlled circulation boilers have been used in a few naval ships during the past few years. In general, however, they are still considered more or less experimental for naval use.

Arrangement of Steam and Water Spaces

Natural circulation boilers are classified as drum-type boilers or as header-type boilers, depending upon the arrangement of the steam and water spaces. Drum-type boilers have one or more water drums (and usually one or more water headers as well). Header-type boilers have no water drum; instead, the tubes enter a great many water headers.

What is a header, and what is the difference between a header and a drum? The term HEADER is commonly used in engineering to describe any tube, chamber, drum, and similar piece to which a series of tubes or pipes are connected in such a way as to permit the flow of fluid from one tube (or group of tubes) to another. Essentially, a header is a type of manifold. As far as boilers are concerned, the only distinction between a drum and a header is the distinction of size. Drums are larger than headers, but both serve basically the same purpose.

Drum-type boilers are further classified according to the overall configuration of the boiler, with particular regard to the shape formed by the steam and water spaces. For example, double-furnace boilers are often called "M-type boilers" because the arrangement of tubes is roughly M-shaped. Single-furnace boilers are often called "D-type boilers" because the tubes form (roughly) the letter D.[3]

Number of Furnaces

All boilers that are now commonly used in the propulsion plants of naval ships may be classified as being either single-furnace boilers or double-furnace boilers. The D-type boiler is a single-furnace boiler; the M-type boiler is a double-furnace (or divided-furnace) boiler.

Furnace Pressure

Recent developments in naval boilers make it convenient to classify boilers on the basis of the

---

[3] An interesting variation in this terminology occurred when the single-furnace or D-type boiler became standard for steam-driven destroyer escorts and thus subsequently became known as a "DE-type boiler." The term "DE-type boiler" is still used rather freely; its use should be discouraged, however, as this general type of boiler is now installed on many ships other than destroyer escorts.

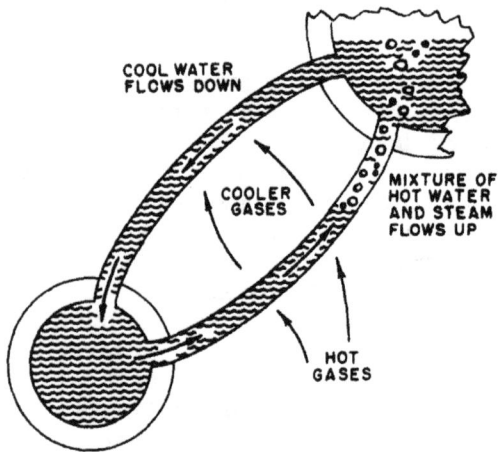

139.17

Figure 10-2.—Natural circulation
(accelerated type).

air pressure used in the furnace. Most boilers now in use in naval propulsion plants operate with a slight air pressure (seldom over 10 psig) in the boiler furnace. This slight pressure, which results from the use of forced draft blowers to supply combustion air to the boilers, is not sufficient to warrant calling these boilers "pressurized-furnace boilers." However, a new type of boiler has recently appeared on the scene and is being installed in some ships. This new boiler is truly a pressurized-furnace boiler, since the furnace is maintained under a positive air pressure of approximately 65 psia (about 50 psig) when the boiler is operating at full power. The air pressure in the furnace is maintained by a special air compressor. Hence we must now make a distinction between this new pressurized-furnace boiler, on the one hand, and all other naval propulsion boilers, on the other hand, with respect to the pressure maintained in the furnace.

Type of Superheater

On almost all boilers currently used in the propulsion plants of naval ships, the superheater tubes are protected from radiant heat by water screen tubes. The water screen tubes absorb the intense radiant heat of the furnace, and the superheater tubes are heated by convection currents rather than by radiation. Hence, the superheaters are referred to as convection-type superheaters.

On a few older ships, the superheater tubes are not screened by water screen tubes but are exposed directly to the radiant heat of the furnace. Superheaters of this kind are called radiant-type superheaters. Although radiant-type superheaters are rarely used at present, it is possible that they may come into use again in future boiler designs.

Control of Superheat

A boiler which provides some means of controlling the degree of superheat independently of the rate of steam generation is said to have controlled superheat. A boiler in which such separate control is not possible is said to have uncontrolled superheat.

Until recently, the term superheat control boiler was used to identify a double-furnace boiler and the term uncontrolled superheat boiler (or no control superheat boiler) was used to identify a single-furnace boiler. Most double-furnace boilers now in use do, in fact, have controlled superheat, and most single-furnace boilers do not have controlled superheat. However, recent developments in boiler design make superheat control independent of the number of furnaces in the boiler. Single-furnace boilers WITH controlled superheat and double-furnace boilers WITHOUT controlled superheat are both possible. The time has come, therefore, to stop relating the number of furnaces in a boiler to the control (or lack of control) of superheat.

Operating Pressure

For some purposes it is convenient to classify boilers according to operating pressure. Most classifications of this type are approximate rather than exact. Header-type boilers and some older drum-type boilers are often called "400-psi boilers" even though the operating pressures may range from 300 psi (or even lower) to about 450 psi. The term "600-psi boiler" is often applied to various double-furnace and single-furnace boilers with operating pressures ranging from about 435 psi to about 700 psi.

The term "high pressure boiler" is at present used rather loosely to identify any boiler that operates at substantially higher pressure than the so-called "600-psi boilers." In general, we will consider any boiler that operates at 751 psi or above as a high pressure boiler. A good many boilers recently installed on naval ships operate at approximately 1200 psi; for some purposes, it is convenient to group these boilers together and refer to them as "1200-psi boilers."

As may be seen, classifying boilers by operating pressure is not very precise, since actual operating pressures may vary widely within one group. Also, any classification based on operating pressure may easily become obsolete. What is called a high pressure boiler today might well be called a low pressure boiler tomorrow.

BOILER COMPONENTS

Most propulsion boilers now used by the Navy have essentially the same components: steam and water drums, generating and circulating tubes, superheaters, economizers, fuel oil burners, furnaces, casings, supports, and a number of accessories and fittings required for boiler operation and control. The basic components of boilers are described here. In later sections of this chapter we will see how the components are arranged to form various common types of naval propulsion boilers.

Drums and Headers

Drum-type boilers are installed in the ship in such a way that the long axis of the boiler drums will run fore and aft rather than athwartships, so that the water will not surge from one end of the drum to the other as the ship rolls.

The steam drum is located at the top of the boiler. It is cylindrical in shape, except that on some boilers, it may be slightly flattened along its lower curved surface. The steam drum receives feed water and serves as a place for the accumulation of the saturated steam that is generated in the tubes. The tubes enter the steam drum below the normal water level of the drum. The steam and water mixture from the tubes goes through separators which separate the water from the steam.

Figure 10-3 shows the way in which a steam drum is constructed. Two sheets of steel are rolled or bent to the required semicircular shape and then welded together. The upper sheet is called the wrapper sheet; the lower sheet is called the tube sheet. Notice that the tube sheet is thicker than the wrapper sheet. The extra thickness is required in the tube sheet to ensure adequate strength of the tube sheet after the holes for the generating tubes have been drilled. The ends of the drum are enclosed with drumheads which are welded to the shell, as shown in figure 10-4. One drumhead contains a manhole which permits access to the drum for inspection, cleaning, and repair.

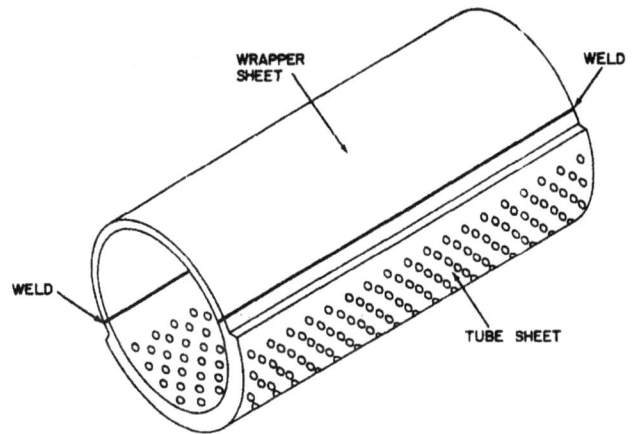

38.19

Figure 10-3.—Boiler steam drum.

The steam drum either contains or is connected to many of the important fittings and instruments required for the operation and control of the boiler. These fittings and controls are discussed separately in chapter 11 of this text.

Water drums and water headers equalize the distribution of water to the generating tubes and provide a place for the accumulation of loose scale and other solid matter that may be present in the boiler water. In drum-type boilers, the water drums and water headers are at the

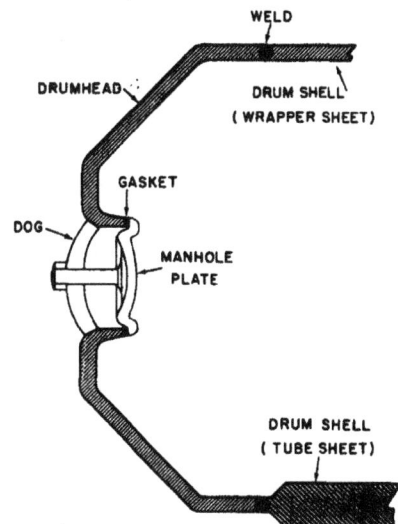

38.20

Figure 10-4.—Drumhead secured to steam drum shell.

bottom of the boiler. Water drums are usually round in cross section; headers may be round, oval, or square. Headers are provided with access openings of the type shown in figure 10-5. Water drums are usually made with manholes similar to the manholes in steam drums.

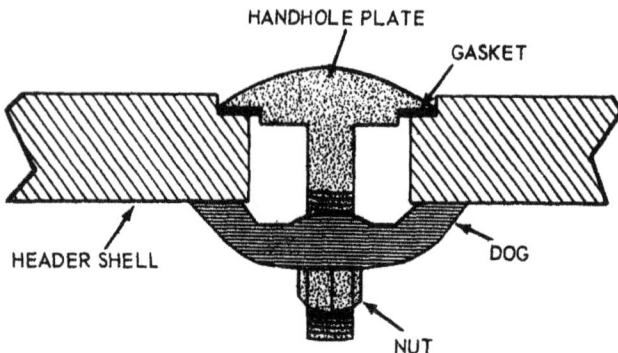

38.21

Figure 10-5.—Header handhole and handhole plate.

Generating and Circulating Tubes

Most of the tubes in a boiler are generating or circulating tubes. There are four main kinds of generating and circulating tubes: (1) generating tubes in the main generating tube bank; (2) water wall tubes, (3) water screen tubes, and (4) downcomers. The tubes are made of steel similar to the steel used for the drums and headers. Most tubes in the main generating bank are about 1 inch or 1 1/4 inches in outside diameter. Water wall tubes, water screen tubes, and the two or three rows of generating tubes next to the furnace are generally a little larger. Downcomers are larger still, being on the average about 3 to 11 inches in outside diameter.

Since the steam drum is at the top of the boiler and the water drums and headers are at the bottom, it is obvious that the generating and circulating tubes must be installed more or less vertically. Each tube enters the steam drum and the water drum (or water header) at right angles to the drum surfaces. This means that all tubes in any one row are curved in exactly the same way, but the curvature of different rows is not the same. Tubes are installed normal to the drum surfaces in order to allow the maximum number of tube holes to be drilled in the tube sheets with a minimum weakening of the drums. However, nonnormal installation is permitted if

certain advantages can be achieved in design characteristics.

What purpose do all these generating and circulating tubes serve? The generating tubes are the ones in which most of the saturated steam is generated. The water wall tubes serve primarily to protect the furnace refractories, thus allowing higher heat release rates than would be possible without this protection. However, the water wall tubes are also generating tubes at high firing rates. Water screen tubes protect the superheater from direct radiant heat. Water screen tubes, like water wall tubes, are generating tubes at high firing rates. Downcomers are installed between the inner and outer casings of the boiler to carry the downward flow of relatively cool water and thus maintain the boiler circulation. Downcomers are not designed to be generating tubes under any conditions.

In addition to the four main types of generating and circulating tubes just mentioned, there are a few large superheater support tubes which, in addition to providing partial support for the steam drum and for the superheater, serve as downcomers at low firing rates and as generating tubes at high firing rates.

Since a modern boiler is likely to contain between 1000 and 2000 tubes, some system of tube identification is essential. Generating and circulating tubes are identified by LETTERING the rows of tubes and NUMBERING the individual tubes in each row. A tube row runs from the front of the boiler to the rear of the boiler. The row of tubes next to the furnace is row A, the next is row B, the next is row C, and so forth. If there are more than 26 rows in a tube bank, the rows after Z are lettered AA, BB, CC, DD, EE, and so forth. Each tube in each row is then designated by a number, beginning with 1 at the front of the boiler and numbering back toward the rear.

The letter which identifies a tube row is often preceded by an R or an L, particularly in the case of water screen tubes, superheater support tubes, and furnace division wall tubes. When an R or an L is used AFTER the regular letter and number identification of a tube, it may indicate either that the tube is bent for a right-hand or left-hand boiler or that the tube is studded or finned on the right-hand side or on the left-hand side.

Figure 10-6 shows a Boiler Tube Renewal Sheet for a double-furnace boiler and illustrates the method used to identify tubes.

The water wall tube row is not identified by a letter in figure 10-6. The tubes in this row are often identified by the letter W and a following letter which indicates the type of tube or its position in the row. Still another letter (an R or an L) may be added to indicate that the tube is studded or finned on the right-hand side or on the left-hand side.

The tubes which screen the superheater from direct radiant heat are identified in figure 10-6 as the LA, LB, and LC rows. Within each row, the individual tubes are numbered: LA-1, LA-2, LA-3, and so forth.

Figure 10-6 identifies the superheater support and drum support tubes as the LD row. Note that these are NOT superheater tubes. The superheater tubes in this boiler, as in most boilers, are installed horizontally. The superheater support tubes and the drum support tubes are installed vertically; they are identified as LD-1, LD-2, LD-3, and so forth.

The first row of division wall tubes is identified in figure 10-6 as the LE row. The second row of division wall tubes may be identified as the LF row or as the D row. Identification of tubes in this row is usually made by using the row identification (LF, D, or whatever row identification is used for the particular boiler) followed by a letter to indicate the type of tube or its position in the row; still another letter (an R or an L) may be added to indicate that the tube is studded or finned on the right-hand side or on the left-hand side.

Tubes in the main generating bank are identified by lettering the rows and numbering the individual tubes, as shown in figure 6-6. The two rows nearest the saturated-side furnace are slightly larger than the rest of the generating tubes; they serve as water screen tubes. These two rows are often called the RA and the RB rows. Individual tubes in these rows are identified by number in the same way that the rest of the generating tubes are identified.

Note that the superheater tubes are also identified in figure 10-6. Identification of superheater tubes is discussed in the section that deals with superheaters.

The discussion of boiler tube identification given here is based on one particular type of boiler—that is, a double-furnace boiler. The same general principles of tube identification apply to most other drum-type boilers now in naval use, but the details of tube identification are necessarily different in different types of boilers.

Superheaters

Most propulsion boilers now in naval service have convection-type superheaters, with water screen tubes installed between the superheater and the furnace to absorb the intense radiant heat and thus protect the superheater.

Most convection-type superheaters have U-shaped tupes which are installed horizontally in the boiler and two headers which are installed more or less vertically at the rear of the boiler. One end of each U-shaped tube enters one superheater header, and the other end enters the other header. The superheater headers are divided internally by one or more division plates which act as baffles to direct the flow of steam. In some cases the superheater headers are divided externally as well as internally.

Figure 10-7 illustrates some convection-type superheater arrangements that are used on double-furnace boilers. Part A is a plan view of the superheater tubes, showing how the tubes enter the headers. Part B shows a superheater in which each header is divided into two sections, and illustrates the flow of steam through the superheater. Part C illustrates the flow of steam through a superheater in which one header has one internal division and the other header has two internal divisions. As may be seen from figure 10-7, the steam makes several passes through the furnaces. The number of passes is determined by the number of header divisions and by the relative locations of the steam inlet and the steam outlet.

The superheater tubes are installed so that their U-shaped ends project forward toward the front of the boiler. In a double-furnace boiler, the superheater tubes project forward into a space between the water screen tubes and some tubes called furnace division wall tubes. The superheater tubes and the surrounding water screen and division wall tubes are thus together the dividing line between the superheater-side furnace and the saturated-side furnace. In a single-furnace boiler, the superheater tubes project forward into a space in the main bank of generating tubes. The tubes between the superheater tubes and the furnace serve as water screen tubes.

Some recent boilers have walk-in or cavity-type superheaters. In this type of superheater,

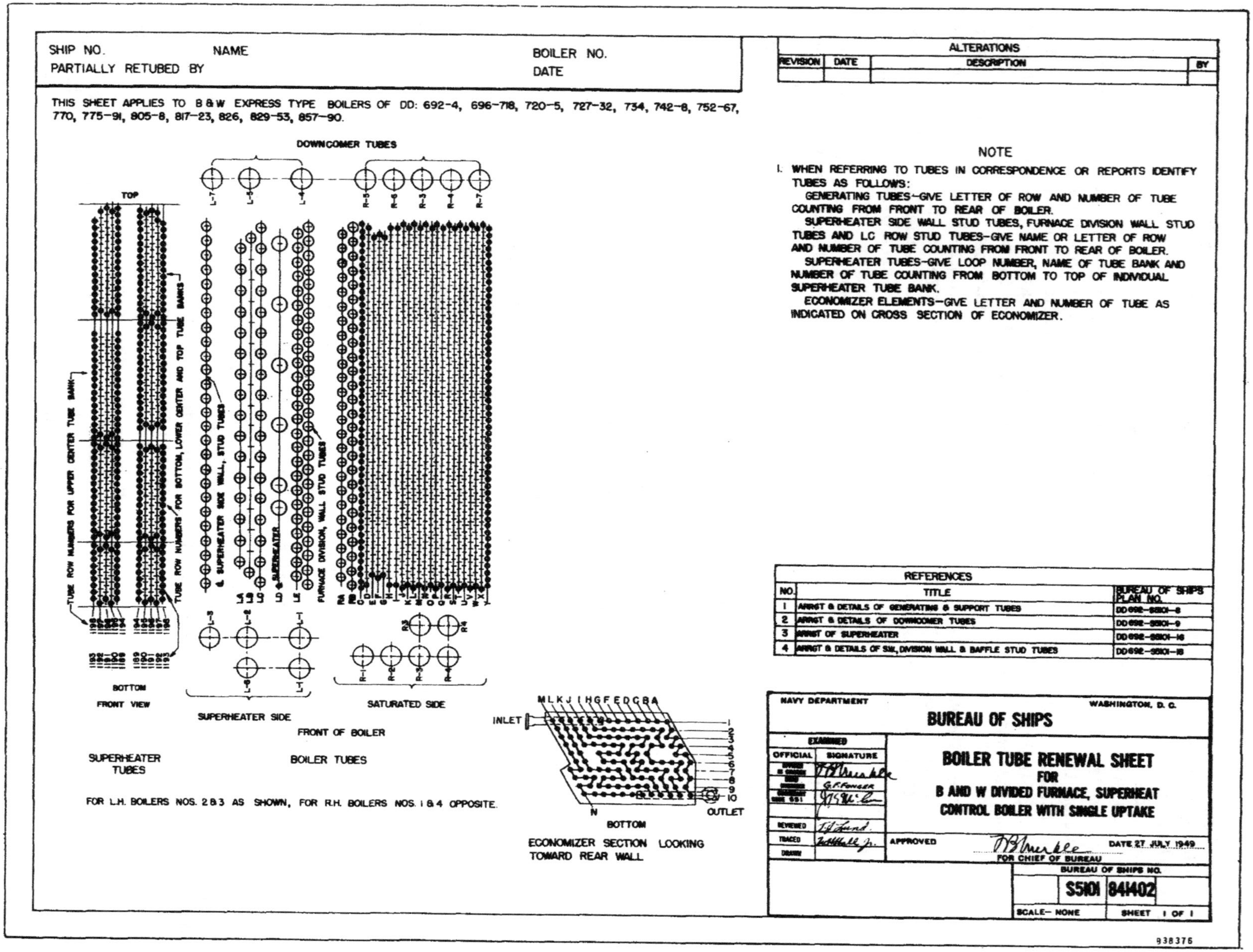

98.168

Figure 10-6.—Boiler Tube Renewal Sheet for a Babcock & Wilcox double furnace boiler.

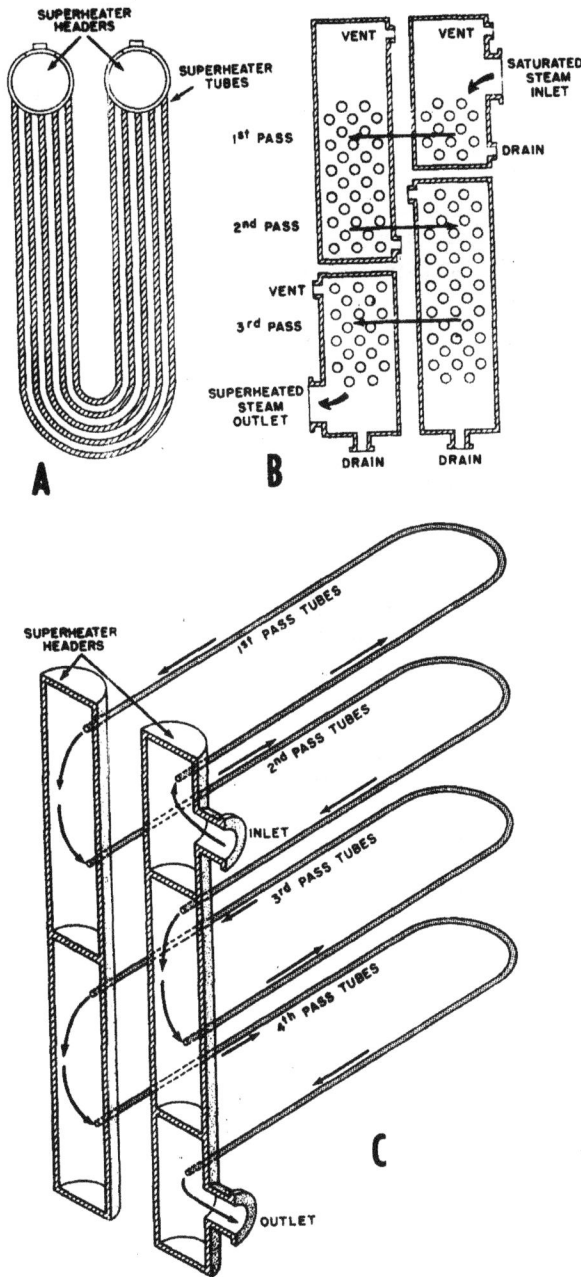

38.23

Figure 10-7.—Convection-type superheater
(double-furnace boiler).

an access space or cavity is provided in the middle of the superheater tube bank. The cavity, which runs the full length and height of the superheater, greatly increases the accessibility of the superheater for cleaning, maintenance, and repair. Some of the walk-in superheaters

have U-shaped tubes. Others, such as the one shown in figure 10-8, have W-shaped tubes.

A few boilers of recent design have vertical, rather than horizontal, convection-type superheaters. In these boilers, the U-bend superheater tubes are installed almost vertically, with the U-bends near the top of the boiler; the tubes are approximately parallel to the main bank of generating tubes and the water screen tubes. Two superheater headers are near the bottom of the boiler, running horizontally from the front of the boiler to the rear.

Superheater tubes are generally identified by loop number, name of tube bank, and number of tube within the bank. In the case of horizontal superheater tubes, as shown in figure 10-6, you count from the bottom toward the top to get the tube number. In the case of vertical superheater tubes, you count from the front of the boiler toward the rear.

Desuperheaters

On boilers with noncontrolled superheaters, all steam is superheated, but a small amount of steam is redirected through a desuperheater line, the desuperheater can be located in either the water drum or the steam drum, most generally the desuperheater will be found in the steam drum below the normal water level. The purpose of the desuperheater is to lower the super-heated steam temperature back to or close to saturated steam temperature for the proper steam lubrication of the auxiliary machinery. The desuperheater is most generally an "S" shaped tube bundle that is flanged to the superheater outlet on the inlet side and the auxiliary steam stop on the outlet side.

Economizers

An economizer is installed on practically every boiler used in naval propulsion plants. The economizer is an arrangement of tubes installed in the uptake space from the furnace; thus the economizer tubes are heated by the rising gases of combustion. All feed water flows through the economizer tubes before entering the steam drum, and the feed water is warmed by heat which would otherwise be wasted as combustion gases pass up the stack. In general, boilers operating at high pressures and temperatures have larger economizer surfaces than boilers operating at low pressure and temperatures.

REAR OF BOILER

SUPPORT CHANNELS

ACCESS
PANEL

ACCESS DOOR

PLAN VIEW

SUPPORT
CHANNEL

ACCESS
DOOR

VERTICAL CENTERLINE
OF WATER DRUM

ELEVATION

38.27

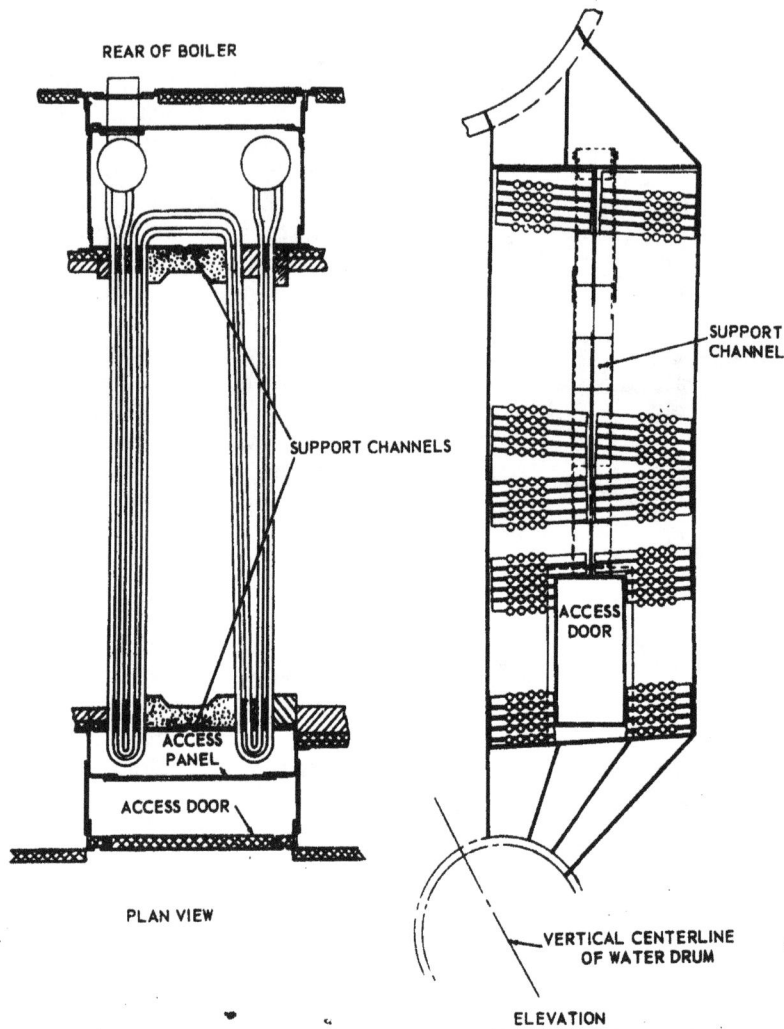

Figure 10-8.—Arrangement of W-tube walk-in superheater.

Economizer tubes may be of various shapes. Most commonly, perhaps, they are a continuous loop of U-shaped elements from inlet to outlet header. Almost all economizer tubes have some sort of metal projections from the outer tube surface. These projections, which are of aluminum, steel, or other metal, are shaped in various ways. Figure 10-9 shows a U-bend economizer tube with aluminum gill rings that are circular in cross section. Other types of projections in use include rectangular fins and star-shaped disks. In all cases, the projections serve to extend the heat transfer surface of the economizer tubes on which they are installed.

38.28
Figure 10-9.—U-bend economizer tube with aluminum gill rings.

Fuel Oil Burners

Almost all fuel oil burners used on naval propulsion boilers are mounted on the boiler front. Special openings called burner cone openings are provided in the furnace front for the burners.

The two main parts of a fuel oil burner are the atomizer assembly and the air register assembly. The atomizers divide the fuel oil into very fine particles; the air registers permit combustion air to enter the furnace in such a way that it mixes thoroughly with the finely divided oil. In addition to the atomizer assembly and the air register assembly, a fuel oil burner includes various valves, fittings, connections, and (on new construction) burner safety devices which prevent spillage of oil when an atomizer assembly is removed from the burner while the burner root valve is still open.

ATOMIZERS.—Three main kinds of atomizers are now in use on naval boilers. Straight-through-flow atomizers are used on most boilers. Return-flow atomizers are used on many of the newer ships, particularly those equipped with automatic combustion controls. Steam-assist atomizers are used on boilers in some of the newest ships.

A fuel oil burner with a straight-through-flow atomizer is shown in cross section in figure 10-10. Figure 10-11 shows how burners of this type look when installed at the boiler front.

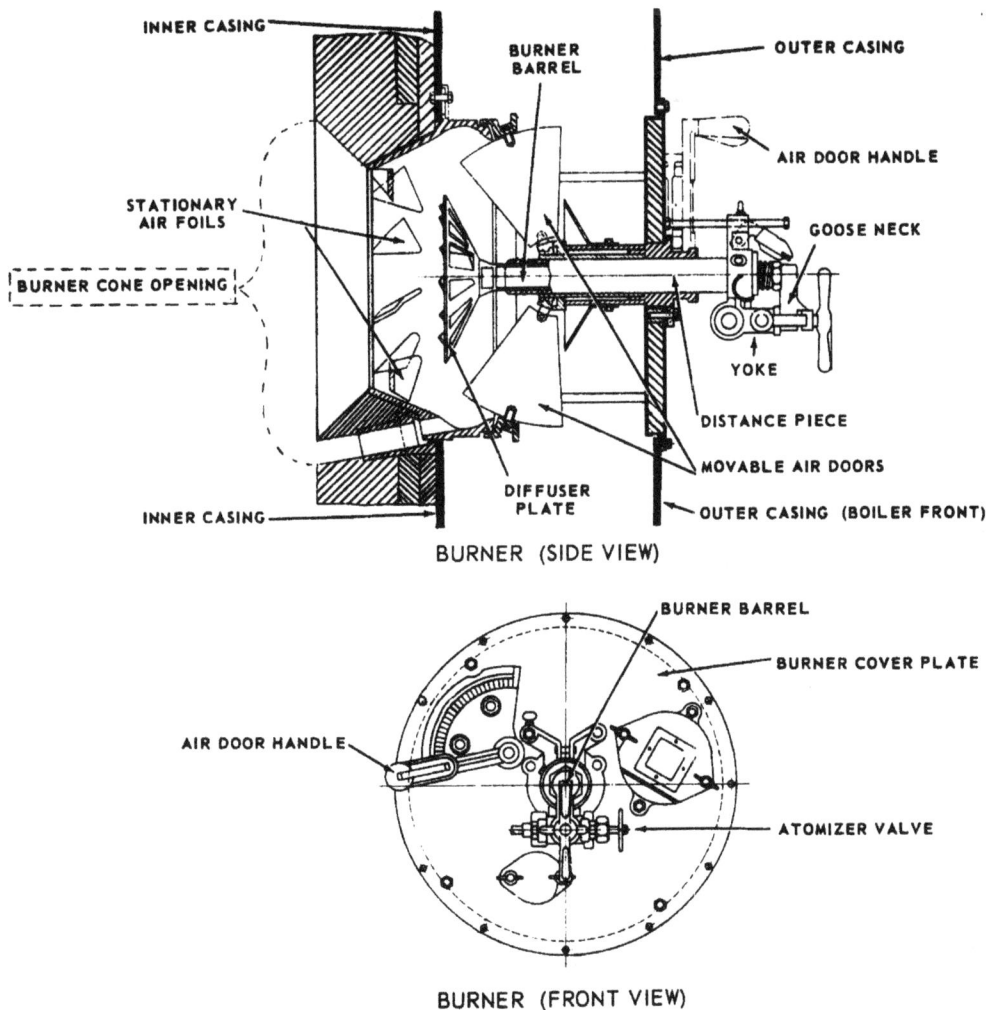

BURNER (SIDE VIEW)

BURNER (FRONT VIEW)

38.69

Figure 10-10.—Cross-sectional view of fuel oil burner with straight-through-flow atomizer.

In a straight-through-flow atomizer, all oil pumped to the atomizer is burned in the boiler furnace. The fuel oil is forced through the atomizer barrel at a pressure between 125 and 300 psi. With this type of atomizer, the firing rate is controlled by changing the number of burners in use, the fuel oil pressure, and the size of the sprayer plates.

A straight-through-flow atomizer assembly consists of a goose neck, a burner barrel (also called an atomizer barrel), a nozzle, a sprayer plate, and a tip. These parts are shown in figure 10-12. The fuel oil goes through the nozzle, which directs the oil to the grooves of the sprayer plate. These grooves are shaped so as to give the oil a high rotational velocity as it discharges into a small cylindrical whirling chamber in the center of the sprayer plate. The whirling chamber is coned out at the end and has an orifice at the apex of the cone. As the oil leaves the chamber by way of the orifice, it is broken up into very fine particles which form a cone-shaped foglike spray. A strong blast of air, which has been given a whirling motion in its passage through the burner register, catches the oil fog and mixes with it. The mixture of air and oil enters the furnace and combustion takes place.

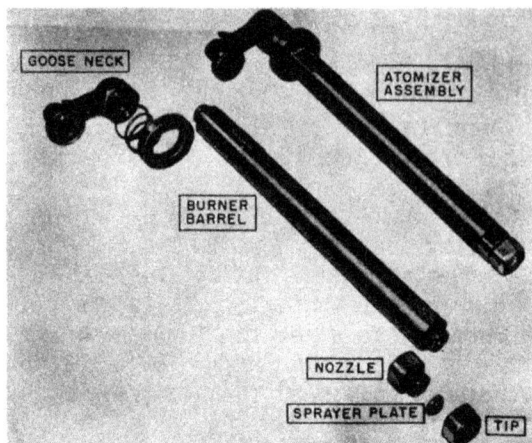

38.71

Figure 10-12.—Parts of a straight-through-flow atomizer assembly.

The sprayer plates most commonly used are called standard sprayer plates. Two types of standard sprayer plates are shown in figure 10-13. Standard sprayer plates may be either flat-faced or dished and rounded, and they may have four, six, or eight oil grooves. Standard sprayer plates with four grooves are most common.

Now that the Defense Department has authorized the conversion to a new distillate fuel (NSDF), sprayer plates for burning this type of fuel will be of the 6, 4, and 3 slot type. Sizes will also be changed as the viscosity of the new fuel is less than that of Navy Special fuel oil (NSFO). Therefore, each class of ship will need the correct size sprayer plates to permit them to burn the correct amount of fuel for full power and overload conditions.

In a return-flow (also called a variable-capacity) atomizer, part of the oil supplied to the atomizer is burned in the boiler furnace and part is returned. Several types of return-flow atomizers are in use. One type (Todd) is designed to operate with a constant fuel oil supply pressure of 300 psig and a minimum return pressure of 25 psig. Another return-flow atomizer (Babcock & Wilcox) operates with a variable fuel oil supply pressure (up to 1000 psi) and a variable return pressure. Still another type (also Babcock & Wilcox) operates with a constant supply pressure of 1000 psi and a variable return pressure.

The return-flow atomizer shown in figure 10-14 operates with a constant fuel oil supply

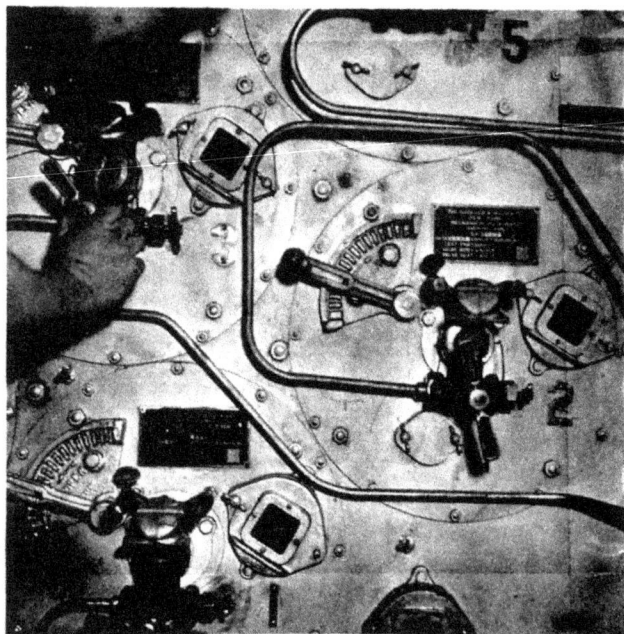

38.70

Figure 10-11.—Fuel oil burners installed on boiler front.

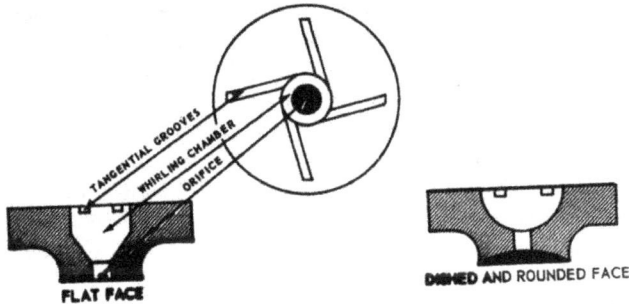

38.74

Figure 10-13.—Two kinds of standard sprayer plates.

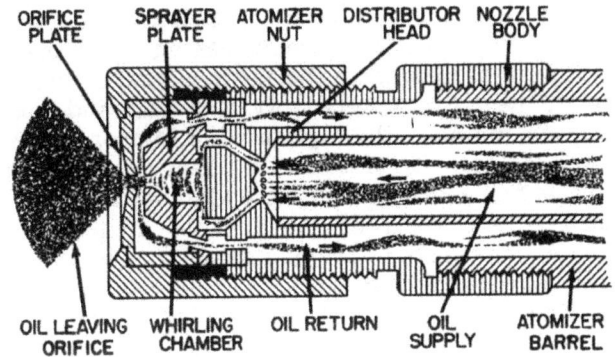

38.75

Figure 10-14.—Return-flow atomizer.

pressure. The amount of oil burned in the furnace is controlled by regulating the oil return pressure. The supply oil enters through the tube-like opening down the middle of the atomizer barrel and passes through the sprayer plate. The tangential slots or grooves in the sprayer plate cause the oil to enter the whirling chamber with a rotary motion. As the oil reaches the return annulus, centrifugal force causes a certain amount of the oil to enter the return annulis. The amount of oil thus returned is determined by the back pressure in the return line; the back pressure is in turn determined by the extent to which the return line control valve is open. The oil which is not returned emerges from the orifice in the form of a hollow conical spray of atomized oil. The amount of oil burned is the difference between the amount of oil supplied and the amount returned.

The straight-through-flow atomizers and the return-flow atomizers just described are both considered to be mechanical atomizers of the pressure type. The steam-assist atomizer, now in use on some new ships, operates on different principles. The fuel oil enters a steam-assist atomizer at relatively low pressure and is very finely atomized by a jet of steam. Combustion air is supplied by forced draft blowers, just as it is in other installations.

A steam-assist atomizer has two supply lines coming into it, one for fuel oil and one for steam. These two lines make the atomizer look a good deal like a return-flow atomizer. However, the steam-assist atomizer does not return any fuel oil; instead, all oil supplied to the atomizer is burned in the boiler furnace. Sprayer plates and other parts are somewhat differently shaped in steam-assist atomizers than they are in

straight-through-flow atomizers and return-flow atomizers.

One reason why steam-assist atomizers have not been used for naval propulsion boilers until quite recently is that they use a considerable amount of steam which cannot be recovered and returned to the feed system. However, they have some advantages that tend to make up for this disadvantage. A major advantage is that the firing range of steam-assist atomizers is much greater than the firing range of other types of atomizers. This characteristic makes the steam-assist atomizer particularly useful for naval service, since it means that large changes of load can be made merely by varying the fuel oil supply pressure, without cutting burners in and out. The fuel oil supply pressure can be varied between 8 and 350 psi.

AIR REGISTERS.—The main parts of an air register are (1) the movable air doors, (2) the diffuser, and (3) the stationary air foils. These parts are shown in figure 10-10. The movable air doors allow operating personnel to open and close the register. When the air doors are open, air rushes in and is given a whirling motion by the diffuser plate. The diffuser thus serves to make the air mix evenly with the oil, and also to prevent flame being blown back from the atomizer. The stationary air foils guide the major quantity of air and cause it to mix with the larger oil spray beyond the diffuser.

Furnaces and Refractories

A boiler furnace is a space provided for the mixing of air and fuel and for the combustion of the fuel. A boiler furnace consists of a more or less rectangular steel casing which is lined on the floor, front wall, side walls, and rear wall with refractory material. The refractory lining serves to protect the furnace casing and to prevent loss of heat from the furnace. Refractories retain heat for a relatively long time and thus help to maintain the high furnace temperatures required for complete and efficient combustion of the fuel. Refractories are also used to form baffles which direct the flow of combustion gases and protect drums, headers, and tubes from excessive heat.

There are many different kinds of refractory materials. The particular use of each type is determined by the chemical and physical characteristics of the material in relation to the required conditions of service. Refractories commonly used in the furnaces of naval propulsion boilers include firebrick, insulating brick, insulating block, plastic fireclay, plastic chrome ore, chrome castable refractory, high temperature castable refractory, air-setting mortar, and burner refractory tile.

Casings, Uptakes, and Smokepipes

In modern boiler installations, each boiler is enclosed in two steel casings. The inner casing is lined with refractory materials, and the enclosed space constitutes the furnace. The outer casing extends around most of the inner casing, with an air space in between. Air from the forced draft blowers is forced into the space between the inner and the outer casings, and from there it flows through the air registers and into the furnace.

The inner casing encloses most of the boiler up to the uptakes. The uptakes join the boiler to the smokepipe. As a rule, the uptakes from two or more boilers connect with one smokepipe.

Both the inner and the outer casings of boilers are made of steel panels. The panels may be flanged and bolted together, with gaskets being used at the joints to make an airtight seal, or they may be welded together. The casings are made in small sections so they can be removed for the inspection and repair of boiler parts.

Saddles and Supports

Each water drum and water header rests upon two saddles, one at the front of the drum or header and one at the rear. The upper flanges of the saddle are curved to fit the curvature of the drum or header, and are welded to the drum

or header. The bottom flanges, which are flat, rest on huge beams built up from the ship's structure. The bottom flange of one saddle is bolted rigidly to its support. The bottom flange of the other saddle is also bolted to its support, but the bolt holes are elongated in a fore-and-aft direction. As the drum expands or contracts because of temperature changes, the saddle which is not rigidly fastened to the support accommodates to the changing length of the drum by sliding backward or forward over the support. The flanges which are not rigidly fastened are known as boiler sliding feet.

Airheaters

Some boilers of recent design have steam-coil airheaters to preheat the combustion air before it enters the furnace. A typical steam-coil airheater consists of two coil blocks, each coil block having three sections of heating coils in a single casing. Each individual section has rows of copper-nickel alloy tubes, helically wound with copper fins. Airheaters used in the past on some older naval ships were installed in the uptakes and the combustion air was preheated by the combustion gases; these airheaters thus utilized heat which would otherwise have been wasted. The use of these older airheaters was discontinued in naval ships because the saving of heat was not considered sufficient to justify the added space and weight requirements. The new steam-coil airheaters use auxiliary exhaust steam as the heating agent; they are installed near the point where the combustion air enters the double casing.

Fittings, Instruments, and Controls

The major boiler components just described could not function without a number of fittings, instruments, and control devices. These additional boiler parts are merely mentioned here for the sake of completeness; they are taken up in detail in chapter 11 of this text.

Internal fittings installed in the steam drum may include equipment for distributing the incoming feed water, for separating and drying the steam, for giving surface blows to remove solid matter from the water, for directing the flow of steam and water within the steam drum, and for injecting chemicals for boiler water treatment. In addition, many boilers have desuperheaters for desuperheating the steam needed for auxiliary purposes.

External fittings and instruments used on naval boilers may include drains and vents; sampling connections, feed stop and check valves; steam stop valves; safety valves; soot blowers, watergage glasses and remote water level indicators; pressure and temperature gages; superheater temperature alarms; superheater steam flow indicators; smoke indicators; and various items used for the automatic control of combustion and water level.

TYPES OF PROPULSION BOILERS

Now that we have examined the basic components used in most naval propulsion boilers, let us put these components together, so to speak, to see how they are arranged to form the types of boilers now used in the propulsion plants of naval ships. The order of presentation is more or less historical, starting with the header-type boiler (which is probably the oldest boiler design still in service), going on to double-furnace boilers and to both older and newer types of single-furnace boilers, and ending with the recently installed pressurized-furnace boiler.

Header-Type Boilers

Sectional header boilers, commonly called header-type boilers, are installed in many auxiliary ships. The basic design of this type of boiler is shown in figures 10-15 and 10-16.

Header-type boilers normally operate at 450 to 465 psig and are designed for a maximum superheater outlet temperature of 740° to 750° F. In capacity, they range from about 25,000 to about 40,000 pounds of steam per hour.

Header-type boilers are sometimes referred to as cross-drum boilers because many of them were designed to be installed with the steam drum athwartships rather than fore and aft. However, some header-type boilers are not of the cross-drum type.

Header-type boilers are also referred to occasionally as side-fired boilers. This term is used to indicate the location of the burners with respect to the position of the steam drum. However, the term "side-fired" tends to be misleading because the surface of a boiler along which the burners are installed is generally regarded as the front of the boiler. In this discussion, we will take as the front of the boiler the surface along which the burners are installed. From this point of view, then, the steam drum is

installed lengthwise along the top of the boiler front.

The header-type boiler gets its name from the header sections which are connected by the generating tubes. There may be 12, 14, or 16 of these header sections, depending upon the size of the boiler. Half of the header sections are installed under the steam drum, at the front of the boiler. The other half are installed at the rear of the boiler, at a somewhat higher level. The header sections are installed at a slight angle from the vertical, leaning somewhat toward the front of the boiler. The angle of inclination of the headers allows the straight generating tubes (which enter the headers normal to the header surfaces) to slope slightly upward from the front of the boiler toward the rear, thus allowing free natural circulation within the boiler.

The header sections installed under the steam drum at the front of the boiler are known as downtake headers. Each downtake header is connected to the steam drum by a short downtake nipple. The lower end of each downtake header is connected to the junction header (sometimes called the mud drum) by a short nipple.

The header sections installed at the rear of the boiler are known as uptake headers. Each uptake header is connected to the steam drum by a large circulator tube which enters the steam drum slightly above the normal water level.

As shown in figures 10-15 and 10-16, the generating tubes in this type of boiler are straight rather than curved. The generating tubes connect the downtake headers at the front of the boiler with the uptake headers at the rear of the boiler.

The superheater consists of U-bend tubes, an upper superheater header, and a lower superheater header. The superheater tubes are installed at right angles to the generating tubes, between the main bank of generating tubes and the water screen tubes.

The steam drum of a header-type boiler usually has a manhole at each end. The steam drum contains the internal fittings, including a desuperheater.

The furnace of a header-type boiler has four vertical walls and a flat floor. The side walls are water cooled, being covered by water wall tubes which form a part of the circulation system of the boiler. There are two water wall downtake headers, one at each corner of the boiler front, installed vertically in the space

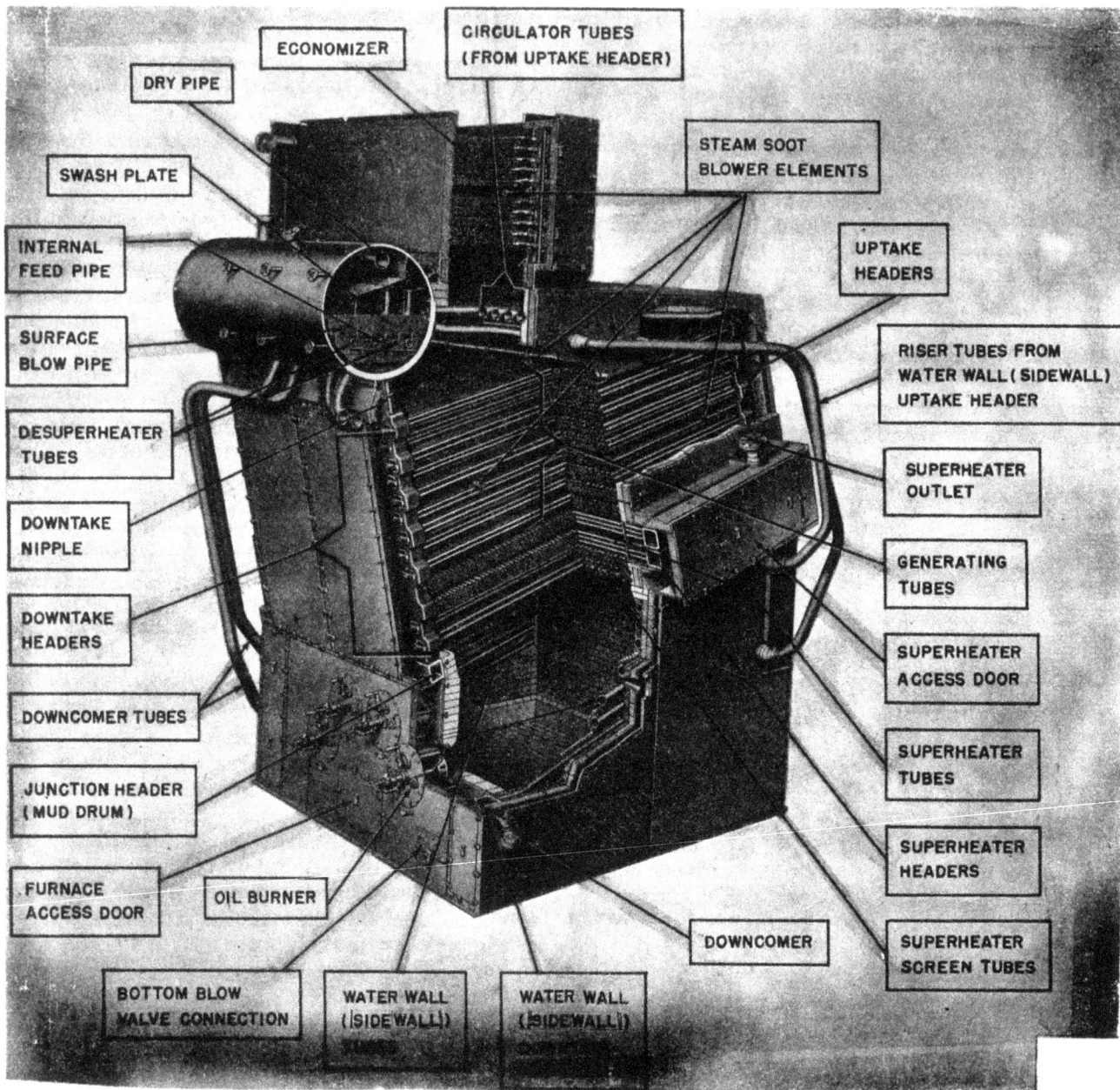

38.33X

Figure 10-15.—Cutaway view of header-type boiler.

between the inner and the outer casing. Two vertical water wall uptake headers are similarly installed at the two rear corners of the boiler. The water wall tubes are rolled into a downtake header at the front and an uptake header at the rear; they are arranged on the same slope as the generating tubes.

Water is supplied to the water wall downtake headers from the junction header. Steam and

water rise through the water wall tubes to the uptake headers, and then through the riser tubes that connect the uptake headers to the steam drum.

As may be seen in figures 10-15 and 10-16, an economizer is located behind the steam drum, in the way of the combustion gas exit.

The boiler is completely enclosed in an insulated steel casing, and an outer casing is

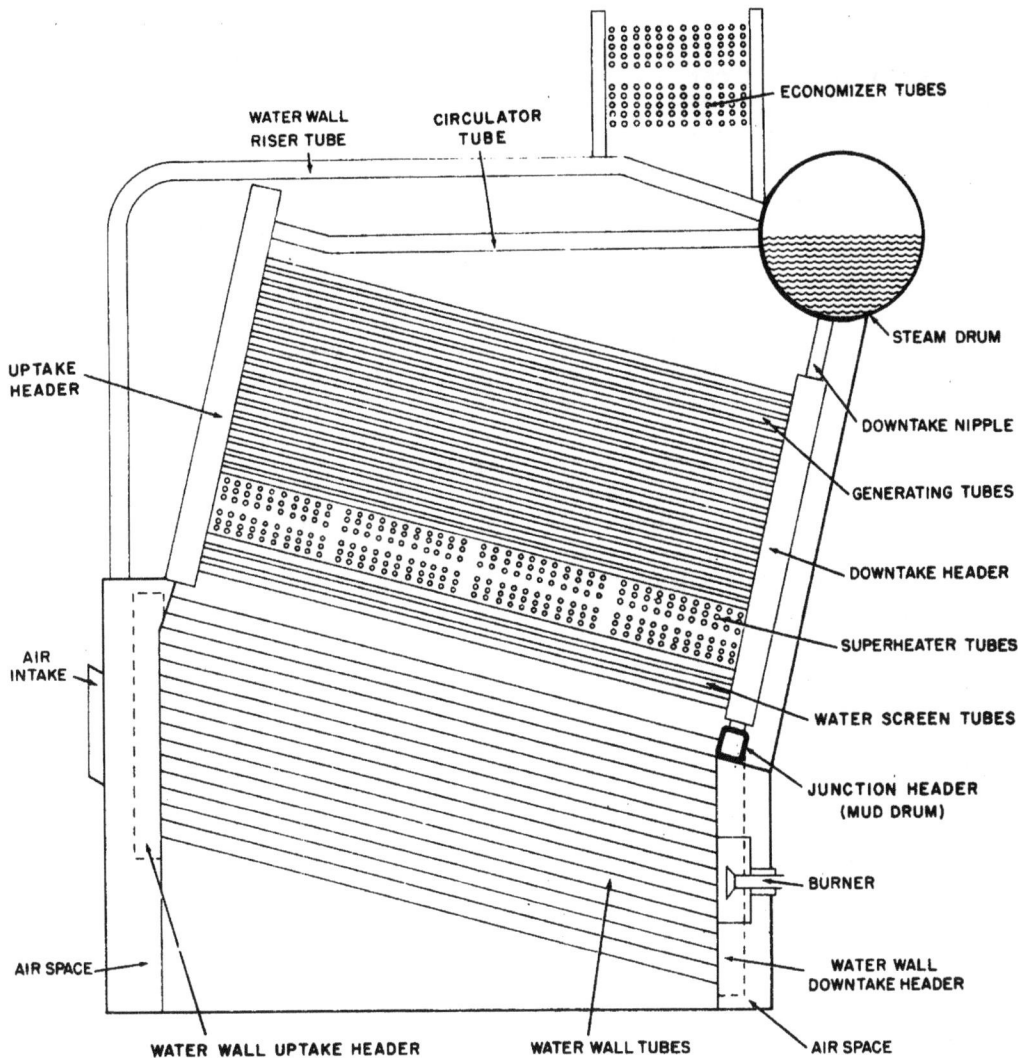

38.34

Figure 10-16.—Side view of header-type boiler.

installed in such a way as to form an air chamber between the inner and outer casings. The air inlet is at the rear of the boiler; an air duct beneath the furnace floor connects the front air chamber and the rear air chamber. The double-cased air chambers at the sides of the boiler are connected directly to the cold air inlet so that an air pressure is maintained in these side chambers at all rates of operation. Removable casing panels are located at various points to permit access for cleaning, inspection, and repair.

In summary, we may consider the header-type boiler as one which, on the basis of the classification methods given earlier in this chapter, has the following characteristics: It is a water-tube boiler with natural circulation of the free (not accelerated) type. It has sectional headers instead of water drums, and so is called a "header-type" boiler instead of a drum-type boiler. It has only one furnace—but the term "single-furnace boiler" is never applied to header-type boilers, possibly because such identification has not been needed. It is not a pressurized-furnace boiler. It does not have controlled superheat. It operates at a pressure of 450 to 465 psig; however, header-type boilers are quite often referred to as "400-psi boilers."

Double-Furnace Boilers

Double-furnace boilers (also called M-type boilers) are installed on most older destroyers and on many other combatant ships. These boilers are designed to carry a steam drum pressure of approximately 615 psig and to generate saturated steam at approximately 490° F. The saturated steam for auxiliaries goes directly from the steam drum to the auxiliary steam system; all other steam goes through the superheater. Double-furnace boilers are designed in various sizes and capacities to suit different installations. They range in capacity from about 100,000 to about 250,000 pounds of steam per hour at full power.

Figure 10-17 shows the general arrangement of a double-furnace boiler. The same type of boiler is shown in sectional view in figure 10-18 and in cutaway view in figure 10-19.

One of the two furnaces in this boiler is used for generating saturated steam; the other is used for superheating the saturated steam. Because each of the two furnaces can be fired separately, thus allowing control of superheated steam temperature over a wide range of operating conditions, the double-furnace boiler has long been called a "superheat control boiler." As noted previously, however, the control of superheat is not necessarily related to the number of furnaces. Therefore we will refer to this boiler as a double-furnace boiler, rather than as a superheat control boiler, even though the boiler shown does in fact have controlled superheat.

Since each furnace has its own burners, the degree of superheat can be controlled by proportioning the amount of fuel burned in the superheater-side furnace to the amount burned in the saturated-side furnace. When burners are lighted only on the saturated side, saturated steam is generated; when burners are lighted on the superheater side as well as on the saturated side, the saturated steam flowing through the superheater becomes superheated. The degree of superheat depends primarily upon (1) the firing rate on the superheater side, and (2) the rate of steam flow through the superheater. However, the rate of steam flow through the superheater is basically dependent upon the firing rate on the saturated side. Therefore we come back again to the idea that the degree of superheat depends primarily upon the ratio of the amount of oil burned in the superheater side to the amount burned in the saturated side.

The flow of combustion gases in the double-furnace boiler is partly controlled by gas baffles on one row of water screen tubes and on one row of division wall tubes, as shown in figure 10-18. The gas baffles on the water screen tubes direct the combustion gases toward the superheater tubes and also deflect the combustion gases away from the steam drum and the water screen header. The baffles on the division wall tubes by the saturated-side furnace keep the saturated-side combustion gases from flowing toward the superheater tubes, thus protecting the superheater when the superheater side is not lighted off. In addition, the baffles on the division wall tubes deflect combustion gases from the superheater side up toward the top of the saturated side, thus allowing the gases to pass toward the uptake without disturbing the fires in the saturated-side furnace.

The double-furnace boiler has a steam drum, one water drum, one water screen header, and one water wall header. All these drums and headers run from the front of the boiler to the rear of the boiler. Most of the saturated steam is generated in the main bank of generating tubes on the uptake side of the boiler; most of these tubes are 1 inch in outside diameter, but a few rows of 2-inch tubes are installed on the side of the tube bank nearest the furnace. The evaporation rate is much higher in the 1-inch tubes than in the 2-inch tubes, since the ratio of heat-transfer surface to the volume of contained water

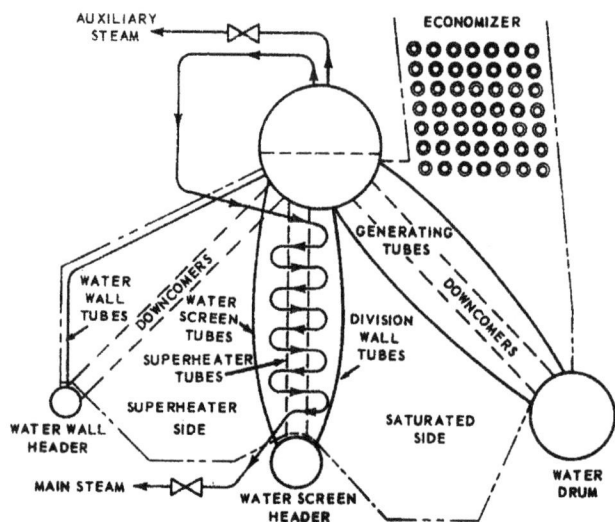

38.36

Figure 10-17.—General arrangement of double-furnace boiler.

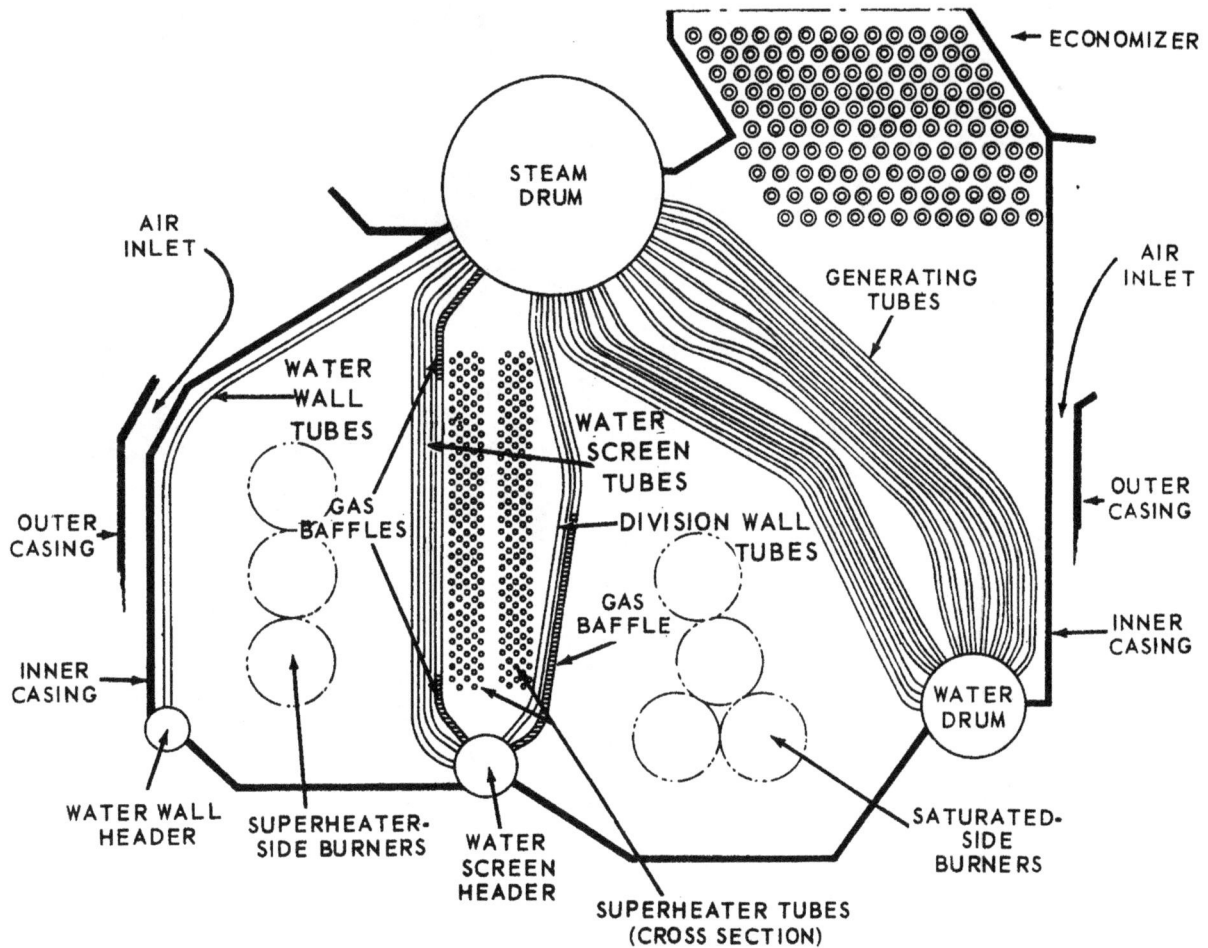

38.37

Figure 10-18.—Double-furnace boiler (sectional view, looking toward rear wall).

is much greater in the smaller tubes. The larger tubes are used in the rows next to the furnace because it is necessary at this point to provide a flow of cooling water and steam sufficient to protect the smaller tubes from the intense radiant heat of the furnaces.

Double-furnace boilers have anywhere from 15 to 50 downcomers, which vary in size from about 3 inches in outside diameter to about 7 inches OD. The downcomers are installed between the inner and the outer casings, as may be seen in figure 10-19.

The use of large-tube downcomers and small generating tubes results in extremely rapid circulation of water. Only a few seconds are required for the water to enter the steam drum as feed water, flow through the downcomers, circulate through the water drum or header, rise in the generating tubes, and return to the steam

drum as a mixture of water and steam. Some notion of the extreme rapidity of circulation may be obtained from the fact that water in the downcomers may flow at velocities of from 3 to 7 feet per second.

The economizer on a double-furnace boiler is usually larger than the economizer on a header-type boiler. As a rule, the economizer on a double-furnace boiler has about 60 U-shaped economizer tubes.

On the basis of the classification methods given earlier in this chapter, we may consider the double-furnace boiler as one which has the following characteristics: It is a water-tube boiler with natural circulation of the accelerated type. It is a drum-type (rather than a header-type) boiler. It has tubes which are arranged roughly in the shape of the letter M—hence it is often called an M-type boiler. It has two

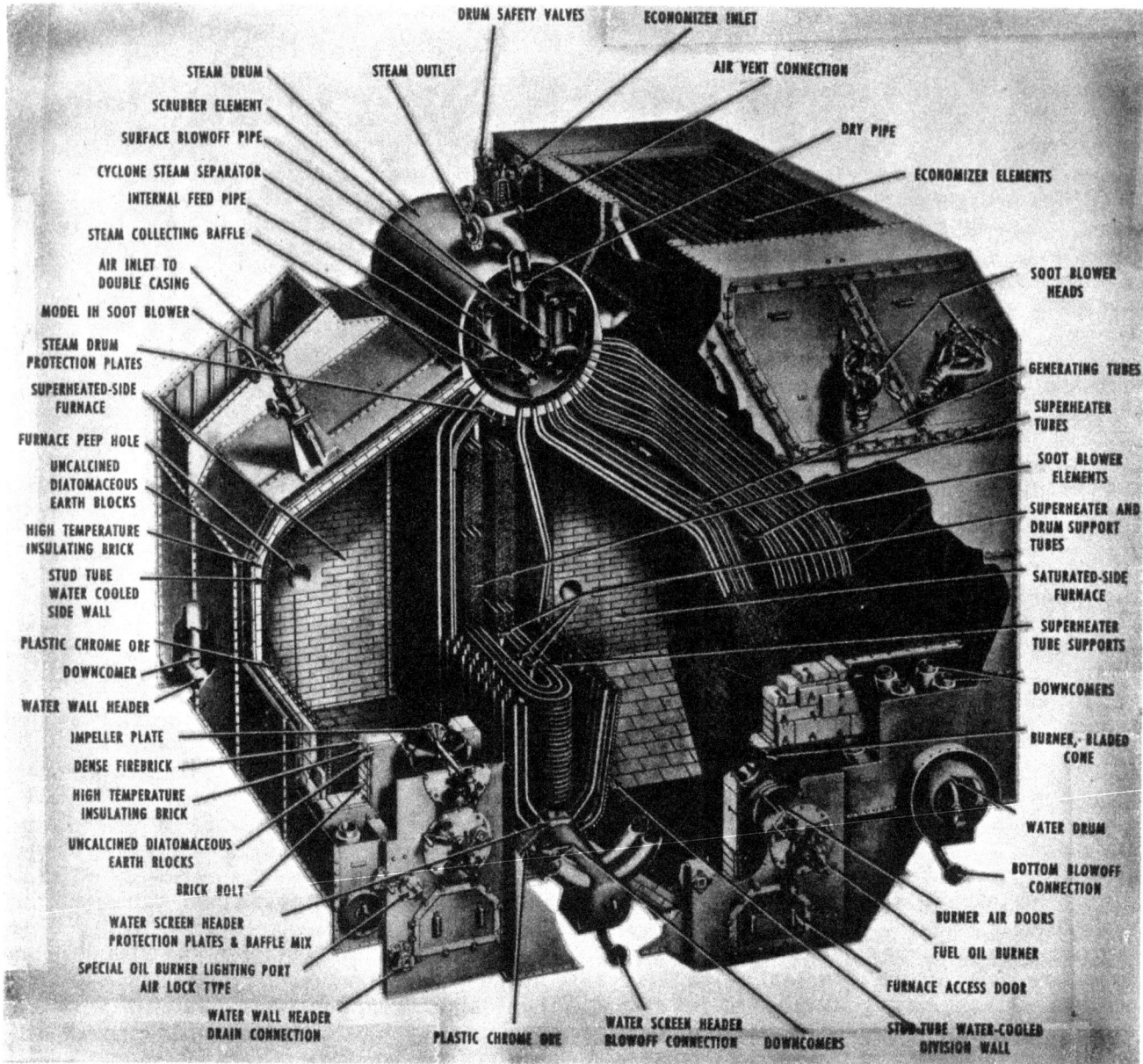

38.38X

Figure 10-19.—Double-furnace boiler (cutaway view).

furnaces—one for the saturated side and one for the superheater side. It has controlled superheat. It operates at a pressure of about 615 psig, and is often called a "600-psi boiler."

The most important advantage of the double-furnace boiler arises from the fact that the separate firing of the superheater side allows positive control of the degree of superheat. In the double-furnace boiler, it is theoretically possible to maintain the maximum designed temperature at the superheater outlet under widely varying conditions of load. In a single-furnace boiler, where one source of heat is used both for generating the steam and for superheating it, the degree of superheat increases as the rate of steam generation increases; and hence the maximum designed temperature at the superheater outlet is normally reached only at full power.

Most double-furnace boilers are designed to carry a superheater outlet temperature of 850° F; this is about 100° F higher than the superheater outlet temperature in a comparable single-furnace boiler, given the same quality of materials for boilers, piping, and turbines. The reason why a higher superheater outlet temperature can be used in a double-furnace boiler than in a comparable single-furnace boiler is that allowance must be made, in the single-furnace boiler, for the maximum superheater temperatures which might occur under adverse conditions of load.

In spite of the advantages resulting from the control of superheat, double-furnace boilers are no longer being installed in naval combatant ships. Experience with these boilers has revealed certain disadvantages which at the present time appear to outweigh the advantages of controlled superheat. Some of the disadvantages are:

1. In practice, it is not possible to maintain maximum designed superheat at low steaming rates. Only the steam for the main turbines and the turbogenerators goes through the superheater; at low firing rates, therefore, the steam flow through the superheater is generally not sufficient to permit a high firing rate on the superheater side. Thus under some conditions the steam supplied to the propulsion turbines and to the turbogenerators may be saturated or only very slightly superheated. As a consequence, therefore, the double-furnace boiler is actually less efficient than the single-furnace boiler at low firing rates.

2. The double-furnace boiler is more difficult to operate than the single-furnace boiler, and requires more personnel for its operation. Once there is any appreciable load on the boiler, the high air pressure in the double casings and in the furnace make it difficult and even dangerous to light burners on the superheater side. In order to avoid this difficulty, operating personnel would have to be able to predict the need for superheat and light off the burners on the superheater side before the air pressure had become so high. Obviously, such prediction is not always possible.

3. The double-furnace boiler is heavier, larger, and generally more complex than a single-furnace boiler of equal capacity.

Single-Furnace Boilers

The older single-furnace boilers that were installed on many World War II ships differ in several important respects from the newer single-furnace boilers that have been installed on ships built since World War II.

A single-furnace boiler of the older type is shown schematically in figure 10-20 and in cutaway view in figure 10-21. This boiler produces about 60,000 pounds of steam per hour at full power. At full power the steam drum pressure is about 460 psig, the superheater outlet pressure is about 435 psig, and the superheater outlet temperature is about 750° F.

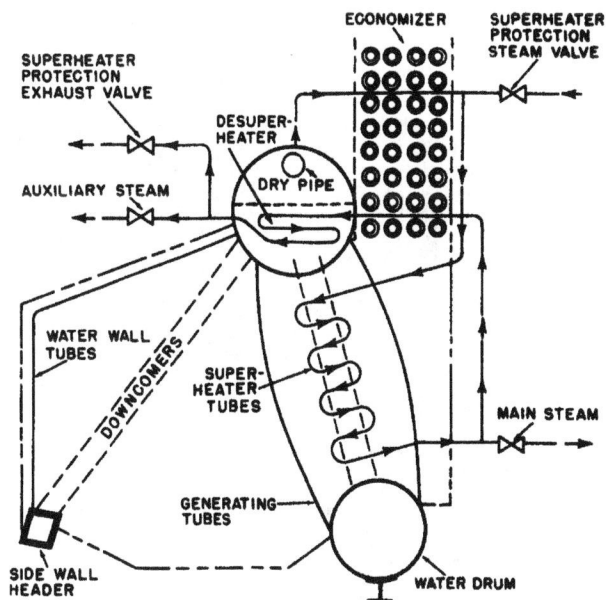

38.39

Figure 10-20.—General arrangement of older single-furnace boiler.

This boiler does not have controlled superheat. When the boiler is lighted off, both the generating tubes and the superheater tubes are heated. In order to protect the superheater tubes from overheating, all steam generated in the boiler must be led through the superheater. The saturated steam goes from the dry pipe in the steam drum to the superheater inlet; it goes through the superheater tubes, out the superheater outlet, and into the main steam line.

Auxiliary steam must go through the superheater (in order to provide a sufficient steam flow to protect the superheater) but must then be desuperheated. Desuperheating is accomplished by passing some of the superheated steam through a desuperheater, which is basically a coil of piping submerged in the water in

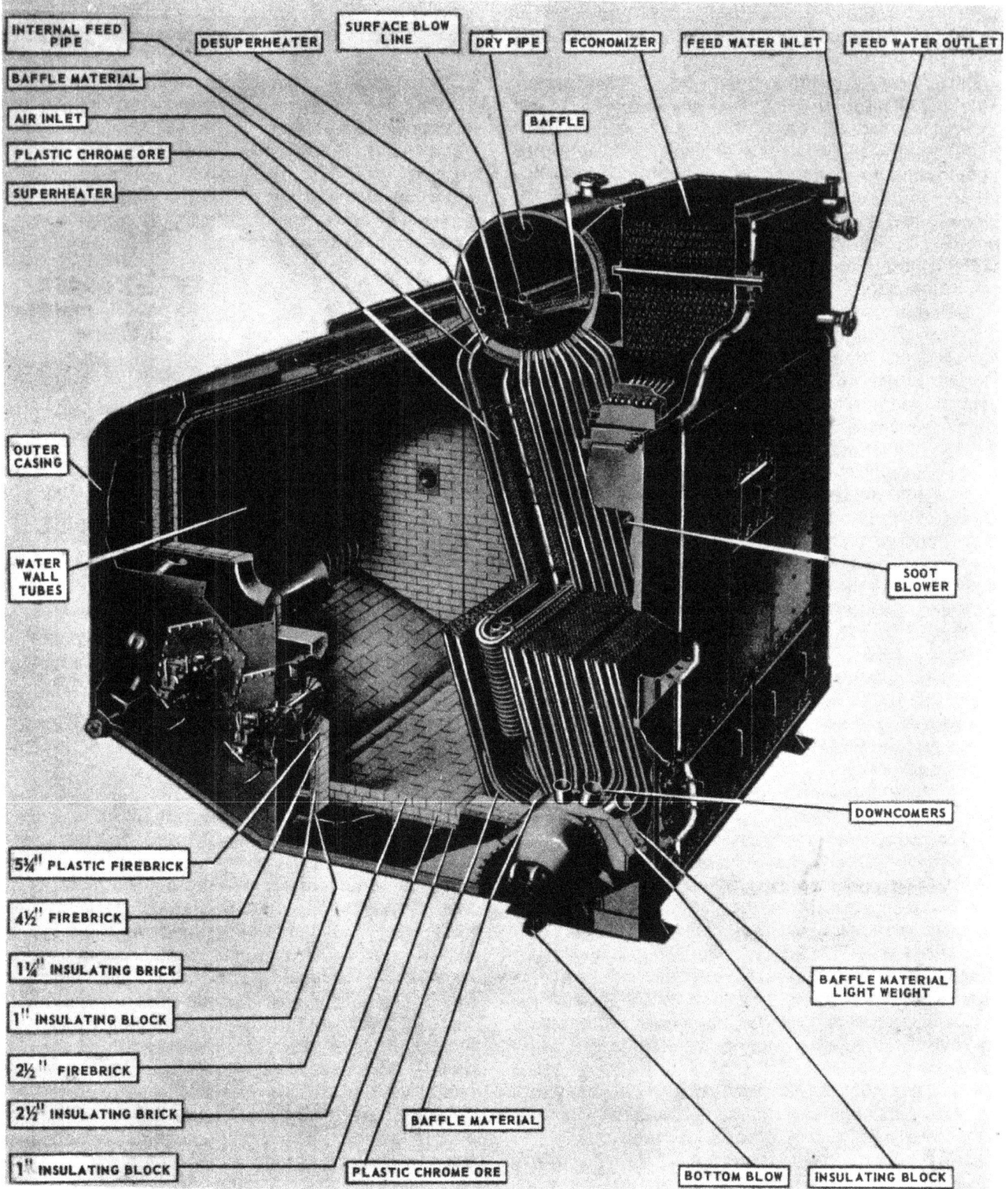

INTERNAL FEED PIPE
DESUPERHEATER
SURFACE BLOW LINE
DRY PIPE
ECONOMIZER
FEED WATER INLET
FEED WATER OUTLET

BAFFLE MATERIAL
AIR INLET
PLASTIC CHROME ORE
SUPERHEATER

BAFFLE

OUTER CASING

WATER WALL TUBES

SOOT BLOWER

5¼" PLASTIC FIREBRICK
4½" FIREBRICK
1¼" INSULATING BRICK
1" INSULATING BLOCK
2½" FIREBRICK
2½" INSULATING BRICK
1" INSULATING BLOCK

DOWNCOMERS

BAFFLE MATERIAL LIGHT WEIGHT

BAFFLE MATERIAL
PLASTIC CHROME ORE
BOTTOM BLOW
INSULATING BLOCK

Figure 10-21.—Cutaway view of older single-furnace boiler.

38.40

252

the steam drum. Heat transfer takes place from the steam in the desuperheater to the water in the steam drum. The desuperheated steam which passes out of the desuperheater and into the auxiliary steam line is once again at (or very close to) saturation temperature.

Thus far we have considered the flow of steam as it occurs after the boiler has been cut in on the steam line. But what happens when a cold boiler is lighted off? How can the superheater tubes be protected from the heat of the furnace after fires are lighted but before sufficient steam has been generated to ensure a safe flow through the superheater?

Various methods are used to protect the superheater during this critical period immediately after lighting off. Very low firing rates are used, and the boiler is warmed up slowly until an adequate flow of steam has been established. Many—but not all—boilers of this type have connections through which protective steam can be supplied from another boiler on the same ship or from some outside source such as a naval shipyard or a tender. As shown in figure 10-20, this steam comes in (under pressure) through the superheater protection steam valve. It enters the superheater inlet, passes through the superheater tubes, goes out the superheater outlet, passes through the desuperheater, and then goes into the auxiliary exhaust line by way of the superheater protection exhaust valve.

On single-furnace boilers which do not have a protective steam system for use during the lighting off period, even greater care must be taken to establish a steam flow through the superheater. In general, the steam flow is established by venting the superheater drains to the bilges while warming up the boiler very slowly.

On the basis of the classification methods given earlier in this chapter, we may consider this older single-furnace boiler as one which has the following characteristics: It is a water-tube boiler with natural circulation of the accelerated type. It is a drum-type (rather than a header-type) boiler. It has tubes which are arranged roughly in the shape of the letter D— hence it is often called a D-type boiler. It has only one furnace. It does not have controlled superheat. It is often classified as a "600-psi boiler," although it actually operates at about 435 psig.

As previously noted, the degree of superheat obtained in a single-furnace boiler of the type being considered is primarily dependent upon the firing rate. However, a number of design features and operational considerations also affect the temperature of the steam at the superheater outlet.

Design features that affect the degree of superheat include (1) the type of superheater installed—that is, whether heated by convection, by radiation, or by both; (2) the location of the superheater with respect to the burners; (3) the extent to which the superheater is protected by water screen tubes; (4) the area of superheater heat-transfer surface; (5) the number of passes made by the steam in going through the superheater; (6) the location of gas baffles; and (7) the volume and shape of the furnace.

Operational factors that affect the degree of superheat include (1) the rate of combustion; (2) the temperature of the feed water; (3) the amount of excess air passing through the furnace; (4) the amount of moisture contained in the steam entering the superheater; (5) the condition of the superheater tube surfaces; and (6) the condition of the water screen tube surfaces. Since these factors may affect the degree of superheat in ways which are not immediately apparent, let us examine them in more detail.

How does the rate of combustion affect the degree of superheat? To begin with, we might imagine a simple relationship in which the degree of superheat goes up directly as the rate of combustion is increased. Such a simple relationship does, in fact, exist—but only up to a certain point. Throughout most of the operating range of this boiler, the degree of superheat goes up quite steadily and regularly as the rate of combustion goes up. Near full power, however, the degree of superheat drops slightly even though the rate of combustion is still going up. Why does this happen? Primarily because the increased firing rate results in an increased generating rate, which in turn results in an increased steam flow through the superheater. The rate of heat absorption increases more rapidly than the rate of steam flow until the boiler is operating at very nearly full power; at this point the rate of steam flow increases more rapidly than the rate of heat absorption. Therefore the superheater outlet temperature drops slightly.

Suppose that the boiler is being fired at a constant rate and that the steam is being used at a constant rate. If we increase the temperature of the incoming feed water, what happens to the superheat? Does it increase, decrease, or remain the same? Surprisingly, the degree of superheat decreases if the feed temperature is increased, more saturated steam is generated from the burning of the same amount of fuel. The increased quantity of saturated steam causes an increase in the rate of flow through the superheater. Since there is no increase in the amount of heat available for transfer to the superheater, the degree of superheat drops slightly.

Under conditions of constant load and a constant rate of combustion, what happens to the superheat if the amount of excess air[4] is increased? To see why an increase in excess air results in an increase in temperature at the superheater outlet, we must take it step by step:

1. An increase in excess air decreases the average temperature in the furnace.

2. With the furnace temperature lowered, there is less temperature difference between the gases of combustion and the water in the boiler tubes.

3. Because of the smaller temperature difference, the rate of heat transfer is reduced.

4. Because of the decreased rate of heat transfer, the evaporation rate is reduced.

5. The lower evaporation rate causes a reduction in the rate of steam flow through the superheater, with a consequent rise in the superheater outlet temperature.

In addition to this series of events, another factor also tends to increase the superheater outlet temperature when the amount of excess air is increased. Large amounts of excess air tend to cause combustion to occur in the tube bank rather than in the furnace itself; as a result, the temperature in the area around the superheater tubes is higher than usual and the superheater outlet temperature is higher.

Any appreciable amount of moisture in the steam entering the superheater causes a very noticeable drop in superheat. This occurs because steam cannot be superheated as long as it is in contact with the water from which it is being generated. If moisture enters the superheater, therefore, a good deal of heat must be used to dry the steam before the temperature of the steam can rise.

The condition of the superheater tube surfaces has an important effect on superheater

outlet temperature. If the tubes have soot on the outside or scale on the inside, heat transfer will be retarded and the degree of superheat will be decreased.

If the water screen tubes have soot on the outside or scale on the inside, heat transfer to the water in these tubes will be retarded. Therefore there will be more heat available for transfer to the superheater as the gases of combustion flow through the tube bank. Consequently, the superheater outlet temperature will rise.

The single-furnace boiler is lighter and smaller, for any given output of steam, than the double-furnace boiler. Because the single-furnace boiler supplies superheated steam at low steaming rates, the overall plant efficiency is better with this type of boiler than with the double-furnace boiler. The single-furnace boiler has the further advantage of simplicity of operation and maintenance. Although the single-furnace boiler considered here does not have controlled superheat, this lack is less important than might have been supposed, since some of the theoretical advantages of controlled superheat have not been entirely realized in practice.

The basic design of the single-furnace boiler has been used increasingly. Except for experimental boilers, no double-furnace boilers have been installed on combatant ships since World War II. The newer single-furnace boilers operate at approximately 600 psi or at approximately 1200 psi. Operating temperature at the superheater outlet is quite commonly 950°F for the 1200-psi boilers; this is 100°F higher than the operating temperature of most double-furnace boilers, and 200°F higher than the operating temperature of the older single-furnace boilers.

One of the most noticeable differences between the older and the newer single-furnace boilers is the change in furnace design. Higher heat release rates are possible in the newer boilers. Although these newer single-furnace boilers are not the type that we refer to as "pressurized-furnace" boilers, they do often use a slightly higher combustion air pressure than the older single-furnace boilers. The use of higher air pressure causes an increase in the velocity of the combustion gases, and the increased velocity results in a higher rate of heat transfer to the generating tubes. Because of the increased heat release rates, a newer single-furnace boiler is likely to have a water-cooled roof and water-cooled rear walls as well as water-cooled side walls.

[4]The term "excess air" is used to indicate any quantity of combustion air in excess of that which is theoretically required for the complete combustion of the fuel. Some excess air is necessary for efficient combustion, but too much excess air is wasteful, as discussed in a later section of this chapter.

In design details and in general configuration, the newer single-furnace boilers vary somewhat among themselves. Figure 10-22 shows a 1200-psi boiler of the type installed on some post World War II destroyers. Except for the additional water-cooled surfaces, this boiler is very much like the older single-furnace boilers. In contrast, figure 10-23 shows a type of single-furnace boiler that has been installed on some recent ships. The superheater tubes are installed vertically, rather than horizontally, between generating tubes and water screen tubes. Note, also, that there is a separate water screen header for the water screen tubes; this feature is quite unusual in single-furnace boilers, though standard for double-furnace boilers.

New Types of Propulsion Boilers

The field of boiler design is by no means static. Although one trend predominates—that of using higher pressures and temperatures—there are almost innumerable ways in which the higher pressures and temperatures can be achieved. New types of boilers are constantly being developed and tested, and existing boiler designs are

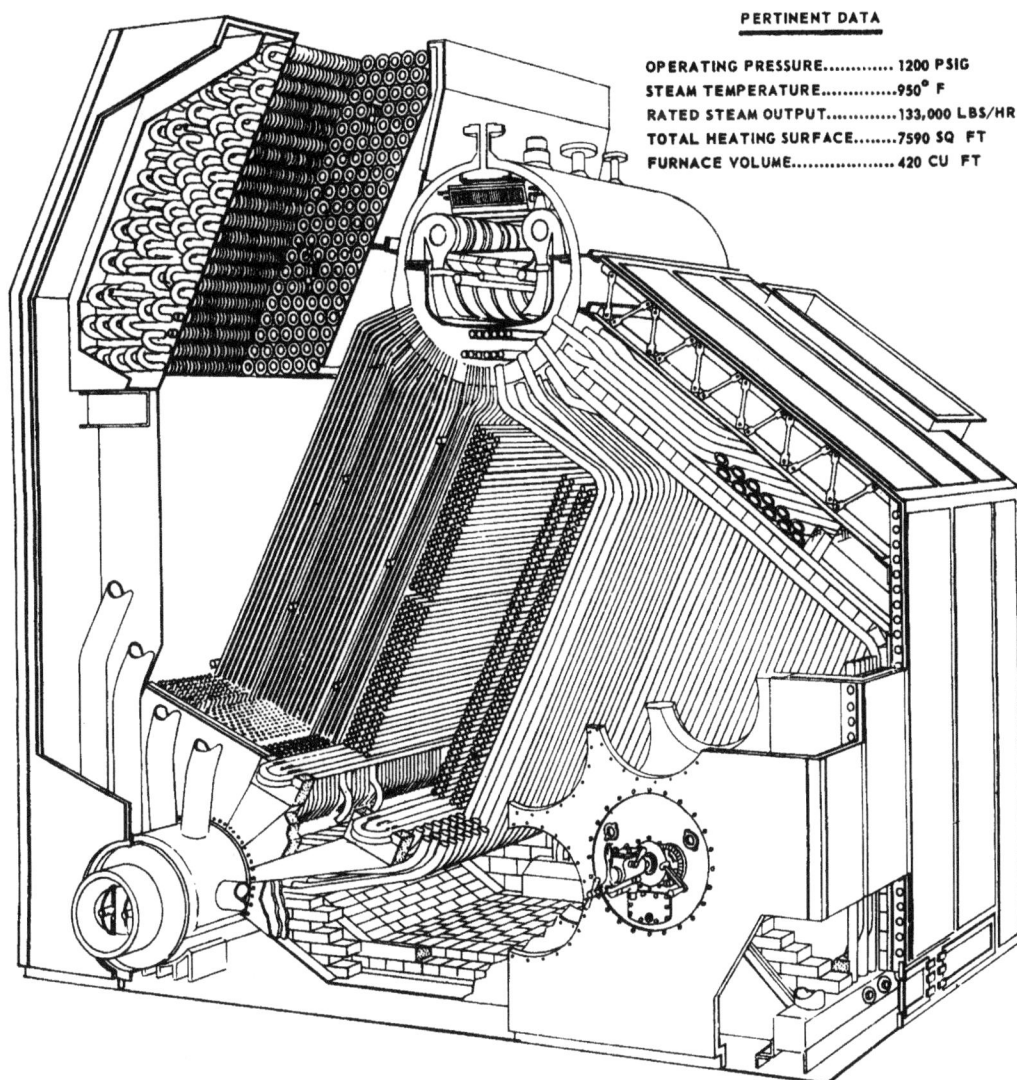

PERTINENT DATA

OPERATING PRESSURE............. 1200 PSIG
STEAM TEMPERATURE...............950° F
RATED STEAM OUTPUT.............133,000 LBS/HR
TOTAL HEATING SURFACE........7590 SQ FT
FURNACE VOLUME..................420 CU FT

38.41

Figure 10-22.—Newer 1200-psi single-furnace boiler for post World War II destroyer.

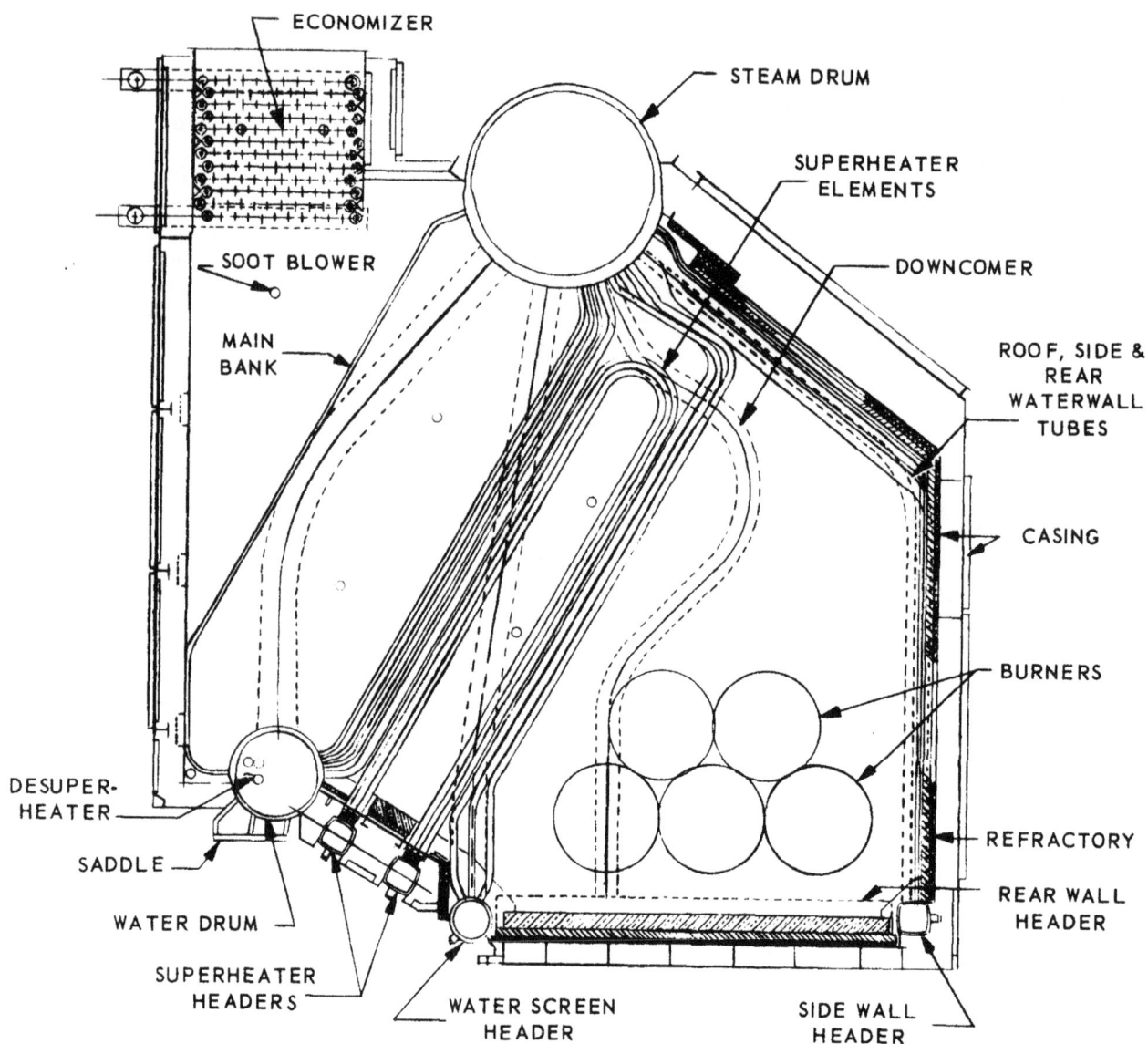

147.84

Figure 10-23.—Newer single-furnace boiler with vertical superheater.

subject to modification and improvement. The new types of boilers discussed here do not by any means exhaust the field of new designs; indeed, it must be emphasized that a wide diversity of design is still possible in this field.

TOP-FIRED BOILERS.—A new boiler design which is at present being used on some auxiliary ships is the top-fired boiler. In this boiler, the fuel oil burners are located at the top of the boiler and are fired downward. The top-fired boiler utilizes certain new construction techniques, including welded walls. The boiler is of the natural circulation type, with a completely water-cooled furnace. The only refractory material that is exposed to the gases of combustion is the refractory that is installed in corners and in a small area around the burners. The top-fired boiler has an in-line generating tube bank and a vertical superheater. It is expected that the top-fired boiler will be much cleaner and thus require less maintenance than older boilers of more conventional design.

CONTROLLED CIRCULATION BOILERS.—Controlled (or forced) circulation boilers have

256

been used for some time in stationary power plants, in locomotives, and in some merchant ships. Only a few controlled circulation boilers have been installed in the propulsion plants of naval ships, and of this few the majority were subsequently removed and replaced by conventional single-furnace boilers with accelerated natural circulation. In theory, however, controlled circulation has some very marked advantages over natural circulation, and it is entirely possible that improved designs of controlled circulation boilers may be developed for future use in naval propulsion plants.

In natural circulation boilers, circulation occurs because the ascending mixture of water and steam is lighter (less dense) than the descending body of relatively cool and steam-free water. As boiler pressure increases, however, there is less difference between the density of steam and the density of water. At pressures over 1000 psi, the density of steam differs so little from the density of water that natural circulation is harder to achieve than it is at lower pressures. At high pressures, controlled circulation boilers have a distinct advantage because their circulation is controlled by pumps and is independent of differences in density. Because controlled circulation boilers can be designed without regard for differences in density, they can be arranged in practically any way that is required for a particular type of installation. Thus a greater flexibility of arrangement is possible and the boilers may be designed for compactness, savings in space and weight requirements, and maximum heat absorption.

There are two main kinds of controlled circulation boilers. One type is known as a once-through or forced flow boiler; the other type is usually called a controlled circulation or a forced recirculation boiler. In both types, external pumps are used to force the water through the boiler circuits; the essential difference between the two kinds lies in the amount of water supplied to the boiler.

In a once-through forced circulation boiler, all (or very nearly all) of the water pumped to the boiler is converted to steam the first time through, without any recirculation. This type of boiler has no steam drum, but has instead a small separating chamber. Water is pumped into the economizer circuit and from there to the generating circuit, the amount of flow being controlled so as to allow practically all of the water to be converted into steam in the generating circuit. The very small amount of water

that is not converted to steam in the generating circuit is separated from the steam in the separating chamber. The water is discharged from the separating chamber to the feed pump suction, if it is suitable for use; if it contains solid matter, it is discharged through the blowdown pipe. Meanwhile, the steam from the separating chamber flows on through the superheater circuit, where it is superheated before it enters the main steam line.

Figure 10-24 shows the boiler circuits of a controlled circulation (or forced recirculation) boiler. In this boiler, more water is pumped through the circuits than is converted into steam. The excess water is taken from the steam drum and is pumped through the boiler circuits again by means of a circulating pump. This type of boiler has a conventional steam drum which contains a feed pipe, steam separators and dryers, a desuperheater, and other fittings. The boiler has an economizer, three generating circuits, and a superheater. Circulating pumps, fitted as integral parts of the boiler, provide positive circulation to all steam generating surfaces.

Both types of controlled circulation boilers have far smaller water capacity than do natural circulation boilers, and therefore have much more rapid response to changes in load. For this reason, automatic controls are required on these boilers to ensure rapid and sensitive response to fuel and feed water requirements.

PRESSURIZED-FURNACE BOILERS.—A boiler recently developed for use in naval propulsion plants is variously known as a pressurized-furnace boiler, a pressure-fired boiler, a supercharged boiler, or a supercharged steam generating system.

A pressurized-furnace boiler is shown schematically in figure 10-25 and in cutaway view in figure 10-26. As may be seen, the boiler is quite unlike other operational boiler types in general configuration. The pressurized furnace is more or less cylindrical in shape, with the long axis of the cylinder running vertically. The boiler drum is mounted horizontally, some distance above the pressurized furnace. The drum is connected to the steam and water elements in the furnace by risers and downcomers, all of which are external to the casing. Some boilers of this type are side-fired. Others (including the one shown) are top-fired; as may be seen in figures 10-25 and 10-26, the burners are at the top

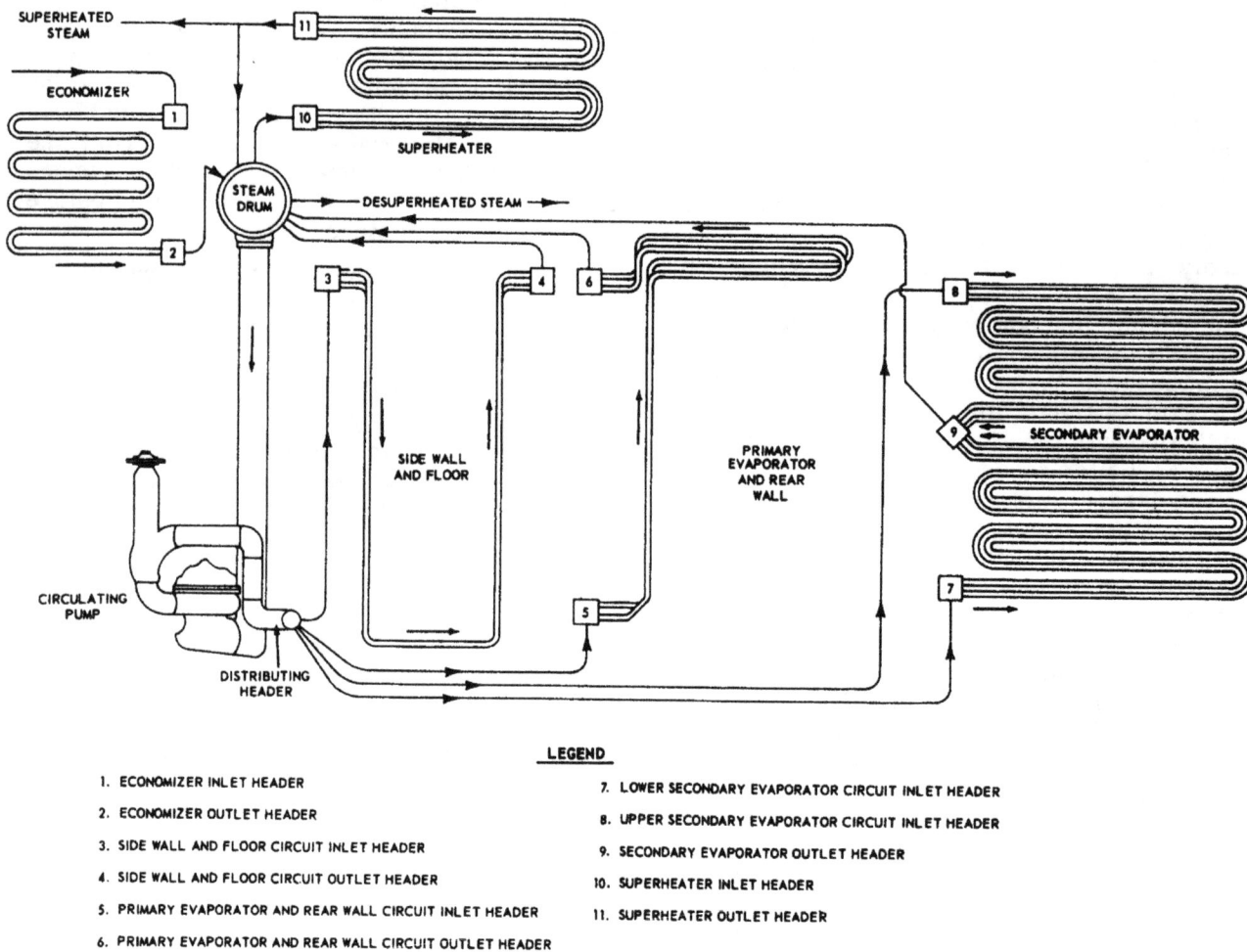

LEGEND

1. ECONOMIZER INLET HEADER
2. ECONOMIZER OUTLET HEADER
3. SIDE WALL AND FLOOR CIRCUIT INLET HEADER
4. SIDE WALL AND FLOOR CIRCUIT OUTLET HEADER
5. PRIMARY EVAPORATOR AND REAR WALL CIRCUIT INLET HEADER
6. PRIMARY EVAPORATOR AND REAR WALL CIRCUIT OUTLET HEADER

7. LOWER SECONDARY EVAPORATOR CIRCUIT INLET HEADER
8. UPPER SECONDARY EVAPORATOR CIRCUIT INLET HEADER
9. SECONDARY EVAPORATOR OUTLET HEADER
10. SUPERHEATER INLET HEADER
11. SUPERHEATER OUTLET HEADER

147.85

Figure 10-24.—Schematic diagram of controlled circulation boiler.

of the pressurized furnace, firing downward into the furnace.

The burners, specially designed for the pressurized-furnace boiler, are quite unlike any we have thus far considered. The burners are designed to burn distillate fuel rather than Navy Special fuel oil. There are no air register doors. There are three burners per boiler, and each burner includes a special type of straight mechanical atomizer (not return-flow) which utilizes three sprayer plates at the same time. All three burners are operated simultaneously, and all three sprayer plates remain in place in each atomizer. The sprayer plates operate in sequence to meet changing conditions of load. The design of these burners allows an enormously wide range of operation without cutting burners

in or out and without even changing sprayer plates.

The generating tubes run vertically inside the pressurized furnace. The superheater is an annular pancake arrangement inserted into the bottom of the pressure vessel. The superheater is designed to be removed without disturbing the main components of the boiler.

The air compressor which supplies the combustion air under pressure is driven by a gas turbine. The air compressor and the gas turbine together are referred to as the supercharger. Part of the energy needed for driving the gas turbine is obtained from the combustion gases leaving the boiler furnace. The combustion gases expand through the gas turbine, and some of the heat is converted into work. This is the same

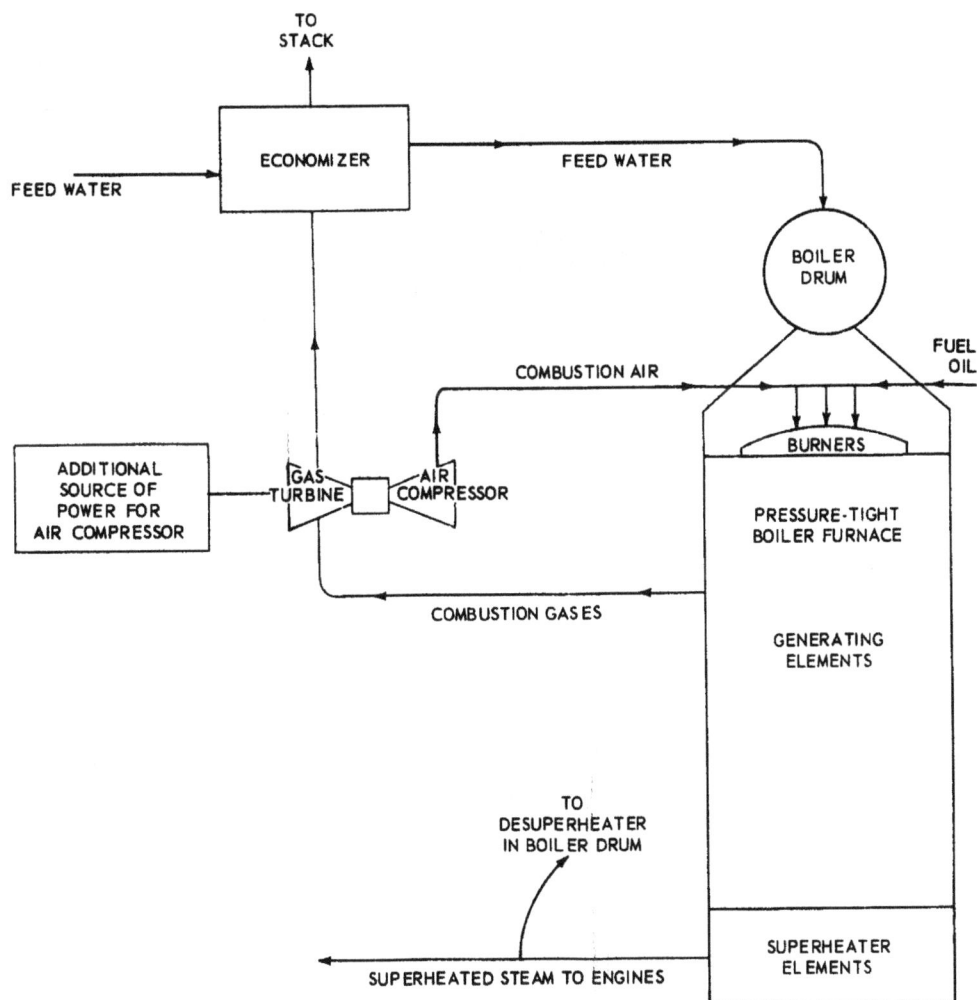

147.86

Figure 10-25.—Schematic view of pressurized-furnace boiler.

kind of energy transformation that occurs in a steam turbine; the difference is that hot combustion gases, rather than steam, carry the energy to the gas turbine. After the combustion gases leave the gas turbine, some of the remaining heat may be used to heat feed water as it flows through an economizer.

There are no forced draft blowers in pressurized-furnace boiler installations. The supercharger takes the place of the forced draft blowers, thus greatly increasing plant efficiency. The steam saved by the use of a supercharger instead of forced draft blowers may amount to as much as 8 or 10 percent of boiler capacity.

Altogether, a pressurized-furnace boiler is not much more than half the size and half the weight of a conventional boiler of equal steam

capacity. A large part of this saving of space and weight occurs because the increased pressure[5] on the combustion gas side causes a very great increase in the rate of heat transfer to the water in the tubes. Thus a smaller generating surface is required to generate the same amount of steam. Another cause of space and weight saving is that the general design of the pressurized-furnace boiler eliminates the need for much of the refractory material that is required in other

—————————

[5] Forced draft blowers for conventional boiler installations furnish air pressures ranging from 0 to 10 psig. In a pressurized-furnace boiler, the air compressor supplies combustion air at pressures ranging from 30 to 90 psig.

SATURATED STEAM
LINE TO SUPERHEATER

STEAM DRUM

AIR INLET TO BOILER

COMPRESSOR
AIR INLET

GAS TURBINE
OUTLET

RISER TUBE

BURNER

IGNITOR

UPPER
BOILER
HEADER

BOILER
CASING
RELIEF
VALUE
CONNECTION

SUPERCHARGER
SET

DOWNCOMER

GAS INLET
TO GAS TURBINE

FURNACE
TUBES

MAIN
CONVECTION
BANK

EXPANSION
JOINT

FURNACE
SCREEN

SUPERHEATER
OUTLET

SUPERHEATER

BOILER
FOUNDATIONS

CONVECTION
HEADER

FURNACE
HEADER

SUPERHEATER
HEADERS

SUPERHEATER
INLET

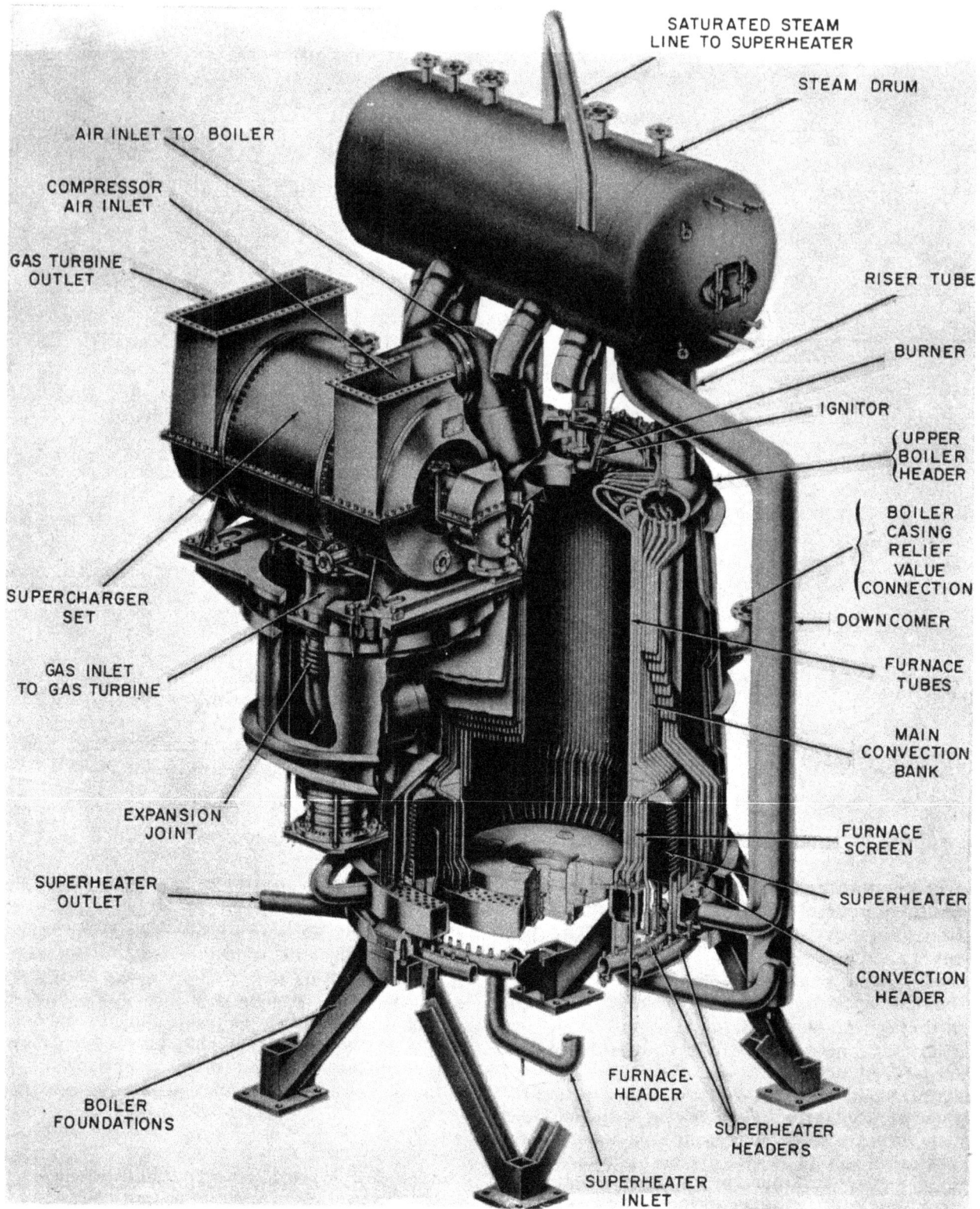

Figure 10-26.—Cutaway view of pressurized-furnace boiler.   139.19

boilers. A pressurized-furnace boiler may require only about 2000 pounds of refractory, as against the 21,000 pounds or more usually required in a conventional boiler of equal capacity.

Increased efficiency, a substantial saving in space and weight requirements, a substantial reduction in ship's force maintenance requirements, shorter boiler start-up time, and better maneuverability and control are the major advantages of the pressurized-furnace boiler. Although some operational and maintenance problems do exist with this boiler, it appears likely that most of them can eventually be solved by increased training of personnel, increased precision in the erection of the boilers, and perhaps continued refinements of design and construction.

BOILER WATER REQUIREMENTS

Modern naval boilers cannot be operated safely and efficiently without careful control of boiler water quality. If boiler water conditions are not just precisely right, the high operating pressures and temperatures of modern boilers will lead to rapid deterioration of the boiler metal, with the possibility of serious casualties to boiler pressure parts.

Although our ultimate concern is with the water actually in the boiler, we cannot consider boiler water alone. We must also consider the water in the rest of the system, since we are dealing with a closed cycle in which water is heated, steam is generated, steam is condensed, and water is returned to the boiler. Because the cycle is continuous and closed, the same water remains in the system except for the water that is lost by boiler blowdown[6] and the very small amount of water that escapes, either as steam or as water, and is replaced by makeup feed.

---

[6] There are two kinds of boiler blowdown: surface blowdown and bottom blowdown. Surface blowdown is used to remove foam and other light contaminants from the surface of the water in the steam drum. Bottom blowdown is used to remove sludge and other material that tends to settle in the lower parts of the boiler. Both surface blowdown and bottom blowdown may be used to remove a portion of the boiler water so that it can be replaced with purer makeup feed, thereby lowering the chloride content of the boiler water. Surface blows may be given while the boiler is steaming; bottom blows must not be given until some time after the boiler has been secured. The valves and piping used for making surface and bottom blows are discussed in chapter 11 of this text.

Although we must remember the continuous or cyclical nature of the shipboard steam plant, we must also distinguish between the water at different points in the system. This distinction is necessary because different standards are prescribed for the water at different points. To identify the water at various points in the steam - water cycle, the following terms are used:

Distillate or sea water distillate is the fresh water that is discharged from the ship's distilling plants. This water is stored in fresh water or feed water tanks. All water in the steam - water cycle begins originally as distillate.

Makeup feed is distillate used as replacement for any water that is lost or removed from the closed steam - water cycle.

Condensate is the water that results from the condensation of steam in the main and auxiliary condensers. This water is called condensate until it reaches the deaerating feed tank.

Boiler feed or feed water is the water in the system between the deaerating feed tank and the boiler.

Deaerated feed water is feed water that has passed through deaerating feed tank and has had the dissolved or entrained oxygen removed from it.

Boiler water is the water actually contained within a boiler at any given moment.

Sea water, the source of practically all fresh water used aboard ship, contains about 35,000 parts per million (ppm) of sea salts. This is equivalent to roughly 70 pounds of sea salts per ton of water. When sea water is evaporated and the vapor is condensed in the distilling plant, the resulting distillate contains about 1.75 ppm of sea salts, or roughly 70 pounds per 20,000 tons. In other words, distillate is actually diluted sea water—sea water that is diluted to about 1/20,000 of its original concentration. It is not "pure water." In considering water problems and water treatment, it is essential to remember that the basic impurity of sea water distillate would make water treatment necessary even if no other impurities entered the water from other sources. The salts that are present in sea water—and, therefore, to a lesser extent in distillate—are chiefly compounds of sodium, calcium, and magnesium.

Although makeup feed enters the tanks as distillate, the makeup feed usually contains a slightly higher proportion of impurities than the distillate. The difference is accounted for by slight seepage or other contamination of the water after it has remained in the tanks for some time.

Just as distillate is diluted sea water, so steam condensate is basically a diluted form of boiler water. The amount of solid matter carried over with the steam varies considerably, depending upon the design of the boiler, the condition of the boiler, the nature of the water treatment, the manner in which the boiler is operated, and other factors. In general, condensate contains from 1.7 to 3.5 ppm of solid matter, or roughly 70 pounds per 20,000 to 10,000 tons. Condensate may pick up additional contamination in various ways. Salt water leaks in the condenser increase the amount of sea salts present in the condensate. Oil leaks in the fuel oil heaters may contaminate the condensate. Corrosion products from steam and condensate lines may also be present in condensate. Under ideal conditions, condensate should be no more contaminated than sea water distillate; under many actual conditions, it is more contaminated.

The solid content of the water (Boiler feed water) in the system between the deaerating feed tank and the boiler is essentially the same as the solid content of the condensate. The main difference between condensate and deaerated boiler feed is that most of the dissolved gases are removed from the water in the deaerating feed tank.

Practically all of the impurities that are present in feed water, including those originally present in the sea water distillate and those that are picked up later, will eventually find their way to the boiler. As steam is generated and leaves the boiler, the concentration of impurities in the remaining boiler water becomes greater and greater. In other words, the boiler and the condenser together act as a sort of distilling plant, redistilling the water received from the ship's evaporators. In consequence, the boiler water would become more and more contaminated if steps were not taken to deal with the increasing contamination.

As an example, suppose that a boiler holds 10,000 pounds of water at steaming level, and suppose that steam is being generated at the rate of 50,000 pounds per hour. After an hour of operation there would be approximately five times as much solid matter in the boiler water as there was in the entering feed water. Now if we continued to steam this boiler for another 2000 to 4000 hours without using blowdown and without using any kind of boiler water

treatment, the boiler water would contain just about the same concentration of sea salts as the original sea water from which the distillate was made. In addition, the boiler water would contain increasingly large quantities of corrosion products and other foreign matter picked up in the steam and condensate systems.

If we continued to steam the boiler with the water in this condition, the boiler would deteriorate rapidly. To prevent such deterioration, it is necessary to do the following things:

1. Maintain the incoming feed water at the highest possible level of purity and as free as possible of dissolved oxygen.

2. Use chemical treatment of the boiler water to counteract the effects of some of the impurities that are bound to be present.

3. Use blowdown at regular intervals to remove some of the more heavily contaminated water so that it may be replaced by purer feed water.

Although there are many sources of boiler water contamination, the contaminating materials tend to produce three main problems when they are concentrated or accumulated in the boiler water. Therefore, boiler water treatment is aimed at controlling the three problems of (1) waterside deposits, (2) waterside corrosion, and (3) carryover.

Waterside deposits interfere with heat transfer and thus cause overheating of the boiler metal. The general manner in which a waterside deposit causes overheating of a boiler tube is shown in figure 10-27. In a boiler operating at 600 psi, the temperature inside a generating tube may be approximately 500° F and the temperature of the outside of the tube may be approximately 100° F higher.[7] Where a waterside deposit exists, however, the tube cannot transfer the heat as rapidly as it receives it. As shown in figure 10-27, the inside of the tube has reached a temperature of 800° F at the point where the waterside deposit is thickest. The tube metal is overheated to such an extent that it becomes plastic and blows out into a bubble or blister under boiler pressure.

Waterside deposits that must be guarded against include sludge, oil, scale, corrosion

---

[7]The temperatures used in this example do not apply to all situations in which a boiler tube is overheated. The exact temperatures of the inside and outside of the tube would depend upon the operating pressure of the boiler, the location of the tube in the boiler, the nature of the deposit, and various other factors.

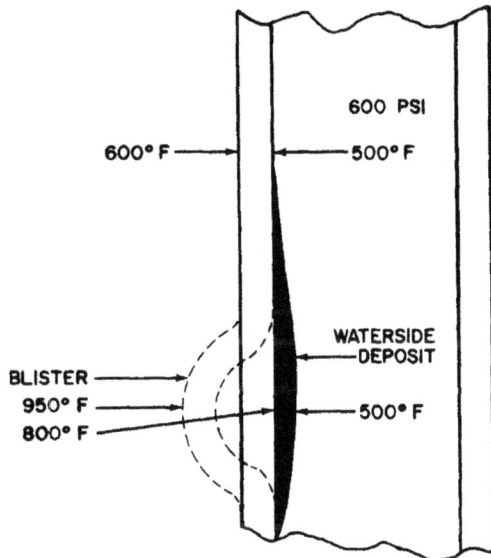

38.131

Figure 10-27.—Effect of waterside deposit on
boiler tube.

38.138

Figure 10-28.—Localized pit in boiler tube
caused by dissolved oxygen in the
boiler water.

deposits, and products formed as the result of
chemical reactions of the tube metal.

The term "waterside corrosion" is used to
include both localized pitting and general corrosion. Most waterside corrosion is electro-
chemical in nature. There are always some
slight variations (both chemical and physical)
in the surface of any boiler metal. These small
chemical and physical variations in the metal
surface cause slight differences in electrical
potential between one area of a tube and another
area. Some areas are anodes (positive terminals) and others are cathodes (negative terminals). Iron from the boiler tube tends to go into
solution more rapidly at the anode areas than
at other points on the boiler tube. Electrolytic
action cannot be completely prevented in any
boiler, but it can be kept to a minimum by
maintaining the boiler water at the proper alkalinity and by keeping the dissolved oxygen content of the boiler water as low as possible.

The presence of dissolved oxygen in the
boiler water contributes greatly to the type of
corrosion in which electrolytic action makes
pits or holes of the type shown in figure 10-28.
A pit of this type actually indicates an anodic
area in which iron from the boiler tube has
gone into solution in the boiler water.

General corrosion occurs when conditions
favor the formation of many small anodes and

cathodes on the surface of the boiler metal. As
corrosion proceeds, the anodes and cathodes
constantly change location. Therefore, there is
a general loss of metal over the entire surface.
General corrosion may occur if the chloride
content of the boiler water is too high or if the
alkalinity is either too low or too high.

The third major problem that results from
boiler water contamination is carryover. Under
some circumstances, very small particles of
moisture (almost like a fine mist) are carried
over with the steam. Under other circumstances,
large gulps or slugs of water are carried over.
The term priming is generally used to describe
the carryover of large quantities of water. Both
kinds of carryover are dangerous and both can
cause severe damage to superheaters, steam
lines, turbines, and valves. Whatever moisture
or water is carried over with the steam brings
with it the solid matter that is dissolved or
suspended in the water. This solid matter tends
to be deposited on turbine blades and in superheater tubes and valves. Figure 10-29 shows a
superheater tube in which solid matter has been
deposited as a result of carryover. Priming, or

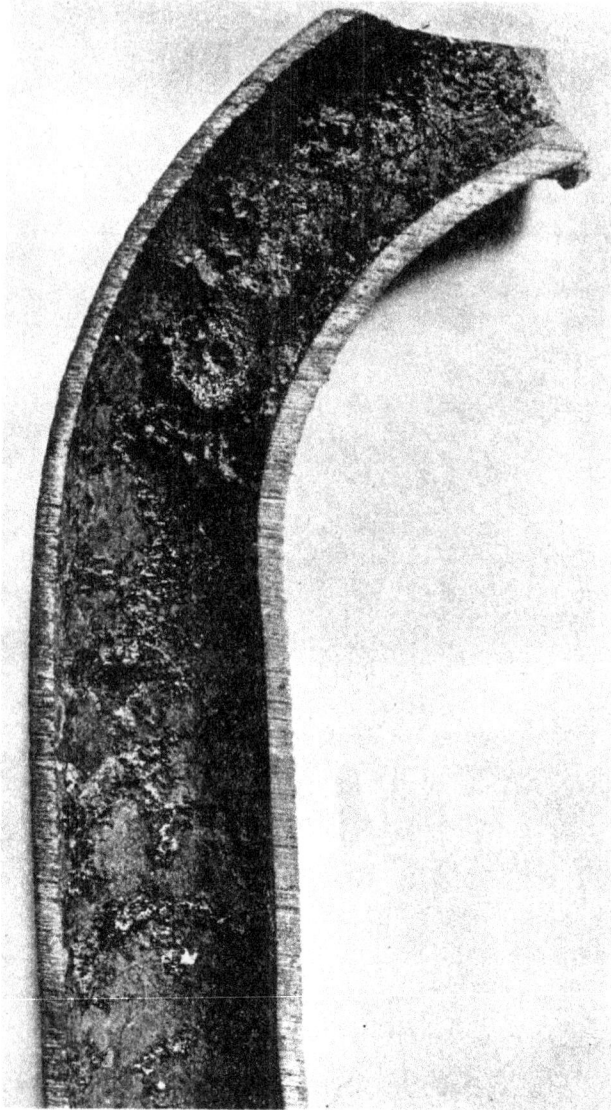

38.140

Figure 10-29.—Evidence of carryover in
superheater tube.

the carryover of large slugs of water, is particularly dangerous because it can do such severe
damage to machinery. For example, priming can
actually rip turbine blades from their wheels.

One cause of carryover is foaming of the
boiler water. Foaming occurs when the water
contains too much dissolved or suspended solid
matter. The solids tend to stabilize the bubbles
and cause them to pile up instead of bursting.
If a great deal of solid matter is present in the
boiler water, a considerable amount of foam

will pile up. Under these conditions, carryover
is almost sure to occur.

In order to counteract the effects of the impurities in boiler water, it is necessary to have
a precise knowledge of the actual condition of
the water. This knowledge is obtained by frequent tests of the boiler water and of the feed
water. Boiler water tests include chloride tests,
hardness tests, alkalinity tests, pH tests, phosphate tests, and electrical conductivity tests
which indicate the dissolved solid content of the
boiler water.[8] Feed water is tested routinely
for chloride, hardness, and dissolved oxygen;
alkalinity, pH, phosphate, and electrical conductivity tests are not normally made on feed water.
The frequency of boiler water and feed water
tests is specified by the Naval Ship Systems
Command. Also, the allowable limits of contamination are specified by the Naval Ship Systems Command. In general, the requirements
for purity of boiler water become more stringent
with increasing boiler pressure.

Water tests aboard ship are made by the oil
and water king (usually a Boilerman), although
certain aspects of the preparation and handling
of the chemicals may require the supervision
of an officer. The tests require some knowledge
of chemistry and a high degree of precision in
preparing, using, and measuring the chemicals.
Therefore, only personnel holding a current
certification resulting from successful completion of a NavShips boiler water/feed water test
and treatment training course may test and
treat boiler water and feed water on propulsion
boilers.

Some of the water tests made aboard ship
give a direct indication of just what contaminating substance is present, and in just what amount
it is present. In other cases, it is more important to know what effects the contaminating
substances have upon the water than it is to
know what the substances are or exactly how
much of each is present. Therefore, some water
tests are designed to measure properties the
water acquires because of the presence of various impurities.

The term <u>chloride content</u> really refers to
the concentration of the chloride ion, rather
than to the concentration of any one sea salt.

---

[8]Note, however, that no one ship makes all of these
tests of boiler water. The types of boiler water tests
required on any particular ship depend upon the method
of boiler water treatment authorized for that ship.

264

Because the concentration of chloride ions is relatively constant in sea water, the chloride content is used as a measure of the amount of solid matter that is derived through sea water contamination. The results of the chloride test are used as one indication of the need for blowdown. Chloride content is expressed in equivalents per million (emp).[9]

Hardness is a property that water acquires because of the presence of certain dissolved salts. Water in which soap does not readily form a lather is said to be hard.

Alkalinity is a property that the water acquires because of the presence of certain impurities. On ships that make alkalinity tests, the results are expressed in epm.

Some ships are required to determine the pH value, rather than the alkalinity, of the boiler water. The pH unit does not measure alkalinity directly; however, it is related to alkalinity in such a way that a pH number gives an indication of the acidity or alkalinity of the water. The pH scale of numbers runs from 0 to 14. On this scale, pH 7 is the neutral point. Solutions having pH values above 7 are defined as alkaline solutions. Solutions having pH values below 7 are defined as acid solutions.

Boiler water that is treated with phosphates must be tested for phosphate content. Boiler water that is treated with standard Navy boiler compound is not tested for phosphates. Phosphate content is expressed in parts per million (ppm). When the phosphate content of boiler water is maintained within the specified limits, the hardness of the water should be zero. Therefore, hardness tests are not required for boiler water when phosphate water treatment is used.

The test for chloride content indicates something about the amount of solid matter that is present in the boiler water, but it indicates only the solid matter that is there because of sea water contamination. It does not indicate anything about other solid matter that may be dissolved in the boiler water. A more accurate indication of the total amount of dissolved solids

can be obtained by measuring the electrical conductivity of the boiler water, since this is related to the total dissolved solid content. All ships are now furnished with special electrical conductivity meters for measuring the conductivity of the boiler water. The total dissolved solid content is expressed in micromhos, a unit of electrical conductivity.

As a regular routine, the test for dissolved oxygen is made only on feed water, although occasional testing of water in other parts of the system is recommended. A chemical test for dissolved oxygen is made aboard ship. Since this test cannot detect dissolved oxygen in concentrations of less than 0.02 ppm, more sensitive laboratory tests are sometimes made as a check on the operation of the deaerating feed tanks.

When tests of the boiler water show that the water is not within the prescribed limits, chemical treatment and blowdown are instituted. Several methods of chemical treatment are now authorized. Each method is designed to completely eliminate hardness and to maintain the alkalinity (or the pH value) within the prescribed limits. The method of boiler water treatment specified for each ship is the method that will best perform these two functions and, at the same time, take account of the total concentration of solids that can be tolerated in the particular type of boiler. The type of water treatment authorized for any particular ship is specified by the Naval Ship Systems Command; it is not a matter of choice by ship's personnel.

Chemical treatment of the boiler water increases, rather than decreases, the need for blowdown. The chemical treatment counteracts the effects of many of the impurities in the boiler water, but at the same time it increases the total amount of solid matter in the boiler water and thus increases the need for blowdown. Each steam boiler must be given a surface blow at least once a day, and more often if the water tests indicate the need. Bottom blows are given at least once a week, usually about an hour after the boiler has been secured. Bottom blows must not be given while a boiler is steaming. Special instructions for boiler blowdown are issued to certain categories of ships.

COMBUSTION REQUIREMENTS

Certain requirements must be met before combustion can occur in the boiler furnace. The fuel must be heated to the temperature that will

[9] Equivalents per million can be defined as the number of equivalent parts of a substance per million parts of some other substance. The word "equivalent" here refers to the chemical equivalent weight of a substance. For example, if a substance has a chemical equivalent weight of 35.5, a solution containing 35.5 parts per million is described as having a concentration of 1 epm.

give it the proper viscosity for atomization. It should be NOTED, however, that with the conversion to the new distillate fuel (NSDF), the fuel will not need to be heated as the viscosity is much lower than the fuel oil (NSFO) now being used. The fuel must be forced into the furnace under pressure through the atomizers which divide the fuel into very fine particles. Meanwhile, combustion air must be forced into the furnace and admitted in such a way that the air will mix thoroughly with the finely divided fuel. And finally, it is necessary to supply enough heat so that the fuel will ignite and continue to burn.

Combustion is a chemical process which results in the rapid release of energy in the form of heat and light. When a fuel burns, the chemical reactions between the combustible elements in the fuel and the oxygen in the air result in new compounds. The combustible components of fuel are mainly carbon and hydrogen, which are present largely in the form of hydrocarbons. Sulfur, oxygen, nitrogen, and a small amount of moisture are also present in fuel.

In almost all burning processes, the principal reactions are the combination of the carbon and the hydrogen in the fuel with the oxygen in the air to form carbon dioxide and a relatively small amount of water vapor. In the absence of sufficient air to form carbon dioxide, carbon monoxide will be formed. A reaction of lesser importance is the combination of sulfur and oxygen to form sulfur dioxide.

Atmospheric air is the source of oxygen for the combustion reactions occurring in a boiler furnace. Air is a mixture of oxygen, nitrogen, and small amounts of carbon dioxide, water vapor, and inert gases. The approximate composition of air, by weight and by volume, is as follows:

Element	Weight (Percent)	Volume (Percent)
Oxygen	23.15	20.91
Nitrogen, etc.	76.85	79.09

At the proper temperature, the oxygen in the air combines chemically with the combustible substances in the fuel. The nitrogen, which is 76.85 percent by weight of all air entering the furnace, serves no useful purpose in combustion but is rather a direct source of heat loss, since it absorbs heat in passing through the furnace and carries off a considerable amount of heat as it goes out the stack.

When a combustion reaction occurs, a definite amount of heat is liberated. The total amount of heat released by the combustion of a fuel is the sum of the heat released by each element in the fuel. The amount of heat liberated in the burning of each of the principal elements in fuel oil is as follows:

Element	Chemical Symbol	Heat Released By Combustion (BTU per lb)
Hydrogen (to water)	H_2	62,000
Carbon (to carbon monoxide)	C	4,440
Carbon (to carbon dioxide)	C	14,540
Sulfur (to sulfur dioxide)	S	4,050

Notice that much more heat is liberated when carbon is burned to carbon dioxide than when it is burned to carbon monoxide, the difference being 10,100 Btu per pound. In burning to carbon monoxide, the carbon is not completely oxidized; in burning to carbon dioxide, the carbon combines with all the oxygen possible, and thus oxidation is complete.

Thus far in this discussion, we have assumed that the oxygen necessary for combustion was present in the exact amount required for the complete combustion of all the combustible elements in the fuel. However, it is not a simple matter to introduce just exactly the required amount of oxygen—no more, no less—into the boiler furnace.

Since atmospheric air is the source of oxygen for the combustion process that occurs in the boiler furnace, let us first calculate the amount of air that would be needed to furnish 1 pound of oxygen. By weight, the composition of air is 23.15 percent oxygen and 76.85 percent nitrogen (disregarding the very small quantities of other gases present in air). To supply 1 pound of oxygen for combustion, therefore, it is necessary to supply 1/0.2315 or 4.32 pounds of air.

Since nitrogen constitutes 76.85 percent of the air (by weight), the amount of nitrogen in this 4.32 pounds of air will be 0.7685 x 4.32 or 3.32 pounds. As mentioned before, the nitrogen serves no useful purpose in combustion and is a direct source of heat loss.

Calculations will show that approximately 14 pounds of air will furnish the oxygen theoretically required for the complete combustion of 1 pound of fuel. In actual practice, of course, the amount of air necessary to ensure complete combustion must be somewhat in excess of that theoretically required. About 10 to 15 percent excess air is usually sufficient to ensure proper combustion. Too much excess air serves no useful purpose, but merely absorbs and carries off heat.

When fuel is burned in the boiler furnace, the difference between the HEAT INPUT and the HEAT ABSORBED represents the HEAT LOSS. Heat losses may be unavoidable, avoidable, or—in some cases—avoidable only to a limited extent. Most heat losses may be accounted for, but some losses cannot normally be accounted for.

All fuel contains a small amount of moisture which must be evaporated and superheated to the furnace temperature. Since the expenditure of heat for this purpose constitutes a heat loss in terms of boiler efficiency, every precaution should be taken to prevent contamination of the fuel oil with water.

All fuel contains some hydrogen which, when combined with oxygen by the process of combustion, forms water vapor. This water vapor must be evaporated and superheated, and in both processes it absorbs heat. Consequently, although the heat of combustion of hydrogen is very great, a small heat loss occurs because the water vapor formed as a result of the combustion of hydrogen must be evaporated and superheated.

Since atmospheric air is the source of the oxygen utilized for combustion in the boiler furnace, there is bound to be some moisture in the combustion air. This moisture must be evaporated and superheated, and therefore constitutes a heat loss.

The heat loss due to heat being carried away by combustion gases is the greatest of all the heat losses that occur in a boiler. Although much of this heat loss is unavoidable, some may be prevented by keeping all heat-transfer surfaces clean and by using no more excess air than is actually required for combustion.

Another heat loss occurs because of incomplete combustion of the fuel. When the carbon in the fuel is burned to carbon monoxide, instead of carbon dioxide, there is a tremendous heat loss of 10,100 Btu per pound. This should be considered an avoidable loss, since the admission of a sufficient amount of excess air will ensure complete combustion.

Heat losses that cannot be measured or that are impracticable to measure are (1) losses due to unburned hydrocarbons, gaseous or solid; (2) losses due to radiation; and (3) other losses not normally accounted for.

FIREROOM OPERATIONS

Although a complete discussion of fireroom operations is beyond the scope of this text, some understanding of the major factors involved in boiler operation may be useful.

Basically, the fireroom force must control three inputs—feed water, fuel, and combustion air—in order to provide one output, steam. Under steady steaming conditions, when steam demands are relatively constant for long periods of time, there is no great difficulty about providing a uniform flow of steam to the propulsion turbines. But one of the special requirements of naval ships is that they must be able to maneuver and to change speed quickly, and this requirement imposes upon the fireroom force the responsibility for making very rapid increases and decreases in the amount of steam furnished to the engineroom. Under conditions of rapid change, boiler operation is a teamwork job that requires great skill and alertness and smooth coordination of efforts by several men.

For manual operation of the boilers, a normal fireroom watch consists of one petty officer in charge of the watch; one checkman for each operating boiler; one burnerman for each operating boiler front; one blowerman for each operating boiler; and one or more men to act as messengers and to check the operation of the auxiliary machinery. When automatic boiler controls are installed, boilers may be operated with fewer men on watch when the controls are being used.

When a boiler is being operated manually, the checkman controls the water level in the boiler by manual operation of the feed stop and check valves. The checkman stands at the upper level, near the feed stop and check valves, and near the boiler gage glass. The checkman admits water to the boiler as necessary to maintain the water at or very near the designed water level. The check watch requires the utmost vigilance and reliability; if any one job in the fireroom can be said to be more important than any other, the checkman's job is the one.

One of the greatest difficulties in maintaining the water level arises from the fact that the boiler water swells and shrinks as the firing rate is changed. As the firing rate is increased, there is an increase in the volume of the boiler water. This increase, which is known as swell, occurs because there is an increase in the number and size of the steam bubbles in the water. As the firing rate is decreased, there is a decrease in the volume of the water. This decrease, which is known as shrink, occurs because there are fewer steam bubbles and they are of smaller size. Thus, for any given weight of boiler water, the volume varies with the rate of combustion.

The problem of swell and shrink becomes even more complex when we remember that the evaporation rate also increases as the firing rate increases and decreases as the firing rate decreases. When the firing rate is increased, therefore, the checkman must remember to feed more water to the boiler, even though the water level has already risen momentarily because of swell. On the other hand, the checkman must remember to feed less water to the boiler when the firing rate is decreased, even though the water level has already dropped. Because these actions may appear to be contrary to common sense to a person who does not understand the concept of swell and shrink, a good deal of training is usually required before a man can be considered qualified to stand a check watch.

The control of combustion involves the control of fuel and the control of combustion air. There are three ways in which the firing rate may be increased or decreased in order to meet changes in steam demand: (1) by increasing or decreasing the fuel pressure, (2) by increasing or decreasing the number of burners in use, and (3) by changing the size of the sprayer plates in the atomizer assemblies. With every change, the amount of combustion air supplied to the boiler must also be changed in order to maintain the proper relationship between fuel and combustion air. The burnerman and the blowerman must therefore work very closely together in order to provide efficient combustion in the boiler furnace.

The burnerman cuts burners in and out and adjusts the oil pressure as necessary to keep the steam pressure at the required value. The burnerman is guided by the steam drum pressure gage. Also, he watches the annunciator which shows the signals going from the bridge to the engineroom, and in this way he can tell what steam demands are going to be made.

On a double-furnace boiler, there are two burnermen—one for the saturated side and one for the superheater side. The burnerman on the superheater side cuts burners in and out and adjusts fuel pressure to keep the superheater outlet temperature at the required value. The burnerman on the superheater side is guided by the distant-reading thermometer which indicates the temperature of the steam at the superheater outlet. In addition, he must keep a close check on the actions of the saturated-side burnerman so that he will always know how many burners are in use on the saturated side.

When two boilers are furnishing steam to the same engine, the burnermen of both boilers must work together to see that the load is equally divided between the two boilers.

The blowerman is responsible for operating the forced draft blowers that supply combustion air to the boiler. Although the air pressure in the double casings is affected by the number of registers in use and by the extent to which each register is open, it is chiefly determined by the manner in which the forced draft blowers are operated. The opening, setting, or adjusting of the air registers is the burnerman's job; the control of the forced draft blowers is the blowerman's job. As may be apparent, the burnerman and the blowerman must each know what the other man is doing at all times. The blowerman must always increase the air pressure before the burnerman increases the rate of combustion, and the burnerman must always decrease the rate of combustion before the blowerman decreases the air pressure.

If a boiler is not being supplied with sufficient air for combustion, everyone in the fireroom will know about it immediately. The boiler will begin to pant and vibrate, and the fireroom force will receive complaints of "heavy black smoke" from the bridge. If the boiler is being supplied with too much air—that is, more excess air than is required for efficient combustion—the fireroom force may or may not know about it immediately. White smoke coming from the smokepipe is always an indication of large amounts of excess air. However, a perfectly clear smokepipe may be deceiving; it may mean that the boiler is operating with only a small amount of excess air, but it may also mean that as much as 300 percent excess air is causing enormous heat losses. The blowerman must learn by experience how much air pressure should be shown on the air pressure gage for all the various combinations of different numbers of burners, different sizes of sprayer plates, and different fuel pressures.

The number of men assigned to operate the fireroom auxiliary machinery varies from one ship to another, depending upon the size of the ship and the number of men available. Some ships may have two or more men assigned to this duty; on other ships, the work may be done by the petty officer in charge of the watch or by the messenger. The burnerman and the blowerman may also take care of some of the auxiliaries. The checkman must never be given any duties other than his primary ones of watching and maintaining the water level.

All fireroom operations are supervised and coordinated by the petty officer in charge of the watch. The petty officer in charge of the watch supervises all lighting off, operating, and securing procedures. He keeps the engineroom and the engineering officer of the watch informed of operating conditions when necessary. He must be constantly alert to the slightest indication of trouble and must be constantly prepared to deal with any casualty that may occur. The petty officer in charge of the watch is responsible for making sure that all safety precautions are being observed and that unsafe operating conditions are not allowed to exist.

FIREROOM EFFICIENCY

The military value of a naval vessel depends in large measure upon her cruising radius, which, in turn, depends upon the efficiency with which the engineering plant is operated. Perhaps the largest single factor in determining the efficiency of the engineering plant is the efficiency with which the boilers are operated. Greater savings in fuel, with consequent increase in steaming radius of the ship, may often be made in the fireroom than in all the rest of the engineering plant put together.

The capacity of a boiler is defined as the maximum rate at which the boiler can generate steam. The rate of steam generation is usually expressed in terms of pounds of water evaporated per hour. You should know something of the limitations upon boiler capacity, the significance of full-power and overload ratings, and the procedure for checking on boiler loads.

The capacity of any boiler is limited by three factors that have to do both with the design of the boiler and with its operation. These limitations, which are known as end points, are (1) the end point for combustion, (2) the end point for moisture carryover, and (3) the end point for water circulation.

Boilers are so designed that the end point for combustion should occur at a lower rate of steam generation than the end point for moisture carryover, and the end point for moisture carryover at a lower rate than the end point for water circulation. Since the end point for combustion occurs first, it is the only end point that is likely to be reached in a properly designed and properly operated boiler. However, it should be understood that it is quite possible to reach the end points for moisture carryover and water circulation before reaching the end point for combustion, by using larger sprayer plates than those recommended by the manufacturer or by the Bureau of Ships. In such a case, the boiler might suffer great damage before the end point for combustion was reached.

End Point for Combustion

The process of burning fuel in a boiler furnace involves forcing the fuel into the furnace at the proper viscosity through atomizers which break up the oil into a foglike spray, and forcing air into the furnace in such a way that it mixes thoroughly with the oil spray. The amount of fuel that can be burned is limited primarily by the actual capacity of the equipment that supplies the fuel (including the capacity of the sprayer plates), by the amount of air that can be forced into the furnace, and by the ability of the burner apparatus to mix this air with the fuel. The volume and shape of the furnace are also limiting factors.

The end point for combustion for a boiler is reached when the capacity of the sprayer plates, at the designed pressure for the system, is reached or when the maximum amount of air that can be forced into the furnace is insufficient for complete combustion of the fuel. If the end point for combustion is actually reached because of insufficient air, the smoke in the uptakes will be black because it will contain particles of unburned fuel. However, this condition should be rare, since the end point for combustion is artificially limited by sprayer plate capacity when the fuel is supplied at the burner manifold at designed operating pressure. As noted before, this artificial limitation upon combustion in the boiler furnace is the factor that would cause the end point for combustion to occur before either of the other two end points.

End Point for Moisture Carryover

The rate of steam generation should never be increased to the point at which an excessive amount of moisture is carried over in the steam. In general, naval specifications limit the allowable moisture content of steam leaving the saturated steam outlet to 1/4 of 1 percent.

As you know, excessive carryover can be extremely damaging to piping, valves, and turbines, as well as to the superheater of the boiler. It is not only the moisture itself that is damaging but also the insoluble matter that may be carried in the moisture. This insoluble matter can form scale on superheater tubes, turbine blades, piping and fittings; in some cases, it may be sufficient to cause unbalance of rotating parts.

As the evaporation rate is increased, the amount of moisture carryover tends to increase also, due to the increased release of steam bubbles. Because modern naval boilers are designed for high evaporation rates, steam separators and various baffle arrangements are used in the steam drum to separate moisture from the steam.

End Point for Water Circulation

In natural circulation boilers, circulation is dependent upon the difference between the density of the ascending mixture of hot water and steam and the density of the descending body of relatively cool water. As the firing rate is increased, the amount of heat transferred to the tubes is also increased. A greater number of tubes carry the upward flow of water and steam, and fewer tubes are left for the downward flow of water. Without downcomers to ensure a downward flow of water, a point would eventually be reached at which the downward flow would be insufficient to balance the upward flow of water and steam, and some tubes would become overheated and burn out. This condition would determine the end point for water circulation.

The use of downcomers ensures that the end point for water circulation will not be reached merely because the firing rate is increased.

Other factors that influence the circulation in a natural circulation boiler are the location of the burners, the arrangement of baffles in the tube banks, and the arrangement of tubes in the tube banks.

Full-power and overload ratings for the boilers in each ship are specified in the manufacturer's technical manual. The total quantity of steam required to develop contract shaft horsepower of the ship, divided by the number of boilers installed, gives boiler full-power capacity. Boiler overload capacity is usually 120 percent of boiler full-power capacity. For some boilers, a specific assigned maximum firing rate is designated.

A boiler should not be forced beyond full-power capacity—that is, it should not be steamed at a rate greater than that required to obtain full-power speed with all the ship's boilers in use. A boiler should never be steamed beyond its overload capacity, or fired beyond the assigned maximum firing rate, except in dire emergency.

Checking Boiler Efficiency

In order to check on boiler efficiency it is necessary to compare the amount of fuel actually burned in a boiler with the amount that should be burned. This check is usually made during economy runs and during full-power runs. As a rule, 4 hours are allowed for each run. During the run, fuel consumption is measured at intervals of precisely 1 hour. This measure, when corrected for meter error and verified by tank soundings, gives the amount of fuel that actually used.

The amount of fuel that should be used under specified conditions may be taken from tables or curves supplied in the manufacturer's technical manual for the boilers or from the ship's fuel performance tables. Since these two sources give different figures for the amount of oil that should be burned under various conditions, it is necessary to make a clear distinction between them. The differences, incidentally, arise from the fact that there are two basic approaches to the problem of checking on fuel consumption. When you are concerned only with boiler performance, you use the tables and charts from the manufacturer's technical manual; when you are concerned with plant performance with respect to fuel consumption, you use the ship's fuel performance tables.

BOILER CASUALTY CONTROL

There are many fireroom casualties which require a knowledge of preventive measures and corrective measures. Some are major, some are minor; but all can be serious. In the event of a casualty, the principal doctrine to be impressed upon operating personnel is the prevention of additional or major casualties. Under normal operating conditions, the safety of personnel and machinery should be given first consideration. Therefore, it is necessary to know instantly and accurately what to do for each casualty. Stopping to find out exactly what must be done for each casualty could mean loss of life, extensive damage to machinery, and even complete failure of the engineering plant. A fundamental principle of engineering casualty control is split-plant operation. The purpose of split-plant design is to minimize the damage that might result from any one casualty which affects propulsion power, steering, and electrical power generation.

Although speed in controlling a casualty is essential, action should never be taken without accurate information; otherwise the casualty may be mishandled, and further damage to the machinery may result. Cross-connecting and intact engineering plant with a partly damaged one must be delayed until it is certain that such action will not jeopardize the intact one.

Cross-connecting valves are provided for the main and auxiliary steam systems and other engineering systems so that any boiler or group of boilers, either forward or aft, may supply steam to each engineroom. These systems are discussed in chapter 9 of this manual showing the construction of the split-plant design on some types of ships.

The discussion of fireroom casualties in this chapter is intended to give you an overall view of how casualties should be handled. For further information on casualty control, study the Naval Ships Technical Manual, Chapter 9880, and the casualty control instructions issued for each type of ship.

Most of the casualties discussed in this chapter are usually treated in a step-by-step procedure, but it is beyond the scope of this chapter to give each step in handling each casualty. In the step-by-step procedure one step is performed, then another, then another, and so forth. In handling actual casualties, however, this step-by-step approach will probably have to be modified. Different circumstances may require a different sequence of steps for control of a casualty. Also, in handling real casualties several steps will have to be performed at the same time. For example, main control must be notified of any casualty to the boilers or to associated equipment. If "Notify main control" is listed as the third step in controlling a particular casualty, does this mean that the main control is not notified until the first two steps have been completed? Not at all. Notifying main control is a step that can usually be taken at the same time other steps are being taken. It is probably helpful to learn the steps for controlling casualties in the order in which they are given; but do not overlook the fact that the steps may have to be performed simultaneously.

FEED WATER CASUALTIES

Casualties in the control of water level include low water, high water, feed pump casualties, loss of feed suction, and low feed pressure. These casualties are some of the most serious ones.

Low water is one of the most serious of all fireroom casualties. Low water may be caused by failure of the feed pumps, ruptures in the feed discharge line, defective check valves, low water in the feed tank, or other defects.

However, the most frequent cause of low water is inattention on the part of the checkman and the PO in charge of the watch, or the diversion of their attention to other duties. The checkman's sole responsibility is to keep the water in the boiler at a proper level.

Low water is extremely damaging to the boiler and may endanger the lives of fireroom personnel. When the furnace is hot and there is insufficient water to absorb the heat, the heating surfaces are likely to be distorted, the brickwork damaged, and the boiler casing warped by the excessive heat. In addition, serious steam and water leaks may occur as a result of low water.

Disappearance of the water level from the water gage glasses must be treated as a casualty requiring the immediate securing of the boiler!

It should be noted that when the water level falls low enough to uncover portions of the tubes, the heat transfer surface is reduced. As a rule, therefore, the steam pressure will drop. Ordinarily a drop in steam pressure is the result of an increased demand for steam, and the natural tendency is to cut in more burners to fulfill the demand. If the drop in steam pressure is caused by low water, however, increasing the firing rate will result in serious damage to the boiler and possibly in injury to fireroom personnel. The possibility that a drop in steam pressure indicates low water must always be kept in mind! Always check the level in the water gage glasses before cutting in additional burners, when steam pressure has dropped for no apparent reason.

High water is another serious casualty that is most frequently caused by the inattention of the checkman and the PO in charge of the watch. If the water level in the gage glass goes above the highest visible part, the boiler must be secured immediately.

By careful observation, it is sometimes possible to distinguish between an empty gage glass and a full one by the presence or absence of condensate trickling down the inside of the glass. The presence of condensate indicates, of course, an empty glass—that is, a low water casualty. However, the boiler must be secured whether the water is high or low. After the boiler has been secured, the location of the water level can be determined by using the gage glass cutout valves and drain valves.

Failure of a feed system pump can have drastic consequences. Unless the pump casualty is corrected immediately, the pump failure will lead to low water in the boiler. In addition to the obvious dangers associated with low water, there are some which are equally serious but not so obvious. For example, low water causes complete or partial loss of steam pressure. When steam pressure is lost or greatly reduced, you will lose the services of vital auxiliary machinery—pumps, blowers, and so forth. It is essential, therefore, that feed pump casualties be handled rapidly and correctly.

If the main feed pump discharge pressure is too low, the first three things to be checked are (1) the feed booster pump discharge pressure, (2) the level and pressure in the deaerating feed tank, and (3) the feed stop and check valves on idle boilers. A failure of the feed booster pump will, of course, cause loss of suction and, therefore, loss of discharge pressure of the main feed pump. If the feed stop and check valves on idle boilers have accidentally been left open, the main feed pump discharge pressure may be low merely because water has been pumped to an idle boiler, as well as to the steaming boiler.

Some of the most likely causes of failure of the main feed pump are (1) malfunction of the constant-pressure pump governor, (2) an air-bound or vapor-bound condition of the main feed pump, (3) faulty pump clearances, and (4) malfunction or improper setting of the speed-limiting governor.

In many installations, the feed booster pump and the main feed pump are in the engineroom. In other installations, the feed booster pump is in the engineroom but the main feed pump is in the fireroom. In this latter type of installation, failure of the feed booster pump will be indicated to the fireroom force by loss of main feed pump discharge pressure and by the sounding of the low pressure feed alarm that is usually fitted where this type of machinery arrangement exists. The casualty to the feed booster pump will be dealt with by engineroom personnel, if the pump is in the engineroom; but fireroom personnel must take immediate action to maintain a supply of feed water to the boiler.

If the engineroom is unable to remedy the situation immediately, start the emergency feed pump on cold suction. The emergency feed pump can take a hot suction from the feed booster pump, or a cold suction from the reserve feed tanks. In standby condition, this pump should always be lined up on cold suction.

If the main feed pump fails and there is no standby pump available, start the emergency feed pump on hot suction and continue to feed the boiler. If the feed also fails then it will be necessary to start the emergency feed pump on cold suction.

If the emergency feed pump fails, the procedures for handling the casualty will vary according to the situation existing at the time of the failure.

In many ships, the emergency feed pump is normally used for in-port operation, with the main feed pump in standby condition and the feed booster pump providing a hot suction for the emergency feed pump. Under these conditions, emergency feed pump failure can be handled by notifying the engineroom so that the main feed pump can be put on the line and used to feed the boiler.

A more difficult problem will arise if the emergency feed pump fails when it is being used because of a previous casualty to the feed booster pump or to the main feed pump. Under these conditions, it may be possible to deal with the situation by cross-connecting and using a pump in some other space to supply feed to the boiler. If the operating conditions do not allow this solution of the problem, it will be necessary to secure the boiler immediately in order to prevent a low water casualty.

FUEL SYSTEM CASUALTIES

Casualties to any part of the fuel oil system are serious and must be remedied at once. Common casualties include (1) oil in the fuel oil heater drains, (2) water in the fuel oil, (3) loss of fuel oil suction, (4) failure of the fuel oil service pump, and (5) fuel oil leaks. It should be noted that these casualties to the fuel oil system are for ships burning NSFO. The procedures for ships burning other types of fuel will differ to some extent, but not in all cases.

Oil leakage from the fuel oil heaters into the drains may cause oil contamination of the drain lines, the reserve feed tanks, the deaerating feed tank, and the feed system piping and pumps. The presence of oil in any part of the feed system is dangerous because of the possibility that the oil will eventually reach the boilers, where it will cause steaming difficulties and serious damage to the boilers.

Fuel oil heater drains must be inspected hourly for the presence of oil.

The presence of an appreciable amount of water in the fuel oil is indicated by hissing and sputtering of the fires and atomizers and by racing of the fuel oil service pump. The situation must be remedied at once; otherwise, choked atomizers, loss of fires, flarebacks, and refractory damage may result.

A loss of fuel oil suction usually indicates that the oil in the service suction tank has dropped below the level of the fuel oil service pump suction line. This causes a mixture of air and oil to be pumped to the atomizers. The atomizers begin to hiss and the fuel oil service pump begins to race. It must be strongly emphasized that the loss of fuel oil suction can cause serious results. Related casualties may include loss of auxiliary steam and electric power, with the complete loss of all electrically driven and steam-driven machinery.

Failure of the fuel oil service pump can cause the same progressive series of casualties as those which result from loss of fuel oil suction.

Fuel oil leaks are very serious, no matter how small they may be. Fuel oil vapors are very explosive. Any oil spillage or leakage must be wiped up immediately.

FLAREBACKS

A flareback is likely to occur whenever the pressure in the furnace momentarily exceeds the pressure in the boiler air casing. Flarebacks are caused by an inadequate air supply for the amount of oil being supplied, or by a delay in lighting the mixture of air and oil.

Situations which commonly lead to flarebacks include: (1) attempting to light off or to relight burners from hot brickwork; (2) gunfire or bombing which creates a partial vacuum at the blower intake, thus reducing the air pressure supplied by the blowers; (3) forced draft blower failure; (4) accumulation of unburned fuel oil or combustible gases in furnaces, tube banks, uptakes, or air casings; and (5) any event which first extinguishes the burners and then allows unburned fuel oil to spray out into the hot furnace. An example of this last situation might be a temporary interruption of the fuel supply which would cause the burners to go out; when the fuel oil supply returns to normal, the heat of the furnace might not be sufficient to relight the burners immediately. In a few seconds, however, the fuel oil sprayed into the furnace would be vaporized, and a flareback or even an explosion might result.

SUPERHEATER CASUALTIES

If the distant-reading superheater thermometer does not register a normal increase in temperature when the superheater is first lighted off, the trouble may be either lack of steam flow or failure of the distant-reading thermometer. Lack of steam flow must be considered as a possible cause even if the superheater steam flow indicator (if installed) shows that there is a flow. If the thermometer does not register a normal increase in temperature, secure all superheater burners.

When operating with superheat, it is essential to keep a constant check on the flow of steam through the superheater and on the superheater outlet temperature. Any deviation from normal conditions must be corrected without delay.

It is important to remember that a casualty to some other part of the engineering plant may reduce or entirely stop the flow of steam through the superheater, and so cause a superheater casualty, unless appropriate action is taken to prevent damage. For example, a casualty to the main engines might call for a sudden large reduction or even a complete stoppage of steam flow. Even if the superheater burners are secured, there will still be a need for steam flow to protect the superheater from the heater of the furnace. In this event, or whenever a greater flow is required than can be obtained by ordinary means, lift the superheater safety valves by hand to ensure a positive flow of steam through the superheater.

When the superheater thermal alarm sounds, the superheater fires must be immediately decreased to bring the temperature below alarm temperature. Do not decrease the temperature further than necessary. It is very seldom necessary to secure all superheater burners in order to bring the temperature down to the prescribed point.

CASUALTIES TO REFRACTORIES

If brick or plastic falls out of the furnace walls and goes unnoticed, burned casings may result. If brick or plastic falls out of a furnace wall, if practicable, secure all the burners. If it is not practicable to secure all the burners, secure those burners which are adjacent to the damaged section. NOTE: It may be necessary to continue operating the boiler until another boiler can be brought in on the line.

CASUALTIES TO BOILER
PRESSURE PARTS

When boiler pressure parts, such as tubes, carry away or rupture, escaping steam may cause serious injury to personnel and damage to the boiler. It is urgent that the boiler be secured, relieved of its pressure, and cooled until no more steam is generated. If a boiler pressure part carries away or ruptures, take steps immediately upon discovery of the casualty, to minimize and localize the damage as much as circumstances will allow.

Gage glasses are connected to the water and steam spaces of the steam drum. If a water gage glass carries away, the mixture of steam and water escaping from the gage connections may seriously burn personnel in the area. A ball check valve in the high pressure gage line functions when the flow is excessive. In addition, the hazard of flying particles of glass makes this casualty very serious. The particles of glass could lodge in your eyes and blind you, or they could lodge elsewhere in your body and cause serious injury. If a gage glass casualty occurs, throw a large sheet of asbestos cloth, rubber matting, or similar material over the glass. Then take immediate action to secure the gage glass.

PRECAUTIONS TO PREVENT FIRES

The following precautions must be taken to prevent fires:

1. Do not allow oil to accumulate in any place. Particular care must be taken to guard against oil accumulation in drip pans under pumps, in bilges, in the furnaces, on the floor plates, and in the bottom of air-encased boilers. Should leakage from the oil system to the fireroom occur at any time, immediate action should be taken to shut off the oil supply by means of quick-closing valves and to stop the oil pump.

2. Absolutely tight joints in all oil lines are essential to safety. Immediate steps must be taken to stop leaks whenever they are discovered. Flange safety shields should be installed on all flanges in fuel oil service lines to prevent spraying oil on adjacent hot surfaces.

3. No lights should be permitted in the fireroom except electric lights (fitted with steam-tight globes, or lenses, and wire guards), and permanently fitted smoke indicator and water gage lights. If work is being done in the vicinity of flammable vapors, or if rust-preventive compound or metal-conditioning compound is being used, all portable lights should be of the explosion proof type.

BOILER MAINTENANCE

The engineer officer must keep himself fully acquainted with the general condition of each boiler and the manner in which each is being operated and maintained. He must satisfy himself, by periodic inspections, that the exterior and interior surfaces of the boiler are clean; that the refractory linings adequately protect the casing, drums, and headers; that the integrity of the pressure parts are being maintained; and that the operating condition of the burners, safety valves, operating instruments, and other boiler appurtenances are satisfactory.

The engineer officer must assure himself that the idle boilers are properly secured at all times, and while steaming, the fuel oil used is free of sea water, and the feed water is within prescribed limits, free of salts, entrained oxygen, and oil.

All parts of the boiler must be carefully examined whenever they are exposed for cleaning and overhauling, and the conditions observed must be described in the boiler record sheet and the engineering log. All unusual cases of damage or deterioration discovered at any time should be reported to the type commander, stating in detail the extent of injury sustained, remedies applied, and the causes, if determined. If considered of sufficient importance, or technical assistance is desired from the Naval Ship Systems Command, a copy of the correspondence should be forwarded to the Naval Ship Systems Command.

The requirements for fireroom maintenance and repair are established by the Planned Maintenance Subsystem; information on this system is contained in the Maintenance and Material Management (3-M) Manual, OPNAV 43P2 Revised edition. All fireroom maintenance shall be conducted in accordance with this system.

CHAPTER 11

BOILER FITTINGS AND CONTROLS

The fittings, instruments, and controls used on naval boilers are sufficiently numerous and important to warrant separate discussion. The term boiler fittings is used to describe a number of attachments which are installed in or closely connected to the boiler and which are required for the operation of the boiler. Boiler fittings are generally divided into two classes. Internal fittings (also called internals) are those installed inside the steam and water spaces of the boiler; external fittings are those installed outside the steam and water spaces. Boiler instruments such as pressure gages and temperature gages are usually regarded as external boiler fittings. Boiler controls are special systems which automatically control the fuel oil, combustion air, and feed water inputs in order to regulate the steam output of the boiler.

INTERNAL FITTINGS

The internal fittings installed in the steam drum usually include equipment for distributing the incoming feed water, for giving surface blows, and for directing the flow of steam and water within the steam drum. In addition, boilers which do not have controlled superheat have desuperheaters for desuperheating steam needed for auxiliary purposes; the desuperheater is most commonly installed in the steam drum, but is installed in the water drum in some of the newer boilers. Internal fittings in some boilers also include equipment for injecting chemicals for boiler water treatment.

The specific design and arrangement of boiler internal fittings varies somewhat from one type of boiler (and from one boiler manufacturer) to another. The arrangement of internals in several boilers is therefore described here.

Figure 11-1 illustrates a typical arrangement of internal fittings installed in the steam drum of a header-type boiler. This illustration also shows many of the external connections. Feed water enters through the feed inlet (A) and flows to the internal feed pipe (B). The feed pipe is capped at one end. The horizontal part of the feed pipe runs about 80 percent of the length of the drum, well below the normal water level. The feed pipe is perforated along the upper side so that the feed water will be evenly distributed along the length of the pipe.

The dry pipe (C) is suspended near the top of the steam drum, along the centerline of the drum. Both ends of the dry pipe are closed. Steam enters by way of perforations in the upper surface of the dry pipe. Thus the steam must change direction in order to enter the dry pipe. Since some moisture is lost whenever steam changes direction, the dry pipe acts as a device to separate steam and moisture. Steam leaves the dry pipe through the main steam outlet (D) and from there goes to the superheater. A few perforations in the bottom of the dry pipe allow water droplets to drain back down to the water in the steam drum.

A longitudinal baffle (N) also helps to separate moisture from the steam before the steam enters the dry pipe. The baffle is installed in such a way as to allow steam to flow to the dry pipe but to keep moisture (and any solid matter that might be carried over with the moisture) from entering the dry pipe.

The surface blow-off pipe (E) is used to remove grease, scum, and light solids from the boiler water and to reduce the salinity of the boiler water while the boiler is steaming. The surface blow-off pipe is installed near the center of the steam drum, with the upper surface of the pipe slightly below the normal water level of the drum. The pipe runs almost the entire length of the drum. Holes are drilled along the top centerline of the pipe. One end of the pipe is blanked off. The other end is connected through

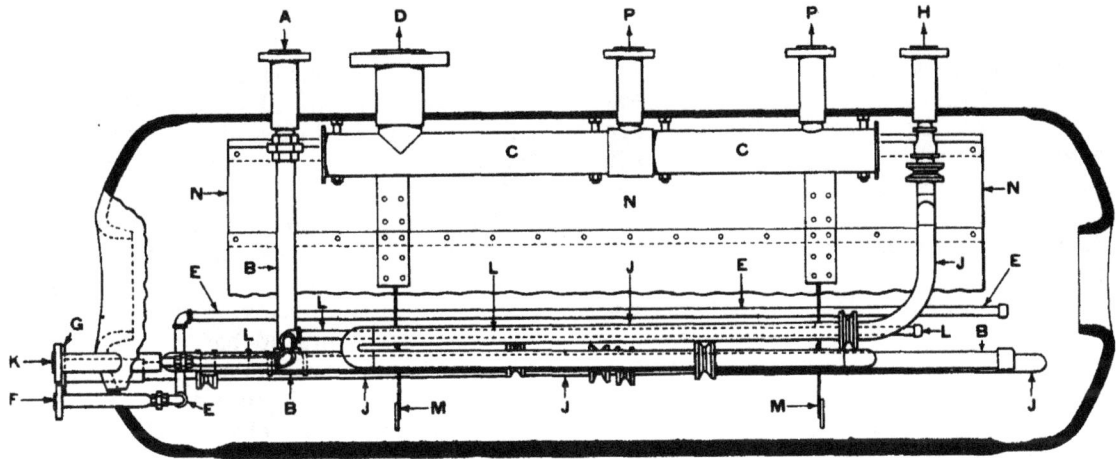

LONGITUDINAL SECTIONAL ELEVATION

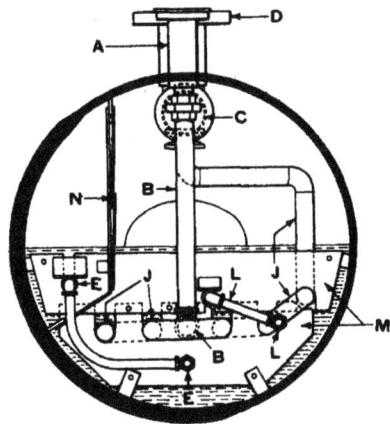

CROSS SECTIONAL ELEVATION

DESIGNATION OF SYMBOLS

A - FEED INLET
B - FEED PIPE
C - DRY PIPE
D - MAIN STEAM OUTLET
E - SURFACE BLOW-OFF PIPE
F - SURFACE BLOW-OFF NOZZLE
G - DESUPERHEATER INLET
H - DESUPERHEATER OUTLET
J - DESUPERHEATER
K - CHEMICAL FEED INLET
L - CHEMICAL FEED PIPE
M - SWASH PLATES
N - BAFFLE
P - SAFETY VALVE NOZZLES

38.42

Figure 11-1.—Arrangement of internal fittings in
header-type boiler.

the drumhead to the surface blow-off nozzle (F). When the surface blow valve is opened, the pressure in the drum forces the water above the blow-off pipe to go into the pipe through the holes on the top surface; the water from the surface blow-off pipe then leaves the boiler by way of the surface blow valve.

The desuperheater (J) is an assembly of pipe lengths and return bends located below the water level of the drum. The superheater steam enters the desuperheater through the desuperheater inlet (G), gives up its superheat to the water in the steam drum, and then—once again at or very close to saturation temperature—passes through

the desuperheater outlet (H) before entering the auxiliary steam line.

The chemical feed inlet (K) is connected to the internal chemical feed pipe (L). The chemical feed pipe has holes drilled in it to allow even distribution of chemicals used for boiler water treatment.

Swash plates (M) are used to reduce the surging or swashing of water from one end of the drum to the other as the ship moves. In addition, the swash plates act as supports for the internal feed pipe and for the desuperheater.

The safety valve nozzles (P) are not normally considered internal fittings. These nozzles connect the safety valves (not shown) to the steam drum.

The arrangement of internal fittings in a double-furnace boiler with controlled superheat is shown in figure 11-2. The dry pipe, the internal feed pipe, and the surface blow line are about the same as the corresponding fittings in the header-type boiler; but some of the other fittings are different.

The double-furnace boiler has no desuperheater, since auxiliary steam is taken directly from the steam drum without passing first through the superheater.

Swash plates are not required in the water spaces of the double-furnace boiler because this type of boiler is always installed with the long axis of the steam drum fore-and-aft rather than athwartships. Surging of water, therefore, is not a particular problem.

The double-furnace boiler does not have a separate chemical feed pipe. Instead, chemicals for boiler water treatment come into the steam drum with the feed water and are therefore distributed with the feed water through the holes in the internal feed pipe.

Perhaps the greatest difference in the internal fittings of the double-furnace boiler and the header-type boiler is in the equipment provided for the separation of moisture from the steam. In the header-type boiler, a steam baffle helps to separate the moisture from the steam before

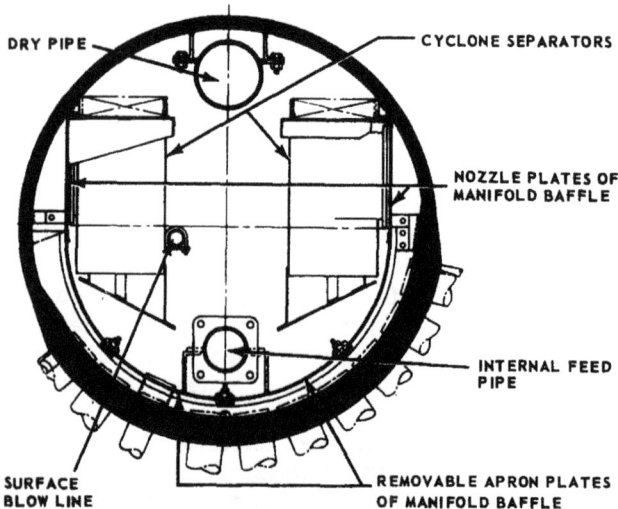

38.43
Figure 11-2.—Arrangement of internal fittings in double-furnace boiler.

the steam enters the dry pipe. In the double-furnace boiler, cyclone steam separators are used instead of a steam baffle. These separators utilize centrifugal force to separate water and steam. There are usually 18 of these separators installed in the steam drum of a double-furnace boiler; half of them are installed on one side of the drum and half on the other side.

The cyclone steam separators are attached to a manifold baffle which extends from just forward of the generating tubes to just aft of them. To avoid interference with boiler circulation, the manifold baffle does not reach as far as the downcomers. The manifold baffle curves around inside the lower half of the steam drum, passing just below the internal feed pipe and leaving a space of about 3 inches between the baffle and the steam drum. The baffle is attached to the drum by means of two flat bars which hang from the drum, one on each side. The bars extend the full length of the baffle. Each bar contains ports or openings, and a cyclone steam separator is placed over each port.

The general arrangement of the manifold baffle and the cyclone steam separators may be seen in figure 11-2. Now let us examine figure 11-3 and trace the flow of steam and water through the steam drum. The generating tubes discharge a mixture of steam and water into the space between the manifold baffle and the steam drum. From this space, the only passage available for the steam and water is through the ports which open to the cyclone separators. As the mixture passes through the separators, the steam passes upward and the water is discharged downward.

The cyclone steam separator is shown in cutaway view in figure 11-4 and in plain view in figure 11-5. The mixture of steam and water enters the separator through the inlet connection, at a tangent to the separator body. Because of its angle of entrance, the mixture of steam and water acquires a rotary motion. As the mixture whirls around, centrifugal force separates the water from the steam. The water, being heavier, is thrown out toward the sides of the separator. The steam, being lighter, tends to remain near the center. An internal baffle further helps to deflect the steam to the center and the water to the outside. The steam then rises through the center of the separator and passes through the scrubber element. The scrubber consists of closely spaced corrugated steel plates. As the steam passes through the

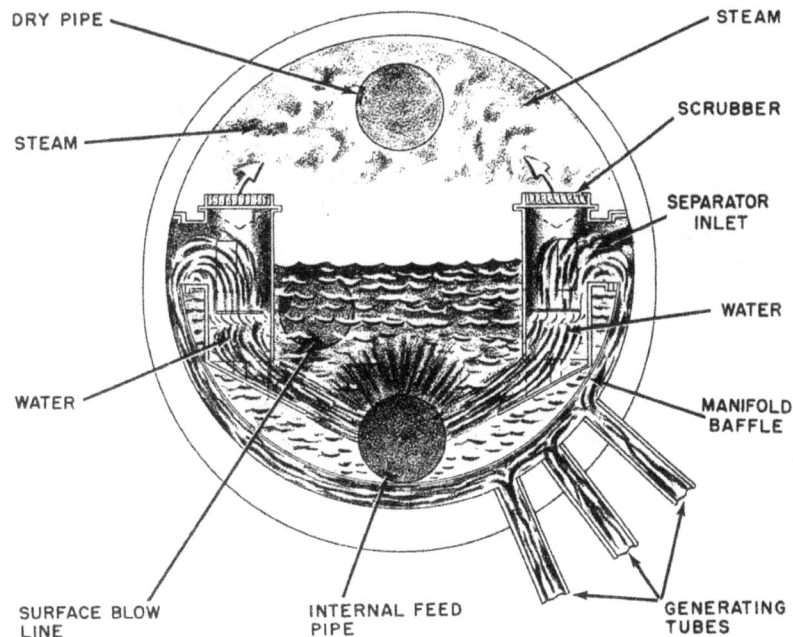

38.44

Figure 11-3.—Flow of steam and water in steam drum
of double-furnace boiler.

scrubber, its direction is changed frequently, and with each change of direction some moisture is lost. The steam passes out of the scrubber element into the top part of the steam drum and then enters the dry pipe.

While the steam is rising, the water is falling to the base of the separator. Stationary curved vanes in the bottom of the separator serve to maintain the rotary motion of the water until the water is finally discharged from the bottom of the separator. Since the vanes are located around the periphery of a flat plate, the water passes from the separator only around the outer edge. The flat plate also serves to keep the steam, which is in the center of the separator, from being carried downward with the water.

A flat baffle plate is fitted at the base of each of the two end separators on each side. This baffle plate guides the water that is being discharged from the separator to the center of the drum, where it mixes thoroughly with the rest of the water in the drum. Without such a baffle, the water from these end separators would tend to flow directly to the downcomers and, since it is hotter than the rest of the water in the steam drum, it would tend to disrupt the boiler circulation.

Figure 11-6 illustrates the arrangement of internal fittings in an older single-furnace boiler. The internal fittings for this boiler are quite similar to those found in the double-furnace boiler, except that the single-furnace boiler has a desuperheater (item 5 in fig. 11-6).

The internal fittings in the steam drum require routine maintenance and upkeep. All maintenance requirements shall be conducted in accordance with the Planned Maintenance Subsystem of the 3-M System.

When a boiler is opened for cleaning of watersides, the internal fittings must be removed. The fittings are bolted into place; removing the bolts allows you to remove the fittings. Be sure that all bolts and tools used in removing the bolts are strictly accounted for. When removing internal fittings, be sure to identify them so that you will be able to reinstall them correctly.

After removing the fittings from the steam drum, thoroughly wirebrush and clean them. Check the dry pipe, the internal feed line, and the surface blow line to be sure that all the holes are free and clear of obstructions. In addition, inspect the inside of the feed line for oil accumulations; if you find any sign of oil notify the CPO in charge of the fireroom.

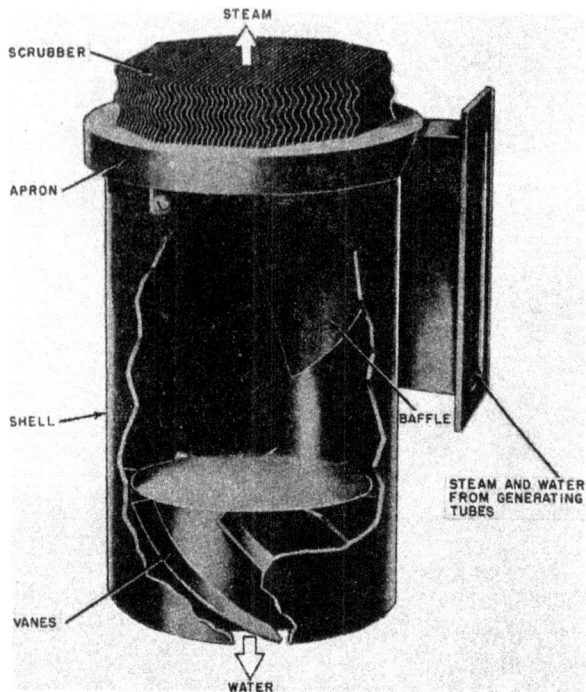

38.45

Figure 11-4.—Cutaway view of cyclone steam separator.

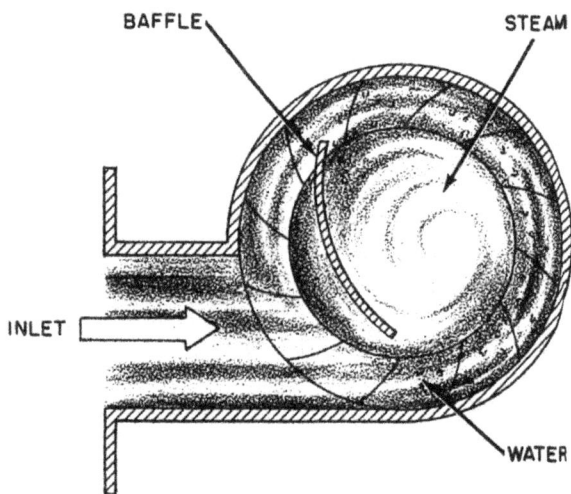

38.46

Figure 11-5.—Plan view of cyclone steam separator.

Before reinstalling the fittings, wirebrush, clean, and hose down the steam drum. Be sure that it is clean and free of any oil or other accumulation.

When replacing the fittings in the steam drum, be sure that all the bolts are drawn tight. The desuperheater flanges must be thoroughly cleaned before the desuperheater is fitted into place and bolted. New gaskets must be installed. The flanges must be drawn up evenly and tightly to prevent any leakage from the flanged joints.

The internal fittings used in newer single-furnance boilers differ from those used in older boilers and also differ among themselves. Figure 11-7 shows the arrangement of fittings used in the steam drum of a boiler on a DLG 9-15 class ship. A little study of this illustration shows several new or different features.

To begin with, notice that there is no desuperheater, even though this is a single-furnace boiler. The boiler does have a desuperheater, but it is installed in the water drum rather than in the steam drum.

Another interesting feature illustrated in figure 11-7 is the vortex eliminator. A vortex eliminator consists of a series of grid-like plates arranged in a semicircular shape to conform to the shape of the lower half of the steam drum. One vortex eliminator is located at the front of the steam drum (as shown in fig. 11-7) and another is located at the rear of the drum. In each case, the eliminator is fitted over the necks of the downcomers. The purpose of the vortex eliminators is to reduce the swirling motion of the water as it enters the downcomers.

The arrangement of internal fittings in the steam drum of a boiler on one of the newer destroyer escorts is shown in figure 11-8. Notice that there are two feed pipes, each of which runs lengthwise in the drum. The discharge holes are drilled along the inner side of each feed pipe so that the incoming feed water is discharged horizontally toward the middle of the drum. Notice also that the internals in this steam drum include horizontal steam separators rather than cyclone steam separators.

A horizontal steam separator is shown in figure 11-9. These separators are installed in much the same way as the cyclone steam separators—that is, one row of separators is installed along each side inside the steam drum. Each horizontal steam separator has a machined flange which is bolted to a matching flange attached to a girth baffle. The mixture of steam

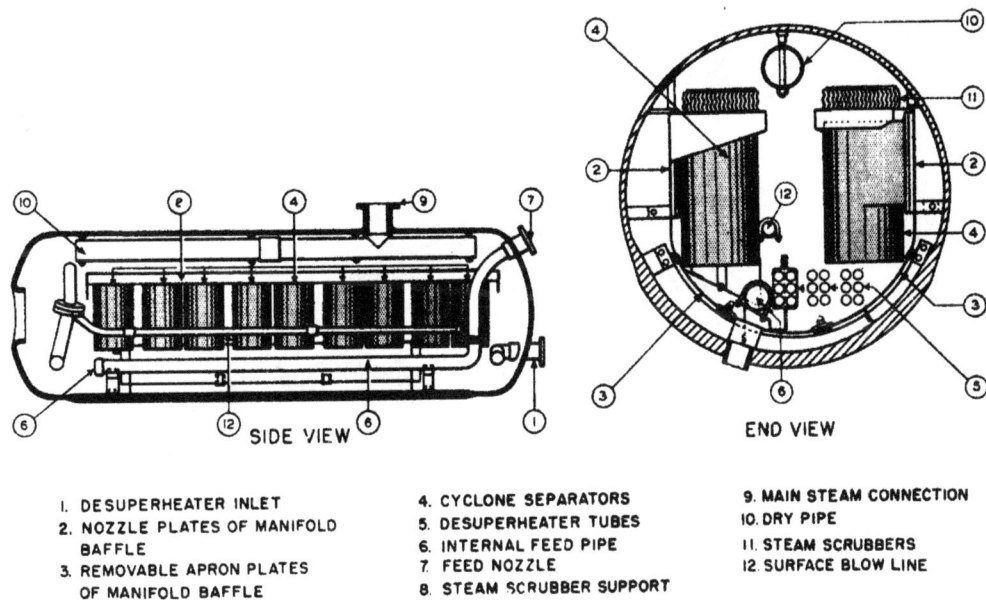

1. DESUPERHEATER INLET
2. NOZZLE PLATES OF MANIFOLD BAFFLE
3. REMOVABLE APRON PLATES OF MANIFOLD BAFFLE
4. CYCLONE SEPARATORS
5. DESUPERHEATER TUBES
6. INTERNAL FEED PIPE
7. FEED NOZZLE
8. STEAM SCRUBBER SUPPORT
9. MAIN STEAM CONNECTION
10. DRY PIPE
11. STEAM SCRUBBERS
12. SURFACE BLOW LINE

38.47X

Figure 11-6.—Arrangement of internal fittings in older single-furnace boiler.

and water enters the separator through a tapered inlet section on the side nearest the shell of the steam drum and follows a curving path along the curve of the separator. The steam leaves through the outlet orifices at each side of the separator. The water, being heavier, continues along the curve of the separator and is discharged through drain holes in the drain baffle. The knife edge on the drain baffle is there to minimize turbulence. A second drain baffle curves down below the knife edge and drains off any water that might pass over the knife edge.

The steam discharged from the horizontal separators is channeled directly to the chevron dryers which are installed near the top of the drum, as shown in figure 11-8. A number of these chevron dryers are installed along the length of the steam drum. From the dryers, the steam enters the rectangular dry box which is fitted against the top of the steam drum. The dry box acts as a chamber for the collection of dry steam.

The chemical feed pipe shown in figure 6-8 is connected to a nozzle on the end of the drum. The chemical feed pipe and nozzle are used to inject chemicals for boiler water treatment while the boiler is in operation; they are also used to draw samples of boiler water for testing.

EXTERNAL FITTINGS AND CONNECTIONS

External fittings and connections commonly used on naval boilers include drains and vents, sampling connections, feed stop and check valves, steam stop valves, safety valves, soot blowers, blow valves, water gage glasses, remote water level indicators, superheater steam flow indicators, pressure and temperature gages, superheater temperature alarms, smoke indicators, oil drip detector periscopes, single-element feed water regulators, and other devices that are closely connected to the boiler but not installed in the steam and water spaces.

Any listing of boiler external fittings and connections tends to sound like a catalog of miscellaneous and unrelated hardware. Actually, however, all of the external fittings and connections serve purposes that are related to boiler operation. Some of the fittings and connections allow you to control the flow of feed water and steam. Others serve as safety devices. Still others allow you to perform operational procedures—removing soot from the firesides, for example, or giving surface blows—that are necessary for efficient functioning of the boiler. The instruments attached to or installed

MAIN STEAM OUTLET NOZZLE

DRY PIPE

HANGER BOLT AND
BOLT ANCHOR

STEAM DRUM BAFFLE
(INTEGRAL PART OF DRUM)

SCRUBBER

SCREEN PLATE

CYCLONE
STEAM
SEPARATOR

SURFACE
BLOW NOZZLE

SURFACE BLOW PIPE

SUPPORT BAR

BAFFLE

FRONT VORTEX
ELIMINATOR

FRONT DOWNCOMERS
(TO WATER DRUM)

FRONT DOWNCOMER
(TO SIDEWALL HEADER)

BAFFLE

CHEMICAL FEED PIPE

FEEDWATER PIPE

38.231

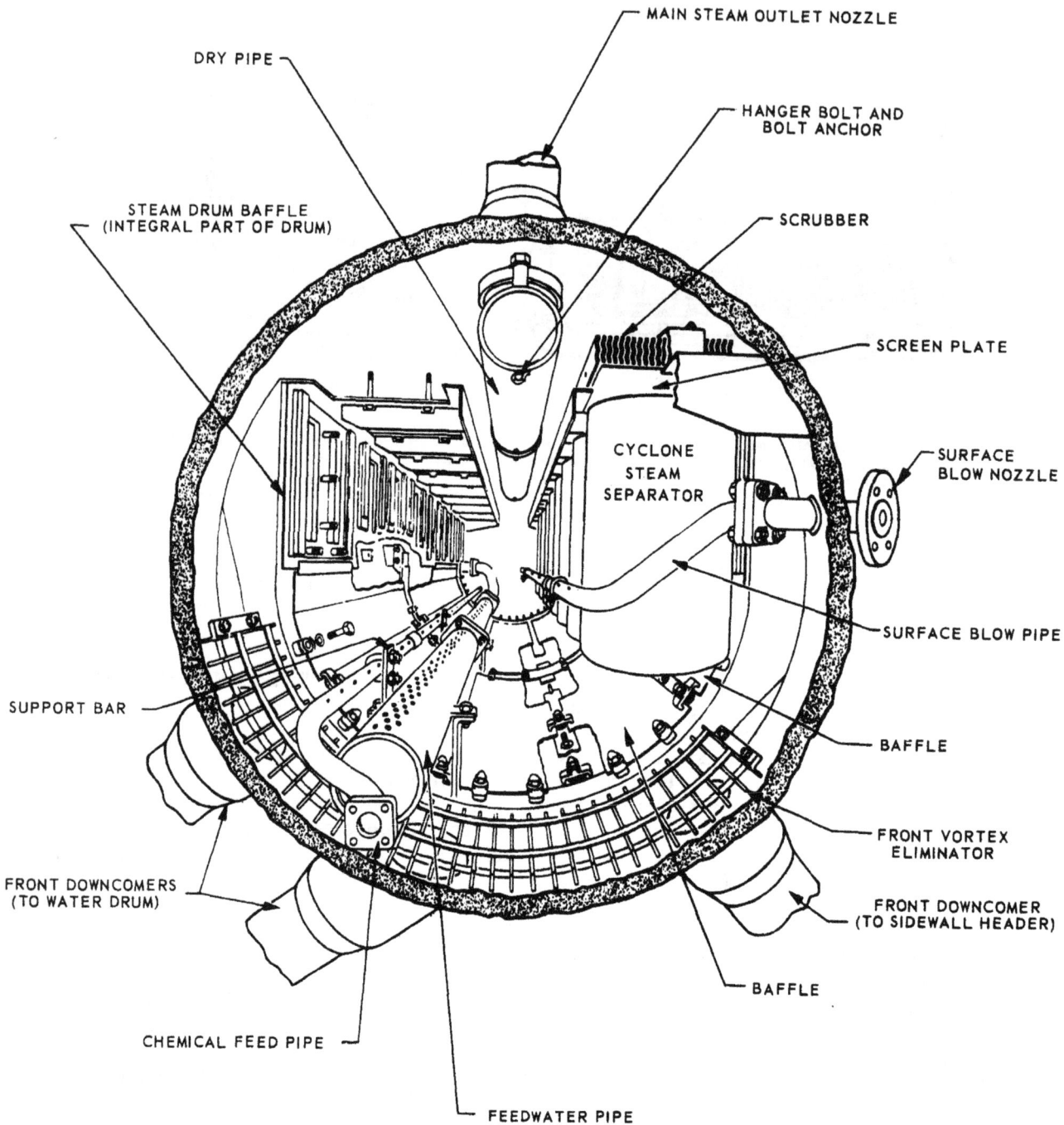

Figure 11-7.—Arrangement of internal fittings in newer
single-furnace boiler (DLG 9-15 class ship).

near the boiler give you essential information concerning the conditions existing inside the boiler. To understand the purposes of the external fittings and connections, then, it is necessary to see how each item is related to boiler operation.

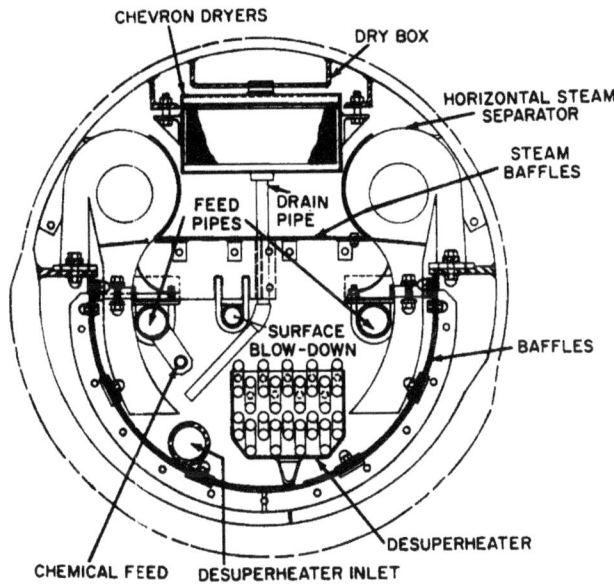

CHEVRON DRYERS
DRY BOX
HORIZONTAL STEAM SEPARATOR
STEAM BAFFLES
FEED PIPES
DRAIN PIPE
SURFACE BLOW-DOWN
BAFFLES
DESUPERHEATER
CHEMICAL FEED
DESUPERHEATER INLET

38.48

Figure 11-8.—Arrangement of internal fittings in newer single-furnace boiler (DE 1006 class ship).

Figures 11-10, 11-11, 11-12, and 11-13 show the locations of many external fittings and connections on a recent 1200-psi single-furnace boiler. As you study the following information on external fittings and connections, you may find it helpful to refer to these figures to see where the various units are installed on or connected to the boiler. Remember, however, that the illustrations shown here are for one particular boiler and that differences in boiler design lead to differences in the type and location of external fittings and connections. Drawings showing the location of external fittings and connections are usually included in the manufacturer's technical manuals for the boilers on each ship.

The maintenance of external fittings is of vital importance to the proper operation of the boiler. Therefore, all maintenance shall be conducted in accordance with the Planned Maintenance Subsystem of the 3-M System.

Drains and Vents

The main part of the boiler, the economizer, and the superheater—in short, all the steam and water sections of the boiler—must be provided with drains and vents.

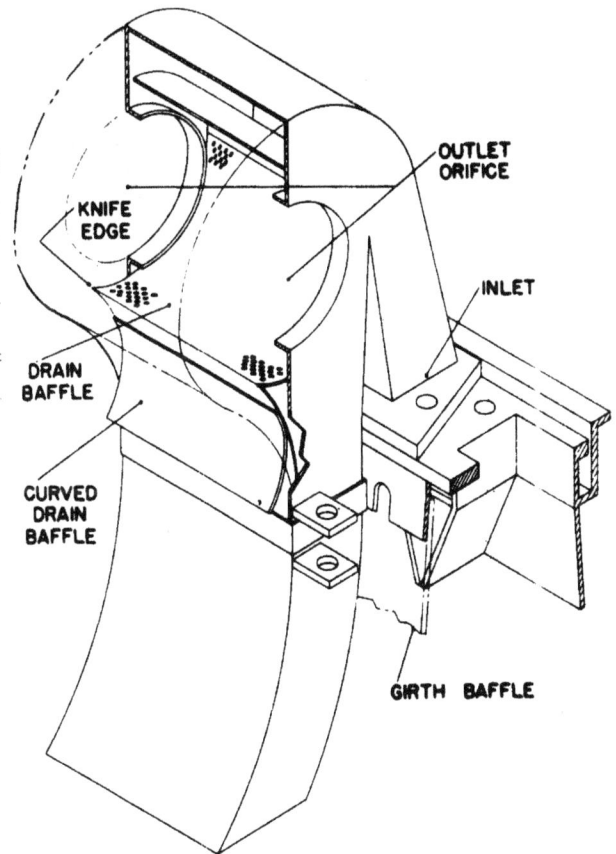

OUTLET ORIFICE
KNIFE EDGE
INLET
DRAIN BAFFLE
CURVED DRAIN BAFFLE
GIRTH BAFFLE

38.49

Figure 11-9.—Horizontal steam separator.

The main part of the boiler may be drained through the bottom blow valves (described later in this chapter) and through water wall header drain valves. It is vented through the aircock, which is a high pressure globe valve[1] installed at the highest point of the steam drum. The aircock allows air to escape when the boiler is being filled and when steam is first forming; it also allows air to enter the steam drum when the boiler is being emptied.

The economizer is vented through a vent valve on the economizer inlet piping. It is drained through a drain line from the economizer outlet header. Another drain line, this one coming from the drain pan installed below

_____
[1]Basic types of valves are discussed in chapter 14 of this text.

283

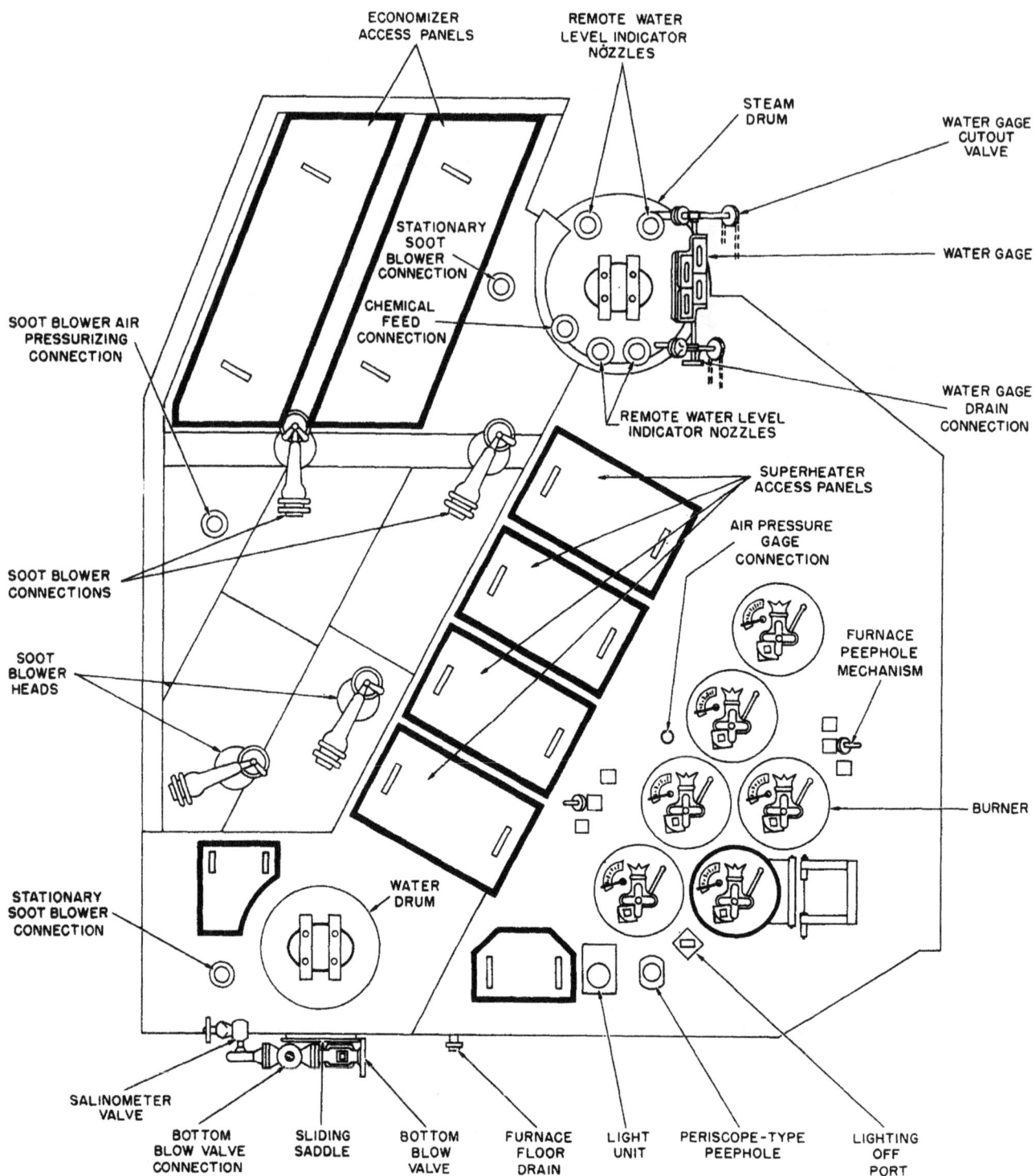

38.232

Figure 11-10.—External fittings and connections on
1200-psi single-furnace boiler (front view).

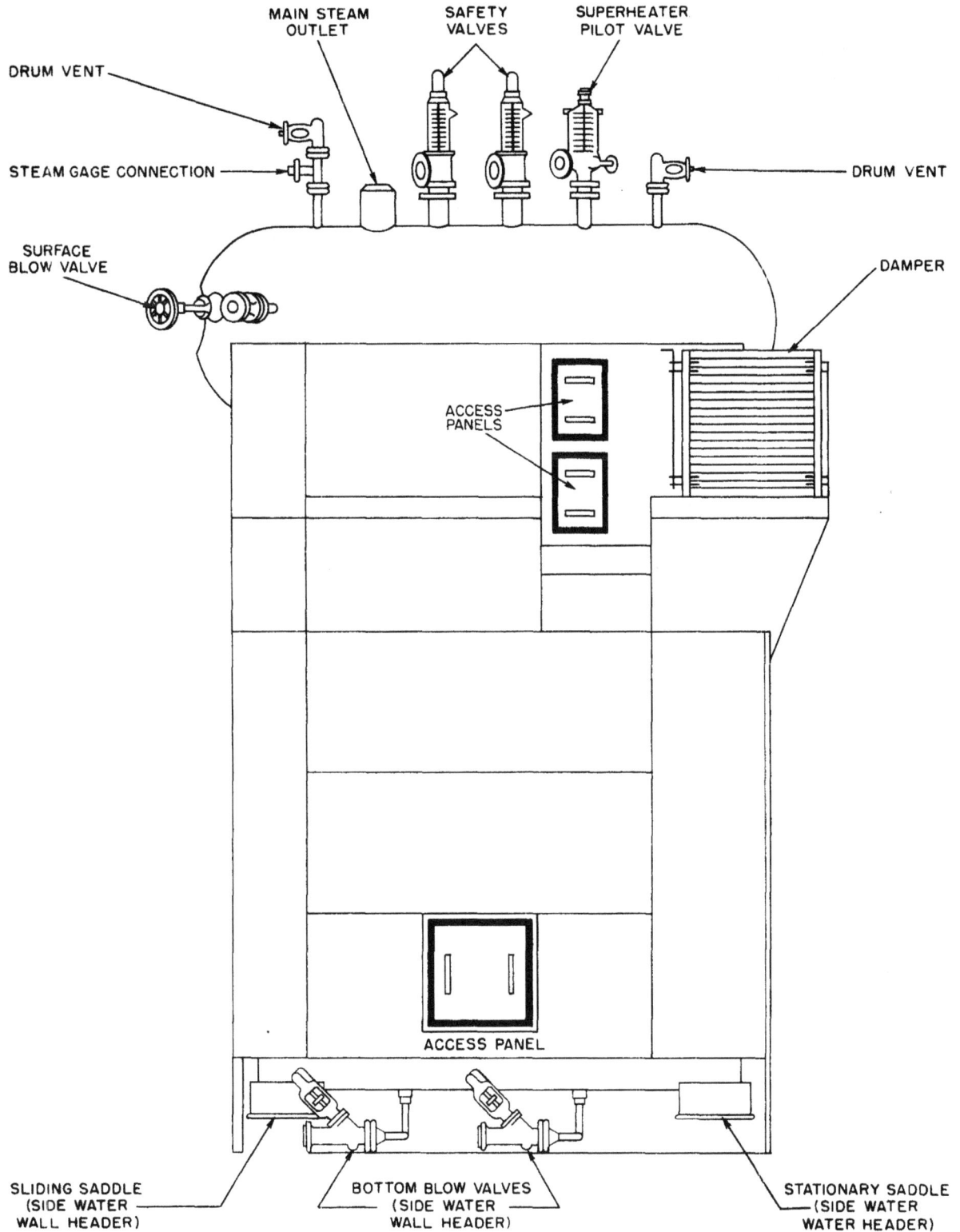

Figure 11-11.—External fittings and connections on 1200-psi single-furnace boiler (furnace side view).

38.233

ECONOMIZER
VENT. CONN.

ECONOMIZER
VESTIBULE PEEPHOLE

ECONOMIZER
INLET CONN.

STATIONARY
SOOT BLOWER
CONNECTIONS

SOOT BLOWER
HEAD

DAMPER

CONN. FOR AIR FLOW
MEASURING DEVICE

SOOT BLOWER HEAD

ECONOMIZER
VESTIBULE
PEEPHOLE

ECONOMIZER
OUTLET CONN.

CONN. FOR AIR FLOW
MEASURING DEVICE

BLANK FLANGE
(CHEMICAL CLEANING
AND BOILER TEST
CONNECTION)

ACCESS
PANEL

ECONOMIZER DRAIN
OUTLET CONN.

MAIN FEED PIPE CONN.

ACCESS PANELS

38.234

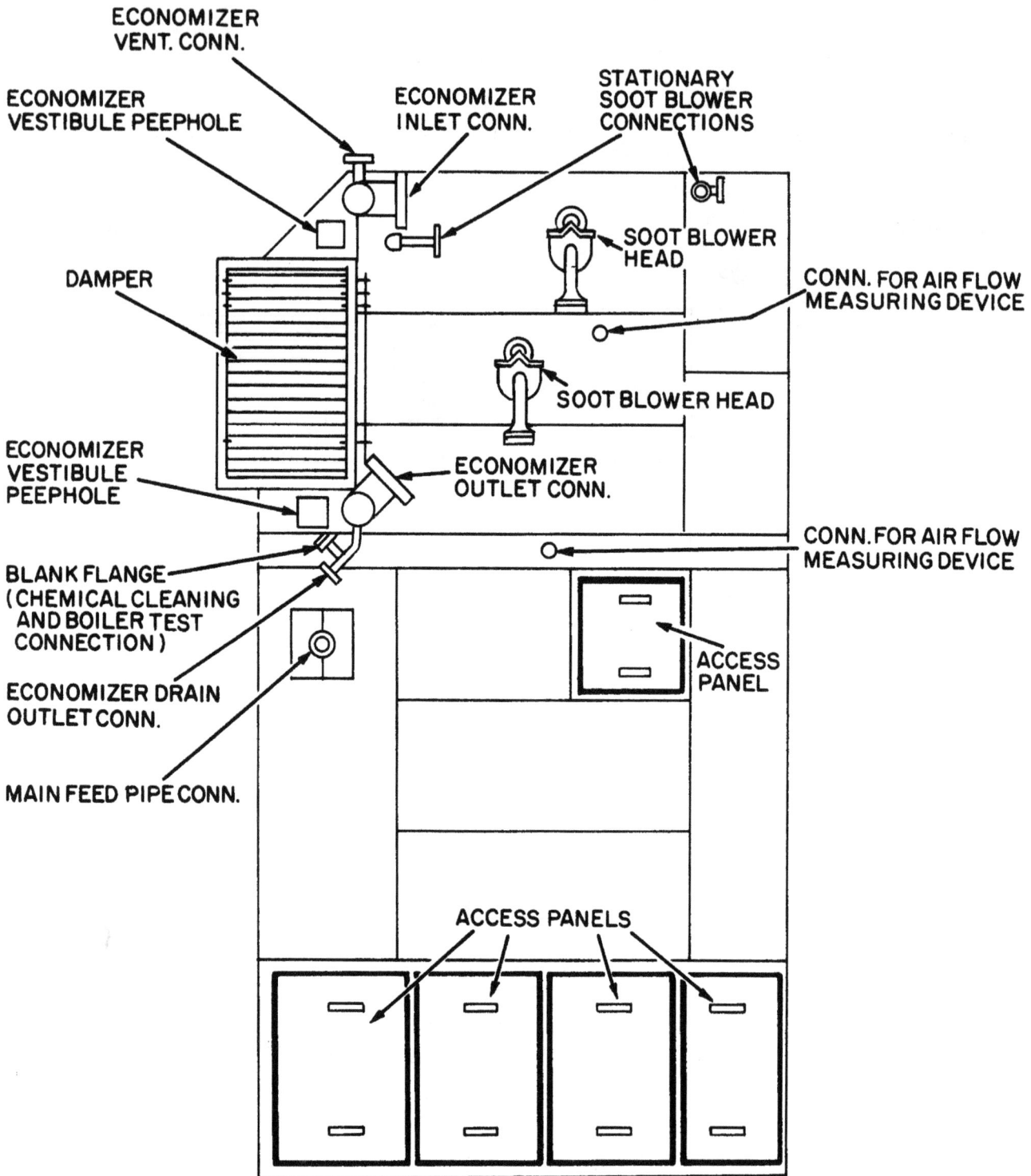

Figure 11-12.—External fittings and connections on 1200-psi single-furnace
boiler (economizer side view).

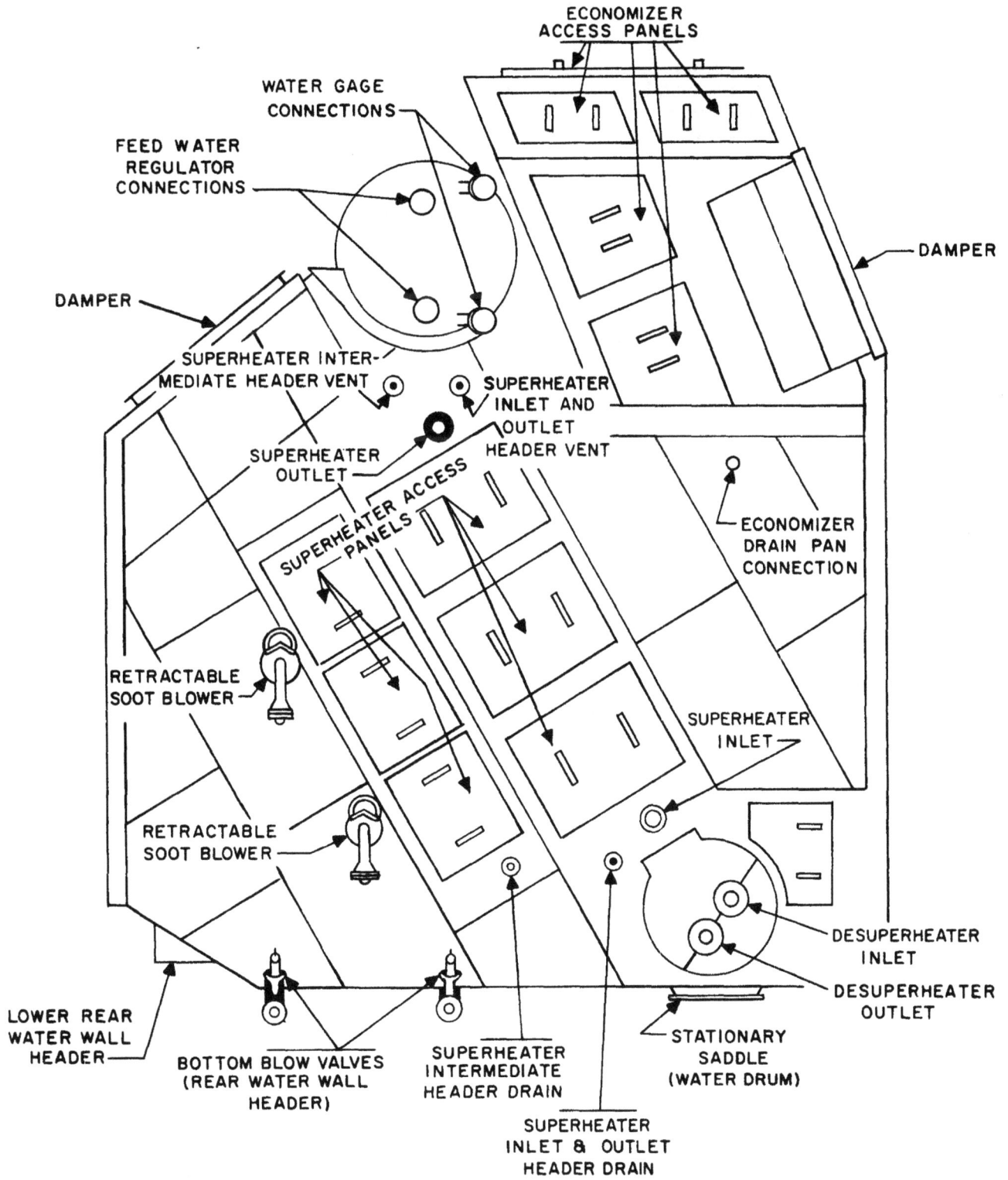

ECONOMIZER
ACCESS PANELS

WATER GAGE
CONNECTIONS

FEED WATER
REGULATOR
CONNECTIONS

DAMPER

DAMPER

SUPERHEATER INTER-
MEDIATE HEADER VENT

SUPERHEATER
INLET AND
OUTLET
HEADER VENT

SUPERHEATER
OUTLET

SUPERHEATER ACCESS
PANELS

ECONOMIZER
DRAIN PAN
CONNECTION

RETRACTABLE
SOOT BLOWER

SUPERHEATER
INLET

RETRACTABLE
SOOT BLOWER

DESUPERHEATER
INLET

DESUPERHEATER
OUTLET

LOWER REAR
WATER WALL
HEADER

BOTTOM BLOW VALVES
(REAR WATER WALL
HEADER)

SUPERHEATER
INTERMEDIATE
HEADER DRAIN

STATIONARY
SADDLE
(WATER DRUM)

SUPERHEATER
INLET & OUTLET
HEADER DRAIN

38.235

Figure 11-13.—External fittings and connections on 1200-psi single-furnace
boiler (rear view).

287

the headers, serves as a telltale[2] in the event of handhole leakage in the economizer.

Superheater vents are installed at or near the top of each superheater header or header section; superheater drains are installed at or near the bottom of each header or header section. Thus each pass of the superheater is vented and drained.

Superheater drains discharge through gravity (open-funnel) drains to the fresh water drain collecting system while steam is being raised in the boiler. After a specified pressure has been reached, the superheater drains are shifted to discharge through steam traps[3] to the high pressure drain system. The steam traps allow continuous drainage of the superheater without excessive loss of steam or pressure.

Figure 11-14 illustrates diagrammatically the arrangement of superheater vents and drains on a newer type of single-furnace boiler. This illustration also shows the superheater protection steam connections. Note, also, the water-drum installation of the desuperheater.

Sampling Connections

It is difficult to say just where the connection for drawing test samples of boiler water may be located, since this connection is found in different places on different types of boilers. On some boilers the sampling connection is located at the rear of the water drum. On others it comes off of the bottom blow line between the water drum and the bottom blow valve, either at the front of the boiler or at about the middle of the water drum. On boilers that have a chemical feed pipe in the steam drum, the test samples may be drawn through the nozzle connection of the chemical feed pipe. On some ships the surface blow line connection is used to take boiler water samples.

It is also difficult to say just what the sampling connection may be called. On some drawings it is identified as a test cock; on others as a salinity cock; on others as a salinometer valve; and on still others as a water test sample connection.

A sample cooler is fitted to the outlet side of the sampling connection. The cooler brings the temperature of the sample water down below the boiling point at atmospheric pressure and thus keeps the water from flashing into steam as it is drawn from the higher pressure of the boiler to the lower pressure of the fireroom.

Feed Stop and Check Valves

Manually operated feed stop and check valves are installed in the feed line to each boiler.[4] Feed stop and check valves are operated manually, with a separate handwheel for each valve. In addition, the feed check valve has remote operating gear so that it can be operated from the firing aisle. In normal operation, the stop valve is kept fully open and the check valve is used to regulate the supply of feed water to the boiler. When automatic feed water controls are in use, both the feed stop valve and the feed check valve are kept fully open so that they will not interfere with the automatic feeding of the boiler. (Similarly, the automatic feed regulating valve is kept fully open when the boiler is being fed manually through the feed check valve.

The feed stop and check valves shown in figure 11-15 are combined in one manifold casting. Note, however, that there are two separate valves. In some installations the two valves are housed in separate flanged castings which are bolted together. No matter what type of installation is used, the feed stop valve is always installed between the feed check valve and the economizer inlet.

Steam Stop Valves

Main steam stop valves are used to cut boilers in on the main steam line and to disconnect them from the line. The main steam stop valve located just after the superheater outlet is usually called the main steam boiler stop. Figure 11-16 shows an external view of a main steam boiler stop. Figure 11-17 shows a cross-sectional view of a globe-type main steam boiler stop. Gate valves instead of globe valves are used as main steam boiler stops on many newer ships.

---

[2] The term telltale is frequently used in connection with engineering equipment to indicate any device which shows leakage, flow, position, or other conditions.

[3] Steam traps are discussed in chapter 14 of this text.

[4] The stop valve is a regular globe-type stop valve. The so-called "check" valve is actually a stop-check valve which functions either as a stop valve or as a check valve, depending upon the position of the valve stem.

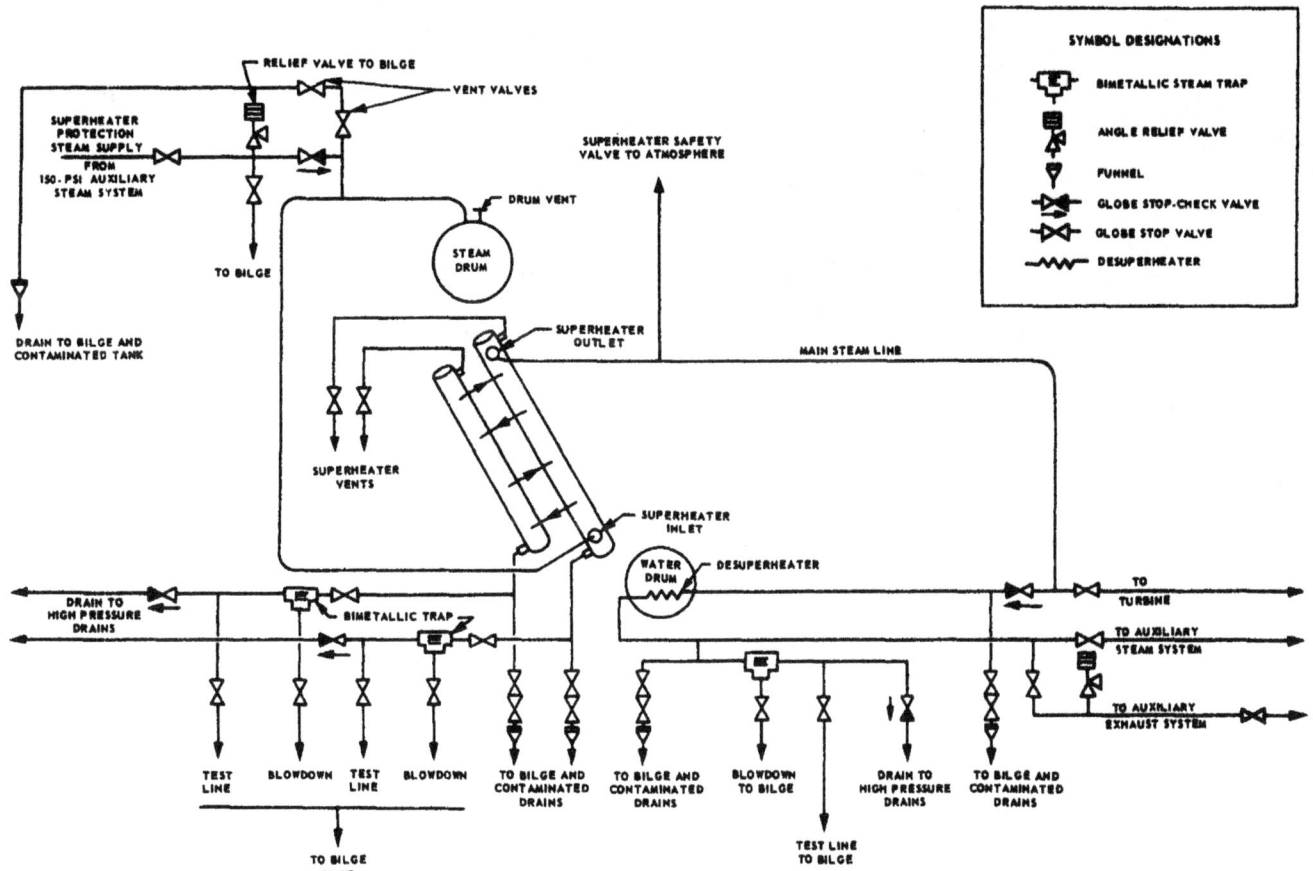

147.88

Figure 11-14.—Diagrammatic arrangement of superheater
vents, drains, and protection steam connections.

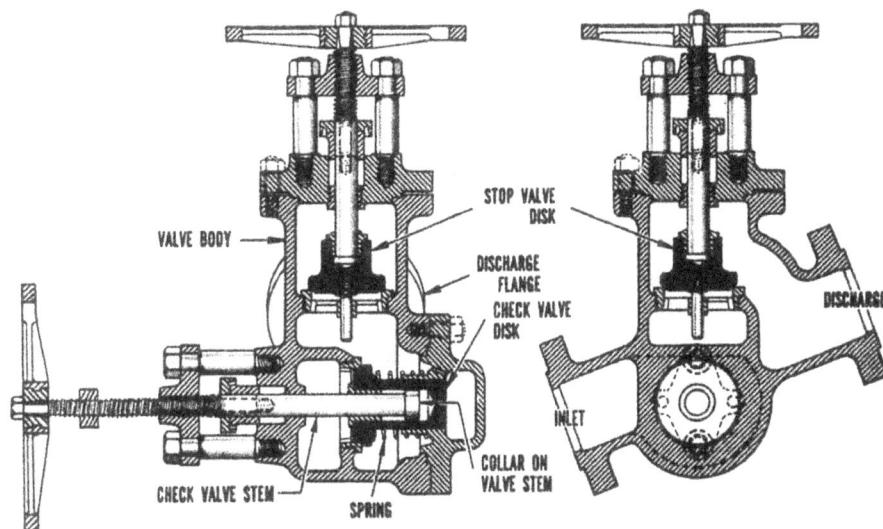

38.51X

Figure 11-15.—Combined feed stop and check valves.

38.52
Figure 11-16.—External view of main steam boiler stop valve.

In use, the main steam boiler stop is always either fully open or fully closed. The valve can be opened and closed manually at the valve itself. In some installations, it can also be closed pneumatically at the valve. The main steam boiler stop can also be operated manually, by remote control cables, from a remote operating station; as a rule, the valve can only be closed (not opened) from the remote station. For manual operation, the toggle operating gear shown in figures 11-16 and 11-17 provides the mechanical advantage required for closing the valve against boiler pressure.

Two-valve protection for each boiler is required on all ships built to U. S. Navy specifications. A second steam stop valve is therefore provided in the main steam line just beyond the main steam boiler stop.

Auxiliary steam stop valves are smaller than main steam stop valves but are otherwise similar. Special turbogenerator steam stop valves control the admission of steam to the turbogenerator line.

Safety Valves

Each boiler is fitted with safety valves which allow steam to escape from the boiler when the pressure rises above specified limits. The capacity of the safety valves installed on a boiler must be great enough to reduce the steam drum pressure to a specified safe point when the boiler is being operated at maximum firing rate with all steam stop valves completely closed. Safety valves are installed on the steam drum and at the superheater outlet.

Several different kinds of safety valves are used on naval boilers, but all are designed to open completely (pop) when a specified pressure is reached and to remain open until a specified pressure drop (blowdown) has occurred. Safety valves must close tightly, without chattering, and must remain tightly closed after seating.

There is an important difference between boiler safety valves and ordinary relief valves. The amount of pressure required to lift a relief valve increases as the valve lifts, since the resistance of the spring increases in proportion to the amount of compression. Therefore a relief valve opens slightly at a specified pressure, discharges a small amount of fluid, and closes at a pressure which is very close to the pressure that causes it to open. Such an arrangement will not do for boiler safety valves. If the valves were set to lift for anything close to boiler pressure, the valves would be constantly opening and closing, pounding the seats and disks and causing early failure of the valves. Furthermore, relief valves would not discharge the large amount of steam that must be discharged to bring the boiler pressure down to a safe point, since the relief valves would reseat very soon after they opened.

To overcome this difficulty, boiler safety valves are designed to open completely at the specified pressure. In all types of boiler safety valves, the initial lift of the disk is caused by static pressure of the steam, just as it would be in a relief valve. But just as soon as the safety valve begins to open, a projecting lip or ring of larger area is exposed for the steam pressure to act upon. The increase in force that results from the steam pressure acting

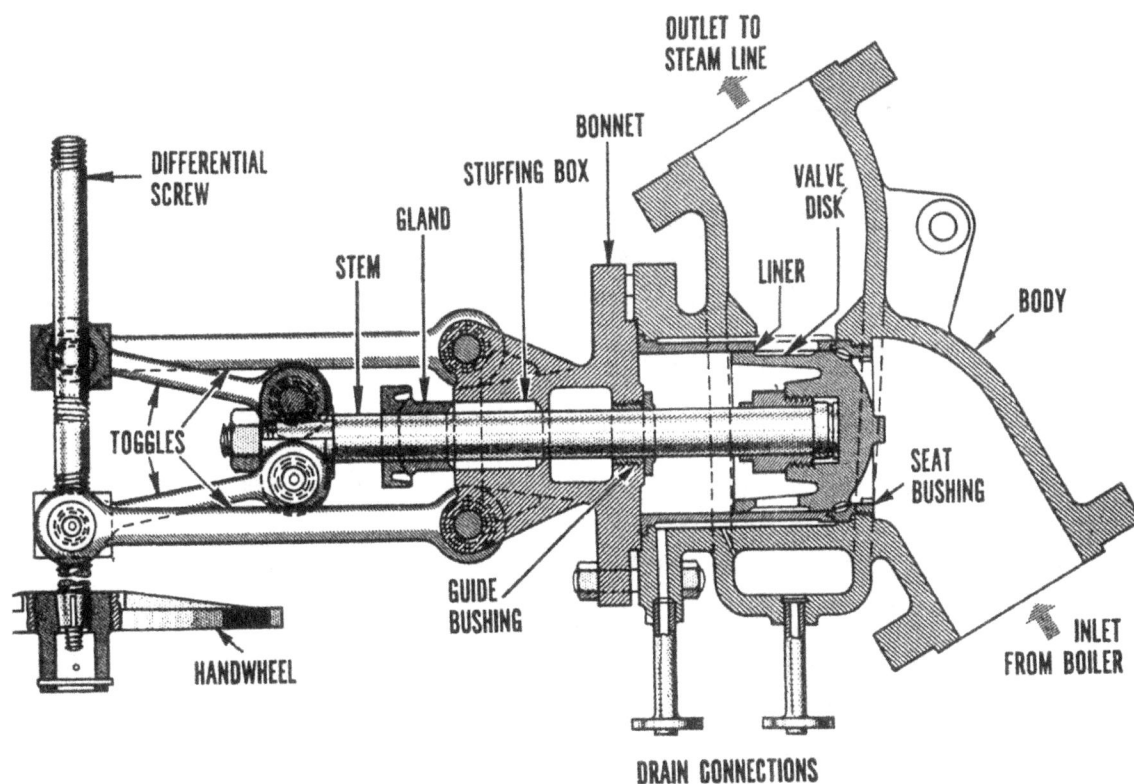

38.53X

Figure 11-17.—Cross-sectional view of main steam boiler stop valve.

upon this larger area overcomes the resistance of the spring, and the valve "pops"—that is, it opens quickly and fully. Because of the larger area now presented, the valve cannot reseat until the pressure has become considerably smaller than the pressure which caused the safety valve to open.

A steam drum safety valve of the huddling chamber type is shown in figure 11-18. As the static pressure of the steam in the steam drum causes the valve to open, the huddling chamber (which is formed by the position of the adjusting ring) fills with steam. The steam in the huddling chamber builds up a static pressure that acts upon the extra area provided by the projecting lip of the feather. The resulting increase in force overcomes the resistance of the spring, and the valve pops. After the specified blowdown has occurred, the valve closes cleanly, with a slight snap. The amount of tension on the spring determines the pressure at which

the valve will pop. The position of the adjusting ring determines the shape of the huddling chamber and thereby determines the amount of blowdown that must occur before the valve will reseat.

A steam drum safety valve of the nozzle reaction type is shown in figure 11-19. The initial lift of the valve occurs when the static pressure of the steam in the drum acts upon the disk insert with force sufficient to overcome the tension of the spring. As the disk insert lifts, the escaping steam strikes the nozzle ring and changes direction. The resulting force of reaction causes the disk to lift higher, up to above 60 percent of rated capacity. Full capacity is reached as the result of a secondary, progressively increasing lift which occurs as an upper adjusting ring is exposed. The ring deflects the steam downward, and the resulting force of reaction causes the disk to lift still higher. Blowdown adjustment in this

291

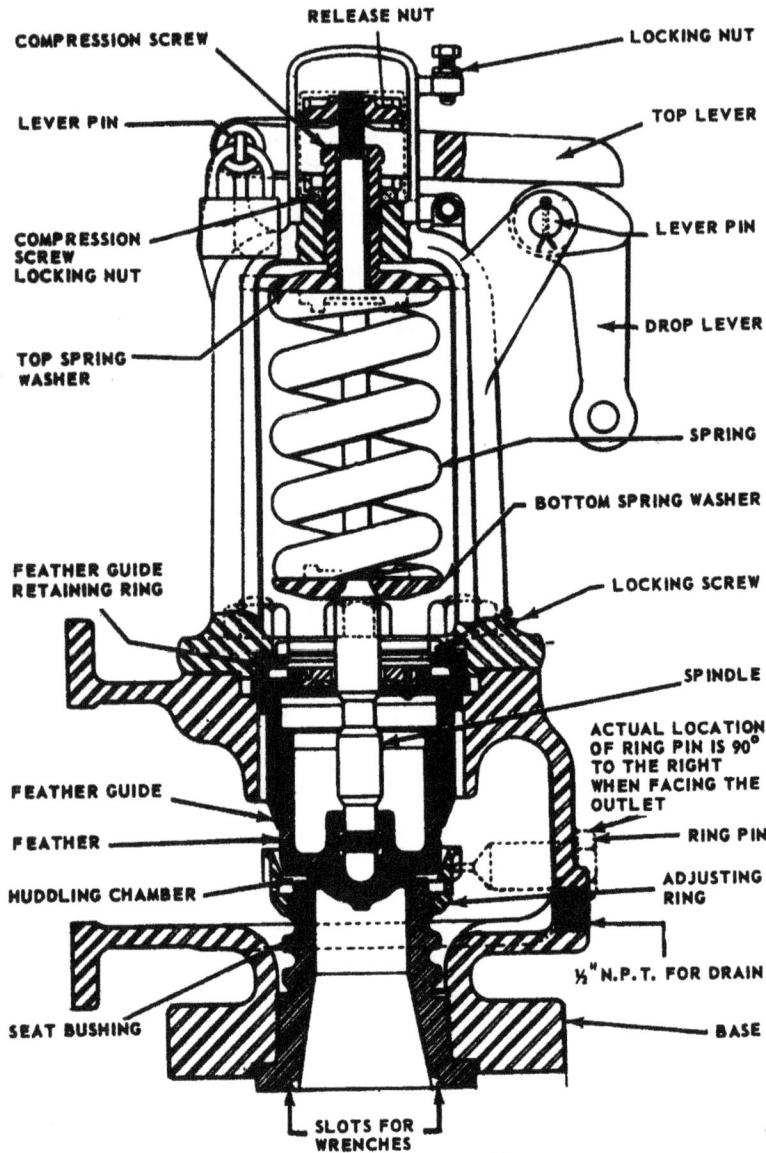

98.80

Figure 11-18.—Steam drum safety valve (huddling chamber type).

type of valve is made by raising or lowering the adjusting ring and by raising or lowering the nozzle ring.

Safety valves are always installed at the superheater outlet as well as on the steam drum. Superheater safety valves are set to lift just below, at, or just above the pressure which lifts the steam drum safety valves, in order to ensure an adequate flow of steam through the superheater when the steam drum safety valves are lifted.

Most double-furnace boilers are fitted with a pressure-pilot operated superheater outlet safety valve assembly of the type shown in figure 11-20. This assembly consists of three connected valves: a small spring-loaded safety valve installed on the steam drum; an actuating valve installed on the steam drum; and an actuated (or unloading) valve installed on the superheater. The stem of the drum valve is connected mechanically, by means of a lever, to the stem of the actuating valve. The actuating valve is

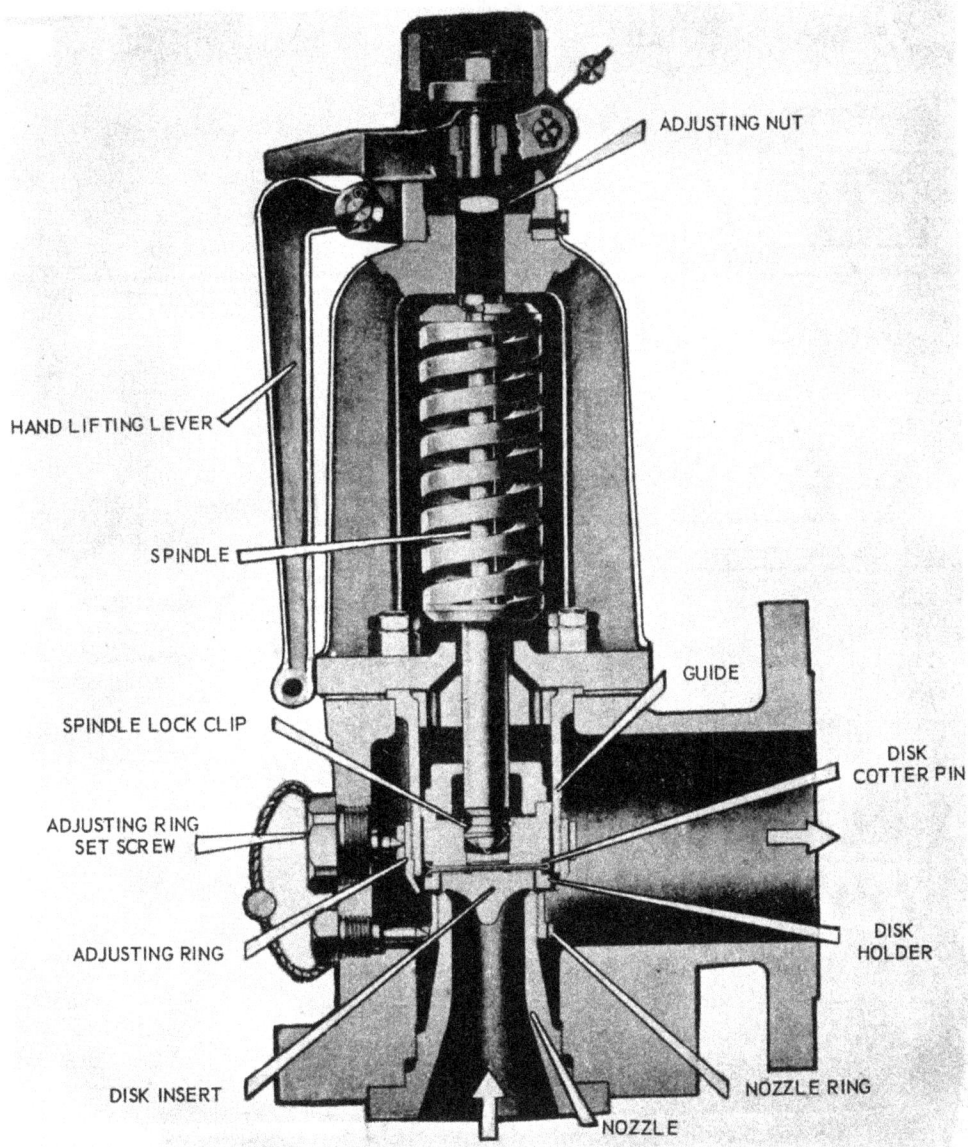

29.219

Figure 11-19.—Steam drum safety valve (nozzle reaction type).

connected by piping to the space above the disk in the superheater actuated (or unloading) valve. The unloading valve has no spring and relies solely on pressure differential for its operation. Normally there is a static pressure above the disk of the unloading valve, since a small orifice (or in some designs a small clearance around the disk) allows some steam at superheater outlet pressure to enter the space above the disk.

When the drum valve is opened by steam drum pressure, the mechanical connection between the drum valve and the actuating valve causes the actuating valve to open also. With the actuating valve open, steam flows from the space above the disk of the superheater unloading valve, through the actuating line, through the actuating valve, to atmosphere. The sudden relief of pressure above the disk of the unloading

293

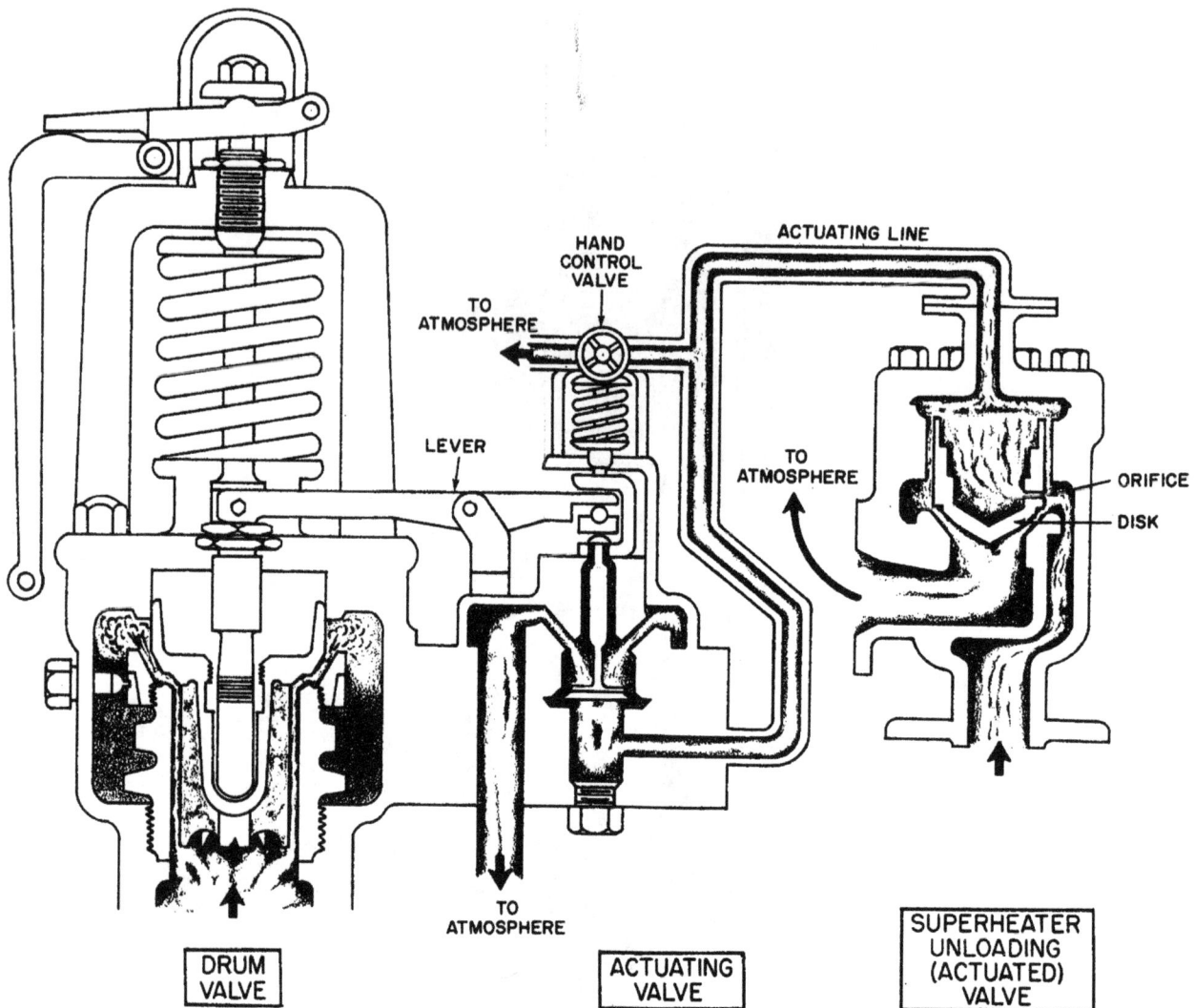

98.81

Figure 11-20.—Pressure pilot-operated superheater outlet safety valve assembly.

valve causes that valve to open fully, so that steam is discharged from the unloading valve to atmosphere.

Many 1200-psi single-furnace boilers, and also some 600-psi single-furnace ones, are equipped with Crosby two-valve superheater outlet safety valve assemblies of the type shown in figure 11-21. In an assembly of this type, both of the valves are spring loaded. The pilot valve on the steam drum and the super-heater valve at the superheater outlet are connected by a pressure transmitting line that runs from the discharge side of the drum pilot valve to the underside of the piston that is

attached to the spindle of the superheater safety valve. The superheater valve is set to pop at a pressure about 2 percent higher than the pressure which causes the drum pilot valve to pop. When the drum pilot valve pops, the steam pressure is transmitted immediately through the pressure line to the piston, and the superheater valve is thus actuated. If for any reason the drum pilot valve should fail to open, the superheater valve would open at a slightly higher pressure.

The Consolidated three-valve superheater outlet safety valve assembly shown in figure 11-22 is used on a number of 1200-psi

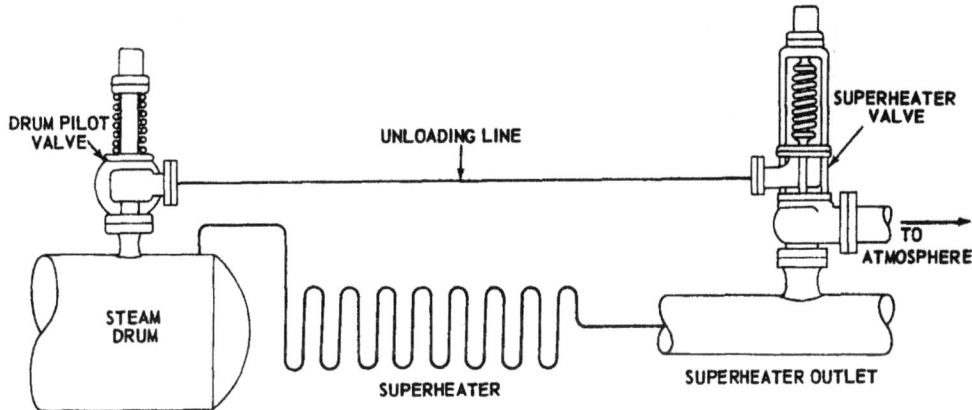

98.82

Figure 11-21.—Crosby two-valve superheater outlet safety valve assembly.

single-furnace boilers. The assembly consists of a pilot valve, an actuating valve, and an unloading valve.

The spring-loaded pilot valve is mounted on the top centerline of the steam drum. The actuating valve and the unloading valve are assembled as a unit and mounted on the piping at the superheater outlet; they are connected to each other by a rocker arm. The actuating valve has a cylinder with a piston inside it. The unloading valve has a piston-type disk, without a stem which is held in line by the cylinder in which it works. The unloading valve is pressure loaded, not spring loaded.

Steam from the superheater outlet enters the unloading valve cylinder and gathers around the valve disk above the seat. The steam bleeds through small ports to the space above the disk. When the actuating valve is closed, the steam above the disk of the unloading valve cannot escape, so the pressure above the disk equalizes with the pressure below the disk—that is, the pressure above the disk is equal to superheater outlet pressure.

The cause of safety valve lifting in this assembly is excessive pressure in the steam drum, not excessive pressure in the superheater. When the pilot valve on the steam drum opens, pressure is transmitted from the pilot valve to the cylinder of the actuating valve. Pressure in the actuating valve cylinder is applied under the piston, causing the spring to compress. The rocker arm moves upward at the end over the actuating valve and downward at the end over the unloading valve, thus opening the actuating valve.

When the actuating valve opens, pressure bleeds off to atmosphere. Since the space above the unloading valve disk is connected to the actuating valve, relief of pressure in the actuating valve also causes relief of pressure above the disk in the unloading valve. The unloading valve therefore opens, allowing steam to flow from the superheater to atmosphere. When the pilot valve reseats, the actuating valve also reseats. As steam bleeds through the ports to the space above the disk in the unloading valve, pressure builds up and rapidly equals the pressure below the disk. The unloading valve closes. In summary, then, the superheater unloading valve always opens immediately after the steam drum pilot valve opens and closes immediately after the pilot valve closes.

Soot Blowers

Soot blowers are installed on each boiler for the purpose of removing soot from the firesides while the boiler is steaming. Soot blowers are used only on steaming boilers, not on idle boilers. Each steaming boiler utilizes its own steam (superheated) to supply its own soot blowers.

The soot blowers must be used frequently, regularly, and in proper sequence in order to prevent the accumulation of heavy deposits of soot which would interfere with heat transfer and which would constitute a fire hazard. The process of using the soot blowers is usually called "blowing tubes."

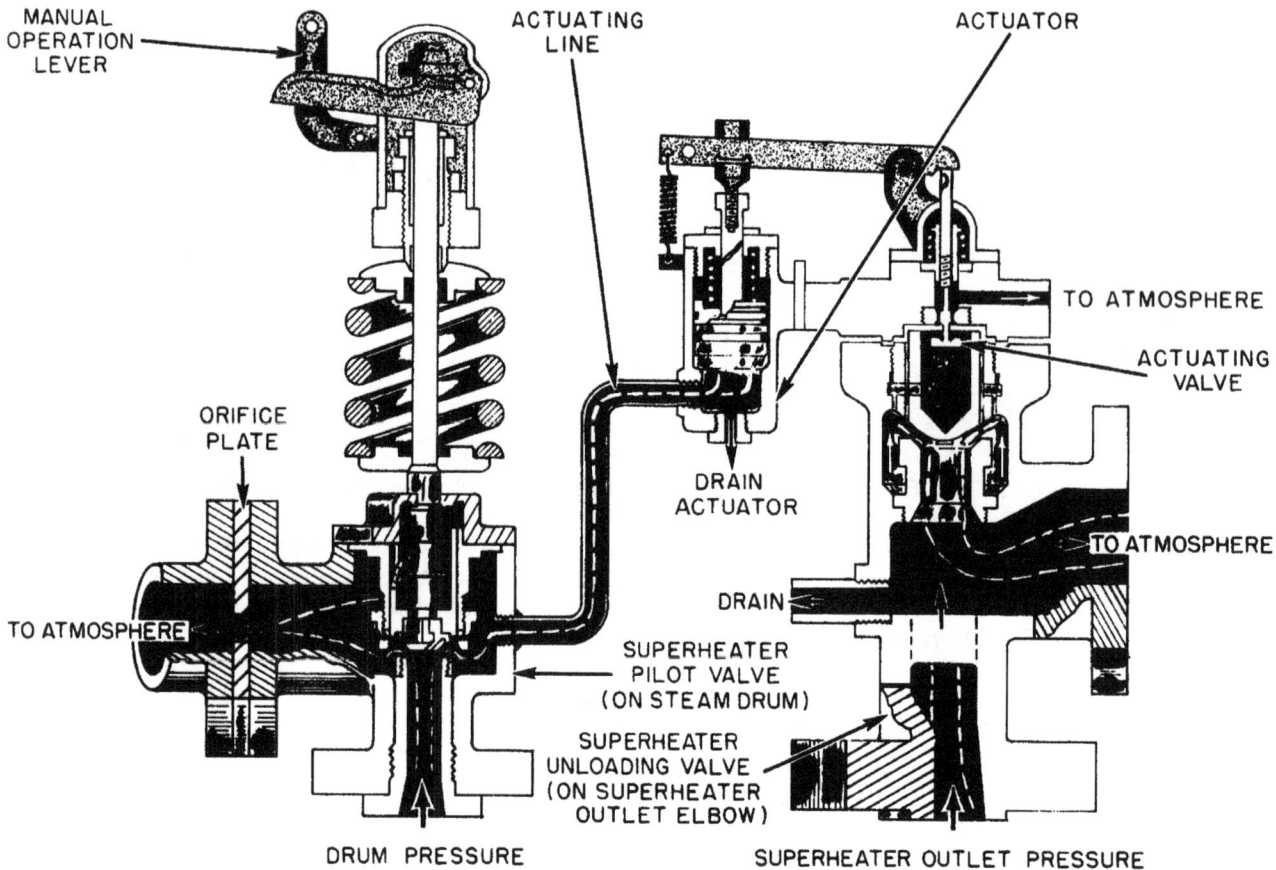

104.22
Figure 11-22.—Three-valve superheater outlet safety valve assembly (Consolidated).

Before instructing fireroom personnel to blow tubes, the engineering officer of the watch must obtain permission from the officer of the deck. Under some conditions, tubes cannot be blown without covering the upper decks with soot; hence the need for obtaining permission from the officer of the deck.

Soot blowers are installed on the boiler with their nozzles projecting into the furnace between the boiler tubes or adjacent to them. The soot blowers are arranged so that operation in the proper sequence will sweep the soot progressively toward the uptakes.

The number of soot blowers installed, the way in which they are arranged, and the blowing arcs for each unit differ from one type of boiler to another. Figure 11-23 shows the arrangement of soot blowers used on one of the newer 1200-psi single-furnace boilers; one of the soot blowers has a blowing arc of 220° and the others have blowing arcs of 360° F, on this particular boiler.

One common type of soot blower is shown in figure 11-24. The part of the soot blower that may be seen from the outside of the boiler is called the head. The soot blower element, a long pipe with nozzle outlets, is the part of each soot blower that projects into the tube banks of the boiler. The soot blower shown in figure 11-24 is operated by an endless chain; when the chain is pulled, the element is rotated and superheated steam is admitted through the steam valve. The steam discharges at high velocity from the nozzles in the elements. The nozzles direct the jets of steam so that they sweep over the tubes. The soot is thus loosened so that it can be blown out of the boiler.

Some soot blowers are operated by turning a crank or handwheel, instead of by an endless chain. On some recent ships, the soot blowers are operated by pushbuttons. One pushbutton is provided for each unit. Pressing the pushbutton admits air to an air motor which drives the units.

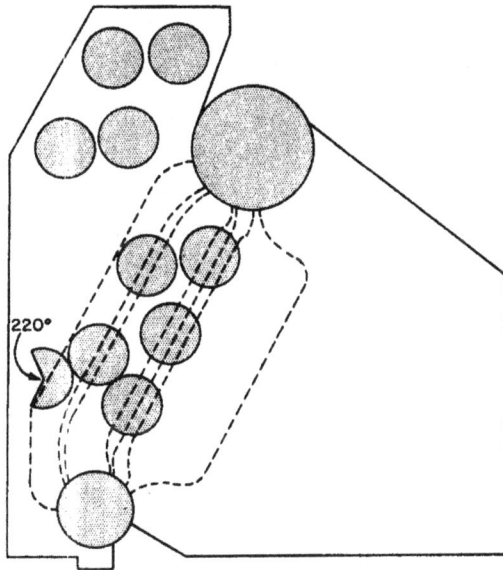

38.57

Figure 11-23.—Arrangement of soot blowers on a 1200 psi single-furnace boiler.

Some soot blowers on some boilers are of the retractable type—that is, the element does not remain in the furnace all the time but instead can be retracted into a housing or shroud between the inner and outer casings.

The scavenging air connection shown in figure 11-24 supplies air to the soot blower element and thus keeps combustion gases from backing up into the soot blower head and piping. A hole in the outer casing allows air to enter the other end of the scavenging air line; thus, scavenging air is blown through the soot blower whenever the forced draft blowers are in operation. A check valve is installed in the scavenging air piping, very near the soot blower head; this valve closes whenever steam is admitted to the soot blower element.

Blow Valves

Some solid matter is always present in boiler water. Most of the solid matter is heavier than water and therefore settles in the water drums and headers. Solid matter that is lighter than water rises and forms a scum on the surface of the water in the steam drum. Since most of the solid matter is not carried over with the steam, the concentration of solids remaining in the boiler water gradually increases as the boiler steams.

For the sake of efficiency and for the protection of the boiler pressure parts, it is necessary to remove some of this solid matter from time to time. Blow valves and blow lines are used for this purpose.

Light solids and scum are removed from the surface of the water in the steam drum by means of the surface blow line—which, as we have already seen, is an internal boiler fitting. Heavy solids and sludge are removed by using the bottom blow valves which are fitted to each water drum and header. Both surface blow valves and bottom blow valves on modern naval boilers are globe-type stop valves.

Both the surface blow and the bottom blow valves discharge to a system of piping called the boiler blow piping. The boiler blow piping system is common to all boilers in any one fireroom. Guarding valves are installed in the line as a protection against leakage from a steaming boiler into the blow piping and against leakage from the blow piping back into a dead boiler. A guarding valve installed at the outboard bulkhead of the fireroom gives protection against salt water leakage into the blow piping. After passing through this guarding valve, the water is discharged through an overboard discharge valve (sometimes called a skin valve) which leads overboard below the ship's waterline. Figure 11-25 shows the general arrangement of boiler blow piping for one of the newer single-furnace boilers.

Water Gage Glasses

Every boiler must be equipped with at least two independent devices for showing the water level in the steam drum, and at least one of these devices must be a water gage glass. Some boilers have more than two devices for indicating water level. Various combinations of water level indicating devices are used on naval boilers. Perhaps the most common arrangement on older boilers is two water gage glasses, one 10 inches long and one 18 inches long. Newer boilers may have two water gage glasses and one remote water level indicator or they may have one water gage glass and two remote water level indicators.

Several types of water gage glasses are used on naval boilers. The older water gage glasses differ in some ways from the ones installed on the newer 1200-psi boilers, and gages made by different manufacturers may

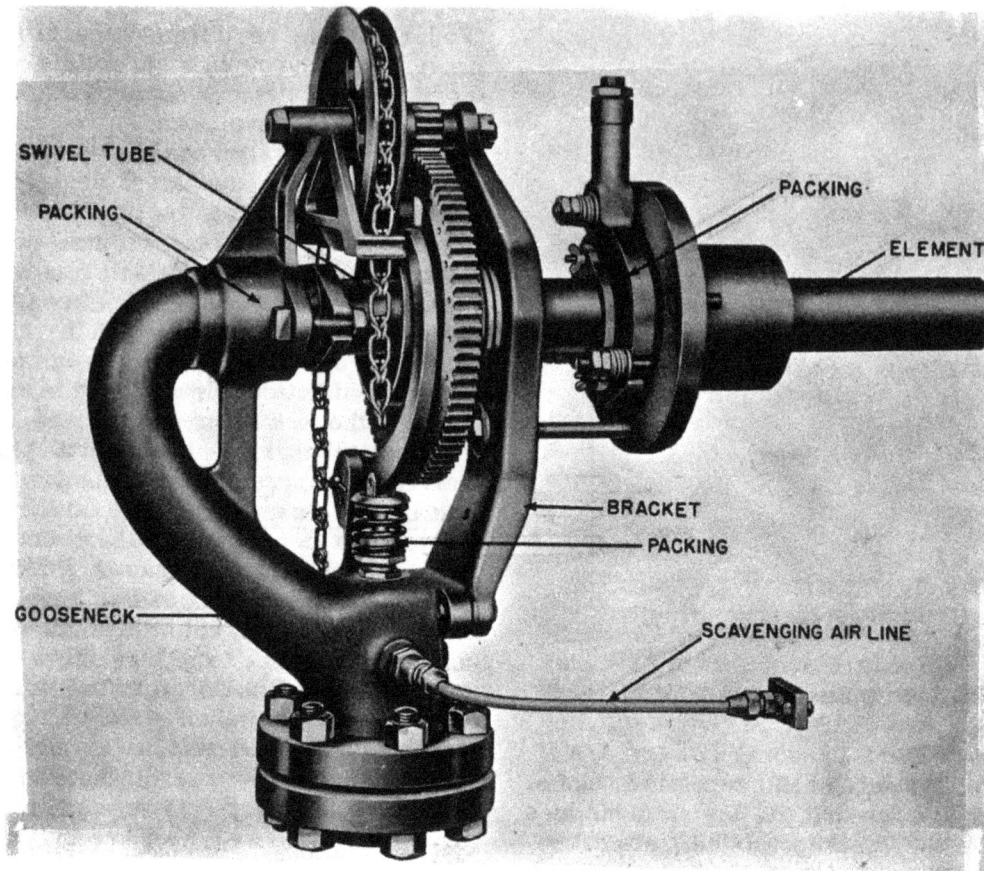

38.54

Figure 11-24.—External view of multi-nozzle soot blower.

vary somewhat in design details. Detailed information on the water gage glasses installed on any particular boiler may be obtained from the manufacturer's technical manual.

An older type of water gage glass is shown in figure 11-26. Figure 11-27 illustrates the construction of this gage. The frame is the centerpiece of the assembly. Hollow stems at each end of the frame connect with the cutout valves at top and bottom, thus allowing water and steam to enter the gage. Two glass plates or strips, ground to flat parallel faces, are used in each water gage. The glass strips are backed up by thin sheets of mica which separate the glass from the high temperature water and steam. The mica sheets serve two purposes. First, they keep the glass from becoming etched by the action of the hot water and steam. And second, they prevent shattering of the glass in case of breakage.

The entire assembly of frame, glass strips, and mica sheets is supported between two steel cover plates which are held together by studs. Asbestos gaskets, 1/32 inch in thickness, are used on each side of the frame. Asbestos cushions, 1/16 inch in thickness, are used between the cover plates and the glass strips. The arrangement of cushions and gaskets is shown in figure 11-27.

The drain connection shown in figure 11-26 permits the water gage to be blown down and also permits it to be drained. The regulator connections shown at the top and bottom of this gage are not found on all gages; they are installed only on boilers which were not orignally designed to use feed water regulators but which were later fitted with the regulators.

A more recent type of water gage glass is shown in figure 11-28. The gage is assembled springs, as shown in the illustration. This type of assembly makes it unnecessary to retorque

298

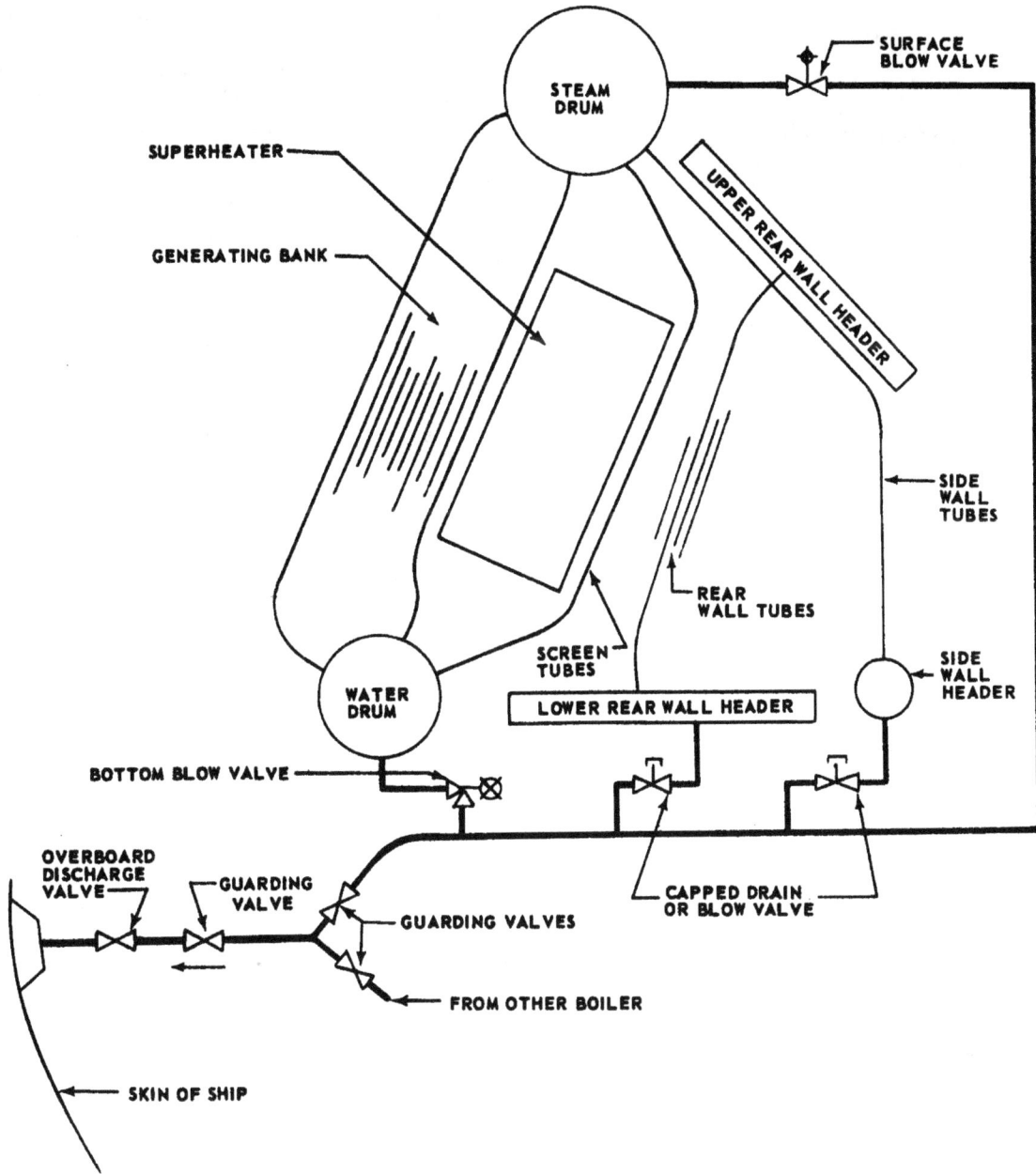

38.58

Figure 11-25.—Boiler blow piping for single-furnace boiler.

the studs after the gage has warmed up. (Notice the numbering of studs in figure 11-28; the numbers indicate the proper sequence of tightening the studs when assembling the gage.)

On some boilers of recent design, a bi-color water gage is used. Bi-color gages show RED in the portion of the glass which is filled with steam and GREEN in the portion of the glass which is filled with WATER. The bi-color

gage glass works on the simple optical principle that a ray of light bends (or refracts) a different amount when it passes through steam than when it passes through water.

Each water gage is connected to the steam drum through two cutout valves, one at the top and one at the bottom. The bottom cutout valve connection contains a ball-check valve. The ball rests on a holder. As long as there is

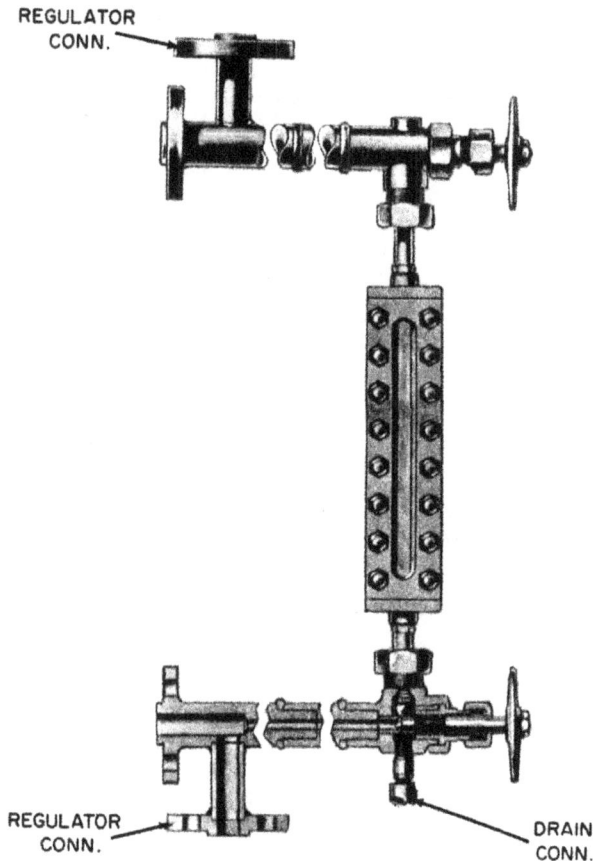

38.59
Figure 11-26.—Older type of water gage glass.

equal pressure on each side of the ball, the ball remains on its holder. But if the water gage breaks, the sudden rush of water through the bottom connection forces the ball upward onto its seat and thus prevents further escape of hot water. No check valve is installed in the top cutout connection.

Most boilers are designed to carry the normal water level at the middle of the steam drum, and most water gages are mounted in such a way that the normal water level shows at the midpoint of the water gages or, in the case of staggered gages, at the midpoint between the bottom of the lower gage and the top of the higher gage. However, this general rule does not apply to some boilers. If the designed normal water level is NOT intended to be shown at the midpoint of the water gages, the location of the normal water level should be marked on the gages.

The water level is considered to be within allowable limits as long as it can be seen in

one or more water gage glasses. However, the water level must always be maintained as close to the normal level as possible. As long as the water level is visible in one gage, you can bring the water level back to normal by increasing or decreasing the amount of water fed to the boiler. If the water level cannot be seen at all, the situation must be treated as an emergency requiring the immediate securing of the boiler.

Water gages must be blown down before the boiler is cut in on the line, at the end of each watch, and at any time when there is the slightest doubt about the water level in the boiler. Frequent blowing down is necessary because the gage connections are easily clogged with dirt, scale, or other solid matter. Failure to blow through the water gages could lead to false indications of water level.

Superheater Steam Flow Indicators

Many boilers—particularly double-furnace boilers—are equipped with superheater steam flow indicators. Where installed, these indicators must be kept in good operating condition at all times. The superheater outlet thermometers indicate temperature at the superheater outlet, not inside the superheater; and, on a double-furnace boiler, it is possible to have no flow through the superheater while the thermometers are giving perfectly normal readings at the outlet. In other words, there is no way to be sure a superheater is not being overheated unless both the superheater outlet thermometers and the superheater steam flow indicator are in good working condition.

Superheater steam flow indicators measure the steam pressure differential between the superheater inlet and the superheater outlet. Since the pressure drop across the superheater is proportional to the rate of steam flow through the superheater, the pressure drop can be used as an indication of the rate of steam flow. Superheater steam flow indicators are usually calibrated in inches of water, since they measure a relatively small pressure differential.

Two types of superheater steam flow indicators are in common naval use. Although both respond to the pressure differential between the superheater inlet and the superheater outlet, they differ in the mechanism by which this pressure difference is measured

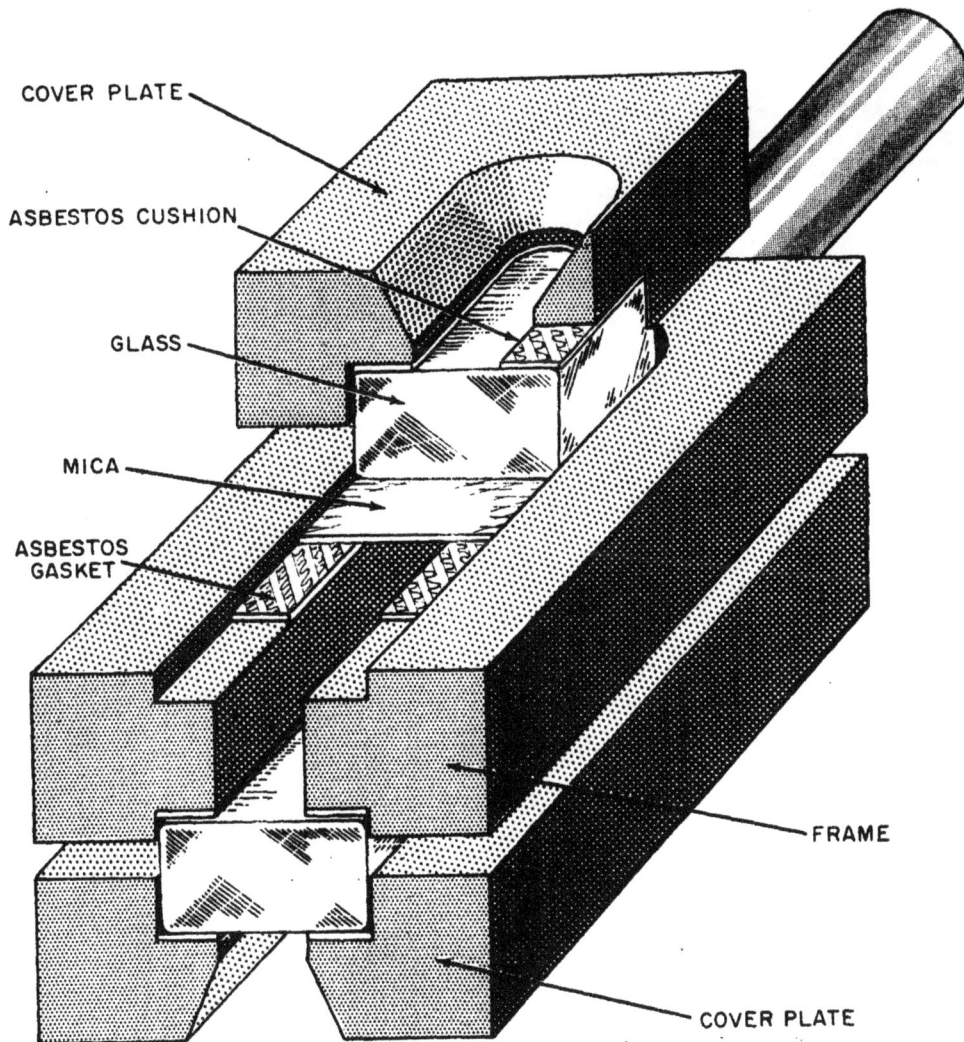

38.238

Figure 11-27.—Construction of older type of water gage glass.

measured and transmitted to an indicating dial.

The Yarway superheater steam flow indicator responds to the difference between superheater inlet pressure and superheater outlet pressure as these two pressures act upon separate columns or heads of water. The general arrangement of the Yarway superheater steam flow indicator is shown in figure 11-29.

A constant water level is maintained in the two head chambers (reservoirs). Both head chambers are in one casting, but the upper head chamber is located slightly above the lower head chamber in order to provide a slight constant pressure differential which serves to stabilize the zero reading. Steam from the superheater inlet is led to the upper head chamber, and steam from the superheater outlet is led to the lower head chamber. Exposed piping connects each head chamber with the indicating unit.

The interior of the indicating unit is shown in figures 11-30 and 11-31. As may be seen, the water from the upper head chamber enters the indicating unit on one side of the diaphragm, and the water from the lower head chamber enters on the other side. The pressure from the upper head chamber is greater than the pressure from the lower head chamber, so the diaphragm is moved accordingly. The diaphragm is connected by a pin linkage to a deflection plate which moves in sensitive response to the movement of the diaphragm.

301

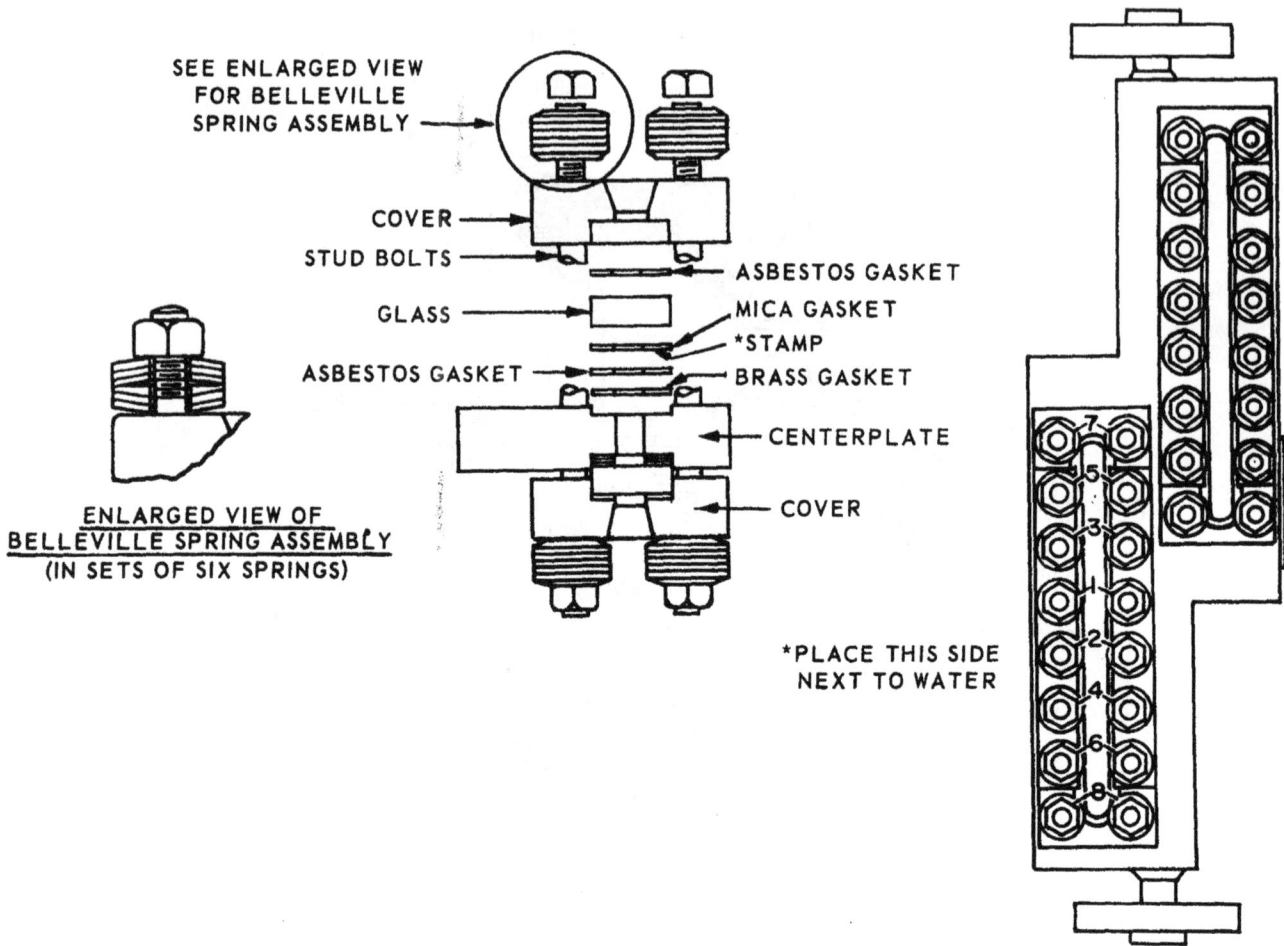

SEE ENLARGED VIEW
FOR BELLEVILLE
SPRING ASSEMBLY

COVER
STUD BOLTS
GLASS
ASBESTOS GASKET

ENLARGED VIEW OF
BELLEVILLE SPRING ASSEMBLY
(IN SETS OF SIX SPRINGS)

ASBESTOS GASKET
MICA GASKET
*STAMP
BRASS GASKET
CENTERPLATE
COVER

*PLACE THIS SIDE
NEXT TO WATER

38.239

Figure 11-28.—Recent type of water gage glass.

A permanent horseshoe magnet is rigidly mounted on that side of the deflection plate which is free to move. The poles of the magnet straddle a tubular well in which a spiral-shaped strip armature is mounted on jeweled bearings. A counterbalanced pointer is attached to the end of the armature mounting shaft.

When the deflection plate moves in response to variations in pressure, the magnet is made to move along the axis of the well. As the magnet moves the spiral-shaped armature rolls in order to keep in alignment with the magnetic field between the poles of the magnet. Thus a rotary motion is imparted to the armature mounting shaft; and the rotation of the shaft causes the pointer to move. The pointer moves over a brightly illuminated vertical dial which is divided into green and red zones to represent safe and unsafe operating conditions.

The Jerguson superheater steam flow indicator, shown schematically in figure 11-32, consists of three main parts: (1) a datum chamber assembly, (2) a valve manifold, and (3) an indicating unit.

The Jerguson indicator is essentially a mercury-filled manometer with a stainless steel float in one leg. As the pressure differential between the superheater inlet and the superheater outlet varies, the mercury level in the instrument changes and thereby actuates the indicating pointer. The movement of the float in the manometer is transmitted to the pointer on the scale by means of a magnetic coupling drive. The coupling consists of an internal magnetic armature on the end of the float shaft and an external yoke with magnetically energized arms. The yoke, forming a part of the pointer system, pivots on precision bearings in order

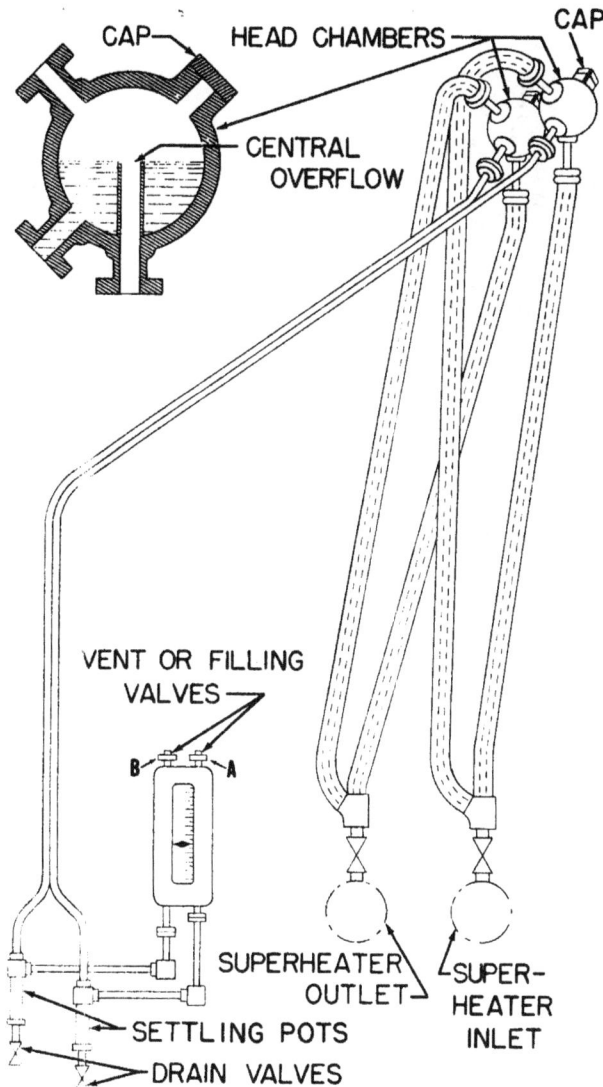

98.90

Figure 11-29.—General arrangement of Yarway superheater steam flow indicator.

to maintain alignment with the internal armature attached to the float shaft; thus the yoke pivots as the armature moves up and down.

The datum chamber is connected by piping to the superheater inlet and outlet connections and thus provides the means for impressing differential pressure on the instrument. As noted previously, the pressure difference between superheater inlet and superheater outlet is used as a measure of the rate of steam flow through the superheater.

Remote Water Level Indicators

Remote water level indicators are used on most ships to provide a means whereby the boiler water level may be observed from the lower level of the fireroom. The two types of remote water level indicators discussed here are in common use on combatant ships; other types may be found on auxiliary ships.

The Yarway remote water level indicator consists of three parts: (1) a constant-head chamber which is mounted on the steam drum at or near the vertical centerline of the drumhead; (2) a graduated indicator which is usually mounted on an instrument panel; and (3) two reference legs that connect the constant-head chamber to the indicator. The reference legs are marked A and B in figure 11-33, which shows the general arrangement of a Yarway remote water level indicator.

A constant water level is maintained in leg A, since the water level in the constant-head chamber does not vary. The level in leg B is free to fluctuate with changes in the steam drum water level. The upper hemisphere of the constant-head chamber is connected to the steam drum at a point above the highest water level to be indicated; because of this connection, boiler pressure is exerted equally upon the water in the two legs. The variable leg B is connected to the steam drum at a point below the water level to be indicated; because of this connection, the water level in leg B is equalized with the water level in the steam drum.

As may be seen in figure 11-33, each leg is connected by piping to the indicator. In the indicator, the two columns of water terminate upon opposite sides of a diaphragm. The indicating unit is almost identical with the indicating unit of the Yarway superheater steam flow indicator, previously described.

The general arrangement of a Jerguson remote water level indicator is shown in figure 11-34. As may be seen, the operating principles of this device are very similar to the operating principles of the Jerguson steam flow indicator, previously discussed.

Superheater Temperature Alarms

Superheater temperature alarms are installed on most boilers to warn operating personnel of dangerously high temperatures in the superheater. One type of superheater temperature alarm is shown in figure 11-35. The bulb,

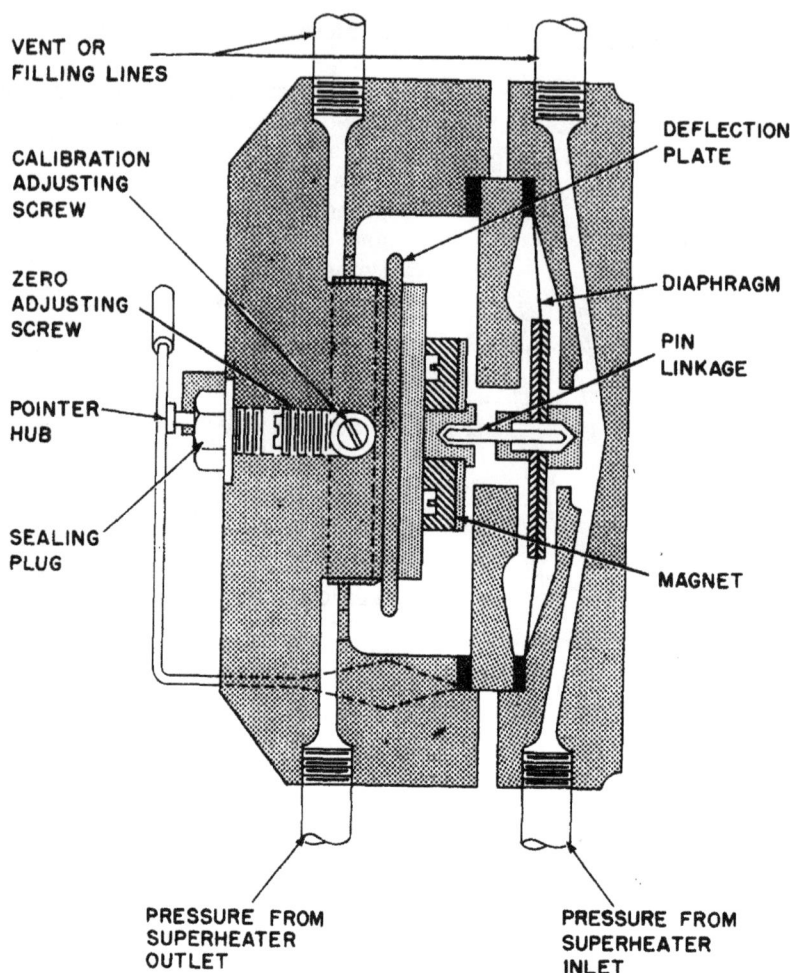

VENT OR FILLING LINES

CALIBRATION ADJUSTING SCREW

ZERO ADJUSTING SCREW

POINTER HUB

SEALING PLUG

DEFLECTION PLATE

DIAPHRAGM

PIN LINKAGE

MAGNET

PRESSURE FROM SUPERHEATER OUTLET

PRESSURE FROM SUPERHEATER INLET

98.91

Figure 11-30.—Cross section of Yarway indicating unit (front view).

the capillary tube, and the spiral-shaped Bourdon element are filled with mercury. As the temperature rises, the mercury expands and causes the Bourdon tube to move so that an attached cantilever arm is moved toward an electric microswitch. When the temperature reaches a predetermined point, the cantilever arm closes the microswitch, actuating a warning light and a warning howler.

Smoke Indicators

Naval boilers are fitted with smoke indicators (sometimes called smoke periscopes) which permit visual observation of the gases of combustion as they pass through the uptakes. Most single-furnace boilers have one smoke indicator installed in the uptake. Double-furnace boilers have two smoke indicators,

one for observing the combustion gases coming from the saturated side and the other for observing the combustion gases coming from the superheater side.

A smoke indicator is shown in figure 11-36. A light bulb is installed in a lamp unit at the rear of the boiler. At the front of the boiler, in direct line of sight with the lamp, is a reflector unit which reflects the image to a second mirror. The second mirror is located so that it may be seen from the fireroom.

Oil Drip Detector Periscopes

Some boilers are equipped with oil drip detector periscopes which permit inspection of the floor between the inner and outer boiler castings, to see if oil has accumulated there.

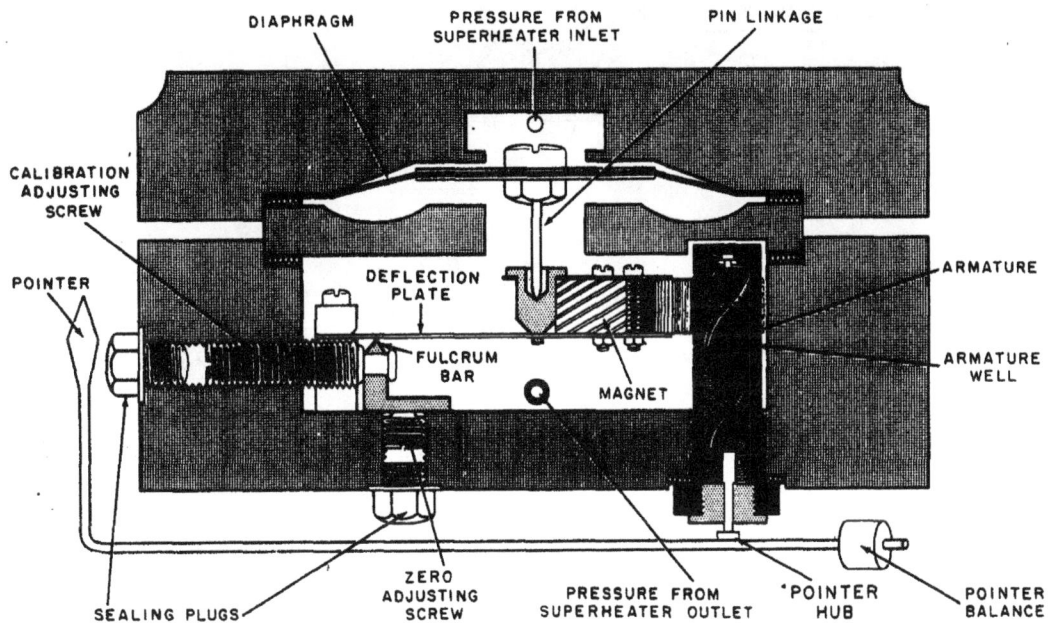

98.92

Figure 11-31.—Cross section of Yarway indicating unit
(plan view).

The oil drip detector periscope operates on much the same principle as the smoke periscope.

Pressure and Temperature Gages

Boiler operation requires constant awareness of the pressures and temperatures existing at certain locations within the boiler and in associated machinery and systems. Operating personnel depend upon a variety of pressure and temperature gages to provide them with the necessary information. Pressure gages are installed on or near each boiler to indicate steam drum pressure, superheater outlet pressure, auxiliary steam pressure, auxiliary exhaust pressure, feed water pressure, steam pressure to the forced draft blowers, air pressure in the double castings, and fuel oil pressure. Temperature gages are installed to indicate superheated steam temperature, desuperheated steam temperature (if the boiler has a desuperheater), feed water temperature at the economizer inlet and outlet, fuel oil temperature at the fuel oil manifold before the boiler, and—in some ships—uptake temperature.

In some firerooms, the gages that indicate steam drum pressure, superheater outlet pressure, superheater outlet temperature, and combustion air pressure are installed on the boiler front. As a rule, however, the indicating units of all pressure gages are mounted on a boiler gage board which is easily visible from the firing aisle. Distant-reading thermometers are also installed with the indicating unit mounted on the boiler gage board. In some installations, a common gage board is used for all the boilers in one space, instead of having separate gage boards for each boiler.

Most of the pressure gages used in connection with boilers are of the Bourdon-tube type, although some diaphragm-type gages and some manometers are also used in the fireroom. The temperature gauges most commonly used in the fireroom are direct-reading liquid-in-glass thermometers and distant-reading Bourdon-tube thermometers. The basic operating principles of these pressure and temperature gages are discussed in chapter 7 of this text.

Single-Element Feed Water Regulators

Single-element automatic feed water regulators are installed on many boilers which are not equipped with complete automatic feed water and combustion control systems. Single-element regulators, unlike the multi-element control

305

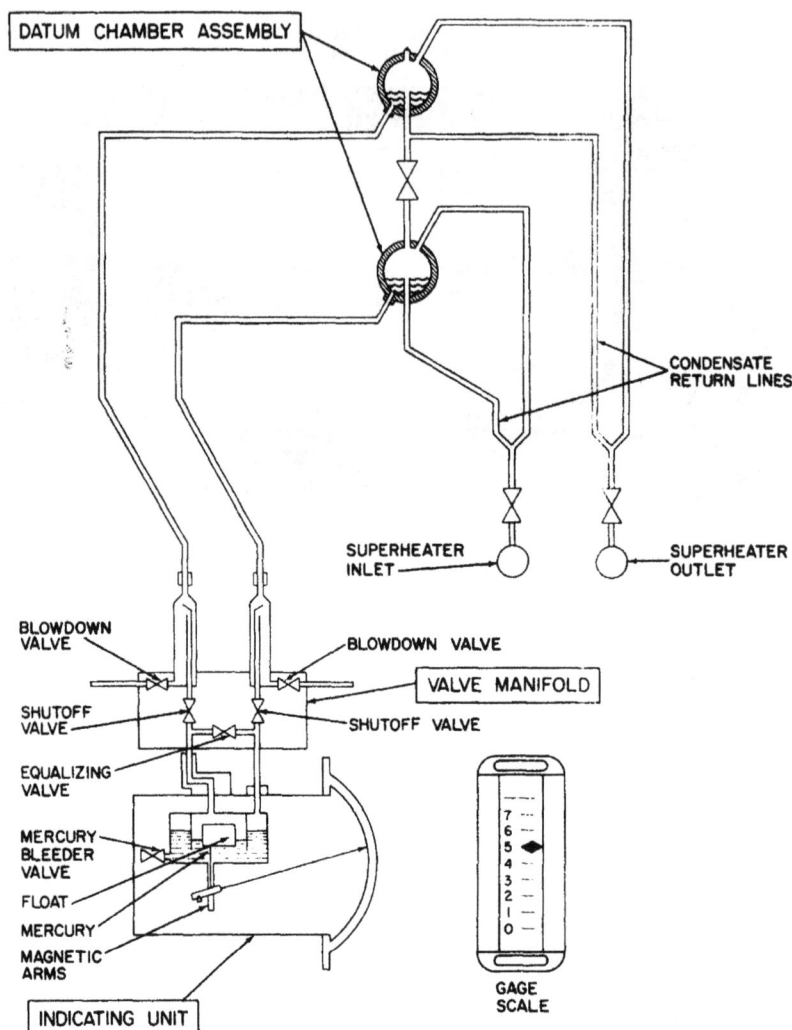

98.96

Figure 11-32.—Jerguson superheater steam flow indicator.

systems, are controlled by one variable only—namely, the water level variation in the steam drum. Hence single-element regulators are not able to compensate for swell and shrink.

Single-element regulators are intended primarily for keeping the boilers supplied with feed water in battle or under other conditions when manual feeding of the boilers might become difficult or impossible. These regulators must be cut in immediately when General Quarters is sounded. They may also be used at other times, and in fact should be used frequently enough to keep them in good working order and ready for use under emergency conditions.

Single-element regulators can control the water level within acceptable limits under relatively steady steaming conditions, but not under severe maneuvering conditions. Since complete reliance cannot be placed on the single-element regulator, a checkman must remain on station and be ready to take manual control if necessary when the single-element regulator is in use in other than emergency conditions.

A single-element automatic feed water regulator is shown in figure 11-37. An inner tube, enclosed by a generator, is connected to the boiler steam drum through valves A and B. Thus the water level in the inner tube is dependent upon the water level in the steam drum.

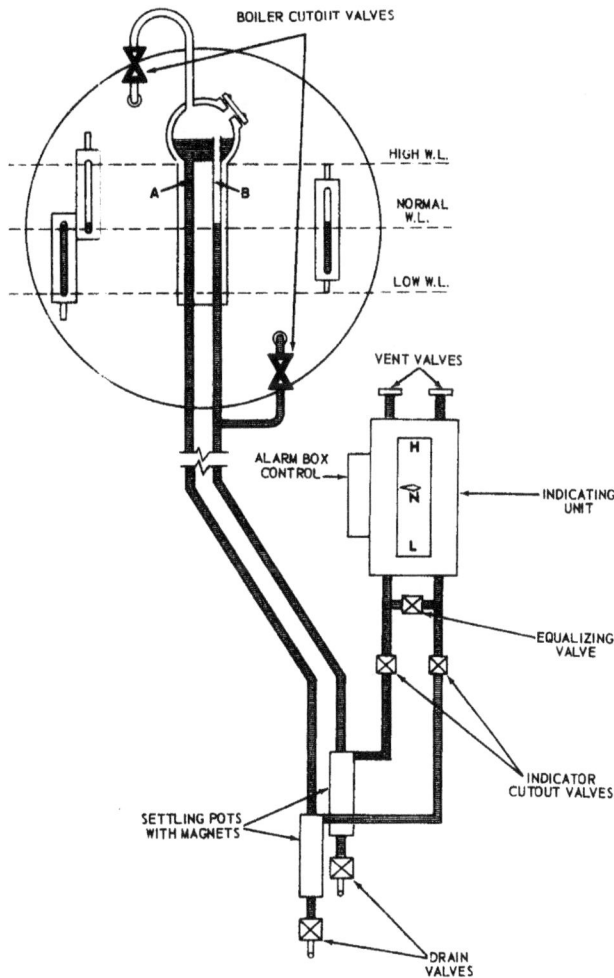

98.97

Figure 11-33.—General arrangement of Yarway remote water level indicator.

Cooling fins attached to the pipe that carries water from the steam drum to the inner tube ensure that the water in the inner tube will be at a slightly lower temperature than the steam in the inner tube. Cooling fins on the generator ensure that the water and steam in the generator will be cooler than the water and steam in the inner tube. For a number of reasons, the transfer of heat is more rapid from the steam in the inner tube to the steam in the generator than it is from the water in the inner tube to the water in the generator.

As the water level in the steam drum drops, causing a corresponding drop in the water level in the inner tube, more of the generator is exposed to the steam in the inner tube. This causes more water in the generator to flash into steam,

thus increasing the pressure on the water in the closed regulator system and expanding the bellows, which are normally compressed by spring pressure. Expansion of the bellows opens the feed-regulating valve in the feed line and allows more water to flow to the boiler. When the water level in the steam drum rises, the reverse process occurs and the feed-regulating valve tends to close.

BOILER CONTROLS

Automatic boiler controls consisting of independent combustion control and feed water control systems have been installed on a number of naval ships. All indications point to an increasing use of automatic controls, particularly as boilers are designed for higher operating pressures and temperatures. Many high pressure boilers require such rapid and sensitive response to feed water, fuel, and combustion air demands that the use of automatic controls is almost a necessity.

The function of an automatic combustion control system is to maintain the fuel input and the combustion air input to the boiler in accordance with the demand for steam and to proportion the amount of air to the amount of fuel in such a way as to provide maximum combustion efficiency. The feed water control system functions to provide the required boiler feed and maintain the steam drum water level at or near normal position (middle of the water gage glass at all steaming rates.

An installed control system is quite often simpler in theory than one would suppose when first viewing the complex assortment of components and tubing. To begin with, then, let us look at the basic principles of automatic control and see how they apply to a very simple control system.

Any control system, simple or complex, must perform four functions. It must:

Measure something on the output side of a process;

Compare the measured value with the desired value;

Compute the amount and direction of change required to bring the measured output value back to the desired output value;

Correct something on the input side of the process so that the output side of the process will be brought back to the desired value.

Measurement, comparison, computation, and correction—these are the basic operations

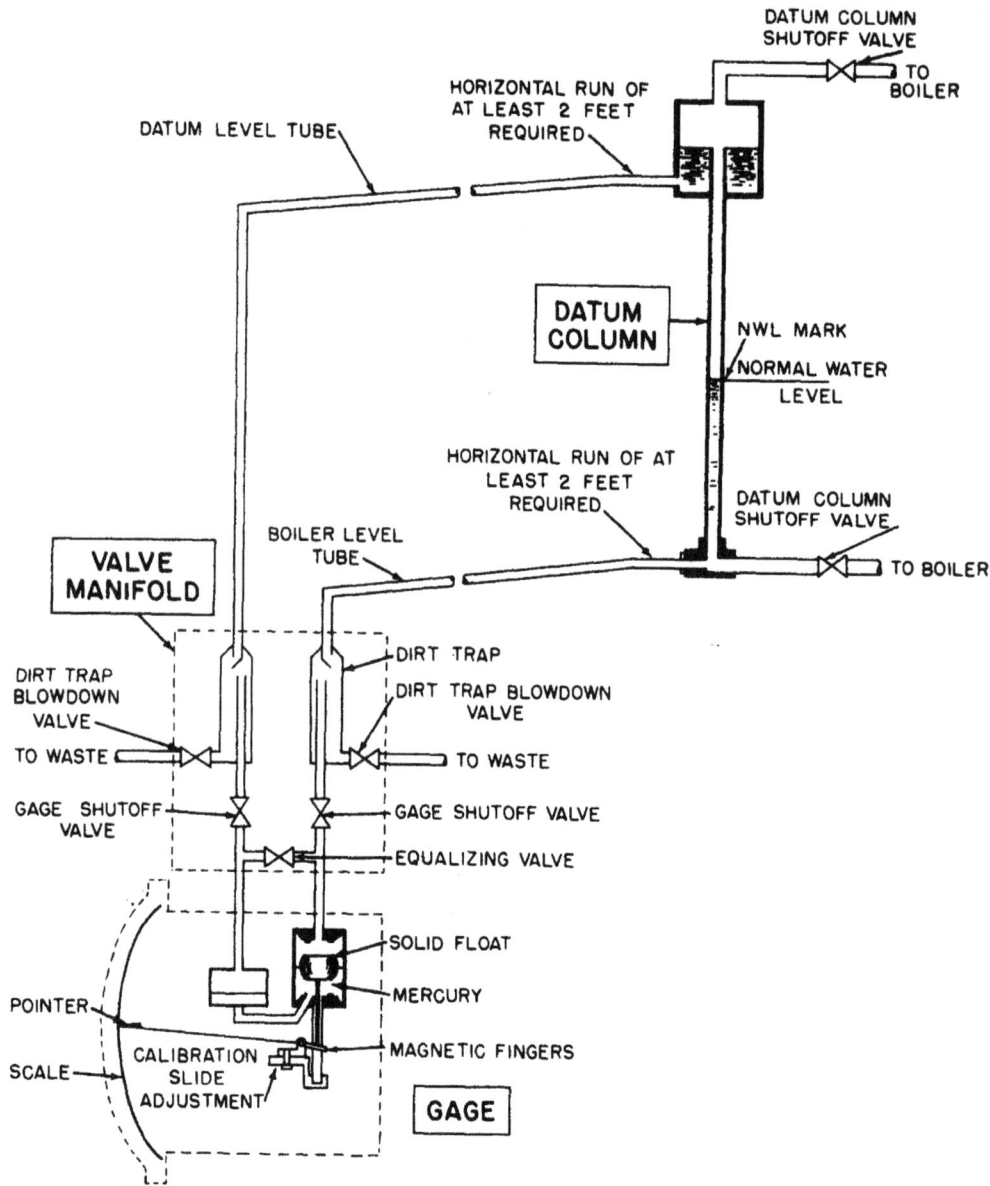

98.98

Figure 11-34.—Jerguson remote water level indicator.

performed by a control system. Taken together, they constitute a closed loop of action and counteraction by which some quantity or condition is measured and controlled. The closed control loop is often called a feedback loop, since it requires a feedback signal from something on the output side of the process to something on the input side of the process. The closed loop concept is illustrated in figure 11-38.

Now let us examine a simple automatic control system such as the one shown in figure 11-39. The process being controlled is a heat exchange process in which steam is used to heat cold water. The two inputs, steam and cold water, produce one output—hot water at some desired temperature. Such a process could be controlled by various kinds of automatic control systems—electrical, hydraulic, pneumatic,

308

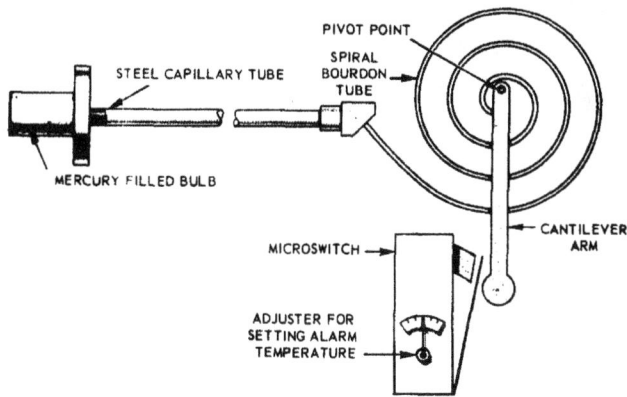

61.31X

Figure 11-35.—Superheater temperature alarm.

or some combination of these. A simple pneumatic control system has been selected because boiler control systems are, almost without exception, pneumatic.

To perform the control functions of measurement, comparison, computation, and correction, the automatic control system must have a measuring means, a controlling means, and some arrangement for comparison and computation.

The measuring means in the system shown in figure 11-39 consists of a thermometer bulb, a Bourdon tube, and connecting capillary tubing. A change in the controlled variable—that is, the temperature of the hot water—leads to a change in the pressure transmitted to the Bourdon tube and thus leads to a change in the position of the Bourdon tube. Through a series of mechanical linkages, the position of the Bourdon tube affects

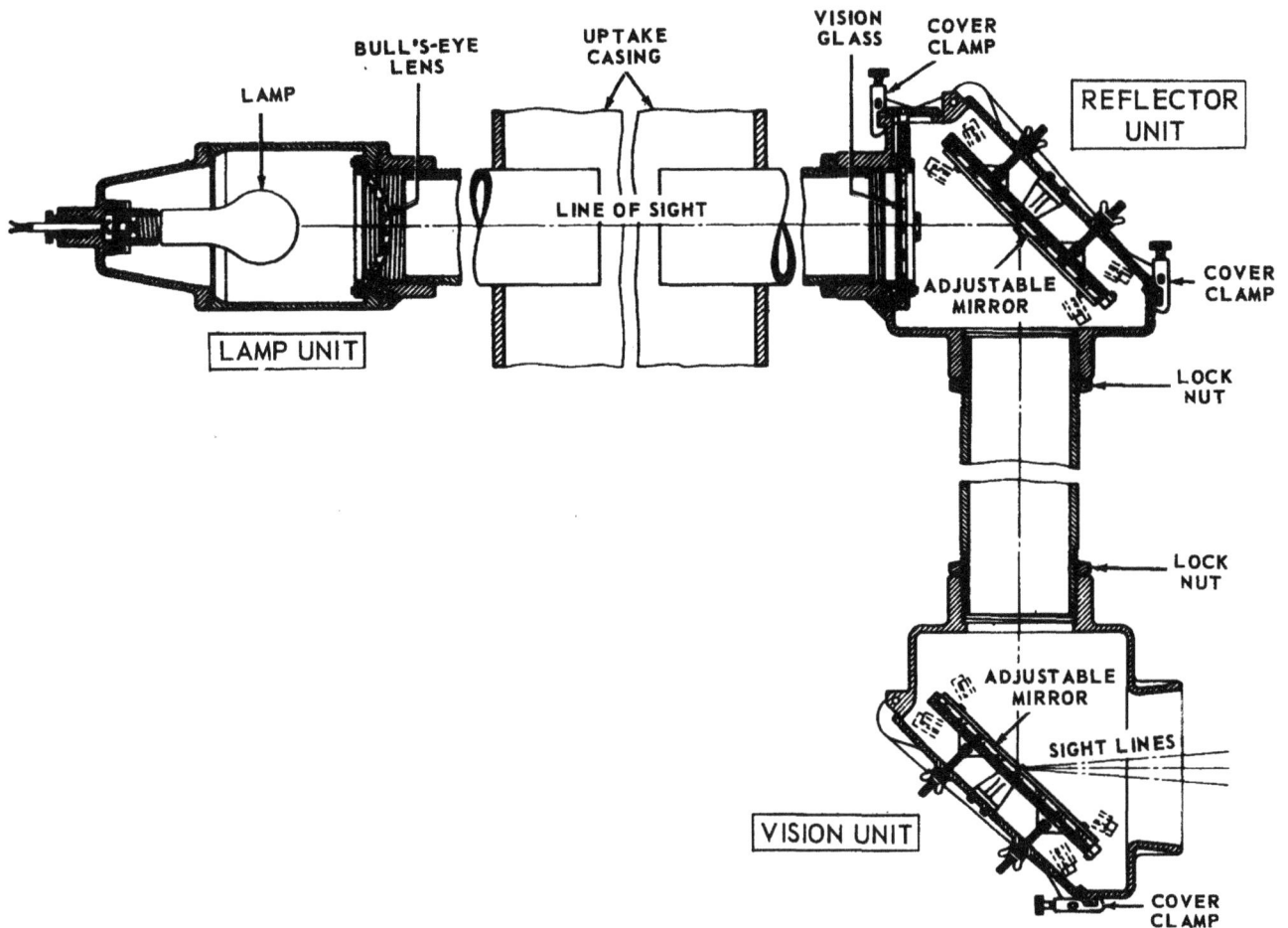

38.60

Figure 11-36.—Smoke indicator.

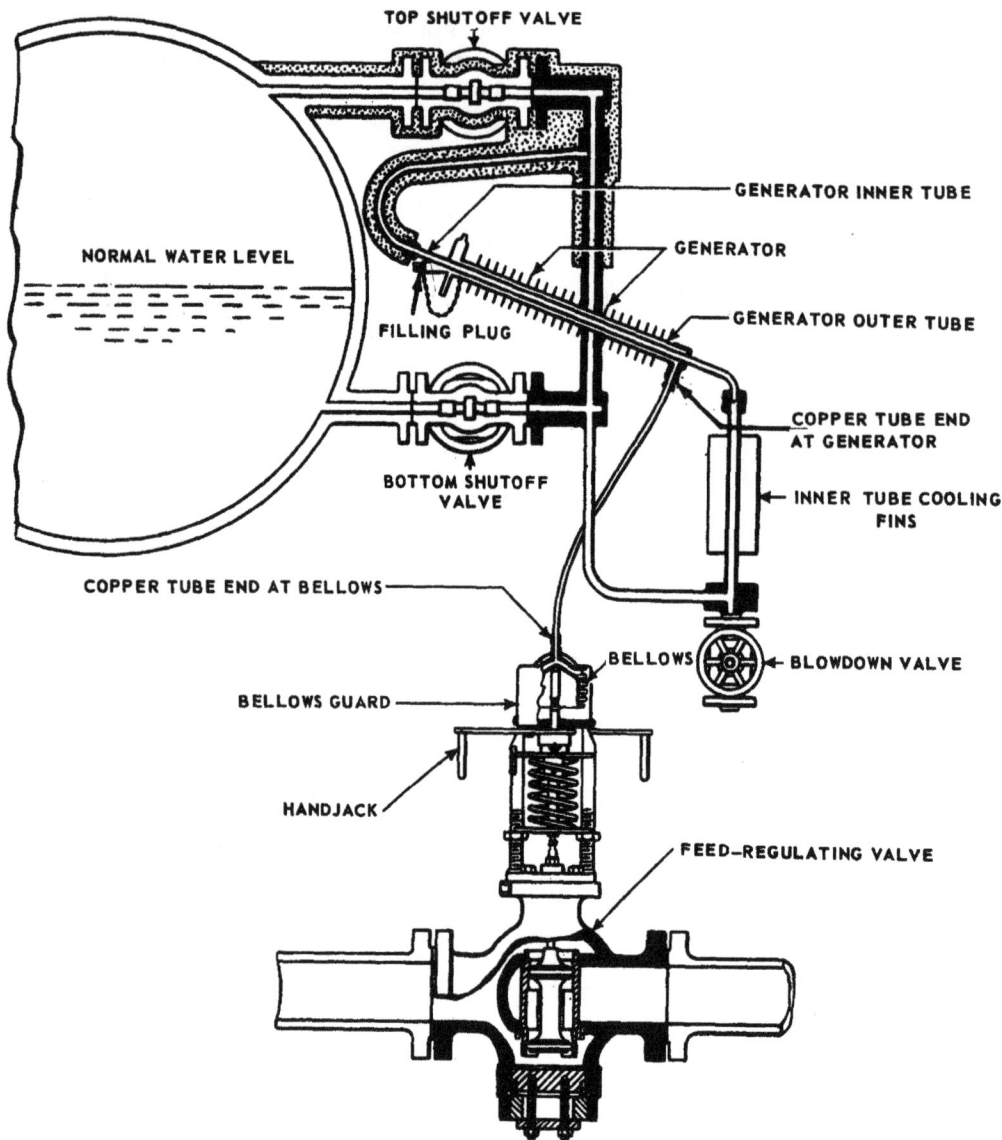

38.184

Figure 11-37.—Single-element feed water regulator.

the position of the vane in a nozzle-and-vane assembly located in the transmitter.

A set point knob is linked in some way to the nozzle of the nozzle-and-vane assembly, so that the setting of the set point knob affects the position of the nozzle. The set point knob is positioned to represent the desired temperature of the hot water output.

The nozzle-and-vane assembly is the comparing and computing device in this pneumatic transmitter. Since the position of the Bourdon tube affects the position of the vane and the setting of the set point knob affects the position of the nozzle, the distance between the nozzle and the vane at all times represents a <u>comparison</u> of the actual measured value (Bourdon tube) and the desired value (set point knob). The distance between the tip of the nozzle and the vane is responsible for the <u>computation</u> of the amount and direction of change that must be made in the position of the steam valve, since:

(1) the rate of air flow from the nozzle depends upon the distance between the nozzle and the vane; and

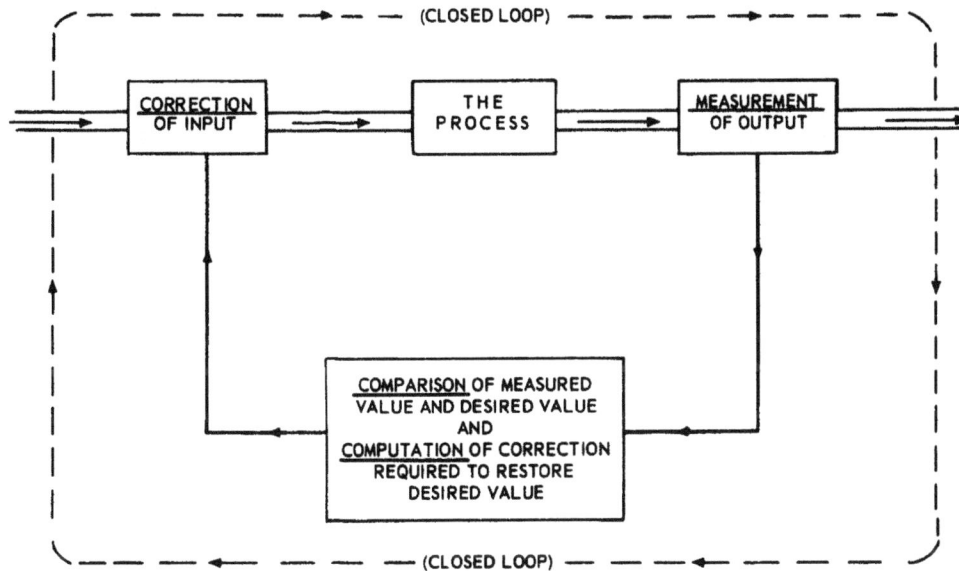

98.100

Figure 11-38.—Closed control loop.

(2) the rate of air flow from the nozzle determines the intensity of the pneumatic pressure imposed upon the valve motor operator.

If the vane and the nozzle are relatively far apart, a good deal of air will flow out of the nozzle and there will be relatively low air pressure acting on the motor operator. If the vane and the nozzle are closer together, the flow of air from the nozzle will be retarded and higher pressure will act upon the motor operator. The level of air pressure acting on the motor operator determines the position of the steam valve, since the motor operator positions the valve in accordance with the air pressure received from the transmitter. The motor operator and the steam valve together thus form the controlling means of this system.

As previously noted, practically all boiler control systems aboard ship are of the pneumatic type, depending upon compressed air for their operation. Compressed air is used as the controlling or balancing force for the operation of the many pneumatic transmitters and relays in the system and, at a higher pressure, as the source of power to operate some or all of the control drives and control valves that control the flow of fuel, combustion air, and feed water. In some older ships, the compressed air supply is obtained from the ship's service compressed air system, through reducing valves. In newer

ships, a separate combustion control compressed air system, with its own air compressor, is installed for the boiler controls.

In any pneumatic boiler control system, a great many units are used to develop, transmit, and receive pneumatic "messages" or "signals" in the form of variable air pressures. The pneumatic units are interconnected by copper tubing. A typical pneumatic unit in a boiler control system operates by receiving one or more pneumatic pressures from one or more sources (frequently from other pneumatic units), altering or combining the pressure or pressures, and then sending a new pneumatic pressure to another pneumatic unit in the system.

The actual mechanisms which develop, transmit, and receive pneumatic signals vary, depending upon the manufacturer and upon the function of the units. The nozzle-and-vane assembly shown in figure 11-39 is only one of a number of pneumatic devices that could be used to accomplish the functions of comparison and compution. Bellows, escapement valves, and various other devices are used in pneumatic control systems to compare the measured value with the desired value and to compute the amount of correction required. Similarly, the measuring means and the controlling means shown in figure 11-39 are commonly used in pneumatic control systems, but they are not the only possible devices for such applications.

311

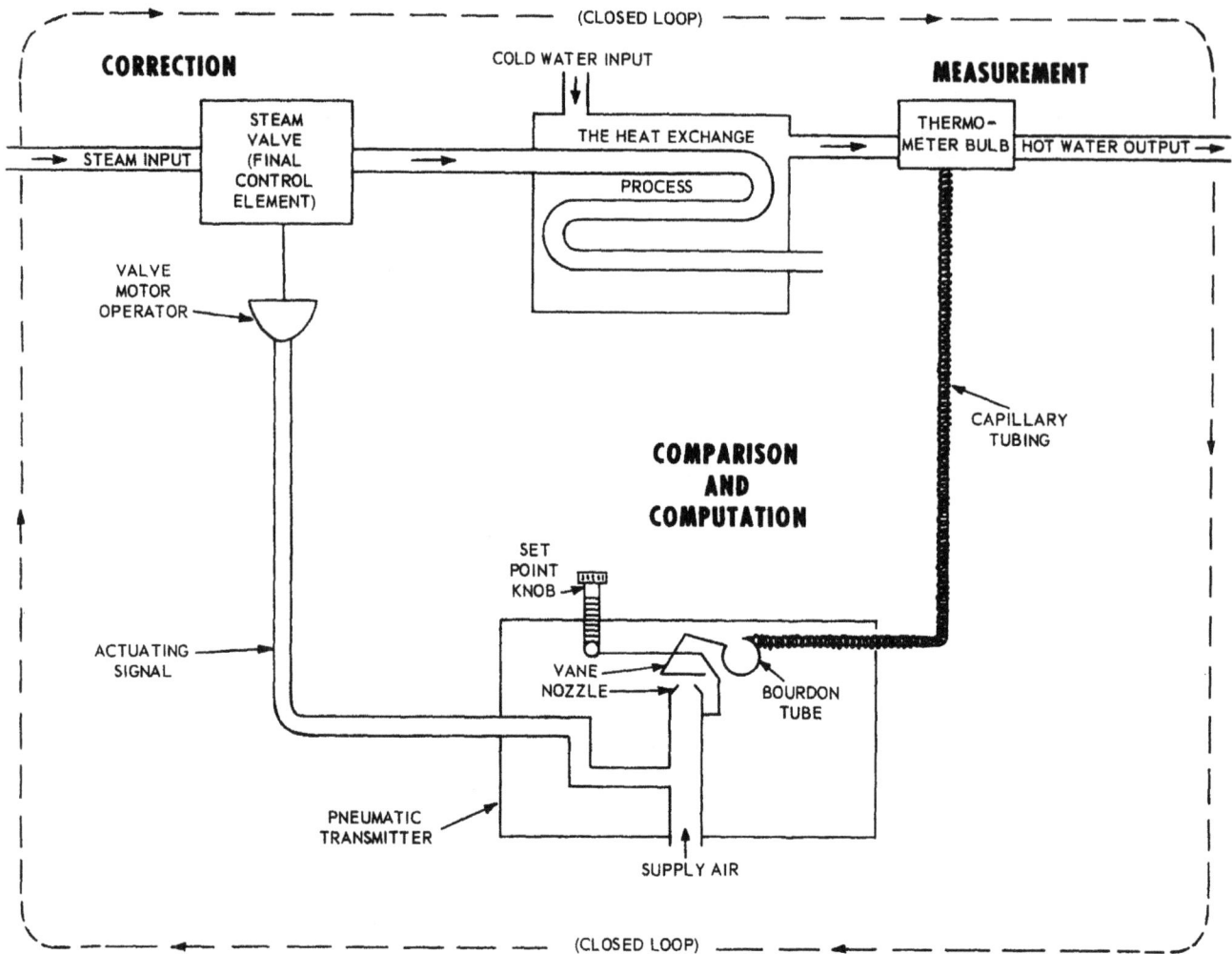

Figure 11-39.—Simple automatic control system for control
of heat exchange process.

The remainder of this discussion deals with Bailey boiler controls which have been installed on several ships that have 1200-psi boilers. It should be noted that boiler control systems installed in naval ships are by no means identical. The systems made by different manufacturers (including Bailey, General Regulator, and Hagan) are different in many respects. Although the systems made by any one manufacturer tend to utilize the same kind of components, the layout of the systems and the variables involved may vary considerably. Hence the following discussion of Bailey boiler controls should be regarded as an example rather than as a standard.

In Bailey boiler control systems, we must distinguish three kinds of air pressure. Loading pressure is the term used to describe the pneumatic signal pressure between two pneumatic units of the control system, except when the pressure is imposed upon a control valve or a control drive. Control pressure is the term used to describe the pneumatic pressure going to the diaphragm of a control valve or to the piston of a control drive. For example, we would call the pneumatic output of a steam pressure transmitter loading pressure because it goes to a pneumatic relay; if the pressure were imposed directly upon a control drive or a control valve, instead of upon the intervening relay, the pressure would be called control pressure. Each pneumatic unit in the system requires a supply of compressed air so that it can develop the appropriate loading pressures or control pressures; the air pressure supplied for this purpose is called supply pressure.

It is not necessary to take up the operating principles of the various pneumatic units, provided we remember their basic function: to develop, transmit, and receive pneumatic signals in the form of variable air pressures. We should also have some idea of the specific functions served by the various kinds of pneumatic units listed below.

Transmitters. In general, a transmitter may be defined as an instrument that produces a pneumatic signal (in the form of variable air pressure) proportional to one of the basic variables in the controlled process.

Relays. A relay is a pneumatic device that receives one or more pneumatic signals, alters or combines signals in various ways, and produces an output signal which goes to one or more other pneumatic units. There are several different kinds of relays: ratio relays, Standatrols, rate relays, selective relays, and limiting relays. The specific functions of these units will become apparent later, as we trace the sequence of events in the boiler control system.

Control Drives. A power unit that mechanically positions valves or dampers in accordance with the amount of control pressure received is called a control drive.

Control Valve. A control valve is a valve used to control the flow of fluid in a line. The control valve is positioned by a control drive in accordance with control pressure. In other words, a control valve is the final control element.

Selector Valves. A selector valve is a pneumatic instrument that provides selection of manual or automatic control of the system components that follow it. A selector valve also provides a means for manual control of the system.

Figure 11-40 shows the control relationships in the combustion control system and the feed water control system. The relationship of the major components is illustrated schematically in figure 11-41. Using this schematic diagram as a guide, we will trace the sequence of events in order to arrive at an understanding of the basic control relationships. Notice that each unit in figure 11-41 is identified by a Bailey number (and in some cases by a name). The numbers and names are given in the legend for figure 11-41 and are used in the following discussion.

It is important to remember that a pneumatic unit may have more than one pneumatic signal coming into it and that it may transmit a pneumatic signal to more than one unit. In describing the sequence of events, it is sometimes necessary to ignore some signals while following others through to their final conclusion. But the system functions as a whole, not as a series of isolated or separate events. This means that a great many signals are being transmitted and received at any given time and that a number of actions are taking place simultaneously.

Combustion Control System.—The combustion control system maintains the energy input to the boiler equal to the energy output by regulating combustion air flow and fuel flow so that the main steam line pressure is maintained at 1200 psig. In other words, the controlled variable is steam pressure, the desired value is 1200 psig, and the manipulated variables are fuel flow and combustion air flow. Combustion air flow and fuel flow are readjusted in accordance with steam demand, as indicated by the measurement of steam flow. The actual measured steam flow thus provides the system with an additional feedback signal.

There are five initial signals in the combustion control system: steam pressure, fuel supply flow, fuel return flow, combustion air flow, and steam flow. Each of these variables is measured, and pneumatic transmitters develop loading pressures that correspond to the measured values of the variables. The two fuel flow signals are combined in a fuel flow differential relay, as described later; in one sense, therefore, it is possible to say that this system has four basic signals instead of five.

The combustion control system is set to maintain the superheater outlet steam pressure at 1200 psig, with variations not exceeding ±0.25 percent of the set pressure at all steaming rates. Steam pressure transmitters (C1a) measure steam pressure from the superheater outlet of each of the two boilers and establish output loading pressure signals that are directly proportional to the measured steam pressure. For the range of steam pressures being measured (900 to 1500 psig), the output loading pressure of the steam pressure transmitter is 3 to 27 psig; for the set steam pressure of 1200 psig, the steam pressure transmitter develops and transmits a pneumatic loading pressure of 15 psig. In other words, the loading pressure varies directly with the applied steam pressure between

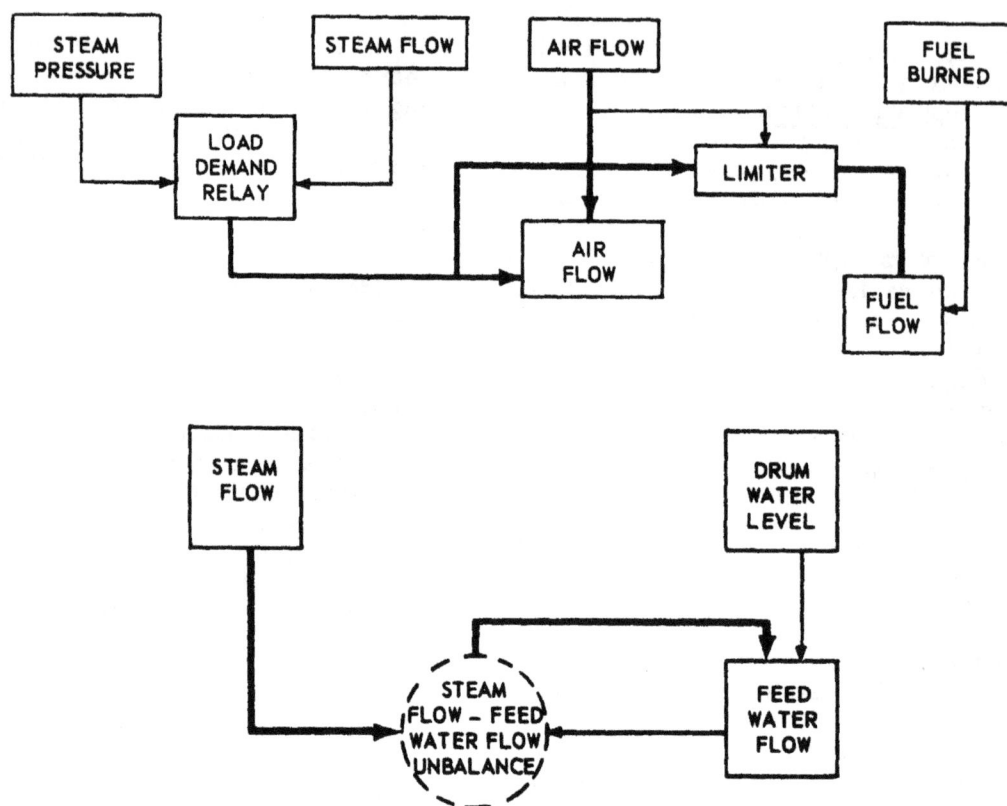

98.108:.109

Figure 11-40.—Control relationships, Bailey combustion control system and feed water control system.

the minimum and the maximum points of the range.

The output signal from each of the steam pressure transmitters is applied to the steam pressure selective relay (C1b). The selective relay selects and transmits the higher of the two signals to the steam demand relay (C4a1). This relay is so adjusted that its output remains constant when the steam pressure is constant at 1200 psig.

While this is going on, steam flow from each boiler is being measured by steam flow transmitters (F2). The steam flow transmitters measure the pressure drop across a restriction in the steam line and extract the square root of this pressure drop. The output loading pressures from the steam flow transmitters are proportional to the square root of the pressure drop—or, in other words, proportional to the rate of flow. The output loading pressure of a steam flow transmitter ranges from 3 psig to 27 psig as the steam flow ranges from minimum

to maximum. The loading pressure from each of the steam flow transmitters is applied to the steam flow selective relay (F2b). The selective relay then transmits the higher of the two loading pressures to the steam demand relay (C4a1).

The output steam demand signal from the steam demand relay (C4a1) passes through the boiler master selector valve (C5a). The boiler master selector valve has no effect on the signal during automatic operation. After passing through the boiler master selector valve, the output loading pressure from the steam demand relay is applied to the combustion air Standatrol (C4b) and to the fuel limiting relay (C15). A measured combustion air flow signal is transmitted from the air flow transmitter (C3) through the excess air remote adjustable relay (C9), where the signal can be manually adjusted for excess air requirements for low-load steaming, maneuvering, or soot blowing. The air flow signal is then applied to the combustion air Standatrol (C4b).

314

LEGEND FOR FIGURE 11-41.

ITEM NO.	ITEM	ITEM NO.	ITEM
C1a.	Superheater outlet pressure transmitter	C15a	Bias hand relay
C1b.	Steam pressure selective relay	C15b	Rate relay
		F1	Feed water flow transmitter
C2a.	Supply fuel flow area meter transmitter	F1a.	Volume chamber with bleed valve
C2b.	Return fuel flow area meter transmitter	F1b.	Feed water flow nozzle
		F2	Steam flow transmitter
C2c.	Fuel flow differential relay	F2a.	Steam flow nozzle
C3	Air flow transmitter	F2b.	Steam flow selective relay
C3a.	Volume chamber	F2c.	Volume chamber with bleed valve
C4a1	Steam demand relay		
C4a2	Fuel flow – air flow Standatrol	F3a.	Drum water level indicating transmitter
C4a3	Transient compensating relay	F3b.	Drum water level indicator
C4b.	Combustion air Standatrol	F4-1	Steam flow – water flow differential relay
C5a.	Boiler master selector valve		
C5b.	Blower selector valve	F4-2	Feed water Standatrol
C5b1	Air pushbutton (for blower)	F5	Feed water selector valve
C5b2	Air pushbutton (for blower)	F6a.	Dual operator feed water flow control valve
C5c.	Fuel selector valve		
C6	Blower speed control drive	FF3	Blower indicating light
C6a.	3-way air trapping valve	IG2	Fuel flow – air flow indicating gage
C7	Blower damper control drive		
C7a.	3-way air trapping valve	IG4	Fuel temperature gage
C8	Fuel control valve	IG5-1	Steam drum pressure gage
C8a.	3-way air trapping valve	IG5-4	Steam drum pressure gage
C9	Excess air remote adjustable relay	IG6-1	Fuel supply pressure indicator
C9a.	Pressure gage	IG6-2	Fuel return pressure indicator
C9b.	Excess air bias (low range)		
C10a	Fuel cutoff valve and operator	IG6a	Fuel separating chamber
		IG7	Superheater outlet pressure gage
C13.	Minimum back pressure valve	IG11	Drum water level indicator
C15.	Fuel limiting relay	R1	Recorder

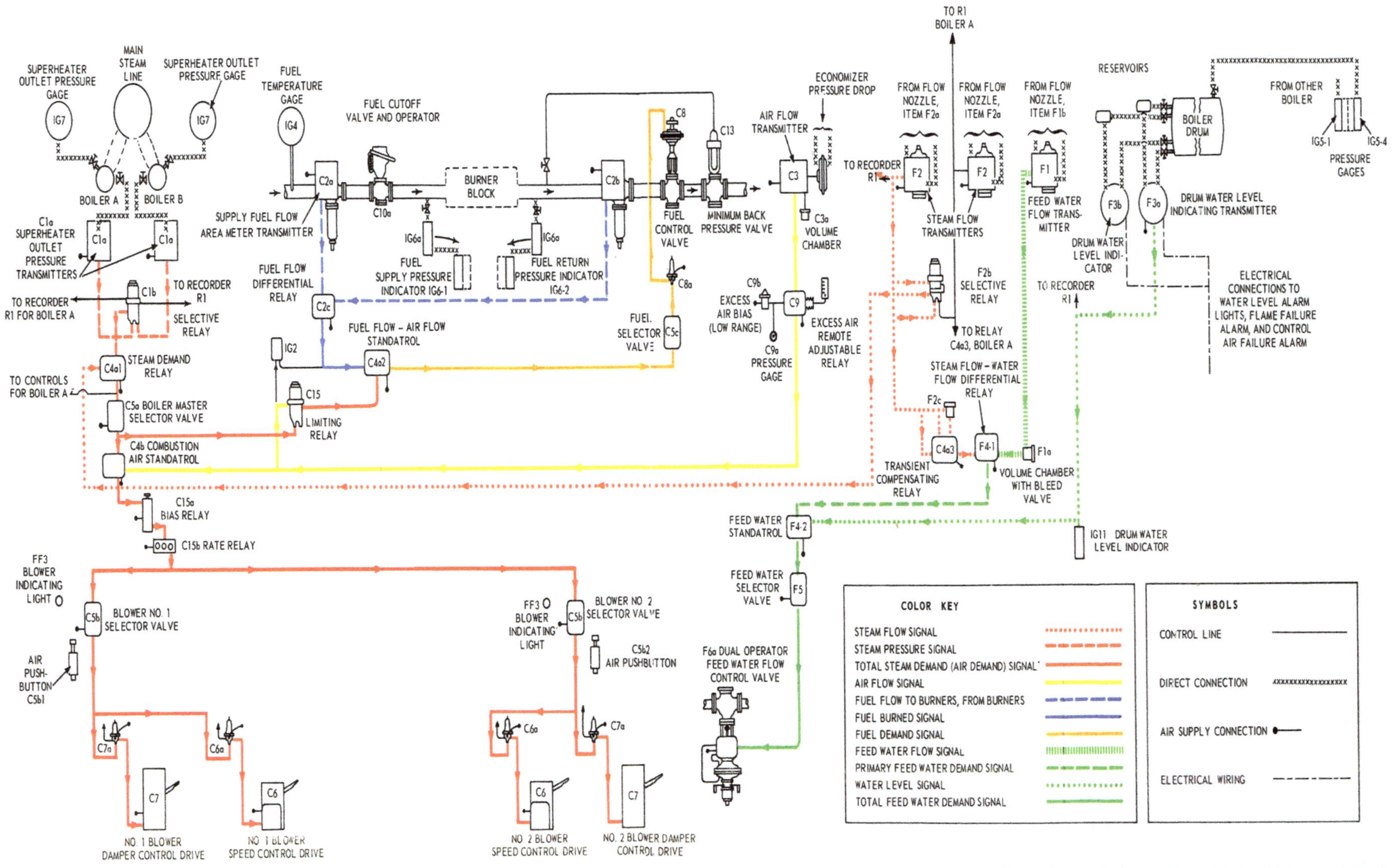

Figure 11-61. — Schematic diagrams of Bailey combustion and feed water control system (Facing page 314).

314b

Under steady boiler loads, the output of the combustion air Standatrol is steady at some value which will maintain combustion air to the furnace at the rate required to maintain the superheater outlet steam pressure at 1200 psig. For each boiler, the combustion air demand signal from C4b is applied, through a bias relay (C15a) and a rate relay (C15b), to the two forced draft blower selector valves (C5b).

The bias relay acts to maintain the minimum air flow demand signal at a value consistent with minimum blower speed and damper position. The rate relay acts in combination with the bias relay to accelerate any changes in the input steam demand signal by providing an exaggerated loading pressure. The rate relay may also be adjusted to decrease the effects of changes in the steam demand signal. The exaggerated signal of the rate relay is slowly returned to normal through the action of a bleed valve within the rate relay.

Each blower selector valve (C5b) transmits a penumatic pressure through the 3-way air trapping valves (C6a and C7a) to the blower speed control drive (C6) and to the blower damper control drive (C7). Note that the pneumatic pressure transmitted by the selector valve is control pressure rather than loading pressure, since it goes to a control drive. The control pressure causes the blower speed control drive and the blower damper control drive to be positioned in accordance with the demand for combustion air. The blower selector valves (C5b) are provided with bias control knobs which can be used to equalize the distribution of combustion air when both blowers are in operation.

The 3-way air trapping valves (C6a and C7a) function to close the forced draft blower dampers to their mechanical bottom stops and to reduce blower speed to the minimum required for stable combustion, in the event of loss of control air supply.

The output from the air flow transmitter (C3) is also applied to the fuel limiting relay (C15). The output of the fuel limiting relay, representing fuel demand, is applied to the fuel flow-air flow Standatrol (C4a2). In the fuel flow-air flow Standatrol, the signal from the fuel limiting relay (C15) is balanced against a signal representing the amount of fuel burned; this "fuel burned" signal comes to the fuel flow-air flow Standatrol (C4a2) from the fuel flow differential relay (C2c). The output signal of the fuel flow-air flow Standatrol (C4a2) is applied to the fuel control valve (C8) in the return fuel

line from the burners. This control pressure from the Standatrol C4a2 positions the fuel control valve so that the required amount of fuel will be burned in order to maintain steam pressure at 1200 psig at the superheater outlet. Notice that the amount of fuel burned is controlled by limiting the return flow of fuel; the supply pressure in the line to the burners is fixed.

Thus, far, we have been considering the combustion control system as it operates when the steam demand (steam flow from the boiler) remains constant. Now let us see what happens when there is an increase in steam demand. For simplicity, the various changes that occur are presented as a numbered list. Remember, however, that some changes may be occurring at the same time as others.

1. Steam flow increases, so there is an increased steam flow signal from the steam flow transmitters (F2) to the steam demand relay (C4a1).

2. Steam pressure drops below 1200 psig, so there is a decrease in the steam pressure signals from the steam pressure transmitters (C1a).

3. The steam demand relay (C4a1) is connected so that an increased signal from the steam flow transmitter and a decreased signal from the steam pressure transmitter result in an increased output loading pressure from C4a1. This increased loading pressure from C4a1 goes to the combustion air Standatrol (C4b) and to the fuel limiting relay (C15).

4. The increased loading pressure from C4a1 to the combustion air Standatrol (C4b) causes an increase in the output loading pressure from the combustion air Standatrol; the ultimate effect of this increase is to increase the control pressure to the blower damper control drives and to the blower speed control drives. The blowers speed up and the dampers open wider. Actually, during this period in which the unbalance is just beginning to be corrected, the blowers speed up enough to allow a temporary "overfiring" rate so that the steam pressure can quickly be restored to normal.

5. As the blowers begin to pick up speed, the measured air flow signal from the air flow transmitter (C3) to the combustion air Standatrol (C4b) and to the fuel limiting relay (C15) also increases.

6. In the fuel limiting relay, the fuel demand signal is held back to a value which corresponds to the value of the measured air

flow signal. Even if the steam demand signal from C4a1 is higher than the measured air flow signal from C3, the output of C15 cannot exceed the air flow signal during the period in which the firing rate is increasing.

7. The output signal of the fuel limiting relay (C15) is applied to the fuel flow-air flow Standatrol (C4a2). The fuel control signal—that is, the output of the fuel flow-air flow Standatrol—begins to increase, thus closing the fuel control valve and increasing the fuel supply to the burners. Since the rate of increase in fuel flow is caused by steam demand but limited by the measured air flow, the system can never supply too much fuel to the burners for the amount of combustion air being supplied.

8. As the steam pressure at the superheater outlet returns to the set pressure of 1200 psig, the steam pressure transmitter signal also rises. Increasing signals from the steam pressure transmitters result in a decreasing loading pressure from the steam demand relay (C4a1) and decreasing control signal from the combustion air Standatrol (C4b) to the forced draft blower and damper drives (C6 and C7, respectively). This reduces the temporary "over-firing" rate previously mentioned. When the measured air flow signal from the air flow transmitter (C3) reaches a value which returns the combustion air Standatrol to balance, the output pressure of C4b stabilizes at a value which will maintain this air flow. The output pressure of the fuel flow-air flow Standatrol (C4a2) stabilizes in a similar way to maintain the same rate of fuel flow to the burners. At this time, the main line steam pressure has returned to 1200 psig and the air flow and fuel flow are adjusted so as to maintain this pressure under the new (and higher) steam demand conditions.

When there is a decrease in steam demand, the system functions to slow down the forced draft blowers, partially close the blower dampers, and open the fuel control valve so that the supply of fuel to the burners will be decreased. After seeing how the system operates when the steam demand is constant and when the steam demand is increasing, it should not be too difficult to trace the signals and events that occur when the steam demand is decreasing.

Feed Water Control System.—While the combustion control system is functioning to control combustion air and fuel, the feed water system is functioning to control the amount of feed water going to the boiler.

There are three elements in the feed water control system: steam flow, feed water flow, and boiler drum water level. The feed water flow transmitter (F1) and the steam flow transmitter (F2) act together to provide a proportioning control—that is, to provide a flow of feed water that is proportional to the flow of steam. The drum water level indicating transmitter (F3a) introduces a secondary signal that continuously adjusts the position of the feed water flow control valve (F6a) in order to maintain the desired water level in the boiler steam drum.

The feed water flow transmitter (F1) develops a pneumatic signal that is proportional to feed water flow. This signal is applied, through a volume chamber (F1a), to the steam flow-water flow differential relay (F4-1). The other input to relay F4-1 is the output pressure from the steam flow transmitter (F2). The output pressure from F2 is applied to F4-1 through the transient compensating relay (C4a3); under conditions of steady steam demand, the output signal of C4a3 exactly duplicates the output signal of the steam flow transmitter (F2), but when there is a change in steam demand the output signal of C4a3 is _not_ the same as the output signal of F2.

The output from the steam flow-water flow differential relay (F4-1) is applied to the feed water Standatrol (F4-2), where it is balanced against a signal from the drum water level indicating transmitter (F3a). When the two inputs to the feed water Standatrol (F4-2) are at their set point values, a constant pneumatic output pressure is transmitted from the Standatrol through the feed water selector valve (F5) to the feed water flow control valve (F6a). A spring adjustment in the feed water Standatrol (F4-2) maintains the steam drum water level at a set height.

When steam demand increases, there is a proportional increase in the loading pressure output of the steam flow transmitter (F2) which is transmitted to the transient compensating relay (C4a3). At the compensating relay, the signal is temporarily reversed—that is, the input signal representing an increase in steam flow becomes an output signal representing a _decrease_ in steam flow. This output signal from C4a3 is applied to the steam flow-water flow differential relay (F4-1), which also receives an input signal from the feed water flow

98.171

System, Subsystem, or Component: Automatic Combustion and Feed Water Control System

Deviation Card Control No.	Maintenance Requirement	M.R. No.	Rate Req'd.	Man Hours	Related Maintenance
MG ZZ8FBEO 45 8954	M1. Lubricate the control drive and linkage.	M-1	ET3	0.2	None
MG ZZ1FTQV 45 8364	Q1. Blow down sensing lines.	Q-1	ET3*	0.1	None
MG ZZDFFC1 45 8955	S1. Clean and inspect air filters.	S-1	ET3 FA	0.4 0.4	None
MG ZZ1FTQV 45 8956	S1. Clean, inspect, and lubricate fuel-air ratio relay.	S-2	ET2 FN	0.4 0.4	None
MG ZZ1FTQ7 45 8957	S1. Test calibration of air flow transmitter.	S-3	ET1*	1.0	None
MG ZZ1FTQ7 65 A523	S1. Test calibration of flow transmitter.	S-4	ET1*	1.0	None
MG ZZ1FTQX 65 A524	S1. Test calibration of drum level transmitter.	S-5	ET1*	1.2	None
MG ZZ1FTQ0 65 A525	S1. Test calibration of ratio totalizer.	S-6	ET1*	1.0	None
MG ZZ8FBEO 45 8961	S1. Test calibration of master sender.	S-7	ET1* ET3	0.5 0.5	None
MG ZZ8FBEO 65 A529	S1. Test the entire air system for leaks.	S-8	ET3 FN	1.0 1.0	None
MG ZZ8FBEO 45 8962	A1. Lubricate fulcrum bearings.	A-1	ET3	0.1	None
MG ZZEFVA5 45 8963	A1. Clean and inspect pressure-reducing valves.	A-2	ET3 FA	0.7 0.7	None
MG ZZ8FBEO 45 8964	A1. Clean and inspect pilot valve on control drive. 2. Renew piston rod packing on control drive.	A-3	ET1* FA	1.2 1.2	None
MG ZZEFVA2 45 8965	A1. Renew valve stem packing on feed water regulator valve.	A-4	ET3	0.6	None
MG ZZEFVA0 45 8966	A1. Clean and inspect needle valve.	A-5	ET1* FA	0.2 0.2	None
MG ZZ8FBEO 45 8967	A1. Clean and inspect escapement valve on master sender.	A-6	ET1* FA	1.2 1.2	S-7

MAINTENANCE INDEX PAGE
OPNAV FORM 4700-3 (4-64)

BUREAU PAGE CONTROL NUMBER F-26/33-65

(Page 1 of 2)

System, Subsystem, or Component: Automatic Combustion and Feed Water Control System

Deviation Card Control No.	Maintenance Requirement	M.R. No.	Rate Req'd.	Man Hours	Related Maintenance
MG ZZ1FTQ7 45 8968	A1. Clean and inspect escapement valve on flow transmitter. 2. Clean and lubricate the cam roller assembly on the transmitter.	A-7	ET1* FA	1.4 1.4	S-3 S-4
MG ZZ1FTQX 45 8969	A1. Clean and inspect escapement valve on drum level transmitter.	A-8	ET1* FA	1.2 1.2	S-5
MG ZZEFVA6 45 8970	A1. Clean and inspect pilot valve.	A-9	ET1* FA	1.2 1.2	None
MG ZZEFVAK 65 A526	C1. Clean and inspect fuel oil control valve. (Perform this MR only if oil pressure at burners fail to follow loading pressure quickly and accurately).	C-1R	ET1* FN	1.0 1.0	None

*Only competent personnel well trained in the function, assembly, and calibration of this equipment, may attempt this MR.

MAINTENANCE INDEX PAGE
OPNAV FORM 4700-3 (4-64)

BUREAU PAGE CONTROL NUMBER F-26/33-65

(Page 2 of 2)

Figure 11-42.—Maintenance index pages.

transmitter (F1). The difference in the two signals put into the differential relay (F4-1) causes a decreased output pressure to be transmitted from the differential relay to the feed water Standatrol (F4-2). The feed water Standatrol therefore sends a decreased signal through the feed water selector valve (F5) to the feed water flow control valve, causing the valve to begin to close.

Let us examine this point more closely. The steam demand has increased but the feed water flow control valve is closing. Why? Because it is necessary to compensate for swell—the momentary increase in the volume of the water that occurs when the firing rate is increased. As swell occurs, the pneumatic signal from the drum water level indicating transmitter (F3a) increases. As a result, the output pressure of the feed water Standatrol (F4-2) begins decreasing even more rapidly, closing down on the feed water flow control valve (F6a) and further restricting the flow of feed water to the boiler.

As the feed water flow decreases, there is a proportional drop in the pneumatic pressure from the feed water flow transmitter. The effects of this pressure decrease are felt slowly, however, because of the restricting action of the bleed valve in volume chamber F1a.

As the steam drum water level begins to drop, there is a proportional decrease in the pneumatic pressure from the drum water level indicating transmitter (F3a). At the same time, the bleed valve in volume chamber F2c is decreasing the pneumatic signal between the steam flow transmitter (F2) and the transient compensating relay (C4a3) and increasing the pressure in another chamber of the compensating relay. The effect of this bleed valve action is to balance the inputs to the two chambers of the compensating relay so that the compensating relay output pressure is now equal to the pressure it is receiving from the steam flow transmitter. In other words, the reversing action of the transient compensating relay (C4a3) has been stopped, and the compensating relay is now transmitting a pneumatic signal that is exactly the same as the new (and higher) steam flow signal it receives.

The increased loading pressure from the transient compensating relay (C4a3), together with the decreased loading pressure from the feed water flow transmitter (F1), increases the output pressure of the steam flow-water flow differential relay (F4-1). The increased output pressure of F4-1 reverses the action of the feed

water Standatrol (F4-2) and causes its output to increase, thus opening the feed water flow control valve wider and allowing more feed water to flow to the boiler.

When the feed water flow is equal to the steam flow, and when the steam drum water level has returned to normal, the system stabilizes and the output of the feed water Standatrol (F4-2) stays at the higher value which will maintain the new and higher rate of feed water flow.

A similar (but of course reversed) series of events occurs when there is a decrease in steam demand. The first effect of the decreased steam demand is a wider opening of the feed water flow control valve to compensate for shrink—that is, the decrease in the volume of the boiler water that occurs when the firing rate is reduced. The final effect is a smaller opening of the feed water flow control valve and a reduced flow of feed water to the boiler.

Maintenance

To ensure trouble-free operation of the control system, it is important that the system be properly maintained and calibrated at all times. Maintenance and calibration should be conducted in accordance with the Planned Maintenance Subsystem of the 3-M System. An example of maintenance actions and tests to be conducted on an Automatic Control System are shown in figure 11-42. Particular emphasis should also be placed on the use of maintenance, repair, calibration procedures found in the applicable manufacturer's technical manual. If each component is kept in a properly maintained and adjusted condition, the need for a general overhaul or major recalibration of the control system will be minimized.

NOTE: When checking the adjustments and calibration of any component of the control system the settings should not be changed except under the supervision of "QUALIFIED" maintenance personnel. It is also extremely important, when making adjustments to the control system, that the person doing the work know the effect adjustments have on the operation of the entire control system. In other words only "QUALIFIED" personnel should be allowed to perform maintenance, repair, and calibration of any automatic control system components.

CHAPTER 12

PROPULSION STEAM TURBINES

In beginning the study of steam turbines, the first point to be noted is that we have now reached the part of the thermodynamic cycle in which the actual conversion of thermal energy to mechanical energy takes place.

We know by simple observation of pressures and temperatures that the steam leaving a turbine has far less thermal energy than it had when it entered the turbine. By observation, again, we know that work is performed as the steam passes through the turbine, the work being evidenced by the turning of a shaft and the movement of the ship through the water. Since we know that energy can be transformed but can be neither created nor destroyed, the decrease of thermal energy and the appearance of work cannot be regarded as separate events. Rather, we must infer that thermal energy has been transformed into work—that is, mechanical energy in transition.

Disregarding such irreversible energy losses as those caused by friction and by heat flow to objects outside the system, it can be shown that two energy transformations are involved. First, there is the thermodynamic process by which thermal energy is transformed into mechanical kinetic energy as the steam flows through one or more nozzles. Second, there is the mechanical process by which mechanical kinetic energy is transformed into work as the steam impinges upon projecting blades of the turbine, thereby turning the turbine rotor.

In order to understand the process by which thermal energy is converted into mechanical kinetic energy, we must have some understanding of the process that takes place as steam flows through a nozzle. The second energy transformation, from kinetic energy to work, is best understood by considering some basic principles of turbine design.

STEAM FLOW THROUGH NOZZLES

The basic purpose of a nozzle is to convert the thermal energy of the steam into mechanical kinetic energy. Essentially, this is accomplished by shaping the nozzle in such a way as to cause an increase in the velocity of the steam as it expands from a high pressure area to a low pressure area. The nozzle also serves to direct the steam so that it will flow in the right direction to impinge upon the turbine blades.

Within certain limitations, the velocity of steam flow through any restricted channel such as a nozzle depends upon the difference between the pressure at the inlet of the nozzle and the pressure at the region around the outlet of the nozzle. Let us begin by assuming equal pressure at inlet and outlet. No flow exists in this static condition. Now, if we maintain the pressure at the inlet side but gradually reduce the pressure at the outlet area, the steam will begin to flow and its velocity will increase as the outlet pressure is reduced. However, if we continue to reduce the outlet pressure, we will reach a point at which the velocity of steam is equal to the velocity of sound in steam. At this point, a further reduction in pressure at the outlet region will not produce any further increase in velocity at the entrance to the nozzle, nor will it produce any further increase in the rate of steam flow.

The ratio of outlet pressure to inlet pressure at which the acoustic velocity (also called the critical flow) is reached is known as the acoustic pressure ratio or the critical pressure ratio. This ratio is about 0.55 for superheated steam. In other words, the velocity of flow through nozzles is a function of the pressure differential across the nozzle, and steam velocity will increase as the outlet pressure decreases (in relation to the inlet pressure). However, no further increase in steam velocity will occur when the outlet pressure is reduced below 55 percent of the inlet pressure.

When the pressure at the outlet area of a nozzle is designed to be higher than the critical pressure, a simple convergent (parallel-wall) nozzle may be used. In this type of nozzle, shown in figure 12-1, the cross-sectional area at the outlet is the same as the cross-sectional area at the throat. This type of nozzle is often referred to as a nonexpanding nozzle because no expansion of steam takes place beyond the throat of the nozzle.

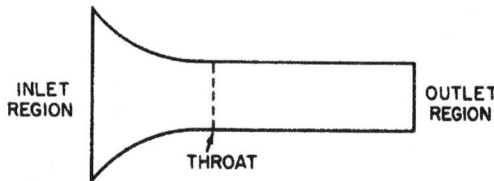

147.90

Figure 12-1.—Simple convergent nozzle.

When the pressure at the outlet area of a nozzle is designed to be lower than the critical pressure, a convergent-divergent nozzle is used to control the turbulence that occurs when steam expansion takes place below the critical pressure ratio. In this type of nozzle, shown in figure 12-2, the cross-sectional area of the nozzle gradually increases from throat to outlet. The critical pressure is reached in the throat of the nozzle, but the gradual expansion from throat to outlet allows the steam to emerge finally in a steady stream or jet. Because expansion takes place from the throat to the outlet, this type of nozzle is often called an expanding nozzle.

The decrease in thermal energy of the steam passing through a nozzle must equal the increase in kinetic energy (disregarding irreversible losses). The decrease in thermal energy may be expressed in terms of enthalpy as

$$h_1 - h_2$$

where

h_1 = enthalpy of the entering steam, in BTU per pound

h_2 = enthalpy of the steam leaving the nozzle, in BTU per pound

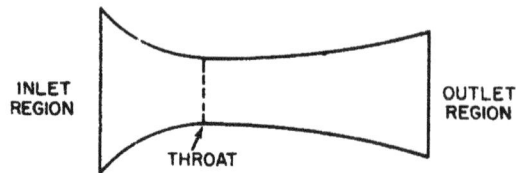

147.91

Figure 12-2.—Convergent-divergent nozzle.

The kinetic energy of the steam jet leaving the nozzle may be determined by using the equation for mechanical kinetic energy:

$$KE = \frac{WV^2}{2g}$$

where

KE = mechanical kinetic energy, in foot-pounds

W = weight of the flowing substance, in pounds per second

V = velocity, in feet per second

g = acceleration due to gravity (32.2 feet per second)

Since we have taken the enthalpy per pound of the entering and departing steam, let us assume 1 pound of steam per second flowing from the nozzle. The kinetic energy of this pound of steam will then be expressed by

$$KE = \frac{V_2^2}{2g}$$

where V_2 is the velocity, in feet per second, of the steam leaving the nozzle. We may now equate the expression for the decrease in thermal energy and the expression for the increase in kinetic energy, Thus,

$$\frac{V_2^2}{64.4} = (h_1 - h_2)\ (778)\ \text{ft-lb}$$

Since 1 BTU is equal to 778 foot-pounds, we have multiplied the expression for the decrease in thermal energy by 778. This puts both sides of the equation in terms of foot-pounds.

The kinetic energy of the steam leaving the nozzle is directly proportional to the square of the velocity. By causing an increase in velocity, therefore, the nozzle causes an increase in the kinetic energy of the steam. Thus it is clear that our last equation has actually described the purpose of a nozzle by equating the decrease in thermal energy with the increase in kinetic energy.

BASIC PRINCIPLES OF TURBINE DESIGN

In essence, a turbine may be thought of as a bladed wheel or rotor that turns when a jet of steam from the nozzles impinges upon the blades. The basic parts of a turbine are the rotor, which has blades projecting radially from its periphery; a casing, in which the rotor revolves; and nozzles, through which the steam is expanded and directed. As we have seen, the conversion of thermal energy to mechanical kinetic energy occurs in the nozzles. The second energy conversion—that is, the conversion of kinetic energy to work—occurs on the blades.

The basic distinction to be made between types of turbines has to do with the manner in which the steam causes the turbine rotor to move. When the rotor is moved by a direct push or "impulse" from the steam impinging upon the blades, the turbine is said to be an impulse turbine. When the rotor is moved by the force of reaction, the turbine is said to be a reaction turbine.

Although the distinction between impulse turbines and reaction turbines in a useful one, and one which is followed in this text, it should not be considered as an absolute distinction in real turbines. An impulse turbine utilizes both the impulse of the steam jet and, to a lesser extent, the reactive force that results when the curved blades cause the steam to change direction. A reaction turbine is moved primarily by reactive force, but some motion of the rotor is caused by the impact of the steam against the blades.

Theory of Impulse Turbines

In discussing the manner in which kinetic energy is converted to work on the turbine blades, it is necessary to consider both the absolute velocity of the steam and the relative velocity of the steam—that is, its velocity relative to the moving blades. The following symbols will be used in the remainder of this discussion:

V_1 = absolute velocity of steam at blade entrance

V_2 = absolute velocity of steam at blade exit

R_1 = relative velocity of steam at blade entrance

R_2 = relative velocity of steam at blade exit

V_b = peripheral velocity of blade

Let us consider, first, a theoretical elementary impulse turbine such as the one shown in figure 12-3. The blades of this imaginary turbine are merely flat vanes or plates. As the steam jet flows from the nozzle and impinges upon the vanes, the rotor is moved.

Assuming that there is no friction as the steam flows across the blade, R_1 must be equal to $V_1 - V_b$ and R_2 must also be equal to $V_1 - V_b$, since theoretically there is no change in velocity as the steam flows across the blade.

It will be apparent that, in order to convert all of the kinetic energy into work, it would be necessary to design a blade from which the steam would exit with zero absolute velocity. This blade would be curved in the manner shown in figure 12-4, and the jet of steam would enter the blade tangentially rather than at an angle. As we shall see, the shape of this blade very closely approximates the shape of the blades used in actual impulse turbines; in a real turbine, however, the steam enters the blade at an angle, rather than tangentially.

When this curved blade is used, the direction of the steam is exactly reversed. The relative velocity of the steam at the blade entrance, R_1, is again $V_1 - V_b$ and R_2 is again $V_1 - V_b$. Since the direction of flow is reversed, however, absolute velocity of the steam at blade exit is now

$$V_2 = (V_1 - V_b) - V_b$$
$$= V_1 - 2V_b$$

As previously noted, the absolute velocity of the departing steam (V_2) should ideally be zero. Therefore, by transposition of the above equation,

$$V_1 = 2V_b$$

In other words, maximum work is obtained from a reversing blade when the blade velocity is exactly one-half the absolute velocity of the steam at the blade entrance.[1] The maximum amount of work obtainable from a reversing blade is twice the amount obtainable from the flat vane shown in figure 12-3.

In actual turbines, it is not feasible to utilize the complete reversal of steam in the blades, since to do so would require that the nozzle be placed in a position that would also be swept by the blades—an obvious impossibility. Furthermore, if the steam entered the blade tangentially it would not be carried through the turbine axially (longitudinally). However, it is only the tangential component of the steam velocity that produces work on the turbine blades; hence the nozzle angle is made as small as possible.

As we know, the work done on the blade must equal the total energy entering minus the total energy leaving. The velocity diagram for an impulse turbine shown in figure 12-5 provides a way of determining the work done on the blade in terms of the various velocities and angles.

The tangential component (which is the only component that produces work) of the velocity of the entering steam is

$$V_1 \cos a$$

which is also equal to

$$V_b + R_1 \cos b$$

The tangential component of the velocity of the departing steam is

$$V_2 \cos c$$

which is also equal to

$$V_b - R_2 \cos d$$

Since entering and leaving velocities are opposed to each other if $V_2 \cos c$ is in the same direction as $V_1 \cos a$, and supplementary to each other if $V_2 \cos c$ is in the opposite direction

from $V_1 \cos a$, the resultant velocity may be expressed as

$$V_1 \cos a - V_2 \cos c$$

or, alternatively, as

$$R_1 \cos b + R_2 \cos d$$

Assuming a steam flow of W pounds per second,

$$\frac{W}{g} = \text{mass per second}$$

and

$$\frac{W}{g} V = \text{force}$$

Therefore, the force on the blade is

$$F_b = \frac{W}{g} (R_1 \cos b + R_2 \cos d) \text{ pounds}$$

$$= \frac{W}{g} (V_1 \cos a - V_2 \cos c) \text{ pounds}$$

Work is force through distance. Therefore, the rate of doing work on the blade is

$$Wk_b = F_b V_b$$

$$= \frac{W}{g}(R_1 \cos b + R_2 \cos d) V_b \text{ foot-pounds per second}$$

$$= \frac{W}{g}(V_1 \cos a - V_2 \cos c) V_b \text{ foot-pounds per second}$$

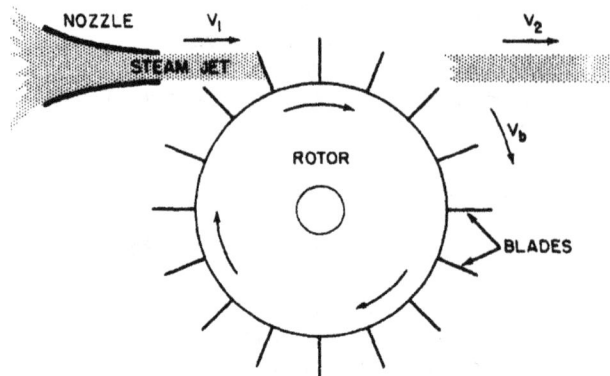

147.92

Figure 12-3.—Elementary impulse turbine.

---

[1]This statement assumes, of course, that the nozzle is tangential to the blades. In actual impulse turbines, the maximum amount of work is done when the blade speed is one-half the cosine of the nozzle angle times the absolute velocity of the entering steam.

The rate of doing work on the blade may also be derived from consideration of the thermal energy and kinetic energy entering the system and leaving the system. Although the actual derivations are not given here, it may be of interest to note the relationships expressed in the following equations:

$$Wk_b = W(h_1 - h_2) + \frac{W(V_1^2 - V_2^2)}{50,000}$$

Btu per second

and

$$Wk_b = \frac{W(R_2^2 - R_1^2) + R_1^2\ W(V_1^2 - V_2^2)}{50,000}$$

Btu per second

The pressure and velocity changes that occur in the nozzle and in the blades of an impulse turbine are shown in figure 12-6. As may be seen, the pressure is the same at the entrance and at the exit of the blade; the only pressure drop occurs in the nozzle. Figure 12-7 shows a section of an impulse turbine rotor, with the blades in place.

Theory of Reaction Turbines

Reaction turbines, as their name implies, are moved by reactive force rather than by a direct push or impulse. Although we commonly think of reactive force as having been

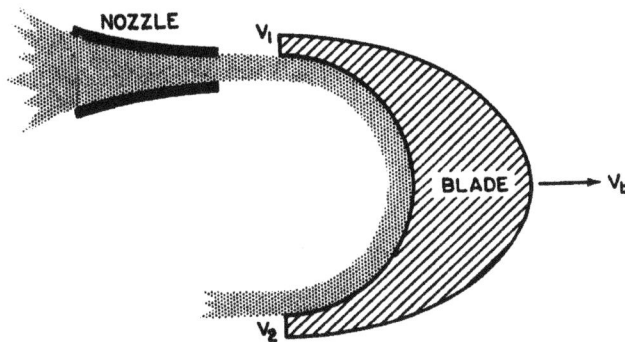

147.93

Figure 12-4.—Curved impulse blade.

"discovered" by Newton, it is interesting to note that the first reaction turbine—and, indeed, perhaps the first steam engine of any kind ever made—was developed by the Greek mathematician Hero about 2000 years ago. This turbine, shown in figure 12-8, consisted of a hollow sphere which carried four bent nozzles. The sphere was free to rotate on the tubes that carried steam from the boiler, below, to the sphere. As the steam flowed out

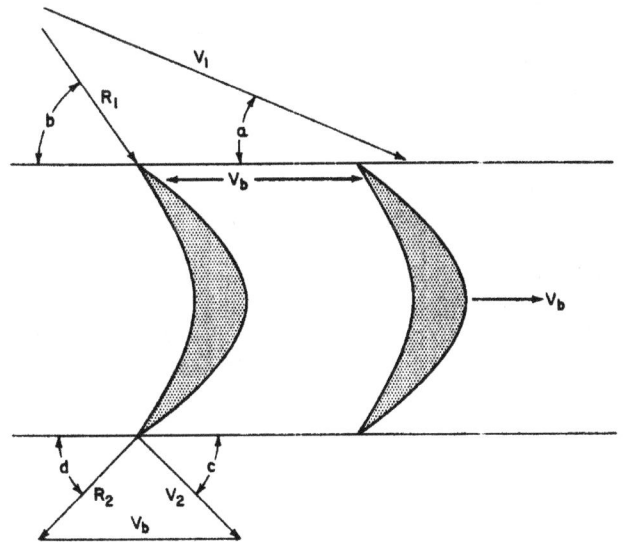

147.94

Figure 12-5.—Velocity diagram for impulse blading.

through the nozzles, the sphere rotated rapidly in a direction opposite to the direction of steam flow.

Reaction turbines used in modern times utilize the reactive force of the steam in quite a different way. In a modern reaction turbine, there are no nozzles as such. Instead, the blades that project radially from the periphery of the rotor are formed and mounted in such a way that the spaces between the blades have, in cross section, the shape of nozzles.[2] Since these

---

[2]The distinction between actual nozzles and the blading which serves the purpose of nozzles in reaction turbines is mechanical rather than functional. The previous discussion of steam flow through nozzles applies equally well to steam flow through the nozzle-shaped spaces between the blades of reaction turbines.

blades are mounted on the revolving rotor, they are called moving blades.

Fixed or stationary blades of the same shape as the moving blades are fastened to the casing in which the rotor revolves; these fixed blades are installed between successive rows of the moving blades. The fixed blades guide the steam into the moving blade system and, since they are also shaped and mounted in such a way as to provide nozzle-shaped spaces between the blades, the fixed blades also act as nozzles. The general arrangement of the fixed and moving blades, together with the pressure and absolute velocity relationships in a reaction turbine, are shown in figure 12-9. Figure 12-10 shows a section of a reaction turbine rotor with one row of moving blades and one row of fixed blades.

A reaction turbine is moved by three main forces: (1) the reactive force produced on the moving blades as the steam increases in velocity as it expands through the nozzle-shaped spaces between the blades; (2) the reactive force produced on the moving blades when the steam changes direction; and (3) the push or "impulse" of the steam impinging upon the blades. Thus, as previously noted, a reaction turbine is moved primarily by reactive force but also to some extent by direct impulse.

From what we have already learned about the function of nozzles, it will be apparent that

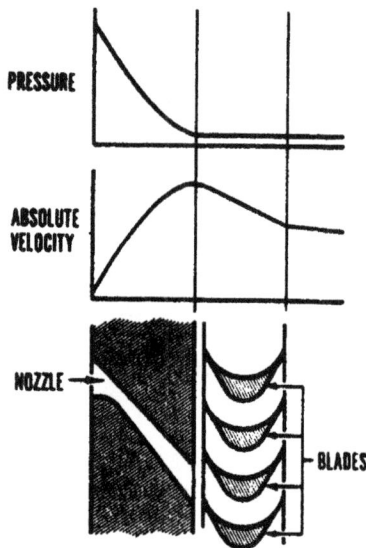

38.76X

Figure 12-6.—Nozzle position and pressure-velocity relationships in an impulse turbine.

thermal energy is converted into mechanical kinetic energy in the blading of a reaction turbine. The second required energy transformation—that is, from kinetic energy to work—also occurs in the blading. A velocity diagram such as was used to analyze the work done on impulse blading may be similarly used to analyze the work done on reaction blading; however, the angles and velocities are different in the two types of blading.

Since the velocity of the steam is increased in the expansion through the moving blades, the initial velocity of the entering steam (V_1) must be lower in a reaction turbine than it would be in an impulse turbine with the same blade speed (V_b); or, alternatively, the reaction turbine must run at a higher speed than a comparable impulse turbine in order to operate at approximately the same efficiency.

TURBINE CLASSIFICATION

As we have seen, turbines are divided into two general groups or classes—impulse turbines and reaction turbines—according to the way in which the steam causes the rotor to move. Turbines may be further classified according to (1) the manner of staging and compounding, and (2) the mode of steam flow through the turbine.

Staging and Compounding

Thus far in this chapter, we have more or less assumed that an impulse turbine had one set of nozzles and one row of blading on the rotor, and that a reaction turbine had one row of fixed blades and one row of moving blades. In reality, however, propulsion steam turbines are not this simple. Instead, they use several rows of blading, arranged in various ways.

It has been shown that the amount of thermal energy which can be utilized in a turbine depends upon the relationship between the velocity of the entering steam (V_1) and the blade speed (V_b). It might seem reasonable, therefore, to think that the work output of the turbine could only be increased by increasing V_1 and V_b in the proper ratio. However, mechanical considerations and problems concerning strength of materials impose certain limits on blade speed. In modern naval ships, the amount of available energy per pound of steam is so great that there is no practicable way of utilizing the major portion of it in one row of blades. When several rows of blades are

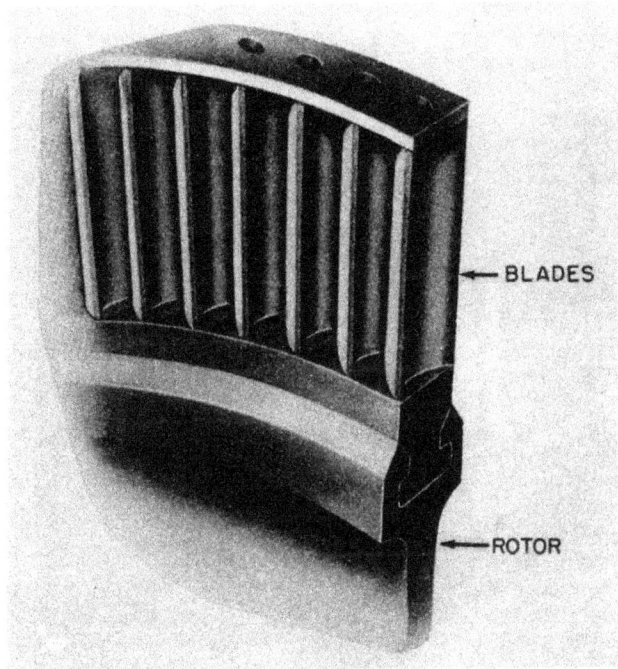

147.95X

Figure 12-7.—Section of impulse turbine rotor (with blades).

139.23

Figure 12-8.—Hero's steam turbine.

used, the steam passes through one row after another, and each row uses part of the energy of the steam.

IMPULSE STAGE.—In an impulse turbine, a stage is defined as one set of nozzles and the succeeding row or rows of moving and fixed blades. Since the only place a pressure drop occurs in an impulse turbine is in the nozzles, another way of defining an impulse stage is to say that it includes the nozzles and blading in which only one pressure drop takes place. A simple impulse stage is often called a Rateau stage. Turbines consisting of a single Rateau stage (fig. 12-11) are not used as propulsion turbines but are frequently used to drive small auxiliary units.

REACTION STAGE.—In reaction turbines, one row of fixed blades and its succeeding row of moving blades are taken as constituting one stage. Since the fixed blades in a reaction turbine are comparable to the nozzles in an impulse turbine, this definition of a reaction stage may seem very similar to the definition of an impulse stage. However, there is this

important difference: a reaction stage includes two pressure drops, whereas an impulse stage includes only one.

VELOCITY-COMPOUNDED IMPULSE TUR-BINE.—One way of increasing the efficiency of an impulse turbine is by velocity-compounding— that is, by adding one or more rows of moving blades to the rotor.[3] Figure 12-12 shows an impulse turbine that has two rows of moving blades on the rotor. This type of turbine is called velocity-compounded because the residual velocity of the steam leaving the first row of moving blades is utilized in the second row of moving blades. If a third row is added, the velocity of the steam leaving the second row is utilized in the third row. The fixed blades,

[3]Velocity-compounding can also be achieved when only one row of moving blades is used, provided the steam is directed in such a way that it passes through the blades more than once. This point is discussed in more detail in chapter 16 of this text, in connection with helical-flow auxiliary turbines.

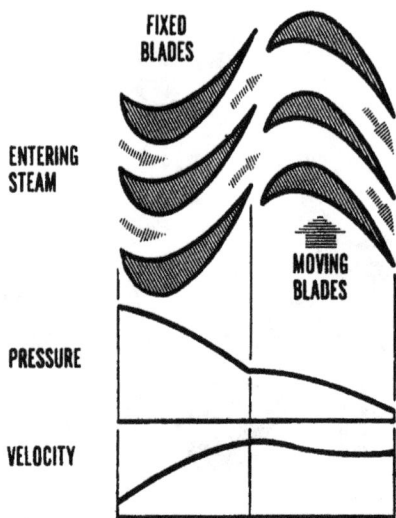

38.77.2X

Figure 12-9.—Arrangement of fixed and moving blades and pressure-velocity relationships in a reaction turbine.

which are fastened to the casing rather than to the rotor, serve to direct the steam from one row of moving blades to another.

As may be seen in figure 12-12, the velocity-compounded impulse turbine has only one pressure drop and therefore, by definition, only one stage. This type of velocity-compounded impulse stage is usually called a Curtis stage.

PRESSURE-COMPOUNDED IMPULSE TURBINE.—Another way to increase the efficiency of an impulse turbine is to arrange two or more simple impulse stages in one casing. The casing is internally divided by nozzle diaphragms. The steam leaving the first stage is expanded again through the first nozzle diaphragm, to the second stage; from the second nozzle diaphragm, to the third stage; and so on. This type of turbine is known as a pressure-compounded turbine because a pressure drop occurs in each stage. Figure 12-13 shows a pressure-compounded impulse turbine with four stages. A pressure-compounded impulse turbine is frequently called a Rateau turbine, since it is essentially a series of simple impulse (Rateau) stages arranged in sequence in one casing.

PRESSURE-VELOCITY-COMPOUNDED IMPULSE TURBINE.—An impulse turbine which consists of one velocity-compounded (Curtis) stage followed by a series of pressure-

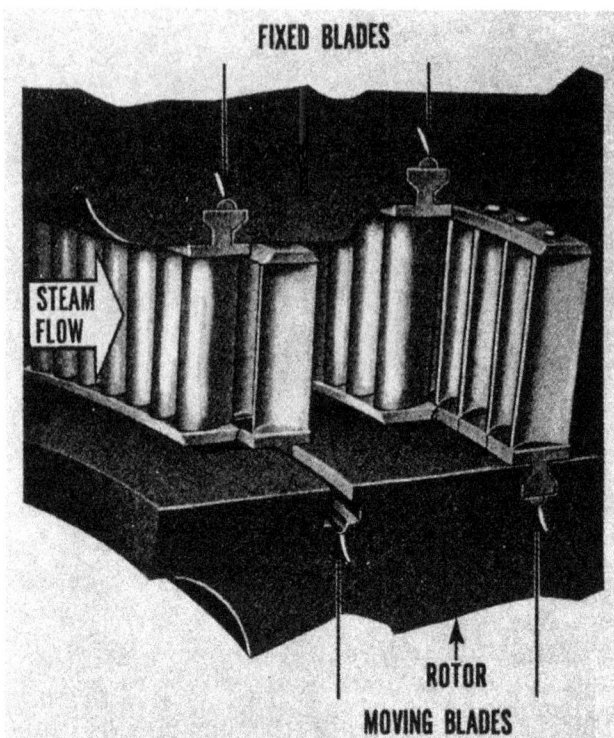

38.771X

Figure 12-10.—Section of reaction turbine rotor, showing fixed and moving blades.

compounded (Rateau) stages is generally referred to as a pressure-velocity-compounded impulse turbine. Turbines of this type are commonly used in the propulsion plants of naval ships.

PRESSURE - COMPOUNDED REACTION TURBINE.—Because the ideal blade speed in a reaction turbine is so high in relation to the velocity of the entering steam (V_1), all reaction turbines are pressure-compounded—that is, they are so arranged that the pressure drop from inlet to exhaust is divided into many steps by means of alternate rows of fixed and moving blades. The pressure drop in each set of fixed and moving blades (i.e., in each stage) is therefore small, thus causing a lowered steam velocity in all stages and consequently a lowered ideal blade velocity for the turbine as a whole.

COMBINATION IMPULSE AND REACTION TURBINE.—A combination impulse and reaction turbine employs a velocity-compounded impulse

(Curtis) stage at the high pressure end of the turbine, followed by impulse staging and then by reaction blading. The impulse blading effects large pressure and temperature drops in the beginning, with a high initial utilization of thermal energy. The reaction blading is more efficient at the low pressure end of the turbine. Hence the combination impulse and reaction turbine is a highly efficient machine that utilizes the advantages of both impulse and reaction blading. Combination impulse and reaction turbines are very commonly used as propulsion turbines.

Mode of Steam Flow

Turbines may be further classified according to the manner in which steam flows through the turbine. The three aspects of steam flow considered here are (1) the direction of flow, (2) the repetition of flow, and (3) the division of flow.

DIRECTION OF STEAM FLOW.—The direction of steam flow through a turbine may be axial, radial, or helical. In general, the direction of flow is determined by the relative positions of nozzles, diaphragms, moving blades, and fixed blades.

Most turbines are of the axial flow type—that is, the steam flows in a direction approximately parallel to the long axis of the turbine shaft. As we have seen, the blades in an axial-flow turbine project outward from the periphery of the rotor.

In a radial-flow turbine, the blades are mounted on the side of the rotor near the periphery. The steam enters in such a way that it flows radially toward the long axis of the shaft. Radial flow is not used for propulsion turbines, but is used for some auxiliary turbines.

In a helical-flow turbine, the steam enters at a tangent to the periphery of the rotor and impinges upon the moving blades. The blades are shaped in such a way that the direction of steam flow is reversed in each blade. Helical flow is not used for propulsion turbines, but is used for some auxiliary turbines.

REPETITION OF STEAM FLOW.—Turbines are classified as single-entry turbines or re-entry turbines, depending on the number of times the steam enters the blades. If the steam

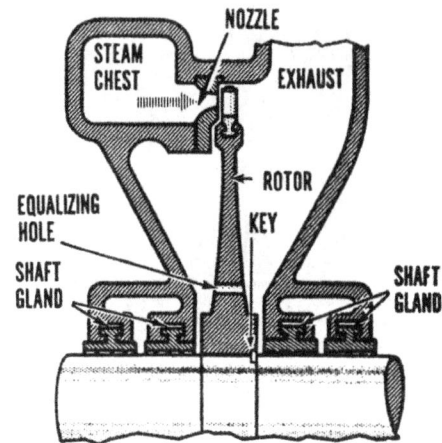

38.78X

Figure 12-11.—Simple impulse turbine (Rateau stage).

passes through the blades only once, the turbine is called a single-entry turbine. All multistage turbines are of the single-entry type.

Re-entry turbines are those in which the steam passes more than once through the blades. Re-entry turbines are used to drive some pumps and forced draft blowers, but are not used as propulsion units.

DIVISION OF STEAM FLOW.—Turbines are classified as single-flow or double-flow, depending upon whether the steam flows in one direction or two. In a single-flow turbine, the steam enters at the inlet or throttle end, flows once through the blading in a more or less axial direction, and emerges at the exhaust end of the turbine. A double-flow turbine consists essentially of two single-flow units mounted on one shaft, in the same casing. The steam enters at the center, between the two units, and flows from the center toward each end of the shaft. The main advantages of the double-flow arrangement are (1) the blades can be shorter than they would have to be in a single-flow turbine of equal capacity, and (2) axial thrust is avoided by having the steam flow in opposite directions. This second point applies primarily to reaction turbines, since impulse turbines develop relatively little axial thrust in any case.

The turbines shown in figures 12-11, 12-12, and 12-13 are single-flow turbines. A double-flow reaction turbine of the type used as the low pressure turbine in some propulsion plants is shown in figure 12-14.

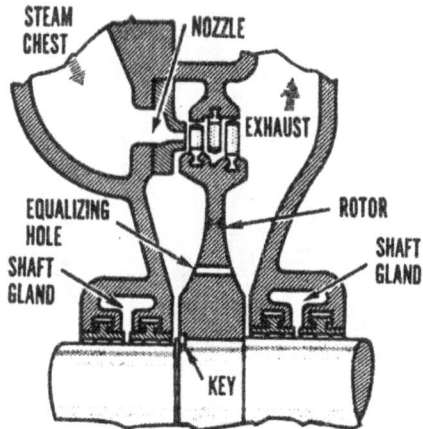

38.79X

Figure 12-12.—Velocity-compounded impulse turbine (one Curtis stage).

TURBINE COMPONENTS AND ACCESSORIES

Propulsion turbine components and accessories include foundations, casings, nozzles (or the equivalent stationary blading), nozzle diaphragms, rotors, blades, bearings, shaft glands, gland seals, oil seal rings, dummy pistons and cylinders (on some reaction turbines), flexible couplings, reduction gears, lubrication systems, and turning gears.

Turbine Foundations

Foundations for propulsion turbines are built up from strength members of the hull so as to provide a rigid supporting base. The after end of the turbine is secured rigidly to the structural foundation. The forward end of the turbine is secured in such a way as to allow a slight freedom of axial movement which allows the turbine to expand and contract slightly with temperature changes.

The freedom of movement at the forward end is accomplished by one of the two methods. Elongated bolt holes or grooved sliding seats may be used to permit the forward end to slide slightly fore-and-aft, as expansions and contraction occur. Or the forward end may be secured to a deep flexible I-beam (fig. 12-15) installed with its longitudinal axis lying athwartship. When the turbine is cold, this I-beam is deflected slightly aft from the vertical position. When the turbine is operating at maximum power, the I-beam is deflected forward. This

arrangement results in minimum stresses in the I-beam over the complete range of turbine expansion. The fixed end of the turbine is aft, so the motion resulting from expansion cannot be transmitted to the reduction gears, where distortion and serious damage would occur if the after end of the turbine were free to move.

Steam lines connected to the turbines are curved, as shown in figure 12-15, to allow for expansion of the steam line and avoid unacceptable strains on turbine casings that could cause distortion or misalignment.

Turbine Casings

Casings for propulsion turbines are divided horizontally to permit access for inspection and repair. Flanged joints on casings are accurately machined to make a steamtight metal-to-metal fit, and the flanges are bolted together. Some high pressure turbine casings are also split vertically to facilitate manufacture, particularly when different alloys are used for the high temperature inlet end and the lower temperature exhaust end. However, these vertical joints are never unbolted and they are usually seal welded.

Each casing has a steam chest to receive the incoming steam and deliver it to the first-stage nozzles or blades. An exhaust chamber receives the steam from the last row of moving blades and delivers it to the exhaust connection. Openings in the casing include drain connections, steam bypass connections, and openings for pressure gages, thermometers, and relief valves.

Nozzles

As previously discussed, the function of a nozzle is to convert the thermal energy of the steam into mechanical kinetic energy. Its secondary function is to direct the steam to the turbine blades. Some turbines have a full arc admission of steam; in this case, the first stage nozzles extend around the entire circle of the first row of blades. Other turbines have partial arc admission; in this case, only a section of the blade circle is covered by the nozzles. In general, the arrangement of nozzles in any turbine depends upon the range of power requirements and upon a number of design factors.

A nozzle is essentially an opening or a passageway for the steam. When we speak of nozzle construction or arrangement, therefore, we are actually concerned with the construction or arrangement of the nozzle blocks in which the openings occur. In most modern turbines, the nozzle blocks are arranged so that the nozzle openings occur in groups, with each group being controlled by a separate nozzle control valve. The quantity of steam delivered to the first stage of the turbine is thus a function of the number of nozzles in use and the pressure differential across the nozzles.

On some auxiliary ships, hand controlled nozzle valves are used in conjunction with a throttle valve to admit steam to the turbine. Any throttling of the inlet steam will reduce efficiency. To avoid throttling losses, all nozzle control valves in use are opened fully before any additional valve is opened. Minor variations in speed within any one nozzle control valve combination are taken care of by the throttle.

On modern combatant ships, the nozzle control valve arrangement shown in figure 12-16 is employed. The throttle valve is omitted and steam enters the turbine through nozzle control valves. Speed control is effected by varying the number of nozzle valves that are opened. The variation in the number of nozzle valves is accomplished through the operation of a lifting beam mechanism. The lifting beam mechanism consists of a steel beam drilled with holes which fit over the nozzle valve stems. The valve stems are of varying lengths and are fitted with shoulders at the upper ends. When the beam is lowered, all valves rest upon their seats. When the beam is raised,

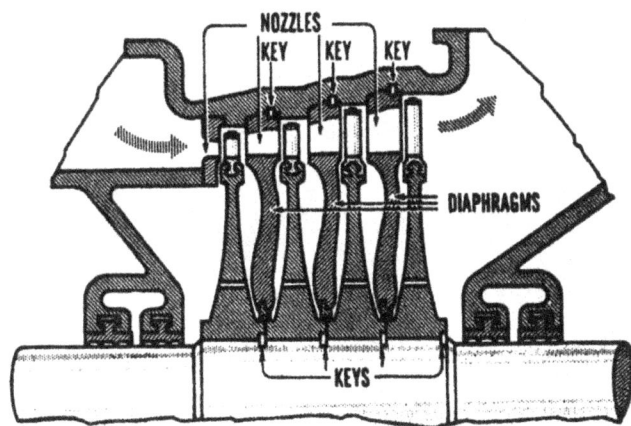

38.80X

Figure 12-13.—Pressure-compounded impulse turbine (Rateau turbine).

the valves open in succession, depending upon their stem length—the shorter ones open first, then the longer ones.

Nozzle Diaphragms

Nozzle diaphragms are installed as part of each stage of a pressure-compounded impulse turbine. The diaphragm serves to hold the nozzles of the stage. Figure 12-17 shows a typical nozzle diaphragm. The nozzle walls are machined, ground, and polished. The nozzles are fitted into a steel plate inner ring. An outer ring fits over the outside of the nozzles. The entire assembly is then welded together. In order to seal against steam leakage, labyrinth packing (discussed later in this chapter) is used between the inner bore of the diaphragm and the rotor.

Turbine Rotors

The turbine rotor carries the moving blades which receive the steam. In some older turbines, the rotors were forged separately, machined, shrunk or pressed onto the shaft, and keyed to the shaft. In most modern turbines, particularly large ones such as those used for ship propulsion, the rotors are forged integrally with the shaft. Figure 12-18 shows an integrally forged turbine rotor to which the blades have not yet been attached.

Turbine Blades

The purpose and function of turbine blading has already been discussed. At this point, it is merely necessary to note that the moving blades are fastened securely and rigidly to the turbine rotor. Figure 12-19 shows several ways of fastening blades to the turbine rotor wheels.

Turbine Bearings

Turbine rotors are supported and kept in position by bearings.[4] The bearings which serve to maintain the correct radial clearance between the rotor and the casing are called radial bearings. Those which serve to limit the axial (longitudinal) movement of the rotor are called thrust bearings.

Propulsion turbines have one radial bearing on each end of the rotor. These bearings are of the type generally known as journal bearings or sleeve bearings. The two metallic surfaces are

---

[4]Bearings are discussed in chapter 5 of this text.

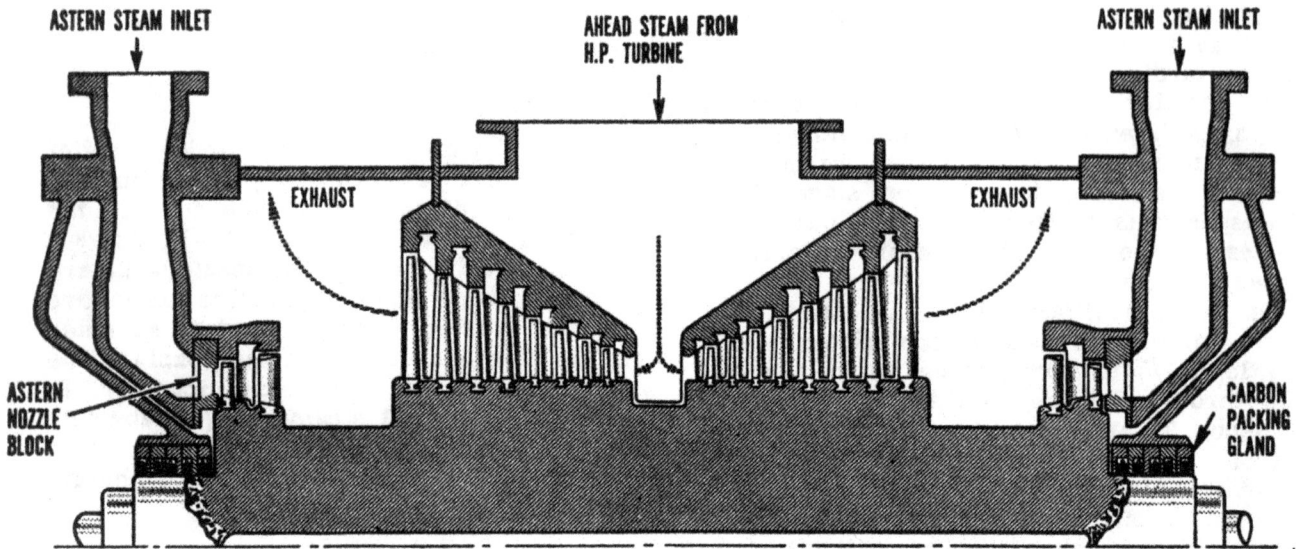

47.8X

Figure 12-14.—Double-flow reaction turbine.

separated only by a fluid film of oil. The effectiveness of oil-film lubrication depends upon a number of factors, including properties of the lubricant (cohesion, adhesion, viscosity, temperature, etc.) and the clearances, alignment, and surface condition of the bearing and the journal. Except for the momentary metal-to-metal contact when the turbine is started, the metallic surfaces of the bearing and the journal are constantly separated by a thin film of oil.[5]

As previously noted, impulse turbines do not, in theory, develop end thrust. In reality, however, a small amount of end thrust is developed which must be absorbed in some way. Kingsbury or pivoted-shoe thrust bearings are usually used on propulsion turbines.

Shaft Glands

Shaft glands are used to minimize steam leakage from the turbine casing (or air leakage into the casing) at the points where the shaft extends through the casing. Two types of packing, carbon packing and labyrinth packing, are used in shaft glands.

Carbon packing is suitable only for relatively low pressures and temperatures. When both types of packing are used in one gland, therefore, as shown in figure 12-20, the labyrinth packing is used at the initial high pressure area and the carbon packing is used at the lower pressure area. Since most modern ships utilize relatively high pressures and temperatures, most modern propulsion turbines are only labyrinth packing.

Labyrinth packing consists of rows of metallic strips or fins. These strips are fastened to the gland liner in such a way as to make a very small clearance between the strips and the shaft. As the steam from the turbine leaks through the small spaces between the packing strips and the shaft, the steam pressure is gradually reduced.

Where carbon packing rings are used, they restrict the passage of steam along the shaft in much the same manner as do the labyrinth packing strips. Carbon packing rings are mounted around the shaft and are held in place by springs. As a rule, three or four carbon rings are used in each gland; each ring is fitted into a separate compartment of the gland housing.

Gland Sealing Systems

On propulsion turbines, the shaft gland packing is not sufficient to entirely stop the

---

[5]Lubricants and the oil-film theory of lubrication are discussed in chapter 6 of this text.

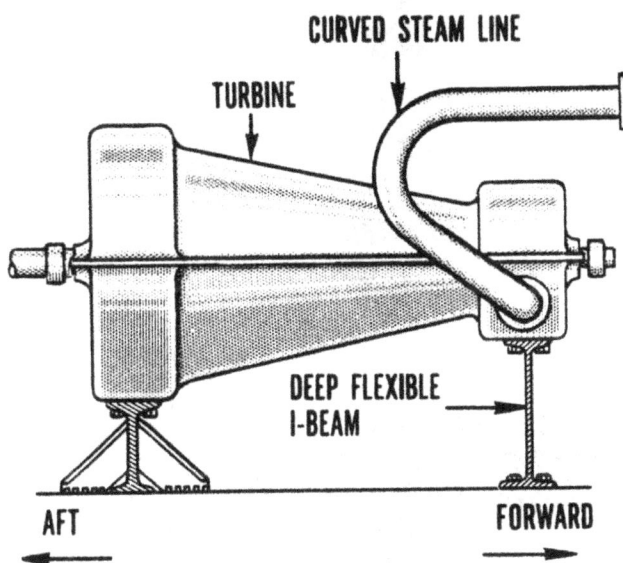

47.12X

Figure 12-15.—Foundation for propulsion turbine.

flow of steam out of the turbine or to entirely prevent the flow of air into the turbine. For this reason, gland sealing steam[6] is brought into the shaft gland in the manner shown in figure 12-20. In this illustration, the gland sealing steam enters a space between the labyrinth packing and the carbon packing. In more recent installations, the sealing steam enters between the segments of the labyrinth packing. The sealing steam enters at a pressure of about 2 psig (17 psia). This pressure is, of course, slightly greater than the atmospheric pressure in the engineroom.

When the pressure of the gland sealing steam is greater than the pressure inside the turbine casing, the sealing steam flows both into the casing and into a line leading to the gland exhaust condenser,[7] excluding all air from the turbine in the process. When a high pressure turbine is operating at high speed, the pressure

of the steam leaking through the shaft gland packing may be slightly higher than the pressure of the gland sealing steam. When this situation prevails, it causes a reversal in the direction of flow of the gland sealing steam. At such times, the gland seal line is closed and the excess steam is led through gland leak-off connections to a later stage of the turbine, to the gland exhaust condenser, or to other glands to be used as gland seal steam. In the illustration (fig. 12-20) the excess steam leaking past the labyrinth packing is being led back into the eighth and twelfth stages of the high pressure turbine.

Dummy Pistons and Cylinders

The steam passing through a multistage impulse turbine does not impart any appreciable axial thrust to the rotor, since the pressure drop actually takes place in the nozzles.[8] In a reaction turbine, however, considerable axial thrust does result from the drop in steam pressure, since a pressure drop occurs in the moving blades as well as in the stationary blades.

In single-flow reaction turbines, this axial thrust is partially counterbalanced by the use of a dummy piston and cylinder arrangement such as that shown in figure 12-21. Space "A" surrounds the inlet area of the turbine rotor and is connected by an equalizing pipe to space "B" which surrounds the outlet area of the rotor. The shoulder on the rotor, shown in figure 12-21, is under full inlet steam pressure, while the corresponding area on the other side of the dummy piston is under exhaust pressure. This difference in pressure causes a thrust toward the high pressure end of the turbine which partially counterbalances the thrust in the opposite direction caused by the pressure drop through the turbine.

Dummy pistons and cylinders are not required in double-flow reaction turbines, since the axial thrust caused by the pressure differential across one-half of the turbine is counterbalanced by the equal and opposite axial thrust in the other half of the turbine.

---

[6]Gland seal and gland exhaust systems are discussed in chapter 9 of this text.

[7]The gland exhaust condenser is discussed in chapter 13 of this text.

[8]An equalizing hole drilled axially through each rotor wheel also helps to minimize thrust in an impulse turbine.

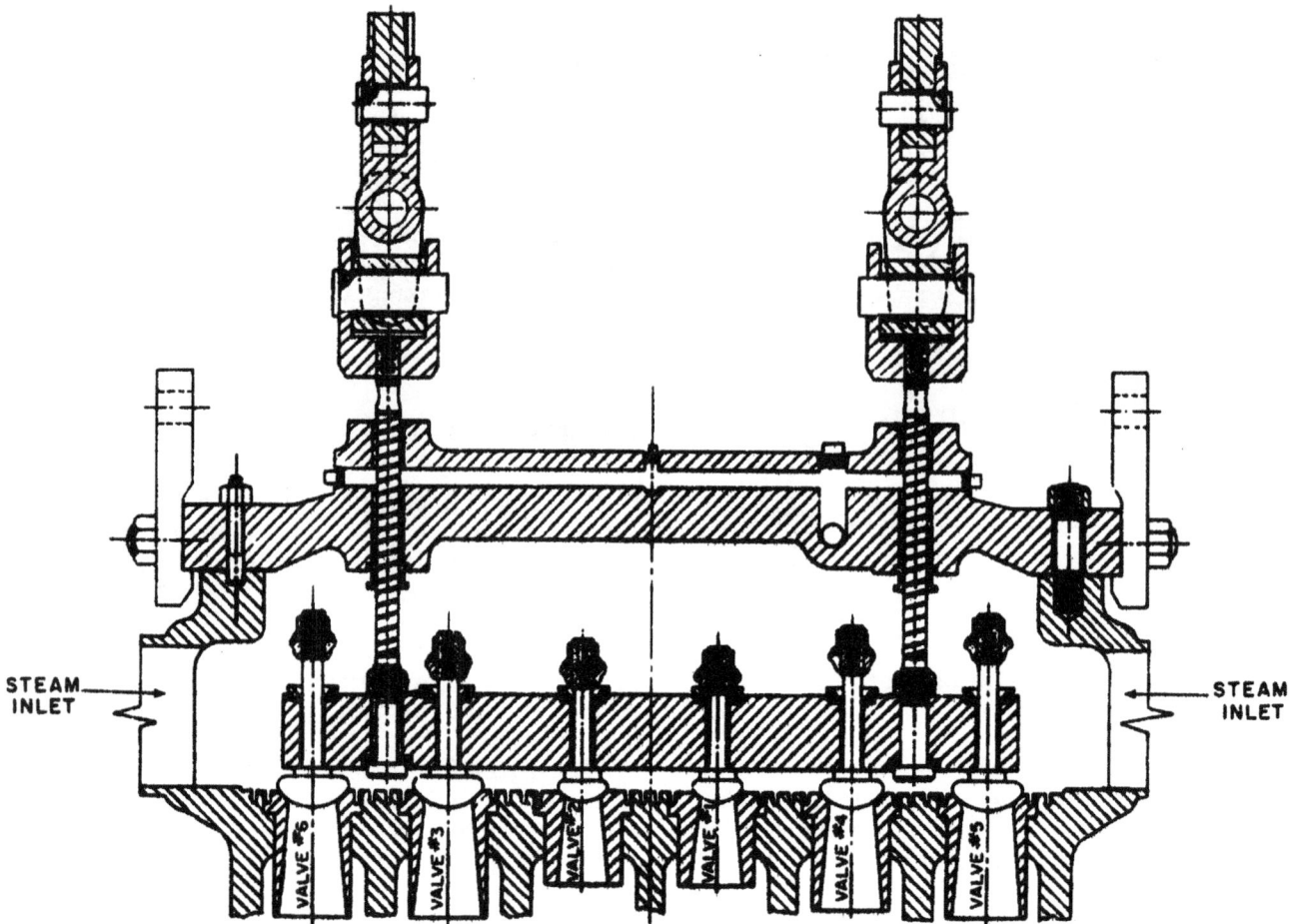

147.96

Figure 12-16.—Arrangement of nozzle control valves.

96.19

Figure 12-17.—Nozzle diaphragm.

Flexible Couplings

Propulsion turbine shafts are connected to the reduction gears by flexible couplings which are designed to take care of very slight misalignment between the two units. Flexible couplings are discussed in chapter 5 of this text.

Reduction Gears

Reduction gears for propulsion turbine installations are described and illustrated in chapter 5 of this text. At this point, it is important merely to note that turbines must operate at relatively high speeds for maximum efficiency, while propellers must operate at lower speeds for maximum efficiency. Reduction gears are used to allow both turbine and propeller to

47.15X

Figure 12-18.—Integrally forged turbine rotor (without blades).

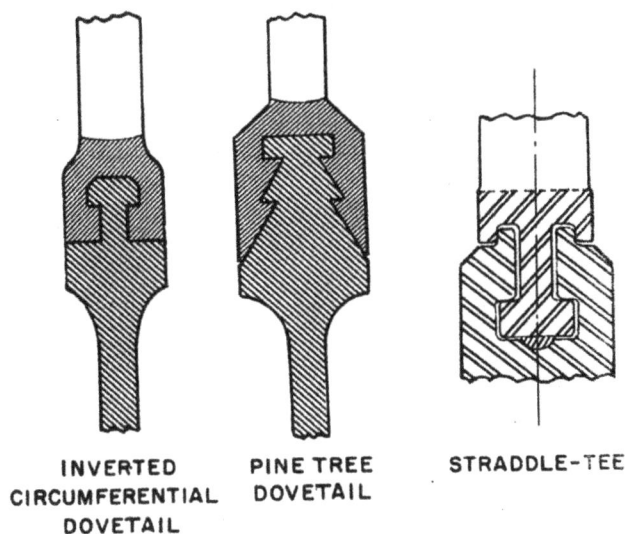

INVERTED
CIRCUMFERENTIAL
DOVETAIL

PINE TREE
DOVETAIL

STRADDLE-TEE

147.97

Figure 12-19.—Methods of fastening blades to turbine rotor wheels.

operate within their most efficient rpm ranges.[9]

Lubrication Systems

Proper lubrication is essential for the operation of any rotating machinery. In particular, the bearings and the reduction gears of turbine installations must be well lubricated at all times.

_____

[9]Reduction gears are not used in ships having turbo-electric drive. In these ships, speed reduction is accomplished electrically.

Main lubricating oil systems are discussed in chapter 9 of this text; the theory of lubrication is discussed in chapter 6.

Turning Gears

All geared turbine installations are equipped with a motor-driven jacking or turning gear. The unit is used for turning the turbine during warming-up and securing periods so that the turbine rotor will heat and cool evenly. The rotor of a hot turbine, or one that is in the process of being warmed up, will become bowed and distorted if left stationary for even a few minutes. The turning gear is also used for turning the turbine in order to bring the reduction gear teeth into view for routine inspection and for making the required daily jacking of the main turbines. The turning gear is mounted on top of and at the after end of the reduction gear casing, as shown in figure 12-22. The brake shown in figure 12-22 is used when it is necessary to lock the shaft after the shaft has been stopped.

STEAM TURBINE PROPULSION PLANTS

The two principal types of steam turbine propulsion plants now in use on naval ships are the geared turbine drive and the turboelectric drive. Direct drive installations, once in common use, are now practically obsolete; however, it is possible that an occasional application for direct drive could again develop in the future.

CLASSIFICATION OF PROPULSION
TURBINE UNITS

Naval propulsion turbines are classified as Class A, Class B, and Class C turbines according to the type ship for which they are designed. Class A turbines are designed for use in submarines. Class B turbines are designed for use in amphibious warfare ships, surface combatant ships, mine warfare ships, and patrol ships. Class C turbines are designed for use in auxiliary ships.

Naval propulsion turbines are also classified according to design features. The six major types are:

1. Type I (single-casing unit).—The Type I propulsion unit consists of one or more ahead elements, each contained in a separate casing and identified as a single-casing turbine. Each

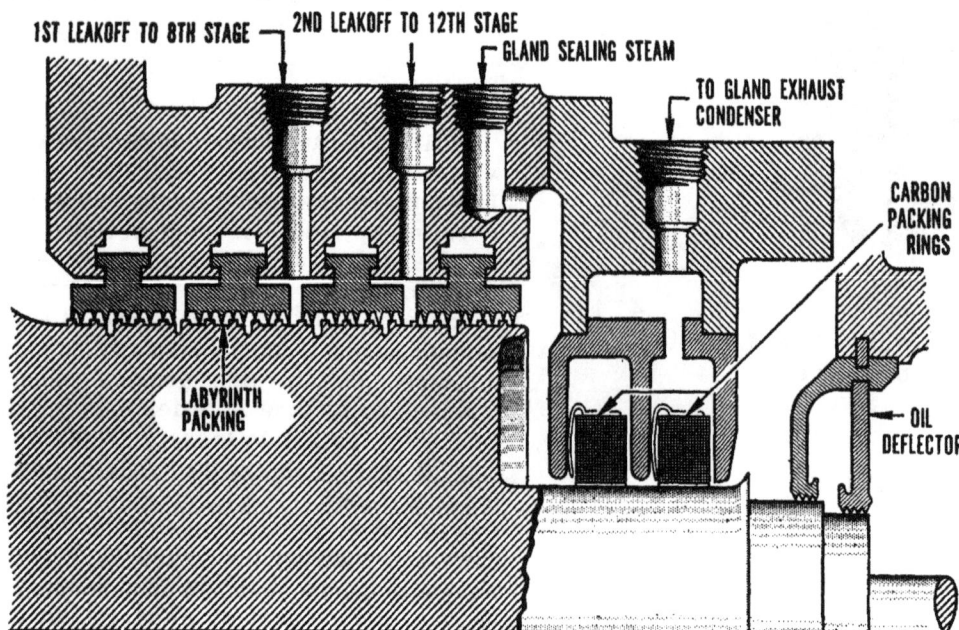

47.19X

Figure 12-20.—Turbine gland.

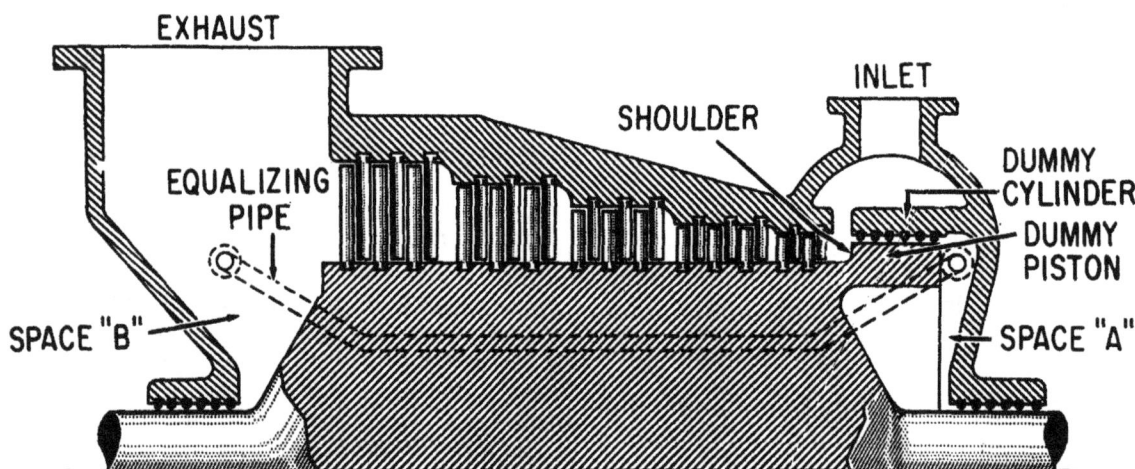

47.5X

Figure 12-21.—Dummy piston and cylinder arrangement.

turbine delivers approximately equal power to a reduction gear.

2. Type II-A (straight-through unit).—The Type II-A propulsion unit is a two element straight-through unit, and consists of two ahead elements, known as a high pressure (HP) element and a low pressure (LP) element. The HP and LP elements are contained in a separate casing and are commonly known as the HP and LP turbines, respectively. The HP and LP turbines deliver power to a single shaft through a gear train and are coupled separately to the reduction gear. Steam is admitted to the HP turbine and flows straight through the turbine

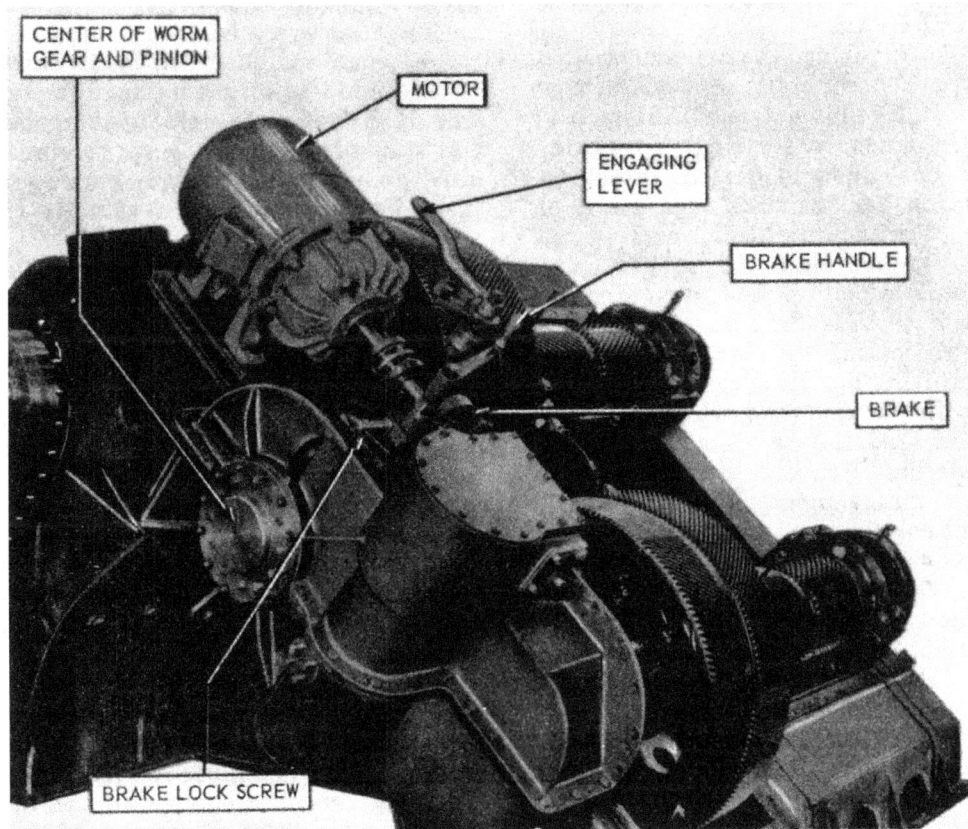

47.35

Figure 12-22.—After end of main reduction gear, showing
turning gear and propeller locking mechanism.

axially without bypassing any stages (there is partial bypassing of the first row of blades at high power), and then is exhausted to the LP turbine through a crossover pipe.

3. Type II-B (EXTERNAL bypass unit).—The Type II-B propulsion unit is similar to the Type II-A, except that provision is made for bypassing of steam around the first stage or first several stages of the HP turbine at powers above the most economical point of operation. Bypass valves are located in the HP turbine steam chest, with the nozzle control valves.

4. Type II-C (INTERNAL bypass unit).—The Type II-C is similar to the Type II-A, except that provision is made for bypassing steam from the first-stage shell around the next several (one or more) stages of the HP turbine at powers above the most economical point of operation. Bypass valves and steam connections are usually integral with the HP turbine casing; however,

some installations have the valves separate, but bolted directly to the casing, with suitable connecting piping between the first-stage shell and valve to the bypass belt.

5. Type III (series-parallel unit).—The Type III propulsion unit consists of three ahead elements, known as the HP element, intermediate pressure (IP) element, and LP element. The HP and IP elements are combined in a single casing, and known as the HP-IP turbine. Steam is admitted to the HP-IP turbine and exhausted to the LP turbine through a crossover pipe. For powers up to the most economical point of operation, only the HP element receives inlet steam, with the IP element being supplied in series with steam from the HP element exhaust. At powers above this point of operation, both elements receive inlet steam in a manner similar to that in a double-flow turbine. During ahead operation no ahead blading is bypassed.

335

Series-parallel units are being used on some of the more recent naval combatant vessels, such as DEs, DDs, and CVAs.

6. Type IV (cruising geared and vented unit).—The Type IV propulsion unit consists of a crusing element, HP element, and LP element— each contained in a separate casing. The cruising turbine is connected in tandem through a cruising reduction gear to the forward end of the HP turbine. The cruising turbine contributes power to the propeller shaft for powers up to the most economical point of operation, and for higher powers it is idled in a partial vacuum and supplied with cooling steam to prevent overheating. For cruising power, steam is admitted to the cruising turbine and then exhausted to the HP turbine inlet, and thence from the HP turbine exhaust into the LP turbine through a crossover pipe. The arrangement of the HP and LP turbine is identical to the Type II units. It is possible to disconnect the cruising turbine to allow for repair. Once disconnected, the HP turbine may be placed in service.

All of these six types of propelling units contain an astern element for backing or reversing. An astern element is located in each end of a double flow LP turbine casing or can be in either end of each single-casing turbine or single flow LP turbine.

Figure 12-23 illustrates the flow of steam in a Type IV propulsion unit when the cruising turbine is in use; figure 12-24 illustrates the flow of steam at higher rates of operation.

Astern elements in noncombatant ships are usually velocity-compounded impulse stages (Curtis stages) mounted in the exhaust end of the ahead turbine. Astern elements in combatant ships are velocity-compounded (Curtis) stages installed at each end of the low pressure turbine. Each astern element has its own steam inlet but the admission of steam to both elements is controlled by one astern throttle. The astern elements exhaust through the low pressure turbine exhaust chamber to the condenser. Figure 12-25 illustrates the flow of steam for astern operation in a Type IV propulsion unit.

Figure 12-26 illustrates a typical high pressure turbine. This turbine has one velocity-compounded impulse stage followed by eleven pressure-compounded impulse stages; hence it is a pressure-velocity-compounded impulse turbine.

A low pressure double-flow turbine is shown in figure 12-27. Note the astern elements. This particular low pressure turbine is a straight reaction turbine. In some ships, double-flow pressure-compounded impulse turbines are used as low pressure turbines.

A typical cruising turbine is shown in figure 12-28. This is an eight-stage impulse turbine. The first stage is a velocity-compounded (Curtis) stage; the remaining seven stages are pressure-compounded (Rateau) states. The turbine is therefore a pressure-velocity-compounded impulse turbine.

Unlike the geared turbine propulsion plants, the turboelectric drive installations have a single turbine unit for each shaft. Figure 12-29 shows the general arrangement of a turboelectric propulsion unit. As may be seen, the plant includes a turbine, a main generator, a propulsion motor, a direct-current generator for supplying excitation current to the generator and the propulsion motor, and a propulsion control board.

Although the speed reduction is brought about electrically, rather than by the use of reduction gears, the speed reduction ratio between turbine and propeller in the turboelectric drive is approximately the same as it is in the geared turbine drive.

One of the outstanding differences between the geared turbine drive and the turboelectric drive is that the turboelectric drive does not have an astern element. In the turboelectric drive, the direction of rotation of the propulsion motor controls the direction of rotation of the propeller. Hence there is no need to reverse turbine rotation for astern operation.

PLANT OPERATION

Operating a ship's propulsion plant requires sound administrative procedures and the cooperation of all engineering departmental personnel. The reliability and the economical operation of the plant is vital to the ship's operational readiness.

A ship must be capable of performing any duty for which it was designed. A ship is considered reliable when it meets all scheduled operations and is in a position to accept unscheduled tasks. In order to do this, the ship's machinery must be kept in good condition so that the various units will operate as designed.

In order to obtain economy, the engineering plant, while meeting prescribed requirements, must be operated so as to use a minimum amount of fuel. The fuel performance ratios are good overall indications of the condition of the engineering plant and the efficiency of

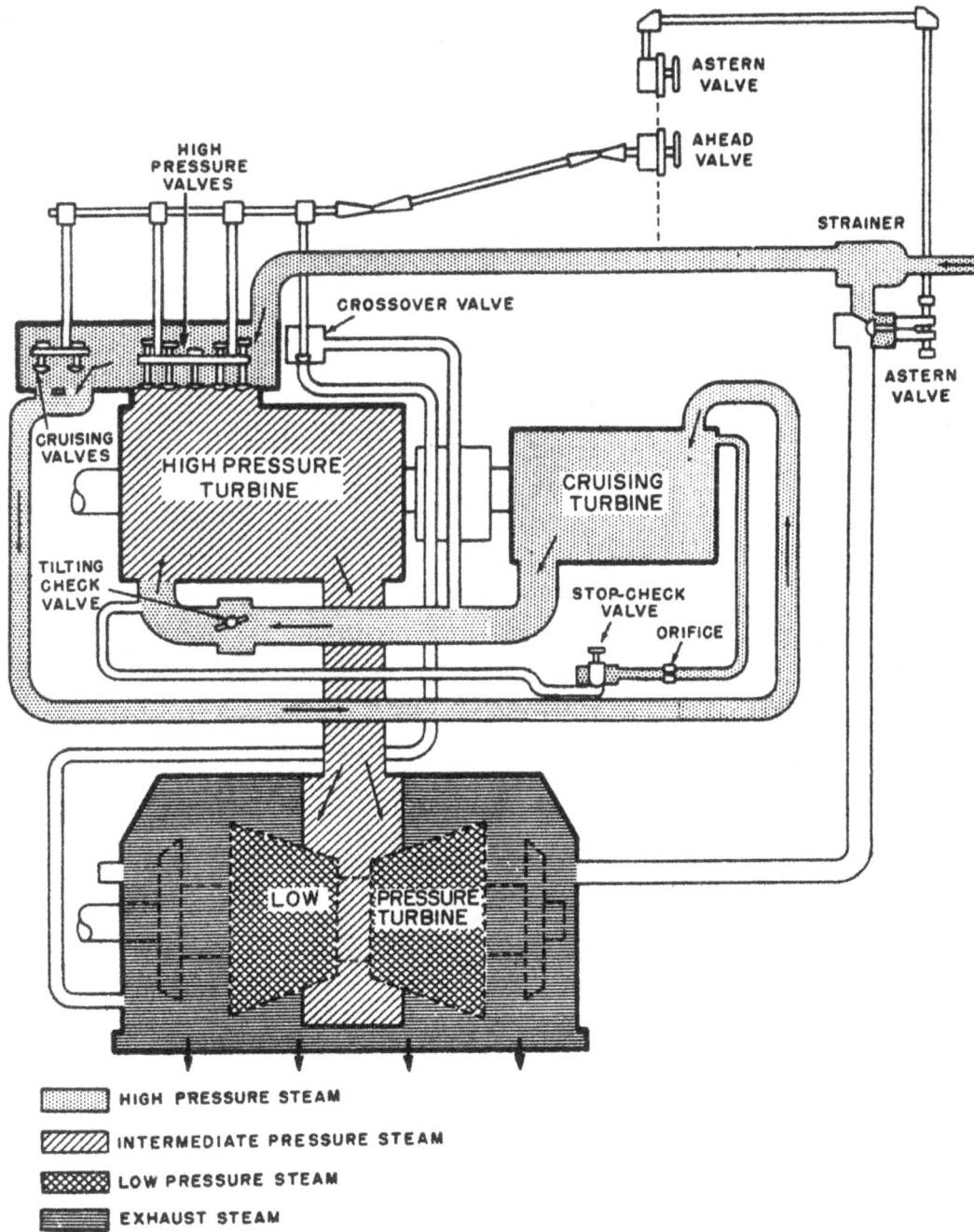

147.98

Figure 12-23.—Flow of steam in Type IV propulsion
unit (cruising turbine in use).

the operating personnel. The fuel performance ratio is the ratio of the amount of fuel oil used as compared to the amount of fuel oil allowed for a certain speed or steaming condition. The fuel performance ratio, which is reported on the Monthly Summary, is a general indication of the ship's readiness to operate economically and within established standards. In determining the economy of a ship's engineering plant, the same consideration is given to

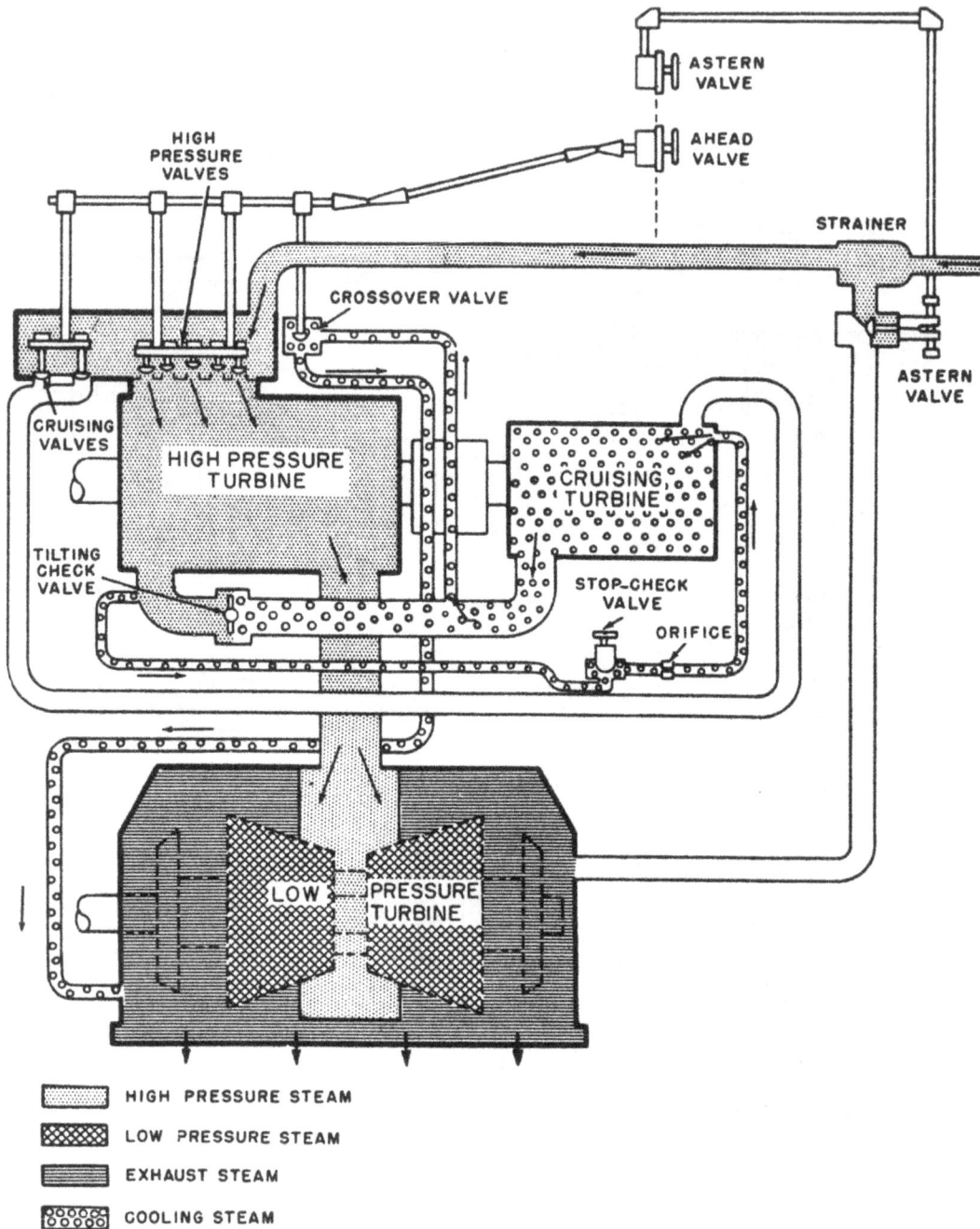

147.99

Figure 12-24.—Flow of steam in Type IV propulsion
unit (cruising turbine not in use).

the amount of water used on board ship. Water consumption is computed in (1) gallons of make-up feed per mile under way, (2) gallons of make-up feed per hour at anchor, and (3) gallons of potable water per man per day.

The increase or decrease in a ship's fuel economy depends largely on the operation of each unit of machinery; economical operation further depends on personnel understanding the function of each unit and knowing how units are

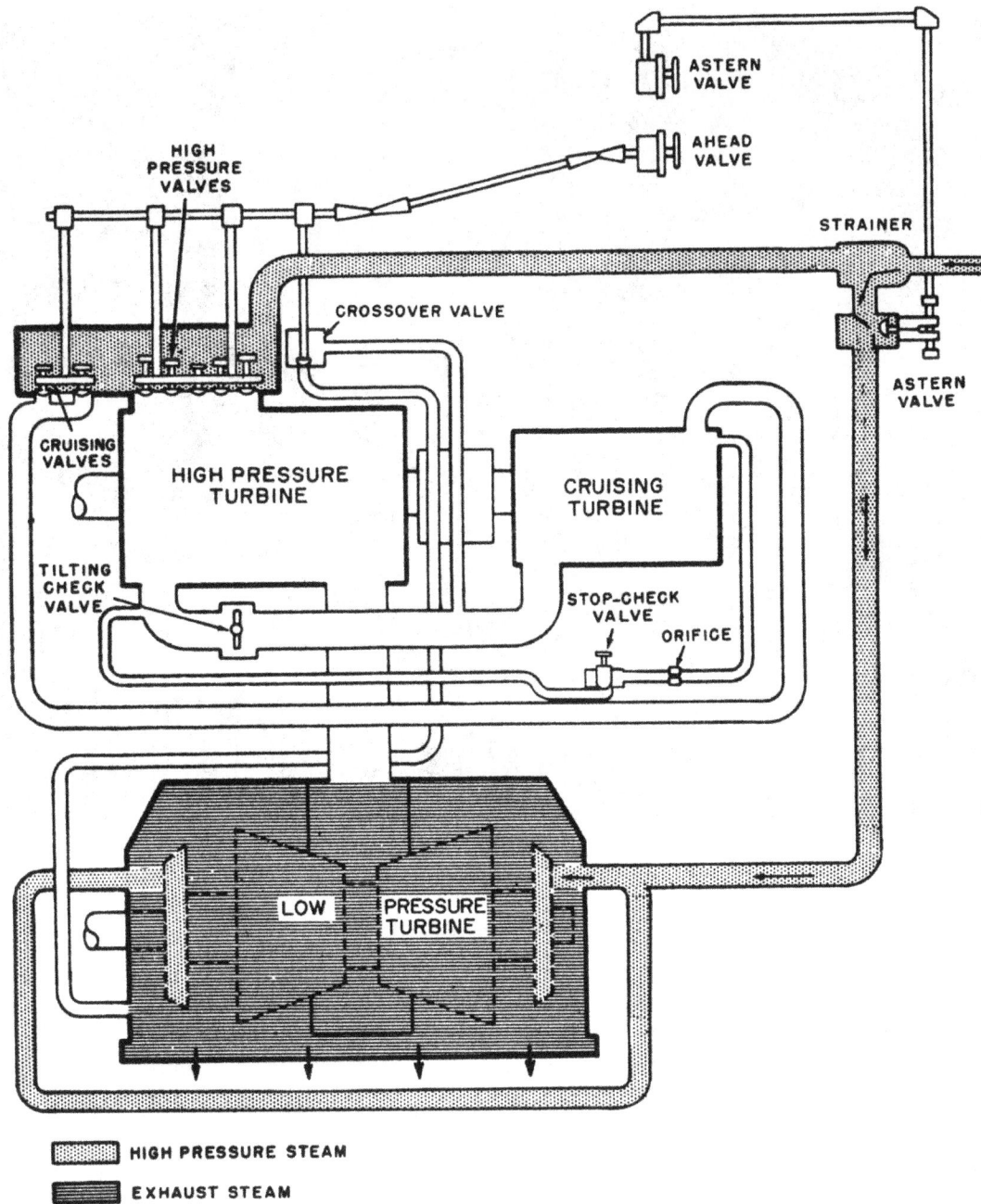

147.100

Figure 12-25.—Flow of steam in Type IV propulsion
unit (astern operation).

used in combination with other units and with
the plant as a whole.

Good engineering practices and safe operation of the plant should never be violated in
the interest of economy—furthermore, factors
affecting the health and comfort of the crew
should meet the standards set by the Navy.

Indoctrination of the ship's crew in methods
of conserving water is of the utmost importance,
and should be given constant consideration.

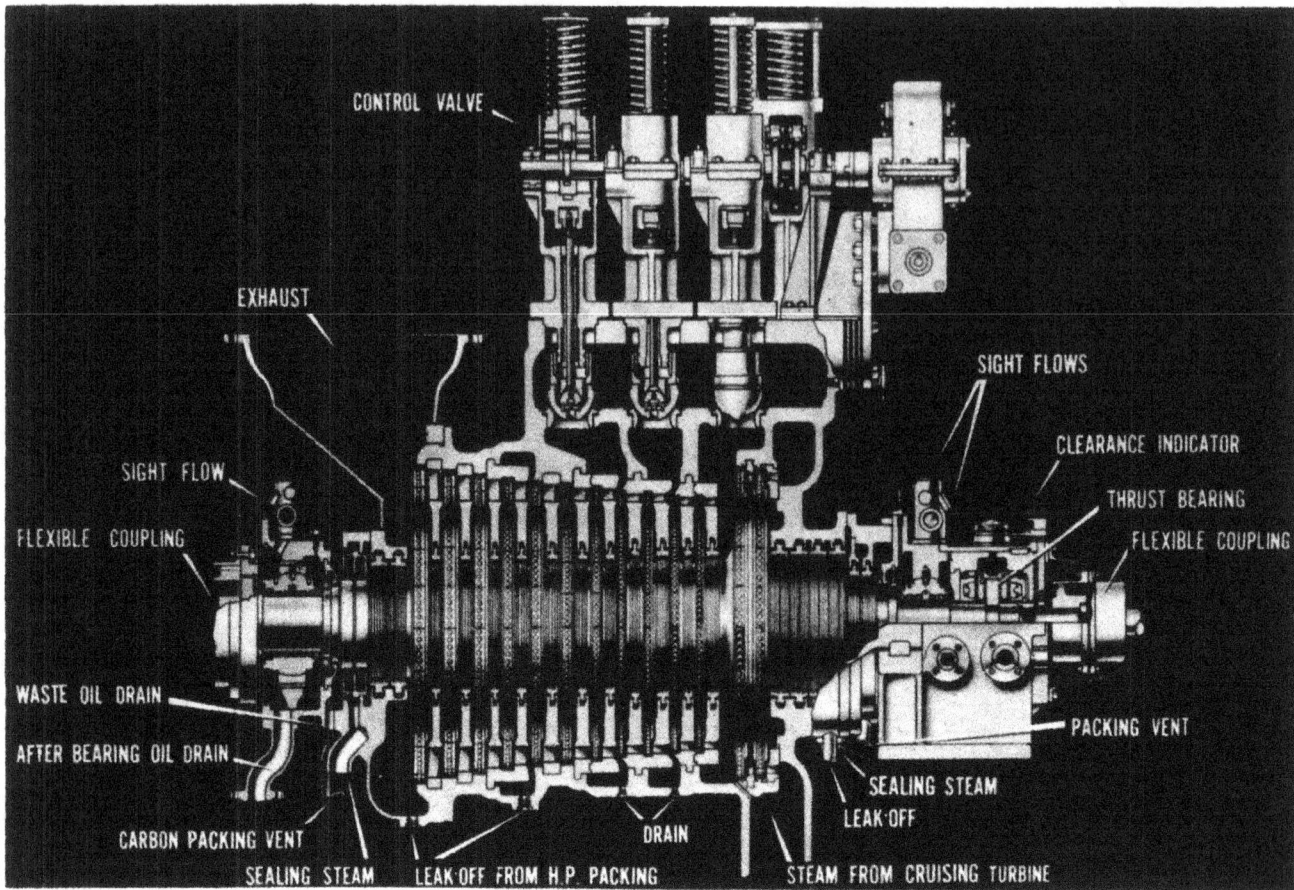

96.10

Figure 12-26.—High pressure turbine.

Aboard naval ships, economy measures cannot be carried to extremes, because there are several safety factors that must be considered. Unless proper safety precautions are taken, reliability may be sacrificed; and in the operation of naval ships, reliability is one of the more important factors. In operating an engineering plant as economically as possible, safety factors and good engineering practice must not be overlooked.

There are several factors that, if given proper consideration, will promote efficient and economical operation of the engineering plant. Some of these factors are: (1) maintaining the designed steam pressure, (2) proper acceleration of the main engines, (3) maintaining high condenser vacuum, (4) guarding against excessive recirculation of condensate, (5) maintenance of proper insulation and lagging, (6) keeping the consumption of feed water and potable water

within reasonable limits, (7) conserving electrical power, (8) using the correct number of boilers for best efficiency at the required load levels, and (9) maintaining minimum excess combustion air to the boilers.

Maintaining a constant steam pressure is important to the overall efficiency of the engineering plant. Wide or frequent fluctuations in the steam pressure or degree of superheat above or below that for which the machinery is designed will result in a considerable loss of economy. Excessively high temperatures will result in severe damage to superheaters, piping, and machinery.

Proper acceleration and deceleration of the main engines are important factors in the economical operation of the engineering plant. A fast acceleration will not only interfere with the safe operation of the boilers but will also result in a large waste of fuel oil. The officer

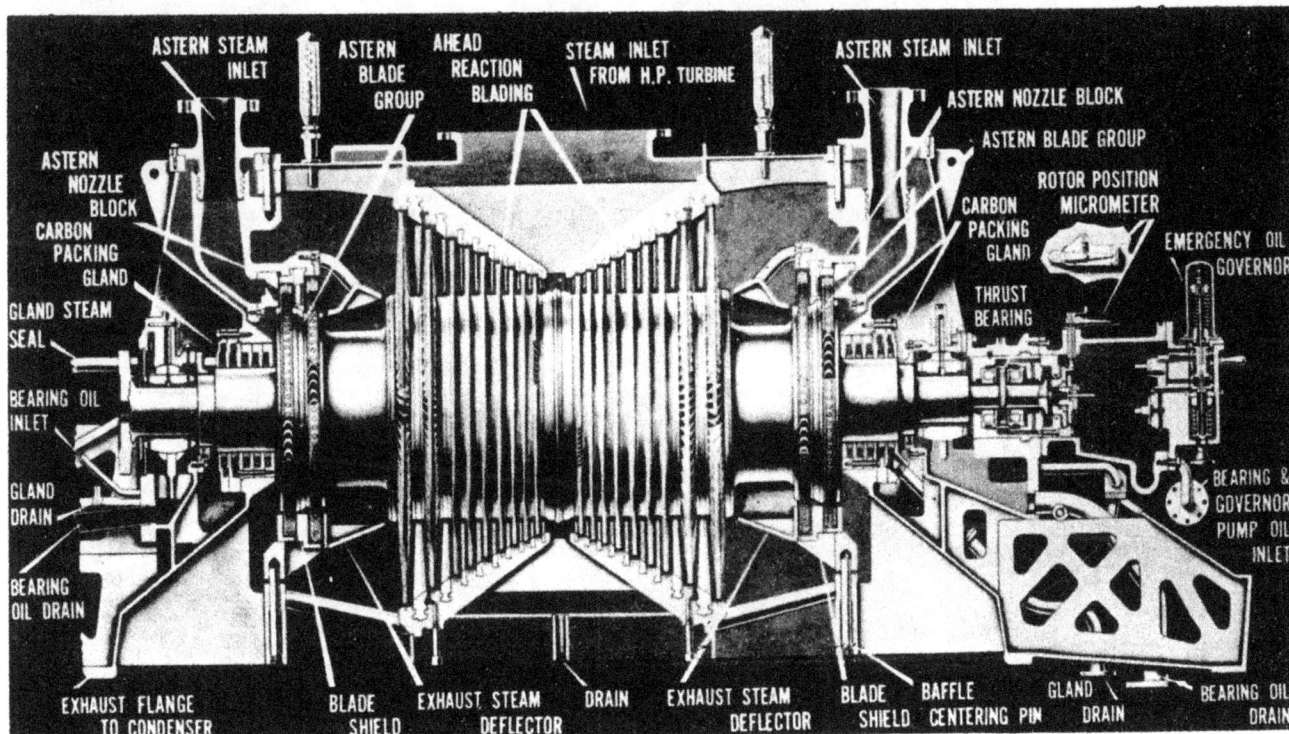

147.101

Figure 12-27.—Low pressure turbine
(with astern elements).

or CPO in charge of an engineroom watch, can contribute a great deal to the economical and safe operation of the boilers if they use the acceleration and deceleration charts provided for that particular propulsion plant.

Acceleration and deceleration charts are posted at each main engine throttle board. These charts give the exact amount of time that the throttleman should use in changing speed. When a speed change is ordered, the throttleman can tell instantly, by checking the chart, the minutes and seconds necessary for him to accelerate or decelerate to the new speed. Main engine control has tachometers indicating the number of rpm each shaft is doing. By means of the tachometers, the engineering officer of the watch can coordinate the rpm of the shafts; if one throttleman accelerates or decelerates too rapidly or too slowly, the engineering officer of the watch can detect the trouble and have it corrected.

Improper acceleration or deceleration wastes fuel, leads to uneconomical operation of the propulsion plant, and may cause operational problems in the fireroom. Each throttleman should have a revolution - pressure table which gives the approximate pressure required in the first stage of the high pressure turbine to develop a certain rpm. By using such a table, together with the acceleration and deceleration charts, the throttleman can make his watchstanding much easier and, at the same time, contribute to economical and efficient operation of the plant. There must always be complete understanding between the engineroom and the bridge as to how many rpm are to be maintained for one-third, two-thirds, standard, and full speed. The throttleman should never relieve the watch without knowing the rpm for these speeds.

For most efficient turbine operation, the highest possible vacuum must be maintained in the condenser. Air must not be allowed to leak into the condenser, exhaust trunks, throttles, lines to air ejectors, gage lines, idle condensate pump packing, makeup feed lines, or any other part of the system under vacuum. A steam pressure of 1/2 to 2 psi must be

147.102

Figure 12-28.—Cruising turbine.

maintained on the glands when the turbine is in operation, and the gland packing must be kept in good condition. An adequate supply of water must be maintained for the makeup feed tank so that air will not be drawn into the condenser.

A combatant ship operates most of the time at speeds far below maximum. At cruising speeds, only a fraction of turbine capacity is required. At low speeds, economy is obtained by one of the following methods: (1) by using cruising turbines which are designed to operate economically at speeds up to about 18 knots, (2) by using cruising stages in the high pressure turbine, and (3) by using turbines which are designed so that they can be operated in series.

On ships that have a cruising turbine, the cruising combination should be used for all underway operations requiring speeds of less than 18 knots. The officer of the watch (or the petty officer of the watch) should obtain permission from the OOD to operate on cruising combination whenever possible.

To prevent casualties to cruising turbines, the protective devices (sentinel valve, direct-reading thermometer, crossover valve lock, and thermal alarm) should be checked continuously.

In order to be a good engineering officer of the watch, he must acquaint himself with all standing orders and operating instructions for his ship. These are made up for each ship and

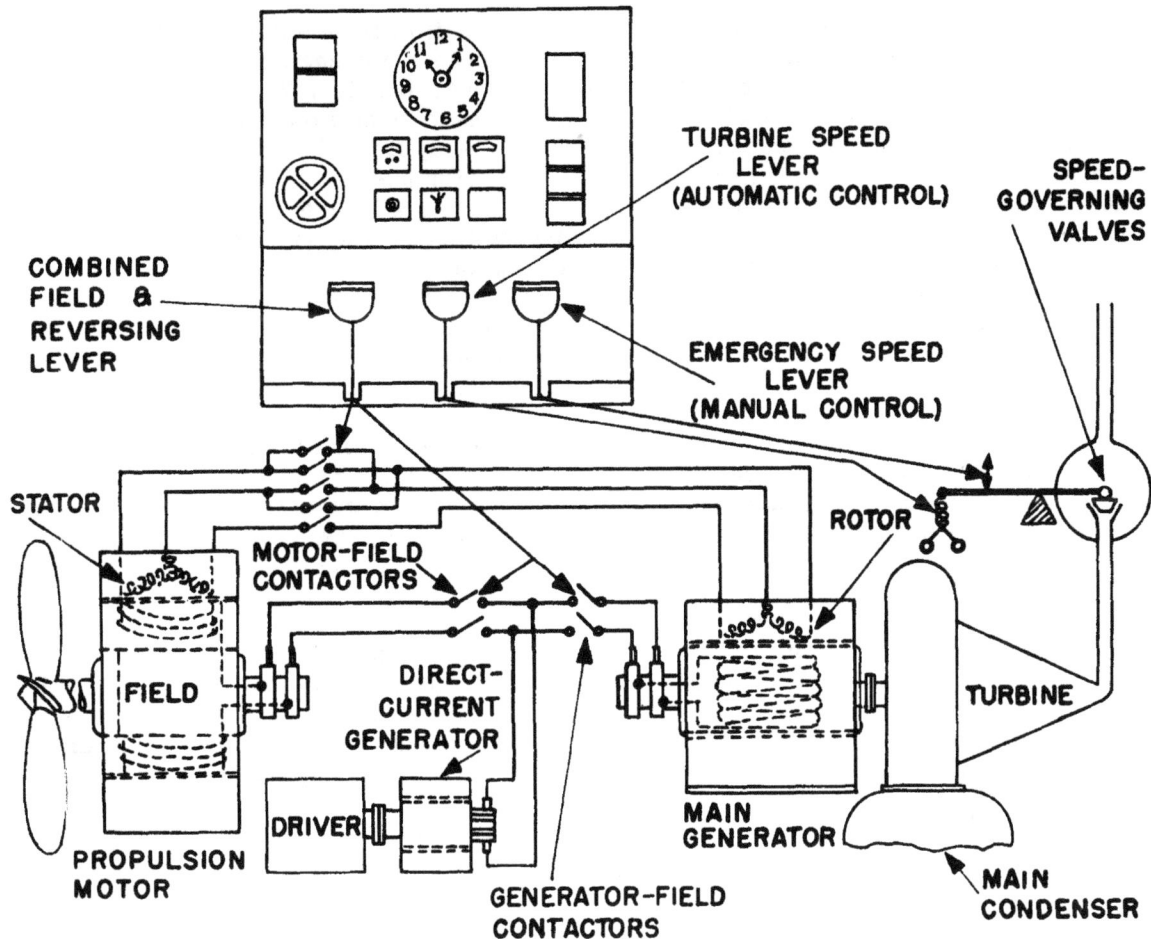

47.2

Figure 12-29.—Diagram of turboelectric drive installation.

show the various plant arrangements (split plant, cross-connected steaming, cruising arrangement, etc.) for the different speeds. Each watchstander must read and understand the steaming orders and any additional orders issued by the engineer officer. At this point the engineering officer of the watch will request permission to light fires under the superheater if two-furnace single-uptake superheater control boilers is installed.

On large combatant ships, there is usually sufficient steam flow (even when steaming for auxiliary purposes) to maintain fires under the superheater side. However, in most installations, and particularly in destroyers, it is usually necessary to be underway and making about 12 knots before the fires can be lighted

under the superheater side of the boiler. When the superheater is operating and the steam flow drops below a safe minimum, the superheater fires must be secured immediately. On destroyers the superheater fires are usually secured when the speed of the ship drops below 10 knots.

From the standpoint of maintenance and repairs to the steam piping, turbine casings, and superheater handhole plates, it is not feasible to put superheaters into operation until it is expected that the ship's speed will be more than 10 knots for a considerable period of time. Furthermore, continually lighting off and securing the superheater fires will cause extensive steam leaks throughout the system subjected to fast changing temperature conditions. These steam leaks will waste more fuel than could be saved by a few minutes of superheat operation.

The no-control, integral superheater boiler creates a different types of problem. The superheater tubes must be protected from the heat of the furnace in the interval during which fires are lighted but the rate of steam generation is still insufficient to ensure a safe flow through the superheater. During operation, there is no problem since all steam passes through the superheater and leaves the boiler at superheat temperature. After the boiler is on the line and furnishing steam, there will be sufficient flow because all steam passes through the superheater.

It is sometimes necessary to light off and put additional boilers on the line, when a ship is underway. With no-control superheat boilers, the steps are much the same as for putting the first boiler or boilers on the line. With superheat control boilers, additional precautions must be taken.

When the steam lines are carrying superheated steam, it would be dangerous to admit saturated steam to the lines. It is not usually possible to establish enough steam flow to light off the superheaters of the incoming boilers, until they are on the line. It is permissible to bring in the incoming boilers, without their superheaters in operation, if the superheater outlet temerature of the steaming boilers is lowered to 600° F. Lowering of the superheat temperature on the steaming boilers should be started in time so that the cutting-in temperature can be reached before the incoming boilers are up to operating pressure. Except in an emergency, the temperature of the superheaters should NOT be lowered or raised at a faster rate than 50° F every 5 minutes.

A number of other items must be checked or inspected at frequent intervals when a ship is underway. Engineeroom personnel must be constantly alert for abnormal pressures, temperatures, sound and vibrations.

The first indication of bearing trouble is usually a rise in temperature. There is no objection to a bearing running warm as long as the temperature is not high enough to cause damage to the bearing. Any RAPID rise in temperature, or any increase over the normal operating temperature, is probably a sign of trouble. The first things to check are the quantity of lube oil and the quality of lube oil. If possible, the amount of oil going to the overheated bearing should be increased and the flow of cooling water through the lube oil cooler should

be increased. If these measures do not reduce the bearing temperature, the unit must be stopped or slowed.

A sight-flow indicator is fitted in the lube oil line of each main engine bearing and each reduction gear bearing. When the plant is in operation, each sight-flow indicator should always show a steady flow of lube oil.

The rotor position indicator for each turbine must be checked every hour and the reading must be logged. Any abnormal reading must be investigated at once.

Once of the first indications of engineroom trouble is an abnormal reading on a thermometer or pressure gage. All gages should be checked frequently.

The oil level in the main engine sump must be checked every hour and logged in the main engine operating record. In addition, other checks should be made in between the required hourly checks. A rise in the oil level may mean that water is entering the lube oil system or that the system is gaining oil in an abnormal manner. A drop in the oil level of the main engine sump may indicate a leak in the lube oil system or incorrect operation of the lube oil purifier.

The water level in each operating deaerating feed tank should be kept between the minimum and the maximum allowable levels. If the water level goes above the maximum, the tank no longer deaerates the water. If the water level is below the minimum, a sudden demand for feed water may empty the deaerating feed tank and cause cumulative casualties to the feed booster pump, the main feed pump, and the boilers.

A salinity indicator is located at or near the throttle board in each engineroom so that engineroom personnel can detect the entrance of salt water into the condensate system. The salinity indicator must be checked constantly. Even a very small amount of salt in the condensate system will very rapidly contaminate a steaming boiler. Any abnormal reading of the salinity indicator must be investigated immediately and the source of contamination must be found and corrected.

PLANT MAINTENANCE

The maintenance of maximum operational reliability and efficiency of steam propulsion plants requires a carefully planned and executed program of inspections and preventive maintenance, in addition to strict adherence to prescribed operating instructions and safety precautions. If proper maintenance procedures are followed, abnormal conditions may be prevented.

Preventive inspection and maintenance are vital to successful casualty control, since these activities minimize the occurrence of casualties by material failures. Continuous and detailed inspection procedures are necessary not only to discover partly damaged parts which may fail at a critical time, but also to eliminate the underlying conditions which lead to early failure (maladjustment, improper lubrication, corrosion, erosion, and other enemies of machinery reliability). Particular and continuous attention must be paid to the following symptoms of malfunctioning:

1. Unusual noises.
2. Vibrations.
3. Abnormal temperatures.
4. Abnormal pressures.
5. Abnormal operating speeds.

Operating personnel should thoroughly familiarize themselves with the specific temperatures, pressures, and operating speeds of equipment required for normal operation, in order that departures from normal operation will be more readily apparent.

If a gage, or other instrument for recording operation conditions of machinery, gives an abnormal reading, the cause must be fully investigated. The installation of a spare instrument, or a calibration test, will quickly indicate whether the abnormal reading is due to instrument error. Any other cause msut be traced to its source.

Because of the safety factor commonly incorporated in pumps and similar equipment, considerable loss of capacity can occur before any external evidence is readily apparent. Changes in the operating speeds from normal for the existing load in the case of pressure-governor-controlled equipment should be viewed with suspicion. Variations from normal pressures, lubricating oil temperatures, and system pressures are indicative of either inefficient operation or poor condition of machinery.

In cases where a material failure occurs in any unit, a prompt inspection should be made of all similar units to determine if there is any danger that a similar failure might occur. Prompt inspection may eliminate a wave of repeated casualties.

Abnormal wear, fatigue, erosion, or corrosion of a particular part may be indicative of a failure to operate the equipment within its designed limits or loading, velocity and lubrication, or it may indicate a design or material deficiency. Unless corrective action can be taken which will ensure that such failures will not occur, special inspections to detect damage should be undertaken as a routine matter.

Strict attention must be paid to the proper lubrication of all equipment, and this includes frequent inspection and sampling to determine that the correct quantity of the proper lubricant is in the unit. It is good practice to make a daily check of samples of lubricating oil in all auxiliaries. Such samples should be allowed to stand long enough for any water to settle. Where auxiliaries have been idle for several hours, particularly overnight, a sufficient sample to remove all settled water should be drained from the lowest part of the oil sump. Replenishment with fresh oil to the normal level should be included in this routine.

The presence of salt water in the oil can be detected by drawing off the settled water by means of a pipette and by running a standard chloride test. A sample of sufficient size for test purposes can be obtained by adding distilled water to the oil sample, shaking vigorously, and then allowing the water to settle before draining off the test sample. Because of its corrosive effects, salt water in the lubricating oil is far more dangerous to a unit than is an equal quantity of fresh water. Salt water is particularly harmful to units containing oil-lubricated ball bearings.

An an example, the maintenance requirements which shall be conducted in accordance with the 3-M System is shown in figure 12-30, (Maintenance Index Page).

CASUALTY CONTROL

The mission of engineering casualty control is to maintain all engineering services in a state of maximum reliability, under all conditions. To carry out this mission, it is necessary for the personnel concerned to know the action necessary to prevent, minimize, and correct the effects of operational and battle casualties on the machinery and the electrical and piping installations of their ship. The prime objective of casualty control is to maintain a ship as a whole in such a condition that it will function effectively as a fighting unit. This requires effective maintenance of propulsion machinery, electrical systems, interior and exterior communications, fire control, electronic services, ship control, firemain supply, and miscellaneous

System, Subsystem, or Component					Reference Publications				
High-Pressure, Low-Pressure and Cruising Turbines									

	Bureau Card Control No.				Maintenance Requirement	M.R. No.	Rate Req'd.	Man Hours	Related Maintenance
MB	ZZZFTR6	B5	3037	D	1. Circulate oil through lube oil system. 2. Jack over idle turbines and reduction gears.	D-1	MM3	0.5	None
MB	ZZZFVA0	55	9752	M	1. Lubricate and operate all valve operating linkage.	M-1	MM3	0.5	None
MB	ZZZFTR6	A4	5409	Q	1. Take depth micrometer readings on the journal bearings of the main propulsion turbines.	Q-1	MM1	0.5	None
MB	ZZZFTR6	25	7724	Q	1. Lift the turbine sentinel relief valves by hand.	Q-2	MM3	0.2	None
MB	ZZZFTR6	65	A197	S	1. Measure thrust clearances of main propulsion turbines.	S-1	MM1 MM2 FN	8.0 8.0 8.0	None
MB	ZZZFTR5	94	5415	A	1. Inspect interior of turbine casings.	A-1	MMC MM2 FN	1.0 1.0 1.0	D-1
MB	ZZZFTR6	A4	5510	C	1. Clean, inspect, and preserve exterior of turbine casing.	C-1	MM2 FN	1.0 1.0	None
MB	ZZZFTR6	49	4052	C	1. Sound and tighten foundation bolts.	C-2	MM3	0.5	None
MB	ZXVFVA1	65	7727	C	1. Remove and test turbine sentinel relief valves.	C-3	MM3	2.5	None
MB	ZZZFTR6	65	A198	C	1. Inspect turbine thrust bearings.	C-4	MMC MM2 FN	4.0 8.0 8.0	S-1
MB	ZZZFTR6	65	A199	C	1. Inspect main propulsion bearings, journals, and oil deflectors. Measure clearances.	C-5	MMC MM2 FN	5.0 10.0 10.0	S-1 C-4
MB	ZZZFTR6	65	A500	C	1. Measure nozzle clearances of main propulsion turbines.	C-6	MMC MM2 FN	1.0 2.0 2.0	S-1 C-4
MB	ZZSFST5	74	4687	C	1. Clean and inspect main steam strainer.	C-7	MM1 FN	2.0 2.0 unit	None
MB	ZZZFTR6	65	A501	C	1. Inspect shaft packing and journals. Measure clearances.	C-8	MMC MM2 FN	3.0 5.0 5.0	C-5

MAINTENANCE INDEX PAGE
OPNAV FORM 4700-3 (4-64)

BUREAU PAGE CONTROL NUMBER   E-1/55-A5

98.171

Figure 12-30.—Maintenance Index Page.

services such as heating, air conditioning, and compressed air systems. Failure of any of these services will affect a ship's ability to fulfill its primary objective, either directly by reducing its power, or indirectly by creating conditions which lower personnel morale and efficiency. A secondary objective—which contributes considerably to the successful accomplishment of the first—is the minimization of personnel casualties and of secondary damage to vital machinery.

The details on specific casualties are beyond the scope of this manual. Detailed information on casualty control can be obtained from the Engineering Casualty Control Manual, the Damage Control Book, the Ship's Organization Book, and the Damage Control Bills. These publications may vary on different ships, but in all cases they give the organization and the procedures to be followed in case of engineering casualties, damage to the ship, and other emergency conditions.

The basic factors influencing the effectiveness of engineering casualty control are much broader than the immediate actions taken at the time of the casualty. Engineering casualty control reaches its peak efficiency by a combination of sound design, careful inspection, thorough plant maintenance (including preventive maintenance), and effective personnel organization and training. CASUALTY PREVENTION IS THE MOST EFFECTIVE FORM OF CASUALTY CONTROL.

CHAPTER 13

CONDENSERS AND OTHER HEAT EXCHANGERS

This chapter deals with the major pieces of heat transfer apparatus found in the condensate and feed system of the conventional steam turbine propulsion plant. Heat exchangers discussed here include the main condenser, the air ejector condenser, the gland exhaust condenser, the vent condenser, the deaerating feed tank, and the auxiliary condenser. The arrangement of piping that connects these units is discussed in chapter 9 of this text.

MAIN CONDENSER

The main condenser is the heat exchanger in which exhaust steam from the propulsion turbines is condensed as it comes in contact with tubes through which cool sea water is flowing. The main condenser is the heat receiver of the thermodynamic cycle—that is, it is the low temperature heat sink to which some heat must be rejected. The main condenser is also the means by which feed water is recovered and returned to the feed system. If we imagine a shipboard propulsion plant in which there is no main condenser and the turbines exhaust to atmosphere, and if we consider the vast quantities of fresh water that would be required to support even one boiler generating 150,000 pounds of steam per hour, it is immediately apparent that the main condenser serves a vital function in recovering feed water.

The main condenser is maintained under a vacuum of approximately 25 to 28.5 inches of mercury. The designed vacuum varies according to the design of the turbine installation and according to such operational factors as the load on the condenser, the temperature of the outside sea water, and the tightness of the condenser. The designed full-power vacuum for any particular turbine installation may be obtained from the machinery specifications for the plant. Some turbines are designed for a full-power exhaust vacuum of 27.5 inches of mercury when the circulating water injection temperature is 75° F; others are designed for a full-power exhaust vacuum of 25 inches of mercury with a circulating water injection temperature of 75° F.

It is often said that an engine can do a greater amount of useful work if it exhausts to a low pressure space than if it exhausts against a high pressure. This statement is undeniably true, but for the condensing steam power plant it may be somewhat misleading because of its emphasis on pressure. The pressure is important because it determines the temperature at which the steam condenses. As noted in chapter 8 of this text, an increase in the temperature difference between the source (boiler) and the receiver (condenser) increases the thermodynamic efficiency of the cycle. By maintaining the condenser under vacuum, we lower the condensing temperature, increase the temperature difference between source and receiver, and increase the thermodynamic efficiency of the cycle.

Given a tight condenser and an adequate supply of cooling water, the basic cause of the vacuum in the condenser is the condensation of the steam. This is true because the specific volume of steam is enormously greater than the specific volume of water. Since the condenser is filled with air when the plant is cold, and since some air finds its way into the condenser during the course of plant operation, the condensation of steam is not sufficient to establish the initial vacuum nor to maintain the required vacuum under all conditions. In modern shipboard steam plants, air ejectors are used to remove air and other noncondensable gases from the condenser. The condensation of steam is thus the major cause of the vacuum, but the air ejectors are required to help establish the initial vacuum and then to assist in maintaining vacuum while the plant is operating.

When the temperature of the outside sea water is relatively high, the condenser tubes are relatively warm and heat transfer is retarded. For this reason, a ship operating in warm tropical waters cannot develop as high a vacuum in the condenser as the same ship could develop when operating in colder waters.

Two basic rules that apply to the operation of single-pass main condensers should be kept in mind. The first is that the OVERBOARD TEMPERATURE should be about 10° higher than the INJECTION TEMPERATURE. The second rule is that the condensate discharge temperature should be within a few degrees of the temperature corresponding to the vacuum in the condenser. The accompanying chart lists vacuums (based on a 30.00-inch barometer) and corresponding temperatures.

Inches of Mercury	Corresponding Temperature (°F)
29.6	53
29.4	64
29.2	72
29.0	79
28.8	85
28.6	90
28.4	94
28.2	98
28.0	101
27.8	104
27.6	107

A main condenser is shown in cutaway view in figure 13-1. A slightly different main condenser is shown in outline drawing in figure 13-2. The operating principles of the two condensers are identical except for minor details.

In any main condenser, there are two separate circuits. The first is the vapor-condensate circuit in which the exhaust steam enters the condenser at the top of the shell and is condensed as it comes in contact with the outer surfaces of the condenser tubes. The condensate then falls to the bottom of the condenser, drains into a space called the hot well, and is removed by the condensate pump. Air and other noncondensable gases that enter with the exhaust steam or that otherwise find their way into the condenser are drawn off by the air ejector through the air ejector suction opening in the shell of the condenser, above the condensate level.

The second circuit is the circulating water circuit. During normal ahead operation, a scoop injection system[1] provides automatic flow of sea water through the condenser. The scoop, which is open to the sea, directs the sea water into the injection piping; from there, the water flows into an inlet water chest, flows once through the tubes, goes into a discharge water chest, and then goes overboard through a main overboard sea chest. A main circulating pump provides positive circulation of sea water through the condenser at times when the scoop injection system is not effective—when the ship is stopped, backing down, or moving ahead at very low speeds.

All main condensers that have scoop injection are of the straight-tube, single-pass type. A main condenser may contain from 2000 to 10,000 copper-nickel alloy tubes. The length of the tubes and the number of tubes depend upon the size of the condenser; and this, in turn, depends upon the capacity requirements. The tube ends are expanded into a tube sheet at the inlet end and expanded or packed into a tube sheet at the outlet end. The tube sheets serve as partitions between the vapor-condensate circuit and the circulating water (sea water) circuit.

Various methods of construction are used to provide for relative expansion and contraction of the shell and the tubes in main condensers. Packing the tubes at the outlet end sometimes makes sufficient provision for expansion and contraction. Where the tubes are expanded into each tube sheet, the shell may have an expansion joint. Expansion joints are also provided in the scoop injection line and in the overboard discharge line. Additional means such as flexible support feet or lubricated sliding feet are provided to compensate for expansion and contraction differentials between the shell and the condenser supporting structure.

As shown in figure 13-3, a central steam lane extends from the top of the condenser all the way through the tube bundle, down to the hot well. The exhaust steam which reaches the hot well through this steam lane tends to be drawn under the tube bundle toward the sides of the condenser shell, in the general direction of the air cooling sections, thus sweeping out any air which would otherwise tend to collect in the hot well. Part of the steam which is drawn through the hot well

---

[1] A major advantage of scoop injection is that it provides a flow of cooling water at a rate which is controlled by the speed of the ship and hence is automatically correct for various conditions. Scoop injection is standard for naval combatant ships and for many of the newer auxiliary ships.

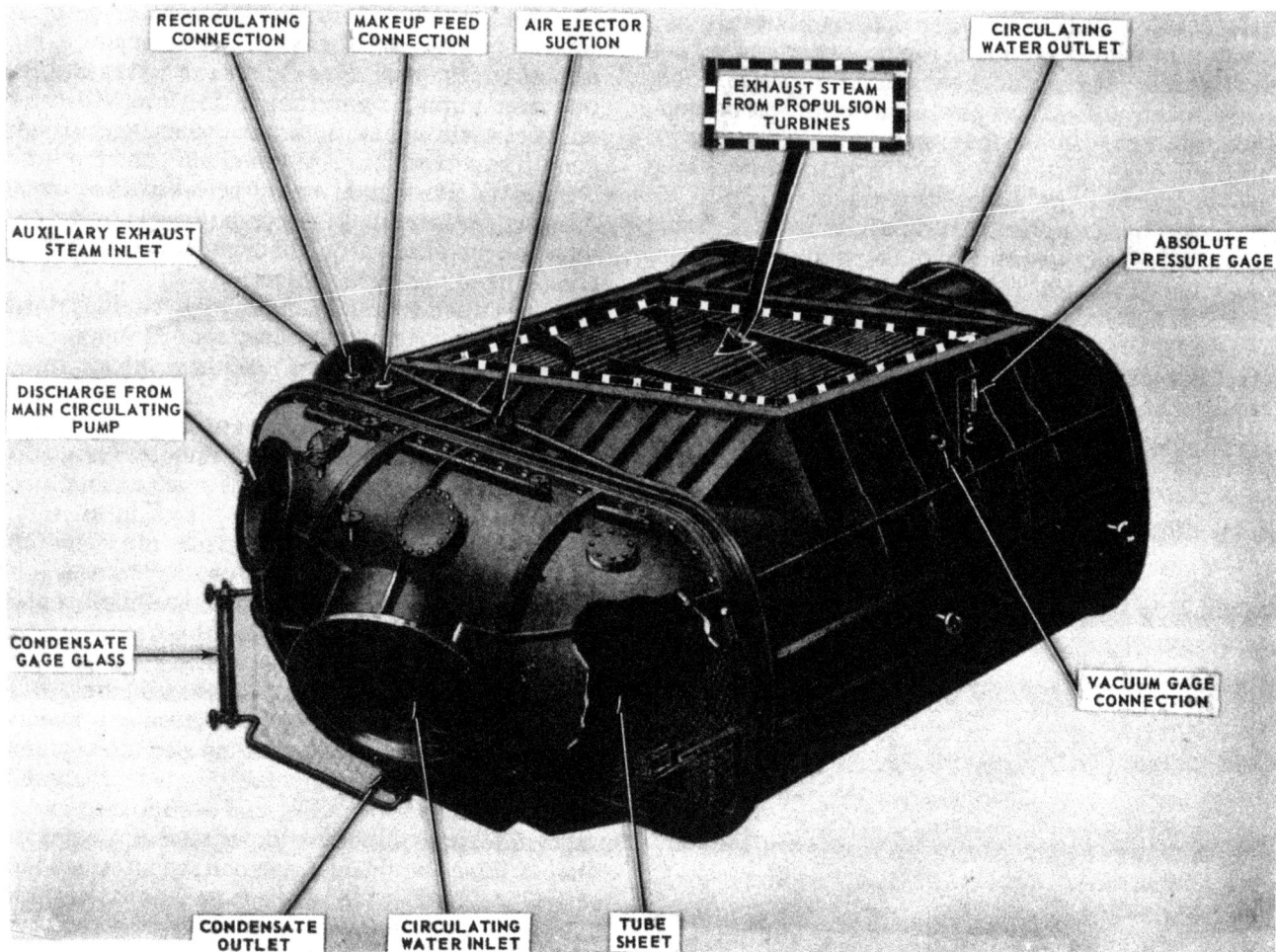

47.70X

Figure 13-1.—Cutaway view of main condenser.

under the tube bundle is condensed by the condensate dripping from the condenser tubes. In this process, the condensate (which has been subcooled by its contact with the cold tubes) tends to become reheated to a temperature which approaches the condensing temperature corresponding to the vacuum maintained in the hot well. The difference between the temperature of the condensate discharge and the condensing temperature corresponding to the vacuum maintained at the exhaust steam inlet to the condenser is called the condensate depression. One measure of the efficiency of design and operation of any condenser is its ability to maintain the condensate depression at a reasonably low value under all normal conditions of operation. Excessive condensate depression decreases the operating efficiency of the plant because the subcooled con-

densate must be reheated in the feed system, with a consequent expenditure of steam. Excessive condensate depression also allows an increased absorption of air by the condensate, and this air must be removed in order to prevent oxygen corrosion of piping and boilers.

Main condensers have various internal baffle arrangements for the purpose of separating air and steam so that the air ejectors will not be overloaded by having to pump large quantities of steam along with the air. Air cooling sections and air baffles may be seen in figure 13-3.

In some installations the condenser is hung from the low pressure turbine in such a way that the turbine supports the condenser. Where this type of installation is used, sway braces are used to connect the lower part of the condenser shell with the ship's structure. Spring supports are

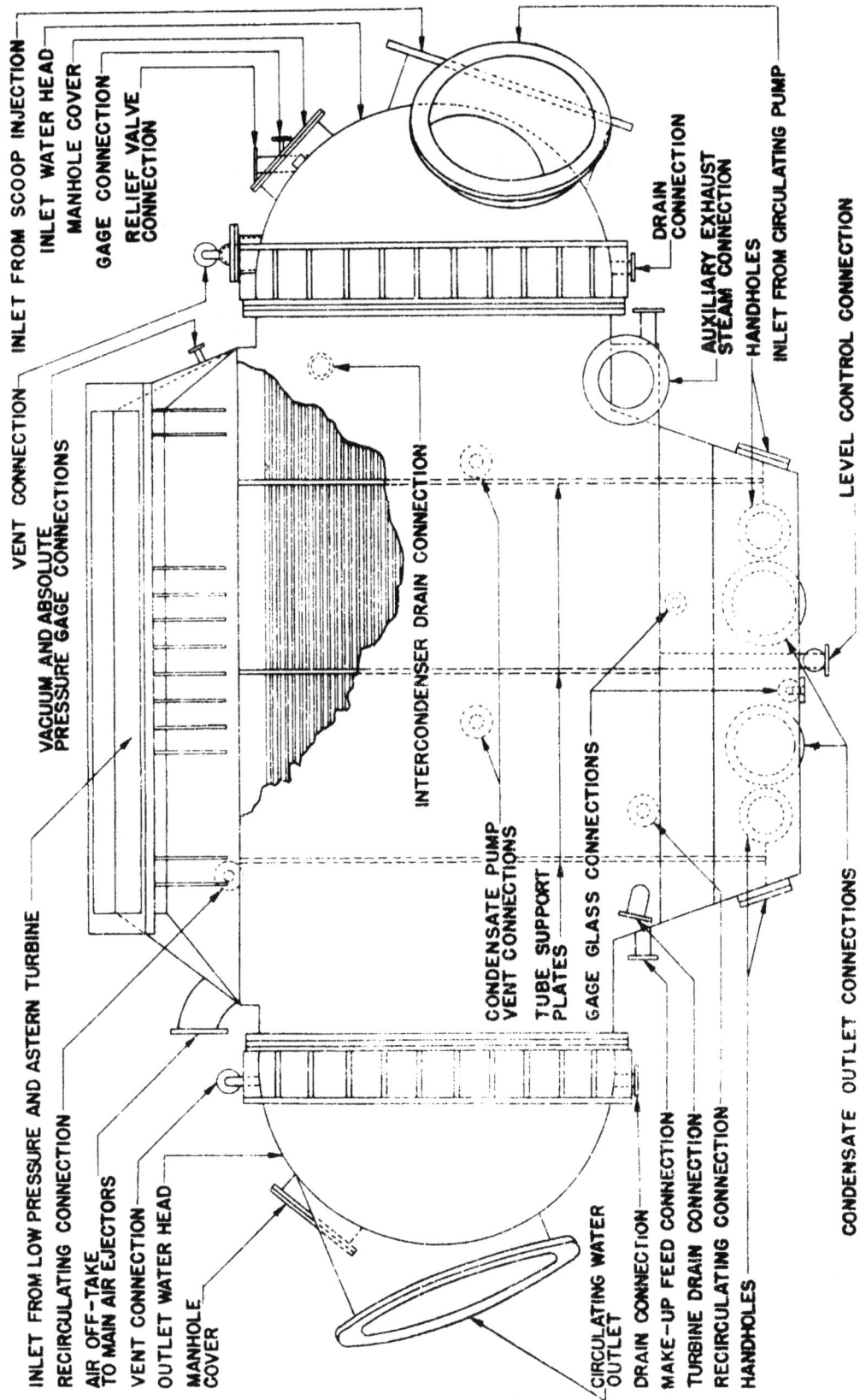

147.103

INLET FROM SCOOP INJECTION

INLET WATER HEAD

MANHOLE COVER

GAGE CONNECTION

RELIEF VALVE CONNECTION

DRAIN CONNECTION

AUXILIARY EXHAUST STEAM CONNECTION

HANDHOLES

INLET FROM CIRCULATING PUMP

VENT CONNECTION

INLET CONNECTION

LEVEL CONTROL CONNECTION

VACUUM AND ABSOLUTE PRESSURE GAGE CONNECTIONS

INTERCONDENSER DRAIN CONNECTION

INLET FROM LOW PRESSURE AND ASTERN TURBINE

RECIRCULATING CONNECTION

AIR OFF—TAKE TO MAIN AIR EJECTORS

VENT CONNECTION

OUTLET WATER HEAD

MANHOLE COVER

CONDENSATE PUMP VENT CONNECTIONS

TUBE SUPPORT PLATES

GAGE GLASS CONNECTIONS

CONDENSATE OUTLET CONNECTIONS

CIRCULATING WATER OUTLET

DRAIN CONNECTION

MAKE—UP FEED CONNECTION

TURBINE DRAIN CONNECTION

RECIRCULATING CONNECTION

HANDHOLES

Figure 13-2.—Outline drawing of main condenser.

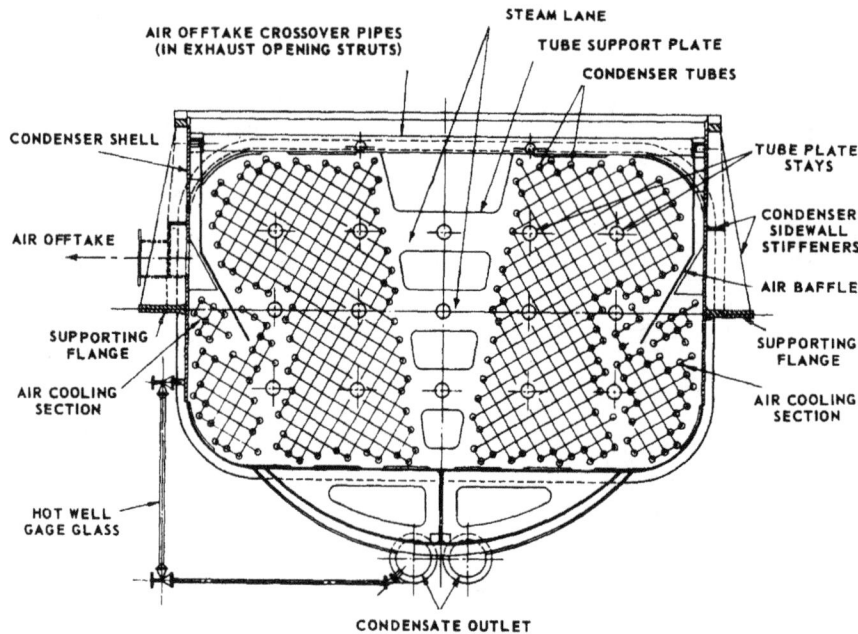

98.33

Figure 13-3.—Cross-sectional view of main condenser.

sometimes used to support part of the weight of the condenser so that it will not have to be entirely supported by the turbine.

Condenser performance may be evaluated by a simple energy balance which takes account of all energy entering and leaving the condenser. In theory, the entering side of the balance should include (1) the mechanical kinetic energy of the entering steam, (2) the thermal energy of the entering steam, (3) the mechanical kinetic energy of the entering sea water, and (4) the thermal energy of the entering sea water. In theory, again, the leaving side of the balance should include (1) the mechanical kinetic energy of the leaving condensate, (2) the thermal energy of the leaving condensate, (3) the mechanical kinetic energy of the leaving sea water, and (4) the thermal energy of the leaving sea water. In considering real condensers, however, the entering and leaving mechanical kinetic energies of the sea water tend to be small and tend to cancel each other out, the mechanical kinetic energy of the entering steam is so small as to be negligible, and the mechanical kinetic energy of the leaving condensate is small enough to disregard. With all of these relatively insignificant quantities omitted, the entering side of the balance includes only the thermal energy of the entering steam and the thermal energy of the entering sea water, and the leaving side includes only the thermal energy of the

leaving condensate and the thermal energy of the leaving sea water.

AIR EJECTOR ASSEMBLIES

The function of air ejectors is to remove air and other noncondensable gases from the condenser. An air ejector is a type of jet pump, having no moving parts. The flow through the air ejector is maintained by a jet of high velocity steam passing through a nozzle. The steam is taken from the 150-psi auxiliary steam system on most ships.

The air ejector assembly (fig. 13-4) used to remove air from the main condenser usually consists of a first-stage air ejector, an inter condenser, a second-stage air ejector, and an after condenser. The two air ejectors operate in series. The first-stage air ejector raises the pressure from about 1.5 inches of mercury absolute (condenser pressure) to about 7 inches of mercury absolute; the second-stage air ejector raises the pressure from 7 inches of mercury absolute to about 32 inches of mercury absolute (about 1 psig).

The first-stage air ejector takes suction on the main condenser and discharges the steam-air mixture to the inter condenser, where the steam content of the mixture is condensed. The resulting condensate drops to the bottom of the inter

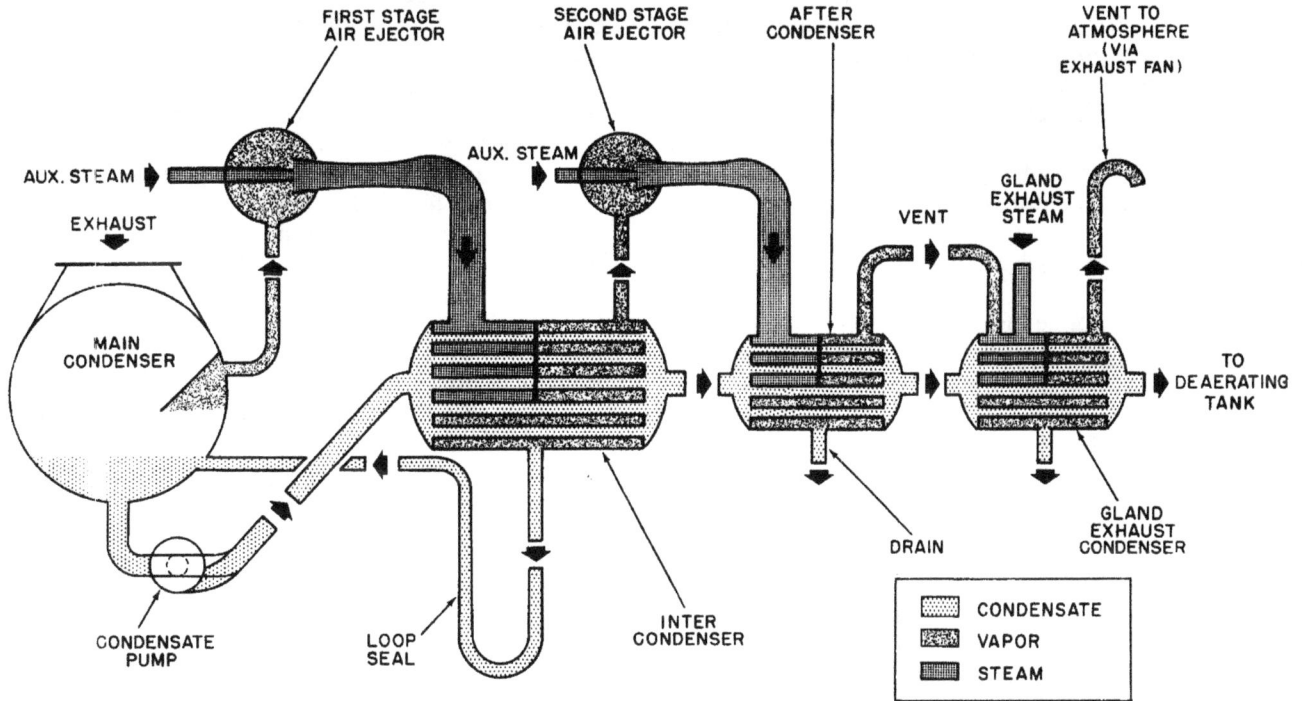

Figure 13-4.—Two-stage air ejector assembly.

47.77X

condenser shell, and from there it drains to the condenser through a U-shaped loop seal line. The air passes to the suction of the second-stage air ejector, where another jet of steam entrains the air and carries it to the after condenser. In the after condenser, the steam is condensed and returned to the condensate system by way of the fresh water drain collecting tank, and the air is vented to atmosphere.

Note that the air ejectors remove air only from the condenser, not from the condensate which passes through the tubes of the inter and after condensers. The condensate merely serves as the cooling medium in these condensers, just as it next serves this purpose in the gland exhaust condenser and in the vent condenser.

GLAND EXHAUST CONDENSER

The gland exhaust condenser receives a steam-air mixture from the propulsion turbine glands. The steam is condensed and returned to the condensate system by way of the fresh water drain collecting tank, and the air is discharged to atmosphere. The atmospheric vent is usually connected to the suction of a small motor-driven fan (gland exhauster), which provides a positive

discharge through piping to the atmosphere above decks. This is necessary to avoid filling the engineroom with steam should the air ejector cooling water supply fail, thereby allowing the steam to pass through the inter condenser and after condenser without being condensed. The cooling medium in the gland exhaust condenser, as in the air ejector condensers, is condensate from the main condenser, on its way to the deaerating feed tank.

In most installations, the gland exhaust condenser appears to be part of the air ejector assembly, since it is attached to the after condenser. However, the gland exhaust condenser is functionally a separate unit even though it is physically attached to the air ejector after condenser.

In serving as the cooling medium in the air ejector condensers and in the gland exhaust condenser, the condensate picks up a certain amount of heat. To some extent this is desirable, since it saves heat which would otherwise be wasted and it reduces the amount of steam required to heat the condensate in the deaerating feed tank. However, overheating of the condensate could result in inefficient operation of the air ejectors and consequent loss of vacuum in the main condenser. To avoid this difficulty, provision is made for returning some of the condensate to the main condenser when the condensate reaches a certain temperature. As a rule, the recirculating line branches off the condensate line just after the gland exhaust condenser. In most installations, the recirculating valve in this recirculating line is thermostatically operated.

VENT CONDENSER

The vent condenser is actually a part of the deaerating feed tank, being installed in the tank near the top. It is described separately here because it is functionally quite separate from the deaerating feed tank.

In the vent condenser, as in the air ejector condensers and the gland exhaust condenser, condensate on its way from the main condenser to the deaerating feed tank is used to cool and condense the steam from a steam-air mixture. The vent condenser receives steam and air from the deaerating feed tank. The steam condenses into water, which falls toward the bottom of the tank. The air goes to the gland exhaust condenser and is vented to atmosphere. The condensate which is used as the cooling medium in the vent condenser is sprayed out into the deaerating feed tank and is deaerated before being used as boiler feed.

DEAERATING FEED TANK

The deaerating feed tank serves to heat, deaerate, and store feed water. The water is heated by direct contact with auxiliary exhaust steam which enters the tank at a pressure just slightly greater than the pressure in the tank. The deaerating feed tank is usually designed to operate at a pressure of about 15 psig and to heat the water to between 240° and 250° F.

One type of deaerating feed tank is shown in figure 13-5. Condensate enters the tank through the tubes of the vent condenser and is forced out through a number of spray valves in a spray head. The spray valves discharge the condensate in a fine spray throughout the steam-filled upper section of the deaerating feed tank. The very small droplets of water are heated, scrubbed, and partially deaerated by the relatively air-free steam. As the steam gives up its heat to the water, much of the steam is condensed into water. The droplets of water (including both the entering condensate sprayed out from the vent condenser and the steam condensed in the deaerating feed

tank) are collected in a cone-shaped baffle which leads them through a central port, to the deaerating unit.

Steam enters the deaerating unit, picks up the partially deaerated water, and throws it tangentially outward through the curving baffles of the deaerating unit. In this process, the water is even more finely divided and is throughly scrubbed by the incoming steam. Thus the last traces of dissolved oxygen are removed from the water. Since the water enters the deaerating unit at saturation temperature, having already been heated by the steam in the upper part of the deaerating feed tank, the incoming steam does not condense to any marked degree in the deaerating unit. Therefore all (or practically all) of the incoming steam is available for breaking up, scrubbing, and deaerating the water.

The thoroughly deaerated water falls into the storage space at the bottom of the tank, where it remains under a blanket of air-free steam until it is pumped to the boilers. Meanwhile, the mixture of steam plus air and other noncondensable gases travels over the spray head (where much of the steam is condensed as it heats the incoming condensate) and over the tubes of the vent condenser (where more steam is condensed into water which then goes into the deaerating unit). The air and other noncondensable gases, together with a little remaining steam, go to the gland exhaust condenser.

As shown in figure 13-5, the deaerating feed tank has a recirculating connection that allows water to be sent back to the condenser from the deaerating feed tank. The recirculating line is used to provide a high enough condensate level in the condenser so that the condensate pump can take suction. The recirculating line is also used at slow speeds and when the plant is first started up to ensure a sufficient supply of cooling condensate to the air ejector condensers and to the gland exhaust condenser and to keep the deaerating feed tank at the prescribed temperature.

The deaerated feed water from the deaerating feed tank is pumped to the boiler by the feed booster pump and the main feed pump. The feed booster pump takes suction from the bottom of the deaerating feed tank and discharges to the suction side of the main feed pump. The feed booster pump provides a positive suction pressure for the main feed pump and thus prevents the hot water from flashing into steam at the main feed pump suction. The main feed pump operates at variable speed in order to maintain a constant discharge pressure under all

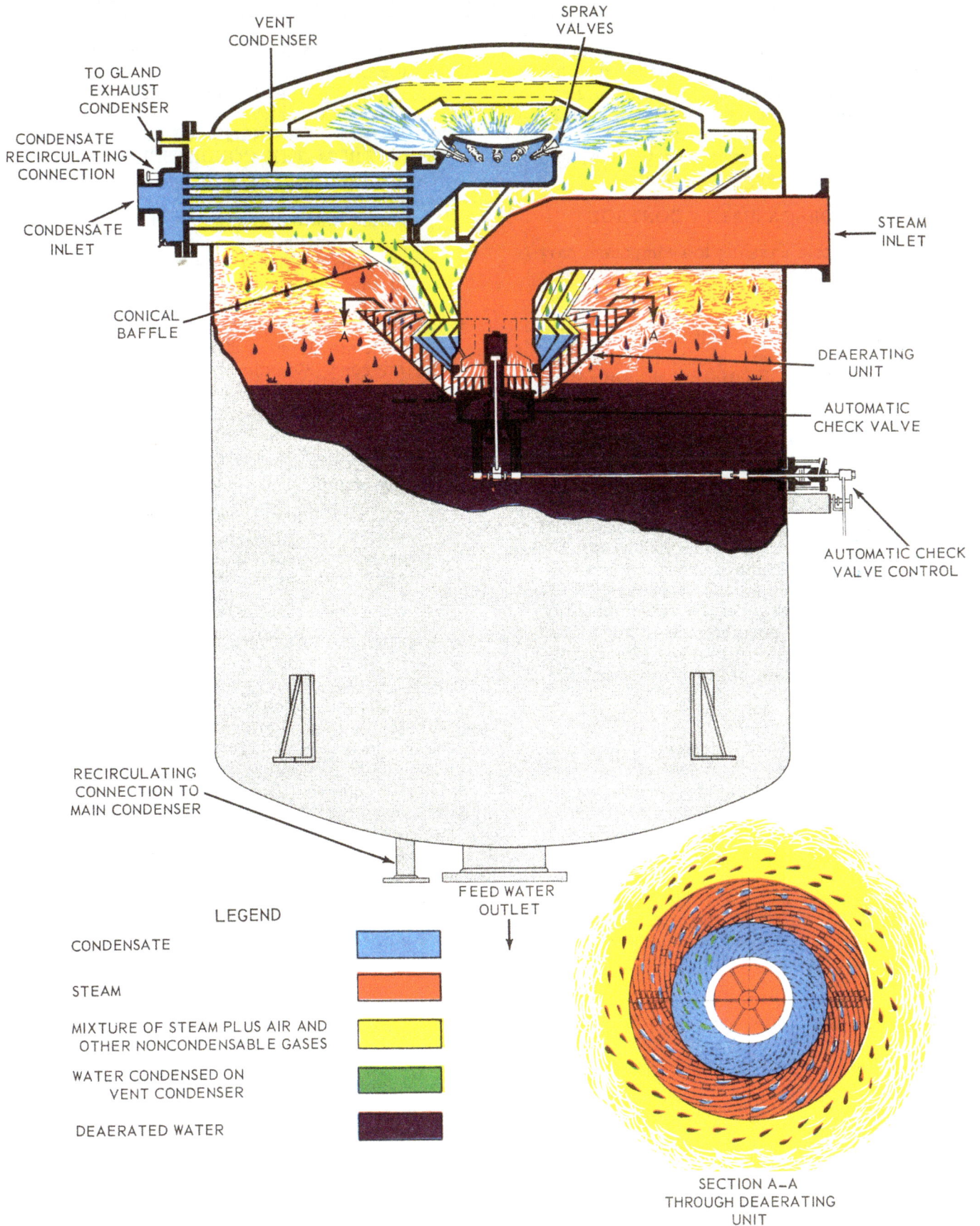

VENT CONDENSER

SPRAY VALVES

TO GLAND EXHAUST CONDENSER

CONDENSATE RECIRCULATING CONNECTION

CONDENSATE INLET

CONICAL BAFFLE

STEAM INLET

DEAERATING UNIT

AUTOMATIC CHECK VALVE

AUTOMATIC CHECK VALVE CONTROL

RECIRCULATING CONNECTION TO MAIN CONDENSER

FEED WATER OUTLET

LEGEND

CONDENSATE

STEAM

MIXTURE OF STEAM PLUS AIR AND OTHER NONCONDENSABLE GASES

WATER CONDENSED ON VENT CONDENSER

DEAERATED WATER

SECTION A—A THROUGH DEAERATING UNIT

Figure 13-5.—Deaerating feed tank.

38.17

355

conditions of load. The discharge pressure of the main feed pump is considerably higher than the pressure carried in the boiler steam drum. Provision is made for the recirculation of water from the main feed pump discharge back to the deaerating feed tank, in order to protect the pump from overheating at very low capacity.

SAFETY AND CASUALTY CONTROL

In the event of a casualty to a component part of the propulsion plant, the principal doctrine to be impressed upon operating personnel is the prevention of additional or major casualties. Under normal operating conditions, the safety of personnel and machinery should be given first consideration. Where practicable, the propulsion plant should be kept in operation by means of standby pumps, auxiliary machinery, and piping systems. The important thing is to prevent minor casualties from becoming major casualties, even if it means suspending the operation of the propulsion plant. It is better to stop the main engines for a few minutes than to put them completely out of commission, so that major repairs are required to place them back into operation. In case a casualty occurs, the officer or CPO in charge of the watch should be notified as soon as possible; he in turn must notify the OOD if there will be any effect on the ship's speed or on the ability to answer bells.

LOSS OF VACUUM

The major causes of a loss of vacuum are: excessive air leakage into the vacuum system, improper functioning of the air-removal equipment, improper drainage of condensate from the condenser, insufficient flow of circulating water, and high injection temperature.

1. EXCESSIVE AIR LEAKAGE INTO THE VACUUM SYSTEM may be caused by:
   a. Insufficient gland sealing steam.
   b. Vent valve on idle condensate pump open.
   c. Loop-seal filling valve open.
   d. Bypass valve on drain tank open.
   e. Drain tank float valve stuck open.
   f. Taking make-up feed from empty feed bottom.
   g. Leakage of flanges, fittings, or valve stem packings under vacuum.
2. IMPROPER FUNCTIONING OF THE AIR-REMOVAL EQUIPMENT may be due to:
   a. Insufficient steam to the air ejectors.

b. Foreign matter lodged in the air ejector nozzle(s).
   c. Erosion of the air ejector nozzle, over a period of time.
3. IMPROPER DRAINAGE OF CONDENSATE FROM THE CONDENSER may be caused by:
   a. Low speed of condensate pump, indicating malfunctioning of the pump's speed-limiting governor.
   b. Condensate pump air-bound because of the vent connection from the first stage being closed or not opened wide.
4. INSUFFICIENT FLOW OF CIRCULATING WATER may be caused by:
   a. Improper adjustment of the overboard discharge valve (the main injection valve being wide open whenever the condenser is under vacuum).
   b. Inadequate speed of the main circulating pump.
   c. Plugged tubes, resulting from mud, shells, small fish, or kelp being trapped against the injection strainer bars or in the inlet water chest.
   d. Air trapped in condenser.
5. HIGH INJECTION TEMPERATURE

Basically, the injection temperature limits the maximum vacuum (minimum absolute pressure) obtainable in a specific plant, assuming the condenser, associated equipment, and piping under vacuum to be clean and properly operated.

Whenever there is a loss of vaccum, the first step in correcting the trouble is to locate the cause of the casualty. The major possible causes are so numerous that no attempt will be made to list the proper action required for each one. The required action may be very simple, such as closing the loop seal filling valve, or it may be much more complicated, such as leaning the main condenser or replacing air ejector nozzles.

SALT WATER LEAKAGE INTO CONDENSER

If a condenser salinity indicator shows a rise in the chloride content, the source of the contamination must be determined immediately. To locate these sources, test the fresh water from different units in the system by checking the proper salinity indicators (if installed) and by making chemical chloride tests. There are four major causes of a salty condenser:

1. Leaky tube(s) in the condenser.
2. Make-up feed tank salted up.
3. Low pressure drain tank salted up.
4. Leaky feed suction and drain lines which run through the bilges.

Each of the aforementioned possibilities must be investigated to determine the source of the contamination and its elimination.

If it is determined that there is a minor leak in the condenser and the ship's prospective arrival time is less than 24 hours, the affected plant will probably be continued in operation. Isolate the condensate system, and limit the number of boilers on the engine involved. When operating under these conditions it will be necessary to blow down the boiler(s) as necessary, to keep the boiler salinity within the specified limit. However, if the leak is serious, secure the plant and locate the leaks.

If leaky tubes are found, they must be plugged so that the condenser can be kept in service. Plugs, which are furnished by the manufacturer, should be driven into the tube ends with light hammer blows. If it becomes necessary to plug tube sheet holes after a tube has been removed, a short section of tube should be expanded into the tube hole before the tube plug is inserted; this will protect the tube holes from damage.

Plugged tubes should be renewed during the next shipyard availability if the water chests are removed for other work; or if more than 10 percent of the tubes are plugged, a retubing request should be submitted, via the type commander, to NavShips.

For the procedure in locating and plugging leaking condenser, tubes, refer to either the manufacturer's technical manual for the specific equipment or chapter 9460 of NavShips Technical Manual.

AIR EJECTOR ASSEMBLY CONTROL AND SAFETY

In order to provide for continuous operation, two sets of nozzles and diffusers are furnished for each stage of the air ejectors. Only one set is necessary for operation of the plant; the other set is maintained ready for use in case of damage or unsatisfactory operation of the set in use. The sets can be used simultaneously when excessive air leakage into the condenser necessitates additional pumping capacity.

Before starting a steam air ejector, the steam line should be drained of all moisture; moisture in the steam will cut the nozzles, and slugs of water will cause unstable operation.

Before cutting steam into the air ejectors, make sure that sufficient cooling water is flowing through the condenser and that the condenser has been properly vented.

The loop seal line must be kept airtight, an air leak may cause all water to drain out of the seal.

If it is necessary to operate both sets of air ejectors to maintain proper condenser vacuum, air leakage is indicated. It is more desirable to eliminate the air leakage than to operate two sets of air ejectors.

Unstable operation of an air ejector may be caused by any of the following: the steam pressure may be lower than the designed amount, the steam temperature and quality may be different than design condition, there may be scale on the nozzle surface, the position of the steam nozzle may not be right in relation to the diffuser, or the condenser drains may be stopped up.

Difficulties due to low pressure are generally caused by improper functioning or improper adjustment of the steam reducing valve supplying motive steam to the air ejector assembly. It is essential that DRY steam at FULL operating pressure be supplied to the air ejector nozzles.

Erosion of fouling of air ejector nozzles is evidence that wet steam is being admitted to the unit. Faulty nozzles make it impossible to operate the ejector under high vacuum. In some instances, the nozzles may be clogged with grease, boiler compound, or some other deposit which will decrease the jet efficiency.

DEAERATING FEED TANK CONTROL AND SAFETY

During normal operation, the only control necessary is maintaining the proper water level. (On some of the newer ships, this is done with automatic control valves.) If the water level is too high, the tank cannot properly remove the air and noncondensable gases from the feed water. A low water level may endanger the main feed booster pumps, the main feed pumps, and the boilers.

Deaerating feed tanks remove gases from the feed water by using the principle that the solubility of gases in feed water approaches zero when the water temperature approaches the boiling point. During operation, steam and water are mixed by spraying the water so that it comes in contact with steam from the auxiliary exhaust line. The quantity of steam must always be proportional to the quantity of water, otherwise, faulty operation or a casualty will result.

Overfilling the deaerating tank may upset the steam-water balance and cool the water to such an extent that ineffective deaeration will take place. Overfilling the deaerating tank also wastes heat and fuel. The excess water, which will have to run down to the condenser, will be cooled—and when it reenters the deaerating tank, more steam will be required to reheat it. If an excessive amount of cold water enters the deaerating feed tank, the temperature drop in the tank will cause a corresponding drop in pressure. As the deaerating feed tank pressure drops, more auxiliary exhaust steam enters the tank. This reduces the auxiliary exhaust line pressure, which causes the augmenting valve (150 psi line to auxiliary exhaust line) to open and bleed live steam into the deaerating feed tank.

When an excessive amount of cold water suddenly enters the deaerating feed tank, a serious casualty may result. The large amount of cold water will cool (quench) the upper area of the deaerating feed tank and condense the steam so fast that the pressure is reduced throughout the deaerating feed tank. This permits the hot condensate in the lower portion of the deaerating feed tank and feed booster pump to boil or flash into vapor causing the booster pump to lose suction until the pressure is restored and the boiling of the condensate ceases. With a loss of feed booster pump pressure, the main feed pump suction is reduced or lost entirely, causing serious damage to the feed pump and loss of feed water supply to the boiler(s). Some of the newer ships have safety devices installed on the main feed pumps which will stop the main feed pump when a partial or total loss of main feed booster pressure occurs.

The mixture of condensate, drains, and make-up feed water, constituting the inlet water to the deaerating tank, enters through the tubes of the vent condenser. The condensate pump discharge pressure forces the water through the spray valves of the spray head and discharges it in a fine spray throughout the steam filled top or preheater section of the deaerating feed tank.

If a spray nozzle sticks open, or if a spray nozzle spring is broken, the flow from the nozzle will not be in the form of a spray and the result will be ineffective deaeration. This condition cannot be discovered except by analysis of the feed water leaving the deaerating feed tank, or by inspecting the spray nozzles.

Inspection of the spray nozzles should be scheduled at frequent intervals.

In most deaerating feed tanks, the manhole provides access for the inspection of spray nozzles; other tanks are so designed that the spray nozzle chamber and the vent condenser must be removed in order to inspect the nozzles.

Complete information on constructing and using a test rig for spray valves can be found in chapter 9560 of NavShips Technical Manual.

SAFETY PRECAUTIONS FOR CONDENSERS

When opening a main condenser for cleaning or inspection, or when testing a main condenser, there are several safety precautions that must be observed. The following procedures and precautions, when carried out properly, will help prevent casualties to personnel and machinery:

1. Before the salt water side of a condenser is opened, all sea connections, including the main injection valve, circulating pump suction valve, and main overboard valve, are to be closed tightly and secured against accidental opening with wire, and tagged, DO NOT OPEN, and signed by the person tagging the valve. This is necessary to avoid the possibility of flooding an engineroom. Safety gates, where provided, are to be installed.

2. On condenser having electrically operated injection and overboard valves, the electrical circuits serving these motors are to be opened and tagged to prevent accidentally energizing these circuits.

3. Before a manhole or handhole plate is removed, drain the salt water side of the condenser by using the drain valve provided in the inlet water box. This is done to make sure that all sea connections are tightly closed.

4. If practicable, inspection plates are to be replaced and secured before work is discontinued each day.

5. Never subject condensers to a test pressure in excess of 15 psig.

6. When testing for leaks, do not stop because one leak is found. The entire surface of both tube sheets must be checked, as other leaks may exist. Determine whether each leak is in the tube joint or in the tube wall, so that the proper repairs can be made.

7. There is always a possibility that hydrogen or other gases may be present in the steam

98.176

Card 1

SYSTEM	COMPONENT	M.R. NUMBER	
Propulsion	Deaerating Feed Tank	E-3 S-1	RATES: MM1 6.0, FN 6.0

SUB-SYSTEM	RELATED M.R.	
Feed Water and Condensate	None	TOTAL M/M 12.0 / ELAPSED TIME: 6.0

PAGE 1 OF 3 | MJ | ZZ2F | TB2 | 44 | 2651 | S

M.R. DESCRIPTION
1. Test spray valves.
2. Inspect and clean the internal components of the tank.
3. Inspect the operation of the automatic check valve.

SAFETY PRECAUTIONS
1. Observe standard safety precautions.
2. Isolate the tank from all systems. Wire valves shut and tag "Do Not Open."
3. Ensure tank is well ventilated before entering.
4. Remove all rags and tools before closing tank.

TOOLS, PARTS, MATERIALS, TEST EQUIPMENT
1. Scissors
2. Water hose
3. Wire brush
4. Safety tags
5. Portable light
6. Clean rags
7. 2-lb Hammer
8. 8" Screwdriver
9. 14" Monkey wrench
10. Wire, 24 gauge
11. 15/16" Wrench (2)
12. Portable blower
13. 8" Adjustable wrench
14. 1/16" Sheet copper
15. 5/8" Gasket punch
16. 1/16" Gasket material, Symbol 2150
17. Monel safety locking wire

PROCEDURE
Preliminary
a. Isolate the tank from all systems. Wire valves shut, tag "Do Not Open."

1. Test Spray Valves.
a. Open manhole cover from the tank.
b. Ventilate the tank at least five minutes before entering.
c. Remove the inspection cover from the spray valves.
d. Remove the spray valves and gaskets.
e. Transport the spray valves to the test activity.
f. Pick up the spray valves from the test activity.

(Cont'd on Page 2)

LOCATION
DATE 18 October 1965

MAINTENANCE REQUIREMENTS CARD
OPNAV FORM 4700-1 (REV. 7/65)

Card 2

SYSTEM	COMPONENT	M.R. NUMBER	
Propulsion	Main Condenser	E-4 C-1	RATES: MMC 2.0, MM1 14.0, FN 14.0

SUB-SYSTEM	RELATED M.R.	
Main Condensers and Air Ejectors	None	TOTAL M/H 30.0 / ELAPSED TIME: 14.0

PAGE 1 OF 2 | MC | ZZZF | CM1 | 84 | 5075 | C

M.R. DESCRIPTION
1. Clean the steam side of condenser.
2. Perform hydrostatic test on steam side of condenser.

SAFETY PRECAUTIONS
1. Observe standard safety precautions.
2. Wire injection and overboard valves shut and tag "Do Not Open."
3. Never exceed 15 psi when performing hydrostatic test.

TOOLS, PARTS, MATERIALS, TEST EQUIPMENT
1. Bucket
2. Air hose
3. Safety tags
4. Wire, 24 gauge
5. Boiler compound
6. Flexible steam hose
7. 6" Slip-joint pliers
8. 0-20 psi Pressure gauge
9. 15/16", 1" Combination wrenches

PROCEDURE
Preliminary
a. Wire injection and overboard valves shut and tag "Do Not Open."

1. Clean the Steam Side of Condenser.
a. Drain salt water side and remove all foreign matter.
b. Shut all valves in lines related to the condenser; wire and tag "Do Not Open."
c. Start filling steam side of condenser through the recirculating line.
d. Start introducing boiler compound mixed with feed water into the condenser, one pound at a time.

NOTE 1: For every 100 gallons of water in condenser, introduce 10 pounds of boiler compound mixed with feed water.

e. Continue the filling process until the top row of tubes in the condenser is covered.

(Cont'd On Page 2)

LOCATION
DATE 1 October 1965

MAINTENANCE REQUIREMENTS CARD
OPNAV FORM 4700-1 (REV. 7 65)

Figure 13-6.—Maintenance Requirement Cards

or the salt water side of a condenser. No open flame or tool which might cause a spark should be brought close to a newly opened condenser. Personnel are not to be permitted to enter a newly opened condenser until it has been thoroughly blown out with steam or air.

8. The salt water side of a condenser must be drained before flooding of the steam side and must be kept drained until the steam side is emptied.

9. The relief valve (set at 15 psig) mounted on the inlet water chest is to be lifted by hand whenever condensers are secured.

10. If a loss of vacuum is accompanied by a hot or flooded condenser, the units exhausting into the condenser must be slowed or stopped until the casualty is corrected. Condensate must not be allowed to collect in condenser and overflow into the turbines or engines.

11. Condenser shell relief valves are to be lifted by hand before a condenser is put into service.

12. No permanent connection which could subject the salt water side to a pressure in excess of 15 psig, is to be retained between any condenser and a water system.

13. No permanent connection which could allow salt water to enter the steam side of the condenser, is to be retained.

14. Test the main circulating pump bilge suction, when so directed by the engineer officer. To conduct this test, it is generally necessary only to start the main circulating pump, open the bilge suction line stop or check valve, and then close down on the sea suction line valve to about 3/4 closed, or until the maximum bilge suction capacity is obtained.

MAINTENANCE

Condensers, heat exchangers and associated equipment should be periodically tested and inspected to ensure that they are operating efficiently. Preventive maintenance is much more economical than corrective maintenance. All preventive maintenance should be conducted in accordance with the 3-M System (PMS Subsystem). As an example, figure 13-6 shows two maintenance requirement cards, one for a main condenser and the other for a deaerating feed tank. Note: These cards contain specific information for conducting the specified preventive maintenance actions.

PART IV—AUXILIARY MACHINERY
AND EQUIPMENT

The chapters included in this part of the text deal with the auxiliary machinery and equipment of the shipboard engineering plant. Some of the units and plants described here are directly related to the operation of the major units of propulsion machinery; others may be regarded as supporting systems. Major emphasis in this part of the text is on the auxiliary machinery and equipment found on conventional steam-driven ships; however, a substantial amount of the information given here applies also to ships with other kinds of propulsion plants.

It should be noted that diesel engines, gasoline engines, and gas turbine engines-all of which may be used to drive auxiliary machinery and equipment-are not included here. These units are discussed in part V of this text.

361

CHAPTER 14

PIPING, FITTINGS, AND VALVES

This chapter deals with pipe, tubing, fittings, valves, and related components that make up the shipboard piping systems used for the transfer of fluids. The general arrangement and layout of the major engineering piping systems is discussed in chapter 9 of this text; in the present chapter, we are concerned with certain practical aspects of piping system design and with the actual piping system components—pipe, tubing, fittings, and valves.

DESIGN CONSIDERATIONS

Each piping system and all its components must be designed to meet the particular conditions of service that will be encountered in actual use. The nature of the contained fluid, the operating pressures and temperatures of the system, the amount of fluid that must be delivered, and the required rate of delivery are some of the factors that determine the materials used, the types of valves and fittings used, the thickness of the pipe or tubing, and many other details. Piping systems that must be subjected to temperature changes are designed to allow for expansion and contraction. Special problems that might arise—water hammer, turbulence, vibration, erosion, corrosion, and creep,[1] for example—are also considered in the design of piping systems.

The requirements governing the design and arrangement of components for shipboard piping

systems are covered in detail by contract specifications and by a number of plans and drawings. The information given here is not intended as a detailed listing but merely as a general guide to the design requirements of shipboard piping systems.

All shipboard piping is installed in such a way that it will not interfere with the operation of the ship's machinery or with the operation of doors, hatches, scuttles, or openings covered by removable plates. As far as possible, piping is installed so that it will not interfere with the maintenance and repair of machinery or of the ship's structure. If piping must be installed in the way of machinery or equipment which requires periodic dismantling for overhaul, or if it must be installed in the way of other piping systems or electrical systems, the piping is designed for easy removal. Piping that is vital to the propulsion of the ship is not installed where it would have to be dismantled in order to permit routine maintenance on machinery or other systems. Piping is not normally installed in such a way as to pass through voids, fuel oil tanks, ballast tanks, feed tanks, and similar spaces.

Valves, unions, and flanges are carefully located to permit isolation of sections of piping with the least possible interference to the continued operation of the rest of the system. The type of valve used in any particular location is specified on the basis of the service conditions to be encountered. For example, gate valves are widely used in locations where the turbulent flow characteristics of other types of valves might be detrimental to the components of the system.

Unnecessary high points and low points are avoided in piping systems. Where high points and low points are unavoidable, vents, drains, or other devices are installed to ensure proper

---

[1]The term creep is used to describe a special kind of plastic deformation that occurs very slowly, at high temperatures, in metals under constant stress. Because creep occurs very slowly—so slowly, in fact, that years may be required to complete a single creep test—the importance of this type of plastic deformation was not recognized in many fields of engineering until fairly recently. Creep-resisting steel is now used in most modern naval boilers and for most high temperature piping.

functioning of the system and equipment served by the system.

Various joints are used in shipboard piping systems. The joints used in any system depend upon the piping service, the pipe size, and the construction period of the ship. Older naval ships have threaded flanges in low pressure piping; rolled-in joints for steel piping that is too large for the threaded flanges; and spelter-brazed flanges for copper and brass piping. On new construction, welded joints are used to the maximum practicable extent in systems that are fabricated of carbon steel, alloy steel, or other weldable material. On both older and newer ships, flanged joints made up with special gaskets are in use.

Components welded in a piping system must be accessible for repair, reseating, and overhaul while in place; they are so located that they can be removed, preheated, rewelded, and stress relieved when major repairs or replacements are necessary. Complex assemblies— for example, assemblies of valves, strainers, and traps in high pressure drain systems—are designed to be removable as a group if they cannot be repaired while in place and if they require frequent overhaul.

Flanged and union joints are placed where they will be least affected by piping system stresses. In general, this means that joints are not located at bends or offsets in the piping.

Valves are designed so that they can be operated with the minimum practicable amount of force and with the maximum practicable convenience. If a man must stand on slippery deck plates to turn a valve handwheel, or if he must reach over his head or around a corner, he cannot apply the same amount of torque that he could apply to a more conveniently located handwheel. Thus the location of the handwheels is an important design consideration. Toggle mechanisms or other mechanical advantage devices are used where the amount of torque required to turn a handwheel is more than could normally be applied by one man. If mechanical advantage devices are not sufficient to produce easy operation of the valve, power operation is used.

If accidental opening or closing of a valve could endanger personnel or jeopardize the safety of the ship, locking devices are used. Any locking device installed on a valve must be designed so that it can be easily operated by authorized personnel; but it must be complex enough to discourage casual or indiscriminate operation by other persons.

Supports used in shipboard piping systems must be strong enough to support the weight of the piping, its contained fluid, and its insulation and lagging. Supports must carry the loads imposed by expansion and contraction of the piping and by the working of the ship, and they must be able to support the piping with complete safety. Supports are designed to permit the movement of the piping necessary for flexibility of the system. A sufficient number of supports are used to prevent excessive vibration of the system under all conditions of operation, but the supports must not cause excessive constraint of the piping. Supports are used for heavy valves and fittings so that the weight of the valves and fittings will not be entirely supported by the pipe.

PIPE AND TUBING

Piping is defined as an assembly of pipe or tubing, valves, fittings, and related components forming a whole or a part of a system for transferring fluids.

It is somewhat more difficult to define pipe and tubing. In commercial usage, there is no clear distinction between pipe and tubing, since the correct designation for each tubular product is established by the manufacturer. If the manufacturer calls a product pipe, it is pipe; if he calls it tubing, it is tubing. In the Navy, however, a distinction is made between pipe and tubing. This distinction is based on the way the tubular product is identified as to size.

There are three important dimensions of any tubular product: outside diameter (OD), inside diameter (ID), and wall thickness. A tubular product is called tubing if its size is identified by actual measured outside diameter (OD) and by actual measured wall thickness. A tubular product is called pipe if its size is identified by a nominal dimension called iron pipe size (IPS) and by reference to a wall thickness schedule of piping.

The size identification of tubing is simple enough, since it consists of actual measured dimensions; but the terms used for identifying pipe sizes may require some explanation. A nominal dimension such as iron pipe size is close to—but not necessarily identical with—an actual measured dimension. For example, a pipe with a nominal pipe size of 3 inches has an actual measured outside diameter of 3.50 inches, and a pipe with a nominal pipe size of 2 inches has an actual measured outside diameter of 2.375 inches. In the larger sizes (above 12 inches) the nominal pipe size and the actual measured outside diameter are the same. For example, a pipe with a nominal pipe size of 14 inches has an actual measured outside diameter of 14 inches. Nominal dimensions are used in order to simplify the standardization of pipe fittings and pipe taps and threading dies.

The wall thickness of pipe is identified by reference to wall thickness schedules established by the American Standards Association. For example, a reference to schedule 40 for a steel pipe with a nominal pipe size of 3 inches indicates that the wall thickness of the pipe is 0.216 inch. A reference to schedule 80 for a steel pipe of the same nominal pipe size indicates that the wall thickness of this pipe is 0.300 inch. A reference to schedule 40 for steel pipe of nominal pipe size 4 inches indicates that the wall thickness of this pipe is 0.237 inch. As may be noted from these examples, a wall thickness schedule identification does not identify any one particular wall thickness unless the nominal pipe size is also specified.

The examples used here are given merely to illustrate the meaning of wall thickness schedule designations. Many other values can be found in pipe tables given in engineering handbooks and piping handbooks.

Pipe was formerly identified as standard (Std), extra strong (XS), and double extra strong (XXS). These designations, which are still used to some extent, also refer to wall thickness. However, pipe is manufactured in a number of different wall thicknesses, and some pipe does not fit into the standard, extra strong, and double extra strong classifications. The wall thickness schedules are being used increasingly to identify the wall thickness of pipe because they provide for the identification of a larger number of wall thicknesses than can be identified under the standard, extra strong, and double extra strong classifications.

It should be noted that pipe and tubing is occasionally identified in ways other than the standard ways described here. For example, some tubing is identified by inside diameter (ID) rather than by outside diameter (OD), and some pipe is identified by nominal pipe size, OD, ID, and actual measured wall thickness.

A great many different kinds of pipe and tubing are used in shipboard piping systems. A few shipboard applications that may be of particular interest are noted in the following paragraphs.

Seamless chromium-molybdenum alloy steel pipe is used for some high pressure, high temperature systems. The upper limit for the piping is 1500 psig and 1050° F.

Seamless carbon steel tubing is used in oil, steam, and feed water lines operating at 775° F and below. Different types of this tubing are available; the type used in any particular system depends upon the working pressure of the system.

Seamless carbon-molybdenum alloy steel tubing is used for feed water discharge piping, boiler pressure superheated steam lines, and boiler pressure saturated steam lines. Several types of this tubing are available; the type used in any particular case depends upon the boiler operating pressure and the superheater outlet temperature. The upper pressure and temperature limits for any class of this tubing are 1500 psig and 875° F.

Seamless chromium-molybdenum alloy steel tubing is used for high pressure, high temperature steam service on newer ships. This type of alloy steel tubing is available with different percentages of chromium and molybdenum, with upper limits of 1500 psig and 1050° F.

Welded carbon steel tubing is used in some water, steam, and oil lines where the temperature does not exceed 450° F. There are several types of this tubing; each type is specified for certain services and certain service conditions.

Nonferrous pipe and nonferrous tubing are used for many shipboard systems. Nonferrous metals are used chiefly where their special properties of corrosion resistance and high heat conductivity are required. Various types of seamless copper tubing are used for refrigeration lines, plumbing and heating systems, lubrication systems, and other shipboard systems. Copper-nickel alloy tubing is widely used aboard ship. Seamless brass tubing is used in systems which must resist the corrosive action of salt water and other fluids; it is available in types and sizes suitable for operating pressures up to 4000 psig. Seamless aluminum tubing is used for dry lines in sprinkling systems and for some bilge and sanitary drain systems.

Many other kinds of pipe and tubing besides the kinds mentioned here are used in shipboard piping systems. It is important to remember that design considerations govern the selection of any particular pipe or tubing for a particular system. Although many kinds of pipe and tubing look almost exactly alike from the outside, they may respond very differently to pressures, temperatures, and other service conditions. Therefore, each kind of pipe and tubing can be used only for the specified applications.

PIPE FITTINGS

Pipe or tubing alone does not constitute a piping system. To make the pipe or tubing into a system, it is necessary to have a variety of fittings, connections, and accessories by which the sections of pipe or tubing can be properly joined and the flow of the transferred fluid may be controlled. The following sections of this chapter deal with some of the pipe fittings most commonly used in shipboard piping systems; these fittings include unions, flanges, expansion joints, flareless fluid connections, steam traps, strainers, and valves.

UNIONS

Union fittings are provided in piping systems to allow the piping to be taken down for repairs and alterations. Unions are available in many different materials and designs to withstand a wide range of pressures and temperatures. Figure 14-1 shows some commonly used types of unions.

FLANGES

Flanges are used in piping systems to allow easy removal of piping and other equipment. The materials used and the design of the flanges are governed by the requirements of service. Flanges in steel piping systems are usually welded to the pipe or tubing. Flanges in nonferrous systems are usually brazed to the pipe or tubing.

EXPANSION JOINTS

Expansion joints are used in some piping systems to allow the piping to expand and contract with temperature changes, without damage to the piping. Two basic types of expansion joints are used in shipboard piping systems: sliding-type joints and flexing-type joints.

Sliding-type expansion joints include sleeve joints, rotary joints, ball and socket joints, and joints made up of some combination of these types. The amount of axial and rotary motion that can be abosrbed by any particular type

of sliding expansion joint depends upon the specific design of the joint.

Flexing-type expansion joints are those in which motion is absorbed by the flexing action of a bellows or some similar device. There are various kinds of flexing-type expansion joints, each kind being designed to suit the requirements of the particular system in which it is installed. Figure 14-2 illustrates the general principle of a bellows-type expansion joint.

Expansion joints are not always used in piping systems, even when allowance must be made for expansion and contraction of the piping. The same effect can be achieved by using directional changes and expansion bends or loops.

FLARELESS FLUID CONNECTIONS

A special flareless fluid connection has recently been developed for connecting sections of tubing in some high pressure shipboard systems. This fitting, which is generally known as the bite-type fitting, is very useful for certain applications because it is smaller and lighter in weight than the conventional fittings previously used to join tubing. The bite-type fitting is used on certain selected systems where the tubing is between 1/8 and 2 inches in outside diameter.

The bite-type fitting, shown in figure 14-3, consists of a body, a ferrule or sleeve that grips the tubing, and a nut. The fitting is not used in places where there is insufficient space for proper tightening of the nut, in places where piping or equipment would have to be removed in order to gain access to the fitting, or in places where the tubing cannot be easily deflected for ready assembly or breakdown of the joint. The fitting is sometimes used on gage board or instrument panel tubing, provided the gage board or panel is designed to be removed as a unit when repairs are required.

STEAM TRAPS

Steam traps are installed in steam lines to drain condensate from the lines without allowing the escape of steam. There are many different designs of steam traps, some being suitable for high pressure use and others being suitable for low pressure use. In general, a steam trap consists of a valve and some device or arrangement that will cause the valve to open and close as necessary to drain the condensate from the lines without allowing the escape of steam. Steam traps are installed at

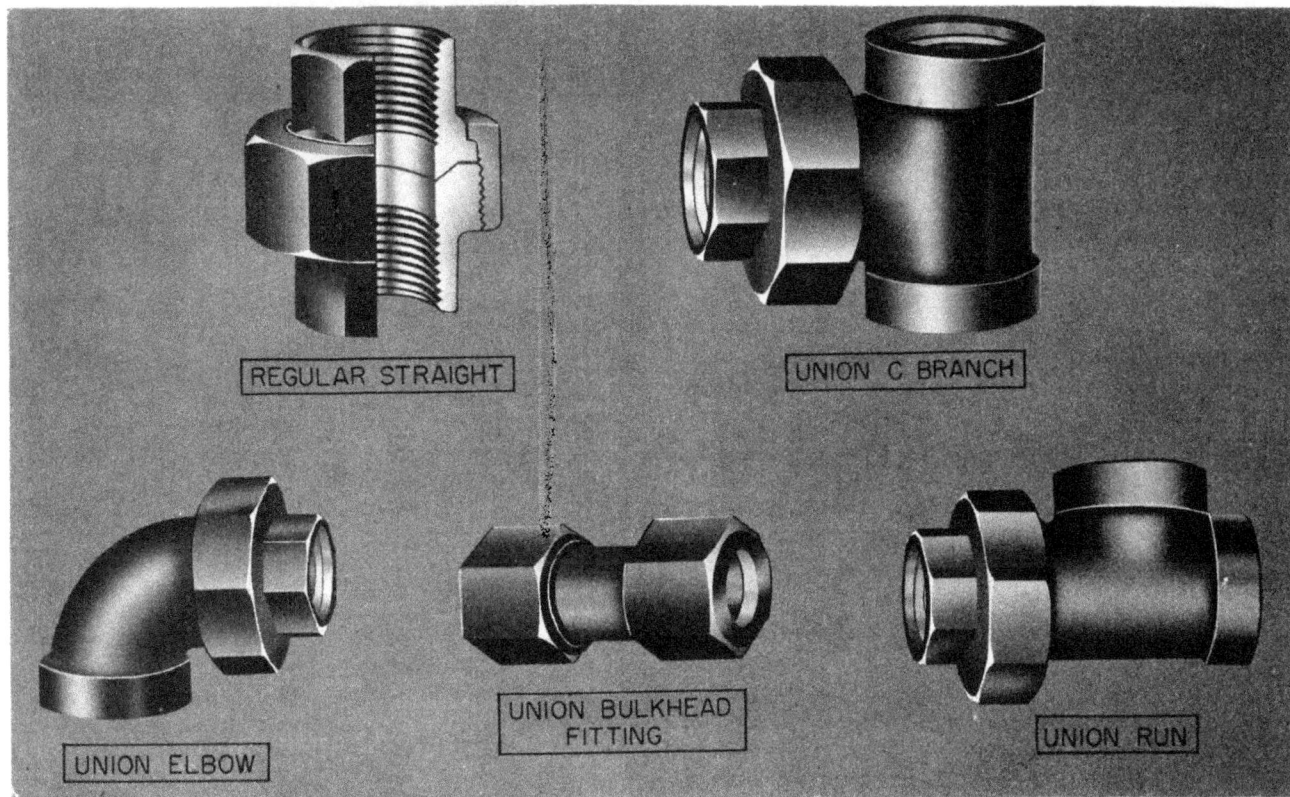

11.313

Figure 14-1.—Unions.

low points in the system or machinery to be drained. Some types of steam traps that are used in the Navy are described here.

MECHANICAL STEAM TRAPS.—Mechanical steam traps in common use include ball float traps and bucket-type traps.

A ball float steam trap is shown in figure 14-4. The valve of this trap is connected to the float in such a way that the valve opens when the float rises. When the trap is in operation, the steam and any water that may be mixed with it flows into the float chamber. As the water level rises, it lifts the float and this in turn lifts the valve plug and opens the valve. The condensate drains out and the float moves down to a lower position, closing the valve. The condensate that passes out of the trap is returned to the feed system.

A bucket-type steam trap is shown in figure 14-5. As condensate enters the trap body, the bucket floats. The valve is connected to the bucket in such a way that the valve closes as the

bucket rises. As condensate continues to flow into the trap body, the valve remains closed until the bucket is full. When the bucket is full, it sinks and thus opens the valve. The valve remains open until enough condensate has passed out to allow the bucket to float, thus closing the valve.

THERMOSTATIC STEAM TRAPS.— There are several kinds of thermostatic steam traps in use. In general, these traps are more compact and have fewer moving parts than most mechanical steam traps.

A bellows-type thermostatic steam trap is shown in figure 14-6. The operation of this trap is controlled by the expansion of the vapor of a volatile liquid which is enclosed in a bellows-type element. Steam enters the trap body and heats the volatile liquid in the sealed bellows, thus causing expansion of the bellows. The valve is attached to the bellows in such a way that the valve closes when the bellows expands. The valve remains closed, trapping steam in the valve body. As the steam cools and condenses,

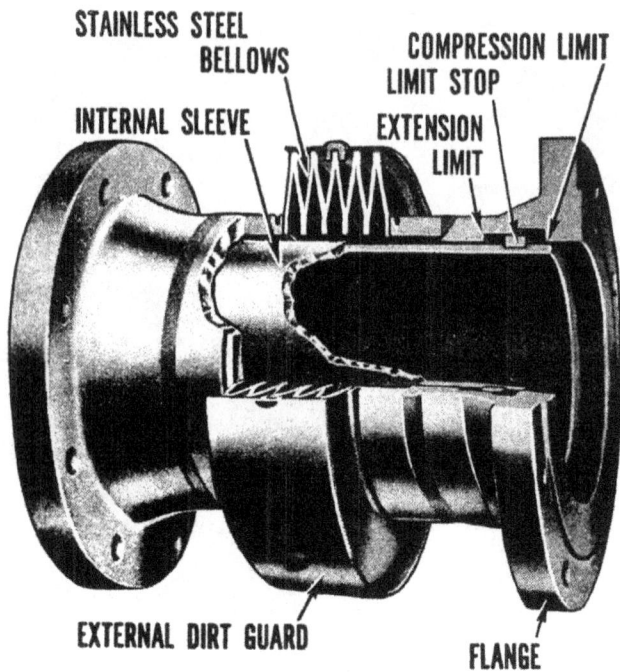

STAINLESS STEEL BELLOWS
COMPRESSION LIMIT LIMIT STOP
INTERNAL SLEEVE
EXTENSION LIMIT
EXTERNAL DIRT GUARD
FLANGE

38.127X

Figure 14-2.—Bellows-type expansion joint.

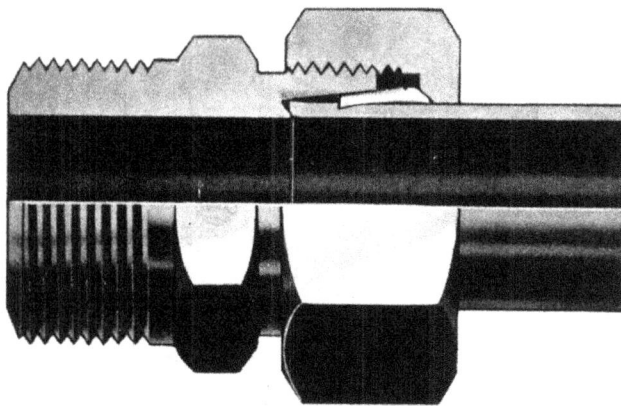

AIR VENT   CONDENSATE INLET
VALVE CONTROL
CONDENSATE DISCHARGE
BALL FLOAT
VALVE
SALINITY CELL CONN.

11.325D

Figure 14-4.—Ball float steam trap.

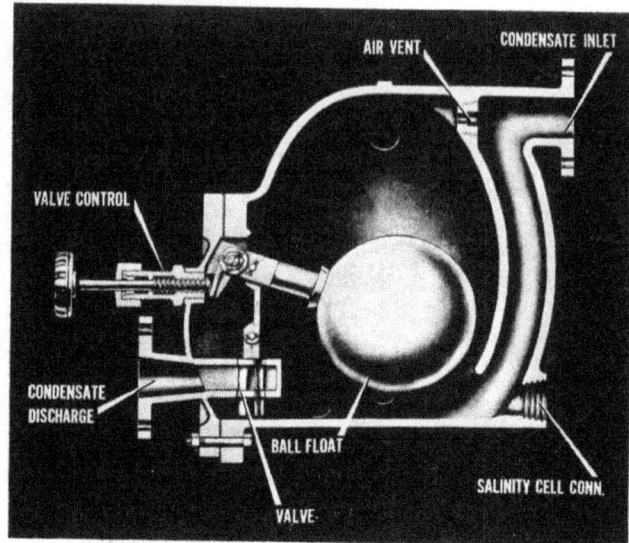

11.312

Figure 14-3.—Flareless fluid connection (bite-type fitting).

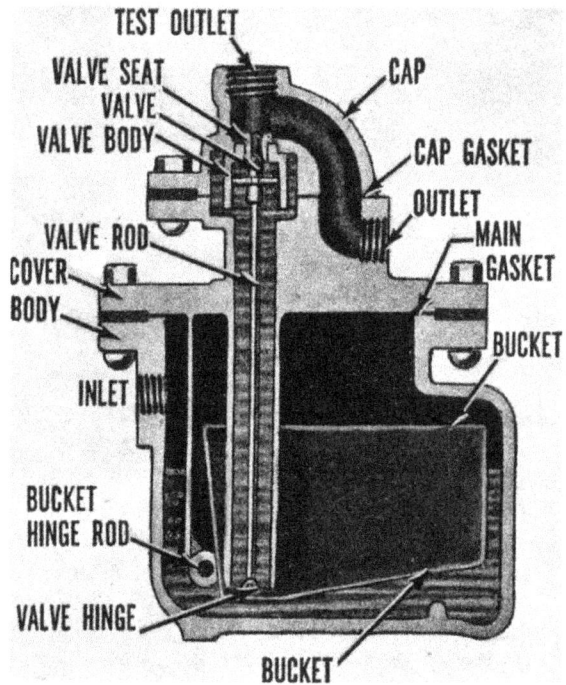

TEST OUTLET
VALVE SEAT
VALVE
VALVE BODY
CAP
CAP GASKET
OUTLET
MAIN GASKET
VALVE ROD
COVER
BODY
BUCKET
INLET
BUCKET HINGE ROD
VALVE HINGE
BUCKET

11.325X

Figure 14-5.—Bucket-type steam trap.

the bellows cools and contracts, thereby opening the valve and allowing the condensate to drain.

IMPULSE STEAM TRAPS.—Impulse steam traps of the type shown in figure 14-7 are used in some steam drain collecting systems aboard ship. Steam and condensate pass through a strainer before entering the trap. A circular

baffle keeps the entering steam and condensate from impinging on the cylinder or on the disk.

The impulse type of steam trap depends for its operation on the fact that hot water under pressure tends to flash into steam when the pressure is reduced. In order to understand how

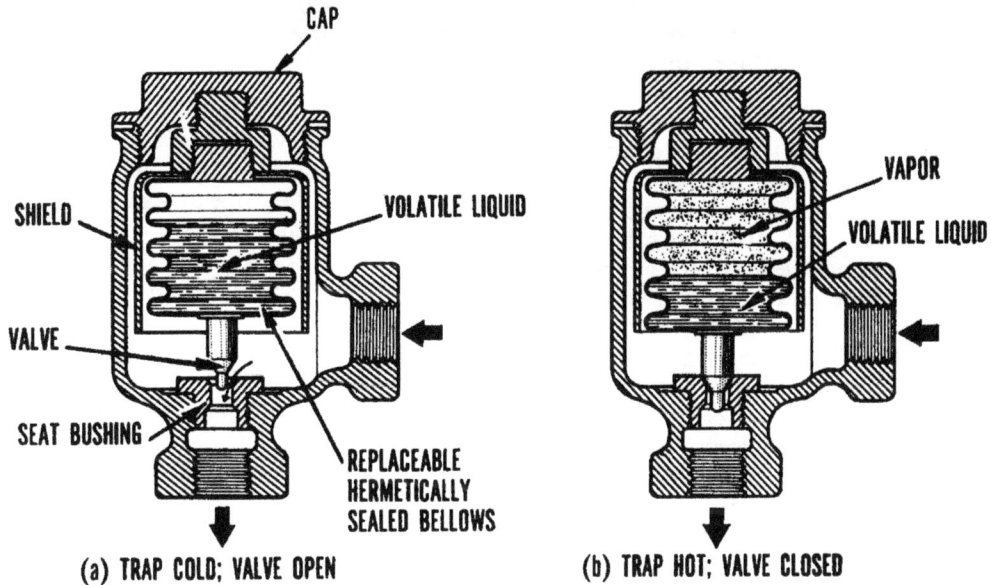

(a) TRAP COLD; VALVE OPEN

(b) TRAP HOT; VALVE CLOSED

11.326X

Figure 14-6.—Thermostatic steam trap.

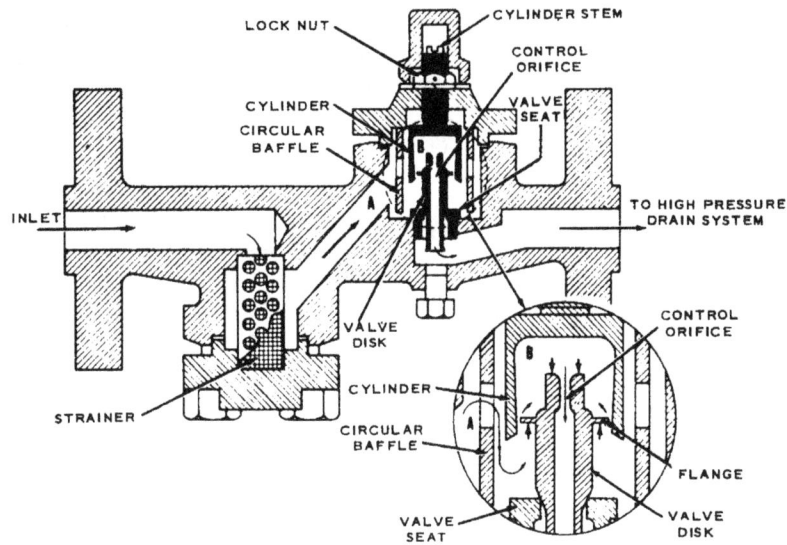

38.126

Figure 14-7.—Impulse steam trap.

this principle is utilized, we will consider the arrangement of parts shown in figure 14-7 and see what happens to the flow of condensate under various conditions.

The only moving part in the steam trap is the disk. This disk is rather unusual in design. Near the top of the disk there is a flange that acts as a piston. As may be seen in figure 14-7, the working surface above the flange is larger than the working surface below the flange; the importance of having this larger effective area above the flange will presently become apparent.

A control orifice runs through the disk from top to bottom, being considerably smaller at the top than at the bottom. The bottom part of the disk extends through and beyond the orifice in the seat. The upper part of the disk (including the flange) is inside a cylinder. The cylinder

369

tapers inward, so the amount of clearance between the flange and the cylinder varies according to the position of the valve. When the valve is open, the clearance is greater than when the valve is closed.

When the trap is first cut in, pressure from the inlet (chamber A) acts against the underside of the flange and lifts the disk off the valve seat. Condensate is thus allowed to pass out through the orifice in the seat; and, at the same time, a small amount of condensate (called control flow) flows up past the flange and into chamber B. The control flow discharges through the control orifice, into the outlet side of the trap, and the pressure in chamber B remains lower than the pressure in chamber A.

As the line warms up, the temperature of the condensate flowing through the trap increases. The reverse taper of the cylinder varies the amount of flow around the flange until a balanced position is reached in which the total force exerted above the flange is equal to the total force exerted below the flange. It is important to note that there is still a pressure difference between chamber A and chamber B. The force is equalized because the effective area above the flange is larger than the effective area below the flange. The difference in working area is such that the valve maintains an open, balanced position when the pressure in chamber B is 86 percent of the pressure in chamber A.

As the temperature of the condensate approaches its boiling point, some of the control flow going to chamber B flashes into steam as it enters the low pressure area. Since the steam has a much greater volume than the water from which it is generated, pressure builds up in the space above the flange (chamber B). When the pressure in this space is 86 percent of the inlet pressure (chamber A), the force exerted on the top of the flange pushes the entire disk downward and so closes the valve.

With the valve closed, the only flow through the trap is past the flange and through the control orifice. When the temperature of the condensate entering the trap drops slightly, condensate enters chamber B without flashing into steam. Pressure in chamber B is thus reduced to the point where the valve opens and allows condensate to flow through the orifice in the valve seat. Thus the entire cycle is repeated continuously.

With a normal condensate load, the valve opens and closes at frequent intervals, discharging a small amount of condensate at each opening. With a heavy condensate load, the valve remains wide open and allows a continuous discharge of condensate.

ORIFICE - TYPE STEAM TRAPS.—Aboard ship, continuous-flow steam traps of the orifice type are used in some constant service steam systems, oil heating steam systems, ventilation preheaters, and other systems or services in which condensate forms at a fairly constant rate. Orifice-type steam traps are not suitable for services in which the condensate formation is not continuous.

There are several variations of the orifice-type steam trap, but all types have one thing in common—they contain no moving parts. One or more restricted passageways or orifices allow condensate to trickle through but do not allow steam to flow through. Some orifice-type steam traps have baffles as well as orifices.

BIMETALLIC STEAM TRAPS.—Bimetallic steam traps of the type shown in figure 14-8 are used on many ships to drain condensate from main steam lines, auxiliary steam lines, and other steam lines. The main working parts of this steam trap are a segmented bimetallic element and a ball-type check valve.

The bimetallic element consists of several bimetallic strips[2] fastened together in a segmented fashion, as shown in figure 14-8. One end of the bimetallic element is fastened rigidly to a part of the trap body; the other end, which is free to move, is fastened to the top of the stem of the ball-type check valve.

Line pressure acting on the check valve tends to keep the valve open. When steam enters the trap body, the bimetallic element expands unequally because of the differential response to temperature of the two metals; the bimetallic element deflects upward at its free end, thus moving the valve stem upward and closing the valve. As the steam cools and condenses, the bimetallic element moves downward, toward the horizontal position, thus opening the valve and allowing some condensate to flow out through the valve. As the flow of condensate begins, a greater area of the ball is exposed to the higher pressure above the seat. The valve now opens wide and allows a full capacity flow of condensate.

---

[2]The principle of bimetallic expansion is discussed in chapter 7 of this text.

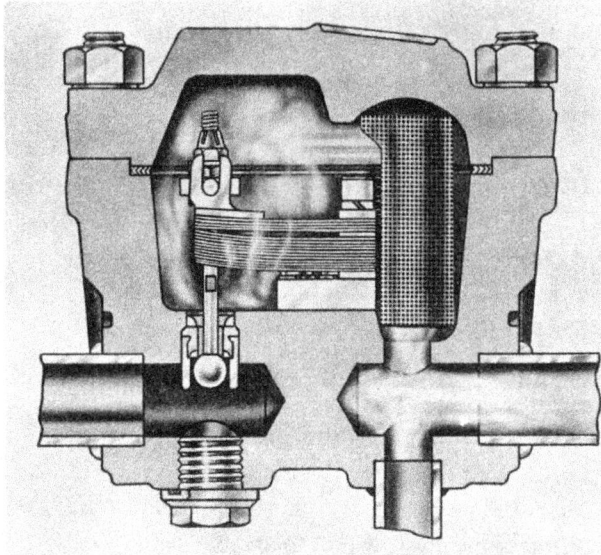

147.104X
Figure 14-8.—Bimetallic steam trap.

STRAINERS AND FILTERS

Strainers are fitted in practically all piping lines to prevent the passage of grit, scale, dirt, and other foreign matter which could obstruct pump suction valves, throttle valves, or other machinery parts.

Figure 14-9 illustrates three common types of strainers. Part A shows a bilge suction strainer located in the bilge pump suction line between the suction manifold and the pump. Any debris which enters the piping is collected in the strainer basket. The basket can be removed for cleaning by loosening the strongback screws, removing the cover, and lifting the basket out by its handle. Part B of figure 14-9 shows a duplex oil strainer of the type commonly used in fuel oil and lubricating oil lines, where it is essential to maintain an uninterrupted flow of oil. The flow may be diverted from one basket to the other, while one is being cleaned. Part C of figure 14-9 shows a manifold steam strainer. This type of strainer is desirable where space is limited, since it eliminates the use of separate strainers and their fittings. The cover is located so that the strainer basket can be removed for cleaning.

Metal-edge filters are used in the lubrication systems of many auxiliary units. A metal-edge filter consists of a series of metal plates or disks. Turning a handle moves the plates or disks across each other in such a way as to remove any particles that have collected on the metal surfaces. Some metal-edge type filters have magnets to aid in removing fine particles of magnetic materials.

VALVES

Every piping system must have some means of controlling the amount and direction of the flow of the contained fluid through the lines. The control of fluid flow is accomplished by the installation of valves.

Valves are usually made of bronze, brass, iron, or steel. Steel valves are either cast or forged, and are made of either plain steel or alloy steel. Alloy steel valves are used in high pressure, high temperature systems; the disks and seats of these valves are usually surfaced with Stellite, an extremely hard chromium-cobalt alloy.

Bronze and brass valves are not used in high temperature systems. Also, they are not used in systems in which they would be exposed to severe conditions of pressure, vibration, or shock. Bronze valves are widely used in salt water systems. The seats and disks of bronze valves used for sea water service are often made of Monel, a metal that is highly resistant to corrosion and erosion.

Many different types of valves are used to control the flow of liquids and gases. The basic valve types can be divided into two groups, stop valves and check valves. Stop valves are those which are used to shut off—or partially shut off—the flow of fluid. Stop valves are controlled by the movement of the valve stem. Check valves are those which are used to permit the flow of fluid in only one direction. Check valves are designed to be controlled by the movement of the fluid itself.

Stop valves include globe valves, gate valves, plug valves, piston valves, needle valves, and butterfly valves. Check valves include ball-check valves, swing-check valves, and lift-check valves.

Combination stop-check valves are valves which function either as stop valves or as check valves, depending upon the position of the valve stem.

In addition to the basic types of valves, a good many special valves which cannot really be classified either as stop valves or as check valves are found in the engineering spaces. Many of these special valves serve to control

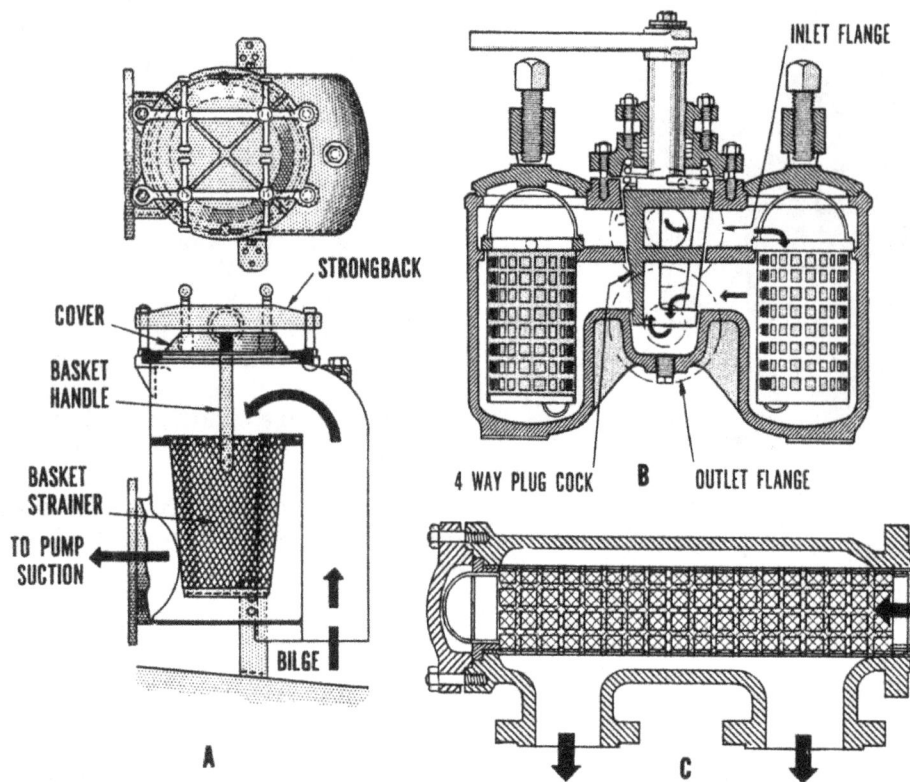

11.329X

Figure 14-9.—A. Bilge suction strainer. B. Duplex oil strainer.
C. Manifold steam strainer.

the pressure of fluids and are therefore general-ly called pressure-control valves. Others are identified by names which indicate their general function—as, for example, thermostatic recir-culating valves. The following sections deal first with the basic types of stop valves and check valves and then with some of the more complex special kinds of valves.

Globe Valves

Globe valves are one of the commonest types of stop valves. Globe valves get their name from the globular shape of their bodies. It is important to note, however, that other types of valves may also have globe-shaped bodies; hence it is not always possible to identify a globe valve merely by external appearance. The inter-nal structure of the valve, rather than the exter-nal shape, is what distinguishes one type of valve from another.

The disk of a globe valve is attached to the valve stem. The disk seats against a seating ring or a seating surface and thus shuts off the flow of fluid. When the disk is moved off the seating surface, fluid can pass through the valve. Globe valves may be used partially open as well as fully open or fully closed.

Globe valve inlet and outlet openings are arranged in several ways, to suit varying re-quirements of flow. Figure 14-10 shows three common types of globe valve bodies. In the straight type, the fluid inlet and outlet openings are in line with each other. In the angle type, the inlet and outlet openings are at an angle to each other. An angle-type globe valve is used where a stop valve is needed at a 90° turn in a line. The cross type of globe valve has three openings rather than two; it is often used in connection with bypass piping.

A globe-type stop valve is shown in cross-sectional view in figure 14-11. Figure 14-12 shows a cutaway view of a similar (but not identical) globe valve.

Globe valves are commonly used in steam, air, oil, and water lines. On many ships, the surface blow valves, the bottom blow valves,

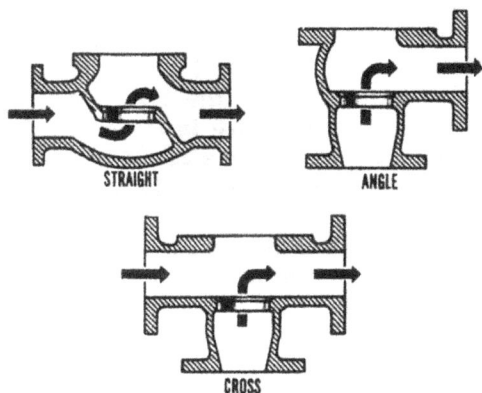

11.316X

Figure 14-10.—Types of globe valve bodies.

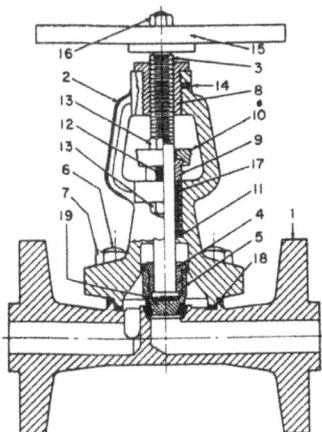

LIST OF PARTS	
PART NO.	NAME OF PART
1	VALVE BODY
2	BONNET
3	STEM
4	DISK NUT
5	DISK
6	BONNET STUD
7	BONNET STUD NUT
8	BONNET BUSHING
9	GLAND
10	GLAND FLANGE
11	PACKING STOP RING
12	GLAND STUD
13	GLAND STUD NUT
14	SET SCREW
15	HANDWHEEL
16	HANDWHEEL NUT
17	PACKING
18	BONNET GASKET
19	DISK WASHER

38.117

Figure 14-11.—Cross-sectional view of globe stop valve.

the boiler stops, the feed stop valve, and many guarding valves and line cutout valves are of this type. Globe valves are also used as stop valves on the suction side of many pumps, as recirculating valves, and as throttle valves.

Gate Valves

Gate valves are used when a straight-line flow of fluid with a minimum amount of restriction is required. Gate valves are frequently used in water lines; for example, firemain cutout valves are usually gate valves. They are also used in steam lines, particularly on the newer ships. In fact, there appears to be an increasing trend toward the use of gate valves in many systems on new construction.

The part of a gate valve that serves the same purpose as the disk of a globe valve is called a gate. The gate is usually wedge-shaped, but some gates are of uniform thickness. When the gate is wide open, the opening through the valve is the same size as the pipe in which the valve is installed. Therefore, there is very little resistance to flow and very little pressure drop through this type of valve. Gate valves are not suitable for use as throttling valves, since the regulation of flow would be difficult and since the flow of fluid against a partially opened gate would cause extensive damage to the valve.

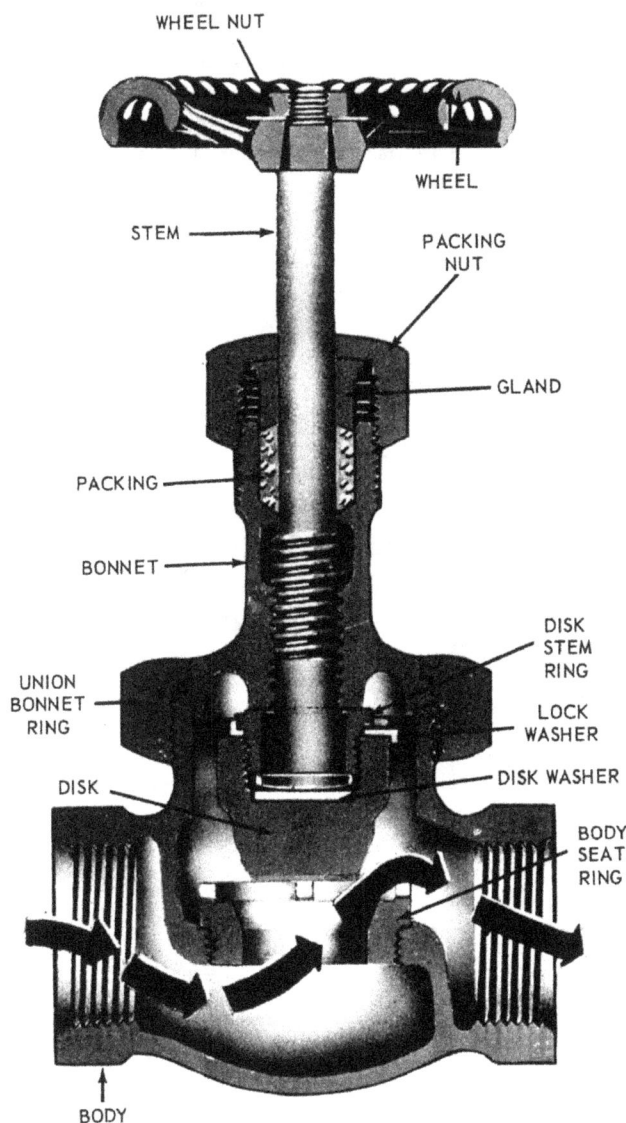

11.317.1X

Figure 14-12.—Cutaway view of globe stop valve.

373

As shown in figures 14-13 and 14-14, the gate is connected to the valve stem. Turning the handwheel positions the valve gate. Some gate valves have nonrising stems—that is, the stem is threaded on the lower end and the gate is threaded on the inside so that the gate travels up the stem when the valve is being opened. Gate valves with nonrising stems are shown in figure 14-13. This type of valve usually has a pointer or a gage to indicate whether the valve is in the open position or in the closed position. Some gate valves have rising stems—that is, both the gate and the stem move upward when the valve is opened. In some rising stem valves, the stem projects above the handwheel when the valve is opened; in other rising stem valves, the stem does not project above the handwheel. A pointer or a gage is required to indicate the position of the valve if the stem does not project above the handweel when the valve is in the open position.

Plug Valves

The body of a plug valve is shaped in such a way that it will hold a cylindrical or tapered plug. Holes or ports in the body line up with the pipe in which the valve is installed. A solid cylindrical plug (or in some cases a plug shaped like a truncated cone) fits snugly into the hollow of the body. The plug is attached to a handle, by means of which the plug can be turned within the body. A passageway is bored through the plug. When the valve is in the open position, the passage in the plug lines up with the inlet and outlet ports of the body, thus allowing fluid to flow through the valve. When the plug is turned

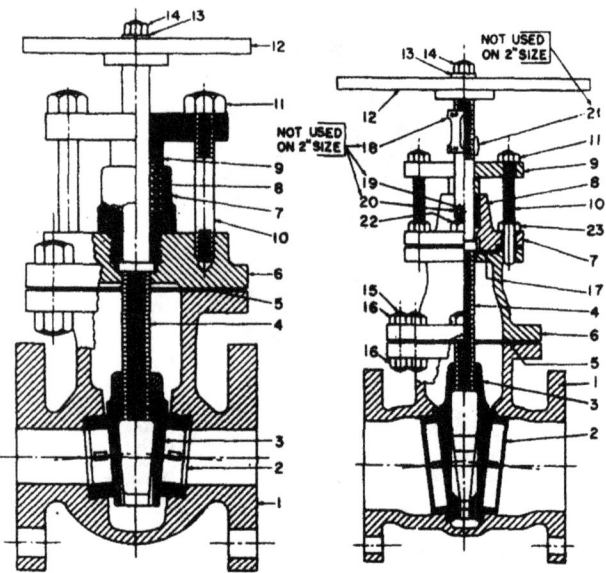

LIST OF PARTS			
PART NO.	NAME OF PART	PART NO.	NAME OF PART
1	BODY	13	HANDWHEEL WASHER
2	SEAT RING	14	HANDWHEEL NUT
3	GATE	15	BONNET STUD
4	STEM	16	BONNET STUD NUT
5	BONNET GASKET	17	STUFFING BOX GASKET
6	BONNET	18	INDICATOR PLATE
7	STUFFING BOX	19	LOCK WASHER
8	PACKING	20	INDICATOR PLATE SCREW
9	GLAND	21	INDICATOR NUT
10	GLAND STUD	22	STUFFING BOX STUD
11	GLAND STUD NUT	23	STUFFING BOX STUD NUT
12	HANDWHEEL		

38.118
Figure 14-13.—Cross-sectional views of gate stop valves (nonrising stem type).

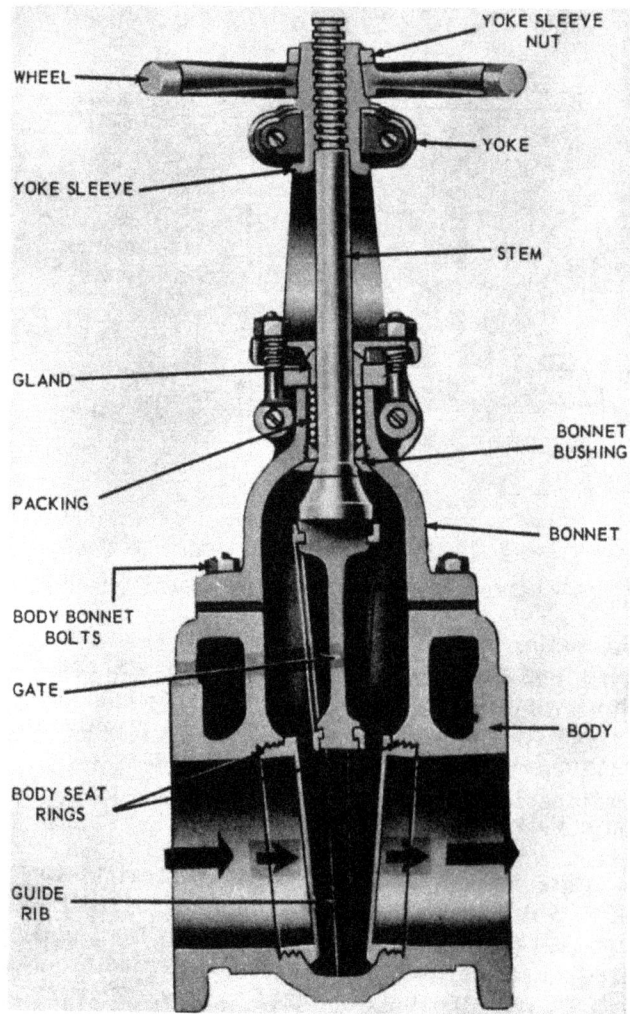

11.317.2X
Figure 14-14.—Cutaway view of gate stop valve (rising stem type).

in the body, the solid part of the plug blocks the ports and thus prevents the flow of fluid.

Plug valves are quite commonly used in connection with auxiliary machinery. The petcocks that are used as vents on lubricating oil coolers for auxiliary machinery are usually plug valves. The three-way and four-way cocks that allow selective routing of various fluids are usually variations of the plug valve. The shutoff device that allows fuel oil or lubricating oil to be diverted from one basket to another of a duplex strainer is often a modified plug valve.

Piston Valves

A piston valve is a stop valve that may be thought of as a combination of a gate valve and a plug valve. The piston valve consists basically of a cylindrical piston operating in a hollow cylinder. The piston is attached to the valve stem, and the valve stem is attached to a handwheel. When the handwheel is turned, the piston is raised or lowered within the hollow cylinder. The cylinder has ports in its walls. When the piston is raised, the ports are uncovered and fluid is allowed to pass through the valve.

Needle Valves

Needle valves are stop valves that are used for making relatively fine adjustments in the amount of fluid that is allowed to pass through an opening. The distinguishing characteristic of a needle valve is the long, tapering, needle-like point on the end of the valve stem. This "needle" acts as the valve disk. The longer part of the needle is smaller than the orifice in the valve seat, and therefore passes through it before the needle seats. This arrangement permits a very gradual increase or decrease in the size of the opening and thus allows a more precise control of flow than could be obtained with an ordinary globe valve.

Needle valves are used as overload nozzles on some auxiliary turbines. Needle valves are often used as component parts of other more complicated valves. For example, they are used in some types of reducing valves. Most constant-pressure pump governors[3] have needle valves to minimize the effects of fluctuations in pump discharge pressure. Needle valves

are also used in some components of automatic boiler control systems.

Butterfly Valves

The butterfly valve, shown in figure 14-15, is being used increasingly in naval ships. The butterfly valve has some definite advantages for certain services. It is light in weight, it takes up less space than a gate valve or a globe valve of the same capacity, and it is relatively quick acting. The butterfly valve provides a positive shutoff and may be used as a throttling valve set in any position from full open to full closed.

Butterfly valves vary somewhat in design and construction. However, a butterfly-type disk and a positive means of sealing are common to all butterfly valves.

The butterfly valve described and illustrated here consists of a body, a resilient seat, a butterfly-type disk, a stem, packing, a notched positioning plate, and a handle. The resilient seat is under compression when it is mounted in the valve body, thus making a seal around the periphery of the disk and both upper and lower points where the stem passes through the seat. Packing is provided to form a positive seal around the stem if the seal formed by the seat should become damaged.

To close a butterfly valve, it is only necessary to turn the handle a quarter of a turn in order to rotate the disk 90 degrees. The resilient seat exerts positive pressure against the disk, ensuring a tight shutoff.

Butterfly valves may be designed to meet a variety of requirements. The shipboard systems in which these valves are now being used include fresh water, salt water, JP-5, Navy special fuel, diesel oil, and lubricating oil.

Check Valves

Check valves are designed to permit flow through a line in one direction only. There are almost innumerable examples of check valves throughout the engineering plant. Check valves are used in open funnel drains, in fuel oil heater drains, and in various other drains. They are used in connection with many pumps, and in any line in which it is important to prevent the back flow of fluid.

The port in a check valve may be closed by a disk, a ball, or a plunger. The valve opens when the pressure on the inlet side is greater,

---

[3]Constant-pressure pump governors are discussed in chapter 16 of this text.

11.318

Figure 14-15.—Butterfly valve.

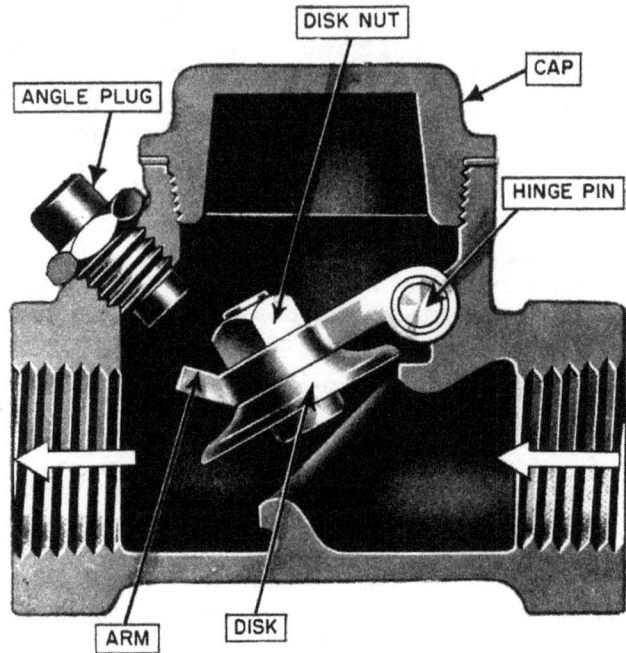

11.319X

Figure 14-16.—Swing-check valve.

and closes when the pressure on the outlet side is greater. All check valves open and close automatically.

A swing-check valve is illustrated in figure 14-16. Figures 14-17 and 14-18 show a lift-check valve. As may be seen, the disk of the swing-check valve moves through an arc, while the disk of a lift-check valve moves up and down in response to changes in the pressure of the incoming fluid. A good example of a ball-check valve is the discharge valve in the bimetallic steam trap, described and illustrated earlier in this chapter.

Stop-Check Valves

As we have seen, most valves can be classified as being either stop valves or check valves. Some valves, however, function either as stop valves or as check valves, depending upon the position of the valve stem. These valves are known as stop-check valves.

Stop-check valves are shown in cross section in figure 14-19. As may be seen, this type of valve looks very much like a lift-check valve. However, the valve stem is long enough so that when it is screwed all the way down it holds the disk firmly against the seat, thus preventing any flow of fluid. In this position, the valve acts as a stop valve. When the stem is raised, the side can be opened by pressure on the inlet side. In this position, the valve acts as a check valve, allowing the flow of fluid in only one direction. The maximum lift of the disk is controlled by the position of the valve stem. Therefore, the position of the valve stem limits the amount of fluid passing through the valve even when the valve is operating as a check valve.

Stop-check valves are widely used throughout the engineering plant. One of the best examples is the so-called boiler feed check valve, which is actually a stop-check valve rather than a true check valve. Stop-check valves are used in many drain lines; on the discharge side of many pumps; and as exhaust valves on auxiliary machinery.

376

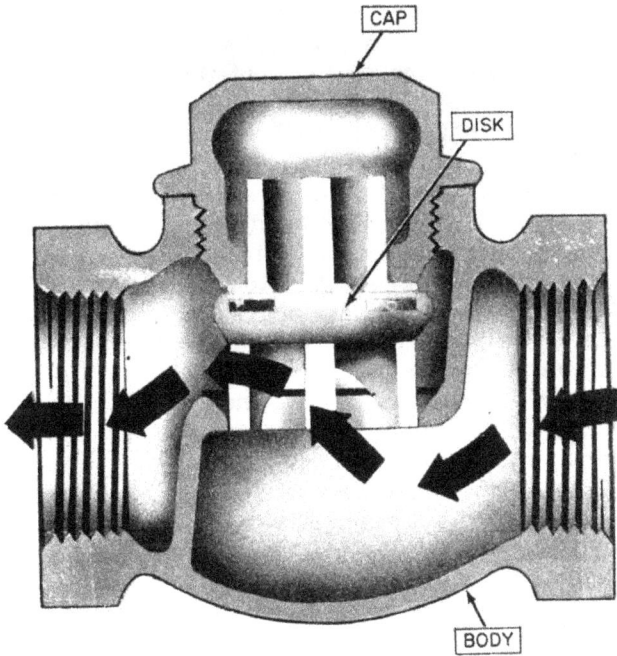

11.320

Figure 14-17.—Cutaway view of lift-check valve.

Pressure-Control Valves

Pressure-control valves are used to relieve or prevent excessive pressure, to reduce pressure, and to control or regulate pressure. Several common types of pressure-control valves are described in the following paragraphs.[4]

RELIEF VALVES.—Relief valves are designed to open automatically when the pressure in the line or in the machinery unit becomes too high. There are several different types of relief valves, but most of them have a disk or a ball which acts against a coil spring. The spring pushes downward against the disk or ball and so tends to keep the valve closed. When the pressure in the line or in the unit is great enough to overcome the resistance of the spring, the disk or ball is forced upward and the valve is thereby opened. After the pressure has been relieved by the escape of fluid through the relief

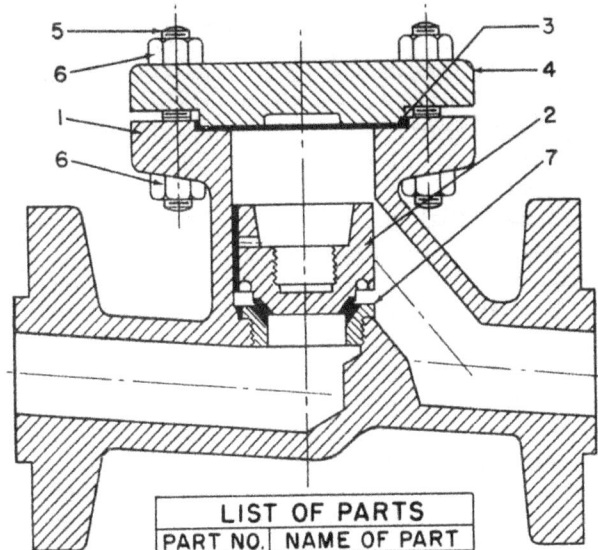

LIST OF PARTS	
PART NO.	NAME OF PART
1	BODY
2	DISK
3	GASKET
4	CAP
5	CAP STUD
6	CAP STUD NUT
7	SEAT RING

38.119

Figure 14-18.—Cross-sectional view of lift-check valve.

valve, the spring again exerts enough force to close the valve.

Relief valves are installed in steam, water, oil, and air lines, and on various units of auxiliary machinery. One or more adjusting nuts at the top of the valve provide a means by which the relief valve setting may be changed when necessary. However, unauthorized changes to relief valve settings must never be permitted.

SENTINEL VALVES.—Small spring-loaded sentinel valves are sometimes attached to the inlet chamber of a relief valve, to give warning of dangerous pressures. Sentinel valves operate on the same general principles as relief valves.

REDUCING VALVES.—Reducing valves are automatic valves used to reduce the supply pressure to a specified lower discharge pressure. A reducing valve can be set for any desired discharge pressure, within the design limits of the valve. After the valve has been set, the reduced pressure will be maintained regardless of changes in the supply pressure (as long as

---

[4]Boiler safety valves, discussed in chapter 11 of this text, might also be considered as a special kind of pressure-control valves.

the supply pressure is at least as high as the desired delivery pressure) and regardless of the amount of reduced pressure fluid that is used.

There are several kinds of spring-loaded reducing valves. The one shown in figure 14-20 is used for steam service, but is very similar to spring-loaded reducing valves used for other services.

The principal parts of the valve are (1) the main valve, an upward-seating valve which has a piston on top of its valve stem; (2) an upward-seating auxiliary (or controlling) valve; (3) a controlling diaphragm; and (4) an adjusting spring.

High pressure steam (or other fluid) enters the valve on the inlet side and acts against the main valve disk, tending to close the main valve. However, high pressure steam is also led through ports to the auxiliary valve, which controls the admission of high pressure steam to the top of the main valve piston. The piston has a larger surface area than the main valve disk; therefore,

a relatively small amount of high pressure steam acting on the top of the main valve piston will tend to open the main valve, and so allow steam at reduced pressure to flow out the discharge side.

But what makes the auxiliary valve open to allow high pressure steam to get to the top of the main valve piston? The controlling diaphragm transmits a pressure downward upon the auxiliary valve stem, and thus tends to open the valve. However, reduced pressure steam is led back to the chamber beneath the diaphragm; this steam exerts a pressure upward on the diaphragm, which tends to close the auxiliary valve. The position of the auxiliary valve, therefore, is determined by the position of the controlling diaphragm.

The position of the diaphragm at any given moment is determined by the relative strength of two opposing forces: (1) the downward force exerted by the adjusting spring, and (2) the upward force exerted on the underside of the diaphragm

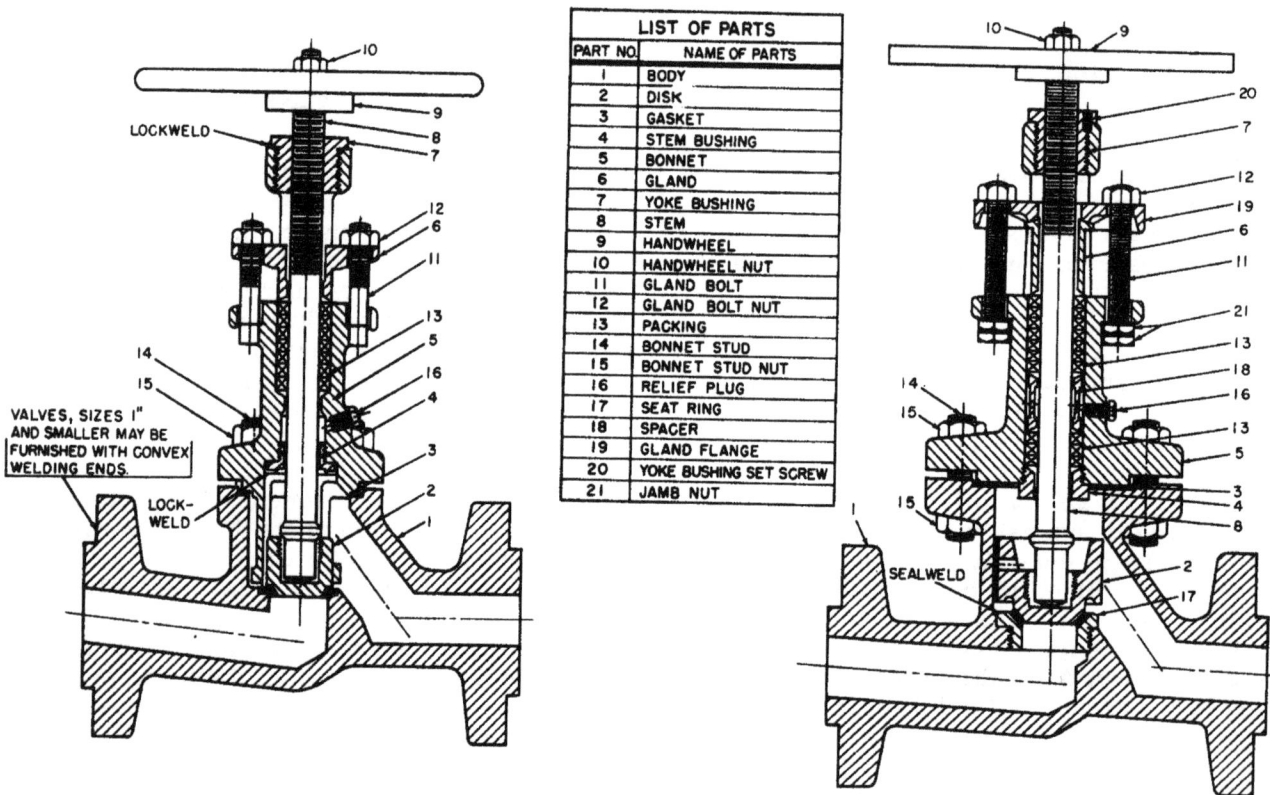

LIST OF PARTS	
PART NO.	NAME OF PARTS
1	BODY
2	DISK
3	GASKET
4	STEM BUSHING
5	BONNET
6	GLAND
7	YOKE BUSHING
8	STEM
9	HANDWHEEL
10	HANDWHEEL NUT
11	GLAND BOLT
12	GLAND BOLT NUT
13	PACKING
14	BONNET STUD
15	BONNET STUD NUT
16	RELIEF PLUG
17	SEAT RING
18	SPACER
19	GLAND FLANGE
20	YOKE BUSHING SET SCREW
21	JAMB NUT

Figure 14-19.—Stop-check valves.

11.321

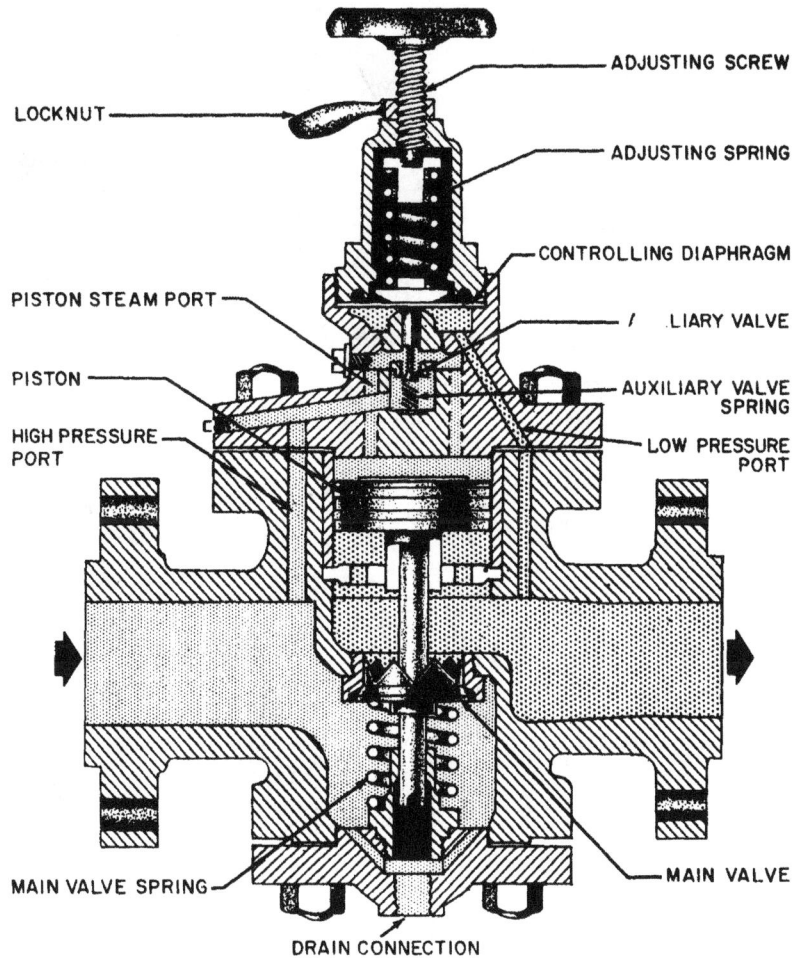

LOCKNUT

ADJUSTING SCREW

ADJUSTING SPRING

CONTROLLING DIAPHRAGM

PISTON STEAM PORT

AUXILIARY VALVE

PISTON

AUXILIARY VALVE SPRING

HIGH PRESSURE PORT

LOW PRESSURE PORT

MAIN VALVE SPRING

MAIN VALVE

DRAIN CONNECTION

47.59X

Figure 14-20.—Spring-loaded reducing valve.

by the reduced pressure steam. These two forces are continually seeking to reach a state of balance; and, because of this, the discharge pressure of the steam is kept constant as long as the amount of steam used is kept within the capacity of the valve.

There are two types of gas-loaded (or pneumatic pressure controlled) reducing valves. One type, shown in figure 14-21, is designed to regulate pressure in low temperature air, water, oil, or other fluids. The other type, shown in figure 14-22, is designed to regulate pressure in high temperature steam, hot water, or other fluids. Both types of valves operate on the principle that the pressure of an enclosed gas varies inversely as its volume.

We will consider first the valve for low temperature service (fig. 14-21). In this valve, a

relatively small change in the large volume within the dome loading chamber produces only a slight pressure variation, while the slightest variation in the small volume within the actuating chamber creates an enormous change in pressure. The restricting orifice connecting these two chambers governs the rate of pressure equalization by retarding the flow of gas from one chamber to another.

The dome loading chamber is charged with air or some other compressible gas at a pressure equal to the desired reduced pressure. When the chamber is loaded, and the loading valve is closed, the dome will retain its charge almost indefinitely. When the regulator is in operation, the trapped pressure within the dome passes into the actuating chamber through the small separation plate orifice and moves the large flexible

379

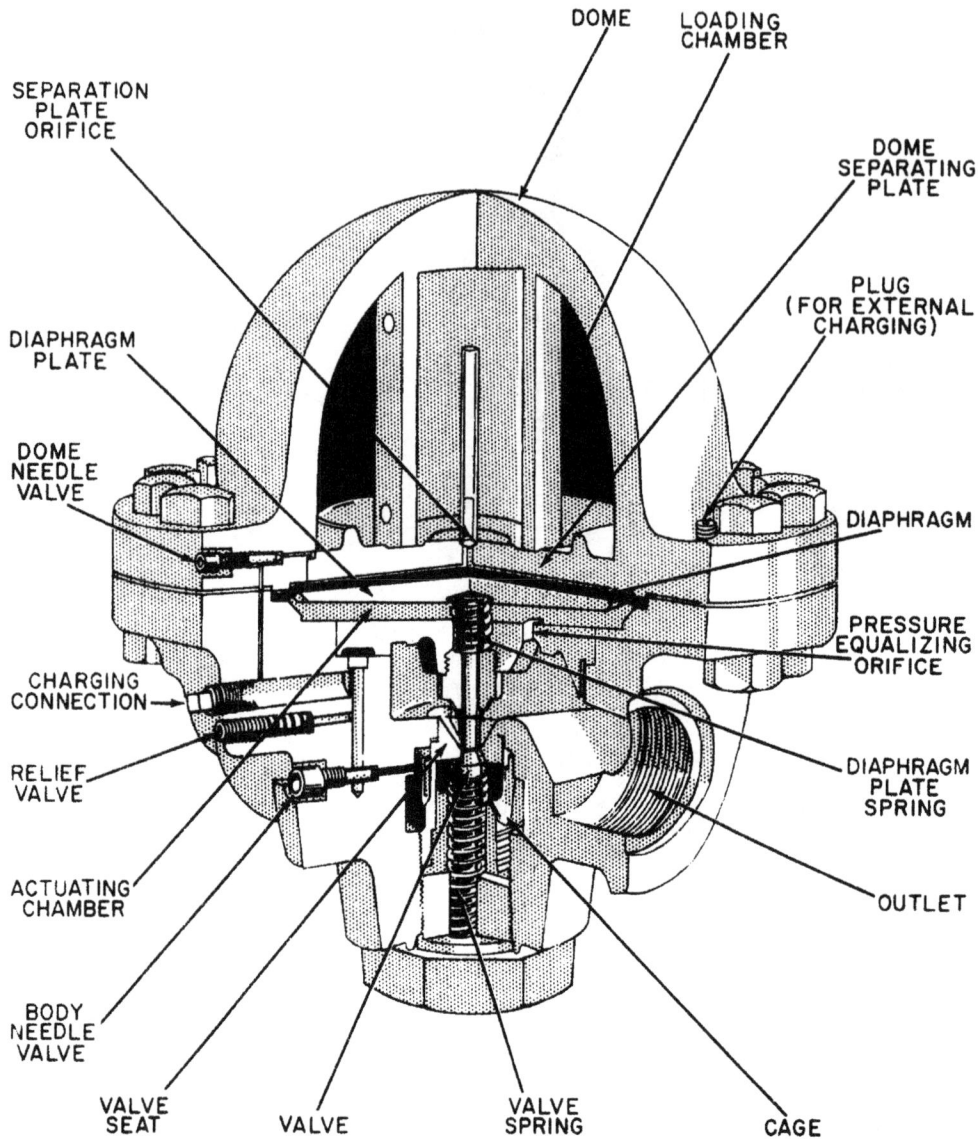

SEPARATION PLATE ORIFICE

DIAPHRAGM PLATE

DOME NEEDLE VALVE

CHARGING CONNECTION →

RELIEF VALVE

ACTUATING CHAMBER

BODY NEEDLE VALVE

VALVE SEAT

VALVE

VALVE SPRING

DOME

LOADING CHAMBER

DOME SEPARATING PLATE

PLUG (FOR EXTERNAL CHARGING)

DIAPHRAGM

PRESSURE EQUALIZING ORIFICE

DIAPHRAGM PLATE SPRING

OUTLET

CAGE

47.60

Figure 14-21.—Pneumatic pressure controlled reducing valve for low temperature service

diaphragm. This action forces the reverse-acting valve off its seat. The pressure entering the regulator is then permitted to flow through the open valve into the reduced pressure line. A large pressure equalizing orifice transmits this pressure directly to the underside of the diaphragm. When the delivered pressure approximates the loading pressure in the dome, and the unbalanced forces are equalized, the valve is closed. With the slightest drop in delivered pressure, the pressure charge in the

dome instantly forces the valve open, thus allowing air to pass through the valve and maintain the outlet pressure relatively constant.

The pneumatic pressure controlled reducing valve for high temperature service (fig. 14-22) operates in much the same way as the valve for low temperature service, except that the valve for high temperature service is designed in such a way as to keep heat from the hot fluid from affecting the gas in the loading chamber. The

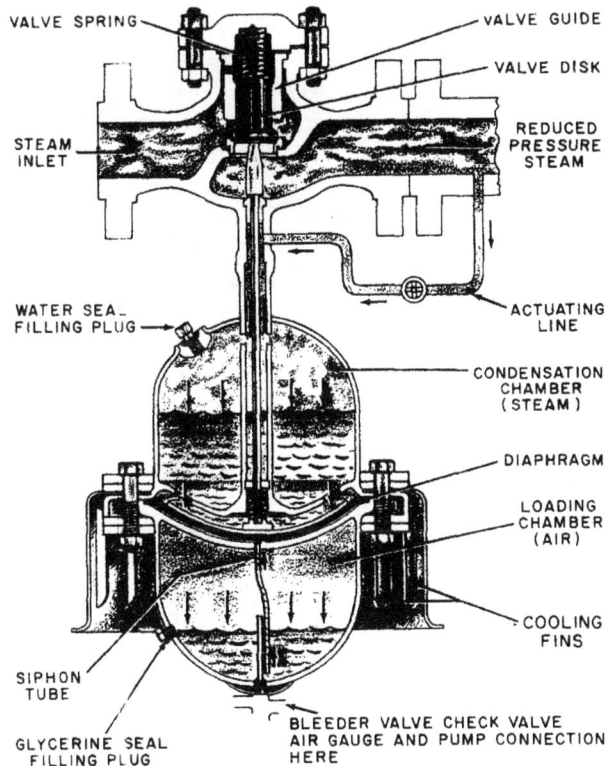

11.323

Figure 14-22.—Pneumatic pressure controlled reducing valve for high temperature service.

loading chamber is surrounded by a finned hood which conducts heat away to atmosphere.

A rubber diaphragm is installed in the middle of the dome. The bottom of the diaphragm is separated from the bottom half of the dome by a fixed steel plate. The area immediately above the diaphragm communicates with the upper part of the dome through holes in the shrouding. The upper half of the dome carries a level of water for sealing; the lower half of the dome carries a level of glycerine for sealing. The area above the glycerine is charged with air, which exerts a downward pressure on the glycerine and forces some of it to go up the tube toward the diaphragm. This pressure causes the diaphragm to move upward; and, since the stem of the valve is in contact with the diaphragm, the upward movement of the diaphragm causes the valve to open. When the valve is open, steam can pass through it.

From the outlet connection, an actuating line leads back to the upper part of the dome, as shown in the illustration. Steam at the reduced pressure is thus allowed to exert a force on the top of the water seal; this force is transmitted through the water and tends to move the diaphragm downward. When the pressure of the steam from the actuating line exceeds the loading air pressure in the lower half of the dome, the diaphragm moves downward sufficiently to close the valve. The closing of the valve reduces the pressure of the steam on the discharge side of the valve. When the pressure on the outlet side of the valve is equal to the air pressure in the lower half of the dome, the valve takes a balanced position which allows the passage of sufficient steam to maintain that pressure.

If the load increases, tending to take more steam away from the valve, the outlet pressure will be momentarily reduced. Thus, the pressure of steam on top of the diaphragm becomes less than the pressure of air below the diaphragm, and the valve then opens wider to restore the pressure to normal. If the load is reduced, this causes a momentary increase in outlet pressure; and this in turn increases the pressure on top of the diaphragm, making it greater than the air pressure below the diaphragm. The diaphragm is therefore displaced downward, and the outlet pressure is again restored to normal.

DIAPHRAGM CONTROL VALVES WITH AIR-OPERATED CONTROL PILOTS.—Diaphragm control valves with air-operated control pilots are being used increasingly on newer ships for various pressure-control applications. These valves and pilots are available in several basic designs to meet different requirements. They may be used to reduce pressure, to augment pressure, or to provide continuous regulation of pressure, depending upon the requirements of the system in which they are installed. Valves and pilots of very similar design can also be used for other services such as liquid level control and temperature control. However, the discussion here is limited to the valves and pilots that are used for pressure-control applications.

The air-operated control pilot may be either direct acting or reverse acting. A direct-acting air-operated control pilot is shown in figure 14-23. In this type of pilot, the controlled pressure—that is, the pressure from the discharge side of the diaphragm control valve—acts on top of a diaphragm in the control pilot. This pressure is balanced by the pressure exerted by the pilot adjusting spring. If the controlled pressure increases and overcomes the pressure exerted by the pilot adjusting spring, the pilot valve stem is forced down. This action causes the pilot valve to open and so to increase the amount of operating

CONTROLLED PRESSURE

LESLIE PRESSURE CONTROL PILOT

AIR SUPPLY

TO DIAPHRAGM
CONTROL VALVE

38.121X

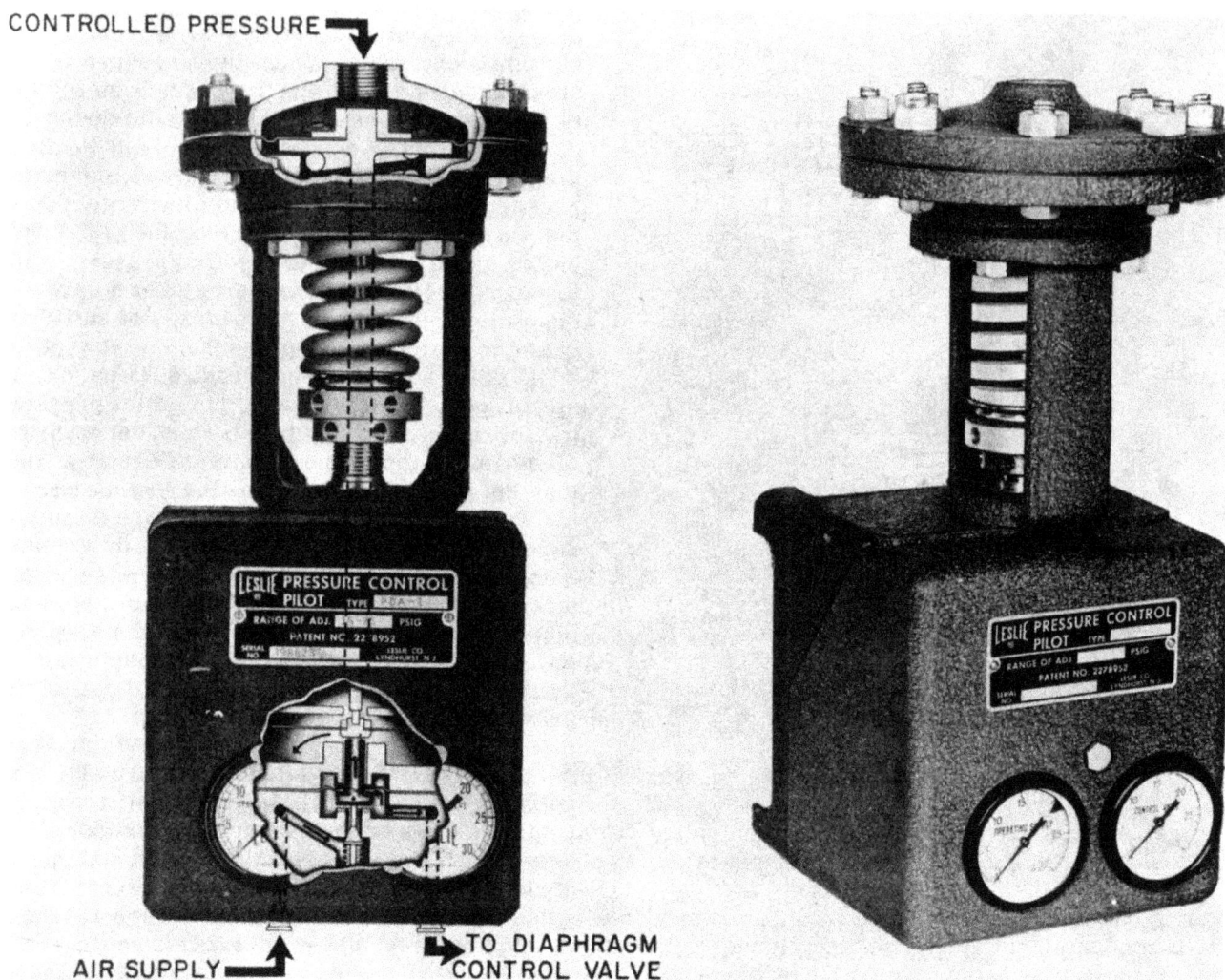

Figure 14-23.—Air-operated control pilot.

air pressure going from the pilot to the diaphram control valve. A reverse-acting pilot has a lever which reverses the pilot action. In a reverse-acting pilot, therefore, an increase in controlling pressure produces a decrease in operating air pressure.

In the diaphragm control valve, operating air from the pilot acts on the valve diaphragm. The superstructure which contains the diaphragm is direct acting in some valves and reverse acting in others. If the superstructure is direct acting, the operating air pressure from the control pilot is applied to the top of the valve diaphragm. If the superstructure is reverse acting, the operating air pressure from the pilot is applied to the underside of the valve diaphragm.

Figure 14-24 shows a very simple type of direct-acting diaphragm control valve, with operating air pressure from the control pilot applied to the top of the valve diaphragm. Since this is a downward seating valve, any increase in operating air pressure pushes the valve stem down and tends to close the valve.

Now let us look at figure 14-25. This is also a direct-acting valve, with operating air pressure from the control pilot applied to the top of the valve diaphragm. But the valve shown in figure 14-25 is more complicated than the one shown in figure 14-24. The valve shown in figure 14-25 is an upward seating valve, rather than a downward seating valve. Therefore, any increase

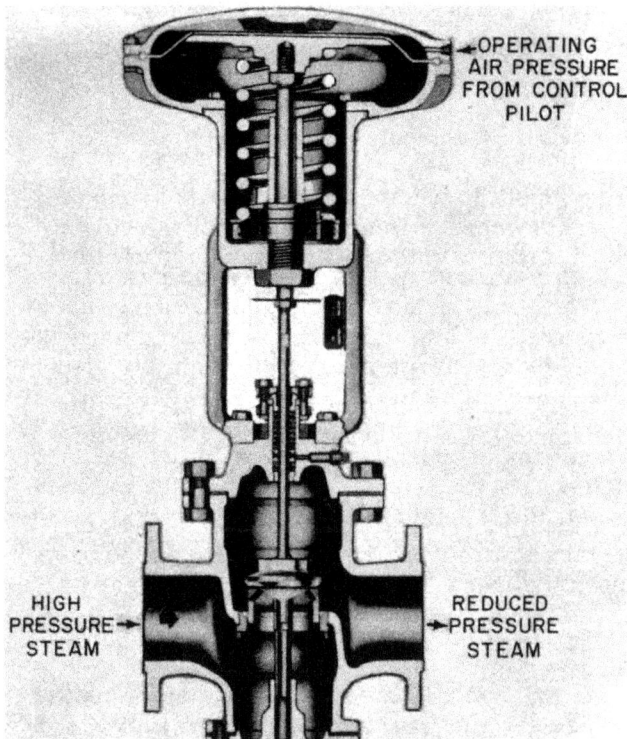

38.123X

Figure 14-24.—Direct-acting downward seating diaphragm control valve.

in operating air pressure from the control pilot tends to open this valve rather than close it.

As we have seen, the air-operated control pilot may be either direct acting or reverse acting, the superstructure of the diaphragm control valve may be either direct acting or reverse acting, and the diaphragm control valve may be either upward seating or downward seating. These three factors, as well as the purpose of the installation, determine how the diaphragm control valve and its air-operated control pilot are installed in relation to each other.

To see how these factors are related, let us consider an installation in which a diaphragm control valve and its air-operated control pilot are to be used to supply reduced pressure steam. Figure 14-26 shows one arrangement that might be used. We will assume that the service requirements indicate the need for a direct-acting upward seating diaphragm control valve. What kind of a control pilot—direct acting or reverse acting would have to be used in this installation?

Suppose that we try it first with a direct-acting control pilot. As the controlled pressure (discharge pressure from the diaphragm control valve) increases, increased pressure would be applied to the diaphragm of the direct-acting control pilot. The valve stem would be pushed down and the valve in the control pilot would be opened, thus sending an increased amount of operating air pressure from the control pilot to the top of the diaphragm control valve. The increased operating air pressure acting on the diaphragm of the valve would push the stem down and—since this is an upward seating valve—this action would open the diaphragm control valve still wider. Obviously, this will not work. For this application, an increase in controlled pressure must result in a decrease in operating air pressure. Therefore, we should have chosen a reverse-acting control pilot rather than a direct-acting one for this particular pressure-reducing application.

It is left as an exercise to the student to trace the sequence of events as they would occur with a reverse-acting control pilot installed in the arrangement shown in figure 14-26.

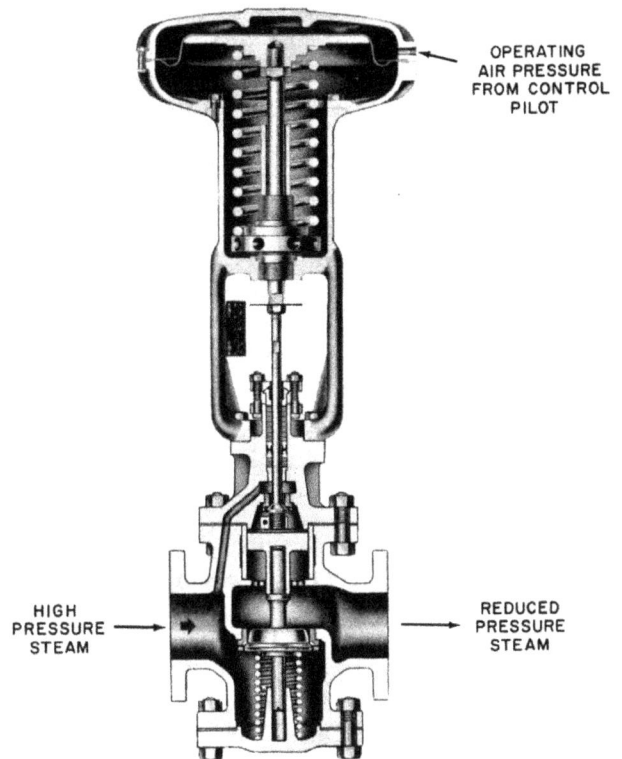

38.124X

Figure 14-25.—Direct-acting upward seating diaphragm control valve.

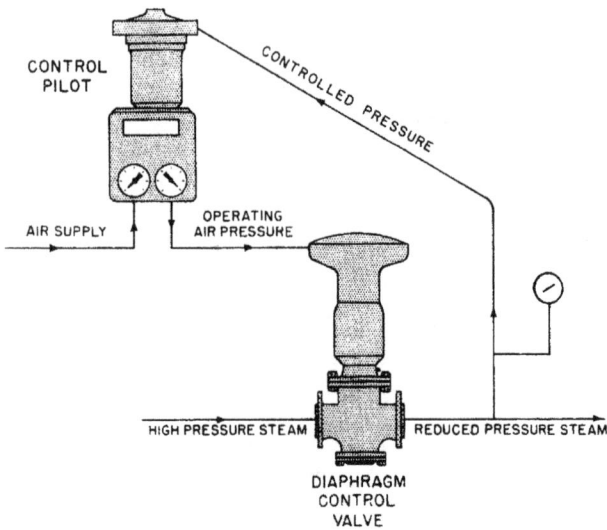

38.125
Figure 14-26.—Arrangement of control pilot and diaphragm control valve for supplying reduced pressure steam.

UNLOADING VALVES.— An automatic unloading valve (also called a dumping valve) is installed at each main and auxiliary condenser. The function of the unloading valves is to discharge steam from the auxiliary exhaust line to the condensers whenever the auxiliary exhaust line pressure exceeds the design operating pressure.

An automatic unloading valve is shown in figure 14-27. Auxiliary exhaust steam is led through valve A to the top of the actuating valve diaphragm. The actuating valve is double seated, and one side is open when the other is closed. When the auxiliary exhaust line pressure is less than the pressure for which the unloading valve is set, the upper seat is closed and the lower seat is open. The valve is thus held by the diaphragm spring. Steam passes into the line through valve B and goes under the unloading valve diaphragm. The pressure acting on this diaphragm holds the unloading valve up and closed. If the auxiliary exhaust pressure exceeds the pressure of the actuating valve diaphragm spring, the diaphragm is forced downward and the lower seat closes while the upper seat opens. This makes a direct connection between the top and the bottom of the unloading valve diaphragm through the actuating valve. The equalized pressure on the diaphragm allows the auxiliary exhaust pressure to force the unloading valve down and steam is thus unloaded

to the condenser. The unloading pressure can be adjusted by turning an adjusting screw, thereby changing the force exerted on the actuating valve diaphragm.

Thermostatic Recirculating Valves

Thermostatic recirculating valves are used in systems where it is necessary to recirculate a fluid in order to maintain the temperature within certain limits. Thermostatic recirculating valves are designed to operate automatically.

The thermostatic recirculating valve shown in figure 14-28 is used to recirculate condensate from the discharge side of the main air ejector condenser to the main condenser.[5] The valve is actuated by the temperature of the condensate. When the condensate temperature becomes higher than the temperature for which the valve is set, the thermostatic bellows expands and automatically opens the valve, allowing condensate to be sent back to the condenser.

Valve Manifolds

A valve manifold is used when it is necessary to take suction from one of several sources and to discharge to another unit or several units of the same or a separate group. One example of a manifold is shown in figure 14-29. This manifold is used in the fuel oil filling and transfer system, where provision must be made for the transfer of oil from any tank to any other tank, to the fuel oil service system, or to another ship. The manifold valves are frequently of the stop-check type.

Remote Operating Gear

Remote operating gear is installed to provide a means of operating certain valves from distant stations. Remote operating gear may be mechanical, hydraulic, pneumatic, or electric.

Some remote operating gear for valves is used in the normal operation of the valves. For example, the propulsion turbine throttle valves are opened and closed by a series of reach rods and gears. In the fireroom, remote operating gear is used to operate the forced draft blowers, to adjust the constant-pressure pump governor on the fuel oil service pump, and to lift safety valves by hand.

_____
[5]Air ejector assemblies are discussed in chapter 13 of this text.

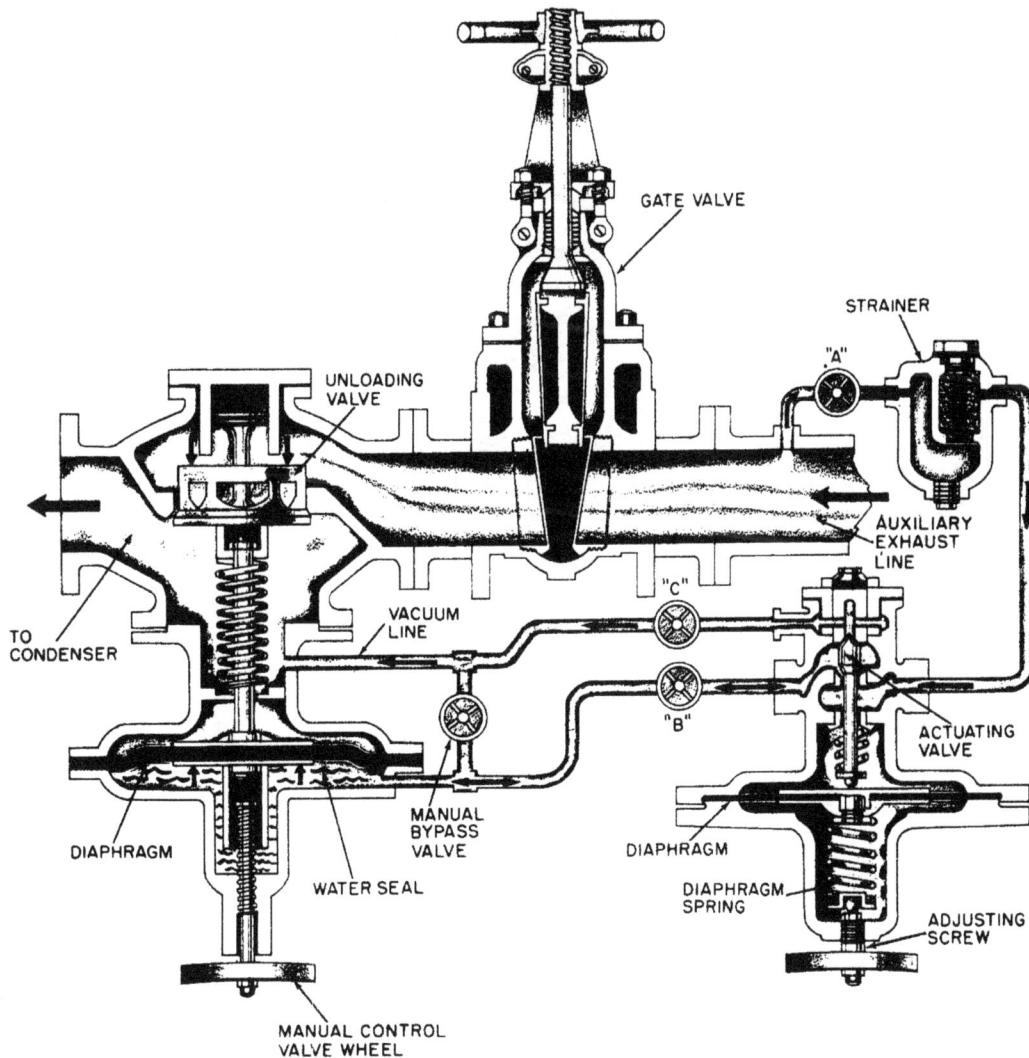

47.61X

Figure 14-27.—Automatic unloading valve.

Other remote operating gear is installed as emergency equipment. For example, the boiler steam stops have remote operating gear which can usually be operated from the main deck or from the second deck, near the general quarters station for damage control personnel of Repair 5 (propulsion repair). This remote operating gear is frequently mechanical or pneumatic, although hydraulic remote operating gear is used for these valves on some ships. The pneumatic and hydraulic types of remote operating gear for these valves allow the valves to be closed but not opened from the remote station.

Another example of emergency remote operating gear is the gear that allows the quick-closing fuel oil valve to be closed from a remote station in the fireroom escape trunk or from the deck above and near the access to the space. Still other examples of emergency remote operating gear include gear for operating some cross-connection valves, main drainage valves, and main condenser injection and overboard discharge valves.

Remote operating gear for valves includes a valve position indicator to show whether the valve is open or closed.

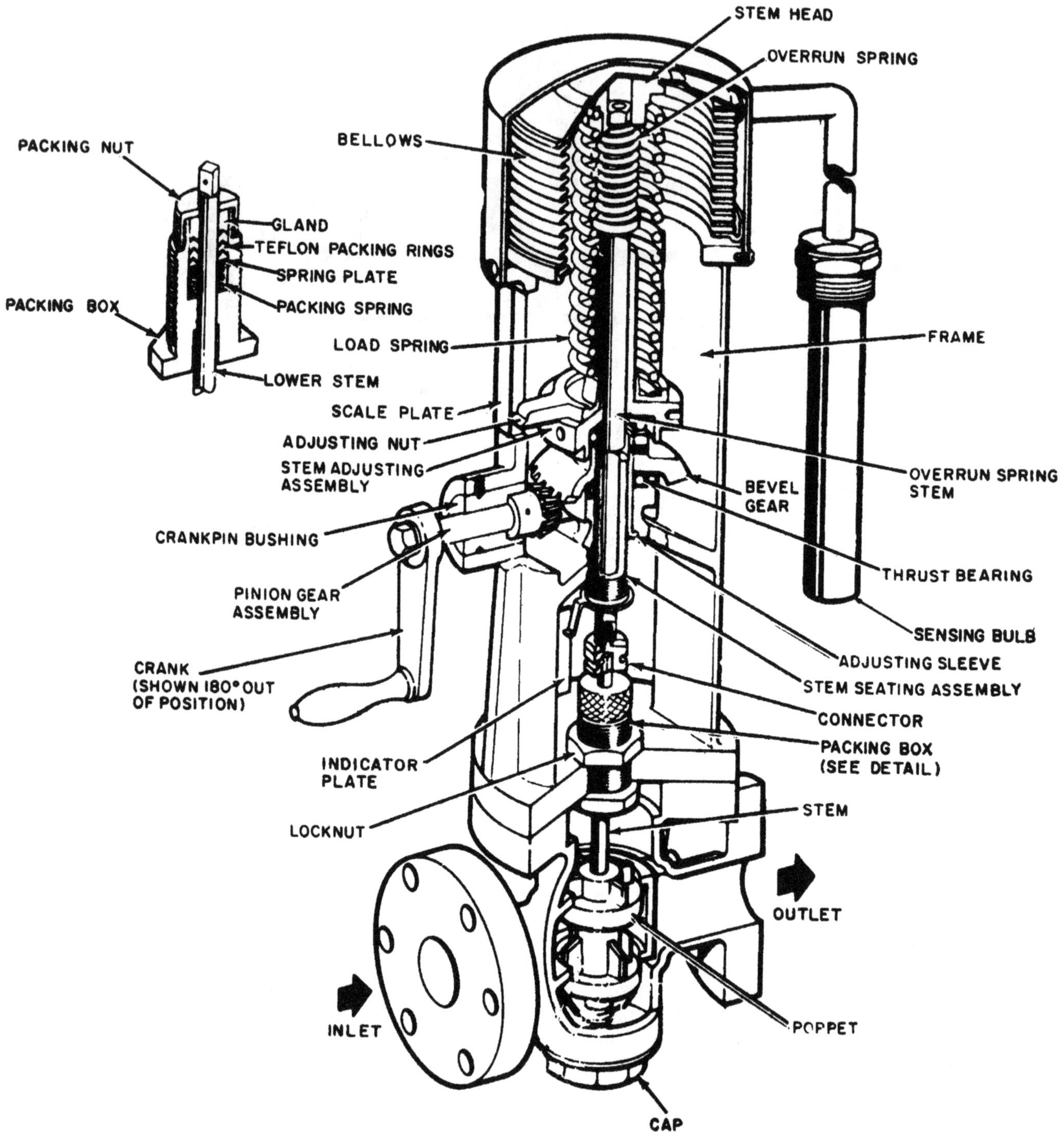

PACKING NUT

GLAND
TEFLON PACKING RINGS
SPRING PLATE
PACKING SPRING

PACKING BOX

LOWER STEM

STEM HEAD

OVERRUN SPRING

BELLOWS

LOAD SPRING

SCALE PLATE

ADJUSTING NUT
STEM ADJUSTING
ASSEMBLY

CRANKPIN BUSHING

PINION GEAR
ASSEMBLY

CRANK
(SHOWN 180° OUT
OF POSITION)

INDICATOR
PLATE

LOCKNUT

INLET

FRAME

OVERRUN SPRING
STEM

BEVEL
GEAR

THRUST BEARING

SENSING BULB

ADJUSTING SLEEVE

STEM SEATING ASSEMBLY

CONNECTOR

PACKING BOX
(SEE DETAIL)

STEM

OUTLET

POPPET

CAP

Figure 14-28.—Thermostatic recirculating valve.

47.175

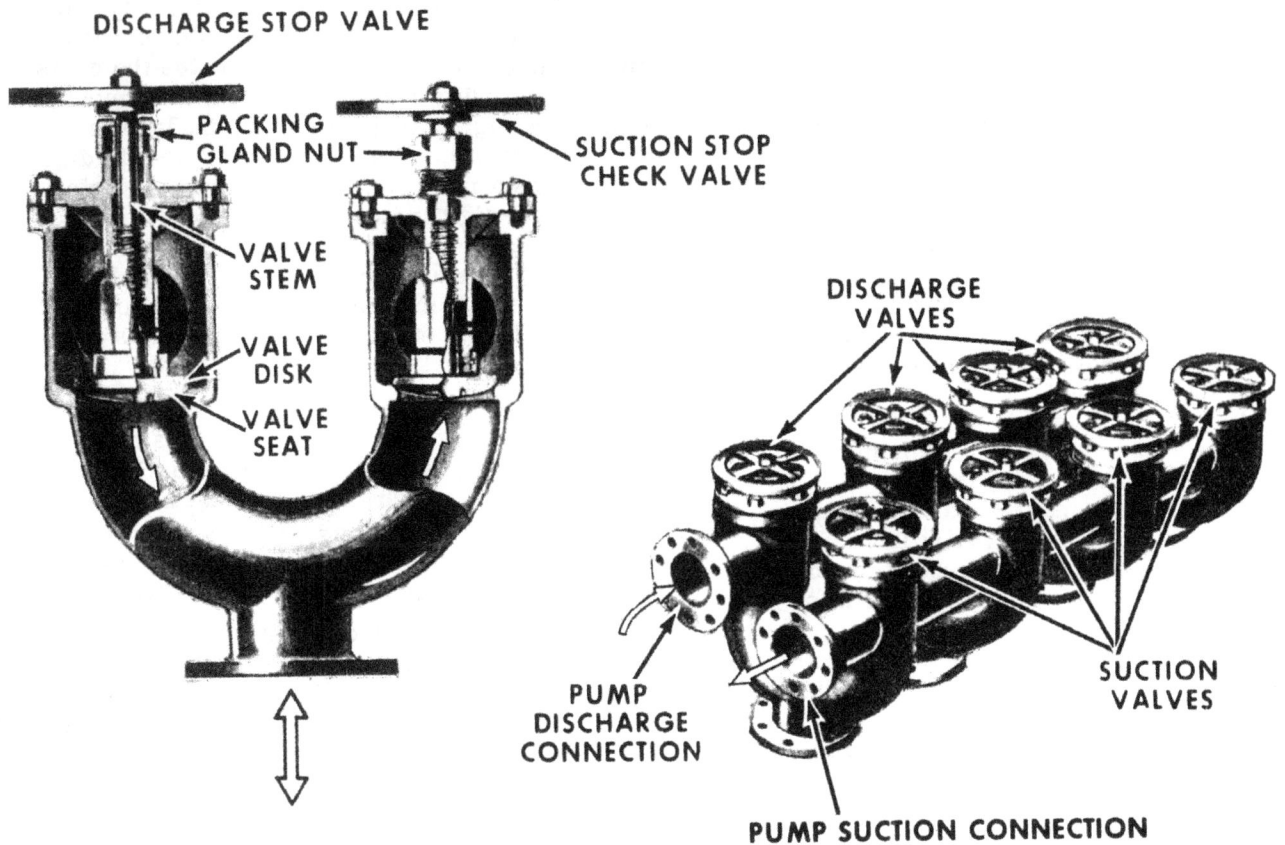

DISCHARGE STOP VALVE

PACKING GLAND NUT

SUCTION STOP CHECK VALVE

VALVE STEM

VALVE DISK

VALVE SEAT

DISCHARGE VALVES

PUMP DISCHARGE CONNECTION

SUCTION VALVES

PUMP SUCTION CONNECTION

47.63X

Figure 14-29.—Valve manifold.

IDENTIFICATION OF VALVES, FITTINGS, FLANGES, AND UNIONS

Most valves, fittings, flanges, and unions used on naval ships are marked with identification symbols of various kinds. The few valves and fittings that are made on board repair ships or tenders or at naval shipyards are usually marked with symbols indicating the manufacturing activity, the size, the melt or casting number, and the material. They may also be marked with an arrow to indicate the direction of flow.

Commercially manufactured valves, fittings, flanges, and unions may be identified according to the requirements of the applicable specifications. However, many valves, fittings, flanges, and unions are now identified according to a standard marking system developed by the Manufacturers Standardization Society (MSS) of the valves and fittings industry. Identification markings in this system usually include the manufacturer's name or trademark, the pressure and service for which the product is intended, and the size (in inches). When appropriate, material identification, limiting temperatures, and other identifying data are included.

The MSS standard identification markings are generally cast, forged, stamped, or etched on the exterior surface of the product. In some cases, however, the markings are applied to an identification plate rather than to the actual surface of the product.

The service designation in the MSS system of marking usually includes a letter to indicate the type of service and numerals to indicate the pressure rating in psi. The letters used in service designations are:

A . . . . air
G . . . . gas
L . . . . liquid
O . . . . oil
W . . . . water
D-W-V . . . . drainage, waste, and vent

When the primary service rating is for steam, and when no other service is indicated, the service designation may consist of numerals only. For example, the number 600 marked on the body of a valve would indicate that the valve is suitable for steam service at 600 psi. If the valve is designed for water at 600 psi, the service designation would be 600 W. Service designations are also used in combination; for example, 3000 WOG indicates a product suitable for water, oil, or gas service at 3000 psi.

Some abbreviations that are commonly used for material identification in the MSS system include the following:

```
AL . . . . . . . . . . . . Aluminum
B . . . . . . . . . . . . Bronze
CS . . . . . . . . . . . Carbon Steel
CI . . . . . . . . . . . Cast Iron
HF . . . . . . . . . . . Cobalt-chromium-
                          tungsten alloy (hard
                          facing)
CU NI . . . . . . . . . Copper-nickel alloy
NI CU . . . . . . . . . Nickel-copper alloy
SM . . . . . . . . . . . Soft metal (lead, Bab-
                          bitt, copper, etc.)
CR 13 . . . . . . . . . 13-percent chrom-
                          ium steel
18 8 . . . . . . . . . . 18-8 stainless steel
18 8SMO . . . . . . . 18-8 stainless steel
                          with molybdenum
SH . . . . . . . . . . . Surface-hardened
                          steel (Nitralloy, etc.)
```

Some examples of MSS standard identification marking symbols are given in figure 14-30.

PACKING, GASKETS, AND INSULATION

Packing and gasket materials are required to seal joints in steam, water, gas, air, oil, and other lines and to seal connections which slide or rotate under operating conditions. There are many types and forms of packing and gasket material available commercially. To simplify the selection of packing and gasket materials commonly used in the naval service, engineering personnel use a packing and gasket chart[6] showing the symbol numbers and the recommended applications of all types and kinds of packing and gasket material.

---
[6] The chart is identified as NavShips Mechanical Standard Drawing B-153.

The symbol numbers used to identify each type of packing and gasket consists of a four-digit number. The first digit indicates the class of service with respect to fixed and moving joints; the digit 1 indicates a moving joint (moving rods, shafts, valve stems, etc.) and the digit 2 indicates a fixed joint (flanges, bonnets, etc.). The second digit indicates the material of which the packing or gasket is primarily composed— asbestos, vegetable fiber, rubber, metal, etc. The third and fourth digits indicate the different styles or forms of the packing or gaskets made from the material.

Pressure, temperature, and other service conditions impose definite restrictions upon the application of the various kinds of packing and gasket materials. Great care must be taken to see that the proper materials are selected for each application, particularly when high pressures or high temperatures are involved.

Insulation is used on most shipboard piping systems. Insulation is actually a composite covering which includes (1) the insulating material itself, (2) the lagging or covering, and (3) the fastenings which are used to hold the insulation and lagging in place. In some instances, the insulation is covered by material which serves both as lagging and as a fastening device.

Insulating materials commonly used in the Navy include magnesia, calcium silicate, diatomaceous silica, asbestos felt, mineral wool, fibrous glass, and high temperature insulating cement. Cork, although light in weight and easy to handle, is not fire-retardant and in burning it gives off a dense, suffocating smoke; hence cork is used only in certain applications and only after it has been treated with a fire-resistant compound.

Lagging may consist of cloth, tape, or sheet metal. Lagging serves to protect the relatively soft insulating material from damage, to give added support to insulation that may be subjected to heavy or continuous vibration, and to provide a smooth surface that may be painted.

Lagging is secured in place by sewing or by using fire-resistant adhesives, insulating cement, or sealing compounds. The method used to fasten the lagging in place depends upon the type of insulation used, the type of lagging used, and the service requirements of the piping or surfaces to be insulated.

3-INCH CAST STEEL SCREWED FITTING SUITABLE FOR WATER, OIL, OR GAS SERVICE AT 1000 PSI: Manufacturer's identification . . A B CO Service designation . . . . . . 1000 WOG Material designation. . . . . . STEEL Size. 3	2-INCH CAST IRON FLANGED FITTING FOR USE IN REFRIGERATION SYSTEM: Manufacturer's identification . . A B CO Service designation . . . . . . 300 GL Temperature designation . . . . 300 F Size. 2
CAST BRASS FITTING FOR DRAINAGE, WASTE, AND VENT SERVICE: Manufacturer's identification . . A B CO Service designation . . . . . . D-W-V	2-INCH BRONZE VALVE RECOMMENDED BY THE MANUFACTURER FOR 200 PSI STEAM SERVICE: Manufacturer's identification . . A B CO Service designation . . . . . . 200 Size. 2

4-INCH STEEL VALVE WITH 13 PERCENT CHROMIUM STEEL VALVE STEM, DISK, AND SEAT, SUITABLE FOR 1500 PSI STEAM SERVICE AT TEMPERATURE OF NO MORE THAN 850° F:

VALVE BODY MARKING:

Manufacturer's identification . . A B CO
Service designation 1500
Material designation. . . . . . STEEL
Size. 4

IDENTIFICATION PLATE MARKING:

Manufacturer's identification . . A B CO
Service designation 1500
Limiting temperature . . . . . MAX 850 F
Body material designation . . . STEEL
Valve stem material designation . STEM CR 13
Valve disk material designation . DISC CR 13
Valve seat material designation . SEAT CR 13
Size. 4

38.128

Figure 14-30.—Examples of MSS standard identification markings for valves, fittings, flanges, and unions.

SAFETY PRECAUTIONS

Most piping system repairs involve breaking joints in the piping system. Before breaking joints in any shipboard piping system, be sure the following precautions are observed:

1. Be sure there is no pressure on the line. This is important in practically all systems, but it is of vital importance in steam lines, hot water lines, and any salt water lines that may have a direct connection with the sea. It is not enough to merely close the valves; the valves

must be locked or wired shut and must be tagged so that they will not be opened accidentally.

2. Be sure that the line is completely drained.

3. In breaking a flanged joint, leave two diametrically opposite securing nuts in place while loosening the others. Then slack off on the last two nuts. When you are sure beyond the slightest doubt that the line is clear, remove the nuts and break the joint.

4. Take all appropriate precautions to prevent fire and explosion when cutting into lines or breaking joints in systems that have contained flammable fluids.

5. Observe all safety precautions required in connection with welding, brazing, or other processes used in repairing the piping.

6. Before repaired piping is put back into service, various tests and inspections may be required. Specific requirements for tests and inspections may be given along with instructions for the repair job; or you may find test and inspection requirements indicated on the plans or blueprints. If specific instructions are not given, consult chapter 9480 of the Naval Ships Technical Manual.

CHAPTER 15

PUMPS AND FORCED DRAFT BLOWERS

This chapter deals with shipboard pumps and with the forced draft blowers used aboard many surface ships to supply combustion air to the propulsion boilers. In general, we are concerned here with the driven end of the units rather than with the driving end; the auxiliary steam turbines used to drive many pumps and blowers are discussed in chapter 16 of this text, and the electric motors used to drive others are discussed in chapter 20.

PUMPS

As we saw in chapter 8, the pump is one of the five basic elements in any thermodynamic cycle. The function of the pump is to move the working substance from the low pressure side of the system to the high pressure side. In the conventional steam turbine propulsion plant, "the pump" of the thermodynamic cycle is actually three pumps—the condensate pump, the feed booster pump, and the main feed pump.

In addition to these three pumps which are a part of the basic thermodynamic cycle, there are, of course, a large number of pumps used for other purposes aboard ship. Pumps supply sea water to the firemains, circulate cooling water for condensers and coolers, empty the bilges, transfer fuel oil, discharge fuel oil to the burners, supply lubricating oil to main and auxiliary machinery, supply sea water to the distilling plant, pump the distillate into storage tanks, supply liquid under pressure for use in hydraulically operated equipment, and provide a variety of other vital services.

Pumps are used to move any substance which flows or which can be made to flow. Most commonly, pumps are used to move water, oil, and other liquids. However, air, steam, and other gases are also fluid and can be moved with pumps, as can such substances as molten metal, sludge, and mud.

A pump is essentially a device which utilizes an external source of power to apply a force to a fluid in order to move the fluid from one place to another. A pump develops no energy of its own; it merely transforms energy from the external source (steam turbine, electric motor, etc.) into mechanical kinetic energy, which is manifested by the motion of the fluid. This kinetic energy is then utilized to do work—for example, to raise a liquid from one level to another, as when water is raised from a well; to transport a liquid through a pipe, as when oil is carried through an oil pipeline; to move a liquid against some resistance, as when water is pumped to a boiler under pressure; or to force a liquid through a hydraulic system, against various resistances, for the purpose of doing work at some point.

Principles and Definitions

Before considering specific designs of shipboard pumps, it may be helpful to examine briefly certain basic concepts and to define some of the terms commonly used in connection with pumps.

FORCE - PRESSURE - AREA RELATIONSHIPS.—When we strike the end of a bar, the main force of the blow is carried straight through to the other end. This happens because the bar is rigid. The direction of the blow almost entirely determines the direction of the transmitted force. The more rigid the bar, the less force is lost inside the bar or transmitted outward at right angles to the direction of the blow.

When we apply pressure to the end of a column of confined liquid, however, the pressure is transmitted not only straight through to the other end but also equally and undiminished in every direction. Figure 15-1 illustrates the difference between pressure applied to a rigid bar and pressure applied to a column of contained liquid.

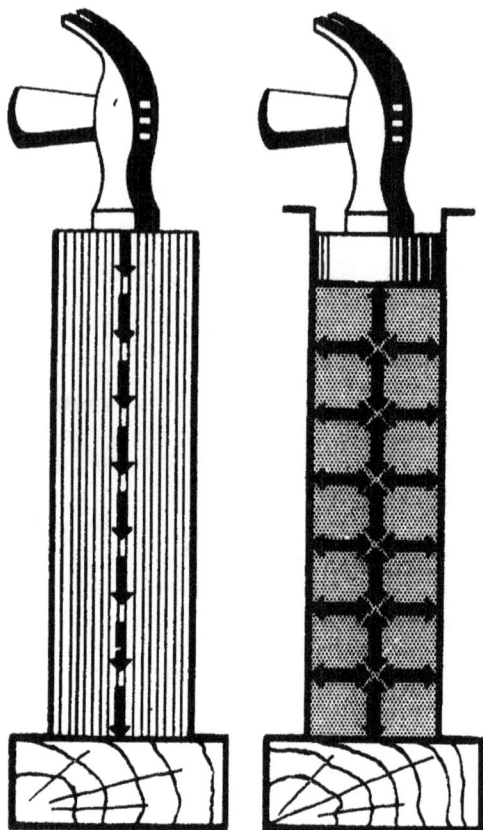

5.180
Figure 15-1.—Results of pressure applied to a rigid bar (left) and to a column of liquid (right).

The principle that pressure is transmitted equally and undiminished in all directions through a contained liquid is known as Pascal's principle. This principle may be regarded as the basic law or foundation of the science of hydraulics.

An important corollary of Pascal's principle is that the transmission of pressure through a liquid is not altered by the shape of the container. This idea is illustrated in figure 15-2. If the pressure due to the weight of the liquid is 8 psi at any one point on the horizontal line H, it is 8 psi at every point along line H. The pressure due to the weight of the liquid at any level thus depends upon the vertical distance from the chosen level to the surface of the liquid. The vertical distance between two horizontal levels in a liquid is known as the head of the liquid. (Since various kinds of head enter into pump calculations, the term head is more fully discussed later.)

Pressure is defined as force per unit area. Alternatively, we may say that force is equal to pressure times area. Figure 15-3 shows how a force of 20 pounds acting on a piston with an area of 2 square inches can produce a force of 200 pounds on a piston with an area of 20 square inches. The system would, of course, work the same in reverse. If we consider piston 2 as the input piston and piston 1 as the output piston, then the output force would be 1/10 the input force.

We are now in a position to state a general rule: If two pistons are used in a hydraulic system, the force acting on each will be directly proportional to its area, and the magnitude of each force will be the product of the pressure and the area.

The second basic rule for two pistons in a hydraulic system such as the one shown in figure 15-3 may be stated as follows: The distance moved by each piston is inversely proportional to the area of the piston. Thus if piston 1 in figure 15-3 is pushed down 1 inch, piston 2 will be raised 1/10 inch.

Consideration of the two basic rules just stated leads us to another basic rule: The input force multiplied by the distance through which it moves is exactly equal to the output force multiplied by the distance through which it moves (disregarding energy losses due to friction). In essence, this rule is merely another statement of the general energy equation—that is, energy in= energy out.

PUMP CAPACITY.—The capacity of a pump is the amount of liquid the pump can handle in a given period of time. For marine applications, the capacity of a pump is usually stated in gallons per minute (gpm).

PRESSURE HEAD.—The power required to drive a pump is a function of pump capacity and of the total head against which the pump operates. Previously we defined head quite simply as the vertical distance between two horizontal levels in a liquid. Since a pump may be installed above, at, or below the surface of the source of supply, it is obvious that other factors must enter into the discussion of pressure head as applied to pumps.

When the pump is installed at the same level as the free surface of the source of supply, no new considerations need apply since the pump merely acts on the liquid like any other applied force.

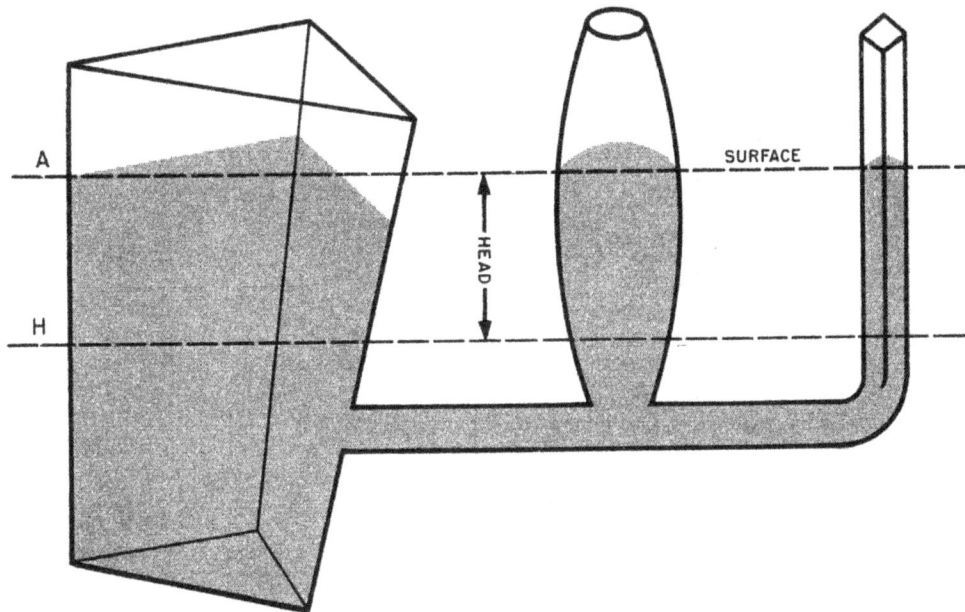

147.105

Figure 15-2.—Pressure in liquid is not affected by the shape of the vessel.

When the pump is installed below this level, as shown in figure 15-4, a certain amount of energy in the form of gravity head will already be available when the liquid enters the pump. In other words, there is a static pressure head, A, on the suction side of the pump. This head is part of the total input head necessary to produce the output head, F, that is required to raise the liquid to the top, T, of the discharge reservoir.

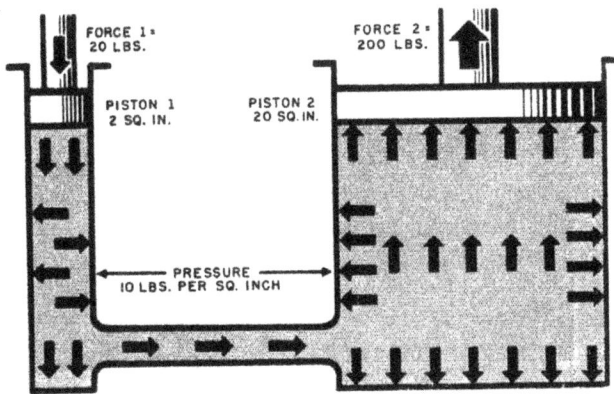

4.7

Figure 15-3.—Relationship of force, pressure, and area in a simple hydraulic system.

The action of the pump produces the total head differential, B, which can be broken down into friction loss, C, and net static discharge head, D. Since D is the vertical distance from the surface of the supply liquid to the surface of the liquid in the discharge reservoir, it is clear that our previous definition of head would apply only to D. E, the total static discharge head, is the vertical distance from the center of the pump to the surface of the liquid in the discharge reservoir; thus, E is equal to D plus A.

As may be seen in figure 15-4, atmospheric pressure is acting upon the free surface of the supply liquid and upon the free surface of the liquid in the discharge reservoir. Since atmospheric pressure is exerted equally on both sides of the pump, in this system, the two heads created by atmospheric pressure cancel out.

Now consider the case of a pump that is installed a vertical distance A above the free surface S of the supply liquid (fig. 15-5). In this case, energy must be supplied merely to get the liquid into the pump (static suction lift, A). In addition, energy must be supplied to produce the static discharge head, E, if the liquid is to be raised to the top of the discharge reservoir, T. B, the total head differential produced by pump action, is here the total energy input. It is divided into A on the suction side—the head required to raise

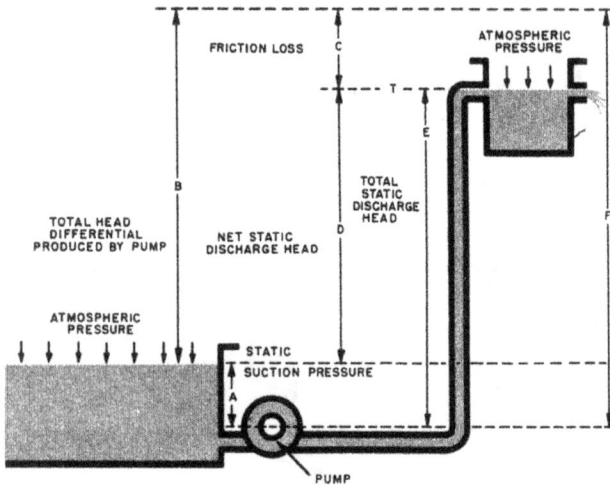

147.106

Figure 15-4.—Pressure head (pump installed below surface of supply liquid).

in a closed system such as the one shown in figure 15-6.

In this system, a pump is being used to drive a work piston back and forth inside a cylinder. We will assume that the pump must develop a pressure equivalent to head $\underline{H}$ in order to drive the piston back and forth against the resistance offered. Under this assumption, the total head differential, $\underline{B}$, must be produced by the pump after the system has begun to operate. $\underline{B}$ is the sum of the friction head (or friction losses), $\underline{C}$, and the static discharge head, $\underline{D}$. Since $\underline{D}$ is equal to $\underline{H}$, $\underline{D}$ therefore produces the pressure required to do the work.

Since the liquid returns to its original level and the system is closed throughout, there will be a siphon effect in the return pipe which will exactly balance the static suction lift, $\underline{A}$. Therefore, atmospheric pressure plays a part in the operation of this system only when the system is being started up, before the entire system has been filled with liquid.

At this point, we may pause and consider the various ways in which the term head has been used, and attempt to formulate a definition. From previous discussion, we may infer that head is (1) measured in feet; (2) somehow related to pressure; and (3) taken as some kind of a measure of energy. But what is it?

the liquid to the pump— and $\underline{C}$ (friction loss) plus $\underline{E}$ (static discharge head) on the discharge side. Atmospheric pressure cancels out here, as before, except that in this case the atmospheric pressure at $\underline{S}$ is required to lift the liquid to the pump.

In both of the cases just described, we have dealt with systems which were open to the atmosphere. Now let us examine the head relationships

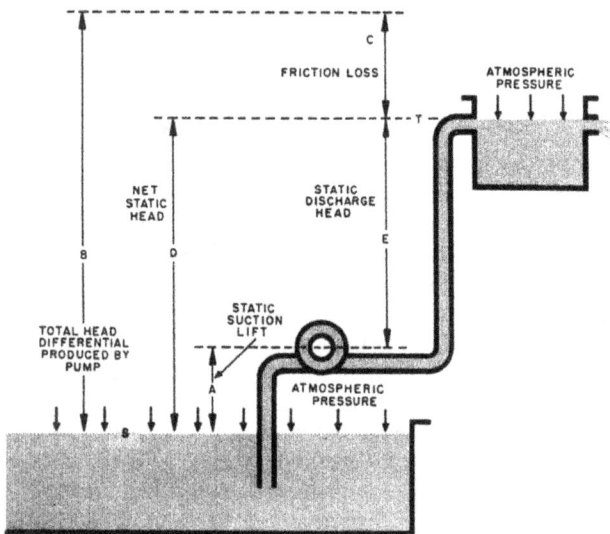

147.107

Figure 15-5.—Pressure head (pump installed above surface of supply liquid).

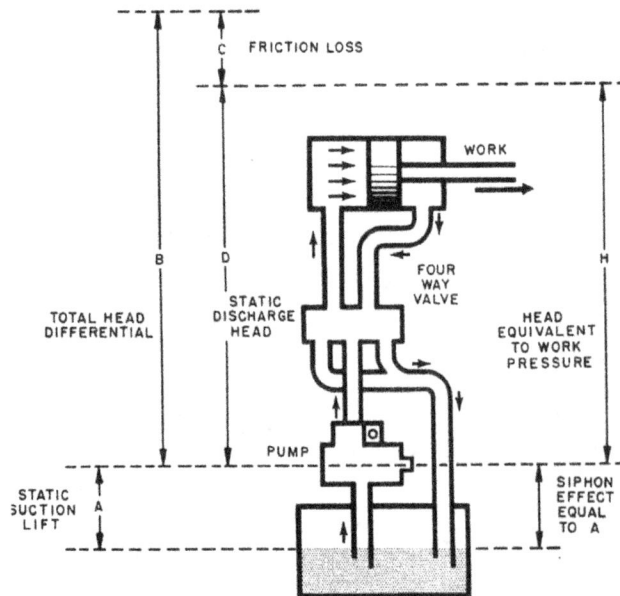

147.108

Figure 15-6.—Pressure head (closed system).

Basically, head is a measure of the pressure exerted by a column or body of liquid because of the weight of the liquid. In the case of water, we find that a column of fresh water 2.309 feet high exerts a pressure of 1 pound per square inch. When we refer to a head of water of 2.309 feet, we know that the water is exerting a pressure of 1 psi because of its own weight. Thus, a reference to a head of so many feet of water does imply a reference to the pressure exerted by that water.

The situation is somewhat different when we have a horizontal pipe through which water is being pumped. In this case, the head is calculated as the vertical distance that would correspond to the pressure. If the pressure in the horizontal pipe is 1 psi, then the head on the liquid in the pipe is 2.309 feet. Further calculations show that a head of 1 foot corresponds to a pressure of 0.433 psi.

The relationship between head and energy can be clarified by considering that (1) work is a form of energy—mechanical energy in transition; (2) work is the product of a force times the distance through which it acts; and (3) for liquids, the work performed is equal to the volume of liquid moved times the head against which it is moved. Thus the head relationships actually indicate some of the energy relationships for a given quantity of liquid.

VELOCITY HEAD. — The head required to impart velocity to a liquid is known as velocity head. It is equivalent to the distance through which the liquid would have to fall in order to acquire the same velocity. If we know the velocity of the liquid, we can compute the velocity head by the formula

$$H_v = \frac{V^2}{2g}$$

where

H_v = velocity head, in feet

V = velocity of liquid, in feet per second

g = acceleration due to gravity (32.2 feet per second per second)

In a sense, velocity head is obtained at the expense of pressure head. Whenever a liquid is given a velocity, some part of the original static pressure head must be used to impart this veloc-ity. However, velocity head does not represent a total loss, since at least a portion of the velocity head can always be reconverted to static pressure head.

FRICTION HEAD.—The force or pressure required to overcome friction is also obtained at the expense of the static pressure head. Unlike velocity head, however, friction head cannot be "recovered" or reconverted to static pressure head, since fluid friction results in the conversion of mechanical kinetic energy to thermal energy. Since this thermal energy is usually wasted, friction head must be considered as a total loss from the system.

BERNOULLI'S THEOREM.—At any point in a system, the static pressure head will always be the original static pressure head minus the velocity head and minus the friction head. Since both velocity head and friction head represent energy which comes from the original static pressure head, the sum of the static pressure head, the velocity head, and the friction head at any point in a system must add up to the original static pressure head. This general principle, which is known as Bernoulli's theorem, may also be expressed as

$$Z_1 + \frac{P_1}{D} + \frac{V_1^2}{2g} = Z_2 + \frac{P_2}{D} + \frac{V_2^2}{2g} + [J(U_2 - U_1) - Wk - JQ]$$

where

Z = elevation, in feet
P = absolute pressure, in pounds per square foot
D = density of liquid, in pounds per cubic foot
V = velocity, in feet per second
g = acceleration due to gravity (32.2 feet per second per second)
J = the mechanical equivalent of heat, 778 foot-pounds per Btu
U = internal energy, in Btu
Wk = work, in foot-pounds
Q = heat transferred, in Btu

When written in this form, Bernoulli's theorem may be readily recognized as a special statement of the general energy equation. The bracketed term represents energy in transition as work, energy in transition as heat, and the increase in internal energy of the fluid arising from friction and turbulence. In some cases of fluid flow, all elements in the bracketed term are of such small magnitude that they may be safely disregarded.

Consideration of Bernoulli's theorem indicates that the term pressure head, as used in connection with pumps and other hydraulic equipment, is actually a measure of mechanical potential energy; that velocity head is a measure of mechanical kinetic energy; and that friction head is a measure of the energy which departs from the system as thermal energy in the form of heat or of the energy which remains in the liquid, generally unusable, in the form of internal energy.

Types of Pumps

Pumps are so widely used for such varied services that the number of different designs is almost overwhelming. As a general rule, however, it may be stated that all pumps are designed to move fluid substances from one point to another by pushing, pulling, or throwing, or by some combination of these three methods.

Every pump has a power end and a fluid end. The power end may be a steam turbine, a reciprocating steam engine, a steam jet, or an electric motor. In steam-driven pumps, the power end is often called the steam end. The fluid end is usually called the pump end. However, it may be called the liquid end, the water end, the oil end, or some other term to indicate the nature of the fluid substance being pumped.

Pumps are classified in a number of different ways according to various design and operational features. Perhaps the basic distinction is between positive-displacement pumps and continuous-flow pumps. Pumps may also be classified according to the type of movement that causes the pumping action; by this classification, we have reciprocating, rotary, centrifugal, propeller, and jet pumps. Another classification may be made according to speed; some pumps run at variable speed, others at constant speed. Some pumps have a variable capacity, others discharge at a constant rate. Some pumps are self-priming, others require a positive pressure on the suction side before they can begin to operate. These and other distinctions are noted as appropriate in the following discussion of specific types of pumps.

RECIPROCATING PUMPS.—A reciprocating pump moves water or other liquid by means of a plunger or piston that reciprocates inside a cylinder. Reciprocating pumps are positive-displacement pumps; each stroke displaces a certain definite quantity of liquid, regardless of the resistance against which the pump is operating.

The two main parts of a reciprocating pump are the water end, which consists of a piston and cylinder arrangement and appropriate suction and discharge valves, and the steam end[1], which consists of another piston and cylinder and appropriate valves for the admission and release of steam.

Reciprocating pumps in naval service are usually classified as:

1. Direct-acting or indirect-acting.
2. Simplex (single) or duplex (double).
3. Single-acting or double-acting.
4. High pressure or low pressure.
5. Vertical or horizontal.

The reciprocating pump shown in figure 15-7 is a direct-acting, simplex, double-acting, high pressure, vertical pump. Now let us see what all these terms mean, with reference to the pump shown in the illustration.

The pump is direct-acting because the pump rod is a direct extension of the piston rod; thus the piston in the power end is directly connected to the plunger in the liquid end. Most reciprocating pumps used in the Navy are direct-acting. An indirect-acting pump may be driven by means of a beam or linkage which is connected to and motivated by the steam piston rod of a separate reciprocating engine; or it may be driven by a crank and connecting rod mechanism which is operated by a steam turbine or an electric motor. An indirect-acting pump might appear to have only one end—that is, the pump end. However, this pump, as all others, must have a power end as well; the separate engine, turbine, or motor which drives the pump is the actual power end of the pump.

The pump shown in figure 15-7 is called a single or simplex pump because it has only one liquid cylinder. Simplex pumps may be either direct-acting or indirect-acting. A double or

_____

[1] Practically all reciprocating pumps in naval use are steam driven. However, a few low pressure, motor-driven reciprocating pumps are used for fresh water, sanitary, bilge, ballast, and fuel oil transfer services. These pumps are generally horizontal. When driven by an electric motor, reciprocating pump is usually referred to as a power pump.

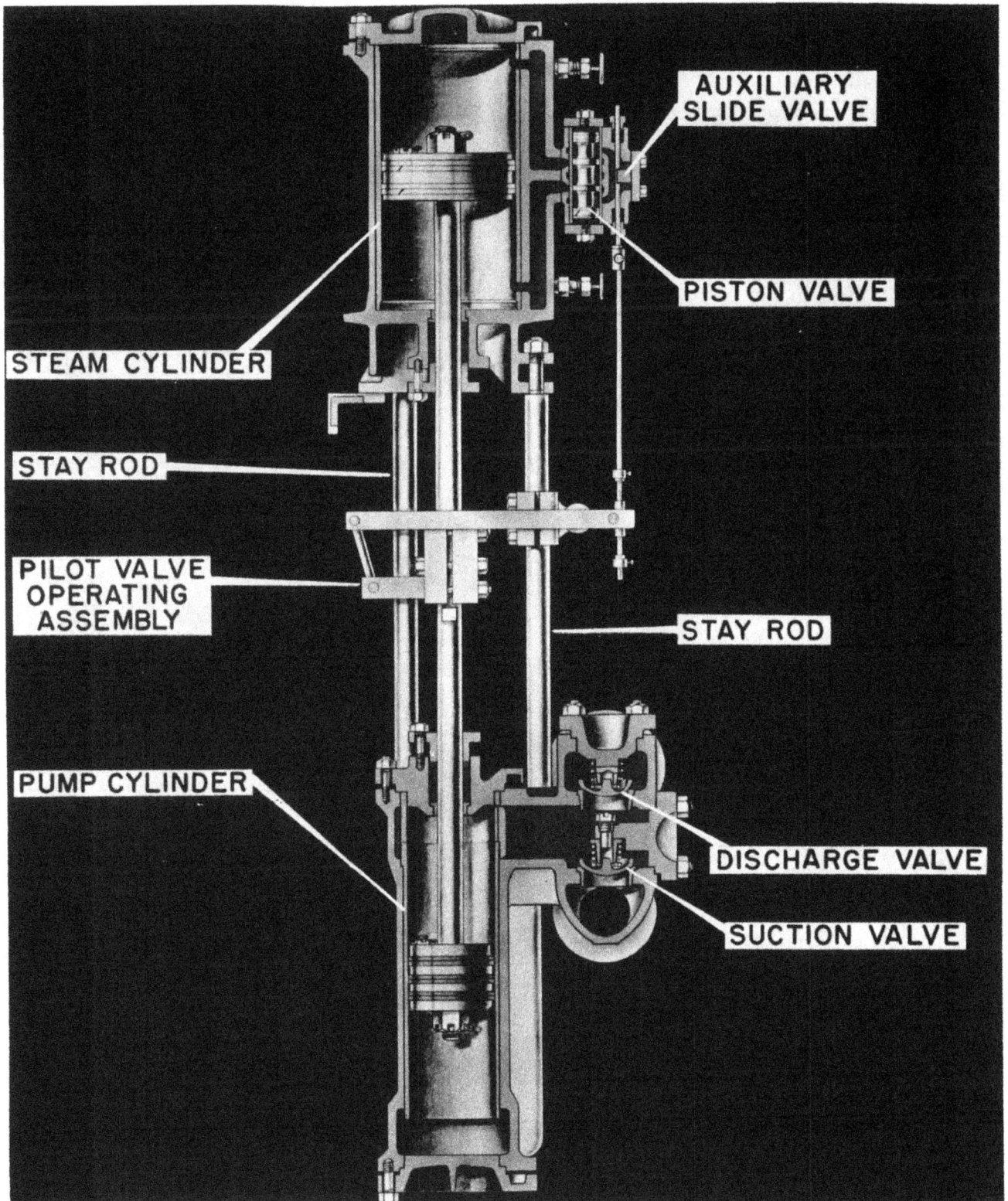

38.98

Figure 15-7.—Reciprocating pump.

duplex pump is an assembly of two single pumps placed side by side on the same foundation; the two steam cylinders are cast in a single block and the two liquid cylinders are cast in another block. Duplex reciprocating pumps are seldom found in modern combatant ships but were once commonly used in the Navy.

In a single-acting pump, the liquid is drawn into the liquid cylinder on the first or suction stroke and is forced out of the cylinder on the return or discharge stroke. In a double-acting pump, each stroke serves both to draw in liquid and to discharge liquid. As one end of the cylinder is filled, the other end is emptied; on the return stroke, the end which was just emptied is filled and the end which was just filled is emptied. The pump shown in figure 15-7 is double-acting, as are most of the reciprocating pumps used in the Navy.

The pump shown in figure 15-7 is designed to operate with a discharge pressure which is higher than the pressure of the steam operating the piston in the steam cylinder. In other words, this is a high pressure pump. In a high pressure pump, the steam piston is larger in diameter than the plunger in the liquid cylinder. Since the area of the steam piston is greater than the area of the plunger in the liquid cylinder, the total force exerted by the steam against the steam piston is concentrated on the smaller working area of the plunger in the liquid cylinder; hence the pressure per square inch is greater in the liquid cylinder than in the steam cylinder. A high pressure pump discharges a comparatively small volume of liquid against a high pressure. A low pressure pump, on the other hand, has a comparatively low discharge pressure but a larger volume of discharge. In a low pressure pump the steam piston is smaller than the plunger in the liquid cylinder.

The standard way of designating the size of a reciprocating pump is by giving three dimensions, in the following order: (1) the diameter of the steam piston, (2) the diameter of the pump plunger, and (3) the length of the stroke. For example, a 12" x 11" x 18" reciprocating pump has a steam piston which is 12 inches in diameter, a pump plunger which is 11 inches in diameter, and a stroke of 18 inches. Thus the size designation indicates immediately whether the pump is a high pressure pump or a low pressure pump.

Finally, the pump shown in figure 15-7 is classified as vertical because the steam piston and the pump plunger move up and down. Most reciprocating pumps in naval use are vertical; a few, however, are horizontal, with the piston

moving back and forth rather than up and down.

The remainder of the discussion of reciprocating pumps is concerned primarily with direct-acting, simplex, double-acting, vertical pumps, since most reciprocating pumps used in the Navy are of this type.

The power end of a reciprocating pump consists of a bored cylinder in which the steam piston reciprocates. The steam cylinder is fitted with heads at each end; one head has an opening to accommodate the piston rod. Steam inlet and exhaust ports connect each end of the steam cylinder with the steam chest. Drain valves are installed in the steam cylinder so that water resulting from condensation may be drained off.

Some reciprocating pumps have cushioning valves at each end of the steam cylinder. These valves can be adjusted to trap a certain amount of steam at the end of the cylinder; thus, when the piston reaches the end of its stroke, it is cushioned by the steam and prevented from hitting the end of the cylinder. When the pump is operating at high speed, the cushioning valves are kept almost closed so that a considerable amount of steam will be trapped at each end of the cylinder; at low speed, the cushioning valves are kept almost open. Some reciprocating pumps do not have cushioning valves.

Automatic timing of the admission and release of steam to and from each end of the steam cylinder is accomplished by various types of valve arrangements. Figure 15-8 shows the piston-type valve gear commonly used for this purpose; it consists of a main piston-type slide valve and a pilot slide valve. Since the rod from the pilot valve is connected to the pump rod by a valve-operating assembly, the position of the pilot valve is controlled by the position of the piston in the steam cylinder. The pilot valve furnishes actuating steam to the main piston-type valve, which, in turn, admits steam to the top or to the bottom of the steam cylinder at the proper time.

The valve-operating assembly which connects the pilot valve operating rod and the pump rod is shown in figure 15-9. As the crosshead arm (sometimes called the rocker arm) is moved up and down by the movement of the pump rod, the moving tappet slides up and down on the pilot valve operating rod. The tappet collars are adjusted so that the pump will make the full designed stroke.

The liquid end of a reciprocating pump has a piston and cylinder assembly similar to that of

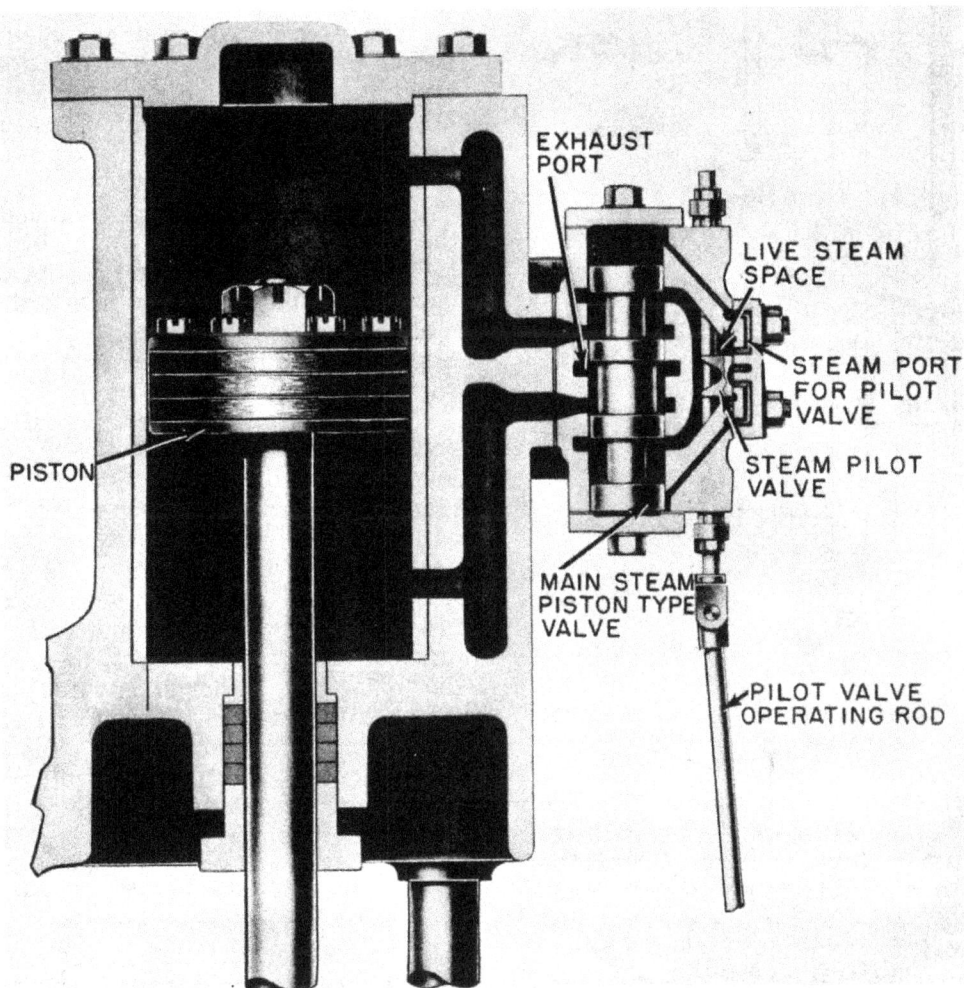

Figure 15-8.—Piston-type valve gear for steam end of reciprocating pump.

the power or steam end. The piston in the liquid end is often called a plunger. A valve chest, sometimes called a water chest, is attached to the liquid cylinder. The valve chest contains two sets of suction and discharge valves, one set to serve the upper end of the liquid cylinder and one to serve the lower end. The valves are so arranged that the pump takes suction from the suction chamber and discharges through the discharge chamber on both the up and down strokes.

An adjustable relief valve is fitted to the discharge chamber to protect the pump and the piping against excessive pressure.

Some reciprocating pumps have an air chamber and a snifter valve installed in the liquid end. The upper part of the air chamber contains air; the lower part contains liquid. On each stroke, the air in the chamber is compressed by the pressure exerted by the plunger. When the plunger stops at the end of a stroke, the air in the chamber expands and allows a gradual, rather than a sudden, drop in the discharge pressure. The air chamber, therefore, smooths out the discharge flow, absorbs shock, and prevents pounding. The snifter valve, if installed, allows a small quantity of air to be drawn in and compressed with each stroke. If no snifter valve is installed, some provision may be made for charging the air chamber with compressed air.

Although reciprocating pumps were once widely used aboard ship for a variety of services, their use on combatant ships is now generally

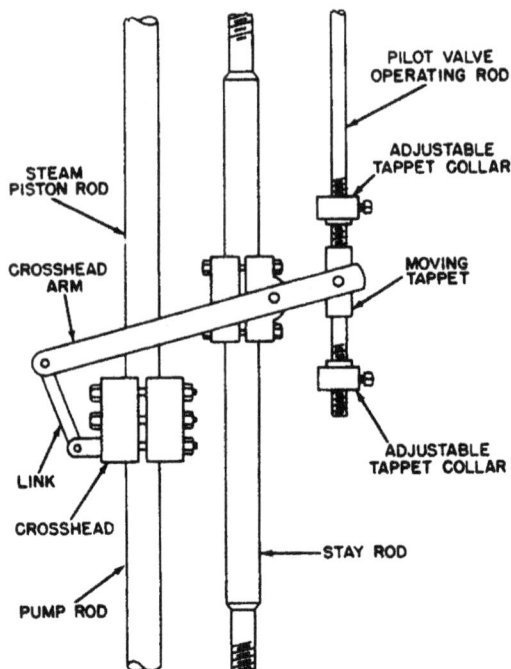

Figure 15-9.—Valve-operating gear of reciprocating pump.

restricted to emergency feed pumps, fire and
bilge pumps, and fuel oil tank stripping and bilge
pumps. On auxiliary ships, reciprocating pumps
are still used for a number of services, including auxiliary feed, standby fuel oil service, fuel
oil transfer, auxiliary circulating and condensate, fire and bilge, ballast, and lube oil transfer.

VARIABLE STROKE PUMPS.— Variable
stroke (also called variable displacement) pumps
are most commonly used on naval ships as part
of an electrohydraulic transmission for anchor
windlasses, cranes, winches, steering gear, and
other equipment. In these applications, the variable stroke pump is sometimes referred to as
the A end and the hydraulic motor which is
driven by the A end is then called the B end. Variable stroke pumps are also used on some ships
as in-port or cruising fuel oil service pumps.

Although variable stroke pumps are often
classifed as rotary pumps, they are actually
reciprocating pumps of a special design. A rotary
motion is imparted to a cylinder barrel or cylinder block in the pump by means of a constant-
speed electric motor; but the actual pumping is
done by a set of pistons reciprocating inside cyl-

indrical openings in the cylinder barrel or cylinder block.

There are two general types of variable
stroke pumps in common use. In the axial-piston
type, the pistons are arranged parallel to each
other and to the pump shaft. In the radial-piston
type, the pistons are arranged radially from the
shaft.

Figure 15-10 shows an exploded view of both
the pump end (A end) and the hydraulic motor
(B end) of an axial-piston type of variable stroke
unit. The pump usually has either seven or
nine[2] single-acting pistons which are evenly
spaced around the cylinder barrel[3] in the manner
shown in figure 15-10.

The piston rods (sometimes called connecting
rods) make a ball-and-socket connection with a
piece called the socket ring. The socket ring
rides on a thrust bearing carried by a casting
called the tilting box or tilting block; thus the
socket ring, which revolves, is actually fitted
into the tilting box, which does not revolve. Figure 15-11 shows diagrammatically the arrangement of the cylinder barrel, the socket ring, and
the tilting box. Although only one piston is shown
in this illustration, the others fit similarly into
the cylinder barrel and into the socket ring.

Figure 15-12 illustrates diagrammatically
the manner in which the position of the tilting box
affects the position of the pistons. (Note that this
is not a continuous cross-sectional view, since
for illustrative purposes two pistons are shown.)
In order to understand how the pumping action
takes place, let us follow one piston as the cylinder barrel and socket ring make one complete
revolution. When the tilting box is set perpendicular to the shaft, as in part A of figure 15-12, the
piston does not move back and forth within its
cylindrical opening as the cylinder barrel and
socket ring revolve. Thus the piston is in the
same position with respect to its own cylindrical
opening when it is at the top position as it is when
the cylinder barrel has completed half a revolution and carried the piston to the bottom position.
Since the piston does not reciprocate, there is no
pumping action when the tilting box is in this
position even though the cylinder barrel and
socket ring are revolving.

---

[2] An uneven number of pistons is always used in order
to avoid pulsations in the discharge flow.

[3] Note that the term cylinder barrel actually refers to
a cylinder block which has cylindrical openings for all
of the pistons.

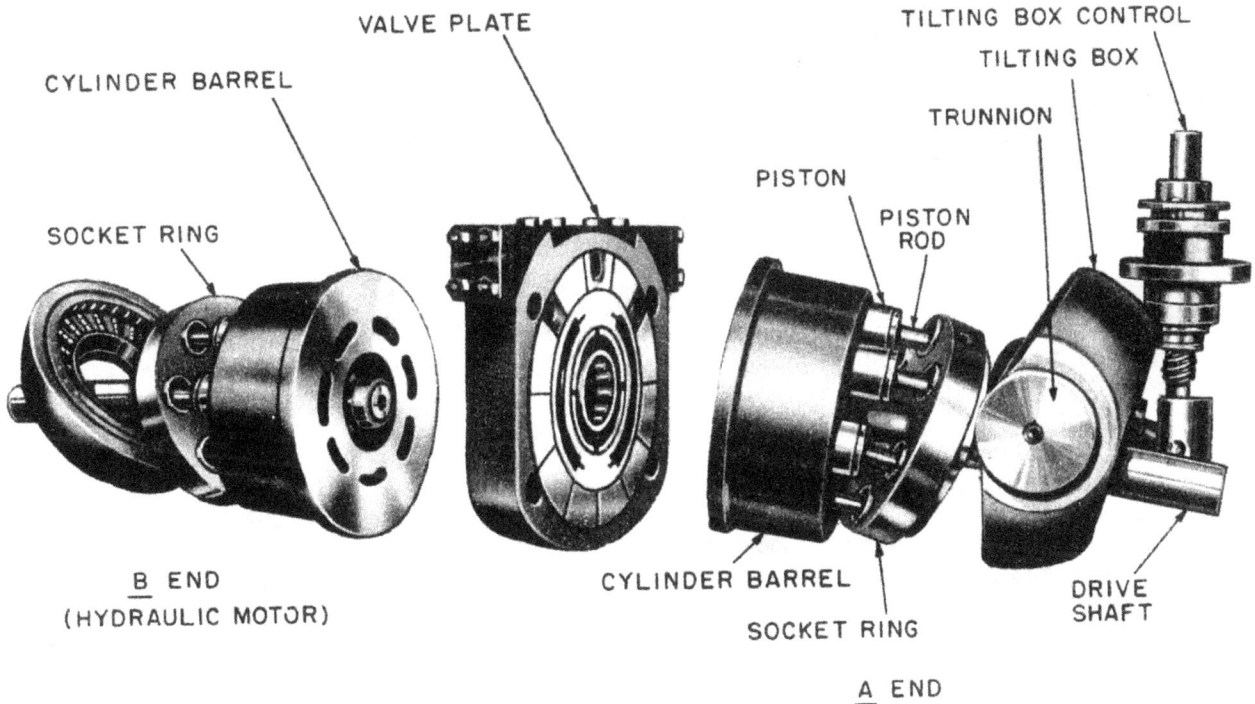

147.109

Figure 15-10.—Exploded view of axial-piston variable stroke pump.

In part B of figure 15-12, the tilting box is set at an angle so that it is farther away from the top of the cylinder barrel and closer to the bottom of the cylinder barrel. As the cylinder barrel and socket ring revolve, the piston is pulled <u>outward</u> as it is carried from the bottom position to the top position, and is pushed <u>inward</u> as it is carried from the top position to the bottom position. Thus

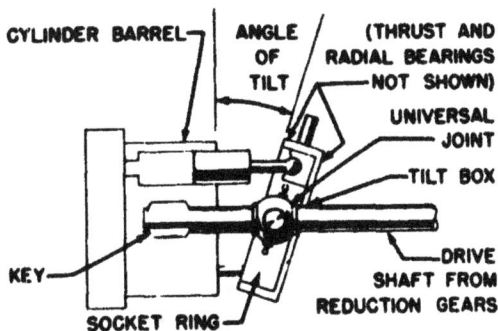

147.110

Figure 15-11.—Diagram showing cylinder barrel, socket ring, and tilting box in axial-piston variable stroke pump.

the piston makes one suction stroke (from the bottom position to the top position) and one discharge stroke (from the top position to the bottom position) for each complete revolution of the cylinder barrel.

In part C of figure 15-12, we see the tilting box set at a somewhat larger angle. Because there is more distance between the cylinder barrel and the socket ring at the top, and less distance between them at the bottom, the piston now moves further on each stroke and thus displaces more liquid on the discharge stroke.

Although we have considered the position of only one piston, it is obvious that the others are being similarly positioned as the cylinder barrel and socket ring revolve. At any given moment, therefore, some pistons are making suction strokes and others are making discharge strokes. In a nine-piston pump, for example, four pistons will be making suction strokes, four will be making discharge strokes, and one will be at the end of its stroke and will therefore be momentarily motionless.

Each cylindrical opening in the cylinder barrel has a port in the face of the cylinder barrel. As we have seen, each port except one will be either a suction port or a discharge port, depend-

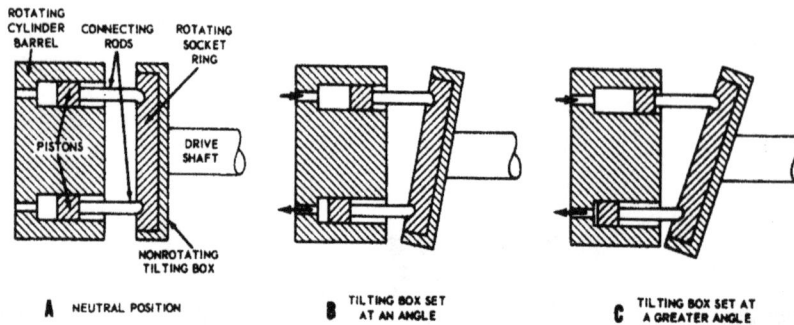

147.111
Figure 15-12.—Diagram showing how tilting box position affects position of pistons.

ing upon the position of the piston in the cylindrical opening. The face of the cylinder barrel bears against the valve plate, a nonrotating piece which has two semicircular ports, one for suction and one for discharge. When a piston is at the top position, at the end of its suction stroke, the port for that piston is over the top land[4] on the valve plate; when a piston is at the bottom position, at the end of the discharge stroke, the port is over the bottom land on the valve plate. Figure 15-13 shows the ports in the face of the cylinder barrel and the ports in the valve plate.

When the A end is used alone as a constant-speed, variable-capacity pump, the tilting box is often so designed that it can be tilted in one direction only. In this case, the flow of the pumped liquid is always in the same direction. When the A end is used as part of an electro-hydraulic system, however, the tilting box is most commonly designed to be tilted in either direction; and in this case the flow of the pumped liquid may be in either direction. Therefore, it should be clear that the position of the tilting box controls both the direction of flow and the amount of flow.

Figure 15-14 shows a cutaway view of an axial-piston variable stroke pump. Note that this particular pump is designed for reversible flow, since the tilting box can be tilted in either direction.

The radial-piston variable stroke pump is similar in general principle to the axial-piston pump just described, but the arrangement of component parts is somewhat different. In the radial-piston pump, the cylinders are arranged radially in a cylinder body that rotates around a nonrotating central cylindrical valve. Each cylinder communicates with horizontal ports in the central cylindrical valve. Plungers or pistons

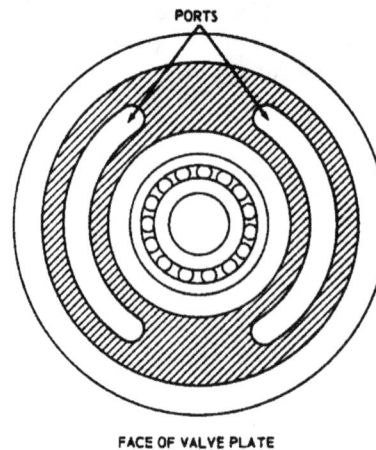

FACE OF VALVE PLATE

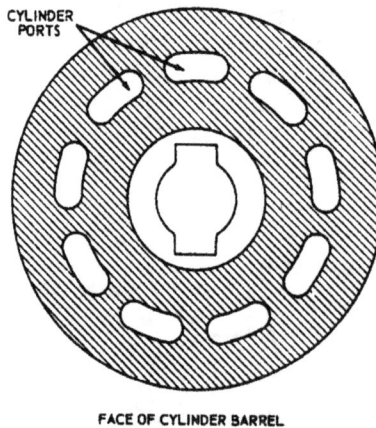

FACE OF CYLINDER BARREL

147.112
Figure 15-13.—Suction and discharge ports in face of cylinder barrel and in valve plate.

which extend outward from each cylinder are pinned at their outer ends to slippers which slide around the inside of a rotating floating ring or housing.

_____
[4] The term land refers to the space between ports.

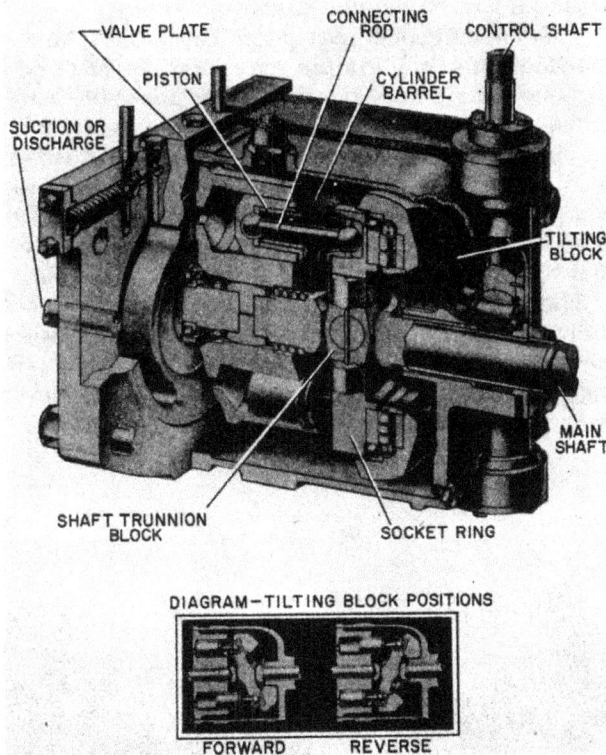

DIAGRAM—TILTING BLOCK POSITIONS

FORWARD   REVERSE

38.103

Figure 15-14.—Cutaway view of axial-piston variable stroke pump.

The floating ring is so constructed that it can be shifted offcenter from the pump shaft. When it is centered, or in the neutral position, the pistons do not reciprocate and the pump does not function, even though the electric motor is still causing the pump to rotate. If the floating ring is forced offcenter to one side, the pistons reciprocate and the pump operates. If the floating ring is forced offcenter to the other side of the pump shaft, the pump also operates but the direction of flow is reversed. Thus both the direction of flow and the amount of flow are determined by the position of the cylinder body relative to the position of the floating ring.

ROTARY PUMPS.—Rotary pumps, like reciprocating pumps, are positive-displacement pumps. The theoretical displacement of a rotary pump is the volume of liquid displaced by the rotating elements on each revolution of the shaft. The capacity of a rotary pump is defined as the quantity of liquid (in gpm) actually delivered under specified conditions. Thus the capacity is

equal to the displacement times the speed (rpm), minus whatever losses may be caused by slippage, suction lift, viscosity of the pumped liquid, amount of entrained or dissolved gases in the liquid, and so forth.

All rotary pumps work by means of rotating parts which trap the liquid at the suction side and force it through the discharge outlet. Gears, screws, lobes, vanes, and cam-and-plunger arrangements are commonly used as the rotating elements in rotary pumps.

Rotary pumps are particularly useful for pumping oil and other heavy, viscous liquids. This type of pump is used for fuel oil service, fuel oil transfer, lubricating oil service, and other similar services. Rotary pumps are also used for pumping nonviscous liquids such as water or gasoline, particularly where the pumping problem involves a high suction lift.

The power end of a rotary pump is usually an electric motor or an auxiliary steam turbine; however, some lubricating oil pumps that supply oil to the propulsion turbine bearings and to the reduction gears are attached to and driven by either the propulsion shaft or the quill shaft of the reduction gear.

Rotary pumps are designed with very small clearances between rotating parts and between rotating parts and stationary parts. The small clearances are necesarry in order to minimize slippage from the discharge side back to the suction side. Rotary pumps are designed to operate at relatively slow speeds in order to maintain these clearances; operation at higher speeds would cause erosion and excessive wear, which in turn would result in increased clearances.

Classification of rotary pumps is generally made on the basis of the type of rotating element. In the following paragraphs the main features of some common types of rotary pumps are discussed.

The simple gear pump (fig. 15-15) has two spur gears which mesh together and revolve in opposite directions. One gear is the driving gear, the other is the driven gear. Clearances between the gear teeth and the casing and between the gear faces and the casing are only a few thousandths of an inch. The action of the unmeshing gears draws the liquid into the suction side of the pump. The liquid is them trapped in the pockets formed by the gear teeth and the casing, so that it must follow along with the teeth. On the discharge side, the liquid is forced out by the meshing of the gears. Simple gear pumps of this

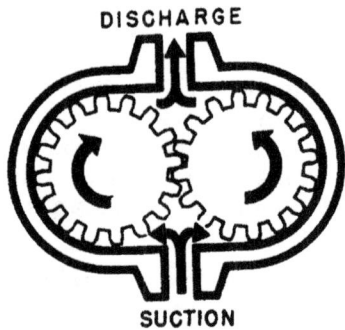

38.108

Figure 15-15.—Simple gear pump.

type are frequently used as lubricating pumps on pumps and other auxiliary machinery.

The herringbone gear pump (fig. 15-16) is a modification of the simple gear pump. In the herringbone gear pump, one discharge phase begins before the previous discharge phase is entirely complete; this overlapping tends to give a steadier discharge than that obtained with a simple gear pump. Herringbone gear pumps are sometimes used for low pressure fuel oil service, lubricating oil service, and diesel oil service.

The helical gear pump (fig. 15-17) is still another modification of the simple gear pump. Because of the helical gear design, the overlapping of successive discharges from spaces between the teeth is even greater than it is in the

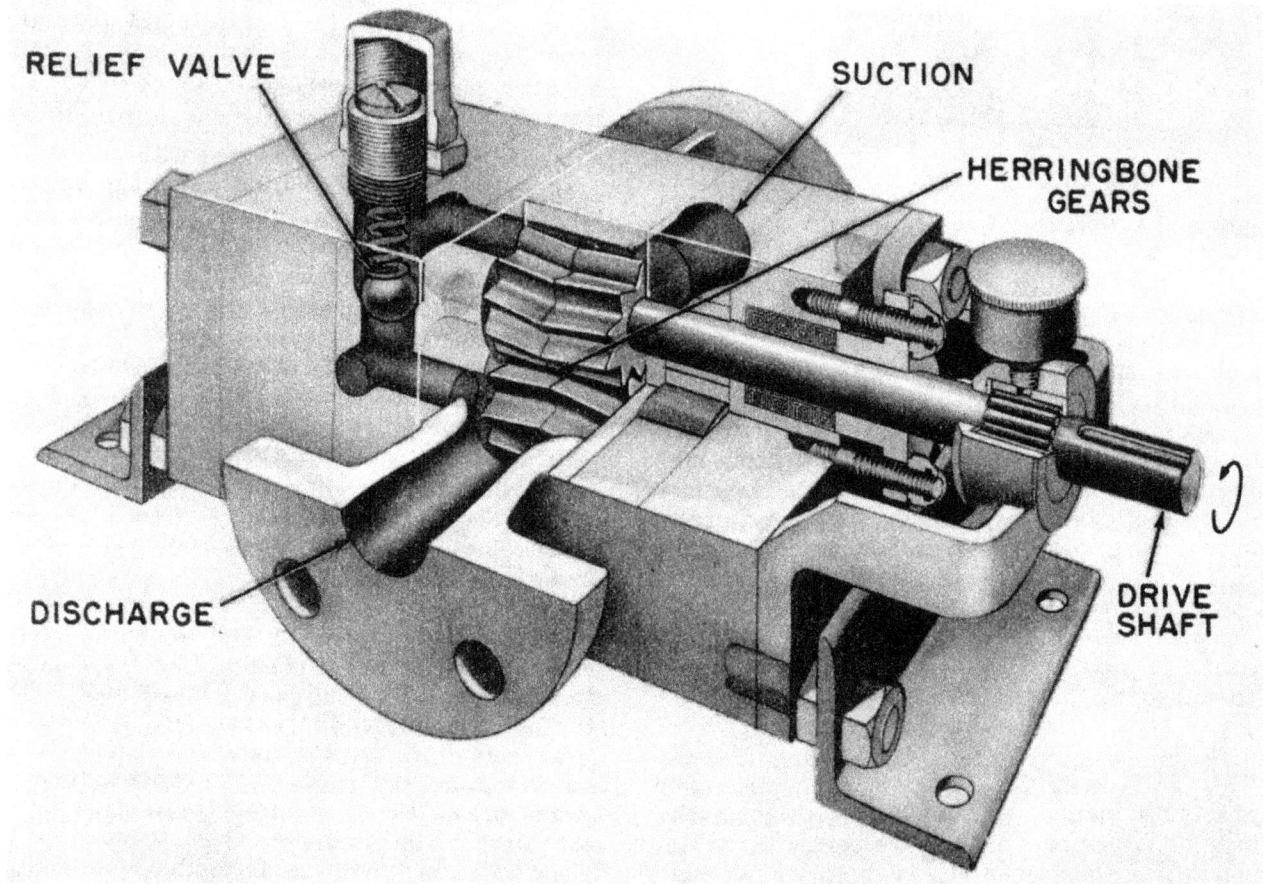

147.113

Figure 15-16.—Herringbone gear pump.

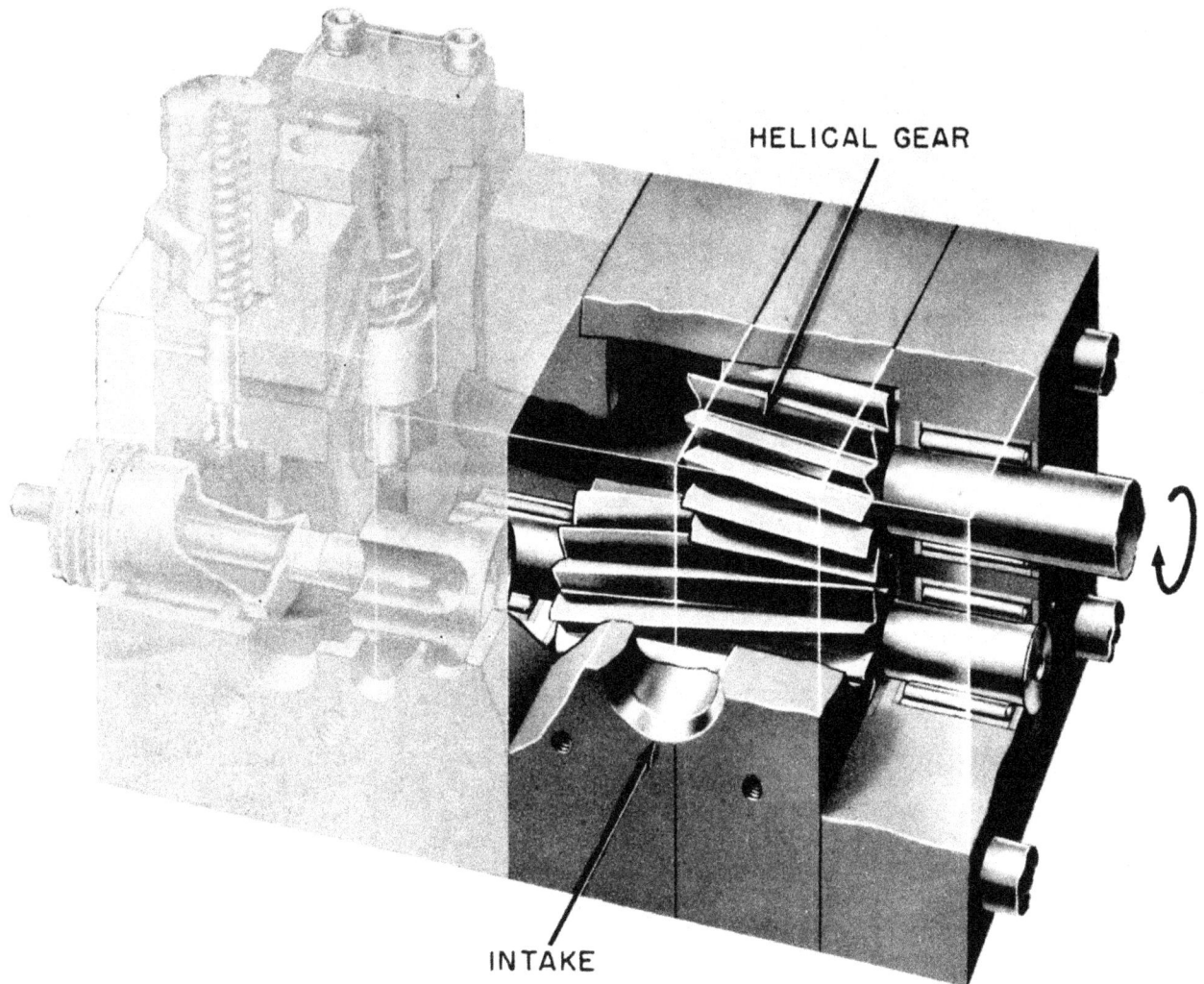

Figure 15-17.—Helical gear pump.

herringbone gear pump; and the discharge flow is, accordingly, even smoother. Since the discharge flow is smooth in the helical gear pump, the gears can be designed with a small number of teeth, thus allowing increased capacity without sacrificing smoothness of flow.

The pumping gears in this type of pump are driven by set of timing and driving gears, which also function to maintain the required close clearances while preventing actual metal-to-metal contact between the pumping gears. As a matter of fact, metallic contact between the teeth of the pumping gears would provide a tighter seal against slippage; but it would cause rapid wear of the teeth because foreign matter in the pumped liquid would be present on the contact surfaces.

Roller bearings at both ends of the gear shafts maintain proper alignment and thus minimize friction losses in the transmission of power. Stuffing boxes are used to prevent leakage at the shafts.

The helical gear pump is used to pump non-viscous liquids and light oils at high speeds and to pump viscous liquids at lower speeds.

Figures 15-18 and 15-19 illustrate two types of lobe pumps. Although these pumps look somewhat like gear pumps, they are not true gear pumps because the rotary elements are not capable of driving each other. One rotor is powered by the drive shaft; the other is driven by a set of timing gears. The lobes are considerably larger

147.115
Figure 15-18.—Lobe pump (heliquad type).

than gear teeth; as a rule, there are only two or three lobes on each rotor.

There are several different types of screw pumps. The main points of difference between the various types are the number of intermeshing screws and the pitch of the screws. A double-screw low pitch pump is shown in figure 15-20, and a triple-screw high-pitch pump in figure 15-21. Both of these pumps are widely used aboard ship to pump fuel oil and lubricating oil. In the double-screw pump, one rotor is driven by the drive shaft and the other by a set of timing gears. In the triple-screw pump, a central power rotor meshes with two idler rotors.

The rotating element in a rotating plunger pump (fig. 15-22) is a plunger which is set off-center on a drive shaft that is rotated by the source of power. The plunger is driven up and down and around the chamber by the rotation of the shaft, in such a way as to make a sliding seal with the walls of the chamber. In moving, the plunger alternately opens and closes a passage to the discharge.

Because of its valveless construction, the rotating plunger pump is suitable for pumping oil that may contain sand or other sediment and for pumping high viscosity liquids. The pump can produce a very high suction lift.

A moving vane pump (fig. 15-23) consists of a cylindrically bored housing with a suction inlet on one side and a discharge outlet on the other

side; a cylindrically shaped rotor of smaller diameter than the cylinder is driven about an axis placed above the centerline of the cylinder in such a way that the clearance between the rotor and the cylinder is small at the top and at a maximum value at the bottom.

The rotor carries vanes which move in and out as the rotor rotates, thus maintaining sealed spaces between the rotor and the cylinder wall. The vanes trap liquid on the suction side and carry it to the discharge side; contraction of the space expels the liquid into the discharge line. The vanes may swing on pivots, as shown in the illustration, or they may slide in slots in the rotor.

The moving vane type of pump is used for lubricating oil service and transfer and, in general, for handling light liquids of medium viscosity.

An internal gear pump is shown in figure 15-24. In the gear pumps previously described, the teeth project radially outward from the center of the gears. In an internal gear system, the teeth of one gear project outward but the teeth of the other project inward toward the center. In an internal gear pump, one gear stands inside the other.

A gear directly attached to the drive shaft of the pump is set offcenter in a circular chamber fitted around its circumference with the spurs of an internal gear. The two gears mesh on one side of the pump chamber, between the suction and the discharge. On the opposite side of the chamber a crescent-shaped form stands in the space between the two gears in such a way that a close clearance exists between each gear and the crescent.

The rotation of the central gear by the shaft causes the outside gear to rotate, since the two

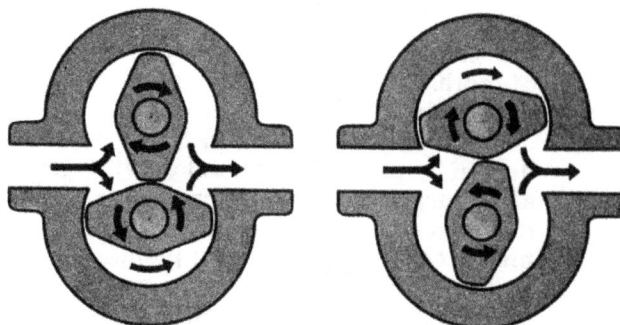

147.116
Figure 15-19.—Two-lobe pump.

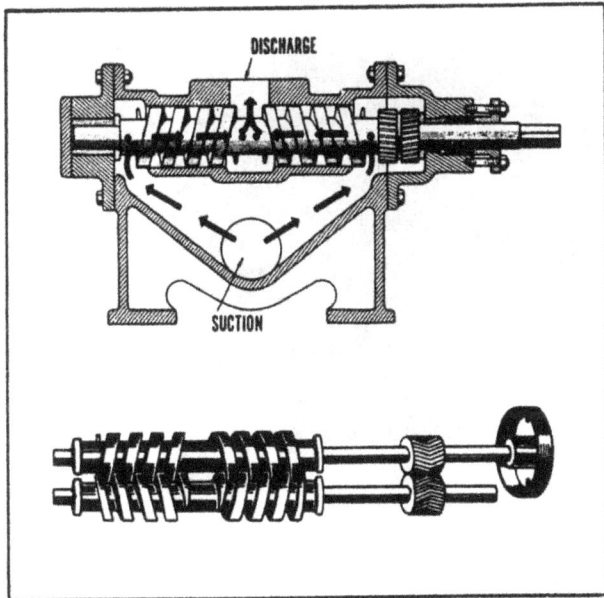

38.105X

Figure 15-20.—Double-screw low-pitch pump.

are in mesh. Everything in the chamber rotates except the crescent, causing liquid to be trapped in the gear spaces as they pass the crescent. The trapped liquid is carried from the suction to the discharge, where it is forced out of the pump by the meshing of the gears. As liquid is carried away from the intake side of the pump, the pressure there is diminished, thus forcing other liquid into the suction side of the pump. The direction of flow in the internal gear pump can be reversed by shifting the position of the crescent 180 degrees.

CENTRIFUGAL PUMPS.—Centrifugal pumps are widely used aboard ship for pumping water and other nonviscous liquids. The centrifugal pump utilizes the throwing force of a rapidly revolving impeller. The liquid is pulled in at the center or eye of the impeller and is discharged at the outer rim of this impeller. By the time the liquid reaches the outer rim of the impeller, it has acquired a considerable velocity. The liquid is then slowed down by being led through a volute or through a series of diffusing passages. As the velocity of the liquid decreases, its pressure increases—or, in other words, some of the mechanical kinetic energy of the liquid is transformed into mechanical potential energy. In the terminology commonly used in discussion of pumps, the velocity head of the liquid is partially converted to pressure head.

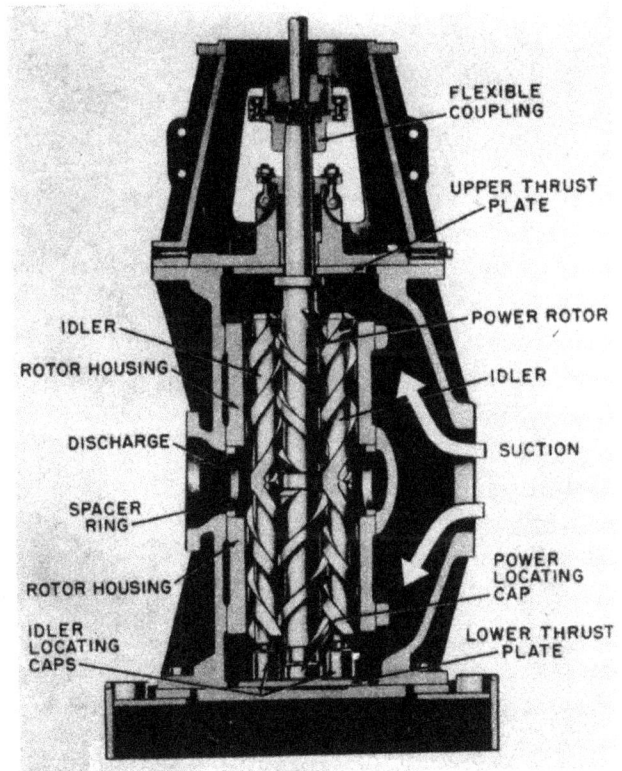

47.80

Figure 15-21.—Triple-screw high-pitch pump.

Centrifugal pumps are not positive-displacement pumps. When a centrifugal pump is operating at a constant speed, the amount of liquid discharged (capacity) varies with the discharge pressure according to the relationships inherent in the particular pump design. The relationships among capacity, total head (pressure), and power are usually expressed by means of a characteristics curve.[5]

Capacity and discharge pressure can be varied by changing the pump speed. However, centrifugal pumps should be operated at or near their rated capacity and discharge pressure whenever possible. Impeller vane angles and the sizes of the pump waterways can be designed for maximum efficiency at only one combination of speed and discharge pressure; under other conditions of operation, the impeller vane angles

_____

[5] Characteristics curves are generally given in the manufacturers' technical manuals or in the outline assembly drawings of pumps.

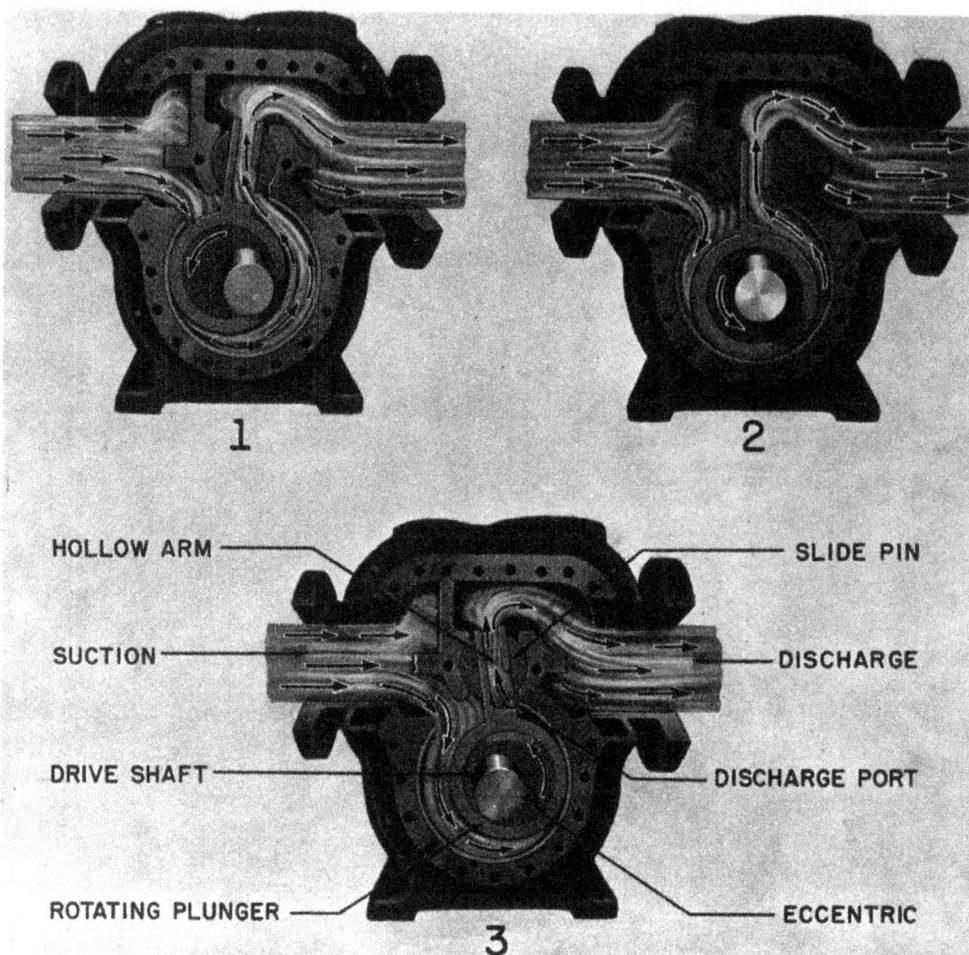

HOLLOW ARM ——————— SLIDE PIN

SUCTION ——————— DISCHARGE

DRIVE SHAFT ——————— DISCHARGE PORT

ROTATING PLUNGER ——————— ECCENTRIC

47.46

Figure 15-22.—Rotating plunger pump.

and the sizes of the waterways will be too large or too small for efficient operation. Therefore, a centrifugal pump cannot operate satisfactorily over long periods of time at excess capacity and low discharge pressure or at reduced capacity and high discharge pressure.

It should be noted that centrifugal pumps are not self-priming. The casing must be flooded before a pump of this type will function. For this reason, most centrifugal pumps are located below the level from which suction is to be taken. Priming can also be effected by using another pump to supply liquid to the pump suction—as, for example, the feed booster pump supplies suction pressure for the main feed pump. Some centrifugal pumps have special priming pumps, air ejectors, or other devices for priming.

Where two or more centrifugal pumps are installed to operate in parallel, it is particularly important to avoid operating the pumps at very low capacity, since it is possible that a unit having a slightly lower discharge pressure might be pushed off the line and thus forced into a shutoff position.

Because of the danger of overheating, centrifugal pumps can operate at zero capacity for only short periods of time. The length of time varies. For example, a fire pump might be able to operate for as long as 15 to 30 minutes before losing suction, but a main feed pump would overheat in a matter of a few seconds if operated at zero capacity.

Most centrifugal pumps—and particularly boiler feed pumps, fire pumps, and others which

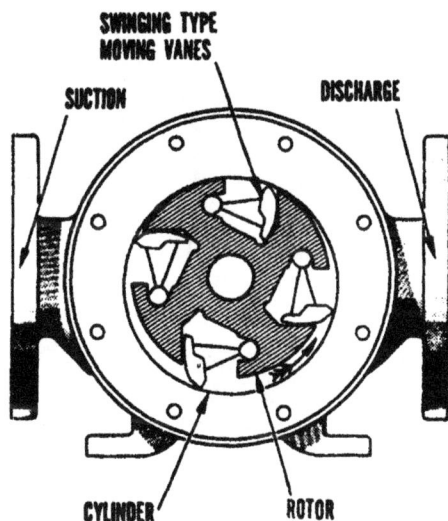

47.47

Figure 15-23.—Moving vane pump.

may be required to operate at low capacity or in shutoff condition for any length of time—are fitted with recirculation lines from the discharge side of the pump back to the source of suction supply. The main feed pump, for example, has a recirculating line going back to the deaerating feed tank. An orifice allows the recirculation of the minimum amount of water required to prevent over-heating of the pump. On boiler feed pumps, the recirculating lines must be kept open whenever the pumps are in operation.

On centrifugal pumps, there must always be a slight leakoff through the packing in the stuffing boxes, in order to keep the packing lubricated and cooled. Stuffing boxes are used either to prevent the gross leakage of liquid from the

pump or to prevent the entrance of air into the pump; the purpose served depends, of course, upon whether the pump is operating with a positive suction head or is taking suction from a vacuum.

If a centrifugal pump is operating with a positive suction head, the pressure inside the pump is sufficient to force a small amount of liquid through the packing when the packing gland is properly set up on. On multistage pumps, it is sometimes necessary to reduce the pressure on one or both of the stuffing boxes. This is accomplished by using a bleedoff line which is tapped in to the stuffing box between the throat bushing and the packing.

If a pump is taking suction at or below atmospheric pressure, a supply of sealing water must be furnished to the packing glands to ensure the exclusion of air. Some of this water must be allowed to leak off through the packing. Most centrifugal pumps use the pumped liquid as the lubricating, cooling, and sealing medium. However, an independent external sealing liquid is used on some pumps.

There are several different designs of centrifugal pumps. The two types most commonly used aboard ship are the volute pump and the volute turbine pump.

The volute pump is shown in figure 15-25. In this pump, the impeller discharges into a volute— that is, a gradually widening channel in the pump casing. As the liquid passes through the volute and into the discharge nozzle, a great part of its kinetic energy is converted into potential energy.

In the volute turbine pump (fig. 15-26) the liquid leaving the impeller is first slowed down by the stationary diffuser vanes which surround the impeller. The liquid is forced through gradually widening passages in the diffuser ring (not shown) and into the volute. Since both the diffuser

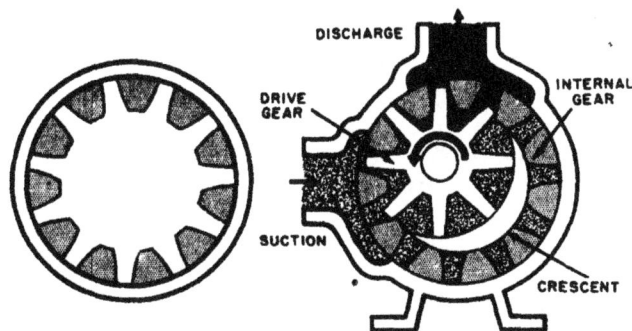

147. 117

Figure 15-24.—Internal gear pump.

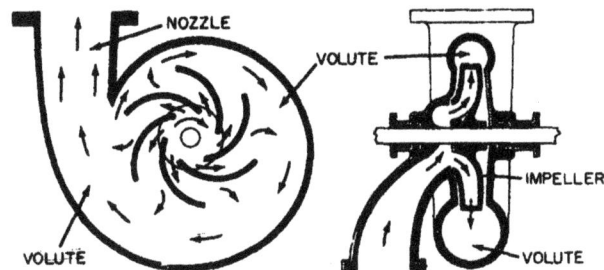

23.18.0

Figure 15-25.—Simple volute pump.

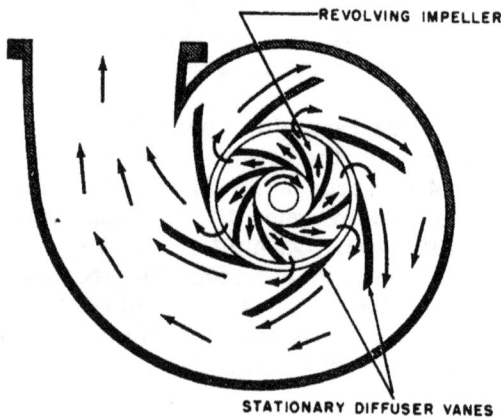

23.19

Figure 15-26.—Volute turbine pump.

vanes and the volute reduce the velocity of the liquid, there is in this type of pump an almost complete conversion of kinetic energy to potential energy.

Centrifugal pumps may be classified in several ways. For example, they may be either single-stage or multistage, a single-stage pump has only one impeller; a multistage pump has two or more impellers housed together in one casing. In a multistage pump, as a rule, each impeller acts separately, discharging to the suction of the next-stage impeller. Centrifugal pumps are also classified as horizontal or vertical, depending upon the position of the pump shaft.

The impellers used in centrifugal pumps may be classified as single suction or double-suction. The single-suction impeller allows liquid to enter the eye of the impeller from one direction only; the double-suction type allows liquid to enter the eye from two directions. Single-suction and double-suction arrangements are shown in figure 15-27. The double-suction arrangement has the advantage of balancing end thrust in one direction by end thrust in the other.

Some of the more important centrifugal pumps used on naval ships are the main feed pump, the feed booster pump, the main and auxiliary condensate pumps, fire pumps, fresh water pumps, and gasoline pumps.

A typical main feed pump is shown in figure 15-28. This is a horizontal, high speed, turbine-driven pump. Main feed pumps on most surface ships operate at a discharge pressure that is 100 to 150 psig above the maximum steam drum pres-

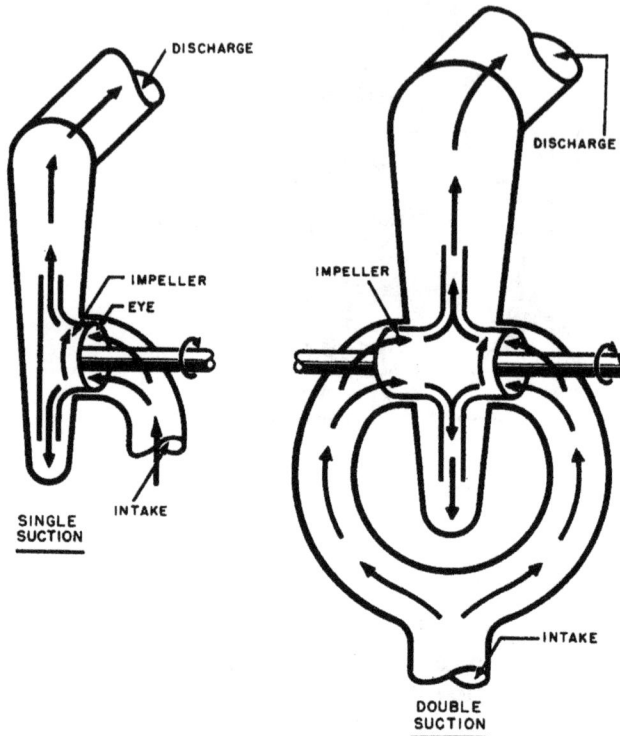

23.25

Figure 15-27.—Single-suction and double-suction arrangements in centrifugal pumps.

sure of the boiler. On ships having 1200-psig boilers, the discharge pressure of the main feed pumps is approximately 200 to 300 psig above the steam drum pressure. A main feed pump must operate at varying speeds to maintain a constant discharge pressure under all conditions of load. A constant-pressure pump governor[6] is used to regulate the admission of steam to the turbine and thus control the discharge pressure of the pump.

PROPELLER PUMPS.—Propeller pumps are used on some ships for pumping water. Although they are often classified as centrifugal pumps, this classification is incorrect because propeller pumps do not actually utilize centrifugal force for their operation.

A propeller pump consists essentially of a propeller fitted into a narrow, tube-like casing. The propeller pumps the liquid by pushing it in a direction parallel to the pump shaft.

[6] Constant-pressure pump governors are discussed in chapter 16 of this text.

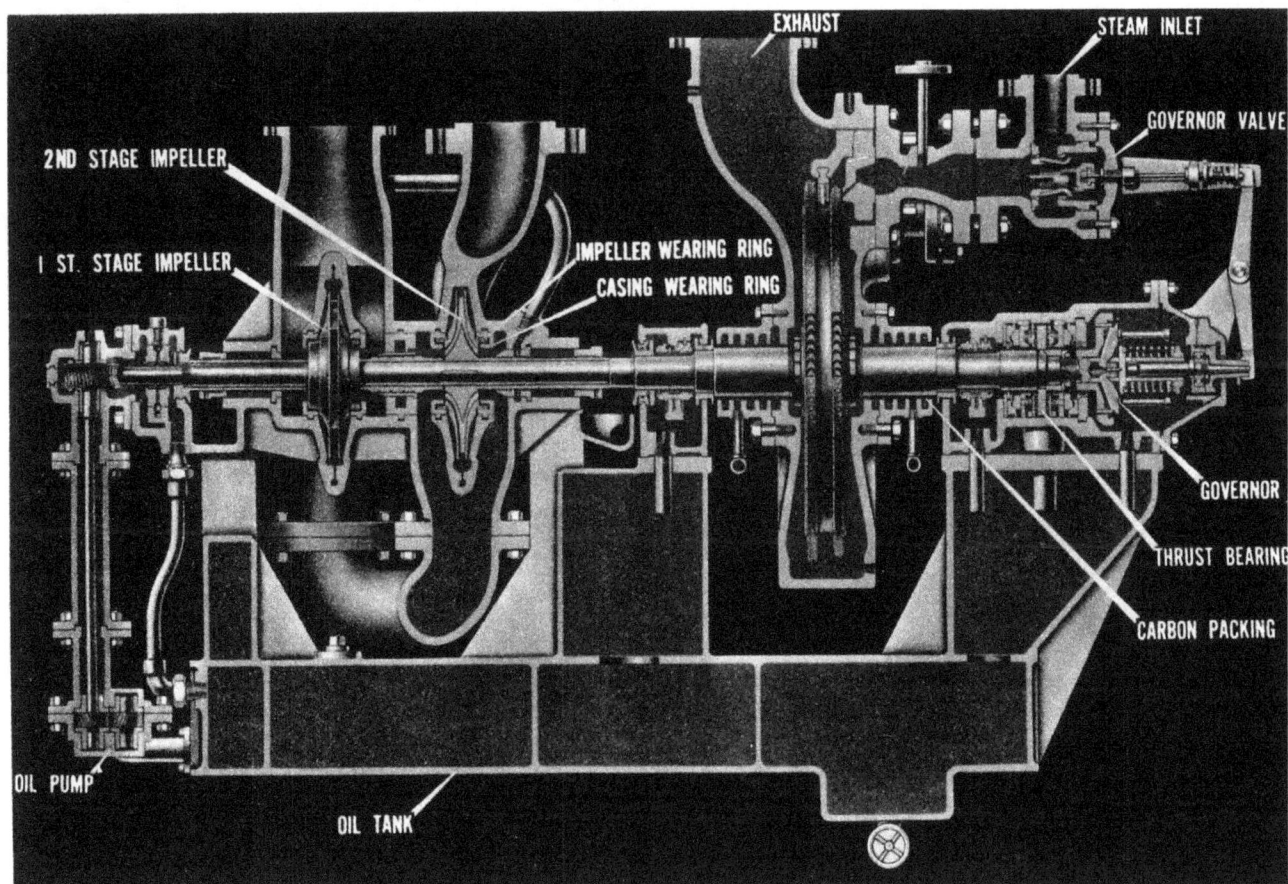

38.109

Figure 15-28.—Main feed pump.

Propeller pumps must be located below or only slightly above the surface of the liquid to be pumped, since they cannot operate with a high suction lift.

MIXED-FLOW PUMPS.—A mixed-flow pump is one in which the pumping action occurs partly by centrifugal force and partly by propeller action. Pumps of this type can be used to handle very viscous liquids or liquids that contain dirt; they are better than either centrifugal pumps or propeller pumps for these services.

JET PUMPS.—Devices which utilize the rapid flow of a fluid to entrain another fluid and thereby move it from one place to another are called jet pumps. Jet pumps are sometimes not considered to be pumps because they have no moving parts. However, in view of our previous definition of a pump as a device which utilizes an external source of power to apply force to a fluid in order to move the fluid from one place to another, it will be apparent that a jet pump is indeed a pump.

Jet pumps are generally considered in two classes: ejectors, which use a jet of steam to entrain air, water, or other fluid; and eductors, which use a flow of water to entrain and thereby pump water. The basic principles of operation of these two devices are identical.

A simple jet pump of the ejector type is shown in figure 15-29. In this pump, steam under pressure enters chamber C through pipe A, which is fitted with a nozzle, B. As the steam flows through the nozzle, the velocity of the steam is increased. The fluid in the chamber at point F, in front of the nozzle, is driven out of the pump through the discharge line, E, by the force of the steam jet. The size of the discharge line

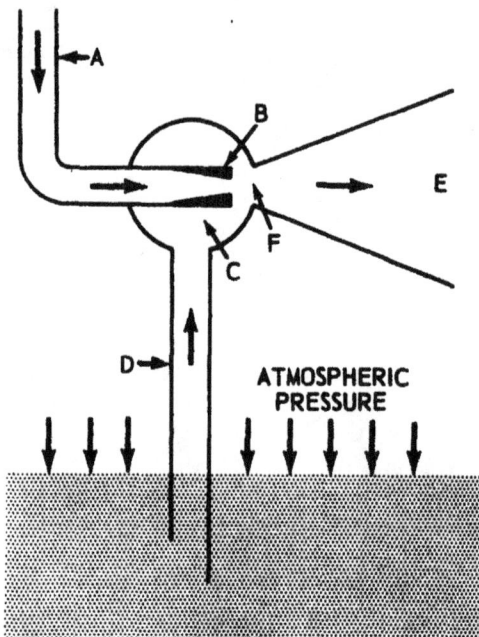

75.283
Figure 15-29.—Jet pump (ejector type).

increases gradually beyond the chamber, in order to decrease the velocity of the discharge and thereby transform some of the velocity head to pressure head. As the steam jet forces some of the fluid from the chamber into the discharge line, pressure in the chamber is lowered and the pressure on the surface of the supply fluid forces fluid up through the inlet, D, into the chamber and out through the discharge line. Thus the pumping action is established.

Jet pumps of the ejector type are occasionally used aboard ship to pump small quantities of drains overboard. Their primary use on naval ships, however, is not in the pumping of water but in the removal of air and other noncondensable gases from main and auxiliary condensers.[7]

An eductor is shown in figure 15-30. As may be seen, the principle of operation is the same as that just described for the ejector type of jet pump; however, water is used instead of steam. On naval ships, eductors are used to pump water from bilges, to dewater compartments, and to

_____
[7] Air ejector assemblies used on condensers are discussed in chapter 13 of this text.

supply a positive pressure head for pumps used in firefighting.

Pump Maintenance

Pumps require a certain amount of routine maintenance and, upon occasion, some repair work. Pumps are so widely used for various services in the Navy that it is necessary to consult the manufacturer's technical manual for details concerning the repair of a specific unit. Routine maintenance, however, is performed in accordance with the Planned Maintenance Subsystem (3-M System) requirements. Figure 15-31 illustrates the planned maintenance requirements for one type of turbine-driven main lube oil pump. Similar requirements are established for all pumps.

Safety Precautions

The following safety precautions must be observed in connection with the operation of pumps:

1. See that all relief valves are tested at the appropriate intervals as required by the Planned Maintenance Subsystem. Be sure that relief valves function at the designated pressure.
2. Never attempt to jack over a pump by hand while the throttle valve to the turbine is open or the power is on. Never jack over a reciprocating pump when the throttle valve or the exhaust valve is open.

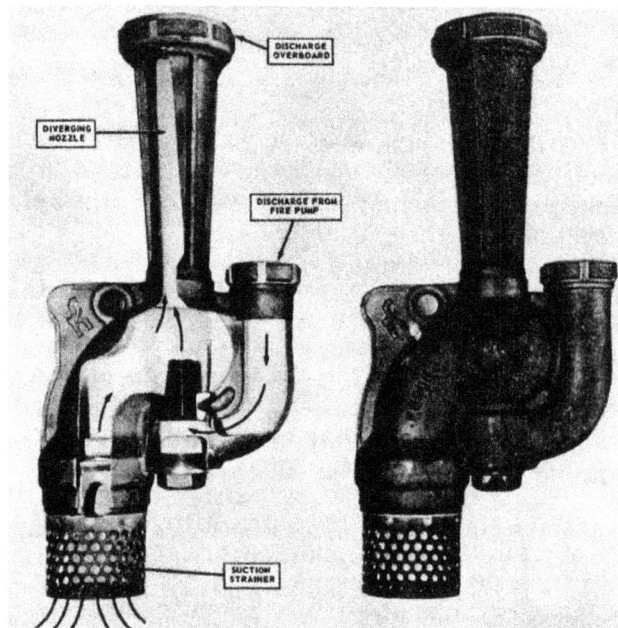

Figure 15-30.—Eductor.   47.48

System, Subsystem, or Component	Reference Publications
Turbine Driven Main Lube Oil Pump	

Bureau Card Control No.	Maintenance Requirement	M.R. No.	Rate Req'd.	Man Hours	Related Maintenance
MM ZVVFPJ4 95 4006 W	1. Sample and inspect lube oil. 2. Lubricate speed limiting governor. 3. Turn idle pump by hand; if free, operate by power.	W-1	FN	0.5	None
MM ZZ1FRP6 95 7768 M	1. Test the speed limiting governor.	M-1	MM2 FN	0.1 0.1	None
MM ZZ2FPJ4 84 4839 Q	1. Clean sump. 2. Clean lube oil filter. 3. Renew oil.	Q-1	MM2 FN	2.0 2.0	None
MM ZVVFPJ4 95 4369 Q	1. Sound and tighten foundation bolts. 2. Lubricate flexible coupling.	Q-2	MM3	0.3	None
MM ZZPFTRO 94 5422 Q	1. Measure turbine thrust clearance.	Q-3	MM2 FN	0.3 0.3	None
MM ZZFFVA1 84 0851 Q	1. Test combination exhaust and relief valve. 2. Test lube oil relief valve.	Q-4	MM2 FN	0.2 0.2	None
MM ZZ1FRP5 A5 4854 Q	1. Clean and inspect pump pressure regulator.	Q-5	MM2	2.0	None
MM ZVVFPJ4 A5 6042 A	1. Inspect shaft journals, thrust collar, and bearings for condition; measure bearing clearance. 2. Inspect and clean steam strainer.	A-1	MM2 FN	3.3 3.3	M-1 Q-1 Q-3
MM ZXFFCW4 95 4370 A	1. Renew stuffing box packing. 2. Inspect flexible coupling.	A-2	MM2	2.0	Q-2
MM ZX6FPJ4 84 6044 A	1. By operational test, inspect internal parts for wear.	A-3	MM2	0.3	None
MM ZZPFTRO 95 6045 C	1. Inspect carbon packing for wear. 2. Inspect turbine exterior.	C-1	MM2 FN	3.0 3.0	None

MAINTENANCE INDEX PAGE
OPNAV FORM 4700-3 (4-64)

BUREAU PAGE CONTROL NUMBER E-9/10-95

98.171

Figure 15-31.—PMS, Maintenance Index Page.

3. Do not attempt to operate a pump while either the speed limiting governor or the constant pressure governor is inoperable. Be sure that the speed limiting governor and the constant pressure governor are properly set.

4. Do not use any boiler feed system pump for any service other than boiler or feed water service, except in an emergency.

FORCED DRAFT BLOWERS

On most steam-driven surface ships, forced draft blowers are used to furnish the large amount of combustion air required for the burning of the fuel oil. A forced draft blower is essentially a very large fan, fastened to a shaft and housed in a metal casing. As a rule, two blowers are furnished for each boiler; they are synchronized for equal distribution of load.

Most forced draft blowers are driven by steam turbines. However, some blowers for in-port use and some main blowers on auxiliary ships are driven by electric motors. Most turbine-driven blowers are direct drive, rather than geared; but some geared turbine drives are used.

On most ships, the forced draft blowers take suction from the space between the inner and outer stack casings and discharge slightly preheated air into a duct that leads to the space between the inner and outer casings of the boiler.

Types of Forced Draft Blowers

Two main types of forced draft blowers are used in naval ships: centrifugal blowers and propeller blowers. The main difference between the two types is in the direction of air flow. The centrifugal blower takes air in axially at the center of the fan and discharges it tangentially off the outer edge of the blades. The propeller blower moves air axially—that is, it propels the air straight ahead in a direction parallel to the axis of the shaft. Most forced draft blowers now in naval use are of the propeller type. However, some older ships and some recent auxiliaries have centrifugal blowers.

Centrifugal blowers may be either vertical or horizontal. In either case, the unit consists of the driving turbine (or other driving unit) at one end of the shaft and the centrifugal fanwheel at the other end of the shaft. Inlet trunks and diffusers are fitted around the blower fanwheel to direct air into the fanwheel and to receive and discharge air from the fan. Centrifugal blowers

are fitted with flaps in the suction ducts. In the event of a casualty to one centrifugal blower, air from another blower blows back toward the damaged blower and closes the flaps.

Both horizontal and vertical propeller blowers are used in naval combatant ships. In general, single-stage horizontal blowers are used on older ships and two-stage or three-stage vertical blowers on ships built since World War II.

Balanced automatic shutters are installed in the discharge ducts between each propeller blower and the boiler casings. These shutters are locked in the closed position whenever the blower is taken out of service so that the blower will not be rotated in reverse.

Figure 15-32 and 15-33 show two views of a single-stage horizontal propeller-type blower. As may be seen, the blower is a complete unit consisting of a driving turbine and a propeller-type fan. The entire unit is mounted on a single bed plate.

The air intake is screened to prevent the entrance of foreign objects. The blower casing

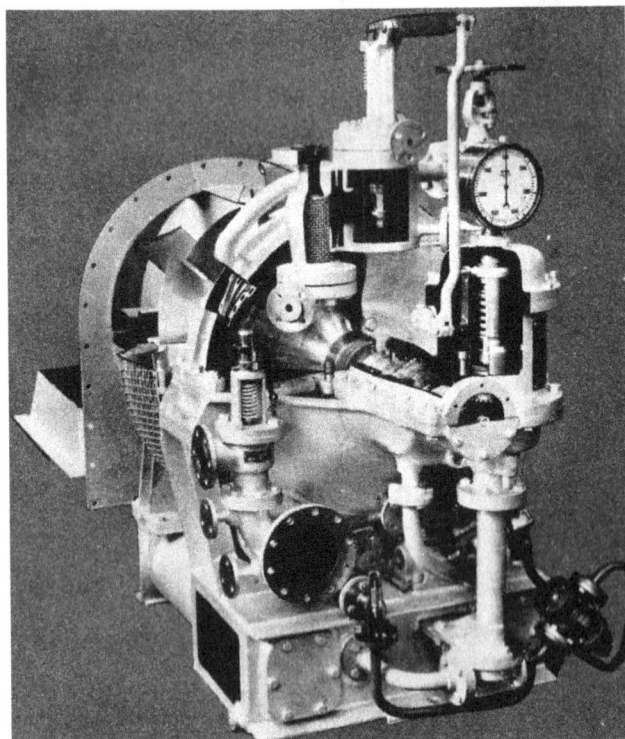

38.111X

Figure 15-32.—Horizontal propeller-type blower (view 1).

38.112X

Figure 15-33.—Horizontal propeller-type blower
(view II).

merges into the discharge duct, and the discharge
duct is joined to the boiler casing. Diffuser vanes
are installed just in front of the blower to prevent
rotation of the air stream as it leaves the blower.
Additional divisions in the curving sections of the
discharge duct also help to control the flow of air.

The shaft that carried both the propeller and
the turbine is a single forging. The propeller is
keyed to the shaft and held to the tapered end of
the shaft by a nut and cotter pin. The entire assembly is supported by two main bearings, one
on each side of the turbine wheel, outside the turbine casing. The main bearing at the governor
end is located in the governor housing, which also
contains the thrust bearing. The speed-limiting
governor spindle and the lubricating oil pump
shaft are driven by the main shaft of the blower,
through a reduction gear.

High powered two-stage or three-stage vertical propeller-type blowers are installed in recent combatant ships. One kind of three-stage
vertical propeller-type blower is shown in figures 15-34 and 15-35. Figure 15-36 shows the
rotating assembly of the same blower. As may
be seen, there are three propellers at the fan
end. Each propeller consists of a solid forged
disk to which are attached a number of forged
blades. The blades have bulb-shaped roots that
are entered in grooves machined across the hub;
the blades are kept firmly in place by locking
devices. Each propeller disk is keyed to the shaft
and secured by locking devices.

38.113X

Figure 15-34.—External view of three-stage
vertical propeller-type blower.

The driving turbine is a velocity-compounded
impulse turbine (Curtis stage) with two rows of
moving blades. The turbine wheel is keyed to the
shaft. The lower face of the turbine wheel bears
against a shoulder on the shaft; a nut screwed
onto the shaft presses against the upper face of
the turbine wheel.

The entire rotating assembly is supported by
two main bearings. One bearing is just below the
propellers and one is just above the thrust bearing in the oil reservoir.

The blower casing is built up of welded plates.
From the upper flange down to a little below the
lowest propeller, the casing is cylindrical in
shape. The shape of the casing changes from cylindrical to cone-shaped and then to square; the
discharge opening of the blower casing is rectangular in shape. Guide vanes in the casing control
the flow of air and also serve to stiffen the

415

PROPELLER, 1st STAGE

PROPELLER, 2nd STAGE

PROPELLER, 3rd STAGE

OIL BAFFLE SEAL RING

OIL GUARD

TURBINE

GLAND SEALS

VISCOSITY PUMP

CENTRIFUGAL PUMP

OIL RESERVOIR

THRUST BEARING

STATIONARY GUIDE VANES

UPPER BEARING

EXTERNAL OIL LINE TO UPPER BEARING

OIL RETURN LINE

STEAM CHEST

NOZZLE VALVE

STATIONARY BLADING

LOWER BEARING

PRESSURE GAGE TACHOMETER

38.114

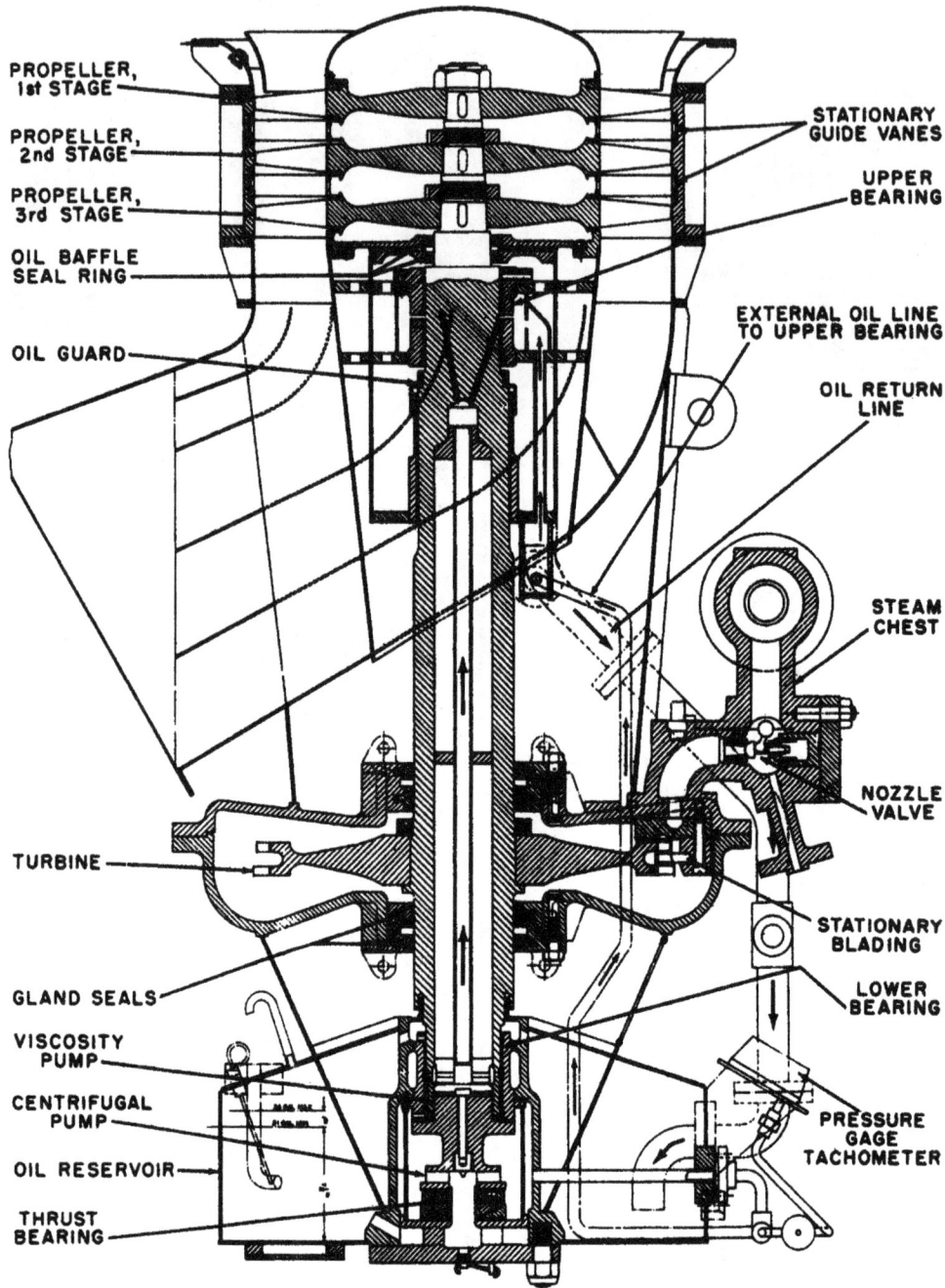

Figure 15-35.—Sectional view of three-stage vertical propeller-type blower.

38.115X

Figure 15-36. — Rotating assembly for three-stage vertical propeller-type blower.

casing. The part of the casing near the propellers is made in sections and is also split vertically to allow removal of the three propellers when necessary. The lower part of the casing, below the air duct, houses the turbine. The lower part of the turbine casing is welded to the oil reservoir structure.

Although all vertical propeller blowers operate on the same principle, and although they may look very much the same from the outside, they are not identical in all details. Perhaps the major differences to be found among vertical forced draft blowers are in connection with the lubrication systems. Some of these differences are noted in the following section.

Blower Lubrication

Because forced draft blowers must operate at very high speeds, correct lubrication of the bearings is absolutely essential. A complete pressure lubrication system for supplying oil to the bearings is an integral part of every forced draft blower. Most forced draft blowers have two radial bearings and one thrust bearing; however, some blowers have two turbine bearings, two fan bearings, and a thrust bearing.

The lubrication system for a horizontal forced draft blower includes a pump, an oil filter, an oil cooler, a filling connection, relief valves, oil level indicators, thermometers, pressure gages, oil sight flow indicators, and the necessary piping. The pump is usually turned by the forced draft blower shaft but is geared down to about one-fourth the speed of the turbine. The lube oil is pumped from the oil reservoir, through the oil filter and the oil cooler, to the bearings. Oil then drains back to the reservoir by gravity.

Simple gear pumps were used in the lube oil systems of older horizontal blowers such as those found on DD 445 and DD 692 classes of destroyers. These pumps were not completely satisfactory for this use, since a simple gear pump does not supply oil to the bearings when the blower is turning in the wrong direction. This is a serious disadvantage in forced draft blower installations because of the possibility that idle blowers may be rotated in reverse when automatic shutters fail to close. To prevent damage to the bearings from this cause, the simple gear pumps in some horizontal blowers were replaced by a special type of gear pump that continues to pump oil, without change in the direction of oil flow, when the direction of blower rotation is reversed. This alteration has been accomplished on many ships that have horizontal forced draft blowers.

Some vertical blowers are fitted with a gear pump and a lubrication system which is generally similar to that just described for horizontal blowers. However, most vertical blowers have quite different lubrication systems.

One type of lubrication system used on some vertical forced draft blowers is shown in figure 15-37. In this system, the gear pump is replaced by a centrifugal pump and a helical-groove viscosity pump. The centrifugal pump impeller is on the lower end of the main shaft, just below the lower main bearing. The viscosity pump (also called a friction pump) is on the shaft, just above the centrifugal pump impeller, inside the lower part of the main bearing. As the main shaft turns, lubricating oil goes to the lower bearing and from there, by way of the hollow shaft, to the upper bearing. In addition, part of the oil is pumped directly to the upper bearing through an external supply line. The oil is returned from the upper bearing to the oil reservoir through an external return line.

In this system, the lubricating oil does not go through the oil strainer or the oil filter on its way to the bearings. Instead, oil from the reservoir is constantly being circulated through an external filter and an external cooler and then back to the reservoir.

The viscosity pump is needed in this system because the pumping action of the centrifugal pump impeller is dependent upon the rpm of the shaft. At low speeds, the centrifugal pump cannot develop enough oil pressure to adequately lubricate the bearings. At high speeds, the centrifugal pump alone would develop more oil

417

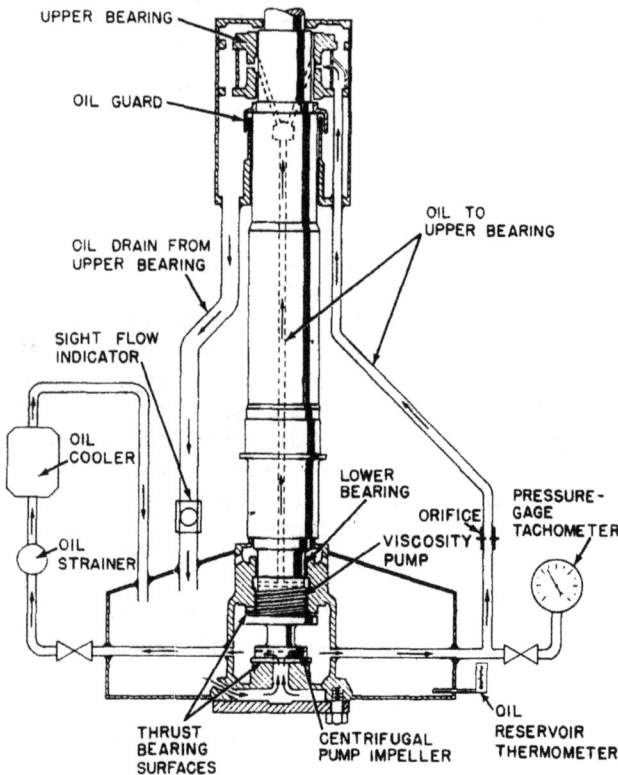

38.116

Figure 15-37.—Lubrication system with hollow-shaft oil supply and external oil supply for vertical forced draft blower.

pressure than is needed for lubrication, and the excessive pressure would tend to cause flooding of the bearings and loss of oil from the lubrication system. The viscosity pump, which is nothing more than a shallow helical thread or groove on the lower part of the shaft, helps to assure sufficient lubrication at low speeds and to prevent the development of excessive oil pressures at high speeds.

The hollow-shaft type of lubrication system just described is still found on some vertical forced draft blowers. The newest vertical blowers, however, do not use a hollow shaft to supply oil to the upper bearings. Instead, oil is pumped to the bearings through an external supply line, passing through an oil filter and an oil cooler on the way to the bearings. Most of these newer vertical blowers have a gear pump and an additional hand pump that is used to establish initial lubrication when the blower is being started. Some of the newer vertical blowers have a centrifugal pump and a viscosity

pump of the type shown in figure 15-36; but these blowers, like the other newer ones, have a completely external oil supply to the bearings instead of a hollow-shaft arrangement.

One feature of some of the newer vertical blowers is an anti-rotation device on the shaft of the lubricating pump. This device prevents windmilling of the blower in a reverse direction in the event of leakage through the automatic shutters. The anti-rotation device is continuously lubricated through a series of passageways which trap some of the leakage from the thrust bearing.

Control of Blower Speed

Forced draft blowers are manually controlled on all naval ships except those that have automatic combustion control systems for the boilers. Speed-limiting governors (discussed in chapter 16 of this text) are fitted to all forced draft blowers, but they function merely as safety devices to prevent the turbine from exceeding the maximum safe operating speed; the speed-limiting governors do not have any control over the turbine at ordinary operating speeds.

Manual control of blower speed is achieved by a valve arrangement that controls the amount of steam admitted to the turbines. In some blowers a full head of steam is admitted to the steam chest; steam is then admitted to the turbine by means of a manually operated lever or handwheel that controls four nozzle valves. The lever or handwheel may be connected by linkage for remote operation. The four valves are so arranged that they open in sequence, rather than all at the same time. The position of the manually operated lever or handwheel determines the number of valves that will open, and thus controls the amount of steam that will be admitted to the driving turbine. The steam chest nozzle valve shafts of all blowers serving one boiler are mechanically coupled so as to provide for synchronized operation of the blowers. If only one blower is to be operated, the root valve of the nonoperating blower must remain closed so that steam will not be admitted to the line.

In other installations, a throttle valve is used to control the admission of steam to the steam chest. From the steam chest, the steam enters the turbine casing through fixed nozzles rather than through nozzle valves. Varying the opening of the throttle valve varies the steam pressure to the steam chest and thus varies the speed of the turbine. The same throttle valve is usually used to control the admission of steam to all blowers serving any one boiler. If only one blower is to be operated, the root valve of the nonoperating blower must be kept closed.

When admission of steam is controlled by the four nozzle valve arrangement, no additional nozzle area is required to bring the blower up to maximum speed. In the other type of installation, a special hand-operated nozzle valve is provided for high speed operation. This nozzle valve, which is sometimes called an overload nozzle valve, is used whenever it is necessary to increase the blower speed beyond that obtainable with the fixed nozzles. As a rule, the use of the overload nozzle valve is required only when steam pressure is below normal.

Checking Blower Speed

Many forced draft blowers are fitted with constant-reading, permanently mounted tachometers for checking on blower speed. Sometimes the tachometer is mounted on top of the governor and is driven by the governor spindle. The governor spindle is driven by the main shaft through a reduction gear, and therefore does not rotate at the same speed as the main shaft. However, the rpm of the governor spindle is proportional to the rpm of the main shaft. The tachometer is calibrated to give readings that indicate the speed of the main shaft rather than the speed of the governor spindle.

Some blowers are equipped with a special kind of tachometer called a pressure-gage tachometer. This instrument, which may be seen in figures 15-35 and 15-37, is actually a pressure gage which is calibrated in both psi and rpm. The pressure-gage tachometer depends for its operation on the fact that the oil pressure built up by the centrifugal lube oil pump has a definite relation to the speed of the pump impeller; and the speed of the impeller, of course, is determined by the speed of the main shaft. Thus the instrument can be calibrated in both psi and rpm.

Some forced draft blowers of recent design are equipped with electric tachometers which have indicating gages at the blower and at the boiler operating station. The electric tachometer (sometimes called a tachometer generator) consists of a stator and a permanent magnet rotor mounted at the bottom of the turbine shaft. The wire from the generator plugs into a connector inside the sump. Another connector is provided outside the sump for attaching the wire from the generator to the transformer box.

Forced Draft Blower Operation

Forced draft blowers supply the air required for combustion of the fuel oil. The amount of air that enters the furnace is determined by the air pressure in the double casings. Although the air pressure is affected by the number of burners in use and by the amount that the air registers are open, it is primarily determined by the speed at which the forced draft blowers are operated. The speed of the forced draft blowers is controlled by manual adjustment of the blower throttle in all installations except those having automatic combustion control systems.

The forced draft blowers should be operated in such a way as to furnish the required amount of air for the complete combustion of the fuel being burned. In actual practice, it is necessary to supply just over 100 percent of the amount of air theoretically required, in order to ensure the complete combustion of the fuel. Higher percentages of excess air are wasteful of fuel, since all air that does not actually enter into a combustion reaction merely absorbs and carries off heat.

On the other hand, an insufficient quantity of air for combustion is also detrimental to boiler efficiency. If there is not enough air for complete combustion, there may be a greater loss in efficiency. Or, if even less air is supplied, some of the carbon will not be burned at all but will pass out the smokestack as black smoke. Insufficient air is also detrimental because it causes the boiler to pant and vibrate; this is one of the major causes of brick work failure in the boilers.

The air pressure in the double casing must be increased BEFORE the rate of combustion is increased, and must be decreased AFTER the rate of combustion is decreased. There is usually little difficulty in teaching fireroom personnel to increase the air pressure before the rate of combustion is increased, since failure to do so results in panting and vibration of the boiler and in heavy smoke. It is more difficult, however, to teach the men to decrease the air pressure after the rate of combustion has been decreased. However, operating the blower at a faster speed than is required for the rate of combustion is definitely a poor practice, since it introduces more excess air than is required for the combustion of the fuel.

Forced Draft Blower Maintenance

Forced draft blowers require relatively little maintenance and repair, provided they are operated and maintained in strict accordance with instructions given in the manufacturer's technical manual. A good deal of the maintenance work required in connection with blowers is related to keeping the lubrication system in proper condition. All maintenance of blowers must be performed in accordance with the requirements of the 3-M System.

The lubricating oil in the reservoir must be kept clean. The reservoir must always be filled to the correct level with oil of the specified weight and grade. Oil samples must be taken routinely at the specified intervals, or more often if you have reason to suspect that the oil is contaminated. When a sample shows an unusual amount of sediment or water, this fact should be reported to the engineering officer or to the PO in charge, who may issue instructions to change the oil. After the oil has been drained, the inside of the reservoir should be wiped clean.

The metal-edge type of filter should be cleaned at least once each watch. This is done by giving the handle one or two complete turns. From time to time the filter should be dismantled and cleaned; this should be done whenever the oil is changed in the reservoir, and more often if necessary.

A common occurrence with vertical blowers is contamination of the oil with fresh water. This happens in an idle blower when leaking steam enters the turbine casing, passes through the upper labyrinth seal, and impinges on the oil slinger, where it condenses and mixes with the oil. This problem can be avoided by keeping the steam valves (especially the exhaust-relief valve) in good repair. If the valves leak, the turbine casing drain of an idle blower must be kept open.

Automatic shutters are not subject to any great amount of wear under normal operating conditions. However, they must be kept well lubricated at all times. Some types of shutters have Zerk-type grease fittings; others have oil holes. Be sure to use the correct lubricant. If the automatic shutters are not properly lubricated, they may stick in the open position and then slam shut with sufficient force to cause

damage to the shutters and to the toggle gear. Broken or sprung parts must be replaced in order to ensure smooth operation of the shutters. Shutters should be inspected frequently to determine whether the leaves operate freely and to be sure that they seal tightly when closed.

Forced draft blowers should not be operated if they are vibrating excessively or making any unusual noise. Vibration may be caused by worn or loose bearings, a bent shaft, loose or broken foundation bolts or rivets, an unbalanced fan, or other defects. All defects should be corrected as soon as possible in order to prevent a complete breakdown of the blower.

Minor repairs to blower fan blades may be made aboard ship in case of emergency, in accordance with procedures specified in chapter 9530 of the Naval Ships Technical Manual. Major repairs to fan blades must NOT be made without specific instructions from the Naval Ship Systems Command. As a matter of routine care, blower fan blades should be wiped down from time to time to remove dirt and dust. (One rapid method of cleaning the blades of a multi-stage blower without any disassembly is discussed in chapter 9530 of the Naval Ships Technical Manual.) Paint must NEVER be applied to a blower fan or to any other rotating part of the unit.

When inspecting and repairing blowers equipped with anti-reverse rotation devices, be careful not to apply reverse torque to the shaft, since this could cause damage to the shaft of the anti-reverse rotation device.

Safety Precautions

Some of the most important safety precautions to be observed in connection with forced draft blowers are:

1. Before starting a blower, always make sure that the fan is free of dirt, tools, rags, and other foreign objects or materials. Check the blower room for loose objects that might be drawn into the fan when the blower is started.

2. Do not try to move automatic shutters by hand if another blower serving the same boiler is already in operation.

3. When only one blower on a boiler is to be operated, make sure that the automatic shutters on the idle blower are closed and locked.

4. Never try to turn a blower by hand when steam is being admitted to the unit.

5. Never tie down the speed-limiting governor. Be sure it is in good operating condition at all times. Be sure it is properly set.

6. Observe all safety precautions required in connection with the operation of the driving turbine.

Casualty Control

If a forced draft blower fails, take the following action:

1. If one blower fails when two blowers are in use, speed up the other blower.

2. If only one blower is in use, secure the burners immediately to prevent a flareback.

3. Start the standby blower and notify the engineroom of the casualty because of a possible need for a reduction in speed.

4. Determine the cause of the failure and remedy the trouble as soon as possible.

The rupture of a forced draft blower lube oil line is a casualty in itself, as well as one which can lead to blower failure. The rupture of a lube oil line is discussed here separately because of the special hazard of fire that is involved. If a forced draft blower lube oil line is ruptured, take the following action immediately.

1. If two blowers are in use, secure the affected blower and speed up the other. If only one blower is in use, secure the burners immediately to prevent a flareback. Secure the affected blower and light off the standby blower.

2. Notify the engineroom because of a possible need for a reduction in speed.

3. Get firefighting equipment to the scene.

4. Wipe up all oil that has spilled out. Flush out and wipe out the bilges as necessary.

5. Repair or renew the ruptured section as soon as possible.

6. Refill the reservoir with clean oil. Test the new section; if it is satisfactory, the blower can be returned to service.

CHAPTER 16

AUXILIARY STEAM TURBINES

Auxiliary steam turbines are used to drive many auxiliary machinery units aboard steam-driven ships. Turbine-driven auxiliaries located in the engineering spaces include ship's service generators, forced draft blowers, air compressors, and a number of pumps such as main condensate pumps, main condenser circulating pumps, main feed pumps, feed booster pumps, fuel oil service pumps, and lubricating oil service pumps.

In many cases, the turbine-driven auxiliaries are duplicated by electrically driven units for in-port or cruising use. Although the motor-driven units have a comparatively high efficiency, their capacity is not sufficient (on some ships, at least) to meet the demands of the engineering plant at high speeds. A further advantage of auxiliary turbines is their greater reliability; in general, there is greater possibility of interruption or loss of electric power supply than of steam supply. In addition, the use of auxiliary turbines improves the overall plant efficiency because exhaust steam from the auxiliary turbines can be utilized in various ways throughout the plant.

The basic principles of steam turbine design, classification, and construction discussed in chapter 12 of this text apply in general to auxiliary turbines as well as to propulsion turbines, except for specific differences noted in the remainder of this chapter.

TYPES OF AUXILIARY TURBINES

Many auxiliary turbines are of the impulse type. Reduction gears are used with most auxiliary turbines[1] to increase efficiency. Since space requirements frequently demand relatively small units, auxiliary turbines are usually designed with comparatively few stages—often only one. This means a large pressure drop and a high steam velocity in each stage. To obtain maximum efficiency, the blade speed must also be high. With auxiliary turbines, as with propulsion turbines, reduction gears serve to reconcile the conflicting speed requirements of the driving and the driven units.

Until about 1950, many generator turbines were designed and installed in such a way that they could be operated on steam from either superheated or saturated steam lines at full boiler pressure. Most of the other auxiliary turbines at this time operated on saturated steam at full boiler pressure. During the early 1950's, a few ships were built in which all auxiliary turbines were designed to operate on steam at full superheat and full boiler pressure. On most oil-fired ships built since 1953, steam at full superheat and full boiler pressure is supplied to the auxiliary turbines for generators, main feed pumps, and forced draft blowers; the other auxiliary turbines on these ships usually operate on steam at reduced temperature and pressure (about 50° F of superheat and 600 psig). On nuclear ships, all turbines (propulsion and auxiliary) are designed to operate on wet steam. The generator turbines usually exhaust to their own separate auxiliary condensers; on recent submarines, however, they exhaust to the main condenser. Most other auxiliary turbines exhaust to the auxiliary exhaust system. The auxiliary exhaust system imposes a back

---

[1] Direct drive, rather than geared drive, units include forced draft blowers, high speed centrifugal pumps, and some recent ship's service turbogenerators.

422

pressure which is approximately 15 psig on oil-fired ships.

Most auxiliary turbines are axial flow units which are quite similar (except for size and number of stages) to the axial-flow propulsion turbines described in chapter 12. However, some auxiliary turbines are designed for helical flow and some for radial flow—types of flow which are seldom if ever used for propulsion turbines.

A helical-flow auxiliary turbine is shown in figure 16-1. In a turbine of this type, steam enters at a tangent to the periphery of the rotor

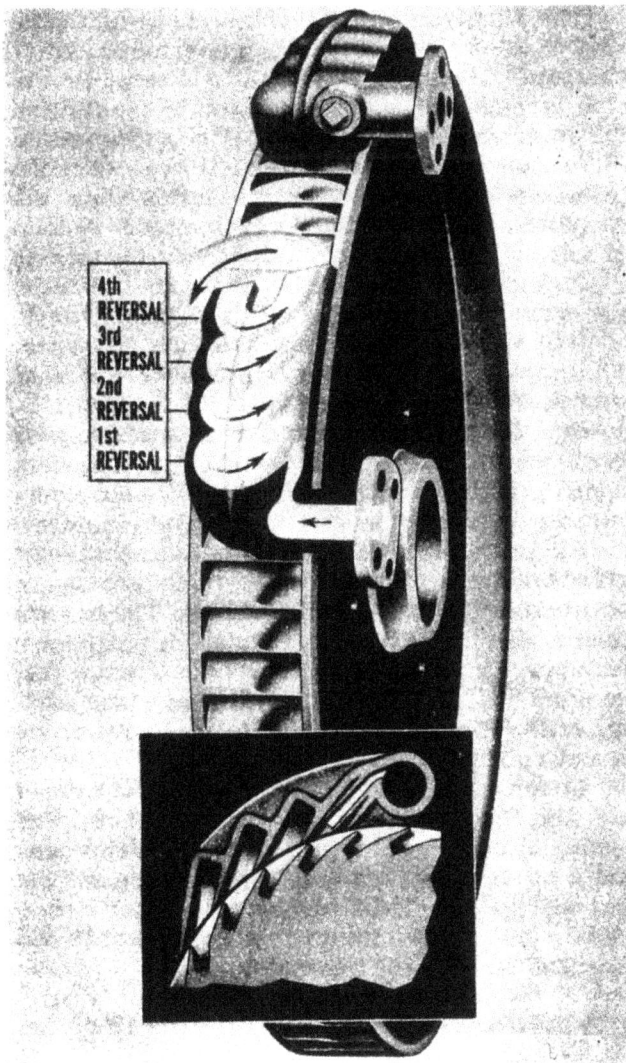

33.45X

Figure 16-1.—Helical-flow turbine.

and impinges upon the moving blades. These blades, which consist of semicircular slots milled obliquely in the wheel periphery, are called buckets. The buckets are shaped in such a way that the direction of steam flow is reversed in each bucket, and the steam is directed into a redirecting bucket or reversing chamber mounted on the inner cylindrical surface of the casing. The direction of the steam is again reversed in the reversing chamber, and the continuous reversal of the direction of flow keeps the steam moving helically.

Several nozzles are usually installed in this type of turbine, and for each nozzle there is an accompanying set of redirecting buckets or reversing chambers. Thus the reversal of steam flow is repeated several times for each nozzle and set of reversing chambers.

Now let us consider the classification of a helical-flow turbine with respect to staging and compounding, as discussed in chapter 12. It is a single-stage turbine because it has only one set of nozzles and therefore only one pressure drop. It is a velocity-compound turbine because the steam passes through the moving blades (buckets) more than once, and the velocity of the steam is therefore utilized more than once. The helical-flow turbine shown in figure 16-1 might be said to correspond roughly to a turbine in which velocity-compounding is achieved by the use of four rows of moving blades.

Helical-flow auxiliary turbines are used for driving some pumps and forced draft blowers.

The arrangement of nozzles and blading that provides radial flow in a turbine is shown in figure 16-2. Turbines of this type are sometimes used for driving auxiliary units such as pumps.

As discussed in chapter 12, turbines may be classified as single-entry or re-entry turbines, depending upon the number of times the steam enters the blading. All multistage (and hence all

38.81X

Figure 16-2.—Radial flow.

423

propulsion) turbines are of the single-entry-type. However, some auxiliary turbines are of the re-entry type.

Re-entry turbines are those in which the steam passes more than once through the blading. Hence the helical-flow turbine just discussed is a re-entry turbine. A different kind of re-entry turbine is shown in figure 16-3. This turbine is similar in principle to the helical-flow turbine,

38.82X

Figure 16-3.—Re-entry turbine with one reversing chamber.

but it has one large reversing chamber instead of a number of redirecting chambers. Re-entry turbines are sometimes made with two reversing chambers instead of one.

The auxiliary turbine shown in figure 16-4 is used to drive main condensate pumps, feed booster pumps, and lubricating oil service pumps on many older destroyers. Note that this is a radial-flow turbine. This same design of turbine is used on some newer ships, but with improved metals designed to withstand higher pressures, higher temperatures, and high-impact (HI) shock.

Another kind of auxiliary turbine is shown in figure 16-5. This turbine, which is used to drive main condenser circulating pumps, is a vertically mounted, axial-flow, velocity-compounded impulse turbine. Although this type of turbine is becoming obsolete, it is still in operation on some older types of ships. The turbine shaft is secured to the vertical shaft of the pump. A thrust bearing, mounted integrally with the upper radial bearing, carries the weight of the rotating ele-

ment and absorbs any downward thrust. A throttle valve and a double-seated balanced inlet valve (normally held wide open by the governor mechanism) admit steam to the turbine.

Figure 16-6 shows an auxiliary turbine used to drive a 400-kilowatt a-c, 50-kilowatt d-c ship's service turbogenerator. The turbine is an axial-flow, pressure-compounded unit. It exhausts to a separate auxiliary condenser which has its own circulating pump, condensate pump, and air ejectors. Cooling water for the condenser is provided by the auxiliary circulating pump, through separate injection and overboard valves. In case of casualty to the auxiliary condenser, the turbine can exhaust to the main condenser when the main plant is in operation.

The turbogenerator turbine shown in figure 16-6 is so designed that it can operate on either saturated steam or superheated steam. Provision is made for supplying steam to the turbine either from the main steam line (superheated) while under way or from the auxiliary steam line (saturated) during in-port operation when the propulsion turbines and the main steam system are secured. The steam is admitted to the turbine through a throttle trip valve to the steam chest, the speed being regulated by a number of nozzle control valves under the control of a governor.

Because the ship's service generator must supply electricity at a constant voltage and frequency, the turbine must run at a constant speed even though the load varies greatly. Constant speed is maintained through the use of a constant-speed governor (discussed later in this chapter).

As may be seen in figure 16-6, the shaft glands of the ship's service generator turbine are supplied with gland sealing steam. The system is much the same as that provided for propulsion turbines. Other auxiliary turbines in naval use do not require an external source of gland sealing steam since they exhaust to pressures above atmospheric pressure.

Generator turbines vary greatly, and are not all like the one shown in figure 16-6. For example, one recent type of turbogenerator consists of seven stages—one Curtis stage and six Rateau stages. This turbine is direct drive, rather than geared; the turbine operates at 12,000 rpm and so does the generator.

AUXILIARY TURBINE LUBRICATION

Auxiliary turbines designed to Navy specifications have pressure lubrication systems to lubricate the radial bearings, reduction gears,

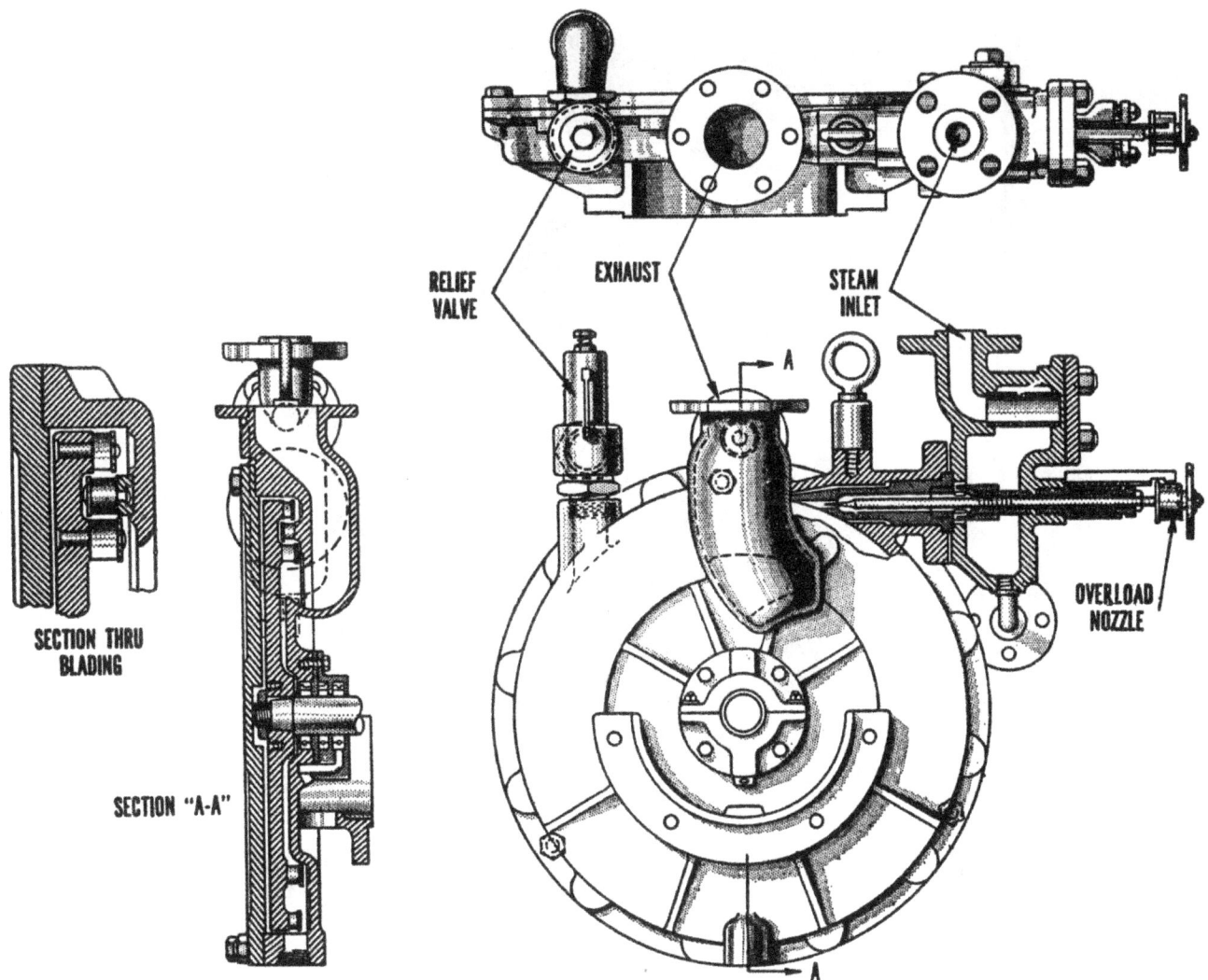

SECTION THRU
BLADING

SECTION "A-A"

RELIEF
VALVE

EXHAUST

STEAM
INLET

OVERLOAD
NOZZLE

47.9X

Figure 16-4.—Auxiliary turbine for main condensate pump, feed booster pump,
and lubricating oil service pump.

and governors.[2] Pressure lubrication systems for auxiliary turbines do not provide lubrication for governor linkages or—except on some turbogenerator sets—for flexible couplings; these parts of the unit must be lubricated separately.

A pressure lubrication system requires a lube oil pump. As a rule, the lube oil pumps used for auxiliary units are positive-displacement pumps of the simple gear type, as discussed in chapter 15 of this text. The lube oil pump is generally installed on the turbine end of a forced draft blower unit, but may be on either the driving or the driven end of pump units. The lube oil pumps for turbogenerators are usually driven by auxiliary gearing connected to the low speed gear shaft. Some forced draft blowers use a centrifugal pump, supplemented by a viscosity pump, for lubrication of the unit; this type of lubrication system is peculiar to forced draft blowers and is

_____

[2] Some very small commercially designed auxiliary turbines have self-oiling bearings instead of pressure lubrication systems. A self-oiling bearing has one or two rings which hang on the turbine shaft and revolve with it, although at a slower rate. On each revolution, the rings dip into an oil reservoir and carry oil around to the upper part of the bearing shell.

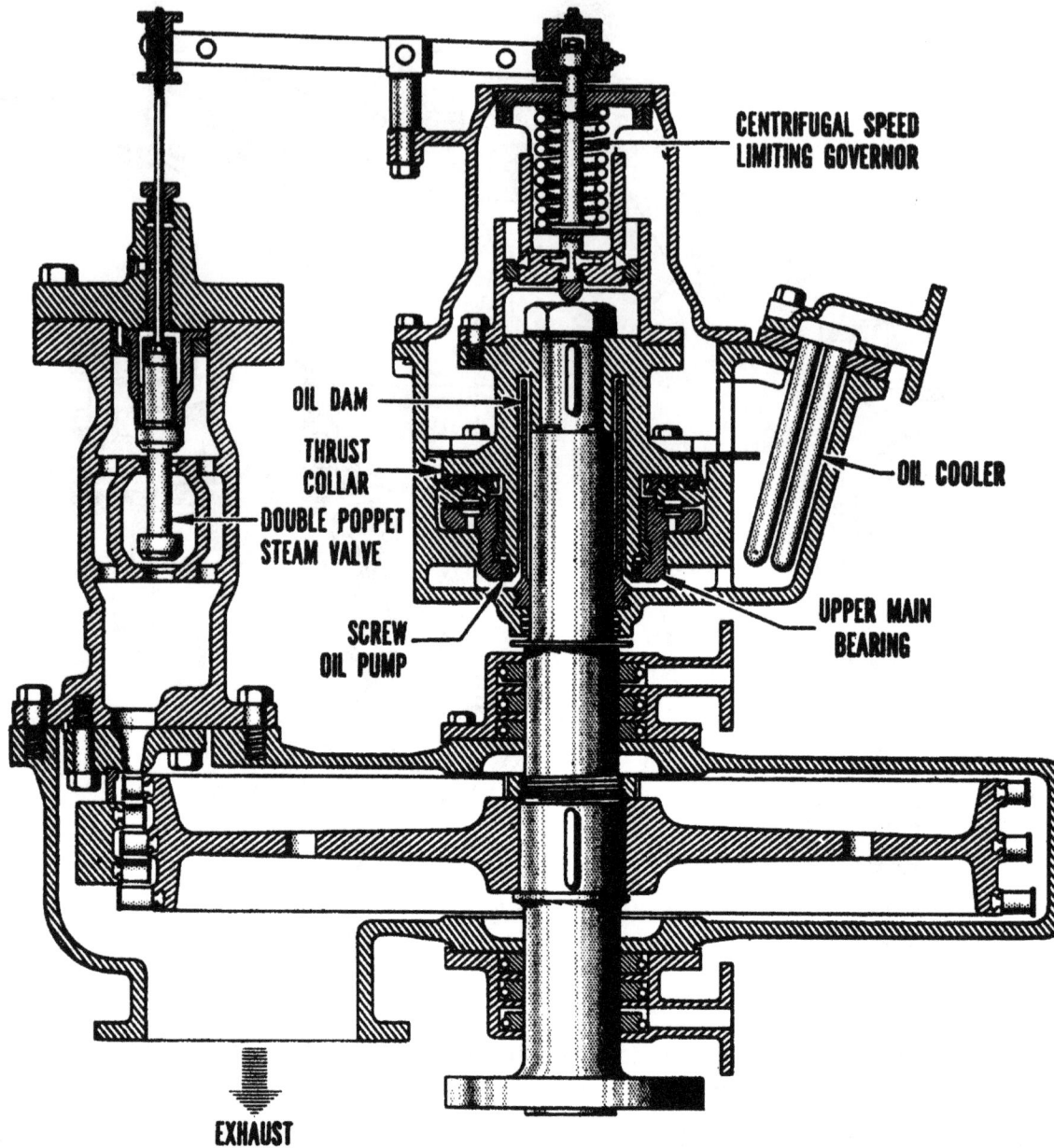

47.10X

Figure 16-5.—Auxiliary turbine for main condenser circulating pump.

therefore discussed in connection with the blowers, in chapter 15 of this text.

The pressure lubrication system shown in figures 16-7 and 16-8 is designed for fuel oil service pumps, fuel oil booster pumps, and lubricating oil service pumps; however, it is similar in principle to the lubricating systems of many other units.

In the system illustrated, the bottom section of the gear casing forms the oil reservoir. The reservoir is filled through an oil filler hole in the top of the casing and emptied through a drain outlet at the base of the casing. The shaft, which carries the gear-type oil pump on one end and the governor on the other end, is geared to the pump shaft. The pump shaft is in turn geared to the turbine shaft.

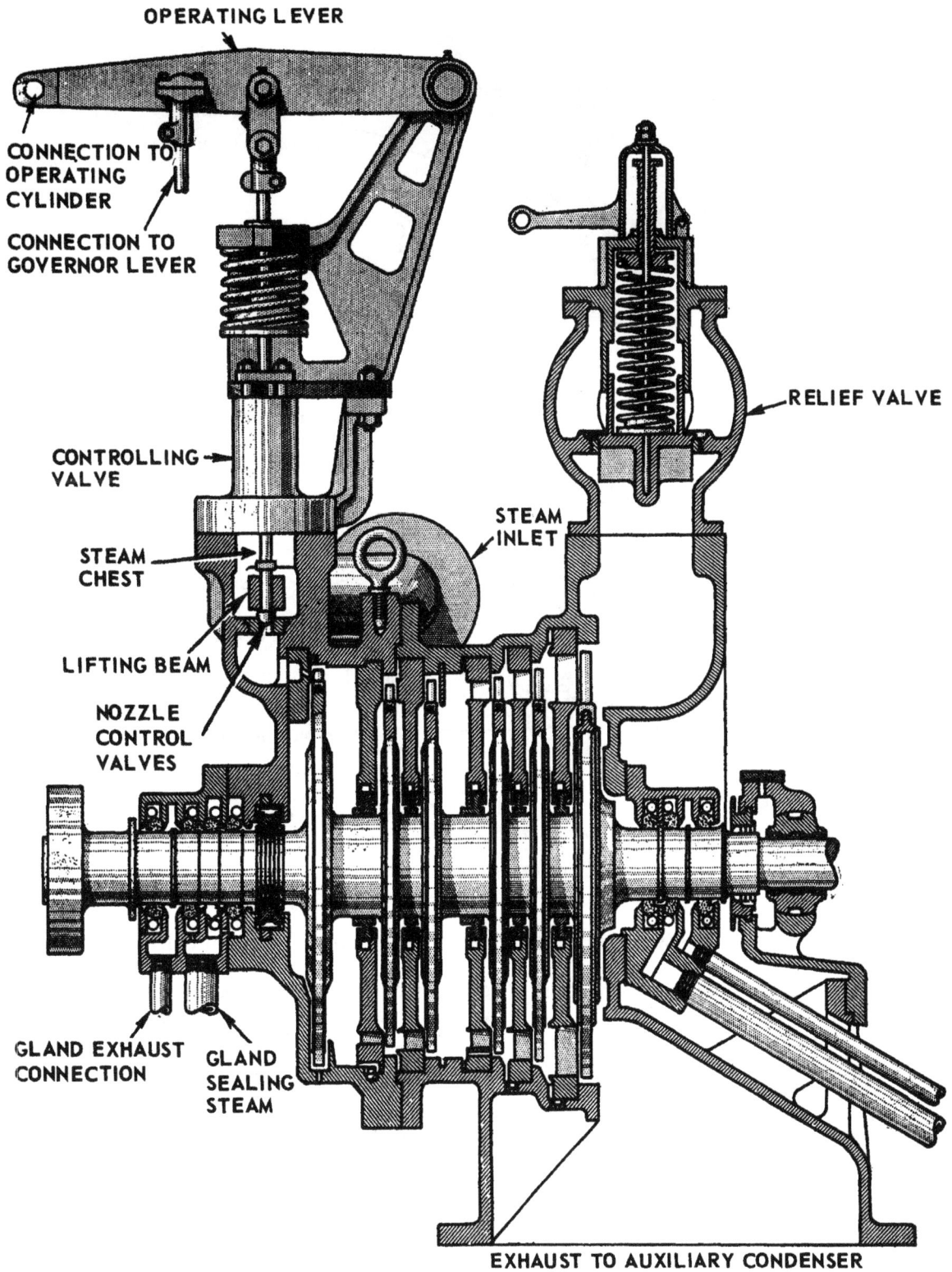

OPERATING LEVER

CONNECTION TO OPERATING CYLINDER

CONNECTION TO GOVERNOR LEVER

CONTROLLING VALVE

STEAM CHEST

LIFTING BEAM

NOZZLE CONTROL VALVES

GLAND EXHAUST CONNECTION

GLAND SEALING STEAM

RELIEF VALVE

STEAM INLET

EXHAUST TO AUXILIARY CONDENSER

47.11X

Figure 16-6.—Auxiliary turbine for ship's service turbogenerator.

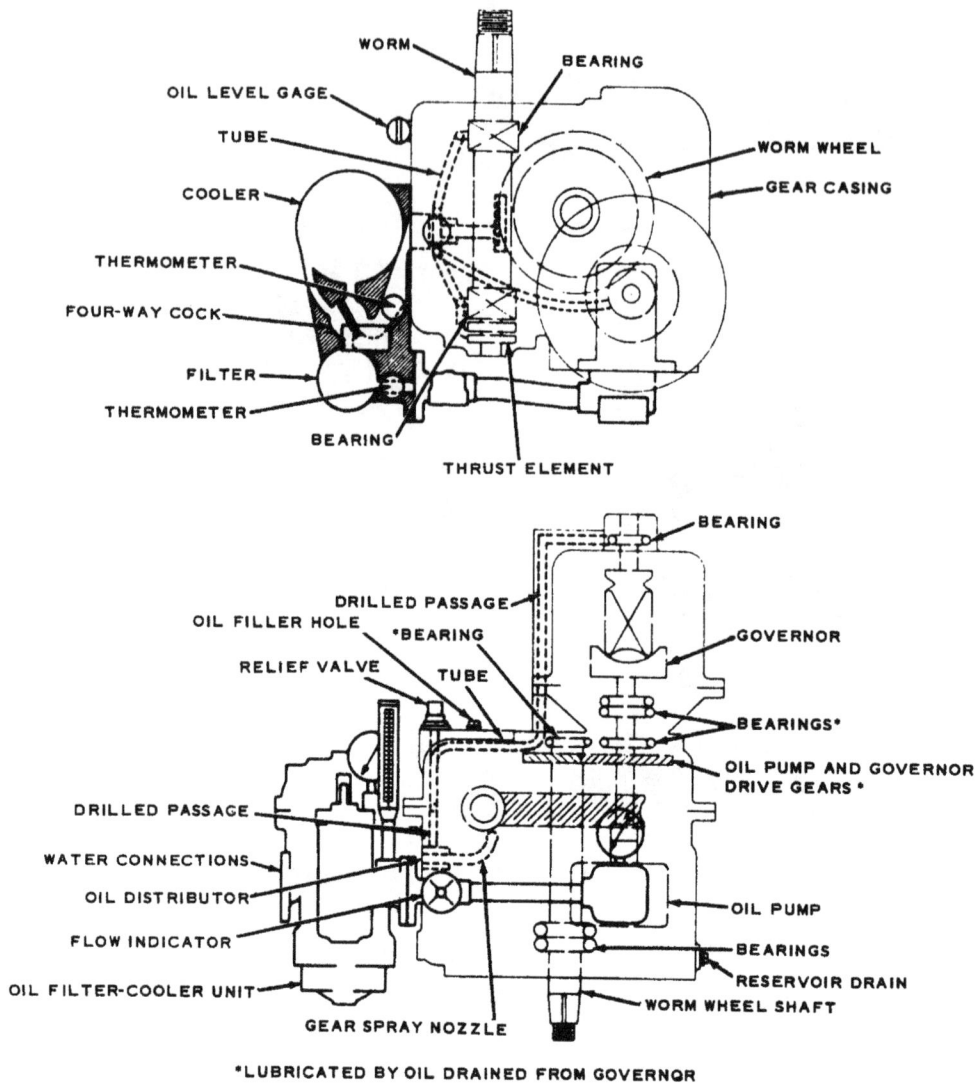

38.96X

Figure 16-7.—Pressure lubrication system for turbine-driven unit.

The lubricating oil passes through an oil sight flow indicator, a metal-edge type of filter, and an oil cooler. Oil is then piped to the bearings on the turbine shaft, to the governor, and to the worm gear on the pump shaft. The bearings and gear on the oil pump and governor shaft are lubricated by oil which drains from the governor and passes back into the oil reservoir. A relief valve is built into the gear casing. This valve serves to protect the system against the development of excessive pressures.

SPEED CONTROL DEVICES

Different types of governors are used for controlling the speed of auxiliary turbines. The discussion here is limited to the constant-speed governor and the constant-pressure pump governor, both of which are in common naval use.[3]

Constant-Speed Governors

The constant-speed governor, sometimes called the speed-regulating governor, is used on

---

[3] Additional governing devices that may be encountered on recent ships include hydraulic or electric load-sensing governors, for turbogenerators, and pneumatic, hydraulic, or electric controls for main feed pumps. On ships having automatic combustion and feed water control systems, the main feed pump controls may be related in some way to the boiler controls.

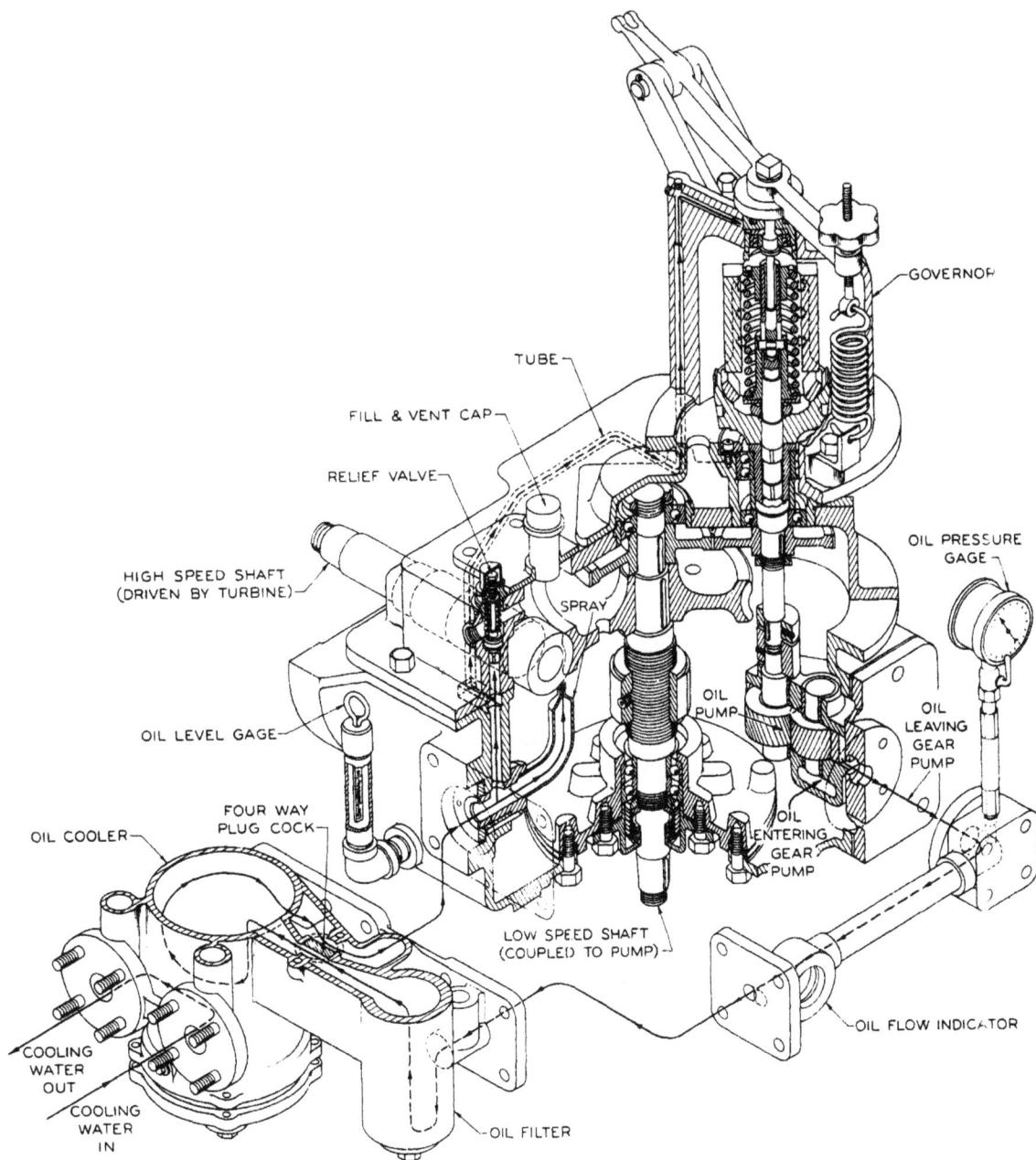

Figure 16-8.—Isometric diagram of pressure lubrication system.

38.97X

constant-speed machines to maintain a constant speed regardless of the load on the turbine. Constant-speed governors are used primarily on generator turbines and on air compressor turbines.

A constant-speed governing system for a ship's service generator turbine is shown in figure 16-9. The constant-speed governor operates a pilot valve which controls the flow of oil to an operating cylinder. The operating cylinder, in turn, controls the extent of the opening or closing of the turbine nozzle valves.

With an increased load on the generator, the turbine tends to slow down. Since the governor is driven by the turbine shaft, through reduction

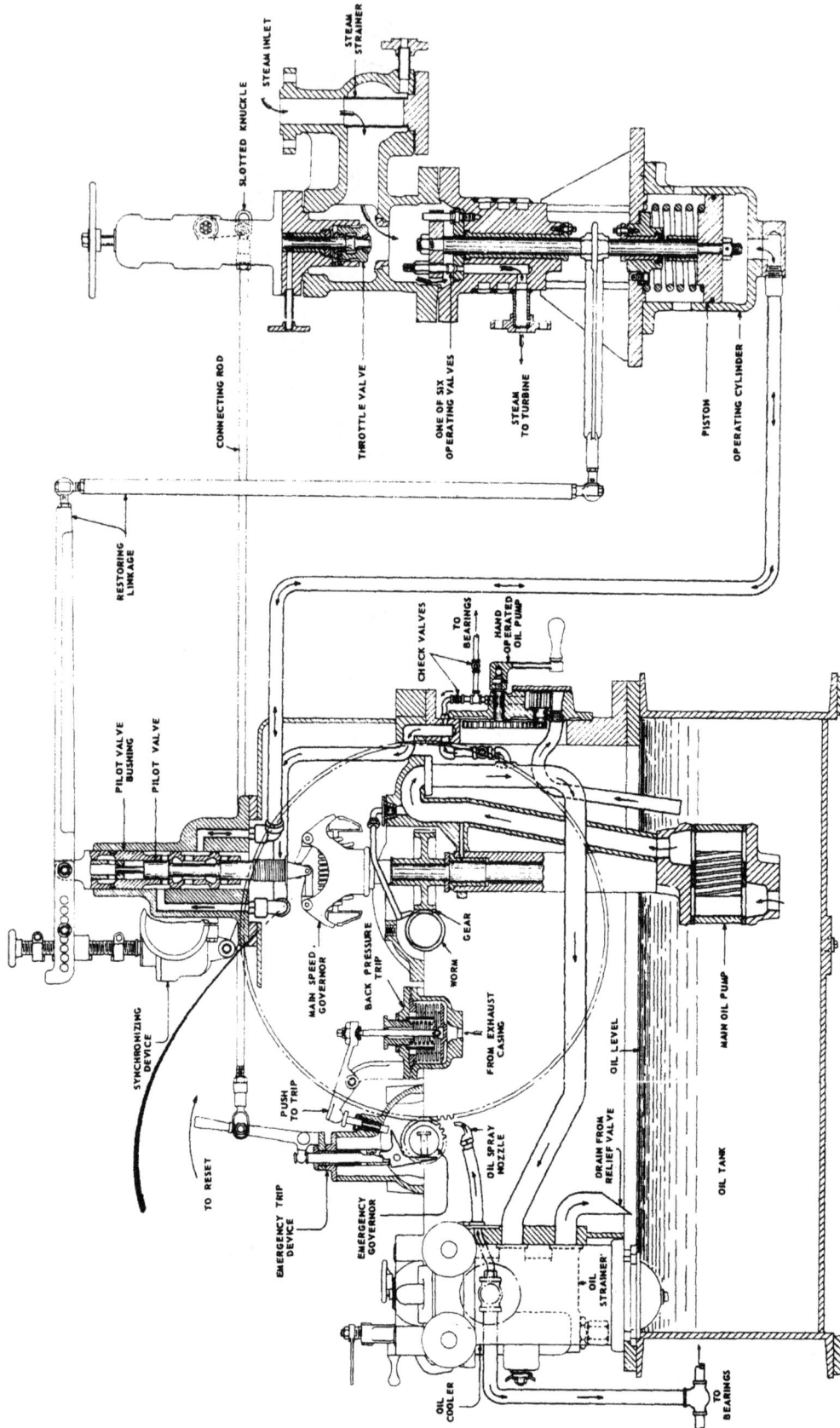

47.21

Figure 16-9.—Constant-speed governing system for ship's service turbogenerator.

gears, the governor also slows. Centrifugal weights on the governor move inward as the speed decreases, and this causes the pilot valve to move upward, permitting oil to enter the operating cylinder. The operating piston rises and, through the controlling valve lever, the lifting beam is raised. The nozzle valves open and admit additional steam to the turbines.

The upward motion of the controlling valve lever causes the governor lever to rise, thus raising the bushing. Upward motion of the bushing tends to close the upper port, shutting off the flow of oil to the operating cylinder; this action stops the upward motion of the operating piston. The purpose of this follow-up motion of the bushing is to regulate the governing action of the pilot valve. Without this feature, the pilot valve would operate with each slight variation in turbine speed and the nozzle valves would be alternately opened wide and closed completely.

A reverse process occurs when the load on the generator decreases. In this case, the turbine speeds up, the governor speeds up, the centrifugal weights move outward, and the pilot valve moves downward, opening the lower ports and allowing oil to flow out of the operating cylinder. The controlling valve lever lowers the lifting beam and thereby reduces the amount of steam delivered to the turbine.

Constant-Pressure Pump Governors

Many turbine-driven pumps are fitted with constant-pressure pump governors. The function of a constant pressure pump governor is to maintain a constant pump discharge pressure under conditions of varying flow. The governor, which is installed in the steam line to the pump, controls the pump discharge pressure by controlling the amount of steam admitted to the driving turbine.

A constant-pressure pump governor for a main feed pump is shown in figure 16-10. The governors used on fuel oil service pumps, lube oil service pumps, fire and flushing pumps, and various other pumps are almost identical. The chief difference between governors used for different services is in the size of the upper diaphragm. A governor used for a pump which operates with a high discharge pressure has a smaller upper diaphragm than one for a pump which operates with a low discharge pressure.

Two opposing forces are involved in the operation of a constant-pressure pump governor.

Fluid from the pump discharge, at discharge pressure, is led through an actuating line to the space below the upper diaphragm. The pump discharge pressure thus exerts an upward force on the upper diaphragm. Opposing this, an adjusting spring exerts a downward force on the upper diaphragm.

When the downward force of the adjusting spring is greater than the upward force of the pump discharge pressure, the spring forces the upper diaphragm and the upper crosshead down. A pair of connecting rods connects the upper crosshead rigidly to the lower crosshead, so the entire assembly of upper and lower crossheads moves together. When the crosshead assembly moves down, it pushes the lower mushroom and the lower diaphragm downward. The lower diaphragm is in contact with the controlling valve. When the lower diaphragm is moved down, the controlling valve is forced down and thus opened.

The controlling valve is supplied with a small amount of steam through a port from the inlet side of the governor. When the controlling valve is open, steam passes to the top of the operating piston. The steam pressure acts on the top of the operating piston, forcing the piston down and opening the main valve. The extent to which the main valve is open controls the amount of steam admitted to the driving turbine. Increasing the opening of the main valve therefore increases the supply of steam to the turbine and so increases the speed of the turbine.

The increased speed of the turbine is reflected in an increased discharge pressure from the pump. This pressure is exerted against the under side of the upper diaphragm. When the pump discharge pressure has increased to the point where the upward force acting on the under side of the upper diaphragm is greater than the downward force exerted by the adjusting spring, the upper diaphragm is moved upward. This action allows a spring to start closing the controlling valve, which in turn allows the main valve spring to start closing the main valve against the now reduced pressure on the operating piston. When the main valve starts to close, the steam supply to the turbine is reduced, the speed of the turbine is reduced, and the pump discharge pressure is reduced.

At first glance, it might seem that the controlling valve and the main valve would be constantly opening and closing and that the pump discharge pressure would be continually varying over a wide range. This does not happen, however, because the governor is designed with an

HANDWHEEL

ADJUSTING SCREW

LOCK NUT

ADJUSTING SPRING

STEAM CHAMBER

DIAPHRAGM DISK
(UPPER MUSHROOM)

CROSSHEAD

UPPER DIAPHRAGM

ACTUATING LINE FROM
DISCHARGE SIDE
OF PUMP

NEEDLE VALVE

INTERMEDIATE DIAPHRAGM

CROSSHEAD CONNECTING ROD

DIAPHRAGM STEM
(LOWER MUSHROOM)

DIAPHRAGM STEM CAP
(INTERMEDIATE MUSHROOM)

DIAPHRAGM STEM GUIDE

LOWER DIAPHRAGM

CONTROLLING VALVE BUSHING

CYLINDER LINER

CONTROLLING VALVE

OPERATING PISTON

CONTROLLING VALVE SPRING

STEAM INLET

STEAM OUTLET
(TO TURBINE)

MAIN VALVE

MAIN VALVE SPRING

STEM (FOR BYPASS)

INDICATOR PLATE

HANDWHEEL (FOR BYPASS)

38.90

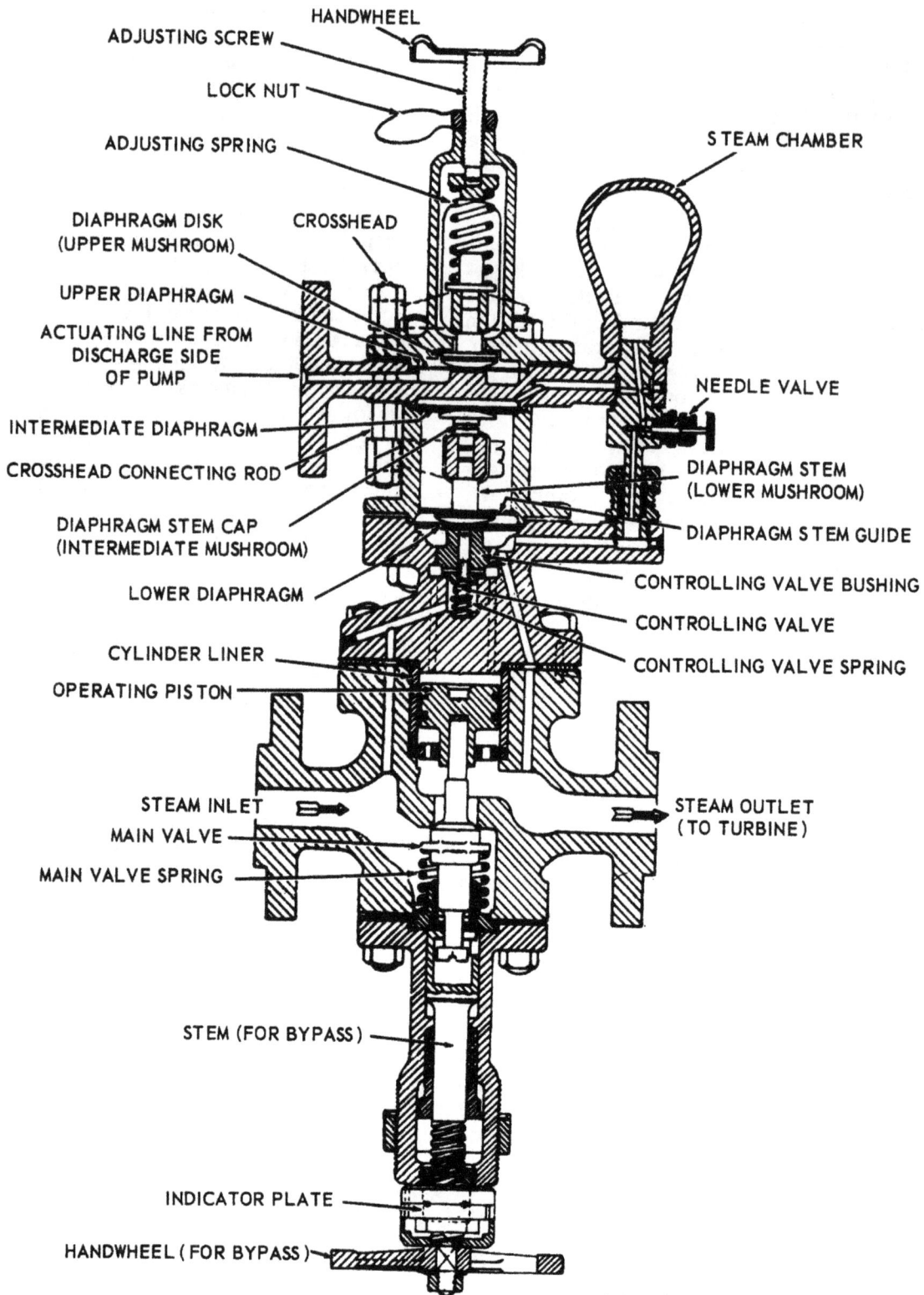

Figure 16-10.—Constant-pressure pump governor for main feed pump.

arrangement which prevents excessive opening or closing of the controlling valve. An intermediate diaphragm bears against an intermediate mushroom which, in turn, bears against the top of the lower crosshead. Steam is led from the governor outlet to the bottom of the lower diaphragm and also, through a needle valve, to the top of the intermediate diaphragm. A steam chamber is provided to assure a continuous supply of steam at the required pressure to the top of the intermediate diaphragm.

Any movement of the crosshead assembly, either up or down, is thus opposed by the force of the steam pressure acting either on the intermediate diaphragm or on the lower diaphragm. The whole arrangement serves to prevent extreme reactions of the controlling valve in response to variations in pump discharge pressure.

Limiting the movement of the controlling valve in the manner just described reduces the amount of hunting the governor must do to find each new position. Under constant-load conditions, the controlling valve takes a position which

causes the main valve to remain open by the required amount. A change in load conditions results in momentary hunting by the governor until it finds the new position required to maintain pump discharge pressure at the new condition of load.

An automatic shutdown device has recently been developed for use on main feed pumps. The purpose of the device is to shut down the main feed pump and so protect it from damage in the event of loss of feed booster pump pressure. The shutdown device consists of an auxiliary pilot valve and a constant-pressure pump governor, arranged as shown in figure 16-11. The governor is the same as the constant-pressure pump governor just described except that it has a special top cap. In the regular governor, the steam for the operating piston is supplied to the controlling valve through a port in the governor valve body. In the automatic shutdown device, the steam for the operating piston is supplied to the controlling valve through the auxiliary pilot valve. The auxiliary pilot valve is actuated by the feed booster

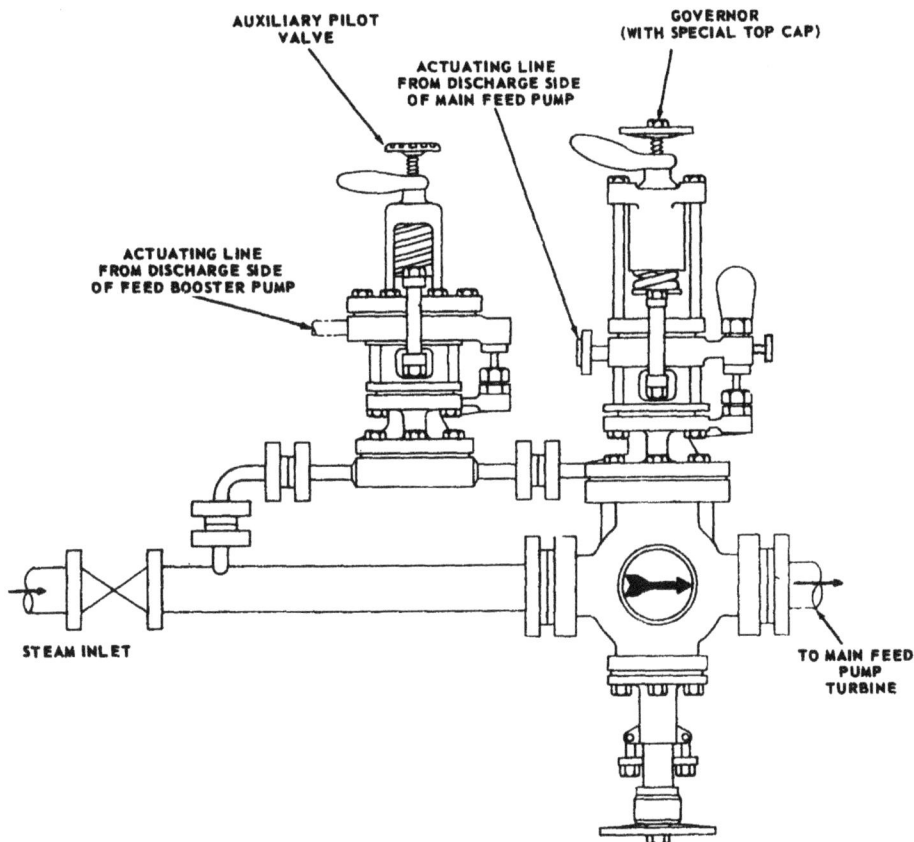

38.91

Figure 16-11.—Automatic shutdown device for main feed pump.

433

pump discharge pressure. When the booster pump discharge pressure is inadequate, the auxiliary pilot valve will not deliver steam to the controlling valve of the governor. Thus inadequate feed booster pump pressure allows the main valve in the governor to close, shutting off the flow of steam to the main feed pump turbine.

SAFETY DEVICES

Safety devices used on auxiliary turbines include speed-limiting governors and several kinds of trips. Safety devices differ from speed control devices in that they have no control over the turbine under normal operating conditions. It is only when some abnormal condition occurs that the safety devices come into use to stop the unit or to control its speed.

Speed-Limiting Governors

The speed-limiting governor is essentially a safety device for variable speed units. It allows the turbine to operate under all conditions from no-load to overload, up to the speed for which the governor is set, but it does not allow operation in excess of 107 percent of rated speed. It is important to note that this type of governor is adjusted to the maximum operating speed of the turbine and therefore has no control over the admission of steam until the upper limit of safe operating speed is reached.

One common type of speed-limiting governor is shown in figure 16-12. This governor is used on main condensate pumps, feed booster pumps, lube oil service pumps, and other auxiliaries in the engineering plant. The particular speed-limiting governor shown here is designed for use on a main condensate pump with a vertical shaft; speed-limiting governors that operate on very much the same principle are used on main feed pumps and other auxiliaries that have horizontal shafts.

The governor shaft is driven directly by an auxiliary shaft in the reduction gear, and rotates at the same speed as the pump shaft. This speed is proportional to—although lower than—the speed of the driving turbine. Two flyweights are pivoted to a yoke on the governor shaft and carry

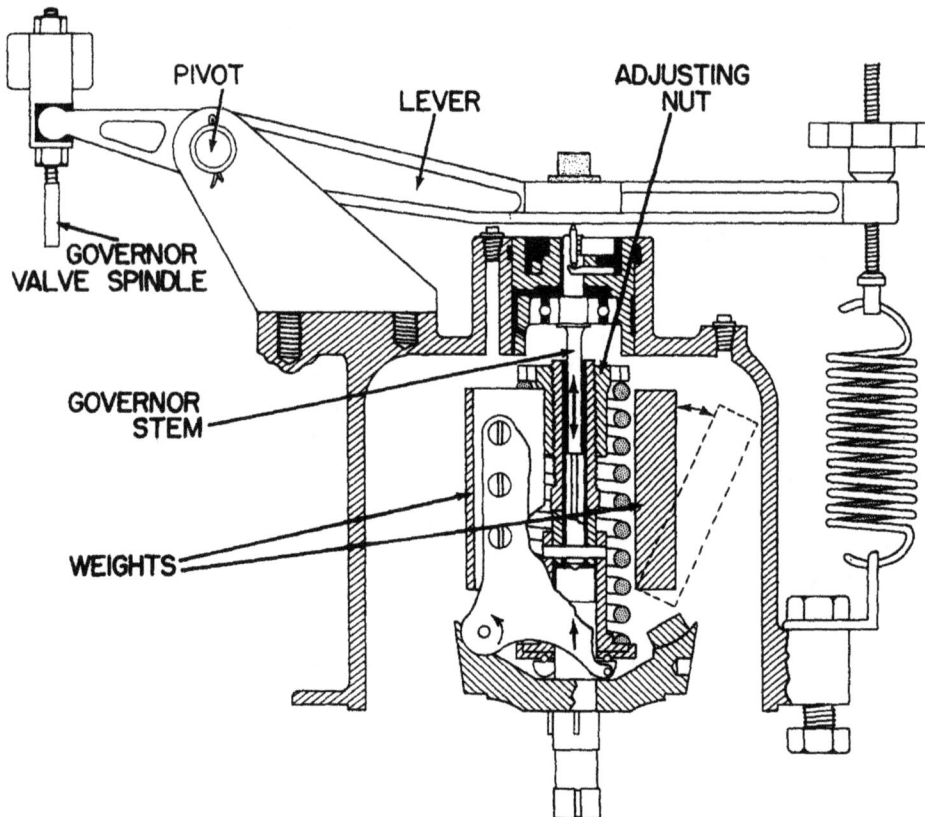

Figure 16-12.—Speed-limiting governor.

47.54

434

arms which bear on a push rod assembly. The push rod assembly is held down by a strong spring.

Because of centrifugal force, the position of the flyweights is at all times a function of turbine speed. As the turbine speed increases, the flyweights move outward and lift the arms. As the speed of the turbine approaches the speed for which the governor is set, the arms lift against the spring tension. If the turbine speed begins to exceed the speed for which the governor is set, the flyweights move even farther out, thereby causing the governor valve to throttle down on the steam.

When the turbine slows down, as from an increase in load, the centrifugal force on the flyweights is diminished and the governor push rod spring acts to pull the flyweights inward. This action rotates the lever about its pivot and opens the governor valve, thus admitting more steam to the turbine. The turbine speed increases until normal operating speed is reached.

The speed-limiting governor acts as a constant-speed governor when the turbine is operating at or near rated speed, although it is designed only as a safety device to prevent overspeeding. The governor has no effect on the speed of the turbine at speeds below about 95 percent of rated speed.

Trips

Several kinds of trips are used as safety devices on auxiliary turbines.

Overspeed trips are used on turbines that have constant-speed governors. The overspeed trip shuts off the supply of steam to the turbine and thus stops the unit when a predetermined speed has been reached. Overspeed trips are usually set to trip out at about 110 percent of normal operating speed. In the past overspeed trips were used primarily on constant-speed turbines and on some commercial-type variable-speed units. Recent specifications require overspeed trips on all naval auxiliary turbines of over 100 horsepower. Figure 16-13 shows the construction of an overspeed trip used on turbogenerator.

Back-pressure trips are installed on turbogenerators to protect the turbine by closing the throttle automatically when the back pressure (exhaust pressure) becomes too high. A back-pressure trip is shown in figure 16-14.

Emergency hand trips are installed on turbogenerators to provide a means for closing the throttle quickly, by hand, in case of damage to

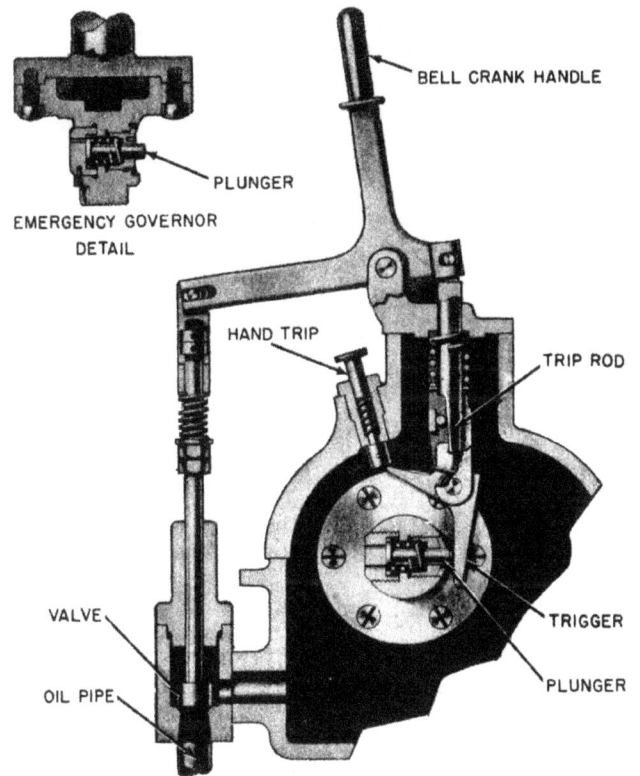

38.93

Figure 16-13.—Overspeed trip for turbogenerator.

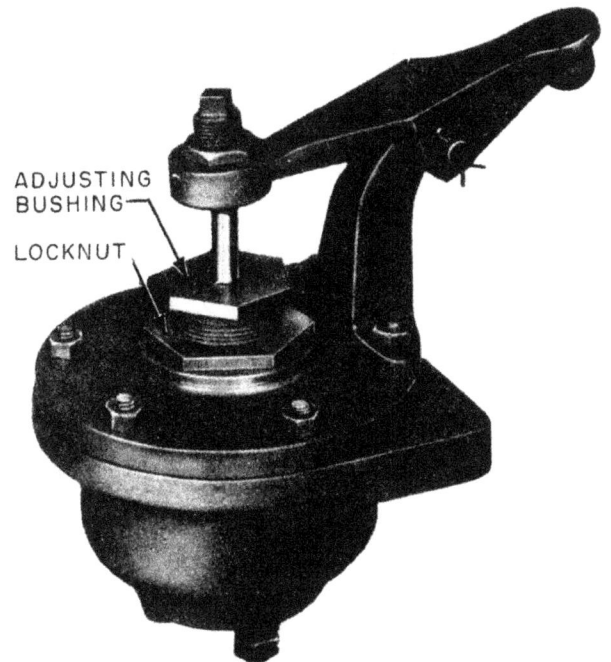

96.25

Figure 16-14.—Back-pressure trip.

	Bureau Card Control No.				Maintenance Requirement	M.R. No.	Rate Req'd.	Man Hours	Related Maintenance
System, Subsystem, or Component Ship Service Turbogenerator					**Reference Publications**				
EA	ZZ4FTRO	35	4842	W	1. Sample and inspect lube oil. 2. Lubricate speed regulating governor linkage. 3. Operate turbine by steam or turn idle turbine by hand while operating hand lube oil pump.	W-1	MM2 FN	0.3 0.3	None
EA	ZZZGGG1	45	4889	W	1. Purify lube oil.	W-2	MM3	1.0	None
EA	ZZEGGG1	75	4432	M	1. Test overspeed trip.	M-1	MM2	0.3	None
EA	ZZEGGG1	45	4035	M	1. Test overspeed trip. 2. Test lube oil pressure alarm.	M-2	MM2 MM3 FN	0.5 0.5 0.5	None
EA	ZZEGGG1	35	4844	Q	1. Clean lube oil sump. 2. Sound and tighten foundation bolts.	Q-1	MM1 FN	3.0 3.0	W-2
EA	ZZEGGG1	75	4845	Q	1. Test back pressure trip. 2. Test lube oil pressure alarm by operation.	Q-2	MM2 MM3	0.5 0.5	None
EA	ZZEGGG1	85	4845	Q	1. Test back pressure trip. 2. Test lube oil pressure alarm by operation.	Q-3	MM2 MM3	0.5 0.5	None
EA	ZZ2GGE5	35	4847	Q	1. Inspect pinion and reduction gears.	Q-4	MMC MM3	0.3 0.3	None
EA	ZZ4FTRO	75	2897	Q	1. Measure turbine thrust clearance.	Q-5	MM2	0.3	None
EA	ZZ4FGE9	75	4714	Q	1. Inspect the high speed pinion shaft gear and intermediate speed shaft gears.	Q-6	MM1 FN	1.0 1.0	None
EA	ZZEGGG1	75	4715	Q	1. Take depth micrometer measurement on the journal bearings.	Q-7	MM1	0.2	None
EA	ZZ4FTRO	75	4038	Q	1. Measure turbine thrust clearance.	Q-8	MM2 FN	0.5 0.5	None
EA	ZZ2GVA1	84	4848	S	1. Operate turbine casing relief valve by hand.	S-1	FN	0.1	None

(Page 1 of 2)

MAINTENANCE INDEX PAGE
OPNAV FORM 4700-3 (4-64)

BUREAU PAGE CONTROL NUMBER E-13/62-65

Figure 16-15-.—Maintenance Index Page.

98.171

either the turbine or the generator. A hand trip may be seen in the illustration of an overspeed trip (fig. 16-13).

Some auxiliary turbines (generator turbines, in particular) are fitted with low oil pressure alarms to warn operating personnel when the lubricating oil pressure becomes dangerously low. When the oil pressure drops below normal, a pressure-actuated switch completes an electrical circuit to sound an audible alarm. Control systems may also be arranged to trip the turbine if lubricating oil pressure drops too low.

TESTING SAFETY DEVICES

Speed limiting governers and safety trips must be tested and maintained in accordance with the requirements set forth by the Naval Ship Systems Command and at the intervals specified by the Planned Maintenance Subsystem of the 3-M System. Testing must be done under the supervision of the engineer officer and the results of the tests must be entered in the engineering log. Figure 16-15 shows some of the required tests and maintenance requirements (Maintenance Index page) for a turbogenerator.

CHAPTER 17

COMPRESSED AIR PLANTS

Compressed air serves many purposes aboard ship, and air outlets are installed in various suitable locations throughout the ship. The uses of compressed air include (but are not limited to) the operation of pneumatic tools and equipment, diesel engine starting and control, torpedo charging, aircraft starting and cooling, air deballasting, and the operation of pneumatic control systems. The systems that supply compressed air for the various shipboard needs are discussed in chapter 9 of this text; in the present chapter we are concerned with the equipment used to compress the air and supply it to the compressed air systems.

Compressed air represents a storage of energy. Work is done on the working fluid (air) so that work can later be done by the working fluid. Air compression may be either an adiabatic or an isothermal process.[1] Adiabatic compression results in a high internal energy level of the air being discharged from the compressor. However, much of the extra energy provided by adiabatic compression may be dissipated by heat losses, since compressed air is usually held in an uninsulated receiver until it is used. Isothermal compression is, in theory, the most economical method of compressing air in that it requires the least work to be done on the working fluid. However, the isothermal compression of air requires a cooling medium to remove heat from the compressor and its contained air during the compression process. The more closely isothermal compression is approached, the greater the cooling effect required; in a compressor of finite size, then, we reach a point at which it is no longer practicable to continue to strive for isothermal compression.

In actual practice, the process of air compression is approximately isothermal when considered from start to finish, and approximately adiabatic when considered within any one stage of the compression process. In order to achieve some benefits from each type of process (isothermal and adiabatic) most air compressors are designed with more than one stage and with a cooling arrangement after each stage. Multistaging and after-stage cooling have the further advantages of preventing the development of excessively high temperatures in the compressor and in the accumulator, reducing the horsepower requirements, condensing some of the entrained moisture, and increasing volumetric efficiency.

The general arrangement of a multistage compressed air plant with after-stage cooling is illustrated in figure 17-1. This illustration shows a reciprocating air compressor with two stages; however, the same general arrangement of parts is found in any type of compressed air plant that utilizes multistaging and after-stage cooling.

The accumulator shown in figure 17-1 is found in all compressed air plants, although the size of the unit varies according to the needs of the system. The accumulator (also called a receiver) helps to eliminate pulsations in the discharge line of the air compressor, acts as a storage tank during intervals when the demand for air exceeds the capacity of the compressor, and allows the compressor to shut down during periods of light load. Overall, the accumulator functions to retard increases and decreases in the pressure of the system, thereby lengthening the start-stop-start cycle of the compressor.

COMPRESSOR CLASSIFICATIONS

Air compressors are classified in various ways. A compressor may be single acting or double acting, single stage or multistage, and horizontal, angle, or vertical, as shown in figure 17-2. A compressor may be designed so that ONLY one stage of compression takes place

---

[1] Thermodynamic processes are discussed in chapter 8 of this text.

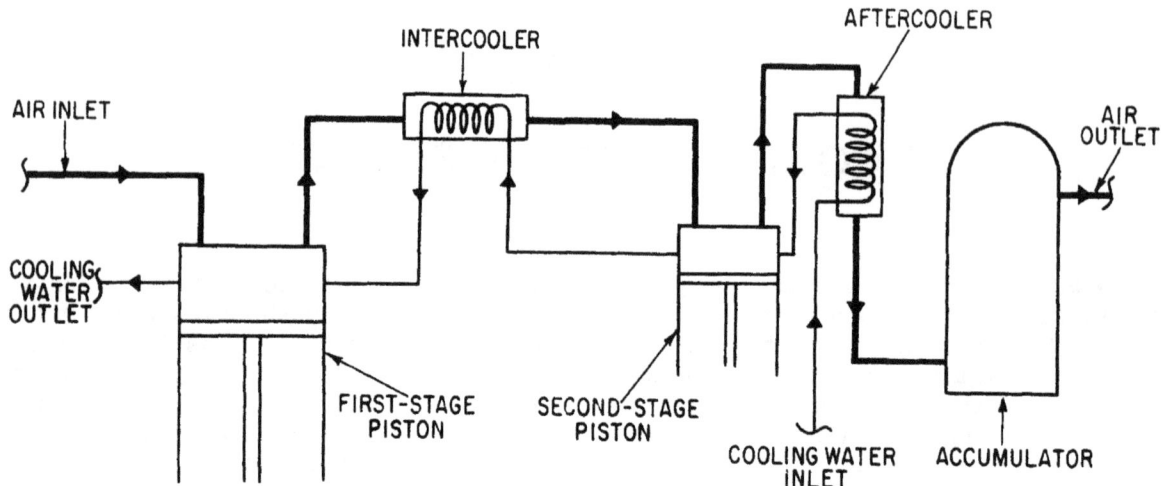

147.118

Figure 17-1.—General arrangement of multistage compressed air plant.

within one compressing element, or so that more than one stage takes place within one compressing element. In general, compressors are classified according to the type of compressing element, the source of driving power, the method by which the driving unit is connected to the compressor, and the pressure developed.

TYPES OF COMPRESSING ELEMENTS.—Air compressor elements may be of the centrifugal, rotary, or reciprocating types. The reciprocating type is generally selected for capacities below 1,000 cfm and for pressures of 100 psi or above, the rotary type for capacities up to 10,000 cfm and for pressures below 100 psi, and the centrifugal type for 10,000 cfm or greater capacities and for up to 100 psi pressures.

Most of the compressors used in the Navy have reciprocating elements (fig. 17-3). In this type of compressor the air is compressed in one or more cylinders, very much like the compression which takes place in an internal combustion engine.

SOURCES OF POWER.—Compressors are driven by electric motors, internal combustion engines, steam turbines, or reciprocating steam engines. Most of the air compressors in naval service are driven by electric motors.

DRIVE CONNECTIONS.—The driving unit may be connected to the compressor by one of

several methods. When the compressor and the driving unit are mounted on the same shaft, they are close coupled. Close coupling is often used for small capacity compressors that are driven by electric motors. Flexible couplings are used to join the driving unit to the compressor where the speed of the compressor and the speed of the driving unit can be the same.

V-belt drives are commonly used with small, low pressure, motor-driven compressors, and with some medium pressure compressors. In a few installations, a rigid coupling is used between the compressor and the electric motor of a motor-driven compressor. In a steam turbine drive, compressors are usually driven through reduction gears.

PRESSURE CLASSIFICATION.—In accordance with General Specifications for Ships of the United States Navy, compressors are classified as low pressure, medium pressure, or high pressure. Low pressure compressors are those which have a discharge pressure of 150 psi or less. Medium pressure compressors are those which have a discharge pressure of 151 psi to 1,000 psi. Compressors which have a discharge pressure above 1,000 psi are classified as high pressure.

Most low pressure air compressors are of the two-stage type with either a vertical V (see fig. 17-3) or a vertical W arrangement of

SUCTION VALVE DISCHARGE VALVE

CYLINDER

PISTON

A

DISCHARGE VALVE

CYLINDER PISTON

SUCTION VALVE

B

DISCHARGE VALVE DISCHARGE VALVE

SUCTION VALVE SUCTION VALVE

CYLINDERS

PISTONS

C

SUCTION VALVES

DISCHARGE VALVES

PISTON CYLINDERS PISTON

D

3RD STAGE 4TH STAGE

1ST STAGE

2ND STAGE

3RD STAGE 4TH STAGE

1ST STAGE 1ST STAGE

2ND STAGE 2ND STAGE

E

47.151

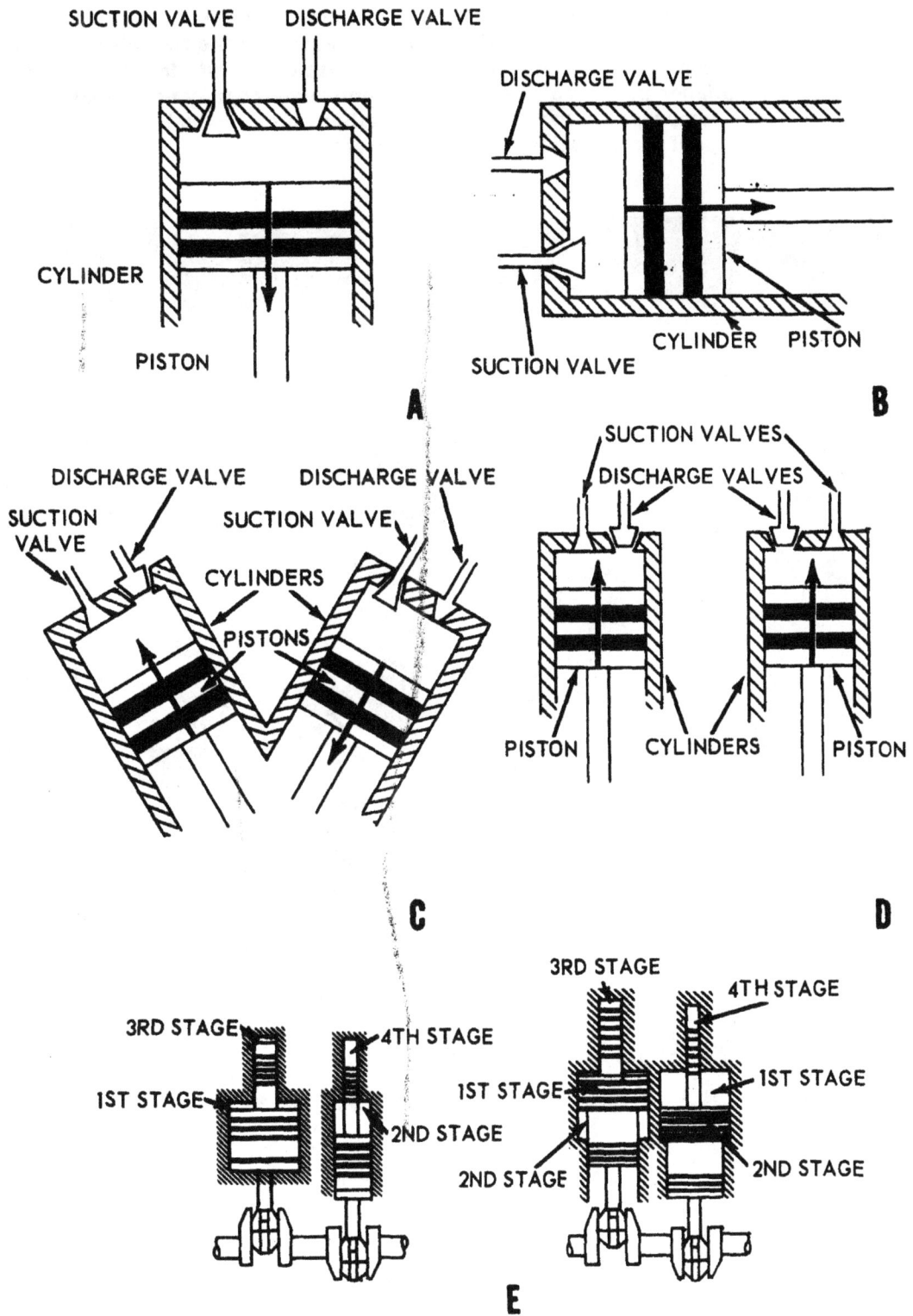

Figure 17-2.—Types of air compressors: A. Vertical. B. Horizontal. C. Angle.
D. Duplex. E. Multistage.

cylinders. Two-stage, V-type low pressure compressors usually have one cylinder for the first (lower pressure) stage of compression, and one cylinder for the second (higher pressure) stage of compression. W-type compressors have two cylinders for the first stage of compression, and one cylinder for the second stage. This arrangement is shown in the two-stage, three-cylinder, radial arrangement in part A of fig. 17-4.

Compressors may be classified according to a number of other design features or operating characteristics.

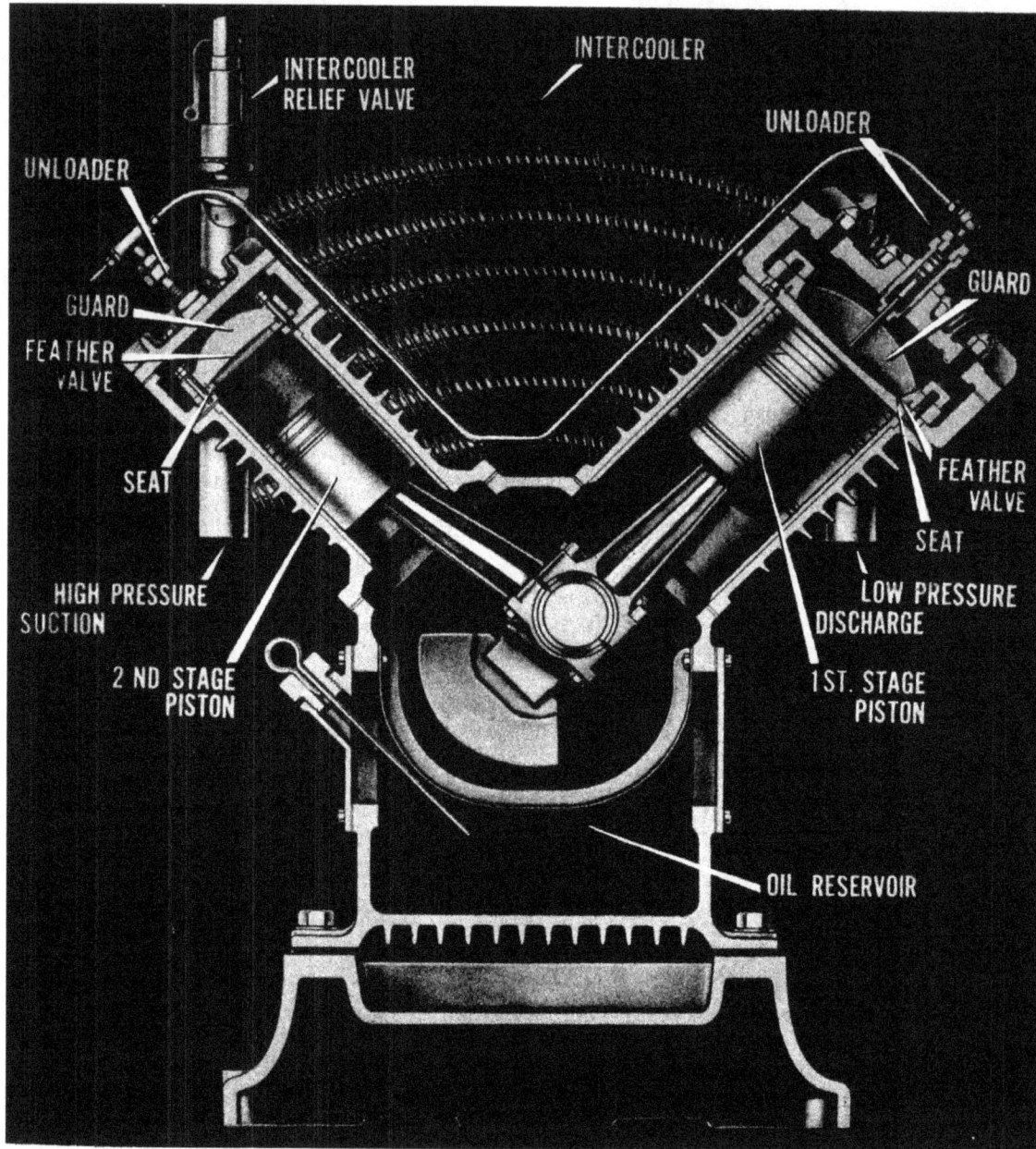

47.152

Figure 17-3.—Reciprocating air compressor (vertical V, two stage, single-acting, low pressure).

RECIPROCATING AIR COMPRESSORS

Reciprocating air compressors are sufficiently similar in design and operation so that the following discussion applies in large part to all reciprocating compressors now in naval use.

Medium pressure air compressors are of the two-stage, vertical, duplex, single-acting type. Many medium pressure compressors have differential pistons; this type of piston has more than one stage of compression during each stroke of the piston. (See fig. 17-4, A.)

Modern air compressors are generally motor-driven (direct or geared), liquid-cooled, four-stage, single-acting units with vertical or horizontal cylinders. Cylinder arrangements for high pressure air compressors installed on Navy ships are illustrated in part B of figure 17-4. Small capacity high pressure air systems may have three-stage compressors. Large capacity, high pressure, air systems may be equipped with five- or six-stage compressors.

Operating Cycle

Let us consider first the operating cycle that occurs during one stage of compression in a single-acting reciprocating air compressor such as the one shown in figure 17-3. The operating cycle consists of two strokes of the piston: a suction stroke and a compression stroke.

The suction stroke begins when the piston moves away from top dead center (TDC). The air under pressure in the clearance space (above the piston) expands rapidly until the pressure falls below the pressure on the opposite side of the air inlet valve. At this point the difference in pressure causes the inlet valve to open and air is admitted to the cylinder. Air continues to flow into the cylinder until the piston reaches bottom dead center (BDC).

The compression stroke starts as the piston moves away from BDC and compression of the air begins. When the pressure in the cylinder

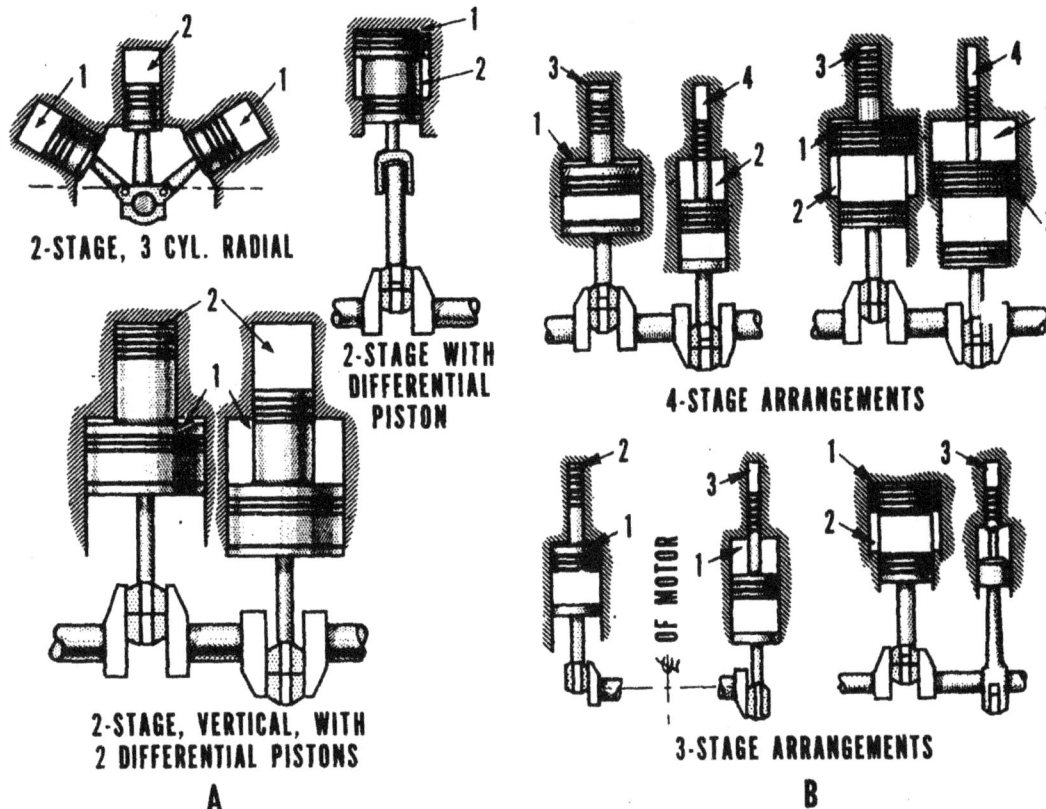

47.153

Figure 17-4.—Air compressor cylinder arrangements. A. Low and medium pressure cylinders. B. High pressure cylinders.

equals the pressure on the opposite side of the air inlet valve, the inlet valve closes. Air is increasingly compressed as the piston moves toward TDC; the pressure in the cylinder finally becomes great enough to force the discharge valve open against the discharge line pressure and the pressure of the valve springs. During the balance of the compression stroke, the air which has been compressed in the cylinder is discharged, at almost constant pressure, through the open discharge valve.

The basic operating cycle just described is repeated a number of times in double acting compressors and in other stages of multistage compressors. In a double-acting compressor, each stroke of the piston is a suction stroke in relation to one end of the cylinder and a compression stroke in relation to the other end of the cylinder. In a double acting compressor, therefore, two basic compression cycles are always in process when the compressor is operating; but each cycle, considered separately, is simply one suction stroke and one compression stroke.

In multistage compressors, the basic compression cycle must occur at least once for each stage of compression. If the compressor is designed with two compressing elements for the first (low pressure) stage, two compression cycles will be in process in the first stage at the same time. If the compressor is designed so that two stages of compression occur at the same time in one compressing element, the two basic cycles (one for each stage) will occur at the same time.

Compressor Components

A reciprocating air compressor consists of a compressor element, a lubrication system, a cooling system, a control system, and an unloading system. In addition to these basic components, the compressor has a system of connecting rods, crankshaft, and flywheel for transmitting the power developed by the driving unit to the air cylinder pistons.

COMPRESSING ELEMENT.—The compressing element of a reciprocating air compressor consists of the air valves, the cylinder, and the piston.

The valves of modern compressors are of the automatic type. The opening and closing of these valves is caused solely by the difference between the pressure of the air in the cylinder and the pressure of the external air on the intake valve or the pressure of the discharged air on the discharge valve. On most compressors, a thin plate, low lift type of valve is used. A valve of this type is shown in figure 17-5.

The design of pistons and cylinders depends primarily upon the number of stages of compression which take place within a cylinder. Common arrangements of pistons and cylinders are shown in a previous illustration (fig. 17-2).

Two types of pistons are in common use. Trunk pistons (fig. 17-6) are driven directly by the connecting rods. Since the upper end of a connecting rod is fitted directly to the piston wrist pin, there is a tendency for a piston to develop a side pressure against the cylinder walls. To distribute the side pressure over a wide area of the cylinder walls or liners, trunk pistons with long skirts are used. This type of piston tends to eliminate cylinder wall wear. Differential pistons (fig. 17-7) are modified trunk pistons having two or more different diameters. These pistons are fitted into special cylinders which are arranged so that more than one stage of compression is served by one piston. The compression for one stage takes place over the piston crown; compression for the other stage or stages takes place in the annular space between the large and small diameters of the piston.

LUBRICATION SYSTEM.—Lubrication of air compressor cylinders is generally accomplished by means of a mechanical force-feed lubricator which is driven from a reciprocating or a rotary part of the compressor. Oil is fed from the lubricator through a separate feed line to each cylinder. A check valve is installed at the end of each feed line to keep the compressed air from forcing the lube oil back to the lubricator. Each feed line is equipped with a sight glass. Lubrication begins automatically as the compressor starts up. The amount of oil that must be fed to the cylinders depends upon the cylinder diameter, the cylinder wall temperature, and the viscosity of the oil.

On small low pressure and medium pressure compressors, the cylinders may be lubricated by the splash method, from dippers on the ends of the connecting rods, instead of by a mechanical force-feed lubricator.

Lubrication of the running gear of most compressors is accomplished by a lube oil pump (usually of the gear type) which is attached to the compressor and driven from the compressor shaft. This pump draws oil from the reservoir, as shown in figure 17-8, and delivers it, through

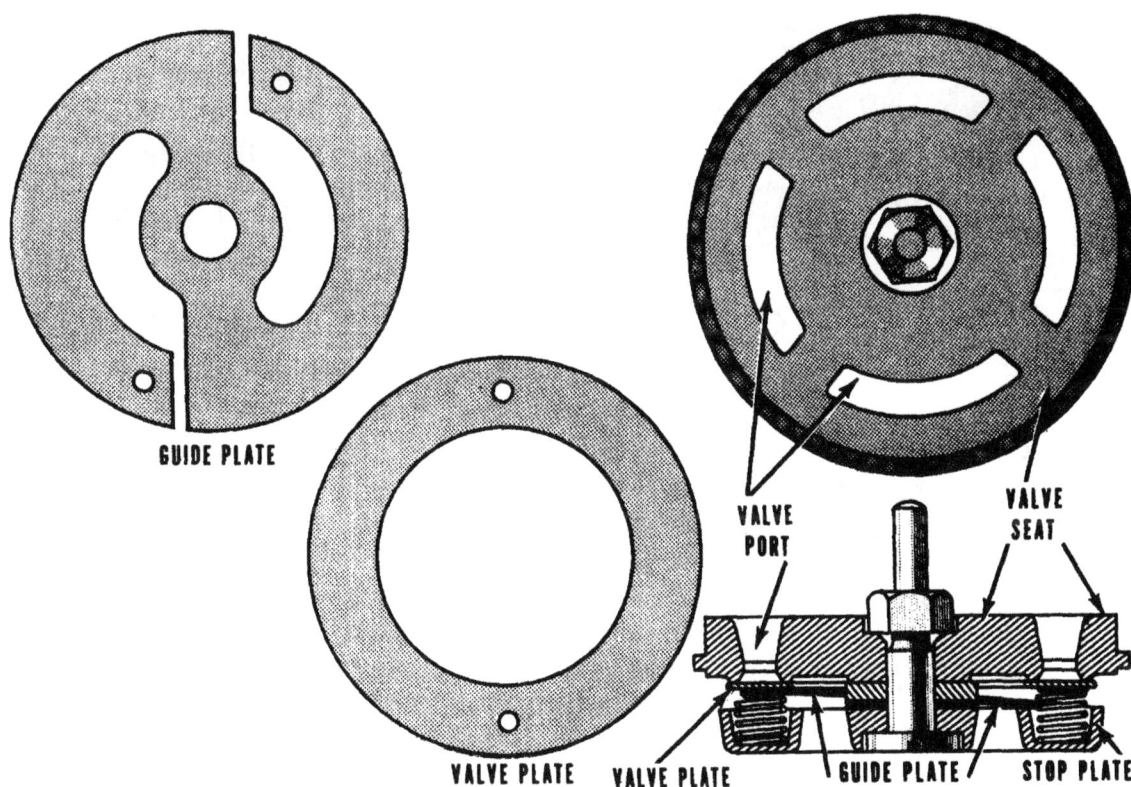

47.154X

Figure 17-5.—Diagram of a thin plate air compressor valve.

a filter, to an oil cooler. From the cooler, the oil is distributed to the top of each main bearing, to spray nozzles for reduction gears, and to outboard bearings. The crankshaft is drilled so that oil fed to the main bearings is picked up at the main bearing journals and carried to the crank pin journals. The connecting rods contain passages which conduct lubricating oil from the crank pin bearings up to the wrist pin bushings. As oil leaks out from the various bearings, it drips back to the reservoir in the base of the compressor and is recirculated. Oil from the outboard bearings is carried back to the base by the drain lines.

The discharge pressure of lubricating oil pumps varies, depending upon the pump design. A relief valve fitted to each pump functions when the discharge pressure exceeds the pressure for the valve is set. When the relief valve opens, excess oil is returned to the reservoir.

COOLING SYSTEM.—Most compressors are cooled by sea water supplied from the ship's fire and flushing system. The cooling water is usually available to each unit through at least two sources. Compressors located outside the larger machinery spaces are generally equipped with an attached circulating water pump as a standby source of cooling water. Some small low pressure compressors are air cooled by a fan mounted on or driven by a compressor shaft.

The path of water in the cooling water system for a four-stage compressor is shown in figure 17-9. The flow paths are not identical in all cooling water systems, but in all systems it is important that the coldest water be available for circulation through the oil cooler. Valves are usually provided so that the flow of water to the cooler can be controlled independently of the rest of the system. Thus the oil temperature can be controlled without affecting other parts of the compressor. Cooling water is then supplied to the intercoolers and the aftercooler and then to the cylinder jackets and heads. A high pressure air compressor may require from 6 to 25 gallons of cooling water per minute, while a medium

443

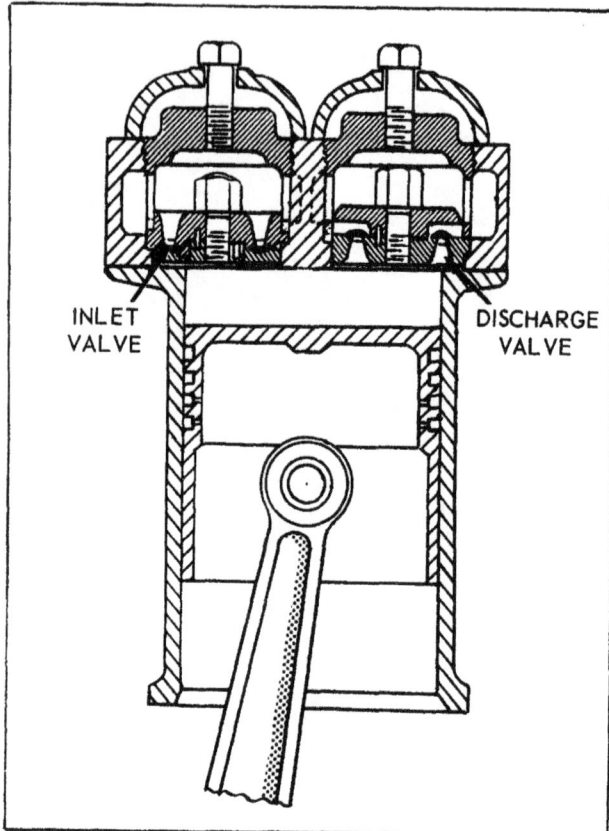

47.155.1

Figure 17-6.—Trunk Piston.

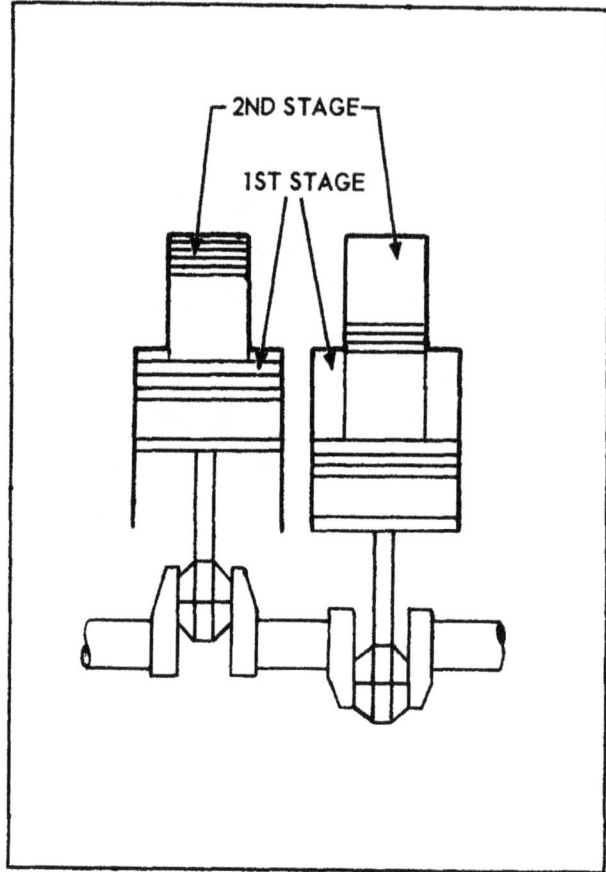

47.155.2

Figure 17-7.—Differential Piston.

pressure air compressor may require from 10 to 20 gallons per minute.

As previously noted, cooling of the air is required for most economical compression. Another reason for cooling the air between stages and after the last stage is to condense any moisture that may be present. The resulting condensate is then drained off. If the moisture is not removed from the air, it will be carried into the accumulator or into the air lines, where it can cause serious trouble.

The intercoolers used between stages and the aftercooler used after the last stage are of the same general construction except that the aftercooler is designed to withstand a higher working pressure than the intercoolers. Water-cooled coolers may be of the straight shell-and-tube type or of the coil type. In coolers designed for air pressures below 250 psig, the air flows either through the tubes or over and around them; in coolers designed for air pressures above 250 psig, the air flows through the tubes. In straight-

tube coolers, baffles are provided to guide the air and water. In coil type coolers, the air passes through the coil and the water flows around the outside. Air-cooled coolers may be of the radiator type or may consist of a bank of finned copper tubes located in the path of a blast of air supplied by the compressor fan.

Both intercoolers and aftercoolers are fitted with moisture separators on the discharge side to remove moisture (and also any oil that may be present) from the air stream. Various designs of moisture separators are in use; the removal of liquid may be accomplished by centrifugal force, impact, or sudden changes in the velocity of the air stream.

CONTROL SYSTEM.—The control system of a reciprocating air compressor may include one or more devices such as automatic temperature shutdown devices, start-stop controls, constant-speed controls, and speed-pressure governors.

444

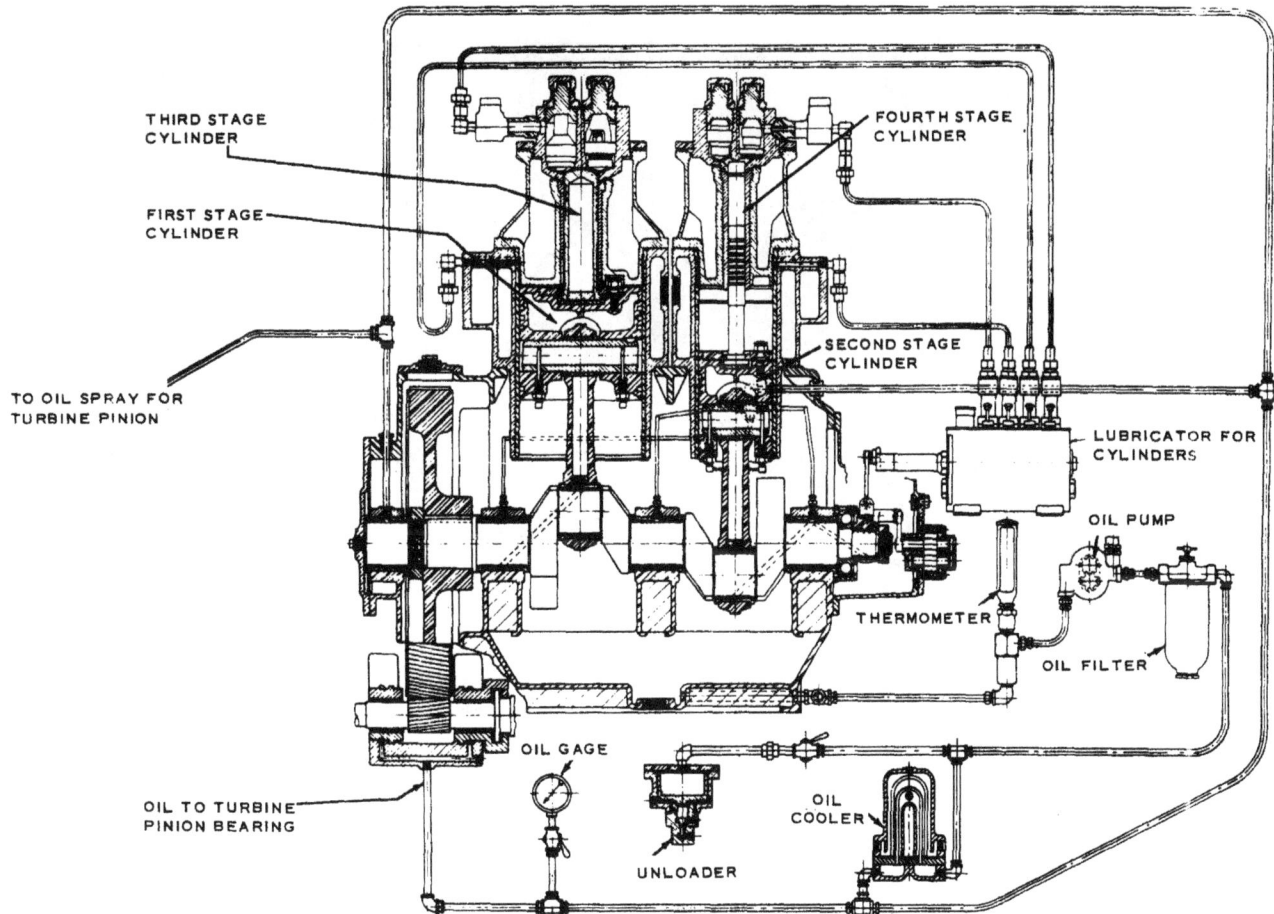

47.156

Figure 17-8.—Lubrication system for turbine-driven high pressure air compressor.

Automatic temperature shutdown devices are fitted to all recent designs of high pressure air compressors. Such a device stops the compressor automatically (and does not allow it to restart automatically) when the cooling water temperature rises above a safe limit. Some compressors are fitted with a device that shuts down the compressor if the temperature of the air leaving any stage exceeds a preset value.

Control or regulating systems for naval air compressors are mainly of the start-stop type. With this type of control, the compressor starts and stops automatically as the accumulator pressure rises or falls to predetermined limits. On electrically driven compressors, the system is very simple: the accumulator pressure operates against a pressure switch that opens when the pressure upon it reaches a given limit and closes when the pressure drops a predetermined amount. On ·steam-driven compressors, the

accumulator pressure is piped to a control valve (also called a pilot, trigger, or auxiliary valve) which, when the designed cutoff pressure is reached, admits air to a plunger connected with the turbine governor valve. This causes steam to be shut off and the compressor to stop. When the pressure falls to a predetermined level, the control valve closes and the air acting upon the plunger is released by leakage or bleeding to atmosphere. The steam is thereby permitted to flow through the governor valve and restart the turbine.

Constant-speed control is a method of controlling the pressure in the air accumulator by controlling the output of the compressor without stopping or changing the speed of the unit. This type of control is used on compressors that have a fairly constant demand for air, where frequent stopping and starting is undesirable.

445

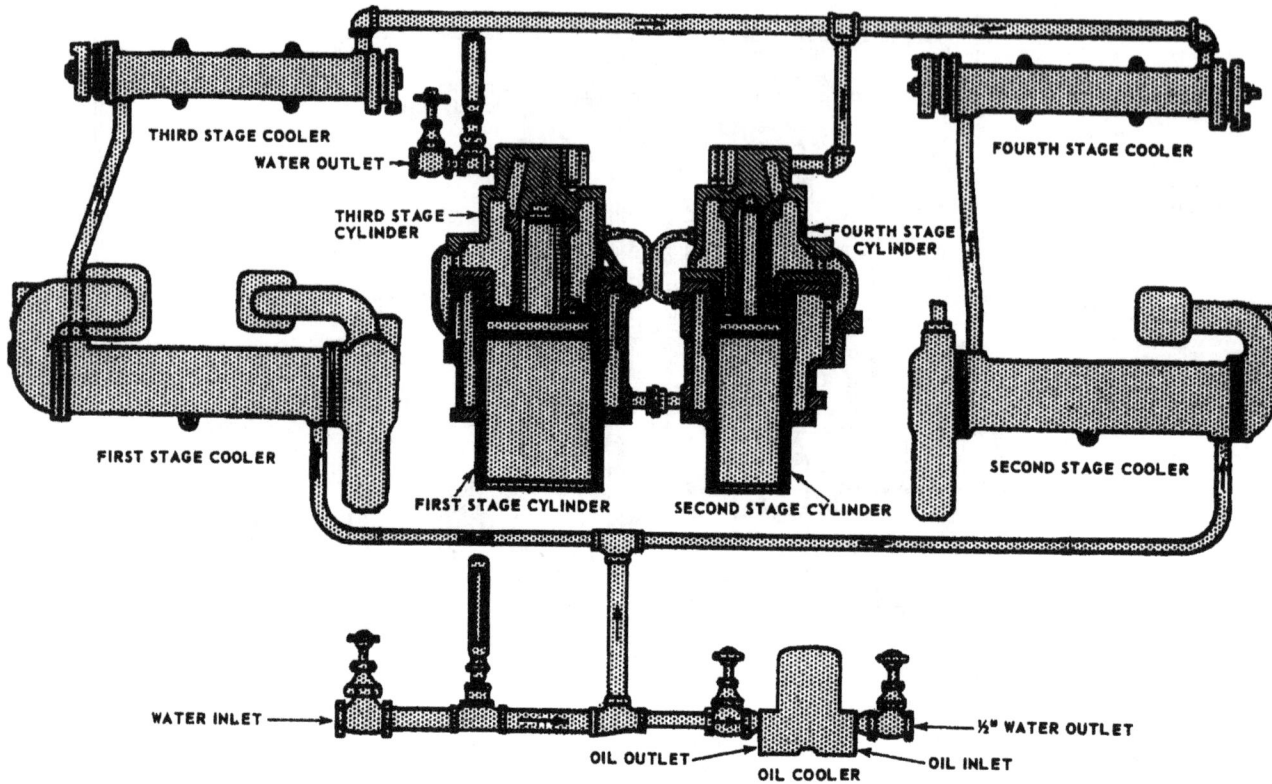

47.157

Figure 17-9.—Cooling water system for multistage air compressor.

Combined speed and pressure governors are usually furnished for compressors which are driven by reciprocating steam engines. Neither the start-stop nor the constant-speed control is entirely satisfactory for this type of compressor.

UNLOADING SYSTEM.—Air compressor unloading systems are installed for the removal of all but the friction loads on the compressors. An unloading system automatically removes the compression load from the compressor while the unit is starting and automatically applies the load after the unit is up to operating speed. For units that have the start-stop type of control, the unloading system is separate from the control system. For compressors equipped with the constant-speed type of control, the unloading system is an integral part of the control system.

A number of different unloading methods are used, including closing or throttling the compressor intake, holding intake valves off their seats, relieving intercoolers to atmosphere, opening a bypass from the discharge to the intake, opening up cylinder clearance pockets, using

miscellaneous constant-speed unloading devices, and using various combinations of these methods.

As an example of a typical compressor unloading device, consider the MAGNETIC TYPE UNLOADER. Figure 17-10 illustrates the unloader valve arrangement. This type of unloader consists of a solenoid-operated valve connected with the motor starter. When the compressor is at rest, the solenoid valve is deenergized, admitting air from the receiver to the unloading mechanism. When the compressor approximates normal speed, the solenoid valve is energized, releasing the pressure from the unloading mechanism and loading the compressor again.

ROTARY-CENTRIFUGAL AIR COMPRESSORS

The one non-reciprocating type of air compressor that is found aboard ship is variously referred to as a rotary compressor, a centrifugal compressor, or a "liquid piston" compressor. Actually, the unit is something of a mixture, operating partly on rotary principles and partly

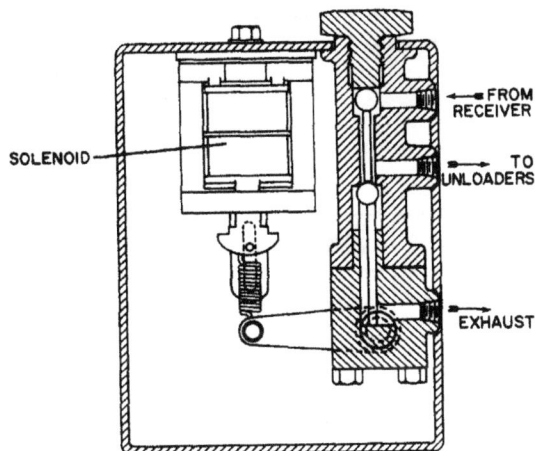

47.158

Figure 17-10.—Magnetic type unloader.

on centrifugal principles; most accurately, perhaps, it might be called a rotary-centrifugal compressor.

The rotary-centrifugal compressor is used to supply low pressure compressed air. Because this compressor is capable of supplying air that is completely free of oil, it is often used as the compressor for pneumatic control systems and for other applications where oil-free air is required.

The rotary-centrifugal compressor, shown in figure 17-11, consists of a round, multi-bladed rotor which revolves freely in an elliptical casing. The elliptical casing is partially filled with high-purity water. The curved rotor blades project radially from the hub. The blades, together with the side shrouds, form a series of pockets or buckets around the periphery. The rotor, which is keyed to the shaft of an electric motor, revolves at a speed high enough to throw the liquid out from the center by centrifugal force, resulting in a solid ring of liquid revolving in the casing at the same speed as the rotor but following the elliptical shape of the casing. This action alternately forces the liquid to enter and recede from the buckets in the rotor at high velocity.

To follow through a complete cycle of operation, let us start at point A. The chamber (1) is full of liquid. The liquid, because of centrifugal force, follows the casing, withdraws from the rotor, and pulls air in through the inlet port. At (2) the liquid has been thrown outward from the chamber in the rotor and has been replaced with atmospheric air. As the rotation continues, the converging wall (3) of the casing forces the liquid back into the rotor chamber, compressing the trapped air and forcing it out through the discharge port. The rotor chamber (4) is now full

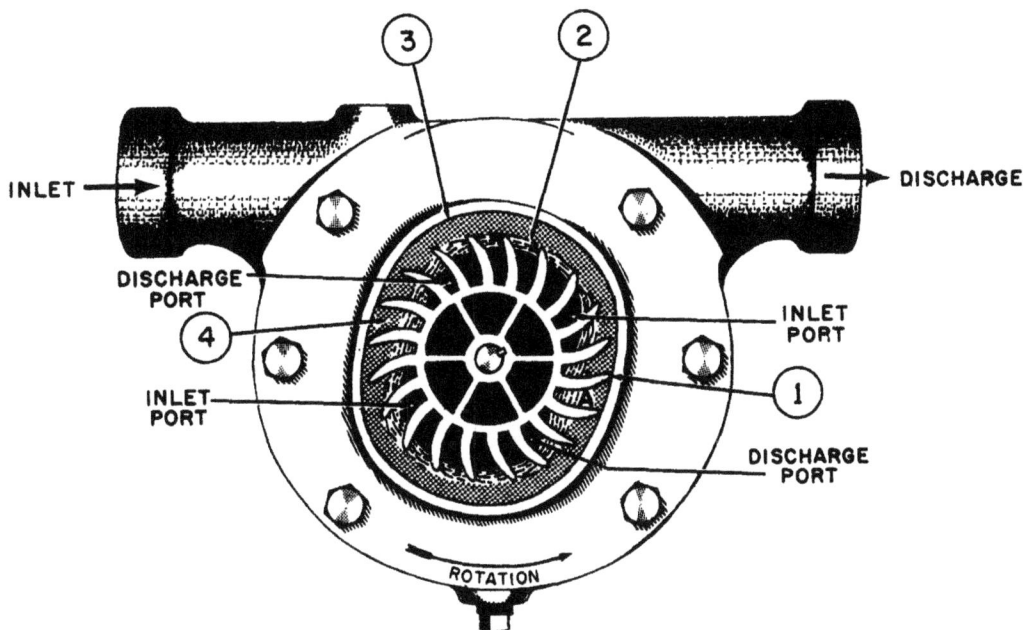

147.119

Figure 17-11.—Rotary-centrifugal compressor.

System, Subsystem, or Component High-Pressure Air Compressor					Reference Publications				

Bureau Card Control No.					Maintenance Requirement	M.R. No.	Rate Req'd.	Man Hours	Related Maintenance
AP	ZZZFCH1	35	4935	W	1. Operate compressor by power. 2. Blow down all air flasks, separators, and filters. 3. Sample and inspect lube oil.	W-1	MM2	0.5	None
AP	ZZPFVA1	84	4936	M	1. Lift relief valves by hand.	M-1	MM3	0.1	None
AP	ZZZFCH1	65	4937	Q	1. Clean suction filter. 2. Test inlet and outlet valves by operation. 3. Test temperature switch by operation. 4. Test automatic start and stop switch by operation.	Q-1	MM2	0.5	None
AP	ZZZFCH1	65	9223	Q	1. Test operation of speed-limiting governor and overspeed trip. 2. Test operation of temperature control.	Q-2	MM2 FN	1.0 1.0	None
AP	ZZZFCH1	65	7659	A	1. Clean lubricator reservoir.	A-1	FN	0.3	None
AP	ZZDFCUO	55	A142	C	1. Clean and test air coolers and oil cooler. 2. Inspect internal parts for wear.	C-1	MM1 FN	24.0 24.0	None
AP	ZZPFVA1	84	4941	C	1. Test relief valves by pressure.	C-2	FN	1.3	None
AP	ZZ6FTRO	25	7652	C	1. Clean, inspect, and preserve exterior of turbine casing.	C-3	MM3	1.0	C-4
AP	ZZ6FTRO	25	7656	C	1. Inspect carbon packing for wear.	C-4	MM1 MM3	4.0 4.0	C-3
AP	ZZZFPGF	B5	2894	C	1. Inspect high-pressure air system for oil contamination.	C-5	MM2 FN	1.5 1.5	W-1

MAINTENANCE INDEX PAGE OPNAV FORM 4700-3 (4-64)

BUREAU PAGE CONTROL NUMBER   A-3/16-95

Figure 17-12.—Maintenance Index Page.

98.171

of liquid and ready to repeat the cycle which takes place twice in each revolution.

A small amount of seal water must be constantly supplied to the compressor to make up for that which is carried over with the compressed air. The water which is carried over with the compressed air is removed in a refrigeration-type dehydrator.

AIR COMPRESSOR MAINTENANCE

Minimum requirements for the performance of inspections and maintenance on high pressure air plants are shown on the maintenance index page figure 17-12.

It is the responsibility of the engineer officer to determine if the condition of the equipment, hours of service, or operating conditions necessitate more frequent inspections and tests. Details for outline tests and inspections may be obtained from the appropriate manufacturer's instruction book or from the Naval Ships Technical Manual.

SAFETY PRECAUTIONS

There are many hazards associated with the process of air compression. Serious explosions have occurred in high pressure air systems because of a diesel effect. [2] Ignition temperatures may result from rapid pressurization of a low pressure dead end portion of the piping system, malfunctioning of compressor aftercoolers, leaky or dirty valves, and many other causes. Every precaution must be taken to have only clean, dry air at the compressor inlet.

Air compressor accidents have also been caused by improper maintenance procedures such as disconnecting parts while they are under pressure, replacing parts with units designed for lower pressures, and installing stop valves or check valves in improper locations. Improper operating procedures have also caused air compressor accidents, with resulting serious injury to personnel and damage to equipment.

In order to minimize the hazards inherent in the process of compression and in the use of compressed air, all safety precautions outlined in the manufacturers' technical manual and in the Naval Ships Technical Manual must be strictly observed.

---

[2] A diesel engine operates by taking in air, compressing it, and then injecting fuel into the cylinders, where the fuel is ignited by the heat of compression. The same effect (normally called the diesel effect) can occur in hydropneumatic machinery and in air, oxygen, or other gas systems, if even a very small amount of "fuel"—a smear of oil, for example, or a single cotton thread—is present to be ignited by the heat of compression.

CHAPTER 18

DISTILLING PLANTS

Naval ships must be self-sustaining as far as the production of fresh water is concerned. The large quantities of fresh water required aboard ship for boiler feed, drinking, cooking, bathing, and washing make it impracticable to provide storage tanks large enough for more than a few days' supply. Therefore, all naval ships depend upon distilling plants to meet the requirements for large quantities of fresh water of extremely high chemical and biological purity.

PRINCIPLES OF DISTILLATION

All shipboard distilling plants not only perform the same basic function but also perform this function in much the same way. The distillation process consists of heating sea water to the boiling point and condensing the vapor to obtain fresh water (distillate). The distillation process for a shipboard plant is illustrated very simply in figure 18-1.

At a given pressure, the rate at which sea water is evaporated in a distilling plant is

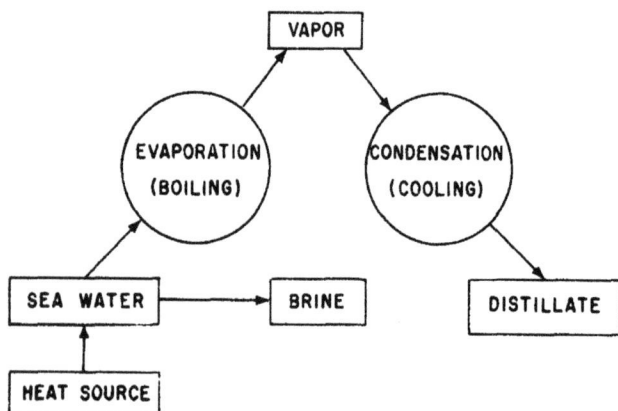

75.284
Figure 18-1.—Simplified diagram of shipboard distillation process.

dependent upon the rate at which heat is transmitted to the water. The rate of heat transfer to the water is dependent upon a number of factors; of major importance are the temperature difference between the substance giving up heat and the substance receiving heat, the available surface area through which heat may flow, and the coefficient of heat transfer of the substances and materials involved in the various heat exchangers that constitute the distilling plant. Additional factors such as the velocity of flow of the fluids and the cleanliness of the heat transfer surfaces also have a marked effect upon heat transfer in a distilling plant.

Since a shipboard distilling plant consists of a number of heat exchangers, each serving one or more specified purposes, the plant as a whole provides an excellent illustration of many thermodynamic processes and concepts. Practical manifestations of heat transfer—including heating, cooling, and change of phase—abound in the distilling plant, and the significance of the pressure-temperature relationships of liquids and their vapors is clearly evident. [1]

The sea water which is the raw material of the distilling plant is a water solution of various minerals and salts. In addition to the dissolved material, sea water also contains suspended matter such as vegetable and animal growths and bacteria and other micro-organisms. Under proper operating conditions, naval distilling plants are capable of producing fresh water which contains only minute traces of the chemical and

---

[1] Much of the information given in chapter 8 of this text has direct and immediate application to the study of distilling plants. Applicable portions of chapter 8 should be reviewed, if necessary, as a basis for the study of distilling plants.

biological contaminants which are found naturally in sea water. [2]

One of the problems that arises in the distillation of sea water occurs because some of the salts present in sea water are negatively soluble—that is, they are less soluble in hot water than they are in cold water. A negatively soluble salt remains in solution at low temperatures but precipitates out of solution at higher temperatures. The crystalline precipitation of various sea salts forms scale on heat transfer surfaces and thereby interferes with heat transfer. In naval distilling plants, this problem is partially avoided by designing the plants to operate under vacuum or (in the case of one type of plant) at approximately atmospheric pressure.

The use of low pressures (and therefore low boiling temperatures) has the additional advantage of greater thermodynamic efficiency than can be achieved when higher pressures and temperatures are used. With low pressures and temperatures, less heat is required to make the sea water boil and less heat is lost overboard through the circulating water that cools and condenses the vapor.

DEFINITION OF TERMS

The manner in which the various kinds of distilling plants accomplish the distilling process can best be understood if we first become familiar with certain terms relating to the process. The terms defined here relate basically to all types of distilling plants now in naval use. Additional terms that apply specifically to a particular type of distilling unit are defined as necessary in subsequent discussion.

Distillation.—The process of boiling sea water and then cooling and condensing the resulting vapor to produce fresh water.

Evaporation.—The first part of the process of distillation. Evaporation is the process of boiling sea water in order to separate it into fresh water vapor and brine.

---

[2] It should be noted that distilling plants are not effective in removing volatile gases or liquids which have a lower boiling point than water, nor are they effective in killing all micro-organisms. These points are of particular importance when a ship is operating in contaminated or polluted waters, as discussed at the end of this chapter.

Condensation.—The latter part of the process of distillation. Condensation is the process of cooling the vapor to produce usable fresh water.

Feed.—The sea water which is the raw material in the distillation process.

Vapor.—The product of the evaporation of sea water. The terms vapor and fresh water vapor are used interchangeably.

Distillate.—The product resulting from the condensation of the fresh water vapor produced by the evaporation of sea water. Distillate is also referred to as condensate, as fresh water, as fresh water condensate, and as sea water distillate. However, the use of the term condensate should be avoided whenever there is any possibility of confusion between the condensate of the distilling plant and the condensate that results from the condensation of steam in the main and auxiliary condensers. In general, it is best to use the term distillate when referring to the product resulting from the condensation of vapor in the distilling plant.

Salinity.—The concentration of salt in water.

Brine.—Water in which the concentration of salt is higher than it is in sea water.

TYPES OF DISTILLING UNITS

Distilling units installed in naval ships are of two general types. The vapor compression type of unit is used aboard submarines and small diesel-driven surface craft where the daily requirements do not exceed 4000 gallons per day (gpd). The low pressure steam distilling unit is used aboard all steam-driven surface ships and on nuclear submarines. The major difference between the two types of distilling units is in the kind of energy used to operate the unit. Vapor compression units use electrical energy; steam distilling units use auxiliary exhaust steam.

VAPOR COMPRESSION DISTILLING UNITS

A vapor compression distilling unit is shown in cutaway view in figure 18-2 and schematically in figure 18-3. The unit consists of three main components—the evaporator, the compressor, and the heat exchanger—and a number of accessories and auxiliaries.

RELIEF VALVE
PRESSURE GAGE
BYPASS VALVE

DRIVE SHAFT

OIL SIGHT

DRIVE MOTOR

DRIVEN SHAFT

(THREE LOBE)

MANOMETER.

COMPRESSOR
(TWO LOBE)

COMPRESSED
VAPOR PIPE

VAPOR SEPARATOR

FEED INLET

MANHOLE

TUBE SHEET

FUNNEL

INSULATION

VENT

BAFFLES

DOWNTAKE

BRINE OVERFLOW TUBE
ELECTRIC HEATERS

DRAIN

BRINE OVERFLOW OUTLET

75.286

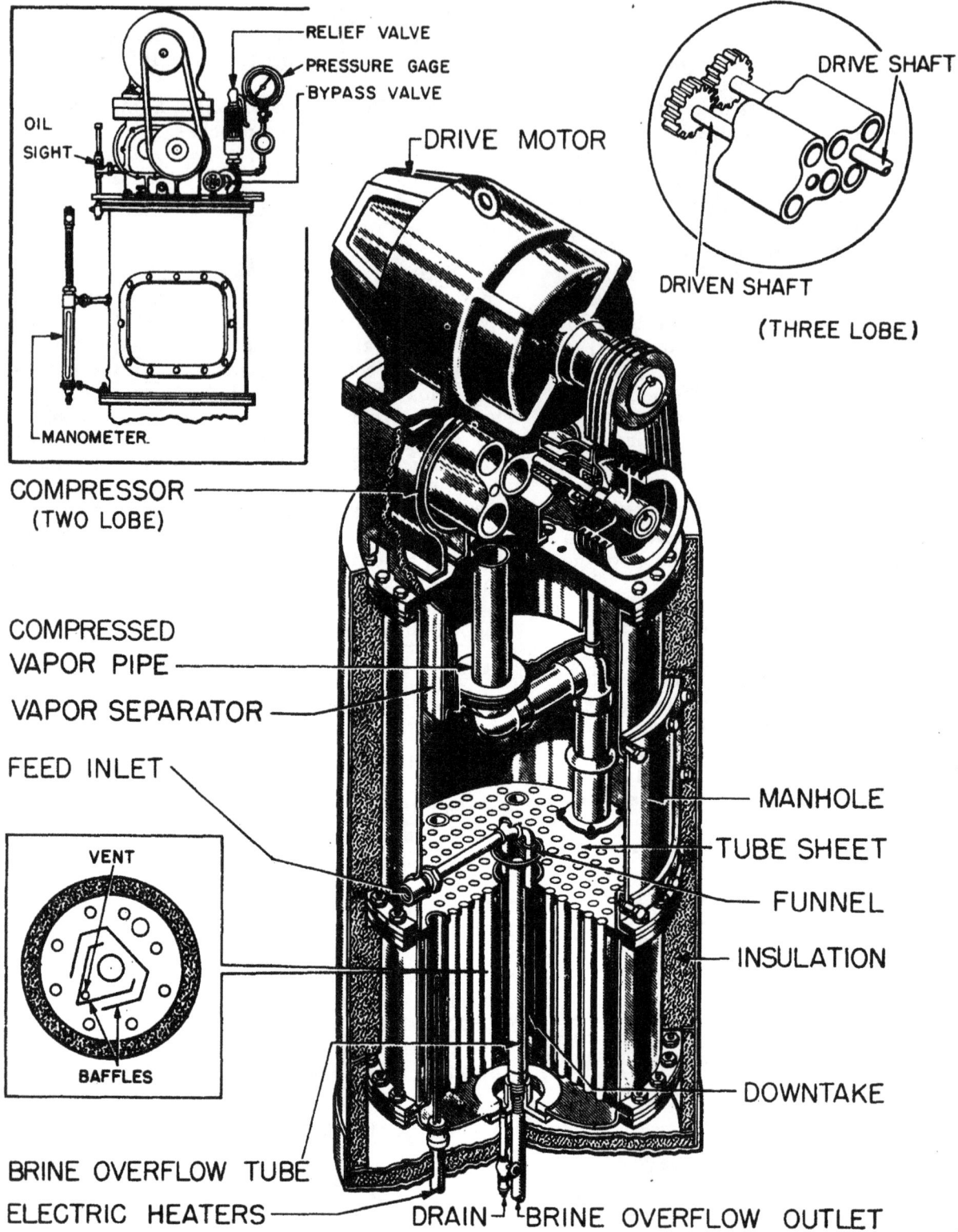

Figure 18-2.—Cutaway view of vapor compression distilling plant.

EVAPORATOR.—The cylindrical shell in which vaporization and condensation occur is commonly called the evaporator. The evaporator consists of two principal elements: the steam chest and the vapor separator.

The steam chest[3] includes all space within the evaporator shell except the space that is occupied by the vapor separator. The steam chest is considered to have an evaporating side and a condensing side. The evaporating side includes the space within the tubes of the tube bundle (which is located in the lower part of the evaporator shell) and the space which communicates with the inside of the tubes. The condensing side includes the space which surrounds the external surfaces of the tubes; this space communicates with the discharge side of the compressor by means of a pipe, as shown in figure 18-3.

The tube bundle is enclosed in a shell. At top and bottom of the bundle the tube ends are expanded into tube sheets. Most of the tubes are small; but a few, set near the periphery, are larger. Each of the larger tubes contains an electric heater. As the sea water feed flows through these larger tubes, it is heated to the boiling point by the heaters.

The feed inlet pipe extends horizontally to the center of the evaporator, where it branches into a Y. The two ends of the feed pipe turn downward into the downtake, as shown in figure 18-2. Sea water feed enters the evaporator through the horizontal inlet pipe, pours into the downtake, and passes down to the bottom head of the evaporator shell; from there, the feed flows upward through the tubes.

A funnel is installed inside the downtake, at the top. The top of the funnel is about 2 inches above the top of the evaporator tubes, and the brine level in the evaporator shell is thus maintained at this height. About one-half to two-thirds of the feed is vaporized; the remaining brine overflows continuously into the funnel and then into the brine overflow tube which is installed inside the downtake. The overflow tube leads the brine out through the bottom of the evaporator shell, to the heat exchanger. In the heat exchanger, the brine gives up its heat and raises the temperature of the incoming feed.

The vapor separator is an internal compartment located at the top of the evaporator shell.

The separator consists of two cylindrical baffles. One cylinder extends downward from the upper head plate of the evaporator; the other extends upward, and is fitted around the upper cylinder to form a baffle. The floor of the separator is formed by the bottom of the outer cylinder. The space between the two cylinders provides a passage for the vapor flowing from the evaporating side of the steam chest to the suction side of the compressor.

The vapor from the boiling sea water rises up through the space between the shell wall and the outer cylinder of the separator; it then flows downward through the space between the cylinders of the separator and enters the separator chamber. From the separator chamber, the vapor travels upward to the intake side of the compressor.

In the course of this roundabout passage through the vapor separator, the vapor is separated from any entrained particles of water. The water drops to the floor of the separator and is continuously drained away. This water has a high salt concentration, and must be continuously drained in order to keep it from entering the compressor and thus getting into the condensing side of the evaporator, where it would contaminate the distillate.

VAPOR COMPRESSOR.— The vapor which flows upward from the separator is compressed by a positive-displacement compressor. The type of compressor discussed here has two three-lobe rotors of the type shown in the insert on figure 18-2 and in figure 18-3. Two-lobe compressors of the type shown in the main part of figure 18-2 were an earlier design.

The two rotors are enclosed in a compact housing which is mounted on the evaporator. The three lobes on each rotor are designed to produce a continuous and uniform flow of vapor. The vapor enters the compressor housing at the bottom and then passes upward between the inner and outer walls of the housing to the rotor chamber, where it fills the space between the rotor lobes. The vapor is then carried around the cylindrical sides of the housing, and a pressure is developed at the bottom as the lobes roll together. Clearances are provided so that the rotor lobes do not actually touch each other and do not touch the housing.

The shaft of one rotor is fitted with a drive pulley on one end and a gear on the other end. This gear meshes with a gear on the shaft of the other rotor, to provide the necessary drive for the second rotor.

---

[3] "Steam chest" is a somewhat misleading term for a unit which is not operated by steam and which has no steam coming into it from an external source. Although called a steam chest, it might more accurately be thought of as a "vapor chest."

PRESSURE GAGE
RELIEF VALVE
BYPASS VALVE

DESUPERHEATER DRIP
COMPRESSOR

VAPOR SEPARATOR
MANOMETER
VAPOR SEPARATOR DRAIN
CHECK VALVE
FROM DESUPERHEATER TANK

VENT LINE
3/64" ORIFICE
STEAM CHEST VENT

FEED INLET
CONDENSATE OUTLET

BRINE OVERFLOW TUBE

DRAIN
ELECTRIC HEATER

STEAM TRAP

HEAT EXCHANGER

VENT OUTLET

BRINE OVERFLOW OUTLET

FEED WATER   CONDENSATE
VAPOR AT ATMOSPHERIC PRESSURE   CONCENTRATED BRINE
COMPRESSED VAPOR   → FLOW PATHS

75.287

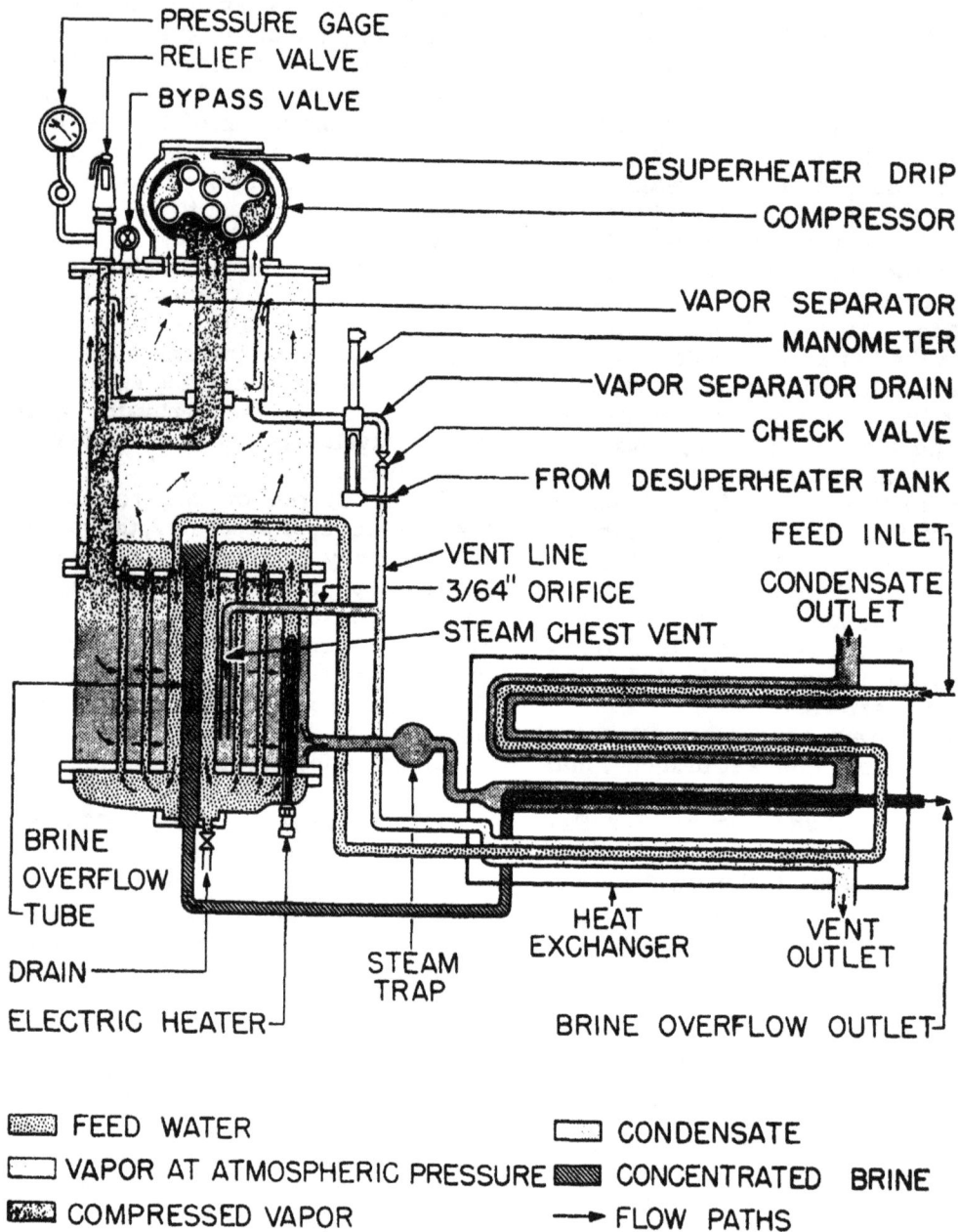

Figure 18-3.—Schematic view of vapor compression distilling plant.

HEAT EXCHANGER.—The heat exchanger preheats the incoming sea water feed by two heat exchange processes. In one process, the sea water feed is heated by the distillate which is being discharged from the distilling unit to the ship's tanks. In the other process, the sea water feed is heated by the brine overflow which is being discharged overboard or to the brine collecting tank.

The heat exchanger is a horizontal double-tube unit. Either sea water or brine flows through the inner tubes, while distillate flows through the space between the inner and the outer tubes.

Figure 18-4 shows the construction of the heat exchanger and also illustrates the flow paths.

454

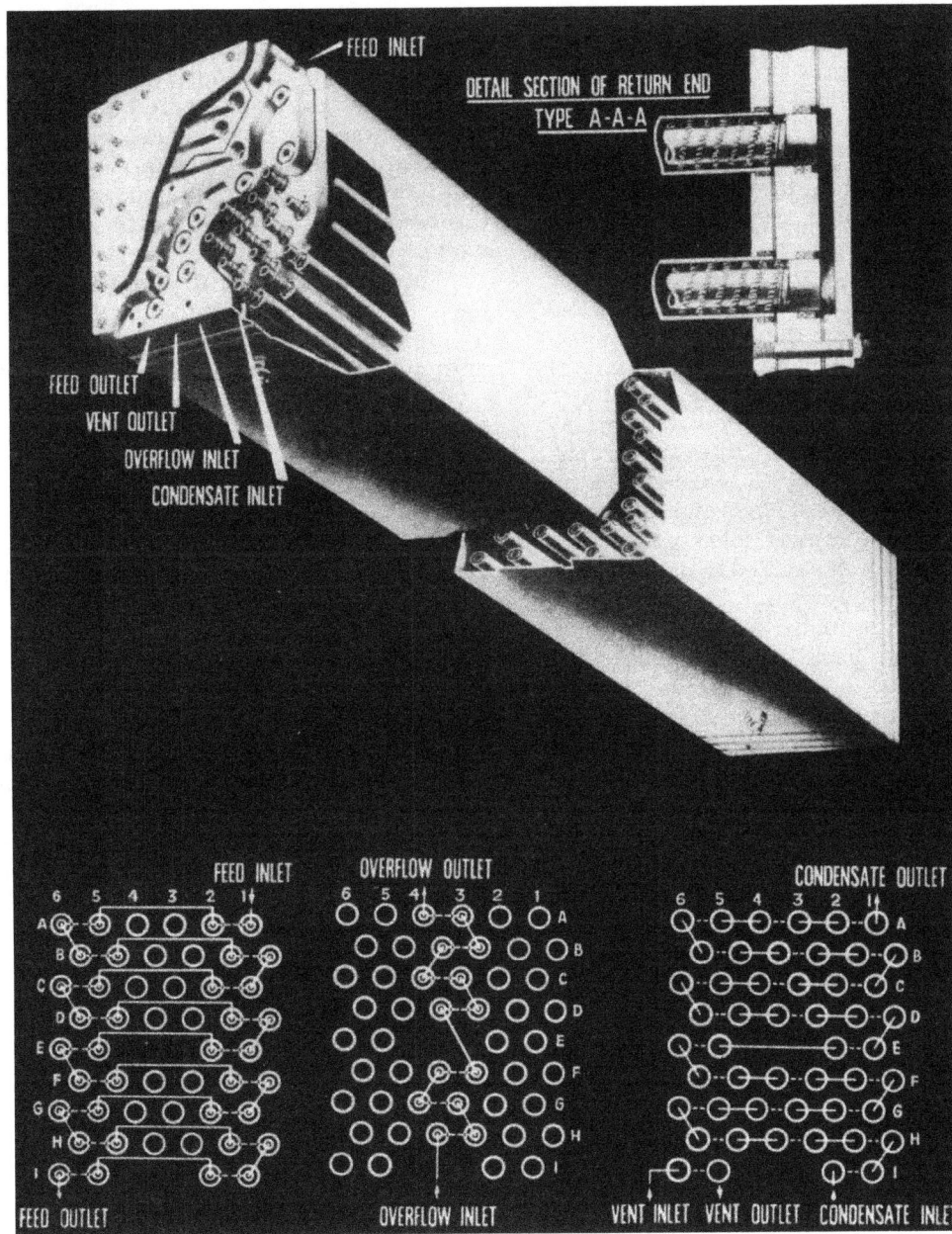

75.288

Figure 18-4.—Heat exchanger for a vapor compression distilling plant.

There are four distinct flow paths: feed, brine overflow, condensate (distillate), and vent.

ACCESSORIES AND AUXILIARIES.—A number of accessories and auxiliaries are required for the operation of the vapor compression distilling unit. These include feed, distillate, and brine overflow pumps; feed regulating and flow control valves; relief valves; compressor bypass valves; rotameters; and a variety of pressure and temperature gages.

THE VAPOR COMPRESSION PROCESS.— Now that the principal parts of a vapor

455

compression unit have been described, let us summarize briefly the sequence of events within the unit and consider some of the factors that are important in the vapor compression process of distillation.

The cold sea water feed enters the heat exchanger and is heated there to about 190° or 200° F. From the heat exchanger, the feed goes into the evaporator. Here it flows down the downtake and into the bottom of the evaporator shell, then upward in the tubes. Boiling and evaporation take place in the tubes at atmospheric pressure. About one-half to two-thirds of the incoming feed is evaporated; the remainder flows out through the brine overflow, thus maintaining a constant water level within the evaporator.

The vapor thus generated rises and enters the vapor separator, where any particles of moisture that may be present are separated from the vapor and drained out of the separator. The vapor goes to the suction side of the compressor. In the compressor, distilled water drips onto the rotors and thus desuperheats the vapor as it is compressed. The vapor is compressed to a pressure of about 3 to 5 pounds above atmospheric pressure, and is discharged to the space surrounding the tubes in the steam chest. As the vapor condenses on the outside of the smaller tubes, the distillate drops down and collects on the bottom tube plate. Every time a pound of compressed vapor condenses, approximately a pound of vapor is formed in the evaporator section; the compressor suction is thus kept supplied with the right amount of vapor

The distillate is drawn off through a steam trap and flows into the heat exchanger at a temperature of about 220° F. As it flows through the heat exchanger, the distillate gives up heat to the incoming feed and is cooled to within about 18° F of the cold feed water temperature. Noncondensable gases, together with a small amount of vapor, flow into the vent line and then to the heat exchanger.

Meanwhile, the sea water which is not vaporized in the evaporator is flowing continuously into the funnel, down the brine overflow tube, and into the heat exchanger. The temperature of this brine is about 214° F. In passing through the heat exchanger, the hot brine raises the temperature of the sea water feed that is entering through the heat exchanger.

The entire distillation cycle is started by using the electric heaters to bring the sea water feed temperature up to the boiling point and to

generate enough vapor for compressor operation. After the cycle has been started and the compressor is adequately supplied with vapor, the normal operating cycle begins and the electric heaters are used henceforth only to provide the heat necessary to make up for heat losses. After the unit has become fully operational, then, the heat input from the heaters is only a small part of the total heat input.

The major part of the heat input comes from the compression work that is done on the vapor by the compressor. The major energy transformations involved in normal operation are thus from electrical energy (put in at the compressor motor) to mechanical energy (work done by the compressor on the vapor) to thermal energy. The thermal energy thus supplied is used to boil the sea water feed and keep the process going.

The compression process serves another vital function in the vapor compression distilling unit. Since the boiling point of sea water is several degrees higher than the boiling point of fresh water at any given pressure, the boiling sea water in the evaporator is actually above 212° F and would therefore be too hot to condense the fresh water vapor if the vapor were at the same pressure as the boiling sea water. By compressing the vapor, the boiling point of the vapor is raised above the boiling point of the sea water at atmospheric pressure. Therefore the compressed vapor can be condensed on the outside of the tubes in which sea water feed is being boiled. This process would not be possible without the pressure difference between the evaporating side and the condensing side of the unit, and this pressure difference is created by the compression of the vapor.

STEAM DISTILLING UNITS

Steam distilling plants now in naval use are practically all of the low pressure type. They are "low pressure" units from two points of view. First, they utilize low pressure steam (auxiliary exhaust steam) as the source of energy; and second, they operate at less than atmospheric pressure. There are three major types of low pressure steam distilling units: submerged tube units, flash-type units, and vertical basket units.

Submerged Tube Units

Submerged tube distilling units range from 4000 to 50,000 gallons per day in capacity. There are three kinds of submerged tube distilling units: (1) the Soloshell double-effect unit, (2) the two-shell double-effect unit, and (3) the three-shell triple-effect unit.

The difference between double-effect units and triple-effect units is merely in the number of stages of evaporation. Two stages of evaporation occur in a double-effect unit, and three in a triple-effect unit.

SOLOSHELL DOUBLE-EFFECT UNITS.— Most Soloshell double-effect units have capacities of 12,000 gallons per day or less. However, some Soloshell units of 20,000 gpd capacity are in use.

A Soloshell double-effect unit is shown schematically in figure 18-5 and in cutaway view in figure 18-6. The unit consists of a single cylindrical shell which is mounted with the long axis in a horizontal position. A longitudinal vertical partition plate divides the shell into a first-effect shell and a second-effect shell. The first-effect shell contains the first effect tube bundle, a vapor separator, and the vapor feed heater. The second-effect shell contains the second-effect tube bundle, a vapor separator, and the distilling condenser. A distillate cooler, not a part of the main cylindrical shell, is mounted at any convenient location, as piping arrangements permit. Another separate unit, the air ejector condenser, is mounted on brackets on the outside of the evaporator shell. The air ejector takes suction on the second-effect part of the shell, maintaining it under a vacuum of approximately 26 inches of mercury. A lesser vacuum—about 16 inches of mercury—is maintained in the first-effect shell.

Steam for the distilling unit is obtained from the auxiliary exhaust line through a regulating valve. This valve is adjusted to maintain a constant steam pressure of 1 to 5 psig in the line between the regulating valve and a control orifice. The size of the opening in the control orifice determines the amount of steam admitted to the distilling unit and hence controls the output of distilled water.

When the steam pressure is reduced by the regulating valve, the steam becomes superheated. Since superheat has the undesirable effect of increasing the rate of scale formation, provision is made for desuperheating the steam. This is done by spraying hot water into the steam line between the control orifice and the point where the steam enters the first-effect shell. The hot water for desuperheating the steam is taken from the first-effect drain pump discharge.

After being desuperheated, the steam passes into the first-effect tube nest, where it heats the sea water feed that surrounds the first-effect tubes. The sea water boils, generating steam which is called vapor to distinguish it from the steam which is the external source of energy for the unit. The condensate that results from the condensation of the supply steam is discharged by the first-effect drain pump to the low pressure drain system or to the condensate system and is thus eventually used again in the boiler feed system.

Although the vapor generated in the first-effect shell is pure water vapor, it does contain small particles of liquid feed. As the vapor rises, a series of baffles above the surface of the water begins the process of separating the vapor and the water particles.

After passing through the baffles, the vapor enters the vapor separator. As the vapor passes around the hooked edges of the baffles and vanes in the separator, it is forced to change direction several times; and with each change of direction some water particles are separated from the vapor. The hooked edges trap particles of water and drain them away, discharging them back into the feed at a distance from the vapor separator.

After passing through the first-effect vapor separator, the vapor goes to the vapor feed heater. Sea water feed passes through the tubes of the vapor feed heater, and part of the vapor is condensed as it flows over the tubes of the heater. This distillate, together with the remaining uncondensed vapor, goes through an external crossover pipe and enters the tube nest of the second-effect shell. The remaining vapor is now condensed as it gives up the rest of its latent heat to the sea water feed in the second-effect shell.

Since the pressure in the second-effect shell is considerably less than the pressure in the first-effect shell, the introduction of the vapor and the distillate from the first-effect shell causes the sea water feed in the second-effect shell to boil and vaporize.

The vapor thus generated in the second-effect shell passes through baffles just above the surface of the water and then goes to the second-effect vapor separator. From the vapor separator, it passes to the distilling condenser. The condensing tubes nearest the incoming vapor are utilized as a feed heating section; the vapor condenses on the outside of the tubes and thus heats the incoming sea water feed which is circulating through the tubes. The remainder of the vapor is

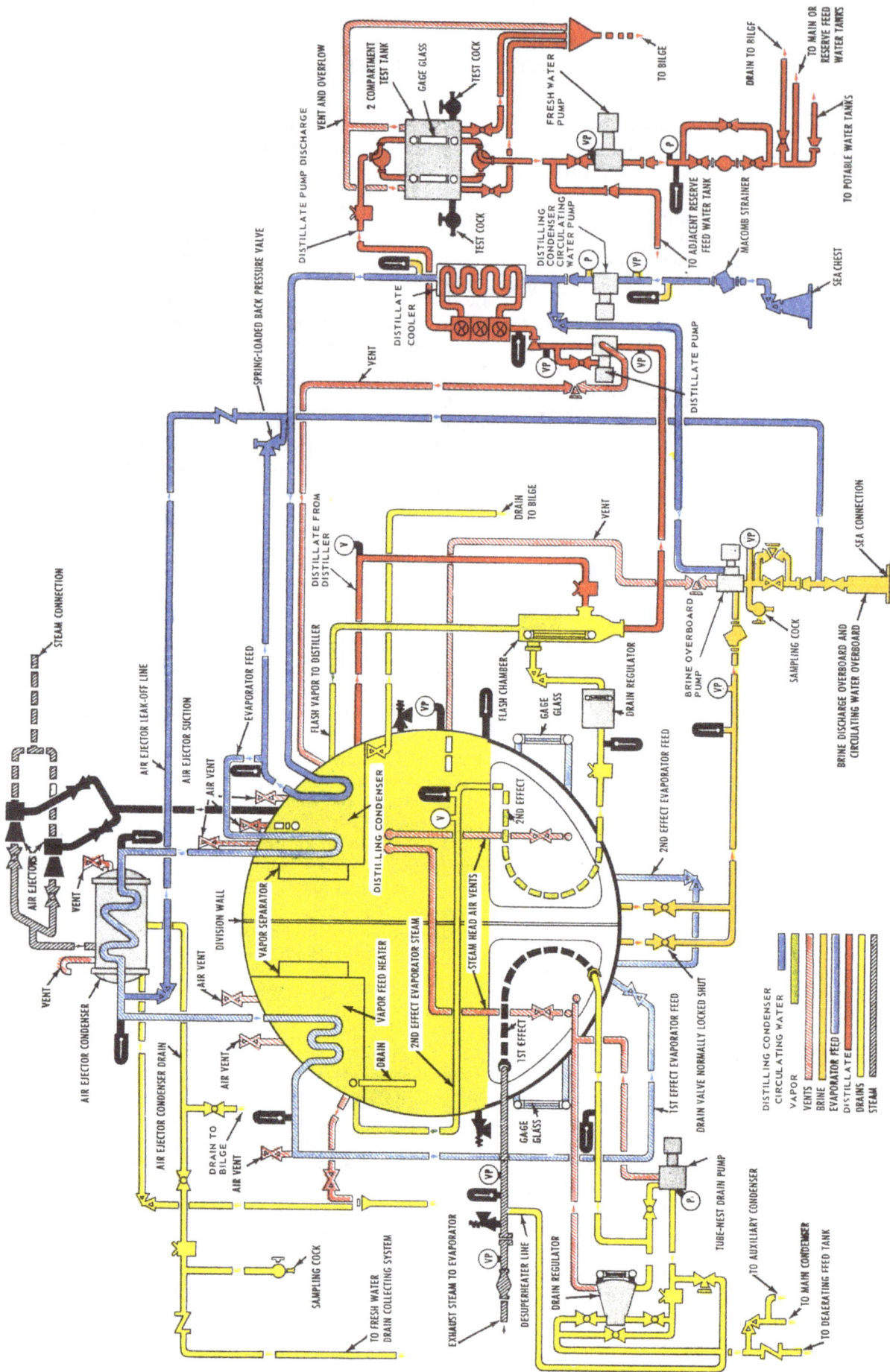

Figure 18-5.—Schematic diagram of Soloshell double-effect distilling plant.

47.114

458

condensed in the condensing section and is discharged to the test tanks as distillate.

The first-effect distillate which was used in the second-effect tubes to boil and vaporize the feed in the second-effect shell is discharged through the second-effect tube nest drain regulator and is led to the distilling condenser by way of a flash chamber. The flash chamber is essentially a receptacle within which the vapor, liberated when the second-effect drains are reduced to a pressure and temperature corresponding to the distilling condenser vacuum, is separated from the condensate and directed to the distilling condenser. As may be seen in figure 18-6, the flash chamber is located just outside of the second-effect shell.

The distilling condenser circulating water pump takes suction from the sea and discharges the sea water through the shell of the distillate cooler (which is external to the unit) and then through the tubes of the distilling condenser. Some of the cooling water is then discharged overboard; but a portion (which is now called evaporator feed) goes through the feed heating

section of the distilling condenser, through the air ejector condenser, and through the first-effect vapor feed heater before it is discharged to the first-effect shell. These paths of the distilling condenser circulating water and the evaporator feed may be traced in figure 18-5.

As previously described, some of the sea water feed in the first-effect shell is boiled and vaporized by the supply steam. The remaining portion becomes more dense and has a higher salinity than the original sea water feed; this denser, saltier water is called brine to distinguish it from sea water. After a certain amount of sea water feed has been vaporized in the first-effect shell, the remaining brine is led to the second-effect shell through a pipe that has a manually controlled feed regulating valve installed in it. When the feed regulating valve is open, the higher pressure in the first-effect shell causes the brine to flow from the first-effect shell to the second-effect shell. After the brine has been used as feed to generate vapor in the second-effect shell, the remaining brine is discharged overboard by the brine overboard discharge pump.

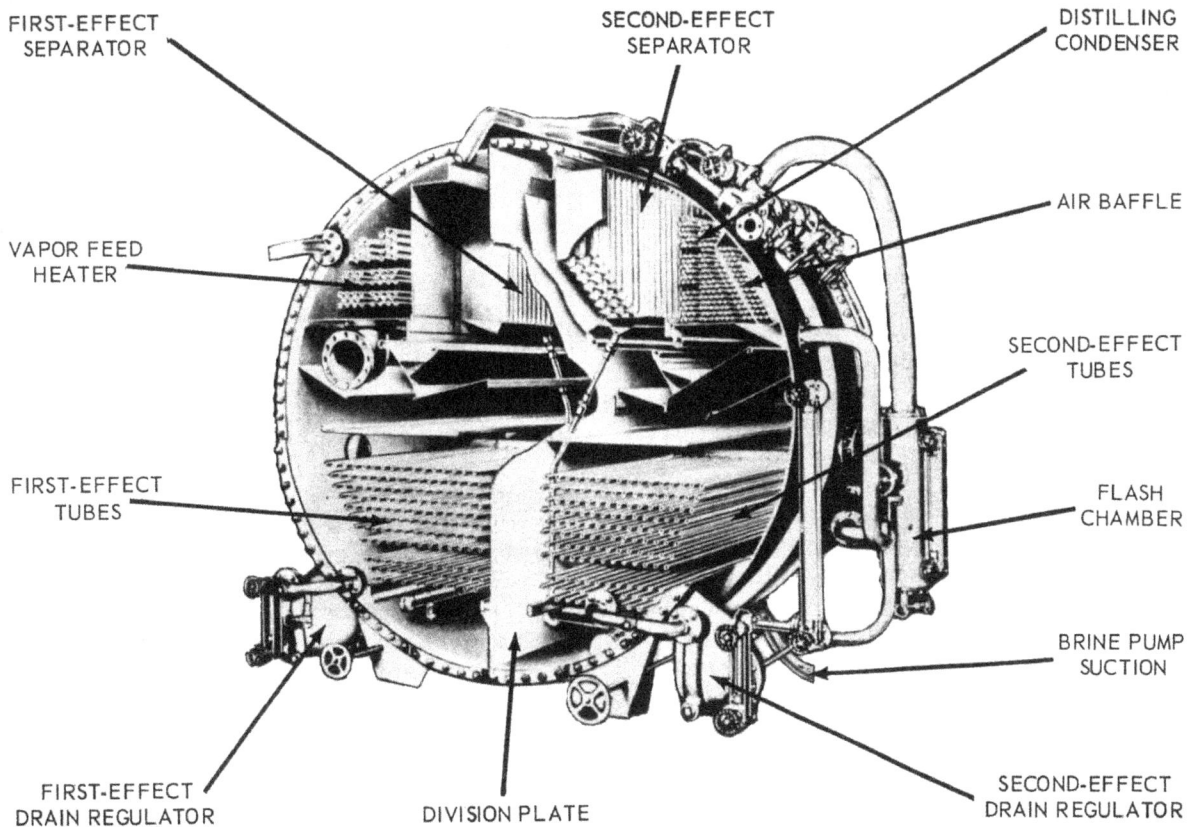

Figure 18-6.—Cutaway view of Soloshell double-effect distilling plant.

47.117

459

TWO-SHELL DOUBLE-EFFECT UNITS.—
Two-shell double-effect units of 20,000 gpd capacity are used on some ships. A typical unit of this kind is shown in figure 18-7. As may be seen, the unit consists of two cylindrical evaporator shells, mounted horizontally, with the long axes of the shells parallel. The first-effect vapor feed heater is built into the upper part of the first-effect shell. The distilling condenser and the distillate cooler are built into separate shells, which are usually mounted between the two evaporator shells. The air ejector condenser is also a separate unit, though it is mounted on one of the shells.

The operation of the two-shell double-effect unit is almost precisely the same as the operation of the Soloshell double-effect unit. The flow paths of steam, condensate, sea water, brine, vapor, and distillate may be traced out on figure 18-7.

THREE-SHELL TRIPLE-EFFECT UNITS.—
Three-shell triple-effect distilling units are similar to the double-effect units previously discussed except that the triple-effect units have an intermediate evaporating stage.

A triple-effect distilling unit is shown schematically in figure 18-8. Although there are several kinds of triple-effect units, the general relationships shown in this illustration hold for any triple-effect plant.

A standard 20,000 gpd triple-effect unit consists of three horizontal cylindrical shells, set side by side with their axes parallel. The first- and second-effect vapor feed heaters are built into the front end of the second- and third-effect evaporator shells. The distilling condenser is contained within the third-effect shell. The air ejector condenser and the distillate cooler are in separate shells and are mounted on the third-effect shell.

Another 20,000 gpd triple-effect design consists essentially of three horizontal shells bolted together end to end, with vertical partition plates between each shell to separate the effects. Vapor separators in independent shells are installed in the vapor piping between effects and between the third effect and the distilling condenser. The first- and second-effect vapor feed heaters are in separate shells and are mounted in the piping at the inlet to the second-effect and third-effect tube nests, respectively. The two sections of the distilling condenser and the distillate cooler are built into a single shell and independently mounted as space and piping arrangements may

permit. The air ejector condenser is also a separately mounted unit.

A standard 30,000 gpd triple-effect unit is also in use. This is similar to the standard 20,000 gpd unit except that the 30,000 gpd unit is larger.

There are two types of 40,000 gpd triple-effect units that may be regarded as standard, since both are widely used in naval ships. The first type uses the same arrangement as the standard 20,000 gpd triple-effect unit but has the larger components needed for the increased capacity. The second type consists of three horizontal shells, usually mounted side by side, with axes parallel. In this design, both vapor feed heaters and distilling condensers are built as three independent units, each mounted separately outside the evaporator shells. The air ejector condenser and the distillate cooler are also in independent shells and are separately mounted outside the evaporator shells.

Triple-effect units operate in virtually the same way as the Soloshell and the two-shell double-effect units previously described, except that the comparable actions in a triple-effect unit are spread out through more equipment and through one more effect. In a triple-effect unit, the sea water feed is piped to the first effect shell, then to the second-effect shell, and then to the third-effect shell. Steam from the auxiliary exhaust line is used to vaporize the feed in the first-effect shell; in the second-effect and third-effect shells, the vapor is generated by the heat given up by vapor generated in the previous shell. In the triple-effect units, as in the double-effect units, this sequence of events is possible because the vacuum is greatest in the shell of the final effect and least in the shell of the first effect.

Flash-Type Distilling Units

Some recent ships are equipped with flash-type distilling units. Although these units differ somewhat in design from the submerged tube units, certain operating principles are common to both types. In particular, both the flash-type and the submerged tube type of unit depends upon pressure differentials between the stages (or effects) to generate vapor from the sea water feed.

Flash-type units consist of two or more stages. Two-stage units of 12,000 gpd capacity are installed on some recent destroyer type ships. Five-stage units of 50,000 gpd capacity are installed on some recent carriers.

Each stage of a flash-type unit has a flash chamber, a feed box, a vapor separator, and a

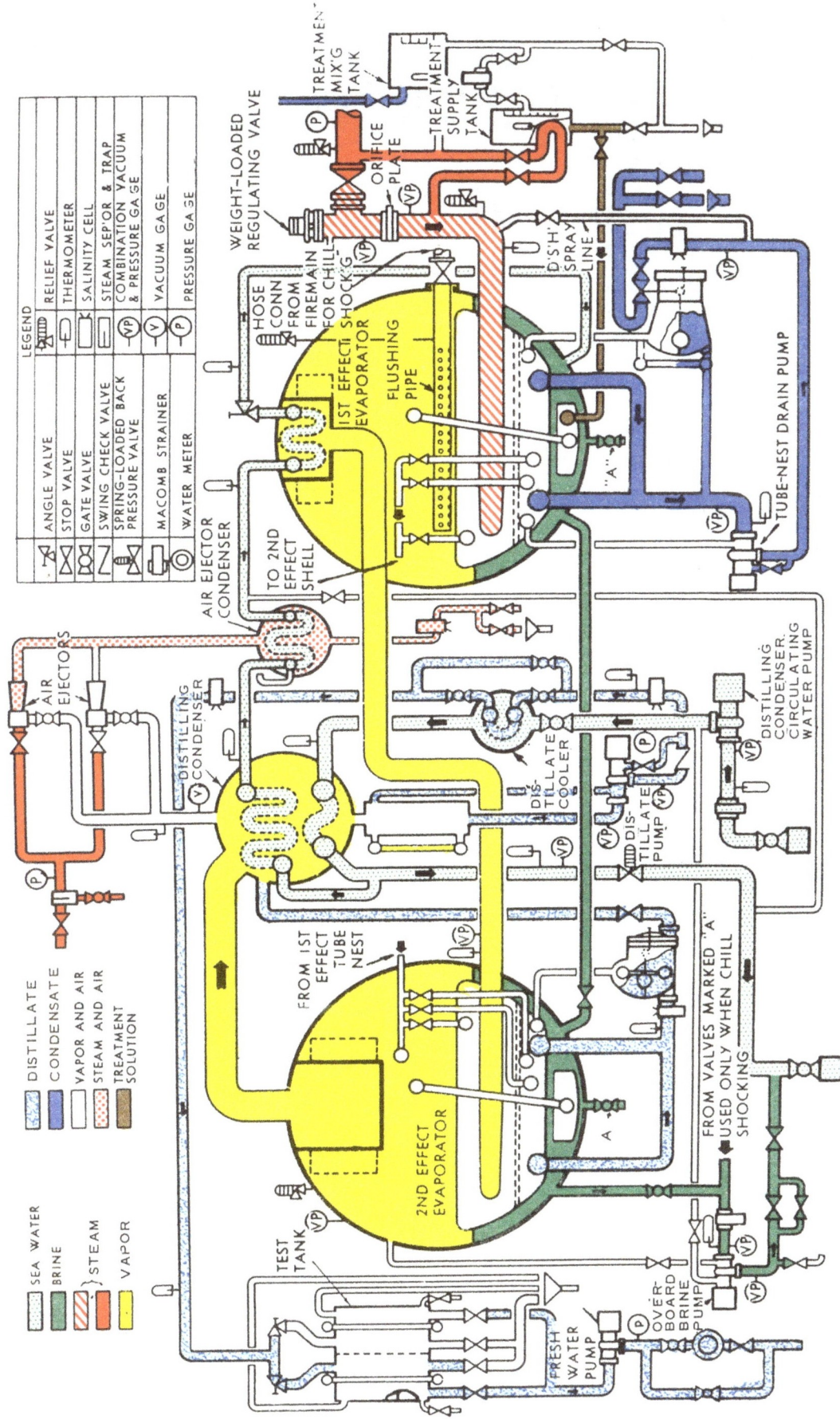

47.116

LEGEND

⋈	ANGLE VALVE	⋈	RELIEF VALVE
⋈	STOP VALVE		THERMOMETER
⋈	GATE VALVE		SALINITY CELL
	SWING CHECK VALVE		STEAM SEP'OR & TRAP
	SPRING-LOADED BACK PRESSURE VALVE	(VP)	COMBINATION VACUUM & PRESSURE GAGE
	MACOMB STRAINER	(V)	VACUUM GAGE
	WATER METER	(P)	PRESSURE GAGE

SEA WATER

BRINE

STEAM

VAPOR

DISTILLATE

CONDENSATE

VAPOR AND AIR

STEAM AND AIR

TREATMENT SOLUTION

TREATMENT MIX'G TANK

TREATMENT SUPPLY TANK

ORIFICE PLATE

WEIGHT-LOADED REGULATING VALVE

HOSE CONN FROM FIREMAIN FOR CHILL SHOCK'G

FLUSHING PIPE

1ST EFFECT EVAPORATOR

TO 2ND EFFECT SHELL

AIR EJECTOR CONDENSER

D'SH' SPRAY LINE

TUBE-NEST DRAIN PUMP

"A"

AIR EJECTORS

DISTILLING CONDENSER

DIS-TILLATE COOLER

DISTILLING CONDENSER CIRCULATING WATER PUMP

DIS-TILLATE PUMP

FROM 1ST EFFECT TUBE NEST

2ND EFFECT EVAPORATOR

A

FROM VALVES MARKED "A" USED ONLY WHEN CHILL SHOCKING

TEST TANK

FRESH WATER PUMP

OVER-BOARD BRINE PUMP

Figure 18-7.—Schematic diagram of two-shell double-effect distilling plant.

461

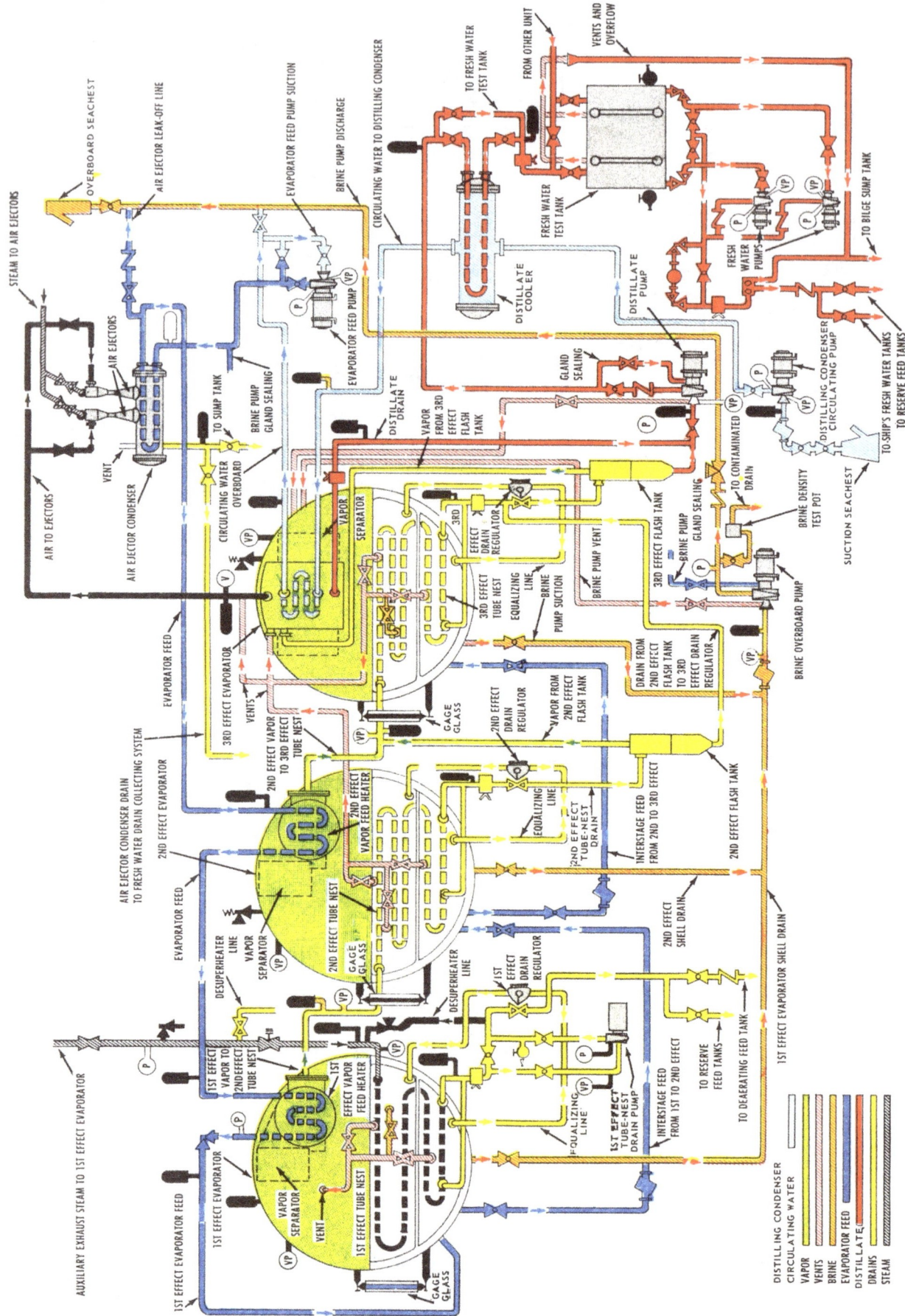

Figure 18-8.—Schematic drawing of a triple-effect distilling plant.

47.115

distilling condenser. A two-stage or three-stage air ejector, a distillate cooler, and a feed water heater are also provided. Feed water passes through the tubes of the distillate cooler, the stage distilling condenser, and the air ejector condenser. In each of these heat exchangers the feed picks up heat. The final heating is done by low pressure steam admitted to the shell of the feed water heater. From this heater the feed water enters the first-stage feed box and comes out through orifices into the flash chamber. As the heated feed water enters the chamber, a portion flashes or vaporizes because the pressure in the chamber is lower than the saturation pressure corresponding to the temperature of the hot feed. The vapor condenses on the tubes of the first-stage distilling condenser. The feed which does not vaporize in the first chamber passes to the second chamber. The process is repeated in each stage and the brine remaining in the last stage is removed by the brine overboard pump. Vapor formed in each stage passes through a vapor separator and into the stage distilling condenser, where it is condensed into distillate. The distillate passes through a loop seal on its way to the distilling condenser of the next stage. The distillate pump removes the distillate from the last stage and discharges it through the distillate cooler and the solenoid-operated dump valve to the ship's tanks.

The general arrangement of a two-stage flash-type unit is shown in figure 18-9; a five-stage unit is shown in figure 18-10. The major circuits are shown in each illustration.

Vertical Basket Distilling Units

Some recent ships are equipped with vertical basket distilling units. A unit of this type is shown in figure 18-11. The unit shown has two effects; however, some units of this type have more than two effects.

The vertical basket unit consists of two or more evaporators, a distiller condenser, vapor feed heaters, a distillate cooler, and air ejectors. The major difference between a vertical basket unit and a submerged tube unit is in the design of the evaporators. In the vertical basket unit, each evaporator consists of a vertical shell in which a deeply corrugated vertical basket is installed. Figure 18-12 shows a sectional view of the evaporator and basket.

Low pressure steam is admitted to the inside of the first-effect basket. This steam boils the feed water in the space between the outside of the basket and the shell of the evaporator. The condensate resulting from the condensation of steam drains downward and is returned to the boiler feed system. The vapor generated from the boiling sea water feed passes through the cyclonic separator above the evaporation section, where most of the entrained liquid particles are removed from the vapor by centrifugal force. The vapor continues on through the second vapor separator (called the "snail"), where the remaining water droplets are separated from the vapor. The liquid particles from both of these separators drain downward and become part of the brine drains.

The vapor generated in the first-effect shell passes from the steam dome of the first-effect shell. It goes through the vapor feed water heater and then enters the steam chest and evaporator basket of the second-effect shell. The first-effect vapor boils the second-effect feed and thus causes the generation of second-effect vapor. The second-effect vapor goes through the cyclonic separator and the snail in the second-effect shell. From the steam dome, this vapor then goes to the distilling condenser, where the vapor is condensed on the outside of the tubes. The second-effect distillate drains down and collects in the flash tank.

As the first-effect vapor is being used to boil the second-effect feed, some of the vapor condenses. This distillate drains downward into the second-effect steam chest and is discharged to the flash tank at the bottom of the distilling condenser, where it mixes with the distillate formed from the second-effect vapor. The distillate is removed from the flash tank by the distillate pump and is discharged through the distillate cooler and the solenoid-operated dump valve to the ship's tanks. Should the salinity of the distillate exceed 0.065 epm, the dump valve automatically dumps the distillate to the bilges.

Sea water flows through the tubes of the distillate cooler and the distilling condenser, creating a suction for the brine pump and maintaining a back pressure for the feed system. About 25 percent of the sea water passes through supplementary heating sections in the distilling condenser to the air ejector condenser, and feeds the evaporator shells in parallel. As the sea water passes through the air ejector condenser, it condenses the air ejector steam; the resulting condensate drains to an atmospheric drain tank.

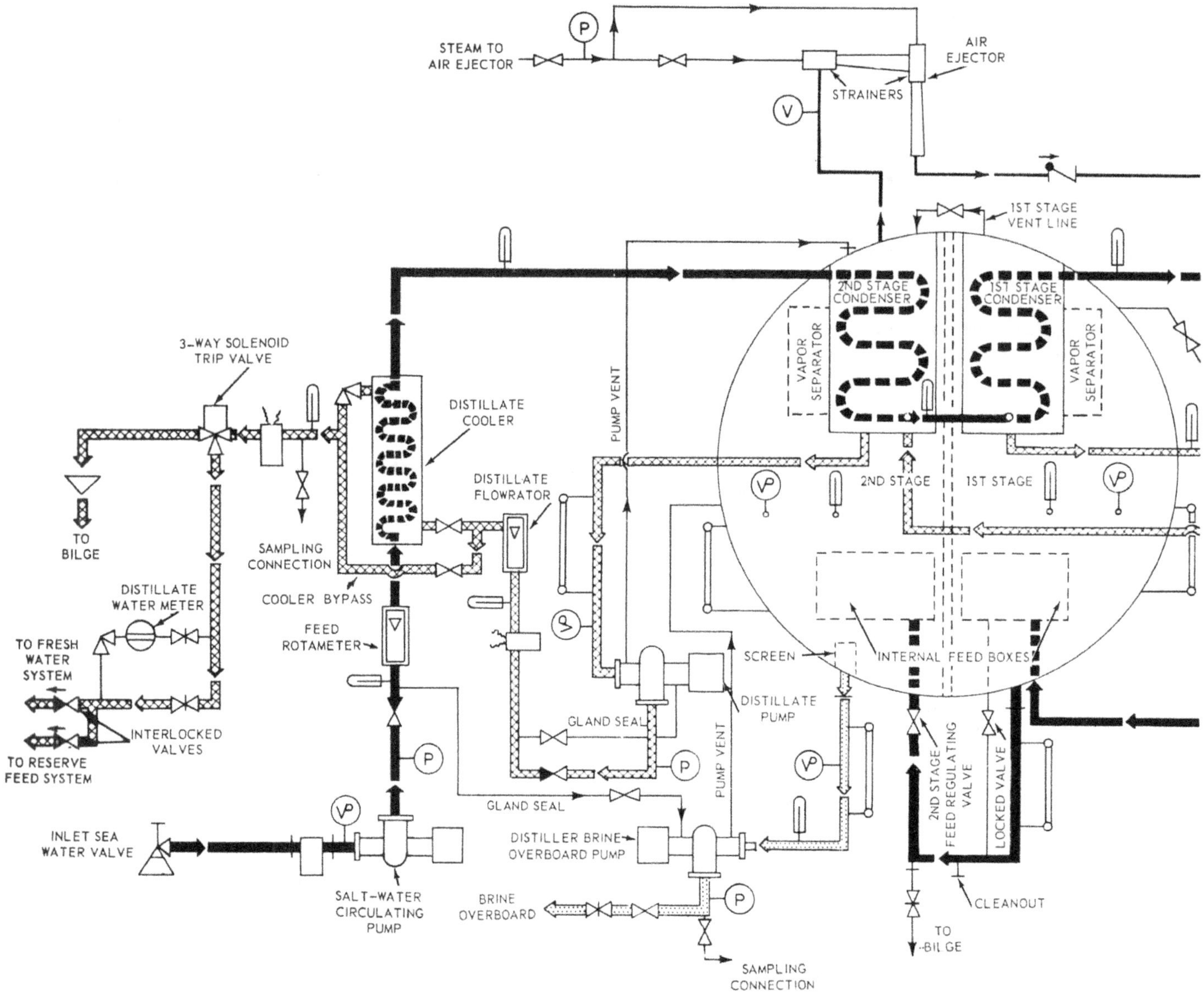

47.131

Figure 18-9.—General arrangement of two-stage flash-type distilling plant.

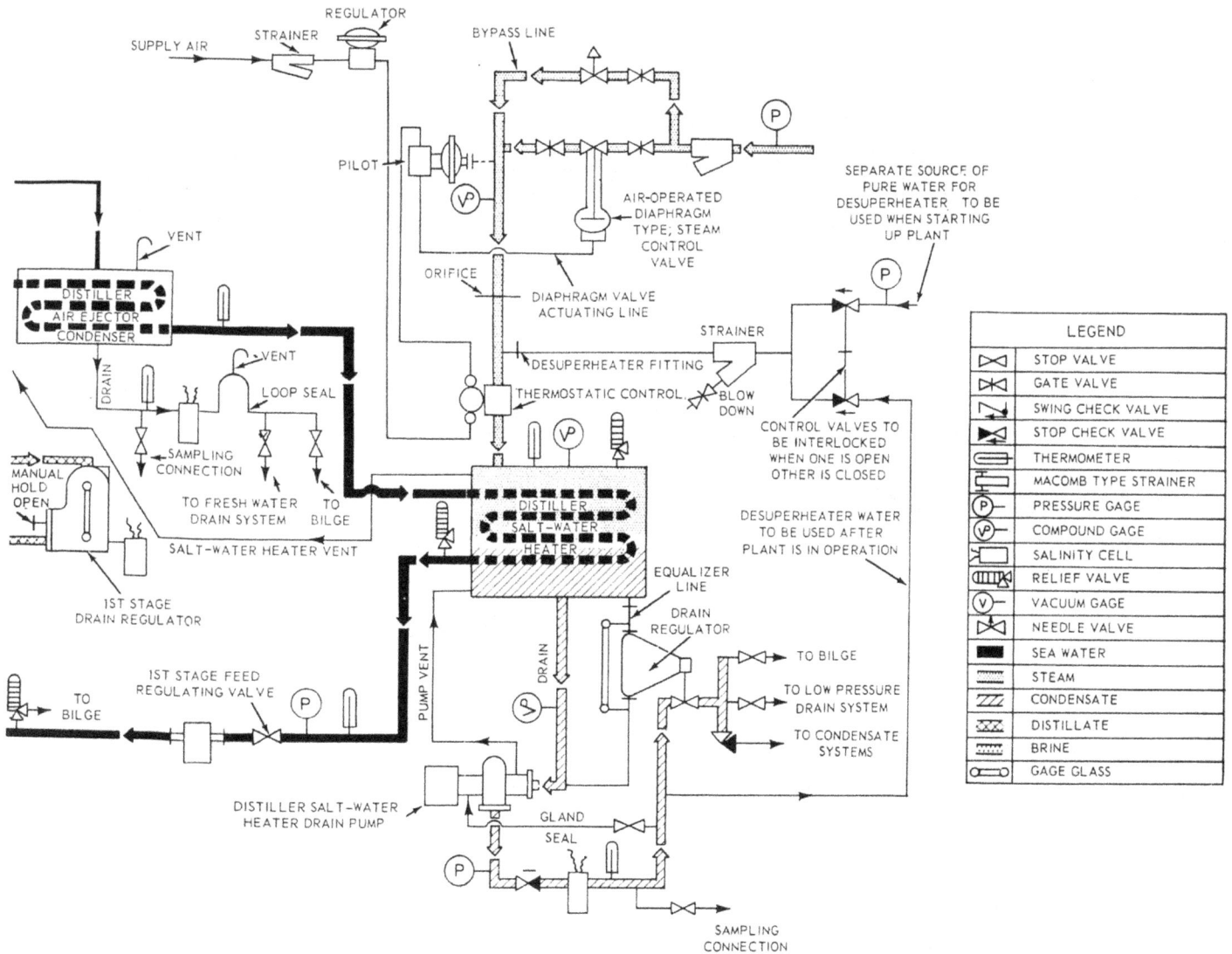

47.131

Figure 18-9.—General arrangement of two-stage flash-type distilling plant--continued.

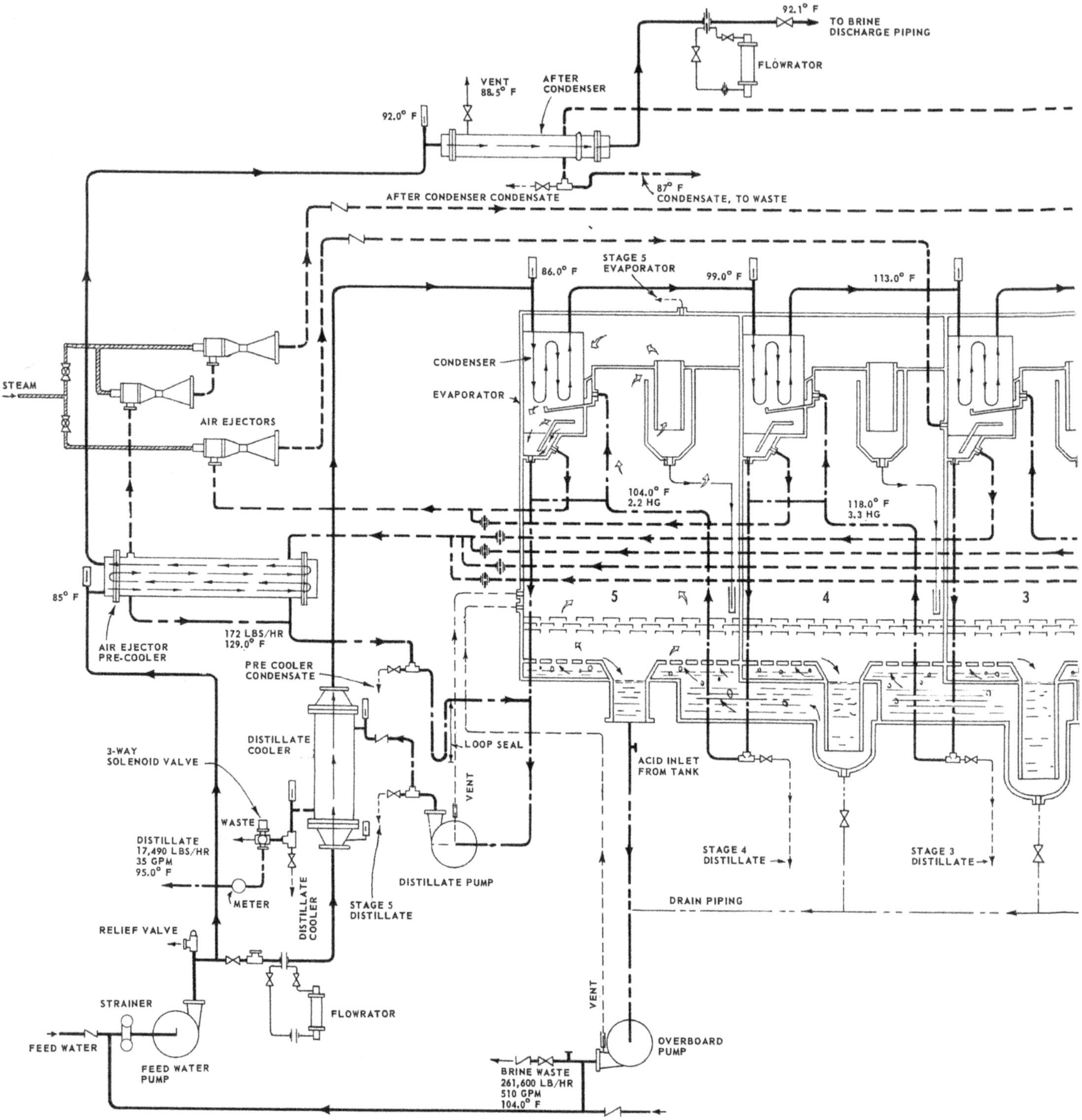

Figure 18-10.—General arrangement of five-stage flash-type distilling plant.

96.30

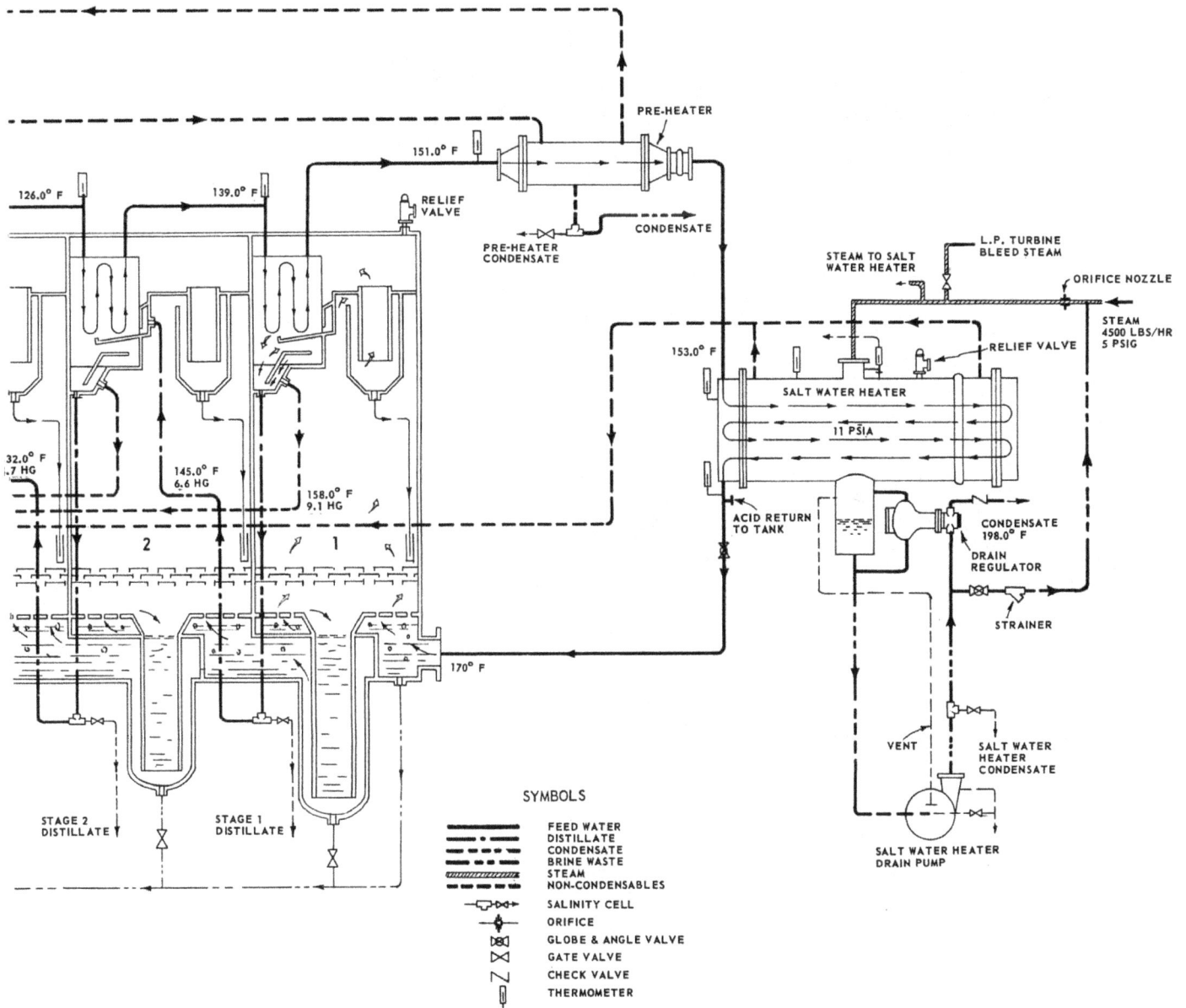

96.30

Figure 18-10.—General arrangement of five-stage flash-type plant--continued

47.126

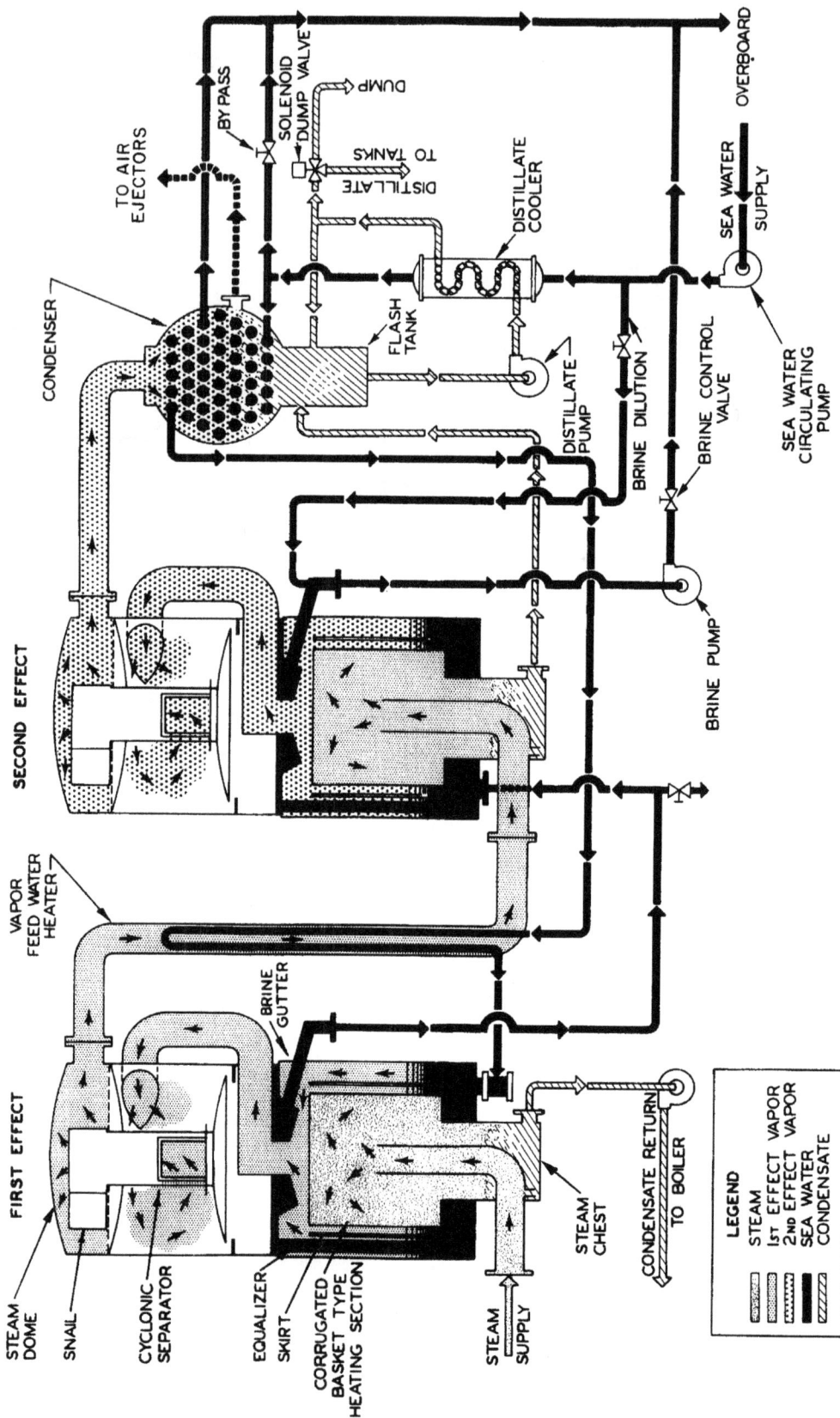

Figure 18-11.—Vertical basket double-effect distilling plant.

468

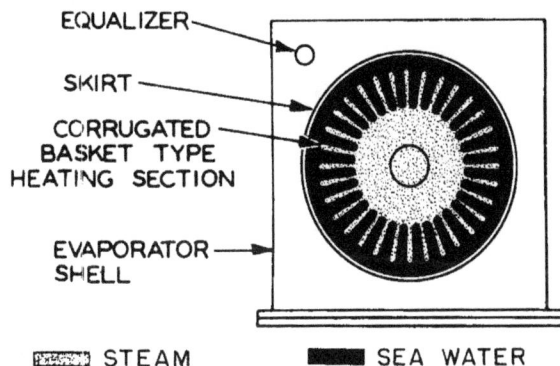

47.127

Figure 18-12.—Sectional view of evaporator and basket in vertical basket distilling plant.

DISTILLING PLANT OPERATION

Although a detailed discussion of distilling plant operation is beyond the scope of this text, certain operational considerations should be noted. The factors mentioned here apply primarily (although not exclusively) to low pressure steam distilling units.

Naval distilling plants are designed to produce distillate of very high quality. The chloride content of distillate discharged to the ship's tanks must not exceed 0.065 equivalents per million. Any distilling unit which cannot produce distillate of this quality is not considered to be operating properly.

Steady operating conditions are essential to the satisfactory operation of a distilling unit. Fluctuations in the pressure and temperature of the first-effect generating steam will cause fluctuations of pressure and temperature throughout the entire unit. Such fluctuations may cause priming, with increased salinity of the distillate, and may also cause erratic operation of the feed and brine pump. Rapid fluctuations of pressure in the last effect tend to cause priming.

To achieve satisfactory operation of a distilling unit, it is necessary to maintain the designed vacuum in all effects. When the unit is operated at less than the designed vacuum, the heat level rises throughout the unit and there is an increased tendency toward scale formation. Scale formation is highly undesirable, since scale interferes with heat transfer and thus reduces the capacity of the unit. Excessive scale formation may also impair the quality of the distillate.

Various methods have been used to retard scale formation in distilling units. In the past, a common method was the continuous injection of a solution of Navy boiler compound and cornstarch into the distilling unit. The boiler compound tends to minimize the formation of scale, and the cornstarch tends to minimize priming.

The boiler compound and cornstarch method of treatment is not fully effective in preventing scale formation, however, and daily removal of scale is required when this method is used. The removal of scale is accomplished by a procedure called chill shocking. For chill shocking, the unit is secured and pumped dry while it is still hot. Then cold sea water is introduced, and the resulting thermal shock causes scale to flake off and fall to the bottom of the tube nest. The unit is then pumped dry, the loose scale is removed, and the unit is filled with water and started up again.

Chill shocking is an effective way of removing scale, but it is somewhat laborious and time-consuming. A particular disadvantage of the chill shocking process is that it requires each operating distilling unit to be out of production for an hour or more each day. This can lead to serious water shortages under some circumstances.

A new chemical compound called HAGEVAP has been adopted as the standard compound for evaporator feed treatment. This compound has proved superior to the boiler compound and cornstarch previously used, and is now authorized for use in submerged tube, vertical basket, and five-stage flash-type distilling units; it does not appear to be necessary for two-stage flash-type units. Where the HAGEVAP treatment is used, chill shocking is not necessary because there is no scale formation (or practically none). The use of this compound requires the installation of certain equipment, including special pumps, tanks, and piping; authorization for such installation has been issued, and the alteration has been or will soon be made for all classes of ships having low pressure steam distilling plants (other than two-stage flash-type units).

The concentration of brine (or brine density, as it is called) has a direct bearing on the quality of the distillate. If the brine concentration is too low, there will be a loss in capacity and economy.

If the brine concentration is too high, there will be an increase in the rate of scaling of the evaporator tube surfaces, and the quality of the distillate may be impaired. The density of the brine overboard discharge should normally be maintained just under 1.5/32, and should never exceed this figure. Since the average sea water contains about 1 part of dissolved sea salts to 32 parts of water (by weight), the brine density should be just under 1 1/2 times that of the average sea water. Brine density is measured with a special kind of hydrometer [4] which is called a salinometer. Salinometers as shown in figure 18-13 are calibrated in thirty-seconds, on four separate scales which indicate the salinity of the brine at four different temperatures (110°, 115°, 120°, and 125° F).

Special restrictions are placed upon the operation of distilling units when the ship is operating in contaminated waters. Because most distilling plants operate at low pressures (and therefore low temperatures) the distillate is not sterilized by the boiling process in the evaporators and may contain dangerous micro-organisms or other matter harmful to health. All water in harbors, rivers, inlets, bays, landlocked waters, and the open sea within 10 miles of the entrance to such waters must be considered contaminated unless a specific determination to the contrary is made. In other areas, contamination may be declared to exist by the fleet surgeon or his representatives, as local conditions may warrant. When the ship is operating in contaminated waters, the distilling units must be operated in strict accordance with special procedures established by the Naval Ship Systems Command.

[4] Hydrometers are discussed in chapter 7 of this text.

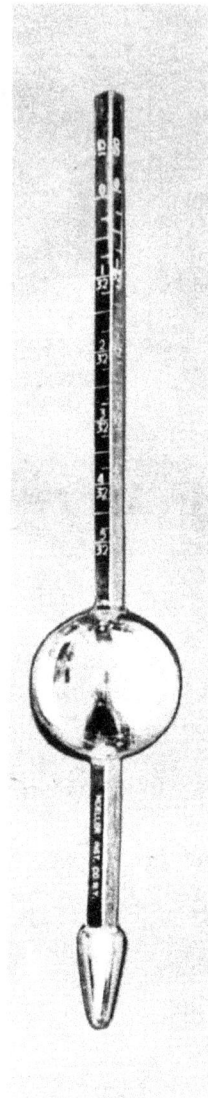

47.136X

Figure 18-13.—Salinometer.

470

CHAPTER 19

REFRIGERATION AND AIR CONDITIONING PLANTS

Refrigeration equipment is used aboard ship for a number of purposes, including the refrigeration of ship's stores, the refrigeration of cargo, the cooling of water, and the conditioning of air for certain spaces. The distinction between refrigeration and air conditioning should be noted. Refrigeration is only a cooling process; air conditioning is a process of treating air so as to simultaneously control its temperature, humidity, cleanliness, and distribution to meet the requirements of the conditioned spaces.

REFRIGERATION

The purpose of refrigeration is to cool spaces, objects, or materials and to maintain them at temperatures below the temperature of the surrounding atmosphere. In order to produce a refrigeration effect, it is merely necessary to expose the material to be cooled to a colder object or environment and allow heat to flow in its "natural" direction—that is, from the warmer material to the colder material. For example, a pan of hot water placed on a cake of ice will be cooled by the flow of heat from the hot water to the ice. We can maintain this refrigeration effect as long as the ice lasts. But no matter how much ice we have, we cannot produce a refrigeration effect any greater than the cooling of the water to $32°$ F. We cannot, for example, cause the water to freeze by this method, since freezing would require the removal of the latent heat of fusion from the water after it had been cooled to $32°$ F; and for this process we would need a temperature difference that does not exist when both the water and the ice are at $32°$ F. When the purpose of refrigeration is the production of ice or the maintenance of temperatures lower than $32°$ F at atmospheric pressure, it is obvious that ice is not a suitable refrigerant.

Refrigeration is a process involving the flow of heat, and is therefore a thermodynamic process. From previous discussion in this text, we may surmise that a closed cycle[1] would be most practicable for a large-scale refrigeration system. When we try to visualize such a cycle, however, it may appear at first glance that the cycle will have to run backwards. Thus far in this text, we have been primarily concerned with a closed cycle in which thermal energy (in the form of heat) is converted into mechanical energy (in the form of work). Now, instead of wanting to convert heat into work, we want to remove heat from a body and we want to continue to remove heat from this body even after its temperature has been lowered below that of its surroundings, in order to maintain the body at its lowered temperature. In other words, we want to extract heat from a cold body and discharge it to a warm area.

The question is: How can this be done, since we know from the second law of thermodynamics that heat cannot, of itself, flow from a colder body or region to a warmer one? It is entirely possible to extract heat from a body at a low temperature and discharge it to a body or region at a higher temperature, provided a suitable expenditure of energy is made to accomplish this. The energy supplied to the refrigeration cycle for this purpose is in the form of work (mechanical energy) done on the working fluid (refrigerant)

---

[1]Thermodynamic cycles are discussed in chapter 8 of this text. A closed cycle is one in which the working fluid never leaves the system except through accidental leakage. Instead, the working fluid undergoes a series of processes which are of such a nature that the fluid is returned periodically to its initial state and is then used again.

by a compressor.[2] In a refrigeration cycle, the refrigerant must alternate between low temperatures and high temperatures. When the refrigerant is at a low temperature, heat flows from the space or object to be cooled to the refrigerant. When the refrigerant is at a high temperature, heat flows from the refrigerant to a condenser. The energy supplied as work is used to raise the temperature of the refrigerant to a high enough value so that the refrigerant will be able to reject heat to the condenser. This point is discussed in more detail later in this chapter, but should be noted now since it is basic to the understanding of a mechanical refrigeration cycle.

Because the energy transformations in a refrigeration cycle occur in an order that is precisely the reverse of the sequence in a power cycle, the refrigeration cycle is sometimes said to be one in which heat is pumped "uphill." This view of a refrigeration cycle is entirely legitimate, provided the "reverse order" of energy transformations does not imply actual thermodynamic reversibility. True thermodynamic reversibility is here, as elsewhere in the observable world, considered to be an impossibility. A refrigeration cycle does not give us something for nothing. Instead, we must put energy into the cycle in order to extract heat at a low temperature and discharge it at a higher temperature.

DEFINITION OF TERMS

Some of the standard terms used in the discussion of refrigeration are defined in this section. A few of these terms have been defined in chapter 8 of this text but are briefly noted here because of their importance in the study of refrigeration.

UNIT OF HEAT.—The British thermal unit (Btu) is the standard unit of heat measurement used in refrigeration, as in most other engineering applications. By definition, 1 Btu is equal to 778.26 foot-pounds.

SPECIFIC HEAT.—The specific heat of a substance is the quantity of heat required to raise the temperature of unit mass of the substance 1 degree. In British systems of measurement, specific heat is expressed in Btu per pound per degree Fahrenheit.

SENSIBLE HEAT.—Sensible heat is the term used to identify heat that is reflected in a change of temperature.

LATENT HEAT OF VAPORIZATION.—The heat required to change a liquid to a gas (or, on the other hand, the heat which must be removed from a gas in order to condense it to a liquid) without any change in temperature is called the latent heat of vaporization.

LATENT HEAT OF FUSION.—The heat which must be removed from a liquid in order to change it into a solid (or, on the other hand, the amount of heat which must be added to a solid to change it to a liquid) without any change in temperature is called the latent heat of fusion.

REFRIGERATING EFFECT.—Since the heat removed from an object that is being refrigerated is absorbed by the refrigerant, the refrigerating effect of a refrigeration cycle is defined as the heat gain per pound of refrigerant.

REFRIGERATION TON.—The unit which measures the amount of heat removal and thereby indicates the capacity of a refrigeration system is known as the refrigeration ton. The refrigeration ton is based on the cooling effect of 1 ton (2000 pounds) of ice at 32° F melting in 24 hours. The latent heat of fusion of ice (or water) is approximately 144 Btu. Therefore, the number of Btu required to melt one ton of ice is 144 x 2000, or 288,000 Btu. The standard refrigeration ton is defined as the transfer of 288,000 Btu in 24 hours. On an hourly basis, the refrigeration ton is 12,000 Btu per hour (288,000 divided by 24 equals 12,000).

It should be noted that the refrigeration ton is not necessarily a measure of the ice-making capacity of a machine, since the amount of ice that can be made depends upon the initial temperature of the water and other factors.

COEFFICIENT OF PERFORMANCE.—The coefficient of performance of a refrigeration cycle is essentially comparable to the thermal

---

[2] A compressor provides the required energy in a vapor-compression refrigeration cycle, which is the cycle most commonly used in naval refrigeration plants. Other kinds of refrigeration cycles use other forms of energy to accomplish the same purpose— namely, to raise the temperature of the refrigerant after it has absorbed heat from the space or object to be cooled.

efficiency of a power cycle. The thermal efficiency of a power cycle is given by the equation

$$\text{thermal efficiency} = \frac{\text{work output}}{\text{heat input}}$$

Since thermal efficiency is a function of absolute temperature alone in the Carnot cycle, the equation may also be given as

$$\text{thermal efficiency} = \frac{T_s - T_r}{T_s}$$

where T_s is the absolute temperature at the heat source and T_r is the absolute temperature at the heat receiver.

For the refrigeration cycle, the coefficient of performance is given by the equation

$$\text{coefficient of performance} = \frac{\text{refrigerating effect}}{\text{work input}}$$

which, as in the power cycle, can be shown to be a function of absolute temperature alone.

THE R-12 PLANT

The refrigeration system most commonly used in the Navy utilizes R-12 as the refrigerant.[3] Chemically, R-12 is dichlorodifluoromethane (CCL_2F_2). The boiling point of R-12 is so low that the subtance cannot exist as a liquid unless it is confined and put under pressure; for example, R-12 boils at -21° F at atmospheric pressure, at 0° F at 9.17 psig, at 50 °F at 46.69 psig, and at 100 °F at 116.9 psig. Because of its low boiling point, R-12 is well suited for use in refrigeration systems designed for only moderate pressures. It also has the advantage of being practically nontoxic, nonflammable, nonexplosive, and noncorrosive; and it does not poison or contaminate foods.

The R-12 refrigeration system is classified as a mechanical system of the vapor-compression type. It is a mechanical system because the energy input is in the form of mechanical energy (work). It is a vapor-compression system because compression of the vaporized refrigerant is the process which allows the refrigerant

---

[3] In accordance with recent policy, refrigerants used in the Navy are no longer identified by trade names. Instead, they are identified by the letter R followed by the appropriate number, or else they are identified simply as "refrigerants." For example, the refrigerant formerly known as "Freon 12" is now identified either as R-12 or simply as a refrigerant.

to discharge heat at a relatively high temperature.

The R-12 Cycle

The basic cycle of an R-12 refrigeration cycle is shown shcematically in figure 19-1. As an introduction to the system, it will be helpful to trace the refrigerant through the entire cycle, noting especially the points at which the refrigerant changes from liquid to vapor and from vapor to liquid, and noting also the concomitant flow of heat in one direction or another.

As shown in figure 19-1, the cycle has two pressure sides. The low pressure side extends from the orifice of the thermostatic expansion valve up to and including the intake side of the compressor cylinders. The high pressure side extends from the discharge side of the compressor to the thermostatic expansion valve. The condensing and evaporating pressures and temperatures indicated in figure 19-1 are not standard for all refrigeration plants, since pressures and temperatures are established as part of the design of any refrigeration system. It should be noted, also, that the pressures and temperatures shown in figure 19-1 are theoretical rather than actual values, even for this particular system. If the system were in actual operation, the pressures and temperatures would vary slightly because they are dependent upon the temperature of the cooling water entering the condenser, the amount of heat absorbed by the refrigerant in the evaporator, and other factors.

Liquid R-12 enters the thermostatic expansion valve at high pressure, from the high pressure side of the system. The refrigerant leaves the outlet of the expansion valve at a much lower pressure and enters the low pressure side of the system. Because of the relatively low pressure, the liquid refrigerant begins to boil and to flash into vapor.

From the thermostatic expansion valve, the refrigerant passes into the cooling coil (evaporator). The boiling point of the refrigerant under the low pressure in the evaporator is extremely low—much lower than the temperature of the spaces in which the cooling coil is installed. As the liquid boils and vaporizes, it picks up its latent heat of vaporization from the surroundings, thereby cooling the space. The refrigerant continues to absorb heat until all the liquid has been vaporized and the vapor has become slightly superheated. As a rule, the amount of superheat is about 10° F.

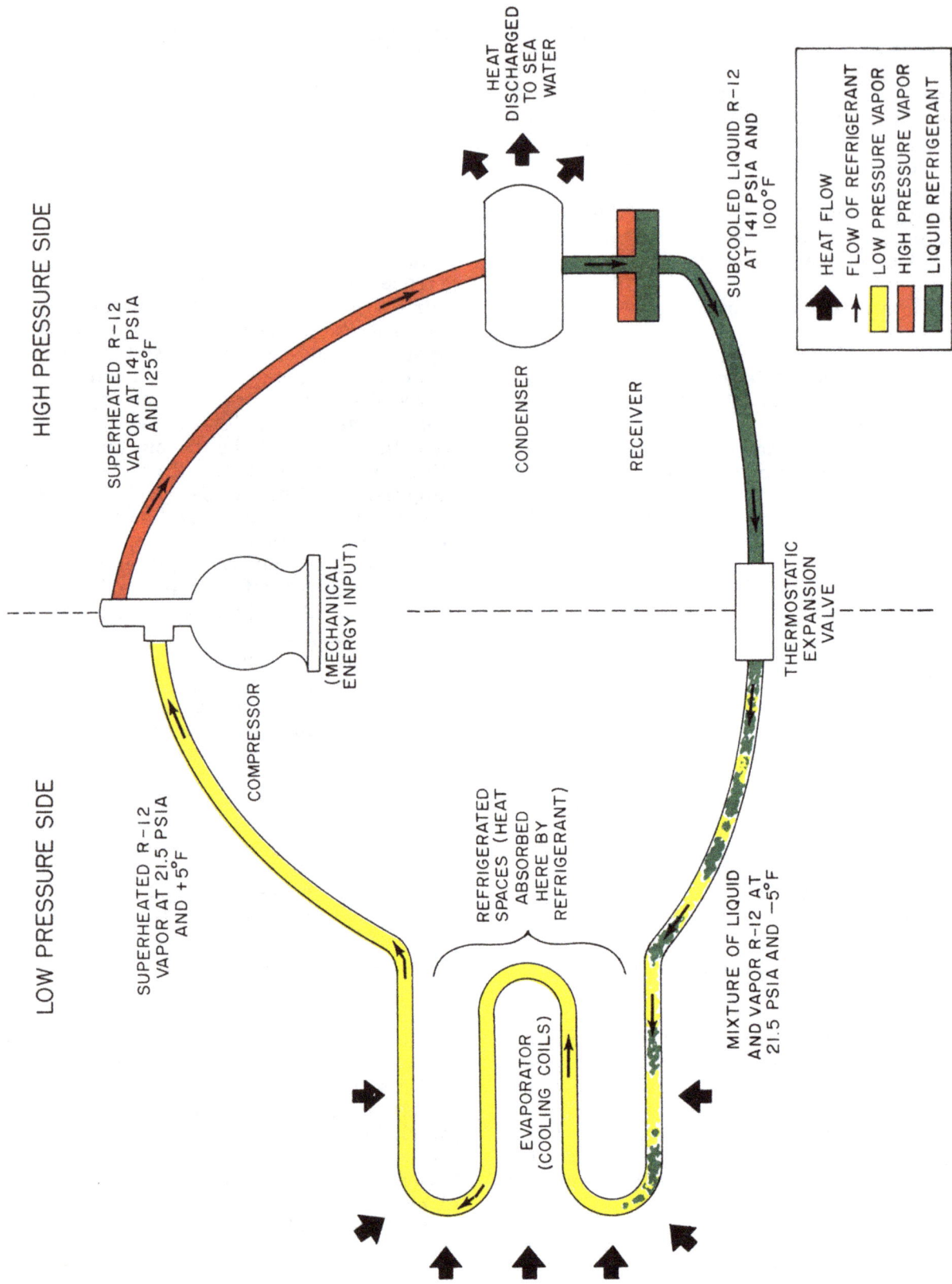

HIGH PRESSURE SIDE

LOW PRESSURE SIDE

HEAT DISCHARGED TO SEA WATER

SUPERHEATED R-12 VAPOR AT 141 PSIA AND 125°F

SUPERHEATED R-12 VAPOR AT 21.5 PSIA AND +5°F

(MECHANICAL ENERGY INPUT)

COMPRESSOR

CONDENSER

RECEIVER

SUBCOOLED LIQUID R-12 AT 141 PSIA AND 100°F

THERMOSTATIC EXPANSION VALVE

REFRIGERATED SPACES (HEAT ABSORBED HERE BY REFRIGERANT)

EVAPORATOR (COOLING COILS)

MIXTURE OF LIQUID AND VAPOR R-12 AT 21.5 PSIA AND −5°F

HEAT FLOW

FLOW OF REFRIGERANT

LOW PRESSURE VAPOR

HIGH PRESSURE VAPOR

LIQUID REFRIGERANT

47.90

Figure 19-1.—Schematic representation of R-12 refrigeration cycle.

474

The refrigerant leaves the evaporator as a low pressure superheated vapor, having absorbed heat and thus cooled the space. The remainder of the cycle is concerned with disposing of this heat and getting the refrigerant back into a liquid state so that it can again vaporize in the evaporator and thus again absorb heat.

The low pressure superheated vapor is drawn out of the evaporator to the suction side of the compressor. The compressor is the unit which keeps the refrigerant circulating through the system. In the compressor cylinders, the refrigerant is compressed from a low pressure vapor to a high pressure vapor, and its temperature rises accordingly.

The high pressure R-12 vapor is discharged from the compressor to the condenser. Here the refrigerant condenses, giving up its superheat, its latent heat of vaporization, and its heat of compression to the cooling sea water which flows through the condenser tubes. The refrigerant, still at high pressure, is now a liquid again.

From the condenser, the refrigerant flows into a receiver, which serves as a storage place for the liquid refrigerant. From the receiver, the refrigerant goes to the thermostatic expansion valve and the cycle begins again.

From this brief summary of an R-12 vapor-compression refrigeration system, it may be seen that the cycle is indeed one in which heat is "pumped uphill" as a result of the arrangements which cause the refrigerant to go through successive phases of expansion, evaporation, compression, and condensation.

Major Components

The major components of a shipboard R-12 refrigeration plant are shown diagrammatically in figure 19-2. The primary parts of the system are the thermostatic expansion valve, the evaporator, the compressor, the condenser, and the receiver. Additional equipment required to complete the plant includes piping, pressure gages, thermometers, various types of control switches and control valves, strainers, relief valves, sight flow indicators, dehydrators, and charging connections. Figure 19-3 shows most of the components on the high pressure side of an R-12 system, as actually installed aboard ship.

In the following discussion of the major components of an R-12 system, we will treat the system as though it had only one evaporator, one compressor, and one condenser. As may be

seen from figure 19-2, however, a shipboard refrigeration system may (and, indeed, usually does) include more than one evaporator and may include additional compressor and condenser units to provide operational flexibility and to protect against loss of refrigerating capacity.

THERMOSTATIC EXPANSION VALVE.—The thermostatic expansion valve, shown in figure 19-4, is essentially a reducing valve between the high pressure side and the low pressure side of the system. The valve is designed to proportion the rate at which the refrigerant enters the cooling coil to the rate of evaporation of the liquid refrigerant in the coil; the amount depends, of course, on the amount of heat being removed from the refrigerated space.

A thermal bulb for the thermostatic expansion valve is clamped to the cooling coil, near the outlet. The bulb contains R-12. Control tubing connects the bulb with the area above the diaphragm in the thermostatic expansion valve. When the temperature at the bulb rises, the R-12 expands and transmits a pressure to the diaphragm; this causes the diaphragm to be moved downward, thus opening the valve and allowing more refrigerant to enter the cooling coil. When the temperature at the bulb falls, the pressure above the diaphragm is decreased and the valve tends to close. Thus the temperature near the evaporator outlet controls the operation of the thermostatic expansion valve.

EVAPORATOR.—The evaporator consists of a coil of copper tubing installed in the space to be refrigerated. Figure 19-5 shows some of this tubing. The liquid R-12 enters the tubing at a very much reduced pressure and the boiling point is therefore very much lowered. In passing through the expansion valve, going from the high pressure side of the system to the low pressure side, some of the refrigerant boils and vaporizes because of the reduced pressure and some of the remaining liquid refrigerant is thereby cooled to its boiling point. Then, as the refrigerant passes through the evaporator, the heat flowing to the evaporator from the surrounding air causes the rest of the liquid refrigerant to boil and vaporize.

After the refrigerant has absorbed its latent heat of vaporization and all the liquid has been vaporized, the refrigerant continues to absorb heat until it has acquired about 10° F of superheat. The amount of superheat is determined by the amount of liquid refrigerant admitted to the

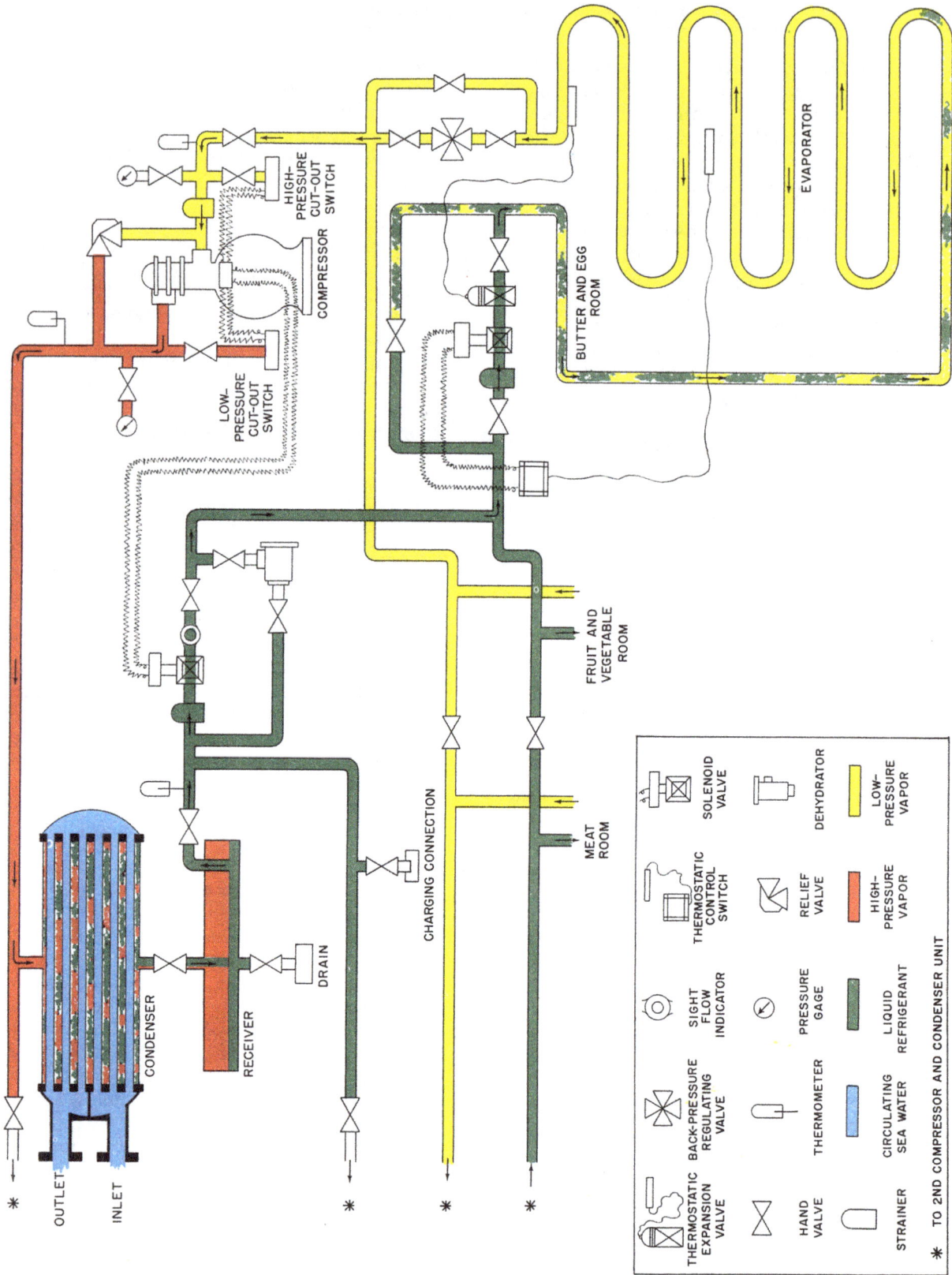

47.91

Figure 19-2.—Diagram of an R-12 refrigeration system.

476

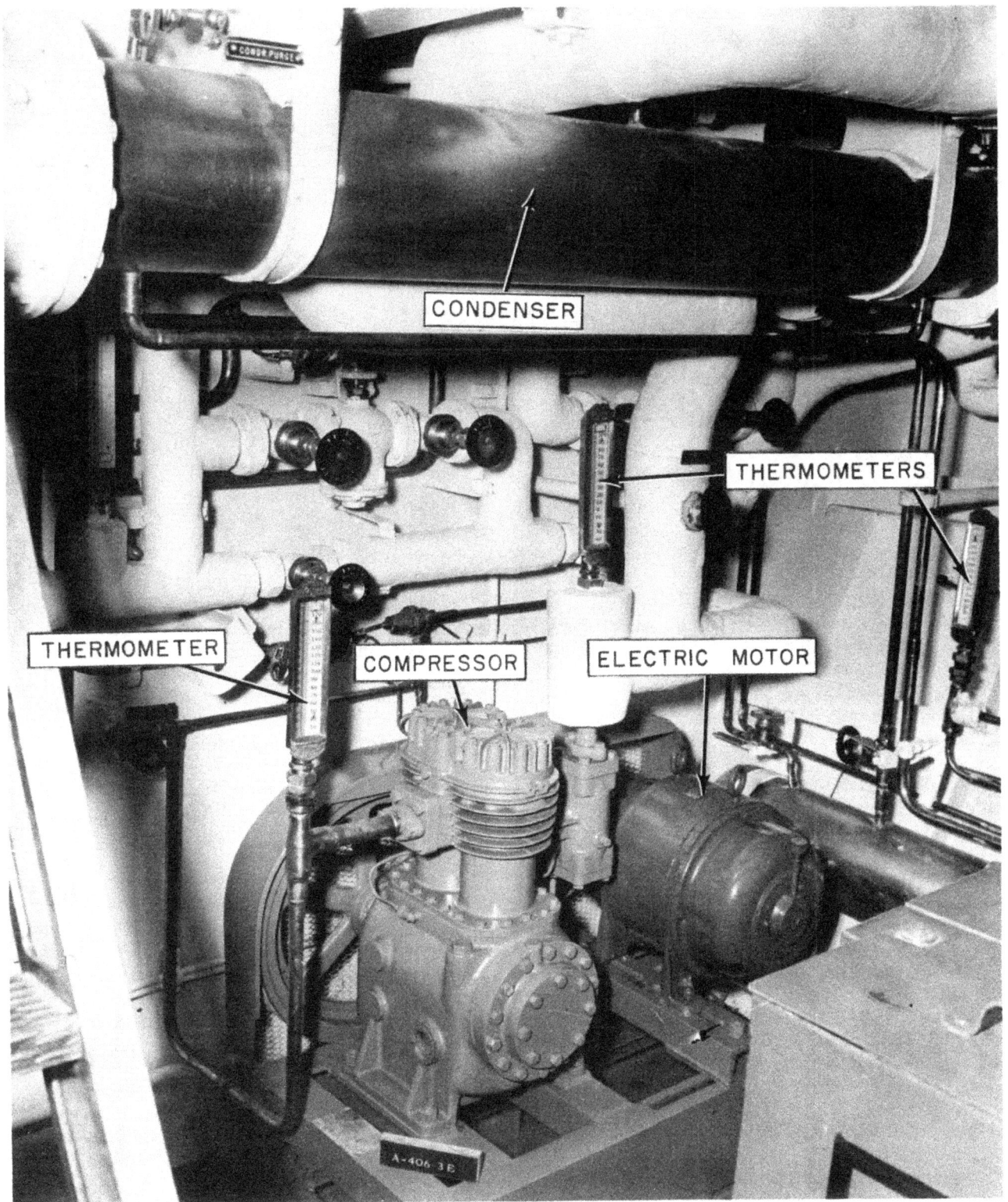

47.92

Figure 19-3.—High pressure side of R-12 installation aboard ship.

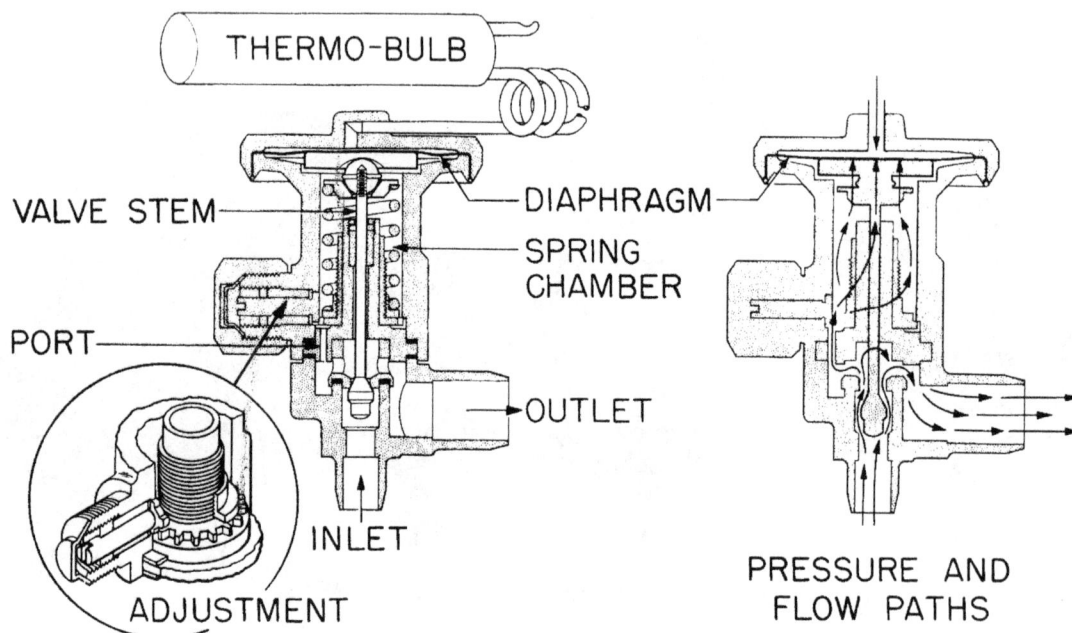

168.1

Figure 19-4.—Thermostatic expansion valve.

evaporator; and this, in turn, is controlled by the spring adjustment of the thermostatic expansion valve. About 10° F of superheat is considered desirable because it increases the efficiency of the plant and because it ensures the evaporation of all liquid, thus preventing liquid carryover into the compressor.

COMPRESSOR.—In a vapor-compression refrigeration system, the compressor is the unit that pumps heat "uphill" from the cold side to the hot side of the system.

The heat absorbed by the refrigerant in the evaporator must be removed before the refrigerant can again absorb latent heat in the evaporator. The only way in which the vaporized refrigerant can be made to give up the latent heat of vaporization that it absorbed in the evaporator is by condensation. In view of the relatively high temperature of the available cooling medium (sea water), the only way to make the vapor condense is by first compressing it.

The vapor drawn into the compressor is at very low pressure and very low temperature. In the compressor, both the pressure and the temperature are raised. Since an increase in pressure causes a proportional rise in temperature, and since the condensation point of a vapor is determined by the pressure, raising the pressure

of the vaporized refrigerant provides a condensation temperature high enough to permit the use of sea water as a cooling and condensing medium. In other words, the compressor raises the pressure of the vaporized refrigerant sufficiently high to permit heat transfer and condensation to take place in the condenser.

In addition to this primary function, the compressor also serves to keep the refrigerant circulating and to maintain the required pressure differential between the high pressure side and the low pressure side of the system.

Many different types of compressors are used in refrigeration systems. Figure 19-6 shows a motor-driven, single-acting, two-cylinder reciprocating compressor of a type commonly used in naval shipboard refrigeration plants.

CONDENSER.—The compressor discharges the high pressure, high temperature refrigerant vapor to the condenser, where it flows around the tubes through which sea water is being pumped. As the vapor gives up its superheat to the circulating sea water, the temperature of the vapor drops to the condensation point. As soon as the temperature of the vapor drops to its condensing point at the existing pressure, the vapor condenses and in the process gives up the latent heat of vaporization that it picked up in the evaporator. The refrigerant, now in liquid form,

47.93

Figure 19-5.—Evaporator tubing.

is subcooled slightly below its boiling point at this pressure to ensure that it will not flash into vapor.

A water-cooled condenser for an R-12 refrigeration system is shown in figure 19-7. Circulating water is obtained through a branch connection from the firemain or by means of an individual pump taking suction from the sea. A water regulating valve (not shown) is usually installed to control the flow of cooling water through the condenser. The purge connection shown in figure 19-6 is on the refrigerant side; it is used to remove air and other noncondensable gases that are lighter than the R-12 vapor.

Most condensers used in naval refrigeration plants are water cooled. However, some small units have air-cooled condensers. These consist of tubing with external fins to increase the heat transfer surface. Most air-cooled condensers have fans to ensure positive circulation of air around the condenser tubes.

RECEIVER.—The receiver, shown in figure 19-8, acts as a temporary storage space and surge tank for the liquid refrigerant which flows from the condenser. The receiver also serves as a vapor seal to prevent the entrance of vapor into the liquid line to the thermostatic expansion valve.

ACCESSORIES AND CONTROLS.—In addition to the five major components just described, a refrigeration system requires a number of controls and accessories. The most important of these are discussed briefly in the following paragraphs.

A dehydrator (or dryer) is placed in the liquid refrigerant line between the receiver and the thermostatic expansion valve. In older installations, such as the one shown in figure 19-2, bypass valves allow the dehydrator to be cut in or out of the system. In newer installations, the dehydrator is installed in the liquid refrigerant line without any bypass arrangement. A refrigerant dehydrator is shown in figure 19-9.

A solenoid valve is installed in the liquid line leading to each evaporator. Figure 19-10 shows a solenoid valve and the thermostatic control switch that operates it. The thermostatic control switch is connected by long flexible capillary tubing to a thermal bulb which is located in the refrigerated space. When the temperature in the refrigerated space drops to the desired point, the thermal bulb causes the thermostatic control switch to open, thereby closing the solenoid valve and shutting off all flow of liquid refrigerant to the thermostatic expansion valve. When the temperature in the refrigerated space rises above the desired point, the thermostatic control switch closes, the solenoid valve opens, and liquid refrigerant once again flows to the thermostatic expansion valve.

The solenoid valve and its related thermostatic control switch serve to maintain the proper temperature in the refrigerated space. However, we may wonder why the solenoid valve is necessary, since the thermostatic expansion valve controls the amount of refrigerant admitted to the evaporator. Actually, the solenoid valve is not necessary in systems having only one evaporator. In systems having more than one evaporator, where there is wide variation in load, the solenoid valve provides the additional control required to prevent spaces from becoming too cold at light loads.

In addition to the solenoid valve installed in the line to each evaporator, a large refrigeration plant usually has a main liquid line solenoid valve installed just after the receiver. If the compressor stops for any reason except normal suction pressure control, the main liquid line solenoid valve closes and prevents liquid refrigerant from flooding the evaporator and flowing to the compressor suction. Great damage to the compressor can result if liquid is allowed to enter the compressor suction.

Whenever several refrigerated spaces of varying temperatures are to be maintained by one compressor, an evaporator pressure

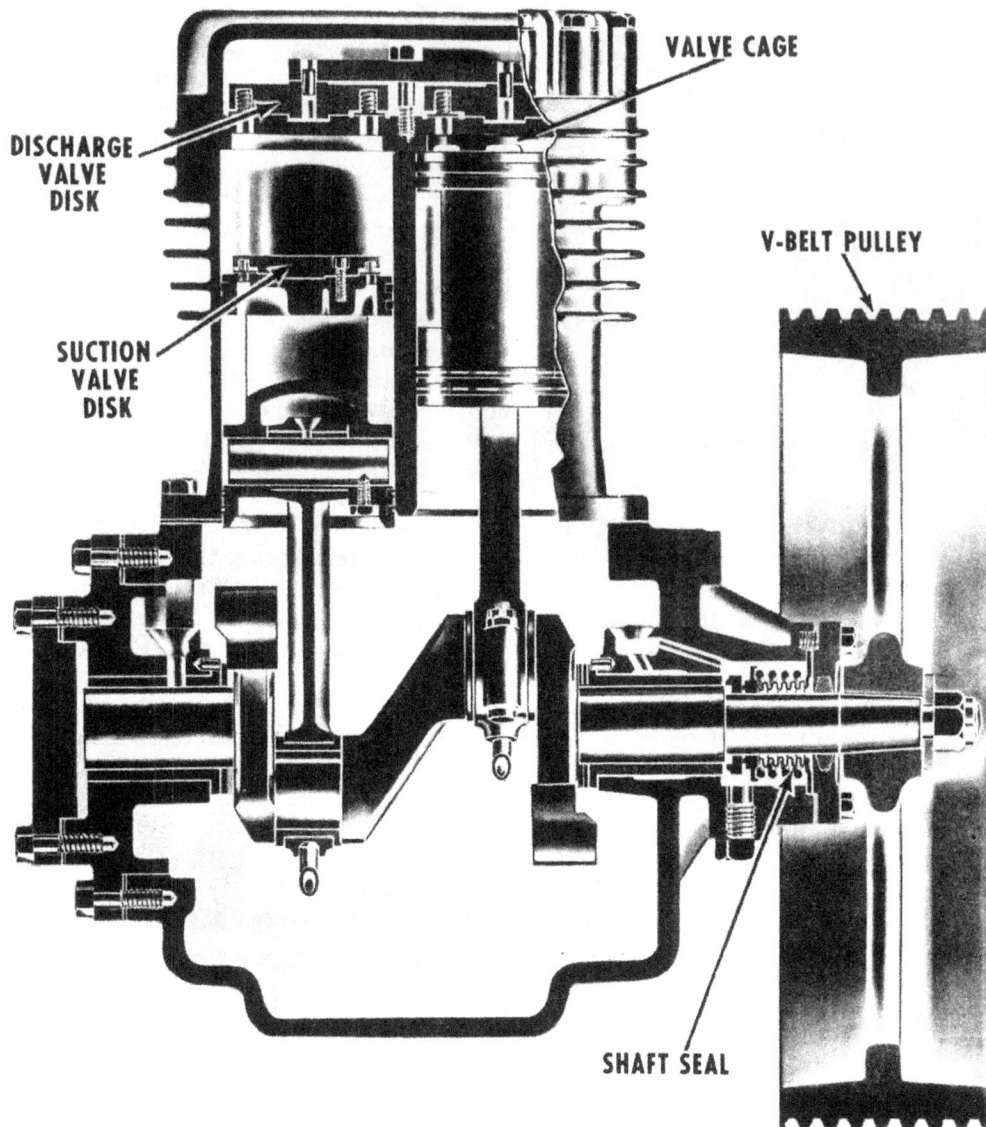

47.94

Figure 19-6.—Reciprocating compressor for R-12 refrigeration plant.

regulating valve is installed at the outlet of each evaporator except the evaporator in the space in which the lowest temperature is to be maintained. The evaporator pressure regulating valve is set to keep the pressure in the coil from falling below the pressure corresponding to the lowest temperature desired in that space.

The low pressure cutout switch is the control that causes the compressor to go on or off as required for the normal operation of the refrigeration plant. This switch is located on the suction side of the compressor and is actuated by pressure changes in the suction line. When the solenoid valves in the lines to the various evaporators are closed, so that the flow of refrigerant to the evaporators is stopped, the pressure of the vapor in the compressor suction line drops quickly. When the suction pressure has dropped to the desired pressure, the low pressure cutout switch causes the compressor motor to stop. When the temperature in the refrigerated space has risen enough to operate one or more of the solenoid valves, refrigerant is again admitted to the cooling coils and the compressor suction pressure builds up again. At the desired pressure,

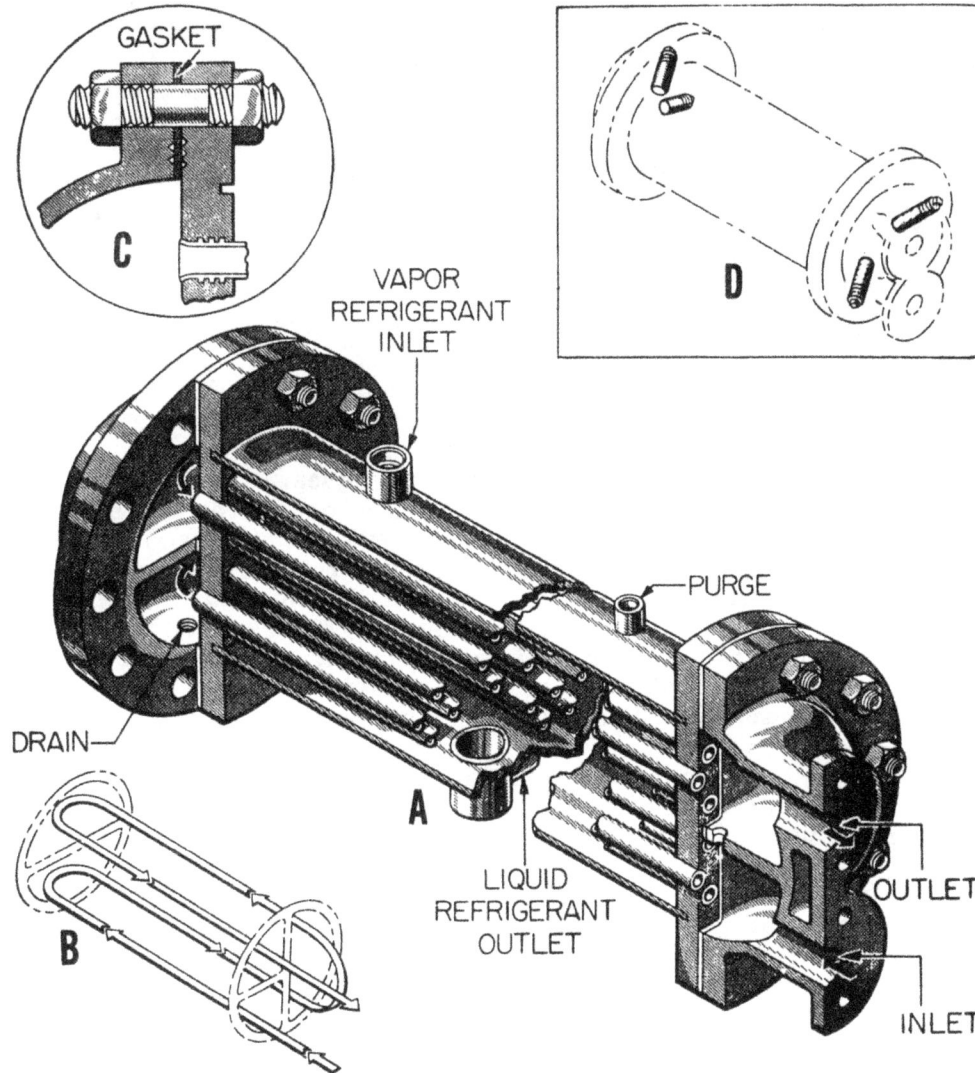

47.95

Figure 19-7.—Water-cooled condenser for R-12 refrigeration system. A. Cutaway view.
B. Water-flow diagram. C. Arrangement of head joints. D. Position of zincs.

the low pressure cutout switch closes, starting the compressor again and repeating the cycle.

A high pressure cutout switch is connected to the compressor discharge line to protect the high side of the system against excessive pressures. This switch is very similar to the low pressure cutout switch; however, the low pressure cutout switch is designed to close when the suction pressure reaches its upper normal limit, whereas the high pressure cutout switch is designed to open when the discharge pressure is too high. The high pressure cutout switch is normally set to stop the compressor

when the pressure reaches 160 psi and to start it again when the pressure drops to 140 psi. As previously noted, the low pressure cutout switch is the compressor control for normal operation of the plant; the high pressure cutout switch, on the other hand, is a safety device only and does not have control of compressor operation under normal conditions.

A spring-loaded relief valve is installed in the compressor discharge line as an additional precaution against excessive pressures. The relief valve is set to open at about 225 psi; therefore, it functions only in case of failure or

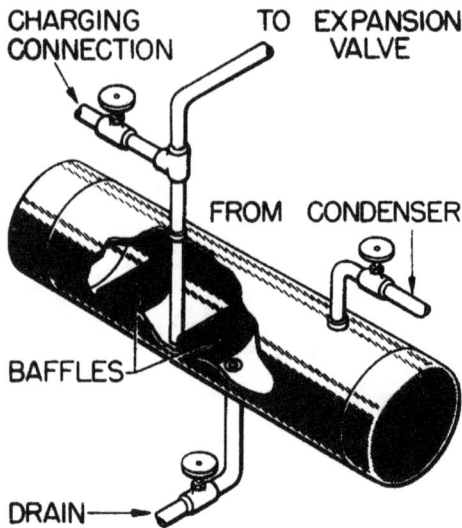

47.96
Figure 19-8.—Receiver for R-12
refrigeration system.

improper setting of the high pressure cutout switch. If the relief valve opens, it discharges high pressure vapor to the suction side of the compressor.

A water regulating valve, as shown in figure 19-11, is usually installed to control the quantity of circulating water flowing to the refrigerant condenser. The valve is located either at the inlet to the condenser or at the outlet from the condenser. The valve is actuated by the refrigerant pressure in the compressor discharge line; this pressure acts upon a diaphragm or a bellows arrangement which transmits motion to the valve stem. As the temperature of the circulating water increases, the temperature of the refrigerant vapor increases; this causes the pressure of the refrigerant to increase, and thereby raises the condensation point. When this occurs, the increased pressure of the refrigerant causes the water regulating valve to open wider, thus automatically permitting more circulating water to flow through the condenser. When the condenser is cooler than necessary, the water regulating valve allows less water to flow through the condenser. Thus the flow of cooling water through the condenser is automatically maintained at the rate actually required to condense the refrigerant under varying conditions of load and temperature.

A water failure switch is provided to stop the compressor in the event of failure of the circulating water supply. This is a pressure-

actuated switch, generally similar to the low pressure cutout switch and the high pressure cutout switch previously described. If the water failure switch should fail to function, the refrigerant pressure in the condenser would quickly build up to the point where the high pressure cutout switch would function.

Because of the solvent action of R-12, any particles of grit, scale, dirt, and metal that the system may contain are very readily circulated through the refrigerant lines. To avoid damage to the compressor from such foreign matter, a strainer is installed in the compressor suction connection. In addition, a liquid strainer is installed in the liquid line leading to each evaporator; these strainers serve to protect the solenoid valves and the thermostatic expansion valves.

A number of pressure gages and thermometers are used in refrigeration systems. A compound refrigerant gage is shown in figure 9-12. The temperature markings on this gage show the boiling point (or condensing point) of the refrigerant at each pressure; the gage cannot measure temperature directly. The dark pointer (which is actually red in color) is a stationary pointer that can be set manually to indicate the maximum working pressure. Other pressure gages and thermometers include a water pressure gage, installed in the circulating water line to the condenser, and standard thermometers of appropriate range, installed in the refrigerant lines.

Refrigerant piping is normally made of copper. Copper is particularly good for this purpose because it does not become corroded by the refrigerant, the internal surface is smooth enough to minimize friction, and the tubing is easily shaped to meet installation requirements.

AIR CONDITIONING

Air conditioning is a field that deals with the design, construction, and operation of equipment used in establishing and maintaining desirable indoor air conditions. It is the science of maintaining the atmosphere of an enclosure at any required temperature, humidity, and purity. As such, air conditioning involves the cooling, heating, dehumidifying, ventilating, and purifying of air.

Aboard ship, air conditioning serves to keep the ship's crew comfortable, alert, and physically fit. The temperature, humidity, cleanliness, quantity, and distribution of the conditioned air supply is a matter of vital concern.

COVER PLATE

SPRING

CARTRIDGE

FITTING SEAL PLUG

SHELL

DEHYDRANT

DISPERSION TUBE

END CAP

47.97

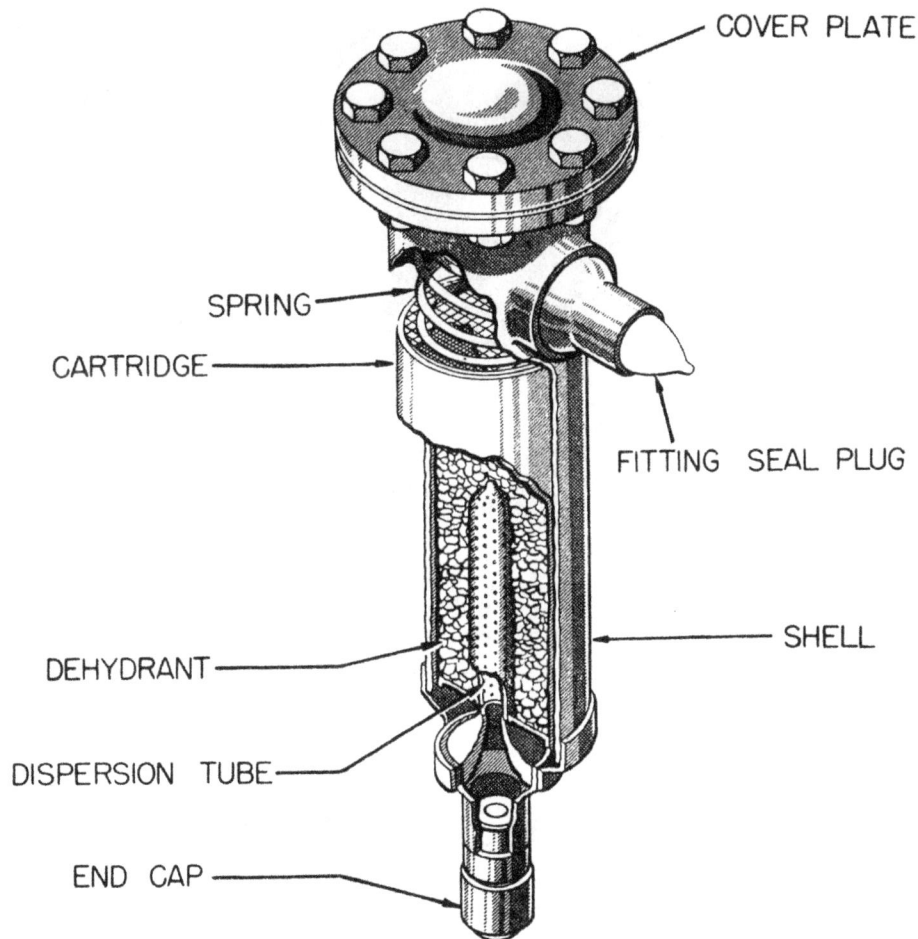

Figure 19-9.—Refrigerant Dehydrator.

The comfort and efficiency of the crew is not the only immediate reason for shipboard air conditioning. Mechanical cooling, heating, or ventilating must be provided for a number of spaces for a variety of reasons. Ammunition spaces must be kept below a certain temperature in order to prevent deterioration of the ammunition; gas storage spaces must be kept cool in order to prevent the buildup of excessive pressures in containers; electrical and electronic equipment must be maintained at certain temperatures, with controlled humidity, in air that is relatively free of dust and dirt.

PRINCIPLES OF AIR CONDITIONING

To achieve the objectives of air conditioning, it is necessary to take account of a number of factors. The principal factors that are important in connection with air conditioning are discussed in the following sections.

HUMIDITY.—The vapor content of the atmosphere is referred to as humidity. Excessive humidity and too little humidity both lead to discomfort and impaired efficiency; hence the measurement and control of the moisture content of the air is an important phase of air conditioning.

The air holds varying amounts of water vapor, depending upon the temperature of the air; the higher the temperature, the greater the amount of moisture the air can hold. For every temperature there is a definite limit as to the amount of moisture the air is capable of holding. When air attains the maximum amount of moisture which it can hold at a specified temperature, the air is said to be saturated.

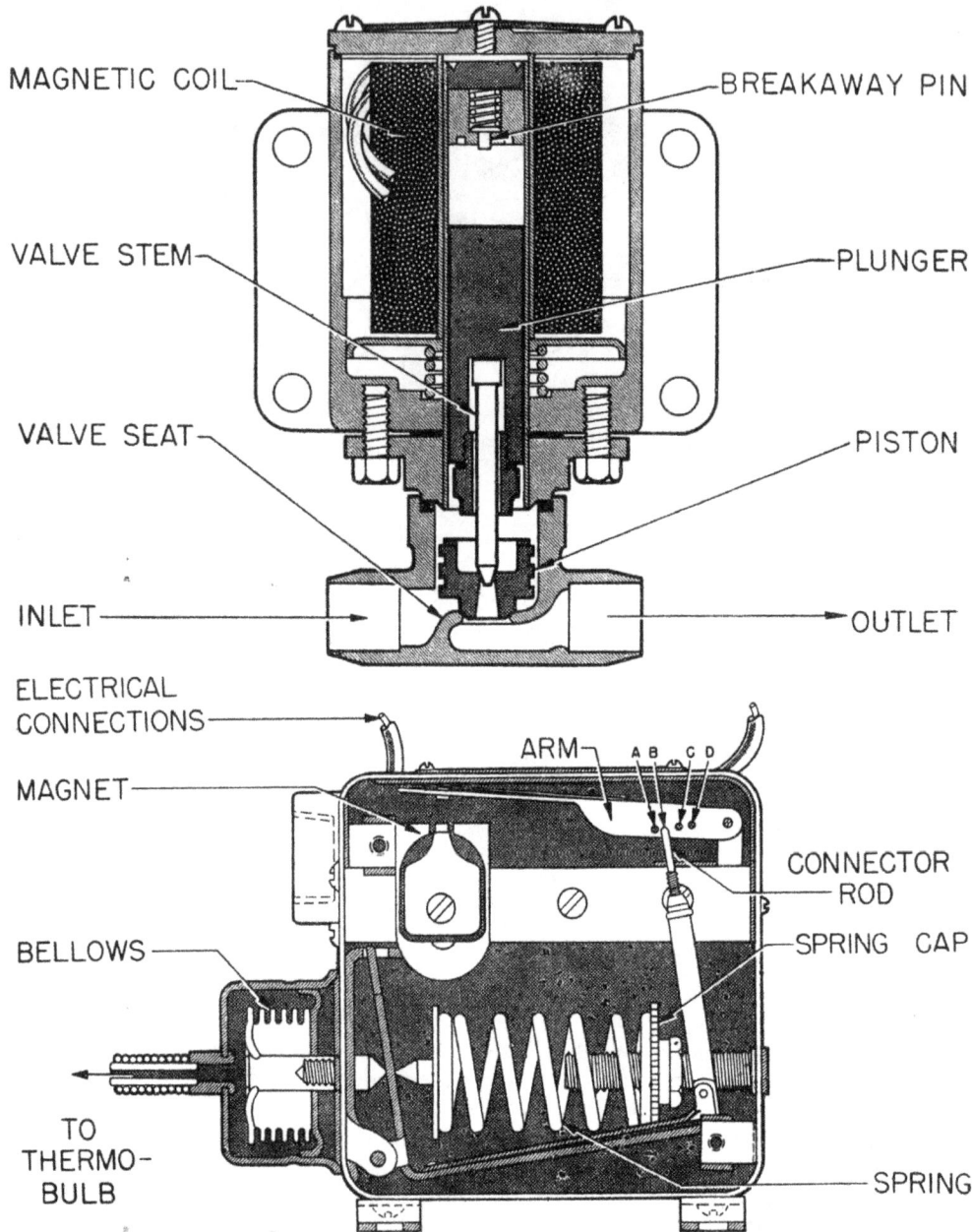

47.98

Figure 19-10.—Solenoid valve (top) and thermostatic control switch (bottom).

The saturation point is usually called the dew point. If the temperature of saturated air falls below its dew point, some of the water vapor in the air must condense into water. The dew that is visible in early morning after a drop in temperature is the result of such condensation. The sweating of cold water pipes is the result of water vapor from the air condensing on the cold surfaces of the pipes.

The amount of water vapor in the air is expressed in terms of the weight of the water vapor. This weight is usually given in grains (7000 grains = 1 pound). Absolute humidity is the weight of water vapor (in grains) per cubic foot of air. Specific humidity is the weight of water vapor (in grains) per pound of air. It should be noted that the weight refers only to the weight of the moisture which is present in the vapor state; it does not include moisture that may be present in the liquid state.

Relative humidity is the ratio of the weight of water vapor in a sample of air to the weight of water vapor which that same sample of air would hold if saturated at the existing temperature. This ratio is usually stated as a percentage. For example, when air is fully saturated, its relative humidity is 100 percent. When air contains no moisture at all, its relative humidity is zero percent. When air is half saturated—that is, holding half as much moisture as it is capable of holding at the existing temperature—its relative humidity is 50 percent.

Relative humidity, rather than absolute humidity or specific humidity, is the factor that affects comfort. This is true because it is the relative humidity that affects evaporation. Moisture tends to travel from regions of greater wetness to regions of lesser wetness. If the air above a liquid is saturated, the liquid and the vapor are in equilibrium contact and no further evaporation can take place. If the air above the liquid is only partly saturated, some evaporation can take place.

A specific example may illustrate the difference between absolute or specific humidity and relative humidity. If the specific humidity of the air is 120 grains per pound and the temperature of the air is 76° F, the relative humidity is nearly 90 percent—that is, the air is nearly saturated. With a relative humidity of 90 percent, the body may perspire freely but the perspiration does not evaporate rapidly; hence there is a general feeling of discomfort.

If the temperature of the air is 86° F, however, with the specific humidity remaining constant at 120 grains per pound, the relative humidity is only 64 percent. Although the amount of moisture in the air is the same as before, the relatively humidity is lower because at 86° F the air is capable of holding more water vapor than it can hold at 76° F. The body can therefore evaporate excess moisture and the general feeling of comfort is much greater even though the temperature is 10 degrees higher.

TEMPERATURE.—When testing the effectiveness of air conditioning equipment and when checking the humidity of spaces, two different

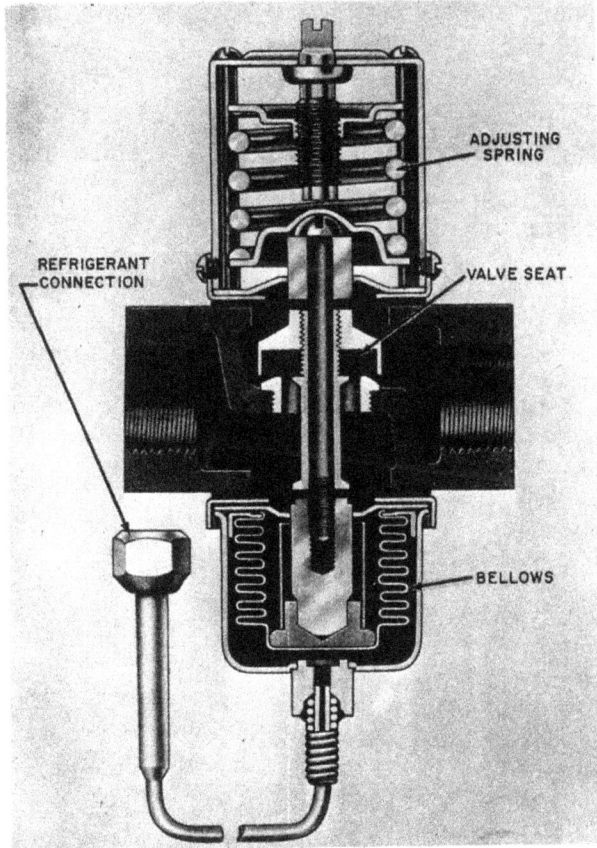

168.4X

Figure 19-11.—Water regulating valve (cross section.)

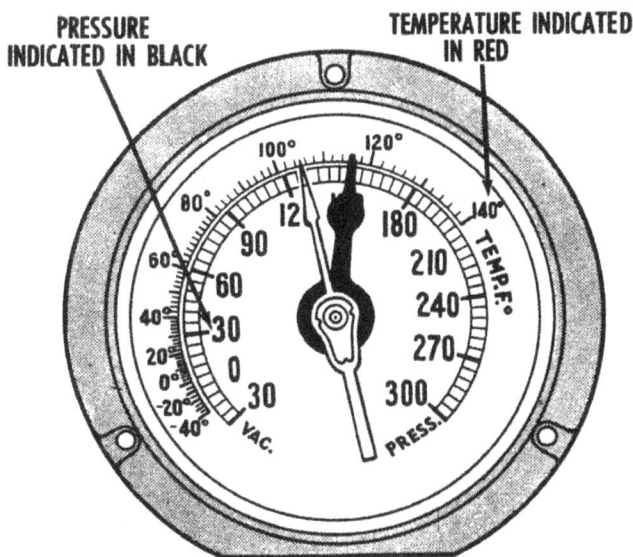

47.102

Figure 19-12.—Compound R-12 pressure gage.

temperatures are usually considered. These are the dry-bulb temperature and the wet-bulb temperature.

The dry-bulb temperature is the temperature of the air as measured by an ordinary dry-bulb thermometer. The dry-bulb temperature reflects the sensible heat of the air.

The wet-bulb temperature is the temperature of the air as measured by a wet-bulb thermometer. A wet-bulb thermometer is an ordinary thermometer with a loosely woven cloth sleeve or wick placed around the bulb and then wet with water. The water in the sleeve or wick is made to evaporate by a current of air at high velocity. The evaporation lowers the temperature of the wet-bulb thermometer. The difference between the dry-bulb temperature and the wet-bulb temperature is called the wet-bulb depression. When the air is saturated, so that evaporation cannot take place, the dry-bulb temperature is the same as the wet-bulb temperature; the condition of saturation is unusual, however, and a wet-bulb depression is normally to be expected.

The wet-bulb thermometer and the dry-bulb thermometer are usually mounted side by side on a frame. A handle or a short chain is attached to the frame so that the thermometers may be whirled in the air, thus providing an air current of high velocity to facilitate evaporation. Such a device is known as a sling psychrometer. (See fig. 19-13.) Motorized psychrometers are provided with a small motor-driven fan and dry cell batteries. Motorized psychrometers are gradually replacing the sling psychrometers. An exposed view of a hand electric psychrometer is shown in figure 19-14. With either type of psychrometer, the wet-bulb temperature must be observed at intervals as the water is being evaporated. The point at which there is no further drop in temperature on the wet-bulb thermometer is the wet-bulb temperature of the space.

As may be inferred from this discussion, the wet-bulb depression is an indication that latent heat of vaporization has been used to vaporize the water in the sleeve or wick around the wet-bulb thermometer.

When the air contains some moisture but is not saturated, the dew-point temperature is lower than the dry-bulb temperature and the wet-bulb temperature is between the dew-point and the dry-bulb temperatures. As the amount of moisture in the air increases, the difference between the dry-bulb temperature and the wet-bulb temperature becomes less and less. When the air is saturated, the dew-point temperature, the dry-bulb temperature, and the wet-bulb temperature are identical.

AIR MOTION.—In perfectly still air, a layer of air adjacent to the body absorbs the sensible heat given off by the body and increases in temperature. This layer of air also takes up the

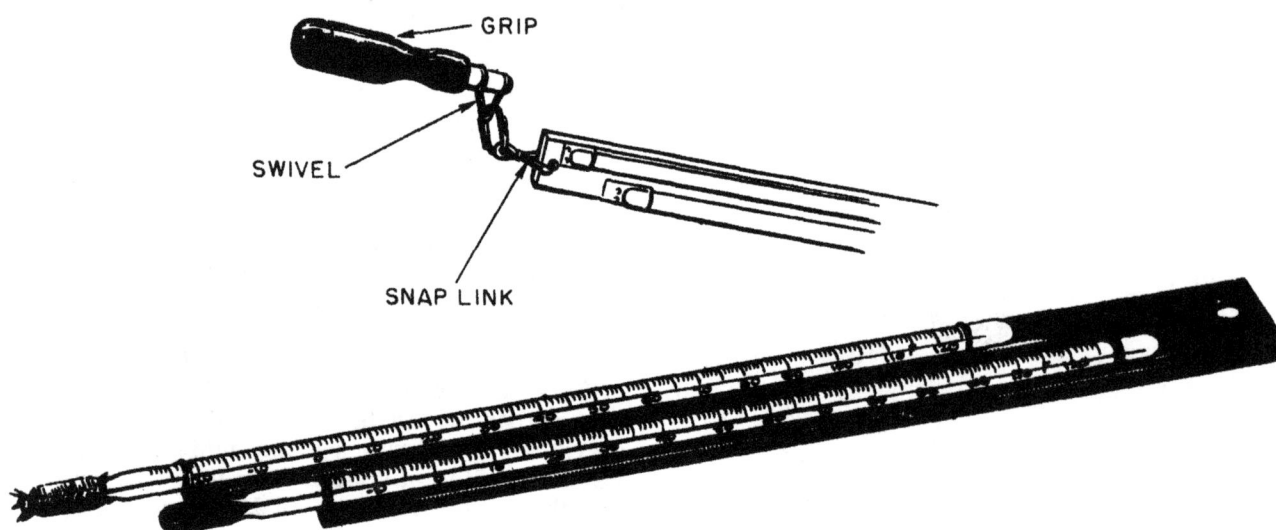

5.65

Figure 19-13.—A standard sling psychrometer.

486

as a higher or lower temperature in conjunction with compensating relative humidity and air motion.

The term used to identify the net effect of these three factors is effective temperature. The effective temperature cannot be measured with any instrument, but can be found on a special psychrometric chart when the dry-bulb temperature, the wet-bulb temperature, and the air velocity are known.

Although all of the combinations of temperature, relative humidity, and air motion of a particular effective temperature may produce the same feeling of warmth or coolness, they are not all equally comfortable or healthful. For best health and comfort, a relative humidity of 40 to 50 percent in cold weather and 50 to 60 percent in warm weather is desirable. An overall range of 30 to 70 percent is acceptable.

MECHANICAL COOLING

Mechanical cooling equipment is provided on ships to cool and dehumidify practically all parts of the ship except the machinery spaces. In general, three types of mechanical cooling equipment are used aboard naval ships: refrigerant circulating systems, chilled water circulating systems, and self-contained air conditioners.

Refrigerant Circulating Systems

A refrigerant circulating system is shown in figure 19-15. As may be seen, this system is essentially a refrigeration system consisting of a compressor, a condenser, cooling coils, a fan, an air filter, and the necessary controls.

Hot moist air from the space to be cooled is drawn through a duct, where it mixes with fresh air drawn from outside. The fan blows the air over the cooling coil and the refrigerant inside the coil cools the surface of the coil. Heat flows from the air to the coil and excess moisture in the air is condensed on the coil. The moisture drips off into a pan below the coil and is carried off by drain piping. The cool dry air leaving the coil is blown into the compartment to be cooled, where it absorbs the excess heat and moisture from the air already in the space The air is then returned to the cooling coil and the cycle is repeated. Air is exhausted from the space being cooled in order to allow fresh air to be drawn into the space. The cooling coils are installed in the ventilation

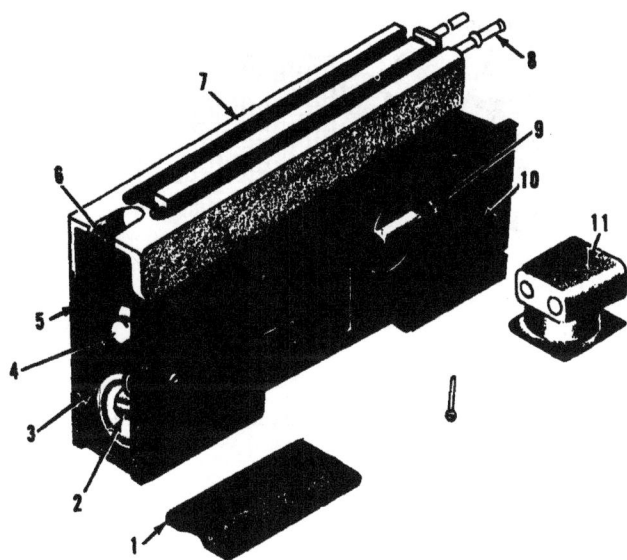

1. Sliding door.
2. Spring contact.
3. Battery compartment.
4. Water bottle.
5. Bottle compartment.
6. Hinge pin.
7. Thermometer holder.
8. Wet-bulb wick.
9. Knob
10. Exhaust parts.
11. Sliding air intake.

168.23
Figure 19-14.—Exposed view of hand electric psychrometer.

water vapor given off by the body and increases in relative humidity. The body is thus surrounded by an envelope of air which is at a higher temperature and higher relative humidity than the ambient air, and the amount of heat that the body can lose to this envelope of motionless air is considerably less than that which it can lose to the ambient air. If the air is set in motion, the motionless envelope of air is broken up and replaced by ambient air, thereby increasing the heat loss from the body. When the increased heat loss improves the heat balance of the body, we are likely to speak cheerfully of feeling a "breeze," but when the increase is excessive, we speak less cheerfully of feeling a "draft."

SENSATION OF COMFORT.—From the previous discussion, it is evident that the three factors of temperature, relative humidity, and air motion are closely interrelated and that all three factors have a definite effect upon comfort and efficiency. In fact, a given combination of temperature, relative humidity, and air motion produces the same feeling of warmth or coolness

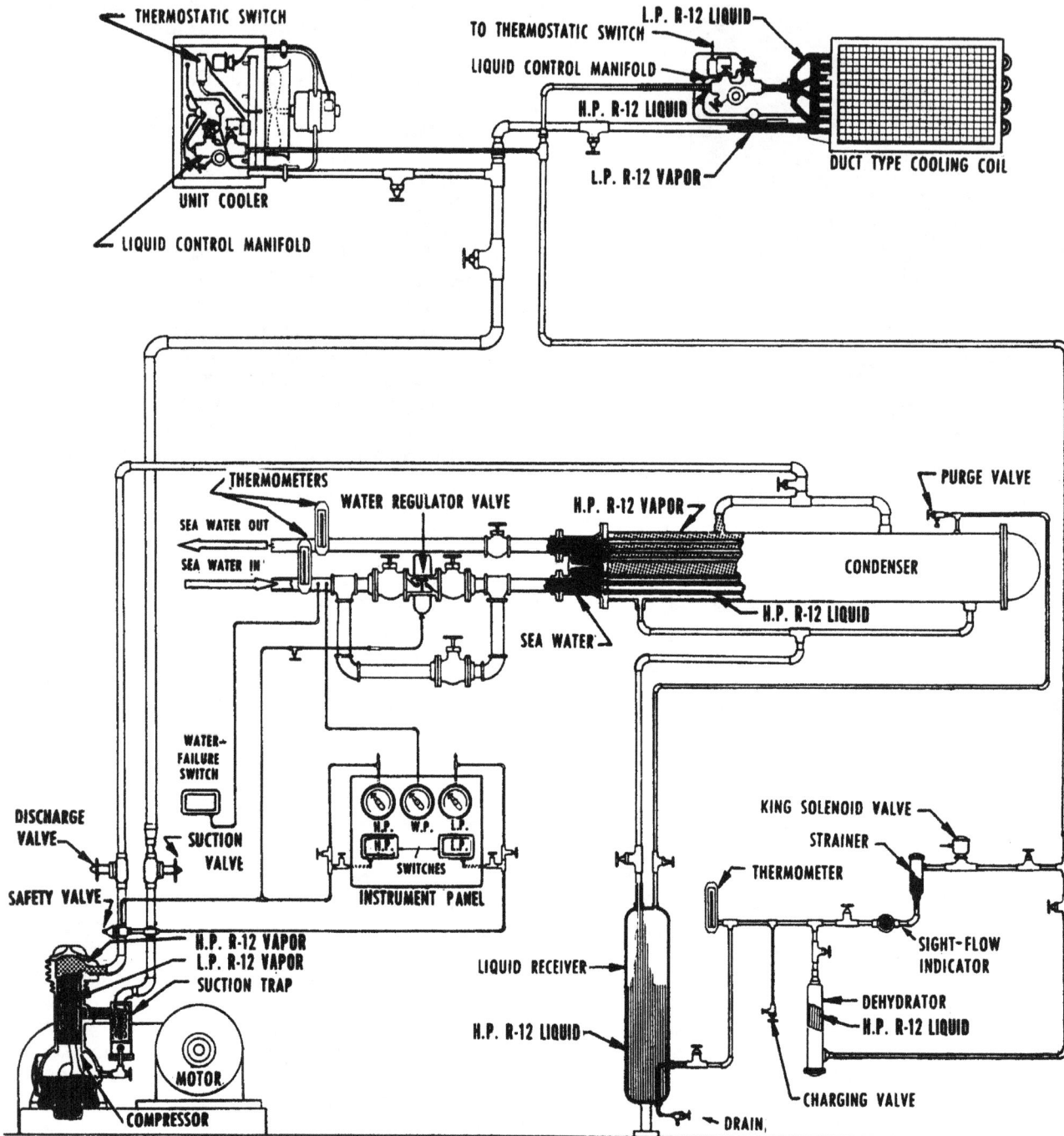

47.109

Figure 19-15.—Refrigerant circulating type of mechanical cooling system.

ducts leading to the spaces to be cooled. The refrigerant used in this system is usually R-12.

Chilled Water Circulating Systems

Two types of chilled water circulating systems are used for mechanical cooling aboard ship. Both systems utilize chilled water as the secondary refrigerant, but one type uses R-12 as the primary refrigerant and the other uses R-11. R-12 systems use reciprocating compressors; R-11 systems use centrifugal compressors.

488

Both types of chilled water circulating systems operate on the same general principle. The secondary refrigerant (chilled water) is circulated to the various cooling coils. Heat from the spaces being cooled is absorbed by the chilled water and is removed from the water by the primary refrigerant in a water chiller.

Figure 19-16 illustrates the flow of primary refrigerant, secondary refrigerant, and condenser water in an R-11 chilled water circulating system which has a single-stage centrifugal compressor. The primary refrigerant vapor goes from the evaporator to the compressor, where it is compressed. It is then discharged to the condenser. In the condenser, the primary refrigerant vapor condenses, giving up its superheat, its latent heat of vaporization, and its heat of compression to the cooling water that flows through the condenser tubes. The liquid primary refrigerant then passes through a high pressure float valve to the cooler.

The secondary refrigerant picks up heat in the coils of the air conditioned space and carries this heat to the cooler. The function of the cooler is to transfer the heat from the secondary refrigerant to the primary refrigerant which surrounds the tubes of the cooler. As this heat is transferred, the liquid primary refrigerant absorbs its latent heat of vaporization, boils, and vaporizes. The quantity of liquid refrigerant thus evaporated varies directly with the amount of heat picked up by the secondary refrigerant. The vaporized primary refrigerant goes to the compressor, and the same sequence of events is repeated in a cyclical manner.

Figure 19-17 illustrates an R-11 chilled water circulating system with a two-state centrifugal compressor. The refrigerant vapor coming from the cooler goes into an opening around the hub of the first wheel of the centrifugal compressor. The blades in the rapidly rotating wheel impart velocity to the vapor. The vapor is then directed to the hub of the second

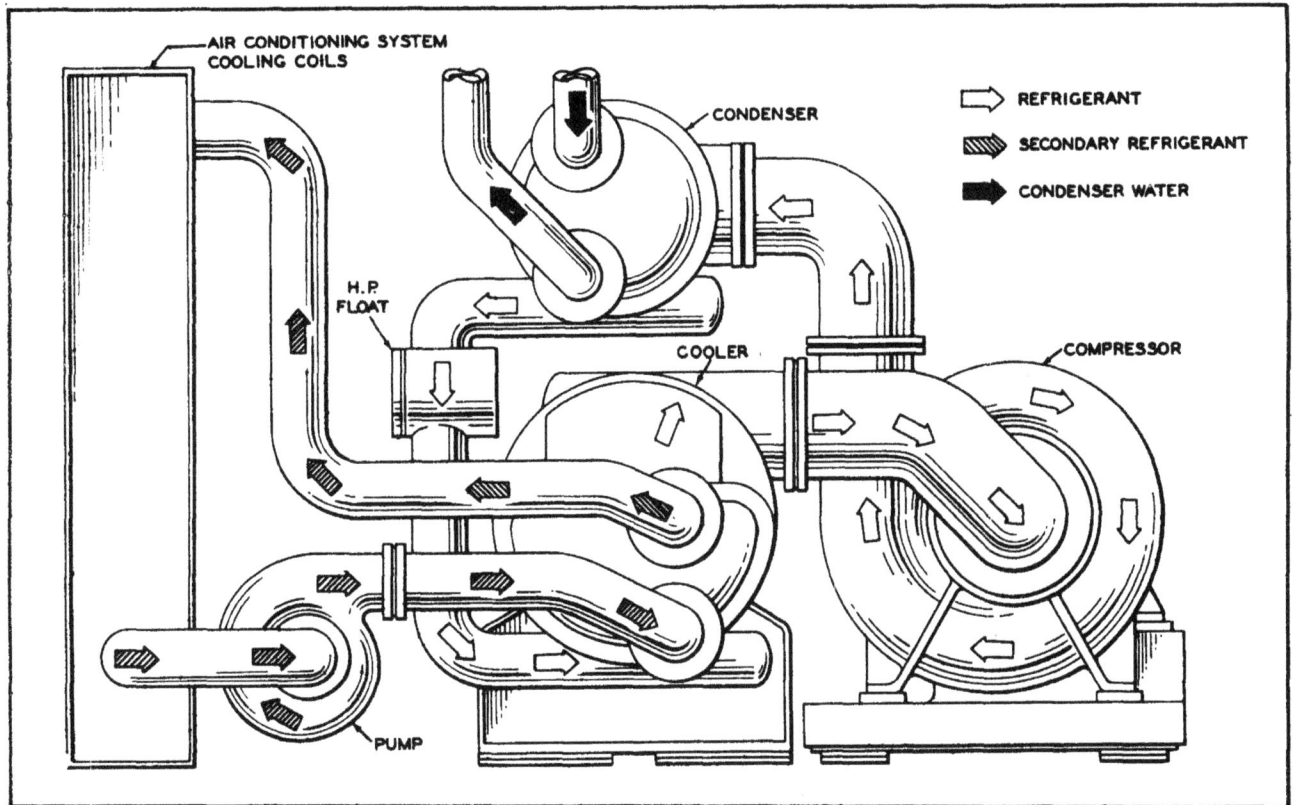

147.120

Figure 19-16.—Flow diagram, chilled water circulating system with single-stage centrifugal compressor.

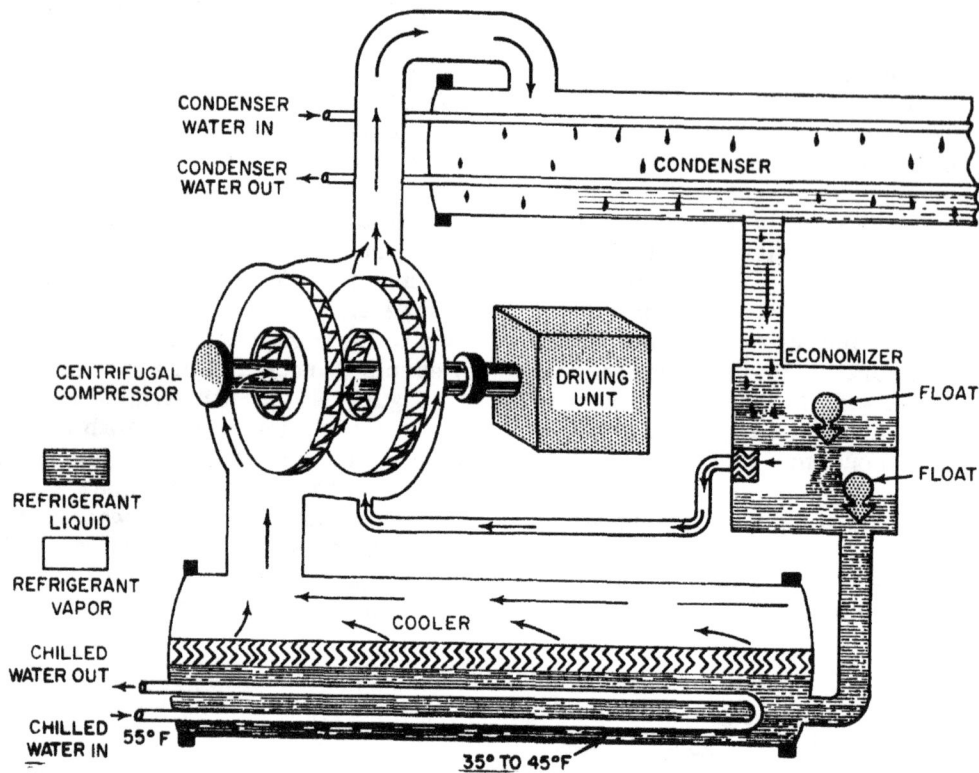

47.110

Figure 19-17.—Chilled water circulating system with two-stage centrifugal compressor.

wheel, where it is compressed and discharged to the condenser.

Between the condenser and the cooler, the liquid refrigerant passes through an economizer. A float in the upper chamber of the economizer allows the passage of refrigerant into the lower chamber. By connecting the economizer to the second stage of the compressor, the pressure in the lower chamber is greatly reduced. The reduced pressure causes some of the liquid refrigerant to flash into vapor, thus cooling the remainder of the refrigerant. Thus the economizer acts as an interstage flash cooler and increases the efficiency of the plant. A float in the lower chamber of the economizer allows the passage of the refrigerant into the cooler. In the cooler, the liquid refrigerant absorbs heat from the water and changes from a liquid to a vapor.

Self-Contained Air Conditioners

Self-contained air conditioners are installed on some ships that were originally built without mechanical cooling systems. A self-contained air conditioner is built with the entire unit in one metal cabinet. The compressing element in the unit is usually of the hermetically sealed type, with the motor and the compressor contained in a welded steel shell. Some self-contained air conditioners utilize a thermostatic expansion valve similar to the type used in large refrigeration plants; others utilize capillary tubes to ensure an even flow of refrigerant through the cooling coil.

HEATING AND VENTILATION

Aboard ship, heating is accomplished by means of steam heaters installed in the ventilation ducts and by means of space heaters. On steam-driven ships, the steam for the heaters is supplied at reduced pressure from an auxiliary steam system. On diesel-driven ships, the steam is supplied by an auxiliary boiler. Some electric heaters are also used aboard ship; these are used primarily for heating spaces which are located at a considerable distance from the steam piping system.

Ventilation is accomplished chiefly by means of fans which supply and exhaust through ventilation duct systems. Most fans used in duct systems are of the axial-flow type, but some centrifugal fans are used. Bracket fans are used to provide local circulation in certain spaces. Portable fans are used for such purposes as temporary ventilation of compartments after painting, exhausting toxic gases from closed spaces and tanks, and cooling hot areas around machinery while repairs are being made.

SAFETY PRECAUTIONS

Refrigerants are furnished in cylinders for use in shipboard refrigeration and air conditioning systems. The following precautions must be observed by personnel handling, using, and storing these cylinders:

1. Never drop cylinders nor permit them to strike each other violently.

2. Never use a lifting magnet or a sling (rope or chain) when handling cylinders. A crane may be used if a safe cradle or platform is provided to hold the cylinders.

3. Caps provided for valve protection must be kept on cylinders except when the cylinders are being used.

4. Whenever refrigerant is discharged from a cylinder, the cylinder should be weighed immediately and the weight of the refrigerant remaining in the cylinder should be recorded.

5. Never attempt to mix gases in a cylinder.

6. NEVER put the wrong refrigerant into a refrigeration system! No refrigerant except the one for which the system was designed should ever be introduced into the system. In some cases, putting the wrong refrigerant into a system may cause a violent explosion.

7. When a cylinder has been emptied, close the cylinder valve immediately to prevent the entrance of air, moisture, or dirt. Also, be sure to replace the valve protection cap.

8. Never use cylinders for any purpose other than their intended purpose. DO NOT use them as rollers, supports, etc.

9. DO NOT tamper with the safety devices in the valves or cylinders.

10. Open cylinder valves slowly. Never use wrenches or other tools except those provided by the manufacturer.

11. Make sure that the threads on regulators or other connections are the same as those on the cylinder valve outlets. Never force connections that do not fit.

12. Regulators and pressure gages provided for use with a particular gas must NOT be used on cylinders containing other gases.

13. Never attempt to repair or alter cylinders or valves.

14. Never fill R-12 cylinders beyond 80 percent of capacity.

15. Whenever possible, store cylinders in a cool, dry place, in an upright position. If the cylinders are exposed to excessive heat, a dangerous increase in pressure will occur. If cylinders must be stored in the open, take care that they are protected against extremes of weather. NEVER allow a cylinder to be subjected to a temperature above 125°F.

16. NEVER allow R-12 to come in contact with a flame or red-hot metal! When exposed to excessively high temperatures, R-12 breaks down into PHOSGENE gas, an extremely poisonous substance. Because R-12 is such a powerful freezing agent that even a very small amount can freeze the delicate tissues of the eyes, causing permanent damage; it is essential that goggles be worn by all personnel who may be exposed to a refrigerant, particularly in its liquid form. If refrigerant does get in the eyes, the person suffering the injury should receive medical treatment immediately in order to avoid permanent damage to the eyes. In the meantime, put drops of clean olive oil, mineral oil, or other nonirritating oil in the eyes, and make sure that the person does not rub his eyes. CAUTION: Do not use anything except clean, nonirritating oil for this type of eye injury. (NOTE: If large leaks are indicated, the soap method should be used to detect leaks; for minute leaks, the halide torch should be employed.)

If R-12 comes in contact with the skin, it may cause frostbite. This injury should be treated as any other case of frostbite. Immerse the affected part in a warm bath for about 10 minutes, then dry carefully. DO NOT rub or massage the affected area.

R-12 is considered a fluid of low toxicity. However, in closed spaces, high concentrations displace the oxygen in the air and thus do not sustain life. If a person should be overcome by R-12 remove him IMMEDIATELY to a well-ventilated place and get medical attention at the earliest opportunity. Watch his breathing. If the person is not breathing, give artificial respiration.

CHAPTER 20

SHIPBOARD ELECTRICAL SYSTEMS

Shipboard electrical systems include a great variety of equipment which provides numerous services indispensable to the operation of a modern naval ship. These systems distribute power throughout the ship for offensive and defensive weapons, the ship's movement, and shipboard habitability. Since the systems and equipment utilizing electric power are often under the cognizance of a division other than the electrical division, a joint responsibility frequently exists for the operation, maintenance, and repair of electrical systems and equipment.

This chapter provides some information on basic electrical theory and gives a brief description of shipboard electrical systems and equipment.

BASIC ELECTRICAL THEORY

The word electric is derived from the Greek word meaning amber. The ancient Greeks used the word to describe the strong forces of attraction and repulsion that were exhibited by amber after it had been rubbed with a cloth. Since scientists are still unable to define electricity clearly, and since many of the phenomena which occur cannot be completely explained, theories can only be postulated from the reactions observed.

Through research and experiment, scientists have observed and described many predictable characteristics of electricity and have postulated certain rules which are often called "laws." These laws of electricity, together with the electron theory, are the basis for our present concepts of electricity.

ELECTRON THEORY

Every atom is primarily an electrical system with high speed planetary electrons orbiting around its nucleus. The electron, whose negative charge forms a natural unit of electricity, is bound to the atom by the positive charge within the nucleus.

The electrons in the outer orbits of certain elements are easily separated from the positive nuclei of their parent atoms. Should an outside force be applied, one of these loosely bound electrons will be released from the parent atom, thus becoming a free electron, and travel to another atom. It is on this ability of an electron to move about from one atom to another that the electron theory is based.

Elements such as silver, copper, gold, and aluminum have many loosely bound electrons and are considered to be good conductors of electricity. In materials used as insulators, electron flow from one atom to another is relatively non-existent, since the planetary electrons in the outer orbital shells are more tightly bound to their parent nuclei.

Ordinarily an atom is most likely to be in that state in which the internal energy is at a minimum, having a neutral electrical charge. However, if an atom absorbs sufficient energy from an outside source, loosely bound electrons in the outer orbital shells will leave the atom. An atom that has lost or gained one or more electrons is said to be ionized. If an atom loses electrons it becomes positively charged and is referred to as a positive ion; if an atom gains electrons it is referred to as a negative ion and is said to have a negative charge. A positive ion will attract any free electron in its surroundings in order to reach a neutral state.

STATIC ELECTRICITY

When two bodies have unlike charges, one positive and the other negative, an electrical force is exerted between the two. This force is called a static charge or an electrostatic force.

A static charge can easily be produced by the force of friction when two materials are rubbed together. If the materials used are both good conductors, it is difficult to obtain a detectable charge because equalizing currents will flow easily in and between the conducting materials. However, if the materials used are poor conductors (insulators), little equalizing current can flow and an electrostatic charge is built up.

Charged Bodies

One of the fundamental laws of electricity is that like charges repel each other and unlike charges attract each other. A positive charge and a negative charge, being unlike, tend to move toward each other; thus in the atom the negative electrons are held in their orbital shells by the positive attraction of the nucleus.

The law of charged bodies may be demonstrated by a simple experiment using two pith (paper pulp) balls suspended near one another by threads, as shown in figure 20-1, and a hard rubber rod. If the hard rubber rod is rubbed to give it a negative charge and then held against the right-hand ball in part (A), the rod will impart a negative charge to the ball. The right-hand ball will be charged negatively with respect to the left-hand ball, and when released the two balls will be drawn together. When the two balls touch, they will remain in contact with each other until the left-hand ball acquires a portion of the negative charge, at which time they will swing apart as shown in part (C) of figure 20-1. Should positive charges be placed on both balls, as shown in part (B) of figure 20-1, the balls would also repel each other.

Coulomb's Law of Charges

The amount of attracting or repelling force which acts between two electrically charged bodies in free space depends upon the magnitude of their charges and the distance between them. This relationship between charged bodies was first discovered by a French scientist named Coulomb. Coulomb's law of charges states that charged bodies attract or repel each other with a force that is directly proportional to the product of their charges and inversely proportional to the square of the distance between them.

The practical unit of charge a body has is expressed in coulombs. One coulomb is the charge carried by approximately 6 x 10^{18} electrons.

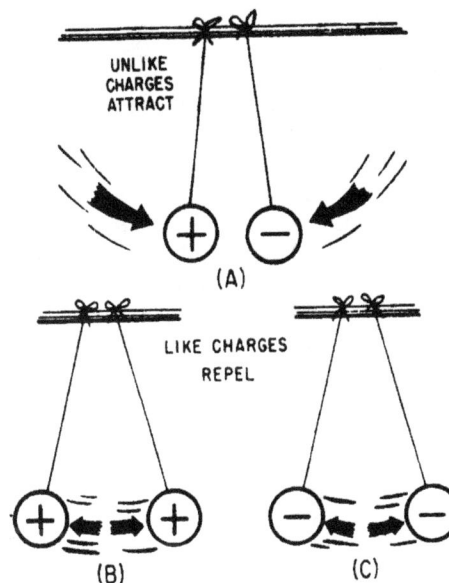

147.121

Figure 20-1.—Reaction between charged bodies.

Electric Current Flow

A difference of potential exists between two bodies having opposite electrostatic charges. If a path is provided between the two bodies, electrons will flow from the negatively charged body to the positively charged body until the charges have equalized and the difference of potential no longer exists. This movement of electrons is called electric current. The rate of flow is measured in amperes. One ampere may be defined as the flow of one coulomb per second past a fixed point in a conductor.

The force or difference in potential which causes electrons to flow from one charged body to another is called electromotive force (emf). Electromotive force is measured in volts. One volt may be defined as the potential difference between two points when one joule of work is required to move a one-coulomb charge between these points.

MAGNETISM

The relationship between magnetism and electricity was first shown in 1819 when the Danish scientist Oersted observed that a small compass needle was deflected when it was passed near a wire carrying a current. About 12 years later, Michael Faraday discovered that moving a conductor in a magnetic field would produce an electric current.

Magnets may be found in the natural state in the form of an iron oxide, but the majority are produced by artificial means. Artificial magnets may be either permanent magnets or temporary magnets, depending upon their ability to retain magnetic strength after the magnetizing force has been removed.

Permanent magnets are bars of hardened steel or other alloy which have been permanently magnetized. Permanent magnets are used extensively in electrical instruments, meters, telephone receivers, and magnetos.

Electromagnets are temporary magnets composed of soft-iron cores around which are wound coils of insulated wire. Electromagnets are used in electric motors, generators, and transformers. When an electric current flows through the coil, the core becomes magnetized.

Magnetism is a field of force exerted in space. A magnetic field consisting of imaginary lines along which the magnetic force acts surrounds each magnet. A visual representation of a magnetic field can be obtained by placing a plate of glass over a magnet and sprinkling iron filings onto the glass. The filings arrange themselves in a pattern of definite paths between the poles, along the magnetic lines of force, as shown in figure 20-2.

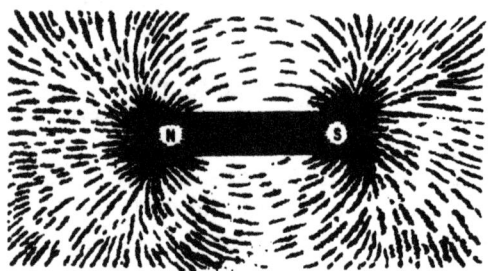

41.4

Figure 20-2.—Magnetic field pattern around a magnet.

Magnetic flux is the entire quantity of lines in a magnetic field, with gauss being the unit measurement of its density. One gauss is equal to one line of force per square centimeter of magnetic field.

PRODUCING A VOLTAGE

There are six commonly used methods of producing a voltage. Magnetism and chemical action are the two methods most commonly used aboard ship; hence the present discussion is limited to these two methods. It should be noted, however, that a voltage can also be produced by friction, pressure, light, and heat.[1]

Voltage Produced By Chemical Action

Chemical energy is transformed into electrical energy within the cells of a battery. Shipboard uses of electricity from this source include power supply for emergency lighting (with dry cell batteries) and the starting of small engines (with wet cell batteries).

The most common dry cell battery consists of a cylindrical zinc container, a carbon electrode, and an electrolyte of ammonium chloride and water in paste form. The zinc container is the negative electrode of the cell; it is lined with a nonconducting material to insulate it from the electrolyte. When a circuit is formed, the current flows from the negative zinc electrode to the positive carbon electrode.

In a common wet cell storage battery, the electrodes and the electrolyte are altered by the chemical action that takes place when the cell delivers current. Such a battery may be restored to its original condition by forcing an electric current through it in the opposite direction to that of discharge.

The most common wet cell storage battery in use is the lead-acid battery having an emf of 2.2 volts per cell. In the fully charged state, the positive plates are pure lead peroxide and the negative plates are pure lead immersed in a dilute sulfuric acid electrolyte.

When a circuit is formed, the chemical action between the ionized electrolyte and dissimilar metal plates converts chemical energy to electrical energy. As the storage battery discharges, the sulfuric acid is depleted by being gradually converted to water, while both positive and negative plates are converted to lead sulfate. This chemical reaction is represented by the following equation, the reversibility of which is dependent upon electrical energy being added during the charging cycle.

[1] One device for producing a voltage by heat is the thermocouple, discussed in chapter 7 of this text.

$$Pb + PbO_2 + 2H_2SO_4 \underset{\text{CHARGING}}{\overset{\text{DISCHARGING}}{\rightleftarrows}} 2PbSO_4 + 2H_2O$$

The capacity of a battery is measured in ampere-hours. The capacity is equal to the product of the current (in amperes) and the time (in hours) during which the battery is supplying this current to a given load. The capacity depends upon many factors, the most important of which are (1) the area of the plates in contact with the electrolyte, (2) the quantity and specific gravity of the electrolyte, (3) the general condition of the battery, and (4) the final limiting voltage.

Voltage Produced by Magnetism

One of the most useful and widely employed applications of magnets is in the production of vast quantities of electric power from mechanical sources. The mechanical power may be provided by a number of different devices, including gasoline engines, diesel engines, water turbines, steam turbines, and gas turbines. The final conversion of these energies to electricity is done by generators employing the principle of electromagnetic induction.

There are three conditions which must exist before a voltage can be produced by electromagnetic induction. First, we must have a magnetic field; second, a conductor; and third, relative motion between the field and the conductor. In accordance with these conditions, when a conductor is moved across a magnetic field so as to cut the lines of force, electrons within the conductor are forced to move; thus a voltage is produced.

Producing a voltage by magnetic induction is illustrated in figure 20-3. If the ends of a conductor are connected to a low-reading voltmeter or galvanometer and the conductor is moved rapidly down through a magnetic field, there is a momentary reading on the meter. When the conductor is moved up through the field, the meter deflects in the opposite direction. If the conductor is held stationary and the magnet is moved so that the field cuts across the conductor, the meter is deflected in the same manner as when the conductor was moved and the field was stationary.

The voltage developed across the conductor terminals by electromagnetic induction is known as an induced emf, and the resulting current that flows is called induced current. The induced

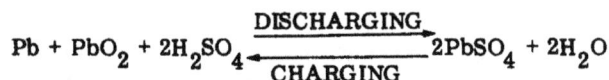

CONDUCTOR MOVED DOWN

CONDUCTOR MOVED UP

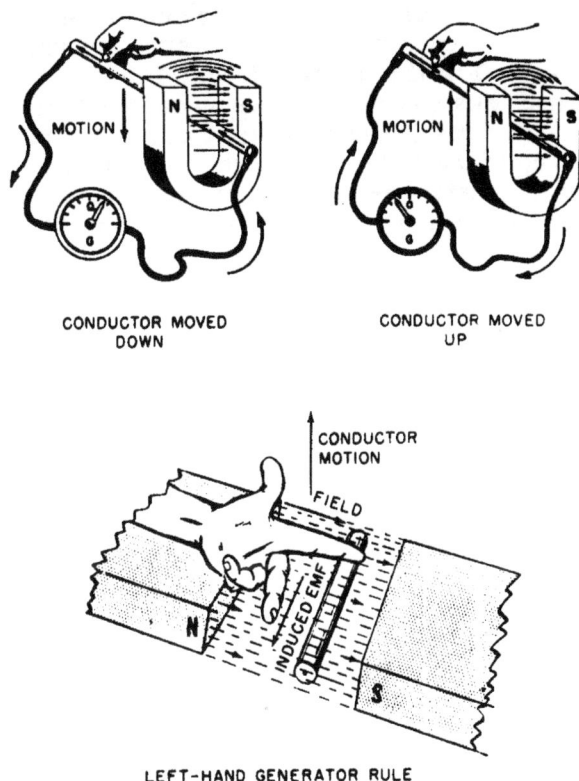

LEFT-HAND GENERATOR RULE

12.143

Figure 20-3.—Left-hand generator rule.

emf exists only so long as relative motion occurs between the conductor and the field.

There is a definite relationship between the direction of flux, the direction of motion of the conductor, and the direction of the induced emf. When two of these directions are known, the third can be found by applying the left-hand rule for generators. To find the direction of the emf induced in a conductor, extend the thumb, the index finger, and the second finger of the left hand at right angles to each other, as shown in figure 20-3. Point the index finger in the direction of the flux (toward the south pole) and the thumb in the direction in which the conductor is moving in respect to the fields. The second finger then points in the direction in which the induced emf will cause the electrons to flow.

DIRECT-CURRENT CIRCUITS

An electric circuit is a complete path through which electrons can flow from the negative terminal of the voltage source, through the connecting wires (conductors), through the load, and back to the positive terminal of the voltage source (fig. 20-4). The resistance[2] of a circuit (opposition to current flow) controls the amount of current flow through the circuit. The unit of electrical resistance, the ohm (symbol Ω), is named after the German physicist Georg Simon Ohm, who in the 19th century proved by experiment the constant proportionality between current and voltage in the simple electric circuit.

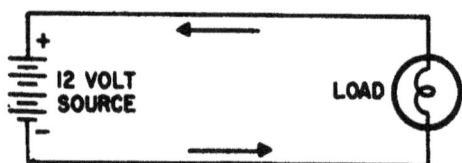

Figure 20-4.—Simple electric circuit.

OHM'S LAW

Ohm's law is fundamentally linear and therefore simple. It is exact and applies to d-c circuits and devices in its basic form; in a modified form it may also be applied to a-c circuits.

Ohm's law may be stated in words as: the intensity of the current (in amperes) in any electric circuit is equal to the difference in potential (in volts) across the circuit divided by the resistance (in ohms) of the circuit. Expressed as an equation, Ohm's law becomes

$$I = \frac{E}{R}$$

where

I = intensity of current (in amperes)
E = difference in potential (in volts)
R = resistance (in ohms)

If any two of these quantities are known, the third may be found by applying the equation.

In addition to the volt, the ampere, and the ohm, the unit of power frequently appears in electric circuit calculations. In a d-c electric circuit, power is equal to the product of the voltage and the current. Expressing the power in watts (P), the current in amperes (I), and the emf in volts (E), the equation is

$$P = IE$$

The various implications of Ohm's law may be derived from the algebraic transposition of the units I, E, R, and P. A summary of the 12 basic formulas which may be derived from transposing these units is given in figure 20-5. The unit in each quadrant of the smaller circle is equivalent to the quantities in the same quadrant of the larger circle.

Series Circuits

The analysis of a series circuit to determine values for voltage, current, resistance, and power is relatively simple. It is necessary only to draw or to visualize the circuit, to list the known values, and to determine the unknown values by means of Ohm's law and Kirchhoff's law of voltages.

Kirchhoff's law of voltages states that the algebraic sum of all the voltages in any complete electric circuit is equal to zero. In other words, the sum of all positive voltages must be equal to the sum of all negative voltages. For any given voltage rise there must be an equal voltage drop somewhere in the circuit. The voltage rise (potential source) is usually regarded

---

[2]All conductors have some resistance, and therefore a circuit made up of nothing but conductors would have some resistance, however small it might be. In circuits containing long conductors, through which an appreciable amount of current is drawn, the resistance of the conductors becomes important. For the purposes of this chapter, however, the resistance of the conducting wires is neglected.

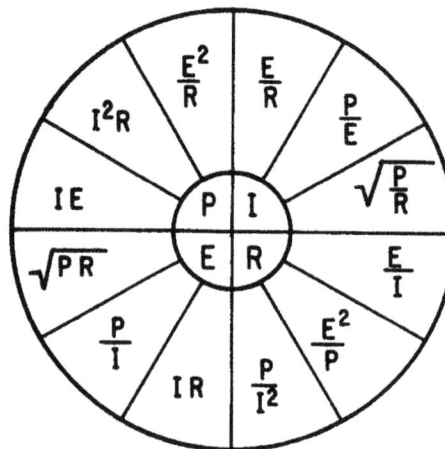

Figure 20-5.—Summary of basic Ohm's law formulas.

as the power supply, such as a battery. The voltage drop is usually regarded as the load, such as a resistor. The voltage drop may be distributed across a number of resistive elements, such as a string of lamps or several resistors. However, according to Kirchhoff's law, the sum of their individual voltage drops must always equal the voltage rise supplied by the power source.

The statement of Kirchhoff's law can be translated into an equation, from which many unknown circuit factors may be determined. (See fig. 20-6.) Note that the source voltage E_s is equal to the sum of the three load voltages E_1, E_2, and E_3. In equation form,

$$E_s = E_1 + E_2 + E_3$$

The following procedure may be used to solve problems applicable to figure 20-6:

1. Note the polarity of the source emf (E_s) and indicate the electron flow around the circuit. Electron flow is out from the negative terminal of the source, through the load, and back to the positive terminal of the source. In the example being considered, the arrows indicate electron flow in a clockwise direction around the circuit.

2. To apply Kirchhoff's law it is necessary to establish a voltage equation. The equation is developed by tracing around the circuit and noting the voltage absorbed (that is, the voltage drop) across each part of the circuit, and expressing the sum of these voltages according to the voltage law. It is important that the trace be made around a closed circuit, and that it encircle the circuit only once. Thus, a point is arbitrarily selected at which to start the trace. The trace is then made and, upon completion,

the terminal point coincides with the starting point.

3. Sources of emf are preceded by a plus sign if, in tracing through the source, the first terminal encountered is positive; if the first terminal is negative, the emf is preceded by a minus sign.

4. Voltage drops along wires and across resistors (loads) are preceded by a minus sign if the trace is in the assumed direction of electron flow; if in the opposite direction, the sign is plus.

5. If the assumed direction of electron flow is incorrect, the error is indicated by a minus sign preceding the current, as obtained in solving for circuit current. The magnitude of the current is not affected.

The preceding rules may be applied to the example of figure 20-6 as follows:

1. The left terminal of the battery is negative, the right terminal is positive, and electron flow is clockwise around the circuit.

2. The trace may arbitrarily be started at the positive terminal of the source and continued clockwise through the source to its negative terminal. From this point the trace is continued around the circuit to a, b, c, d, and back to the positive terminal, thus completing the trace once around the entire closed circuit.

3. The first term of the voltage equation is $+E_s$.

4. The second, third, and fourth terms are, respectively, $-E_1$, $-E_2$, $-E_3$. Their algebraic sum is equated to zero, as follows:

$$E_s - E_1 - E_2 - E_3 = 0$$

Transposing the voltage equation and solving for E_s,

$$E_s = E_1 + E_2 + E_3$$

Since $E = IR$, from Ohm's law, the voltage drop across each resistor may be expressed in terms of the current and resistance of the individual resistor, as follows:

$$E_s = IR_1 + IR_2 + IR_3$$

where R_1, R_2, and R_3 are the resistances of resistors R1, R2, and R3, respectively. E_s is the source voltage and I is the circuit current.

E_s may be expressed in terms of the circuit current and total resistance as IR_t. Substituting IR_t for E_s, the voltage equation becomes

$$IR_t = IR_1 + IR_2 + IR_3$$

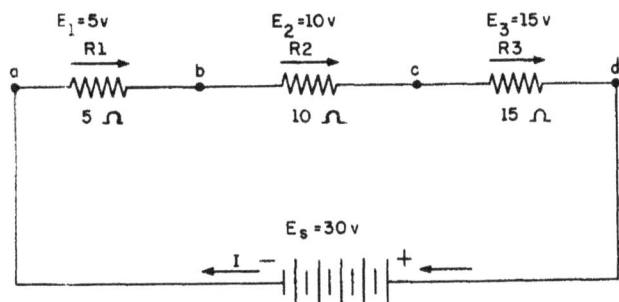

13.15

Figure 20-6.—Series circuit for demonstrating Kirchhoff's law of voltages.

Since there is only one path for current in the series circuit, the total current is the same in all parts of the circuit. Dividing both sides of the voltage equation by the common factor I, an expression is derived for the total resistance of the circuit in terms of the resistances of the individual devices:

$$R_t = R_1 + R_2 + R_3$$

Therefore, in series circuits the total resistance is the sum of the resistances of the individual parts of the circuit.

In the example of figure 20-6, the total resistance is $5 + 10 + 15 = 30$ ohms. The total current may be found by applying the equation

$$I_t = \frac{E_t}{R_t} = \frac{30}{30} = 1 \text{ ampere}$$

The power absorbed by resistor R_1 is $I^2 R_1$, or $1^2 \times 5 = 5$ watts. Similarly, the power absorbed by R_2 is $1^2 \times 10 = 10$ watts, and the power absorbed by R_3 is $1^2 \times 15 = 15$ watts. The total power absorbed is the arithmetic sum of the power of each resistor, or $5 + 10 + 15 = 30$ watts. The value is also calculated by $P_t = E_t I_t = 30 \times 1 = 30$ watts.

Parallel Circuits

The parallel circuit differs from the simple series circuit in that two or more resistors, or loads, are connected directly to the same source of voltage. There is accordingly more than one path that the electrons can take. The more paths (or resistors) that are added in parallel, the less opposition there is to the flow of electrons from the source. This condition is opposite to the effect that is produced in the series circuit where added resistors increase the opposition to the electron flow.

As may be seen from figure 20-7, the same voltage is applied across each of the parallel resistors. In this case the voltage applied across the resistors is the same as the source voltage, E_s.

Current flows from the negative terminal of the source to point a where it divides and passes through the three resistors to point b and back to the positive terminal of the voltage source. The amount of current flowing through each individual branch depends on the source voltage and on the resistance of that branch—that is, the lower the

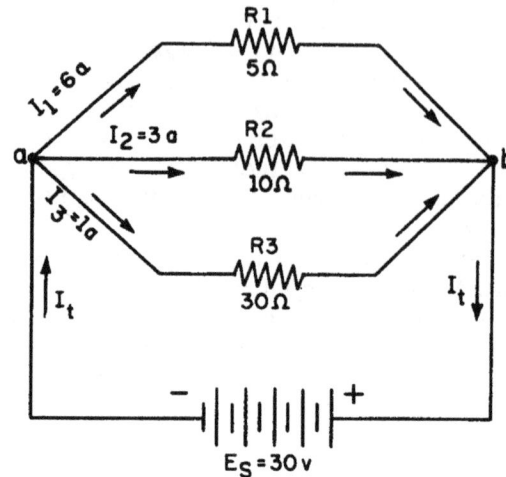

13.16

Figure 20-7.—Parallel electric circuit.

resistance of the branch, the higher will be the current through that branch. The individual currents can be found by the application of Ohm's law to the individual resistors. Thus,

$$I_1 = \frac{E_s}{R_1} = \frac{30}{5} = 6 \text{ amperes}$$

and

$$I_2 = \frac{E_s}{R_2} = \frac{30}{10} = 3 \text{ amperes}$$

and

$$I_3 = \frac{E_s}{R_3} = \frac{30}{30} = 1 \text{ ampere}$$

The total current, I_t, of the parallel circuit is equal to the sum of the currents through the individual branches. This, in slightly different words, is Kirchhoff's law. In this case, the total current is

$$I_t = I_1 + I_2 + I_3 = 6 + 3 + 1 = 10 \text{ amperes}$$

In order to find the equivalent, or total, resistance (R_t) of the combination shown in figure 20-7, Ohm's law is used to find each of the currents (I_t, I_1, I_2, and I_3) in the preceding formula. The total current is equal to the sum of the branch currents. Thus,

$$\frac{E_s}{R_t} = \frac{E_s}{R_1} + \frac{E_s}{R_2} + \frac{E_s}{R_3}$$

or

$$\frac{E_s}{R_t} = E_s\left(\frac{1}{R_1} + \frac{1}{R_2} + \frac{1}{R_3}\right)$$

Both sides of this equation may be divided by E_s without changing the value of the equation; therefore,

$$\frac{1}{R_t} = \frac{1}{R_1} + \frac{1}{R_2} + \frac{1}{R_3}$$

By means of the preceding equation the total resistance of the circuit shown in figure 20-7 may be determined. Thus

$$\frac{1}{R_t} = \frac{1}{5} + \frac{1}{10} + \frac{1}{30}$$

and

$$\frac{1}{R_t} = \frac{10}{30}$$

Taking the reciprocals of both sides,

$$R_t = \frac{30}{10} = 3 \text{ ohms}$$

A useful rule to remember in computing the equivalent resistance of a d-c parallel circuit is that the total resistance is always less than the smallest resistance in any of the branches.
In addition to adding the individual branch currents to obtain the total current in a parallel circuit, the total current may be found directly by dividing the applied voltage by the equivalent resistance, R_t. For example, in figure 20-7:

$$I_t = \frac{E_s}{R_t} = \frac{30}{3} = 10 \text{ amperes}$$

Three or more resistors may be connected in series and parallel combinations to form a compound circuit. One basic series-parallel circuit composed of three resistors is shown in figure 20-8.

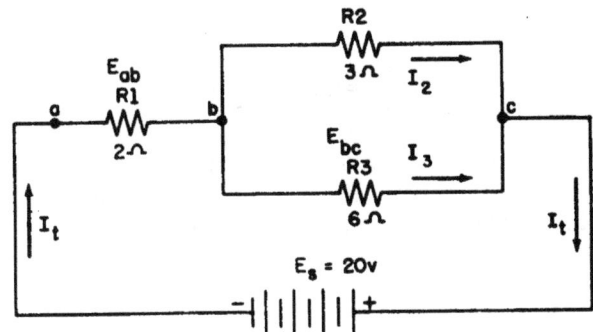

R1 IN SERIES WITH PARALLEL COMBINATION OF R2 AND R3

27.237

Figure 20-8.—Compound electric circuit.

The total resistance, R_t, of figure 20-8 is determined in two steps. First, the resistance $R_{2,3}$ of the parallel combination of R2 and R3 is determined as

$$R_{2,3} = \frac{R_2 R_3}{R_2 \ R_3} = \frac{3 \times 6}{3 + 6} = \frac{18}{9} = 2 \text{ ohms}$$

The sum of $R_{2,3}$ and R_1 (that is, R_t) is

$$R_t = R_{2,3} + R_1 = 2 + 2 = 4 \text{ ohms}$$

If the total resistance, R_t, and the source voltage, E_s, are known, the total current, I_t, may be determined by Ohm's law. Thus, in figure 20-8,

$$E_{ab} = I_t R_1 = 5 \times 2 = 10 \text{ volts}$$

and

$$E_{bc} = I_t R_{2,3} = 5 \times 2 = 10 \text{ volts}$$

According to Kirchhoff's voltage law, the sum of the voltage drops around the closed circuit is equal to the source voltage. Thus,

$$E_{ab} + E_{bc} = E_s$$

or

$$10 + 10 = 20 \text{ volts}$$

If the voltage drop E_{bc} across $R_{2,3}$—that is, the drop between points b and c—is known, the current through the individual branches may be determined as

$$I_2 = \frac{E_{bc}}{R_2} = \frac{10}{3} = 3.333 \text{ amperes}$$

and

$$I_3 = \frac{E_{bc}}{R_3} = \frac{10}{6} = 1.666 \text{ amperes}$$

According to Kirchhoff's current law, the sum of the currents flowing in the individual parallel branches is equal to the total current. Thus,

$$I_2 + I_3 = I_t$$

or

$$3.333 + 1.666 = 5 \text{ amperes (approx.)}$$

The total current flows through R1; at point b it divides between the two branches in inverse proportion to the resistance of the branches. Twice as much goes through R2 as through R3 because R2 has one-half the resistance of R3. Thus, 3.333 (or two-thirds of 5) amperes flow through R2; and 1.666 (or one-third of 5) amperes flow through R3.

WHEATSTONE BRIDGE

A type of circuit that is widely used for precision measurements of resistance is the Wheatstone bridge. The circuit diagram of a Wheatstone bridge is shown in figure 20-9. R1, R2, and R3 are precision variable resistors, and R_x is the resistor whose unknown value is to be determined. The galvanometer, G, is inserted across terminals b and d to indicate the condition of balance. When the bridge is properly balanced, there is no difference in potential across terminals b and d and the galvanometer deflection, when the switch is closed, will be zero. Should the bridge become unbalanced due to a change in resistance of R_x, the difference of potential between terminals b and d will cause a deflection in the galvanometer.

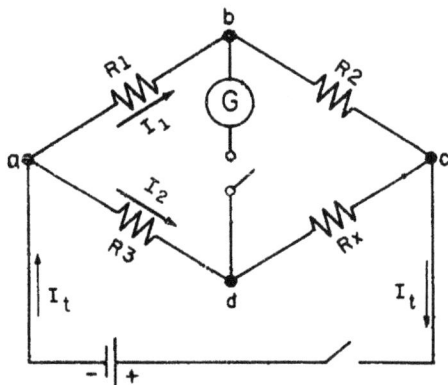

When this type of circuit is used as a component of a resistance thermometer, R_x is the temperature-sensing element. The resistance of R_x varies directly with the temperature; thus a change in temperature results in an unbalanced bridge and a deflection of the galvanometer.

DIRECT-CURRENT GENERATORS

A d-c generator is a rotating machine that converts mechanical energy into electrical energy. This conversion is accomplished by rotating an armature, which carries conductors, in a magnetic field, thus inducing an emf in the conductors.

A d-c generator (fig. 20-10) consists essentially of a steel frame or yoke containing the pole pieces and field windings; an armature consisting of a group of copper conductors mounted in a slotted cylindrical core; a commutator for maintaining the current in one direction through the external circuit; and brushes with brush holders to carry the current from the commutator to the external load circuit.

The frame, in addition to providing mechanical support for the pole pieces, serves as a portion of the magnetic circuit in that it provides a path for the magnetic flux between the poles.

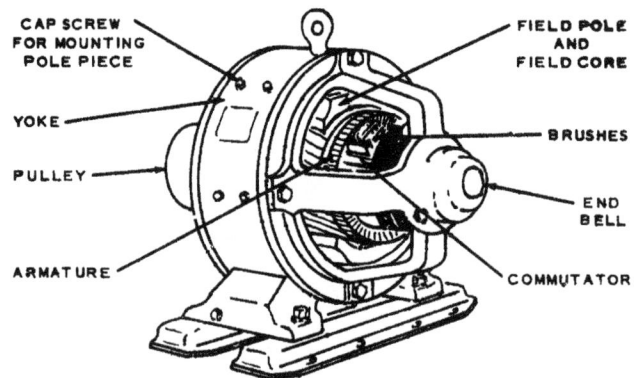

12.251
Figure 20-9.—Wheatstone bridge circuit diagram.

73.161
Figure 20-10.—A d-c generator.

GENERATING A VOLTAGE

The field windings of a d-c generator receive current either from an external d-c source or directly across the armature, thus becoming electromagnets. They are connected so that they produce alternate north and south poles and, when energized, they establish magnetic flux in the field yoke, pole pieces, air gap, and armature core, as shown in figure 20-11.

The armature is mounted on a shaft and is rotated through the field by an outside energy source (prime mover). Thus we have a magnetic field, a conductor, and relative motion between the two—which, it will be remembered, are the three essentials for producing a voltage by magnetism. If the output of the armature is connected across the field windings, the voltage and the field current at start will be small because of the small residual flux in the field poles. However, as the generator continues to run, the small voltage across the armature will circulate a small current through the field coils and the field will become stronger. In a self-excited generator, this action causes the generator voltage to rise quickly to the proper value and the machine is said to "build up" its voltage.

The simplest generator armature winding is a loop or single coil. Rotating this loop in a magnetic field will induce an emf whose strength is dependent upon the strength of the magnetic field and the speed of rotation of the conductor.

A single-coil generator with each coil terminal connected to a bar of a two-segment metal ring is shown in figure 20-12. The two

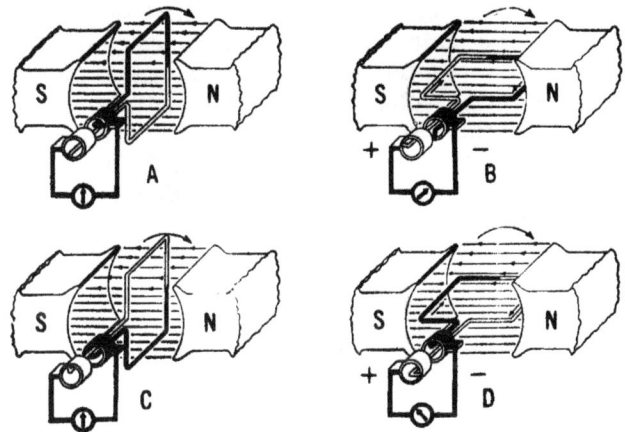

41.10

Figure 20-12.—Single-coil generator with commutator.

segments of the split ring are insulated from each other and the shaft, thus forming a simple commutator which mechanically reverses the armature coil connections to the external circuit at the same instant that the direction of generated voltage reverses in the armature coil.

The emf developed across the brushes is pulsating and unidirectional. Figure 20-13 is a graph of the pulsating emf for one revolution of a single-loop armature in a 2-pole generator. A pulsating direct voltage of this characteristic (called ripple) is unsuitable for most applications. In practical generators, more coils and more commutator bars are used to produce an output voltage waveform with less ripple. Figure

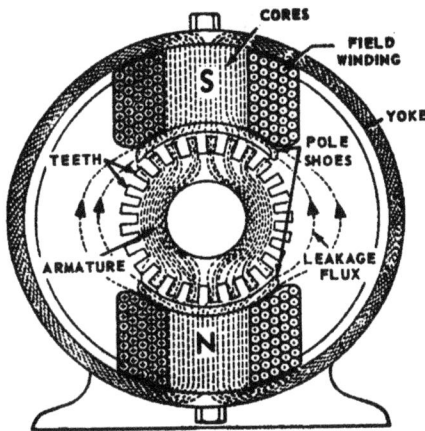

27.248.1

Figure 20-11.—Magnetic circuit of a 2-pole generator.

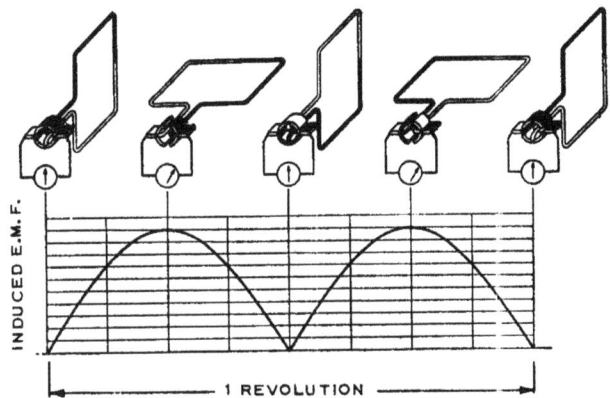

41.10

Figure 20-13.—Pulsating voltage from a single-coil armature.

20-14 shows the reduction in ripple obtained by the use of two coils instead of one. Since there are now four commutator segments and only two brushes, the voltage cannot fall any lower than point A; therefore, the ripple is limited by the rise and fall between points A and B. By adding still more armature coils, the ripple can be reduced still more.

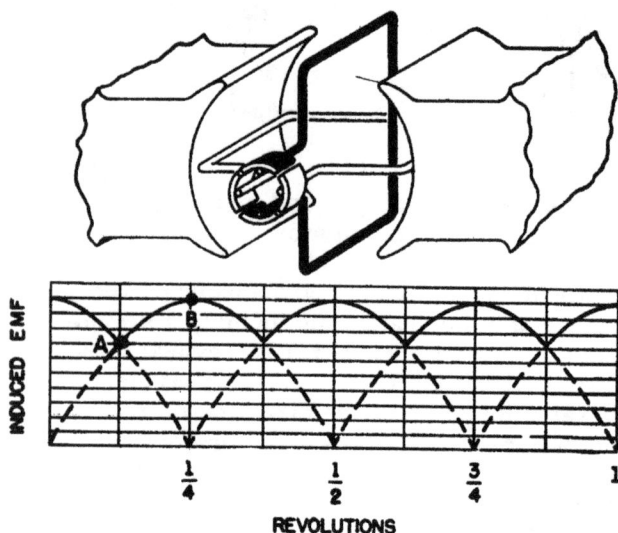

41.9

Figure 20-14.—Voltage from a two-coil armature.

TYPES OF D-C GENERATORS

D-c generators are usually classified according to the manner in which the field windings are connected to the armature circuit (fig. 20-15).

A separately excited d-c generator is indicated in part A of figure 20-15. In this machine the field windings are energized from a d-c source other than its own armature.

Self-excited d-c generators may be of three types, as indicated in part B of figure 20-15. A shunt generator has its field windings connected parallel with the armature, whereas the field windings of a series generator are connected in series with the armature. The compound d-c generator employs both shunt and series field windings.

The d-c generator most widely used in the Navy is the stabilized shunt generator, which employs a light series field winding on the same poles with the shunt field windings. This

type of generator has good voltage regulation characteristics and at the same time ensures good parallel operation.

VOLTAGE CONTROL

Voltage control is either manual or automatic. In most cases, the process involves changing the resistance of the field circuit, thus controlling the field current which permits control of the terminal voltage. The major difference between the various voltage regulator systems is merely the method by which the field circuit resistance is controlled.

DIRECT-CURRENT MOTORS

The construction of a d-c motor is essentially the same as that of a d-c generator. The d-c generator converts mechanical energy into electrical energy, and the d-c motor converts the electrical energy into mechanical energy. A d-c generator may be made to function as a motor by applying a suitable source of direct voltage across the normal output electrical terminals.

There are various types of d-c motors, depending upon the way in which the field coils are connected. Each type has characteristics that are advantageous under given load conditions.

Shunt motors have the field coils connected in parallel with the armature circuit. This type of motor, with constant potential applied, develops variable torque at an essentially constant speed, even under changing load conditions. Such loads are found in drives for such machine shop equipment as lathes, milling machines, drills, planers, and shapers.

Series motors have the field coils connected in series with the armature circuit. This type of motor, with constant potential applied, develops variable torque but its speed varies widely under changing load conditions. The speed of a series motor is low under heavy loads but becomes excessively high under light loads. Series motors are commonly used to drive electric cranes, hoists, and winches.

Compound motors are a compromise between shunt and series motors, having one set of field coils in parallel with the armature circuit and another set of field coils in series with the armature circuit. The compound motor develops an increased starting torque over the shunt motor and has less variation in speed than the series motor.

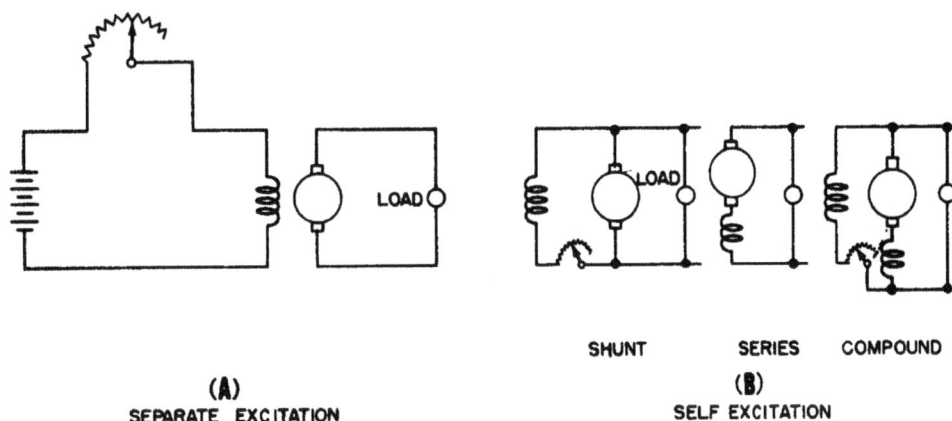

(A)
SEPARATE EXCITATION

SHUNT　SERIES　COMPOUND
(B)
SELF EXCITATION

147.122

Figure 20-15.—Types of d-c generators.

The operation of a d-c motor depends on the principle that a current-carrying conductor placed in, and at right angles to, a magnetic field tends to move at right angles to the direction of the field. A convenient method of determining the direction of motion of a current-carrying conductor in a magnetic field is by use of the right-hand motor rule for electron flow (fig. 20-16). Extend the thumb, index finger, and second finger of the right hand at right angles to each other, with the index finger pointed in the direction of the flux (toward the south pole) and the second finger pointed in the direction of electron flow. The thumb then points in the direction of motion of the conductor with respect to the field.

ALTERNATING-CURRENT THEORY

Just as a current flowing in a conductor produces a magnetic field around the conductor, the reverse of this process is true. A voltage can be generated in a circuit by moving a conductor so that it cuts across lines of magnetic force or, conversely, by moving the lines of force so that they cut across the conductor. An a-c generator utilizes this principle of electromagnetic induction to convert mechanical energy into electrical energy.

In the case of alternating current, electrons move first in one direction and then in the other. Thus the direction of the current reverses periodically and the magnitude of the voltage is constantly changing. This variation in current is represented graphically in sine waveform in figure 20-17.

The vertical projection (dotted line in fig. 20-17) of a rotating vector may be used to represent the voltage at any instant. Vector E_M represents the maximum voltage induced in a conductor rotating at uniform speed in a 2-pole field (points 3 and 9). The vector is rotated counterclockwise through one complete revolution (360°). The point of the vector describes a circle. A line drawn from the point of the vector perpendicular to the horizontal diameter of the circle is the vertical projection of the vector.

The circle also describes the path of the conductor rotating in the bi-polar field. The vertical projection of the vector represents the voltage generated in the conductor at any instant corresponding to the position of the rotating vector as indicated by angle θ. Angle θ represents selected instants at which the generated voltage is plotted. The sine curve plotted at the

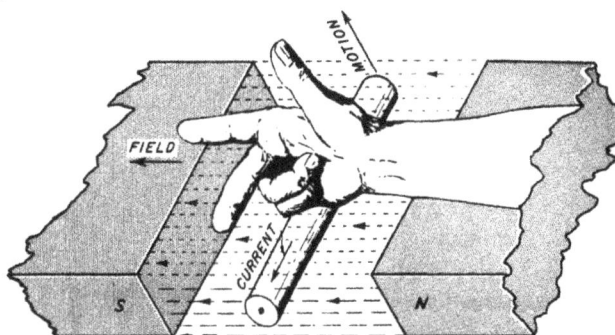

12.143
Figure 20-16.—Right-hand motor rule for electron flow.

503

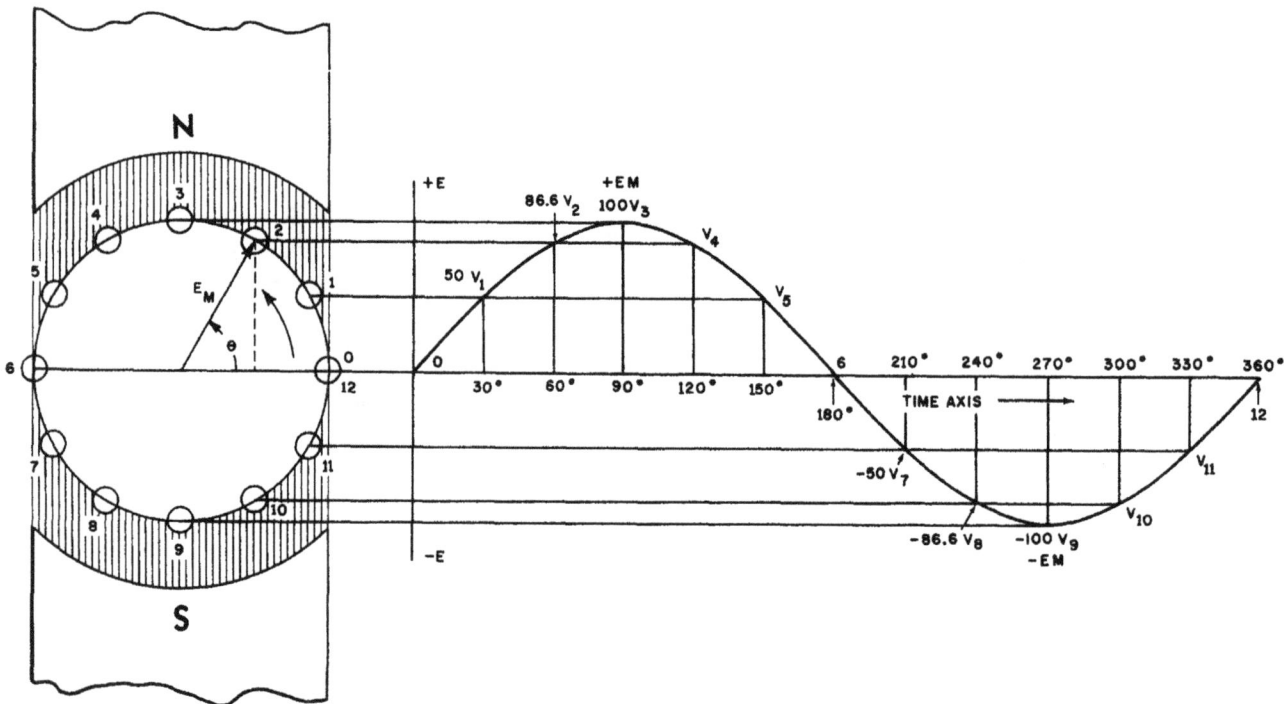

41.19

Figure 20-17.—Generation of sine-wave voltage.

right of the figure represents successive values of the a-c voltage induced in the conductor as it moves at uniform speed through the 2-pole field, because the instantaneous values of rotationally induced voltage are proportional to the sine of the angle θ that the rotating vector makes with the horizontal.

The sine wave in figure 20-17 represents one complete revolution of the armature or one voltage cycle. The <u>frequency</u> of a-c voltage is measured in cycles per second (cps) and may be determined by the following formula:

$$f = \frac{P \times rpm}{120}$$

where

    f = frequency (in cps; according to the National Bureau of Standards Special Publication 304, frequency in cycles per second in the International Systems of Units is expressed as Hertz (H_z). One hertz equals one cycle per second.)

rpm ■ revolutions per minute

    P ■ number of poles in the generator

A generator made to deliver 60 cps, and having two field poles, would need an armature designed to rotate at 3600 rpm.

PROPERTIES OF A-C CIRCUITS

Resistance, the opposition to current flow, has the same effect in an a-c circuit as it does in a d-c circuit. However, in the application of Ohm's law to a-c circuits, other properties must be taken into consideration.

<u>Inductance</u> is that property which opposes any change in the current flow and <u>capacitance</u> is that property which opposes any change in voltage. Since a-c current is constantly changing in magnitude and direction, the properties of inductance and capacitance are always present.

The amount of opposition to current flow in an inductive circuit is referred to as its inductive reactance, X_L. The value of inductive reactance (in ohms) depends on the inductance of the circuit and the frequency of the applied voltage. Expressed in equation form,

$$X_L = 2\pi fL$$

where

X_L = inductive reactance, in ohms
π = 3.1416
f = frequency, in cycles per second
L = inductance, in henrys

The current flowing in a capacitive circuit is directly proportional to the capacitance and to the rate at which the applied voltage is changing. The rate at which the voltage changes is determined by the frequency. The value of the capacitive reactance, X_C, is inversely proportional to the capacitance of the circuit and the frequency of the applied voltage. Thus,

$$X_C = \frac{1}{2\pi f C}$$

where

X_C = capacitive reactance, in ohms
π = 3.1416
f = frequency, in cycles per second
C = capacitance, in farads

The effects of capacitance and inductance in an a-c circuit are exactly opposite. Inductive reactance causes the current to lag the applied voltage and capacitive reactance causes the current to lead the applied voltage. These effects tend to neutralize each other, and the combined reactance is the difference between the individual reactances.

The total opposition offered to the flow of current in an a-c circuit is the impedance, Z. The impedance of a circuit, expressed in ohms, is composed of the capacitive reactance, the inductive reactance, and the resistance.

The effects of capacitive reactance, inductive reactance, and resistance in an a-c circuit can be shown graphically by the use of vectors. For example, consider the series circuit shown in part A of figure 20-18.

The vector representation of the reactances is shown in part B of figure 20-18. Because the inductive reactance and the capacitive reactance are exactly opposite, they are subtracted directly and the difference shown in part C of figure 20-18 as capacitive reactance. The resultant is found vectorially by constructing a parallelogram, as shown in part D of figure 20-18. The resultant vector is also the hypotenuse of a right triangle; therefore,

$$Z = \sqrt{R^2 + (X_C - X_L)^2}$$

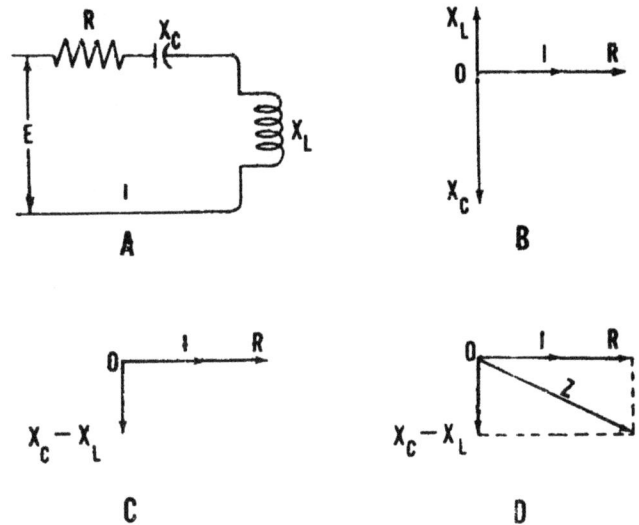

Figure 20-18.—Vector solution of an a-c circuit.

In accordance with Ohm's law for a-c circuits, the effective current through a circuit is directly proportional to the effective voltage and inversely proportional to the impedance. Thus,

$$I = \frac{E}{Z}$$

where

I = current, in amperes
E = emf, in volts
Z = impedance, in ohms

A-C GENERATORS

Most of the electric power for use aboard ship and ashore is generated by alternating-current generators.

A-c generators are made in many different sizes, depending upon their intended use. For example, any one of the generators at Boulder Dam can produce millions of volt-amperes, while generators used on aircraft produce only a few thousand volt-amperes.

Regardless of size, however, all generators operate on the same basic principle: a magnetic field cutting through conductors, or conductors passing through a magnetic field. Thus all generators will have at least two distinct sets of conductors. They are (1) a group of conductors

505

in which the output voltage is generated, and (2) a group of conductors through which direct current is passed to obtain an electromagnetic field of fixed direction. The conductors in which the output voltage is generated are always referred to as the armature windings. The conductors in which the electromagnetic field originates are always referred to as the field windings.

In addition to the armature and field, there must also be relative motion between the two. To provide this relative motion, a-c generators are built in two major assemblies—the stator and the rotor. The rotor rotates inside the stator. The rotor may be driven by any one of a number of commonly used prime movers, including steam turbines, gas turbines, and internal combustion engines.

TYPES OF A-C GENERATORS

In the revolving-armature a-c generator, the stator provides a stationary electromagnetic field. The rotor, acting as the armature, revolves in the field, cutting the lines of force, producing the desired output voltage. In this generator, the armature output is taken through sliprings and thus retains its alternating characteristics.

For a number of reasons, the revolving-armature a-c generator is seldom used. Its primary limitation is the fact that its output power is conducted through sliding contacts (sliprings and brushes). These contacts are subject to frictional wear and sparking. In addition, they are exposed, and thus liable to arc-over at high voltages. Consequently, revolving-armature generators are limited to applications of low power and low voltage.

The revolving-field a-c generator (fig. 20-19) is by far the most commonly used type. In this type of generator, direct current from a separate source is passed through windings on the rotor by means of sliprings and brushes. This maintains a rotating electromagnetic field of fixed polarity (similar to a rotating bar magnet). The rotating magnetic field, following the rotor, extends outward and cuts through the armature windings embedded in the surrounding stator. As the rotor turns, alternating voltages are induced in the windings, since magnetic fields of first one polarity and then the other cut through them. Since the output power is taken from stationary windings, the output may be connected through fixed terminals directly

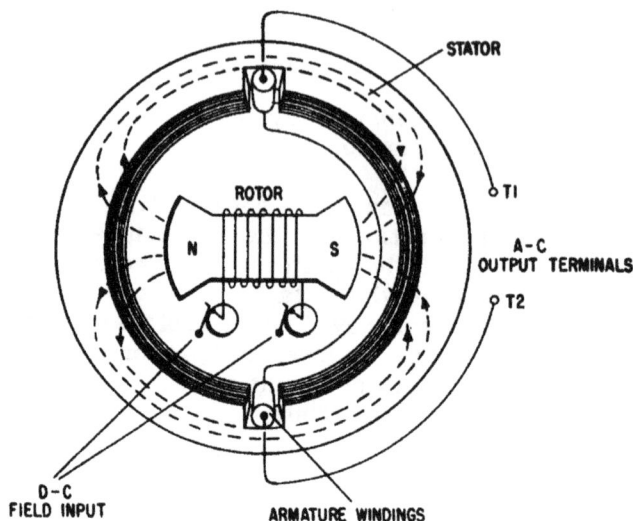

147.124

Figure 20-19.—Essential parts of a rotating-field a-c generator.

to the external loads, as through terminals T1 and T2 in figure 20-19. This is advantageous because there are no sliding contacts and the whole output circuit is continuously insulated, thus minimizing the danger of arc-over.

Sliprings and brushes are still used on the rotor to supply direct current to the field; they are adequate for this purpose because the power level in the field is much lower than in the armature circuit.

THREE-PHASE GENERATORS

The three-phase a-c generator has three single-phase windings spaced so that the voltage induced in each winding is 120° out of phase with the voltages in the other two windings. A schematic diagram of a three-phase stator showing all the coils becomes complex and is difficult to understand. A simplified schematic diagram, showing all the windings of a single phase as one winding, is given in figure 20-20. The rotor is omitted for the sake of simplicity. The waveforms of voltage are shown to the right of the schematic. The three voltages are 120° apart and are similar to the voltages that would be generated by three single-phase a-c generators whose voltages are out of phase by angles of 120°. The three phases are independent of each other.

Rather than have six leads come out of the three-phase alternator, one of the leads from

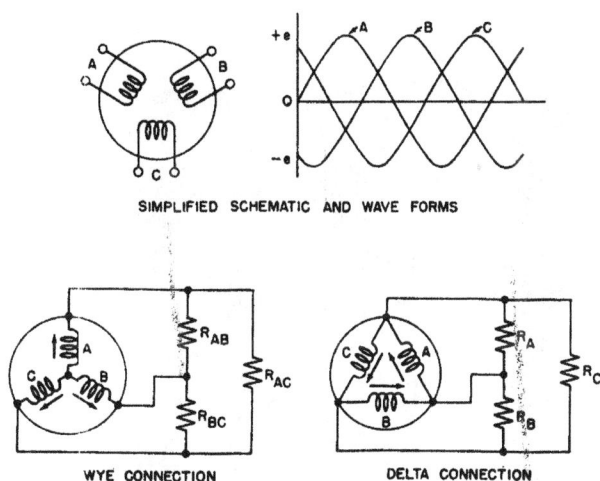

SIMPLIFIED SCHEMATIC AND WAVE FORMS

WYE CONNECTION   DELTA CONNECTION

27.244

Figure 20-20.—Three-phase a-c generator.

each phase may be connected to form a common junction. The stator is then called a wye-connected or star-connected stator. The common lead may or may not be brought out of the machine. If it is brought out, it is called the neutral. The simplified schematic diagram (fig. 20-20) shows a wye-connected stator with the common lead not brought out. Each load is connected across two phases in series. Thus, R_{ab} is connected across phases A and B in series, R_{ac} is connected across phases A and C in series, and R_{bc} is connected across phases B and C in series. Thus the voltage across each load is larger than the voltage across a single phase. The total voltage, or line voltage, across any two phases is the vector sum of the individual phase voltages. For balanced conditions, the line voltage is 1.73 times the phase voltage. Since there is only one path for current in a line wire and the phase to which it is connected, the line current is equal to the phase current.

A three-phase stator can also be connected so that the phases are connected end to end, as shown in figure 20-20. This arrangement is called a delta connection. In the delta connection, the line voltages are equal to the phase voltages. The line currents are equal to the vector sum of the phase currents. The line current is equal to 1.73 times the phase current, when the loads are balanced.

VOLTAGE REGULATION

When the load on an a-c generator is changed, the terminal voltage varies with the load. The

amount of variation depends on the design of the generator and on the amount of reactance from the inductive or capacitive loads. Under practical shipboard operating conditions, the load varies widely with the starting and stopping of motors.

The only practicable way to regulate the voltage output of an a-c generator is to control the strength of the rotating magnetic field. The strength of the electromagnetic field may be varied by changing the amount of current flowing through the coil, which is done by connecting a rheostat in series with the coil. Thus, voltage regulation in an a-c generator is accomplished by varying the field current. This allows a relatively large a-c voltage to be controlled by a much smaller d-c voltage and current.

Since manual adjustment of a-c voltage is not practicable when the load fluctuates rapidly, automatic voltage regulators are used. The construction and operating principles of voltage regulators varies; however, the essential function of any voltage regulator is to use the a-c output voltage, which the regulator is designed to control, as a sensing influence to control the amount of current the exciter supplies to its own control field.

TRANSFORMERS

A transformer (fig. 20-21) is an a-c device that has no moving parts and that transfers energy from one circuit to another by electromagnetic induction. The energy is always transferred without a change in frequency but usually with changes in voltage and current. A step-up transformer receives electrical energy at one voltage and delivers it at a higher voltage. A stepdown transformer receives electrical energy at one voltage and delivers it at a lower voltage. Transformers are not used on direct current.

The conventional constant-potential transformer is designed to operate with the primary connected across a constant-potential source and to provide a secondary voltage that is substantially constant from no load to full load.

Various types of small single-phase transformers are used on shipboard equipment. In many installations, transformers are used on switchboards to step down the voltage for indicating lights. Low-voltage transformers are included in some motor control panels to supply control circuits or to operate overload relays. Other common uses include low-voltage supply

147.125

Figure 20-21.—Single-phase transformer.

for gunfiring circuits, special signal lights, and high-voltage ignition circuits.

The typical transformer has two windings which are electrically insulated from each other. These windings are wound on a common magnetic circuit made of laminated sheet steel. The principal parts are the <u>core</u>, which provides a circuit of low reluctance for the magnetic flux; the <u>primary winding</u>, which receives the energy from the a-c source; and the <u>secondary winding</u>, which receives the energy by mutual induction from the primary and delivers it to the load.

When a transformer is used to step up the voltage, the low-voltage winding is the primary. When a transformer is used to step down the voltage, the high-voltage winding is the primary. The primary is always connected to the source of the power; the secondary is always connected to the load. It is common practice to refer to the windings as the primary and the secondary,

rather than as the high-voltage and the low-voltage windings.

The operation of the transformer is based on the principle that electrical energy can be transferred efficiently by mutual induction from one winding to another. When the primary winding is energized from an a-c source, an alternating magnetic flux is established in the transformer core. This flux links the turns of both primary and secondary, thereby inducing voltages in them. Because the same flux cuts both windings, the same voltage is induced in each turn of both windings. Hence the total induced voltage in each winding is proportional to the number of turns in that winding. That is,

$$\frac{E_1}{E_2} = \frac{N_1}{N_2}$$

where E_1 and E_2 are the induced voltages in the primary and secondary windings, respectively, and N_1 and N_2 are the number of turns in the primary and secondary windings, respectively. In ordinary transformers, the induced primary voltage is almost equal to the applied primary voltage; hence, the applied primary voltage and the secondary induced voltage are approximately proportional to the respective number of turns in the two windings.

A-C MOTORS

A-c motors are manufactured in many different sizes, shapes, and ratings for use in a wide variety of applications. Since this discussion cannot possibly cover all aspects of all kinds of a-c motors, it will be limited to the polyphase induction motor. Information on other types of motors may be found in Basic Electricity, NavPers 10086-A, and in various manufacturers' technical manuals.

The induction motor is a widely used type of a-c motor because it is simple, rugged, and inexpensive. It consists essentially of a stator and a rotor; it can be designed to suit most applications requiring constant speed and variable torque.

The stator of a polyphase induction motor consists of a laminated steel ring with slots on the inside circumference. The stator winding is similar to the a-c generator stator winding and is generally of the two-layer distributed preformed type. Stator phase windings are symmetrically placed on the stator and may be either wye connected or delta connected.

Most induction motors used by the Navy have a cage-type rotor (fig. 20-22) consisting of a laminated cylindrical core with parallel slots in the outside circumference to hold the windings in place. The rotor winding is constructed of individual short circuited bars connected to end rings.

In induction motors, the rotor currents are supplied by electromagnetic induction. The stator windings contain two or more out-of-time-phase currents which produce corresponding magnemotive forces which establish a rotating magnetic field across the air gap. This magnetic field rotates continuously at constant speed, regardless of the load on the motor. The stator winding corresponds to the primary winding of a transformer.

The induction motor derives its name from the fact that mutual induction (transformer action) takes place between the stator and the rotor under operating conditions. The magnetic

77.77
Figure 20-22.—Cage-type induction motor rotor.

revolving field produced by the stator cuts across the rotor conductors, thus inducing a voltage in the conductors which causes rotor current to flow. Hence, motor torque is developed by the interaction of the rotor current and the magnetic revolving field.

POWER DISTRIBUTION SYSTEM

The power distribution system is the connecting link between the generators that supply electric power and the electrical equipment that utilizes this power to furnish the various services necessary to operate the ship. The power distribution system includes the ship's service power distribution system, the emergency power distribution system, and the casualty power distribution system.

Most a-c power distribution systems on naval ships are 450-volt, three-phase, 60-cycle, three-wire systems. The lighting distribution systems are 115-volt, three-phase, 60-cycle, three-wire systems supplied from the power circuits through transformer banks. On some ships, the weapons systems, some I.C. circuits, and aircraft starting circuits receive electrical power from a 400-cps system.

SHIP'S SERVICE POWER

The ship's service power distribution system is the electrical system that normally supplies electric power to the ship's equipment and machinery. The switchboards and associated generators are located in separate engineering spaces to minimize the possibility that a single hit will damage more than one switchboard.

The ship's service generators and distribution switchboards are interconnected by bus ties so that any switchboard can be connected to feed power from its generator to one or more of the other switchboards. The bus ties also connect two or more switchboards so that the generator plants can be operated in parallel (or the switchboards can be isolated for split-plant operation).

In large installations, power distribution to loads is from the generator and distribution switchboards or switchgear groups to load centers, to distribution panels, and to the loads, or directly from the load centers to some loads.

On some ships, such as large aircraft carriers, a system of zone control of the ship's service and emergency power distribution is provided. Essentially, the system establishes a number of vertical zones, each of which contains one or more load center switchboards supplied through bus feeders from the ship's service switchgear group. A load center switchboard supplies power to the electrical loads within the electrical zone in which it is located. Thus, zone control is provided for all power within the electrical zone. The emergency switchboards may supply more than one zone, depending on the number of emergency generators installed. Figure 20-23 shows the ship's service and emergency power distribution system in a large aircraft carrier.

In smaller installations (fig. 20-24) the distribution panels are fed directly from the generator and distribution switchboards. The distribution panels and load centers (if any) are located centrally with respect to the loads they feed to simplify installation. This arrangement also requires less weight, space, and equipment than if each load were connected to a switchboard.

At least two independent sources of power are provided for selected vital loads through automatic bus transfer equipment. The normal and alternate feeders to a common load run from different ship's service switchboards and are located below the waterline on opposite sides of the ship to minimize the possibility that both will be damaged by a single hit.

The lighting circuits are supplied from the secondaries of 450/115-volt transformer banks connected to the ship's service power system. In large ships, the transformer banks are installed in the vicinity of the lighting distribution panels, at some distance from the generator and distribution switchboards. In small ships, the transformer banks are located near the

generator and distribution switchboards and energize the switchboard buses that supply the lighting circuits.

EMERGENCY POWER

The emergency power distribution system is provided to supply an immediate and automatic source of electric power to a limited number of selected vital loads in the event of failure of the ship's service power distribution system. The emergency power system, which is separate and distinct from the ship's service power distribution system, includes one or more emergency distribution switchboards. Each emergency switchboard, supplied by its associated emergency generator, has feeders which run to the bus transfer equipment at the distribution panels or loads for which emergency power is provided.

The emergency generators and switchboards are located in separate spaces from those containing the ship's service generators and distribution switchboards. As previously noted, the normal and alternate ship's service feeders are located below the waterline on opposite sides of the ship. The emergency feeders are located near the centerline and higher in the ship (above the waterline). This arrangement provides for horizontal separation between the normal and alternate ship's service feeders and vertical separation between these feeders and the emergency feeders, thereby minimizing the possibility of damaging all three types of feeders simultaneously.

The emergency switchboard is connected by feeders to at least one and usually to two different ship's service switchboards. One of these switchboards is the preferred source of ship's service power for the emergency switchboard and the other is the alternate source. The emergency switchboard and distribution system are normally energized from the preferred source of ship's service power. If both the preferred and the alternate sources of ship's service power fail, the diesel-driven emergency generator starts automatically and the emergency switchboard is automatically transferred to the emergency generator.

When the voltage is restored on either the preferred or the alternate source of the ship's service power, the emergency switchboard is automatically retransferred to the source that is available (or to the preferred source, if voltage is restored on both the preferred and the alternate sources). The emergency generator

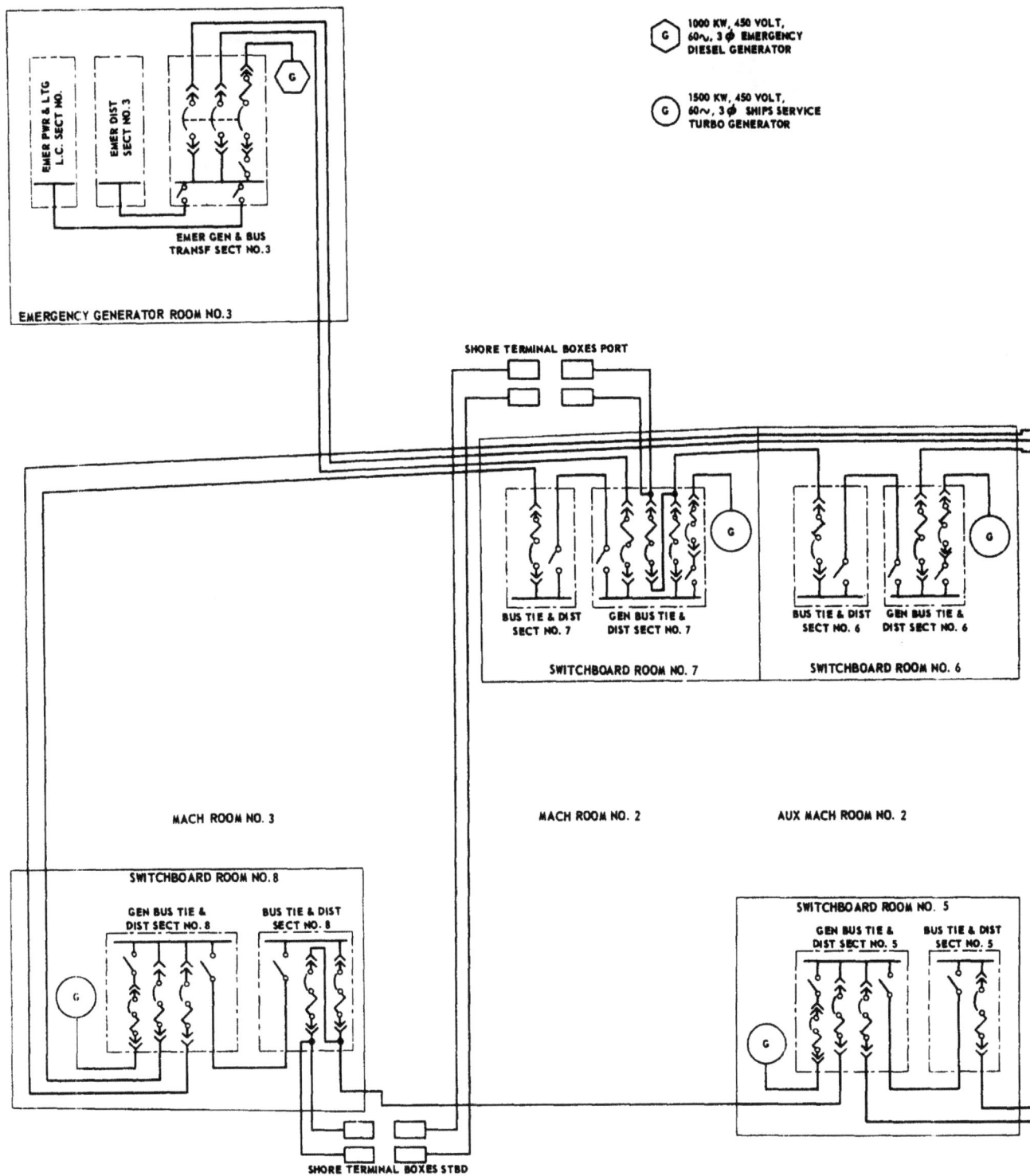

Figure 20-23.—Ship's service and emergency power distribution system in a large carrier.

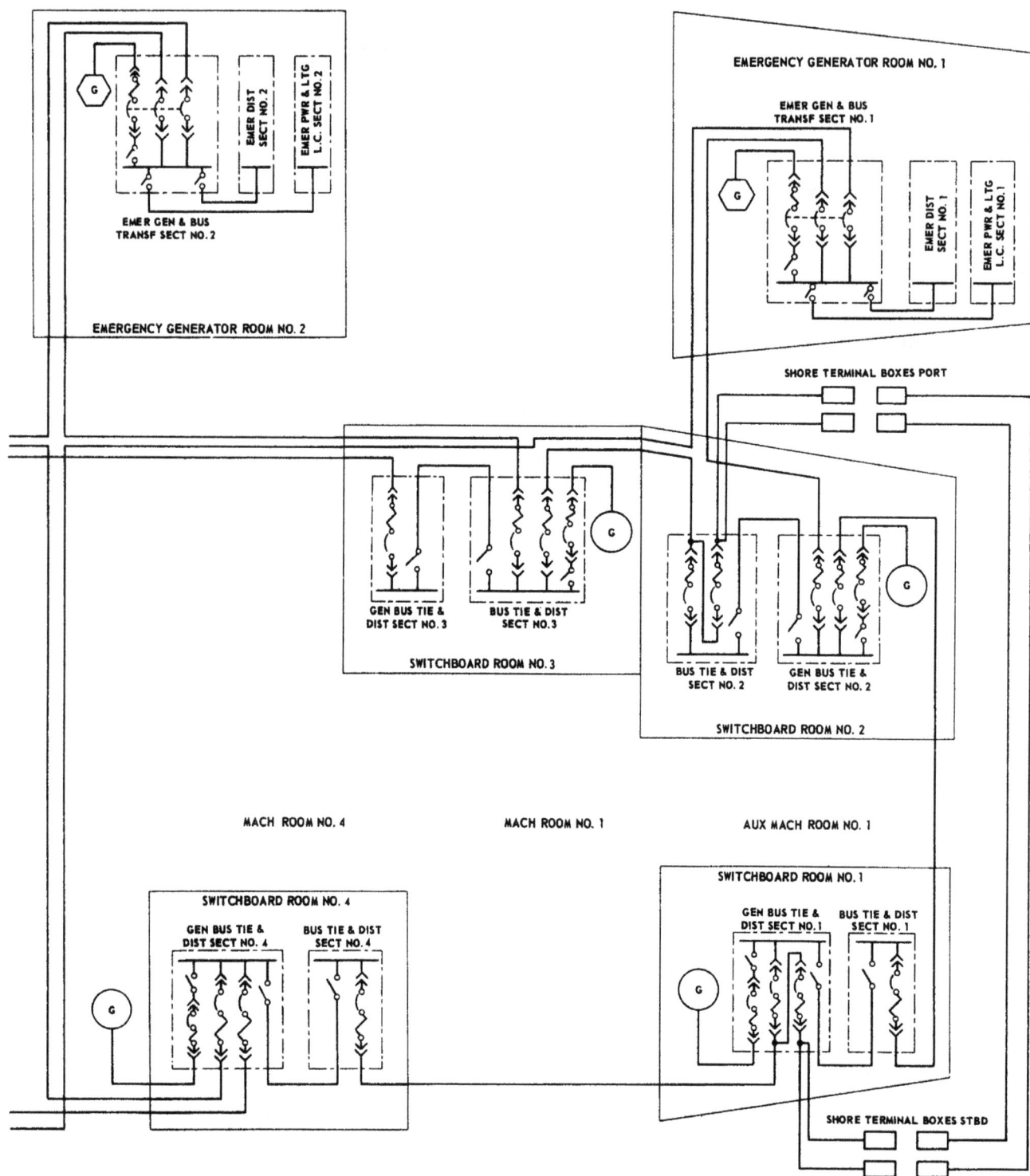

65.54.2

Figure 20-23.—Ship's service and emergency power distribution system
in a large carrier.—Continued.

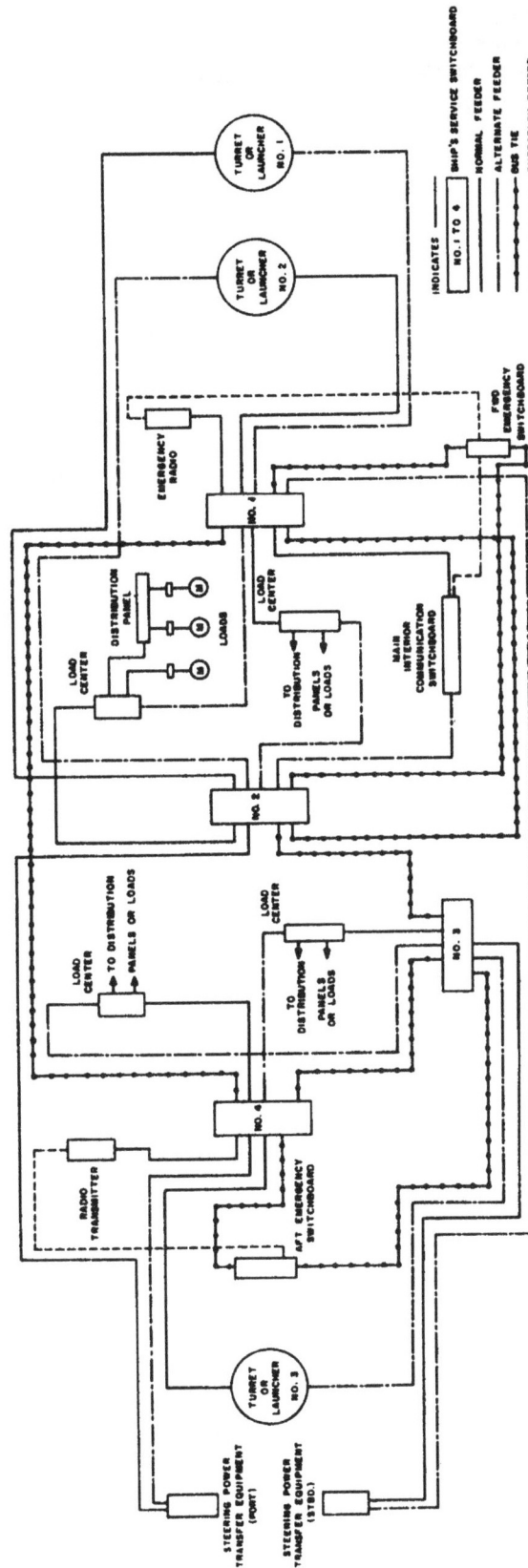

Figure 20-24. — Power distribution in a destroyer.

77. 146X

must be manually shut down. Hence, the emergency switchboard and distribution system are always energized either by a ship's service generator or by the emergency generator. Therefore, the emergency distribution system can always supply power to a vital load if both the normal and the alternate sources of the ship's service power to this load fail. The emergency generator is not started if the emergency switchboard can receive power from a ship's service generator.

A feedback tie from the emergency switchboard to the ship's service switchboard (fig. 20-24) is provided on most ships. The feedback tie permits a selected portion of the ship's service switchboard load to be supplied from the emergency generator. This feature facilitates starting up the machinery after major alterations and repairs and provides power to operate necessary auxiliaries and lighting during repair periods when shore power and ship's service power are not available.

CASUALTY POWER

The casualty power distribution system is provided for making temporary connections to supply electric power to certain vital auxiliaries if the permanently installed ship's service and emergency distribution systems are damaged. The casualty power system is not intended to supply power to all the electrical equipment in the ship but is confined to the facilities necessary to keep the ship afloat and to get it away from a danger area. The system also supplies a limited amount of armament, such as antiaircraft guns and their directors, that may be necessary to protect the ship when in a damaged condition. The casualty power system for rigging temporary circuits is separate and distinct from the electrical damage control equipment, which consists of tools and appliances for cutting cables and making splices for temporary repairs to the permanently installed ship's service and emergency distribution systems.

The casualty power system includes portable cables, bulkhead terminals, risers, switchboard terminals, and portable switches. Portable cables in suitable lengths are stowed in convenient locations throughout the ship. The bulkhead terminals are installed in watertight bulkheads so that the horizontal runs of cables can be connected on the opposite sides to transmit power through the bulkheads without the loss

of watertight integrity. The risers are permanently installed vertical cables for transmitting power through the decks without impairing the watertight integrity of the ship. A riser consists of a cable that extends from one deck to another with a riser terminal connected to each end for attaching portable cables.

CONTROL AND SAFETY DEVICES

The distribution of electric power requires the use of many devices to control the current and to protect the circuits and equipment.

Control devices are those electrical accessories which govern, in some predetermined way, the power delivered to any electrical load. In its simplest form, the control applies voltage to (or removes it from) a single load. In more complex control systems, the initial switch may set into action other control devices that govern the motor speeds, the compartment temperatures, the depth of liquid in a tank, the aiming and firing of guns, or the direction of guided missiles.

Switchboards make use of hand-operated (manual) switches as well as electrically operated controls. Manually operated switches are those familiar electrical items which can be operated by motions of the hand, as with a pushing, pulling, or twisting motion. The type of action required to operate the manually operated switch is indicated by the names of the controls—push-button switch, pull-chain switch, or rotary switch.

Automatic switches are devices which perform their function of control through the repeated closing and opening of their contacts, without requiring a human operator. Limit switches and float switches are representative automatic switches.

The Navy uses many different types of switches and controllers, which range from the very simple to the very complex. A typical a-c across-the-line magnetic controller is shown in figure 20-25.

The simplest protective device is a fuse, consisting of a metal alloy strip or wire and terminals for electrically connecting the fuse into the circuit. The most important characteristic of a fuse is its current-versus-time or "blowing" ability. Three time ranges for existence of overloads can be broadly defined as fast (5 microseconds through 1/2 second), medium (1/2 second to 5 seconds), and delayed (5 to 25 seconds).

77.73X

Figure 20-25.—A-c magnetic controller.

Normally, when a circuit is overloaded or when a fault develops, the fuse element melts and opens the circuit that it is protecting. However, all fuse openings are not the result of current overload or circuit faults. Abnormal production of heat, aging of the fuse element, poor contact due to loose connections, oxides or other corrosion products forming within the fuse holder, and unusually high ambient temperatures will alter the heating conditions and the time required for the element to melt.

A more complex type of protective device is the circuit breaker. In addition to acting as protective devices, circuit breakers perform the function of normal switching and are used to isolate a defective circuit while repairs are being made.

Circuit breakers are available in many types; some may be operated both manually and electrically, while others are restricted to one mode of operation. Figure 20-26 shows a circuit breaker which may be operated either manually or electrically. When operated electrically, the operation is usually in conjunction with a pilot device such as a relay or switch. Electrically operated circuit breakers employ an electromagnet, used as a solenoid, to trip a release mechanism that causes the breaker contacts to open. The energy to open the breaker is derived from a coiled spring, and the electromagnet is controlled by the contacts in a pilot device.

Circuit breakers designed for high currents have a double-contact arrangement, consisting of the main bridging contacts and the arcing contacts. When the circuit opens, the main contacts open first, allowing the current to flow through the arc contacts and thus preventing burning of the main contacts. When the arc contacts are open, they pass under the front of the arc runner, causing a magnetic field to be set up which blows the arc up into the arc quencher and quickly opens the circuit.

SYNCHROS AND SERVOMECHANISMS

Synchros, as identified by the Armed Forces, are a-c electromagnetic devices which are used primarily for the transfer of angular-position data. Synchros are, in effect, single-phase transformers in which the primary-to-secondary coupling may be varied by physically changing the relative orientation of these two windings.

27.73
Figure 20-26.—Circuit breaker.

Synchro systems are used throughout the Navy to provide a means of transmitting the position of a remotely located device to one or more indicators located away from the transmitting area.

Part A of figure 20-27 shows a simple synchro system. When the handwheel is turned, an electrical signal is generated by the synchro transmitter and is transmitted through interconnecting leads to the synchro receivers. The synchro receivers will always turn the same amount and direction and at the same speed as the synchro transmitter.

Part B of figure 20-27 shows the same type of system using mechanical linkage. As may be readily seen, mechanical systems are impracticable because of the need for associated belts, pulleys, gears, and rotating shafts.

Synchro systems are widely used for input control of electromechanical devices (servomechanisms) that position an object in accordance with a variable signal. The essential components of a servomechanism system are the input controller and the output controller.

The input controller provides the means, either mechanical or electrical, whereby the

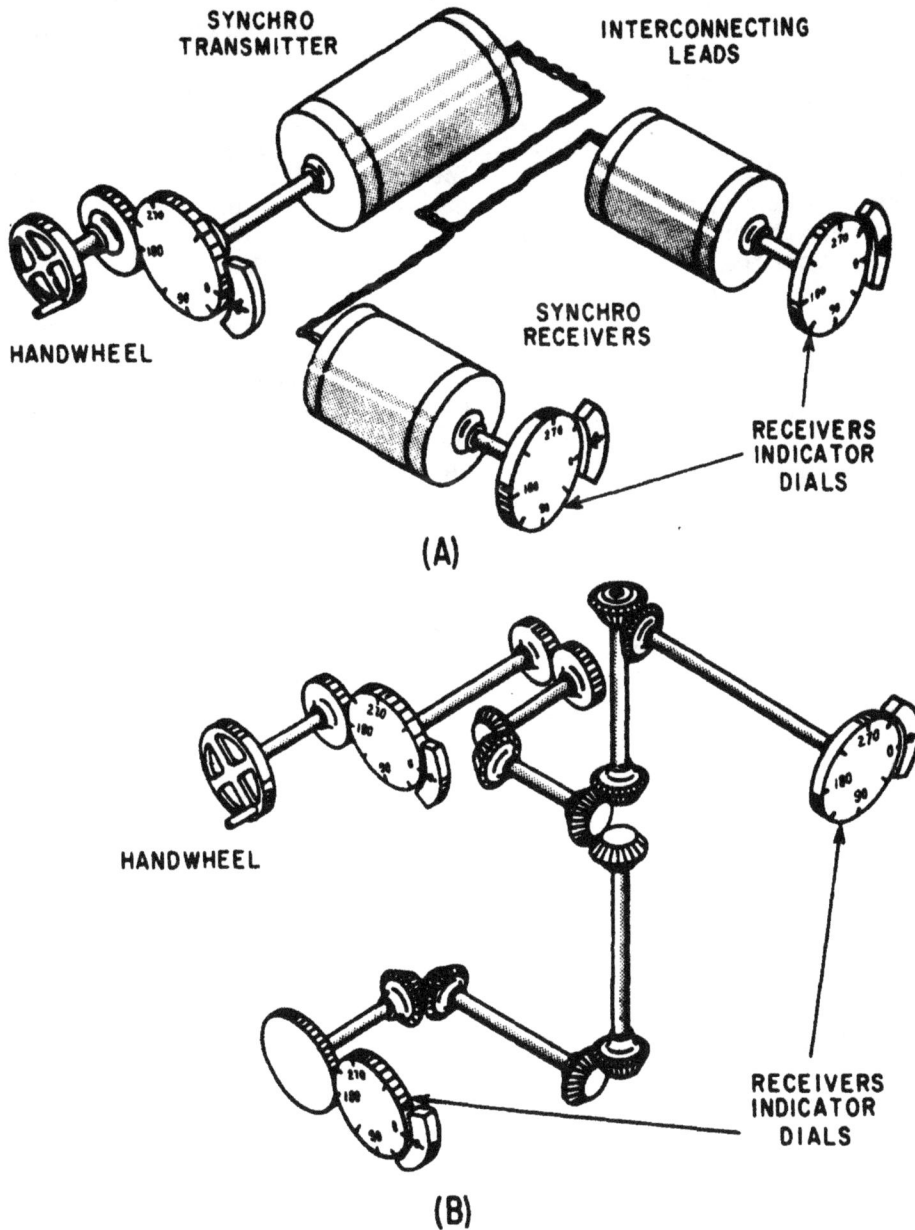

Figure 20-27.—Simple synchro system.

72.38

human operator may actuate or operate a remotely located load.

The output controller of a servomechanism system is the component (or components) in which power amplification and conversion occur. This power is usually amplified by vacuum-tube or magnetic amplifiers and then converted by the servomotor into mechanical motion of

the direction required to produce the desired function.

Figure 20-28 shows a simplified block diagram of a servomechanism. When the shaft of the input controller is rotated in either direction, a voltage is induced in the rotor of the control transformer. This voltage is fed to the amplifier, where it is sent through the necessary stages of amplification, and drives the servo-motor.

DEGAUSSING INSTALLATIONS

A ship is a magnet because of the presence of magnetic material in its hull and machinery. A ship is therefore surrounded by a magnetic field which is strong near the ship and weak at a considerable distance from the ship. As a ship passes over a point on the surface of the earth, the magnetic field of the ship is superimposed upon the magnetic field of the earth, thus tending to distort the earth's field around the ship. If the ship is close to a magnetic mine or torpedo, the distortion caused by the ship's field will activate the firing mechanism to detonate the mine or torpedo.

Degaussing equipment is installed aboard ship to neutralize the disturbance of the earth's magnetic field caused by the ship, and thus to reduce the possibility of detonating a magnetic mine or torpedo. A shipboard degaussing installation consists of one or more coils of electric cable in specific locations inside the ship's hull, a d-c power source to energize these coils, and a means of controlling the magnitude and polarity of the current through the coils. Compass-compensating equipment, consisting of compensating coils and control boxes, is also installed as a part of the degaussing system, to compensate for the deviation effect of the degaussing coils on the ship's magnetic compasses.

Naval ships are tested periodically at magnetic range stations to determine the configuration of the ship's magnetic field. Sensitive

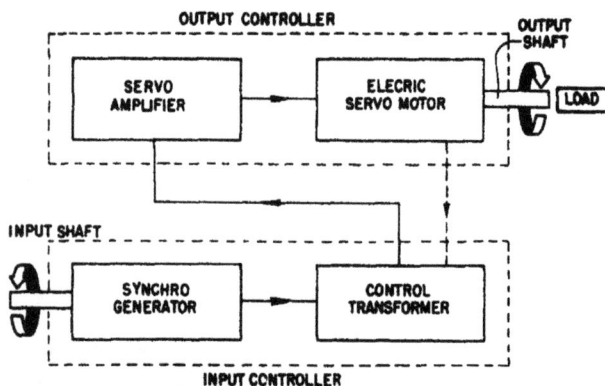

72.46

Figure 20-28.—Simplified block diagram of a servomechanism.

measuring coils, located at or near the bottom of the channel, and recording equipment respond to the signals induced in the coils as the ship passes over them. These measurements indicate the distortion of the earth's magnetic field caused by the ship and are used to determine the values of current needed in the ship's degaussing coils to neutralize this distortion.

GYROCOMPASSES

The gyrocompass system provides a means of indicating the ship's course at various stations throughout the ship. There are various types of gyrocompasses in use; however, all depend upon gyroscopic principles and the rotation of the earth for their operation.

The gyroscope is a heavy wheel, or rotor, suspended so that it has the freedom to spin, the freedom to turn, and the freedom to tilt. These three degrees of freedom permit the rotor to assume any position within the supporting frame.

When a gyroscope rotor is spinning rapidly, the gyroscope develops two properties it does not have when the rotor is at rest. These two properties, which make it possible to develop the gyroscope into a gyrocompass, are (1) rigidity of plane, and (2) precession.

Rigidity of plane results from the fact that the rotating wheel of a gyroscope has high angular momentum and kinetic energy. When the rotor is set spinning with its axle pointed in one direction, it will continue to spin with its axle pointing in the same direction, no matter how the frame of the gyroscope is tilted or turned.

Any force that attempts to change the angle of the plane of rotation of a gyroscope with respect to its earlier position produces a movement known as precession. Precession takes place whenever any torque tends to tilt the axle of a spinning gyroscope rotor.

A gyrocompass is simply a gyroscope with a means of exerting a force at right angles to the end of the axle whenever the axle tilts with respect to the surface of the earth. Because of the rotation of the earth, the axle tilts whenever it is not on the meridian. The axle is precessed automatically into a north-south direction.

SAFETY PRECAUTIONS

Because of the possibility of injury to personnel, the danger of fire, and the possibility of damage to material, all repair and maintenance work on electrical equipment should be performed only by duly authorized and assigned persons.

When any electrical equipment is to be overhauled or repaired, the main supply switches or cutout switches in each circuit from which power could possibly be fed should be secured in the open position and tagged. The tag should read: "This circuit was ordered open for repairs and shall not be closed except by direct order of_____." The name given is usually the name of the person directly in charge of the repairs. After the work has been completed, the tag or tags should be removed by the same person.

The covers of fuse boxes and junction boxes should be kept securely closed except when work is being done. Safety devices such as interlocks, overload relays, and fuses should never be altered or disconnected except for replacements. Safety or protective devices must never be changed or modified in any way without specific authorization.

Fuses should be removed and replaced only after the circuit has been deenergized. When a fuse blows, it should be replaced only with a fuse of the correct current and voltage ratings. When possible, circuit should be carefully checked before the replacement is made, since the burned-out fuse is often the result of a circuit fault.

CHAPTER 21

OTHER AUXILIARY EQUIPMENT

In addition to the shipboard auxiliary machinery described in previous chapters of this text, there are a number of other units of machinery that are essential to the operation of a ship and which are directly or indirectly of concern to engineering department personnel. Such auxiliary machinery includes steering gears and their remote control equipment, elevators, winches, capstans, windlasses, and catapults. Some of this machinery may be located within the engineering spaces of the ship; but many of the units are located outside the engineering spaces and are sometimes referred to as outside machinery.

ELECTROHYDRAULIC TRANSMISSION

Some shipboard auxiliary machinery must operate at variable speeds over a considerable range. In addition, there must be close control of speed between minimum and maximum limits. Many auxiliary machines operate with a high starting torque and must be capable of accelerating to maximum speed very quickly. To meet these requirements, the electrohydraulic transmission is used on naval ships. Since the electrohydraulic transmission is utilized in more than one type of auxiliary machine, it is discussed first.

Electrohydraulic transmissions are used for driving or controlling machinery such as steering gears, gun turrets, anchor windlasses, boat and airplane handling equipment, capstans, hoists, and certain shipboard valves. Some electrohydraulic transmissions are designed to deliver rotary motion; others are designed to deliver reciprocating motion.

An electrohydraulic transmission designed to deliver rotary motion to an auxiliary machine consists basically of an electric motor, a hydraulic pump, a hydraulic motor, and piping to allow the flow of fluid from and to the pump through the motor. The pump is of the variable displacement reversible type[1]; it is sometimes called the A-end of the transmission. A complex machinery system may include one or more pumps and one or more motors. The hydraulic motor of an electrohydraulic transmission is similar in design to the pump, except that the motor is usually of fixed displacement. Occasionally, to provide very wide speed variation, the motor may also be of the variable displacement type. A hydraulic motor for use in an electrohydraulic transmission is shown in figure 21-1.

The components, control equipment, and piping system for an electrohydraulic transmission used to drive a winch are illustrated in figure 21-2. The transmission illustrated is typical of those designed to deliver rotary motion to various types of shipboard auxiliary machinery.

The pump is driven at constant speed by an electric motor. The hydraulic motor is driven by the fluid under pressure, and the auxiliary machine is driven by the mechanical output of the motor. By controlling the variable output of the pump, the direction and speed of rotation of the motor can be controlled; therefore, the direction and speed of motion utilized in the operation of the auxiliary machine can be controlled.

The flow of fluid under pressure from the pump to the motor exerts force on the faces of the pistons open to the valve port receiving the fluid under pressure. This force on the piston results in a thrust component along the axis of rotation of the socket ring and a turning component at right angles to the thrust component.

---

[1]Operating principles of variable displacement pumps are discussed in chapter 15 of this text.

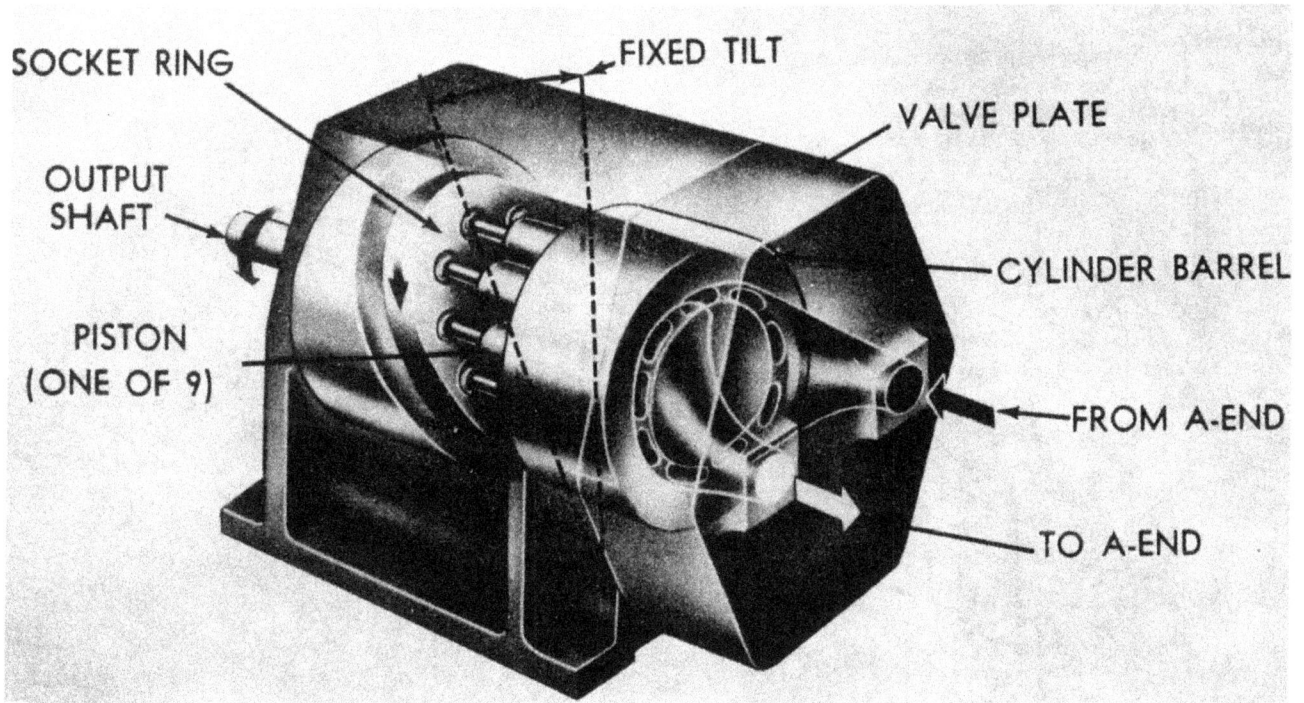

110.23

Figure 21-1.—Hydraulic motor for electrohydraulic transmission.

(See fig. 21-3.) the turning component rotates the socket ring. The rotation of the socket ring causes the cylinder barrel and output shaft of the motor to rotate, and thereby provides the rotary motion utilized to drive a machine.

When reciprocating motion is required, as in the case of a steering gear (fig. 21-4), the motor of an electrohydraulic transmission is replaced by a piston or plunger. The complete hydraulic assembly of which the plunger is a part is commonly called a ram. The force of the hydraulic fluid from the pump causes the movement of the piston or plunger. The tilting box in the pump can be controlled either locally (as on the anchor windlass) or by remote control (as on the steering gear).

STEERING GEARS

The steering gears installed on naval ships are of two types: electromechanical and electrohydraulic. Most modern naval ships have steering gears of the electrohydraulic type; however, electromechanical steering gears are also described briefly.

ELECTROMECHANICAL STEERING GEAR

Electric motors were first introduced as prime movers of steering gears on combatant ships to serve in case of failure of the steam steering engines, at one time the only prime mover used. Electric motors were used later as the primary source of power, with steam as a reserve. Steam is no longer used as a prime mover for steering gears. The use of electromechanical steering gear is now limited to small noncombatant vessels.

The principles of operation are about the same for all designs of electromechanical steering gear. Any differences that exist are chiefly in the manner in which the driving motor is connected to the tiller and the method by which the motor is controlled. The motor may drive the tiller by means of gears and a quadrant, a right- and left-hand screw assembly (fig. 21-5), or by means of wire rope from a drum. In the gear and quadrant type, the steering engine is located in the steering gear room; in the wire rope and drum type, it may be installed in a nearby machinery space.

521

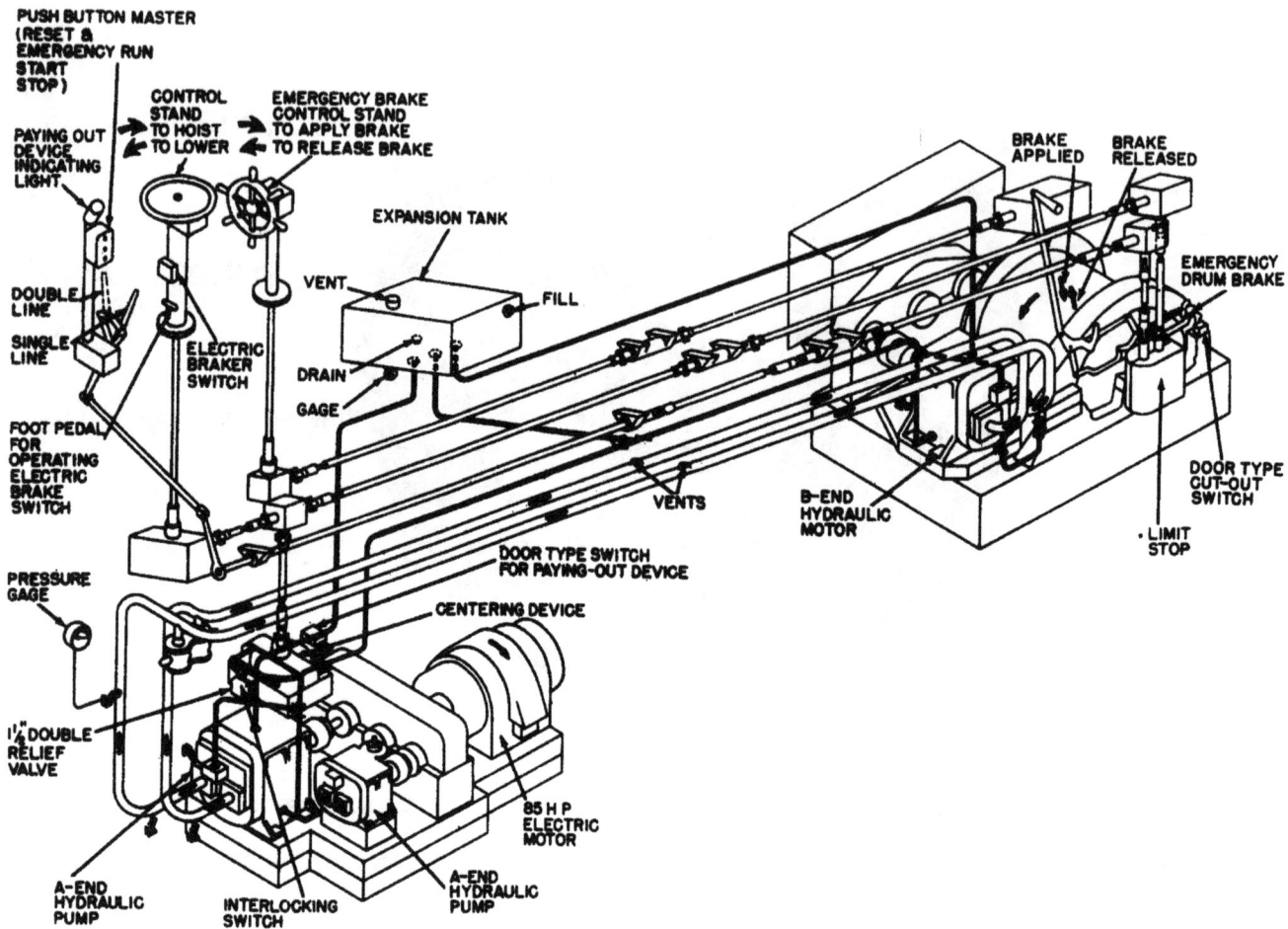

147.128

Figure 21-2.—Electrohydraulic transmission.

The steering motor control may be either of the follow-up or the nonfollow-up type. In the nonfollow-up type, the motor is controlled by a master controller at the steering station. When the master controller is brought to neutral, dynamic braking action takes place to slow down the motor; the motor is finally brought to rest and held by a magnetic brake.

In the follow-up type of control, the follow-up feature is incorporated in a contact ring assembly in the steering stand. The rings make contact with rollers which control the circuits to contractors on the control panel. Movement of the steering wheel rotates the contact rollers in the proper direction to start the motor. Motion of the steering motor is transmitted through shafting to the contact ring assembly, which follows up the motion of the rollers. By this action the rudder is moved an amount proportional to the rotation of of the steering wheel.

Most wire rope and drum type steering gears utilize a follow-up control arrangement. The follow-up motion is transmitted from the steering gear to the steering stand by means of shafting, bevel gears, and flexible couplings.

In most electromechanical installations, the shafting connecting the steering engine to the wheel is utilized not only to provide follow-up control to the steering stand but also to provide a means for steering by hand from the pilot house if power is lost.

ELECTROHYDRAULIC STEERING GEAR

Steering gear installations on most modern naval ships are of the electrohydraulic type.

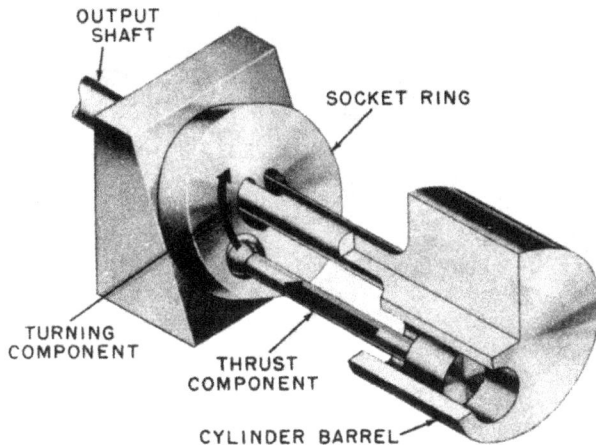

83.82

Figure 21-3.—Thrust and turning components in hydraulic motor operation.

The development of this type of steering gear was prompted primarily by the large momentary electric power requirements for electromechanical steering gears—particularly for ships of large displacement and high speed, with attendant increased rudder torques.

Electrohydraulic steering gears in use include various types of equipment. Some shipboard installations have double hydraulic rams and cylinders; others have single-ram arrangements. (See figs. 21-4 and 21-6.)

STEERING GEAR ARRANGEMENTS.—Only the pump of the previously described electrohydraulic transmission is used in electrohydraulic steering gears. Axial-piston variable displacement pumps are used in most installations; radial-piston pumps are used in some.

The pumps are connected by piping to the ram cylinders of the steering gear. Two pipes from each pump are united at a main transfer valve. The transfer valve is a multiported valve which permits the ram cylinders to be connected to either pump while the pipes from the other pump are connected for bypassing.

Various methods are used for connecting the hydraulic rams to the tiller. The arrangements depend on the design and on the space available for the installation. Two common arrangements are shown in figures 21-4 and 21-6.

Typical cruiser steering gear installations include two rams set fore and aft, one on either side of the rudder stock. The rams operate the rudder through a double yoke tiller fitted with sliding blocks.

The gear illustrated in figure 21-6 is typical of those installed on destroyers. The installation includes a single ram set athwartship. The ram operates the rudder through a single yoke tiller fitted with a sliding block.

Some cruiser steering gear installations have two rams set fore and aft but located forward from the rudder stock (fig. 21-4). The rams are connected to the tiller by connecting links and pins. Some ships are equipped with twin rudders and an independent steering gear for each rudder. Carriers and auxiliary ships may have any one of the above-mentioned steering gear arrangements.

PRINCIPLES OF OPERATION.—Regardless of the type of equipment (double ram or single ram, axial pump or radial pump) included in electrohydraulic steering installations, the principles of operation are basically the same. The discharge volume and direction of flow from the variable displacement pumps are controlled by the operation of the tilting block in the pump. This control is accomplished mechanically by means of trick wheels in the steering gear room, and by remote control from one or more steering stations.

Any movement, right or left, of the control from any of the various steering stations places the hydraulic pump on stroke and causes the pump to supply liquid under pressure to the hydraulic rams, resulting in a corresponding right or left movement of the rudder. This rudder movement actuates the follow-up gear which in turn immediately acts to return the pump control to neutral but does not accomplish this until the assigned rudder position has been attained. The rudder is held in the assigned position by a hydraulic lock until another movement is originated at the steering station.

EMERGENCY STEERING SYSTEMS.—All naval combatant and auxiliary ships equipped with electrohydraulic steering gears are also equipped with an auxiliary steering gear. This emergency steering system generally consists of a relief and shuttle valve, hand-operated hydraulic pump, and the piping, valves, and fittings necessary to complete the system. The emergency equipment is installed in or near the steering gear compartment.

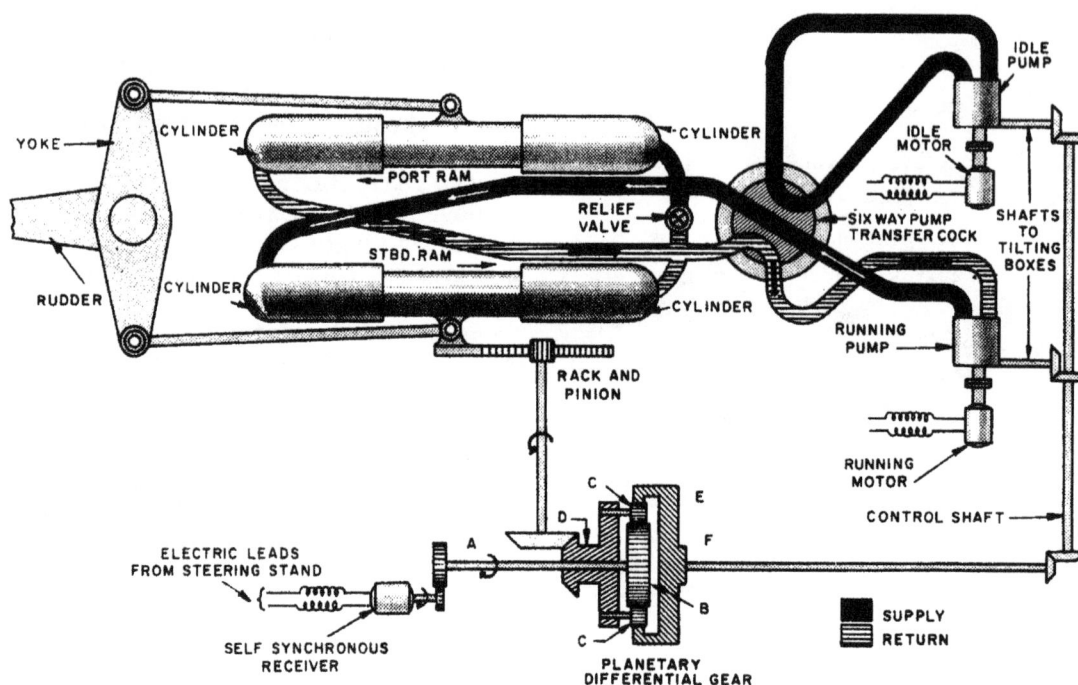

47.139X

Figure 21-4.—Arrangement of a double-ram electrohydraulic steering gear.

To prevent the pressure developed by the hand pump from causing motoring or leakage through the main hydraulic units, the piping from the emergency pump to the main hydraulic system is so arranged that the high-pressure stop valves may be closed. The emergency pump is usually connected to the main hydraulic system in a manner whereby all ram cylinders will be in use. Since it is necessary to block off the emergency system under normal steering gear operation, the emergency lines are usually connected to the drain valves to eliminate the necessity of additional high pressure valves.

Some ships are equipped with a dual emergency, submersible steering system. The purpose of this system is to provide emergency steering by means of either electric motor or hand power in event of failure of the main system. This emergency gear, with driving motor integrally mounted, is located in the steering gear compartment of the ship. Hand operation is accomplished by use of a remotely located crank stand connected to the unit by shafting.

REMOTE CONTROLS — Electrohydraulic steering gears may be controlled from remote steering stations (1) electrically by either a pilot motor and its controller or by a synchronous transmission; (2) hydraulically by means of a telemotor system; or (3) mechanically by means of wire rope. (See subsequent section on remote control systems.)

Only a few pilot motor control systems are in use; the majority of naval ships utilize either synchronous transmission or hydraulic telemotor systems.

In control arrangements for electrohydraulic steering gears, the trick wheel and the receiver of the control system, either synchronous receiver, hydraulic telemotor receiver, or pilot motor, are geared to and actuate the pump control cam through one end of a differential to put the pump on stroke. The follow-up acts through the opposite end of the differential to reverse the movement of the cam and to take the pump off stroke. The differential control unit and cam are so arranged that the control unit may lead the rudder by the full amount of rudder travel.

REMOTE CONTROL SYSTEMS
FOR STEERING GEAR

Control of the steering gear from the steering wheel on the bridge may be accomplished by any of the following remote control systems.

524

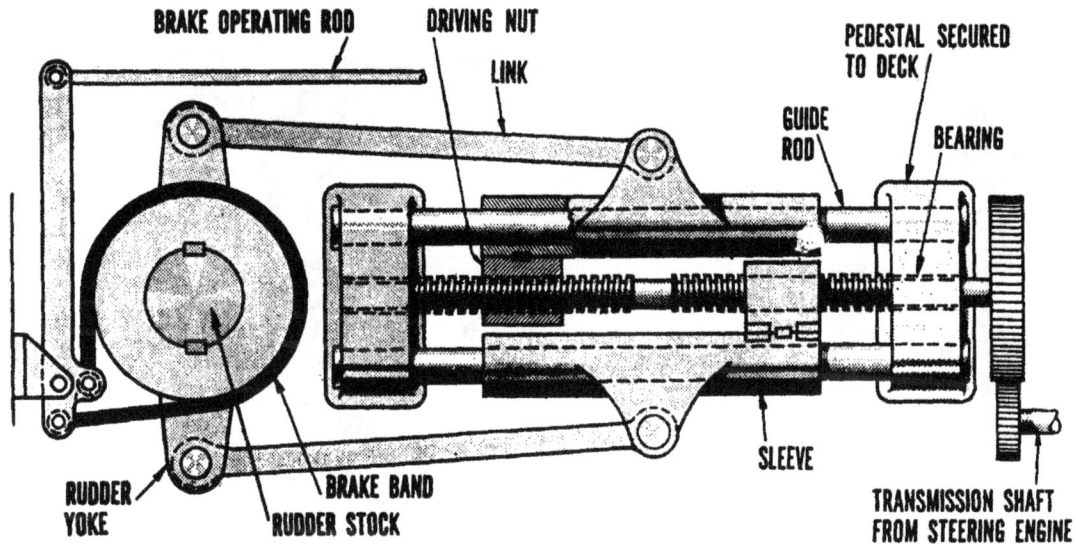

47.140X

Figure 21-5. —Right-and-left screw steering gear.

ELECTRICAL SYSTEMS. —Steering gear control systems of the electrical type are divided into two general types—the direct current pilot motor type and the alternating current synchronous transmission type. The direct current pilot motor type is no longer used on new construction.

The direct current pilot motor type of remote control consists of a small reversible direct current motor which is connected through the differential gear to the control shaft of a variable displacement hydraulic pump. The control of the pilot motor is effected by means of a magnetic contactor control panel located adjacent to the motor and through master controllers located at remote control stations. The motor is equipped with a magnetic brake which promptly stops and holds the motor when the master controller is returned to the neutral position.

The alternating current synchronous transmission type of remote control consists of interchangeable receiving and transmitting units which are, in reality, small wound rotor induction motors with interconnected three-phase rotor windings; their stator windings are connected to the same alternating current supply. When the transmitter rotor is turned, the receiver rotor turns at the same speed and in the same direction.

The transmitters are located in steering stands at remote control stations such as the pilot house, conning tower, central station, etc., and are mechanically connected through gearing to the wheels. A transmitter at one of the remote stations is electrically connected to a receiver in the steering room. Where more than one remote steering station is provided, as on cruisers and carriers, a switch is provided for selecting the desired control station. Indicating lights are provided on the steering stands and at the selector switch to indicate the selected circuit and the power available.

The receiver is connected to the control shaft of the variable displacement hydraulic pump through a differential. On large hydraulic units where the torque required to stroke the pump is greater than the torque that can be exerted by the receiver, the stroke is controlled through an auxiliary hydraulic servosystem. In installations involving the use of a servosystem, the synchronous receiver actuates a pilot valve which controls the flow of oil, under pressure, to and from a power cylinder. The direction and amount of motion of the power piston controls the stroke of the main pump which actuates the rudder.

Electrical control circuits from the transmitter selector switch (from the pilot house selector switch in the case of destroyers and auxiliary vessels) to the steering gear compartment are installed in duplicate. Drum-type selector switches, one in the steering room and one located at the terminus of the duplicate run of control circuits, are provided for selecting the port or starboard cable. When independent synchronous receivers are provided for each

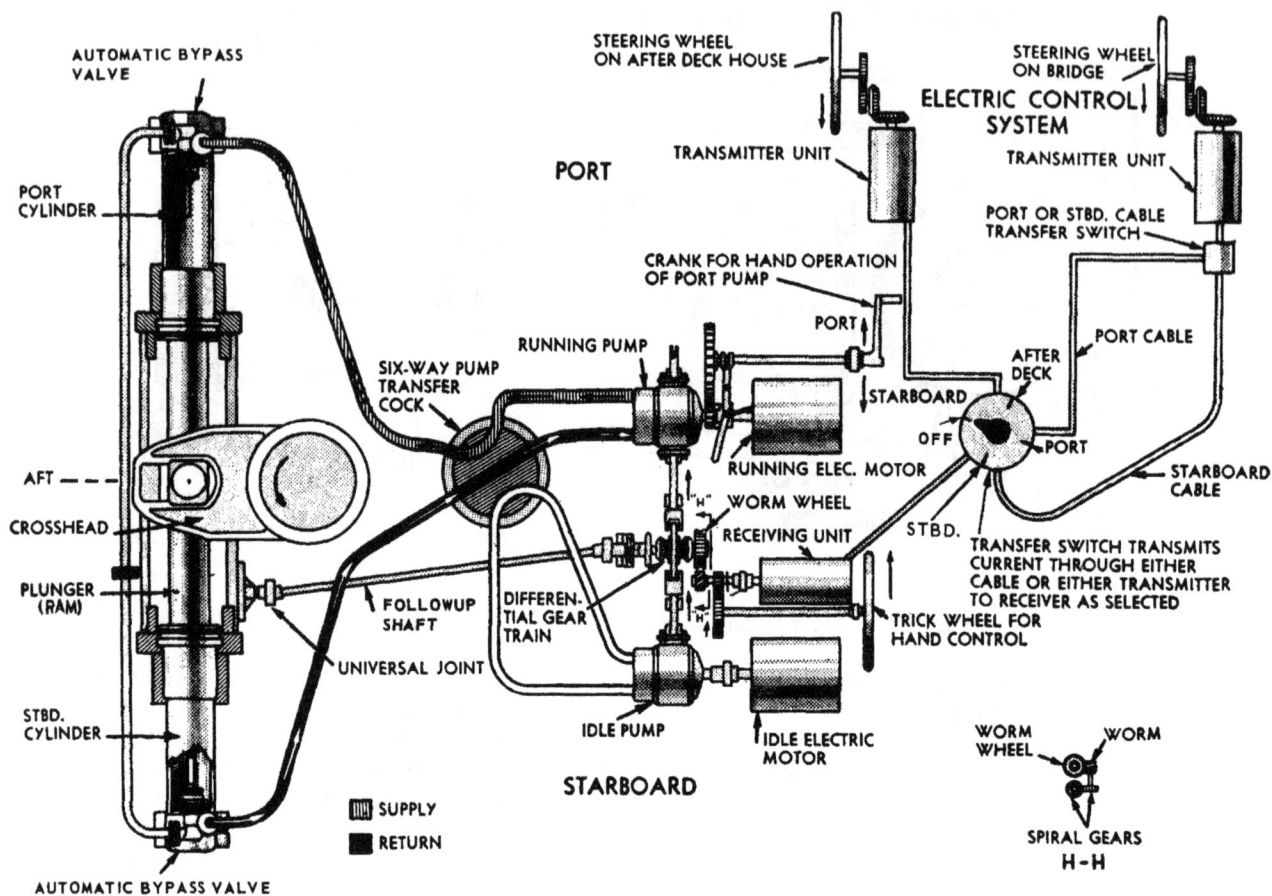

27.83X

Figure 21-6.—Single-ram electrohydraulic steering gear.

steering gear, an additional switch is provided in the steering room for selecting the proper receiver.

HYDRAULIC TELEMOTOR SYSTEM.—A telemotor is a hydraulic device by means of which the motion of the steering gear is controlled from the pilot house. In general, telemotor systems are employed for remote steering control where it is impractical to provide an electrical synchronous transmission system and where the length of runs of shafting or wire rope and the paths for such shafting or ropes would make the use of these types of mechanical controls impracticable.

The hydraulic telemotor system consists of one or more transmitters located at remote steering stations connected by piping to a receiver or receivers located in the steering engineroom. Each transmitter unit is either a

a pair of cylinders or a fixed-delivery piston-type pump connected so that movement of the steering wheel causes fluid to flow through the system, resulting in a corresponding movement of the plungers in the receiver unit. The receiver unit is connected by suitable means to the pump control or valve operating mechanism of the steering engine. The principal components of a hydraulic telemotor control system are shown schematically in figure 21-7.

WIRE ROPE SYSTEM.—This type of remote control is found in some small ships. The steering engine control mechanism is connected to the wheel by wire ropes. The system has the disadvantages of requiring long leads involving large friction loads; of the ropes being vulnerable to gunfire above decks; of impairing watertight integrity by passage of the cables through bulkheads and decks; and of requiring a

PILOT HOUSE STEERING WHEEL

INDICATOR LIGHTS

TO PILOT HOUSE TOP STEERING STATION

DISENGAGE CLUTCH WHEN STEERING FROM PILOT HOUSE TOP

HELM ANGLE INDICATOR

DIMMER KNOB AND SWITCH I

ELECTRIC HELM ANGLE RECEIVER

ENGAGE CLUTCH WHEN STEERING FROM PILOT HOUSE TOP

STEERING CONSOLE
(IN PILOT HOUSE)

ELECTRIC CABLE TO HELM ANGLE TRANSMITTER

FILLING CAP

REPLENISHING TANK

LIQUID LEVEL GAGE
(TRUE LOCATION ON OPPOSITE SIDE)

AIR COCK

BYPASS VALVE LEVER

CAM

RELIEF VALVE

AUTO STEERING INTERLOCK SWITCH

BYPASS VALVE

CHECK VALVE

HYDRAULIC PUMP(CUTAWAY TO SHOW PISTONS, AND TILTED PLATE)

AIR COCK

CROSSHEAD FOR CONNECTION TO STEERING GEAR

PLUNGER

HYDRAULIC CYLINDER

ELECTRIC HELM ANGLE TRANSMITTER

ADJUSTABLE STOP

HELM ANGLE TRANSMITTER RACK AND PINION

ADJUSTABLE STOP

RECEIVER
(IN STEERING GEAR COMPARTMENT)

FILLING OR CHARGING CONNECTION

47.142X

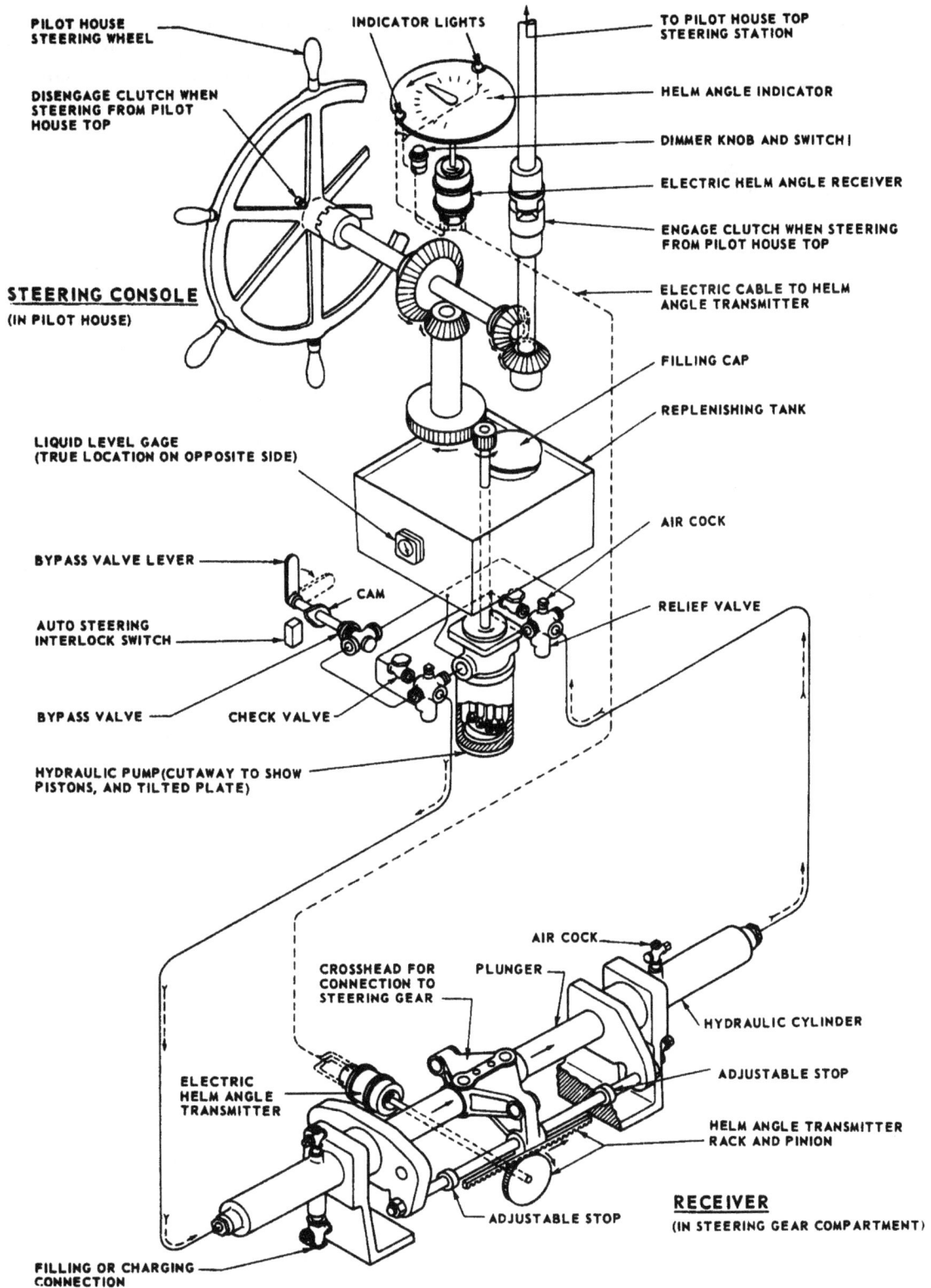

Figure 21-7.—Hydraulic telemotor control system.

comparatively great amount of time for maintenance.

ELECTROHYDRAULIC ELEVATORS

Many naval ships are equipped with electrohydraulic elevators which are used to handle airplanes, bombs, freight, mines, torpedoes, ammunition, and other material. Electrohydraulic elevators may be divided into two general types: the direct plunger lift and the plunger-actuated wire rope lift.

DIRECT PLUNGER LIFT ELEVATORS.— The platform of the direct plunger lift type elevator is raised and lowered by direct connection under the platform, with one or more vertical hydraulic rams. Oil from a high pressure tank is directed into the ram during the hoisting operation. Lowering is accomplished by the oil being discharged from the rams into a low pressure tank. Pressure is maintained in the high pressure tank by means of two electrical variable displacement pumps, which take suction from the low pressure tank. One of the pumps is capable of maintaining elevator operation at reduced speed. Two electric sump pumps keep the volume of oil in the pressure system within specified limits.

Special control valves (operated by pilot valves or a motor) in the pressure and exhaust lines regulate elevator speeds by varying the amount of oil admitted to or discharged from the rams. Positive stops and mechanical locks, interlocked with the elevator control system, enable the platform to be stopped, locked, and held in position at deck level. An equalizer system maintains the platform at uniform level under conditions of unequal loading. Automatic quick-closing valves in the oil line prevent an unrestricted fall of the elevator.

PLUNGER-ACTUATED WIRE ROPE LIFT ELEVATORS.—The primary difference between the direct plunger lift elevator and the plunger-actuated wire rope lift elevator is that the latter type is raised by wire rope fastened to the platform at two or four symmetrically located points. Most hydraulic airplane elevators are of the plunger-actuated wire rope lift type. The wire ropes in an airplane elevator, through a series of sheaves, are actuated by a horizontal hydraulic ram located beneath the hangar deck.

Hydraulic bomb elevators differ from plunger-actuated wire rope lift elevators in that the hoisting wire ropes are wound on drums driven through reduction gears by the hydraulic motor. Raising, lowering, or speed changes are accomplished by varying the stroke of the variable delivery pump through differential gearing. Hydraulic accumulators are not used with hydraulic bomb elevators.

WINCHES

A winch is a deck machine used for hoisting or hauling loads. The main components of a winch are a wire rope drum (or drums), a reduction gear train, and a power unit. Some winches are provided with one or two gypsy heads for handling manila or other fiber lines.

Most ships constructed before World War II were provided with steam-powered winches, a few of which remain in naval service. During World War II, Auxiliary ships were provided with winches powered by either alternating or direct-current electric motors. On modern ships, a-c electric drive winches are used. Figure 21-8 shows an a-c electric motor drive winch.

Where stepless speed control between zero and design maximum is required, a variable speed hydraulic transmission is included between the electric motor and the gear train on the same bed frame. The variable speed hydraulic transmission consists of a variable volume pump connected by high pressure tubing to a hydraulic motor, which is usually of the fixed-displacement type. The fluid output from the pump passes through the motor and returns to the pump in a closed circuit, and the speed of the motor varies as the volume of fluid from the pump varies. The speed and direction of rotation of the motors are obtained through a manually operated lever control at the pump or at a remote station. Figure 21-9 shows a typical electrohydraulic winch.

One type of steam-driven winch is illustrated in figure 21-10. The winch illustrated is equipped with two gypsy heads (1), one mounted at each end of the main drive shaft, and a single hoisting drum (10). The drum is provided with a standard type of brake band (3) with a foot-operated control and ratchet lock. The winch is driven by a two-cylinder, single-expansion, double-acting reciprocating engine. The drive is by means of a train of spur gears. A gear shift is provided to give two drive speeds.

The clutch mechanism consists of a sleeve (11) which is keyed to the crankshaft (12) and

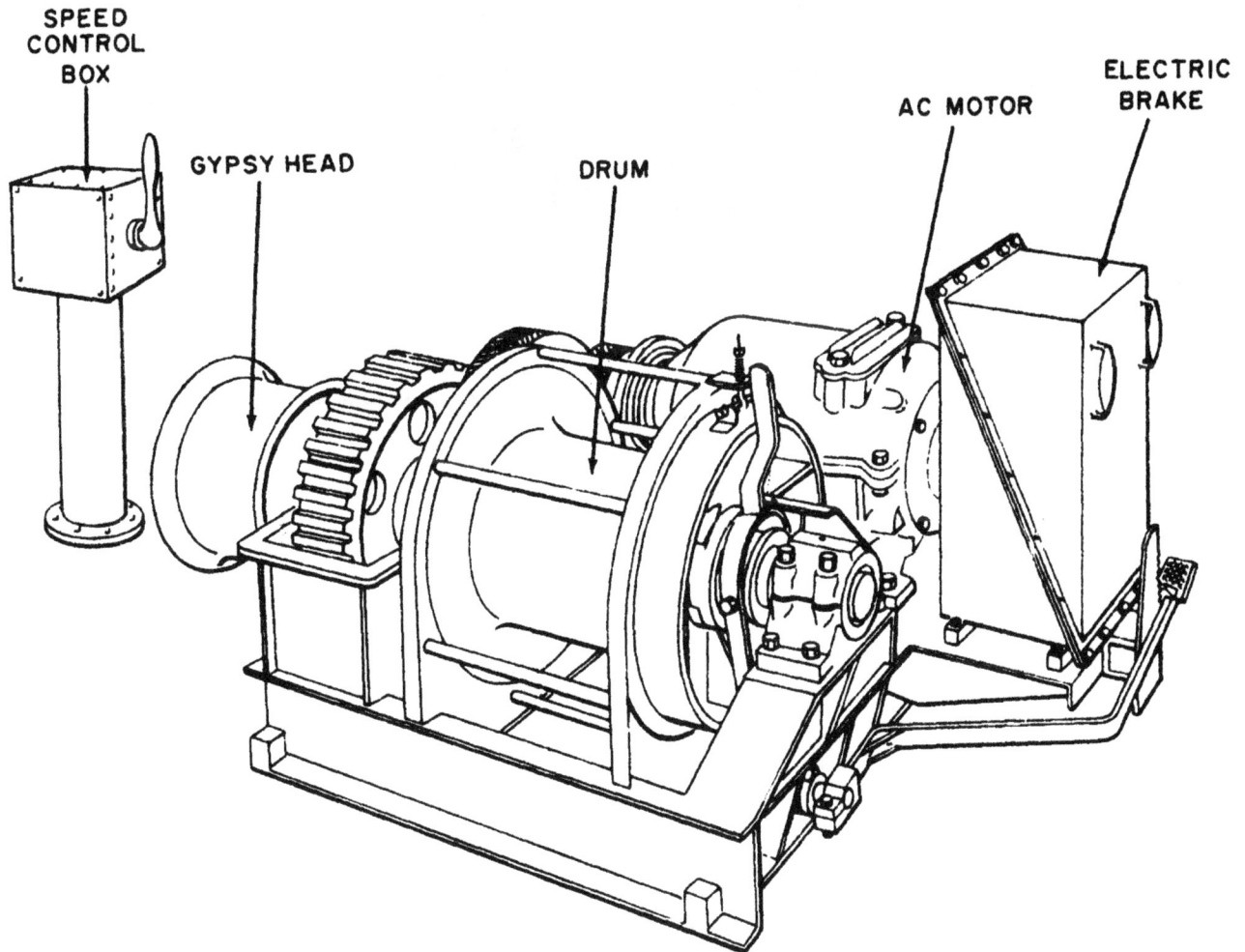

80.149

Figure 21-8.—A-C electric motor drive winch.

provided with a shifter yoke which is operated by a lever (7) located near the reverse valve (9). When this lever is moved to the left, the position shown in part A of figure 21-10, the pinion which is integral with it is engaged with the gear (2) on the intermediate shaft. This is the "compound gear" position and gives a slower drum speed with an increase in available line pull. When the shifter lever (7) is moved to the right, the driving sleeve is shifted to the right. This disengages the pinion from the intermediate shaft gear (2) and engages the square jaw clutch (13) and pinion which is always in mesh with the main drive gear. This gives a direct drive from the crankshaft to the main shaft, and is called the "single gear" position. The "single gear" gives a higher-drum speed for a given

engine speed, but with a decrease in the available line pull.

The speed and direction of rotation of the engine is controlled by a hand lever (14). With the winch in compound gear, the drum turns to lift a load when the hand lever is raised. With the lever in this position, a spool type valve passes steam from the reverse valve (9) through the two top horizontal pipe lines (5) to the cylinders where it drives the engine. The two lower pipes (4) are exhaust lines. When the lever is lowered below the horizontal position, the direction of steam flow is reversed and the engine turns in the opposite direction, thus lowering the load. On some winches, the hand lever is provided with an automatic latch for holding it in the horizontal, or neutral position.

529

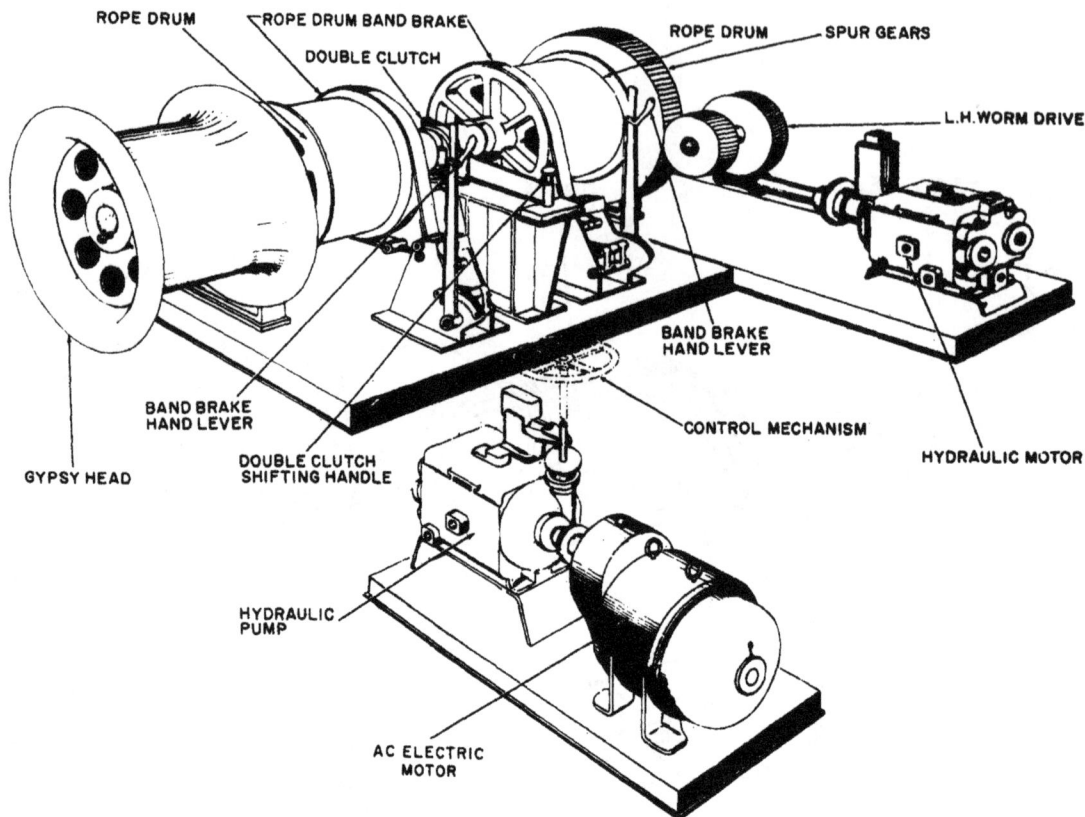

47.145

Figure 21-9.—Typical electrohydraulic winch.

For all other positions, the lever must be held in position for the desired speed.

CAPSTANS

A capstan is a spool-shaped, vertical revolving drum used for heaving in on heavy mooring lines. When a capstan is used to haul a load, as in mooring, several turns of mooring line are placed around the capstan head. A manpower strain is then taken on the free end of the line. Maintaining the strain causes the line to bind on the capstan head, which in turn hauls the load.

A capstan head may be a component part of an anchor windlass. Since the shaft for a capstan head is vertical, a capstan is always free of the fair lead problem which is often present in connection with gypsy heads which are mounted on horizontal shafts. A line leads fair to a capstan from any horizontal direction. Capstans that are not components of anchor windlasses are usually electrically powered.

ANCHOR WINDLASSES

A windlass is a piece of deck machinery used primarily for paying out and heaving in an anchor chain. A wildcat (drum) may be mounted vertically or horizontally at the end of the windlass shaft for handling the anchor chain. The wildcat is usually fitted with whelps to engage the anchor chain. On the windlass there may also be a capstan head or warping head (concave drum) for handling lines. A vertical-shaft anchor windlass with capstan head is shown in figure 21-11.

All anchor windlasses were formerly powered by steam, and some windlasses on auxiliaries still use steam as the source of power. Small combatant ships have electrically powered windlasses; larger combatant ships have vertical-shaft windlasses with electrohydraulic transmission. Hand-operated windlasses are in use, but they are found only on small ships where the weight of the anchor gear is small enough to be handled in a reasonable

530

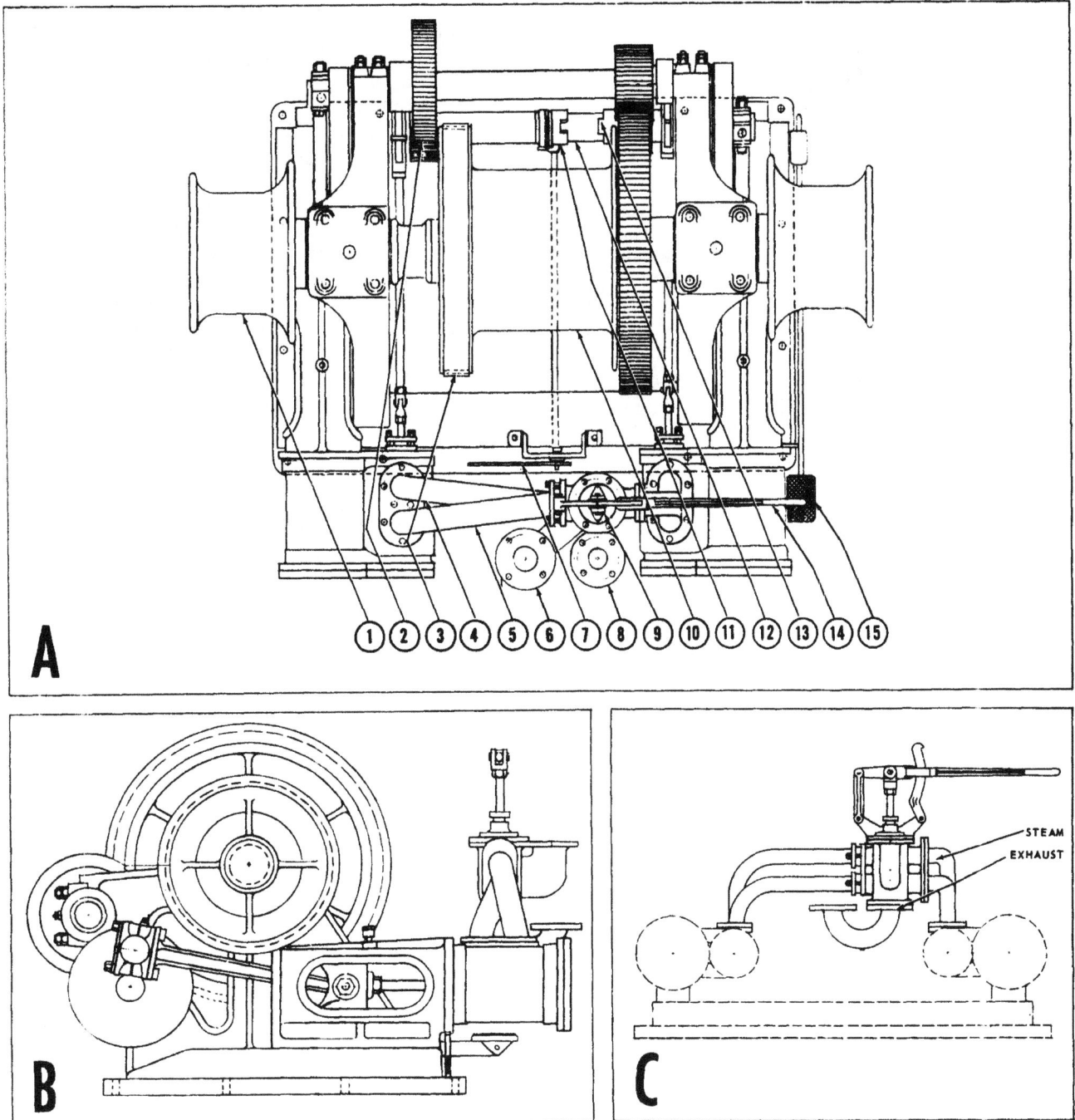

47.144

Figure 21-10.—Steam-operated winch. (A) Plan view. (B) Side view. (C) Reversing throttle valve.

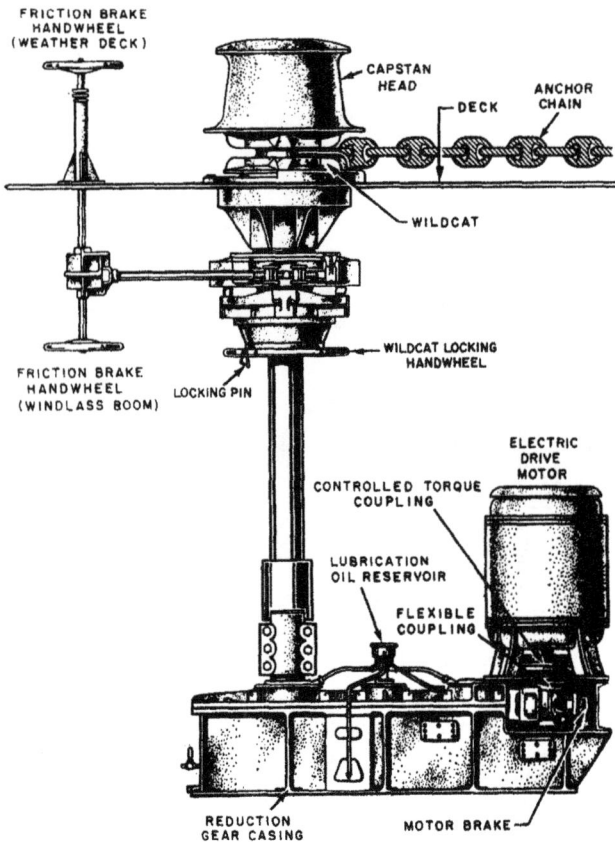

3.224X

Figure 21-11.—Vertical-shaft anchor windlass.

time and without excessive effort on the part of operating personnel.

CARRIER CATAPULTS

The efficiency of an aircraft carrier depends upon the speed of its airplane launching operations. Therefore, a compact and efficient device for getting all airplanes into the air within a short time is needed. This requirement is met by the modern carrier catapult. The catapult permits controlled application of a predetermined amount of power at any desired instant. Through the controlled power of the catapult, the plane on the catapult is safely accelerated from a standstill to flying speed within the limited space available on the flight deck of a carrier.

The type of catapults used during World War II and through the Korean incident were of the pneumatic-hydraulic type. Catapults of this type adequately met launching requirements, but the gradual increase in the weight of newly designed

aircraft and the attendant higher launching speeds continually necessitated the development of larger and heavier catapults. By 1950, the size and weight of the pneumatic-hydraulic type catapult had increased to a point where any further increase would be impracticable. British investigation of steam as the source of power for catapults attracted the attention of U. S. Navy officials; the Navy's powerful steam catapult of today is the result of basic British research.[2] The present discussion deals only with the steam catapult.

The major components of a steam catapult are shown schematically in figure 21-12.

During the operational cycle of the steam catapult, the plane is first spotted astride the catapult slot slightly aft of the shuttle. The airplane is coupled with the shuttle by means of the bridle which slips over the shuttle hook and over the hooks mounted on the underside of the airplane frame. The airplane is anchored to the deck by means of the holdback device which is released at the moment of launch. The grab, attached to the shuttle, pushes forward after the bridle is attached so that the shuttle puts tension on the bridle.

When the airplane is ready to be launched, with its engines running at full power, the launching valves are opened and steam is admitted to the after side of each piston. The resulting accelerating force combined with the engine thrust causes a calibrated "breaking" link in the holdback to part and the grab releases the shuttle. The shuttle and airplane are free to be moved forward by the accelerating force.

At the end of the launching run, the plane is airborne and the bridle is automatically released from the hook. The brake stops the piston-shuttle assembly. The grab, driven by the retracting engine, now moves along the track, hooks the shuttle, and returns it to the launching or battery position.

The principal component of the steam catapult is a cylinder-piston assembly—two power cylinders and two pistons per catapult. The spear-tipped pistons, which in the launching

---

[2] For greater detail on the history and operation of the steam catapult, see The Steam Catapult, NavAer 00-80T-69.

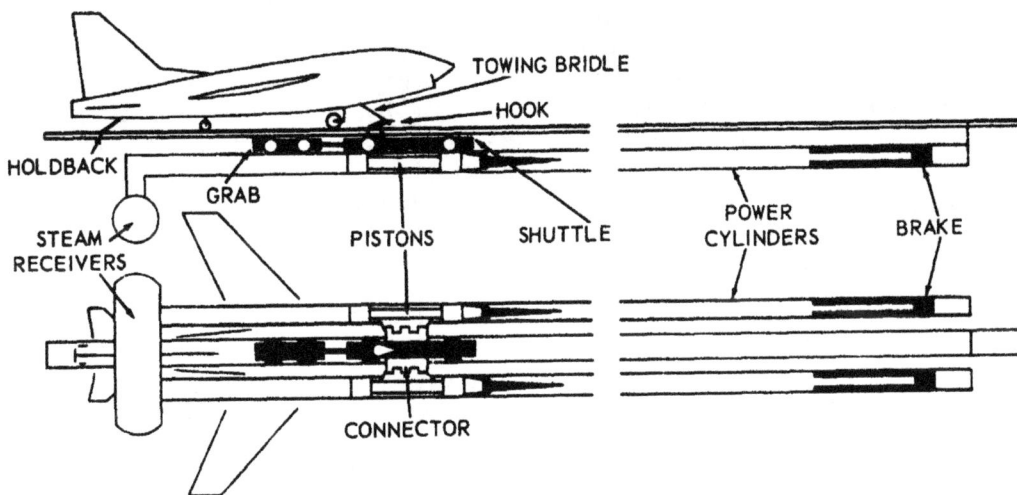

47.146

Figure 21-12.—Major components of a steam catapult.

operation are forced at high speed through the cylinders by steam pressure, are solidly interconnected by means of a connector shaped like an inverted T. The vertical leg of the inverted T extends upward through a slot in the flight deck, and serves as the hook to which the aircraft towing bridle is connected. The piston connector is attached to the shuttle. The shuttle is a small roller-mounted car which moves, during the launch, on tracks installed just under the flight deck. (See fig. 21-13.)

Power to drive the shuttle and its airplane load comes from expanding steam piped to the catapult from the main boilers of the ship. This steam is placed under pressure in large tanks—called accumulators or receivers—located under the launching engine on the hangar deck. From the receivers, the steam is transferred

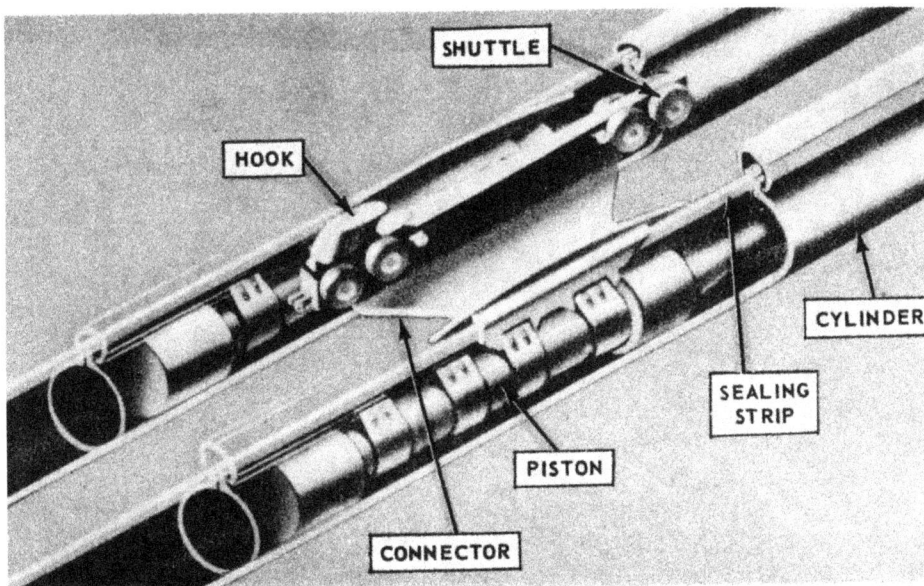

147.131

Figure 21-13.—Shuttle-connector-piston assembly for steam catapult.

533

at the moment of launch into the power cylinders. Steam pressure acts directly on the pistons and propels the piston-shuttle assembly through the cylinders. A sealing strip closes the slot in each cylinder as the pistons are driven forward, thus preventing the escape of steam from the cylinder slots through which the connector moves.

Prior to a launch, the engines of an airplane must be operating at full power. A holdback device is utilized to prevent the airplane from being moved forward by the thrust of its own engines, until the time of launch. The holdback device hooks into a fitting in the flight deck.

The piston-connector-shuttle assembly is stopped at the end of its launching run by a water brake. The brake consists of two cylinders of water located co-axially with the power cylinders at the forward end of the catapult. The spear tips of the pistons ram into the water-filled cylinders. As the spear tips penetrate the water, pressure builds up and stops the assembly. (See fig. 21-14.)

The principal unit in the shuttle retraction and tensioning systems is the grab. This unit is essentially a spring-loaded latch mounted on a wheeled frame just aft of the shuttle. The grab is driven along the shuttle track through a system of cables by hydraulic force. The hydraulic retraction engine consists of two cylinders. In one cylinder, hydraulic pressure is converted into the mechanical motion of a piston rod which is installed in this cylinder. The other cylinder is an accumulator in which hydraulic liquid is stored under pressure. The motion of the piston rod is transmitted to a device called a crosshead to which the drive cables of the grab are attached. (See fig. 21-15.)

Prior to a launch, the grab is moved forward by a hydraulic cylinder-piston assembly. This assembly is located aft of the grab. (See fig. 21-16.) When liquid is introduced into the cylinder, the piston pushes the grab forward. The grab, in turn, exerts force on the shuttle so that it moves forward enough to place tension on the towing bridle which connects the shuttle to the airplane. When the launch is made, the grab releases the shuttle and it is driven through the power cylinders of the catapult by steam pressure. After the launch is made, the grab is

147.132

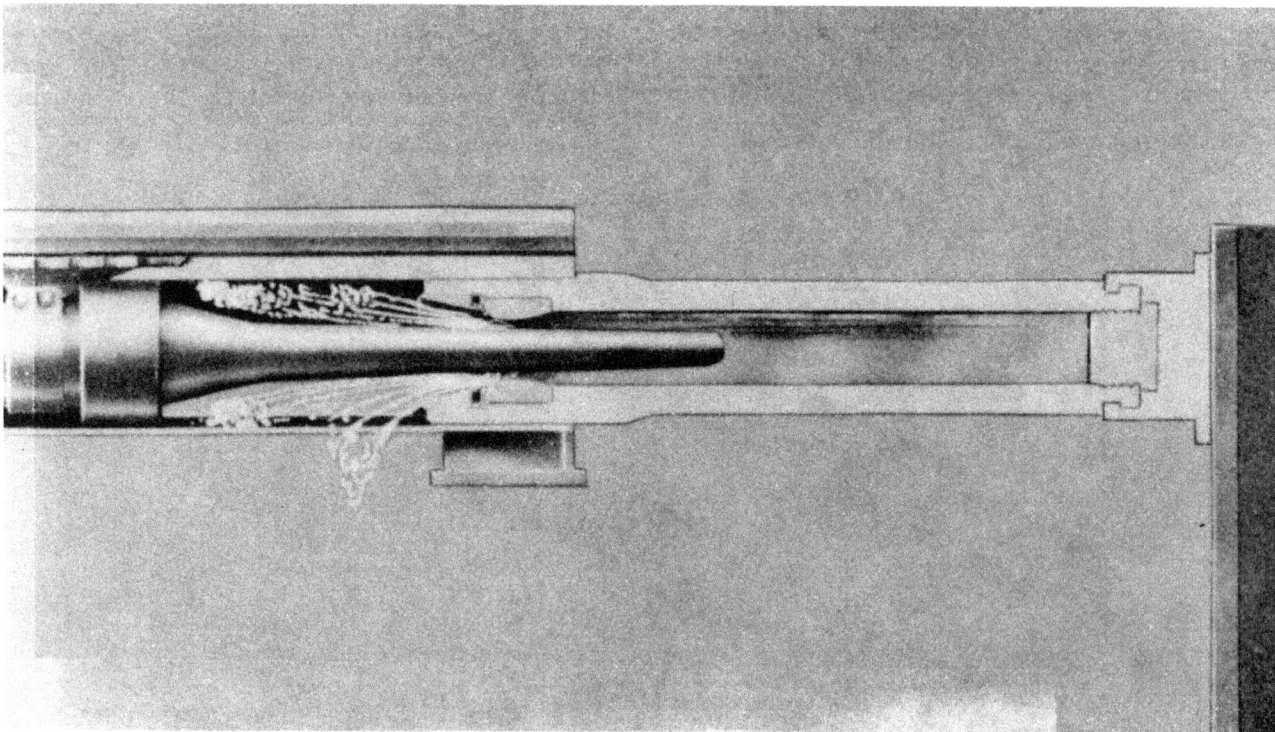

Figure 21-14.—Water brake of a steam catapult.

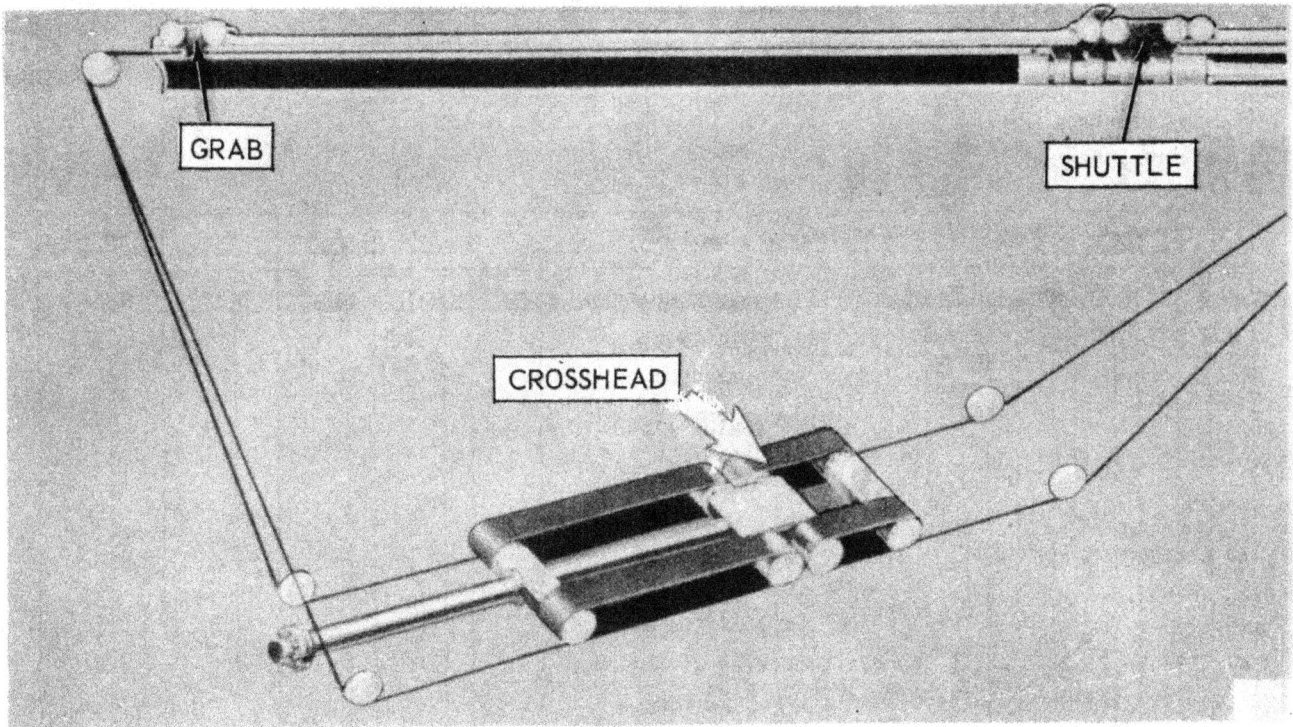

147.133

Figure 21-15.—Cables and crosshead of retraction system for steam catapult.

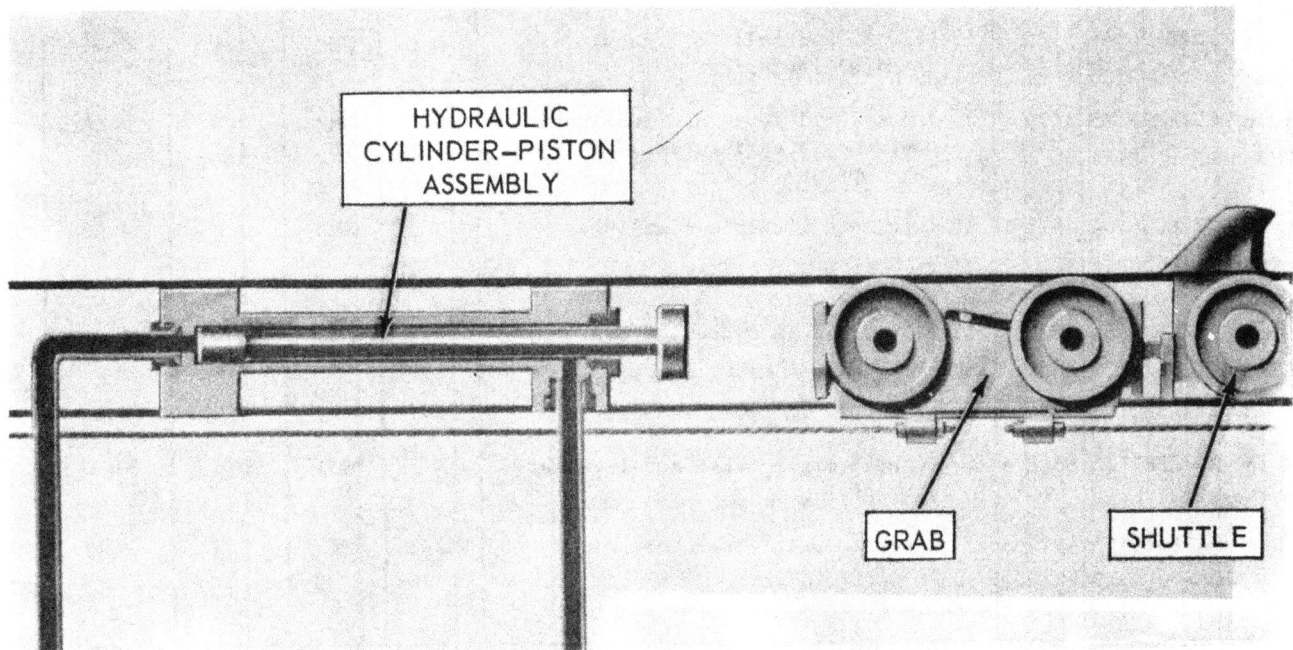

147.134

Figure 21-16.—Hydraulic assembly for moving the grab in the tensioning process.

System, Subsystem, or Component					Reference Publications			
Steering Gear								

	Bureau Card Control No.				Maintenance Requirement	M.R. No.	Rate Req'd.	Man Hours	Related Maintenance
AU	ZZZESRO	75	4065	W	1. Inspect all pins, couplings, and shafts. 2. Inspect agreement of helm angle indicator with mechanical rudder indicators. 3. Inspect ram packing glands for correct tightness. 4. Inspect system for excessive oil leaks.	W-1	MM3 FN	0.5 0.5	None
AU	ZZZESRO	65	A627	W	1. Lubricate ram room machinery. 2. Lubricate pump room machinery. 3. Inspect oil levels.	W-2	MM3	1.0	None
AU	ZZZESRO	25	2383	M	1. Inspect rudder stock packing. 2. Clean packing gland.	M-1	FN	0.5	None
AU	ZZZESRO	65	A628	M	1. Lubricate pump room machinery.	M-2	MM3	0.5	None
AU	ZZZESRO	72	6669	Q	1. Sound and tighten foundation bolts.	Q-1	FN	0.6	None
AU	ZZZESRO	45	9347	Q	1. Drain hydraulic oil filter on fill and drain pump.	Q-2	FN	0.2	None
AU	ZZZESRO	45	8150	Q	1. Provide hydraulic oil sample for chemical analysis.	Q-3	FN	0.1	None
AU	ZZZESRF	65	A629	Q	1. Inspect oil level and lubricate emergency steering gear. 2. Test operate unit.	Q-4	MM2 FN	0.5 0.5	None
AU	ZZ1ECW4	84	4915	S	1. Lubricate flexible couplings.	S-1	MM3	0.8	None
AU	ZZZESRO	65	A630	S	1. Lubricate follow-up gears.	S-2	FN	0.1	None
AU	ZZZESRO	65	A631	S	1. Filter oil in hydraulic system.	S-3	MM3 FN	3.0 3.0	Q-2
AU	ZZZESRO	25	7756	A	1. Renew oil in speed reducers.	A-1	MM3	1.0	None
AU	ZZZESRF	85	6368	A	1. Inspect motor brake and lubricate linkage.	A-2	MM3	0.3	None
AU	ZZZESRO	65	4919	A	1. Conduct operational test of steering gear.	A-3	MM1	0.3	None

(Page 1 of 2)

MAINTENANCE INDEX PAGE
OPNAV FORM 4700-3 (4 64)

BUREAU PAGE CONTROL NUMBER   A-1/56-65

98.171

Figure 21-17.—Planned Maintenance Index Page, steering gear.

Page 2 of 2	AZ	ZZZE	HGO	75	9966	M

Procedure (Cont'd)

b. Ensure ladder is firmly anchored in second deck pit under platform.

c. Apply grease with pressure gun to guide roller housing and roller shaft bearings.

d. Wipe off all excess grease.

WARNING: Before energizing electrical circuit, ensure there are no personnel in second deck pit under the platform.

e. Remove tag and energize circuit.

SYSTEM	COMPONENT	M. R NUMBER	
Auxiliary	Platform Hoisting Parts	A-77	M-2

SUB-SYSTEM	MAINT. SIGNI. FICANT NO.	RATES	M H
Aircraft Elevators	RELATED M.R. None	MM2	1.0
		FN	1.0

TOTAL M.H 2.0
ELAPSED TIME: 1.0

Page 1 of 2	AZ	ZZZE	HGO	75	9966	M

M. R. DESCRIPTION

1. Lubricate platform guide rails.
2. Lubricate platform hoisting cables and sheaves.
3. Lubricate platform guide shoe housing and roller shaft.

SAFETY PRECAUTIONS

1. Observe standard safety precautions.
2. Ensure all personnel are clear of platform edge while platform is in motion.
3. While performing MR 3, raise platform above main deck level, de-energize circuit, and tag "Out OF Service."

TOOLS, PARTS, MATERIALS, TEST EQUIPMENT

1. 5 lbs Grease, MIL-G-18709A
2. Clean empty bucket
3. Clean rags
4. Flashlight
5. Putty knife
6. Grease gun
7. Ladder
8. Safety tag

PROCEDURE

1. Lubricate Platform Guide Rails.

WARNING: Ensure all personnel are clear of platform edge while platform is in motion.

a. Scrape off all old grease from guide rails.
b. Apply grease with putty knife to inside surface of guide rails while elevator platform is traveling up and down at slow speed.

2. Lubricate Platform Hoisting Cables and Sheaves.

a. Swab grease to all platform hoisting cables. Use pressure gun to grease upper sheaves and bearings.

3. Lubricate Platform Guide Shoe Housing and Roller Shaft.

a. Raise platform above main deck level and de-energize electrical circuit. Grease roller shaft housing under platform.

(Cont'd on Page 2)

LOCATION

Figure 21-18.—Maintenance Requirement Cards.

98.176

SYSTEM	COMPONENT	M. R. NUMBER	
Auxiliary	Anchor Windlass	A-5	W-1

SUB-SYSTEM	RELATED M. R.	RATES	M/H
Winches, Capstans, Cranes and Anchor Handling	None	MM3	0.3

M. R. DESCRIPTION

1. Inspect oil levels.
2. Test operate windlass.

TOTAL M/H
0.3
ELAPSED TIME:
0.3

SAFETY PRECAUTIONS

1. Observe standard safety precautions.
2. De-energize circuit and tag "Out Of Service."
3. Ensure wildcat is disengaged before starting windlass.

TOOLS, PARTS, MATERIALS, TEST EQUIPMENT

1. Oil, Symbol 2190 TEP
2. Oil, Symbol 2135 H
3. Oil, Symbol 2110 H
4. Flashlight
5. Rags
6. Funnel
7. 8" Adjustable wrench
8. Safety tag

PROCEDURE

Preliminary

a. De-energize circuit and tag "Out Of Service."

1. Inspect Oil Levels.
   a. Inspect oil level in main gearcase, power unit gear reducer, and hydraulic system storage tank. Proper oil level is at top mark on gauge rod. Replenish gearcase and reducer with oil, Symbol 2190 TEP; replenish hydraulic system with oil, Symbol 2135 H.
   b. Inspect oil level in power brake storage tank. Proper level is at center line on gauge. Replenish with oil, Symbol 2110 H.

WARNING: Ensure wildcat is disengaged before starting windlass.

2. Test Operate Windlass.
   a. Remove safety tag and energize circuit.
   b. Operate unit; inspect for proper operation.

PAGE 1 OF 1 | AW | ZZZE | WC1 | 65 | A658 | W

LOCATION	DATE	
	1 July 1965	W

Figure 21-18. —Maintenance Requirement Cards—continued.   98.176

driven along the track by the retraction engine, hooks onto the shuttle, and returns it to the launching position.

AUXILIARY EQUIPMENT OPERATION, MAINTENANCE AND SAFETY

The operation and safety pertaining to auxiliary equipment should be in accordance with NavShips Technical Manual and/or the instructions posted on or near each individual piece of equipment. All maintenance actions, tests, and inspections should be accomplished in accordance with the 3-M System (PMS Subsystem). Figure 21-17 (Maintenance Index Page) shows the minimum maintenance requirements for a steering gear. Examples of maintenance requirements for two types of auxiliary equipment are shown in figure 21-18. Note that the Maintenance Requirement Cards list safety precautions to be observed, the tools, parts, materials, and test equipment required, and give the procedures to follow when performing the specified maintenance.

PART V—OTHER TYPES OF PROPULSION PLANTS

The conventional steam turbine propulsion plant, although widely used, is by no means the only propulsion plant in naval use. Chapter 22 deals with internal combustion engines of the reciprocating type—diesel and gasoline. Chapter 23 discusses the increasingly important gas turbine engine. Chapter 24 takes up the nuclear power plant—a plant which utilizes the steam turbine as a prime mover but which employs the nuclear reactor rather than the conventional boiler as a source of heat for the generation of steam. Chapter 25 provides a brief survey of new developments in naval engineering and indicates some of the areas in which future developments may change the nature of our present shipboard engineering plants.

CHAPTER 22

DIESEL AND GASOLINE ENGINES

Much of the machinery and equipment discussed in the preceding chapters utilizes steam as the working fluid in the process of converting thermal energy to mechanical energy. This chapter deals with internal combustion engines, in which air (or a mixture of air and fuel) serves as the working fluid. The internal combustion engines considered are those to which the thermodynamic cycles of the open and heated-engine types[1] apply. In engines which operate on these cycles, the working fluid is taken into the engine, heat is added to the fluid, the energy available in the fluid is utilized, and then the fluid is discarded. During the process, thermal energy is converted to mechanical energy. The purpose of this chapter is to present the basic theory and the fundamental principles underlying the energy conversion in internal combustion engines, and the functions of the engine parts, accessories, and systems essential for the conversion. No attempt is made to describe design, construction, models, etc., except as necessary to make the theory of operation and the function of components readily understandable.

Internal combustion engines are used extensively in the Navy, serving as propulsion units in a variety of installations such as ships, boats, airplanes, and automotive vehicles. Engines of the internal combustion type are also used as prime movers for auxiliary machinery. Internal combustion engines in a majority of the shipboard installations are of the reciprocating type. In relatively recent years, engines of the gas turbine type have been placed in Navy service as power plants. Gas turbine engines are discussed in chapter 23 of this text.

---

[1]Thermodynamic cycles are discussed in chapter 8 of this text.

RECIPROCATING ENGINES

Most of the internal combustion engines in marine installations of the Navy are of the reciprocating type. This classification is based on the fact that the cylinders in which the energy conversion takes place are fitted with pistons, which employ a reciprocating motion. Internal combustion engines of the reciprocating type are commonly identified as diesel and gasoline engines. The general trend in navy service is to install diesel engines rather than gasoline engines unless special conditions favor the use of the latter.

Most of the information on reciprocating engines in this chapter applies to diesel and gasoline engines. These engines differ, however, in some respects; the principal differences which exist are noted and discussed.

Basic Principles

The operation of an internal combustion engine of the reciprocating type involves the admission of fuel and air into a combustion space and the compression and ignition of the charge. The resulting combustion releases gases and increases the temperature within the space. As temperature increases, pressure increases and forces the piston to move. This movement is transmitted through a chain of parts to a shaft. The resulting rotary motion of the shaft is utilized for work; thus, heat energy is transformed into mechanical energy. In order for the process to be continuous, the expanded gases must be removed from the combustion space, a new charge admitted, and then the process repeated.

In the study of engine operating principles, starting with the admission of air and fuel and following through to the removal of the expanded gases, it will be noted that a series of events takes place. The term cycle identifies the sequence of events that takes place in the cylinder of an engine for each power impulse transmitted to the crankshaft. These events always occur in the same order each time the cycle is repeated. The number of events occurring in a cycle of operation will depend upon the engine type—diesel or gasoline. The difference in the events occurring in the cycle of operation for these engines is shown in the following table.

The events and their sequence in a cycle operation for a:	
DIESEL ENGINE	GASOLINE ENGINE
INTAKE of air COMPRESSION of air INJECTION of fuel IGNITION AND COMBUSTION of charge EXPANSION of gases REMOVAL of waste	INTAKE of fuel and air COMPRESSION of fuel-air mixture. IGNITION and COMBUSTION of charge EXPANSION of gases REMOVAL of waste

The principal difference, as shown in the table, in the cycles of operation for diesel and gasoline engines involves the admission of fuel and air to the cylinder. While this takes place as one event in the operating cycle of a gasoline engine, it involves two events in diesel engines. Thus, insofar as events are concerned, there are six main events taking place in the diesel cycle of operation and five in the cycle of a gasoline engine. This is pointed out in order to emphasize the fact that the events which take place and the piston strokes which occur during a cycle of operation are not identical. Even though the events of a cycle are closely related to piston position and movement, all of the events will take place during the cycle regardless of the number of piston strokes involved. The relationship of events and piston strokes is discussed later under a separate heading.

The mechanics of engine operation is sometimes referred to as the mechanical or operating cycle of an engine; while the heat process which produces the forces that move engine parts may be referred to as the combustion cycle. A cycle of each type is included in a cycle of engine operation.

Mechanical Cycles

In the preceding section, the events taking place in a cycle of engine operation were emphasized. Little was said about piston strokes except that a complete sequence of events would occur during a cycle regardless of the number of strokes made by the piston. The number of piston strokes occurring during any one series of events is limited to either two or four, depending upon the design of the engine; thus, the 4-stroke cycle and the 2-stroke cycle. These cycles are known as the mechanical cycles of operation.

Four- and Two-Stroke Cycles.—Both types of mechanical cycles are used in diesel and gasoline engines. However, most large gasoline engines in Navy service operate on the 4-stoke cycle; a greater number of diesel engines operate on the 2-stroke than on the 4-stroke cycle. The relationship of the events and piston strokes occurring in a cycle of operation involves some of the differences between the 2-stroke cycle and the 4-stroke cycle.

RELATIONSHIP OF EVENTS AND STROKES IN A CYCLE.—A piston stroke is the distance a piston moves between limits of travel. The cycle of operation is an engine operating on the 4-stroke cycle involves four piston strokes—intake, compression, power, and exhaust. In the case of the 2-stroke cycle, only two strokes apply—power and compression.

A check of the previous table listing the series of events which take place during the cycles of operation of diesel and gasoline engines will show that the strokes are named to correspond to some of the events. However, since six events are listed for diesel engines and five events for gasoline engines, it is evident that more than one event takes place during some of the strokes, especially in the case of the 2-stroke cycle. Even though this is the case, it is common practice to identify some of the events as strokes of the piston. This is because such events as intake, compression, power and exhaust in a 4-stroke cycle involve at least a major portion of a stroke and, in some cases, more than one stroke. The same is true of power and compression events and strokes in a 2-stroke cycle. Such association of events and strokes overlooks other events taking place during a cycle of operation. This oversight sometimes leads to confusion when the operating principles of an engine are being considered.

This discussion points out the relationship of events to strokes by covering the number of events occurring during a specific stroke, the duration of an event with respect to a piston stroke, and the cases where one event overlaps another. The relationship of events to strokes can be shown best by making use of graphic representation of the changing situation occurring in a cylinder during a cycle of operation. Figure 22-1 illustrates these changes for a 4-stroke cycle diesel engine.

The relationship of events to strokes is more readily understood, if the movements of a piston and its crankshaft are considered first. In part A of figure 22-1, the reciprocating motion and stroke of a piston are indicated and the rotary motion of the crank during two piston strokes is shown. The positions of the piston and crank at the start and end of a stroke are marked "top" and "bottom, " respectively. If these positions and movements are marked on a circle (part B, fig 22-1) the piston position, when at the top of a stroke, is located at the top of a circle. When the piston is at the bottom of a stroke, the piston position is located at the bottom center of the circle. Note in parts A and B of figure 22-1 that the top center and bottom center identify points where changes in direction of motion take place. In other words, when the piston is at top center, upward motion has stopped and downward motion is ready to start or, with respect to motion, the piston is "dead."

The points which designate changes in direction of motion for a piston and crank are commonly called top dead center (TDC) and bottom dead center (BDC).

If the circle illustrated in B is broken at various points and "spread out" (part C, fig. 22-1), the events of a cycle and their relationship to the strokes and how some of the events of the cycle overlap can be shown. TDC and BDC should be kept in mind since they identify the start and end of a stroke and they are the points from which the start and end of events are established.

By following the strokes and events as illustrated, it can be noted that the intake event starts before TDC, or before the actual down stroke (intake) starts, and continues on past BDC, or beyond the end of the stroke. The compression event starts when the intake event ends, but the

upstroke (compression) has been in process since BDC. The injection and ignition events overlap with the latter part of the compression event, which ends at TDC. The burning of the fuel continues a few degrees past TDC. The power event or expansion of gases ends several degrees before the down (power) stroke ends at BDC. The exhaust event starts when the power event ends and continues through the complete upstroke (exhaust) and past TDC. Note the overlap of the exhaust event with the intake event of the next cycle. The details on why certain events overlap and why some events are shorter or longer with respect to strokes will be covered later in this chapter.

From the preceding discussion, it can be seen why the term "stroke" is sometimes used to identify an event which occurs in a cycle of operation. However, it is best to keep in mind that a stoke involves 180° of crankshaft rotation (or piston movement between dead centers) while the corresponding event may take place during a greater or lesser number of degrees of shaft rotation.

The relationship of events to strokes in a 2-stroke cycle diesel engine is shown in figure 22-2. Comparison of figures 22-1 and 22-2 reveals a number of differences between the two types of mechanical or operating cycles. These differences are not too difficult to understand if one keeps in mind that four piston strokes and 720° of crankshaft rotation are involved in the 4-stroke cycle while only half as many strokes and degrees are involved in a 2-stroke cycle. Reference to the cross-sectional illustrations (fig. 22-2) will aid in associating the event with the relative position of the piston. Even though the two piston strokes are frequently referred to as power and compression, they are identified as the "down stroke" (TDC to BDC) and "up stroke" (BDC to TDC) in this discussion in order to avoid confusion when reference is made to an event.

Starting with the admission of air, (1) figure 22-2, we find that the piston is in the lower half of the down stroke and that the exhaust event (6) is in process. The exhaust event started (6') a number of degrees before intake, both starting several degrees before the piston reached BDC. The overlap of these events is necessary in order that the incoming air (1') can aid in clearing the cylinder of exhaust gases. Note that the exhaust event stops a few degrees before the intake event stops, but several degrees after the upstroke of the piston has started. (The exhaust

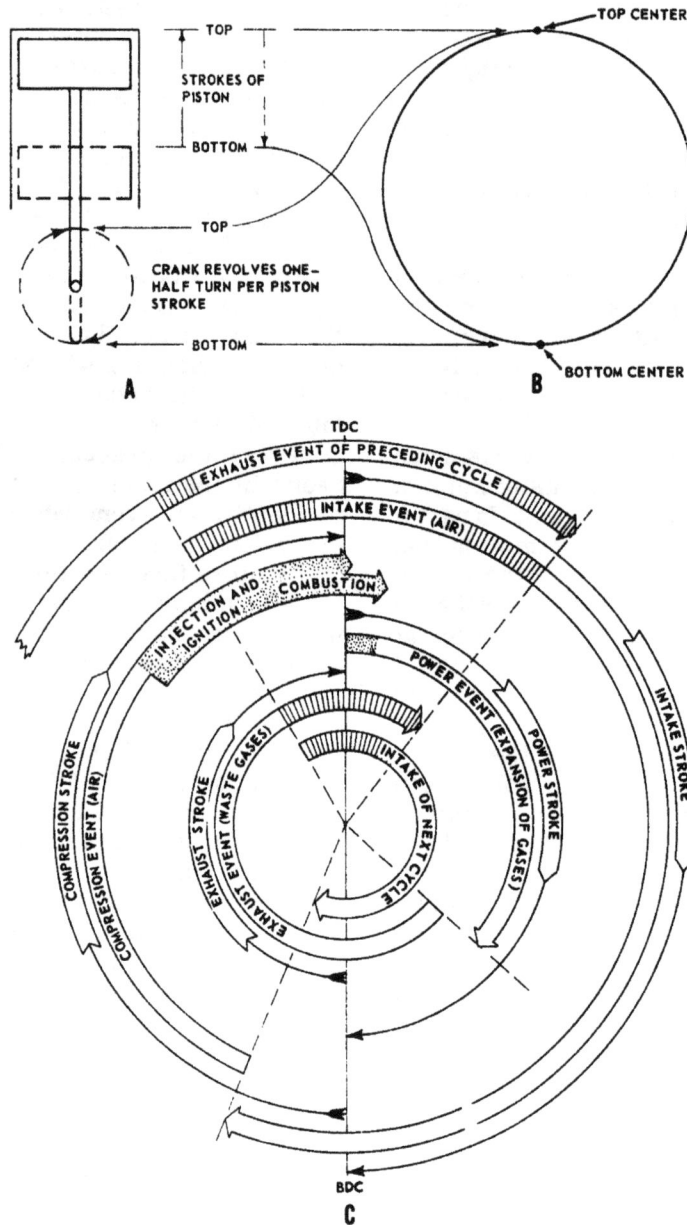

Figure 22-1.—Relationship of events and strokes in a 4-stroke cycle diesel engine.

event in some 2-stroke cycle diesel engines ends a few degrees after the intake event ends.) When the scavenging event ends, the cylinder is charged with the air which is to be compressed. The compression event (2) and (2') takes place during the major portion of the upstroke. The injection event (3) and (3') and ignition and combustion (4) and (4') occur during the latter part of the upstroke. (The point at which the injection

ends varies with engines. In some cases, it ends before TDC; in others, a few degrees after TDC.) The intense heat generated during the compression of the air ignites the fuel-air mixture and the pressure resulting from combustion forces the piston down. The expansion (5 and 5') of the gases continues through a major portion of the down stroke. After the force of the gases has been expended, the exhaust valve opens (6') and

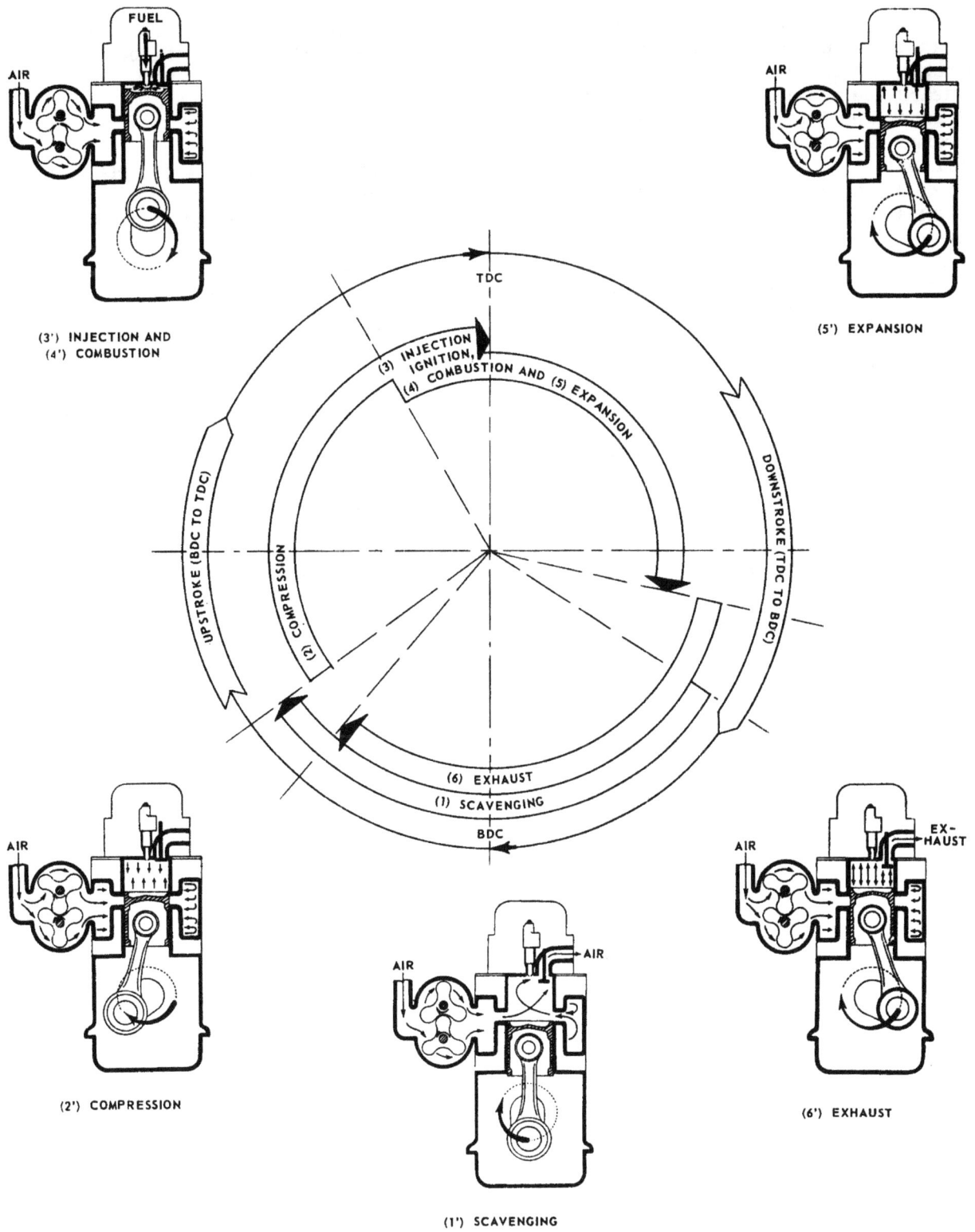

(3') INJECTION AND
(4') COMBUSTION

(5') EXPANSION

(2') COMPRESSION

(1') SCAVENGING

(6') EXHAUST

TDC

(3) INJECTION IGNITION, (4) COMBUSTION AND (5) EXPANSION

DOWNSTROKE (TDC TO BDC)

UPSTROKE (BDC TO TDC)

(2) COMPRESSION

(6) EXHAUST

(1) SCAVENGING

BDC

54.20AX

Figure 22-2.—Strokes and events of a 2-stroke cycle diesel engine.

permits the burned gases to enter the exhaust manifold. As the piston moves downward, the intake ports are uncovered (1') and the incoming air clears the cylinder of the remaining exhaust gases and fills the cylinder with a fresh air charge (1); thus, the cycle of operation has started again.

Now what is the difference between the 2- and 4-stroke cycles? From the standpoint of the mechanics of operation, the principal difference is in the number of piston strokes taking place during the cycle of events. A more significant difference is the fact that a 2-stroke cycle engine delivers twice as many power impulses to the crankshaft for every 720° of shaft rotation. (See fig. 22-3.)

Diagrams showing the mechanical cycles of operation in gasoline engines would be somewhat similar to those described for diesel engines except that there would be one less event taking place during the gasoline engine cycle. Since air and fuel are admitted to the cylinder of a gasoline engine as a mixture during the intake event, the injection event does not apply.

The figures shown here representing the cycles of operation are for illustrative purposes only. The exact number of degrees before or after TDC or BDC that an event starts and ends will vary between engines. Information on such details should be obtained from appropriate technical manuals dealing with the specific engine in question.

Combustion Cycles

To this point, the strokes of a piston and the related events taking place during a cycle of operation have been given greater consideration than the heat process involved in the cycle. However, the mechanics of engine operation cannot be discussed without dealing with heat. Such terms as ignition, combustion, and expansion of gases, all indicate that heat is essential to a cycle of engine operation. So far, particular differences between diesel and gasoline engines have not been pointed out, except the number of events occurring during the cycle of operation. Whether a diesel engine or a gasoline engine, the 2- or the 4-stroke cycle may apply. Then, one of the principal differences between these types of engines must involve the heat process utilized to produce the forces which make the engine operate. The heat processes are sometimes called combustion or heat cycles.

The three most common combustion cycles associated with reciprocating internal combustion engines are the Otto cycle, the true diesel cycle, and the modified diesel cycle.

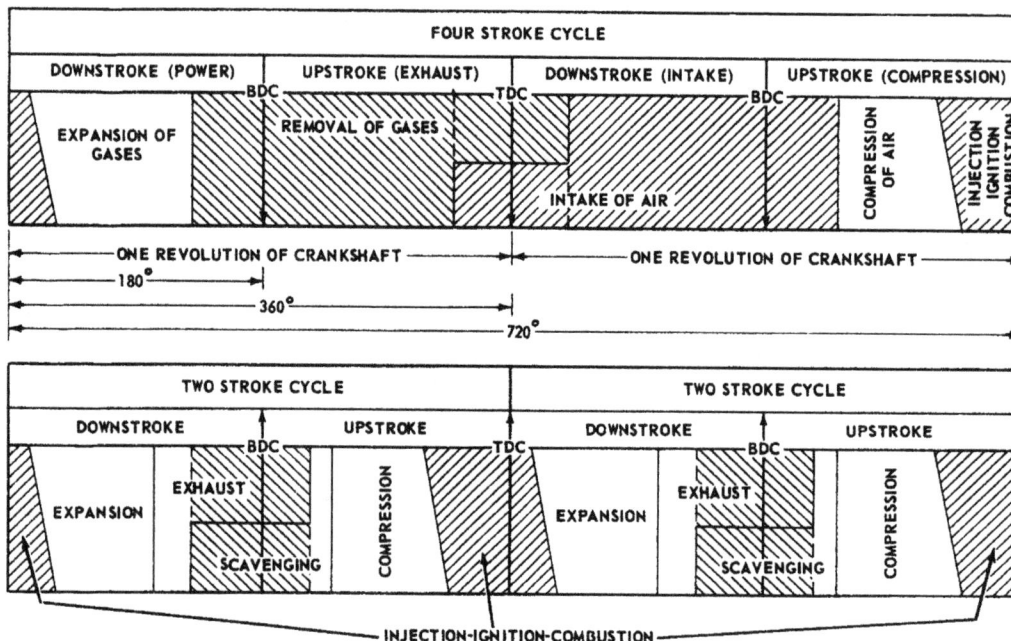

54.19:.20X

Figure 22-3.—Comparison of the 2- and 4-stroke cycles.

Reference to combustion cycles suggests another important difference between gasoline and diesel engines—compression pressure. This factor is directly related to the combustion process utilized in an engine. Diesel engines have a much higher compression pressure than gasoline engines. The higher compression pressure in diesels explains the difference in the methods of ignition used in gasoline and diesel engines. Compressing the gases within a cylinder raises the temperature of the confined gases. The greater the compression, the higher the temperature. In a gasoline engine, the compression temperature is always lower than the point where the fuel would ignite spontaneously. Thus, the heat required to ignite the fuel must come from an external source—spark ignition. On the other hand, the compression temperature in a diesel engine is far above the ignition point of the fuel oil; therefore, ignition takes place as a result heat generated by compression of the air within the cylinder—compression ignition.

The difference in the methods of ignition indicates that there is a basic difference in the combustion cycles upon which diesel and gasoline engines operate. This difference involves the behavior of the combustion gases under varying conditions of pressure, temperature, and volume. Since this is the case, the relationship of these factors is considered before the combustion cycles.

RELATIONSHIP OF TEMPERATURE, PRESSURE, AND VOLUME.—The relationship of these three conditions as found in an engine can be illustrated by considering what takes place in a cylinder fitted with a reciprocating piston. (See fig. 22-4.)

Instruments are provided which indicate the pressure within the cylinder and the temperature inside and outside the cylinder. Consider that the air in the cylinder is at atmospheric pressure and that the temperatures, inside and outside the cylinder, are about 70°F. (See fig. 22-4A.)

If the cylinder is an airtight container and a force pushes the piston toward the top of the cylinder, the entrapped charge will be compressed. As the compression progresses, the volume of the air decreases, the pressure increases, and the temperature rises (see B and C). These changing conditions continue as the piston moves and when the piston nears TDC (see D) we find that there has been a marked decrease in volume and that both pressure and temperature are much greater than at the beginning of compression. Note that pressure has gone from 0 to 470 psi and temperature has increased from 70° to about 1000° F. These changing conditions indicate that mechanical energy, in the form of work done on the piston, has been transformed into heat energy in the compressed air. The temperature of the air has been raised sufficiently to cause ignition of fuel injected into the cylinder.

Further changes take place after ignition. Since ignition occurs shortly before TDC, there is little change in volume until the piston passes TDC. However, there is a sharp increase in pressure and temperature shortly after ignition takes place. The increased pressure forces the piston downward. As the piston moves downward, the gases expand, or increase in volume, and pressure and temperature decrease rapidly. The changes in volume, pressure, and temperature, described and illustrated here, are representative of the changing conditions within the cylinder of a modern diesel engine.

The changes in volume and pressure in an engine cylinder can be illustrated by diagrams similar to those shown in figure 22-5. Such diagrams are made by devices which measure and record the pressures at various piston positions during a cycle of engine operation. Diagrams which show the relationship between pressures and corresponding piston positions are called pressure-volume diagrams or indicator cards. Examples of theoretical and actual pressure-volume diagrams are used in this chapter with the description of combustion cycles.

On diagrams which provide a graphic representation of cylinder pressure as related to volume, the vertical line P on the diagram (fig. 22-5) represents pressure and the horizontal line V represents volume. When a diagram is used as an indicator card, the pressure line is marked off in units of pressure and the volume line is marked off in inches. Thus, the volume line could be used to show the length of the piston stroke which is proportional to volume. The distance between adjacent letters on each of the diagrams represents an event of a combustion cycle—that is, compression of air, burning of the charge, expansion of gas, and removal of gases.

The diagrams shown in figure 22-5 provide a means by which the Otto and true diesel combustion cycles can be compared. Reference to the diagrams during the following discussion of these combustion cycles will aid in identifying

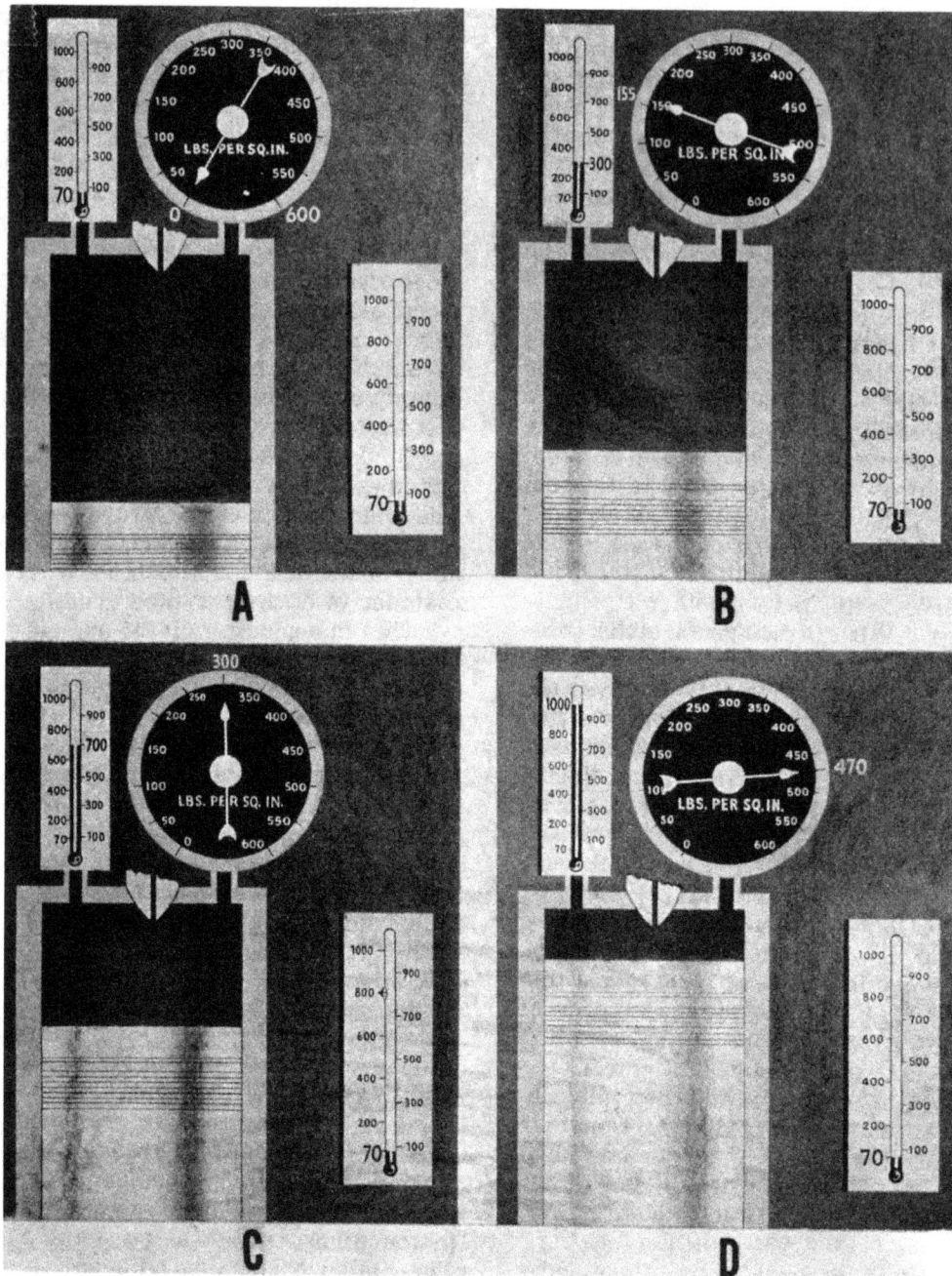

75.4

Figure 22-4.—Volume, temperature, and pressure relationships in a cylinder.

the principal differences existing between the cycles. The diagrams shown are theoretical pressure-volume diagrams. Diagrams representing conditions in operating engines are given later. Information obtained from actual indicator diagrams may be used in checking engine performance.

OTTO (CONSTANT-VOLUME) CYCLE.—In theory, this combustion cycle is one in which

550

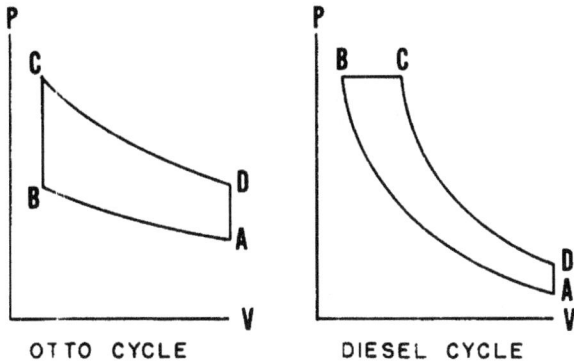

75.5

Figure 22-5.—Pressure-volume diagrams
for theoretical combustion cycles.

combustion, induced by spark ignition, occurs at constant volume. The Otto cycle and its principles serve as the basis for modern gasoline engine designs.

Compression (see line A-B, figure 22-5) of the charge in the cylinder is adiabatic. Spark ignition occurs at B, and, due to the volatility of the mixture, combustion practically amounts to an explosion. Combustion, represented by line BC, occurs (theoretically) just as the piston reaches TDC. During combustion, there is no piston travel; thus there is no change in the volume of the gas in the cylinder. This accounts for the descriptive term, constant volume. During combustion, there is a rapid rise of temperature followed by a pressure increase which performs the work during the expansion phase, represented by line CD. The removal of gases, represented by line DA, is at constant volume.

TRUE DIESEL (CONSTANT-PRESSURE) CYCLE.—This cycle may be defined as one in which combustion, induced by compression ignition, theoretically occurs at a constant pressure. Adiabatic compression (represented by line AB, fig. 22-5) of the air increases its temperature to a point where ignition occurs automatically when the fuel is injected. Fuel injection and combustion are so controlled as to give constant-pressure combustion (represented by line BC). This is followed by adiabatic expansion (represented by line CD) and constant volume (represented by the line DA).

In the true diesel cycle, the burning of the mixture of fuel and compressed air is a relatively slow process when compared with the quick, explosive-type combustion process of the Otto cycle. The injected fuel penetrates the

compressed air, some of the fuel ignites, then the rest of the charge burns. The expansion of the gases keeps pace with the change in volume caused by piston travel; thus combustion is said to occur at constant pressure (represented by line BC).

MODIFIED COMBUSTION CYCLES.—The preceding discussion covers the theoretical combustion cycles which serve as the basis for modern engines. In actual operation, modern engines operate on modifications of the theoretical cycles. However, characteristics of the true cycles are incorporated in the cycles of modern engines. This is pointed out in the following discussion of examples representing the actual cycles of operation in gasoline and diesel engines.

The following examples are based on the 4-stroke mechanical cycle since the majority of gasoline engines use this type of cycle; thus, a means of comparing the cycles found in both gasoline and diesel engines is provided. Differences existing in diesel engines operating on the 2-stroke cycle are pointed out.

The illustrations in figures 22-6 and 22-7 represent the changing conditions in a cylinder during engine operation. Some of the events are exaggerated in order to show more clearly the change which takes place and, at the same time, to show how the theoretical and actual cycles differ.

The compression ratio situation and a pressure-volume diagram for a 4-stroke Otto cycle is shown in figure 22-6. Illustration A shows the piston on BDC at the start of an upstroke. (In a 4-stroke cycle engine, this stroke could be either that identified as the compression stroke or the exhaust stroke.) Notice that in moving from BDC to TDC (illustration B), the piston travels 5/6 of the total distance ab. In other words, the volume has been decreased to 1/6 of the volume when the piston was at BDC. Thus, the compression ratio is 6 to 1.

Illustration C shows the changes in volume and pressure during one complete 4-stroke cycle. Note that the lines representing the combustion and exhaust phases are not straight as they were in the theoretical diagram. As in the diagram of the theoretical cycle, the vertical line at the left represents cylinder pressure in psi. Atmospheric pressure is represented by a horizontal line called the atmospheric pressure line. Pressures below this line are less than atmospheric pressures, while pressures above

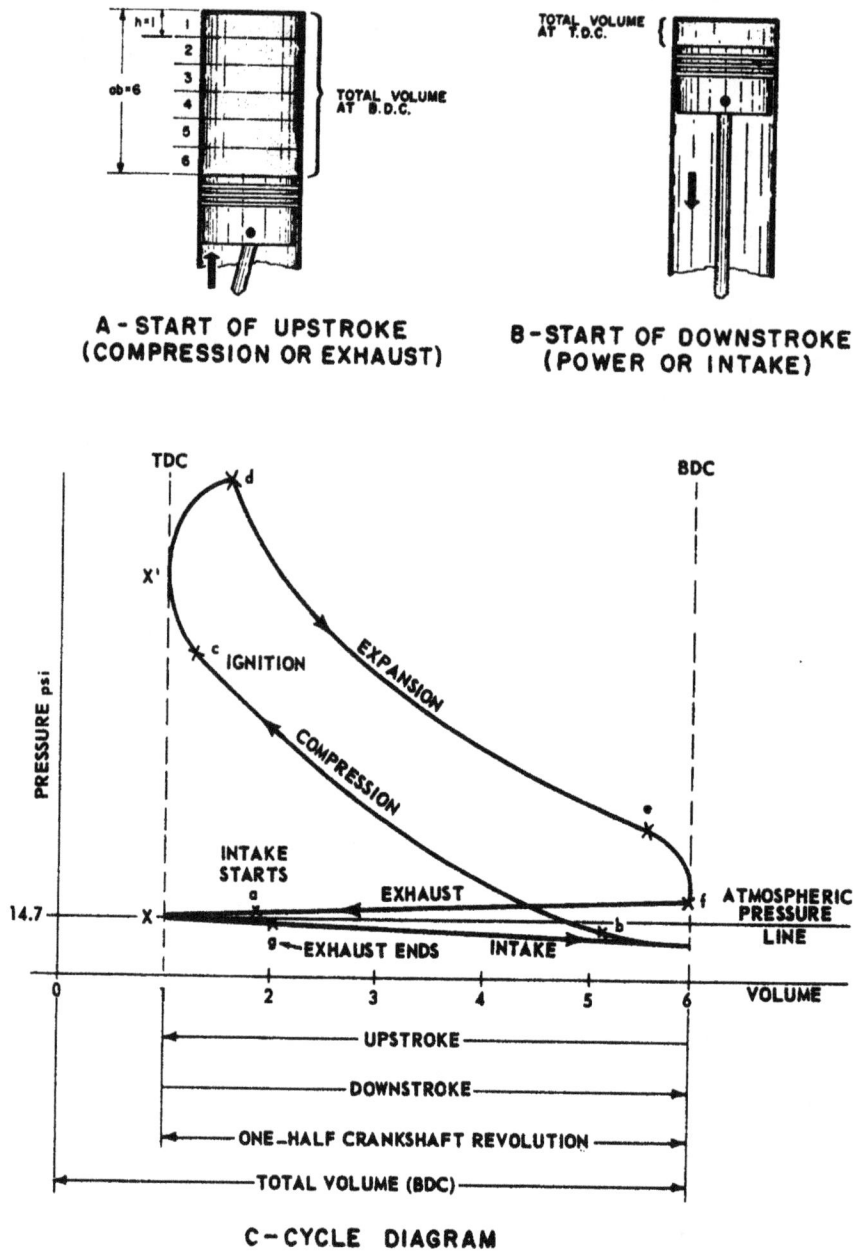

A – START OF UPSTROKE
(COMPRESSION OR EXHAUST)

B – START OF DOWNSTROKE
(POWER OR INTAKE)

C – CYCLE DIAGRAM

54.19B

Figure 22-6.—Pressure-volume diagram, Otto 4-stroke cycle.

the line represent compression. The bottom horizontal line provides a means of representing cylinder volume and piston movement. The volume line has been divided into six parts which correspond to the divisions of volume shown in illustration A. Since piston movement and volume are proportional, the distance between O and 6 indicated the volume when the piston is at BDC, and the distance from O to 1 the volume with the piston at TDC. Thus, the

distance from 1 to 6 corresponds to total piston travel and units of the distance may be used to identify changes in volume resulting from the reciprocating motion of the piston.

The curved lines of illustration C represent the changes of both pressure and volume which take place during the four piston strokes of the cycle. To conform to the discussion on the relationship of strokes and events (see fig. 22-1), the cycle of operation starts with intake.

552

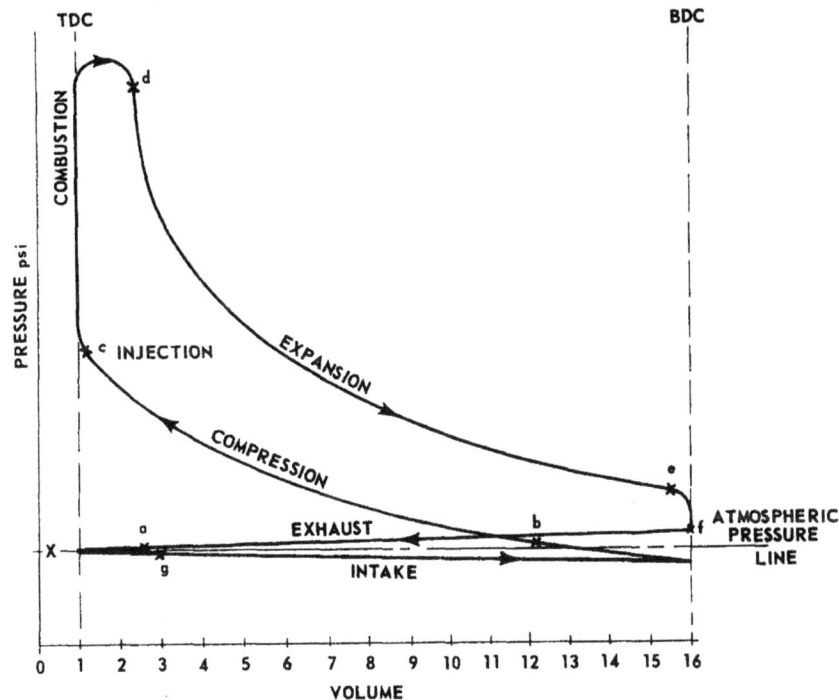

54.19C

Figure 22-7.—Pressure-volume diagram, diesel 4-stroke cycle.

In the case of Otto cycle, this event includes the admission of fuel and air. As indicated earlier, the intake event starts before TDC, or at point a, illustration C. Note that pressure is decreasing and after the piston reaches TDC and starts down, a vacuum is created which facilites the flow of the fuel-air mixture into the cylinder. The intake event continues a few degrees past BDC, ending at point b. Since the piston is now on an upstroke, compression takes place and continues until the piston reaches TDC. Note the increase in pressure (x to x') and the decrease in volume (f to x). Spark ignition at c starts combustion which takes place very rapidly. There is some change in volume since the phase starts before and ends after TDC.

There is a sharp increase in pressure during the combustion phase. The relative amount is shown by the curve cd. The increase in pressure provides the force necessary to drive the piston down again. The gases continue to expand as the piston moves toward BDC, and the pressure decreases as the volume increases, from d to e. The exhaust event starts a few degrees before BDC, at e, and the pressure drops rapidly until the piston reaches BDC. As the

piston moves toward TDC, there is a slight drop in pressure as the waste gases are discharged. The exhaust event continues a few degrees past TDC to point g so that the incoming charge aids in removing the remaining waste gases.

The modified diesel combustion cycle is one in which the combustion phase, induced by compression ignition, begins on a constant-volume basis and ends on a constant-pressure basis. In other words, the modified cycle is a combination of the Otto and true diesel cycles. The modified cycle is used as the basis for the design of practically all modern diesel engines.

An example of a pressure-volume diagram for a modified 4-stroke cycle diesel engine is shown in figure 22-7. Note that the volume line is divided into 16 units, indicating a 16 to 1 compression ratio. The higher compression ratio accounts for the increased temperature necessary to ignite the charge. By comparing this illustration with illustration C of figure 22-6, it will be found that the phases of the diesel cycle are relatively the same as those of the Otto cycle, except for the combustion phase. Fuel is injected at point c and combustion is represented by line cd. While combustion in the

Otto cycle is practically at constant-volume throughout the phase, combustion in the modified diesel cycle takes place with volume practically constant for a short time, during which period there is a sharp increase in pressure, until the piston reaches a point slightly past TDC. Then, combustion continues at a relatively constant pressure, dropping slightly as combustion ends at d. For these reasons, the combustion cycle in modern diesel engines is sometimes referred to as the constant-volume constant-pressure cycle.

Pressure-volume diagrams for gasoline and diesel engines operating on the 2-stroke cycle would be similar to those just discussed, except that separate exhaust and intake curves would not exist. They do not exist because intake and exhaust occur during a relatively short interval of time near BDC and do not involve full strokes of the piston as in the case of the 4-stroke cycle. Thus, a pressure-volume diagram for a 2-stroke modified diesel cycle would be similar to a diagram formed by f-b-c-d-e-f of figure 22-7. The exhaust and intake phases would take place between e and b with some overlap of the events. (See fig. 22-2.)

The preceding discussion has pointed out some of the main differences between engines which operate on the Otto cycle and those which operate on the modified diesel cycle. In brief, these differences involve (1) the mixing of fuel and air, (2) compression ratio, (3) ignition, and (4) the combustion process.

Action of Combustion Gases on Pistons

Engines are classified in many ways. Mention has already been made of some classifications such as those based on (1) the fuels used (diesel fuel and gasoline), (2) the ignition methods (spark and compression), (3) the combustion cycles (Otto and diesel), and (4) the mechanical cycles (2-stroke and 4-stroke). Engines may also be classified on the basis of cylinder arrangements (V, in-line, opposed, etc.), the cooling media (liquid and air), and the valve arrangements (L-head, valve-in head, etc.). The manner in which the pressure of combustion gases acts upon the piston to move it in the cylinder of an engine is also used as a method of classifying engines.

The classification of engines according to combustion-gas action is based upon a consideration of whether the pressure created by the combustion gases acts upon one or two surfaces of a single piston or against single surfaces of

two separate and opposed pistons. The two types of engines under this classification are commonly referred to as single-acting and opposed-piston engines.

SINGLE-ACTING ENGINES.—Engines of this type are those which have one piston per cylinder and in which the pressure of combustion gases acts only on one surface of the piston. This is a feature of design rather than principle, for the basic principles of operation apply whether an engine is single acting, opposed piston, or double acting.

The pistons in most single-acting engines are of the trunk type (length greater than diameter). The barrel or wall of a piston of this type has one end closed (crown) and one end open (skirt end). Only the crown of a trunk piston serves as part of the combustion space surface. Therefore, the pressure of combustion can act only against the crown; thus, with respect to the surfaces of a piston, pressure is single acting. Most modern gasoline engines as well as many of the diesel engines used by the Navy are single acting.

OPPOSED-PISTON ENGINES.—With respect to combustion-gas action, the term opposed piston is used to identify those engines which have two pistons and one combustion space in each cylinder. The pistons are arranged in "opposed" positions—that is, crown to crown, with the combustion space in between. (See fig. 22-8.) When combustion takes place, the gases act against the crowns of both pistons, driving them in opposite directions. Thus, the term "opposed" not only signifies that, with respect to pressure and piston surfaces, the gases act in "opposite" direction, but also classifies piston arrangement within the cylinder.

In modern engines which have the opposed-piston arrangement, two crankshaft (upper and lower) are required for transmission of power. Both shafts contribute to the power output of the engine. They may be connected in one of two ways; chains as well as gears have been used for the connection between shafts. However, in most opposed-piston engines common to Navy service, the crankshafts are connected by a vertical gear drive. (See fig. 22-8.)

The cylinders of opposed-piston engines have scavenging air ports located near the top. These ports are opened and closed by the upper piston. Exhaust ports located near the bottom of the cylinder are closed and opened by the lower piston.

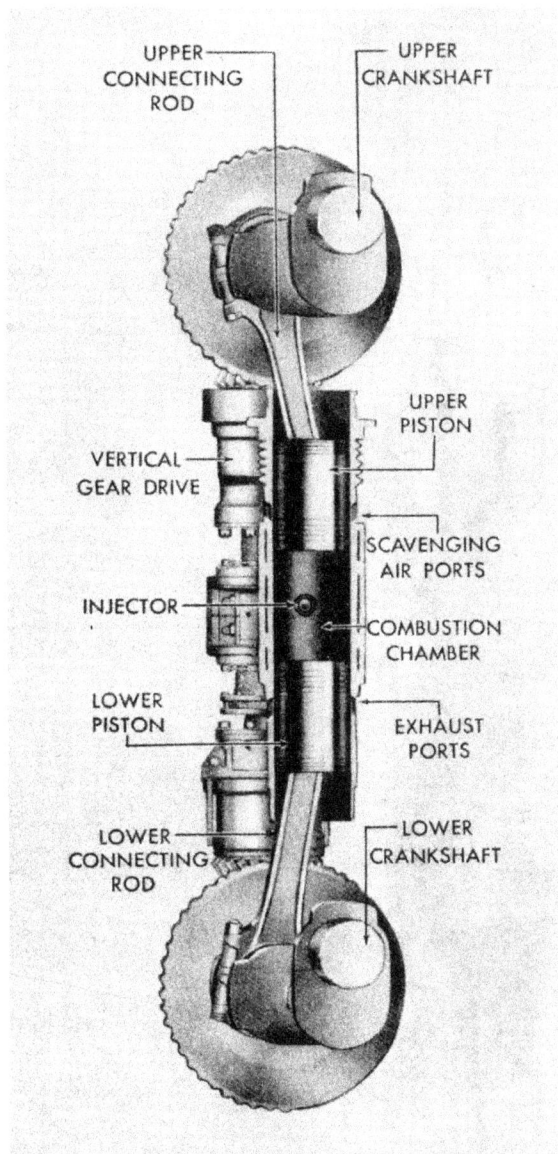

75.8

Figure 22-8.—Cylinder and related parts-
opposed-piston engine.

Movement of the opposed pistons is such that
the crowns are closest together near the center
of the cylinder. When at this position, the pistons
are not at the true piston dead centers. This is
because the lower crankshaft operates a few
degrees in advance of the upper shaft. The
number of degrees that a crank on the lower
shaft travels in advance of a corresponding
crank on the upper shaft is called lower crank
lead. This is illustrated in figure 22-9.

Opposed-piston engines used by the Navy
operate on the 2-stroke cycle. In engines of the
opposed-piston type, as in 2-stroke cycle single-
acting engines, there is an overlap of the various
events occurring during a cycle of operation. Injection and the burning of the fuel start during the
latter part of the compression event and extend
into the power phase. There is also an overlap of
the exhaust and scavenging periods. The events
in the cycle of operation of an opposed-piston,
2-stroke cycle diesel engine are shown in figure
22-10.

Modern engines of the opposed-piston design
have a number of advantages over single-acting
engines of comparable rating. Some of these
advantages are: less weight per horsepower
developed; lack of cylinder heads and valve
mechanisms (and the cooling and lubricating
problems connected with them); and fewer
moving parts.

Functions of Reciprocating
Engine Components

The design of most internal combustion engines of the reciprocating type follows much the
same general pattern. Though engines are not
all exactly alike, there are certain features common to all, and the principal components of most
engines are similarly arranged. Since the general structure of gasoline engines is basically the
same as that of diesel engines, the following
discussion of the engine components applies
generally to both types of engines. However, differences do exist and these will be pointed out
whereever applicable.

The principal components of an internal
combustion engine may be divided into two
principal groups—parts and systems. The main
parts of an internal combustion engine may be
further divided into structural parts and moving
parts. Structural parts, for the purpose of this
discussion, include those which, with respect
to engine operation, do not involve motion;
namely, the structural frame and its components
and related parts. The other group of engine
parts includes those which involve motion. Many
of the principal parts which are mounted within
the main structure of an engine are moving parts.
Moving parts are considered as those which convert the power developed by combustion in the
cylinder to the mechanical energy that is available for useful work at the output shaft.

The systems commonly associated with the
engine proper are those necessary to make

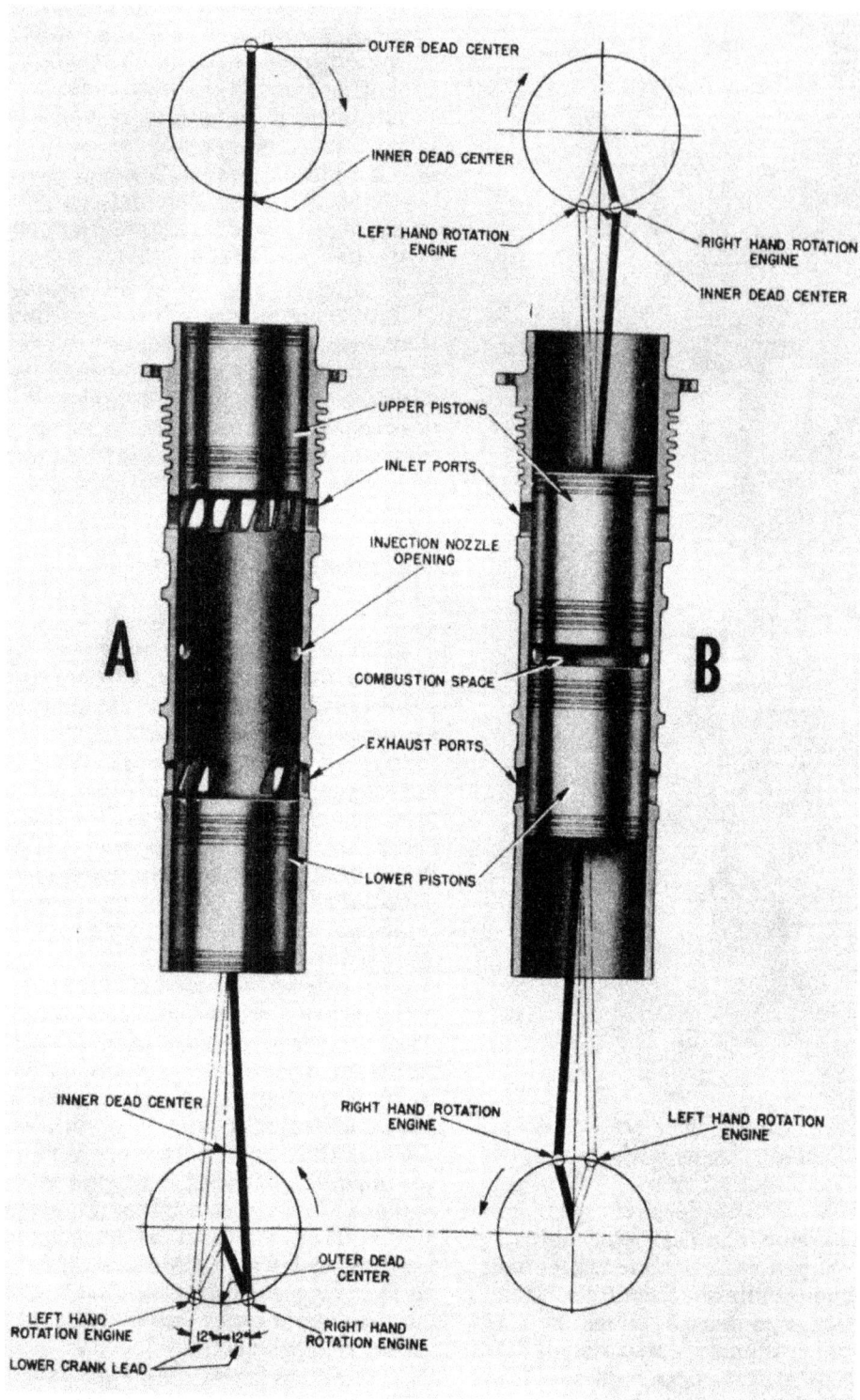

Figure 22-9.—Lower crank lead, opposed-piston engine.

75.9X

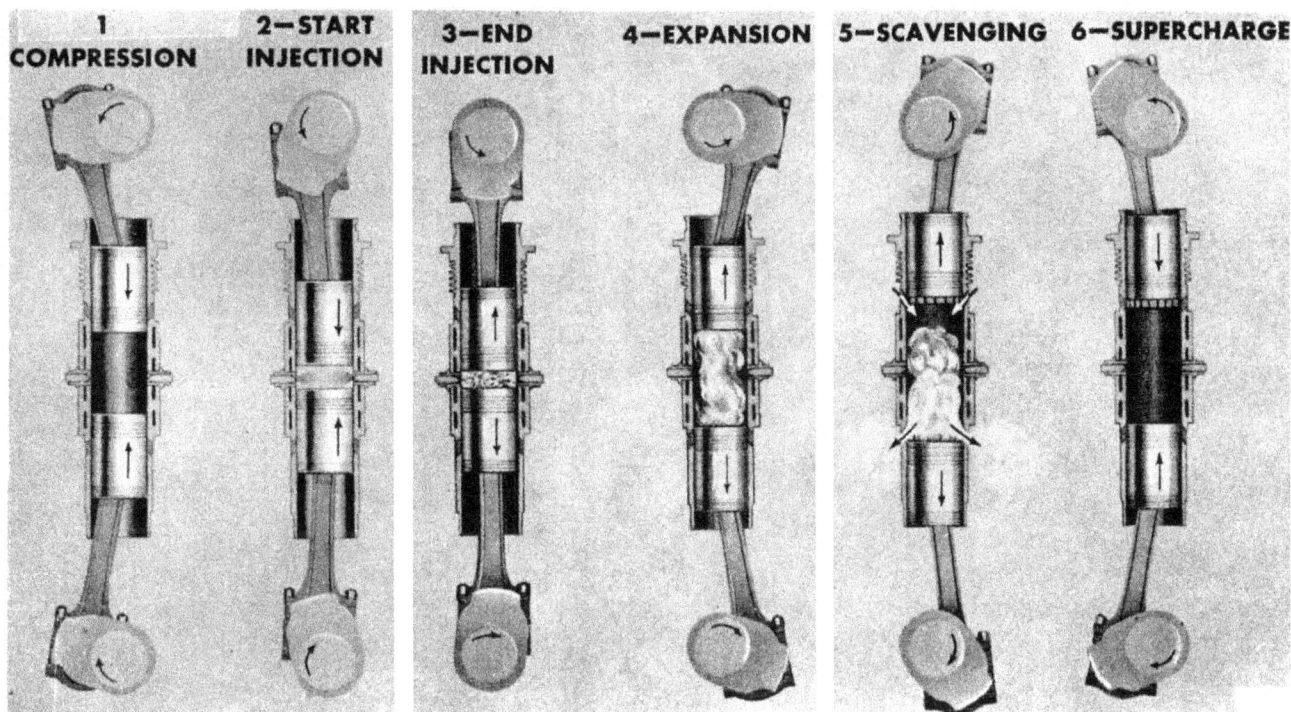

75.10

Figure 22-10.—Events in operating cycle of an opposed-piston engine.

combustion possible, and those which minimize and dissipate heat created by combustion and friction. Since combustion requires air, fuel, and heat (ignition), systems providing each may be found on some engines. However, since a diesel engine generates its own heat for combustion within the cylinders, no separate ignition system is required for engines of this type. The problem of heat, created as a result of combustion and friction, is taken care of by two separate systems—cooling and lubrication. The functions of the parts and systems of engines which operate on the principles already described are discussed briefly in the following paragraphs.

MAIN STRUCTURAL PARTS.—The main purpose of the structural parts of an engine is to maintain the moving parts in their proper relative position. This is necessary if the gas pressure produced by combustion is to fulfill its function.

The term frame is sometimes used to identify a single part of an engine; in other cases, it identifies several stationary parts fastened together to support most of the moving engine parts and engine accessories. For the purpose

of this discussion, the latter meaning will be used. As the load-carrying part of the engine, the frame of the modern engine may include such parts as the cylinder block, crankcase, bedplate or base, sump or oil pan, and end plates.

The part of the engine frame which supports the engine's cylinder liners and head or heads is generally referred to as the cylinder block. The blocks for most large engines are of the welded-steel type construction. Blocks of small high-speed engines may be of the en bloc construction. In this type construction, the block is cast in one piece. Two types of cylinder blocks coming to Navy service are shown in figures 22-11 and 22-12. The block shown in figure 22-11 is representative of blocks designed for some large engines with in-line cylinder arrangement. The block illustrated in figure 22-12 is representative of blocks constructed for some engines with V-type cylinder arrangement.

The engine frame part which serves as a housing for the crankshaft is commonly called the crankcase. In some engines, the crankcase is an integral part of the cylinder block (see fig. 22-11), requiring an oil pan, sump, or base

557

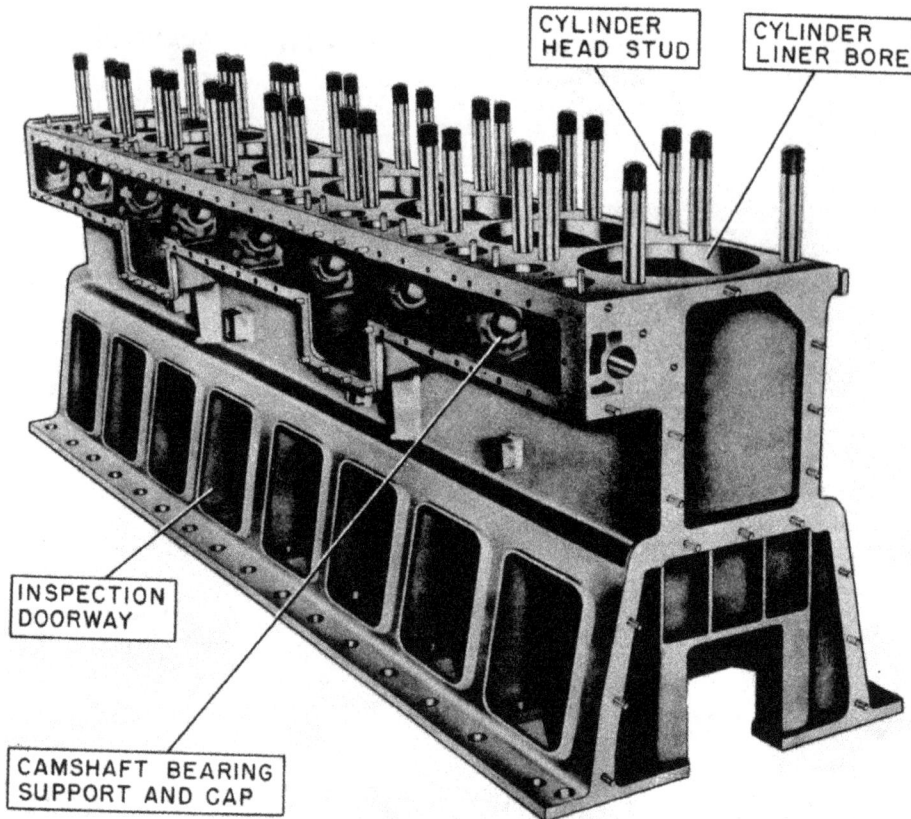

Figure 22-11.—Cylinder block with in-line cylinder arrangement.

75.14X

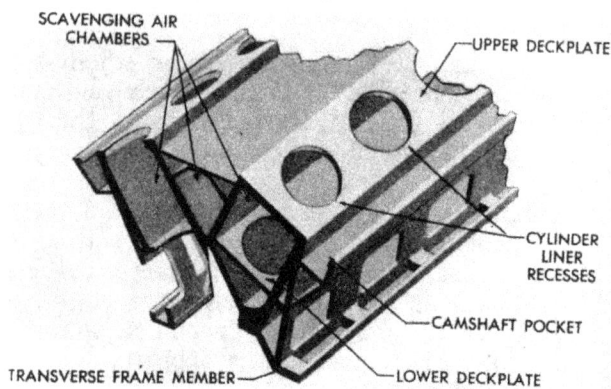

75.16

Figure 22-12.—An example of a V-type cylinder block construction.

to complete the housing. In others the crankcase is a separate part and is bolted to the block.

In large engines of early design, the support for the main bearings was provided by a bedplate. The bedplate was bolted to the crankcase and an oil pan was bolted to the bedplate when a separate oil pan was used. In some large engines of more modern design the support for main bearings is provided by a part called the base. Figure 22-13 illustrates such a base, which is used with the block shown in figure 22-11. This type base serves as a combination bedplate and oil plan. This base requires the engine block to complete the frame for the main engine bearings. Some crankcases are designed so that the crankshaft and the main bearings are mounted and secured completely within the crankcase.

Since lubrication is essential for proper engine operation, a reservoir for collecting and holding the engine's lubricating oil is a necessary part of the engine structure. The reservoir may be called a sump or an oil pan, depending upon its design, and is usually attached directly to

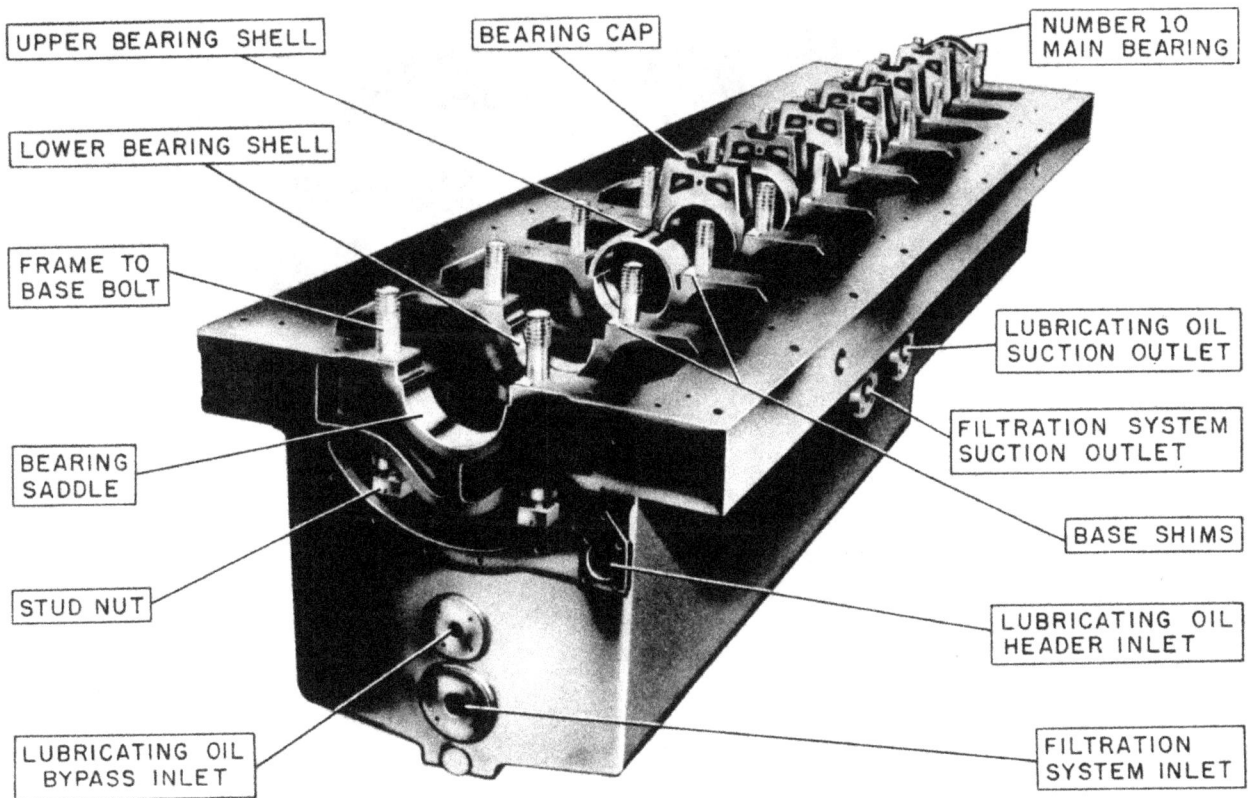

Figure 22-13.—Engine base.

75.20X

the engine. However, in some engines, the oil reservoir may be located at some point relatively remote from the engine; such engines may be referred to as dry sump engines.

In the engine base shown in figure 22-13, oil sump is an integral part of the base or crankcase, which has functions other than just being an oil reservoir. Many of the smaller engines do not have a separate base or crankcase; instead, they have an oil pan, which is secured directly to the bottom of the block. In most cases, an oil pan serves only as the lower portion of the crankshaft housing and as the oil reservoir.

Some engines have flat steel plates attached to each end of the cylinder block. End plates add rigidity to the block and provide a surface to which may be bolted housings for such parts as gears, blowers, pumps, and generators.

Many engines, especially the larger ones, have access openings in some part of the engine frame. (See fig. 22-11.) These openings permit access to the cylinder liners, main and connecting rod bearings, injector control shafts, and various other internal engine parts. Access doors (sometimes called covers or plates) for the openings are usually secured with handwheel or nut-operated clamps and are fitted with gaskets to keep dirt and foreign material out of the engine's interior.

The cylinder assembly completes the structural framework of an engine. As one of the main stationary parts of an engine, the cylinder assembly, along with various related working parts, serves to confine and release the gases. For the purpose of this discussion, the cylinder assembly will be considered as consisting of the head, the liner, the studs, and the gasket. (See fig. 22-14.)

The design of the parts of the cylinder assembly varies considerably from one type of engine to another. Regardless of differences in design, however, the basic components of all cylinder assemblies function, along with related moving parts, to provide a gas- and liquid-tight space.

559

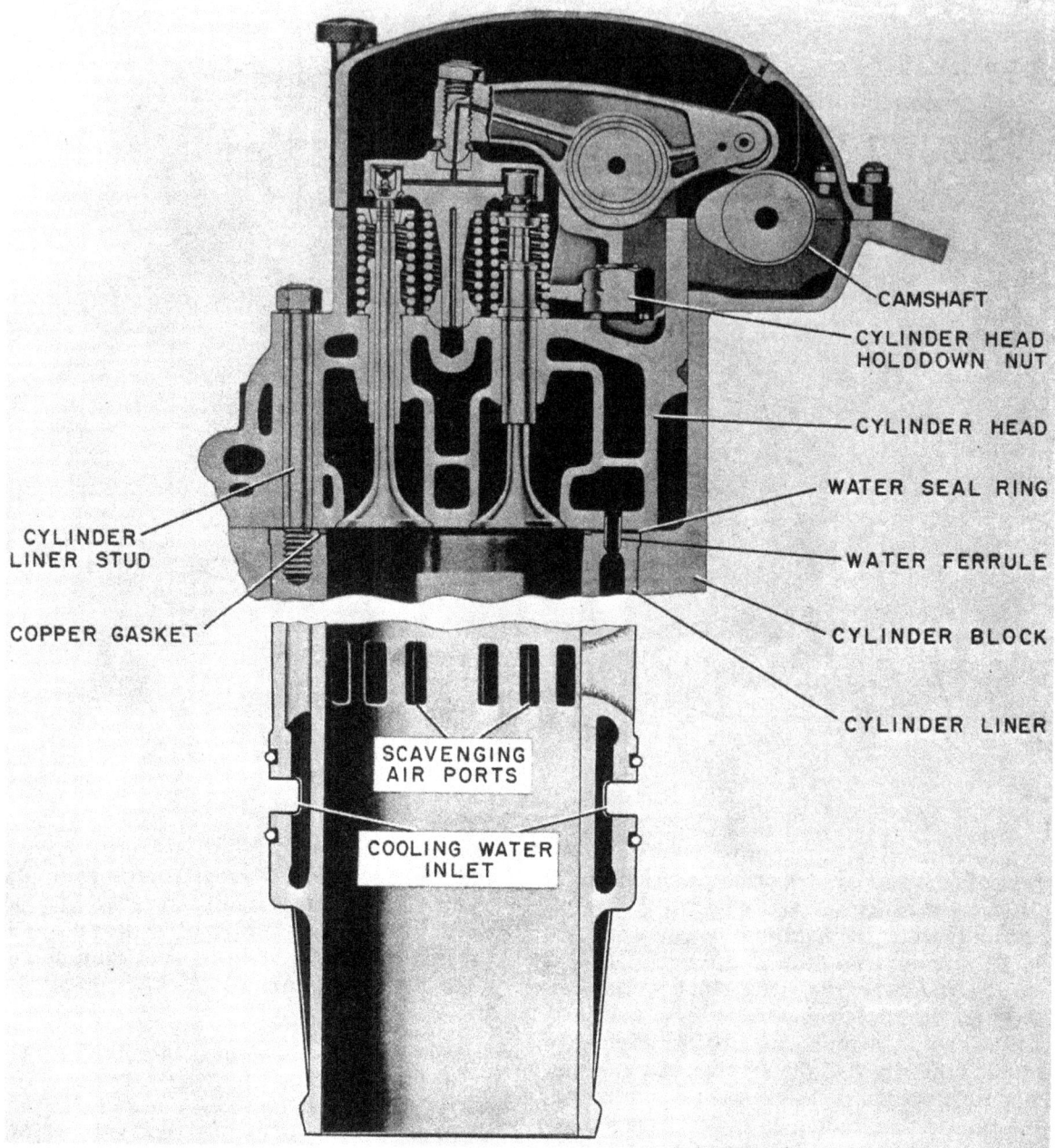

75.24X

Figure 22-14.—Principal stationary parts of a cylinder assembly.

The barrel or bore in which an engine piston moves back and forth may be an integral part of the cylinder block or it may be a separate sleeve or liner. The first type, common in gasoline engines, has the disadvantage of not being replaceable. Practically all diesel engines are constructed with replaceable cylinder liners.

Six cylinder liners of the replaceable type are shown in figure 22-15. These liners illustrate some of the differences in the design of liners and the relative size of the engines represented.

The liners or bores of an internal combustion engine must be sealed tightly to form the combustion chambers. In most Navy engines, except

560

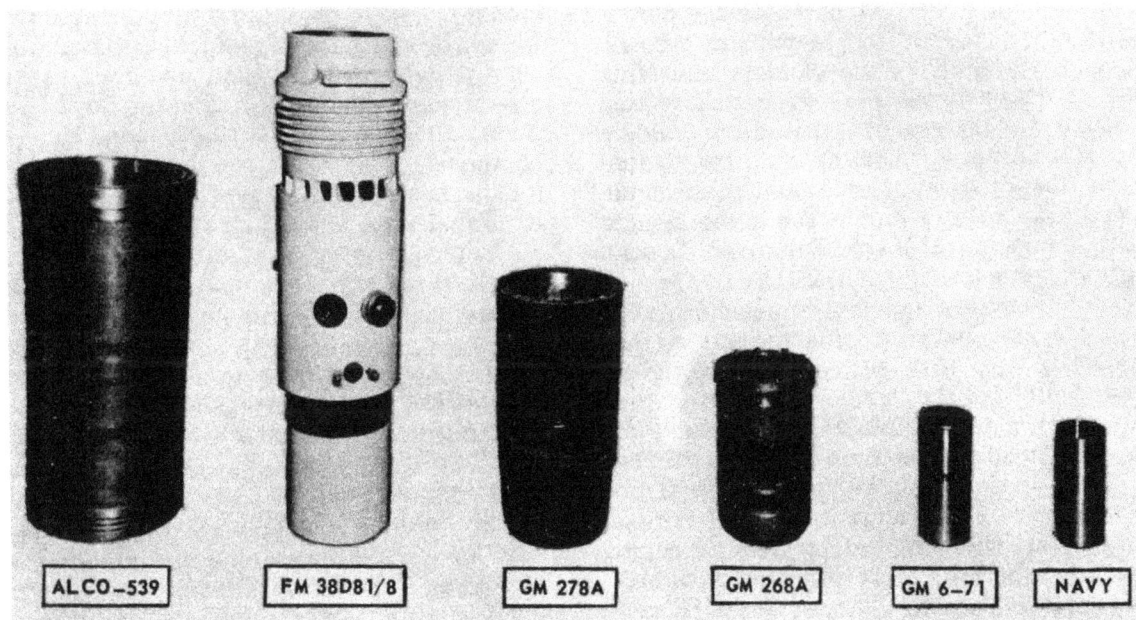

75.25

Figure 22-15.—Cylinder liners of diesel engines.

for engines of the opposed-piston type, the space at the combustion end of a cylinder is formed and sealed by a cylinder head which is a separate unit from the block. (See fig. 22-14.)

A number of engine parts which are essential to engine operation may be found in or attached to the cylinder head. The cylinder head may house intake and exhaust valves, valve guides and valve seats, or only exhaust valves and related parts. Rocker arm assemblies are frequently attached to the cylinder head. The fuel injection valve is almost universally in the cylinder head or heads of a diesel engine, while the spark plugs are always in the cylinder head of gasoline engines. Cylinder heads of a diesel engine may also be fitted with air starting valves, indicator and blow down valves, and safety valves.

The number of cylinder heads found on engines varies considerably. Small engines of the in-line cylinder arrangement utilize one head for all cylinders. A single head serves for all cylinders in each bank of some V-type engines. Large diesel engines generally have one cylinder head for each cylinder. Some engines use one head for each pair of cylinders.

In most cases, the seal between the cylinder head and the block depends principally upon the studs and gaskets. The studs, or stud bolts, secure the cylinder head to the cylinder block. A gasket between the head and the block is compressed to form a seal when the head is properly tightened down. In some cases, gaskets are not used between the cylinder head and block; the mating surfaces of the head and block are accurately machined to form a seal between the two parts.

PRINCIPAL MOVING PARTS.—In order that the power developed by combustion can be converted to mechanical energy, it is necessary for reciprocating motion to be changed to rotating motion. The moving parts included in the conversion process, from combustion to energy output, may be divided into the following three major groups: (1) the parts which have only reciprocating motion (pistons), (2) the parts which have both reciprocating and rotating motion (connecting rods), and (3) the parts which have only rotating motion (crankshafts and camshafts).

The first two major groups of moving parts may be further grouped under the single heading of piston and rod assemblies. Such an assembly may include a piston, piston rings, piston pin, connecting rod, and related bearings.

As one of the principal parts in the power transmitting assembly, the piston must be so designed and must be made of such materials that it can withstand the extreme heat and pressure of combustion. Pistons must also be light enough to keep inertia loads on related parts to a minimum. The piston aids in the sealing of the cylinder to prevent the escape of gas and transmits some of the heat through the piston rings to the cylinder wall. In addition to serving as the unit which transmits the force of combustion to the connecting rod and conducts the heat of combustion to the cylinder wall, a piston serves as a valve in opening and closing the ports of a two-stroke cycle engine. The nomenclature for the parts of a typical trunk type piston is given in figure 22-16.

Piston rings are particularly vital to engine operation in that they must effectively perform three functions: seal the cylinder, distribute and control lubricating oil on the cylinder wall, and transfer heat from the piston to the cylinder wall. All rings on a piston perform the latter function, but two general types of rings—compression and oil—are required to perform the first two functions. There are numerous types of rings in each of these groups, contructed in different ways for particular purposes. Some of the variations in ring design are illustrated in figure 22-17.

In trunk-type piston assemblies, the connection between the piston and the connecting rod is usually the piston pin (sometimes referred to as the wrist pin) and its bearings. These parts must be of especially strong construction because the power developed in the cylinder is transmitted from the piston through the pin to the connecting rod. The pin is the pivot point where the straight-line or reciprocating motion of the piston changes to the reciprocating and rotating motion of the connecting rod. Thus, the principal forces to which a pin is subjected are the forces created by combustion and the side thrust created by the change in direction of motion. (See fig. 22-18.)

The connecting link between the piston and crankshaft or the crankshaft and the crosshead of an engine is the connecting rod. In order that the forces created by combustion can be transmitted to the crankshaft, the rod changes the reciprocating motion of the piston to the rotating motion of the crankshaft.

Most marine engines in Navy service use the trunk-type piston connected directly to the connecting rod.

The camshaft is a shaft with eccentric projection, called cams, designed to control the operation of valves, usually through various intermediate parts as described later in this chapter. Originally cams were made as separate pieces and fastened to the camshaft. However, in most modern engines the cams are forged or cast as an integral part of the camshaft.

To reduce wear and to help them withstand the shock action to which they are subjected, camshafts are made of low-carbon alloy steel with the cam and journal surfaces carburized before the final grinding is done.

The cams are arranged on the shaft to provide the proper firing order of the cylinders served. The shape of the cam determines the point of opening and closing, the speed of opening and closing, and the amount of the valve lift. If one cylinder is properly time, the remaining cylinders are automatically in time. All cylinders will be affected if there is a change in timing.

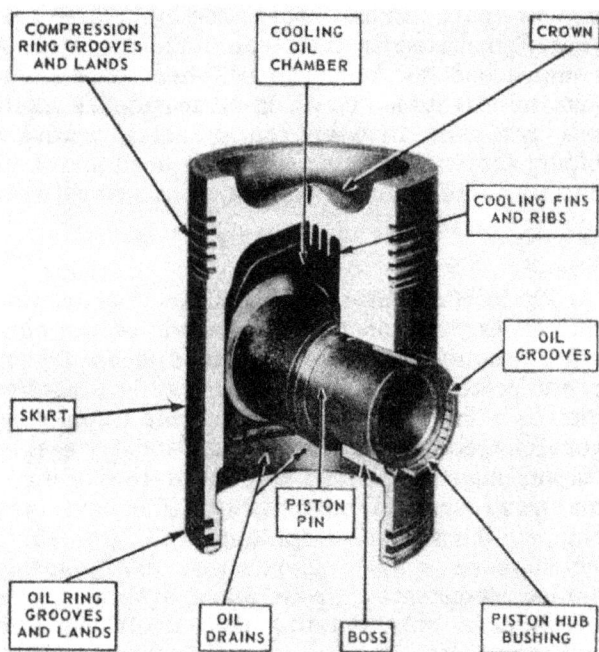

75.47
Figure 22-16.—Piston nomemclature.

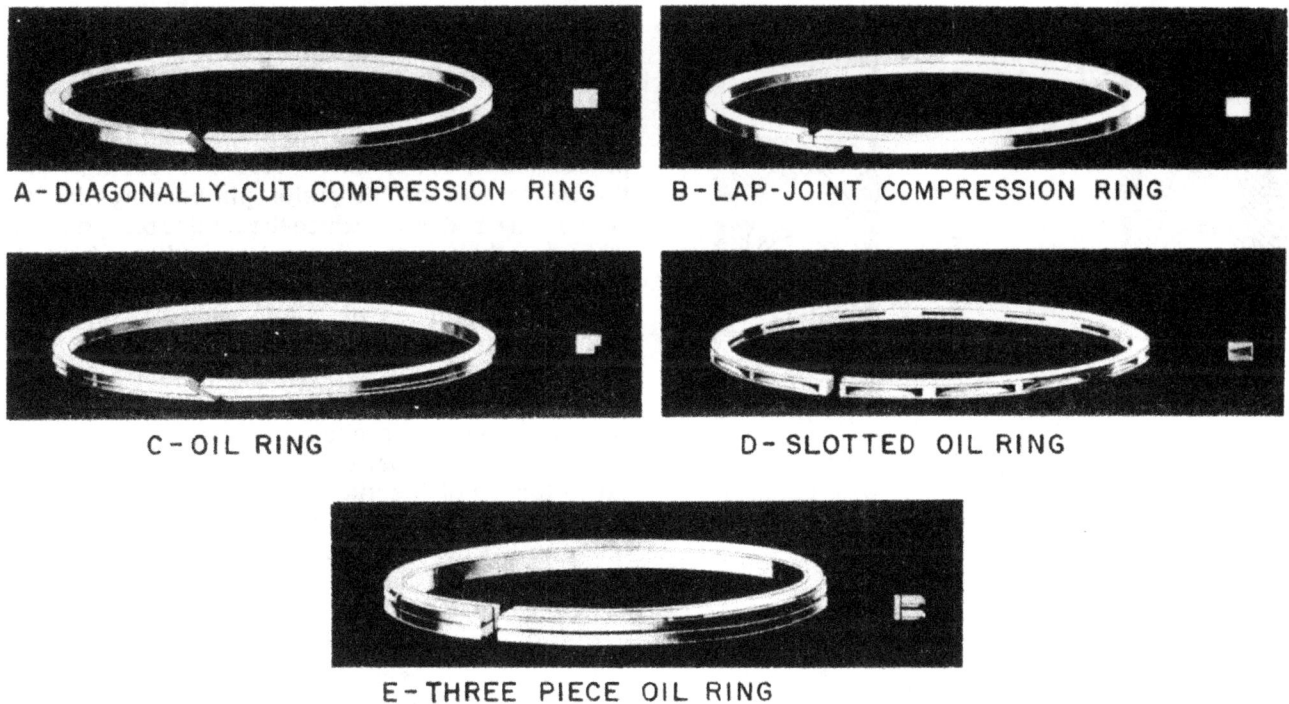

A - DIAGONALLY-CUT COMPRESSION RING

B - LAP-JOINT COMPRESSION RING

C - OIL RING

D - SLOTTED OIL RING

E - THREE PIECE OIL RING

75.51

Figure 22-17.—Types of piston rings.

The camshaft is driven by the crankshaft by various means, the most common being by gears or by a chain and sprocket. The camshaft for a 4-stroke cycle engine must turn at one-half of the crankshaft speed; while in the 2-stroke cycle engine, it turns at the same speed as the crankshaft.

The location of the crankshaft in various engines differs. Camshaft location depends on the arrangement of the valve mechanism. The location of a camshaft is shown in figure 22-14.

One of the principal engine parts which has only rotating motion is the crankshaft. As one of the largest and most important moving parts in an engine, the crankshaft changes the movement of the piston and the connecting rod into the rotating motion required to drive such items as reduction gears, propeller shafts, generators, pumps, etc. As a result of its function, the crankshaft is subjected to all the forces developed in an engine.

While crankshafts of a few larger engines are of the built-up type (forged in separate sections and flanged together), the crankshafts of most modern engines are of the one-piece type construction. A shaft of this type is shown in figure 22-19. The parts of a crankshaft may be identified by various terms; however, those shown in figure 22-19 are common in the technical manuals for most of the engines used by the Navy.

The speed of rotation of the crankshaft increases each time the shaft receives a power impulse from one of the pistons; and it then gradually decreases until another power impulse is received. These fluctuations in speed (their number depending upon the number of cylinders firing in one crankshaft revolution) would result in an undesirable situation with respect to the driven mechanism as well as the engine; therefore, some means must be provided to stabilize shaft rotation. In some engines this is accomplished by installing a flywheel on the crankshaft; in others, the motion of such engine parts as the crankpins, webs, lower ends of connecting rods, and such driven units as the clutch, generator, etc., serve the purpose. The need for a flywheel decreases as the number of cylinders firing in one revolution of the crankshaft and the mass of the moving parts attached to the crankshaft increases.

A flywheel stores up energy during the power event and releases it during the remaining events of the operating cycle. In other words, when the

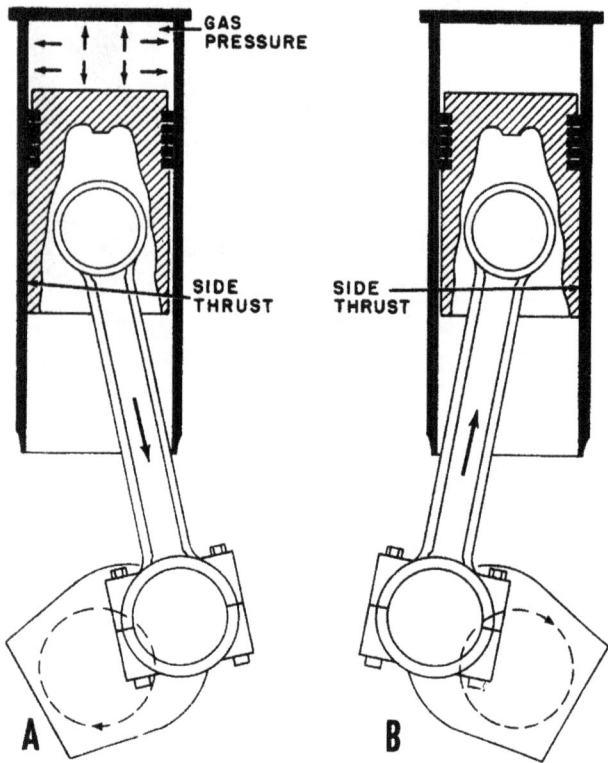

75.57

Figure 22-18.—Side thrust of a trunk-type piston, single-acting engine.

speed of the shaft tends to increase, the flywheel absorbs energy, and when the speed tends to decrease, the flywheel gives up energy to the shaft in an effort to keep shaft rotation uniform. In doing this, a flywheel (1) keeps variations in speed within desired limits at all loads; (2) limits the increase or decrease in speed during sudden changes of load; (3) aids in forcing the piston through the compression event when an engine is running at low or idling speed; and (4) helps bring the engine up to speed when it is being cranked.

An important group of engine parts consists of the bearings. Some bearings remain stationary in performing their function while others move. One principal group of stationary bearings in an engine is that which supports the crankshaft. These bearings are generally called main engine bearings. (See fig. 22-13). Main bearings in most engines are of the sliding contact, or plain type, consisting of two halves or shells.

Main bearings are subjected to a fluctuating load. This is also true of the crankpin bearings and the piston-pin bearings. However, the manner in which main journal bearings are loaded depends upon the type of engine in which they are used.

In a 2-stroke cycle engine, a load is always placed on the lower half of the main bearings and the lower half of the piston pin bearings in the connecting rod; meanwhile the load is placed upon the upper half of the connecting rod bearings at the crankshaft end of the rod. This is true because the forces of combustion are greater than the inertia forces created by the moving parts.

In a 4-stroke cycle engine, the load is applied first on one bearing shell and then on the other. The reversal of pressure is the result of the large forces of inertia imposed during the intake and exhaust strokes. In other words, inertia tends to lift the crankshaft in its bearings during the intake and exhaust strokes.

There is a definite reversal of load application on the main bearings of a double-acting engine. In this case, the reversal is caused by combustion taking place first on one end of the piston and then on the other.

The bearings used in connection with piston pins are of three types: the integral bearing, the sleeve bearing or bushing, and the needle type roller bearing. The bearings in the bosses (hubs) of most pistons are of the sleeve bushing type. However, in a few cases, the boss bearings are an integral part of the piston. In such cases, the bearing surface is precision bored directly in the bosses. Pistons fitted with stationary piston pins require no bearing surfaces in the bosses.

Even though the piston pins in most engines are equipped with bushing type bearings, some have been fitted with bearings of the needle roller type.

The types of bearings used for main bearings and in connection with piston-pin assemblies are representative of those used at other points in an engine where bearing surfaces are required.

All of the parts which make a complete engine have by no means been covered in the preceding section of this chapter. Since many engine parts and accessories are commonly associated with the systems of an engine, functions of some of the principal components not covered to this point are considered with the applicable system which they affect.

ENGINE AIR SYSTEMS.—Parts and accessories which supply the cylinders of an engine with air for combustion, and remove the waste gases after combustion and the power events

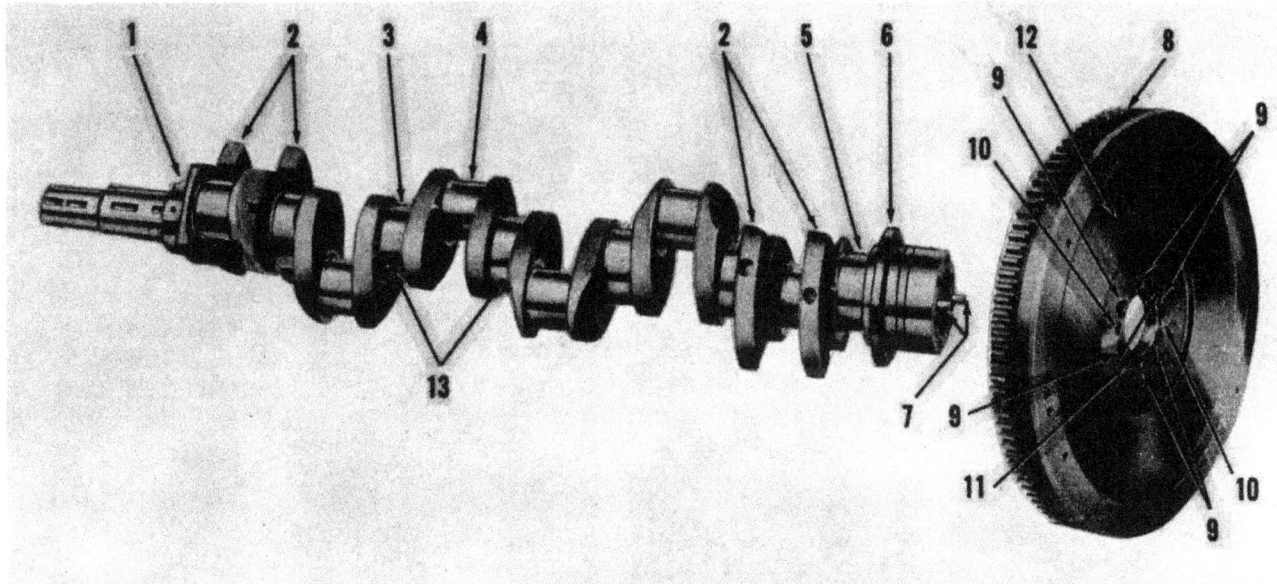

1. MAIN BEARING JOURNAL-FRONT
2. COUNTERWEIGHT
3. MAIN BEARING JOURNAL-
   INTERMEDIATE
4. CONNECTING ROD JOURNAL-NO. 3

5. MAIN BEARING JOURNAL-REAR
6. BOLTING FLANGE-TIMING GEAR
7. DOWEL-FLYWHEEL
8. RING GEAR
9. RETAINING BOLT HOLE

10. DOWEL HOLE
11. PULLER SCREW HOLE
12. FLYWHEEL
13. LUBRICATING OIL HOLES

75.81X

Figure 22-19.—One-piece six-throw crankshaft with flywheel.

are finished, are commonly referred to as the intake and exhaust systems. These systems are closely related and, in some cases, are referred to as the air systems of an engine. A cross-sectional view of the air systems of one type of high-speed diesel engine is shown in figure 22-20.

The following information on air systems deals primarily with the systems of diesel engines; nevertheless, much of the information dealing with the parts of diesel engine air systems is also applicable to most of the parts in similar systems of gasoline engines. However, the intake event in the cycle of operation of a gasoline engine includes the admission of air and fuel as a mixture to the cylinder. For this reason, the intake system of a gasoline engine differs, in some respects, from that of a diesel engine. (See subsequent section on fuel systems.)

A discussion of the air systems of deisel engines frequently involves the use of two terms which identify processes related to the functions of the intake and exhaust systems. These terms— scavenging and supercharging—and the processes they identify are common to many modern diesel engines.

In the intake systems of all modern 2-stroke cycle engines and some 4-stroke cycle engines, a device, usually a blower, is installed to increase the flow of air into the cylinders. This is accomplished by the blower compressing the air and forcing it into an air box or manifold (reservoir) which surrounds or is attached to the cylinders of an engine. Thus, an increased amount of air under constant pressure is available as required during the cycle of operation.

The increased amount of air available as a result of blower action is used to fill the cylinder with a fresh charge of air and, during the process, aids in clearing the cylinder of the gases of combustion. This process is called scavenging. Thus, the intake system of some engines, especially those operating on the 2-stroke cycle, is sometimes called the scavenging system. The air forced into the cylinder

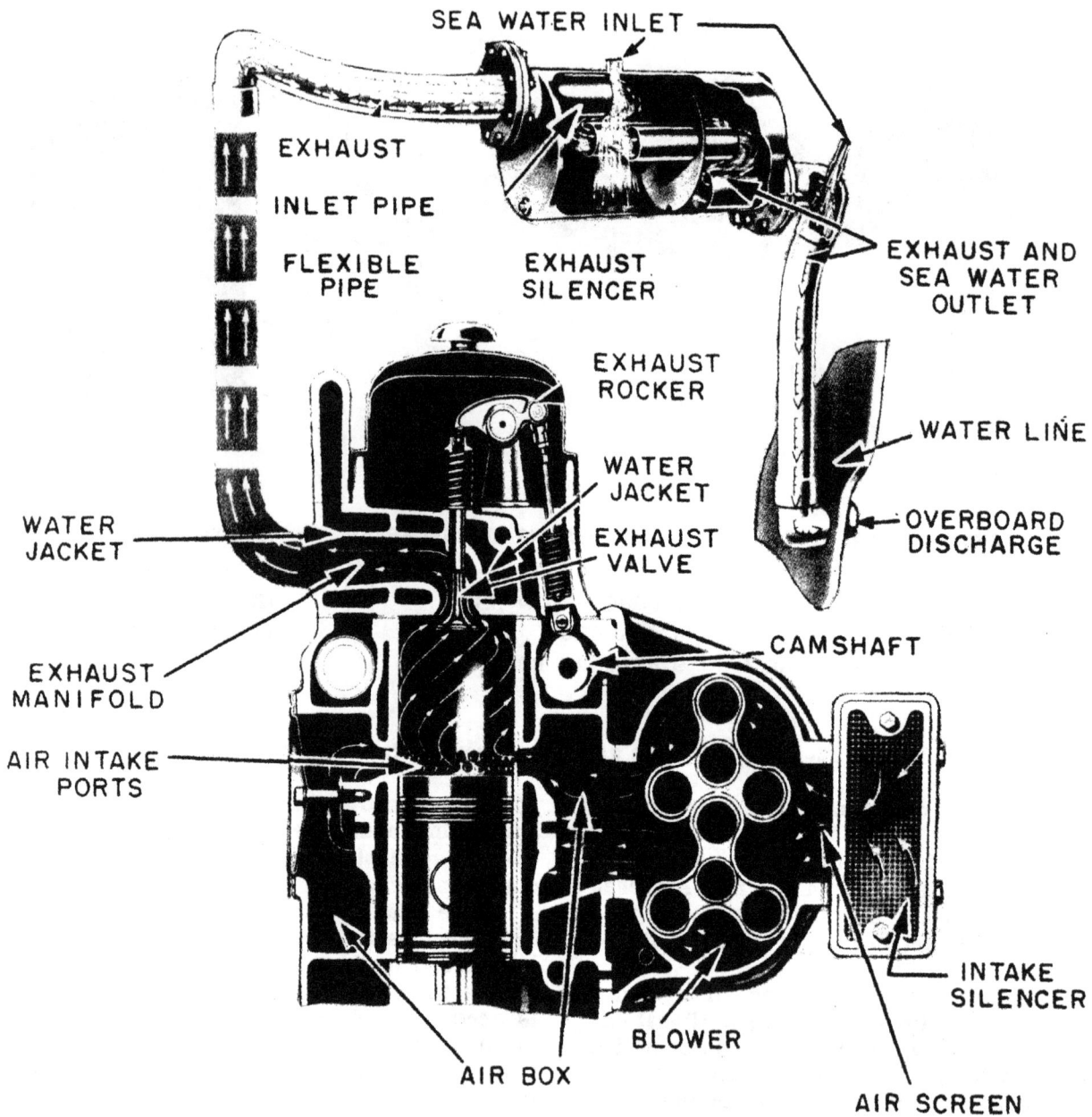

75.151X

Figure 22-20.—Air systems of a 2-stroke cycle engine.

is called <u>scavenge</u> (<u>or scavenging</u>) <u>air</u> and the ports through which it enters are called <u>scavenge ports</u>.

The process of scavenging must be accomplished in a relatively short portion of the operating cycle; however, the duration of the process differs in 2- and 4-stroke cycle engines. In a 2-stroke cycle engine, the process takes place during the later part of the downstroke (expansion) and the early part of the upstroke (compression). In a 4-stroke cycle engine, scavenging takes place when the piston is nearing and passing TDC during the latter part of an upstroke (exhaust and the early part of a downstroke (intake). The intake and exhaust openings are both open during this interval of time. The overlap of intake and exhaust permits the air from the blower to pass through the cylinder into the exhaust manifold, cleaning out the exhaust gases from the cylinder and, at the same time, cooling the hot engine parts.

Scavenging air must be so directed, when it enters the cylinder of an engine, that the waste gases are removed from the remote parts of the cylinder. The two principal methods by which this is accomplished are sometimes referred to as port scavenging and valve scavenging. Port scavenging may be of the direct (or cross-flow) loop (or return), or uniflow type. (See fig. 22-21.)

An increase in air flow into cylinders of an engine can be used to increase power output, in addition to being used for scavenging. Since the power of an engine is developed by the burning of fuel, an increase of power requires more fuel; the increased fuel, in turn, requires more air, since each pound of fuel requires a certain amount of air for combustion. Supplying more air to the combustion spaces that can be supplied through the action of atmospheric pressure and piston action (in 4-stroke cycle engines) or

scavenging air (in 2-stroke cycle engines) is called supercharging.

In some 2-stroke cycle diesel engines, the cylinders are supercharged during the air intake simply by increasing the amount and pressure of scavenge air. The same blower is used for supercharging and scavenging. Whereas scavenging is accomplished by admitting air under low pressure into the cylinder while the exhaust valves or ports are open, supercharging is done with the exhaust ports or valves closed. This latter arrangement enables the blower to force air under pressure into the cylinder and thereby increase the amount of air available for combustion. The increase in pressure resulting from the compressing action of the blower will depend upon the engine involved, but it is usually low, ranging from 1 to 5 psi. With this increase in pressure, and the amount of air available for combustion, there is a corresponding increase in the air-fuel ratio and in combustion efficiency within the cylinder. In other words, a given size engine which is supercharged can develop more power than the same size engine which is not supercharged.

Supercharging a 4-stroke cycle diesel engine requires the addition of a blower to the intake system since the operations of exhaust and intake in an unsupercharged engine are performed

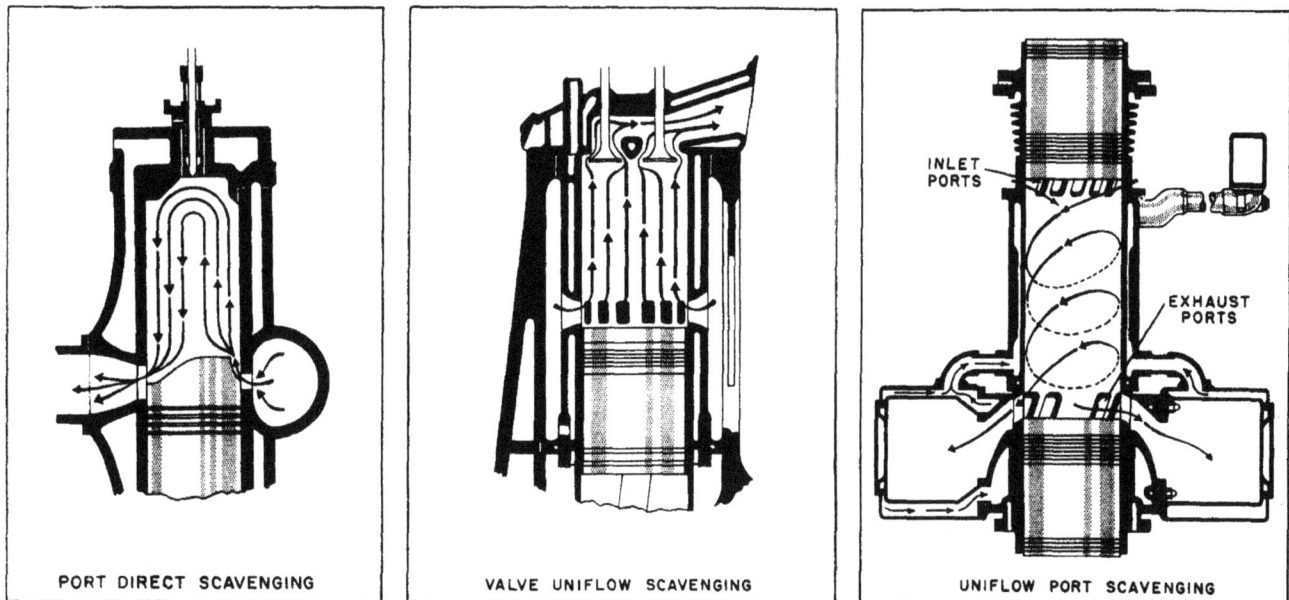

PORT DIRECT SCAVENGING   VALVE UNIFLOW SCAVENGING   UNIFLOW PORT SCAVENGING

75.152

Figure 22-21.—Methods of scavenging—diesel engines.

567

by the action of the piston. The timing of the valves in a supercharged 4-stroke cycle engine is also different from that in a similar engine which is not supercharged. In the supercharged engine the intake-valve opening is advanced and the exhaust-valve closing is retarded so that there is considerable overlap of the intake and exhaust events. This overlap increases power, the amount of the increase depending upon the supercharging pressure. The increased overlap of the valve openings in a supercharged 4-stroke cycle engine also permits the air pressure created by the blower to be used in removing gases from the cylinder during the exhaust event. How the opening and the closing of the intake and exhaust valves or ports affect both scavenging and supercharging, and the differences in these processes as they occur in supercharged 2- and 4-stroke cycle engines, can be seen by studying the diagrams in figure 22-22.

As in the case of the diagrams used in connection with the discussion of engine operating principles, the circular pattern in figure 22-22 represents crankshaft rotation. Some of the events occurring in the cycles are shown in terms of degrees of shaft rotation. However, the numbers (of degrees) shown on the diagrams are for purposes of illustration and comparison only. When these diagrams are being studied, it must be kept in mind that the crankshaft of a 4-stroke cycle engine makes two complete revolutions in one cycle of operation while the shaft in a 2-stroke cycle engine makes only one revolution per cycle. It should also be remembered that the exhaust and intake events in a 2-stroke cycle engine do not involve complete piston strokes as they do in a 4-stroke cycle engine.

Even though the primary purpose of a diesel engine intake system is to supply the air required for combustion, the system generally has to perform one or more additional functions. In most cases, the system cleans the air and reduces the noise created by the air as it enters the engine. In order to accomplish the functions of intake, an intake system may include an air silencer, an air cleaner and screen, an air box or header, intake valves or ports, a blower, an air heater, and an air cooler. All of these parts are not common to every intake system. An intake system in which only a silencer, a screen, a blower, an air box, and intake ports provide a clean supply of air, with minimum noise, to the combustion spaces is shown in figure 22-20.

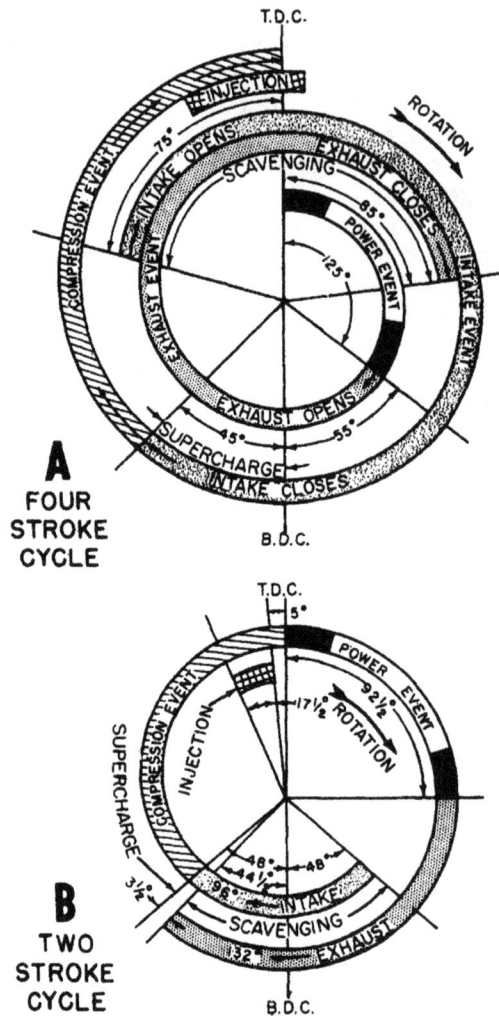

A
FOUR STROKE CYCLE

B
TWO STROKE CYCLE

54.19:.20B

Figure 22-22.—Scavenging and supercharging in diesel engines.

The system which functions primarily to convey gases away from the cylinders of an engine is called the exhaust system. In addition to this principal function, an exhaust system may be designed to perform one or more of the following functions: muffle exhaust noise, quench sparks, remove solid material from exhaust gases, and furnish energy to a turbine-driven supercharger. The principal parts which may be used in combination to accomplish the functions of an engine exhaust system are shown in figures 22-20 and 22-23.

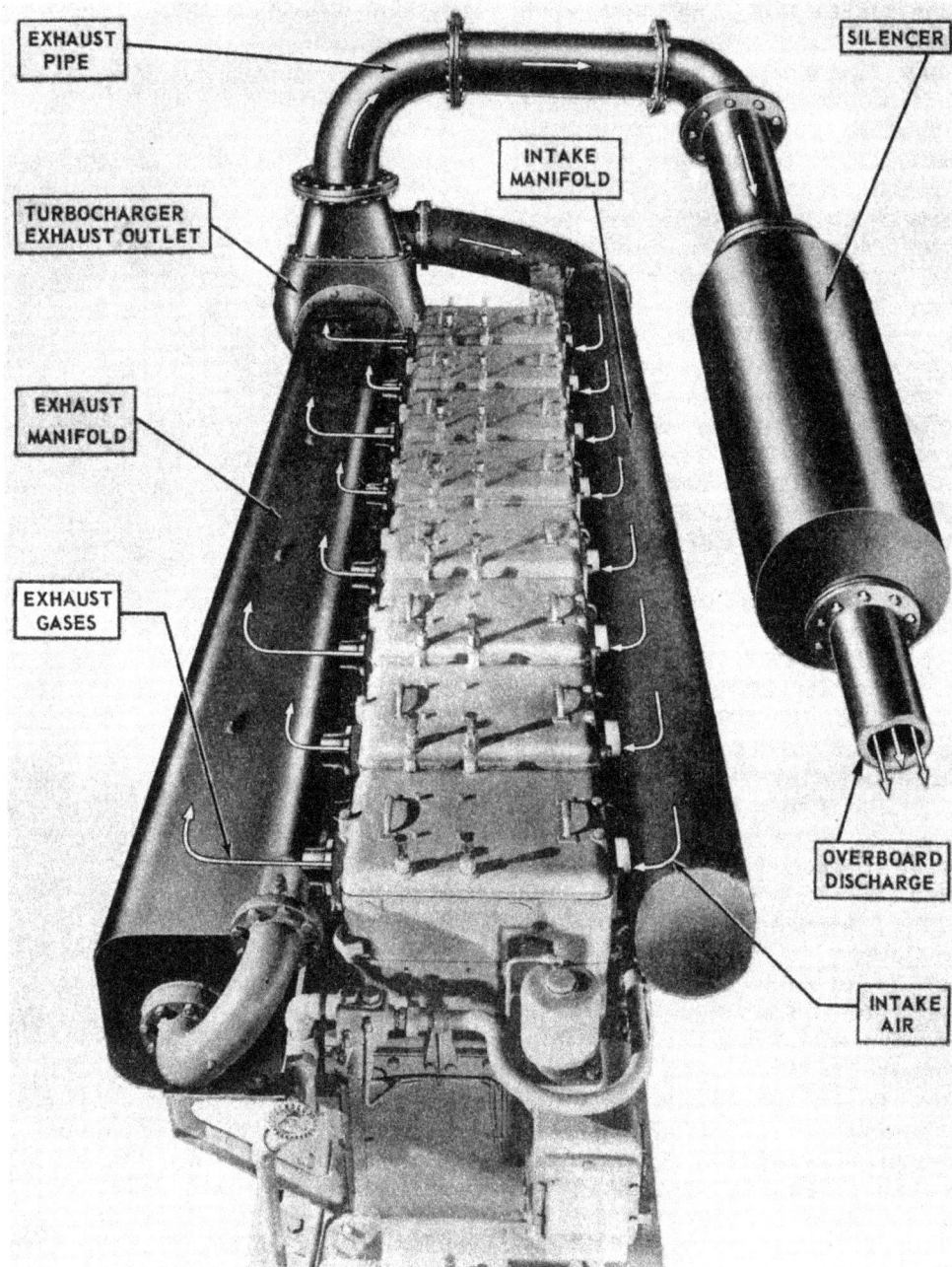

75.167**X**

Figure 22-23.—Intake and exhaust systems.

OPERATING MECHANISMS FOR SYSTEM PARTS AND ACCESSORIES.—To this point, consideration has been given only to the main engine parts—stationary and moving—and to two of the systems common to internal combustion engines. At various points in this chapter, reference has been made to the operation of some of the engine parts. For example, it has been pointed out that the valves open and close at the proper time in the operating cycle and that the impellers or lobes of a blower rotate to compress intake air. However, little consideration has been given to the source of power or to the mechanisms which cause these parts to operate.

In many cases, the mechanism which transmits power for the operation of the engine valves and blower may also transmit power to parts and accessories which are components of various engine systems. For example, such items as the governor; fuel, lubricating, and water pumps; and overspeed trips, are, in some engines, operated by the same mechanism. Since mechanisms which transmit power to operate specific parts and accessories may be related to more than one engine system, such operating mechanisms are considered here before the remaining engine systems are discussed.

The parts which make up the operating mechanisms of an engine may be divided into two groups: the group which forms the drive mechanisms and the group which forms the actuating mechanisms. The source of power for the operating mechanisms of an engine is the crankshaft.

As used in this chapter, the term drive mechanism identifies the group of parts which takes power from the crankshaft and transmits that power to various engine parts and accessories. In engines, the drive mechanisms does not change the type of motion, but it may change the direction of motion. For example, the impellers or lobes of a blower are driven or operated as a result of rotary motion which is taken from the crankshaft and transmitted to the impellers or lobes by the drive mechanism, an arrangement of gears and shafts. While the type of motion (rotary) remains the same, the direction of motion of one impeller or lobe is opposite to that of the other impeller or lobe as a result of the gear arrangements within the drive mechanism.

A drive mechanism may be of the gear, chain or belt type. Of these, the gear type is the most common; however, some engines are equipped with chain assemblies. A combination of gears and chains is used as the driving mechanism in some engines.

Some engines have a single drive mechanism which transmits power for the operation of engine parts and accessories. In other cases, there may be two or more separate mechanisms. When separate assemblies are used, the one which transmits power for the operation of the accessories is called the accessory drive. Some engines have more than one accessory drive. A separate drive mechanism which is used to

transmit power for the operation of engine valves is generally called the camshaft drive or timing mechanism.

The camshaft drive, as the name implies, transmits power to the camshaft of the engine. The shaft, in turn, transmits the power through a combination of parts which causes the engine valves to operate. Since the valves of an engine must open and close at the proper moment (with respect to the position of the piston) and remain in the open and closed positions for definite periods of time, a fixed relationship must be maintained between the rotational speeds of the crankshaft and the camshaft. Camshaft drives are designed to maintain the proper relationship between the speeds of the two shafts. In maintaining this relationship, the drive causes the camshaft to rotate at crankshaft speed in a 2-stroke cycle engine; and at one-half crankshaft speed in a 4-stroke cycle engine.

The term actuating mechanism, as used in this chapter, identifies that combination of parts which receives power from the drive mechanism and transmits the power to the engine valves. In order for the valves (intake, exhaust, fuel injection, air starter) to operate, there must be a change in the type of motion. In other words, the rotary motion of the crankshaft and drive mechanism must be changed to a reciprocating motion. The group of parts which, by changing the type of motion, causes the valves of an engine to operate is generally referred to as the valve actuating mechanism. A valve-actuating mechanism may include the cams, cam followers, push rods, rocker arms, and valve springs. In some engines, the camshaft is so located that the need for push rods is eliminated. In such cases, the cam follower is a part of the rocker arm. (Some actuating mechanisms are designed to transform reciprocating motion into rotary motion, but in internal combustion engines most actuating mechanisms change rotary motion into reciprocating motion.)

There is considerable variation in the design and arrangement of the parts of operating mechanisms found in different engines. The size of an engine, the cycle of operation, the cylinder arrangement, and other factors govern the design and arrangement of the components as well as the design and arrangement of the mechanisms. Three types of operating mechanisms are shown in figures 22-24, 22-25, and 22-26.

The mechanisms which supply power for the operation of the valves and accessories of gasoline engines are basically the same as those

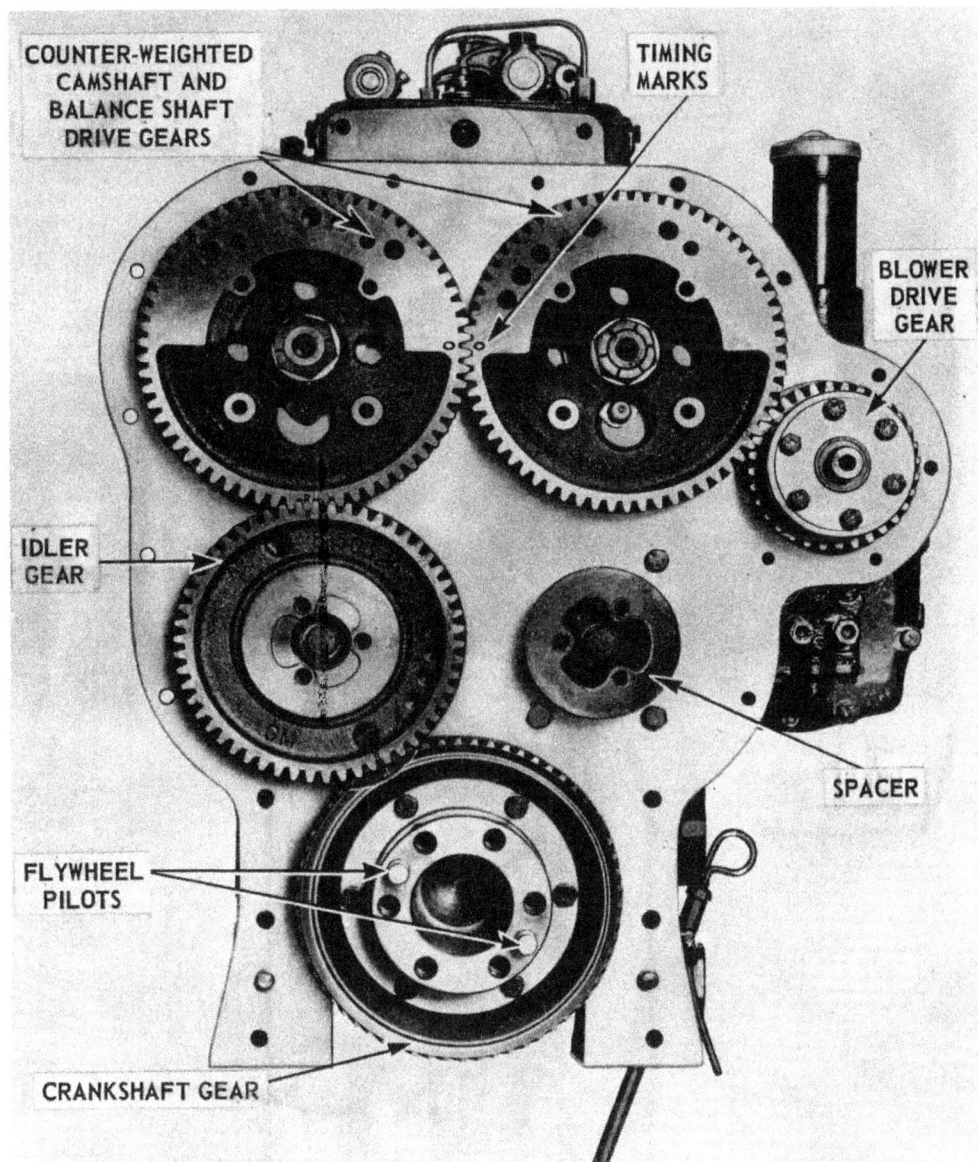

75.98X

Figure 22-24.—Camshaft and accessory drive.

found in diesel engines. Some manufacturers utilize mechanisms consisting primarily of chain assemblies, while others use gears as the primary means of transmitting power to engine parts. Combination gear-chain drive assemblies are used on some gasoline engines.

ENGINE FUEL SYSTEMS.—The method of getting fuel into the cylinder is one of the major differences between gasoline and diesel engines. As pointed out earlier, fuel for gasoline engines

is mixed with air outside the cylinder and the mixture is then drawn into the cylinder and compressed. On the other hand, fuel for diesel engines is injected or sprayed into the combustion space after the air is already compressed. The equipment which supplies fuel to the cylinders of a gasoline engine would necessarily be different from that of a diesel engine.

There are several types of fuel injection systems in use. The function of each type is, however, the same. The primary function of a

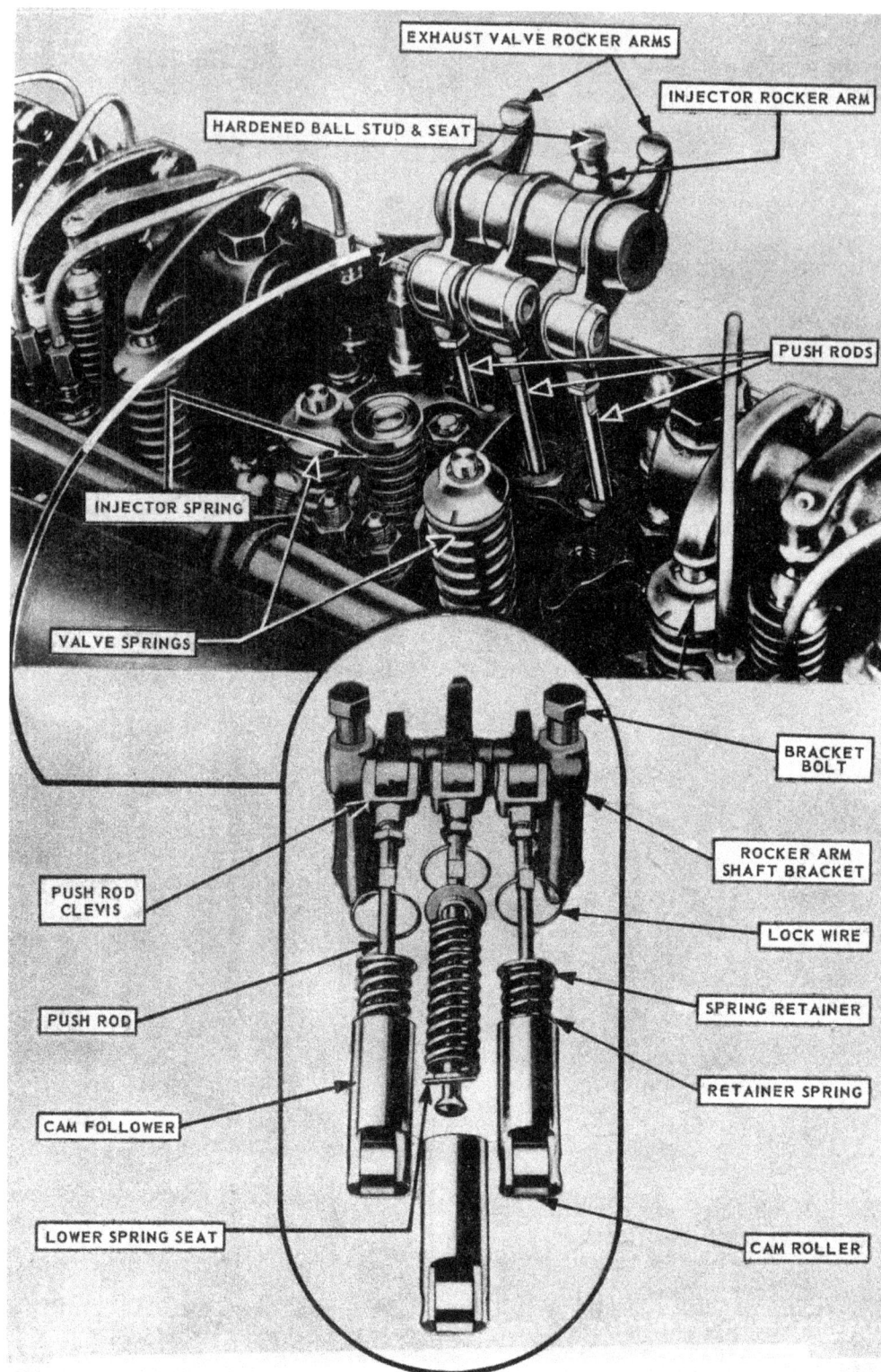

Figure 22-25.—Valve-actuating mechanism.

75.101X

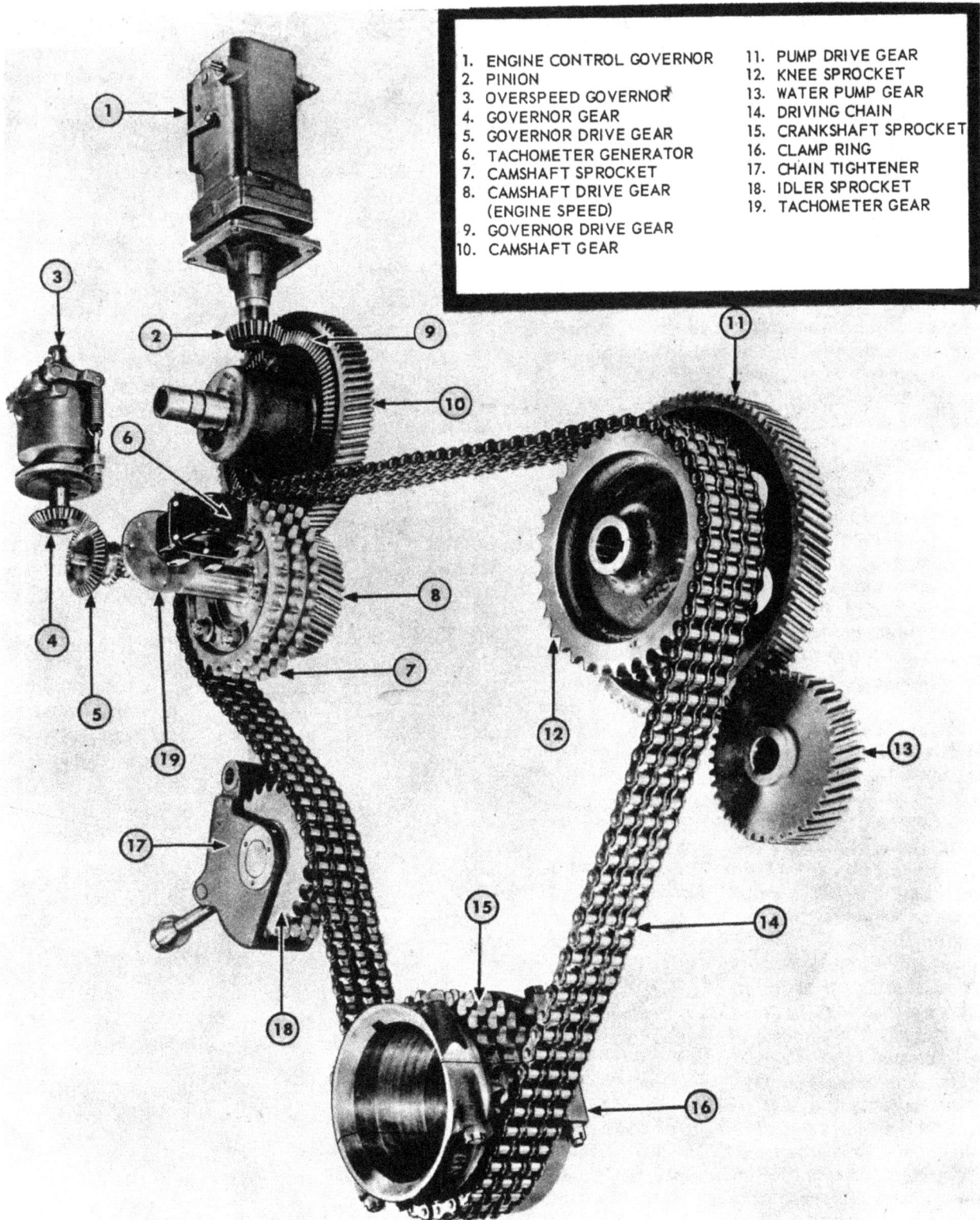

1. ENGINE CONTROL GOVERNOR
2. PINION
3. OVERSPEED GOVERNOR
4. GOVERNOR GEAR
5. GOVERNOR DRIVE GEAR
6. TACHOMETER GENERATOR
7. CAMSHAFT SPROCKET
8. CAMSHAFT DRIVE GEAR (ENGINE SPEED)
9. GOVERNOR DRIVE GEAR
10. CAMSHAFT GEAR
11. PUMP DRIVE GEAR
12. KNEE SPROCKET
13. WATER PUMP GEAR
14. DRIVING CHAIN
15. CRANKSHAFT SPROCKET
16. CLAMP RING
17. CHAIN TIGHTENER
18. IDLER SPROCKET
19. TACHOMETER GEAR

Figure 2-26.—Camshaft and accessory drive mechanism.

75.111X

573

fuel injection system is to deliver fuel to the cylinders, under specified conditions. The conditions must be in accordance with the power requirements of the engine.

The first condition to be met is that of the injection equipment. The quantity of fuel injected determines the amount of energy available, through combustion, to the engine. Smooth engine operation and even distribution of the load between the cylinders depend upon the same volume of fuel being admitted to a particular cylinder each time it fires, and upon equal volumes of fuel being delivered to all cylinders of the engine. The measuring device of a fuel injection system must also be designed to vary the amount of fuel being delivered as changes in load and speed vary.

In addition to measuring the amount of fuel injected, the system must properly time injection to ensure efficient combustion, so maximum energy can be obtained from the fuel. Early injection tends to develop excessive cylinder pressures; and extremely early injection will cause knocking. Late injection tends to decrease power output; and, if extremely late, it will cause incomplete combustion. In many engines, fuel injection equipment is designed to vary the time of injection, as speed or load varies.

A fuel system must also control the rate of injection. The rate at which fuel is injected determines the rate of combustion. The rate of injection at the start should be low enough that excessive fuel does not accumulate in the cylinder during the initial ignition delay (before combustion begins). Injection should proceed at such a rate that the rise in combustion pressure is not excessive, yet the rate of injection must be such that fuel is introduced as rapidly as is permissible in order to obtain complete combustion. An incorrect rate of injection will affect engine operation in the same way as improper timing. If the rate of injection is too high, the results will be similar to those caused by an excessively early injection; if the rate is too low, the results will be similar to those caused by an excessively late injection.

A fuel injection system must increase the pressure of the fuel sufficiently to overcome compression pressures and to ensure proper distribution of the fuel injected into the combustion space. Proper distribution is essential if the fuel is to mix thoroughly with the air and burn efficiently. While pressure is a prime contributing factor, the distribution of the fuel

is influenced in part, by "atomization" and "penetration" of the fuel. As used in connection with fuel injection, atomization means the breaking up of the fuel, as it enters the cylinder, into small particles which form a mist-like spray. Penetration is the distance through which the fuel particles are carried by the kinetic energy imparted to them as they leave the injector or nozzle.

Atomization is obtained when the liquid fuel, under high pressure, passes through the small opening or openings in the injector or nozzle. As the fuel enters the combustion space, high velocity is developed because the pressure in the cylinder is lower than the fuel pressure. The friction created as the fuel passes through the air at high velocity causes the fuel to break up into small particles. Penetration of the fuel particles depends chiefly upon the viscosity of the fuel, the fuel-injection pressure, and the size of the opening through which the fuel enters the cylinder.

Fuel must be atomized into particles sufficiently small so as to produce a satisfactory ignition delay period. However, if the atomization process reduces the size of the fuel particles too much, they will lack penetration; the smaller the particles the less the penetration. Lack of sufficient penetration results in the small particles of fuel igniting before they have been properly distributed. Since penetration and atomization tend to oppose each other, a compromise in the degree of each is necessary in the design of fuel injection equipment if uniform fuel distribution is to be obtained. The pressure required for efficient injection, and, in turn, proper distribution, is dependent upon the compression pressure in the cylinder, the size of the opening through which the fuel enters the combustion space, the shape of the combustion space, and the amount of turbulence created in the combustion space.

The fuel system of a gasoline engine is basically similar to that of a diesel engine, except that a carburetor is used instead of injection equipment. While injection equipment handles fuel only, the carburetor handles both air and fuel. The carburetor must meet requirements similar to those of an injection system except that in the carburetor air is also involved. In brief, the carburetor must accurately meter fuel and air, and in varying percentages, according to engine requirements. The carburetor also functions to vaporize the fuel charge and then mix it with the air, in the proper ratio.

The amount of fuel mixed with the air must be carefully regulated, and must change with the engine's different speeds and loads. The amount of fuel required by an engine which is warming-up is different from the amount required by an engine which has reached operating temperature. Special fuel adjustment is needed for rapid acceleration. All of these varying requirements are met automatically by the modern carburetor.

Engine Ignition Systems.—The methods by which the fuel mixture is ignited in the cylinders of diesel and gasoline engines differ as much as the methods of obtaining a combustible mixture in the cylinders of the two engines. An ignition system, as such, is not commonly associated with diesel engines. There is no one group of parts in a diesel engine which functions only to cause ignition, as there is in a gasoline engine. However, a diesel engine does have an "ignition system"; otherwise, combustion would not take place in the cylinders.

In a diesel engine, the parts which may be considered as forming the ignition system are the piston, the cylinder liner, and the cylinder head. These parts are not commonly thought of as forming an ignition system since they are generally associated with other functions such as forming the combustion space and transmitting power. Nevertheless, ignition in a diesel engine depends upon the piston, the cylinder, and the head. These parts not only form the space where combustion takes place but also provide the means by which the air is compressed to generate the heat necessary for self-ignition of the combustible mixture. In other words, both the source (air) of ignition heat and its generation (compression) are wholly within a diesel engine.

This is not true of a gasoline engine because the combustion cycles of the two types of engines are different. In a gasoline engine, even though the piston, the cylinder, and the head form the combustion space, as in a diesel engine, the heat necessary for ignition is caused by energy from a source external to the combustion space. The completion of the ignition process, involving the transformation of mechanical energy into electrical energy and then into heat energy, requires several parts, each performing a specific function. The parts which make the transformation of energy and the system which they form are commonly thought of when reference is made to an ignition system.

The spark which causes the ignition of the explosive mixture in the cylinders of a gasoline engine is produced when electricity is forced across a gap formed by two electrodes in the combustion chamber. The electrical ignition system furnishes the spark periodically to each cylinder, at a predetermined position of piston travel. In order to accomplish this function, an electrical ignition system must have, first of all, either a source of electrical energy or a means of developing electrical energy. In some cases, a storage battery is used as the source of energy; in other cases, a magneto generates electricity for the ignition system. The voltage from either a battery or a magneto is not sufficiently high enough to overcome the resistance created by pressure in the combustion chamber and to cause the proper spark in the gap formed by the two electrodes in the combustive chamber. Therefore, it is essential that an ignition system include a device which increases the voltage of the electricity supplied to the system sufficiently to cause a "hot" spark in the gap of the spark plug. The device which performs this function is generally called an ignition coil or induction coil.

Since a spark must occur momentarily in each cylinder at a specific time, an ignition system must include a device which controls the timing of the flow of electricity to each cylinder. This control is accomplished by interrupting the flow of electricity from the source to the voltage-increasing device (ignition coil). The interruption of the flow of electricity also plays an important part in the process of increasing voltage. The interrupting device is generally called the breaker assembly. A device which will distribute electricity to the different cylinders in the proper firing order is also necessary. The part which performs this function is called the distributing mechanism. Spark plugs to provide the gaps and wiring and switches to connect the parts of the system are essential to complete an ignition system.

All ignition systems are basically the same, except for the source of electrical energy. The source of energy is frequently used as a basis for classifying ignition systems; thus the battery-ignition system and the magneto-ignition system.

Engine Cooling Systems.—A great amount of heat is generated within an engine during operation. Combustion produces the greater portion of this heat; however, compression of gases within the cylinders and friction between moving parts add to the total amount of heat developed within an engine. Since the temperature of combustion alone is about twice that at which iron melts, it is apparent that, without some means of dissipating heat, an engine would operate for only a very limited time. Without proper temperature control, the lubricating-oil film between moving parts would be destroyed, proper clearance between parts could not be maintained, and metals would tend to fail.

Of the total heat supplied to the cylinder of an engine by the burning fuel, only one-third, approximately, is transformed into useful work; an equal amount is lost to the exhaust gases. This leaves approximately 30 to 35 percent of the heat of combustion which must be removed in order to prevent damage to engine parts. The greater portion of the heat which may produce harmful results is transferred from the engine through the medium of water; lubricating oil, air, and fuel are also utilized to aid in the cooling of an engine. All methods of heat transfer are utilized in keeping engine parts and fluids (air, water, fuel, and lubricating oil) at safe operating temperatures.

In a marine engine, the cooling system may be of the open or closed type. In the open system, the engine is cooled directly by salt water.

In the closed system, fresh water (or an antifreeze solution) is circulated through the engine. The fresh water is then cooled by salt water. In marine installations, the closed system is the type commonly used; however, some older marine installations use a system of the open type. The cooling systems of diesel and gasoline engines are similar mechanically and in function performed.

The cooling system of an engine may include such parts as pumps, coolers, engine passages, water manifolds, valves, expansion tank, piping, strainers, connections, and instruments. The schematic diagrams in figure 22-27 and 22-28 show the parts and the path of water flow in the fresh- and sea-water circuits of one arrangement of a closed cooling system.

Even though there are many types and models of engines used by the Navy, the cooling systems of most of these engines include the same basic parts. Design and location of parts, however, may differ considerably from one engine to another.

ENGINE LUBRICATING SYSTEMS.—It is essential to the operation of an engine that the contacting surfaces of all moving parts of an

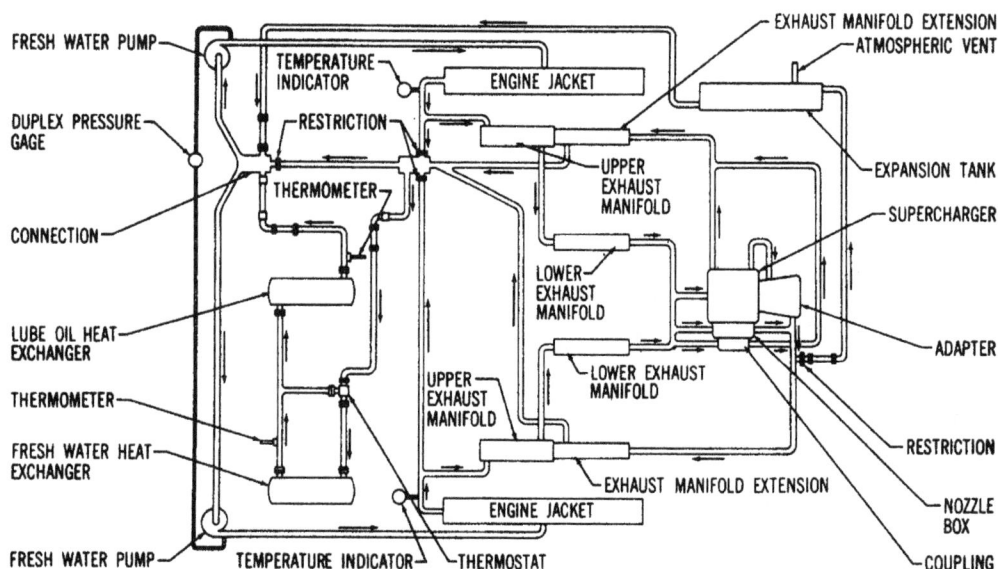

75.208X

Figure 22-27.—Fresh water circuit of a closed cooling system.

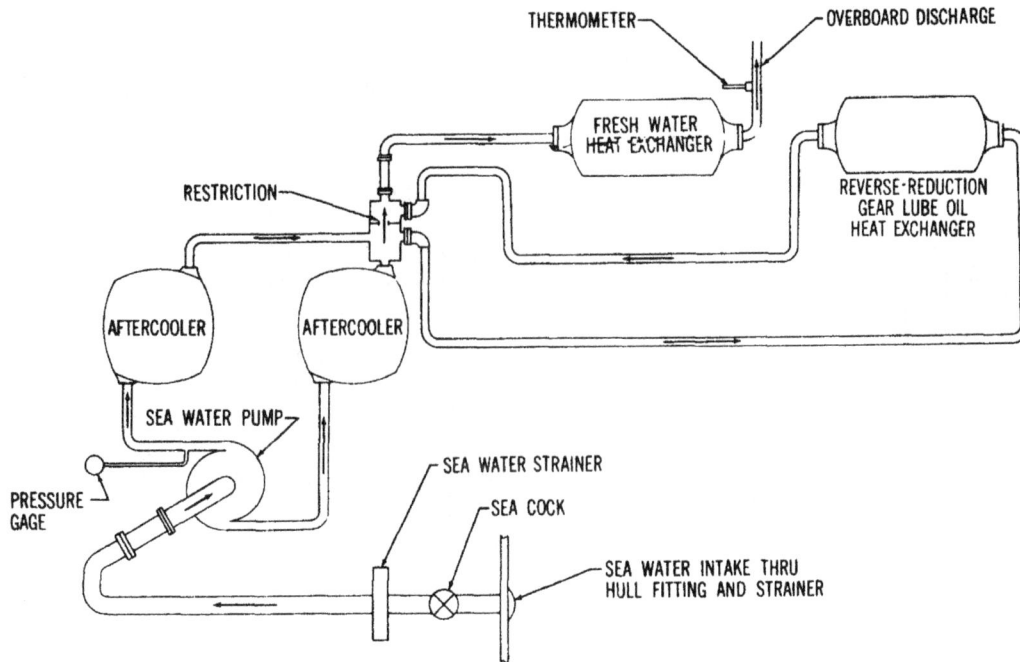

75.209X

Figure 22-28.—Salt water circuit of a closed cooling system.

engine be kept free from abrasion and that there be a minimum of friction and wear. If sliding contact is made by two dry metal surfaces under pressure, excessive friction, heat, and wear result. Friction, heat, and wear can be greatly reduced if metal-to-metal contact is prevented by keeping a clean film of lubricant between the metal surfaces.

Lubrication and the system which supplies lubricating oil to engine parts that involve sliding or rolling contact are as important to successful engine operation as air, fuel, and heat are to combustion. It is important not only that the proper type of lubricant be used, but also that the lubricant be supplied to the engine parts in the proper quantities, at the proper temperature, and that provisions be made to remove any impurities which enter the system. The engine lubricating oil system is designed to fulfill the above requirements.

The lubricating system of an engine may be thought of as consisting of two main divisions, that external to the engine and that within the engine. The internal division, or engine part, of the system consists principally of passages and piping; the external part of the system includes several components which aid in supplying the oil in the proper quantity, at the

proper temperature, and free of impurities. In order to meet these requirements, the lubricating systems of many engines include, external to the engine, such parts as tanks and sumps, pumps, coolers, strainers and filters, and purifiers. These parts and their relative location for one type of engine are shown in figure 22-29.

The engine system which supplies the oil required to perform the functions of lubrication is of the pressure type in practically all modern internal combustion engines. Even though many variations exist in the details of engine lubricating systems, the parts of such a system and its operation are basically the same, whether the system is in a diesel or a gasoline engine. Any variance between the systems of the two types of engines is generally due to differences in engine design and in opinions of manufacturers as to the best location of the component parts of the system. In many cases, similar types of components are used in the systems of diesel and gasoline engines.

TRANSMISSION OF ENGINE POWER

The fundamental characteristics of an internal combustion engine make it necessary, in many cases, for the drive mechanism to

577

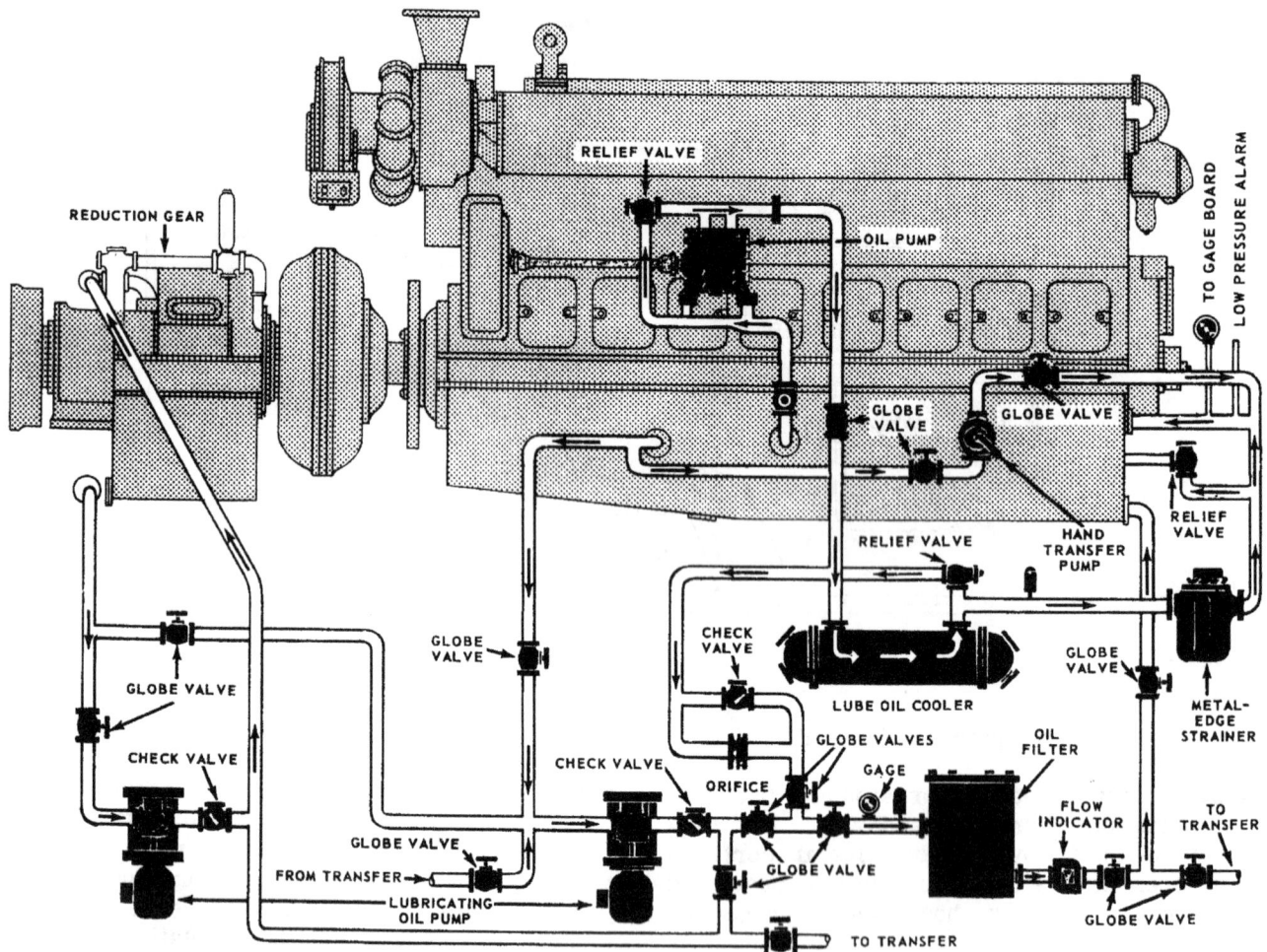

75.198X

Figure 22-29.—Sump-type lubricating oil filtering system.

change both the speed and the direction of shaft rotation in the driven mechanism. There are various methods by which required changes of speed and directions may be made during the transmission of power from the driving unit to the driven unit. In most installations the job is accomplished by a drive mechanism consisting principally of gears and shafts.

The process of transmitting engine power to a point where it can be used in performing useful work involves a number of factors. Two of these factors are torque and speed.

The force which tends to cause a rotational movement of an object is called torque or "twist". The crankshaft of an engine supplies a twisting force to the gears and shafts which transmit power to the driven unit. Gears are

used to increase or decrease torque. If the right combination of gears is installed between the engine and the driven unit, the torque or "twist" will be sufficient to operate the driven unit.

If maximum efficiency is to be obtained, an engine must operate at a certain speed. In order to obtain efficient engine operation, it might be necessary in some installations for the engine to operate at a higher speed than that required for efficient operation of the driven unit. In other cases the speed of the engine may have to be lower than the speed of the driven unit. Through a combination of gears, the speed of the driven unit can be increased or decreased so that the proper speed ratio exists between the units.

Types of Drive Mechanisms

The term <u>indirect drive</u> describes a drive mechanism which changes speed and torque. Drives of this type are common to many marine engine installations. Where the speed and the torque of an engine need not be changed in order to drive a machine satisfactorily, the mechanism used is a <u>direct drive</u>. Drives of this type are commonly used where the engine ırnishes power for the operation of auxiliaries such as generators and pumps.

INDIRECT DRIVES.—The drive mechanism of most engine-powered ships and many boats are of the indirect type. With indirect drive, the power developed by the engine(s) is transmitted to the propeller(s) indirectly, through an intermediate mechanism which reduces the shaft speed. Speed reduction may be accomplished mechanically (by a combination of gears) or by electrical means.

<u>Mechanical drives</u> include devices which reduce the shaft speed of the driven unit, provide a means for reversing the direction of shaft rotation in the driven unit, and permit quick-disconnect of the driving unit from the driven unit.

The combination of gears which effects the speed reduction is called a reduction gear. In most diesel engine installations, the reduction ratio does not exceed 3 to 1; there are some units, however, which have reductions as high as 6 to 1.

The propelling equipment of a boat or a ship must be capable of providing backing-down power as well as forward motive power. There are a few ships and boats in which backing down is accomplished by reversing the pitch of the propeller; in most ships, however, backing down is accomplished by reversing the direction of rotation of the propeller shaft. In mechanical drives, reversing the direction of rotation of the propeller shaft may be accomplished in one of two ways: by reversing the direction of engine rotation, or by the use of reverse gears. Of these two methods, the use of reverse gears is more commonly employed in modern installations.

More than reducing speed and reversing the direction of shaft rotation is required of the drive mechanism of a ship or a boat. It is frequently necessary to allow an engine to operate without power being transmitted to the propeller. For this reason, the drive mechanism of a ship or boat must include a means of disconnecting the engine from the propeller shaft. Devices used for this purpose are called <u>clutches</u> and <u>couplings</u>.

The arrangement of the components in an indirect drive varies, depending upon the type and size of the installation. In some small installations, the clutch or coupling, the reverse gear, and the reduction gear may be combined in a single unit; in other installations, the clutch or coupling and the reverse gear may be in one housing and the reduction gear in a separate housing attached to the reverse-gear housing. Drive mechanisms arranged in either manner are usually called <u>transmissions</u>. The arrangement of the components in two different types of transmissions are shown in figures 22-30 and 22-31.

In the transmission shown in figure 22-30 the housing is divided into two sections by the bearing carrier. The clutch or coupling assembly is in the forward section, and the gear assembly is in the after section of the housing. In the transmission shown in figure 22-31, note that the clutch assembly and the reverse gear assembly are in one housing, while the reduction gear unit is in a separate housing (attached to the clutch and the reverse gear housing).

In large engine installations, the clutch or coupling and the reverse gear may be combined; or they may be separate units, located between the engine and a separate reduction gear; or the clutch or coupling may be separate and the reverse gear and the reduction gear may be combined. An assembly of the last type is shown in figure 22-32.

In most geared-drive, multiple-propeller ships, the propulsion units are independent of each other. An example of this type of arrangement is illustrated in figure 22-33.

In some installations, the drive mechanism is arranged so that two or more engines drive a single propeller. This is accomplished by having the driving gear which is on, or connected to, the crankshaft of each engine transmit power to the driven gear on the propeller shaft. In one type of installation, each of two propellers is driven by four diesel engines. The arrangement of the engines, the location of the reduction gear, and the direction of rotation of the crankshaft and the propeller shaft in one type of "quad" power unit are illustrated in figure 22-34.

The drive mechanism illustrated includes four clutch assemblies (one mounted to each engine flywheel) and one gear box. The box

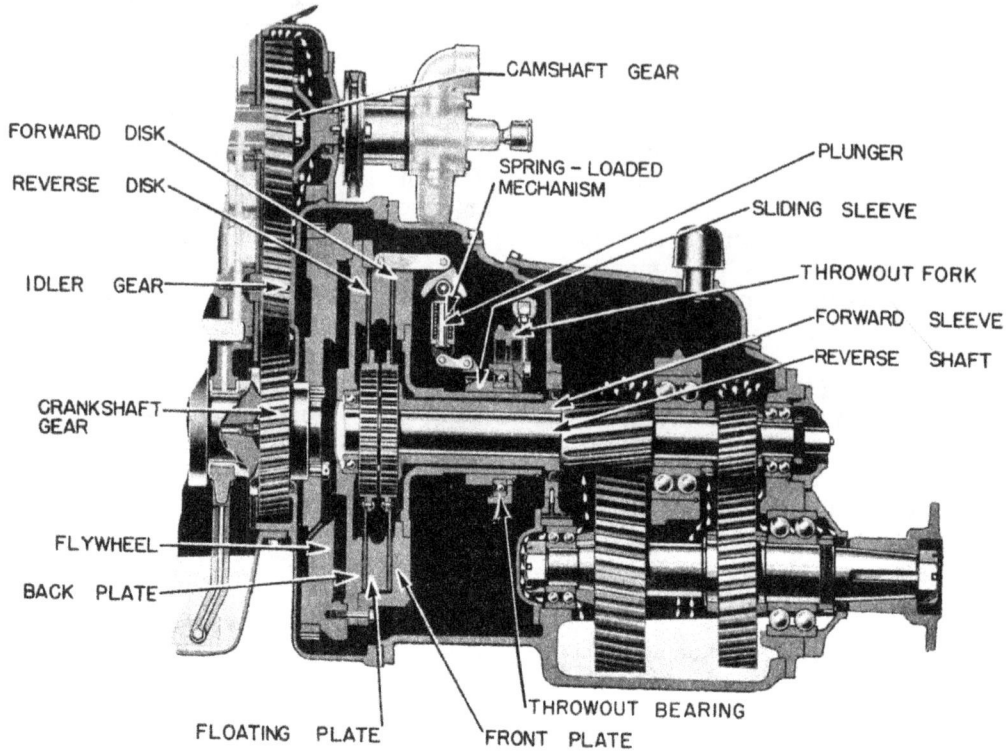

Figure 22-30.—Transmission with independent oil system.

75.245

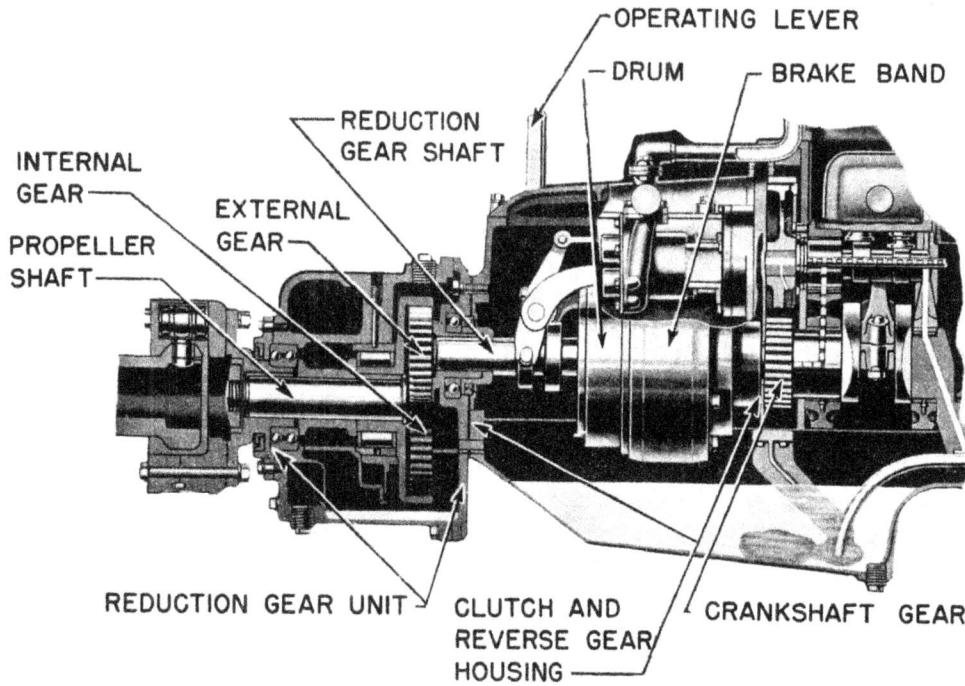

Figure 22-31.—Clutch and reverse gear assembly with attached reduction gear unit.

75.246

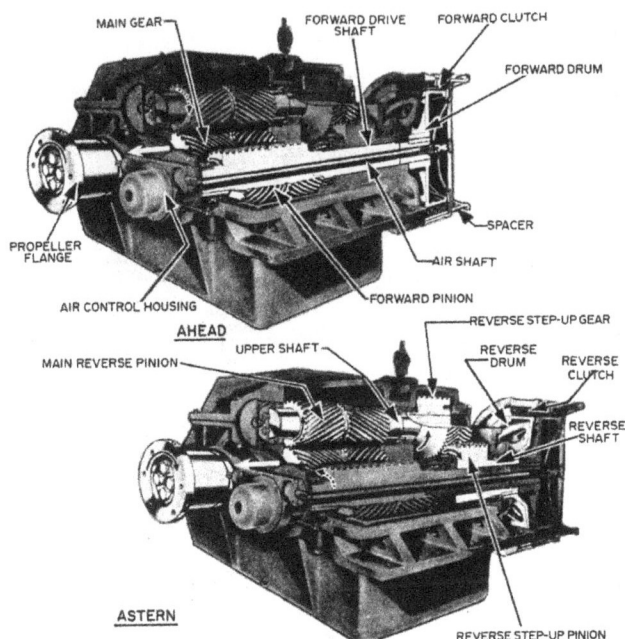

75.247

Figure 22-32.—Clutch and reverse reduction gear assembly.

contains two drive pinions and the main drive gear. Each pinion is driven by the clutch or coupling shafts of two engines, through splines in the pinion hubs. The pinions drive the single main gear, which is connected to the propeller shaft.

Electric drives are used in the propulsion plants of some diesel-driven ships. With electric drive, there is no mechanical connection between the engine(s) and the propeller(s). In such plants, the diesel engines are connected directly to generators. The electricity produced by such an engine-driven generator is transmitted, through cables, to a motor. The motor is connected to the propeller shaft directly, or indirectly through a reduction gear. When a reduction gear is included in a diesel-electric drive, the gear is located between the motor and the propeller.

The generator and the motor of a diesel-electric drive may be of the alternating current (a-c) type or of the direct current (d-c) type; almost all diesel-electric drives in the Navy, however, are of the direct current type. Since the speed of a d-c motor varies directly with the voltage furnished by the generator, the control system of an electric drive is so arranged that the generator voltage can be changed at

any time. An increase or decrease in generator voltage is used as a means of controlling the speed of the propeller. Changes in generator voltage may be brought about by electrical means, by changes in engine speed, and by a combination of these methods. The controls of an electric drive may be in a location remote from the engine, such as the pilot house.

In an electric drive, reversing the direction of rotation of the propeller is not accomplished by the use of a reverse gear. The electrical system is arranged so that the flow of current through the motor can be reversed. This reversal of current flow causes the motor to revolve in the opposite direction. Thus, the direction of rotation of the motor and of the propeller can be controlled by manipulating the electrical controls.

DIRECT DRIVES.—In some marine engine installations, power from the engine is transmitted to the driven unit without a change in shaft speed; that is, by a direct drive. In a direct drive, the connection between the engine and the driven unit may consist of a "solid" coupling, a flexible coupling, or a combination of both. A clutch may or may not be included in a direct drive, depending upon the type of installation. In some installations, a reverse gear is included.

Solid couplings vary considerably in design. Some solid couplings consist of two flanges bolted solidly together. In other direct drives, the driven unit is attached directly to the engine crankshaft by a nut.

Solid couplings offer a positive means of transmitting torque from the crankshaft of an engine; however, a solid connection does not allow for any misalignment nor does it absorb any of the torsional vibrations transmitted from the engine crankshaft or shaft vibrations.

Since solid coupling will not absorb vibration and will not permit any misalignment, most direct drives consist of a flange-type coupling which is used in connection with a flexible coupling. Connections of the flexible type are common to the drives of many auxiliaries, such as engine-generator sets. Flexible couplings are also used in indirect drives to connect the engine to the drive mechanism.

The two solid halves of a flexible coupling are joined by a flexible element. The flexible element is made of rubber, neoprene, or steel springs. Two views of one type of flexible coupling are shown in figure 22-35.

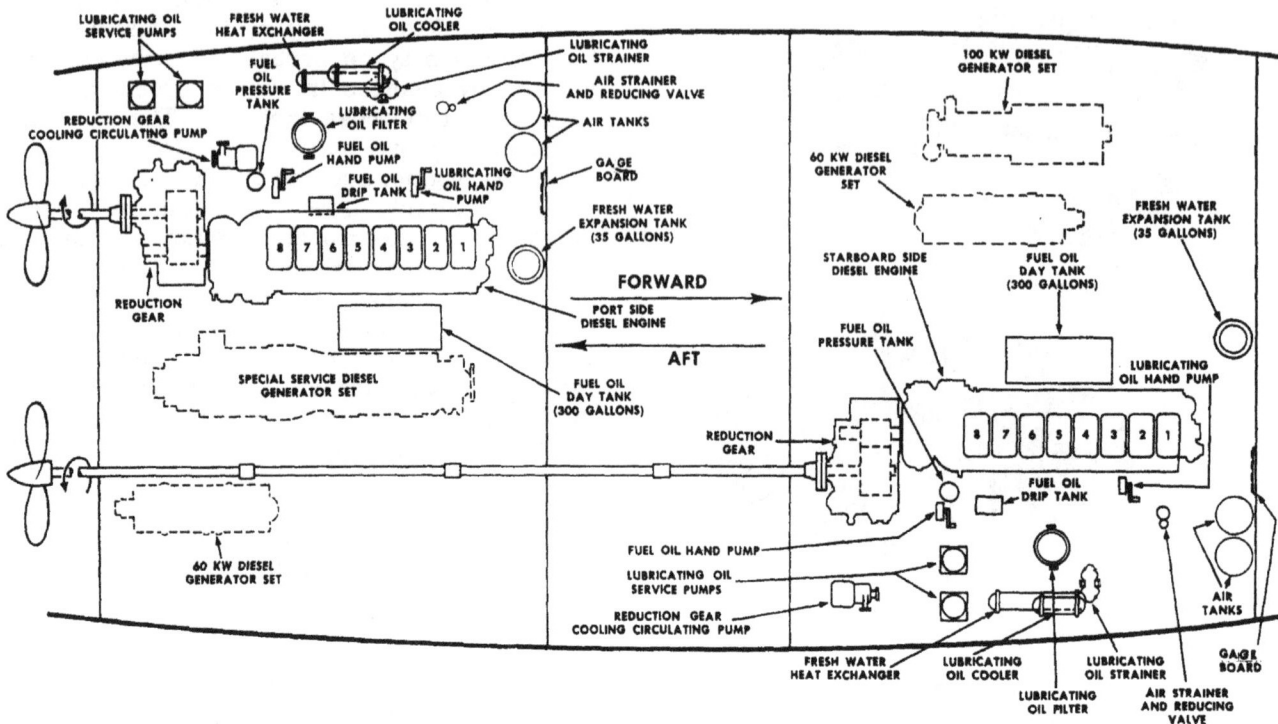

Figure 22-33.—Example of independent propulsion units.

The coupling illustrated has radial spring packs as the flexible element. The power from the engine is transmitted from the inner ring, or spring holder, of the coupling, through a number of spring packs to the outer spring holder, or driven member. A large driving disk connects the outer spring holder to the flange on the driven shaft. The pilot on the end of the crankshaft fits into a bronze, bushed bearing on the outer driving disk to center the driven shaft. The ring gear of the jacking mechanism is pressed onto the rim of the outer spring holder.

The inner driving disk, through which the camshaft gear train is driven, is fastened to the outer spring holder. A splined ring gear is bolted to the inner driving disk. This helical, internal gear fits on the outer part of the crankshaft gear and forms an elastic drive, through the crankshaft gear which rides on the crankshaft. The splined ring gear is split and the two parts are bolted together with a spacer block at each split-joint.

The parts of the coupling shown in figure 22-35 are lubricated by oil flowing from the

bearing bore of the crankshaft gear through the pilot bearing.

CLUTCHES, REVERSE GEARS, AND REDUCTION GEARS

Clutches may be used on direct-driven propulsion Navy engines to provide a means of disconnecting the engine from the propeller shaft. In small engines, clutches are usually combined with reverse gears and used for maneuvering the ship. In large engines, special types of clutches are used to obtain special coupling or control characteristics, and to prevent torsional vibration.

Reverse gears are used on marine engines to reverse the direction of rotation of the propeller shaft, when maneuvering the ship, without changing the direction of rotation of the engine. They are used principally on relatively small engines. If a high-output engine has a reverse gear, the gear is used for low-speed operation only, and does not have full-load and full-speed capacity. For maneuvering ships with large

582

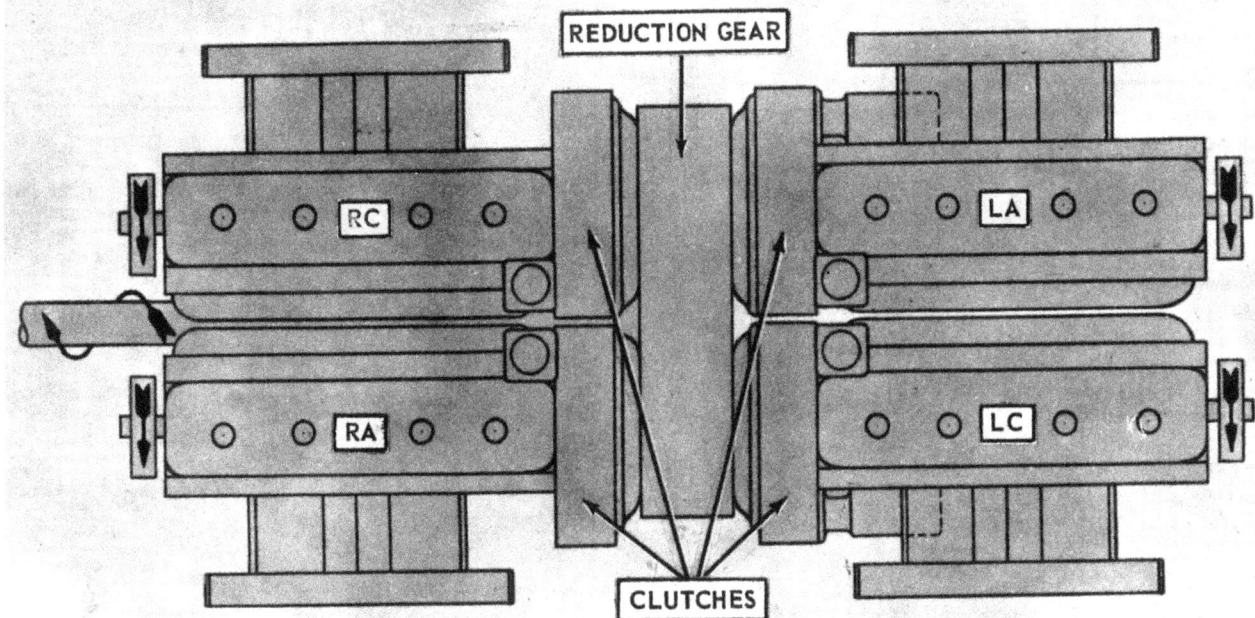

75.249X

Figure 22-34.—Four engines (quad unit) arranged to drive one propeller.

direct-propulsion engines, the engines are reversed.

Reduction gears are used to obtain low propeller-shaft speed with a high engine speed. When accomplishing this, the gears correlate two conflicting requirements of a marine engine installation. These opposing requirements are: (1) for minimum weight and size for a given power output, engines must have a relatively high rotative speed; and (2) for maximum efficiency, propellers must rotate at a relatively low speed, particularly where high thrust capacity is desired.

Friction Clutches and Gear Assemblies

Friction clutches are commonly used with smaller, high-speed engines, up to 500 hp. However, certain friction clutches, in combination with a jaw-type clutch, are used with engines up to 1400 hp; and pneumatic clutches, with a cylindrical friction surface, with engines up to 2000 hp.

Friction clutches are of two general styles; the disk and the band styles. In addition, friction clutches can be classified into dry and wet types, depending upon whether the friction surfaces operate with or without a lubricant. The designs of both types are similar, except that the wet clutches require a large friction area because of the reduced friction coefficient between the lubricated surfaces. The advantages of wet clutches are smoother operation and less wear of the friction surfaces. Wear results from slippage between the surfaces not only during engagement and disengagement, but also, to a certain extent, during the operation of the mechanism. Some wet-type clutches are filled with oil periodically; in other clutches the oil, being a part of the engine-lubricating system, is circulated continuously. Such a friction clutch incorporates provisions which will prevent worn-off particles from being carried by the circulating lubricating oil to the bearings, gears, etc.

The friction surfaces are generally constructed of different materials, one being of cast iron or steel; the other is lined with some asbestos-base composition, or sintered iron or bronze for dry clutches, and bronze, cast iron, or steel for wet clutches. Cast-iron surfaces are preferred because of their better bearing qualities and greater resistance to scoring or scuffing.

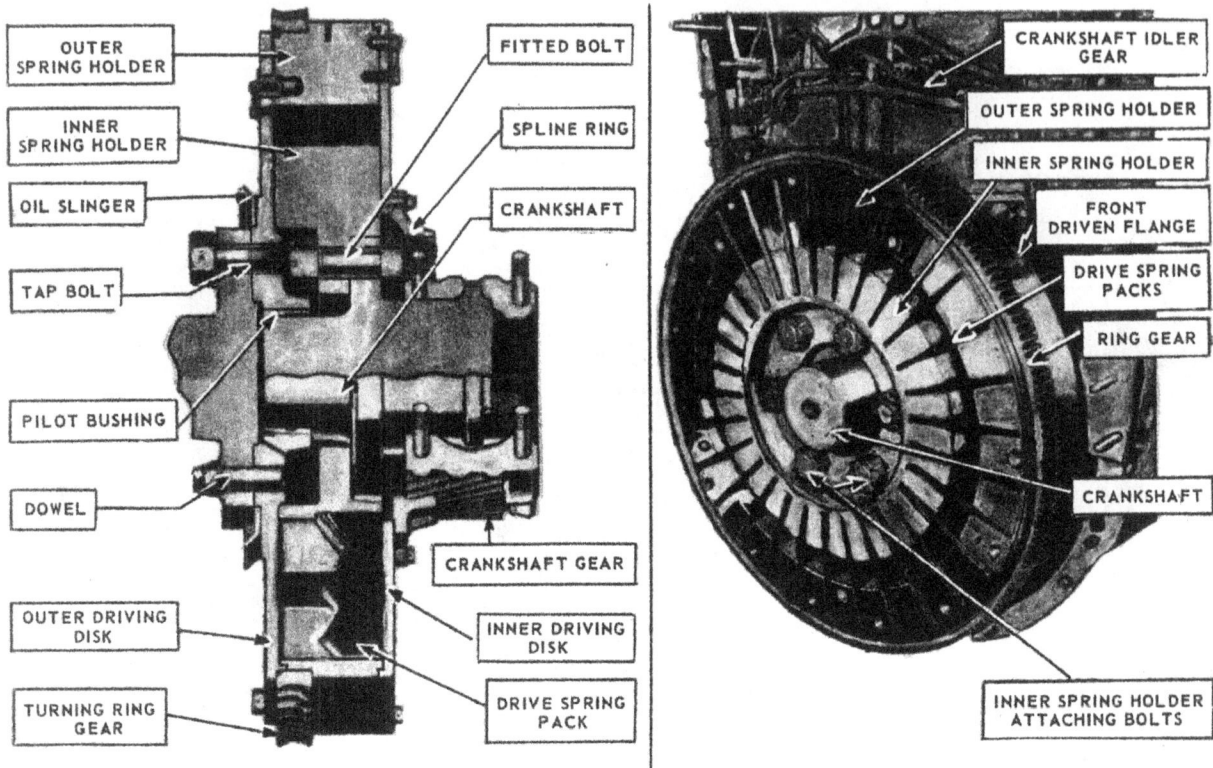

75.251X

Figure 22-35.—Flexible coupling.

Sintered blocks are made of finely powdered iron or bronze particles, molded in forms to the desired shape, under high temperature and pressure.

As far as engagement of the friction clutches is concerned, the application of force-producing friction can be obtained either by mechanically jamming the friction surfaces together by some toggle-action linkage, or through stiff springs (coil, leaf, or flat-disk type). Air pressure is also used to engage friction clutches.

TWIN-DISK CLUTCH AND GEAR MECHANISM.—One of the several types of transmissions used by the Navy is the twin disk transmission mechanism, shown in figure 22-30. Gray Marine high-speed diesel engines are generally equipped with a combination clutch, and reverse and reduction gear unit—all contained in a single housing, at the after end of the engine. A sectional view of this mechanism is shown in figure 22-30.

The clutch assembly of the twin disk transmission mechanism is contained in the part of the housing nearest the engine. It is a dry-type, twin-disk clutch with two driving disks. Each disk is connected, through shafting, to a separate reduction gear train in the after part of the housing. One disk and reduction train is for reverse rotation of the shaft and propeller, the other disk and reduction train for forward rotation. The forward and reverse gear trains for Gray Marine engines are illustrated in figure 22-36. In figures 22-30 and 22-36, it will be observed that the gear trains are different in the two illustrations; however, the operation of the mechanisms shown is basically the same.

Since the gears for forward and reverse rotation of the twin-disk clutch and gear mechanism remain in mesh at all times, there is no shifting of gears. In shifting the mechanism, only the floating plate, located between the forward and reverse disks is shifted. The shifting mechanism is a sliding sleeve, which does not rotate, but has a loose sliding fit around the hollow forward shaft. A throwout fork (yoke) engages a pair of shifter blocks pinned on either side of the sliding sleeve.

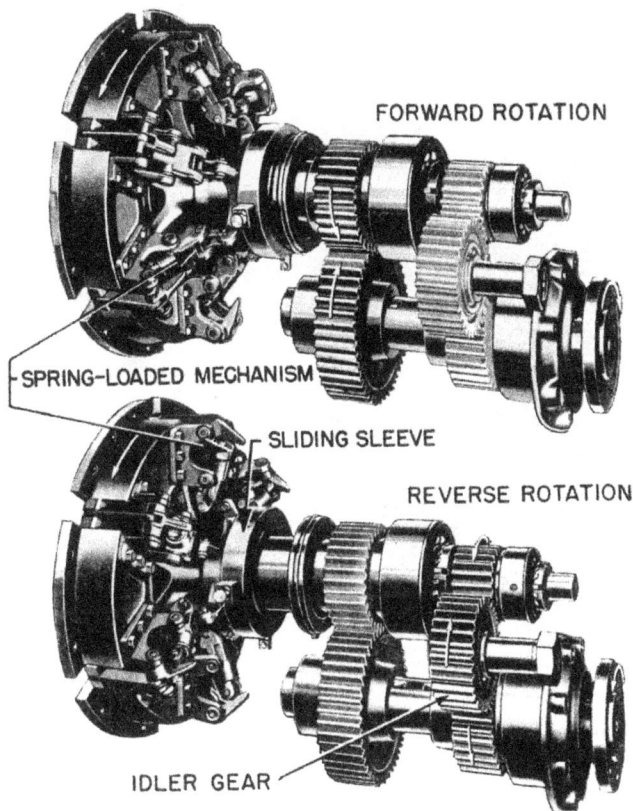

FORWARD ROTATION

SPRING-LOADED MECHANISM

SLIDING SLEEVE

REVERSE ROTATION

IDLER GEAR

75.252

Figure 22-36.—Forward and reverse gear trains for Gray Marine engines.

The clutch operating lever moves the throw-out fork, which in turn shifts the sliding sleeve lengthwise along the forward shaft. When the operating lever is placed forward, the sliding sleeve is forced backward. In this position the linkages of the spring-loaded mechanism pull the floating pressure plate against the forward disk, and cause forward rotation. When the operating lever is pulled back as far as it can go, the sliding sleeve is pushed forward. In this position, the floating pressure plate engages the reverse disk and back plate for reverse rotation.

The clutch has a positive neutral which is set by placing the operating lever in a middle position. Then the sliding sleeve is also in a middle position, and the floating plate rotates freely between the two clutch disks. (The only control that the operator has is to cause the floating plate to bear heavily against either the forward disk or the reverse disk, or to put the floating plate in the positive neutral position

so that it rotates freely between the two disks.)

The reversing gear unit is lubricated separately from the engine by its own splash system. The oil level of the gear housing should never be kept over the high mark because too much oil will cause overheating of the gear unit. The oil is cooled by air which is blown through the baffled top cover by the rotating clutch. Grease fittings are installed for bearings not lubricated by the oil.

JOE'S DOUBLE CLUTCH REVERSE GEAR.— A gear mechanism found on many power boats is Joe's double clutch reverse gear. The installation of a typical Joe's reverse gear and clutch assembly for the Navy type DC engine is shown in figure 22-37. The drive from the engine crankshaft is taken into the clutch and reverse gear housing by an extension of the crankshaft drive gear. The crankshaft rotation is transmitted to the reduction gear shaft through the clutch and the reverse gear unit.

If one could open the clutch and reverse gear housing and watch the reverse gear drum and the reduction gear shaft while the engine is running, the following operation would be observed:

When the operating lever is thrown forward, the drum and reduction gear shaft rotate in the same direction as the engine crankshaft. This causes forward rotation of the propeller.

In the intermediate position of the operating lever, the drum rotates but the reduction gear shaft remains stationary. This is the neutral setting.

Forward rotation is obtained by dual clutch action while reverse rotation is obtained through the operation of the planetary gears. The unit consists of a housing enclosing a split conical clutch and a multi-plate friction clutch and gearing. Additional components include the collar and yoke and an outer brake band with an operating toggle mechanism. Movement of the sliding collar selects the direction of rotation.

When the operating lever is placed in the forward position, the linkage between the lever and the collar and yoke assembly slides the collar lengthwise to the left along the reduction gear shaft. This motion operates the toggle assembly which, in turn, drives the three plungers to the right, pressing them hard against the disk clutch.

When the plungers are driven hard against the disk clutch, the disks are locked together by friction. This locks the drum housing to the

585

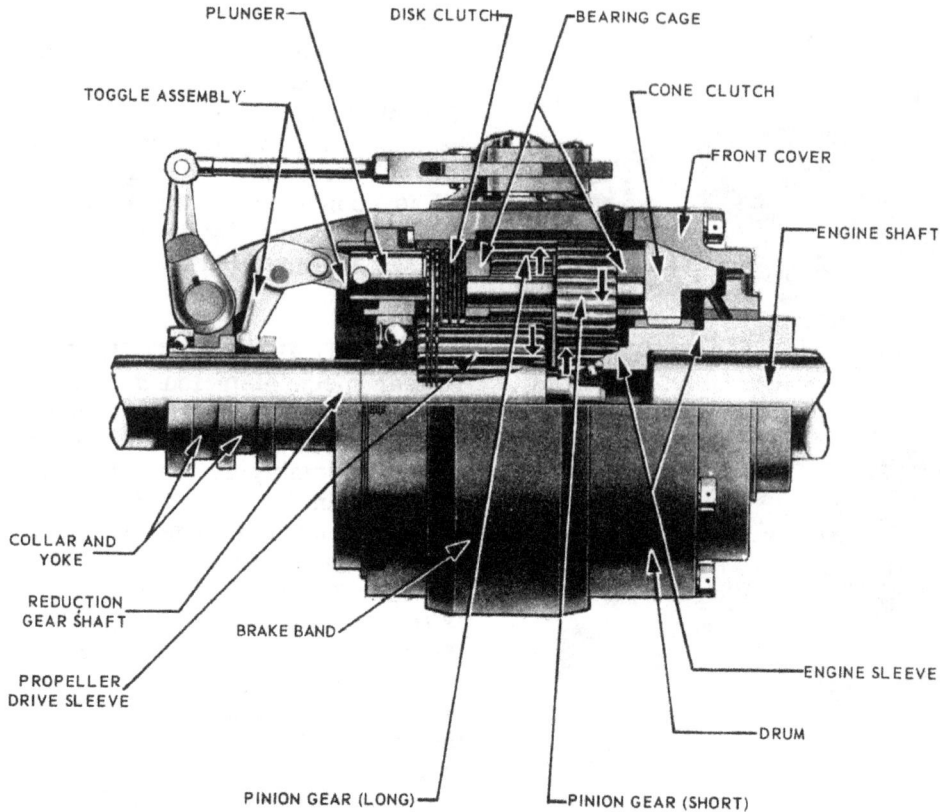

75.254

Figure 22-37.—Cutaway view of Joe's clutch and reverse gear.

propeller drive sleeve. In addition, the force of the plungers on the disk clutch is transmitted to the bearing cage, which is a cylinder containing the reverse gear pinions. The bearing cage, in turn, is pressed against the cone clutch. Thus, the cone clutch is forced against its seat in the front cover of the gear box, clamping the clutch to the front cover by friction. Since the cone clutch is in mesh on its inner surface with the engine sleeve, which is in turn keyed to the engine shaft, the front cover is now locked to the engine shaft. The front cover must rotate with the engine shaft, in the same direction.

Now, since the front cover is bolted to the drum housing, which is locked to the propeller drive sleeve by the disk clutch, there is a complete lock from the engine shaft to the reduction gear shaft. The entire assembly rotates as a unit in the same direction as the engine shaft; this motion gives the propeller a forward rotation.

When the operating lever is thrown into the reverse position, the plungers are withdrawn, and both clutches are disengaged. At the same time the brake band is tightened around the drum, holding the drum stationary. The bearing cage is locked to the drum. The cone clutch rotates freely out of contact with the front cover. Then the motion from the engine shaft to the reduction gear shaft is transmitted through the inner gear assembly.

The reverse gear pinions are held in the bearing cage, which is stationary for reverse rotation. There are three short pinions, each in mesh with the small inner gear of the engine sleeve. The three short pinions mesh with the three long pinions, each of which also meshes with the propeller drive sleeve gear. Engine rotation is transmitted from the engine sleeve to the short pinions, to the long pinions, and to the propeller drive sleeve. These pinions (gear train) cause the reduction gear shaft to rotate opposite to the engine rotation (see arrows in fig. 22-37), and give the propeller a reverse rotation.

Note that in figure 22-38, which shows the mechanism more clearly, the gears are set for reverse rotation, and the brake band is clamped to the drum. The parts which are shaded are held stationary by the brake band, and the remaining internal parts, which are not shaded, rotate. (The rotation of the engine shaft and engine sleeve is transmitted directly to the cone clutch and the short pinions. The cone clutch rotates freely out of contact with the stationary front cover. The short pinions drive the long pinions, which drive the propeller drive sleeve. The latter unit is keyed to and drives the reduction gear shaft, which rotates opposite to the engine shaft.)

The reduction gear unit is bolted to the reverse gear housing, as shown in figure 22-31. It consists merely of an external gear, mounted on the reduction gear shaft, and in mesh with a larger internal gear, mounted on the propeller shaft. Power is transferred, at a reduced speed, from the smaller drive gear to the larger internal gear.

Lubrication of the clutch and reverse gear mechanism is accomplished by means of a drilled passage in the crankshaft which supplies oil, as a spray, to the gears and other moving parts. This oil returns to the engine sump by gravity.

Lubrication of the reduction gear unit is accomplished by an external line from the engine's main oil gallery. Oil is sprayed over the gears and moving parts to lubricate and cool them. Excess oil either drains back to the engine sump by gravity, or, where the unit is

below the engine, returns to the sump by means of a scavenging pump.

Airflex Clutch and Gear Assembly.—On the larger diesel-propelled ships, the clutch, reverse and reduction gear unit has to transmit an enormous amount of power. To maintain the weight and size of the mechanism as low as possible, special clutches have been designed for large diesel installations. One of these is the airflex clutch and gear assembly used with some General Motors engines on LST's.

A typical airflex clutch and gear assembly, for ahead and astern rotation, is shown in figure 22-32. There are two clutches, one for forward rotation and one for reverse rotation. The clutches are bolted to the engine flywheel by means of a steel spacer, so that they both rotate with the engine at all times, and at engine speed. Each clutch has a flexible tire (or gland) on the inner side of a steel shell. Before the tires are inflated, they will rotate out of contact with the drums, which are keyed to the forward and reverse drive shafts. When air under pressure (100 psi) is sent into one of the tires, the inside diameter of the clutch decreases. This causes the friction blocks on the inner tire surface to come in contact with the clutch drum, locking the drive shaft with the engine.

The parts of the airflex clutch which give the propeller ahead rotation are illustrated in the upper view of figure 22-32. The clutch tire nearest the engine (forward clutch) is inflated to contact and drive the forward drum with the engine. The forward drum is keyed to the forward

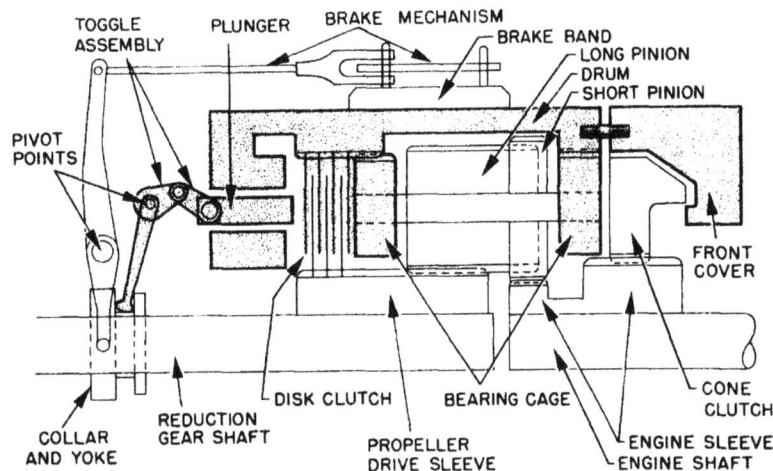

Figure 22-38.—Schematic diagram of Joe's reverse gear assembly.

75.255

drive shaft, which carries the double helical forward pinion at the after end of the gear box. The forward pinion is in constant mesh with the double helical main gear, which is keyed on the propeller shaft. By following through the gear train, you can see that, for ahead motion, the propeller rotates in a direction opposite to the engine's rotation.

The parts of the airflex clutch which give the propeller astern rotation are illustrated in the lower view of figure 22-32. The reverse clutch is inflated to engage the reverse drum, which is then driven by the engine. The reverse drum is keyed to the short reverse shaft, which surrounds the forward drive shaft. A large reverse step-up pinion transmits the motion to the large reverse step-up gear on the upper shaft. The upper shaft rotation is opposite to the engine's rotation. The main reverse pinion on the upper shaft is in constant mesh with the main gear. By tracing through the gear train, it may be seen that, for reverse rotation, the propeller rotates in the same direction as the engine.

The diameter of the main gear of the airflex clutch is approximately 2 1/2 times as great as that of the forward and reverse pinions. Thus, there is a speed reduction of 2 1/2 to 1 from either pinion to the propeller shaft.

Since the forward and main reverse pinions are in constant mesh with the main gear, the set that is not clutched in will rotate as idlers driven from the main gear. The idling gears rotate in a direction opposite to their rotation when carrying the load. For example, with the forward clutch engaged, the main reverse pinion rotates in a direction opposite to its rotation for astern motion (note the dotted arrow in the upper view of figure 22-32. Since the drums rotate in opposite directions, a control mechanism is installed to prevent the engagement of both clutches simultaneously.

The airflex clutch is controlled by an operating lever which works the air control housing, located at the after end of the forward pinion shaft. The control mechanism, shown with the airflex clutches in figure 22-39, directs the high pressure air into the proper paths to inflate the clutch glands (tires). The air shaft, which connects the control mechanism to the clutches, passes through the forward drive shaft.

The supply air enters the control housing through the air check valve and must pass through the small air orifice. The purpose of the restricted orifice is to delay the inflation of the clutch to be engaged, when shifting from one direction of rotation to the other. The delay is necessary to allow the other clutch to be fully deflated and out of contact with its drum before the inflating clutch can make contact with its drum.

The supply air goes to the rotary air joint in which a hollow carbon cylinder is held to the valve shaft by spring tension. This prevents leakage between the stationary carbon seal and the rotating air valve shaft. The air goes from the rotary joint to the four-way air valve. The sliding-sleeve assembly of the four-way valve can be shifted endwise along the valve shaft by operating the control lever.

When the shifter arm on the control lever slides the valve assembly away from the engine, air is directed to the forward clutch. The four-way valve makes the connection between the air supply and the forward clutch, as follows: there are eight neutral ports which connect the central air supply passage in the valve shaft with the sealed air chamber in the sliding member. In the neutral position of the four-way valve, as shown in figure 22-39, the air chamber is a dead end for the supply air. In the forward position of the valve, the sliding member uncovers eight forward ports, which connect with the forward passages conducting the air to the forward clutch. The air now flows through the neutral ports, air chamber, forward ports, and forward passages to inflate the forward clutch gland. As long as the valve is in the forward position, the forward clutch will remain inflated and the entire forward air system will remain at a pressure of 100 psi.

LUBRICATION.—On most large gear units, a separate lubrication system is used. One lubrication system is shown in figure 22-40. Oil is picked up from the gear box by an electric-driven gear-type lubricating oil pump and is sent through a strainer and cooler. After being cleaned and cooled, the oil is returned to the gear box to cool and lubricate the gears. In twin installations, such as shown in figure 22-40 a separate pump is used for each unit and a standby pump is interconnected for emergency use.

Hydraulic Clutches or Couplings

The fluid clutch (coupling) is widely used on Navy ships. The use of hydraulic coupling

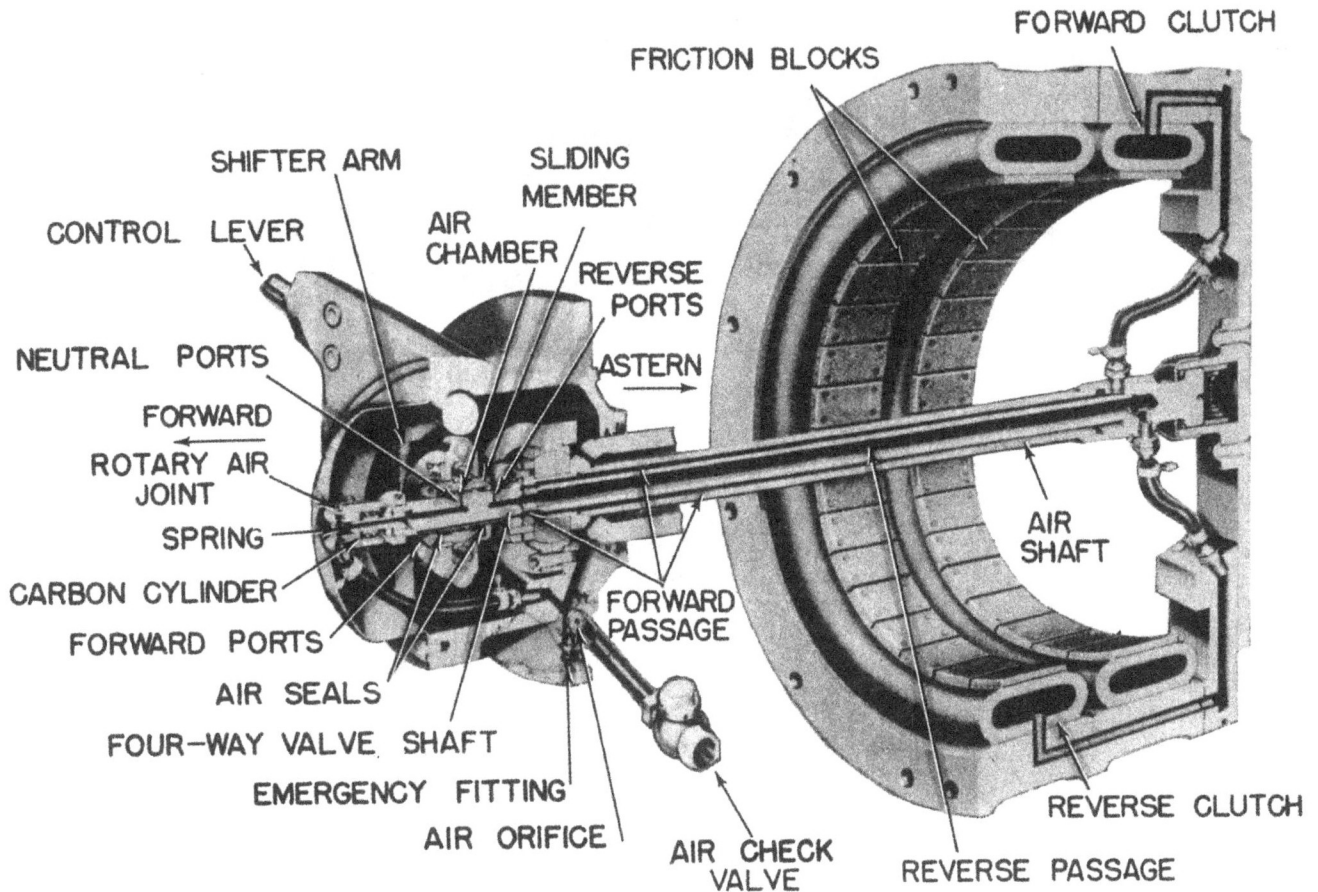

75.257

Figure 22-39.—Airflex clutches and control valves.

eliminates the need for a mechanical connection between the engine and the reduction gears. Couplings of this type operate with a minimum of slippage.

Some slippage is necessary for operation of the hydraulic coupling, since torque is transmitted because of the principle of relative motion between the two rotors. The power loss resulting from the small amount of slippage is transformed into heat which is absorbed by the oil in the system.

Compared with mechanical clutches, hydraulic clutches have a number of advantages. There is no mechanical connection between the driving and driven elements of the hydraulic coupling. Power is transmitted through the coupling very efficiently (97 percent) without transmitting torsional vibrations, or load shocks, from the engine to the reduction gears. This protects the engine, the gears, and the shafting from sudden

loads which may occur as a result of piston seizure or fouling of the propeller. The power is transmitted entirely by the circulation of a driving fluid (oil) between radial passages in a pair of rotors. In addition, the assembly of the hydraulic coupling will absorb or allow for slight misalignment.

The two rotors and the oil-sealing cover of a typical hydraulic coupling are shown in figure 22-41. The primary rotor (impeller) is attached to the engine crankshaft. The secondary rotor (runner) is attached to the reduction gear pinion shaft. The cover is bolted to the secondary rotor and surrounds the primary rotor. Each rotor is shaped like a half-doughnut with radial partitions. A shallow trough is welded into the partitions around the inner surface of the rotor. The radial passages tunnel under this trough (as indicated by the white arrows in fig. 22-41).

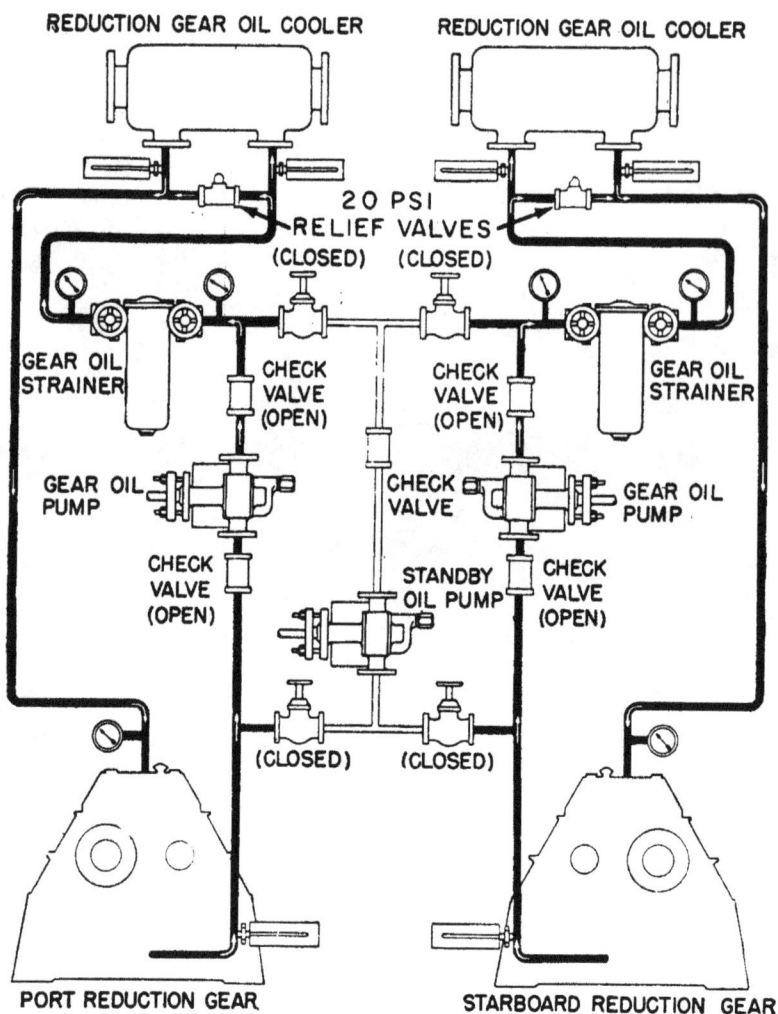

75.258

Figure 22-40.—Schematic diagram of reverse gear lubrication system.

When the coupling is assembled, the two rotors are placed facing each other to complete the doughnut (fig. 22-42). The rotors do not quite touch each other, the clearance between them being 1/4 to 5/8 inch, depending on the size of the coupling. The curved radial passages of the two rotors are opposite each other, so that the outer passages combine to make a circular passage except for the small gaps between the rotors.

In the hydraulic coupling assembly, shown in figure 22-42, the driving shaft is secured to the engine crankshaft and the driven shaft goes to the reduction gear box. The oil inlet admits oil directly to the rotor cavities, which become completely filled. The rotor housing is bolted to the secondary rotor and has an oil-sealed joint with the driving shaft. A ring valve, going entirely around the rotor housing, can be operated by the ring valve mechanism to open or close a series of emptying holes (fig. 22-42) housing. When the ring valve is opened, the oil will fly out from the rotor housing into the coupling housing, draining the coupling completely in two or three seconds. Even when the ring valve is closed, some oil leaks out into the coupling housing, and additional oil enters through the inlet. From the coupling housing, the oil is drawn by a pump to a cooler, then sent back to the coupling.

Another coupling assembly used on several Navy ships is the hydraulic coupling with piston-type quick-dumping valves, shown in figure 22-43. In this coupling, in which the operation is

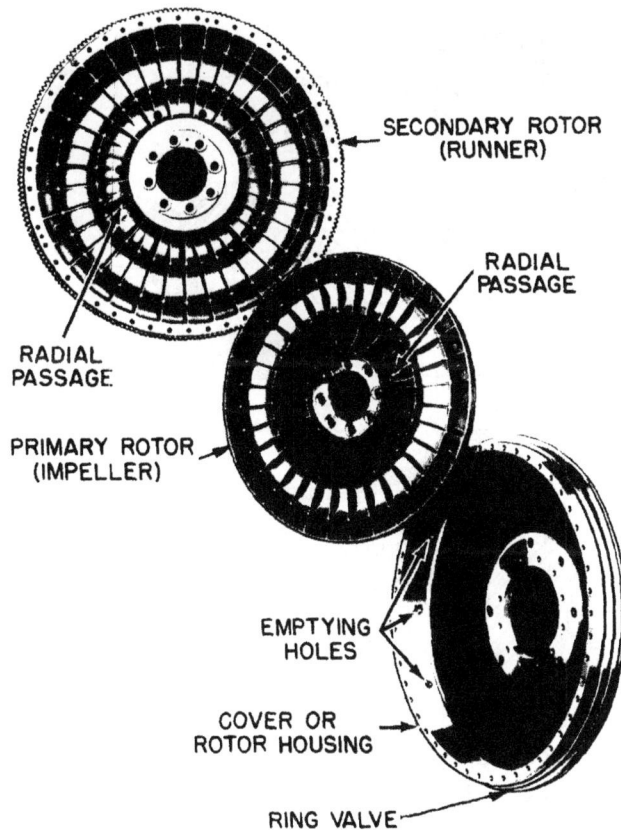

5.33

Figure 22-41.—Runner, impeller, and cover of
hydraulic coupling.

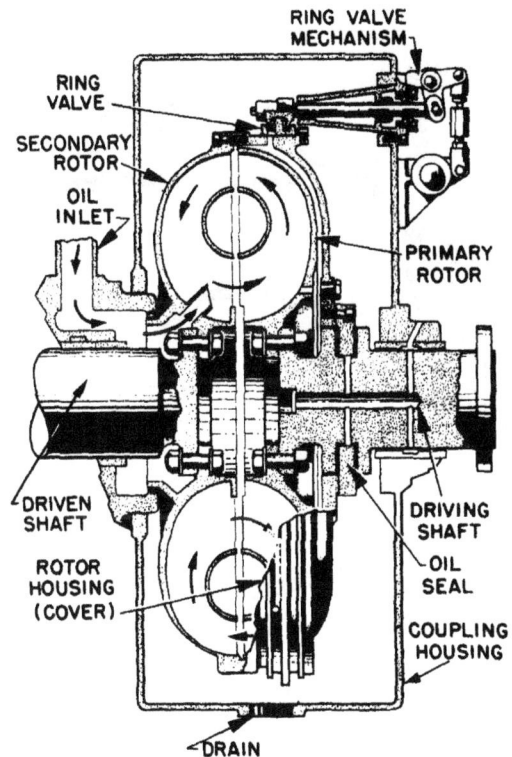

75.260

Figure 22-42.—Hydraulic coupling assembly.

similar to the one described above, a series
of piston valves, around the periphery of the
rotor housing, are normally held in the closed
position by springs. By means of air oil pressure admitted to the valves, as shown in figure
22-43, the pistons are moved axially so as to
uncover drain ports, allowing the coupling to
empty. Where extremely rapid declutching is
not required, the piston-valve coupling offers
the advantages of greater simplicity and lower
cost than the ring-valve coupling.

Another type of self-contained unit for certain diesel engine drives is the scoop control
coupling, shown in figure 22-44. In couplings
of this type, the oil is picked up by one of two
scoop tubes (one tube for each direction of
rotation), mounted on the external manifold.
Each scoop tube contains two passages: a
smaller one (outermost) handles the normal
flow of oil for cooling and lubrication, and a
larger one which rapidly transfers oil from the
reservoir directly to the working circuit.

The scoop tubes are operated from the control stand through a system of linkages. As
one tube moves outward from the shaft centerline and into the oil annulus, the other is being
retracted.

Four spring-loaded centrifugal valves are
mounted on the primary rotor. These valves
are arranged to open progressively as the speed
of the primary rotor decreases. The arrangement provides the necessary oil flow for cooling as it is required. Quick-emptying piston
valves are provided to give rapid emptying of
the circuit when the scoop tube is withdrawn
from contact with the rotating oil annulus.

Under normal circulating conditions, oil fed
into the collector ring passes into the piston
valve control tubes. These tubes and connecting
passages conduct oil to the outer end of the
pistons. The centrifugal force of the oil in the
control tube holds the piston against the valve
port, thus sealing off the circuit. When the
scoop tube is withdrawn from the oil annulus
in the reservoir, the circulation of oil will be

591

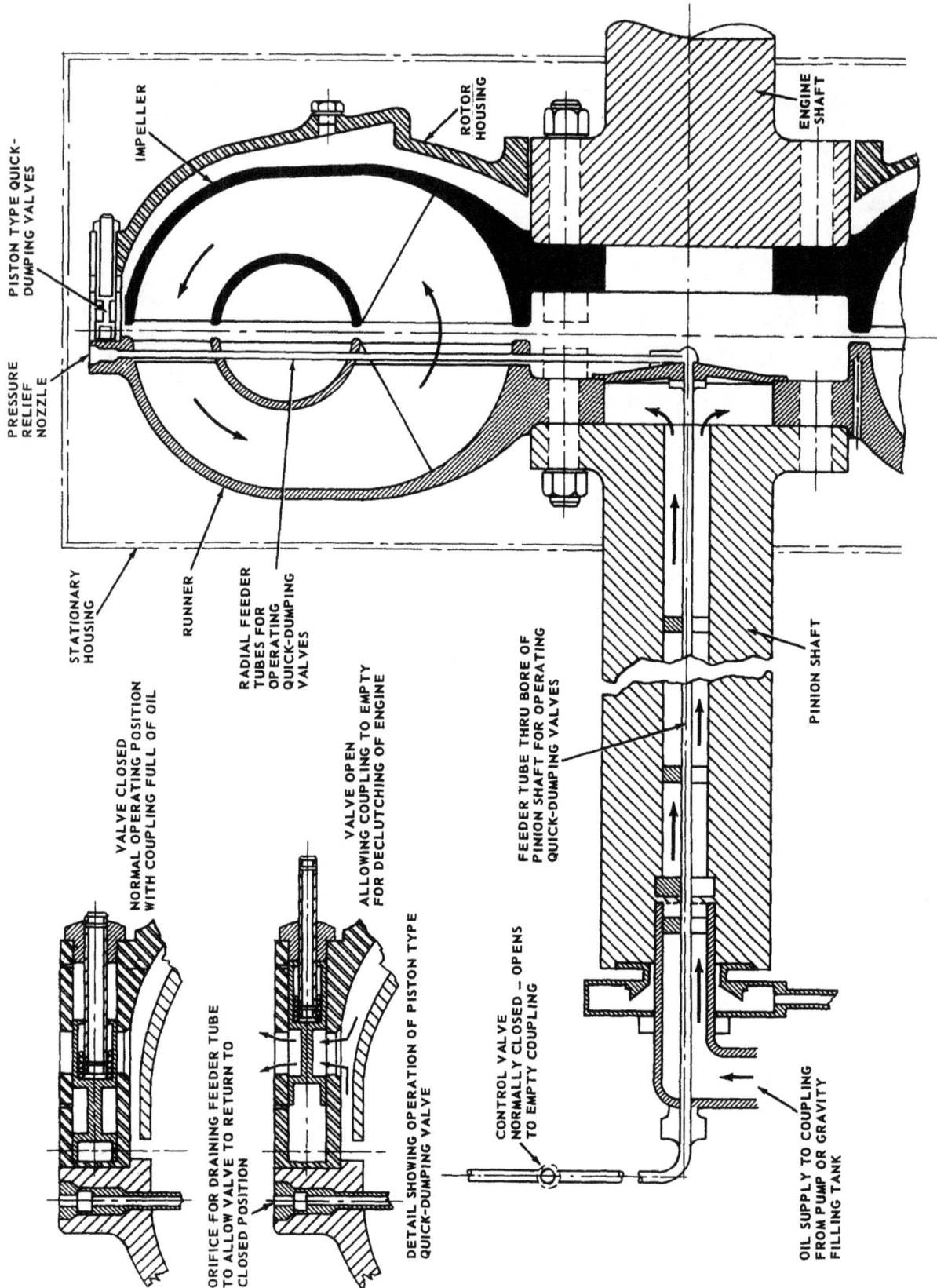

IMPELLER

ROTOR HOUSING

ENGINE SHAFT

PISTON TYPE QUICK-DUMPING VALVES

PRESSURE RELIEF NOZZLE

STATIONARY HOUSING

RUNNER

RADIAL FEEDER TUBES FOR OPERATING QUICK-DUMPING VALVES

PINION SHAFT

FEEDER TUBE THRU BORE OF PINION SHAFT FOR OPERATING QUICK-DUMPING VALVES

VALVE CLOSED
NORMAL OPERATING POSITION WITH COUPLING FULL OF OIL

ORIFICE FOR DRAINING FEEDER TUBE TO ALLOW VALVE TO RETURN TO CLOSED POSITION

VALVE OPEN
ALLOWING COUPLING TO EMPTY FOR DECLUTCHING OF ENGINE

DETAIL SHOWING OPERATION OF PISTON TYPE QUICK-DUMPING VALVE

CONTROL VALVE NORMALLY CLOSED – OPENS TO EMPTY COUPLING

OIL SUPPLY TO COUPLING FROM PUMP OR GRAVITY FILLING TANK

75.261X

Figure 22-43.—Hydraulic coupling with piston-type quick-dumping valves.

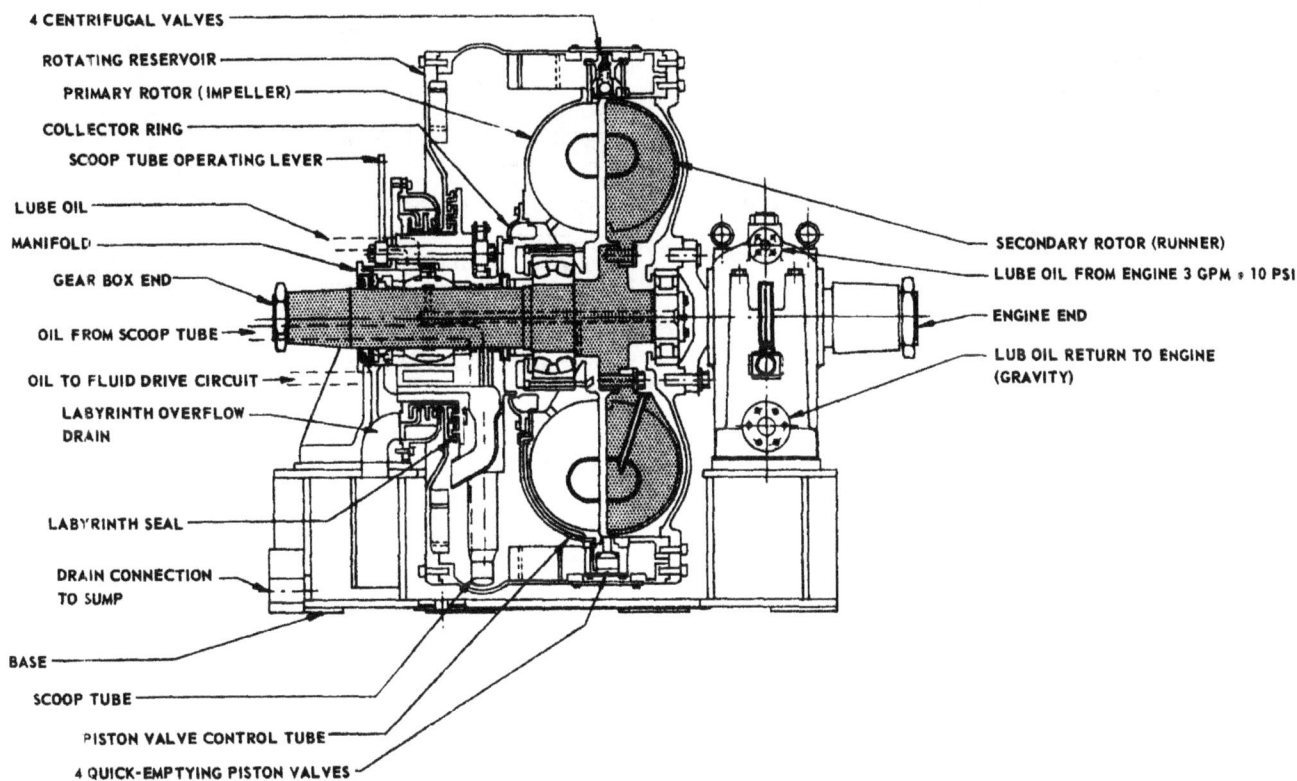

75.262

Figure 22-44.—Scoop control hydraulic coupling.

interrupted and the oil in the control tubes will be discharged through the orifice in the outer end of the piston housing. This releases the pressure on the piston and allows it to move outward, thus opening the port for rapid discharge of oil. Resumption of oil flow from the scoop tube will fill the control tubes; and the pressure will move the piston to the closed position.

When the engine is started and the coupling is filled with oil, the primary rotor turns with the engine crankshaft. As the primary rotor turns, the oil in its radial passages flows outward, under centrifugal force. (See arrows in fig. 22-42.) This forces oil across the gap at the outer edge of the rotor and into the radial passages of the secondary rotor, where the oil flows inward. The oil in the primary rotor is not only flowing outward, but is also rotating. As the oil flows over and into the secondary rotor, it strikes the radial blades in the rotor.

The secondary rotor soon begins to rotate and pick up speed, but it will always rotate more slowly than the primary rotor because of

drag on the secondary shaft. Therefore, the centrifugal force of the oil in the primary rotor will always be greater than that of the oil in the secondary rotor. This causes a constant flow from the primary rotor to the secondary rotor at the outer ends of the radial passages, and from the secondary rotor to the primary rotor at the inner ends.

The power loss in the hydraulic clutch is small (3 percent) and is caused by friction in the fluid itself. This means that approximately 97 percent of the power delivered to the primary rotor is transmitted to the reduction gear. The loss power is transformed into heat that is absorbed by the oil—which is the reason for sending part of the oil through a cooler at all times.

MAINTENANCE

Keeping an internal combustion engine (diesel or gasoline) in good operating condition demands a well-planned procedure of periodic inspection, adjustments, maintenance, and repair. If inspections are made regularly, many maladjustments can be detected and corrected before a serious casualty results. A planned maintenance program will help to prevent major casualties and the occurrence of many operating troubles.

There may be times when service requirements interfere with a planned maintenance program. In this event, routine maintenance must be performed as soon as possible after the specified interval of time has elapsed. Necessary corrective measures should be accomplished as soon as possible; if repair jobs are allowed to accumulate, the result may be hurried and incomplete work.

Since the Navy uses so many models of internal combustion engines, it is impossible to specify any detailed overhaul procedure that is adaptable to all models. However, there are several general rules which apply to all engines. They are:

1. Detailed repair procedures are listed in manufacturers' instruction manuals and maintenance pamphlets. Study the appropriate manuals and pamphlets before attempting any repair work. Pay particular attention to tolerances, limits, and adjustments.

2. The highest degree of cleanliness must be observed in handling engine parts during overhaul.

3. Before starting repair work, be sure that all required tools and replacements for known defective parts are available.

4. Detailed records of repairs should be kept. Such records should include the measurements of parts, hours in use, and new parts installed. An analysis of such records will indicate the hours of operation that may be expected from the various engine parts. This knowledge is helpful as an aid in determining when a part should be renewed in order to avoid a failure.

5. Detailed information on preventive maintenance is contained in the PMS Manual for the engineering department. All preventive maintenance should be accomplished in accordance with the (3-M System) Planned Maintenance Subsystem which is based upon the proper utilization of the PMS manuals, Maintenance

SYSTEM Propulsion	COMPONENT Propulsion Diesel T3 - T4 MAINT. SIGNIFICANT NO.	M.R. NUMBER E1D R-6		
SUB-SYSTEM Propulsion Units	RELATED M.R. R-5	RATES	EN2 EN3 FN	M/H 2.0 2.0 2.0

M.R. DESCRIPTION AFTER 750 HOURS OF OPERATION:
1. Inspect cap screws on cylinder retainers for tightness.
2. Inspect valve tappets for proper clearance.

TOTAL M/H:
6.0
ELAPSED TIME:
2.0

SAFETY PRECAUTIONS
1. Air starting valve wired shut and tagged, "DO NOT OPEN".
2. Injection pump must be in stop position when jacking engine by hand.

TOOLS, PARTS, MATERIALS, TEST EQUIPMENT
1. Cover gasket H-75 PAM 10866   6. Jacking bar
2. Torque wrench   7. Wire and tag
3. Socket set   8. NAVSHIPS 341-3393
4. Feeler gauge
5. Blowdown valve wrench

PROCEDURE
NOTE: To be done when engine is hot.

1. a. Remove valve cover.
   b. Remove vibration damper cover plate.
   c. Install barring socket plate.
   d. Open blow down valves.
   e. Using torque wrench and special torqueing tool, inspect cap screws in cylinder retainer for tightness (216 inch pounds torque).
   f. Continue checking and adjusting torque until a complete round is made with no movement of the screws.
NOTE: Working clockwise, check cap screw tightness in following order 1, 5, 9, 7, 4, 2, 6, 10, 8, 3.

2. a. With barring socket in place, jack engine as needed to measure each valve tappet clearance.
       Maximum clearance 0.15   Minimum clearance .007
   b. Remove barring socket plate and re-install the cap screws in vibration damper.
   c. Re-install vibration damper cover plate.
   d. Start engine and observe for diesel oil leak. (cont)

LOCATION

Page 1 of 2 MB ZZIP EE1 C3 1283 R

75.262

Figure 22-45.—Maintenance requirement card.

Requirement Cards (MRCs), and schedules for the accomplishment of planned maintenance actions. An MRC Card is shown in figure 22-45.

It should be noted that the PMS does not cover certain operating checks and inspections that are required as a normal part of the regular watchstanding routine. For example, you will not find such things as hourly pressure and temperature checks or routine oil level checks listed as maintenance requirements under the PMS. Even though these routine operating checks are not listed as PMS requirements, you must of course still perform them in accordance with all applicable watchstander's instructions.

CHAPTER 23

GAS TURBINES

The gas turbine engine, long regarded as a promising but experimental prime mover, has in recent years been developed to the point where it is entirely practicable for ship propulsion and for a number of auxiliary applications. Gas turbine engines are currently installed as prime movers on minesweepers, landing craft, PT boats, air-sea rescue boats, hydrofoils, hydroskimmers, and other craft. In addition, the gas turbine engine is finding increasing application as the driving unit for ship's service generators, pumps, and other auxiliary units.

Although the gas turbine engine as a type need no longer be regarded as experimental, many specific models of gas turbine engines are still at least partially experimental and subject to further change and development. The discussion in this chapter therefore deals primarily with the general principles of gas turbine engines rather than with specific models. Detailed information on any specific model may be obtained from the manufacturer's technical manual furnished with the equipment.

BASIC PRINCIPLES

The gas turbine engine bears some resemblance to an internal combustion engine of the reciprocating type and some resemblance to a steam turbine. However, a brief consideration of the basic principles of a gas turbine engine reveals several ways in which the gas turbine engine is quite unlike either the reciprocating internal combustion engine or the steam turbine.

Let us look first at the thermodynamic cycles[1] of the three engine types. The reciprocating internal combustion engine[2] has an open, heated-engine cycle and the steam turbine[3] has a closed, unheated-engine cycle. In contrast, the gas turbine has an open, unheated-engine cycle—a combination we have not previously encountered in our study of naval machinery. The gas turbine cycle is open because it includes the atmosphere; it is an unheated-engine cycle because the working substance is heated in a device which is separate from the engine.

Another way in which the three types of engines differ is in the working substance. The working fluid in a steam turbine installation is steam. In both the reciprocating internal combustion engine and the gas turbine engine, the working fluid may be considered as being the hot gases of combustion that result from the burning of fuel in air. However, there are very important differences in the way the working fluid is used in the reciprocating internal combustion engine and in the gas turbine engine.

Still other differences in the three types of engines become apparent when we consider the arrangement and relationship of component parts and the processes that occur during the cycle. From our study of previous chapters of this text, we are already familiar with the functional arrangement of parts in steam turbine installations and in reciprocating internal combustion engines. Now let us look at the relationship of the major components in a basic gas turbine engine, as illustrated schematically in figure 23-1.

In the steam turbine installation, the processes of combustion and steam generation take

---

[1] Thermodynamic cycles are discussed in chapter 8 of this text.

[2] Internal combustion engines are discussed in chapter 22.

[3] Steam turbines are discussed in chapter 12 and in chapter 16.

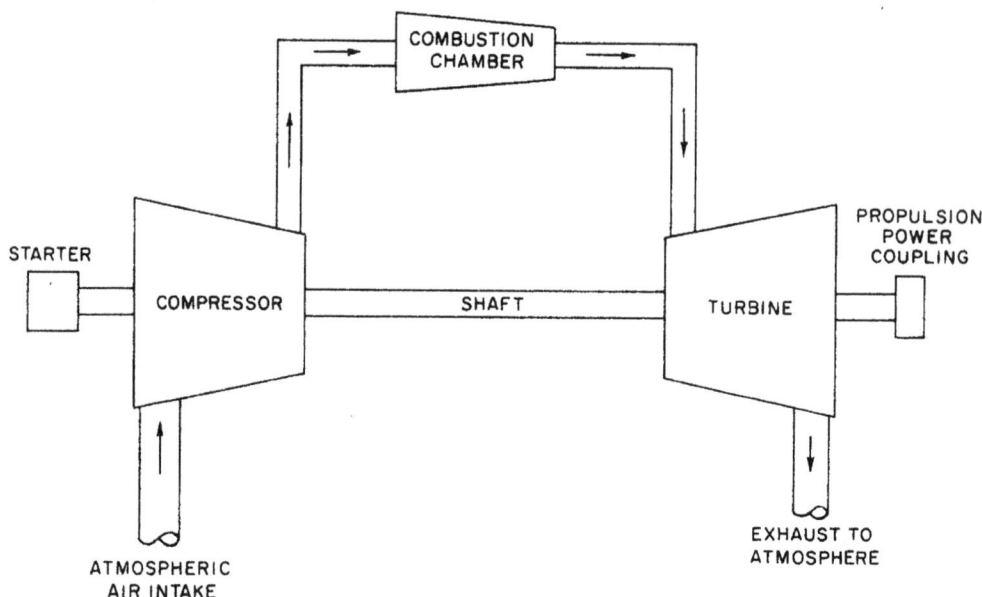

147.135

Figure 23-1.—Schematic diagram showing relationship of parts in single-shaft gas turbine engine.

place in the boiler, while the process by which the thermal energy of the steam is converted into mechanical work takes place in the turbine. In the reciprocating internal combustion engine, three processes—the compression of atmospheric air, the combustion of a fuel-air mixture, and the conversion of heat to work—all take place in one unit, the cylinder. The gas turbine engine is similar to the reciprocating internal combustion engine in that the same three processes—compression, combustion, and conversion of heat to work—occur; but it is unlike the reciprocating internal combustion engine in that these three processes take place in three separate units rather than in one unit. In the gas turbine engine, the compression of atmospheric air is accomplished in the compressor; the combustion of fuel is accomplished in the combustion chamber; and the conversion of heat to work is accomplished in the turbine.

Many different types and models of gas turbine engines are in use. The gas turbine engine shown in figure 23-1 is called a single-shaft type because one shaft from the turbine rotor drives the compressor and an extension of this same shaft drives the load.

The gas turbine engine shown in figure 23-2 is called a split-shaft type. This engine is considered to be split into two sections: the gas-

producing section, or gas generator, and the power turbine section. The gas-generator section, in which a stream of expanding gases is created as a result of continuous combustion, includes the compressor, the combustion chamber (or chambers), and the gas-generator turbine. The power turbine section consists of a power turbine and the power output shaft. In this type of gas turbine engine, there is no mechanical connection between the gas-generator turbine and the power turbine. When the engine is operating, the two turbines produce basically the same effect as that produced by a hydraulic torque converter. The split shaft gas turbine engine is well suited for use as a propulsion unit where loads vary, since the gas-generator section can be operated at a steady and continuous speed while the power turbine section is free to vary with the load. Starting effort required for a split-shaft gas turbine engine is far less than that required for a single-shaft gas turbine engine connected to the reduction gear, propulsion shaft, and propeller.

In the twin-spool gas turbine engine (fig. 23-3) the air compressor is split into two sections or stages and each stage is driven by a separate turbine element. The low pressure turbine element drives the low pressure compressor element and the high pressure turbine element drives the high pressure compressor

596

GAS-GENERATOR SECTION | POWER TURBINE SECTION

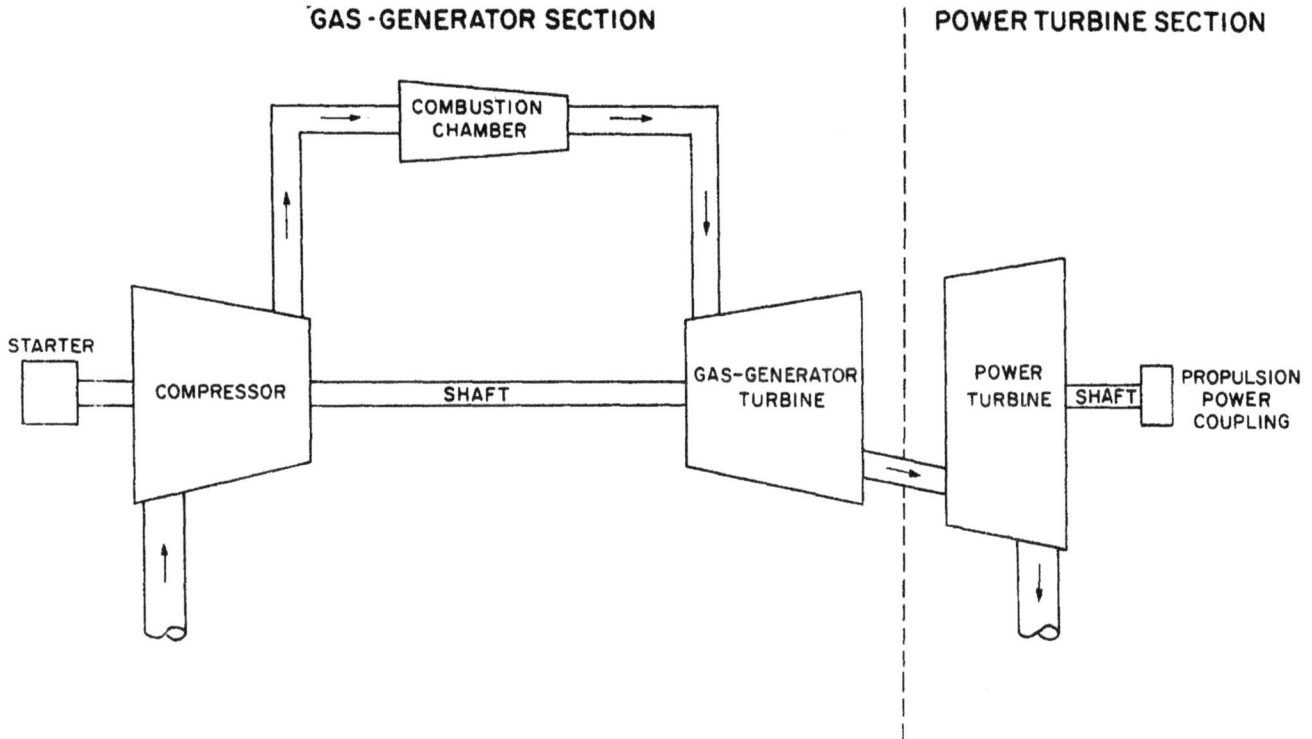

147.136

Figure 23-2.—Schematic diagram showing relationship of parts in split-shaft gas turbine engine.

element. Like the split-shaft type, the twin-spool gas turbine engine is usually divided into a gas-generator section and a power turbine section. However, some twin-spool gas turbine engines are so arranged that the low pressure turbine element drives the low pressure compressor element and the power output shaft.

The basic cycle of the gas turbine engine is one of isentropic compression, constant-pressure heat addition, isentropic expansion, and constant-pressure heat rejection. As the hot combustion gases are expanded through the turbine, converting thermal energy into mechanical work, some of the turbine work is used to drive the compressor and the remainder is used to drive the load. The power output from the turbine is steady and continuous and, after the initial start, self-sustaining.

Although this chapter deals only with gas turbine engines which operate on the simple open cycle, it should be mentioned that other cycles are also of interest to designers of gas turbine engines. Among the cycles that have been considered (and to some extent used) are the closed cycle, the semi-open cycle, and various modifications of the simple open cycle. In one such modification, known as the regenerated open cycle, the hot exhaust gases from the turbine are passed through a heat exchanger in which they give up some heat to the air between the compressor discharge and the inlet to the combustion chamber. The utilization of this heat decreases the amount of fuel required and thereby increases the efficiency of the cycle.

FUNCTIONS OF COMPONENTS

As we have seen, the three major components of a gas turbine engine are the compressor, the combustion chamber, and the turbine. In addition, the engine requires a number of other components, accessories, and systems in order to operate as a complete unit. The functions of the gas turbine engine components are described in the following sections.

Compressor

The compressor takes in atmospheric air and compresses it to a pressure of several

597

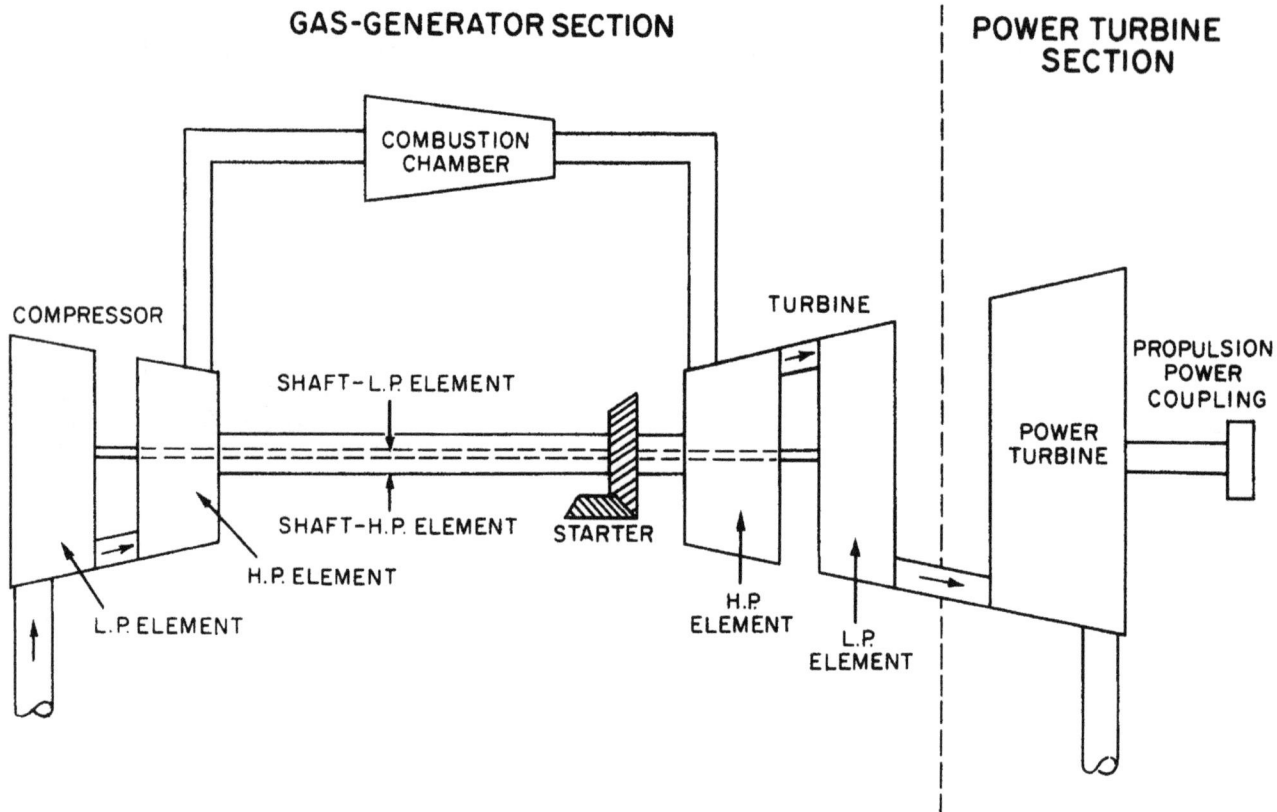

147.137

Figure 23-3.—Schematic diagram showing relationship of
parts in twin-spool gas turbine engine.

atmospheres. Part of the compressed air, called
primary air, enters directly into the combustion
chamber where it is mixed with the atomized
fuel so that the mixture can be ignited and burned.
The remainder of the air, called secondary air,
is mixed with the gases of combustion. The purpose of the secondary air is to cool the combustion gases down to the desired turbine inlet
temperature.

Both axial-flow compressors and centrifugal
(radial-flow) compressors are currently used
in gas turbine engines. There are several possible configurations of these basic types, some
of which are in use and some of which are in
experimental phases of development.

In the axial-flow compressor the air is compressed as it flows axially along the shaft.
An axial-flow compressor of good design may
achieve efficiencies in the range of 82 to 88
percent at compressor pressure ratios up to
8:1. At higher pressure ratios, the efficiency
tends to decrease. Axial flow compressors may

be of the single-spool type, previously illustrated
in figures 23-1 and 23-2, or of the twin-spool
type, as shown in figure 23-3. The gas turbine
engine shown in figure 23-4 has an axial-flow
compressor of the single-spool type. Figure
23-5 shows an axial-flow single-spool compressor removed from its engine. Where twin-
spool axial-flow compressors are used, a separate turbine drives each spool, as shown in
figure 23-3.

The centrifugal (radial-flow) compressor
picks up the entering air and accelerates it
outward by means of centrifugal force. The
centrifugal compressor (fig. 23-6) may achieve
efficiencies of 80 to 84 percent at pressure
ratios of 2.5 to 4 and efficiencies of 76 to 81
percent at pressure ratios of 4 to 10.

The advantages of the axial-flow compressor
include high peak efficiencies; a relatively small
frontal area for any given air flow; and only
negligible losses between stages, even when
a large number of stages are used. The

598

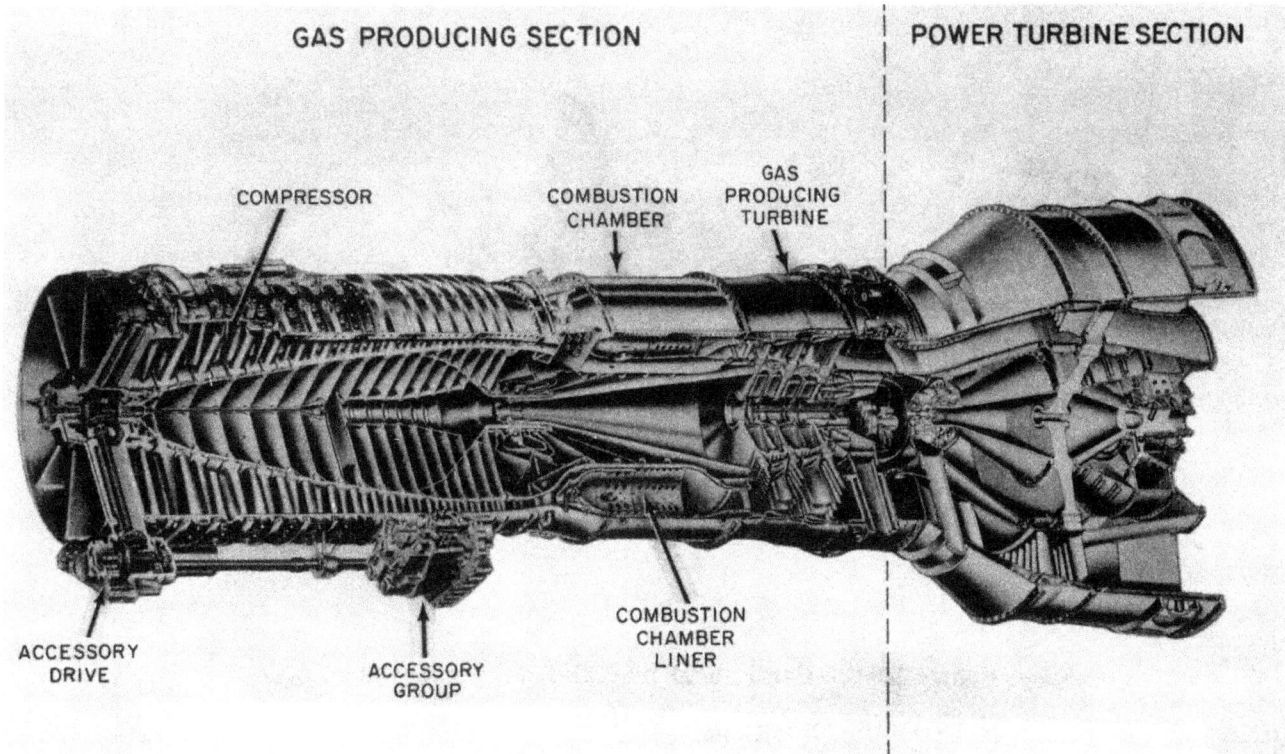

GAS PRODUCING SECTION   POWER TURBINE SECTION

147.138

Figure 23-4.—Gas turbine engine with single-spool axial-flow compressor.

disadvantages of the axial-flow compressor include difficulty of manufacture, high initial cost, and relatively great weight.

147.139

Figure 23-5.—Single-spool axial-flow compressor.

The advantages of the centrifugal compressor include a high pressure rise per stage, simplicity of manufacture, low initial cost, and relatively light weight. The disadvantages of the centrifugal compressor include the need for a relatively large frontal area for a given air flow and the difficulty of using two or more stages because of losses that would occur between the stages.

Combustion Chamber

The combustion chamber is the component in which the fuel-air mixture is burned. The combustion chamber consists of a casing, a perforated inner shell, a fuel nozzle, and a device for initial ignition. The number of combustion chambers used in a gas turbine engine varies widely; as few as one and as many as sixteen combustion chambers have been used in one gas turbine engine.

The combustion chamber is the most efficient component of a gas turbine engine. Efficiencies between 95 and 98 percent can be obtained over a wide operating range. To produce such efficiencies, combustion chambers are designed to operate with low pressure losses, high combustion

599

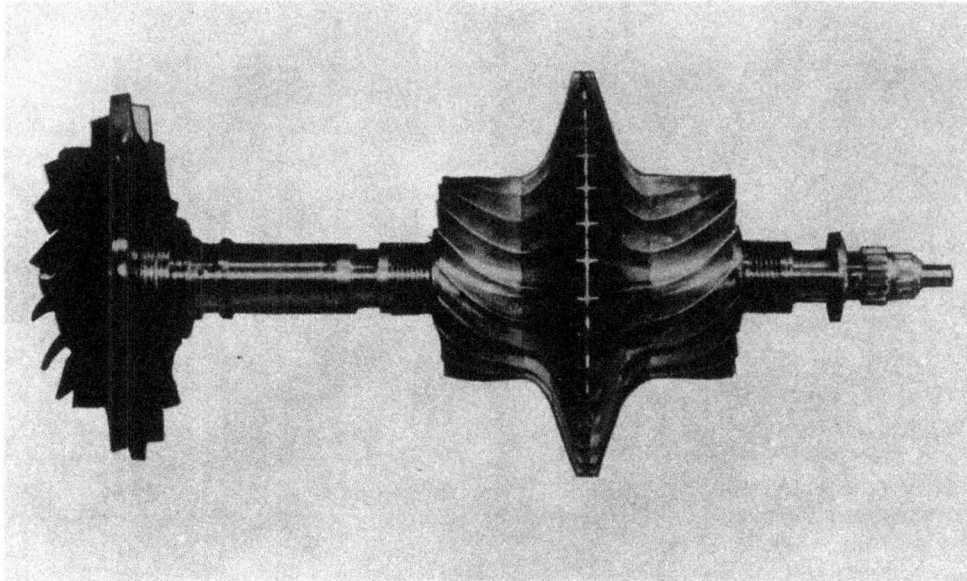

147.140

Figure 23-6.—Centrifugal (radial-flow) compressor.

efficiency, and good flame stability. Additional requirements for the combustion chamber include low rates of carbon formation, light weight, reliability, reasonable length of life, and the ability to mix cold air with the hot combustion gases in such a way as to give uniform temperature distribution to the turbine blades.

Only a small part (perhaps one-fourth) of the air which enters the combustion chamber area is burned with the fuel. The remainder of the air is used to keep the temperature of the combustion gases low enough so that the turbine nozzles and blades will not be overheated and thereby damaged.

The basic types of combustion chambers in current use are the tubular or can-type chamber, the annular chamber, the can-annular chamber, and the elbow chamber.

The tubular or can-type chamber, shown in figures 23-7 and 23-8, is used with both axial-flow and centrifugal compressors. The can-type combustion chamber consists of an outer case or housing within which is a perforated, stainless steel, highly heat resistant combustion chamber liner. The combustion chamber housing is divided to facilitate liner replacement. Each can-type chamber has its own individual air inlet duct.

Interconnector tubes (flame tubes) are a necessary part of can-type combustion chambers.

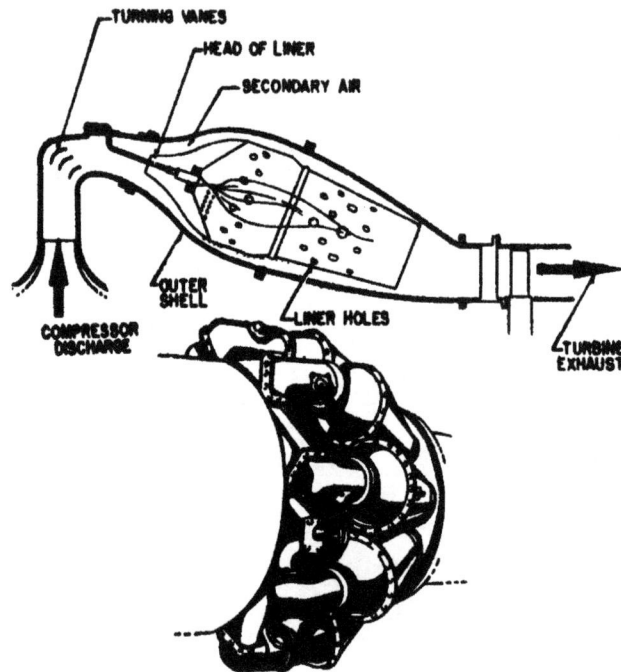

147.141

Figure 23-7.—Tubular or can-type combustion chamber.

600

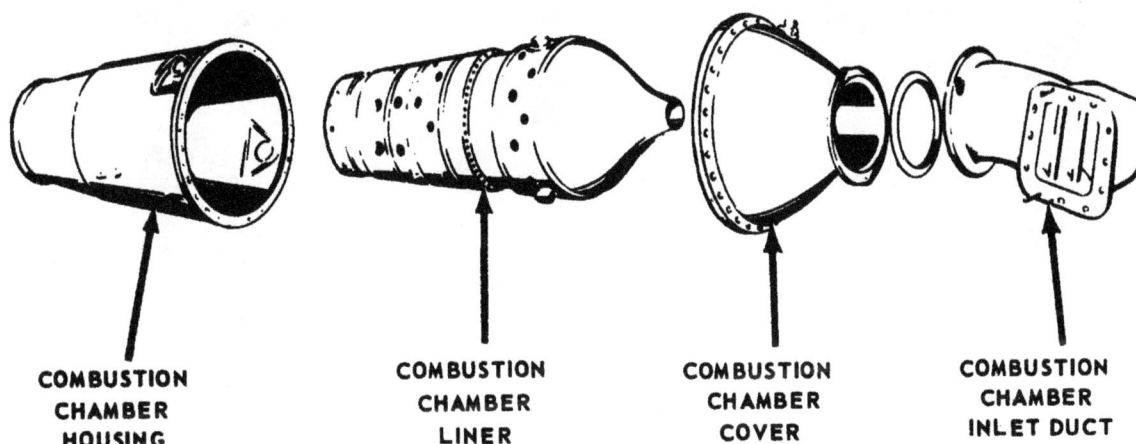

COMBUSTION CHAMBER HOUSING   COMBUSTION CHAMBER LINER   COMBUSTION CHAMBER COVER   COMBUSTION CHAMBER INLET DUCT

Figure 23-8.—Elements of tubular or can-type combustion chamber.

Since each of the combustion chambers has its own separate burner, each one operating independently of the others, there must be some way to spread the flames during starting. This requirement is met by interconnecting all the chambers so that, as the flame is started by the spark ignition plugs in the lower chambers, the flame passes through the interconnector tubes and ignites the combustible mixture in the adjacent chambers. Once ignition is obtained, the spark igniters are automatically cut off.

The annular combustion chamber is more efficient than the can-type chamber. Although details of design may vary, the annular combustion chamber consists essentially of a single chamber which completely surrounds the engine. Fuel enters the combustion chamber through a series of nozzles which are mounted equidistant from each other on a ring at the front end of the combustion chamber; because of this arrangement, the flame is distributed evenly around the entire circumference of the combustion chamber. Diffusion of air and an efficient flame pattern are maintained by means of rows of holes which are punched in the outer liner or basket of the combustion chamber.

The can-annular combustion chamber combines features of both the can-type chamber and the annular chamber. The can-annular chamber allows an annular discharge from the compressors, from which the air flows to individual burners where the fuel is injected and burned. The can-annular combustion chambers are arranged radially around the axis of the engine—the axis in this instance being the rotor

shaft housing. Figure 23-9 shows the arrangement of can-annular combustion chambers.

The can-annular combustion chambers are enclosed by a removable steel shroud which covers the entire burner section. This feature makes the burners readily accessible for any required maintenance.

The can-annular combustion chambers are interconnected by means of projecting flame tubes. These flame tubes facilitate starting, as previously described in connection with the can-type combustion chamber.

Each of the can-annular combustion chambers contains a central, bullet-shaped, perforated liner. The size and shape of the holes are designed to admit the correct quantity of air at the required velocity and angle. Cutouts are provided in two of the bottom chambers for the installation of the spark igniters.

Each can-annular combustion chamber receives fuel through duplex nozzles installed at the forward end of the chamber. Guide vanes around the fuel nozzles direct the primary air and cause it to enter the combustion chamber with a swirling motion which mixes the fuel and air and thus leads to even and complete combustion.

Turbine

In theory, design, and operating characteristics, the turbines used in gas turbine engines are quite similar to the turbines used in a steam plant. The gas turbine differs from the steam turbine chiefly in the type of blading

147.143

Figure 23-9.—Can-annular combustion chamber.

material used, the means provided for cooling the bearings and highly stressed parts, and the higher ratio of blade length to wheel diameter which is required to accommodate the large gas flow.

The turbine section of a gas turbine engine is located directly behind the combustion chamber outlet. The turbine consists of two basic elements, the <u>stator</u> and the <u>rotor</u>. Part of a stator element is shown in figure 23-10; a rotor element is shown in figure 23-11.

The stator element is referred to by various names, including <u>turbine nozzle vanes</u> and <u>turbine guide vanes</u>. The vanes of the stator element serve the same purpose as the nozzles in an impulse steam turbine or the stationary blading in a reaction steam turbine—that is, they convert thermal energy into mechanical kinetic energy. The vanes of the stator element are contoured and set at such an angle that they form a number of small nozzles which discharge the gas as extremely high speed jets. As in the case of the nozzles (or stationary blading) of steam turbines, the increase in

velocity may be equated with the decrease in thermal energy. The vanes of the stator element direct the flow of gas to the rotor blades at the required angle while the turbine wheel is rotating.

The rotor element of the turbine consists of a shaft and a bladed wheel or disk. The wheel is attached to the main power transmitting shaft of the gas turbine engine. The jets of combustion gas leaving the vanes of the stator element act upon the turbine blades and cause the turbine wheel to rotate at a very high rate of speed. The high rotational speed imposes severe centrifugal loads on the turbine wheel, and at the same time the very high temperatures result in a lowering of the strength of the material. Consequently, the engine speed and temperature must be controlled to keep turbine operation within safe limits. Even so, the operating life of the turbine blading is accepted as the governing factor in determining the life of the gas turbine engine.

The turbine may be of the single-rotor type or of the multiple-rotor type. Either single-rotor

147.144

Figure 23-10.—Stator element of turbine
assembly.

147.145

Figure 23-11.—Rotor element of turbine
assembly.

or multiple-rotor turbines may be used with
either centrifugal or axial-flow compressors.
In the single-rotor type of turbine, the power
is developed by one rotor and all engine-driven
parts are driven by this single wheel. In the
multiple-rotor type, the power is developed by
two or more rotors. It is possible for one or
more rotors to drive the compressor and the
accessories, while one or more other rotors
are used for the power output.

Main Bearings

The main bearings in a gas turbine engine
serve the critical function of supporting the
compressor, the turbine, and the engine shaft.
The number and position of main bearings required for proper support vary according to
the length and stiffness of the shaft, with both
length and stiffness being affected by the type
of compressor used in the engine. In general,
a gas turbine engine requires at least three
main bearings and may require six or even
more.

Several types of main bearings are used
in gas turbine engines. Ball and roller bearings
have been quite commonly used in the past, and
they are still used in many aircraft gas turbine
engines. Sleeve bearings, split-sleeve bearings,
floating-sleeve bearings, and slipper bearings
are commonly used in gas turbine engines designed for marine propulsion.

The slipper or pivoted-shoe type of bearing
has recently attracted considerable attention and
is being used increasingly for main bearings on
gas turbine engines and other high speed engines. This type of bearing is designed with
relatively large radial clearances. Since a
rotating object tends to rotate about its true
balance center when it is not restrained by
bearings or supports, the large radial clearances in the slipper bearings allow a kind of
self-balancing action or automatic compensation

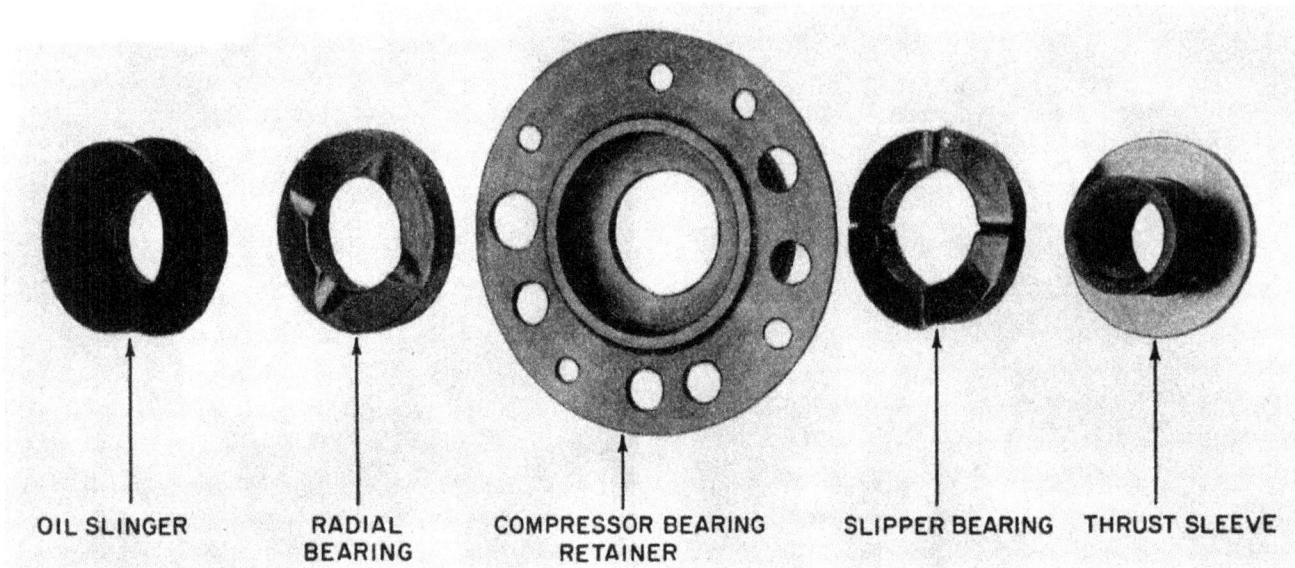

147.146

Figure 23-12.—Slipper bearing.

for balance errors to take place when the engine is operating.

One type of slipper bearing is shown in figure 23-12. In this type of bearing, the slipper consists of four pivoted-shoe segments, similar to the pivoted shoes used in Kingsbury bearings. The segments are held together loosely by a wire spring.

Another type of slipper bearing is shown in figure 23-13. In this type of bearing, the slipper consists of six segments which are fastened in place by dowel pins. This type of bearing is sometimes called a fixed-pivot slipper bearing.

Accessory Drives

Because the turbine and the compressor are on the same rotating shaft, a popular misconception is that the gas turbine engine has only one moving part. This is not the case, however. A gas turbine engine requires a starting device (which is usually a moving part), some kind of control mechanism, and power take-offs.

The accessory drive section of the gas turbine engine takes care of these various accessory functions. The primary purpose of the accessory drive section is to provide space for the mounting of the accessories required for the operation and control of the engine. Secondary purposes include acting as an oil reservoir and/or oil sump and providing for

147.147

Figure 23-13.—Slipper bearing (fixed-pivot type).

and housing accessory drive gears and reduction gears. The accessory drive section of a gas turbine engine is shown in figure 23-14.

The gear train is driven by the engine rotor through an accessory drive shaft gear coupling.

The reduction gearing within the case provides suitable drive speeds for each engine accessory or component. Because the operating rpm of the rotor is so high, the accessory reduction gear ratios are relatively high. The accessory drives

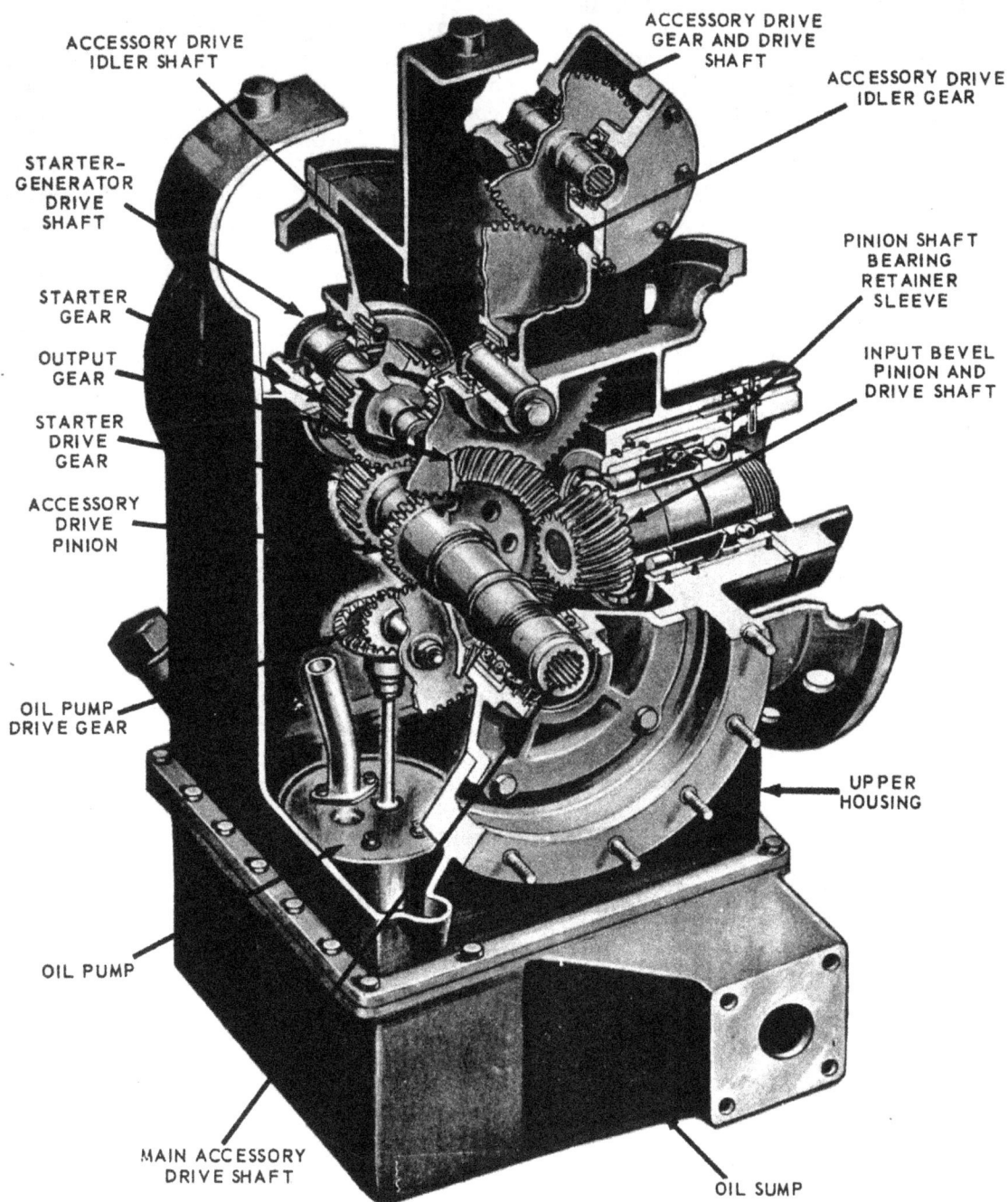

147.148

Figure 23-14.—Cutaway view of accessory drive section.

are supported by ball bearings assembled in the mounting bores of the accessory case.

Accessories always provided in the accessory drive section include the fuel control, with its governing device; the high pressure fuel oil pump or pumps; the oil sump; the oil pressure and scavenging pump or pumps; the auxiliary fuel pump; and a starter. Additional accessories which may be included in the accessory drive section or which may be provided elsewhere include a starting fuel pump, a hydraulic oil pump, a generator, and a tachometer. Some gas turbine engines are equipped with magnetos for ignition and these magnetos are driven from the accessory drive. Most of these accessories are essential for the operation and control of any gas turbine engine; however, the particular combination and arrangement of engine-driven accessories depends upon the use for which the gas turbine engine is designed.

Engine Systems

The major systems of a gas turbine engine are those which supply fuel, lubricating oil, and electricity.

FUEL SYSTEM.—The fuel system supplies the specified fuel for combustion. The components of a fuel system depend to some extent upon the type of gas turbine engine; however, the basic fuel oil system shown in figure 23-15 may be regarded as typical of a simple fuel system. The engine-driven pump receives filtered fuel from a motor-driven supply pump at a constant pressure. The engine-driven fuel pump increases the pressure and forces the fuel through the high pressure filter to the fuel control governor in the fuel control assembly. The fuel control governor provides fuel to the

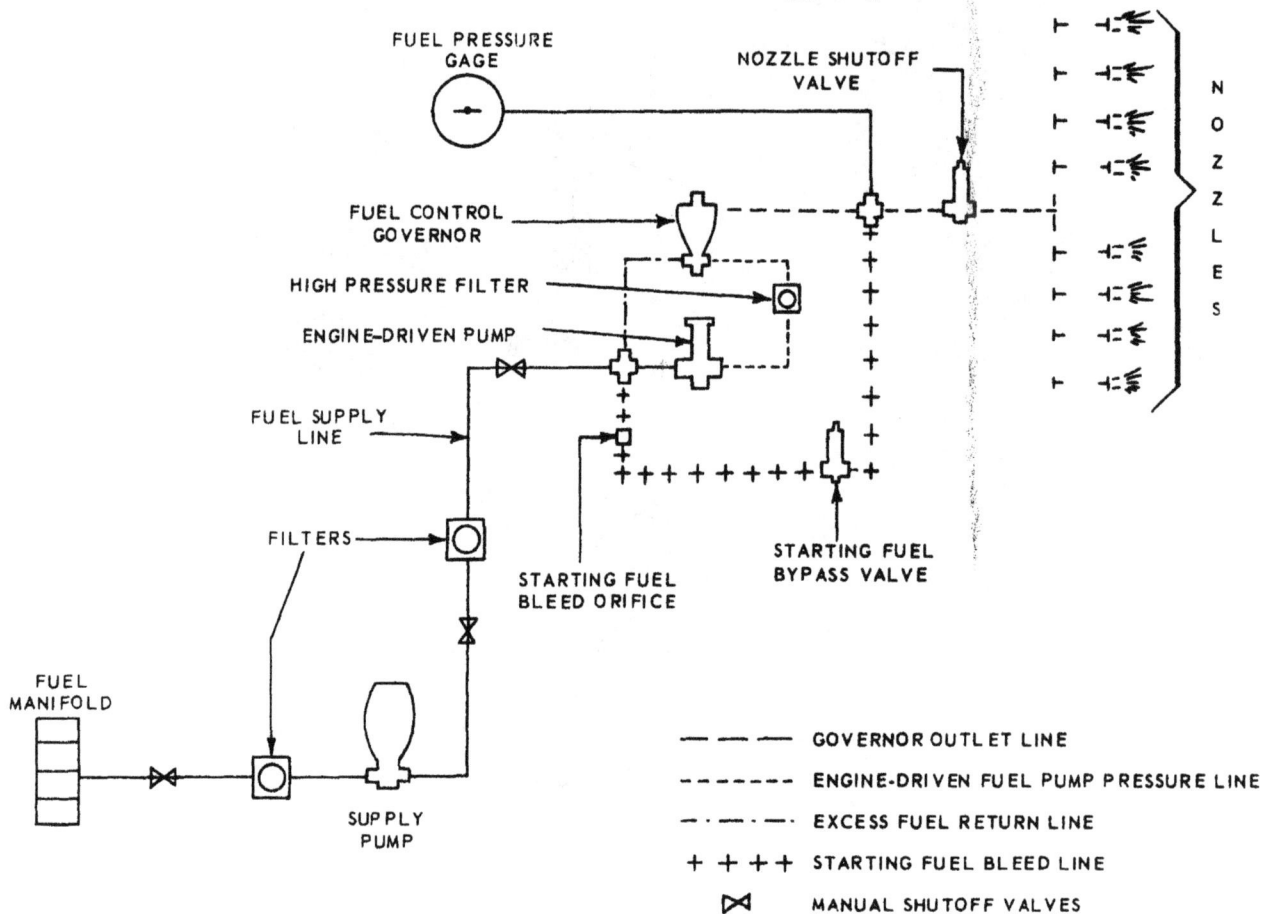

Figure 23-15.—Fuel flow diagram for gas turbine engine.

147.149

nozzles at the pressure and volume required to maintain the desired engine performance. At the same time, the fuel control governor limits fuel flow to maintain operating conditions within safe limits.

The fuel nozzles serve to introduce the fuel into the combustion chamber. The fuel is sprayed into the combustion chamber under pressure, through small orifices in the nozzles. Various kinds of fuel nozzles are in use. Figure 23-16 shows a simplex nozzle and a duplex nozzle. The simplex nozzle was used on some older gas turbine engines. Most recent gas turbine engines use some kind of duplex nozzle. A duplex nozzle requires a dual manifold just ahead of the nozzles and a flow divider (before the manifold) to divide the fuel into primary and secondary streams. The duplex type of nozzle provides a desirable spray pattern for combustion

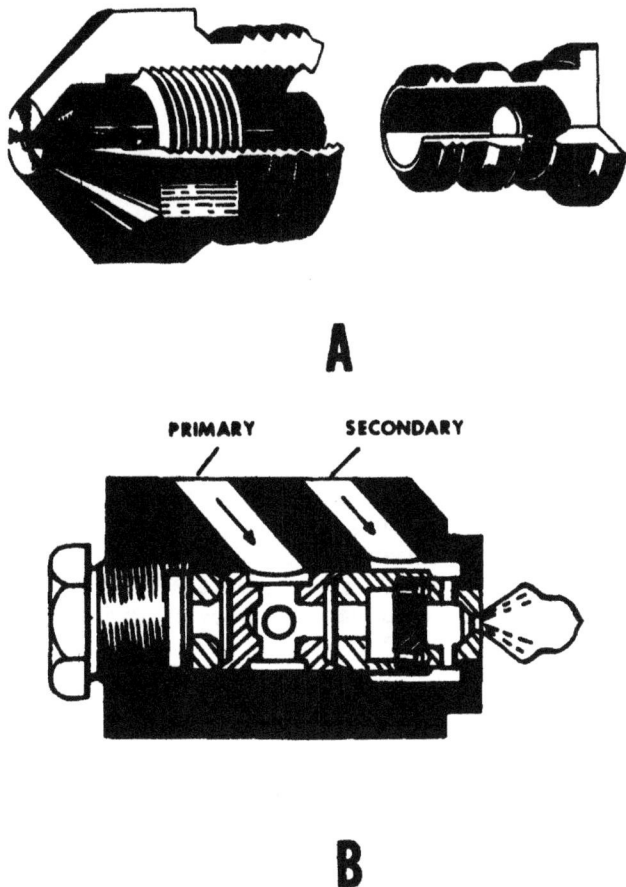

A

B

147.150

Figure 23-16.—Fuel nozzles.
A. Simplex.   B. Duplex.

over twice the range of that provided by the simplex nozzle.

The fuel control assembly is the unit which regulates the turbine rpm by adjusting fuel flow from the high pressure engine-driven pump to the fuel nozzle. The major parts of the fuel control assembly are shown in figure 23-17. Fuel enters the fuel control assembly and is pumped through a filter. High pressure fuel is routed to the differential relief valve, then to the fuel shutoff valve, and finally to the fuel nozzles.

The speed setting lever on the outboard end of the governor is connected to a speed control device on the control console either by a cable or by an electric servomotor.[4] At the fuel control end, the lever is keyed to a pinion. This pinion positions a rack which in turn controls the governor flyweight spring. The mechanism regulates gas producer speed according to the position of the control lever. With the control lever in any particular position, variations from the preset speed are sensed by the governor flyweights and a compensating movement of the fuel control valve results. An externally adjustable needle valve provides a constant minimum fuel flow during deceleration, when the governor valve is closed, to prevent loss of combustion. An acceleration limiter, consisting of a needle valve positioned by a shaft, arm, and bellows, is actuated by compressor discharge pressure. During acceleration, this mechanism controls fuel flow to the point at which the governor flyweight mechanism and its fuel control valve take over.

LUBRICATING SYSTEM.—Because of the high operating rpm and the high operating temperatures encountered in gas turbine engines, proper lubrication is of vital importance. The lubricating system is designed to supply bearings and gears with clean lubricating oil at the desired pressures and temperatures. In some installations, the lubricating system also furnishes oil to various hydraulic systems. Heat absorbed by the lubricating oil is transferred to the cooling medium in a lube oil cooler.

The lubricating system shown in figure 23-18 has a combined hydraulic system—in this case, the hydraulic system is for the operation of a hydraulic clutch in a gas turbine propulsion

_____
[4]Servomechanisms are discussed in chapter 20 of this text.

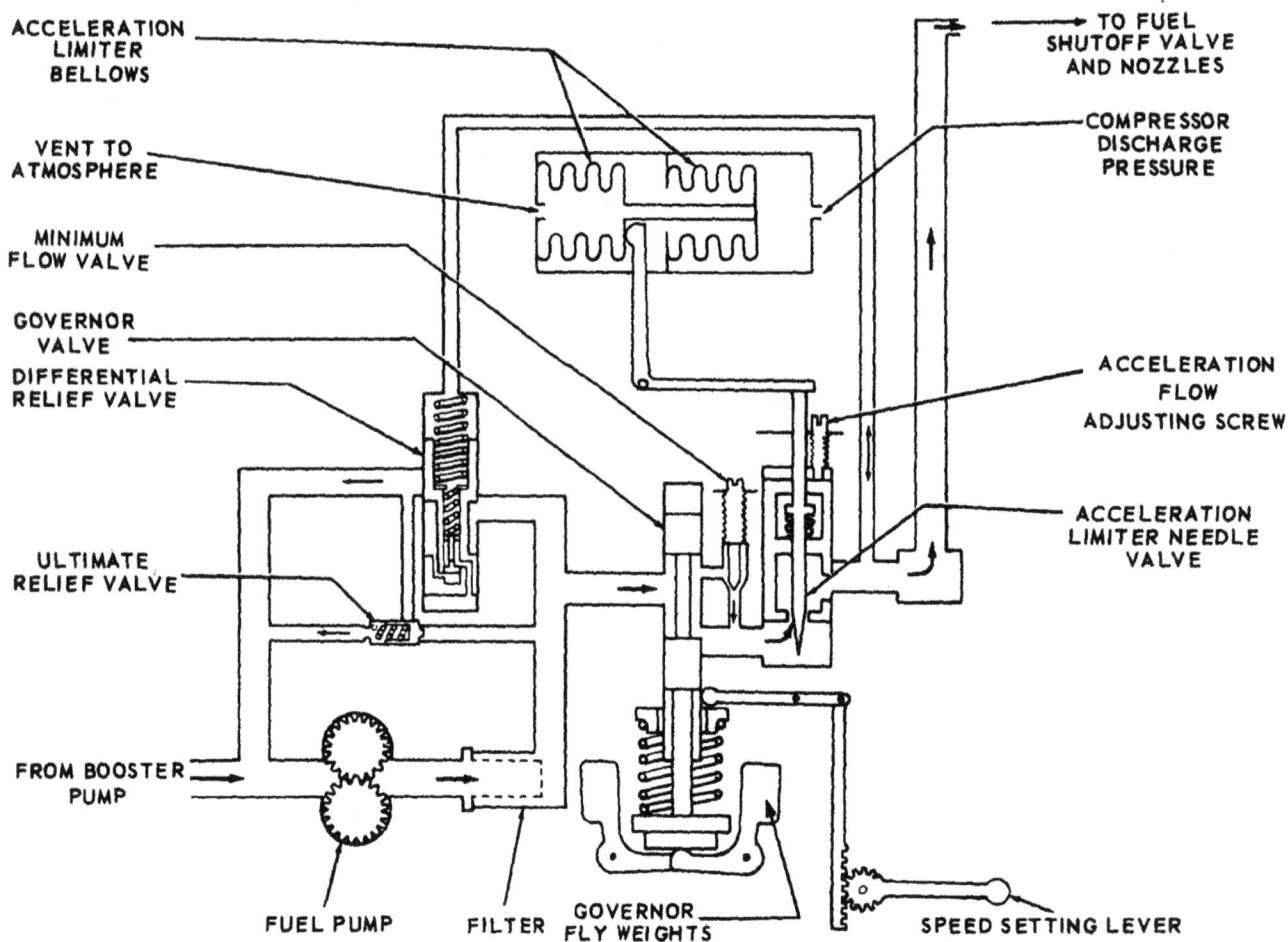

147.151

Figure 23-17.—Schematic diagram of fuel control assembly.

system. The lubricating system illustrated is of the dry-sump type, with a common oil supply from an externally mounted oil tank. The system includes the oil tank, the lubricating oil pump, the hydraulic oil pump, the air inlet scavenging pump, the oil temperature switch, the oil cooler, oil filters, the pressure regulating valve, the diverter valve, and a low pressure switch.

All bearings and gears in the engine, accessory drives, reduction gears, and reverse gear are lubricated and cooled by the lubricating system. Also, as may be observed in figure 23-18, the system illustrated here supplies oil for the lubrication of the fuel control governor.

ELECTRICAL SYSTEM.—In the gas turbine engine, the electrical system (fig. 23-19) is the principal means of automatic control of the engine. Electrical circuits which incorporate speed and pressure sensing switches control the starting and ignition sequence by opening and closing various valves in the fuel system. Engine operating conditions are reported by speed, pressure, and temperature operated switches and temperature bulbs.

The electrical system usually includes a starter, an ignition circuit, a control battery, and relays. In addition, it includes electrical accessories and control components such as the starting and ignition control switches and relays; the panel-mounted instruments and indicator lights for oil temperature, oil pressure, and engine rpm; and engine-mounted reporting devices such as fuel pressure switches, oil temperature switches, oil pressure switches, and thermocouples.

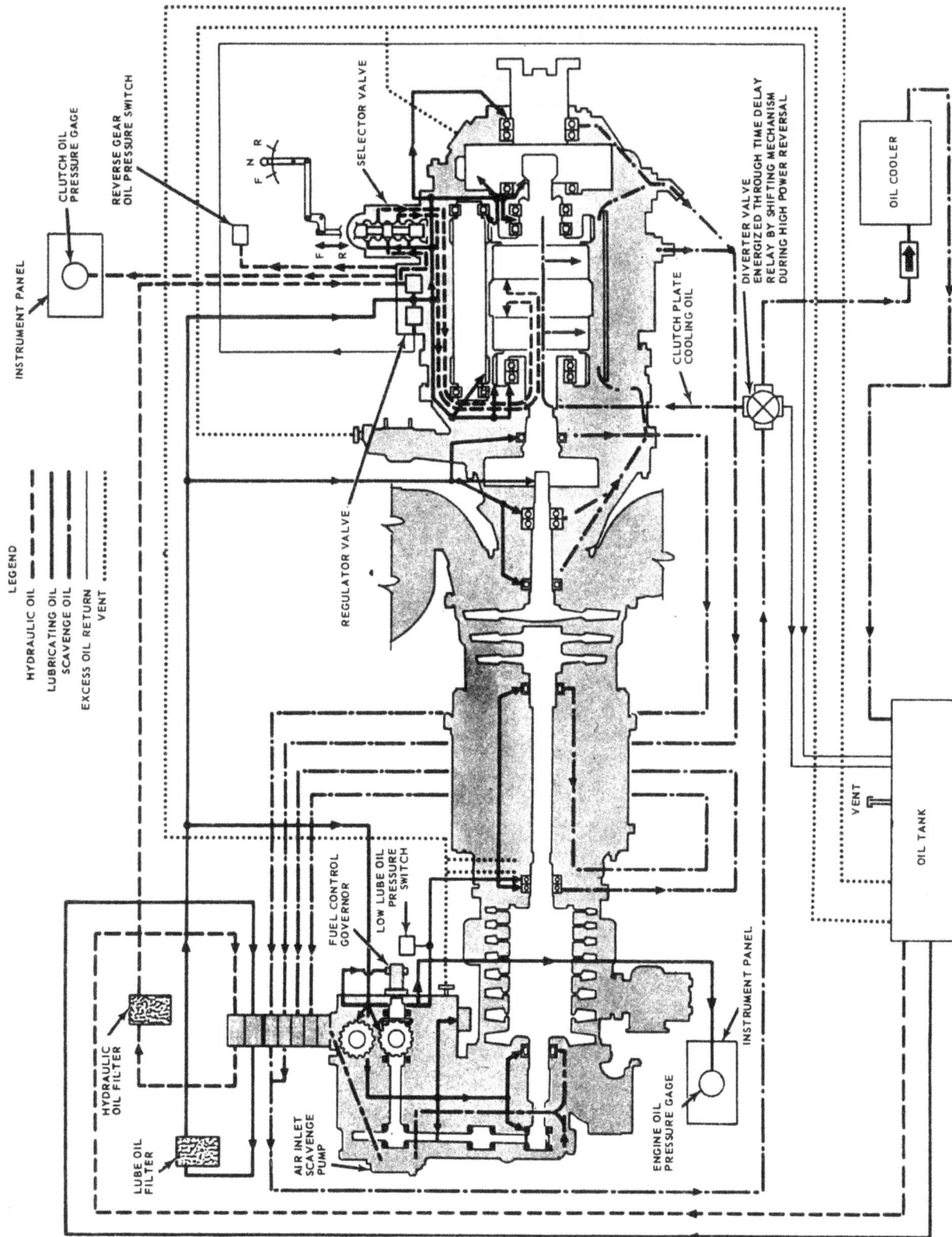

Figure 23-18.—Lubricating and hydraulic oil system for gas turbine engine.

147.152

LEGEND

HYDRAULIC OIL

LUBRICATING OIL

SCAVENGE OIL

EXCESS OIL RETURN

VENT

CLUTCH OIL PRESSURE GAGE

REVERSE GEAR OIL PRESSURE SWITCH

INSTRUMENT PANEL

SELECTOR VALVE

REGULATOR VALVE

DIVERTER VALVE ENERGIZED THROUGH TIME DELAY RELAY BY SHIFTING MECHANISM DURING HIGH POWER REVERSAL

CLUTCH PLATE COOLING OIL

OIL COOLER

HYDRAULIC OIL FILTER

LUBE OIL FILTER

AIR INLET SCAVENGE PUMP

FUEL CONTROL GOVERNOR

LOW LUBE OIL PRESSURE SWITCH

ENGINE OIL PRESSURE GAGE

INSTRUMENT PANEL

VENT

OIL TANK

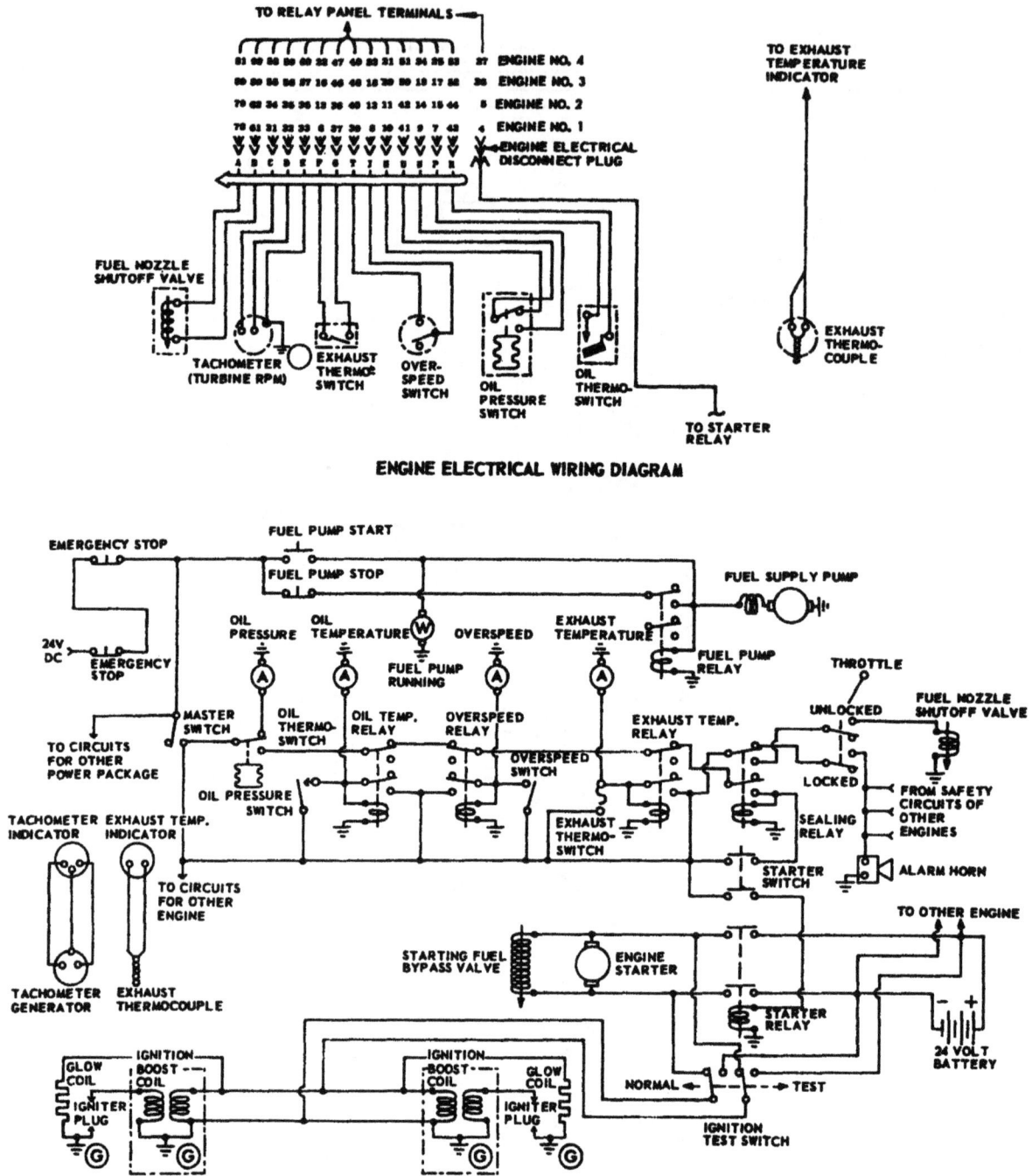

ENGINE ELECTRICAL WIRING DIAGRAM

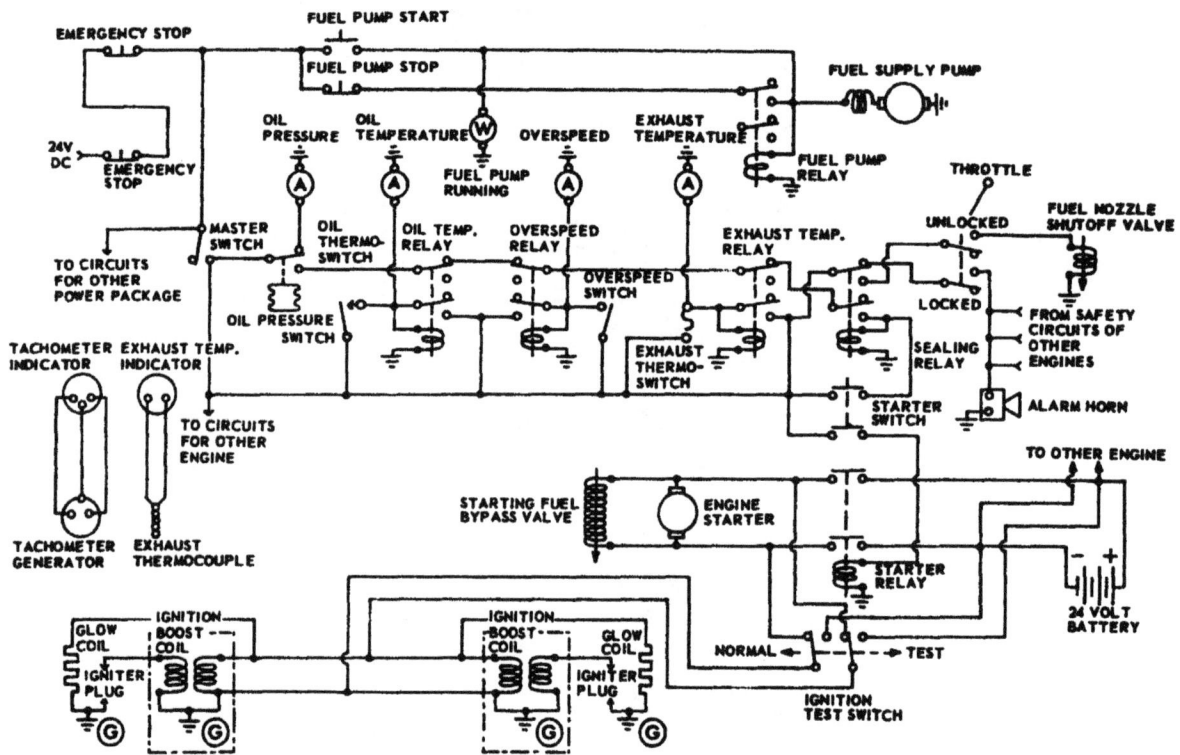

NOTES

GROUND SYMBOLS INDICATE
A COMMON RETURN WIRE
EXCEPT POINTS ⓖ WHICH
ARE ACTUALLY GROUNDED
TO TURBINE.
CIRCUITS SHOWN WITH EN-
GINE NOT RUNNING.

75.275

Figure 23-19.—Electrical system for gas turbine engine.

The starting and ignition circuits receive power from storage batteries. The warning and safety circuits receive power from the ship's power supply panel. Power for the indicating circuits is self generated by thermocouples and other units in the circuits.

Engine Starters

Of the various methods used for starting gas turbine engines, the three most common devices are the air turbine, the hydraulic starting device, and the electric starter-generator.

The air turbine starter is a turbine-air motor with a radial inward-flow turbine wheel assembly and an engaging and disengaging mechanism. Compressed air is supplied to the air turbine from an external source.

The hydraulic motor starter consists of a motor-driven hydraulic pump mounted separately. It supplies high pressure hydraulic oil to the hydraulic motor starter, which is mounted on the accessory pad along with its engaging and disengaging mechanism. The hydraulic motor starter is quite similar to the air turbine starter; however, the hydraulic motor starter is usually used for larger and higher horsepower gas turbine engines.

The electric starter-generator is a shunt-wound d-c generator with compensating windings and a series winding, using a 24-volt battery power source. The generator is usually mounted on the accessory drive pad. The generator is so designed and controlled that it can be used as an engine starter. When the designed engine speed is reached, the starter-generator is automatically switched from a starter to a generator.

TRANSMISSION OF ENGINE POWER

The two main types of gas turbine engine installations used for ship propulsion are (1) the geared drive, and (2) the turboelectric drive.

The fundamental characteristics of the gas turbine engine make it necessary for the drive mechanism to change both the speed and the direction of shaft rotation in the driven mechanism. The process of transmitting engine power to a point where it can be used in performing useful work involves a number of factors, two of which are torque and speed. The gas turbine engine does not produce high torque, but it does produce high speed. Therefore, a gear train is used with most gas turbine engines to lower

speed and increase torque. This is true in both types of installations. In the case of the geared drive installation, the gears are used between the gas turbine engine and the propeller shaft. In the case of the turboelectric drive, the gears are usually used between the gas turbine engine and the generator shaft, to reduce the rpm of the generator to a practicable operating value.

The propelling equipment of a boat or ship must be capable of providing reversing power as well as forward power. In a few ships and boats, reversing is accomplished by the use of controllable pitch propellers.[5] In most vessels, however, reversing is accomplished by the use of reversing gears.

Reducing the speed of rotation and reversing the direction of shaft rotation are not the only requirements of the drive mechanism of a ship or boat. It is also necessary to make some provision for the fact that the engine must be able to operate at times without transmitting power to the propeller shaft. In the electric drive, this is no problem because the transmission of power is controlled electrically. With the gear type of drive, however, it is necessary to include a means of disconnecting the engine from the propeller shaft. Devices used for this purpose are called <u>clutches</u>.

The arrangement of components in a gear-type drive varies, depending upon the type and size of the installation. In some of the small installations, the clutch, the reversing gear, and the reduction gear may be combined in a single unit. This type of arrangement is shown in figure 23-20. In other installations, the clutch and the reversing gear may be in one housing and the reduction gear in a separate housing attached to the reversing gear housing. Drive mechanisms arranged in either manner are called <u>transmissions</u>.

GAS TURBINE ENGINES AND
JET PROPULSION

Thus far, we have considered the gas turbine engine as a prime mover which delivers power in the form of torque on an output shaft. In concluding this chapter, it should be noted that the gas turbine engine also serves as the prime

---

[5]Controllable pitch propellers are discussed in chapter 5 of this text.

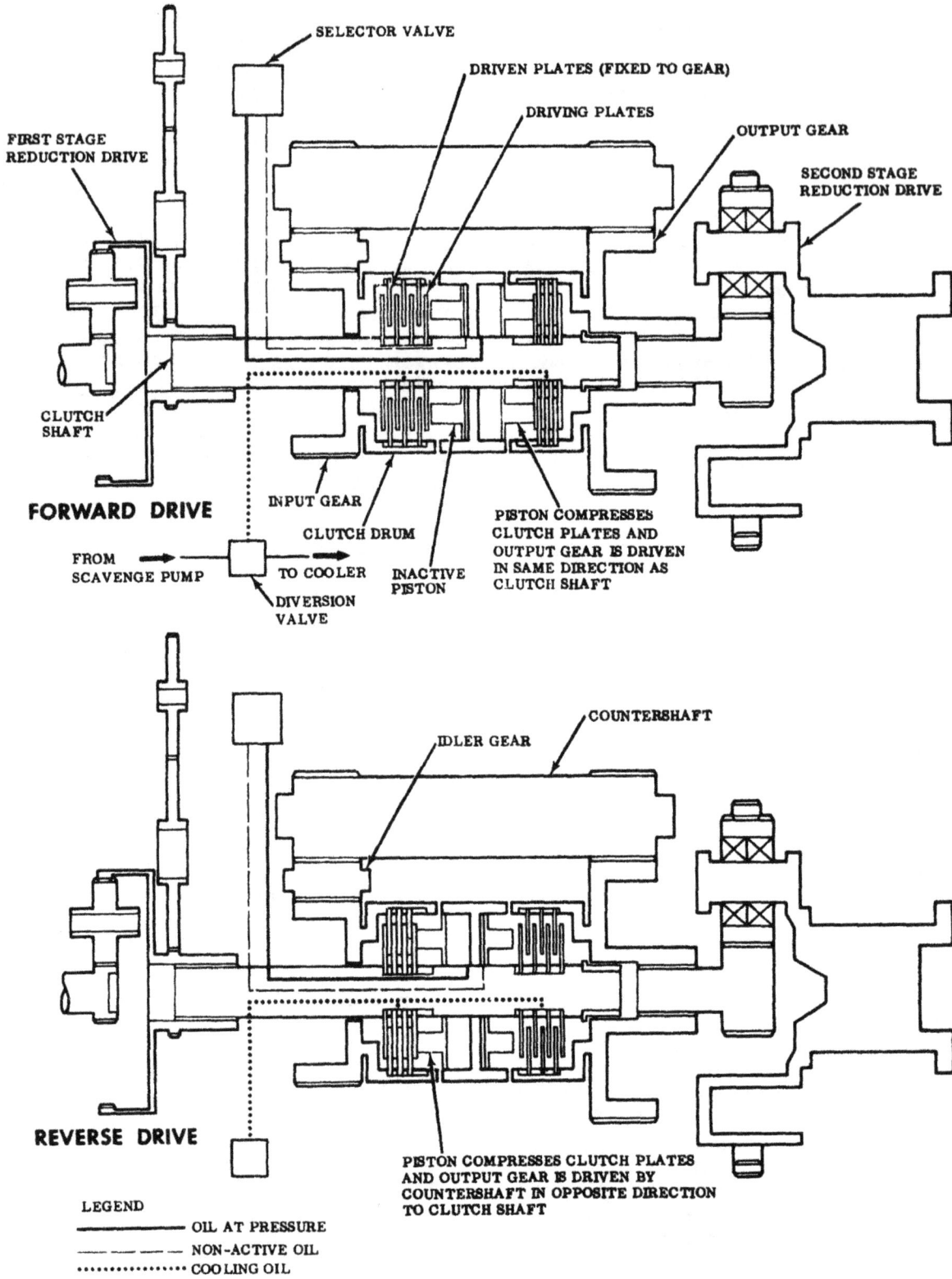

SELECTOR VALVE

DRIVEN PLATES (FIXED TO GEAR)

DRIVING PLATES

OUTPUT GEAR

FIRST STAGE
REDUCTION DRIVE

SECOND STAGE
REDUCTION DRIVE

CLUTCH
SHAFT

FORWARD DRIVE

INPUT GEAR

CLUTCH DRUM

INACTIVE
PISTON

PISTON COMPRESSES
CLUTCH PLATES AND
OUTPUT GEAR IS DRIVEN
IN SAME DIRECTION AS
CLUTCH SHAFT

FROM
SCAVENGE PUMP

TO COOLER

DIVERSION
VALVE

IDLER GEAR

COUNTERSHAFT

REVERSE DRIVE

PISTON COMPRESSES CLUTCH PLATES
AND OUTPUT GEAR IS DRIVEN BY
COUNTERSHAFT IN OPPOSITE DIRECTION
TO CLUTCH SHAFT

LEGEND

———————— OIL AT PRESSURE

—— — —— — NON-ACTIVE OIL

·················· COOLING OIL

147.153

Figure 23-20.—Clutch, reduction gear, and reverse gear arrangement.

mover in the power plants of many military aircraft. When so adapted, the gas turbine engine develops power by converting thermal energy into mechanical kinetic energy in a high velocity gas stream. The highly accelerated gas stream creates <u>thrust</u> which propels the aircraft. This method of creating thrust is called the <u>direct reaction</u> or <u>jet propulsion</u> method.

The concept of thrust is basic to an understanding of jet propulsion. The concept of thrust is based on Newton's third law of motion, which may be stated as follows: <u>For every acting force there is an equal and opposite reacting force.</u> In the case of aircraft in flight, the acting force is the force the engine exerts on the air mass as it flows through the engine. The reacting force (thrust) is the force which the air mass exerts on the components of the engine as the heated air mass is discharged from the jet nozzle at the rear of the airplane. In other words, thrust is not produced by the ejected air mass reacting against the atmosphere; rather, thrust is created within the engine as the air mass flowing through the engine is accelerated and discharged.

Engines which include the gas turbine and which create thrust by the direct reaction method are commonly identified as <u>turbojet engines</u>. Except for a diffuser and a <u>different</u> type exhaust system in engines of the turbojet type, the basic components of the turbojet engine are similar in design and function to the components of any open-cycle gas turbine engine. The function of the diffuser is to decrease the velocity of the inlet air and to increase its pressure before the air enters the compressor. The exhaust system of a turbojet engine consists basically of a cone and a convergent nozzle. The exhaust cone is designed to exhaust to the nozzle the accelerated air mass which the other components of the engine deliver to the cone. As the accelerated air mass flows through the convergent nozzle, its velocity is greatly increased and thrust is created within the engine.

MAINTENANCE

The maintenance of gas turbines is the normal function of operating activities. Cleanliness is one of the most important basic essentials in operation and maintenance of gas turbines. Particular care should be exercised in keeping fuel, air, coolants, lubricants, rotating elements, and combustion chambers clean. Periodic inspection procedures should be followed in order to detect maladjustments, possible failures, and excessive clearances of moving parts. All inspection and maintenance requirements should be accomplished in accordance with the 3-M System (PMS Subsystem).

<u>CAUTION</u>: Never use lead pencils for marking gas turbine hot parts, because the carbon content of the pencil lead will cause stainless steel to become brittle, causing a possible failure of the parts that were marked. A grease pencil should be used in marking gas turbine parts. Do not use steel wool to clean gas turbine parts, unless the wool is stainless steel.

Overhaul periods and procedures are set up by the Naval Ship Systems Command. These periods and procedures are reported to the Fleet through NavShips Technical Manuals and/or direct correspondence. Accurate operating logs should be kept on each engine so the number of hours and operational history on each engine is readily known. These records aid in developing measures which improve engine reliability.

CHAPTER 24

NUCLEAR POWER PLANTS

Nuclear reactors release nuclear energy by the fission process and transform this energy into thermal energy. While we are learning more daily about the phenomena which occur in nuclear reactions, the knowledge already gained has been put to use in both the submarine and the surface fleets. The Navy is now in the second decade of the utilization of nuclear energy for ship propulsion.

Nuclear engineering is a field that is in the stage of rapid development at the present time; therefore the discussion in this chapter is limited to the basic concepts to reactor principles. The discussion of nuclear physics is limited to the fission process, since all power reactors in operation at this time use the fissioning of a heavy element to release nuclear energy.[1]

ADVANTAGES OF NUCLEAR POWER

A major advantage of nuclear power for any naval ship is that less logistic support is required. On ships using conventional petroleum fuels as an energy source, the cruising range and strategic value are limited by the amount of fuel which can be stored in their hulls. A ship of this type must either return to port to take on fuel or refuel for a tanker at sea—an operation which is time consuming and hazardous.

Nuclear-powered ships have virtually unlimited cruising range, since the refueling is done routinely as part of a regular scheduled overhaul. On her first nuclear fuel load, the USS Nautilus steamed 62,562 miles, more than half of this distance fully submerged. The USS Enterprise steamed over 200,000 miles before

being refueled. In 1963, Operation Sea Orbit, a 30,000-mile cruise around the world in 65 days, completely without logistic support of any kind, proved conclusively the strategic and tactical flexibility of a nuclear-powered task force.

There are other (and perhaps less obvious) advantages of nuclear power for aircraft carriers. For one thing, tanks that would otherwise be used to store boiler fuels can be used on nuclear-powered carriers to store additional aircraft fuels, thus giving the ship a greater striking potential. Another advantage is the lack of stacks; since there are no stack gases to cause turbulence in the flight deck atmosphere, the operation of aircraft is less hazardous than on conventionally powered ships.

The fact that a nuclear-powered ship requires no outside source of oxygen from the earth's atmosphere means that the ship can be completely closed off, thereby reducing the hazards of any nuclear attack. This greatly increases the potential of the submarine fleet by giving it the capability of staying submerged for extended periods of time. In 1960 the nuclear-powered submarine USS Triton completed a submerged circumnavigation of the world, traveling a distance of 35,979 miles in 83 days and 10 hours.

NUCLEAR FUNDAMENTALS

At the present time there are 103 known elements of which the smallest particle that can be separated by chemical means is the atom. The Rutherford-Bohr theory of atomic structure (fig. 24-1) describes the atom as being similar to our solar system. At the center of every atom is a nucleus which is comparable with the sun; moving in orbits around the nucleus are a number of particles called electrons. The electrons have a negative charge and are held in orbit by the attraction of the positively charged nucleus.

---

[1] For a discussion of nuclear fusion, see John F. Hogerton, The Atomic Energy Deskbook (New York: Reinhold Publishing Corp., 1963), p. 196.

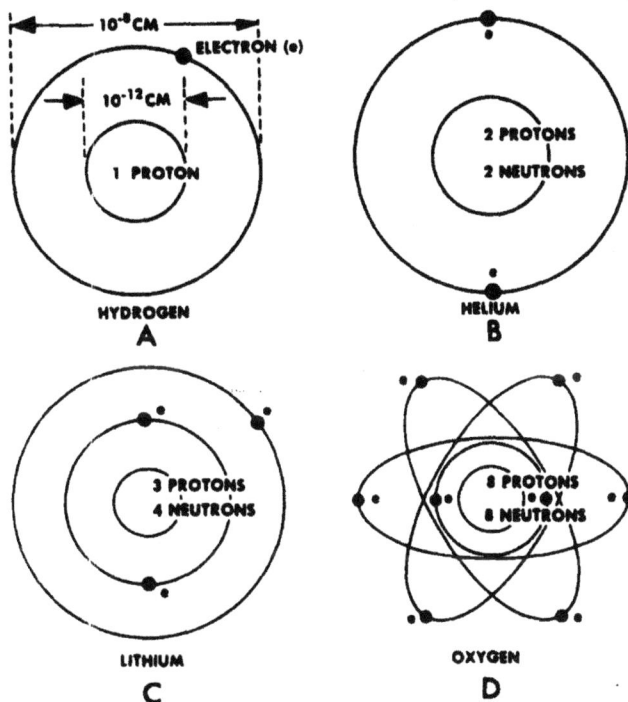

41.2

Figure 24-1.—Rutherford-Bohr models
of simple atoms.

Two elementary particles, protons and neutrons, often referred to as nucleons, compose the
atomic nucleus. The positive charge of atomic
nuclei is attributed to the protons. A proton has
an electrical charge equal and opposite to that of
an electron. A neutron has no charge.

The number of electrons in an atom and
their relative orbital positions predict how an
element will react chemically, whereas the
number of protons in an atom determines which
element it is. An atom which is not ionized
contains an equal number of protons and electrons; thus it is said to be neutral, since the
total atomic charge is zero.

As shown in part A of figure 24-1, the
hydrogen atom has a single proton in the nucleus
and a single orbital electron. Hydrogen, the
lightest element, is said to have a mass of
approximately one. The next heavier atom,
that of helium (part B of fig. 24-1), had a mass
of four relative to hydrogen and was expected
to contain four protons. It was found that the
helium atom has only two protons instead of the
four expected; the remainder of its mass is
attributed to two neutrons located in the nucleus
of the helium atom. The more complex atoms
contain more protons and neutrons in the

Particle	Charge	Mass (amu)
Proton	+1	1. 00758
Neutron	0	1. 00894
Electron	−1	0. 00055

147.154

Figure 24-2.—Characteristics of elementary
atomic particles.

nucleus, with a corresponding increase in the
number of planetary electrons. The planetary
electrons are arranged in orbits or shells of
definite energy levels outside the nucleus.

The characteristics of the elementary atomic
particles are compiled in figure 24-2. Note
that the mass of a proton is much greater than
that of an electron; it takes about 1847 electrons
to weigh as much as one hydrogen proton.

It is possible for atoms of the same element
to have different numbers of neutrons, and
therefore different masses. Atoms which have
the same atomic number (number of protons in
the atom) but different masses are called
isotopes. Different isotopes of the same element
are identified by the atomic mass number,
which is the total number of neutrons and
protons contained within the nucleus of the
atom.

The element hydrogen has three known
isotopes, as shown in figure 24-3. The simplest
and most common known form of hydrogen
consists of 1 proton, which is the nucleus,
and 1 orbital electron. Another form of hydrogen, deuterium, consists of 1 proton and 1
neutron forming the nucleus. The third
form, tritium, consists of 1 proton and 2
neutrons forming the nucleus and 1 orbital
electron.

In scientific notation, the three isotopes
of hydrogen are written as follows:

Common hydrogen . . . . $_1H^1$

Deuterium . . . . $_1H^2$

Tritium . . . . $_1H^3$

In this notation, the subscript preceding
the symbol of the element indicates the atomic
number of the element. The superscript following

615

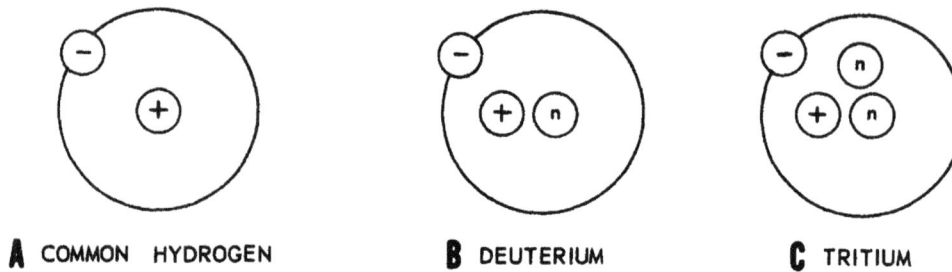

A COMMON HYDROGEN **B** DEUTERIUM **C** TRITIUM

5.40(147A)

Figure 24-3.—Isotopes of hydrogen.

the symbol of the element is the atomic mass number; thus the superscript indicates which isotope of the element is being referred to.

The geneal symbol for any atom is thus

$$_Z X ^A$$

where

    X symbol of the element
    Z atomic number (number of protons)
    A atomic mass number (sum of the number of protons and the number of neutrons)

Of the known 103 elements, there are approximately 1000 isotopes, most of which are radioactive.[2] Figure 24-4 gives the nuclear compositon of various isotopes.

RADIOACTIVITY

All isotopes with atomic number Z greater than 83 are naturally radioactive and many more isotopes can be made artifically radioactive by bombarding with neutrons which upset the neutron-proton ratio of the normally stable nucleus.

Naturally radioactive isotopes undergo radioactive decomposition, thereby forming lighter and more stable nuclei. Radioactive decomposition occurs through the emission of an alpha particle or a beta particle. One or more gamma rays may also be emitted with the alpha or beta particle.

An alpha particle (symbol α) is composed of two protons and two neutrons. It is the nucleus of a helium ($_2He^4$) atom, has an electrical charge of +2, and is very stable. In the decay process to a more stable element,

Element	Symbol	Atomic No. or No. of protons	No. of electrons	No. of neutrons
Hydrogen	$_1H^1$	1	1	0
Hydrogen (Deuterium)	$_1H^2$	1	1	1
Hydrogen (Tritium)	$_1H^3$	1	1	2
Helium	$_2He^3$	2	2	1
Helium	$_2He^4$	2	2	2
Helium	$_2He^5$	2	2	3
Helium	$_2He^6$	2	2	4
Beryllium	$_4Be^9$	4	4	5
Cadmium	$_{48}Cd^{113}$	48	48	65
Polonium	$_{84}Po^{210}$	84	84	126
Radium	$_{88}Ra^{226}$	88	88	138
Uranium	$_{92}U^{234}$	92	92	142
Uranium	$_{92}U^{235}$	92	92	143
Uranium	$_{92}U^{238}$	92	92	146
Uranium	$_{92}U^{239}$	92	92	147
Neptunium	$_{93}Np^{239}$	93	93	146
Plutonium	$_{94}Pu^{239}$	94	94	145

5.36

Figure 24-4.—Nuclear composition of various isotopes.

[2]For a detailed discussion of nuclear stability, see Francis W. Sears and Mark W. Zemansky, University Physics (3d ed.; Reading, Mass.: Addison-Wesley Publishing Company, Inc., 1964), p. 997.

many unstable nuclei emit an alpha particle. The results of alpha emission can be seen from the following equation:

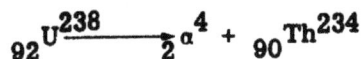

$$_{92}U^{238} \xrightarrow{} {}_2\alpha^4 + {}_{90}Th^{234}$$

In the above equation, the parent isotope of uranium ($_{92}U^{238}$) is a naturally occurring, radioactive isotope which decays by alpha emission. Since the A and Z numbers must balance in a nuclear equation, and since an alpha particle contains two protons, we see that the uranium has changed to an entirely new element.

The radioactive isotope of thorium ($_{90}Th^{234}$) produced in the above reaction further decays by the emission of a beta particle symbol β) as indicated in the following equation:

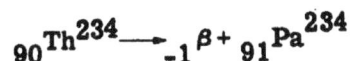

$$_{90}Th^{234} \longrightarrow {}_{-1}\beta + {}_{91}Pa^{234}$$

The beta particle has properties similar to an electron.[3] However, the origin of the beta particle is within the nucleus rather than the orbital shells of an atom. It is postulated that a beta particle is emitted at an extremely high energy level when a neutron within the nucleus decays to a proton and an electron (beta particle). When this phenomenon occurs, the proton stays within the nucleus forming an isotope of a different element having the same mass.

A radioactive isotope may go through several transformations of the above types before reaching a stable state. In the case of $_{92}U^{238}$ there are a total of eight alpha particles and six beta particles emitted prior to reaching a stable isotope of lead ($_{82}Pb^{206}$).

The third manner in which a naturally radioactive isotope may reach a more stable configuration is by the emission of gamma rays (symbol γ). The gamma ray is an electromagnetic type of radiation having frequency,

high energy, and a short wave length. Gamma rays are similar to X-rays in that the properties are the same. The distinguishing factor between the two is the fact that gamma rays are originated in the nucleus of an atom, whereas the X-ray originates from the orbital electrons. In general it can be said that a gamma ray is of higher energy, higher frequency, and shorter wave length than an X-ray.

Frequently an isotope which emits an alpha or beta particle in the decay process will emit one or more gamma rays at the same time, as in the case of $_{27}Co^{60}$, an isotope that decays by beta emission and at the same time emits two gamma rays of different energy levels. Some radioactive isotopes reach a stable state by the emission of gamma rays only. In the latter case, since gamma rays have neither mass nor electrical charge, the A and Z numbers of the isotope remain unchanged but the energy level of the nucleus is reduced.

An important property of any radioactive isotope is the time involved in radioactive decay. To understand the time element, it is necessary to understand the concept of half-life. Half-life may be defined as the time required for one-half of any given number of radioactive atoms to disintegrate, thus reducing the radiation intensity of that particular isotope by one-half. Half lives may vary from microseconds to billions of years. At times an isotope may be said to be "short-lived" or "long-lived", depending upon its peculiar radio-active half-life. Some half-lives of typical elements are:

$$_{92}U^{238} = 4.51 \times 10^9 \text{ years}$$

$$_{92}U^{235} = 7.13 \times 10^8 \text{ years}$$

$$_{88}Ra^{226} = 1620 \text{ years}$$

$$_{53}I^{135} = 6.7 \text{ hours}$$

$$_{84}Po^{214} = 10^{-6} \text{ seconds}$$

---

[3] Francis W. Sears and Mark W. Zemansky, University Physics (3d ed.; Reading, Mass.: Addison-Wesley Publishing Company, Inc., 1964), p.986.

As stated previously, naturally radioactive isotopes decay by the emission of alpha particles, beta particles, gamma rays, or a combination thereof. In the case of induced nuclear reactions there are many other phenomena which may occur, including fission and the emission of neutrons, positrons, nutrions, and other forms of energy.[4]

CONSERVATION OF MASS AND ENERGY

The conservation of energy is discussed in chapter 8 of this text. It now becomes necessary to consider mass and energy as two phases of the same principle. In so doing, the law of conservation becomes:

(mass + energy) before =
(mass + energy) after.

Fundamental to the above and to the entire subject of nuclear power is Einstein's mass-energy equation where the following relation holds:

$$E = mc^2$$

where

E = energy in ergs,
M = mass in grams,
C = velocity of light (3 x 10^{10} cm/sec)

Mass and energy are not conserved separately but can be converted into each other.

Several units and conversion factors which have become conventional to the field of nuclear engineering are listed below.

1 ev (electron-volt = the energy acquired by an electron as it moves through a potential difference of 1 volt

1 Mev (million electron-volts)   = 10^6 ev
            = 1.52 x 10^{-16} Btu

---

[4]For detailed information on nuclear particles, refer to Samuel Glasstone, Sourcebook on Atomic Energy (2d ed.; Princeton: D. Van Nostrand Company, Inc., 1958).

1 amu (atomic mass unit = 1/16 of an oxygen atom (by definition)

1 amu = 1.49 x 10^{-3} erg
       = 1.66 x 10^{-24} gm
       = 931 Mev
       = 1.415 x 10^{-13} Btu

NUCLEAR ENERGY SOURCE

It was previously stated that the atomic mass number is the total number of nucleons within the nucleus. It can also be said that the atomic mass number is the nearest integer (as found by experiment) to the actual mass of an isotope. In nuclear equations, the entire mass must be accounted for; therefore the actual mass must be considered.

The atomic mass of any isotope is somewhat less than indicated by the sum of the individual masses of the protons, neutrons, and orbital electrons which are the components of that isotope. This difference is termed mass defect: it is equivalent to the binding energy of the nucleus. Binding energy may be defined as the amount of energy which was released when a nucleus was formed from its component parts.

The binding energy of any isotope may be found, as in the following example of copper ($_{29}Cu^{63}$) which contains 34 neutrons, 29 protons, and 29 electrons. Using the values given in figure 24-2 we find:

34 x 1.00894 = 34.30496 amu
29 x 1.00785 = 29.21982 amu
29 x 0.00055 =   0.01595 amu
Total of component masses = 63.54073 amu
Less actual mass of atom   = 62.9298 amu
Mass defect   =   0.61093 amu

Converting to energy, we find:

931 Mev/amu x 0.61093 amu = 568.77583 Mev, or 568.8 : 63 = 8.9 Mev/nucleon

The relationship between mass number and the average binding energy per nucleon is shown in figure 24-5.

Since binding energy was released when a nucleus was formed from its component parts, it is necessary to add energy to separate a nucleus. In the fissioning of uranium 235, the additional energy is supplied by bombarding the fissionable fuel with neutrons. The fissionable material absorbs a neutron and is converted

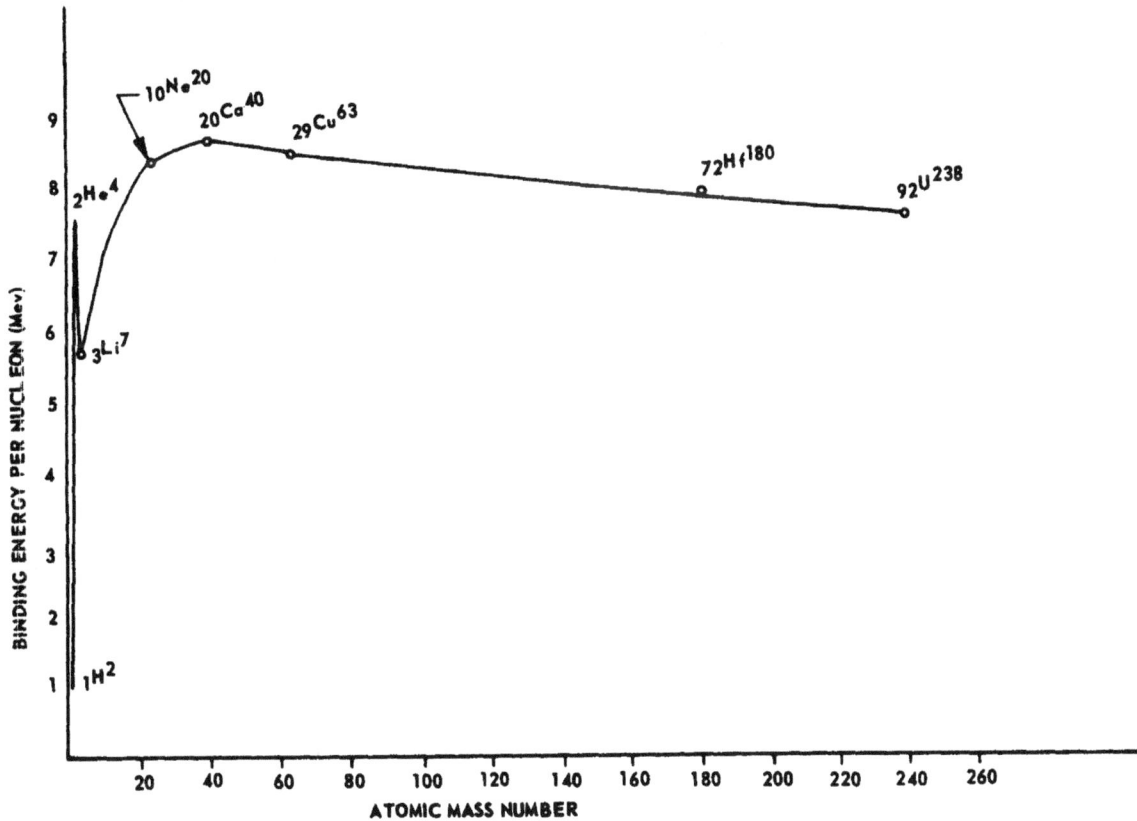

147.155

Figure 25-5.—Relationship between atomic mass number and
average binding energy per nucleon.

into the compound nucleus of uranium-236, which fissions instantaneously.

There are more than 40 different ways a uranium-235 nuclei may fission, resulting in more than 80 different fission products.[5] For the purpose of this discussion, let us consider the most probable fission of a uranium-235 nucleus. In slightly more than 6 percent of the fissions, the uranium-235 nucleus will split into fragments having mass numbers of 95 and 139. The following equation is typical:

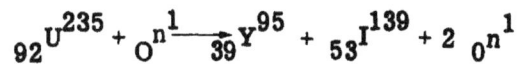

$$_{92}U^{235} + _0n^1 \longrightarrow _{39}Y^{95} + _{53}I^{139} + 2 \, _0n^1$$

where the daughter products, yttrium and iodine, are toh radioactive and decay through beta emission to the stable isotopes of molybdenum ($_{42}Mo^{95}$) and lanthanum ($_{57}La^{139}$), respectively.

One method of determining the energy released from the above reaction is to find the difference in atomic mass units of the daughter products and the original nucleus. It is also necessary that we account for the neutron used to bombard the uranium-235 atom and the two neutrons liberated in the fission process. In

[5] For a detailed discussion on nuclear fission, refer to Samuel Glasstone, Sourcebook on Atomic Energy (2d ed.; Princeton: D. Van Nostrand Company, Inc., 1958).

the investigation of energy released in this reaction we find:

Mass of uranium-235 atom = 235.0439
Mass of neutron = 1.00894
Original mass = 236.05284 amu
Mass of molybdenum-95
    atom = 94.9058
Mass of lanthanum-139
    atom = 138.9061
Mass of 2 neutrons = 2.01788
Total mass of fission
    fragments = 235.82978
Mass defect = 236.05284 - 235.82978 = 0.22306 amu/fission

Hence,

0.22306 amu/fission x 931 Mev/amu
207.7 Mev/fission

Thus we find that from each fission approximately 200 Mev of energy is released, most of which (about 80 percent) appears immediately as kinetic energy of the fission fragments. As the fission fragments slow down, they collide with other atoms and molecules; this results in a transfer of velocity to the surrounding particles. The increased molecular motion is manifested as sensible heat. The remaining energy is realized from the decay of fission fragments by beta particle and gamma ray emission, kinetic energy of fission neutrons, and instantaneous gamma ray energy.

In a nuclear reactor, the two neutrons liberated in the above reaction are available, under certain conditions, to fission other uranium atoms and assist in maintaining the reactor critical. A nuclear reactor is said to be critical if the neutron flux remains constant. Neutron flux is defined as the number of neutrons passing through unit area in unit time. A neutron flux of 10^{13} neutrons per square centimeter per second is not uncommon. If the neutron flux is decreasing, the reactor is said to be subcritical; conversely, a reactor is supercritical if the neutron flux is increasing.

NEUTRON REACTIONS

Neutrons may be classified by their energy levels. A fast neutron has an energy level of greater than 0.1 Mev, an intermediate neutron in the process of slowing down possesses an energy level between 1 ev and 0.1 Mev, a thermal neutron is in thermal equilibrium with its surroundings and has an energy level of less than 1 ev.

Neutrons lose their kinetic energy by interacting with atoms in the surrounding area. The probability of a neutron interacting with one atom is dependent upon the target area presented by that atom for a neutron reaction. This target area (which is the probability of a neutron reaction occurring) is called cross section. The unit of cross section measurement is barns. The size of a barn is 10^{-24} square centimeters. Four of the different cross sections that an element may have for neutron processes are as follows:

Scattering cross section is a measure of the probability of an elastic (billiard ball) collison with a neutron. In this type of collision part of the kinetic energy of the neutron is imparted to the atom and the neutron rebounds after collision. Neutrons are thermalized (reduced to an energy level below 1 ev) by elastic collisions.

Capture cross section is a measure of the probability of the neutron being captured without causing fission.

Fission cross section is a measure of the probability of fission of the atom after neutron capture.

Absorption cross section is a measure of the probability that an atom will absorb a neutron. The absorption cross section is the sum of the capture cross section and the fission cross section.

The cross section for any given element may vary with the energy level of the approaching neutron. In the case of uranium-235, the absorption cross section for a thermal neutron is 100 times the cross section for a fast neutron.

REACTOR PRINCIPLES

A nuclear reactor must contain a critical mass. A critical mass contains sufficient fissionable material to enable the reactor to maintain a self-sustaining chain reaction, thereby keeping the reactor critical. A critical mass is dependent upon the species of fissionable material, its concentration and purity the geometry and size of the reactor, and the matter surrounding the fissionable material.[6]

[6]For a thorough discussion of the aspects of reactor design, see Samuel Glasstone, Sourcebook on Atomic Energy (2d ed.; Princeton: D. Van Nostrand Company, Inc., 1958).

REACTOR FUELS

The form and composition of a reactor fuel may vary both in design and in the fissionable isotope used. Many commercial power reactors use a solid fuel element fabricated in plate form, with the fissionable material being enriched uranium in combination with aluminum, zirconium, or stainless steel. Fuel elements may be arranged in thin sandwich layers, as shown in figure 24-6. This construction provides a relatively large heat transfer area between the fuel elements and the reactor coolant.

The outer cladding on the fuel elements confines the fission fragments within the fuel elements and serves as a heat transfer surface. Cladding materials should be resistant to corrosion, should be able to withstand high temperatures, and should have a small cross section for neutron capture. Three common cladding materials are aluminum, zirconium, and stainless steel. The fuel elements may be assembled in groups, some of which may contain control rods. Several groups of fuel elements placed within a reactor vessel make up the reactor core. It is not necessary that all fuel groups within the reactor contain control rods.

CONTROL RODS

Control rods serve a dual purpose in a reactor. They keep the neutron density (neutron flux) constant within a critical reactor and they provide a means of shutting down the reactor.

The material for a control rod must have a high capture cross section for neutrons and a low fission cross section. Three materials suitable for control rod fabrication are cadmium, boron, and hafnium. Hafnium is particularly suitable for control rods because it has a relatively high capture cross section and because several daughter products after neutron capture are stable isotopes which also have good capture cross sections.

The control rods are withdrawn from the reactor core until criticality is obtained; thereafter very little movement is required. It is important to note at this point that after criticality is reached, movement of control rods does not control the power output of the reactor; it controls only the temperature of the reactor.

Control rod drive mechanisms are so designed that, should an emergency shutdown of the reactor be required, the control rods may be inserted in the core very rapidly. A shutdown of this type is called a scram.

MODERATORS

A moderator is the material used to thermalize the neutrons in a reactor. As previously stated, neutrons are thermalized by elastic collisions; therefore, a good moderator must have a high scattering cross section and a low absorption cross section to reduce the speed of a neutron in a small number of collisions. Nuclei whose mass is close to that of a neutron are the most effective in slowing the neutron; therefore, atoms of low atomic weight generally make the best moderators. Materials which have been used as moderators include light and heavy water, graphite, and beryllium.

Ordinary light water makes a good moderator since the cost is low; however it must be free from impurities which may capture the neutrons or add to the radiological hazards.

REACTOR COOLANTS

The primary purpose of a reactor coolant is to absorb heat from the reactor. The coolant may be either a gas or a liquid; it must possess good heat transfer properties, have good thermal properties, be noncorrosive to the system, be nonhazardous if exposed to radiation, and be of low cost. Coolants which have been used in operational and experimental reactors include light and heavy water, liquid sodium, and carbon dioxide.

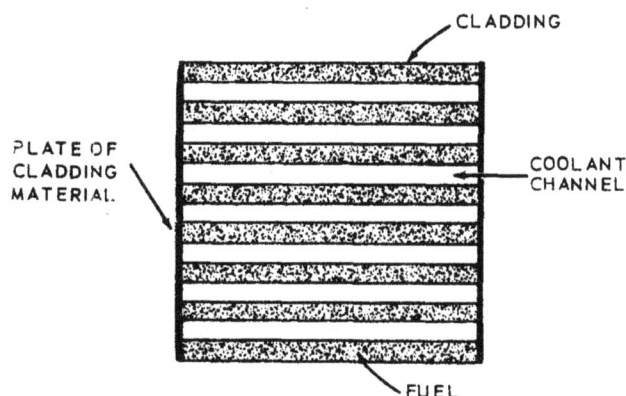

147.156

Figure 24-6.—PWR fuel element.

REFLECTORS

In a reactor of finite size, the leakage of neutrons from the core becomes somewhat of a problem. To minimize the leakage, a reflector is used to assist in keeping the neutrons in the reactor. The use of a reflector reduces both the required size of the reactor and the radiation hazards of escaping neutrons. The characteristics required for a reflector are essentially the same as those required for a moderator.

Since ordinary water of high purity is suitable for moderators, coolants, and reflectors, the inference is that it could serve all three functions in the same reactor. This is indeed the case in many nuclear reactors.

SHIELDING

The shielding of a nuclear reactor serves the dual purpose of (1) reducing the radiation so that it will not interfere with the necessary instrumentation, and (2) protecting operating personnel from radiation.

The type of shielding material used is dependent upon the purpose of the particular reactor and upon the nature of the radioactive particles being attenuated or absorbed.

The shielding against alpha particles is a relatively simple matter. Since an alpha particle has a positive electrical charge of 2, a few centimeters of air is all that is required for attenuation. Any light material such as aluminum or plastics makes a suitable shield for beta particles.

Neutrons and gamma rays have considerable penetrating power; therefore, shielding against them is more difficult. Since neutrons are best attenuated by elastic collisions, any hydrogenous material such as polyethylene or water is suitable as a neutron shield. Sometimes polyethylene with boron is used for neutron shields, as boron has a high neutron capture cross section. Gamma rays are best attenuated by a dense material such as lead.

NUCLEAR REACTORS

The purpose of any power reactor is to provide thermal energy which can be converted to useful work. Several types of experimental and operational reactors have been designed. They include the Pressurized Water Reactor (PWR), the Sodium Cooled Reactor, the Experimental Boiling Water Reactor, the Experimental Breeder Reactor, and the Experimental Gas Cooled Reactor.

The first full-scale nuclear-powered central station in the United States was the Pressurized Water Reactor (PWR) at Shippingport, Pennsylvania. The Shippingport PWR is a thermal, heterogeneous reactor fueled with enriched uranium-235 "seed assemblies" arranged in a square in the center of the core, surrounded by "blanket assemblies" of uranium-238 fuel elements. Figure 24-7 shows a cross-sectional view of the PWR reactor and core. This type of reactor can be called a converter, since the uranium-238 is converted into the fissionable fuel of plutonium-239.

A schematic diagram of PWR and its associated steam plant with power output and flow ratings is shown in figure 24-8. The reactor plant consists of a single reactor with four main coolant loops; the plant is capable of maintaining full power on three loops. Each coolant loop contains a steam generator, a pump, and associated piping.

High purity water at a pressure of 2000 psia serves as both moderator and coolant for the plant. At full power the inlet water temperature to the reactor is 508° F and the outlet temperature is 542° F.

The coolant enters the bottom of the reactor vessel (fig. 24-9) where 90 percent of the water

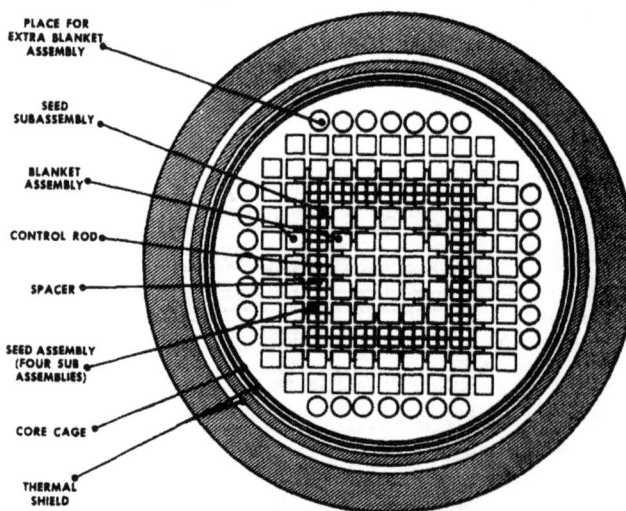

147.157X

Figure 24-7.—Cross-sectional view of PWR reactor and core.

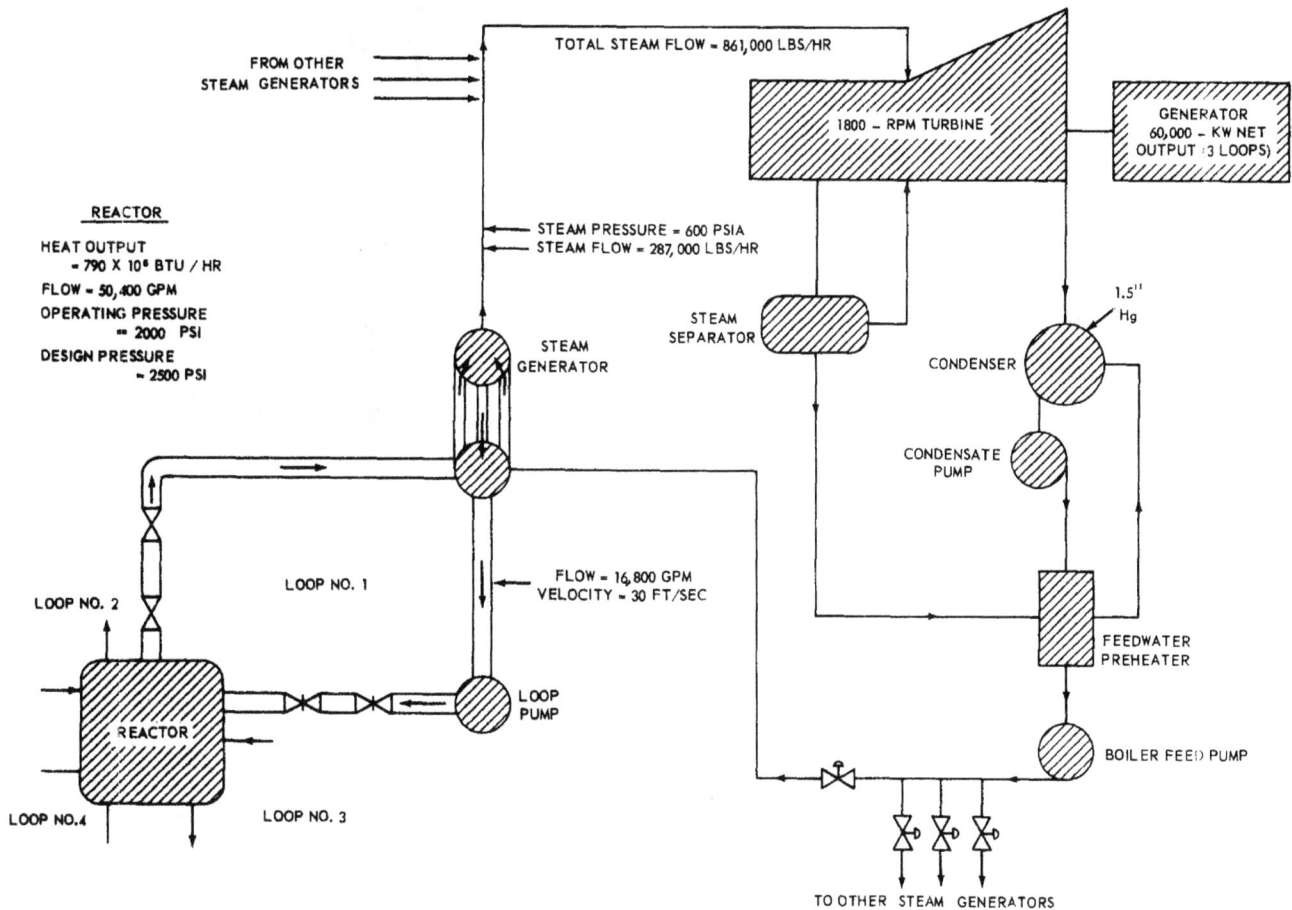

Figure 24-8.—Schematic diagram of PWR plant.

147.158X

flows upward between the fuel plates, with the remainder bypassing the core in order to cool the walls of the reactor vessel and the thermal shield. After having absorbed heat as it goes through the core, the water leaves the top of the reactor vessel through the outlet nozzles and flows through connecting piping to the steam generator.

The steam generator is a shell-and-tube type of heat exchanger with the primary coolant (reactor coolant) flowing through the tubes and the secondary water (boiler water) surrounding the tubes. Heat is transferred to the secondary water in the steam generator, producing high quality saturated steam for the use in the turbines.

The primary coolant flows from the steam generator to a hermetically sealed (canned rotor) pump (fig. 24-10) and is pumped through connecting piping to the bottom of the reactor vessel to complete the primary coolant cycle.

The pressure on the reactor vessel and the main coolant loop is maintained by a pressurizing tank (fig. 24-11) which operates under the saturation conditions of 636° F and 2000 psia. A second function of the pressurizing tank is to act as a surge tank for the primary system. Under no load conditions the inlet, outlet, and average temperatures of the reactor coolant

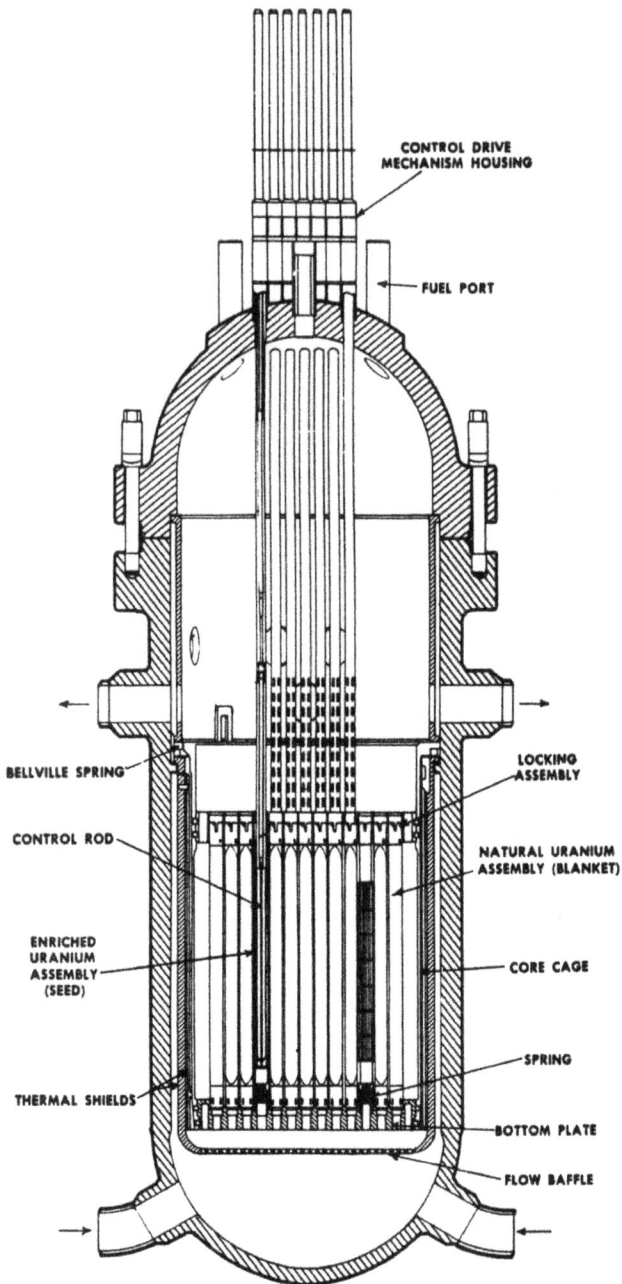

147.159X

Figure 24-9.—Longitudinal section of
PWR reactor.

147.160X

Figure 24-10.—PWR main coolant pump.
(A) External view. (B) Cutaway view.

are nearly equal in value. As the power increases, the average temperature remains constant but the inlet and outlet temperatures diverge. Since the colder leg of the primary coolant is the longer, the net effect in the pressurizer is a decrease in level to make up for the increase in density of the water in the primary loop. The reverse holds true with a decreasing power level. Electrical heaters and a spray valve with a supply of water from the cold leg of the primary coolant assist in maintaining a steam blanket in the upper part of the pressurizer and also assist in maintaining saturation conditions of 2000 psia and 636° F.

PRINCIPLES OF REACTOR CONTROL

Reactor control principles[7] which are of particular interest to this discussion include the negative temperature coefficient, the delayed neutron action, and the poisoning of fuel.

The term "negative temperature coefficient" is used to express the relationship between temperature and reactivity—as the temperature decreases, the reactivity increases. The negative temperature coefficient is a design requirement and is achieved by the proper ratio of elements in the reactor, the geometry of the reactor, and the physical size of the reactor. The negative temperature coefficient makes it possible to keep a power reactor critical with minimum movement of the control rods.

The concept of negative temperature coefficient may be most easily understood by use of an example. Assume that, in the PWR plant shown in figure 24-8, the reactor is critical and the machinery is operating at a given power level. Now, if the valve is opened to increase the turbine speed, the rate of steam flow, and the power level of the reactor, the measurable effect with installed instrumentation is a decrease in the temperature of the primary coolant leaving the steam generator. The decrease in temperature is small but significant in that it results in an increase in density of the coolant. As the density of the

coolant increases, so does the magnitude of the neutron scattering cross section. The higher value of the scattering cross section allows the coolant, in its capacity as moderator, to thermalize neutrons at faster rate, supplying more

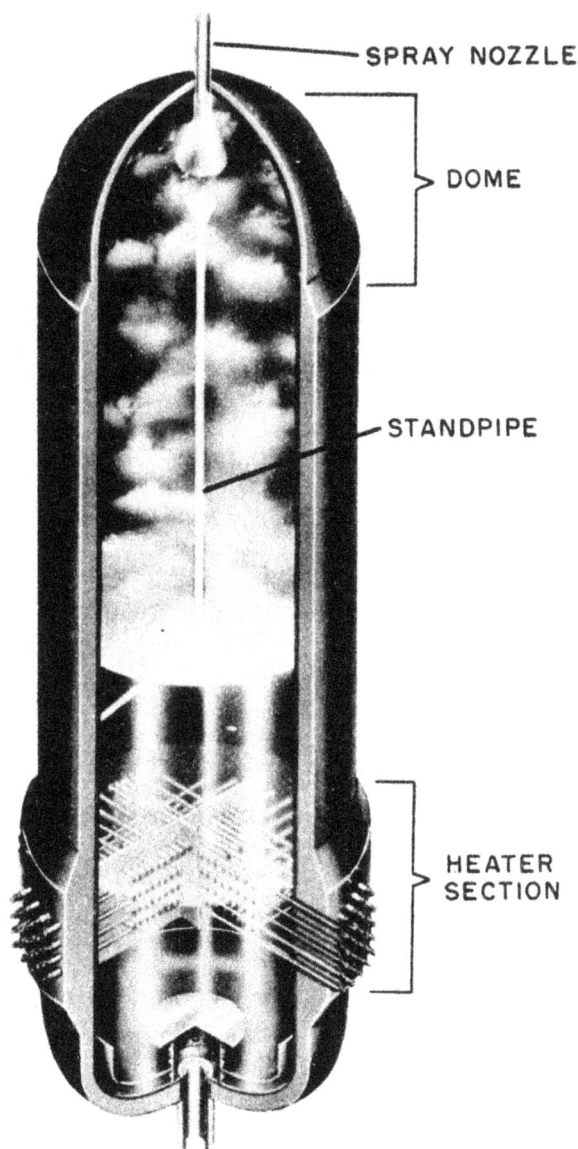

147.161X
Figure 24-11.—Cutaway view of PWR pressurizing tank.

---

[7]John F. Hogerton, The Atomic Energy Deskbook (New York: Reinhold Publishing Corp., 1963), p. 463.

thermal neutrons to be absorbed in the fuel. As more neutrons are absorbed in the fuel, more fissions occur, resulting in a higher power level and more heat being generated by the reactor. The additional heat is removed by the reactor coolant to the secondary water in the steam generator to compensate for the increased steam demand by the turbine. The temperature of the primary coolant leaving the steam generator increases slightly, lowering the scattering cross section of the moderator, and the reactor settles out at a higher power level.

The delayed neutron action is a phenomenon that simplifies reactor control considerably. Each fission in a nuclear reactor releases on the average between two and three neutrons which either leak out of the reactor or are absorbed in reactor materials. If the reactor material which absorbs the neutron happens to be the fissionable fuel, and if the neutron is of proper energy level, another fission is likely to result. The majority of the neutrons released in the fission process appear instantaneously and are termed prompt neutrons; but other neutrons are born after fission and are termed delayed neutrons. The delayed neutrons appear in a time range of seconds to 3 or more minutes after the fission takes place. The weighted mean lifetime of the delayed neutrons is approximately 12 seconds. About 0.75 percent of the neutrons produced in the fission process are delayed neutrons.

Should a reactor become prompt critical (critical on prompt neutrons), it would be very difficult to control and any delayed neutrons would tend to make it supercritical. The delayed neutrons have the effect of increasing the reactor period sufficiently to permit reactor control. Reactor period is the time required to change the power level by a factor of e (the base of the system of natural logarithms).

A nuclear poison is material in the reactor that has a high absorption cross section for neutrons. Some poisons are classed as burnable poisons and are placed in the reactor for the purpose of extending the core life; other poisons are generated in the fission process and have a tendency to be a hindrance to reactor operation.

A burnable poison has a relatively high cross section for neutron absorption but is used up in the early part of the core life. By adding a burnable poison to the reactor, more fuel can be loaded into the core, thus extending the life of the core.

Most of the fission products produced in a reactor have a small absorption cross section. The most important one that does have a high absorption cross section for neutrons is xenon-135; this can become a problem near the end of core life. Xenon-135 is a direct fission product a small percentage of the time but is mostly produced in the decay of iodine-135 as indicated in the following decay chain:

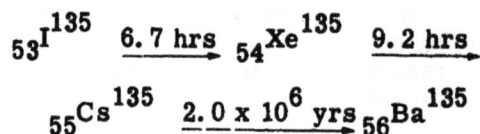

$$_{53}I^{135} \xrightarrow{\text{6. 7 hrs}} {}_{54}Xe^{135} \xrightarrow{\text{9. 2 hrs}}$$

$$_{55}Cs^{135} \xrightarrow{\text{2.0 x 10}^6 \text{ yrs}} {}_{56}Ba^{135}$$

Xenon-135 has a high neutron absorption cross section. In normal operation of the reactor, xenon-135 absorbs a neutron and is transformed to the stable isotope of xenon-136, which presents no poison problem to the reactor. Equilibrium xenon is reached after about 40 hours of steady-state operation. At this point the same amount of xenon-135 is being "burned" by neutron absorption as is being produced by the fission process.

The second, and perhaps the more serious, effect of xenon poisoning occurs near the end of core life. As indicated by the half-lives shown in the xenon decay chain, xenon-135 is produced at a faster rate than it decays. The buildup of xenon-135 in the reactor reaches a maximum about 11 hours after shutdown. Should a scram occur near the end of core life, the xenon buildup may make it impossible to take the reactor critical until the xenon has decayed off. In a situation of this type, the reactor may have to sit idle for as much as two days before it is capable of overriding the poison buildup.

THE NAVAL NUCLEAR POWER PLANT

Since many aspects of the design and operation of naval nuclear propulsion plants involve classified information, the information presented here is necessarily brief and general in nature.

In a nuclear power plant designed for ship propulsion, weight and space limitations and other factors must be taken into consideration in addition to the factors involved in the design of a shore-based power plant.

The thermodynamic cycle of the shipboard nuclear propulsion plant is similar to that of the conventional steam turbine propulsion plant.

Instead of a boiler, however, the nuclear propulsion plant utilizes a pressurized water reactor as the heat source and a steam generator as a heat exchanger to generate the steam used to drive the propulsion turbines.

The steam generator is a heat exchanger in which the primary coolant transfers heat to the secondary system (boiler water) by conduction. The water in the secondary side of the steam generator, being at lower pressure, changes from the physical state of water to the physical state of steam. This steam then flow through piping to the engineroom.

The engineroom equipment consists of propulsion turbines, turbogenerators, condensers, and associated auxiliaries.

PROBLEMS OF NUCLEAR POWER

Although many developmental and engineering problems associated with nuclear power have been solved to some extent, some problems remain. A few problems that are of particular importance in connection with the shipboard nuclear power plant are noted here briefly.

The remote possibility of radiological hazards exists even though the radiation is well contained in the shipboard nuclear reactor. To eliminate or minimize the radiological hazards, a high degree of quality control is essential in the design, construction, and operation of nuclear power plants. The high pressures and temperatures used in nuclear reactors, together with the prolonged periods of continuous operation, pose materials problems. For shipboard use, the great weight of the materials required for shielding presents still other problems.

Although many of these problems may be solved by further technological developments, the problems involved in the selection and training of personnel for nuclear ships appear to be continuing ones. The safe and efficient operation of a shipboard nuclear plant requires highly skilled, responsible personnel who have been thoroughly trained in both the academic and the practical aspects of nuclear propulsion. The selection and training of such personnel is inevitably costly in terms of time and money.

CHAPTER 25

NEW DEVELOPMENTS IN NAVAL ENGINEERING

New developments in naval engineering tend to be closely related to concepts of strategy. In some instances, new concepts of strategy may force the development of new engineering equipment to meet specific needs; in other instances, the development of a new source of power, a new engine, a new hull form, or a new propulsive device opens up new strategic possibilities. In any event, engineering capability is a major limiting factor in strategic planning, since it determines what is possible and what is not possible in the way of ship operation.

In previous chapters of this text, we have been concerned with naval engineering equipment currently installed in naval ships. But it would be unreasonable to assume that present achievements, impressive though they may be, are the last word in naval engineering. Practically everything in the Navy—policies, procedures, publications, systems, and equipment—is subject to rapid change and development, and naval engineering is certainly no exception. The rate of change in technological areas is increasing all the time. The officer who is just beginning his naval career may well, in the course of a few years, see more changes in naval engineering than have been seen in the past half century or more. And, difficult though it may be, every naval officer has a responsibility for keeping up with new developments.

Because of the increasingly rapid rate of technological development, it is no mean feat to keep abreast of changes in engineering equipment. In order to keep up with new developments in naval engineering, it is necessary to read widely in the literature of the field and to develop a special kind of alertness for information that may ultimately have an effect on naval engineering.

In the present chapter, we will depart from our previous framework of the here-and-now and mention briefly a few areas which, at present,

appear to offer some promise for future application in the field of naval engineering. With a few exceptions, the areas noted in this chapter are ones in which some actual work has been done or in which some serious thought has been given to naval engineering applications. It should be emphasized that this chapter is neither a complete survey of new developments nor a crystal-ball type reading of the future. Some of the areas mentioned here may turn out to have little or no application to naval engineering, while areas that are not even mentioned may suddenly come into prominence and importance.

HULL FORMS

Many new developments in naval engineering have been aimed, directly or indirectly, at increasing the speed of ships. One approach to this problem is to increase the size or change the nature of the propulsion machinery—a solution which, for various reasons, is not always feasible. Even when larger or better propulsion machinery is feasible, it is not always a total solution to the problem of increased speed, since at least some of the increased power thus provided is needed to overcome the increased resistance of the ship at the resulting higher speed. In the continuing search for ways to achieve higher operating speeds, therefore, a considerable amount of thought has been given to new hull forms which will reduce the resistance of the ship as it moves through the water.

A surface ship moving through the water is impeded by various resistances,[1] chiefly the frictional resistance of the water and the resistance that results from the generation of wave trains by the ship itself. Overcoming each of

---

[1] Fundamentals of ship resistance are discussed in chapter 5 of this text.

these resistances requires the expenditure of definite, calculable amounts of energy; hence each kind of resistance must be considered in connection with efforts to increase the speed of surface ships.

Among the interesting hull forms that have been developed (or at least considered) with a view to increasing surface ship speed by decreasing one or both of these resistances are the bulbous-bow form, the slender hull form, the semi-submarine form, the hydrofoil, and the various forms devised to utilize an air cushion or an air bubble. Each of these forms has specific advantages and disadvantages; the search is not for one perfect hull form but rather for a variety of hull forms suitable for a variety of functions.

Bulbous-Bow Forms

The bulbous-bow configuration is currently used on most large naval ships.[2] The theory of the bulbous-bow form is that the bulb will generate its own system of waves which will interfere with the systems of waves formed by the ship, thus reducing the resistance that results from the wave-making of the ship.

Although the bulbous-bow configuration is not so very new either in theory or in application, the concept of using much larger bulbs is a fairly recent development. In theory, the wave-making resistance of a ship could be substantially reduced by locating a large bulb just below the surface and just forward of the bow. In reality, there are enormous design difficulties involved, since the bulb must be specifically designed to interfere with complex wave trains generated by the ship.

Some idea of the complexity of the problem may be obtained by tossing pebbles into a pond and observing the waves that are formed. When two pebbles are tossed in together, each pebble generates its own systems of waves. Under some conditions, the systems of waves tend to cancel each other out; under other conditions, they tend to enhance or amplify each other; and under still other conditions, they interfere with each other in a chaotic, unpredictable manner which is essentially useless as far as achieving any cancellation of waves is concerned.

In spite of design difficulties, the use of large bulbs is of value in the design of certain types of ships. Tests of models have shown that a ship with a large bulb at the bow generates smaller and smoother systems of waves than a ship without such a bulb. However, this gain is not all free, since the bulb adds frictional resistance to the ship.

Slender Ships

A slender hull ship is basically similar to a destroyer except that, for any given displacement, the slender ship is about 30 percent longer than the destroyer. The extra length is designed to increase speed by decreasing the wave-making resistance of the ship. A slender ship has less static stability than a comparable destroyer, due to the narrower beam in relation to length. Also, the structural weight of the slender ship must be considerably greater than the structural weight of a comparable destroyer.

Some slender ships have been designed utilizing a large bulb at the bow and another one at the stern. Such ships are of interest because they have improved longitudinal stability characteristics as well as decreased wave-making resistance.

Semi-Submarine

The semi-submarine is still another approach to the problem of increasing speed by using a special hull form. The semi-submarine is shaped somewhat like a very streamlined submarine. The main hull of the semi-submarine runs submerged, while surface-piercing structures or fins (hydrofoils) at the stern increase the dynamic stability characteristics of the vessel and provide a means for handling engine air intake and exhaust. Because the semi-submarine runs submerged, except for the fins (hydrofoils), the craft avoids both storm waves and the self-generated wave-making resistance of surface ships.

Hydrofoils

The hydrofoil has been described as a cross between a high speed boat and an airplane. The craft has two modes of operation; it may run on the surface of the water, as a conventional surface ship, or it may fly on the foils with the hull clear of the water. When flying, the hydrofoil is supported clear of the water by the dynamic lift of the foils.

---

[2] See the discussion of stem and bow structure in chapter 2 of this text.

The primary advantages of the hydrofoil form are high speed and superior seakeeping abilities. The high speed is attainable because the resistances encountered by other ship forms are substantially reduced in the hydrofoil. When the hydrofoil is flying, the hull is entirely above the surface of the water and is therefore not impeded by frictional resistance or wave-making resistance and is not disturbed by waves, swells, or choppy surfaces that slow down other craft.

At present, the hydrofoil form is not considered feasible for very large vessels. However, the future development of propulsion plants with smaller specific weights[3] could well extend the tonnage range of the hydrofoil form. In fact, a "hydrofoil destroyer" has even been proposed.

Several types of hydrofoil systems have been developed. The Navy hydrofoil program has concentrated largely on fully submerged subsurface wings or foils. Control surfaces on the fully submerged foils act like aircraft ailerons and control the course of the craft through the water; the foils are controlled by a height-sensing system which maintains level flight. In another type of control (incidence control) the entire foil is moved instead of flaps. Other hydrofoil systems include skids or other planing devices and surface-piercing foils. Surface-piercing foils have a lifting area that is proportional to the amount of foil immersed.

A number of hull forms have been used for hydrofoils, but most of them are basically adaptations of conventional hull forms. The hull is relatively long and narrow, with the length of the craft being eight to ten times the beam. It has been suggested that the catamaran[4] design may have certain specific advantages for hydrofoils; however, this type of design is not currently used in Navy hydrofoils except on an experimental basis.

Although a wide variety of power plants have been considered for hydrofoil craft, the gas turbine and the diesel engine are the types primarily installed in these craft at present. Some hydrofoils are equipped with both gas turbines and diesel engines.

A major problem in the development of hydrofoils has centered around the transmission of power. When an underwater propeller is used, power must be transmitted downward from the prime mover in the hull to the propeller, which is at a deeper level. Various solutions to this problem have been tried. One solution is to use an inclined shaft and an inclined prime mover which drives into a V-gear. Another solution is to use double right-angled gearing. To date, no solution has been found that is entirely satisfactory.

Propellers have also been a continuing problem with hydrofoils. The cavitation encountered at high speeds has led to numerous propeller failures. One solution to this problem is the supercavitating (SC) propeller discussed later in this chapter. Because of the propeller problem, other propulsion devices such as airscrews and airjet or waterjet propulsion systems have been considered for use with hydrofoils.

The Navy's first operational hydrofoil, USS High Point, PCH 1, is shown in figures 25-1 and 25-2. In flight, this hydrofoil reaches speeds of more than 45 knots.

Hydroskimmers

The hydroskimmer belongs to the general category of craft designed to ride on a bubble or a cushion of air. Vehicles and craft in this category are called ground effect machines (GEM), air cushion vehicles (ACV), or surface effect ships. In general, any GEM that is designed to operate over water is called a hydroskimmer.

The hydroskimmer has been referred to as a "flying washtub,"[5] and the comparison is apt. The basic principles of the hydroskimmer (and, indeed, of all ground effect machines) may be grasped by considering an inverted washtub with a fan mounted inside the tub. When the fan is turned on, the tub begins to rise off the ground as soon as the air pressure inside the tub becomes sufficiently great. This, in essence, is the principle of the hydroskimmer.

In a real hydroskimmer or other GEM, the air escapes uniformly around the bottom edges of the craft, thus providing a cushion of air which lifts the craft evenly above the ground or the water. The air cushion is developed and maintained in various ways, depending upon the basic design of the vehicle.

---

[3]The specific weight of a propulsion plant is the number of pounds of propulsion machinery required per shaft horsepower.

[4]The catamaran or twin-hull design consists of two slender hulls which are joined together above the waterline.

---

[5]Erwin A. Sharp, JOC, USN, "Sailing on a Bubble of Air," All Hands, December 1960, pp. 8-11.

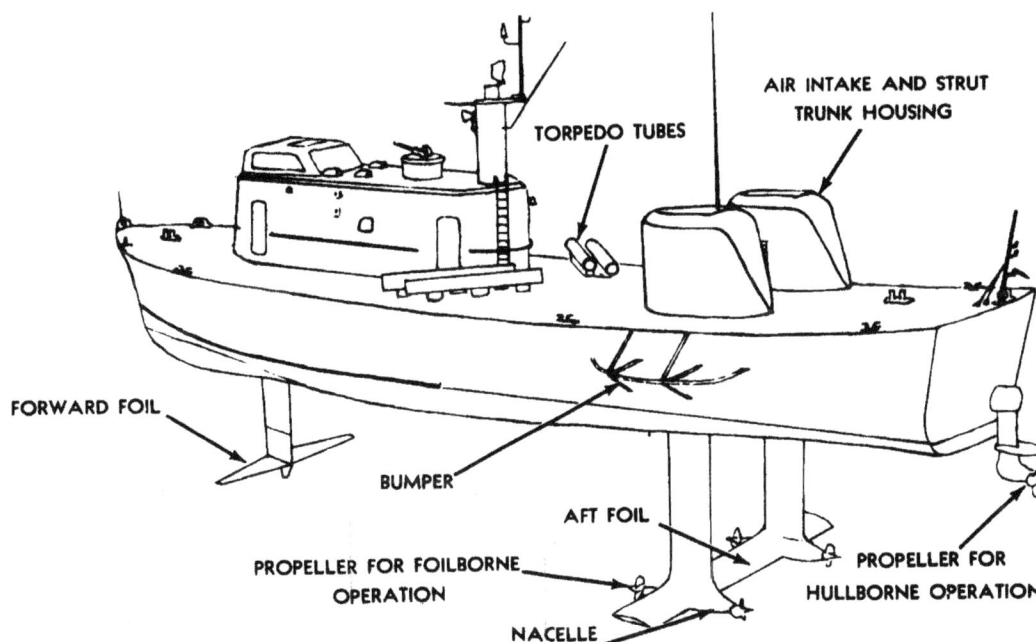

3.88

Figure 25-1.—Hydrofoil (USS High Point, PCH 1)

In the plenum-chamber type, air is forced down from the top of one chamber and allowed to escape around the edges at the bottom.

In the air-curtain type, the interior is divided into sections, with open air ducts between the sections. Air is forced down through the ducts to form a high pressure air cushion for the craft to ride upon. Air jets, pointing downward and inward, are installed around the periphery of the craft; the air from these jets holds the air bubble in place. In some air-curtain designs, the air jets are used only at the front and the rear. Side walls (called skegs) are used to enclose the air bubble at the sides.

In the water-curtain type, a scoop and a water pump are used to form water jets (instead of air jets) around the periphery of the craft. The water jets are even more efficient than the air jets in keeping the air cushion under the craft. Obviously, the water-curtain design is suitable only for craft operating over water.

The air cushion gets the hydroskimmer into the air, but it does not provide it with any means of horizontal propulsion. Both airscrews and water screws have been tried; each type has some advantages, with the choice depending upon the nature of the vehicle and the service conditions.

The hydroskimmer, although still experimental at this stage, gives promise of being a very fast and effective craft for a variety of uses. Speeds of 80 to 120 knots are considered feasible. The hydroskimmer is being considered for such uses as ASW craft, patrol craft, high speed transport, amphibious assault, mine countermeasures craft, and rescue craft. The Navy's 25-ton research hydroskimmer, SKMR-1, is shown in figure 25-3. This vessel was built with the specific aim of providing more knowledge on the preferred shapes, propulsion machinery, and propulsive devices for this type of craft. The air cushion system for SKMR-1 is shown in figure 25-4.

PROPULSION AND STEERING

For more than a hundred years, the underwater screw propeller has been the conventional device for ship propulsion. Although the screw propeller is in no danger of being replaced within the foreseeable future, the increasing emphasis on high speed ships has led to an increasing concern with other types of propulsion devices. Some of the propulsion devices presently under development or consideration are mentioned here.

631

3.88

Figure 25-2.—Hydrofoil in flight.

The supercavitating (SC) propeller is a recent development that may have particular application to hydrofoils and other high speed ships. The supercavitating propeller is intended for use at high forward speeds and high rpm, with a relatively shallow depth of submergence. It is designed to operate under conditions of full cavitation, although it may encounter conditions of partial cavitation when operating at less than designed forward speed and rpm. Under conditions of partial cavitation, the supercavitating propeller may be subject to cavitation erosion similar to that encountered on conventional propellers.

A supercavitating propeller is designed to operate with the suction side (back) of the blades enclosed in a vapor cavity. Because of the special design of the blades, this vapor cavity collapses far enough downstream from the propeller blades to prevent any cavitation effects on the face of the blades. In essence, the basic line of reasoning behind the design of the supercavitating propeller is not to do away with or prevent cavitation but rather to accept it as inevitable at high speeds and to control it. The supercavitating propeller controls cavitation by making sure that the cavity collapse occurs in an area in which it can do relatively little damage.

3.265

Figure 25-3.—Navy research hydroskimmer (SKMR-1).

Another approach to the problem of cavitation and the resulting reduction in thrust is the use of propeller nozzles and shrouds. Nozzles and shrouds reshape the flow of water to the propeller in such a way as to delay cavitation by increasing and "containing" the pressure around the propeller. A similar reshaping of the water flow may be obtained by the use of hydraulic jets.

One "new" propulsion device is actually quite old; although it has not recently had any major application for ship propulsion, it has been used on some torpedoes. This is the contra-rotating propeller, which consists of two screws turning in opposite directions. The advantage of the contra-rotating propeller design is that the after propeller is able to utilize some of the energy from the wake of the forward propeller, thus leading to higher efficiencies than are obtainable with a single screw propeller. The contra-rotating propeller also offers possible weight savings and efficiency improvements in the propulsion machinery.

One variation of the contra-rotating propeller installation is called a tandem propeller installation. In this arrangement, one propeller is installed near the bow and the other near the stern. The propeller blades are mounted on a blade ring, with the blades projecting out through the hull. The pitch of the contra-rotating propeller blades

can be varied in such a way as to provide complete maneuverability, as well as propulsion; hence this device is actually a combination of a propulsion device and a steering device. Although originally proposed for small submersible craft, it is possible that this type of installation may have application for larger ships in the future.

A number of other devices have been suggested which combine the functions of propulsion and steering. Among these are the steering screw, which consists of a propeller mounted on a vertical shaft. The shaft can be rotated through 360° in order to propel the vessel in any desired direction. Other devices which combine the functions of propulsion and steering to some extent are screw propellers arranged as bow thrusters or as stern thrusters. In each case, the screw propeller is mounted on a horizontal shaft. The propeller is located within a tube which runs athwartship through the bow or the stern. The propellers can be reversed to provide thrust in either athwartship direction. The tubes can be arranged to be closed off when not in use.

Airscrew propulsion for ships has been under investigation for the past few years. The primary advantage of the airscrew is that it provides very great maneuverabiltiy, particularly at low speeds. In a Navy test of airscrew propulsion on the liberty ship John L. Sullivan (YAG 37), it was found that maneuverability at low speeds was better with the airscrews than with conventional propulsion and steering. When approaching piers or mooring buoys, the ship was able to maneuver without the assistance of a tug because the airscrews provided the capability for applying propulsive force in any direction.

The disadvantage of airscrew propulsion for ships is that enormous airscrews would be required to propel even a medium-sized ship at any great speed. For certain types of craft, however, it is possible that some combination of water screw and airscrew propulsion may be feasible.

Although none of the combined devices thus far developed have solved all propulsion and steering problems, there is much to recommend the combination approach. Propulsion and steering are very closely related; a truly effective propulsion and maneuvering combination should result in greater simplicity and greater efficiency than is obtainable with two separate devices.

Waterjet propulsion for ships was tried out many years ago but abandoned because of its relatively low efficiency. However, the Navy has recently been investigating the possibility of

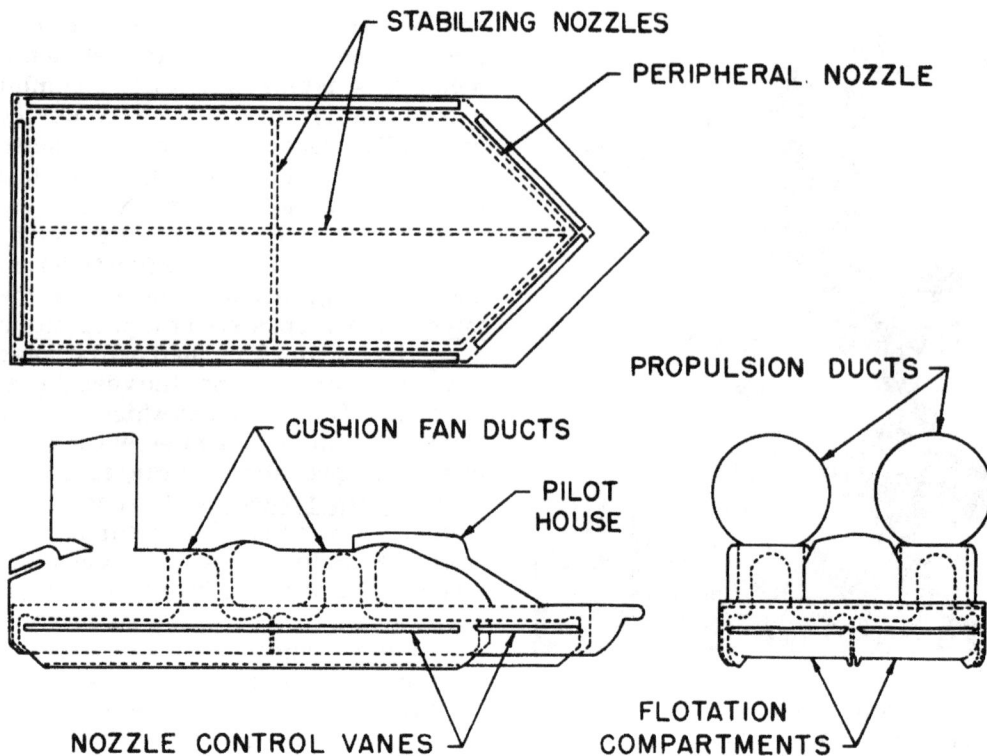

3.265

Figure 25-4.—Air cushion system for SKMR-1).

using the waterjet as a prime mover in small craft such as the one shown in figure 25-5.

The general principle of waterjet propulsion is illustrated in figure 25-6. The "prime mover" consists merely of a water pump. The rotating impellers deliver thrust by accelerating a large volume of water at a high velocity through a nozzle.

The velocity of flow through the nozzle is directly proportional to the flow rate and inversely proportional to the volume of flow times the velocity of flow.

One of the outstanding advantages of the waterjet mode of propulsion is that the waterjet produces much less underwater noise than a conventional propeller-driven craft of the same size and general configuration. Preliminary studies of the waterjet indicate that a comparable conventional propeller-driven craft could be detected approximately ten times farther away than the waterjet.

Quite a different system of jet propulsion has been suggested as a possibility for the propulsion of underwater vehicles. This propulsion system (fig. 25-7) is generally known as an underwater ramjet system. The fuel in the combustion chamber reacts with the water; hence the water that enters the combustion chamber may be regarded as a co-propellant. The products of the reaction are expanded through a nozzle at the after end of the craft, thus propelling the craft forward. Although a propulsion system of this sort would have many advantages for short-range operation, there are some disadvantages; one present disadvantage is that the system is extremely noisy. Also, it is very inefficient except at very high speeds (above 80 knots).

DIRECT ENERGY CONVERSION

The production of power for ship propulsion begins with the conversion of some stored form of energy. In all present propulsion plants, the stored energy that is the original source of power must undergo a series of transformations before it can be utilized to propel the ship. We have two major sources of stored energy: fossil fuel and nuclear fuel. In each case the stored energy must be converted into thermal energy which is then converted into mechanical energy.

During the past few years, a considerable amount of interest has developed concerning

147.162

Figure 25-5.—Small craft with waterjet propulsion.

direct energy conversions that may ultimately have application to the production of power for ship propulsion. This interest arises from two major considerations. First, the Carnot cycle[6] which is the thermodynamic basis for our heat engines is inherently inefficient, with the theoretical maximum efficiency of the cycle being

$$\frac{T_1 - T_2}{T_1}$$

limited to where T_1 is the absolute temperature at which heat flows from the source to the working fluid and T_2 is the absolute temperature at which heat is rejected to the receiver. Since the temperature of the heat receiver (the ocean) cannot be lowered, the only way to improve the efficiency of an actual cycle based on the Carnot cycle is to increase the temperature of T_1. The past few years have seen great advances in the use of higher T_1 temperatures (e.g., boilers operating at higher pressures in order to increase the difference between T_1 and T_2), but materials limitations eventually impose a barrier to progress in this direction.

The second reason for current interest in novel energy conversions is that the actual shipboard cycles in which stored energy is converted to thermal energy which is then converted to work

---

[6]The Carnot cycle is discussed in chapter 8 of this text.

require a great deal of equipment to perform these various energy conversions. Beginning with an inherently inefficient cycle which cannot operate unless a great deal of heat is "wasted" because it must be rejected to a heat receiver, we must accept even greater inefficiency because of the mechanical losses and miscellaneous heat losses that inevitably occur throughout the plant.

There are two major approaches to the problem of direct energy conversion. One approach is to find an energy conversion which is not based on the Carnot cycle and is therefore not limited by the requirement that some heat be rejected to a low temperature heat receiver. The other approach is to utilize a "static" heat engine which is based on the Carnot cycle, and therefore subject to its limitations, but which has no moving parts and therefore no mechanical losses. The fuel cell is an example of a device that bypasses the Carnot cycle to make a direct energy conversion. Thermoelectric converters, thermionic converters, and magnetohydrodynamic generators are examples of static heat engines which, although operating on the Carnot cycle, come very much closer to the maximum theoretical efficiency of the cycle by reducing or eliminating mechanical losses.

Fuel Cells

A fuel cell is a battery-type device in which chemical energy is converted directly into electrical energy. The reaction involves a free energy release, without the rejection of heat to a heat sink; hence the process is independent of the Carnot cycle and free of Carnot cycle limitations.

The major parts of a fuel cell (fig. 25-8) are an anode, a cathode, and an electrolyte. The fuel is fed continuously to the anode, while the oxidant is fed continuously to the cathode. The conversion from chemical energy to electrical energy occurs as electrons, released at the anode, flow to the cathode.

Several types of fuel cells are under investigation and development. Some operate at relatively low pressures and temperatures, others at high pressures and temperatures. A wide variety of fuels have been considered for fuel cells; hydrogen, various hydrocarbons, and methanol appear to offer particular promise for many applications, while a sodium amalgam is being considered for use in certain small fuel cells. The oxidants most commonly used are air and oxygen; however, peroxides, chlorine, and other substances have also been tried. Electrolytes

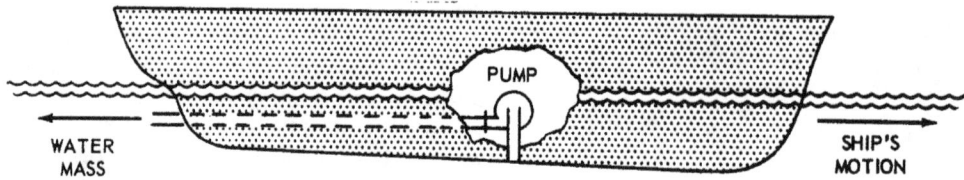

147.163

Figure 25-6.—Principle of waterjet propulsion.

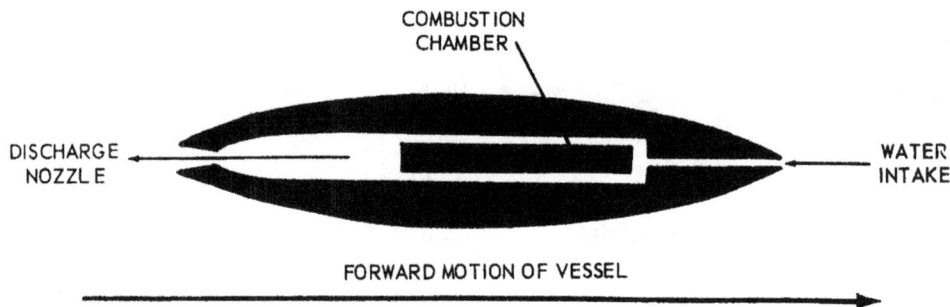

147.164

Figure 25-7.—Underwater ramjet propulsion system.

that have been used in fuel cells include potassium hydroxide, fused salts, and liquid salts.

As may be inferred, the fuel cell is simplicity itself in basic concept, and it offers the promise of enormously greater efficiencies than can ever be obtained through any energy conversion that is based on the Carnot cycle. However, there are many problems to be overcome before the fuel cell can be regarded as a major source of power for applications requiring large power outputs.

Thermoelectric Converters

A thermoelectric converter is a device for converting heat to direct-current electricity. The general arrangement of a thermoelectric converter is illustrated in figure 25-9. As may be seen, the converter is basically a thermocouple device in which an emf is developed through the application of heat to dissimilar materials.

In the thermoelectric converter, one of the dissimilar materials is of the type known as an N material and the other is of the type known as a P material. As far as electron behavior is concerned, N and P materials react differently when heat is applied to one end. In N materials, the application of heat tends to make negatively charged free electrons move toward the cold end

of the piece. In P materials, the application of heat tends to produce positive charges toward the cold end of the piece. When heat is applied to the hot junction of a thermoelectric converter (as shown in fig. 25-9) the cold end of the N material is negative and the cold end of the P material is positive. When the cold ends are connected through a load, the electric circuit is complete.

The efficiency of the thermoelectric converter is limited by Carnot cycle limitations. As with all Carnot cycle energy conversions, the efficiency of the thermoelectric converter increases as the temperature differential increases. It is believed that the maximum attainable efficiency of the thermoelectric converter may be on the order of 25 to 30 percent, although efficiencies thus far achieved are not nearly so high.

Thermionic Converters

Another device for converting heat to direct-current electricity is the thermionic converter. In its simplest form, the thermionic converter is similar to a vacuum tube, consisting of two metal electrodes separated by a space under vacuum. The cathode is heated, and the anode is maintained at a lower temperature. When the cathode is heated, electrons are thermally agitated and driven from the cathode to the anode. Connecting

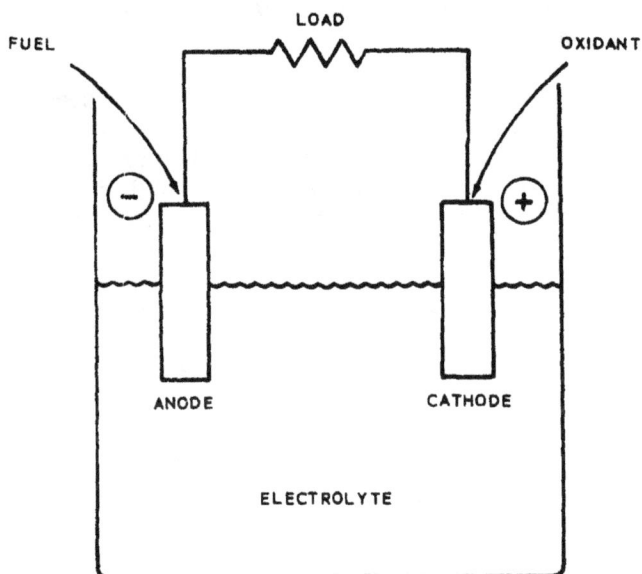

147.165

Figure 25-8.—Fuel cell.

the two electrodes to a load establishes a circuit and allows the flow of electric current.

The thermionic converter, like the thermoelectric converter, is a static heat engine and is limited by Carnot cycle considerations. It is believed, however, that higher efficiencies can be obtained with the thermionic converter than with the thermoelectric converter. One reason for the higher efficiencies is that the thermionic converter typically operates at substantially higher temperatures (and with substantially greater temperature differentials) than the thermoelectric converter. Another reason for the higher efficiencies is that very little thermal energy is lost between the hot and cold terminals because of the vacuum that is maintained between the cathode and the anode.

In itself, however, the use of a vacuum creates some significant problems in connection with the thermionic converter. As electrons pass through a space under vacuum, there is a tendency for a negative "space charge" to be built up. This space charge is merely a cloud of electrons which, after escaping the cathode, have insufficient energy to reach the anode. As the negative space charge builds up, electron flow is greatly diminished. The space charge effect can be reduced by putting the electrodes extremely close together, but this is difficult to do except in very small converters. One recent approach to this problem is to fill the space with positive ions,

thus neutralizing the space charge effect and allowing the unimpeded flow of electrons. Thermionic converters in which the space between electrodes is filled with cesium vapor appear to be promising.

Because a thermionic converter must operate at an extremely high temperature in order to develop any great efficiency, some thought has been given to utilizing the decay heat of radioactive isotopes as the heat source. One device that has been successful in laboratory tests combines reactor fuel and the thermionic converter in one unit, thus in effect producing a "thermionic fuel cell." It is also believed that thermionic conversion may be suitable for solar energy conversion for use in space.

Magnetohydrodynamic Generators

The magnetohydrodynamic (MHD) generator is still another device for converting heat to direct-current electricity. In the MHD generator, the conversion is accomplished by passing a high temperature gas through a magnetic field. Heating the gas to a very high temperature ionizes the gas and makes it electrically conductive. Passing the electrically conductive gas through a fixed magnetic field induces an electrical voltage in accordance with Faraday's law.

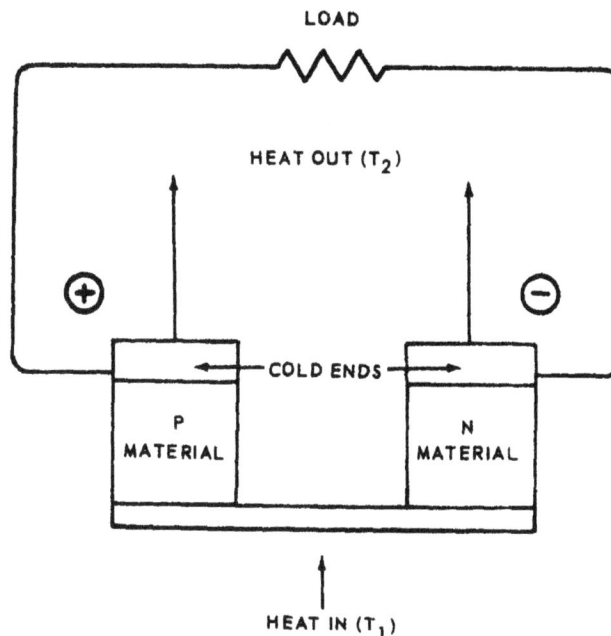

147.166

Figure 25-9.—Thermoelectric converter.

The MHD conversion is similar to the thermoelectric conversion and the thermionic conversion in some respects but quite unlike them in others. All three conversions involve the direct conversion of thermal energy into electrical energy, without the intervening step of conversion to mechanical energy; in this sense, all three may be regarded as "direct energy conversions." But the MHD conversion, unlike the thermoelectric and thermionic conversions, requires a working fluid—namely, a hot ionized gas. In this respect, then, the MHD conversion is somewhat less a "direct energy conversion" than the other two processes.

The major problems in connection with magnetohydrodynamic conversion arise from the fact that extremely high temperatures (in excess of 4000 °F) must be developed in order to produce ionization of the gas. Obviously, such high temperatures pose materials problems. Also, it is difficult to achieve such temperatures on the large scale desired for MHD generators. Nuclear reactors capable of operating at these ultra-high temperatures are under development but are not fully operational. When chemical fuels such as oil or powdered coal are used, the desired temperatures can be obtained only if combustion takes place with almost pure oxygen or if the combustion air is preheated to approximately 2000° F.

In spite of the temperature problem, the magnetohydrodynamic conversion process continues to arouse great interest among scientists and engineers. It should be noted, in fact, that the temperature <u>problem</u> is only one side of the coin. On the other side, the use of such high temperatures leads to the possibility of thermal efficiencies far greater than any that are even theoretically possible with conventional heat engines. It has been estimated that overall efficiencies as high as 50 to 60 percent may be achieved through MHD conversion, if provision is made for utilizing the "waste" heat of the MHD process. The advantage of utilizing the waste heat is enormous, since the ionized gas is at a temperature of 2500° to 3000° F when it is discharged from the generator.

COMBINED POWER PLANTS

In recent years there has been a great deal of interest in the use of combined power plants for ship propulsion. In a combined power plant, two basically different kinds of prime movers are used to furnish propulsive power. Each type of prime mover has its own inherent limitations, as well as its own unique advantages; the purpose of combining two prime movers is to make full use of the special advantages of each and, at the same time, to minimize or bypass its limitations.

Combined power plants are of particular interest for naval ships because of the constant need to reconcile conflicting operational requirements. On the one hand, a naval ship must be able to operate at high speeds when necessary. On the other hand, the ship must be able to cruise economically at lower speeds for extended distances and extended periods of time.[7] If the prime mover is selected specifically for high speed operation, there is normally some sacrifice of cruising radius. If the prime mover is selected specifically for economical operation at cruising speeds, there is normally some sacrifice of speed capability. In most cases, then, the selection of a prime mover represents a compromise between high speed capability and large cruising radius.[8]

For many naval applications, it appears that conflicting operational requirements can be reconciled by combining a base-load plant of moderate weight and high efficiency with a booster plant of very light weight and lesser efficiency. The base-load plant is selected to meet cruising requirements, and should be able to go many hours between overhauls. The booster plant will inevitably require overhauls at much shorter intervals but is capable of providing additional power for high speed operation when necessary.

A combined power plant for ship propulsion usually consists of two prime movers which are mechanically connected by gearing, clutching, or both. In some combination plants, the two prime movers have interrelated thermodynamic cycles in which one prime mover utilizes waste heat from the other. In other combination plants, the thermodynamic cycles of the two prime movers are entirely separate and independent.

A great many combinations of prime movers are possible, though not all combinations are equally feasible or desirable. Also, for any given

[7]More than 80 percent of the total operating time of naval ships is at speeds requiring less than a third of the power available from the installed plant.

[8]This does not necessarily apply to nuclear ships. Obviously, however, the use of nuclear power brings about the necessity for another set of compromises.

combination of prime movers, various arrangements are possible. The three combination plants which at present appear to offer great advantages for naval ships are (1) the combined steam and gas turbine plant, known as COSAG, (2) the combined diesel and gas turbine plant, known as CODAG, and (3) the combined gas turbine and gas turbine plant, known as COGAG. Other combinations, including some that utilize nuclear power, are also under consideration.

Combined steam and gas turbine (COSAG) plants have been installed in some combatant ships of the British Navy and are being investigated by our own Navy. Figure 25-10 illustrates the general arrangement of one COSAG plant installed in a British twin-screw guided missile destroyer; there are two such plants, each serving one propeller. Each shaft set consists of a cross-compound steam turbine of 15,000 shaft horsepower plus two gas turbines of 7500 horsepower each. All prime movers drive into one gear box.

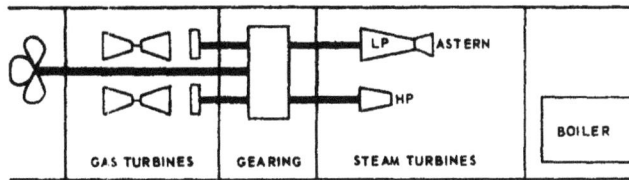

147.167

Figure 25-10.—COSAG plant for one screw of a twin-screw ship.

A slightly different arrangement of a COSAG plant is shown in figure 25-11. This plant, which is installed in a British single-screw frigate, consists of a single casing steam turbine of 12,500 shaft horsepower and one gas turbine of 7500 shaft horsepower.

In both of the COSAG plants illustrated, the steam turbine installation is capable of propelling the ship at approximately 85 percent of full power ship's speed, in the event of complete failure of the gas turbine unit. The ability of the gas turbines to make a rapid start allows the ship to get underway very quickly with the steam plant cold. Maneuverability (including reversing) with the gas turbine is achieved by fairly complex gearing and clutching. Each gear box incorporates a reversing section for the gas turbine; two manual clutches for the gas turbine, one for the boost drive and one for maneuvering; a manual clutch for the steam turbine; synchronizing

clutches which automatically connect the gas turbine drive to the main shaft; and two hydraulic couplings which are used when maneuvering on the gas turbine. With this arrangement, the steam turbines or gas turbines (or both) can be used for propulsion, and maneuvering can be accomplished either on the astern steam turbine or on the gas turbine and the reversing gears.

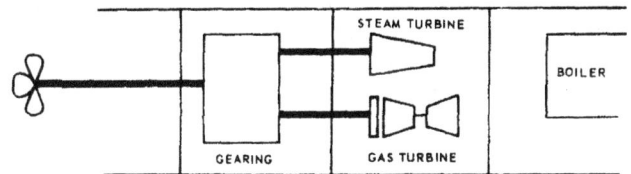

147.168

Figure 25-11.—COSAG plant for one-screw ship.

Combined diesel and gas turbine (CODAG) plants utilize diesels for the base-load plant and gas turbines for the booster plant. The use of multiple diesels and multiple boost gas turbines means that the loss in ship's speed will be very small in the event of failure of any one or even any two units. The ship can get underway very quickly with either diesel or gas turbine power. The efficiency of the diesel is much higher than the efficiency of a steam plant, and (for small sizes of engines) the specific weight is somewhat less. In general, CODAG plants appear to be suitable for ships which have moderate requirements for cruising power.

A combined gas turbine and gas turbine (COGAG) plant has been proposed. Such a plant would combine a long-life, efficient, moderate-weight gas turbine for the base load and a light, aircraft-type gas turbine for the booster load. COGAG plants have not yet been tried out because gas turbines suitable for the base loads are still in developmental stages.

CENTRAL OPERATIONS SYSTEM

Advances in engineering technology, the use of solid state devices, and computer circuitry have made possible significant automation factors in the operation of the naval engineering installations. Much of this automation has been applied in the area of ship control and plant surveillance. A general discussion is presented in this chapter on a portion of an automated engineering plant considered representative of those currently being employed on naval vessels.

The automated engineering plant is designed to bring together in one location all of the major control functions and indications previously located throughout the engineering spaces. In addition, major advances made in the areas of boiler control, turbine control, and plant surveillance have also been incorporated in the control systems. Control systems such as the central operation system provide for direct control of shaft speed and direction at a console located on the bridge. These control systems are located in an enclosed Engineering Operation Station (EOS) located within the machinery plant.

The Central Operations System (COS), shown in figure 25-12, as found in naval vessels, provides for control of the electrical plant, the main turbine, selected auxiliary equipment and surveillance of the entire engineering plant. It utilizes solid state components in the analog and the digital circuitry. Analog components are used in the throttle control systems and in the input to the plant surveillance equipment. Throttle control features are provided by standard operational amplifiers and functional generators used in a closed loop system which maintains operation at a desired point.

Information on plant conditions is provided by digital demand displays, alarm indications, indicating lights, meters and the printout typewriters. This is handled by a time sharing system made up of logic circuitry, and controlled by a synchronous timing generator.

A substantial decrease in the number of watch standers required to operate the engineering installation can be achieved through the use of a system of this type. The bridge throttle control feature provides the OOD with a greater feel of the ship as well as a faster response to desired changes.

ENGINE ROOM CONSOLE

The engine room console (fig. 25-13) is the heart of the central operations system (COS) and is divided into five functional sections. generators, propulsion, boilers, auxiliaries, and data logger. The desk top of the console houses the controls and devices required to be within the operator's reach. The vertical surface above the desk top is used primarily for instrument display and visual indicators. Solid state control modules with printed circut elements are used which can

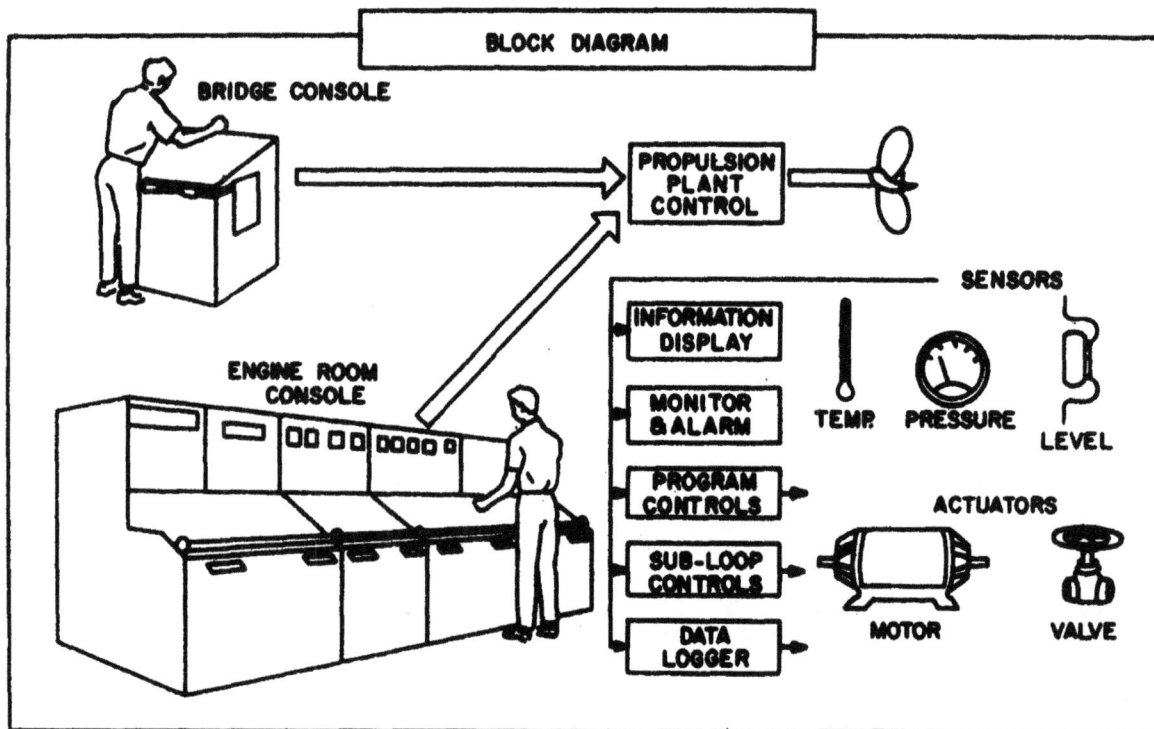

27.344

Figure 25-12.—Central operations system major units.

27.345

Figure 25-13.—Engine Room Console.

be easily removed from the panels by unplugging. Temperatures, pressures, and liquid levels are converted to electrical signals by sensors located at various points throughout the system.

Significant readings on the console are displayed on flush mounted electric meters. Other readings which only need to be checked periodically are read on digital meters (called digital demand display readouts). The console provides monitoring of the boilers and monitoring and control of the main propulsion plant, turbogenerator sets, main condensate pumps, lube oil pumps, fire pumps, and other auxiliary machinery, thus, enabling the engine room operator to observe all important operating functions without leaving his station. At any time a plant status record may be made with the data logger.

The COS continually monitors key temperatures, pressures, levels and motor conditions. If any go beyond operating limits, the system sounds an alarm to alert the operator and the alarm logger automatically records the out of limit conditions. An alarm log review, plant status log and bell log printout may be obtained at any time by pressing a push button. Selected points also have individual alarm lights. A bell logger automatically records engine order telegraph signals and responses, propellor r.p.m., throttle control location, and throttle control wheel position together with time and date.

Generator Section

The generator section of the engine room console (fig. 25-14) provides control and monitoring of the ship service generators. Provisions are included which enable the operator to adjust the frequency and voltage and monitor the output of each generator; monitor the current in the shore power connection; open, close, or monitor the position of generator or bus tie circuit breakers and the shore power circuit breaker; test each switch-board bus for grounds, and control and monitor the space heaters in each generator.

Propulsion Section

The propulsion section of the engine room console (fig. 25-15) contains the throttle controls and transfer switches, engine order telegraph, shaft revolution indicator-transmitter, and the necessary gages and indicators for monitoring the operating conditions of the main turbines, reduction gears, and propeller shaft.

The throttle control handwheel controls the position of pilot valves on hydraulic power actuators. The hydraulic power actuators, in turn, open or close the main steam valves to the ahead and astern turbines, thus, controlling the speed and direction of the propellor shaft. An alternate electrical control and a direct mechanical control of the throttle are also provided. Throttle

27.346

Figure 25-14.—Generator Section.

control may also be shifted to the bridge control console, but may be reclaimed by the engine room personnel at any time.

Boiler Section

The boiler section (fig. 25-16) contains the necessary pressure, temperature, and level indicators and alarms for monitoring boiler operation. Also mounted is an underwater log speed indicator and a sound-powered telephone handset and jack. Actual boiler control is accomplished by a Bailey boiler control console (not shown).

Auxiliaries Section

The auxiliaries section (fig. 25-17) provides for remote operation (start and stop) of such equipment as fire pumps, condensate and circulating pumps, and ventilation systems. It also monitors nonvital systems such as potable water, air conditioning, and refrigeration. This section also contains three digital demand meters. These meters will display, upon demand, any one or any three simultaneously, of approximately 170 different readings relating to the boilers, fuel and lube oil, main condensers, main turbines, ship service generators, and auxiliary machinery. To obtain a reading, the operator looks up the number (address) of the function he wishes to read on the function address nameplate located on the data logger section and turns a thumbwheel switch beside the digital demand meter to this address. The meter will then display the value of the function selected. The same function may be selected and read on all three meters or three

27.347

Figure 25-15.—Propulsion Section.

different functions may be displayed simultaneously.

Data Logger Section

The data logger section (fig. 25-18) consists of plant performance data logging, alarm scanning, and bell logging equipment. The plant performance logging equipment can be set up to print at regular time intervals and on demand, with continuous scanning of all sensor points. Abnormal conditions are printed in red by the performance typewriter (shown on the left in fig. 25-18). The bell log typewriter (on the right) records each engine order telegraph signal along with time and date, location of throttle control, throttle control wheel position, and shaft RPM as stated previously.

27.348

Figure 25-16.—Boiler Section.

BRIDGE CONSOLE

The bridge console provides remote control of the throttle as stated earlier. The throttle control handwheel and other necessary equipment for control of the propulsion plant are mounted on the left section of the console. The ship's helm and other steering and navigation equipment are mounted on the right section as shown in figure 25-19.

SENSORS

The sensing devices used with the automated controls are in most cases improved versions of detectors already receiving wide usage throughout the fleet.

27.349

Figure 25-17.—Auxiliaries Section.

In all cases the manufacturer's technical manuals used with the system give complete installation, operation, and maintenance instructions.

Pressure Sensors

Pressure sensors are used to convert plant pressure to an electrical signal for further transmission to the Engine Room Console. Two of the main types of sensors are the pressure-to-current transmitter and the pressure switch types.

The pressure-to-current transmitter (fig. 25-20) converts the applied pressure to a direct current proportional to the applied pressure. This current can then be applied to a meter for

27.350

Figure 25-18.—Data Logger Section.

remote indication. The meter is calibrated in the desired scale (PSI). These meters are in most applications d.c. microammeters.

The pressure switch type transmitter finds its application in the alarm circuitry on the Engine Room Console. Herein, at a given high or low pressure, electrical contacts within the switch housing are actuated completing or breaking a circuit and actuating an alarm.

Level Sensors

Three types of level sensors are generally employed throughout the system. Since pressure may be a function of level the first two devices used are the two detectors mentioned under pressure; the pressure-to-current for indication and the pressure switch for alarm.

27.351

Figure 25-19.—Bridge Console.

An additional level switch is employed in bilges and unvented tanks. This switch is similar in operation to that described in chapter 3 as a float switch.

Temperature Sensors

Temperature is measured by means of resistance temperature detectors (RTD). The RTD (fig. 25-21) consists of a sensing element incased in a protective tube. Since the electrical resistance of the element changes with temperature changes, the temperature can be determined by measuring the resistance.

For temperatures having a maximum of 600 degrees F., a nickel resistance element is used. These detectors are found to be of the stem-sensitive type, where the sensing element is located within a few inches of the stem, or the tip-sensitive type, where the sensing element is within the tip of the detector tube which must be pressed against the material being measured.

Above 600 degrees a platinum element is employed in the stem-sensitive element type. Currently this is the only application of this type element.

In most applications the RTD is installed in thermo wells, bored and threaded to receive the detector. By the use of this method the unit may be removed from the measured component or piping without disturbing the integrity of the component. Extra protection is also afforded to the element in this manner.

DATA SCANNER SYSTEM

Figure 25-22 is a block diagram of the Data Scanner system. The inputs from the sensing devices are placed into the scanner (block 27) as analog values. The scanner is an electronic selector governed by the Synchronous Timing Generator (22), the Program Contact (23) and the Point Drive (26). When there are no requests for the system such as the Bell Log, Alarm Log, Status Log, or Display Triggers, the scanner continues to check each of the inputs. If there is a request present, the scanner will go directly to the address requested and process that signal before monitoring all of the addresses. The signals are sent to the Isolation Amplifier (28) from the scanner, and after amplification it passes on to the Analog/Digital converter (29). The

647

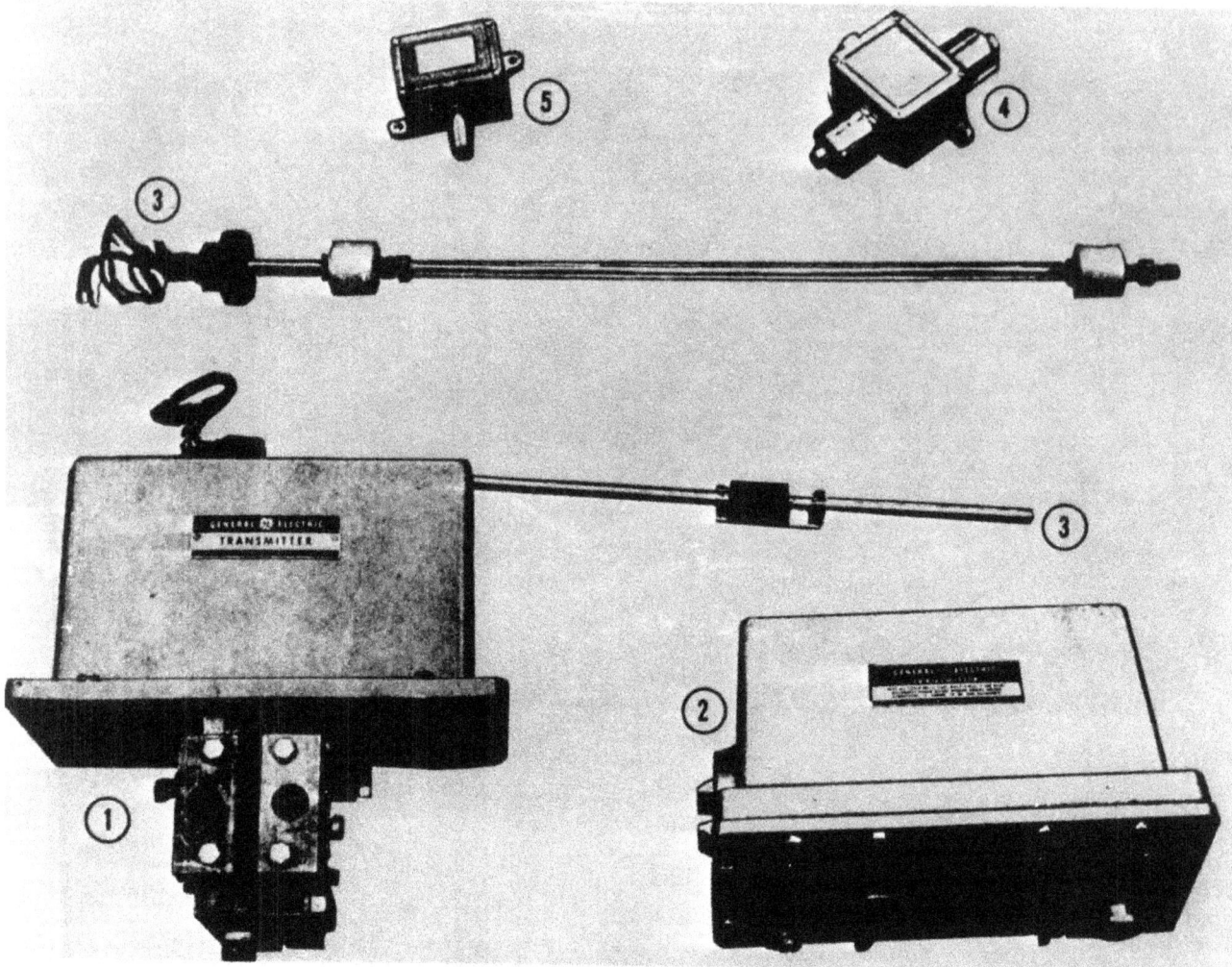

(1) Differential Pressure-to-Current Transmitter
(2) Pressure-to-Current Transmitter
(3) Float Switch
(4) Differential Pressure Switch
(5) Pressure Switch

140.92

Figure 25-20.—Pressure and Level Sensors.

program contact, controlled by the scanner, sets up the comparison values for the signal as well as any adjustments to the signal required during conversion and scaling.

If the information is requested and/or the point is in alarm, the A/D converter then transfers the values via the Word Distributor (24) to the scaling module. If neither of the previously mentioned conditions exist, the scanner executes branch back and picks up the next address and repeats the process.

After the information leaves the A/D converter it is sent to the scaling module which scales all signals into a zero to 1000 scale. This information is in the form of pulses numbering from zero to 1000 according to the value of the input signal to the A/D converter. The address information is then placed in the necessary registers and along with inputs from the Real Time Clock (40) and Digital Inputs (43) and made ready to be sent to the log Printout Buffer (37) and on to the Typewriter Drive (38) for printing.

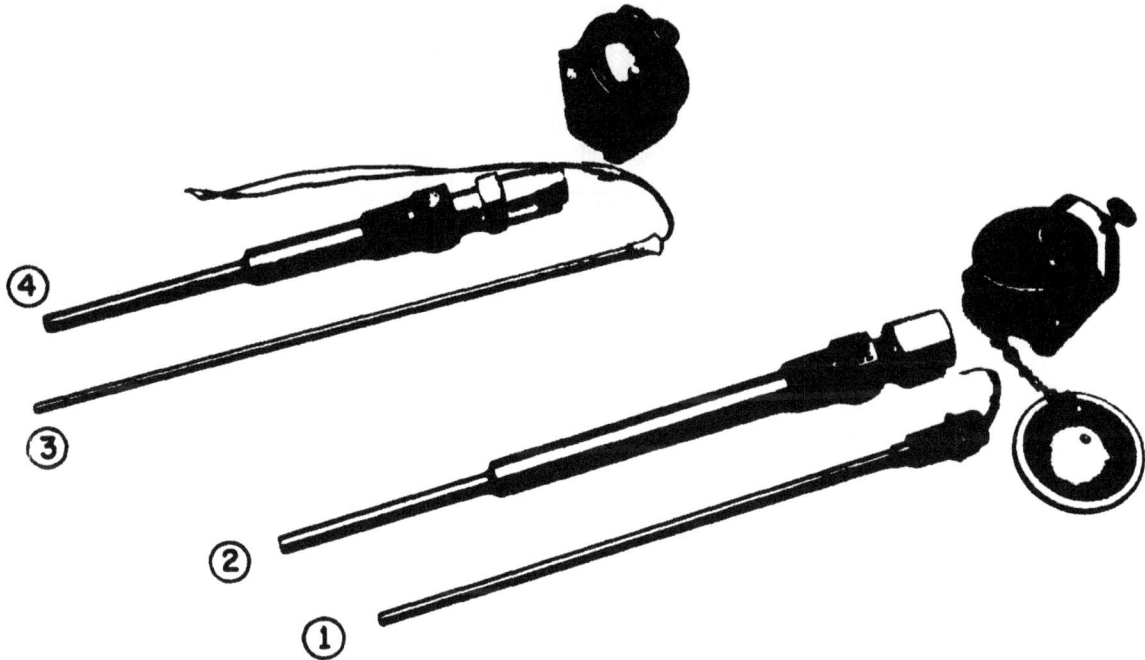

(1) Stem Sensitive RTD
(2) Thermowell with Adapter
(3) Tip Sensitive RTD
(4) Thermowell with Holder

140.93

Figure 25-21.—Resistance Temperature Detectors.

Information leaving the scaling module for display is sent directly to the Digital Display Buffers and Readouts (44) and appear at the readout units on the console face.

The entire operation from pick-up of the input address to activation of the printout units requires a time span of .0376 milliseconds. The assembly will monitor the complete bank of 273 inputs in approximately two seconds providing there are no requests or alarms conditions presented to the system during that time period.

THROTTLE CONTROL

Figure 25-23 is a block diagram of the throttle control system. A reference input signal may be taken from either the bridge or engine room reference handwheel potentiometer and fed to the system. Negative voltages are used for ahead speeds and positive for astern. The signal then passes through a common operational amplifier where it is inverted and then goes to the common circuit for both the ahead and astern turbines. The function generators will accept only a signal of a given polarity. The ahead function generator accepts positive signals and the astern function generator negative signals. The signal to the function generator is also used as a reference signal for the speed feedback system. This circuit compares the reference and speed feedback signals and uses the algebraic sum as the input to the speed error amplifier.

The signal to the function generator is adjusted within the amplifier so that the output is equivalent to the cube of the input. This is done to change the linear movement of the reference to the non-linear characteristics of the throttle valve. Inversion once again takes place in the function generator.

The output of the function generator is matched with the speed error signal and the throttle position signal at the summing junction and the algebraic sum is fed to the summing amplifier. Inversion takes place and the output controls the action of the SCR power package.

The SCR power package will cause the pilot motor to drive in either direction depending upon the input. A positive input will cause the pilot

Figure 25-22.—Block diagram of data scanner system.

140.94

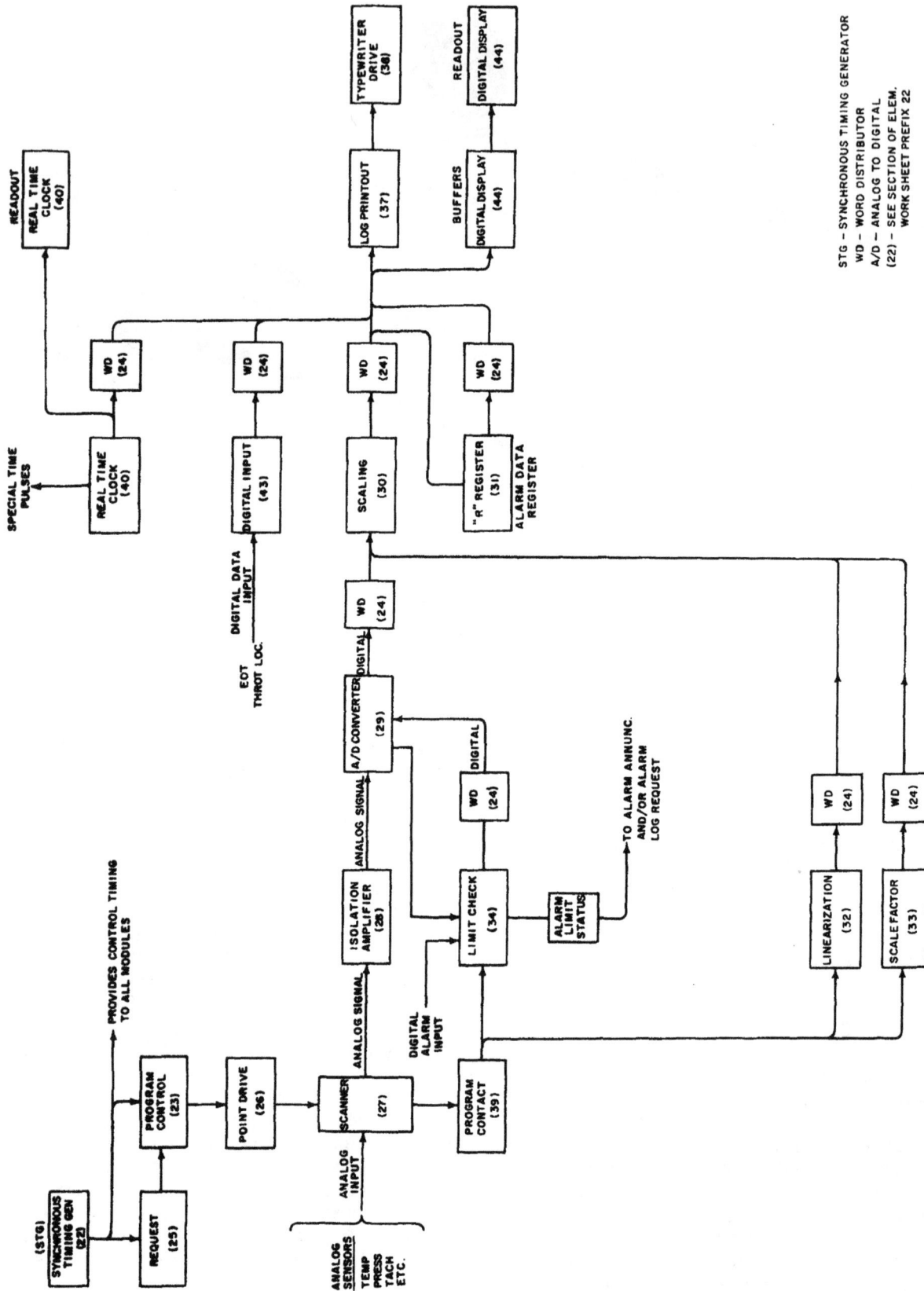

STG – SYNCHRONOUS TIMING GENERATOR
WD – WORD DISTRIBUTOR
A/D – ANALOG TO DIGITAL
(22) – SEE SECTION OF ELEM.
WORK SHEET PREFIX 22

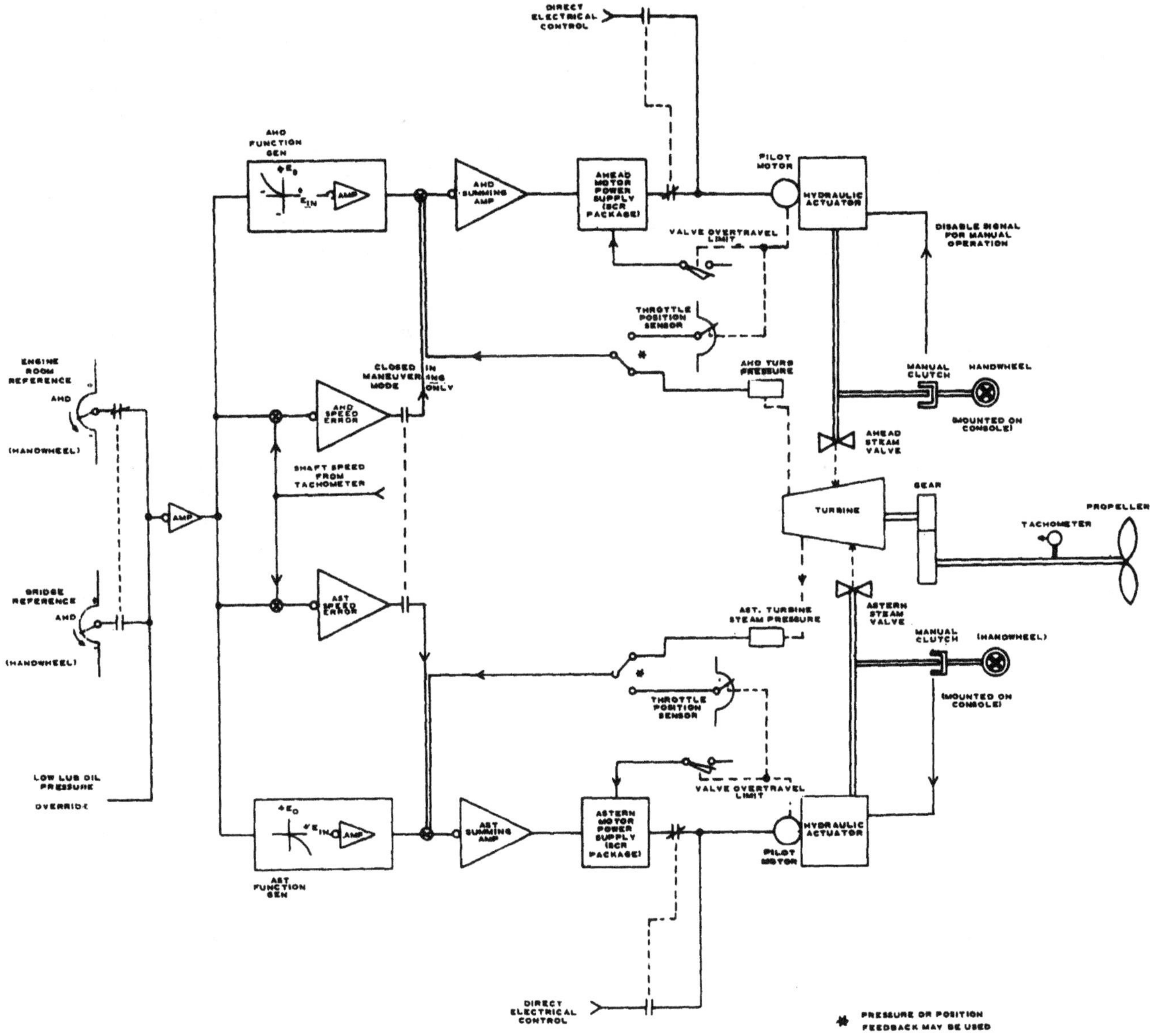

140.95

Figure 25-23.—Block diagram of throttle control system.

motor to drive in a direction to open the throttle valve. A negative input will close the throttle valve. The SCR power package will be inhibited by limit switches if the motor travel exceeds a predetermined point of travel.

The pilot motor positions a pilot valve in the hydraulic actuator which ports oil in the proper direction to correctly position the throttle valve.

A reference signal for throttle position, which is controlled by the pilot motor, is fed back to the summing junction. This section cancels the input signal when the desired valve opening is reached.

During direct electrical control of the throttle, the contacts in the throttle location switch change the circuitry eliminating the regulated signal and setting up the circuitry for signals from the direct throttle switches.

During manual operation, the manual clutch is engaged and the hydraulic actuator is inhibited. In addition, the hydraulic system is vented to prevent a hydraulic lock and permit the movement of the handwheel for manual throttle control.

A tachometer generator on the shaft produces an output signal that is fed back as the speed error signal. This signal produces a rapid response from the system when the engineering plant is in the maneuvering mode. Under normal mode of plant operation the speed feedback signal is not utilized.

The signals for astern throttle movement are handled in the same manner but all of the polarities are reversed.

In the near future we may see a substantial increase in the automation of naval propulsion machinery and auxiliary machinery. At the present time, it is entirely possible to design a completely automated ship. Although complete automation is an unlikely goal for the naval ship, three is little doubt that automation will increase to some extent within the next few years.

FUEL CONVERSION PROGRAM

As of this writing, the Department of Defense has authorized the Navy to shift to an all distillate marine diesel type fuel which will replace the Navy Special Fuel Oil (NSFO) now in use. The shift will take place on a gradual basis over a three year period. This conversion will ease the principal adverse factors associated with the use of Navy Special Fuel Oil such as:

1. Fouling of firesides of boilers by permissible impurities in Navy Special Fuel Oil, principally sulphur, ash, and carbon residue.
2. Decrease in ship readiness associated principally with cleaning of firesides.
3. High corrosion rate of above-deck equipment associated with exposure to products of combustion of Navy Special Fuel Oil.
4. Substantially below average retention rate of Navy enlisted personnel who perform boiler cleaning operations.

Testing planned completion date 1975, is presently taking place with diesel engines and gas turbine propulsion plants; the results of these tests are to be evaluated for the "Single-Fuel" Navy concept, which will permit the Navy to operate either steam diesel, or gas turbine driven propulsion plants with one and the same type fuel.

GENERAL TRENDS

In conclusion, it may be of interest to note some general trends in naval engineering and to hazard a few predictions concerning possible future developments.

First, we may expect continuing refinement and improvement of the machinery and equipment now in use. The steam turbine, the diesel engine, the gas turbine, the nuclear propulsion plant—all are capable of further development and perhaps increased efficiency. We may reasonably look for new designs in boilers, turbines, reducing gears, bearings, propellers, condensers and other heat exchangers, and a wide variety of auxiliary machinery. Some improvements may be aimed at reducing mechanical losses, others at increasing the utilization of power developed by the prime mover, others at reducing noise levels, and still others at minimizing maintenance requirements.

We may look forward to the introduction of new engineering materials-metals and alloys, plastics, ceramics, lubricants, and others. We may watch for—though not necessarily count on— a materials breakthrough that would raise the upper temperature limits of our present machinery. If it is not possible to devise new materials to withstand ultra-high temperatures, we may perhaps look for new designs that will enable us to utilize higher temperatures with some of our present materials. We may also expect new and improved techniques for welding or otherwise joining metals, new methods of metal forming and shaping, new methods of treating metals to obtain desired properties, and new procedures for the nondestructive testing of engineering materials.

In the more distant future, perhaps, we might look for some entirely new concepts of ship propulsion. In particular, we might expect to see ship and machinery designs tending toward the ultimate goal of integrating the prime mover, the propulsive device, the steering device, and the hull form into one coordinated unit. Designers of ships and propulsion machinery have long looked with envy at the fully integrated propulsion systems of many fish, and a good deal of work has been done in analyzing fish propulsion with a view to picking up some usable ideas. One approach that has been suggested is to effect undulation of a flexible hull by pumping water in a sinusoidal path through a series of compartments. Still another approach utilizes a series of undulating plates. Although no type of simulated fish propulsion is even close to being operational at present, these approaches should not be dismissed as frivolous or trivial. A great deal has already been learned through biological and simulation studies of fish propulsion.

Altogether, we may expect the future to bring at least a few surprises, a few practical results from ideas which at the moment might be classified as exotic if not downright ludicrous. There is at present a great proliferation of new ideas in the field of naval engineering, and some of these ideas will doubtless find application in the propulsion plants of the future.

INDEX

☆ U.S. GOVERNMENT PRINTING OFFICE: 1987-730-025/60013